It's Like Having The Entire Radon Industry Delivered To Your Door!

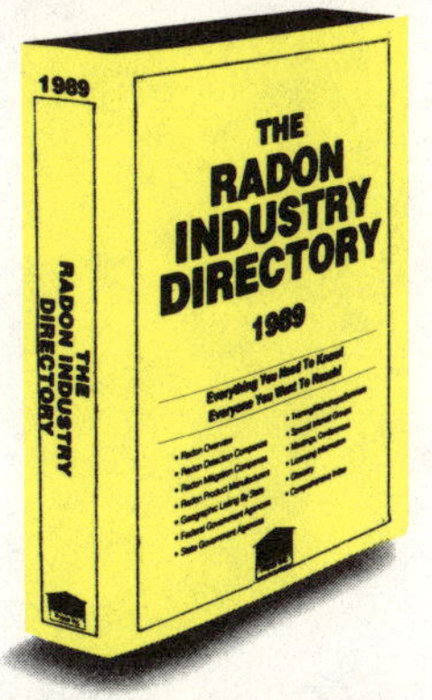

ISBN: 0-929840-00-3 ISSN: 0899-9023

Satisfaction Guaranteed

You must be completely satisfied with The 1989 Radon Industry Directory or simply return it within 10 days of receipt for a prompt refund.

For Faster Service, and Standing Order Information, Call Toll-Free

1-800-548-1567

RADON Press Inc.

ORDER FORM

The RADON INDUSTRY DIRECTORY

Yes! Please send me _____ copies of the Directory for only $75 each — Total Price $________. All invoices payable U.S. dollars.

☐ Bill Me ☐ Bill Company ☐ P.O. Enclosed #_______ ☐ Check Enclosed

☐ Begin a Standing Order (Standing order qualifies for a 10% discount).

Name________________ Title________________

Company________________

Address________________

City________________ State______ Zip________

Phone________________ Ext.________

Just Drop In The Mail. Postage Is Prepaid. DB1001

ORDER FORM

The RADON INDUSTRY DIRECTORY

Yes! Please send me _____ copies of the Directory for only $75 each — Total Price $________. All invoices payable U.S. dollars.

☐ Bill Me ☐ Bill Company ☐ P.O. Enclosed #_______ ☐ Check Enclosed

☐ Begin a Standing Order (Standing order qualifies for a 10% discount).

Name________________ Title________________

Company________________

Address________________

City________________ State______ Zip________

Phone________________ Ext.________

Just Drop In The Mail. Postage Is Prepaid. DB1002

ORDER FORM

The RADON INDUSTRY DIRECTORY

Yes! Please send me _____ copies of the Directory for only $75 each — Total Price $________. All invoices payable U.S. dollars.

☐ Bill Me ☐ Bill Company ☐ P.O. Enclosed #_______ ☐ Check Enclosed

☐ Begin a Standing Order (Standing order qualifies for a 10% discount)

Name________________ Title________________

Company________________

Address________________

City________________ State______ Zip________

Phone________________ Ext.________

Just Drop In The Mail. Postage Is Prepaid. DB1003

NO POSTAGE
NECESSARY
IF MAILED
IN THE
UNITED STATES

BUSINESS REPLY MAIL
FIRST-CLASS MAIL PERMIT NO. 8873 ALEXANDRIA, VIRGINIA

POSTAGE WILL BE PAID BY ADDRESSEE

RADON PRESS, INC.
P.O. BOX 25551
ALEXANDRIA, VA. 22313

NO POSTAGE
NECESSARY
IF MAILED
IN THE
UNITED STATES

BUSINESS REPLY MAIL
FIRST-CLASS MAIL PERMIT NO. 8873 ALEXANDRIA, VIRGINIA

POSTAGE WILL BE PAID BY ADDRESSEE

RADON PRESS, INC.
P.O. BOX 25551
ALEXANDRIA, VA. 22313

NO POSTAGE
NECESSARY
IF MAILED
IN THE
UNITED STATES

BUSINESS REPLY MAIL
FIRST-CLASS MAIL PERMIT NO. 8873 ALEXANDRIA, VIRGINIA

POSTAGE WILL BE PAID BY ADDRESSEE

RADON PRESS, INC.
P.O. BOX 25551
ALEXANDRIA, VA. 22313

THE RADON INDUSTRY DIRECTORY

1989

THE RADON INDUSTRY DIRECTORY

1989

Everything You Need To Know
Everyone You Want To Reach

RADON Press Inc.

Published by Radon Press, Inc.

Printed and bound in the United States of America.
International Standard Book Number 0-929840-00-3.
International Standard Serial Number 0899-9023.

Table of Contents

Foreword

On a cold December day in 1984, Stanley Watras began setting off alarms.

Like his fellow workers at the Limerick Nuclear Generating Station in Pottstown, Pa., Watras, an engineer, had to pass through a radiation monitor before leaving the plant. And as the monitor indicated Watras' repeated exposure to disturbing levels of radiation, Watras and his employers grew increasingly baffled with each passing day. Then early one morning, on a hunch, Watras decided to try the monitor *before* starting work. As the alarms went off, his fears were confirmed.

Ironically, the source of contamination was eventually traced to his own home, several miles away from the plant. Radon, a little known, naturally occurring radioactive gas, had contaminated Watras and his family to a level equivalent to smoking over 238 packs of cigarettes per day.

Unfortunately, Watras' story was not an isolated incident. Years later, the alarms he triggered continue to sound.

Radon, the invisible and odorless radioactive gas, is found at dangerously high levels in millions of homes, offices, schools and public buildings across the nation. Produced by the natural decay of uranium in the earth's crust, radon finds its way into places where we live and work through cracks and holes in foundations, water supplies, and open spaces between soil and floor. Radon decays further into radioactive particles that attach to dust and, when inhaled, lodge in the lungs.

According to health officials, radon is the second leading cause of lung cancer. Between 10,000 to 30,000 lung cancer deaths annually can be attributed to long-term exposure to radon. At present, experts generally consider radon to be our greatest environmental health threat — dwarfing the menace of asbestos, hazardous waste, toxic chemicals, and other recognized environmental pollutants.

As with other national health threats, public awareness of radon has grown rapidly. In response to the heightened concern, Congress has passed legislation supporting the implementation of federal and state radon programs to aid in the detection and mitigation of radon nationwide.

A once embryonic radon services marketplace has grown dramatically into a full-fledged radon industry. Within a few short years, thousands of radon detection firms, mitigation firms, product manufacturers, government agencies, and consultants have mobilized for action. Professionals within the areas of health, real estate, law, medicine, physics, construction, the environment, private industry and numerous other fields have quickly assumed active roles within this new industry.

Today, all facets of this new industry continue to grow at an exponential rate. And as the radon industry scrambles to manage and monitor itself, a need has arisen for consolidating and organizing its various dimensions — to document and systematize all that comprises the radon industry.

The first edition of *The Radon Industry Directory* has been published in response to this demand. It is our sincere hope that this annual directory — by serving as a forum for communication and information — will contribute toward remediating and controlling the devastating health effects of radon for this generation and generations to come.

Larry Siegelman

Publisher

Acknowledgments

A publication of this magnitude cannot be produced without the dedication of all those concerned. First and foremost, the publisher would like to express appreciation to *The Radon Industry Directory* Staff:

Rob Morrow, Editor
Marc O'Brien, Associate Editor
Sara Powell, Research Specialist
Suzanne Wood, Research Specialist
Barbara Bunner, Research Specialist
Russ Jackson, Research Specialist
Bob Peak, Art Director
Larry Kuekes, Software Consultant
Jill Kovacs
Mary Lou Downing Morrow

Their perseverance, enthusiasm, and commitment to excellence guaranteed the success of this publication.

Additionally, the publisher is grateful to several individuals representing a cross-section of knowledge and experience within the radon industry:

Gunnar Bar, Obar Systems
Terry Brennan, Camroden Associates
Bill Brodhead, W.P.B. Associates
Bonnie Chiles, U.S. EPA/Radon Division
Jim Craffey, AARST
Fred Cross, U.S. Department of Energy/Pacific Northwest Laboratories
Jim Davidson, Radon Detection Services
Jed Harrison, U.S. EPA/Radon Division
Craig Kneeland, New York State Energy Office/Radon Program
Milton Lammering, U.S. EPA/Region 8
Jerry Lowry, Lowry Engineering
Richard McGregor, National Health and Welfare/Canada
James Otten, U.S. Geological Survey/Radon Project
Steve Page, U.S. EPA/Radon Division
Melinda Ronca-Battista, U.S. EPA/Radon Division
Susan Rose, Department of Energy/Office of Health and Energy Research
David Saum, Infiltec
Donald Schutz, AARST
Ron Simon, R.F. Simon Company
Michael Terpilak, Radon Detection Services
George Woodruff, U.S. EPA/Radon Division

Their contributions during various stages of the project were invaluable.

Finally, the publisher would like to thank the members of the radon industry. Without their overwhelming response to our questionaires, letters, and phone calls, this directory would not have been possible.

Introduction

This first annual edition of *The Radon Industry Directory* provides a comprehensive source of information for all areas of the radon industry. In addition to its value as a reference for those involved in the radon industry, it also provides quick access to a wide range of resources and services to those outside the industry.

Criteria For Inclusion In The Radon Industry Directory

The Radon Industry Directory contains information on several thousand organizations and key personnel. Directory listings reflect the diverse backgrounds, specialties, and activities of those involved in the radon industry in the United States and Canada. To warrant inclusion in the Directory, an individual or organization had to meet one or more of the following criteria:

- Directly provides or supports radon-related services
- Conducts radon-related research
- Manufactures, distributes, or supplies products used in radon detection or mitigation
- Demonstrates involvement in the influence, formulation, or administration of public policy addressing radon
- Demonstrates involvement in special interest organizations or fields with a vested interest in radon-related issues.

Data Collection

To compile the information in the Directory, a master database was first developed by merging records derived from association, training, and business lists; government documents, correspondence, library research, and personal visits to key agencies and groups. Data request forms were then mailed to the several thousand names contained in the database. Directory announcements in various media and data request forms distributed at conferences, trade shows, and workshops lead to hundreds of additional inquiries.

Returned data request forms were checked against the database and a "last call" mailing was made. Those who failed to respond to the mailings were contacted by telephone. Individuals and organizations that could not be verified or fell short of selection criteria were dropped from the database.

Finally, to further our objective of depth and breadth of coverage, the research staff identified and verified hundreds of additional individuals and organizations who merited inclusion and satisfied selection criteria.

While every effort has been made to ensure accuracy and currency of information in the Directory, the radon industry is growing at a rapid rate and organizations are forming, expanding, or relocating daily. There are bound to be errors and omissions and the publisher would appreciate hearing of them.

The reader should also note that the publisher has reserved the right to delete irrelevant material or questionable self-appraisals, and to edit the style of information supplied by respondents. However, respondents remain fully responsible for the validity of the information they have supplied.

The publisher strongly recommends that proper evaluation of an organization include direct contact and personal assessment of the organization's qualifications and ability to meet your requirements.

Organization

As indicated in the *Table of Contents* information in *The Radon Industry Directory* is grouped into twenty five categories. Each category comprises several sections identified by section numbers. The *How To Use The Directory* section provides a brief description of each section in the Directory. Additionally, all twenty five sections include annotated descriptions of section content, listing criteria, and section organization.

Indexing

The Radon Industry Directory contains two major indexes enabling the reader to easily locate any organization or professional listed in the Directory. Entries in both *Name* and *Company Indexes* reference every company and individual by section and include name, address, and phone number. Additionally, several sections are followed by supplemental indexes to provide further aid in locating listings within that section. Supplemental indexes are indicated in the *Table of Contents.*

Other Radon Services

Radon Press provides several additional publications and services that the reader may find valuable:

•Reader Service Bureau
–Radon Press maintains a Reader Service Bureau to give assistance to individuals requiring information which can't be located in *The Radon Industry Directory* or other Radon Press publications. This service is free of charge.

•The Radon Hotline Letter
– a news hotline addressing radon-related legislative, licensing, and regulatory issues, policy trends, new products, business news and analysis, grant and contract opportunities, conferences, exhibits and other new developments in the radon industry.

•Radon Industry Mailing Lists – computerized lists of companies and personnel active in the radon industry.

Full information on Radon Press publications and services can be obtained by contacting Radon Press, Inc., P.O. Box 25551, Alexandria, VA 22313 or calling (703) 548-2756.

How to Use the Directory

The Radon Industry Directory is designed to provide the reader a quick and easy reference to radon-related information.

The Directory is organized into twenty five subject sections as indicated in the *Table of Contents*. Each section begins with a brief annotation describing its content and arrangement. Index tabs are also provided to assist referencing.

The reader can also locate information by referring to the *Index of Names* and *Index of Organizations* found at the end of the Directory. Each index entry is referenced by section number and contains name, organization, address, and phone information.

Directory Listings

Radon Industry Directory listings provide basic information about organizations in the private sector, government, research, and special interest groups. Listings are arranged in a standardized format. Listings within a section are alphabetized by organization name and include such information as address, telephone number, key personnel, description of activities, qualifications, or other items specific to the section. The omission of specific information from a listing signifies the information is either not applicable to that organization or was unavailable.

Organizations listed in the Directory are alphabetized in a letter-by-letter manner. The articles "an" and "the" are ignored in alphabetizing. Hyphens, slashes, commas, and parentheses are treated similarly to single spaces. When a personal name is the only information provided, the surname is generally listed first.

It should be noted that many organizations function in more than one area of the radon industry. When specifically requested by an organization or by editorial judgment in the interest of accurate representation, duplicate listings are listed in the appropriate sections. The respective section numbers at the end of each entry in the *Name and Organization Indexes* at the back of the Directory make it simple to locate any listing.

Sample Company Listing

The following explains the content and format of a typical listing:

HEALTH PHYSICS ASSOCIATES, LTD.
3312 Commercial Ave.
Northbrook, IL 60062-1909
(312) 480-1900
Ted Fields - President
Locke Bailey - General Manager
William Rivkin - Vice President
Services & Products:
Broad based radiation service company provides radon detection services using charcoal canister and alpha track methods, provides consultation on radiation safety.
Qualifications:
EPA RMP, certified health physicists, technicians registered for radon.
Area Served:
Nationwide.
EPA RMP 5:
CC(P).

Organization Name.
The name of the organization is printed in upper case and boldface type. A federal government listing begins with the primary administrative agency and proceeds hierarchically to the end office to which all information in the listing pertains.

Address.
Many organizations have several branches and offices. These branches are often listed in the Directory.

Telephone Number.
These are usually for the main office of activity. Several organizations maintain national toll-free telephone numbers.

EPA RMP 5.
Refers to the organization's participation in The National Radon Measurement Proficiency (RMP) Program--Round 5 administered by the U.S. Environmental Protection Agency. The codes describe the organization's demonstrated proficiency in measuring indoor radon and radon progeny.

Names/Titles of Key Personnel.
Contains the names and titles of the principal administrative officers and contact persons within the organization. For government agencies and special interest groups, the names of officials directly involved with radon-related activities are usually listed.

Services & Products.
Information about services, products, or activities is supplied by the organization and generally describes the organization's involvement with the radon industry. Where no description is given, the nature of the organization is self-explanatory; in other cases, no description was obtained.

Qualifications.
Specific information on credentials, training, experience, certification, affiliations, education, degrees, specialties, and professional expertise was provided by the organization.

Area Served.
Describes the geographic area within which services are provided.

For additional information on a specific section's content and arrangement, please refer to the annotated description on the first page of that section.

Radon Overview

Definition and Background

Radon is a naturally occurring, chemically inert, odorless, tasteless radioactive gas. It is produced from the radioactive decay of radium, which is formed, through several intermediate steps, from the decay of uranium. Since radium and uranium are ubiquitous elements in rock and soil, radon is also. Because radon is not chemically bound or attached to other materials, it can move easily through very small spaces such as those between particles of soil and rock. Radon-222, produced by the decay of radium-226, is the type of radon of most concern in indoor environments.

A major source of indoor radon in residences is radium in the soil and rock under the building. Secondary sources include groundwater and radium-containing building materials. Radon gas can enter a residence through cracks or openings such as sewer pipe and sump pump openings, cracks in concrete, wall-floor joints or, hollow concrete block walls. The amount of radon entering a residence is dependent on several factors, including the concentration of radium in the surrounding soil or rock and the soil porosity and permeability, and lower air pressure in the residence which may be caused by stack effect or operation of exhaust fans.

Radon has a half-life of approximately four days, decaying into progeny and emitting alpha particles in the process. The four sequential decay products of radon gas are chemically active particles that have half-lives under 30 minutes. Alpha particles in the lung transfer their energy into nearby cells and kill or damage the irradiated cells.

Health Effects

The inhalation of the decay products of radon is one of the most significant sources of risk to human health from indoor air pollution, and has been estimated to contribute to thousands of lung cancer cases per year in the United States.

Long-term exposure to the short-lived radon decay products is associated with an increase in the risk of lung cancer. The decay products of radon tend to adhere to dust particles or other surfaces and, if inhaled, can also adhere to the airways of the lung. Radon gas, on the other hand, is almost chemically inert and an inhaled radon atom is very likely to be exhaled before it decays. Thus, the main risk from radon is lung cancer associated with inhalation of its short-lived decay products, or progeny. Estimates of the risk of lung cancer due to radon exposure are based on studies of workers in underground mines who were exposed to radon progeny. These workers developed lung cancer at a higher rate than that observed in the general population. Although the epidemiological evidence is not directly applicable to people who do not work in mines, it is an adequate basis for concern about indoor radon exposure.

Factors Affecting Radon Concentration

Many factors affect the radon concentration in a residence. These include building type, local geology and climate, and occupant activities. Differences in these factors can cause the annual average indoor radon concentrations to vary greatly from one residence to another. These variations make it difficult to predict which localities or which buildings will have elevated indoor radon concentrations. Important factors affecting radon concentrations are:

- Soil characteristics
  - The potential for indoor radon increases with increasing permeability and radium content of the soil. Clays and waterladen soils, for example, are likely to retard the flow of radon.
- Construction type
  - The type of structure and its design affect the amount of area in contact with the soil, the air exchange rate in the residence, and the number and size of entry points for radon. For example, residences with well ventilated crawlspaces are likely to have less radon than other residences on the same site with tightly enclosed crawlspaces.
- Foundation condition
  - Cracks and openings in the foundation walls and floor can serve as entry points for radon.
- Occupant lifestyle and weather
  - Both factors can affect the ventilation rate for a residence and hence, the degree to which radon or any other indoor pollutant is trapped inside.

Radon Measurement

When measuring indoor radon, it is important to recognize that indoor concentrations can vary hourly, daily, and seasonally. For example, measurements made during a mild spring may underestimate the annual average level because of ventilation from open windows. Similarly, indoor levels during a cold winter may be higher than average because the residence is sealed. Thus, longer term measurements (six months to one year) made in residences during normal use generally provide more reliable estimates of the radon concentration. Shorter term measurements made under controlled conditions by trained personnel can also provide reliable information.

Radon Mitigation

Two approaches to reducing exposure to indoor radon are (1) to prevent radon entry into a residence and (2) to remove radon or radon decay products after entry into a residence. The first approach is the more desirable, especially if accomplished in a manner that does not require maintenance or checks.

The most effective measures include:

- Sealing cracks and openings such as basement drains and utility openings in the foundation with caulking or epoxy sealants.
- Exhausting from crawlspaces or underfloor areas, including sumps or other drain systems.
- Sealing tops of hollow concrete block foundations with concrete.
- Covering exposed earth inside or under the residence with concrete or a tightly sealed polymeric vapor barrier.
- Increasing the ventilation rate. The use of air-to-air heat exchangers will reduce the heating or air conditioning load.

Radon Detection Companies

This section lists companies in the U.S. and Canada that have indicated capabilities for measuring radon and radon progeny. Included here are private businesses, research facilities, laboratories, and product manufacturers that use various methods to measure radon under a wide range of conditions. Many of the companies listed here also provide other radon-related services and can be found in additional sections of the Directory.

The item "EPA RMP 5", when found at the end of a listing, designates the company's successful participation in The National Radon Measurement Proficiency (RMP) Program--Round 5. The "EPA RMP 5", sponsored by the U.S. Environmental Protection Agency, is the most recent of a series of programs in which a radon detection company may voluntarily enroll to demonstrate proficiency in measuring radon.

The codes following "EPA RMP 5" in a particular listing denote one or more of eight radon measurement methods in which a company can be judged proficient. The eight measurement methods are as follows:

AT - Alpha Track Detector
CC - Charcoal Canister
CR - Continuous Radon Monitor
CW - Continuous Working Level Monitor
EP - Electret Passive Environmental Radon Monitor (E-Perm)
GR - Grab Radon Sampling Unit
GW - Grab Working Level Sampling Unit
RP - Radon Progeny Integrated Sampling Unit (RPISU)

The code (P) which may follow a radon measurement method code indicates that the radon detection company is a "Primary Company" for that radon measurement method. A "Primary Company" (P) either maintains radon analysis equipment on its premises or conducts all radon measurements and analysis with its own instrumentation and operators.

Please note that mention of EPA RMP in the "Qualifications" item in a listing denotes only that the radon detection company has participated in at least one of the past five EPA RMP rounds. Also note that absence of an "EPA RMP 5" designation should not be construed as judgment on the competency or proficiency of a particular radon detection company.

An index listing radon detection companies by state served can be found immediately following this section.

A

A & A CONSULTING AND INSPECTION SERVICES, INC.
349 Chenango St.
Binghamton, NY 13901
(607) 723-7968
Larry Richards - President
John Meade - Vice President
Peter Connell - Environmental Consulting
John Link - Industrial Hygienist
Services:
Provides radon detection services, radon screening, house characterization for residential and commercial structures.
Qualifications:
NRA, AARST, staff includes professional engineer, geologist, NYS DOE workshops.
Area Served:
NY.

A & A RADON TESTING SERVICES, INC.
1805 South Bellaire St.
Denver, CO 80222
(303) 753-1044
Robert W. Goard - Senior Partner
Area Served:
Nationwide.
EPA RMP 5:
CC.

A & M ASSOCIATES
1813 Northwest 27th
Portland, OR 97210
(503) 274-2922
Phyllis A. McLaughlin
Area Served:
OR, WA.
EPA RMP 5:
AT, CC.

A & N RADON COMPANY
P.O. Box 188
Keene, NH 03431
(603) 357-0155
Robert Pecor - President
Lea Martin - Vice President
Area Served:
NH.

A & W RADON SPECIALISTS, INC.
12 Beechwood Rd.
Hudson, NH 03051
(603) 883-3883, (800) 727-7236
Robert C. Wilburn - President
Adrian Austin - Vice President
Services:
Provides radon detection services, water testing, and laboratory analysis.
Area Served:
Nationwide.
EPA RMP 5:
CC(P).

A 'LARY' SERVICE
686 Somerset Dr.
Lawrenceville, GA 30245
(404) 441-6819, (404) 963-8218
Susan Lary
Area Served:
GA.
EPA RMP 5:
CC.

A BLACK CROSS SANITATION COMPANY
1024 Cottman Ave.
Philadelphia, PA 19111
(215) 342-5100
Scott Ecksel - Certified Radon Technician

Area Served:
Nationwide.
EPA RMP 5:
AT, CC.

A-1 RELIABLE BASEMENT WATERPROOFING
3008 Gracemore St.
Dayton, OH 45420
(513) 293-6699
Lloyd D. Cohen
Area Served:
OH.
EPA RMP 5:
CR(P).

A-1 SHERLOCK HOME INSPECTORS
6-01 Fairlawn Dr.
Fairlawn, NJ 07410
(201) 796-4942
Steven Luxton - Technical Director
Area Served:
CT, DE, MA, NJ, NY, PA.
EPA RMP 5:
CC, CW(P).

A-1 TERMITE AND PEST CONTROL, INC.
Rte. 11, Box 20
Lenoir, NC 28645
(704) 758-9116, (704) 437-9380
Jack F. Roberts
Area Served:
NC.

A. ARTHUR LIPSHUTZ ASSOCIATES
7650 South Tamiami Trail Suite 9
Sarasota, FL 34231
(813) 923-7300
A. Arthur Lipshutz - Manager
Services:
Provides radon detection and mitigation services for residential and commercial markets, specializes in slab-on-grade construction, consultants to builders and real estate professionals.
Qualifications:
AARST, BBB, Executive Network, three years of radon mitigation experience.
Area Served:
Nationwide.
EPA RMP 5:
CC(P), CR(P).

A.B.E. RADIATION MEASUREMENTS LABORATORY
RD #1, Box 214
Lenhartsville, PA 19534
(215) 756-4153
Anthony LaMastra - President
Services:
Provides radon detection services using charcoal canister and alpha track methods.
Qualifications:
PA certified.
Area Served:
Nationwide.
EPA RMP 5:
AT, CC(P), GW(P), CW(P).

A.B.E. RADON TESTING, INC.
P.O. Box 508
Trexlertown, PA 18087
(215) 481-9916
Dean L. Ritter - President
Carl Distenthal - Certified Health Physicist
Services:
Provides radon detection services using charcoal cannisters and alpha track methods for real estate transactions, home kits, retesting.
Qualifications:
Two years in testing business, applied for EPA 3-day reduction course, PA DER certification.
Area Served:
NJ, PA.
EPA RMP 5:
CC.

A.D.T. INSPECTIONS, INC.
P.O. Box 425
Keyport, NJ 07735
(201) 566-4079
Anthony Tarantola
Area Served:
NJ.
EPA RMP 5:
CC.

A.J. PERRI, INC.
401 Hwy. 35
Red Bank, NJ 07701
(201) 747-3131
Albert J. Perri
Area Served:
NJ.

A.L. BOUCHARD COMPANY
56 Franklin Pl.
Totowa, NJ 07512
(201) 790-6259
A.L. Bouchard - President
Services:
Provides radon detection and mitigation services.
Qualifications:
AARST.
Area Served:
NJ.

A.S. MCCREATH AND SONS, INC.
P.O. Box 1453
Harrisburg, PA 17105-1453
(717) 238-9331
Fred A. Pennington, Jr.
Area Served:
Nationwide.
EPA RMP 5:
CC(P).

A.T.E.I.
809 East 11th St.
Michigan City, IN 46360
(800) 833-0999
George Perry - President
Services:
Provides radon detection and mitigation services for residential and commercial structures.
Area Served:
Nationwide.

AA HOME INSPECTIONS
P.O. Box 248
Springfield, NJ 07081
(201) 635-6060
Paul Uslan - President
Services:
Provides radon detection services for residential and commercial properties, serves as a consultant and expert witness in litigation matters.
Area Served:
NJ.

AA RADCO
1447 Old Sauk Trail
Niles, MI 49120
(616) 684-8115
Glenn Diamond
Area Served:
IL, IN, MI, MN, MS, OH.
EPA RMP 5:
CC.

AAA DISASTER FORCE
18219 West 3rd Ave. Suite 3
Golden, CO 80401
(303) 399-4203
Mary Knudson
Area Served:
CO.
EPA RMP 5:
CC.

AAA NEW ENGLAND RADON SERVICES
P.O. Box 1096
Westbrook, CT 06498
(203) 399-5748
Donald C. Kowalski

Area Served:
CT.

AAA RADON SERVICES, INC.
P.O. Box F
Hellertown, PA 18055
(215) 868-6025, (215) 821-2938
Robert Grosset - President
Services:
Provides radon detection and mitigation services.
Qualifications:
PA DER approved course.
Area Served:
Nationwide.
EPA RMP 5:
CC.

AAA RADON TESTING
26 Flynn Pl.
Wayne, NJ 07470
(201) 772-7166
Nicholas Verhagen - President
Services:
Provides radon detection and mitigation services.
Qualifications:
NJ, PA certification.
Area Served:
NJ, PA.
EPA RMP 5:
CC.

AAAIRSCAN ENGINEERING CORPORATION
8 Eric Dr.
Granby, CT 06035
(203) 653-5839
Kevin O'Rourke
Area Served:
AZ, CA, CO, CT, DC, DE, FL, GA, HI, MA, MD, ME, MN, NC, NE, NH, NJ, NV, NY, PA, RI, SC, VA, VI, VT, WV.
EPA RMP 5:
AT, CC.

AADVANCED AAIR AANALYSIS
2670 Billingsley Rd.
Columbus, OH 43235
(614) 792-3400
James Cutcher - President
Services:
Provides radon detection, mitigation, and reduction repair services for residential, commercial, and industrial sites.
Qualifications:
EPA RMP, ARA, BBB, 18 years of engineering and building experience.
Area Served:
Nationwide.
EPA RMP 5:
CR(P).

AARDEN TESTING AND EVALUATION, INC.
2015 Dodge Ave.
Sarasota, FL 34234
(813) 951-4060
Byron Schenk - President
Richard Bainbridge - Vice President
Services:
Provides radon detection services for soil and water analysis, uses charcoal canister and alpha track methods.
Area Served:
FL.

ABACUS LABS RADON ASSESSMENT AND MITIGATION (ALRAM)
P.O. Box 5490
Portsmouth, NH 03801
(603) 433-3332, (800) 228-3332
Gary W. Seesman
Area Served:
CT, DE, MA, ME, NC, NH, RI, VT.
EPA RMP 5:
CC(P).

ABC INSPECTION COMPANY
9911 Santa Cruz St.
Seabrook, MD 20706
(301) 577-1915
Fred W. Jones
Area Served:
DC, MD, VA.
EPA RMP 5:
CC.

ABIONICS, LTD.
12509 Kings Lake Dr.
Reston, VA 22091
(703) 620-3767
William R. Hancuff
Services:
Provides radon detection services using charcoal canister, E-Perm, and alpha track methods.
Qualifications:
EPA training programs.
Area Served:
DC, MD, VA.
EPA RMP 5:
AT, CC, EP(P).

ABMS, INC.
739 Allston Way
Berkeley, CA 94710
(415) 548-4300
Tim Hussler
Area Served:
CA.

ACADEMY OF IRW, INC.
The Friary, Winchester-on-Severn 1600 Winchester Rd.
Annapolis, MD 21401
(301) 757-7887
Bobby E. Leonard
Area Served:
Nationwide.
EPA RMP 5:
CC(P).

ACCENT ENGINEERING, INC.
5624 Yukon St.
Arvada, CO 80002
(303) 425-8700
Paula Knudson
Area Served:
CO.

ACCORD CORPORATION
84 Schuyler St.
Belmont, NY 14813
(716) 268-7605
Charles Kalthoff - Executive Director
Area Served:
NY.

ACCREDITED WATERPROOFING COMPANY, INC.
2504 Church Rd.
Cherry Hill, NJ 08002
(609) 482-2630, (215) 265-2888
Allan Konkus - Vice President
Services:
Provides radon mitigation services for residential market.
Area Served:
DE, NJ, PA.

ACCU-TEST
1350 Beverly Rd. Suite 115-343
McLean, VA 22101
(703) 448-8088
Bruce Johnson
Brian Keff
Area Served:
DC, DE, MD, NJ, PA, VA.
EPA RMP 5:
AT, CC.

ACCURAD TESTING SERVICE
P.O. Box 8766
Rolling Meadows, IL 60008
(312) 529-6633
Mark A. Doruff
Area Served:
IL.

ACCURATE AIR TESTING
1209 Piney Woods Dr.
Friendswood, TX 77546
(713) 482-5097

Charles T. Gallagher
Area Served:
TX.

ACCURATE HOME INSPECTION, INC.
P.O. Box 131
Holmdel, NJ 07733
(201) 264-5485
Steve Mizrahi
Area Served:
NJ.

ACCURATE HOME SERVICE
928 Weatherbee Pl.
Downers Grove, IL 60516
(312) 971-2662
Sue F. Mitch
Area Served:
IL.

ACCUTEK LABORATORY
11126 Deerfield Rd.
Cincinnati, OH 45242
(513) 984-4112
Timothy G. Barrett
Area Served:
IN, KY, OH.
EPA RMP 5:
CR(P).

ACE PEST CONTROL
Rte. 2, Box 244
Syracuse, IN 46567
(219) 856-2294
Howard E. Long
Area Served:
IN.

ACI CHEMICAL COMPANY, INC.
P. O. Box 306
Walnut Ridge, AR 72476
(501) 886-6625
Charles E. Snapp
Area Served:
AR, MO, TN.
EPA RMP 5:
CC.

ACME EXTERMINATING COMPANY
P.O. Box 85
Gainesville, GA 30503
(404) 536-3036, (404) 532-4351
John E. Hulsey
Area Served:
GA.

ACRES INTERNATIONAL CORPORATION
140 John James Audubon Pkwy.
Buffalo, NY 14228-1180
(716) 689-3737
Larry Zamojski - Engineer
Area Served:
NY.

ACS CONSULTANTS, INC.
129 Maitland Ave.
Paramus, NJ 07652
(201) 845-8278
Anthony Szabo - President
Services:
Provides radon detection services using charcoal canister method.
Qualifications:
AARST.
Area Served:
NJ, NY.
EPA RMP 5:
CC.

ADCO
P.O. Box 1267
Burlington, CT 06013
(203) 589-3333
Aniello Depascale
Area Served:
CT.
EPA RMP 5:
CC.

ADDY'S RADON DETECTION SERVICE
245 Moonlight Dr.
Columbia, SC 29210
(803) 798-5379
H. Paul Addy
Area Served:
SC.
EPA RMP 5:
CC.

ADIRONDACK ENVIRONMENTAL INVESTIGATIONS
P.O. Box 270
Cambridge, NY 12816
(518) 677-3958
John V. Freitas
Qualifications:
AARST.
Area Served:
NY.

ADVANCE PEST CONTROL
P.O. Box 749
Franklin, NC 28734
(704) 524-8061
Franklin Thompson
Area Served:
NC.

ADVANCE TEX CORPORATION
204 Plantation Rd.
Seneca, SC 29678
(803) 654-2300
Thomas Lau
Area Served:
GA, NC, SC.
EPA RMP 5:
CC.

ADVANCED AIR ANALYSIS
61836 30th St.
Lawton, MI 49065
(616) 624-2741
Timothy J. Diebolt
Area Served:
MI.
EPA RMP 5:
CC.

ADVANCED AUTOMATIC ALARM COMPANY, INC.
1722 Mayappleway
Gambrills, MD 21054
(301) 261-0011
Janette E. Hupman
Area Served:
DC, DE, MD, VA, WV.
EPA RMP 5:
CC.

ADVANCED HOME INSPECTION AND TESTING
23 Magna Vista Cr.
Tewksbury, MA 01876
(617) 851-5866
Joseph F. Laliberte
Area Served:
MA, NH.
EPA RMP 5:
CC.

ADVANCED MATERIALS ENTERPRISES, INC.
P.O. Box 121174
Nashville, TN 37212
(615) 377-0624
Ron Quarles
Area Served:
TN.

ADVANCED RADIATION MONITORING SERVICE, INC. (ARMS)
2219 Prosperity
Portage, MI 49081
(616) 323-9483
Philip B. Noack - President
Mary K. Noack - Secretary
Services:
Manufactures and distributes activated charcoal test monitors and radon laboratory equipment, provides radon mitigation services.
Qualifications:

Nuclear medicine technologist, physicists, engineers on staff.
Area Served:
Nationwide.
EPA RMP 5:
CC(P).

ADVANCED RADON DETECTION SERVICES, INC.
P.O. Box 30458
Knoxville, TN 37930
(615) 691-9905
George F. Haynes
Art Roberts, Jr.
Qualifications:
AARST.
Area Served:
TN.

ADVANCED RADON SERVICES
P.O. Box 729
Parker, CO 80134
(303) 841-6702
John A. Turbett
Area Served:
CO.
EPA RMP 5:
CC.

ADVANTAGE HOME INSPECTION
P.O. Box 580
South Dennis, MA 02660
(617) 398-0861
Donald Lovering
Area Served:
MA.

AERO-AQUA
54 Miller St.
Belfast, ME 04915
(207) 338-2809
Tom Reuwer - Field Office Manager
Area Served:
ME.

AERTEC LABORATORIES, INC.
P.O. Box 2268
Kokomo, IN 46904-2268
(317) 453-5173
Arden C. Scruggs
Area Served:
IN.

AFFILIATED BUILDING INSPECTORS, INC.
P.O. Box 15770
Chevy Chase, MD 20815
(301) 986-8866
Larry Wasson - President
Services:
Provides radon detection services using charcoal canister, alpha track, E-Perm and continuous radon monitor methods, distributes E-Perms, charcoal canisters, and alpha track detectors.
Qualifications:
EPA RMP.
Area Served:
DC, DE, MD, NJ, PA, VA.
EPA RMP 5:
AT, CC, EP(P).

AFFTREX, LTD.
3530 Washington Ave.
Finleyville, PA 15332
(412) 348-9211
William H. Button - President
Douglas J. Button - Vice President
Services:
Provides radon detection services using charcoal canister, alpha track, and grab sample methods, provides mitigation services.
Qualifications:
PA DER interim radon testing and mitigation certification.
Area Served:
NY, PA.

AFI ENVIRONMENTAL/DIVISION OF BUFFALO CONSULTANTS, INC.
6103 Robinson Rd.
Lockport, NY 14094
(716) 625-8434
Kathy Fregelette - Senior Technician
William Heitzenrater - President
Kevin Cronin - Senior Geologist
Services:
Initial and follow-up radon detection in residential, commercial and public buildings.
Qualifications:
AARST, NYSDOH radon workshop.
Area Served:
NY.

AGAPE RADON SERVICES, INC.
P.O. Box 969
Alpharetta, GA 30239-0969
(404) 664-4106
Sandra Hightower - President
Services:
Provides full service radon detection and consultation services for homeowners and commercial buildings.
Qualifications:
Thirty years experience in the building industry.
Area Served:
GA.

AGENTERPRISE, INC.
227 North 7th St.
Allentown, PA 18102
(215) 776-0777
John Murphy
Area Served:
NJ, PA.
EPA RMP 5:
CC.

AIDCO MAINE CORPORATION
Grassy Pond House
Rindge, NH 03461
(603) 899-5166
R. Multer - President
Services:
Provides radon detection, mitigation, and on-site consultation services.
Area Served:
CT, DC, DE, FL, GA, IA, IL, KY, MA, MD, ME, MI, NC, NH, NJ, NY, OH, PA, RI, VT, VA, WV.
EPA RMP 5:
CC.

AIR ANALYSIS ASSOCIATES
P.O. Box 240
Unadilla, NY 13849
(607) 369-7689
Elaine J. Reed - President
John R. Isler - Vice President
Linda W. Isler - Secretary/Treasurer
Services:
Provides radon detection services using charcoal canister and alpha track methods.
Qualifications:
EPA RMP.
Area Served:
NY.
EPA RMP 5:
CC.

AIR ANALYSIS ASSOCIATES
2063 Stanley Rd.
Cazenovia, NY 13035
(315) 655-8948
John R. Isler - President
Linda W. Isler - Vice President
Services:
Provides radon detection services.
Area Served:
NY.

AIR CHEK, INC.
180 Glenn Bridge Rd. Box 2000
Arden, NC 28704
(704) 684-0893, (800) 257-2366
B.V. Alvarez - President
C.S. Amos - Vice President
J.L. Alvarez - Director of Research
Ned Doyle
Services:
Manufactures, distributes, and provides

analysis for radon detection kits nationwide, direct to consumers as well as through dealership network.
Area Served:
Nationwide.
EPA RMP 5:
CC(P).

See Ad Page 188

AIR CONCERNS
1919 Oakwell Farms Pkwy. Suite 120
San Antonio, TX 78218
(512) 820-8875
Donna Jordan - Owner
Services:
Provides radon detection services for residential and commercial structures.
Qualifications:
AARST.
Area Served:
TX.

AIR HAZARDS ASSESSMENT, INC.
415 Main St. Suite 3
Laurel, MD 20707
(301) 792-7111
Victor Suben
Area Served:
DC, MD, PA, VA, WA.
EPA RMP 5:
CC.

AIR QUALITY ANALYSTS, INC.
5400 Eisenhower Ave.
Alexandria, VA 22304
(703) 823-7567
Bruce C. Phillips - President
Services:
Distributes residential radon detection devices.
Area Served:
DC, DE, MD, NC, NJ, PA, VA, WV.
EPA RMP 5:
AT, CC.

AIR QUALITY CONTROL, INC.
4 Terry Dr. Suite 3
Newtown, PA 18940
(215) 860-6198
Dan Smith - Branch Manager
Services:
Provides radon detection and mitigation services.
Area Served:
NJ, PA.

AIR QUALITY CONTROL, INC.
585 Phoenixville Pike
Malvern, PA 19335
(215) 889-3100, (800) 654-2144
Dave Smith - Owner
Buzz Clark - Owner
Services:
Provides radon detection and mitigation services.
Area Served:
PA.

AIR QUALITY DETECTION, INC.
11700 Old Columbia Pike Suite 2102
Silver Spring, MD 20904
(301) 622-1661
Barry Rowe
Area Served:
DC, MD, VA.
EPA RMP 5:
CC.

AIR RADON SERVICE, INC.
292 Main St. Suite 8
Hackensack, NJ 07601
(201) 585-2994
Eddie Tseng
Robert Musumeci
Qualifications:
AARST.
Area Served:
NJ.

AIR RESOURCE SPECIALISTS, INC.
1901 Sharp Point Dr. Suite E
Ft. Collins, CO 80525
(303) 484-7941
John Faust
David L. Dietrich
Services:
Provides radon detection services.
Area Served:
CO.

AIR SCIENCES, INC.
508 North Truman Blvd. P.O. Box 320
Crystal City, MO 63019
(314) 933-3915, (800) 242-8886
Kerry Van House - Chief Executive Officer
Ralph Overman - Vice President
Services:
Provides radon detection and mitigation services, distributes charcoal canister home detection kits.
Qualifications:
Staff includes nuclear scientist with over 40 years of radiation work and physicist skilled in radiation measurement.
Area Served:
Nationwide.
EPA RMP 5:
CC(P).

AIR SENSE, INC.
25 William St.
Pawcatuck, CT 06379-2108
(203) 599-5599
Douglas C. Hoagland - President
Services:
Provides radon detection and mitigation services.
Qualifications:
AARST.
Area Served:
CT, MA, ME, NH, RI, VT.

AIR SOLUTIONS
1004 Dennis Ave.
Silver Spring, MD 20901
(301) 681-9242
Kevin St. Clair - President
Services:
Provides radon mitigation services and consultation for residential market.
Qualifications:
Registered architect, participant in Maryland indoor radon evaluation methods course.
Area Served:
DC, MD, VA.
EPA RMP 5:
CC.

AIR TRON MECHANICAL SERVICES
3178 South Halifax St.
Aurora, CO 80013
(303) 699-1836
Rob Brooks - Owner
Area Served:
CO.

AIR VENTILATION SYSTEMS, INC.
3343 Hollins Ferry Rd.
Baltimore, MD 21227
(301) 525-3174
John Guanti
Area Served:
DC, DE, MD, PA, VA.
EPA RMP 5:
CC.

AIR-N-SOL CORPORATION
3 Cedar St. P.O. Box 437
Frenchtown, NJ 08825
(201) 996-2028
Dale L. Johnson - President
Services:
Provides radon detection services for residential and commercial buildings, diagnostic testing and preparation of specifications for radon mitigation.
Qualifications:
NJ DEP, EPA RMP, EPA courses, AARST.
Area Served:
Nationwide.
EPA RMP 5:
AT, CC, EP(P), GR(P), GW(P), CW(P).

AIR-TIGHT OF FARMINGTON
28000 Middlebolt
Farmington Hills, MI 48018
(313) 855-5065
David Kolodziej
Area Served:
MI.

AIR-TIGHT OF GENESSEE
2429 North Dort Hwy.
Flint, MI 48506
(313) 235-5010
David Neblock
Area Served:
MI.
EPA RMP 5:
CC.

AIRCARE SYSTEMS
P.O. Box 4942
Lancaster, PA 17604
(717) 293-9223
John E. Henderson
Area Served:
PA.

AIRMASTER RADON DETECTION, INC.
P.O. Box 3177
Wayne, NJ 07474-3177
(201) 794-3733
William Mirabile - Owner
Services:
Provides radon detection services using charcoal canister method, provides radon mitigation services.
Area Served:
NJ.

AIRMASTERS RADON DIVISION
1746 Boggs Rd.
Forest Hill, MD 21050
(301) 893-1993
Kelly F. McGill - President
Services:
Provides radon detection and mitigation services.
Area Served:
DC, MD, VA.

AIRREGS, INC./AIR POLLUTION REGULATORY SERVICE, INC.
1640 Powers Ferry Rd. Building 17, Suite 210
Atlanta, GA 30067
(404) 984-1919
Robert Geddis - Hygienist
Richard Shutt - Engineer
Services:
Provides radon detection and mitigation services, training to homeowners and realtors.
Qualifications:
EPA RMP, 38 years of combined experience, both professionals have master's degrees.
Area Served:
Nationwide.
EPA RMP 5:
AT, CC.

ALABAMA TESTING LABORATORY
P.O. Box 2337
Gasden, AL 35903
(205) 547-5311
Don Cotton
Area Served:
Nationwide.
EPA RMP 5:
CC.

ALL AMERICAN RADON SERVICES, INC.
P.O. Box 1079
Orange Park, FL 32067
(904) 642-1963
Paul Bliton
J.E. Bliton
Qualifications:
AARST.
Area Served:
FL.

ALL EASTERN HOME INSPECTION COMPANY
309 Bloomfield Ave.
Caldwell, NJ 07006
(201) 228-5548
Peter Kras
William Rodgers
Area Served:
NJ, NY, PA.
EPA RMP 5:
CC.

ALL PRO PEST CONTROL
P.O. Box 1313
Amarillo, TX 79105-1313
(806) 374-7886, (806) 355-6660
Bobbie McReynolds
Area Served:
TX.

ALL SERVICE RADON REDUCERS, INC.
166 Southeast Crain Hwy.
Upper Marlboro, MD 20772
(301) 390-7440
Bill Culwell
Area Served:
DC, MD, VA.
EPA RMP 5:
CC.

ALL TIME PEST CONTROL, INC.
728 East Main St.
Benton Harbor, MI 49022
(616) 925-0607, (616) 944-1907
Maurice Bishop
Area Served:
MI.

ALLIED BIOLOGICAL
240 Main St.
Gladstone, NJ 07934
(201) 234-2476
Bruce O. Ellis - Biologist
Area Served:
NJ.

ALLSAFE ENVIRONMENTAL CORPORATION
P.O. Box 89
Bayville, NJ 08721
(201) 269-6060
Gregory Makowski
Area Served:
Nationwide.
EPA RMP 5:
CC.

ALLSTATE RADON TESTING SERVICE
2231 Pinecliff Dr.
Atlanta, GA 30345
(404) 938-2238
Richard F. Cain
Area Served:
Nationwide.
EPA RMP 5:
CC.

ALPHA BETA GAMMA TECHNOLOGIES, INC.
1829 Cawley Ave.
Bethlehem, PA 18017
(215) 866-8895
Glenn E. Aranyos - Owner
Area Served:
PA.

ALPHA CHECK
7200 Wisconsin Ave.
Bethesda, MD 20814
(301) 856-3637
Brian Koepf
Larry Adair
Services:
Performs radon detection using charcoal canisters, alpha track, E-Perms and continuous radon monitors.
Qualifications:
EPA RMP.
Area Served:
DC, DE, MD, NJ, PA, VA.

EPA RMP 5:
AT, CC.

ALPHA CONCEPTS, INC.
P.O. Box 119
Hopatcong, NJ 07843
(201) 770-3046
Thomas Hatton
Area Served:
Nationwide.
EPA RMP 5:
AT, CC, CR(P).

ALPHA CONTROL, INC.
3509 Cherry Valley Dr.
Olney, MD 20832
(301) 924-0377
Scott Parry, Jr. - Owner
Area Served:
Nationwide.
EPA RMP 5:
AT, CC, EP(P), GR(P).

ALPHA DETECTORS
1040 Serpentine Way
Sandy, UT 84070
(801) 572-4659
Patty Rehn
Area Served:
Nationwide.
EPA RMP 5:
CC.

ALPHA ENERGY LABORATORIES, INC.
2100 Road to Six Flags
Arlington, TX 76011
(817) 792-3434
Mike Buvinghausen
Area Served:
Nationwide.
EPA RMP 5:
CC(P).

ALPHA GUARD/TECTROL
5613 Yarrow St.
Arvada, CO 80002
(303) 421-7955
Greg Piette - Owner
Area Served:
CO.

ALPHA RADIATION ENGINEERING, INC.
P.O. Box 90
Littleton, CO 80160-0090
(303) 721-0068, (303) 721-0872
Howard E. Dixon
Area Served:
AZ, CO, FL, KS, MT, ND, NE, NM, NV, OK, TX, UT, WY.
EPA RMP 5:
AT, CC.

ALPHA RADON ENVIRONMENTAL TESTING
Government Center P.O. Box 737
Portland, ME 04104
(207) 773-8004
David Dalphonse
Area Served:
MA, ME, NH, NY, PA, RI, VT.
EPA RMP 5:
AT, CC(P), CR(P), CW(P).

ALPHA RADON SCREENING AND HOME INSPECTION, INC.
1216 Chinook Way
Boulder, CO 80303
(303) 494-9533
Bill Jewell
Area Served:
CO.

ALPHA RADON TESTING, INC.
1915 Kyle Pl.
Golden Valley, MN 55422
(612) 529-1929
Francine Archambault
Area Served:
Nationwide.
EPA RMP 5:
CC.

ALPHA SERVICES/AIR AND WATER RADON TESTING
296 Grindle Brook Rd.
South Glastonberry, CT 06073
(203) 659-8099
Pietro A. Catizone - Owner
Services:
Provides radon reduction services or do-it-yourself air and water radon test kits.
Area Served:
CT, MA, ME, NH, NJ, NY, PA, RI, VT.
EPA RMP 5:
CC.

ALPHA SPECTRA OF KENTUCKY
11518 Main St.
Louisville, KY 40243
(502) 245-5557
Patrick S. Holmes
Area Served:
KY.
EPA RMP 5:
CC.

ALPHA SPECTRA, INC.
P.O. Box 6772 210 Clayton St.
Denver, CO 80206
(303) 377-2118
Frank Wilkinson III
Area Served:
Nationwide.
EPA RMP 5:
CC(P).

ALPHA-BAN
3719 Michelle Way
Baltimore, MD 21208
(301) 484-7120
David Nevin
Area Served:
DC, DE, MD, PA.
EPA RMP 5:
CC.

ALPHANUCLEAR COMPANY
1125 Derry Rd., E.
Mississauga, ON L5T 1P3
(416) 676-1364
G.M. Denham - President
See Ad Page 197
Services:
Develops, manufactures, and distributes radon/radon progeny measurement instrumentation, provides charcoal canister radon test kits and analysis in Canada.
Qualifications:
Manufacturer of radon detection instrumentation since 1975, supplier to EPA, DOE, and many state agencies.
Area Served:
Nationwide.

ALPHATECH, INC.
P.O. Box 254
Haverford, PA 19041
(215) 525-1490
Evelyn S. Kritchevsky - President
Nancy W. Pusey - Secretary
Services:
Provides radon detection services using charcoal canister, grab sampling, and water (well) for homeowners, realtors, home inspection services.
Qualifications:
Applied for PA license.
Area Served:
Nationwide.
EPA RMP 5:
CC(P), GW(P).

AMBIC BUILDING INSPECTION CONSULTANTS, INC.
1200 Rte. 130
Robbinsville, NJ 08691
(609) 426-1212
W. David Goldstein
Services:
Provides radon detection services.
Qualifications:
AARST.

Area Served:
NJ, PA.

AMC HOME INSPECTION SERVICES
421 West Union Ave.
Bound Brook, NJ 08805
(201) 469-6050
John Hendricks
Area Served:
NJ.
EPA RMP 5:
CC.

AMERICAN BUILDING INSPECTORS, INC.
P.O. Box 1105
Haverford , PA 19083
(215) 449-9361
Ralph D'Angelo
Area Served:
DE, MD, NJ, PA.
EPA RMP 5:
CC.

AMERICAN COMPLIANCE TECHNOLOGIES, INC.
4404 South Florida Ave. Suite 12
Lakeland, FL 33813
(813) 647-1200
Robert O. Kincart - Hazardous Materials
Services:
Comprehensive environmental services programs including mobile laboratory, radon detection and mitigation, real estate inspections and assessments.
Qualifications:
Certified hazardous materials manager, radon chemistry and detection courses.
Area Served:
Nationwide.

AMERICAN HERITAGE CORPORATION
P.O. Box 1145
Paoli, PA 19301
(717) 275-0300, (800) 247-0300
Kenneth G. Enochs - Director
Services:
Provides radon detection and mitigation services, fully equipped vehicles and teams (200 mile radius).
Qualifications:
PA DER, NY DOE.
Area Served:
DE, NJ, NY, PA.

AMERICAN HI-TECH
263-265 Leighton Ave.
Red Bank, NJ 07701
(201) 842-9477, (201) 671-5887
J. Charles Tully

Area Served:
NJ.

AMERICAN HOME INSPECTION SERVICE
684 Main St.
Wakefield, MA 01880
(617) 536-6510
Dolly Thomas - Office Manager
Area Served:
MA.

AMERICAN HOME SERVICE COMPANY
57 South Grand Ave.
Poughkeepsie, NY 12603
(914) 473-1999
Dan Friedman - Home Inspector
Services:
Provides radon detection services for residential and commercial properties prior to purchase using charcoal canister method.
Area Served:
NY.

AMERICAN HOME-TECH, INC.
3415 Northwest 48th Ter.
Gainesville, FL 32606
(904) 375-4323
Bill Zetrouer - President
Services:
Provides radon detection services.
Area Served:
FL.

AMERICAN HOMECHECK, INC.
P.O. Box 565
Wayne, PA 19087
(215) 259-8707
Ralph D'Angelo
Roman Paul
Area Served:
DE, NJ, PA.
EPA RMP 5:
CC.

AMERICAN INDEPENDENT LABS
P.O. Box 1069
Sandpoint, ID 83864
(208) 263-6935
Bruce Newman
Area Served:
ID.

AMERICAN INSPECTION AGENCY
P.O. Box 2444
Lehigh Valley, PA 18001
(215) 759-2444
George H. Harris
Qualifications:
AARST.

Area Served:
PA.

AMERICAN INSPECTIONS SERVICES
R.D. #4 Box 344
Duncansville, PA 16635
(800) 922-5857
Area Served:
PA.

AMERICAN INSULATION, INC.
15 Cherry St. P.O. Box 266
Danvers, MA 01923
(617) 777-0617
David O'Heir - President
Services:
Provides radon mitigation services using insulation, distributes radon test kits.
Area Served:
MA.

AMERICAN MEDICAL MARKETING SERVICES
2429 West Davie Blvd. Suite 17
Ft. Lauderdale, FL 33312
(305) 462-2575
Jim Phillips - Vice President
Services:
Provides radon detection and mitigation services.
Area Served:
FL.

AMERICAN MONITORING AND ENGINEERING SERVICES
200 High Tower Blvd. Suite 205
Pittsburgh, PA 15205
(215) 547-5460, (609) 392-1422
John Phillips - Safety Services
Bryan Fisher - Geologist
Services:
Provides radon detection services using charcoal canister method.
Area Served:
OH, PA.

AMERICAN PEST MANAGEMENT
3929 Martin Way Suite D
Olympia, WA 98506
(206) 491-1809
David Scott - President
Services:
Provides radon detection services for relocation companies and distributes charcoal canisters and alpha track detectors.
Area Served:
WA.

AMERICAN RADON CONSULTANTS, INC.
109 Ladbrook Rd.
Knoxville, TN 37922
(615) 891-9581
Jack Buddenbaum
Area Served:
Nationwide.
EPA RMP 5:
CC.

AMERICAN RADON CONTROL, INC.
2758 S. Langley Ct.
Denver, CO 80210
(303) 758-6580
Dave Brockway
Area Served:
CO.

AMERICAN RADON CORPORATION
8300 North Hayden Rd. Suite 100
Scottsdale, AZ 85258
(602) 967-8029
James Krueger - President
Susan Bollin - Senior Director
Services:
Provides radon detection and mitigation services, provides educational services for real estate and legal professions.
Qualifications:
President is member of ASTM-E6 task group subcommittee and director of AARST.
Area Served:
Nationwide.
EPA RMP 5:
AT, CC, EP(P).

AMERICAN RADON DETECTION COMPANY, INC.
39 Folcroft Ave.
Folcroft, PA 19032
(215) 532-8845
Christopher G. Mellis - President
Services:
Provides radon detection services using charcoal canister method.
Qualifications:
AARST.
Area Served:
PA.

AMERICAN RADON SERVICES, LTD.
303 I.S.I.S. Center Iowa State University
Ames, IA 50011-2130
(515) 296-9926
Kenneth D. Wiggers - President
Services:
Provides radon detection services using E-Perm and charcoal canister methods, provides radon mitigation services.
Area Served:
Nationwide.
EPA RMP 5:
CC(P), EP(P).

AMERICAN RADON TECHNICAL SERVICES, INC.
1941A Nears Pkwy.
Margate, FL 33063
(305) 968-5555
Gordon Marsh - President
Marge Semenuk - Vice President
Services:
Provides radon detection services using charcoal canister and E-Perm methods, provides radon mitigation services.
Qualifications:
General contractor and physicist on staff.
Area Served:
NJ, FL.

AMERICAN RADON TESTERS
5705 Center Ln. Office 1
Falls Church, VA 22041
(703) 239-0227
Michael Grissley
Area Served:
VA.

AMERICAN RADON TESTING LABORATORIES
1015 South Birch St. #610
Glendale, CO 80222
James W. Kruse
Area Served:
CO.

AMERICAN RADON TESTING LABS, INC.
21360 Center Ridge Rd. Suite 307
Rocky River, OH 44116
(216) 331-0200
Timothy Freeman
Area Served:
Nationwide.
EPA RMP 5:
CC.

AMERICAN RADON TESTING SERVICES
2302 Basswood Cr.
Tustin, CA 92680
(714) 838-4990
Mary Jane Henry
Area Served:
CA.

AMERICAN RADON TESTING SERVICES, INC.
P.O. Box 70
Vandalia, OH 45377
(513) 898-1949
John Ross
Area Served:
OH.

AMERICAN RADON TESTING, INC.
P.O. Box 60
Trexlertown, PA 18087
(215) 778-2167
Murray Samuels - Owner
Area Served:
PA.

AMERICAN REAL ESTATE INSPECTIONS
P.O. Box 61
Blairstown, NJ 07825
(201) 850-7435
Peter Russell - President
Services:
Radon detection services for home buyers.
Qualifications:
AARST, participant in NJ DEP voluntary program for radon testing.
Area Served:
NJ.

AMES ASSOCIATES, INC.
6935 Wisconsin Ave.
Chevy Chase, MD 20815
(301) 656-5222
Michael J. Ames
Area Served:
Nationwide.
EPA RMP 5:
CC.

ANALYTICAL MANAGEMENT, INC.
P.O. Box 11279
Lexington, KY 40574
(606) 233-7774
Lee Vittitow
Area Served:
KY.

ANGELO, MARC
110 Woods Ln.
Radnor, PA 19087
(215) 687-6183
Marc Angelo
Area Served:
PA.

ANNAPOLIS PROPERTY SERVICES
1680 Yorktown Ct.
Crofton, MD 21114
(301) 721-7177
Robert S. Jacobson
Area Served:
Nationwide.
EPA RMP 5:
CC.

ANTI-PEST COMPANY
P.O. Box 8082
Longview, TX 75607
(214) 758-0169
James T. Way
Area Served:
TX.

ANTI-PEST COMPANY OF SHREVEPORT
P.O. Box 4266
Shreveport, LA 71104
(318) 221-6181
J.T. Way
Area Served:
LA.

APPALACHIAN ENVIRONMENTAL TESTING, INC.
105 South Union St. Suite 326
Danville, VA 24541
(804) 792-1300
Christopher Halladay - President
Services:
Provides radon detection services using charcoal canister and alpha track detectors, conducts on-site working level measurements and geological/radiometric site evaluations.
Qualifications:
EPA RMP, certified professional geologist.
Area Served:
Nationwide.
EPA RMP 5:
AT, CC.

APPLIED ANALYTICAL TECHNOLOGIES, INC.
P.O. Box 668
Narragansett, RI 02882
(401) 782-8188
Paul B. Champlin
Area Served:
CT, DC, DE, FL, MA, MD, ME, NH, NJ, NY, PA, RI, VT, VA.
EPA RMP 5:
CC.

APPLIED ANTENNA TECHNOLOGIES/RADON MEASUREMENTS DIVISION
2270 Honeycomb Ct.
Duluth, GA 30136
(404) 623-0403
William Barry Stewart
Area Served:
GA.
EPA RMP 5:
GR(P).

APPLIED ENERGY COMPANY
3316 Augusta Ave.
Omaha, NE 68144
(402) 334-5881
Jon Traudt
Area Served:
Nationwide.
EPA RMP 5:
CC.

APPLIED HEALTH PHYSICS, INC.
2986 Industrial Blvd.
Bethel Park, PA 15102
(412) 563-2242, (518) 477-7974
R.G. Gallagher - President
John Douglas - Instrument Manager
John Harshman - Radon Specialist
Services:
Professional radiation safety firm distributes charcoal canisters, provides instrument calibration, repair, and leasing.
Qualifications:
EPA RMP, twenty-five years of experience, NRC and state licensed, all services by board certified registered professional engineers.
Area Served:
Nationwide.
EPA RMP 5:
AT, CC, EP(P).

APPLIED RADON, INC.
52205 Warehime Rd.
Cumberland, OH 43732-9804
(614) 638-2078, (800) 638-2078
Frank J. Donia - Vice President
Nancy A. Donia - President
Services:
Provides radon detection and mitigation services, distributes At-Ease radon monitors, presents seminars to real estate industry.
Qualifications:
ARA, EPA RMP, health physicist on staff.
Area Served:
Nationwide.
EPA RMP 5:
AT, CC, CR(P).

APPLIED TECHNICAL SERVICES, INC.
1190 Atlanta Industrial Dr.
Marietta, GA 30066
(404) 423-1400, (800) 451-3405
Donnissa Duvic - Environmental Science
Services:
Distributes radon test kits including canister, instructions, pre-paid return postage, analysis and report, retail packaging available.
Area Served:
Nationwide.
EPA RMP 5:
AT, CC(P), EP(P).

AQUATEK LABORATORY/DIVISION OF GRAHAM-MASSEY ANALYTICAL LABORATORY
2212 Main St.

Bridgeport, CT 06606
(203) 576-0104
David Graham - Laboratory Director
Area Served:
Nationwide.
EPA RMP 5:
CC(P).

AREA HOME INSPECTIONS, INC.
341 Littleton Rd.
Harvard, MA 01451
(617) 263-7251
Anthony Galeota - President
Services:
Provides radon detection services using charcoal canister method.
Area Served:
MA.

ARGONNE NATIONAL LABORATORY
9700 South Cass Ave.
Argonne, IL 60439
(312) 972-4172
Henry Lucas
Services:
Non-commercial laboratory for research purposes or state agencies' assistance only.
Area Served:
Nationwide.
EPA RMP 5:
GR(P).

ARIX CORPORATION
325 East Southern Ave. Suite 18
Tempe, AZ 85282
(602) 968-0052
Lee Compton
Area Served:
AZ.

ARIX CORPORATION
760 Horizon Dr.
Grand Junction, CO 81506
(303) 242-6203
J. Tell Tappan - President
David Reinersten
Area Served:
Nationwide.
EPA RMP 5:

AT, CC(P), GR(P).
ARIX CORPORATION
P.O. Box 2021
Greeley, CO 80632-2021
(303) 356-3930
William F. McKelvey
Patti Terry
Area Served:
CO.
EPA RMP 5:
CC.

ARIX CORPORATION
2155 North 200 West Suite 100
Provo, UT 84604
(801) 373-2749
Toni Sullivan
Area Served:
UT.
EPA RMP 5:
CC.

ARIX CORPORATION
877 North Eighth, W. Suite 8
Riverton, WY 82501
(307) 858-6506
Melody Reeb
Patty Gartrell
Area Served:
MT, WY.
EPA RMP 5:
CC.

ARIX CORPORATION/A DIVISION OF VERSAR, INC.
1200 East High St.
Pottstown, PA 19464
(215) 327-1140
W. McKelvey - Operations Manager
R. Meier - Radiologic Technician
L. Schlegel - Business Development
Services:
Radon consultants provide mitigation, design, and testing for residential and commercial buildings, and schools in the public and private sectors, provides education and training courses.
Qualifications:
EPA RMP, PA DER testing and mitigation courses, listed by NJ DEP.
Area Served:
Nationwide.
EPA RMP 5:
AT, CC, GR(P).

ARIZONA RADIATION REGULATORY AGENCY/ENVIRONMENTAL SURVEILLANCE LABORATORY
4814 South 40th St.
Phoenix, AZ 85040
(602) 255-4845
Paul J. Weeden - State Contact
Services:
Non-commercial laboratory for research purposes or state agencies' assistance only.
Area Served:
AZ.
EPA RMP 5:
CC(P).

ASAFE-HOUSE, INC.
260 Litchfield Rd.
New Preston, CT 06777
(203) 868-0512
Erhard Werner
Area Served:
CT, MA, NY, RI.
EPA RMP 5:
CC, GR(P).

ASBESTOS AND RADON CONSULTANTS QUALI-PRO-TECH SERVICES, INC.
204 Devon Dr.
Exiom, PA 19341
(215) 363-7268
George W. Woods
Area Served:
Nationwide.
EPA RMP 5:
CC.

ASSAY SERVICES
P.O. Box 1266
Friendswood, TX 77546
(713) 641-0391
Robert D. Gallagher
Area Served:
Nationwide.
EPA RMP 5:
CC(P).

ASSOCIATED BUILDERS OF THE ROCKIES, INC.
P.O. Box 704
Wheat Ridge, CO 80034
(303) 237-3428
Robert C. Petty
Area Served:
CO.
EPA RMP 5:
CC.

ASSOCIATED PROTECTIVE SYSTEMS
2511 Winford Ave.
Nashville, TN 37211
(615) 244-7637, (615) 244-7638
Kenneth E. Hair
Area Served:
TN.

ASSOCIATED RADON TESTING SERVICES, INC.
303 South Ave.
Media, PA 19063
(215) 566-2274
Michael P. Yorke - President
Katee M. Yorke - Vice President
Services:
Provides radon detection services and follow-up testing using charcoal canister, alpha track, continuous radon and continuous working level measurements, provides pre-construction and mitigation consultation.

Qualifications:
Extensive experience in construction and building inspection, PA training seminars, participant in EPA RMP.
Area Served:
DE, MD, NJ, PA.
EPA RMP 5:
AT, CC, CR(P).

ASSOCIATION FOR IMPROVEMENT IN MOHEGAN, INC.
East Main St. P.O. Box 103
Mohegan Lake, NY 10547
(914) 528-3300
William Stevens - Rehabilitation
Services:
Housing preservation company funded by New York State Energy Office to assist residents with referrals for radon detection, mitigation, and education concerning radon. Helps coordinate state radon effort at the local level.
Area Served:
NY.

ASTECO, INC.
P.O. Box 2204
Niagara Univeristy, NY 14109
(716) 297-5992
John E. Larson
Area Served:
NY.

ASTLE CORPORATION
146 Old Farm Rd.
Leominister, MA 01453-2029
(508) 534-0979
William Astle - President
Services:
Provides radon detection and mitigation services, manufactures a patented air-to-air energy recovery device that reclaims 95% of heat lost up flue.
Area Served:
MA.

ATESYS, INC.
1 Pheasant Ln.
Westport, CT 06880
(203) 261-0528, (203) 226-5173
Judith Auslander
Area Served:
CT.

ATLANTA TESTING AND ENGINEERING, INC.
11420 Johns Creek Pkwy.
Duluth, GA 30136
(404) 476-3555
L.T. Gregg - Geologic Services
J.O. Costello - Sr. Env. Scientist
J.J. Holmes - Staff Geologist
Services:
Detection and measurement of radon in soil-gas and groundwater primarily for real estate developers and lending institutions.
Qualifications:
Registered professional geologists supervise all surveys.
Area Served:
Nationwide.

ATLANTIC ELECTRIC
65 W. Highland Ave.
Atlantic Highlands, NJ 07716
(201) 291-3468
Frank Feola - Owner
Area Served:
NJ.

ATLANTIC ENVIRONMENTAL LABORATORY
16 Elm St.
West Haven, CT 06516
(203) 932-5808
Richard Morrow
Area Served:
CT, MA, NH, NJ, NY, RI, VT.
EPA RMP 5:
CC(P).

ATLAS WATERPROOFING COMPANY, INC.
P.O. Box 14953
Reading, PA 19612-4953
(215) 374-6454
Leo Runge - Owner
Area Served:
PA.

ATOMETRICS
911 Veterans Blvd. Suite 202
Metairie, LA 70005
(504) 837-1367
Hugh A. Smith - President
R. G. Wille - Associate Director
Services:
Provides radon detection services, radioactivity analyses, and gramma spectroscopy services.
Area Served:
Nationwide.
EPA RMP 5:
AT, CC(P).

AUDUBON CONTROLS, INC.
P.O. Box 477
Audubon, PA 19403
(215) 566-7089
Silas A. Belden
Area Served:
CT, DE, NJ, NY, PA.
EPA RMP 5:
CR(P).

AUGUSTUS W. SMITH, INC.
1 Short Hills Ave. P.O. Box 164
Short Hills, NJ 07078
(201) 376-7740
Augustus W. Smith
Area Served:
NJ.

B

B & H RADON SERVICES
P.O. Box 247
Silver Spring, PA 17575-0247
(717) 684-2612, (717) 426-3723
James Bradley - Owner
Alfred Henne - Owner
Area Served:
PA.

B & I SERVICES
P.O. Box 291
Edmond, KS 67636
(913) 622-4341
Ben Cooper
Area Served:
KS.

B & W SALES AND MARKETING
P.O. Box 130
Chiefland, FL 32626
(904) 528-3600
Bernard Wilson Cooper - Sales
Area Served:
FL.

B & W SALES AND MARKETING CORPORATION
4045-A Jonesboro Rd.
Forest Park, GA 30050
(404) 366-0957, (800) 331-1449
Robert D. Black - President
Ben Williams - Secretary Treasurer
Area Served:
Fl, GA.

B-DRY SYSTEM OF DELAWARE AND CHESTER COUNTY, INC.
1054 Saunders Ln.
West Chester, PA 19380
(215) 430-0393
Jim Gane - President
Area Served:
PA.

BAIN PEST CONTROL SERVICE, INC.
1320 Middlesex St.
Lowell, MA 01851
(617) 453-7792

Jeffrey L. Bain
Area Served:
MA.

BALTIMORE GAS AND ELECTRIC
P.O. Box 1475
Baltimore, MD 21203
(301) 281-3781
Lawrence Bartal - Environmental Health
Services:
Provides local citizens with information on effect of home weatherization on indoor air quality (including radon).
Area Served:
MD.
EPA RMP 5:
CC(P).

BAR-TEL INFRASCAN
748 Kirkland Cir. F-102
Kirkland, WA 98033
(206) 822-1482
Fred C. Bartel
Area Served:
ID, MT, OR, WA.
EPA RMP 5:
CC.

BARBARA L. SALTHOUSE
251 Crown St.
Stratford, CT 06497
(203) 375-5837
Barbara L. Salthouse
Area Served:
CT.

BARRINGER
15000 West 6th Ave. Suite 300
Golden, CO 80401
(303) 277-1687
B.E. Sabels - Radon Manager
David Lasher
Services:
Distributes charcoal canisters and alpha track detectors, mail order services, mitigation, and consultation services.
Qualifications:
AARST, EPA RMP, general environmental laboratory listed with EPA, Health Department, etc.
Area Served:
Nationwide.
EPA RMP 5:
AT(P), CC(P).

BARRINGER MAGENTA, LTD.
304 Carlingview Dr.
Rexdale, ON M9W 5G2
(416) 675-3870
John Ip - Chemist
Services:
Provides radon detection services using charcoal canisters and alpha track detection methods.
Area Served:
ON.

BASEMENT DE-WATERING SYSTEMS, INC.
156 Curry Hill Rd.
Canonsburg, PA 15317
(412) 745-5959
Robert Johnston - Owner
Services:
Provides radon detection and mitigation services.
Area Served:
Nationwide.

BATTELLE MEMORIAL INSTITUTE/NUCLEAR TECHNOLOGY SECTION
505 King Ave.
Columbus, OH 43201-2693
(614) 879-5192
V. Pasupathi - Section Manager
Mike Stenhouse - Research Scientist
Services:
Contracts with federal government and other oraganizations for radon-related research projects. Radon analysis performed at laboratory.
Area Served:
Nationwide.
EPA RMP 5:
AT, CC(P), CR(P).

BAUBIOLOGIE BY BANTA
P.O. Box 51250 Suite 125
Pacific Grove, CA 93950
(408) 372-8626
John C. Banta
Services:
Provides radon detection services for real estate industry and residential market.
Area Served:
CA.
EPA RMP 5:
CR(P).

BAYSURE BUILDING INSPECTION, INC.
290 King of Prussia Rd.
Radnor, PA 19087
(215) 688-2572
Jack L. Gaskill - President
Services:
Provides radon detection services.
Area Served:
DE, MD, PA.

BBC & M ENGINEERING
195 Chittenden Ave.
Columbus, OH 43201
(614) 294-3745
Richard Brown
Area Served:
OH.
EPA RMP 5:
CC.

BCA LABORATORIES, INC.
1102 South Main St.
Bloomington, IL 61701
(309) 828-7772
Kurt Benckendorf - Lab Manager
Dave Martin - Business Manager
Don Curry
Services:
Provides radon detection services by grab sampling, charcoal canisters and other methods, distributes charcoal canisters and alpha track detectors to the public.
Qualifications:
EPA RMP.
Area Served:
AK, IA, IL, IN, MO, WI.
EPA RMP 5:
GW(P).

BCT
12 Magnolia Ave.
Denville, NJ 07834
(201) 627-2802
Kevin Scollans
Area Served:
NJ.
EPA RMP 5:
CC.

BECKMAN INSTRUMENTS, INC.
2500 Harbor Blvd.
Fullerton, CA 92634
(714) 871-4848
Robert J. Jarnutowski - Marketing
Sheila Brann - Communications
Services:
Manufactures and distributes laboratory instrumentation and supplies, radon detection systems, including samplers and software for report generation.
Area Served:
Nationwide.

See Ad Page 431

BENCHMARK INSPECTION SERVICES
939 South 26th St.
Arlington, VA 22202
(703) 548-5690
A.R. Dahms - Field Inspector
K.F. Dahms - Administrative Asst.
Services:
Provides radon detection services, pre- and

post-mitigation measurement, and consultation.
Qualifications:
Certified radon technologist.
Area Served:
DC, MD, VA.

BERG GROUP
1401 West Hamburg
Baltimore, MD 21230
(301) 837-5525
David Berg - Vice President
Services:
Provides radon detection services using charcoal canister method, provides mitigation services.
Qualifications:
NRA.
Area Served:
MD.

BERGEN HOME INSPECTION SERVICE
838 Palisades Ave.
Teaneck, NJ 07666
(201) 836-3232
Richard Laprete
Allan Bien
Area Served:
NJ, PA.
EPA RMP 5:
CC.

BERNELL PROFESSIONAL SERVICES
361 12th St.
Cresskill, NJ 07626
(201) 568-7083
Bernard Kasmir - President
Services:
Provides radon detection services using alpha track and charcoal canister methods, provides consultation on radon mitigation methods.
Area Served:
NJ.

BET, INC.
222 South Rolff
Davenport, IA 52802
(319) 322-8749
Tom Peters - Secretary
Services:
Provides radon detection services.
Area Served:
IA.

BILL EICH CONSTRUCTION COMPANY
Box 63
Spirit Lake, IA 51360
(712) 366-4438
William Holland
Area Served:
IA.

BIOHAZARDOUS MATERIALS CONSULTING
P.O. Box 4376
Ithaca, NY 14852
(607) 257-5010, (315) 478-0653
Mark Schenerman - Radon Specialist
Kirk Leister - Radon Specialist
Services:
Provides radon detection services using charcoal canister method, provides consultation services for associated contractors to perform radon mitigation procedures.
Qualifications:
NY certified for radon measurement and mitigation.
Area Served:
NY.

BIOMEDICAL TOXICOLOGY ASSOCIATES
P.O. Box 3568
Frederick, MD 21701
(301) 662-0783
W.G. Palmer - Director
Services:
Provides radon detection and consultation services in toxicology and health sciences.
Area Served:
DC, MD, VA, WV.
EPA RMP 5:
AT, CC.

BIOMONITORING SERVICES LABORATORY/TECHNICAL RESOURCES, INC.
3202 Monroe St.
Rockville, MD 20852
(301) 231-5250
Gary Petrazzuolo
Harold Siefried
Area Served:
Nationwide.
EPA RMP 5:
CC.

BIONOMIC INDUSTRIES, INC.
479 Route 17 N.
Mahwah, NJ 07430
(201) 529-1094, (800) 338-3210
John Enhoffer - Vice President
Steve Augulis - Design Engineer
Barbara Purcel - Customer Service
Services:
Provides complete engineered radon mitigation systems for residential, commercial, and industrial buildings, designs for both new and existing construction.
Qualifications:
ARRST, registered with EPA, NJ DEP, and other state organizations.
Area Served:
Nationwide.
EPA RMP 5:
AT, CC, CW(P).

BIOSOURCES, LTD.
18 Sevilla Rd.
Andover, MA 01810
(508) 698-8707
Richard E. Rammacher - Director
Qualifications:
AARST.
Area Served:
MA.

BIOSOURCES, LTD.
7 Bailey St.
Nashua, NH 03063
(603) 880-6004
Paul M. Tyree - Director
Qualifications:
AARST.
Area Served:
MA, NH.

BIOSPHERICS, INC.
4928 Wyaconda Rd.
Rockville, MD 20852
(301) 770-7700
Frank Fitzpatrick
Area Served:
DC, DE, MD, NC, NJ, NY, PA, VA, WV.
EPA RMP 5:
CC.

BLUEGRASS RADON SERVICES
424 Ridgeway Rd.
Lexington, KY 40502
(606) 266-6148
Vickie Jordan
Area Served:
KY.

BONDAR-CLEGG, INC.
12980 West Cedar Dr.
Lakewood, CO 80228
(303) 989-1404
W.G. Sherwood
Services:
Provides radon detection and mitigation services for residential and commercial markets.
Qualifications:
AARST, EPA RMP.
Area Served:
AZ, CA, CO, KS, NE, NM, NV, UT, WY.
EPA RMP 5:
CC.

BORO RADON TESTING SERVICE, INC.
501 Sharp Ave.
Glenolden, PA 19036
(215) 532-8378, (215) 586-5640
Rob Bruno
Len Bruno
Ron Bruno
Area Served:
PA.
EPA RMP 5:
AT, CC.

BOSTON UNIVERSITY MEDICAL CENTER/RADIATION PROTECTION OFFICE
80 East Concord
Boston, MA 02118
(617) 638-7052
Victor Evdokimoff - Director
Services:
Provides radon detection services for Medical Center and Boston University employees. Non-commercial laboratory for research purposes or state agencies' assistance only.
Qualifications:
EPA RMP.
Area Served:
Nationwide.
EPA RMP 5:
CC(P).

BOULDER INSPECTION COMPANY
P.O. Box 3369
Boulder, CO 80307
(303) 444-6657
Ian Rosebrook - Engineer
John Gilbert - Engineer
Don Drake - Engineer
Services:
Provides radon detection services, distributes radon detection products to the public.
Qualifications:
EPA courses in radon detection.
Area Served:
CO.

BOXFORD RADON LABORATORY
52 Wildmeadow Rd.
Boxford, MA 01921
(617) 887-9641
John Fisher - President
Cathy Fisher - Vice President
Area Served:
CT, MA, ME, NH, NY, RI, VT.
EPA RMP 5:
CC(P).

BRADLEY HARRISON ASSOCIATES, INC.
55 Virginia Ave.
Freeport, NY 11520
(516) 546-2283, (516) 546-7760
Bradley Harrison - President
Services:
Consulting firm specializing in environmental matters, waste management, and industrial safety and health.
Area Served:
CT, MA, NY, RI.
EPA RMP 5:
CC.

BRAMA CONSTRUCTION CORPORATION
47 Buena Vista Ave.
Yonkers, NY 10701
(914) 968-4700
Al Diodati - President
Area Served:
NY.

BRANDEIS UNIVERSITY/ROSENTIEL RESEARCH CENTER
415 South St.
Waltham, MA 02154
(617) 736-2461
Daniel Perlman - Senior Scientist
Services:
Developing new passive and active radon detector based on adsorption. Developer of liquid scintillation counting technology in radon detection. Developed first 100 percent humidity independent carbon-based detector. Non-commercial laboratory for research purposes or state agencies' assistance only.
Area Served:
Nationwide.
EPA RMP 5:
CC(P).

BRANDYWINE ENVIRONMENTAL SERVICES
P.O. Box 3343 206-B East Ayre St.
Newport, DE 19804
(302) 995-7565
Thomas S. Anthony - President
Area Served:
DE.

BRENT, DAVID J.
3000 South Jamaica Ct. Suite 130
Aurora, CO 80014
(303) 755-7012
David J. Brent
Area Served:
CO.

BROOKS LABORATORIES, INC.
44 Codfish Ln.
Weston, CT 06883
(203) 226-6384
Pramod Patel
Services:
Provides radon detection, mitigation, and consultation services.
Qualifications:
Industrial hygiene consultants.
Area Served:
CT.

BRUN-REUSS ASSOCIATES
P.O. Box 0842
Lionville, PA 19353-0482
(800) 522-0072
J.F. Brown - President
Area Served:
PA.

BRYANT AND BRYANT SERVICE COMPANY
992 1/2 Davidson Dr.
Nashville, TN 37205
(615) 352-2570
B.J. Purtle
Harold Bryant
Area Served:
Nationwide.
EPA RMP 5:
CC.

BUILDING DIAGNOSTICS, INC.
380 27th St.
Boulder, CO 80303
(303) 494-2431
Mike Holecek
Area Served:
CO.

BUILDING INSPECTION SERVICES, INC.
1675 Winding Rd.
Southampton, PA 18966
(215) 357-8055
Marvin Goldstein - President
Area Served:
PA.

BUILDING INSPECTOR OF AMERICA, THE
243 Reed Rd.
Westport, MA 02790
(508) 771-2477, (508) 636-6767
Tim Borden - President
Area Served:
MA.

BUILDING INSPECTOR OF AMERICA, THE
129C Gaither Dr.
Mt. Laurel, NJ 08054
(609) 273-8333
Jack Moreno
Qualifications:
AARST.
Area Served:
NJ.

BUILDING INSPECTOR OF AMERICA, THE
1 Mason Ct. Box 1654
Hawley, PA 18428
(717) 775-9619
Lane F. Quevedo
Qualifications:
AARST.
Area Served:
PA.

BUILDING INSPECTOR OF AMERICA, THE
112 Jockey Hollow Way
Union, NJ 07083
(201) 473-5656, (201) 579-2734
Joel A. Pagano - President
Services:
Provides radon detection services for home buyers.
Qualifications:
NJ DEP voluntary program for radon testing, EPA RMP.
Area Served:
NJ.

BUILDING INSPECTORS, INC.
282 Harlow St. P.O. Box 2220
Bangor, ME 04401
(207) 941-0870
Larry Lussier
Area Served:
ME.
EPA RMP 5:
CC.

BULAVA ENVIRONMENTAL, INC.
13 Hunt Club Rd.
Belle Mead, NJ 08502
(201) 874-6207
Edward J. Bulava
Area Served:
NJ.

BURENS ENVIRONMENTAL SERVICES CORPORATION
P.O. Box 96
Wallingford, PA 19086
(215) 565-7455
Area Served:
PA.

BURNS AND ASSOCIATES
265 Kearsarge Way
Portsmouth, NH 03801
(603) 431-3400
Robert J. Burns - Principal
Virginia Steckowych - Q/A Manager
Services:
Provides radon air and water detection services, diagnostic work ,and design of mitigation techniques.
Qualifications:
Registered professional engineer, completed AARST Radon Specialist course.
Area Served:
MA, ME, NH, VT.
EPA RMP 5:
AT, CC, EP(P).

C

C & D ASSOCIATES HOME INSPECTIONS
542 Union Ave.
Middlesex, NJ 08846
(201) 469-0263
Lisa Anthony
Area Served:
NJ.
EPA RMP 5:
CC.

C & D HEATING AND CONSTRUCTION
1111 South Sheridan Blvd.
Lakewood, CO 80226
(303) 935-6461
Charles Smith
Sami Todd
Services:
Provides radon detection and mitigation services.
Area Served:
CO.

C SENTINEL ENTERPRISES
426 Ingel Blvd.
Salem, VA 24153
(703) 380-4828
Richard C. Moore
Area Served:
NC, VA, WA.
EPA RMP 5:
CC.

C-P LABORATORY
783 State Rd. Suite 174
Plymouth, MA 02360
(617) 477-8128

Philip DeLizza
Area Served:
MA.

C. KENT CHAMBERLAIN AND ASSOCIATES
6579 West David Dr.
Littleton, CO 80123
(303) 979-3916
C. Kent Chamberlain - President
Services:
Provides radon detection services using charcoal canister and alpha track methods, provides radon mitigation services.
Qualifications:
Principal holds doctorate in geology, certified by two professional societies.
Area Served:
CO.

C.R. ADAMS COMPANY, INC.
17228 West 16th St.
Golden, CO 80401
(303) 279-5967
Charles Adams
Area Served:
CO, ID, KS, MT, ND, NE, NM, NV, SD, UT.
EPA RMP 5:
CC.

CALIFORNIA DEPARTMENT OF HEALTH SERVICES/SANITATION AND RADIATION LABORATORY
2151 Berkeley Way Room 465
Berkeley, CA 94704
(415) 540-2201
George S. Uyesugi - Director
Services:
Non-commercial laboratory for research purposes or state agencies' assistance only.
Area Served:
CA.
EPA RMP 5:
GR(P).

CALOIERO RADON TECHNOLOGIES
RD5, Box 84 North Main St.
Williamstown, NJ 08094
(609) 728-9257
Joseph Caloiero
Services:
Provides radon detection and mitigation services.
Qualifications:
AARST.
Area Served:
NJ.

CAMBRIDGE ASSOCIATES, LTD.
11 Harwood Ln. Trowbridge Complex
Clementon, NJ 08021
(609) 435-2626, (215) 563-6611
Al Vasys
Area Served:
DE, NJ, PA.
EPA RMP 5:
CC.

CAMP DRESSER AND MCKEE, INC.
Plaza 1 Raritan Ctr.
Raritan, NJ 08818
(201) 225-7000
Jay W. Davis
Area Served:
Nationwide.
EPA RMP 5:
CC.

CAMPTON DEVELOPMENT CORPORATION
P.O. Box 354
Campton, NH 03223
(603) 726-3030
Charles F. Orr
Area Served:
MA, ME, NH, RI, VT.
EPA RMP 5:
CC.

CAMRODEN ASSOCIATES
RFD 4, Box 62
Rome, NY 13440
(315) 865-4269
Terry Brennan - President
Services:
Provides radon mitigation services, indoor air quality investigation, and training programs for federal, state, and private sector groups.
Qualifications:
APCA, ASTM, NAHB.
Area Served:
Nationwide.
EPA RMP 5:
AT, CC.

CANADIAN INSTITUTE FOR RADIATION SAFETY/ELLIOTT LAKE LABORATORY
7 Timmins Rd. Suite 7-15
Elliott Lake, ON P5A 2R7
(705) 848-3687
Ernest Becker - Senior Scientist
Services:
• Non-profit agency operates as a registered charity. Has enacted radon-daughter assessments in uranium mines (550 miners). • Provides mitigation and assessment training and measures radon levels in private homes. • Presently developing detection devices. • Environmental monitoring and applied research.
Area Served:
Nationwide.

CANBERRA INDUSTRIES, INC.
1 State St.
Meriden, CT 06450
(203) 238-2351
Tom Perkins - Marketing Specialist
Michela Hirnak - Marketing Specialist
Bruce M. Gillespie
Services:
Manufactures and distributes turnkey systems for radon detection using the charcoal canister method.
Qualifications:
EPA RMP.
Area Served:
Nationwide.
EPA RMP 5:
CC(P).

See Ad Page 186

CANCERSCREEN, INC.
1120 First Colonial Rd.
Virginia Beach, VA 23454
(804) 481-4742
Steven Warden
Area Served:
Nationwide.
EPA RMP 5:
CC.

CANNONDALE HEATING AND AIR CONDITIONING
P.O. Box 458
Wilton, CT 06897
(203) 834-0875
Paul T. Bonamo
Area Served:
CT.

CAPE CHEMICAL COMPANY
33 North Frederick St.
Cape Girardeau, MO 63701
(314) 334-3002
Elizabeth A. Knote
Qualifications:
AARST.
Area Served:
MO.

CAPE COD HOME INSPECTIONS
243 Reed Rd.
Westport, MA 02790
(508) 771-2477, (800) 828-1919
Timothy E. Borden - President
Services:
Provides radon detection services for buyers and sellers of real estate and for homeowners.
Qualifications:
Licensed construction supervisor, MA registered as radon inspector.

Area Served:
MA.

CAPITAL CONSTRUCTION SERVICES
P.O. Box 568
Troy, NY 12180
(518) 434-0105
Merrill Greenberg - Manager
Area Served:
NY.

CAPITAL HOME INSPECTION CORPORATION
P.O. Box 2541
Manassas, VA 22110
(703) 369-4206
Chris Burkert
Area Served:
DC, MD, VA.
EPA RMP 5:
CC.

CAPITAL MATERIALS TESTING, INC.
P.O. Box 81 5 Hemlock St.
Latham, NY 12110
(518) 783-3150, (800) 541-4109
James Swartz - Regional Manager
Bob Field - Technician
Services:
Provides radon detection and mitigation services, distributes radon test kits.
Qualifications:
NY state certified lab, EPA RMP.
Area Served:
NJ, NY, PA.

CARL HENRY AND COMPANY, INC.
Benson East Suite 207B
Jenkintown, PA 19046
(215) 576-7766
Carl Feldman - President
Dorothy Feldman - Secretary/Treasurer
Jack Feldman
Services:
Provides radon detection services using charcoal canisters, continuous monitors and grab sampling methods.
Qualifications:
EPA RMP, AARST.
Area Served:
DE, NJ, PA.
EPA RMP 5:
CC, CR(P).

CARLSON HEATING, COOLING, AND ELECTRIC COMPANY
3340 West Lake Ave.
Glenview, IL 60025-1297
(312) 729-0123
Ken Carlson
Area Served:
IL.
EPA RMP 5:
CC.

CARLSON INSPECTION ASSOCIATES, INC.
51 R Dudley St.
Arlington, MA 02174
(617) 641-0600
Warner Carlson - President
Services:
Sells radon at-home test kits.
Area Served:
MA.

CAROLINA RADON CONSULTANTS
P.O. Box 2000
Etowah, NC 28729
(704) 891-9559
Ned Doyle - President
Area Served:
GA, NC, SC.

CAROLINA RADON CONSULTANTS
P.O. Box 6683
Columbia, SC 29260
(803) 779-2043
Ned Doyle - President
Area Served:
GA, NC, SC.
EPA RMP 5:
CC.

CAROLINA RADON DETECTION
209 Tebblewood Dr.
Simpsonville, SC 29681
(803) 282-0494
Alan Huggins
Services:
Provides radon detection and mitigation services.
Qualifications:
AARST.
Area Served:
SC.

CARPENTER, CHARLES D.
13462 West Florida Dr.
Lakewood, CO 80228
(303) 988-6818
Charles D. Carpenter
Area Served:
CO.

CARROLL COUNTY RADON CONTROL, INC.
Box 027
Westminster, MD 21157
(301) 848-6324
Tim Kyle
Area Served:
MD, PA, VA.
EPA RMP 5:
CC.

CAVALIER CORPORATION
11516 East Sprague
Spokane, WA 99206
(509) 926-6217
Warren Riddle
Sylvia Riddle
Services:
Provides radon detection and mitigation services specializing in real estate, commercial, and public structures, distributes Fantech products.
Qualifications:
EPA training courses.
Area Served:
ID, MT, OR, WA.
EPA RMP 5:
AT, CC.

CAVIN ANALYTICAL CONSULTANTS, INC.
2165K West Park Ct.
Stone Mountain, GA 30087
(404) 498-8295
Donald K. Cavin - Director
Richard H. Payne - Manager
Services:
Provides radon detection services by charcoal canister, grab sampling, and working level measurement.
Qualifications:
NRA, EPA RMP.
Area Served:
Nationwide.
EPA RMP 5:
CC(P), GR(P).

CCI-RADON AND HOME INSPECTION/RADON DETECTION
P.O. Box 1045
Malvern, PA 19355
(215) 889-1554, (215) 363-9015
Charles E. Weber - President
Bruce Jordan - Vice President
Norman Warner - Vice President
Services:
Provides radon detection and mitigation services for home owners and buyers, distributes charcoal canisters and alpha track detectors to retail outlets.
Qualifications:
AARST, EPA RMP PA DER certification.
Area Served:
DE, NJ, PA.
EPA RMP 5:
CC.

CDM FEDERAL PROGRAMS CORPORATION
13135 Lee Jackson Memorial Hwy.
Fairfax, VA 22033
(703) 968-0900
David S. Rosa
Daniel Gilroy
Services:
Non-commercial laboratory for research purposes or state agencies' assistance only.
Area Served:
Nationwide.
EPA RMP 5:
AT, CC, CR(P), GR(P).
CENTARUS CORPORATION
P.O. Box 513
Glasgow, KY 42141
(502) 651-2396
G. Wayne Calvert
Area Served:
Nationwide.
EPA RMP 5:
CC.

CENTER FOR RADON DETECTION
12 Bank St.
Summit, NJ 07901
(201) 277-3200, (201) 992-9702
Rebecca Kirschbaum
John R. Blazo
Area Served:
NJ.
EPA RMP 5:
AT, CC.

CENTER FOR RADON SERVICES
P.O. Box 2250
Cherry Hill, NJ 08034
(609) 424-7191
Michael S. Stocknoff
Paul R. Allen
Services:
Provides radon detection and consulting services.
Qualifications:
Licensed professional engineers on staff, ANS, Health Physics Society.
Area Served:
NJ.
EPA RMP 5:
AT, CC.

CENTRAL HOME INSPECTION, INC.
435 Rte. 18
East Brunswick, NJ 08816
(201) 545-0300
Sam Mayer - President
Services:
Sells radon at-home test kits.
Area Served:
NJ.

CENTRAL JERSEY RADON, INC.
P.O. Box 600
Lincroft, NJ 07738
(201) 741-0400
John A. Murray - President
Services:
Provides radon detection and mitigation services and consultations.
Area Served:
NJ.

CENTRAL RADON TESTING
435 Rte. 18
East Brunswick, NJ 08816
(201) 545-0300
Sam Mayer
Area Served:
NJ.
EPA RMP 5:
CC.

CERT WATER SERVICES, INC.
25 Duval Ln.
Plainville, CT 06062
(203) 793-9142
David Gworek - President
Services:
Provides radon detection services using charcoal canister, alpha track , and water testing methods for residential, commercial and industrial structures, distributes continuous radon monitors.
Qualifications:
President is a registered professional engineer.
Area Served:
CT.
EPA RMP 5:
AT, CC.

CERTIFIED TESTING SERVICES, INC.
474 Easton Rd.
Horsham, PA 19044
(215) 672-8100
James Bucciarelli - President
David Delavan - Mitigation Specialist
Bernard J. Salera - Executive Secretary
Richard Salera
Services:
Provides radon detection and mitigation services, specializes in real estate transactions.
Qualifications:
Full compliance with PA DER interim requirements for certification.
Area Served:
PA.

CERTUS LABORATORIES
P.O. Box 105633
Atlanta, GA 30348
(404) 762-5411
Thomas R. Stephenson - President
Services:
Primary radon testing laboratory for charcoal canisters, educational seminars ongoing.
Qualifications:
EPA RMP, AARST, AGSE.
Area Served:
Nationwide.
EPA RMP 5:
CC(P).

CHAPPARAL ENTERPRISES, INC.
P.O. Box 601
Elberton, GA 30635
(404) 283-7611
R. Jack Spivey
Area Served:
GA.

CHEM-SAFE LABORATORIES
North 131 Grand Ave. P.O. Box 546
Pullman, WA 99163-0546
(509) 334-0922
Cliff Cooper - General Manager
Services:
Manufactures and distributes charcoal canisters and alpha track detectors for the indoor environment, operates a full service laboratory.
Qualifications:
EPA RMP, Institute for Environmental Sciences.
Area Served:
Nationwide.
EPA RMP 5:
CC(P).

See Ad Page 21

CHENANGO ENVIRONMENTAL LAB, INC.
35 State St.
Binghamton, NY 13901
(607) 723-8175
Tim Sawyers - Lab Director
Services:
Provides radon consulting services.
Area Served:
Nationwide.

CHEROKEE DE-WATERING, INC.
Rte. 4, Oak Hill Cr.
Canton, GA 30114
(404) 928-5091
Jackie McCollum
Area Served:
GA.

CHESTER COUNTY DEVELOPMENT SERVICE
420 West Market St.
West Chester, PA 19382

(215) 692-4509
Richard Marion - Vice President
Area Served:
PA.

CHIEF HOME AND BUILDING INSPECTION SERVICES
Box 284
Towaco, NJ 07082
(201) 299-9087
Richard D. Conklin - President
Services:
Provides radon detection services.
Qualifications:
AARST, ASHI.
Area Served:
NJ, NY, PA.
EPA RMP 5:
CC.

CHILTON ASSOCIATES (RADON GROUP)
144 Valley Rd.
Haworth, NJ 07641
(201) 385-1660
George E. Chilton
Area Served:
NJ, NY, PA.
EPA RMP 5:
CC(P).

CLARK UNIVERSITY/ENVIRONMENT, TECHNOLOGY, AND SOCIETY PROGRAM/DEPARTMENT OF PHYSICS
950 Main St.
Worcester, MA 01610
(617)793-7687
Robert L. Goble - Associate Professor
Cristoph Hohenemser - Chairman
Halina Brown - Associate Professor
Services:
Analyzes distribution of indoor radon exposure and radon epidemiology (possibly expanding to environmental epidemiology). Performs some radon measurements and experiments with measurement devices utilizing physics laboratory.
Area Served:
MA.

CLASSIC RENOVATIONS AND BUILDERS
6762 Riverview Rd.
Peninsula, OH 44264
(216) 468-3119
Bill Rickenbacher
Judy Rickenbacher
Services:
Provides radon detection services, distributes radon test kits to real estate industry and retail markets.
Qualifications:
AARST.
Area Served:
OH.

CLEAN AIR ENGINEERING, INC.
207 North Woodwork Ln.
Palatine, IL 60067
(312) 991-3300
Ann Marie Ramsden
Area Served:
Nationwide.
EPA RMP 5:
CC.

CLEAN HARBORS ANALYTICAL SERVICES, INC.
P.O. Box 327 325 Wood Rd.
Braintree, MA 02184
(617) 849-6070
Richard Fix
Services:
Provides radon detection services and operates a primary laboratory.
Area Served:
MA.

CMT, INC.
2813 Rio Vista
Emporia, KS 66801
(316) 342-2760
Michael M. Nichols
Area Served:
Nationwide.
EPA RMP 5:
CC.

COLORADO FRONT RANGE BUILDING INSPECTION SERVICE
P.O. Box 9614
Ft. Collins, CO 80525
(303) 482-1976, (800) 248-1976
Willliam Warren - President
William V. Merker - Professional Engineer
Services:
Provides radon detection services using continuous working level, charcoal canister, and alpha track methods for commercial and residential buildings.
Qualifications:
ASHI
Area Served:
CO, NM, WY.

COLORADO HOME PROTECTION
6318 South Gallop Ct.
Littleton, CO 80120
(303) 797-2007
J. Keith Brown

Area Served:
CO.

COLORADO RADON DETECTION COMPANY
8824 West Ontario Ave.
Littleton, CO 80123
(303) 973-3261
Ronald Deerwater
Area Served:
CO.

COLORADO RADON ENGINEERING, INC.
5104 Greenview Ct.
Ft. Collins, CO 80525
(303) 223-9755
Bill Alexander - President
Jamie Alexander - Vice President
Services:
Provides radon detection services, home inspections, and mitigation consultation services.
Qualifications:
EPA RMP.
Area Served:
CO, WY.
EPA RMP 5:
CC.

COLORADO RADON INSTITUTE
4675 East Louisiana Suite 114
Denver, CO 80222
(303) 756-0701
Robert W. Goard
Area Served:
CO.

COLORADO REAL ESTATE INSPECTION SERVICES
P.O. Box 24962
Denver, CO 80224-0962
(303) 753-0543
R.A. Campbell - President
Services:
Provides radon detection services for residential and commercial properties using the charcoal canister method.
Area Served:
CO.

COLORADO SAFE-N-DRY, INC.
7950 East Prentice Ave. Suite 201
Englewood, CO 80111
(303) 771-2212
Charles Austin - President
Services:
Provides radon detection and mitigation services.
Area Served:
CO.

COLORADO VENTURE MANAGEMENT, INC.
2995 Wilderness Pl.
Boulder, CO 80301
(303) 440-4055
Kyle Lefkoff - General Manager
Services:
Provides radon detection services for residential and commercial properties using the charcoal canister method, provides radon mitigation services.
Area Served:
Nationwide.

COLUMBIA RADIOLOGICAL TESTING LABORATORY
4054 Victoria Ct.
Columbia, MO 65201
(314) 443-3538, (314) 442-6080
William E. Becker
M. Milewski
Area Served:
Nationwide.
EPA RMP 5:
AT, CC(P).

COMFORT SYSTEMS
4100 Nebraska Ave.
Nashville, TN 37209
(615) 298-1855
Randy Robinson
Area Served:
TN.
EPA RMP 5:
CC.

COMMONWEALTH ENGINEERING AND TECHNOLOGY
3113 North Front St.
Harrisburg, PA 17110
(717) 232-2288
Judy F. Musselman
Area Served:
CT, DC, DE, FL, GA, IA, IL, IN, KY, MA, MD, ME, MI, NC, NH, NJ, NY, OH, PA, RI, SC, TN, VT, VA, WV.
EPA RMP 5:
CC.

COMMONWEALTH ENTERPRISES, INC.
P.O. Box S187
St. Davids, PA 19087
(215) 688-7097
William DiAddezio - President
Doris Natale
Services:
Provides radon detection and mitigation services to property owners and buyers privately and through realtors.
Qualifications:
EPA detection and mitigation courses.
Area Served:
PA.

COMMONWEALTH RADON SPECIALISTS
2204 Grundy Rd.
Woodbridge, VA 22191
(703) 494-6241
Robert H. Adkins
Area Served:
Nationwide.
EPA RMP 5:
CC.

COMMUNITY HEALTH ASSOCIATES, INC.
316 Sherman St.
Canton, MA 02021
(617) 828-3123
Thomas Gecewizcz - President
William Thibeault - Vice President
Services:
Provides radon detection and mitigation services.
Area Served:
MA.

COMMUNITY RADON TESTING/ DIVISION OF SANTIAGO ENTERPRISES
4702 Kilburn Ct.
Woodbridge, VA 22193
(703) 690-4282, (800) 451-8164
J.A. Santiago
Area Served:
Nationwide.
EPA RMP 5:
AT, CC.

COMPREHENSIVE MANAGEMENT SERVICES
650 Division St.
Harrisburg, PA 17110
(717) 238-1010
Area Served:
PA.

COMPREHENSIVE RADON SERVICES, INC.
P.O. Box 548
Manchester, NH 03105
(603) 644-7301
Mark Malbon - President
Services:
Provides radon detection services using charcoal canister, alpha track, continuous radon, grab sampling, working level methods, water analysis, and open land testing, provides radon mitigation services, distributes radon mitigation supplies.

Area Served:
NH.
EPA RMP 5:
CC, CR(P), GR(P).

CON-TEST, INC.
126 Shaker Rd.
East Longmeadow, MA 01028
(413) 525-1198
David A. MacLean
Area Served:
Nationwide.
EPA RMP 5:
GW(P).

CONCEPT 90, INC.
12843 West Jewell Dr.
Lakewood, CO 80228
(303) 985-5355
Pramila Sinha
Area Served:
CO.

CONNECTICUT BASEMENT DE-WATERING AND RADON SYSTEMS, INC.
1 Clayton St.
Milford, CT 06460
(203) 874-3491
Larry Janeski - President
Services:
Provides radon detection and mitigation services.
Area Served:
CT.
EPA RMP 5:
CC.

CONNECTICUT ENVIRONMENTAL TESTING
6 East Main St.
Westport, CT 06880
(203) 454-1544
Craig Burry
Area Served:
CT, MA, NH, NY, RI, VT.
EPA RMP 5:
CC.

CONSHOHOCKEN RADON TESTING SERVICES
476 West Elm St.
Conshohocken, PA 19428
(215) 828-2059, (215) 272-3694
Bruce Rakowski
Area Served:
PA.
EPA RMP 5:
CC.

CONSTRUCTION EVALUATION, INC.
102 Hickok Rd.
Greenfield Center, NY 12833
(518) 587-6366
Rich D. Martin - President
Services:
Provides radon detection services.
Area Served:
NY.

CONSULTING CHEMISTS
3501 Lakeshore Dr.
Kingsport, TN 37663
(615) 239-5329, (615) 477-8253
Cephas H. Sloan
Robert C. Mu
Area Served:
NC, TN, VA.
EPA RMP 5:
CC.

CONSUMER ENERGY SERVICES, INC.
12 Westerville Sq. Suite 200
Westerville, OH 43081
(614) 861-0600
Stephen R. Helmuth - President
Services:
Provides radon detection services using continuous monitor method, provides radon mitigation services for the real estate industry and residential markets.
Area Served:
OH.
EPA RMP 5:
CC, CR(P).

CONTEK/CORPORATE HEADQUARTERS
P.O. Box 3065
Newport Beach, CA 92663
(714) 759-9226
David Hardman - Vice President
Services:
Provides radon detection services.
Qualifications:
NRA.
Area Served:
Nationwide.

CONTINENTAL ENVIRONMENT COMPANY, INC.
34 Maple St.
Summit, NJ 07901
(201) 277-2255
Alvin M. Natkin - President
Services:
Provides radon detection and consultation services using charcoal canister, alpha-track detectors, and E-Perm for private and public sectors, real estate transactions, provides radon mitigation services.
Qualifications:
EPA RMP, AARST, ACS, APCA, listed with NJ DEP for testing and mitigation.
Area Served:
CT, DC, DE, FL, MD, NJ, NY, PA, TX, VA.
EPA RMP 5:
AT, CC, EP(P).

CORE LABORATORIES, INC.
420 West First St.
Casper, WY 82601
(307) 235-5741, (800) 666-0306
David L. Demerest - Supervisor
Services:
Provides radon detection services to homeowners.
Qualifications:
Experience in radon testing at uranium mill tailing sites and EPA drinking water testing standards.
Area Served:
Nationwide.

CORNELL COLLEGE
First St. West
Mt. Vernon, IA 52314
(319) 895-8811
Conrad Weiffenbach
Services:
Non-commercial laboratory for research purposes or state agencies' assistance only.
Area Served:
IA.
EPA RMP 5:
AT.

CORNERSTONES
1770 Central St.
Yorktown Heights, NY 10598
(914) 962-7037
Bruce Barber - President
Services:
Provides radon detection services.
Area Served:
NY.

COUNTY COLLEGE OF MORRIS/CHEMISTRY DEPARTMENT
Rte. 10 and Center Grove Rd. A-304
Randolph, NJ 07869
(201) 361-5000
Samuel Helf - Adjunct Professor
Area Served:
NJ.
EPA RMP 5:
CC(P).

COUNTY HOME INSPECTIONS
P.O. Box 391
Wilton, CT 06897
(203) 762-2608

Paul T. Bonamo
Area Served:
CT.
EPA RMP 5:
CC.

COURTNEY CONSULTANTS, INC.
520 Carriage Dr.
Atlanta, GA 30328
(404) 256-2487
F.E. Courtney - President
Peter Courtney - Scientist
Services:
Provides radon detection services using charcoal canister method.
Qualifications:
NRA.
Area Served:
GA.

CRIPPEN CONSULTING ASSOCIATES, INC.
242 North James St. Suite 202
Wilmington, DE 19804
(302) 994-8000
Lee W. Thomas - Vice President
William D. Garlington - Vice President
Services:
Provides radon detection, mitigation and consultation services for residential, commercial, and public buildings.
Qualifications:
Licensed professional engineers and radon certified consultants.
Area Served:
Nationwide.

CTL ENVIRONMENTAL SERVICES
2905 East Century Blvd.
South Gate, CA 90280
(213) 564-2641
Rafik Vartanian
Area Served:
CA.

CTL/THOMPSON, INC.
1971 West 12th Ave.
Denver, CO 80204
(303) 825-0777
Ron McOmber - President
Area Served:
CO.

CUMBERLAND ANALYTICAL LABS, INC.
56 North 2nd St.
Chambersburg, PA 17201
(717) 263-5943
Donald R. Richner, Jr.
Area Served:
PA.
EPA RMP 5:
EP(P).

D

D-TECH, INC.
7 Laurence Pl.
Plymouth Meeting, PA 19462
(215) 825-4832
C. Wilson Kile, Jr.
Area Served:
PA.

DAMES AND MOORE
7101 Wisconsin Ave. Suite 700
Bethesda, MD 20814
(301) 652-2215
Gary Mayer - Principal
Tracey Brak - Auditor
Jack Parrott
Services:
Provides radon detection services including sales analyses as part of comprehensive environmental audits.
Area Served:
AL, CT, DC, DE, FL, GA, IL, IN, KY, LA, MA, MD, ME, MI, MN, MS, NC, NH, NJ, NY, OH, PA, PR, RI, SC, TN, VI, VT, VA, WI, WV.
EPA RMP 5:
CC.

DAN HIMMELFARB COMPANY
7416 Chrome Mine Rd.
Gaithersburg, MD 20879
(301) 744-4004
Daniel J. Himmelfarb
Area Served:
MD.

DAVID A. MASE REAL ESTATE
6515 Main St.
Trumbull, CT 06611
(203) 261-8080
David A. Mase - President
Services:
Provides at-home radon test kits.
Area Served:
CT.

DAVID E. WOLFE CONSULTANTS, INC.
6728 Memorial Hwy. Suite 255
Tampa, FL 33615
(813) 885-5887, (813) 887-8164
David E. Wolfe - President
Services:
Provides radon detection services using charcoal canister, grab sample, and alpha track methods.
Area Served:
FL.

DEAN ARRINGTON ASSOCIATES
89A Haytown Rd.
Lebanon, NJ 08833
(201) 236-9697
Dean Arrington
Area Served:
NJ.
EPA RMP 5:
CC.

DEGENSHEIN DENKER
10 South Broadway
Nyack, NY 10960
Stuart G. Gerger
Area Served:
NY.

DELAWARE VALLEY GEOTECHNICAL ASSOCIATES
P.O. Box 98
Washington Crossing, PA 18977
(215) 357-0907
Rocky Richards - President
Area Served:
PA.

DELAWARE VALLEY RADON INSPECTIONS/ARSENAL BUSINESS CENTER
5301 Tacony St. Box 221
Philadelphia, PA 19137-2309
(215) 289-3300, (215) 947-1169
Ralph R. Freed
Area Served:
NJ, PA.
EPA RMP 5:
CC.

DEMAR PEST CONTROL COMPANY
3522 Kensington Ave.
Philadelphia, PA 19134
(215) 744-0336
E. David DeMar
Area Served:
PA.
EPA RMP 5:
AT, CC.

DESERT-DRY WATERPROOFING
904 Rydal Rd.
Rydal, PA 19046
(215) 886-6140, (703) 490-4010
H. Edward Reiff
Area Served:
PA.

DEVELOPMENT SERVICES, LTD.
420 West Market St.

West Chester, PA 19382
(215) 692-4509
Richard Merion - Vice President
Services:
Provides radon detection services for residential and commercial properties using the charcoal canister method.
Qualifications:
PA license, registered sanatarian..
Area Served:
PA, NJ.

DIAGNOSTICS AND DEVICES, INC.
1201 Mt. Kemble Ave.
Morristown, NJ 07960
(201) 766-7111
J.G. Murnick - President
Services:
Distributes radon test kits for radon in air or water.
Area Served:
DE, MD, NJ, NY, PA.
EPA RMP 5:
CC.

DIAMOND BACK SERVICES, INC.
7396 Reynolds Dr.
Sedalia, CO 80135
(303) 791-1984
Richard Rocquemore
Larry Wujcikowski
Area Served:
CO.

DIAMOND REALTY INSPECTIONS, INC.
211 Monocacy Creek Rd.
Birdsboro, PA 19508
(215) 582-2070
Barry Diamond - President
Services:
Provides radon detection services.
Area Served:
PA.

DITTMAR INSPECTION COMPANY, INC.
20 Redding Pl.
Towaco, NJ 07082
(201) 263-4745
Willy C. Dittmar
Area Served:
NJ.
EPA RMP 5:
CC.

DJ'S RADIATION PROTECTION
Route 2 Box 332B
St. Pauls, NC 28384
(919) 425-6680
Debra J. Jacobs
Area Served:
NC.

DK ROCKWELL
101 East Cornell St.
Somerdale, NJ 08083
(609) 627-2166
Frank J. Tomasetti
Area Served:
NJ, PA.
EPA RMP 5:
CC.

DMC ASSOCIATES
P.O. Box 11
Bergenfield, NJ 07621
(201) 384-6060
Mary Ellen Collins
Services:
Provides radon detection services.
Qualifications:
AARST.
Area Served:
NJ.

DONALD WALKER COMPANY
60591 Mount Vernon
Rochester, MI 48064
(313) 652-1504
Donald G. Walker, Jr.
Area Served:
MI.

DOUGLAS MARTIN AND ASSOCIATES, INC./DMA ANALYSIS GROUP
520 Chestnut St.
Emmaus, PA 18049
(215) 965-2914
JoAnne D. Martin - Senior Scientist
Ken Smith - Health and Safety Services
Richard Martin - President
Services:
Nuclear analysis laboratory provides radon detection services, manufactures and provides gamma spectroscopy analysis of four inch EPA-specification canisters, water analysis.
Qualifications:
AARST, EPA RMP, licensed in NJ, PA, interim/voluntary paticipant in NJ programs.
Area Served:
Nationwide.
EPA RMP 5:
CC(P).

See Ad Page 199

DPR BUILDERS
21 Stacy Dr.
Belle Mead, NJ 08502
(201) 874-3419
Ralph Viola - President
Services:
Provides radon diagnostic, consultation, mitigation (pre- or post-construction), and detection services using charcoal canister method.
Qualifications:
AARST, NJ DEP, licensed electrical contractor, graduate construction engineer.
Area Served:
NJ, NY, PA, VA.

DR. RADON TESTING AND REDUCTION
687 North Cary St.
Brockton, MA 02402
(508) 583-2950
George A. Dyroff
Area Served:
MA.
EPA RMP 5:
CC.

DREWRY HOME INSPECTIONS, INC.
2625 Oakton Glen Dr.
Vienna, VA 22180
(703) 938-1044
David R. Drewry - President
Services:
Provides radon detection, mitigation, and consultation services.
Qualifications:
AARST, EPA training courses.
Area Served:
DC, VA.

DUDLEY, PAUL
140 Railroad Mills Rd.
Pittsford, NY 14534
(716) 385-2368
Paul Dudley - Environmental Consultant
Services:
Provides radon detection and consultation services for residential markets.
Area Served:
NY.

E

E-THREE, INC.
4454 Genesee St.
Buffalo, NY 14225
(716) 631-5858
Edward Nesselbeck - President
Services:
Provides radon detection services.
Area Served:
NY.

E.C. CHAMBERLAIN, RADIOLOGICAL CONSULTANT
3304 Susileen Dr.

Reno, NV 89509
(702) 329-5638, (702) 323-8341
E. Carl Chamberlain
Area Served:
CA, ID, NV, OR, UT.
EPA RMP 5:
AT.

E.D.A., INC.
P.O. Box 4854
Austin, TX 78765
(512) 322-9013
Lorrie Brown
Area Served:
TX.

E.O. CHURCH, INC.
925 East 17th Ave.
Denver, CO 80218
(303) 832-9692
E.O. Church
Susan Jones
Area Served:
CO.
EPA RMP 5:
CC.

EASTERN ANALYTICAL SERVICES
4 Westchester Plaza
Elmsford, NY 10523
(914) 939-6992, (800) 327-0121
Adam Yablonsky - Industrial Hygienist
Paul Stascavage
Services:
Provides radon detection and diagnostic services to residential and commercial clients.
Qualifications:
EPA RMP.
Area Served:
Nationwide.
EPA RMP 5:
CC(P), CW(P).

EASTERN INTEGRATED SERVICES CORPORATION
3490 U.S. Rte. 1 Building 17-2
Princeton , NJ 08540
(609) 520-8344
W.A. Thompson - Co-President
R.A. Pettit - Co-President
Services:
Provides radon detection services.
Qualifications:
NJ DEP voluntary radon testing program.
Area Served:
NJ.
EPA RMP 5:
CC.

EBERLINE INSTRUMENT CORP.
P.O. Box 2108
Santa Fe, NM 87504-2108
(505) 471-3232
Eric L. Geiger - Marketing Manager
Wayne Gemmill - Sales Manager
Michael D. Shepherd - Sales Manager
Services:
Manufactures and distributes measuring instruments for radon concentrations and working levels.
Area Served:
Nationwide.
EPA RMP 5:
CR(P), CW(P).

See Ad Inside Back Cover

ECI
2955 Valmont Suite 300
Boulder, CO 80301
(303) 442-8995
Kevin O'Hornett
Area Served:
Nationwide.
EPA RMP 5:
CC.

ECM ENVIRONMENTAL CONSULTANTS
7425 Hawkins Dr.
Hanover, MD 21076
(301) 761-1136
Ernest C. Michaelson
Services:
Provides radon detection and mitigation services.
Qualifications:
AARST.
Area Served:
MD.

ECODEX
1 Aaron Rd.
Lexington, MA 02173
(617) 862-3440
Steven Wightman - Operations Chief
Margaret Wacks - Quality Control
Services:
Provides radon mitigation services, distributes radon home test kits.
Qualifications:
DOD certified in radiation monitoring, medical technology, AIHA.
Area Served:
Nationwide.
EPA RMP 5:
CC(P).

ECOLOGICAL INTERACTIONS, INC.
96 East Main St.
Rockaway, NJ 07866
(201) 625-7700, (800) 543-2116
F. Michael Fitzpatrick - Vice President
Bernadette E. Secallus - Director
Services:
Provides radon detection and mitigation services, distributes radon test kits.
Qualifications:
AARST board of directors, EPA RMP, NJ DEP approved testing and mitigation course.
Area Served:
CT, DE, FL, MD, NJ, NY, PA, VA.
EPA RMP 5:
AT, CC.

ECONOMY HEATING AND AIR CONDITIONING
519 C Pulaski Hwy.
Joppa, MD 21085
(301) 679-3000
Robert L. Allen
Area Served:
DE, MD, PA, VA, WV.
EPA RMP 5:
CC.

EDA INSTRUMENTS, INC.
9200 East Mineral Ave. Suite 370
Englewood, CO 80112
(303) 790-2541
Jerry McJunkin - General Manager
Services:
Manufactures and distributes radon detection instruments including grab radon, grab working level, continuous radon, continuous working level, and gamma scintillometers.
Area Served:
Nationwide.

EDINBORO SCIENTIFIC COMPANY
P.O. Box 589 Wetsell Ridge Rd.
Edinboro, PA 16412
(814) 734-4648
Thomas A. Walkiewicz - Vice President
Services:
Provides radon detection services, laboratory analysis, consultation in experimental nuclear/radiation physics, distributes charcoal canisters.
Qualifications:
EPA RMP.
Area Served:
Nationwide.
EPA RMP 5:
CC(P).

EES OF MINNESOTA, INC.
600 Fairview Ave., N.
St. Paul, MN 55104
(612) 646-7991
Ronald K. Nagle
Area Served:
IA, MN, ND, SD, WI.

EPA RMP 5:
AT.

EG & G ORTEC
100 Midland Rd.
Oak Ridge, TN 37831
(615) 482-4411, (800) 251-9750
Ken Knight
Services:
Manufactures and distributes radiation detection, measurement and analysis instruments and systems.
Area Served:
Nationwide.
See Ad Page 200
EPA RMP 5:
CC(P).

EGAL, INC.
P.O. Box 14983 11633 West 83 Terr.
Kansas City, KS 66215
(913) 541-1862
Toby B. Glyde - President
Paul Sabados - General Manager
Services:
Provides radon detection services using charcoal canister, continuous working level , and alpha track methods.
Qualifications:
AARST quality assurance program, EPA RMP.
Area Served:
KS, MO.
EPA RMP 5:
CC.

EKS RADTECH, INC.
1000 Herald Sq.
Ashton, PA 19014
(215) 358-9425, (800) 682-5378
John Rooks - General Manager
Services:
Manufactures and distributes humidity independent charcoal canister detectors.
Qualifications:
NJ, PA certified, EPA RMP.
Area Served:
Nationwide.
See Ad Page 201
EPA RMP 5:
CC(P), GR(P).

ELECTRO ANALYTICAL, INC.
7118 Industrial Park Blvd.
Mentor, OH 44060
(216) 951-3514
Joel M. Javitch
Munroe Brown
Joe Turner
Services:
Provides radon detection services.
Qualifications:
AIHA.
Area Served:
Nationwide.
EPA RMP 5:
CC(P).

ELECTRO MECHANICAL CONCEPTS
28 Woodruff Dr.
Carnegie, PA 15106-3054
(412) 276-2272
Matt Kovac
Area Served:
PA.

ELLIOTT AND ASSOCIATES
146 North Drexel Ave.
La Grange, IL 60525
(312) 354-1481
Elliott Wall - President
E. Hughes
Area Served:
IL, IN, MI, MO, MS, WI.
EPA RMP 5:
AT.

EMERGENCY RADON TESTING, INC.
29 Freeman St.
West Orange, NJ 07052
(201) 481-3992, (201) 674-1570
Jack Harvey - President
Services:
Provides radon detection services.
Area Served:
NJ.

EMMI-RADON CHECK LABORATORY
E-57 Rd. Box 93
Kelley, IA 50134-8724
(515) 769-2400
Kenneth D. Wiggers
Area Served:
IA.

EMPAC, INC.
25 Science Pk.
New Haven, CT 06511
(203) 786-5420, (800) 872-3662
Albert W. Doolittle, Jr.
Services:
Provides radon detection services, distributes radon detection products to commercial, municipal, and residential markets.
Qualifications:
Listed with CT DOH-Toxic Hazards.
Area Served:
CT, MA, ME, NH, NJ, NY, RI, VT.
EPA RMP 5:
AT, CC, CR(P), EP(P).

EMPIRE INSPECTION, INC.
1020 Springfield Ave.
Mountainside, NJ 07092
(201) 654-3111
Thomas J. Sanford - President
Services:
Provides radon detection services, 24-hour test for radon progeny, radon gas sampling, distributes charcoal canisters.
Area Served:
CT, NJ, NY, PA.
EPA RMP 5:
AT, CC, CW(P).

EN-THERM, INC.
509 West Fayette St.
Syracuse, NY 13204
(315) 474-6549
Richard Kornbluth - President
Services:
Provides radon detection and mitigation services.
Area Served:
NY.

ENCOSYS CORPORATION OF AMERICA
7202 East 87th St. Suite 109
Indianapolis, IN 46250
(317) 849-6060
Raymond P. Barker - President
Charles J. Lynn - Vice President
Zachary S. Taylor - Mgr. Env. Programs
Services:
Provides radon detection and consulting for real estate industry, private industry, governmental sector and residential markets.
Qualifications:
AARST, ARA, EPA RMP, certified health physicist on staff.
Area Served:
IL, IN, KY, MI, OH.

ENERGY ALTERNATIVES UNLIMITED, INC.
8384 Middlecrest Rd.
Boulder, CO 80302
(303) 444-6063
David Nyquist
Services:
Provides radon detection, mitigation and consultation services for disposal of toxic and hazardous materials.
Qualifications:
Charter member AARST.
Area Served:
CO, NV, UT, WY.

ENERGY CONSERVATION SERVICES
6290 North Vine Valley Rd.
Rushville, NY 14544
(716) 554-3994
Peter Duffy - Principal
Services:
Provides radon detection, mitigation, and

energy consultation services.
Area Served:
NY.

ENERGY INDEPENDENCE/THE RADON RESPONSE TEAM
9940 East 36th St.
Independence, MO 64052
(816) 353-3755
Kenneth C. Riead - President
Services:
Provides radon detection and mitigation services for residential and commercial markets, distributes Vanee line of heat recovery ventilation systems.
Qualifications:
EPA RMP, 12 years of experience in the field.
Area Served:
IL, KS, MN, MO, NE, WI.

ENERGY LABORATORIES, INC.
254 North Center St. P.O. Box 3258
Casper, WY 82602
(307) 235-0515
Donald D. Phillips
Area Served:
AZ, CA, CO, ID, MT, ND, NE, NM, NV, OR, SD, TX, UT, WA, WY.
EPA RMP 5:
CC(P).

ENERGY MARKETING MIDWEST, INC.
East 57 Rd. Box 93
Kelley, IA 50134-8724
(515) 769-2400
Kenneth D. Wiggers
Area Served:
IA, IL, MN, NE, SD.
EPA RMP 5:
CC.

ENERGY PROFESSIONALS, INC.
387 Irwin St.
Lock Haven, PA 17745
(717) 748-5910, (717) 726-3825
Jay Crust - President
Mike Miller - Vice President
Jim Donley - Officer
Services:
Provides radon detection and mitigation services.
Qualifications:
PA state certified in testing and mitigation.
Area Served:
Nationwide.

ENERGY RESOURCE CENTER
P.O. Box 267
Colorado Springs, CO 80901
(719) 636-2447
Lee Salmon - Radon Specialist
Services:
Provides radon detection and mitigation services, project management, independent evaluation, presents seminars, distributes charcoal canisters and alpha tracks to homeowners.
Qualifications:
Graduate, Colorado School of Mines course on radon and indoor environment, 15 years of experience in energy management.
Area Served:
Nationwide.

ENERGY SYSTEMS, INC.
134 Carr Ave.
Cookeville, TN 38501
(615) 528-7247
Dwight G. Moore, Jr.
Area Served:
AL, KY, TN.
EPA RMP 5:
AT, CC, GR(P).

ENERGY UNLIMITED
RFD 1, Box 1090
Mercer, PA 16137
(412) 662-3591
Mitchell Moseley - Partner
Services:
Provides radon detection and mitigation services.
Area Served:
OH, PA.

ENGINEERING ASSOCIATES
4682 Watch Hill Rd.
Manlius, NY 13104
(315) 682-4389
David Miller - Principal
Services:
Provides radon detection and mitigation services.
Qualifications:
Licensed professional engineer, NYSERDA trained in radon detection and mitigation.
Area Served:
NY.

ENRAD, INC./RADON TESTING AND ENGINEERING, INC.
18705 B North Frederick Rd.
Gaithersburg, MD 20879
(301) 948-8040
Robert L. Watters - Manager
Charles L Osterberg
Services:
Primary laboratory analyses of activated charcoal for homeowners, commercial buildings, public buildings, and secondary radon testing companies.
Qualifications:
EPA RMP, certified health physicist.
Area Served:
Nationwide.
EPA RMP 5:
AT, CC(P).

ENSA, INC.
4550 Bailey Ave.
Buffalo, NY 14226
(716) 834-3111
Michael J. Dittenhauser - Senior Technician
Services:
Provides radon detection services.
Area Served:
NY.

ENTESCO
1910 Bobby Ray Rd.
Clover, SC 29710
(803) 222-4130
David Grigg
Area Served:
SC.

ENTROPY ENVIRONMENTALISTS, INC.
P.O. Box 12291
Research Triangle Park, NC 27709
(919) 781-3550
Robin R. Segall
Area Served:
Nationwide.
EPA RMP 5:
CC.

ENV SERVICES, INC.
216 Goddard Blvd. Suite 117
King of Prussia, PA 19406
(215) 337-8222, (800) 345-6094
Scott Berger
Area Served:
Nationwide.
EPA RMP 5:
AT, CC.

ENVIRADON, INC.
P.O. Box 13683
Research Triangle Park, NC 27709
(919) 848-2128
Brad McRee - President
Tim Owens
Sharon Kralick
Services:
Provides radon detection services for home.
Area Served:
NC, SC.

ENVIRADON, INC.
914 Rollingwood Dr.

Mt. Holly, NC 28120
(704) 827-1293
William B. McRee
Area Served:
NC, SC.
EPA RMP 5:
AT, CC.

ENVIRO SCIENCES, INC
1548 Valwood Pkwy. Suite 118
Carrollton, TX 75066
(214) 406-8100 (817) 792-3434
Michael L. Buvinghausen
Area Served:
Nationwide.
EPA RMP 5:
CC(P).

ENVIRO TECH SYSTEMS
430 Fernhill Ave.
Pittsburgh, PA 15226
(412) 561-2100
Jack Stanizzo - Owner
Services:
Provides radon detection and mitigation services, distributes radon mitigation products.
Qualifications:
PA DER, 450+ mitigations completed.
Area Served:
PA.

ENVIRO TECHNOLOGY CORPORATION
945 Spring Rd.
Pelham Manor, NY 10803
Terrence O'Hanlon
Area Served:
Nationwide.
EPA RMP 5:
CC.

ENVIRO TESTING
97 Greenridge St.
Johnson City, NY 13790
(607) 770-9098, (607)797-3665
Richard Tarnowski
Area Served:
NJ, NY, PA.
EPA RMP 5:
AT, CC.

ENVIRO-CARE RADON SERVICES
661 Wyoming St.
Buffalo, NY 14215
(716) 834-1788
Charles Mertz - President
Services:
Provides radon detection and mitigation services.
Area Served:
NY.

ENVIROHAZARDS MANAGEMENT, INC.
P.O. Box 1054
South Orange, NJ 07079
(201) 761-7226
Diane Trainor
Area Served:
Nationwide.
EPA RMP 5:
CC.

ENVIROHEALTH, INC.
1090 General Greene Rd.
Wayne, PA 19087
(215) 964-0668
R.M. Patterson - President
Services:
Provides consultation on radon detection, industrial hygiene, and air pollution technical and management services.
Qualifications:
Certified industrial hygienist, PA certified.
Area Served:
DE, NJ, PA.
EPA RMP 5:
CC.

ENVIRONMENTAL AIR CONTROL, INC.
2321 Northwest 30th Ct.
Fort Lauderdale, FL 33311
(305) 739-5159
Ray Narushko - President
Services:
Provides radon detection and mitigation services.
Area Served:
FL.

ENVIRONMENTAL ASPECS, INC.
1527 North Dale Mabry Suite 103
Lutz, FL 33549
(813) 948-1387
Tony Egitto - Secretary-Treasurer
Services:
Provides radon detection services using charcoal canister and alpha track methods.
Area Served:
FL.

ENVIRONMENTAL ASSOCIATES
1222 Fairfield Ave.
Bridgeport, CT 06605
(203) 368-6064, (203) 368-6032
Usha Maru
Area Served:
CT.

ENVIRONMENTAL AUDIT, INC.
1901 West 47th Pl. Suite 310
Westwood, KS 68205
(913) 384-0000
Charles H. Hajinian
Area Served:
Nationwide.
EPA RMP 5:
CC.

ENVIRONMENTAL CHEMISTRY, INC.
1233 Ogden Suite 504
Denver, CO 80218
(303) 831-0437
John Taylor - President
Susan Romeo - Vice President
Edward Shirley - Secretary
Services:
Provides radon detection services, distributes radon test kits.
Qualifications:
EPA RMP.
Area Served:
Nationwide.
EPA RMP 5:
CC(P).

ENVIRONMENTAL CONSULTANTS ASSOCIATES
14 Ramapo Ln.
Upper Saddle River, NJ 07458
(201) 934-9866
Richard W. Goodwin
Area Served:
CT, FL, NJ, NY, PA.
EPA RMP 5:
CC.

ENVIRONMENTAL CONSULTING LABORATORY
1353 Boston Post Rd. Suite 11
Madison, CT 06443
(203) 245-7039
David C. Barris - President
Services:
Environmental consulting and testing laboratory for industry, commercial businesses and residential buildings.
Qualifications:
EPA RMP, state-approved EPA laboratory.
Area Served:
Nationwide.
EPA RMP 5:
CC(P).

ENVIRONMENTAL CONTROL SYSTEMS, LTD.
15325 Jones Ln.
Gaithersburg, MD 20878
(301) 421-5150
Moe Vakili - Environmental Specialist

Jasmin Karim - Virologist
Services:
Provides radon detection services, environmental research on indoor air quality, radon detection and mitigation.
Qualifications:
ASHRAE, ASME.
Area Served:
MD, NY.

ENVIRONMENTAL CONTROL TECHNOLOGIES
400 Amherst St.
Nashua, NH 03063
(603) 881-7133
Harold W. Eriksen
Area Served:
NH.

ENVIRONMENTAL DATA COMPANY, INC.
2700 Youngfield
Lakewood, CO 80215
(303) 238-1104
Frederick Varani
Area Served:
CO.

ENVIRONMENTAL DETECTION AGENCY, INC.
P.O. Box 4854
Austin, TX 78765
(512) 322-9013
Lorrie Brown - President
Services:
Provides radon detection services for home inspectors, realtors, etc., distributes radon test kits, services in conjunction with the Texas A&M University radon program.
Area Served:
Nationwide.

ENVIRONMENTAL DETECTION ASSOCIATES, INC.
P.O. Box 1006 21 Middlemount Rd.
Pisgah Forest, NC 28768-1006
(704) 877-4300
Wolfgang Selle - President
Harry Hill - Vice President
Services:
Provides radon detection services, distributes radon detection devices, consultation services to architects. Non-commercial laboratory for research purposes or state agencies' assistance only.
Qualifications:
EPA RMP.
Area Served:
Nationwide.
EPA RMP 5:
AT.

ENVIRONMENTAL DETECTION MANAGEMENT
P.O. Box 12316
Research Triangle Park, NC 27709
(919) 469-8005
Wade A. Davis
Area Served:
Nationwide.
EPA RMP 5:
CC.

ENVIRONMENTAL DETECTION SERVICES
1046 West Peakview Ave.
Littleton, CO 80120
(303) 797-7771
Paul E. Johnson
Area Served:
CO.
EPA RMP 5:
CC.

ENVIRONMENTAL ENGINEERING AND TESTING
9185 East Kenyon Ave. Suite 190
Denver, CO 80237
(303) 722-2112
G.P. Logan
Area Served:
AZ, CA, CO, IL, IN, KS, MD, MO, MT, NJ, NV, OR, TX, UT, VA, WY.
EPA RMP 5:
CC.

ENVIRONMENTAL ENGINEERING LABORATORIES, INC.
P.O. Box 55304
Grand Junction, CO 81505
(303) 245-2500
Walid Bou-Matar
Area Served:
AR, CA, CO, MT, NM, NV, OR, UT, WA, WY.
EPA RMP 5:
CC(P).

ENVIRONMENTAL ENTERPRISES, INC.
P.O. Box 16847
Atlanta, GA 30321
(404) 766-5551
Chris Esworthy - President
Services:
Provides radon detection and mitigation services, distributes full line of radon mitigation products including fans, sealers, etc.
Area Served:
Nationwide.
EPA RMP 5:
CC.

ENVIRONMENTAL FIELD SERVICES, INC.
63 Elm St.
Topsham, ME 04086
(207) 725-4112
Theodore M. Reed - President
Michael D. Sauda - General Manager
Todd C. Hobson - General Manager
Valerie A. Caron
Services:
Provides radon detection and consulting services to residential and commercial clients.
Qualifications:
NRA.
Area Served:
CT, MA, ME, NH, NJ, NY, PA, VT.
EPA RMP 5:
EP(P).

ENVIRONMENTAL HAZARD CONTROL
590 Whitney Ave.
New Haven, CT 06511
(203) 787-1871
George Curry
Services:
Provides radon detection services for commercial and private properties.
Qualifications:
NRA.
Area Served:
CT.

ENVIRONMENTAL HEALTH AND SAFETY
P.O. Box 20052 Suite 211
Brandenton, FL 34206
(313) 242-1699
C.R. Hardy, Jr. - President
Area Served:
Nationwide.
EPA RMP 5:
CC.

ENVIRONMENTAL HEALTH CONSULTANTS, INC.
1505 North Prospect
Champaign, IL 61820
(217) 356-4797
M. Dehgan
Area Served:
Nationwide.
EPA RMP 5:
CC(P).

ENVIRONMENTAL HEALTH INSPECTIONS, INC.
1200 Rte. 9 Cross River Mill Suite 5E
Lakewood, NJ 08701
(201) 905-8888

Robert Gogats - President
Mark Guarino - Vice President
Services:
Specializes in mortgage commitment inspections and certifications, radon detection and consultation, well water sampling, inspection and certification.
Qualifications:
Partners in firm are NJ licensed health officers and sanitarians.
Area Served:
Nationwide.
EPA RMP 5:
CC.

ENVIRONMENTAL HEALTH TECHNOLOGIES
566 South Corona St.
Denver, CO 80209
(303) 778-1632
Jerel K. McElroy
Area Served:
Nationwide.
EPA RMP 5:
CC.

ENVIRONMENTAL HEALTH WATCH OF GREATER CLEVELAND
4115 Bridge Ave.
Cleveland, OH 44113
(216) 961-4646
Bob Staib - Education Director
Stuart Greenberg - Research Director
Services:
Previously known as Council on Hazardous Materials. Provides general information on radon hazards, products and services. Provides radon testing services (charcoal and continuous monitor) for the general public. Sponsors "Blueprint for a Healthy House" conference annually and is releasing "Healthy House Catalog" for information on indoor air pollutants.
Area Served:
OH.
EPA RMP 5:
CC, CR(P).

ENVIRONMENTAL HOME SERVICES, INC.
12354 Howard Lodge Dr.
Sykesville, MD 21784
(301) 442-5861
Charles K. Clopein - President
Frank Checkes - Vice President
Services:
Provides radon detection and mitigation services for realtors, builders and homeowners.
Qualifications:
EPA RMP, EPA training courses in radon mitigation methods.
Area Served:
MD.
EPA RMP 5:
CC.

ENVIRONMENTAL INDOOR AIR QUALITY CONSULTANTS, INC. (EIC)
58 Keller Ave.
Lancaster, PA 17601
(717) 393-5774, (800) 367-7223
William H. Miller - Radon Manager
H. Edward Carr, Jr.
Services:
Provides radon detection and mitigation services using ventilation equipment.
Area Served:
PA.

ENVIRONMENTAL MANAGEMENT CONTROL, INC.
2408 Belknap Beach
Louisville, KY 40259
(502) 228-0284
Pat Masterson - President
Area Served:
KY.

ENVIRONMENTAL MANAGEMENT GROUP
9841 Broken Land Pkwy. #117
Columbia, MD 21046
(301) 290-7078
Michael E. Logsdon
Area Served:
AL, DC, DE, KY, MD, NC, NJ, NY, OH, PA, SC, TN, VA, WV.
EPA RMP 5:
CC.

ENVIRONMENTAL OIL, INC.
P.O. Box 315
Syracuse, NY 13209
(315) 471-0503, (800) 262-1012
Daniel Ours - Environmental Technician
Area Served:
CT, MA, NJ, NY, OH, PA, VT.
EPA RMP 5:
AT.

ENVIRONMENTAL POLLUTION CONTROL
195 Cliffwood Dr.
Atlanta, GA 30328
(400) 425-6454
Eugene A. Bryson, Sr.
Area Served:
GA.

ENVIRONMENTAL PRO-TECH SERVICES, INC.
13 Edgar Dr. P.O. Box 599
West Milford, NJ 07480
(201) 728-2666
Peter Dominski - President
Services:
Provides radon detection services and operates a radon testing laboratory.
Area Served:
NJ.

ENVIRONMENTAL PRODUCTS ALLIANCE, LTD.
218 North 4th St., Suite 21 P.O. Box 2156
Bismarck, ND 58502
(701) 258-5050
Dan Schmaltz - Director of Marketing
Services:
Distributes charcoal canister radon test kits.
Area Served:
Nationwide.

ENVIRONMENTAL QUALITY ASSOCIATES, INC.
308 Harrison St.
Fulton, NY 13069
(315) 593-2085, (800) 727-2366
George J. Vargo - President
John A. Solini - Vice President
Services:
Provides radon detection services using charcoal canister and alpha track methods.
Qualifications:
NYS Department of Health approved environmental lab, certified health physicist on staff.
Area Served:
Nationwide.
EPA RMP 5:
CC(P).

ENVIRONMENTAL QUALITY CONTROL, INC.
2130 South Sawtelle Ave.
Los Angeles, CA 90025
(213) 478-6668
Andreas Marx
Area Served:
CA.
EPA RMP 5:
AT, CC.

ENVIRONMENTAL RADIATION SERVICES, INC.
305 West Magnolia St. Suite 373
Fort Collins, CO 80521
(303) 226-5530
Robert V. Huff - President
Philip D. Kearney - Vice President
Services:

Provides radon detection services using charcoal canister method for the real estate industry and home buyers.
Qualifications:
EPA RMP.
Area Served:
Nationwide.
EPA RMP 5:
CC(P).

ENVIRONMENTAL RADIOACTIVITY MEASUREMENTS, INC.
P.O. Box 226
Piermont, NY 10968
(914) 534-8679, (201) 768-7651
Guy Mathieu
Katherine Ellins
William Smethie
Services:
Provides laboratory analyses of air samples for radon (using 3-day integrated samples and grab samples), and water analysis for radon and radium for home owners and building inspectors.
Qualifications:
EPA RMP, AARST.
Area Served:
CT, MD, NJ, NY.
EPA RMP 5:
CR(P), GR(P).

ENVIRONMENTAL RADON MEASUREMENTS
49 North Orange Ave.
Orlando, FL 32801
(407) 648-1969, (800) 237-3065
Terrie Keene
Paul Pagello
Mark Riffe
Area Served:
Nationwide.
EPA RMP 5:
CC(P).

ENVIRONMENTAL RESEARCH ASSOCIATES
P.O. Box 219 490 Darby-Paoli Rd.
Villanova, PA 19085
(215) 687-5629
Michael H. Levin - Director
Services:
Provides radon detection services, research and consulting to real estate, trade, and legal professionals.
Qualifications:
Services since 1970.
Area Served:
Nationwide.

ENVIRONMENTAL RISK CONSULTANTS, INC.
800 Cluster
Evanston, IL 60202
(312) 864-1255
David Slutzky
Area Served:
IL.

ENVIRONMENTAL SAFETY PRODUCTS
8351 Roswell Rd. Suite 325
Atlanta, GA 30350
(404) 889-3117
June E. Peters
Services:
Provides radon detection services using alpha track method.
Area Served:
FL, GA.

ENVIRONMENTAL SCIENCE AND ENGINEERING, INC.
P.O. Box ESE
Gainesville, FL 32602-3053
(904) 332-3318
Christopher P. Amstutz
Les E. Sharpe
Area Served:
Nationwide.
EPA RMP 5:
CC(P).

ENVIRONMENTAL SCIENCE LABORATORY, INC.
26 Pearl St. Suite 110
Bellingham, MA 02019
(508) 966-4344, (800) 343-0707
Bob Juliano - President
Services:
Provides radon detection services using charcoal canister method, distributes radon test kits.
Qualifications:
EPA RMP.
Area Served:
Nationwide.
EPA RMP 5:
CC(P).

ENVIRONMENTAL SCIENCE SERVICES
235 Promenade St.
Providence, RI 02908
(401) 421-0398
Kathleen Grubb
Richard Hittinger
Area Served:
CT, MA, ME, NH, RI, VT.
EPA RMP 5:
CC.

ENVIRONMENTAL SERVICES AND TECHNOLOGY, INC.
22 Oakland Ave.
Atlantic Highlands, NJ 07716
(201) 291-5525
Allan Dean - Radiation Physicist
Services:
Provides radon detection and mitigation services serving residential, commercial, industrial, and government clients.
Qualifications:
On-staff radiation physicist, EPA RMP.
Area Served:
Nationwide.
EPA RMP 5:
AT, CC.

ENVIRONMENTAL SERVICES OF ILLINOIS, INC.
302 West State St.
O'Fallon, IL 62269
(618) 624-8079, (314) 421-4374
Stephen J. Willig - Air Quality Division
Douglas R. Wall - Vice President
Mark Warchol - Project Manager
G.R. Wall - Project Manager
Services:
Provides radon detection and mitigation services.
Qualifications:
IL Department of Safety voluntary certification program.
Area Served:
IL, MO.
EPA RMP 5:
CC.

ENVIRONMENTAL SOLUTIONS, INC.
P.O. Box 11323
St. Louis, MO 63105
(314) 863-9801
William M. Vaughan - Senior Scientist
Trudy Y. Smith - Technical Assistant
Services:
Provides consultation services on air quality and air toxins, distributes radon test kits.
Qualifications:
EPA RMP.
Area Served:
AR, IA, IL, IN, KS, KY, MO, NE, OH, OK, TN.
EPA RMP 5:
CC.

ENVIRONMENTAL SYSTEMS SERVICE
P.O. Box 512
Culpepper, VA 22701
(703) 825-6660
Cheryl B. Ludy
Pamela M. Pruett
Area Served:
CT, DC, DE, FL, GA, IA, IL, IN, KY, MA, MD, ME, MI, NC, NH, NJ, NY, OH, PA, RI, SC, TN, VA, VT, WV.
EPA RMP 5:
CC.

ENVIRONMENTAL SYSTEMS SERVICE, LTD.
P.O. Box 220
College Park, MD 20740
(301) 779-0606
G.D. Fleming
Area Served:
CT, DC, DE, FL, GA, IA, IL, IN, KY, MA, MD, ME, MI, NC, NH, NJ, NY, OH, PA, RI, SC, TN, VT, VA, WV.
EPA RMP 5:
CC.

ENVIRONMENTAL TECHNOLOGIES, INC.
P.O. Box 1300
Monticello, MN 55362
(612) 295-3008
Charles A. Stein
Area Served:
MN.

ENVIRONMENTAL TESTING
133 Defense Hwy. Suite 105
Annapolis, MD 21401
(301) 266-8700
Paul E. Wren
Area Served:
MD.

ENVIRONMENTAL TESTING AND CONSULTING
1316 Gress St.
Manville, NJ 08835
(201) 722-5293
Thomas R. Reilly - President
Services:
Provides radon detection (air and water) services and working level monitor and charcoal canister detectors for home buyers and home owners.
Qualifications:
NJ DEP, EPA RMP.
Area Served:
NJ.
EPA RMP 5:
CC, CW(P).

ENVIRONMENTAL TESTING ASSOCIATES
6021 Sandy Springs Cr. Suite 205
Atlanta, GA 30328
(404) 255-3884
David C. Perkins
Area Served:
Nationwide.
EPA RMP 5:
CC.

ENVIRONMENTAL TESTING ASSOCIATES
240 Harvest Ln.
Altoona, WI 54720
(715) 832-8585
Bill Steiger
Area Served:
Nationwide.
EPA RMP 5:
CC.

ENVIRONMENTAL TESTING ASSOCIATES
9 White Pine Trail
Madison, WI 53717
(608) 833-4073
Don Steiger - President
Services:
Provides radon detection and mitigation services.
Area Served:
Nationwide.
EPA RMP 5:
CC.

ENVIRONMENTAL TESTING ASSOCIATES
2201 East Ridge Rd.
Beloit, WI 53511
(608) 365-2774
Steve Boldt
Area Served:
Nationwide.
EPA RMP 5:
CC.

ENVIRONMENTAL 4TESTING ASSOCIATES, INC.
P.O. Box 142
Media, PA 19050
(215) 565-5665
Area Served:
PA.

ENVIROSCAPE, LTD.
4500 19th St., Suite 172
Boulder, CO 80302
(303) 444-9076
Edward Roth - President
Services:
Provides radon detection services for homes and open land.
Qualifications:
EPA RMP.
Area Served:
CO.

EPA RMP 5:
AT, CC.

ENVIROSCIENCE CONSULTANTS, INC.
66 Cedar St.
Newington, CT 06111
(203) 666-7167
Neal B. Freudan - President
Scott D. Herzog - Vice President
Gregg Stevens - Technician
Services:
Provides radon detection services, full service environmental engineering and industrial hygiene firm.
Qualifications:
Certified industrial hygienist, chemical engineer.
Area Served:
CT, MA, ME, NH, NJ, NY, RI, VT.

ENVIROSERV/MAIN OFFICE
15 Buckley Hill Rd.
Morristown, NJ 07960
(201) 285-1065
Donald Ulbrich - President
Services:
Provides radon detection and consultation services, provides laboratory analysis for other radon detection companies, distributes charcoal canisters.
Area Served:
Nationwide.
EPA RMP 5:
CC(P).

See Ad Page 35

ENVIROSERV/RADON DETECTION OF ALABAMA
751-D Raleigh Ct.
Birmingham, AL 35209
(205) 942-0756
Douglas Ulbrich - Manager
Betty Perez - Assistant Manager
Services:
Provides radon detection services, distributes mail-order charcoal canisters, placements for real estate companies.
Qualifications:
EPA RMP.
Area Served:
Nationwide.

ENVIROSERV/RADON DETECTION OF OHIO
781 Mount Vernon Ave.
Marion, OH 43302
(614) 382-2354
D. Todd Ulbrich
Services:
Provides radon detection services.
Area Served:
OH.

ENVIROTECH
132 Owsley Ave.
Lexington, KY 40502
(606) 269-4228
Joel Beatty
Area Served:
KY.
EPA RMP 5:
AT, CC.

ENVIROTECH SERVICES
201 South Paul
Springfield, IL 62703
(217) 753-1346
Mark E. Harris
Area Served:
IL.

ENVIROTEST
P.O. Box 223
North Ridgeville, OH 44039-0223
(216) 327-9432
Vincent A. Gargaro
Area Served:
Nationwide.
EPA RMP 5:
CC.

ERM, INC.
49 North Orange Ave.
Orlando, FL 32801
(407) 648-1969, (800) 237-3065
Wayne Dean - President
Alan Baker - Marketing Director
Paul Pagello - Laboratory Director
Services:
Provides radon detection and mitigation services to real estate and mortgage lending institutions.
Qualifications:
EPA RMP.
Area Served:
Nationwide.

ERNEST W. BLOCHINGER, INC.
2 Burnbrae Ln.
Sparta, NJ 07871
(201) 729-6638
Ernest W. Blochinger - President
Services:
Provides radon detection services using charcoal canister and grab sampling methods.
Area Served:
NJ.

ERO RESOURCES CORPORATION
2801 Youngfield Suite 121
Golden, CO 80401
(303) 231-9022
Mark DeHaven
Area Served:
CO.

ETE CONSULTING ENGINEERING, INC.
P.O. Box 3471
Oak Ridge, TN 37831
(615) 482-4503, (615) 546-0154
Paris R. Walker, Jr.
Area Served:
TN.

ETS ENVIRONMENTAL TESTING SERVICE
Rte. 2, Box 54
Mechanicsville, MD 20659
(301) 884-4489, (301) 475-2169
Daniel G. Bellere
Area Served:
DC, MD, VA.
EPA RMP 5:
AT, CC.

EVANSHIRE COMPANY, LTD., THE
Rte. 4, Box 57
Rome, NY 13440
(315) 865-5941, (800) 962-5941
Edwin Evans - President
Wade Evans - Vice President
Services:
Provides radon detection services using charcoal canister, alpha track detector, and grab sampling methods, provides radon mitigation services, distributes full line of radon detection and mitigation products, including sealers, sump covers, fans.
Qualifications:
Experience in state and federally funded projects.
Area Served:
Nationwide.

EXCEL INSPECTION SERVICE, INC.
438 Windmill Way
Somerville, NJ 08876
(201) 526-7277
Daniel Loreti
Area Served:
NJ.
EPA RMP 5:
CC.

EXECUTIVE HOME CONSULTANTS
19 Grove St.
Waldwick, NJ 07403
(201) 444-9571
Joseph R. Agner - Structural Engineer
Qualifications:
AARST.

Area Served:
NJ.

EXPEDITERS, INC., THE
404 North Summit Ave.
Oconomowoc, WI 53066
(414) 965-3063
Kay Muchka - Vice President
Robert Muchka - President
Services:
Provides charcoal canister radon detectors, detector placement, processing and written results of radon screening for homeowners.
Qualifications:
EPA RMP.
Area Served:
IL, MI, MN, WI.
EPA RMP 5:
CC.

F

FAYTEK, INC.
P.O. Box 2229
Coeur D'Alene, ID 83814
(208) 667-3263
Jan Elizabeth Fay
Area Served:
CA, CO, CT, FL, ID, MT, NJ, NV, NY, OR, PA, UT, WA, WY.
EPA RMP 5:
AT, CC.

FEDERAL NUCLIDE MONITORS
2023 Park Rd., NW
Washington, DC 20010
(202) 265-0429
Paul A. Danckwerth
Area Served:
CT, DC, FL, MA, MD, ME, NH, NJ, NY, VA.
EPA RMP 5:
CC.

FEDERAL RADON, INC.
206 Third St. P.O. Box 374
Cameron, MO 64429
(816) 632-1318
R.M. Yetter - President
Services:
Provides radon detection services, distributes charcoal canisters to public.
Area Served:
Nationwide.
EPA RMP 5:
CC.

FEDERAL RADON, INC.
93 Parmalee Ave.
Hawthorne, NJ 07506
(201) 427-9391
Ernest J. Revell
Area Served:
Nationwide.
EPA RMP 5:
CC(P).

FEMTO-TECH, INC.
150 C Industry Dr.
Carlisle, OH 45005
(513) 746-4427
E.I. Burgess - President
C.J. Kershner - Vice President
R. Manning - Director of Marketing
Services:
Manufactures and distributes continuous radon monitors.
Qualifications:
ARA
Area Served:
Nationwide.
EPA RMP 5:
CR(P), EP(P).

FINN CONSULTANT ENGINEERS, INC.
227 North St.
Dalton, MA 01226
(413) 684-2019
George Finn - President
Services:
Provides radon detection and mitigation services, consultation in mitigation design.
Area Served:
MA.

FINNEX (RADON)
23 North Ave.
Cincinnati, OH 45215
(513) 821-7019 (915) 757-2068
Pentti Toivonen
Area Served:
IN, KY, NM, OH, TX.
EPA RMP 5:
CC.

FLORIDA DEPARTMENT OF HEALTH AND REHABILITATIVE SERVICES/DEPUTY SECRETARY FOR HEALTH/OFFICE OF RADIATION CONTROL/ENVIRONMENTAL SECTION
P.O. Box 15490
Orlando, FL 32858
(407) 628-2274, (407) 297-2097
Benjamin Prewitt - Health Physicist
Harlan Keaton - Manager
Services:
Non-commercial laboratory for research purposes or state agencies' assistance only.
Area Served:
FL.

EPA RMP 5:
CC(P).

FLORIDA RADON CONSULTANTS, INC.
5448 Hoffner Ave. Suite 305
Orlando, FL 32812
(407) 281-0339
Leonard Bloomfield - President
Services:
Provides radon detection services using charcoal canister method.
Area Served:
FL.

FMT EQUIPMENT
P.O. Box 6198
Hamden, CT 06517
(203) 865-2864
E. Juhani Jaske
Area Served:
CT.
EPA RMP 5:
CR(P).

FORCHETTI PSI
P.O. Box 1234
Morrisville, PA 19067
(215) 493-1646, (609) 695-1229
Vince Forchetti - President
Area Served:
DE, MD, NJ, NY, PA.
EPA RMP 5:
CC.

FORENSIC ANALYTICAL SPECIALTY
745 Lincoln Ave. Suite 4
Alameda, CA 94501
(415) 887-8828, (415) 769-1180
William R. James
Area Served:
CA.

FORESIGHT ENGINEERING
P.O. Box 349
Pittstown, NJ 08867
(201) 996-3090
Charles J. Yanulevich - Manager
Services:
Provides radon detection using charcoal canisters and on-site working level monitors, provides radon mitigation services.
Qualifications:
AARST.
Area Served:
NJ.

FORESIGHT ENGINEERING, INC.
P.O. Box 3111
Wayne, NJ 07474-3111
(201) 942-3061
Ernie Borsellino - Engineer
Services:
Provides radon detection services using charcoal canisters and on-site working level monitors, provides radon mitigation services.
Area Served:
NJ.

FORESIGHT ENGINEERING, INC.
P.O. Box 362
Monasquan, NJ 08736
(201) 223-3316
Ken Lucas - Engineer
Services:
Provides radon detection and mitigation services for residential and commercial properties.
Area Served:
NJ.

FORESIGHT ENGINEERING, INC.
17 Cook Ave. P.O. Box 621
Madison, NJ 07940
(201) 377-0600
Christine Hammitt - President
F.B. Kaufmann - Vice President
T.D. Kaufmann - Treasurer
F.G. Hammitt - Secretary
Services:
Provides radon detection and mitigation services.
Qualifications:
AARST, EPA RMP, NJ professional engineers.
Area Served:
NJ, PA, VA.
EPA RMP 5:
CC.

FORESIGHT ENGINEERING, INC., VIRGINIA
P.O. Box 7333
Arlington, VA 22207-0333
(703) 525-4790
Steven Duarte
Services:
Provides radon detection services using charcoal canisters and working level monitors and mitigation services.
Area Served:
DC, MA, MD, RI, VA, WV.
EPA RMP 5:
CC.

FRANKLIN ENVIRONMENTAL ANALYSIS
716 North Franklin St. Box 1-R
Philadelphia, PA 19123-2721
(215) 592-8322
Cynthia Lawn - Lab Director
Joseph Solge - Quality Assurance
Services:
Provides radon detection services using the charcoal canister method.
Qualifications:
NJ, PA certification pending.
Area Served:
DE, NJ, PA.
EPA RMP 5:
CC(P).

FRANKS, TERRY M.
P.O. Box 5224
Louisville, KY 40205
(603) 437-1425
Terry M. Franks
Area Served:
KY.

FRAZIER COMPANY, THE
15106 Industrial Rd.
Omaha, NE 68144
(402) 330-3400
Dean Frazier - President
Area Served:
IA, NE.
EPA RMP 5:
CC.

FTLC SERVICES, INC.
P.O. Box 563
Amherst, OH 44001
(216) 960-2550
David Eastaugh
John Vegh
Nancy Eastaugh
Cindy Vegh
Qualifications:
AARST.
Area Served:
OH.

FUTURA MEDICAL SYSTEMS, INC.
387 Regency Ridge
Centerville, OH 45459
(513) 436-2215
E. Theodore Agard
Area Served:
OH.

G

G.A.M. BUILDING INSPECTION AND RADON MEASUREMENTS AND CONTROL, INC.
17223 East Purdue Pl.
Aurora, CO 80013
(303) 699-2827
George A. Michael - President
Services:
Provides radon detection and mitigation services for the real estate industry, home

buyers and sellers, and the general public.
Qualifications:
Licensed professional engineer.
Area Served:
CO.
EPA RMP 5:
AT, CC, GW(P).

G.S. TECHNOLOGIES, INC.
108 Hooten Rd.
Mt. Laurel, NJ 08054
(609) 234-2605
Glenn S. Schwartz
Area Served:
NJ.

GALLAGHER ENGINEERING AND INSPECTION
1164 Lincoln Heights Ave.
Ephrata, PA 17522
(717) 733-3734
Michael J. Gallagher
Area Served:
PA.

GALLIPOT, INC.
2401 Pilot Knob Rd.
St. Paul, MN 55120
(612) 681-9517, (800) 423-6967
Michael J. Jones - President
Services:
Manufactures and distributes charcoal canisters, distributes alpha track detectors, provides radon detection services
Area Served:
Nationwide.
EPA RMP 5:
CC(P).

GALSON TECHNICAL SERVICES
6601 Kirkville Rd.
East Syracuse, NY 13057
(315) 432-0506
Marcia Schwartz
Area Served:
NY.

GARCO RESEARCH COMPANY
144A Sutton Rd.
Lebanon, NJ 08833
(201) 832-7895
Gary J. Gilbert - Technical Director
Laslo Veradi - Safety Specialist
Garry M. Annibal - Env. Specialist
Services:
Provides radon detection and mitigation services for homeowners and businesses.
Qualifications:
Certified industrial hygienist on staff.
Area Served:
MD, NJ, NY, PA, VA.
EPA RMP 5:
CC.

GARDEN STATE HOME INSPECTION SERVICE
475 Valley Rd.
West Orange, NJ 07052
(201) 678-3333
Thomas Pasquariello
Area Served:
Nationwide.
EPA RMP 5:
CC.

GARDEN STATE RADON DEWATERING SERVICES, INC.
2 Edgewater Pl. Suite 1A
New Brunswick, NJ 08901
(201) 846-1948
Ron Weshnak - Owner
Area Served:
NJ.

GARDNER, PHILIP
136B Paoli Pike
Malvern, PA 19355
(215) 889-0239
Philip Gardner
Qualifications:
AARST.
Area Served:
PA.

GARNER CONSULTANTS INSPECTION SERVICE
1686 Fifth St.
Trenton, NJ 08638
(609) 883-2021
Larry Swirsky - Manager
Services:
Provides radon detection services using charcoal canister method.
Area Served:
NJ.

GASOMETRY ASSOCIATES
29 Pocahontas Trail
Medford, NJ 08055
(609) 983-3846
R.J. McDonald
Area Served:
NJ.

GAVIN, GERARD M.
214 East Township Line Rd.
Havertown, PA 19083
(215) 789-2718
Gerard M. Gavin
Area Served:
PA.

GEIER CONSTRUCTION COMPANY, INC./DIVISION OF RADON CONTROL
1506 Gridley Ln.
Silver Spring, MD 20902
(301) 649-2141
Robert J. Geier
Area Served:
DC, MD, PA, VA.
EPA RMP 5:
CC, EP(P).

GENE CALVERT AND ASSOCIATES
P.O. Box 217
Waxahachie, TX 75165
(214) 937-2223
Gene Calvert
Area Served:
TX.

GENERAL ENVIRONMENTAL MONITORING
3312 Commercial Ave.
Northbrook, IL 60062-1909
(312) 291-0772
Theodore Fields - Vice President
Services:
Provides radon detection services for the retail trade.
Area Served:
Nationwide.

GENERAL HEALTH PHYSICS, INC.
7217 Lockport Pl.
Lorton, VA 22079
(703) 550-7525
John B. Davis
Area Served:
Nationwide.
EPA RMP 5:
CC(P).

GENERAL PHYSICS CORPORATION
10650 Hickory Ridge Rd.
Columbia, MD 21044
(301) 290-2300, (301) 997-6241
Desmond Chan - Health Physics
Bruce Jacobs
Area Served:
Nationwide.
EPA RMP 5:
EP(P).

GENUS CORPORATION, THE
2400 Morris Ave.
Union, NJ 07083
(201) 687-5000
Fred A. Buro - President
Area Served:
NJ.

GEO-CHEK, INC.
2270 Stony Hill Rd.
Boulder, CO 80303
(303) 494-0576
Phil Herrington
Area Served:
CO, IA, MO.
EPA RMP 5:
CC.

GEO-ENVIRONMENTAL SERVICES, INC.
130 West Wieuca Rd. Suite 108
Atlanta, GA 30342
(404) 256-6437
Ion Petre
Area Served:
Nationwide.
EPA RMP 5:
CC(P), GW(P).

GEO-ENVIRONMENTAL SERVICES, INC.
444 North Frederick Ave. Suite L-148
Gaithersburg, MD 20877-2432
(301) 353-0338
John Razzolini
Area Served:
MD.
EPA RMP 5:
CC.

GEO-RADON SERVICES
P.O. Box 5985
Berkeley, CA 94705
(415) 843-8844
Barbara Moed - Owner
Services:
Provides radon detection and mitigation services, consultation and research on influence of geology and soil properties on indoor radon.
Qualifications:
Eight years of indoor radon research funded by DOE, EPA, BPA, masters' degree in geology, consultant to California radon survey.
Area Served:
Nationwide.

GEOCENTRIC ENGINEERING CORPORATION
P.O. Box 277
Lorton, VA 22070
(703) 550-9253
Paul W. Steinberger
Area Served:
DC, DE, MD, NC, NJ, NY, OH, PA, SC, VA, WV.
EPA RMP 5:
CC.

GEOCHEM, INC.
3200 New Seward Hwy. Suite 201
Anchorage, AK 99503
(907) 272-2121
Joseph Neubauer - President
David Neubauer - Vice President
Donald Gibson - Marketing
Services:
Provides radon detection and mitigation services.
Area Served:
AK.

GEOHAZARDS, INC.
P.O. Box 14956
Gainesville, FL 32604
(904) 371-7243
Douglas L. Smith - President
Services:
Provides radon detection services and consultation on radon mitigation options.
Area Served:
FL.

GEOMET TECHNOLOGIES, INC.
20251 Century Blvd.
Germantown, MD 20874
(301) 428-9898
Niren L. Nagda - Vice President
Bill Mountel - Marketing
Services:
Research on radon detection and mitigation technologies. Presently studying two homes in Maryland to analyze particular mitigation strategies. Large scale surveys in Florida and Massachusetts on charcoal canister and alpha track detection methods.
Area Served:
Nationwide.
EPA RMP 5:
AT.

GEORGIA RADON DETECTION COMPANY
3400 Elm Creek Dr.
Marietta, GA 30064-1132
(404) 429-0384
Fred March
Area Served:
Nationwide.
EPA RMP 5:
CC.

GERALD HARGROVE INSPECTION SERVICE
5308 Meadowlark Ln.
Grovetown, GA 30813
(404) 860-0797
Gerald Hargrove
Area Served:
GA.

GIDDINGS AND ASSOCIATES, INC.
3049 Enterprise Dr.
State College, PA 16801
(814) 238-5927
Todd Giddings
Area Served:
PA.

GITERMAN ASSOCIATES
2912 Xenwood
St. Louis Park, MN 55416
(612) 332-1129
Michael Giterman - President
Area Served:
OH.

GLOBAL BUILDING INSPECTORS
39 East 4th St.
Bayonne, NJ 07002
(201) 858-2386, (800) 321-4677
Robert Petrie
Area Served:
NJ.

GLOBAL RADON TESTING SERVICE TRAINING CENTER
163 Padgett Dr.
Lakeland, FL 33809
(813) 859-5891
Alexander Z. Kobryn - Training Director
Paul Jarrett - Executive Director
Services:
Regional training facility with seminars in radon business, specialist, technician and mitigation.
Qualifications:
State certified general contractor.
Area Served:
Nationwide.
EPA RMP 5:
CC, CR(P).

GLOBAL WHOLESALE
2410 East Lincolnway
Cheyenne, WY 82001
(307) 634-7318
Ron Langley - President
Services:
Provides radon detection services using charcoal canister, alpha track and Honeywell Professional Radon Monitors, distributes Foil Ray products, useful as radon barrier.
Area Served:
CO, NB, WY.

GLOGORE ENGINEERING COMPANY
3116 Stanley Rd.
Akron, OH 44313-3254
(216) 836-1880 (800) 702-9905
Robert Glogore

Area Served:
OH.
EPA RMP 5:
CC.

GODBEER, GODBEER AND ASSOCIATES
P.O. Box 128
New London, NH 03257
(603) 526-9235
Barrie C. Godbeer - President
Area Served:
NH.

GOLDEN CONSTRUCTION SERVICE
20 Distillery Rd.
Warwick, NY 10990
(914) 986-5600
Henry Goldman - Home Inspector
Area Served:
NY.

GOY, TERRY
1466 East Lake Dr.
Chaska, MN 55318
(612) 448-3664
Terry Goy
Qualifications:
AARST, ARA.
Area Served:
MN.

GRANITE STATE ANALYTICAL, INC.
16 Manning St. Unit 107
Derry, NH 03038
(603) 432-3044
Philip West - President
Services:
Distributor of Niton radon monitors.
Area Served:
ME.

GRANITE STATE RADON SERVICES
P.O. Box 864
Windham, NH 03087
(603) 437-1820
Duane Howes - President
Lewis Howes - Vice President
Services:
Provides radon detection services using charcoal canister method.
Qualifications:
EPA RMP.
Area Served:
MA, ME, NH.
EPA RMP 5:
CC.

GRAVES RADON CONTRACTING
P.O. Box 266
Terre Hill, PA 17581
(215) 445-7886
Thomas Graves
Area Served:
PA.

GREAT PLAINS NUCLEAR SERVICES, INC.
4502 Leavenworth St.
Omaha, NE 68106
(402) 553-3776
Sam Augustine
Dave Jacobson
Area Served:
Nationwide.
EPA RMP 5:
CC.

GREAT TRACERS
3 Schoenbeck Rd.
Prospect Heights, IL 60070-1435
(312) 255-0436
David Blustein - Industrial Hygienist
Josh Blustein - President
Teryln Blustein - Quality Assurance
Services:
Provides radon detection services using charcoal canister method, distributes charcoal canisters.
Qualifications:
EPA RMP.
Area Served:
Nationwide.
EPA RMP 5:
CC.

GREEN, GEORGE
RD 2, Box 2621 Big Hollow Rd.
Lake George, NY 12845
(518) 668-3233
George Green - Home Inspector
Services:
Provides radon detection services using charcoal canister method.
Area Served:
NY.

GREGORY K. MARSH AND ASSOCIATES
7700 West 61st Ave. Suite 12
Arvada, CO 80004
(303) 421-3383
Greg Marsh - President
Services:
Provides detection services using diffusion barrier charcoal canister method.
Qualifications:
Bachelors degree in chemistry, certified industrial hygienists on staff with 20 years experience.
Area Served:
CO.

GROUNDWATER TECHNOLOGY, INC.
12 Walker Way
Albany, NY 12205
(518) 456-2444
William Welling - Geologist
Services:
Provides radon detection and mitigation services.
Area Served:
Nationwide.

GS ANALYTICS, INC.
108 Hooten Rd.
Mt. Laurel, NJ 08054
(609) 722-5557
Glenn S. Schwartz - President
Sheryl Schwartz - Vice President
Services:
Provides radon detection, mitigation and consultation services.
Qualifications:
EPA RMP.
Area Served:
NJ.
EPA RMP 5:
AT, CC.

GUNTER EXTERMINATING COMPANY
220 West 72nd St.
Kansas City, MO 64114
(816) 523-0777
Jay G. Besheer
Area Served:
KS, MO.
EPA RMP 5:
CC.

H

H & H DECORATING
610 Great Falls Rd. Suite 100
Rockville, MD 20850
(301) 251-6654
Mae J. Hill
Area Served:
MD.

H & J FREILE HOME INSPECTIONS, INC.
99 Harrison Ave.
North Plainfield, NJ 07060
(201) 561-7147
Joyce Zimko - President
Herb Freile
Qualifications:
AARST.
Area Served:
NJ.

H.M. COOK, JR., PROFESSIONAL ENGINEER
14632 Lufthansa Cr.
Chantilly, VA 22021
(703) 378-0114, (703) 248-3809
H.M. Cook, Jr.
Services:
Provides radon detection services using charcoal canister method.
Area Served:
VA, WV.
EPA RMP 5:
CC.

H.P.J. INSPECTION SERVICE, INC.
1303 Woodbridge Commons Way
Iselin, NJ 08830
(201) 855-9475
Jess D. Greenstein
Services:
Non-commercial laboratory for research purposes or state agencies' assistance only.
Area Served:
NJ.
EPA RMP 5:
CC.

HABITABILITY ENGINEERING CONSULTANTS (HECO)
2108 Southeast 12th St.
Ocala, FL 32671
(904) 622-3873
Daniel H. Ruzic
Services:
Provides radon detection services using charcoal canister and alpha track methods.
Area Served:
FL.
EPA RMP 5:
AT, CC.

HALL-KIMBRELL ENVIRONMENTAL SERVICES
5319 Southwest Westgate
Portland, OR 97221
(503) 292-9406
David Butts - Technical Director
Services:
Provides radon detection, mitigation, and consultation services.
Qualifications:
NRA.
Area Served:
Nationwide.

HANDY ANDY
90 Corona St. #1102
Denver, CO 80218
(303) 733-2903
Ron Alpert
Area Served:
CO.

HARRIMAN ASSOCIATES
292 Court St.
Auburn, ME 04210
(207) 784-5728
William Turner - Engineer
Area Served:
ME.

HARRIS CONSTRUCTION
P.O. Box 1203
Breckenridge, CO 80424
(303) 453-2045
Tony Harris - Owner
Services:
Provides radon detection and mitigation services.
Area Served:
CO.

HARTSELL HOME INSPECTION SERVICE
5501 Executive Center Dr.
Charlotte, NC 28212
(704) 588-3781
Susan Hartsell
Area Served:
NC, SC.
EPA RMP 5:
CC.

HARVARD ENGINEERING SERVICES COMPANY
28 Shaker Rd.
Harvard, MA 01451
(617) 456-3716
James Galeski - President
Lorelei Galeski - Vice President
Services:
Provides radon detection services, design and installation of radon reduction methods.
Qualifications:
Bachelor's degree in mechanical engineering, construction supervisor's license.
Area Served:
Nationwide.

HAWKS INTER-STATE
814 North Main
Wichita, KS 67203
(316) 267-8331, (316) 681-0031
Bill Hawks, Jr.
Area Served:
KS.

HAYNES REALTY
335 Nashville Pike
Gallatin, TN 37066
(615) 452-7500
Hub Reese - Manager
Services:
Provides radon detection using charcoal canister method.
Area Served:
TN.

HAZEN RESEARCH, INC.
4601 Indiana St.
Golden, CO 80203
(303) 279-4501
John C. Jarvis
Area Served:
CO.

HEALTH AND SAFETY ASSOCIATES
35 Grove St.
Medford, MA 02155
(617) 488-7081
Neil A. Gaeta - Health Physicist
Services:
Provides radon detection services, training in radon for realtors and corporate firms and consultation services for the industry.
Qualifications:
Certified health physicists and safety professionals.
Area Served:
Nationwide.
EPA RMP 5:
AT, CC.

HEALTH HAZARDS DETECTION
Route 3, Box 372
Albany, KY 42602
(606) 387-7021
Jim DeForest
Area Served:
Nationwide.
EPA RMP 5:
CC.

HEALTH PHYSICS ASSOCIATES, INC.
RD 1 Box 796
Lenhartsville, PA 19534
(215) 756-4153
Anthony LaMastra - Health Physicist
Area Served:
MD, PA.
EPA RMP 5:
CC.

HEALTH PHYSICS ASSOCIATES, LTD.
3312 Commercial Ave.
Northbook, IL 60062-1909
(312) 480-1900
Ted Fields - President
Locke Bailey - General Manager
William Rivkin - Vice President
Services:
Radiation service company provides radon

detection services using charcoal canister, alpha track, and liquid scintillation methods, instrument calibration using radon chamber, and consultation on radiation safety.
Qualifications:
Certified health physicists, technicians registered for radon.
Area Served:
Nationwide.

HEALTH PHYSICS NORTHWEST, INC.
11535 Southwest 67th Ave.
Tigard, OR 97223
(503) 620-6617, (800) 762-8444
Ross L. Mercer - President
Corwin M. Hymes - Sales Manager
Services:
Radiation service company provides radiation consultation, instrumentation, materials and calibration and repair services to medical, industrial and commercial clients.
Area Served:
Nationwide.

HEALTH PHYSICS SERVICES, INC.
4 Research Pl. Suite 140
Rockville, MD 20850
(301) 670-1818
Catherine Haney - President
Services:
Provides radon detection services, operates a primary laboratory.
Area Served:
MD.

HEALTH SCIENCE SERVICES CORPORATION
P.O. Box 102
Marlboro, NJ 07746
(201) 536-2066
Robert DeAngelis
Area Served:
Nationwide.
EPA RMP 5:
CC(P).

HEIMER INSPECTIONS OF NEW JERSEY, INC.
17 Catherine St.
Carteret, NJ 07008
(201) 541-8995
William J. Powll
Area Served:
NJ.

HELIX ENVIRONMENTAL, INC.
416 Triangle Ave.
Dayton, OH 45419
(513) 298-2990
Ralph A. Froehlich - President
R.C. Froehlich - Asbestos Planner
E.E. Rothwell - Industrial Hygienist
Services:
Occupational and environmental health consultants specializing in the recognition, evaluation, and control of indoor air contaminants.
Qualifications:
CIH, Ohio Asbestos Building Inspectors, Ohio Abatement Supervisors.
Area Served:
Nationwide.
EPA RMP 5:
CR(P).

HERITAGE RADON SERVICES, INC.
6968 Birchwood Rd.
Harrisburg, PA 17112
(717) 599-5060
Richard C. Smith - President
Glenn Livetsberger - Vice President
Services:
Provides radon detection and mitigation services.
Area Served:
Nationwide.
EPA RMP 5:
CC.

HERRICK ENGINEERING, INC.
2004 Belvedere Dr.
Toledo, OH 43614
(419) 385-5655
Robert Herrick - Professional Engineer
Services:
Provides radon detection and consultation services.
Qualifications:
NRA.
Area Served:
IN, MI, OH.

HILLMANN ENVIRONMENTAL COMPANY
1089 Cedar Ave.
Union, NJ 07083
(201) 686-3335
Joseph Hillmann - Vice President
Services:
Provides radon detection services using charcoal canister and alpha track methods.
Qualifications:
NRA.
Area Served:
DC, FL, MD, NJ, VA.

HMA HOME INSPECTION AND WARRANTY SERVICE
13250 Ridge Rd. Unit 7-4
Largo, FL 32257
(813) 587-0776, (800) 330-0776
Janet Karr - Director
Services:
Provides radon detection services using charcoal canister method.
Area Served:
FL.

HOLIDAY TERMITE AND PEST CONTROL
612 3rd St.
Herndon, VA 22070
(703) 471-5533
Frederick R. Gunnels
Area Served:
DC, MD, VA.
EPA RMP 5:
CC.

HOLMES INSPECTION SERVICE
159 Liberty St.
Little Ferry, NJ 07643
(201) 641-8377, (800) 642-4603
Vincent DeLuca
Area Served:
NJ, NY, PA.
EPA RMP 5:
CC, CW(P).

HOME ANALYSTS, INC.
748 Considine Ave.
Cincinnati, OH 45205-2336
(513) 244-2860
Paul L. Nienaber, Jr.
Area Served:
IN, KY, OH.
EPA RMP 5:
CC.

HOME ASSURANCE
17 Birchwood Dr.
Naperville, IL 60540
(312) 961-9154, (312) 355-6792
Vicky Lanksbury
Area Served:
DC, IL, MD, OH.
EPA RMP 5:
CC.

HOME BUYERS EDGE
880 Benton St.
Lakewood, CO 80214
(303) 238-6275
Robert Fisher
Area Served:
CO.

HOME CHECK SERVICES
P.O. Box 98
Front Royal, VA 22630
(703) 635-1274
William H. McSpadden - President
Glen Hickerson - Geologist

Mary D. Sitlon - Geologist
Services:
Provides radon detection, mitigation, and consulting services.
Qualifications:
AARST, EPA contractor's course.
Area Served:
DE, VA.

HOME CHEK, INC.
234 South Cardington
Wichita, KS 67209
(316) 267-2324
Louis Tangney - President
Gary Huntsperger - Treasurer
Services:
Provides radon detection services using charcoal canister method.
Qualifications:
ASHI, CBO, AARST.
Area Served:
KS, OH.

HOME DETOX SPECIALISTS
81 Candy Ln.
Hamburg, NY 14075
(716) 648-3321
Thomas Deere - President
Guy Gessner - Vice President
Services:
Provides radon detection and mitigation services.
Qualifications:
Trained by NY State Health Department.
Area Served:
NY.

HOME ENVIRONMENT COMPANY
P.O. Box 24744
Speedway, IN 46224
(317) 247-1881
Gil Ward - Owner
Services:
Manufactures Radon Sentry Alarm System, alarm signaling radon mitigation system failure, provides radon mitigation services.
Qualifications:
Completed certified course by Radon Detection Services, Inc.
Area Served:
IL, IN, KY, OH.

HOME EXTERMINATING COMPANY
4608 Bucks Schoolhouse Rd.
Baltimore, MD 21237
(301) 661-7911, (800) 492-5544
Craig Strobel - Vice President
Jack Brennan - Manager
Kevin Kordex - Technical Director
James Cotton - Vice President
Services:
Provides radon detection services.
Area Served:
DC, DE, MD, VA.

HOME INSPECTION OF KENTUCKIANA
2416 Frankfort Ave. Suite 206
Louisville, KY 40206
(502) 897-1450
William G. Harpole, Jr.
Area Served:
IN, KY.
EPA RMP 5:
CC.

HOME INSPECTION ONE
P.O. Box 1177 2 Valley Rd.
Denville, NJ 07834
(201) 586-2730
Myron M. Bytz - President
Services:
Provides radon detection services.
Qualifications:
ASHI.
Area Served:
NJ.
EPA RMP 5:
CC.

HOME INSPECTION SERVICE, INC.
1916 Southland Ave.
Highland Park, IL 60035
(312) 831-4545
Jerry T. Comer - President
Services:
Provides radon detection and mitigation services.
Qualifications:
EPA RMP.
Area Served:
Nationwide.
EPA RMP 5:
AT, CC.

HOME INSPECTION SERVICE, INC.
RR 1 Box 225 Walsh Rd.
La Grangeville, NY 12540
(914) 227-5417
John R. Abraham
Area Served:
NY.

HOME INSPECTION SERVICES
P.O. Box 455
Boulder, CO 80306
(303) 443-9015
Paul Shippee - Owner
Services:
Provides radon detection services and consultation to building owners.
Qualifications:
ASHI, licensed building contractor, degree in civil engineering.
Area Served:
CO.

HOME INSPECTIONS BY CHANEL, INC.
20 Redding Pl.
Towaco, NJ 07082
(201) 263-9163
Willy C. Dittmar
Area Served:
NJ.
EPA RMP 5:
CC.

HOME INSPECTIONS OF MICHIGAN
1007 St. Joseph Dr. P. O. Box 936
St. Joseph, MI 49085
(616) 983-4663 , (616) 983-7098
Paul G. Freudenburg
Area Served:
MI.

HOME INSPECTOR OF NORTH AMERICA, THE
301 North Main St.
Randolph, MA 02368
(617) 963-0343
Sumner Levine - President
Robert Bryant
Area Served:
MA.

HOME RADON ASSOCIATES
490 Main St.
Norwich, CT 06360
(203) 889-8104 (203) 887-1528
Anthony Benoit
Area Served:
CT.

HOME RADON DETECTION COMPANY
5824 El Monte
Fairway, KS 66205
(913) 262-1555
Randolph S. Davis - President
Gary Rittenhouse - Field Supervisor
Services:
Provides radon detection services for the real estate industry.
Qualifications:
AARST, EPA RMP, Certified Radon Inspector (by Air-Chek).
Area Served:
Nationwide.
EPA RMP 5:
CC.

HOME RADON DETECTION, INC.
33488 Lone Rock
Wildwood, IL 60030
(312) 223-6517
Richard J. Karpel - President
Services:
Provides radon detection services in homes, commercial and public buildings.
Qualifications:
EPA RMP.
Area Served:
Nationwide.
EPA RMP 5:
CC(P).

HOME RADON SURVEYS
4209 Olley Ln.
Fairfax, VA 22032
(703) 323-7236
John A. Leonard - Quality Assurance Officer
Tom E. Tobin - Manager
J. Douglas Hough - Manager
Services:
Provides radon detection services using charcoal and alpha-track detectors.
Area Served:
Nationwide.
EPA RMP 5:
AT, CC.

HOME SENTRY, INC.
P.O. Box 555
Midland Park, NJ 07432
(201) 652-2497
Christopher Pedersen
Area Served:
Nationwide.
EPA RMP 5:
CC.

HOME TESTING LABS, INC.
299 West Fort Lee Rd.
Bogota, NJ 07603
(201) 343-1199
Thomas Parisi - Director
Services:
Provides consultation for radon mitigation design.
Area Served:
Nationwide.
EPA RMP 5:
CC.

HOME TESTING LABS, INC.
44 Crestview Rd.
Phoenixville, PA 19460
(215) 768-7960
Thomas Parisi - Director
Services:
Provides consultation for radon mitigation design.
Area Served:
PA.

HOME TESTING SERVICE
1 Aaron Rd.
Lexington, MA 02173
(617) 862-3440
Steven D. Wightman - Chief Technologist
Margaret Wacks - Quality Assurance
Services:
Provides radon detection services for home owners and real estate transactions using charcoal canister method.
Qualifications:
Certified asbestos inspector, ASCP, AARST, BBB.
Area Served:
Nationwide.
EPA RMP 5:
CC(P).

HOME-TECH
P.O. Box 423
Arnold, MD 21012
(301) 757-8458
Scott A. Stangle
Area Served:
MD.
EPA RMP 5:
CC.

HOME-TECH ENGINEERING
P.O. Box 43129 51 Upper Montclair Plaza
Upper Montclair, NJ 07043
(201) 744-6674
David A. Osage
Robert Schwalbenberg
Area Served:
Nationwide.
EPA RMP 5:
CC.

HOMEAIR TESTERS
1234 Old Town South Dr.
Indianapolis, IN 46260
(317) 846-5711
Tom Boneham - President
Roger Boneham
Services:
Provides radon detection services.
Qualifications:
AARST.
Area Served:
IL, IN, KY, OH.
EPA RMP 5:
CW(P).

HOMEBUYERS INSPECTION SERVICE OF NORTH WEST FLORIDA, INC.
P.O. Box 9433
Pensacola, FL 32513-9433
(904) 479-3276
Steve Butler - Owner
Services:
Provides radon detection services using charcoal canister and alpha track methods.
Area Served:
AL, FL, MS.

HOMEBUYERS INSPECTION SERVICE OF SOUTH WEST FLORIDA
2062 Stella St.
Ft. Myers, FL 33901
(813) 334-6116
Carl C. Fowler - President
Services:
Provides radon detection services using charcoal canister method.
Area Served:
FL.

HOMEBUYERS INSPECTION SERVICE, INC.
1000 East Chicago Ave.
Naperville, IL 60540
(312) 357-0903
John R. Ball - President
Tony Ross - Inspector
Tom Iversen - Inspector
Services:
Provides radon detection services using charcoal canister method.
Qualifications:
President is registered professional nuclear engineer, AARST, ASHI, EPA RMP.
Area Served:
IL, IN, WI.

HOMEPRO SYSTEMS, INC.
6715 Lowell Ave. Suite 2
McLean, VA 22101
(703) 761-1400
Mike Lennon
Services:
Provides consultation for analysis of structures, feasibility studies, specifications, radon detection and mitigation training.
Area Served:
VA.

HOMERAD
908 Wild Forest Dr.
Gaithersburg, MD 20879
(301) 889-5769
Ara Carapetian
Area Served:
DC, DE, MD, NJ, NY, PA, VA, WV.
EPA RMP 5:
CC.

HOMES PLUS RADON SERVICES, INC.
490 Main St. P.O. Box 654

Norwich, CT 06360
(203) 887-5128
Scott Harrington - President
Tony Benoit - Scientist
Martin Grisham - Marketing
Susan Axline - Assistant Manager
Services:
Provides radon detection services, distributes charcoal canister test kits.
Qualifications:
AARST.
Area Served:
Nationwide.

HOMESAFE RADON ENGINEERING
P.O. Box 6
Livingston, NJ 07039
(201) 992-8060, (201) 342-2246
Henry Smith
Area Served:
NJ.
EPA RMP 5:
CC.

HOMESTEAD INSPECTION SERVICE
7 Sunny View Dr.
Succasunna, NJ 07876
(201) 927-8343
Jim Benson
Area Served:
NJ.
EPA RMP 5:
CC.

HOMESURE INSPECTIONS, INC.
444 North Frederick Ave. Suite L348
Gaithersburg, MD 20877-2432
(301) 428-7672
James Scheible
Qualifications:
AARST.
Area Served:
DC, MD, VA.
EPA RMP 5:
GR(P).

HORIZON ENGINEERING AND ENVIRONMENTAL SERVICES, INC.
599 Waldron Rd.
LaVergne, TN 37086
(615) 793-7993
Randall G. Smith
Area Served:
Nationwide.
EPA RMP 5:
AT, CC.

HOUSE ANALYSIS AND ASSOCIATES
747 Hallowell Dr.
Huntingdon Valley, PA 19006
(215) 947-1169, (215) 947-9209

Fred Powell
Nathaniel Burden, Jr.
James McKay
Services:
Non-commercial laboratory for research purposes or state agencies' assistance only.
Area Served:
Nationwide.
EPA RMP 5:
AT, CR(P), EP(P), GR(P), GW(P).

HOUSE CHECK, INC./MARYLAND
2025 Maryland Ave.
Baltimore, MD 21218
(301) 727-5553, (703) 550-7525
John B. Davis
Area Served:
MD.
EPA RMP 5:
CC.

HOUSE CHEK
P.O. Box 209
Port Monmouth, NJ 07758
(201) 495-0303
Albert J. Reid - Owner
Area Served:
NJ.

HOUSE DETECTIVES, INC.
2927 Champa St.
Denver, CO 80205
(303) 399-6557
Dennis Williams - President
Area Served:
CO.

HOUSE DOCTORS OF LAKEWOOD, INC.
P.O. Box 489
Englewood, CO 80151-0489
(303) 841-2443
Brent Tyson
Area Served:
CO.

HOUSE DOCTORS, INC.
P.O. Box 10070
Colorado Springs, CO 80932
(303) 574-6960
David Kittinger
Area Served:
CO.

HOUSE MASTER OF AMERICA
9989 West 60th Ave. Suite 201A
Arvada, CO 80004
(303) 431-5383
Thomas W. Stroud - President
Services:
Provides radon detection services using charcoal canister method.
Area Served:
CO.

HOUSE-CHECK, INC.
P.O. Box 731
Valley Forge, PA 19482
(215) 933-4431
Roman Korocij - Owner
Area Served:
Nationwide.
EPA RMP 5:
AT, CC.

HOUSEHOLD LABORATORIES, INC.
P. O. Box 2247
Glenview, IL 60025
(312) 724-6330
Pat Schroeder
Area Served:
Nationwide.
EPA RMP 5:
CC(P).

HOUSEHOLD RADON TESTERS, INC.
4 Seabury Ln.
Downingtown, PA 19335
(215) 525-8427, (215) 458-8730
Mike Schneider - President
Area Served:
PA.

HOUSEMASTER
601 N. Olive St.
Media, PA 19063
(302) 658-2880
Valerie Bradley - Manager
Services:
Provides radon detection services using charcoal canister method.
Area Served:
PA.

HOUSEMASTERS OF AMERICA
2935 Hawk
Carson City, NV 89701
(702) 849-2229
Michael Tedesco - Director
Services:
Provides radon detection services using alpha track and charcoal canister methods.
Area Served:
NV.

HOUSING CONSULTANTS, INC.
P.O. Box 4838
Lousiville, KY 40204
(502) 459-0772
Ralph Wirth - President
Area Served:
KY.

HOUSING INSPECTION SERVICES, INC.
P.O. Box 373
Murrysville, PA 15668
(412) 374-1900
John A. Jaccarello - President
James Esperseth - Inspector
Services:
Provides radon detection services for the public, realtors and mortgage lenders.
Qualifications:
EPA training course, other classes, seminars and workshops.
Area Served:
PA.
EPA RMP 5:
CC.

HUDSON VALLEY RADON TESTING CORPORATION
P.O. Box 790
Poughkeepsie, NY 12601
(914) 431-6424, (914) 485-5561
Paul Costello
Robert Hare
Area Served:
Nationwide.
EPA RMP 5:
CC.

HUGHES, RONALD L.
68 Hanshue Rd.
Hummelstown, PA 17036
(717) 534-1168
Ronald L. Hughes
Area Served:
PA.

HUNT AND ASSOCIATES
802 Avenue B
Rome, GA 30161
(404) 232-7653, (404) 232-2034
W.H. Hunt
Area Served:
AL, GA, TN.
EPA RMP 5:
CC.

HUNT'S WATER CONTROL, INC.
607 Tower View Dr.
Taylor Mill, KY 41015
(606) 292-0333
Jim Hunt - President
Services:
Provides radon detection services using charcoal canister and alpha track methods, provides mitigation services.
Area Served:
KY.

HURT AND PROFFITT, PROFESSIONAL ENGINEERS
2524 Langhorne Rd.
Lynchburg, VA 24501
(804) 847-7796
Roy Bradley
Area Served:
VA.
EPA RMP 5:
CC.

HUTCHISON PEST CONTROL
2662 North Freeway
Pueblo, CO 81003
(719) 544-7378, (719) 633-7378
A.E. Chirby
Area Served:
CO.

I

I-T.E.M., LTD.
12850 Middlebrook Rd. Suite 301
Germantown, MD 20874
(301) 353-0585
Corey Thompson - Vice President
Michael Hentgen - Laboratory Services
Services:
Provides radon detection services using charcoal canister and alpha track methods.
Qualifications:
NRA.
Area Served:
Nationwide.

ILLINOIS DEPARTMENT OF NUCLEAR SAFETY/OFFICE OF ENVIRONMENTAL SAFETY
1035 Outer Park Dr.
Springfield, IL 62704
(217) 786-6399
Terry Lash - Department Director
John Cooper - Office Manager
Melanie Ann Hamill - Health Physicist
Richard Allen - Radon Programs
Services:
Non-commercial laboratory for research purposes or state agencies' assistance only.
Area Served:
IL.
EPA RMP 5:
AT, CC.

ILLINOIS STATE GEOLOGICAL SURVEY
Natural Resources Building 615 East Peabody Dr. Room 121
Champaign, IL 61820
(217) 333-5111
Morris W. Leighton - State Geologist
Richard A. Cahill
Services:
Non-commercial laboratory for research purposes or state agencies' assistance only.
Area Served:
IA, IL, IN, WI.
EPA RMP 5:
CC(P).

IMPROVED TECHNOLOGY SYSTEMS
P.O. Box 8244
Canton, OH 44711-8244
(216) 452-1493
William Kuendig
Betsy Kuendig
Area Served:
FL, OH.
EPA RMP 5:
CC.

INDEPENDENT HOME INSPECTION SERVICE, INC.
51 Forest Rd.
Demarest, NJ 07627
(201) 767-3449
Robert S. Zenorini
Qualifications:
AARST.
Area Served:
NJ.

INDOOR POLLUTION CONTROL CENTER
3921 Sawyer Rd.
Sarasota, FL 34233
(813) 923-3641, (800) 226-2636
Ronald V. Miller - President
Robert L. Brogan - Manager
David V. Miller - Vice President
Services:
Provides radon detection and mitigation services.
Qualifications:
EPA RMP, certified by Honeywell, Inc.
Area Served:
FL.
EPA RMP 5:
AT, CC.

INDUSTRIAL OPERATIONS, INC./RADON EVALUATION
3736 South 7th Ave.
Phoenix, AZ 85041
(602) 268-8772
Ray Cunning - President
Richard Kaz - Program Coordinator
Services:
Provides radon detection services using charcoal canister, alpha track, and E Perm methods.
Area Served:
AZ.

INDUSTRIAL RESOURCE DEVELOPMENT COMPANY
2490 General Armistead Ave. Suite 303
Norristown, PA 19403
(215) 630-9150
Maurice Silvestris - Manager
William Minnick - Vice President
Services:
Provides radon detection and mitigation services.
Area Served:
DE, NJ, PA.

INFILTEC, D.C.
1105 Park St., NE
Washington, DC 20002
(202) 547-1611
Marc Messing
Area Served:
DC, MD, VA.
EPA RMP 5:
CC.

INFILTEC/FACTORY
108 South Delphine Ave. P.O. Box 1125
Waynesboro, VA 22980
(703) 943-2776
David Brubaker - Factory Manager
Frankie Patterson - Warehouse Manager
Arthur Saum - Research Manager
Services:
Manufactures and distributes a full line of radon detection and mitigation products including fans, couplings, pressure gauges, alarms, sumps and drain covers, radon test kits, continuous radon monitors, caulks and sealants, manometers, blower doors, etc.
Qualifications:
EPA RMP.
Area Served:
Nationwide.
EPA RMP 5:
CC.

See Ad Page 214

INFILTEC/MAIN OFFICE
P.O. Box 8007
Falls Church, VA 22041
(703) 820-7696
David Saum - President
Marc Messing - Vice President
Services:
Provides radon mitigation, research and development, nationwide training, consulting services, specializes in large buildings.
Qualifications:
EPA RMP, EPA subcontractor, EPA trainer.
Area Served:
Nationwide.
EPA RMP 5:
AT, CC, EP(P), GR(P).

INFOTECH HOME INSPECTION SERVICES
984 Route 9
Parlin, NJ 08859
(201) 727-5335
Ernest W. Feist
Qualifications:
AARST.
Area Served:
NJ.

INSPECT
7737 South Normandy
Burbank, IL 60459
(312) 599-3553
Wayne Maleck
Area Served:
IL.

INSPECT-A-HOME OF UTAH
455 East 400 S.
Salt Lake City, UT 84111
(801) 532-4663
Michael H. Christensen - President
Area Served:
UT.

INSPECT-A-HOME, INC.
37 Janet Ln.
Springfield, NJ 07081
(201) 379-4124
Steven Hockstein
Area Served:
CT, DE, NJ, NY, PA.
EPA RMP 5:
CC.

INSPECTIONS UNLIMITED
514 Ryers Ave.
Cheltenham, PA 19012
(215) 848-0720
Fred Klein
Area Served:
PA.

INSPEX OF BOULDER, INC.
2955 Valmont Suite 300
Boulder, CO 80301
(303) 442-8995
Kevin M. O'Hornett - President
Jim Johansen - Chief Field Technician
Barbara Farnsworth - Office Manager
Services:
Provides radon detection services for home buyers and the real estate industry.
Qualifications:
EPA RMP, AARST, AIHA, CEHA, Colorado School of Mines course on radon and the indoor environment.
Area Served:
Nationwide.
EPA RMP 5:
CC(P), EP(P).

INSUL-TECH, INC.
5559 Westerville Rd.
Westerville, OH 43081
(614) 891-9555
Dan West - President
Services:
Provides radon detection services using charcoal canister, alpha track, and continuous radon monitor methods, provides radon mitigation services.
Qualifications:
ARA, ICAA, NRA, contractor's license.
Area Served:
IN, KY, MI, OH, PA, WA, WV.
EPA RMP 5:
AT, CC, CR(P).

INSULATING SYSTEMS
340 Gargrave Rd.
West Carrollton, OH 45449
(513) 435-0079
James Dunn - President
Services:
Provides radon detection and mitigation services.
Qualifications:
EPA workshops.
Area Served:
OH.
EPA RMP 5:
CC, CR(P).

INSURANCE HEALTH SERVICES
35 Stony Brook Rd.
Westport, CT 06880
(203) 226-3444
Sandy Staw
Area Served:
CT.

INTEGRAL RESOURCES, INC.
1916 East El Parque
Tempe, AZ 85282
(602) 967-8029
James W. Krueger
Area Served:
AZ.

INTEGRATED PEST MANAGEMENT
Box 93, RD #2 White Horse Pike
Hammonton, NJ 08037
(609) 567-0100
David Damiano
Area Served:
NJ.
EPA RMP 5:
CC.

INTERMOUNTAIN RADON SERVICE
P.O. Box 3
Salida, CO 81201
(719) 539-6726
Monty Holmes - President
Services:
Provides radon detection and mitigation services, conducts training seminars for real estate personnel, is currently producing a radon video for Coldwell-Banker.
Area Served:
CO.

INTERNATIONAL EXPLORATIONS
577 Sackettsford Rd.
Warminster, PA 18974
(215) 598-7137
Jacob Jacobsen - President
Area Served:
PA.

INTERNATIONAL EXTERMINATOR CORPORATION
320 South Freeway
Fort Worth, TX 76104
(817) 335-5531, (800) 666-7287
Kenneth Baker - Director
Services:
Provides radon detection and consulting services.
Qualifications:
EPA RMP.
Area Served:
TX.

INTERNATIONAL PEST CONTROL, INC.
P.O. Box 5922
Edison, NJ 08818
(201) 494-9313
Richard Tantleff
Area Served:
NJ.
EPA RMP 5:
CC.

INTERNATIONAL RADON TECHNOLOGIES
2417 Regency Rd.
Lexington, KY 40503
(606) 278-0104, (800) 537-5066
Ed Davis - President
William G. Buchman - Inventor
Services:
Distributes radon-related products including a water reducible, two-part polyamide epoxy coating to block radon passage, manufactures continuous radon progeny monitor and electret cylindrical airborne radon monitor, operates a primary laboratory.

Area Served:
Nationwide.
EPA RMP 5:
CC(P).

J

J & J QUALITY HOME INSPECTIONS
909 East Lynda Ln.
Marietta, GA 30060
(404) 425-8868
Richard Jolley
Area Served:
GA.
EPA RMP 5:
CC.

J. BECK AND ASSOCIATES, INC.
35 Meetinghouse Ln.
Bradford Woods, PA 15015
(412) 935-6711
Joseph R. Beck - President
Services:
Provides radon detection and mitigation services to home owners, real estate companies, and relocation companies.
Area Served:
MD, NJ, OH, PA, WV.
EPA RMP 5:
CR(P).

J.A. ASSOCIATES/PUBLIC HEALTH CONSULTANTS
313 Clear Spring Rd.
Lansdale, PA 19446
(215) 855-5592
John A. Messa, Jr.
Area Served:
AK, CA, CO, DE, FL, HI, KY, MA, MD, ME, NJ, NY, PA, TN, VA, WV.
EPA RMP 5:
CC.

J.J. LAY COMPANY
3130 South Platte River Dr. P.O. Box 1745
Englewood, CO 80150
(303) 761-2021
Area Served:
CO.

J.M. FROHN CORPORATION
35 Bowdoin Dr.
Milford, MA 01757
(508) 478-4342
Jeffrey M. Frohn - President
Services:
Provides radon detection services for home owners/buyers as part of home inspection services.
Qualifications:
Licensed in MA as construction supervisor.
Area Served:
MA.
EPA RMP 5:
CC.

J.M. MONTGOMERY ENGINEERS
555 East Walnut St.
Pasadena, CA 91101
(818) 796-9141
Andrew Eaton - Laboratory Director
Services:
Provides radon detection services and operates a primary laboratory.
Area Served:
Nationwide.

J.M.C. ENTERPRISES
700 Ferndale Rd.
Johnson City, TN 37604
(615) 282-2136
John M. Conley
Area Served:
KY, NC, TN, VA.
EPA RMP 5:
CC.

JB TESTING COMPANY
P.O. Box 1303
Frederick, MD 21701
(301) 662-7043
John W. Brennan
Area Served:
MD.

JBK INSPECTION SERVICE, INC.
225B Cottrell Rd.
Matawan, NJ 07747
(201) 525-9410
Barry Benevento
Area Served:
NJ.
EPA RMP 5:
CC.

JH AND ASSOCIATES, INC.
2524 East 71st St.
Tulsa, OK 74136
(918) 455-4984
Michael M. Nichols
Jack W. Hughey
Area Served:
Nationwide.
EPA RMP 5:
CC.

JMX INDUSTRIES, INC.
3959 Wilshire Blvd. Suite 200/B11
Los Angeles, CA 90010
(213) 871-8054
Myron Hale
Area Served:
CA.

JOHN A. FIXARY, INC.
P.O. Box 490
Riverside, CT 06878
(203) 622-6661
John A. Fixary - President
Services:
Provides radon detection services.
Qualifications:
AARST.
Area Served:
CT.

JOHN PRIKKEL AND ASSOCIATES
P.O. Box 14145
Dayton, OH 45414
(513) 222-7749, (513) 278-3176
John Prikkel - President
Area Served:
Nationwide.
EPA RMP 5:
CR(P).

JOHNSON PLUMBING AND HEATING
232 South Main St.
Chambersburg, PA 17201
(717) 264-5028
Kirk Johnson - Partner
Services:
Provides radon detection and mitigation services.
Area Served:
Nationwide.

JOHNSON'S ENERGY SERVICE
104 Central Dr.
Forest City, IA 50436
(515) 582-2998
Harvey Johnson - Owner
Services:
Provides radon detection and mitigation services.
Qualifications:
Master's degree in radiation biology.
Area Served:
Nationwide.

JOYCE EXTERMINATING COMPANY
Route 2, Box 249
Ramseur, NC 27316
(919) 824-2358
Donald R. Joyce
Area Served:
NC.

JP HOME INSPECTION, INC.
100 Polifly Rd. Suite 4I
Hackensack, NJ 07601
(201) 488-0417

Peter A. Schmidt
Area Served:
NJ.
EPA RMP 5:
CC.

JRM ASSOCIATES
Box 602 North Barnstead Rd.
Center Barnstead, NH 03225
(603) 776-6046
J. Robert Marr, Sr.
Area Served:
MA, ME, NH.
EPA RMP 5:
CC.

JRW INDUSTRIES, INC.
2312 North 2nd St.
Harrisburg, PA 17110
(717) 236-1418
James R. Williams - President
Kevin J. Williams - General Manager
Valeria Hollern - Secretary
Services:
Provides radon detection services, manufactures and distributes radon test kits, charcoal canisters, test stands, warning signs, certification kits, and other accessories.
Qualifications:
EPA RMP, PA DER, AARST.
Area Served:
Nationwide.
EPA RMP 5:
CC.

See Ad Page 236

JUNGBLUT, DAVID W., GEOLOGIST
Route 12 Box 168
Baptistown, NJ 08803
(201) 996-3688
David W. Jungblut
Area Served:
Nationwide.
EPA RMP 5:
CC.

K

K & J SOLAR SERVICE
1142 Fillmore St.
Denver, CO 80206
(303) 393-8372
James Valastro
Area Served:
Nationwide.
EPA RMP 5:
CC.

KALKA MIDWEST CONSTRUCTION, INC.
1710 South Michigan St.
South Bend, IN 46613
(219) 287-3399, (219) 234-9206
James E. Kalka, Jr.
Area Served:
IN.

KANSAS DEPARTMENT OF HEALTH AND ENVIRONMENT/BUREAU OF AIR QUALITY AND RADIATION CONTROL/ENVIRONMENTAL SURVEILLANCE AND EMERGENCY PLANNING SECTION
Forbes Field Building Suite 740
Topeka, KS 66620
(913) 296-1567
Harold Spiker - Chief
Dominic To - Laboratory Director
John Irwin - Manager
Services:
Non-commercial laboratory for research purposes or state agencies' assistance only.
Area Served:
KS.
EPA RMP 5:
CC(P).

KANSAS TESTING SERVICES
957 North Coolidge
Wichita, KS 67203
(316) 267-8378
Peter Austin - Owner
Services:
Provides radon detection services for real estate companies and the public using Honeywell Professional Radon Monitors.
Area Served:
KS.

KBC TECHNOLOGY
Box 170 RR 2
Raymond, IL 62580
(217) 229-4437
Keith Boehler
Area Served:
Nationwide.
EPA RMP 5:
CC.

KELLOGG ENVIRONMENTAL SERVICES, INC.
816 Anthony Dr.
Mechanicsburg, PA 17055
(717) 732-3364, (717) 732-9263
Ed Kellogg - President
Services:
Provides radon detection and mitigation services.
Area Served:
PA.

KEY TECHNOLOGY, INC.
P.O. Box 562
Jonestown, PA 17038
(717) 274-8310, (800) 523-4964
Claude Wiblin - President
Robert Garland - Marketing Director
Robby Marquette - Secretary/Treasurer
Edward Suter - Training Director
Services:
Manufactures and distributes a complete line of radon detection products and distributes radon mitigation products.
Qualifications:
EPA RMP, PA DER testing and mitigation courses.
Area Served:
Nationwide.
EPA RMP 5:
CC(P), GR(P).

See Ad Inside Front Cover

KSE, INC.
P.O. Box 368
Amherst, MA 01004
(413) 549-5506
Charles W. Quinlan - Vice President
G. Ruth Kittrell - Manager
Services:
Radon primary analytical laboratory, air and water radon analyses for consumers, distributors, inspection services and retailers.
Qualifications:
EPA RMP.
Area Served:
Nationwide.
EPA RMP 5:
CC(P).

L

L. CLARK AND SONS, INC.
986 Fisher Rd.
Fitchburg, MA 01420
(508) 343-7849
LeRoy E. Clark - President
Services:
Provides radon mitigation services, installation of radon control systems, monitoring, and consultation services to the home owner/buyer.
Area Served:
DE, NH, NJ, NY, MA, ME, PA, RI, VT.

L.G. PAUL, SR., INC.
333 Union St.
Allentown, PA 18102
(214) 437-3530
L.G. Paul - General Contractor
Area Served:
PA.

LAMSON HOME INSPECTIONS
9 Lougheed Ave.
West Caldwell, NJ 07006

(201) 228-1862
William M. Lamson
Area Served:
NJ.
EPA RMP 5:
CC.

LANDMARK RADON LABORATORIES, LTD.
3521 West Eisenhower Blvd.
Loveland, CO 80537
(303) 667-6286 , (303) 356-6286
Bruce F. Meyer - Architect
Services:
Provides radon detection services and mitigation consultation.
Area Served:
CO, KS, NE, NM, UT, WY.
EPA RMP 5:
CC.

LANIDEX CORPORATION
300 Lanidex Plaza
Parsippany, NJ 07054
(201) 386-5800
Thomas C. Graham - President
Area Served:
NJ.

LAPTEFF ASSOCIATES
P.O. Box 4150
Woodbridge, VA 22194
(703) 491-6700
Alexander Lapteff
Area Served:
Nationwide.
EPA RMP 5:
CC.

LAW ENVIRONMENTAL, INC.
P.O. Box 23184
Tampa, FL 33623
(813) 877-9182
Arthur Benson
Area Served:
AL, FL, GA, LA, NC, SC, TN.
EPA RMP 5:
CC.

LEAFE, EDWARD
212 Wilson St.
Hackensack, NJ 07601
(201) 343-0780
Edward Leafe
Area Served:
NJ.

LEGIS CORPORATION
2211 Saxon
Houston, TX 77018
(713) 680-8863
Richard L. Miller - President
Services:
Microclimate surveys.
Qualifications:
Certified industrial hygienist, certified safety professional.
Area Served:
Nationwide.

LEHIGH VALLEY HOUSE INSPECTION SERVICE
200 Delaware Ln.
Whitehall, PA 18052
(215) 435-0208
Pete Potochney - Director
Services:
Provides radon detection and mitigation services.
Qualifications:
BOCA, CSI, NAHRO.
Area Served:
PA.

LEONARDS HOME INSPECTIONS
6416 Loring Dr.
Columbia, MD 21045
(301) 381-2240
Christopher Leonard
Qualifications:
AARST.
Area Served:
MD.

LEWIS AND KAHLSON RADON CONSULTANTS
459 Dunlin Plaza
Secaucus, NJ 07094
(201) 866-7219
Don Kahlson
Qualifications:
AARST.
Area Served:
NJ.

LEWIS HOME INSPECTION
P.O. Box 6149
Lawrenceville, NJ 08648
(609) 924-1411
Craig Lewis
Qualifications:
AARST.
Area Served:
NJ.

LIFELINE ENVIRONMENTAL TESTING
2522 Stuart Ave.
Richmond, VA 23220
(804) 358-4192
Clark D. Roan
Area Served:
Nationwide.
EPA RMP 5:
CC.

LINCOLN DEVORE, INC.
1000 West Fillmore St.
Colorado Springs, CO 80907
(303) 632-3593
Joseph Winston
Area Served:
CO, KS, NM, UT, WY.
EPA RMP 5:
CC.

LMS ENGINEERS
1 Blue Hill Plaza
Pearl River, NY 10965
(914) 735-8300
D. Groff - Head, Hydrology
Peter McGroddy
Thomas Vanderbeek
Services:
Provides radon detection services.
Area Served:
NY.

LONCHAR, NIKOLA
771 South 20th St.
Philadelphia, PA 19146
(215) 985-1474
Nikola Lonchar
Area Served:
PA.

LOS ANGELES URBAN DEVELOPMENT
1725 Radcliff St.
Los Angeles, CA 90026-1131
(213) 668-2655
W.R. Dowling - CEO
Randy Morceau - Vice President
Services:
Provides radon detection services, computer Bulletin Board includes radon information exchange, accessable 24 hours a day at (213) 666-6639.
Qualifications:
AARST.
Area Served:
CA.

LOWRY ENGINEERING, INC.
P.O. Box 536 Back Troy Rd.
Unity, ME 04988
(207) 948-3790
Sylvia Lowry - President
Jerry Lowry - Resarch Director
Services:
Engineering research and consulting firm specializes in radionuclide analysis and

removal from water supplies, and removal equipment for homes and public water supplies, manufactures and distributes the Stripper Aeration System, removes 99.9% of radon from water.
Qualifications:
Director is a leading expert on radon removal from water supplies, teaches EPA courses on mitigation.
Area Served:
Nationwide.

See Ad Page 211

LV SAFETY AND HEALTH CONSULTANTS
446 Timothy Dr.
Bethlehem, PA 18017
(215) 758-9072
Laslo Varadi - President
Area Served:
PA.

LW ENVIRONMENTAL SERVICES
2111 Jennings St.
San Francisco, CA 94124
(415) 822-4555
Richard S. DellaValle
Qualifications:
AARST.
Area Served:
CA.

LYNN PEST MANAGEMENT COMPANY
P.O. Box 274
Jarrettsville, MD 21084
(301) 557-6839, (800) 327-7134
Jeffrey N. Lynn
Area Served:
DC, DE, MD, NY, PA, VA.
EPA RMP 5:
CC, CR(P).

M

M.A.D. EXTERMINATORS, INC.
3116 Methacton Ave.
Norristown, PA 19403
(215) 539-7647
Mark DeCinque
Qualifications:
AARST.
Area Served:
PA.

M.L. DOWNING AND ASSOCIATES
2806 Farm Rd.
Alexandria, VA 22314
(703) 548-8607
Mary Lou Downing - President
Area Served:
VA.

M.V.E., INC.
1911 Lelaray St.
Colorado Springs, CO 80909
(303) 635-5736
Kent Rockwell
Area Served:
CO, NE, WY.
EPA RMP 5:
AT, CC.

MACK, E.F.
11 South 8th Ave.
Marshalltown, IA 50158
(515) 753-7328
Thomas E. Mack - Partner
Area Served:
IA.

MACLAREN ENGINEERS, INC.
33 Yonge St.
Toronto, ON M5E 1E7
(416) 756-4919
E. Chart
Area Served:
ON.

MACMILLAN RESEARCH, LTD.
1221 Barclay Cir.
Marietta, GA 30061
(404) 427-3101
George Kordares - Director of Chemistry
Services:
Independent testing laboratory provides radon detection and consulting services.
Qualifications:
EPA RMP, 32 years of experience.
Area Served:
AL, FL, GA, MS.

MADRONA DIVERSIFIED, INC.
1938 Fairview Ave., E.
Seattle, WA 98102
(206) 324-0920
Samuel E. Kelly
Area Served:
WA.
EPA RMP 5:
CC.

MAHONEY HEATING AND COOLING
13620 C St.
Omaha, NE 68144
(402) 895-3333
Pat Mahoney - President
Bob Mahoney - General Manager
Services:
Provides radon detection and mitigation services.
Qualifications:
Licensed radon mitigation contractor for NE.
Area Served:
IA, NE.
EPA RMP 5:
CC.

MAINE DEPARTMENT OF HUMAN SERVICES/BUREAU OF HEALTH/PUBLIC HEALTH LABORATORY
State House Station 12
Augusta, ME 04333
(207) 289-2727
Philip Haines - Director
Cheryl L. Baker - Chemist
Services:
Laboratory provides testing services to determine radon levels in homes and buildings. Non-commercial laboratory for research purposes or state agencies' assistance only.
Area Served:
Nationwide.
EPA RMP 5:
CC(P).

MARTIN J. EGAN ASSOCIATES
37 Waldron Ave.
Summit, NJ 07901
(201) 522-0931
Martin J. Egan - Owner
Area Served:
NJ.

MARTIN MARIETTA ENERGY SYSTEMS
P.O. Box 1410 Hobbs Rd.
Paducah, KY 42001
(502) 444-6311
Rickie Byrd
Services:
Non-commercial laboratory for research purposes or state agencies' assistance only.
Area Served:
KY.
EPA RMP 5:
CC(P).

MARYLAND RADON LABORATORY, INC.
P.O. Box 2578
Columbia, MD 21045
(301) 997-2366
Paul Jennemann - President
Services:
Provides radon detection services using charcoal canister, grab sample, and continuous radon monitor methods.
Qualifications:
AARST, NBS, EPA RMP.
Area Served:
Nationwide.

EPA RMP 5:
CC(P), CR(P), GR(P).

MASSACHUSETTS INSTITUTE OF TECHNOLOGY/RADON PROJECT
Building 26 Room 421
Cambridge, MA 02139
(617) 253-4151
Lee Grodzins - Professor
Services:
Non-commercial laboratory for research purposes or state agencies' assistance only.
Area Served:
Nationwide.
EPA RMP 5:
CC(P), CR(P), GR(P).

MASSACHUSETTS RADON ALERT, INC.
20 Robertson St.
Quincy, MA 02169
(617) 471-9403, (617) 331-7900
Joseph Lamparelli - President
Eleanor Lamparelli - Vice President
Services:
Provides radon detection services for commercial and residential structures.
Qualifications:
Texas A&M radon seminar.
Area Served:
CT, MA, NH, RI.

MASTER BUILDING SERVICE
245 Huguenot St. Suite 107
New Rochelle, NY 10801
(914) 235-8282
Marvin Kapus
Area Served:
NY.

MATTER RESEARCH, INC.
1923 New York Ave.
Huntington Station, NY 11746
(516) 549-8777
Richard Heimer - President
Esther Heimer - Executive Director
Services:
Provides radon detection services using charcoal canister method.
Qualifications:
NRA.
Area Served:
NY.

MAYA-TEK, INC.
P.O. Box 1092
Chadds Ford, PA 19317
(215) 358-5453
Mark D. Daniels - President
Len W. Daniels - Specialist
Services:
Provides radon detection and mitigation services.
Area Served:
DE, NJ, PA.

MBI RADON TECHNOLOGY
P.O. Box 21-157
St. Paul, MN 55121
(612) 454-4534
Bob Ronning
Area Served:
Nationwide.
EPA RMP 5:
CC.

MCCRONE ENVIRONMENTAL SERVICES, INC.
1412 Oakbrook Dr. Suite 100
Norcross, GA 30093
(404) 368-9600
Kent Sprague
Area Served:
Nationwide.
EPA RMP 5:
AT, CC.

MCGUIRE BUILDERS, INC.
Rte. 2, Box 1840
Warfordsburg, PA 17267
(717) 294-3108
Robert McGuire - Vice President
Area Served:
PA.

MCHILL, A.D. AND SONS, INC.
1121 Second St. Pike
Richboro, PA 18954
(215) 364-2784
A.D. McHill - General Contractor
Area Served:
PA.

MCWILLIAMS PEST CONTROL
7014 McCallum St.
Philadelphia, PA 19119
(215) 849-2223
Alfred R. McWilliams - Owner
Michael T. Stafford
William Vetter
Qualifications:
AARST.
Area Served:
PA.

MECHANICAL EQUIPMENT SERVICES CORPORATION
3122 49th St.
Des Moines, IA 50310
(515) 276-9131
Charlie Stephenson - Owner
Services:
Provides radon detection and mitigation services.
Area Served:
IA.

MECKLENBURG PURIFIER
P.O. Box 404
Newell, NC 28126
(704) 537-8503
Eric Roberts
Area Served:
NC, SC.
EPA RMP 5:
CC.

MEDLAB ENVIRONMENTAL TESTING, INC.
P.O. Box 2045
Wilmington, DE 19899
(302) 655-5227, (800) 633-5221
James Boylan - Vice President
Sev Sinasian - Laboratory Director
Services:
Provides radon detection services, distributes charcoal canisters to public.
Qualifications:
NRA.
Area Served:
DE, MD, NJ, PA.

MEDRAD, INC.
Kappa Manor II 271 Kappa Dr.
Pittsburgh, PA 15238-2870
(412) 767-4600
John Friel
Services:
Non-commercial laboratory for research purposes or state agencies' assistance only.
Area Served:
Nationwide.
EPA RMP 5:
CR(P).

MENKE CONSULTING, INC.
6070 Routzong Rd.
Greenville, OH 45331
(513) 447-4225
Tom Menke - President
Bill Hinger - Consultant
Services:
Provides radon detection and environmental assessment services for home owners and the real estate industry.
Area Served:
Nationwide.
EPA RMP 5:
CC.

MERIT ENVIRONMENTAL SERVICES, INC.
P.O. Box 3083

Berea, KY 40403
(606) 986-7070
David McFarland - President
Services:
Provides radon detection and mitigation services.
Qualifications:
ARA, professional geologist and certified soil scientist.
Area Served:
AK, AR, FL, GA, IA, IL, IN, KY, MO, NC, OH, TN, VA, WV.
EPA RMP 5:
CC.

METRO RADON MEASUREMENT
1565 Cliff Rd. Suite 3-101
Eagan, MN 55122
(612) 681-1829
Steven Joyce - President
Sandra Joyce - Owner
Services:
Distributes charcoal canisters and alpha track detectors.
Qualifications:
EPA RMP.
Area Served:
Nationwide.
EPA RMP 5:
AT, CC.

METRO SERVICE LABORATORIES, INC.
6309 Fern Valley Pass
Louisville, KY 40228
(502) 964-0865, (800) 678-6878
Gregory A. Mayer
Area Served:
KY.

METROPOLITAN RADON SERVICE
4011 Newbys Bridge Rd.
Chesterfield, VA 23832
(804) 745-4702
Linda Alt - President
Wayne Alt - Vice President
Services:
Provides radon detection and mitigation services.
Qualifications:
EPA RMP.
Area Served:
Nationwide.
EPA RMP 5:
CC.

METROPOLITAN REAL ESTATE ANALYSIS
1810 South Josephine
Denver, CO 80210
(303) 777-7594
Kay Shofner - Owner
Services:
Provides radon detection services and inspections/analysis for real estate industry.
Area Served:
CO.
EPA RMP 5:
CC.

MICROBAC ENVIRONMENTAL LABORATORY
P.O. Box 49
Fayetteville, NC 28303
(919) 864-1920
Ronald W. Sanders
Area Served:
CT, GA, IA, IL, IN, KY, MA, MD, ME, MI, NC, NH, NJ, NY, OH, PA, RI, SC, TN, VA, VT, WV.
EPA RMP 5:
CC.

MICROBAC LABORATORIES, INC./ERIE TESTING DIVISION-MAIN OFFICE
2411 West 26th St.
Erie, PA 16506
(814) 833-4790
James K. Willmot - Operations Manager
Mark Bannister - Project Supervisor
Don Tyler - Radon Analyst
Jeff Porte
Services:
Environmental testing laboratory for radon detection companies, primary laboratory facility, distributes charcoal canisters with 24 hour results.
Qualifications:
EPA RMP.
See Ad Page 55
Area Served:
CT, DC, DE, FL, GA, IL, IN, KY, LA, MA, MD, ME, NC, NH, NJ, NY, OH, PA, RI, SC, TN, VA, VT, WV.
EPA RMP 5:
CC(P).

MICROBAC LABORATORIES, INC./HANOVER DIVISION
701 3rd St.
Hanover, PA 17331
(717) 633-6011
Ronald M. Barber
Area Served:
CT, DC, DE, FL, GA, IA, IL, IN, KY, MA, MD, ME, MI, NC, NH, NJ, NY, OH, PA, RI, SC, TN, VA, VT, WV.
EPA RMP 5:
CC.

MICROBAC LABORATORIES, INC./KENTUCKY TESTING DIVISION
1121 West Broadway
Louisville, KY 40203
(502) 583-5256, (502) 589-6019
Pamela A. Ellis
Area Served:
CT., DC, DE, FL, GA, IL, IN, KY, LA, MA, MD, ME, NC, NH, NY, OH, PA, RI, SC, TN, VA, VT, WV.
EPA RMP 5:
CC.

MICROBAC LABORATORIES, INC./MASSACHUSETTS TESTING DIVISION
202 Bussey St.
Dedham, MA 02026
(617) 326-7117
Andrew W. Niemann - Laboratory Director
Robert McMahon - Assistant Director
Services:
Provides radon detction services.
Qualifications:
EPA RMP.
Area Served:
CT, DC, DE, FL, GA, IA, IL, IN, KY, MA, ME, MI, MD, NH, NJ, NC, NY, OH, PA, RI, SC, TN, VA, VT, WV.
EPA RMP 5:
CC.

MICROBAC LABORATORIES, INC./NEW CASTLE TESTING DIVISION
R. D. #3 Pulaski Rd.
New Castle, PA 16105
(412) 854-4212
Mary V. Wolford
Area Served:
CT, DC, DE, FL, GA, IA, IL, IN, KY, MA, MD, ME, MI, NC, NH, NJ, NY, OH, PA, RI, SC, TN, VA, VT, WV.
EPA RMP 5:
CC.

MICROBAC LABORATORIES, INC./PITTSBURGH TESTING DIVISION
4580 McKnight Rd.
Pittsburgh, PA 15237
(412) 931-5851
William A. Laird - Vice President
Services:
Provides radon detection services using charcoal canister method.
Area Served:
CT, DC, DE, FL, GA, IA, IL, IN, KY, MA, MD, ME, MI, NC, NH, NJ, NY, OH, PA, RI, SC, TN, VA, VT, WV.
EPA RMP 5:
CC.

MICROBAC LABORATORIES, INC./SCHILLER DIVISION
449 Rochester Rd.
Pittsburgh, PA 15237-1733
(412) 369-4830
David J. Danis - Laboratory Director
Area Served:
CT, DC, DE, FL, GA, IA, IL, IN, KY, MA, MD, ME, MI, NC, NH, NJ, NY, OH, PA, RI, SC, TN, VA, VT, WV.
EPA RMP 5:
CC.

MICROBAC LABORATORIES, INC./SEAWAY DIVISION
542-544 Conkey St.
Hammond, IN 46324
(219) 932-1770
Karen A. Erny - Laboratory Director
Services:
Provides radon detection services using charcoal canister method.
Qualifications:
EPA RMP.
Area Served:
CT, DC, DE, FL, GA, IA, IL, IN, KY, MA, MD, ME, MI, NC, NH, NJ, NY, OH, PA, RI, SC, TN, VA, VT, WV.
EPA RMP 5:
CC.

MICROBAC LABORATORIES, INC./STRAND TESTING DIVISION
110 Kinzey St.
Johnstown, PA 15904
(814) 266-9548
Gerard M. Hazic
Area Served:
CT, DC, DE, FL, GA, IL, IN, KY, LA, MA, MD, ME, NC, NH, NJ, NY, OH, PA, RI, SC, TN, VA,VT.
EPA RMP 5:
CC.

MICROBAC MID-ATLANTIC, INC.
5808 Jefferson Ave.
Newport News, VA 23605
(804) 245-8613
J. Trevor Boyce
Area Served:
Nationwide.
EPA RMP 5:
CC.

MID AMERICA RADON TESTING, INC.
P.O. Box 25643
Overland Park, KS 66225
(913) 469-1997
Bill Slentz - President
Services:
Primary lab for radon analysis.
Qualifications:
EPA RMP.
Area Served:
Nationwide.
EPA RMP 5:
CC(P).

MIDWEST HEALTH PROGRAMS, INC.
P.O. Box 3023
Urbana, IL 61801
(217) 367-5293
Hector Mandel - President
Services:
Provides radiation safety technical services.
Qualifications:
Certified health physicist.
Area Served:
IL.

MIDWEST RADON ANALYSIS, INC.
P.O. Box 33093
Cleveland, OH 44133
(216) 582-4848
Ray J. Gannon
Area Served:
Nationwide.
EPA RMP 5:
CC.

MIDWEST RADON INSPECTION, INC.
25005 West 129 Ter.
Olathe, KS 66061
(913) 780-2000
Gary Hodgden - President
Steve Hodgden - Vice President
Services:
Provides radon detection services and mitigation consultation.
Area Served:
KS, MO.

MINCHIK AND MINCHIK
6319 Utah Ave. NW
Washington, DC 20015
(202) 364-8845
Richard Minchik - President

MINUTEMAN HOME INSPECTORS, INC.
90 Conz St.
Northampton, MA 01060
(413) 586-1009, (413) 549-1800
Matthew Zacarian
Qualifications:
AARST.
Area Served:
MA.

MISCAVAGE AND COMPANY
2814 Pearl Ave.
Allentown, PA 18103
(215) 797-3865
Joe Miscavage - General Contractor
Area Served:
PA.

MISSOURI TESTING LABORATORIES, INC.
P.O. Box 1
Trenton, MO 64683
(816) 359-6833
Sid R. Rogers
Area Served:
MO.

MIXLEE ENVIRONMENTAL RESEARCH CORPORATION
684-B Commerce Dr.
Fairfield, CT 06430
(203) 336-1211, (800) 545-1211
Norman Schenck - Vice President
Wendy Bowers
Services:
Provides radon detection services using continuous radon monitors,grab sampling, alpha track, and charcoal canister methods.
Qualifications:
EPA RMP, CT radon testing techniques workshop.
Area Served:
CT, MA, ME, NH, NY, RI, VT.
EPA RMP 5:
AT, CC, CR(P), GR(P), GW(P), CW(P).

MOBILE EMISSIONS TESTING CORPORATION
1780 Maple St. Suite 21
Northfield, IL 60093
(312) 501-2525
Keevin Schier - President
Services:
Provides radon detection services using charcoal canister, alpha track, and E-Perm methods.
Area Served:
IL.

MOBILE RADON DIAGNOSTICS, INC.
1018 Delta Ave. Suite 302
Cincinnati, OH 45208
(513) 321-5300
Richard J. Bolan
Area Served:
IN, KY, OH.
EPA RMP 5:
CW(P).

MONARCH ANALYTICAL LABORATORIES, INC.
P.O. Box 2990
Toledo, OH 43606
(419) 535-1780
Ernest L. Lippert
Area Served:
OH.

MONTGOMERY LABORATORIES
555 East Walnut
Pasadena, CA 91109
(818) 796-9141
Andrew D. Eaton - Laboratory Director
Services:
Analytical laboratory specializes in testing water for radon.
Qualifications:
Numerous state certifications to perform radiological analyses.
Area Served:
Nationwide.

MORRISON WATERPROOFING COMPANY, INC.
7644 L Fullerton Rd.
Springfield, VA 22153
(703) 569-1666
Robert B. Morrison - President
Services:
Provides radon detection services.
Area Served:
VA.

MOSS, ROBERT
19514 Fisher Ave.
Poolesville, MD 20837
(301) 972-8730
Robert Moss
Area Served:
MD.

MR. RENOVATOR
395 South End Ave. Apt. 27P
New York, NY 10280
(212) 321-9129
Arthur M. Diamond
Qualifications:
AARST.
Area Served:
NY.

MRC HOME INSPECTIONS
P.O. Box 157
Wyckoff, NJ 07481
(201) 891-7057
Henry H. Doron
Area Served:
CT, NJ, NY.
EPA RMP 5:
CC.

N

NALEPA, DAN
6 East Courtney St.
Dunkirk, NY 14048
(716) 366-1778
Dan Nalepa
Services:
Provides radon detection services.
Area Served:
NY.
EPA RMP 5:
CC.

NATIONAL ASSOCIATION OF HOME BUILDERS (NAHB)/NATIONAL RESEARCH CENTER
400 Prince Georges Center Blvd.
Upper Marlboro, MD 20772
(301) 249-2000
John Spears - Program Manager
Services:
Non-commercial laboratory for research purposes or state agencies' assistance only.
Area Served:
Nationwide.
EPA RMP 5:
AT, CC, CR(P), EP(P), GR(P).

NATIONAL ENVIRONMENTAL ENGINEERING AND DIAGNOSTICS
Box 519
Rockland, MA 02370
(617) 871-4426
Edmund M. Cannon
Area Served:
MA.

NATIONAL FUEL
2484 Seneca St.
Buffalo, NY 14210
(716) 824-2500
Thaddeus Kolankowski
Area Served:
NY.

NATIONAL FUEL
800 State St.
Erie, PA 16501
(814) 871-8000
Norman E. Filley
Area Served:
PA.

NATIONAL PROPERTY INSPECTIONS
7917 North Anita Dr.
Kansas City, MO 64151
(816) 587-9595
Clyde Yost - Owner
Services:
Provides radon testing services.

Qualifications:
AARST.
Area Served:
KS, MO.
EPA RMP 5:
AT, CC.

NATIONAL PROPERTY INSPECTIONS
1344-S Briar Creek Rd.
Charlotte, NC 28205
(704) 568-6324
Tom Leisinger - Owner
Services:
Provides radon detection services.
Area Served:
NC.

NATIONAL PROPERTY INSPECTIONS
11212 Davenport St. Suite 15
Omaha, NE 68154
(402) 330-1074
Mark Culbertson - Owner
Services:
Provides radon detection services using charcoal canisters and alpha track methods
Qualifications:
AARST, NE DOH, EPA RMP.
Area Served:
IA, NE.
EPA RMP 5:
AT.

NATIONAL PROPERTY INSPECTIONS
247 Sunnyhill Ln.
Oconomowoc, WI 53066
(414) 569-9377
Ken Helbing - President
Services:
Provides radon detection services using activated charcoal canisters.
Area Served:
WI.

NATIONAL PROPERTY INSPECTIONS, INC.
224 South 108th Ave. Suite 2
Omaha, NE 68154
(402) 333-9807, (800) 333-9807
Roland Bates - President
Services:
Provides radon detection services to home buyers and realtors.
Area Served:
Nationwide.

NATIONAL RADON CONTROL, INC.
P.O. Box 5342
North Branch, NJ 08876
(201) 231-0844
David T. DeNeufville
Area Served:
NJ.

NATIONAL RADON LABORATORIES, INC.
1000 Johnson Ferry Rd. Suite B-145
Marietta, GA 30068
(404) 565-3886
C. Fred Stafford
Area Served:
Nationwide.
EPA RMP 5:
CC(P).

NATIONAL RADON LABORATORIES/DIVISION OF ANALYTICAL SERVICES CORPORATION
P.O. Box 10681
Arlington, VA 22210-1681
(703) 525-6672
Arthur J. Pallotta
Area Served:
Nationwide.
EPA RMP 5:
CC(P).

NATIONAL RADON LABORATORIES/DIVISION OF COLORADO SERVICES CORPORATION
14677 East Easter Ave. Suite H
Englewood, CO 80112
(303) 699-7378
Loel Passe, Jr. - President
Paul Seppo - Technician
Steve Shearer - Laboratory Chief
Lance Downing - Technician
Services:
Provides radon detection services, independent laboratory offering testing and evaluation for other service companies using the computer generated multichannel sprectroscopy method.
Qualifications:
AARST.
Area Served:
CO.

NATIONAL RADON SERVICE
P.O. Box 3596
Gaithersburg, MD 20878-0596
(301) 953-4598
Edward Boswell
Area Served:
DC, MD, PA, VA, WV.
EPA RMP 5:
CC.

NATIONAL RADON TEST COMPANY
P.O. Box 50468
Austin, TX 78763
(512) 495-9330
Kevin Callaway
Area Served:
TX.

NATIONAL RADON TESTING CENTER, INC.
18 Marshall St.
Norwalk, CT 06854
(203) 853-4777, (203) 853-1219
Michael Ivancich
Area Served:
Nationwide.
EPA RMP 5:
AT, CC.

NATIONAL RADON TESTING SERVICE
P.O. Box 266
Maple Glen, PA 19002
(215) 628-8738
Area Served:
PA.

NATIONWIDE TESTING
RD 1, Box 266B
Annandale, NJ 08801
(201) 735-2770
Peter W. Georges
Area Served:
Nationwide.
EPA RMP 5:
CC.

NATKIN SERVICE COMPANY
2241 West Altorfer
Peoria, IL 61615
(309) 691-9206, (800) 642-8391
Paul Graves
Area Served:
Nationwide.
EPA RMP 5:
CC.

NATLSCO
Rte. 22 and Kemper Dr.
Long Grove, IL 60049
(312) 540-2488
Joan Wronski
Area Served:
Nationwide.
EPA RMP 5:
CC(P).

NATURAL ENERGY CONSULTANTS
P.O. Box 586
Conifer, CO 80433
(303) 838-5291
Rob Haines - Consultant

Services:
Provides radon detection, mitigation, and consultation services.
Area Served:
Nationwide.

NATURAL FLORIDA RETROFIT, INC.
P.O. Box 301
Montverde, FL 32756
(305) 469-2173, (305) 841-4376
John J. Tooley, Jr.
Area Served:
AL, FL, GA, LA.
EPA RMP 5:
EP(P).

NATURAL PANTRY, THE
300 West 36th Ave.
Anchorage, AK 99503
(907) 563-2727
Richard Solberg
Area Served:
AK.

NATURAL RESOURCE LABORATORY, INC.
1100 Simms St.
Golden, CO 80401
(303) 233-8155
Ronald L. Keil
Area Served:
Nationwide.
EPA RMP 5:
CC(P).

NATURAL RESOURCE TECHNOLOGIES
32131 Steven Way
Conifer, CO 80433
(303) 850-9335
George H. Sylvester
Area Served:
CO, NM, WY.
EPA RMP 5:
CC.

NDL ORGANIZATION, INC., THE
P.O. Box 791
Peekskill, NY 10566
(914) 737-7200
Theodore E. Rahon
Area Served:
Nationwide.
EPA RMP 5:
CC.

NEA, INC.
10950 Southwest 5th St. Suite 380
Beaverton, OR 97005
(503) 643-4661
John A. Cooper
Area Served:
OR.

NEBRASKA WESLEYAN UNIVERSITY/PHYSICS DEPARTMENT/RADON MEASUREMENTS LABORATORY
5000 St. Paul Ave.
Lincoln, NE 68504
(402) 465-2252, (402) 465-2247
Walter French - Professor
Area Served:
CO, IA, KS, MN, ND, NE, SD.
EPA RMP 5:
CC(P).

NELSON, C. STUART
1007 River Rd.
Quarryville, PA 17566
(717) 548-3151
C. Stuart Nelson
Area Served:
PA.

NERING, INC.
P.O. Box 209
Ridgefield, CT 06877
(203) 431-3597
Catherine Hershey
Alexander Nering
Area Served:
Nationwide.
EPA RMP 5:
AT, CC(P), CR(P), GR(P).

NEW ENGLAND RADON SERVICES
95 Vauxhall St.
New London, CT 06320
(203) 442-6675
Martin Grisham - Owner
Services:
Provides radon detection and mitigation consulting services for the do-it-yourself homeowner.
Area Served:
CT, MA, ME, NH, NY, RI, VT.
EPA RMP 5:
CC.

NEW ENGLAND RADON, LTD.
373 Main St.
Salem, NH 03079
(603) 893-4260, (800) 637-2366
Daniel Espinal - Vice President
Services:
Provides radon detection and mitigation services, distributes radon test kits, laboratory analysis.
Qualifications:
EPA RMP.
Area Served:
CT, MA, ME, NH, PR, RI, VT.
EPA RMP 5:
CC(P).

NEW JERSEY DEPARTMENT OF ENVIRONMENTAL PROTECTION/DIVISION OF ENVIRONMENTAL QUALITY/BUREAU OF ENVIRONMENTAL RADIATION
729 Alexander Rd. CN-415
Trenton, NJ 08625
(609) 987-6396, (800) 648 0394
Robert Stern - Chief
Gerald Nicholls - Radiation Programs
Bill Hoyle - Radon Programs
Elaine Makatura - Community Relations
Services:
Non-commercial laboratory for research purposes or state agencies' assistance only.
Area Served:
NJ.
EPA RMP 5:
CC(P), GR(P), GW(P), CW(P).

NEW JERSEY DEPARTMENT OF ENVIRONMENTAL PROTECTION/DIVISION OF ENVIRONMENTAL QUALITY/BUREAU OF RADIATION AND INORGANIC ANALYTICAL SERVICES
380 Scotch Rd. CN-411
Trenton, NJ 08628
(609) 530-4116
Patricia Gardner - Chief
Services:
Non-commercial laboratory for research purposes or state agencies' assistance only.
Area Served:
NJ.
EPA RMP 5:
CC(P).

NEW JERSEY DEPARTMENT OF ENVIRONMENTAL PROTECTION/DIVISION OF SCIENCE AND RESEARCH
401 East State St. 7th Floor, West Wing
Trenton, NJ 08625
(609) 984-6070
Robert Tucker - Director
Barbara Litt - Program Manager, Radon
Services:
Funding development of personal radon monitor to compare human exposure to radon with levels in the home. Testing 150 wells in state to compare level of radon contamination in different strata.
Area Served:
NJ.

NEW JERSEY HOME SERVICE, INC.
16 Westwood Dr., N.
West Orange, NJ 07052
(201) 669-8166
William K.H. Chow
Area Served:
NJ.
EPA RMP 5:
CC.

NEW MEXICO DEPARTMENT OF HEALTH AND ENVIRONMENT/COMMUNITY SERVICES BUREAU
P.O. Box 968
Santa Fe, NM 87504-0968
(505) 827-2948
J. Margo Metzger-Keele - Radon Programs
Services:
Non-commercial laboratory for research purposes or state agencies' assistance only.
Area Served:
NM.
EPA RMP 5:
CC(P).

NEW YORK STATE HEALTH DEPARTMENT/WADSWORTH CENTER FOR LABORATORIES AND RESEARCH/LABORATORY FOR INORGANIC AND NUCLEAR CHEMISTRY
Empire State Plaza Room D366-B
Albany, NY 12201
(518) 474-6649
Christopher D. Parker - Research Scientist
Charles Kunz - Research Scientist
Services:
Non-commercial laboratory for research purposes or state agencies' assistance only.
Area Served:
NY.
EPA RMP 5:
CC(P).

NIGRO CONSTRUCTION COMPANY, INC.
37 James Rd.
Monroe, NY 10950
(914) 783-4254
Dan Nigro - President
Area Served:
NY.

NIMCO/NESSI
P.O. Box 356
Menasha, WI 54952-0356
(414) 725-0357
Mike Thornell - Vice President
Services:
Provides radon detection services using charcoal canister method.
Qualifications:
NRA.
Area Served:
WI.

NITE HAWK ENTERPRISES, INC.
17 Lincoln Ave.
White Plains, NY 10606
(914) 949-3207
Jerry Shepherd
Area Served:
NY.

NITON CORPORATION
P.O. Box 9153
Lexington, MA 02173
(617) 229-4929, (800) 451-6467
Lee Grodzins - President
Ethel Romm - Vice President
Diana B. Doran - Operations Manager
Ann McGuineas - Marketing
Services:
Manufactures and distributes radon detection equipment including ultra-sensitive charcoal canisters and liquid scintillation counting method licensed by Packard Instrument Pico-Rad Sysyem.
Area Served:
Nationwide.
EPA RMP 5:
CC(P), CR(P).

See Ad Page 202

NODAR, INC.
Route 5 Box 333R
Athens, AL 35611
(206) 729-1900, (800) 445-3297
William Young
Area Served:
Nationwide.
EPA RMP 5:
CC, EP.

NORDBERG ASSOCIATES
Clover Hill Farm P.O. Box 520
Alfred, ME 04002
(207) 490-1105
Margit Lassen Nordberg
Area Served:
CT, DE, ME, NH, VT.

NORTH CAROLINA DEPARTMENT OF HUMAN RESOURCES/DIVISION OF FACILITY SERVICES/RADIATION PROTECTION SECTION
701 Barbour Dr.
Raleigh, NC 27603-2008
(919) 733-4283
Dayne H. Brown - Chief
Richard M. Fry - Deputy Chief
Felix Fong - Radiation Surveillance
Harry J. Newman - Radiation Surveillance
Services:
Non-commercial laboratory for research purposes or state agencies' assistance only.
Area Served:
NC.
EPA RMP 5:
CC(P), GR(P).

NORTH EAST ENVIRONMENTAL SERVICES
424 Mine Rd.
Asbury, NJ 08802
(201) 735-6080, (800) 447-2366
David C. Grammer
Area Served:
NJ, VA.
EPA RMP 5:
CC.

NORTHEAST RADON CONSULTANTS
180 Zachary Rd.
Manchester, NH 03103
(603) 622-9391
R.R. Comtois
Area Served:
NH.

NORTHEAST RADON TESTING SERVICE, INC.
P.O. Box 94
Stormville, NY 12582
(914) 221-0236
Karen Bard
Area Served:
Nationwide.
EPA RMP 5:
CC.

NORTHEAST VECTOR MANAGEMENT, INC.
P.O. Box 754
Freehold, NJ 07728
(201) 431-9563
William J. Zawicki
Area Served:
NJ.

NORTHEASTERN RADON SERVICE
21 Rte. 202
Montville, NJ 07045
(201) 402-1030
Bogdan Choruza - President
Services:
Provides radon detection and mitigation services.
Area Served:
NJ.

NORTHERN ENVIRONMENTAL LABORATORY, INC.
RR 1, Box 1810
Dixmont, ME 04932-9740
(207) 234-2387
Ulois Vanags
Area Served:
Nationwide.
EPA RMP 5:
AT, CC(P).

NORTHERN NECK RADON SERVICE
Rte. 1, Box 259-E
Hague, VA 22469
(804) 472-4252
Milford L. Brann - Owner
Jean E. Brann - Manager
Services:
Provides radon detection services using EPA-approved charcoal canister method, provides radon mitigation services.
Area Served:
Nationwide.
EPA RMP 5:
CC.

NORTHWEST FURNACE COMPANY, INC.
3139 5th Ave., S.
Fort Dodge, IA 50105
(515) 944-7262
Rob Kruthoff
Area Served:
IA.

NORTON CONSERVATION COMPANY, INC.
P.O. Box 583
Hackettstown, NJ 07840
(201) 852-6046
Joseph Norton - President
Bryan Tafuri - Pennsylvania Manager
Services:
Provides radon mitigation and energy management services for residential and commercial markets.
Qualifications:
NJ DOE, NJ DEP, PA DER.
Area Served:
Nationwide.
EPA RMP 5:
CC.

NUCHEM CORPORATION
P.O. Box 3369 2900 William Penn Hwy.
Palmer, PA 18043
(215) 252-9050
Leroi R. Yaffey - Owner
Services:
Provides radon detection and mitigation services, operates a primary laboratory.
Area Served:
PA.

NUCLEAR MEASUREMENTS CORPORATION
2460 North Arlington Ave.
Indianapolis, IN 46218
(317) 543-4419, (317) 546-2415
Alan Brent Scott
Area Served:
IN.

NUCLEAR RESEARCH CORPORATION
P.O. Box H
Warrington, PA 18976
(215) 343-5900
Tom T. Tseng
Area Served:
PA.

NUCLEAR SOURCES AND SERVICES, INC.
P.O. Box 34042
Houston, TX 77234
(713) 641-0391
Charles T. Gallagher
Area Served:
Nationwide.
EPA RMP 5:
CC(P).

NUCLEAR TECHNOLOGY, INC./RADON CHEK
20 H. South State St.
Westerville, OH 43081
(614) 899-2435
James M. Copp - President
Daniel L. Bowman - Secretary
Services:
Manufactures and distributes charcoal adsorption canisters.
Qualifications:
ARA, EPA RMP.
Area Served:
Nationwide.
EPA RMP 5:
CC.

NUCLEON LECTERN ASSOCIATES, INC.
1916 Grayslake Dr.
Silver Spring, MD 20906
(301) 774-3301
Michael S. Terpilak - President
Area Served:
MD.

NUCLEUS, THE
P.O. Box 2561
Oak Ridge, TN 37830-2561
(615) 482-4041
Larry Webb - Marketing Manager
Services:
Manufactures and distributes a complete radon monitoring system for charcoal canisters including data analysis software and calibration sources. Non-commercial laboratory for research purposes or state agencies' assistance only.
Area Served:
Nationwide.
EPA RMP 5:
CC(P).

See Ad Page 192

NUWS LABORATORIES, INC.
28 Montgomery St.
Rhinebeck, NY 12572
(914) 876-5666
Henry Mikaelian - President
Services:
Provides radon gas screening and radon in water detection services.
Qualifications:
NY State Energy Office seminar.
Area Served:
NY.

O

O'RIORDAN INSULATION COMPANY
8901 North Nebraska Ave.
Tampa, FL 33604
(813) 932-0008
Joe O'Riordan - Vice President
Services:
Provides radon detection and mitigation services.
Area Served:
FL.

O.K. REMS COMPANY
10 Hillsborough Rd.
Flemington, NJ 08822
(201) 782-3398
Bob Kretvix - Owner
Area Served:
NJ.

O.K. REMS CORPORATION
181 London Rd.
Hebron, CT 06248
(203) 647-7440
Jeffrey C. Olcott
Area Served:
CT.

O.K. REMS CORPORATION
174 Flock Rd.
Mercerville, NJ 08619
(609) 588-9627
Jeffrey C. Olcott - Owner

Area Served:
Nationwide.
EPA RMP 5:
CC, GW(P).

OAK HILL RADON TESTING AND REPAIR
15 1/2 North State St.
Westerville, OH 43081
(614) 895-8365
Jim Lubinsky - President
Services:
Provides radon detection, mitigation and consultation services.
Qualifications:
EPA RMP, ARA.
Area Served:
OH.
EPA RMP 5:
AT, CC, CR(P), EP.

OCCUPATIONAL AND ENVIRONMENTAL HEALTH CONSULTING SERVICES
635 Harding Rd.
Hinsdale, IL 60521
(312) 325-2083
Robert C. Brandys - President
Services:
Provides radon detection and mitigation services to industry, architects, and homeowners.
Qualifications:
Certified industrial hygienist, registered environmental assessor.
Area Served:
Nationwide.

OLD COLONY PEST CONTROL COMPANY
P.O. Box 580
Pembroke, MA 02359
(617) 294-0937
Alan J. Jacobson - General Manager
Area Served:
MA.

ONTARIO RESEARCH FOUNDATION (ORF)
Sheridan Park Research Community
Mississauga, ON L5K 1B3
(416) 822-4111
Peter Piersol - Consultant
Services:
Crown Corporation analyzes results of radon testing and provides results to provincial government.
Area Served:
ON.

ORKIN ENVIRONMENTAL SERVICES
P.O. Box 647
Atlanta, GA 30301
(404) 888-2801
Joseph H. Cline
Area Served:
Nationwide.
EPA RMP 5:
CC.

ORR-DUVALL AND ASSOCIATES
22346 Metamora
Birmingham, MI 48010
(313) 644-0706
Owen DuVall
Area Served:
MI.

P

PACIFIC RADIATION CORPORATION
2945 Stonehill Dr.
Altadena, CA 91001
(818) 798-8100
Dan Gollnick - President
Services:
Provides radon detection service using charcoal canister method.
Qualifications:
EPA RMP, certified health physicist.
Area Served:
AZ, CA, NV.
EPA RMP 5:
CC(P).

PACKARD INSTRUMENT COMPANY
2200 Warrenville Rd.
Downers Grove, IL 60515
(312) 969-6000, (800) 323-1891
Jim Floeckher - Product Manager
Services:
Manufactures and distributes the Pico-Rad complete radon system for air and water, including liquid scintillation counters, software, canisters, etc. Non-commercial laboratory for research purposes or state agencies assistance only.
Qualifications:
EPA RMP.
Area Served:
Nationwide.
EPA RMP 5:
CC(P).

See Ad Page 205

PAGANO ASSSOCIATES, INC.
112 Jockey Hollow Way
Union, NJ 07083
(201) 686-2939
Joel A. Pagano
Area Served:
NJ.

EPA RMP 5:
CC.

PALMER TECHNOLOGY
P.O. Box 1072
Glen Burnie, MD 21061
(301) 766-3176
Jeffrey A. Palmer - President
Laurie E. Palmer - Operations Manager
Donna M. Andrauskas - Operations Manager
Services:
Provides radon detection services for home owners using charcoal canister and alpha track methods.
Qualifications:
Professional engineer, AARST, EPA RMP.
Area Served:
Nationwide.
EPA RMP 5:
CC.

PARADISE COMPANY
109 Greenleaf
Lumberton, TX 77656
(409) 755-3524
Linda McKinney - President
Ed McKinney - Vice President
Dale Vandine - Vice President
Liz McKinney - Secretary
Services:
Provides radon detection, mitigation, and consulting services.
Area Served:
TX.

PARAMOUNT PEST CONTROL
110 Gordon Rd.
Falls Church, VA 22046
(703) 241-9140
Walter Tilley - President
Joe Gallagher - Vice President
Kevin Kordex - Technical Director
Craig Strobel - Vice President
Services:
Provides radon detection services to the real estate industry.
Area Served:
DC, MD, VA.

PELICAN ENVIRONMENTAL CORPORATION
42 Lehigh Rd.
Wellesley, MA 02181
(617) 235-5997
Don Crawshaw - President
Jeff Crawshaw - Engineer
Randy Foster - Mitigator
Services:
Provides radon detection, mitigation and consultation services.

Qualifications:
Master's degree in civil engineering, accredited in MA for testing.
Area Served:
CT, MA, NH, RI.

PENNSYLVANIA BUILDING INSPECTORS, INC.
1273 Wilton Crescent
Yardley, PA 19067
(215) 493-9426
Arthur Brasof
Area Served:
DE, MD, NJ, NY, PA.
EPA RMP 5:
CC.

PENNSYLVANIA RADON CENTER
21 Hillbrook Cr.
Malvern, PA 19355
(215) 647-9999
Area Served:
PA.

PENNSYLVANIA STATE UNIVERSITY, THE/BREAZEALE NUCLEAR REACTOR/LOW LEVEL RADIATION MONITORING LABORATORY
101 Academic Projects Building
University Park, PA 16802
(814) 863-2647
William A. Jester - Professor
Bonnie C. Ford - Laboratory Supervisor
Services:
Laboratory established for monitoring research and service including airborne radon and radon and radium in water.
Area Served:
PA.
EPA RMP 5:
CC(P).

PENRAD, INC.
54 South Market St.
Elizabethtown, PA 17022
(717) 367-8410, (717) 367-3995
Bud Longenecker
Area Served:
PA.

PEPPADYNE RADON DETECTION SERVICE
126 East 9th St. Suite 25
St. Paul, MN 55101
(612) 293-0864
Tyrone Powell
Area Served:
MN.

PERISON, JANICE B.
13 Beaver Branch Rd.
Pennsylvania Furnace, PA 16865
(814) 234-1871
Janice B. Perison
Qualifications:
AARST.
Area Served:
PA.

PERMA TREAT, INC.
P.O. Box 8260
Fredericksburg, VA 22404
(703) 373-6686
Joseph R. Wilson
Area Served:
VA.

PEST CONTROL TECHNICIANS, INC.
P.O. Box 1951
Norristown, PA 19401
(215) 275-7378
Thomas F. Silvestrini - Owner
Area Served:
PA.

PEST DOCTOR
P.O. Box 32
Lithia Springs, GA 30057
(404) 920-2288
Stan Dockery
Area Served:
GA.
EPA RMP 5:
CC.

PETROLEUM ANALYTICAL LABORATORY, INC.
P.O. Drawer AK
Eaton Park, FL 33840
(813) 665-0028
Jerard A. Kent - President
Paul S. Kent - Vice President
John A. Kent - Vice President
Ann Sherrouse - Manager
Services:
Provides radon detection services using charcoal canister method.
Qualifications:
EPA RMP, Certified by state of FL, licensed professional engineer on staff.
Area Served:
Nationwide.
EPA RMP 5:
CC(P).

PHOENIX LABS
L. Rorbes Rd. Route 1, Box 247
Colebrook, NH 03578
(603) 237-5773
Gilbert McCarthy
Area Served:
Nationwide.
EPA RMP 5:
CC(P).

PHOENIX LABS
P.O. Box 472
Hollis, NH 03049
(603) 465-7868
Robert Brodeur
Area Served:
NH.

PLUM GROVE CORPORATION
33 West Queens Way Mall Old Towne
Hampton, VA 23669
(804) 722-9841
Garland H. Harrington - President
Services:
Provides radon detection and mitigation services.
Area Served:
VA.

PMI CONTRACTORS, INC.
P.O. Box 6125
Gulfport, MS 39506
(601) 863-1988, (601) 863-1998
William C. Ladnier
Area Served:
MS.

PNEUMODYNAMICS, INC.
227 Kasson Rd.
Bethlehem, CT 06751
(203) 266-7786
Frederick J. Mendyka, Jr.
Area Served:
CT.
EPA RMP 5:
CC.

POLK COUNTY HEALTH UNIT/RADIOLOGICAL HEALTH SECTION
P.O. Box 1480
Winter Haven, FL 33882-1480
(813) 294-7481 ext. 283
William T. McNally - Director
Area Served:
FL.
EPA RMP 5:
RP(P).

PORTER CONSULTANTS, INC.
125 Argyle Rd.
Ardmore, PA 19003
(215) 896-5353
S.W. Porter, Jr. - Health Physicist
M.C. Roberts - Health Physicist
J. Sykes, III - Health Physicist
Services:
Offers training courses on radon detection

and mitigation, provides radon detection services using charcoal canister, grab sample, alpha track methods, water and soil.
Qualifications:
ABH certified health physicists on staff, certified radon testing specialists, EMP RMP, Vice President of AARST.
Area Served:
DE, NJ, PA.
EPA RMP 5:
AT, CC, GW(P), CW(P).

POTOMAC AIR TESTING
P.O. Box 30091
Bethesda, MD 20814
(301) 657-4391
Robert B. Samworth
Area Served:
Nationwide.
EPA RMP 5:
EP(P).

PRE-PURCHASE HOME INSPECTION
3512 Breeland Ave.
Louisvile, KY 40241
(502) 425-6810
Herbert T. Strausburg
Area Served:
KY.

PRECISION RADON TESTING
302 Leidy Rd.
Souderton, PA 18964
(215) 721-1958, (800) 545-8378
Doris A. Burner - President
Services:
Provides radon detection services using charcoal canisters and alpha track methods.
Qualifications:
EPA RMP, state seminar on radon reduction methods
Area Served:
PA.

PRIGEL RADON TESTING
2866 Travis Ct.
York, PA 17403
(717) 741-3188, (301) 252-1246
Don Prigel
Area Served:
MD, PA.
EPA RMP 5:
CR(P).

PRINCETON HOME INSPECTION SERVICE
197 Wall St.
Princeton, NJ 08540
(609) 921-3775
Clive Usiskin - President
Services:
Provides radon detection and mitigation services.
Area Served:
NJ, PA.
EPA RMP 5:
CC.

PRINCETON PLASMA PHYSICS LABORATORY
P.O. Box 451
Princeton, NJ 08543
(609) 683-3319, (609) 243-3319
Joseph M. Greco
Services:
Non-commercial laboratory for research purposes or state agencies' assistance only.
Area Served:
NJ.
EPA RMP 5:
CC(P).

PRINCETON TESTING LABORATORY
Princeton Service Center 3490 U.S. Rte. 1
Princeton, NJ 08540
(609) 452-2037
Anthony Damato
Area Served:
Nationwide.
EPA RMP 5:
AT, CC.

PRO CHEK, INC.
Box 2000
Arden, NC 28704
(800) 522-2435
B.V. Alvarez
Area Served:
Nationwide.
EPA RMP 5:
CC(P).

PRO-GUARD BUILDING INSPECTIONS, INC.
RD 1, Box 193B
Auburn, PA 17922
(717) 754-7673
William X. Cairnes
Services:
Provides radon mitigation services.
Area Served:
MD, PA.

PRO-TEC HOME INSPECTORS, INC.
P.O. Box 132
Franklin, NJ 07416
(201) 827-4996
Harry Benjamin
Area Served:
NJ.
EPA RMP 5:
CC.

PRO-TECH RADON
459 Forestwood Dr.
Gahanna, OH 43230
(614) 475-0023
Mark Majeski
Area Served:
OH.

PROFESSIONAL HOME ANALYSIS
1025 Northeast Sanchez Ave.
Ocala, FL 32670
(904) 694-3982
G. Douglas Foreman - Owner
Keith Brandies - Analysis
Jeff Foreman - Analysis
Services:
Provides radon detection services.
Area Served:
FL, GA.
EPA RMP 5:
CR(P).

PROFESSIONAL HOME INSPECTIONS, INC.
28790 Pine Dr.
Evergreen, CO 80439
(303) 674-4620
G. Thomas Martin - President
Services:
Provides radon detection services and consultation on mitigation design and building materials.
Qualifications:
AARST, ASHI, Colorado School of Mines course on radon and the indoor environment.
Area Served:
Nationwide.

PROFESSIONAL HOME INSPECTIONS, INC.
R.D. 1, Fenimore Rd. Box 371-A
Medford, NJ 08055
(609) 261-7374
Michael D. Huber
Area Served:
NJ.
EPA RMP 5:
CC.

PROFESSIONAL HOUSE DOCTORS
599 Short Ave.
Marengo, IA 52301
(319) 642-7133
Tom Voightman
Area Served:
IA.

PROFESSIONAL HOUSE DOCTORS, INC.
1406 East 14th St.
Des Moines, IA 50316

(515) 265-6667
Dane J. Shearer - President
Services:
Provides radon detection and mitigation services, manufactures blower doors with 25-6,000 cfm capacity, manufactures structural analyzers providing complete and detailed indoor air and energy analysis, performs mitigation training nationwide.
Area Served:
Nationwide.

PROFESSIONAL RADON ASSESSMENT
405 Valley Dr.
Oliver Springs, TN 37840
(615) 435-6432, (615) 690-2208
James Summers
Area Served:
TN.
EPA RMP 5:
CC.

PROFESSIONAL RADON MEASUREMENTS
P.O. Box 791
Normal, IL 61761
(309) 827-3377, (309)452-9904
John W. Foster
Area Served:
IL.
EPA RMP 5:
GW(P).

PROTECT-A-HOME
P.O. Box 584
Newtown, CT 06470
(203) 426-3000
B. Michael Meehan - Radon Diagnostician
Services:
Provides radon detection and diagnostic services for private and commercial use.
Qualifications:
EPA RMP.
Area Served:
CT, MA, NJ, NY, RI.
EPA RMP 5:
CC, CR(P).

PROTECTION INSPECTION, INC.
Rte. 11 Hill Rd.
Knoxville, TN 37938
(615) 688-9779
Norman R. Kurtz - President
Thomas A. Walkiewicz - Consultant
Services:
Provides radon detection and mitigation services to residential and commercial markets, provides "watchdog" service to prevent tampering during radon testing.
Qualifications:
ASHI, EPA RMP.
Area Served:
GA, KY, NC, TN, VA.
EPA RMP 5:
CC.

PROTECTRON, INC.
229 East Main St. Suite E
Westminster, MD 21157
(301) 857-9116
Ronald G. Harrington - President
Services:
Provides radon detection services using electronic continuous working level dosimetry.
Qualifications:
AARST.
Area Served:
DC, DE, MD, PA, VA, WV.
EPA RMP 5:
CW(P).

PUKAL CONTRACTING COMPANY
390 Moorehead Rd.
Cabot, PA 16023
(412) 352-4030
Richard Pukal - Owner
Area Served:
PA.

PYLON ELECTRONIC DEVELOPMENT COMPANY, LTD.
147 Colonnade Rd.
Nepean, ON K2E 7L9
(613) 226-7920
Dave Hanneson - Sales Manager
Chuck Palmer - Marketing
Aurel Lebel
Services:
Manufactures and distributes radon gas monitors, working level monitors, radon in water analyzers, radon gas calibration sources, scintillation cells.
Qualifications:
EPA RMP.
Area Served:
Nationwide.
EPA RMP 5:
CR(P), GR(P), GW(P), CW(P).

See Ad Page 209

PYRAMID ENVIRONMENTAL SYSTEMS, INC.
P.O. Box 548
Sparta, NJ 07871
(201) 729-9375
Jeffrey R. DeChacon
Area Served:
Nationwide.
EPA RMP 5:
CC, CW(P).

Q

QUALITY AIR RADON REDUCTION
5-70 Oakwood Rd.
Glastonbury, CT 06033
(203) 633-3629
Mike McKeon
Area Served:
CT.
EPA RMP 5:
AT, CC.

QUALITY ASSURANCE, INC.
P.O. Box 4087
Highlands Ranch, CO 80126
(303) 791-2624
Herm Henderson
Area Served:
CO.

QUALITY ENVIRONMENTS, INC.
P.O. Box 3689
Chapel Hill, NC 27515
(919) 493-7033
Sharon Berry - Vice President
Services:
Provides radon detection services and consultation.
Area Served:
NC, SC, TN, VA.
EPA RMP 5:
CC.

QUALITY HOME INSPECTIONS, INC.
81 Main St.
High Bridge, NJ 08829
(201) 638-4994
Robert S. Koczanowski
Area Served:
NJ.
EPA RMP 5:
CC.

QUALITY RADON TESTING CORPORATION
P.O. Box 11980
Chicago, IL 60611
(312) 280-4908
D.R. Thornburg - President
Services:
Provides radon detection services using charcoal canister method.
Area Served:
Nationwide.

QUANTEC
212 Lucia St.
Middlesex, NJ 08846
(201) 469-9461
Robert Sherr

Area Served:
NJ.

QUANTUM INFORMATION, LTD.
P.O. Box 467
Westown, PA 19395-0467
(215) 431-9577
Louis Macauley - President
Services:
Provides radon detection services and operates testing laboratory.
Qualifications:
PA DER, EPA RMP.
Area Served:
Nationwide.
EPA RMP 5:
CC(P).

R

R. BERNT AND SONS
96 Laural Ave. P.O. Box 721
Woodacre, CA 94973
(415) 488-9106
Ron Bernt
Area Served:
CA.

R.A.D. SERVICE AND INSTRUMENTS, LTD.
50 Silver Star Blvd. Unit 208
Scarborough, ON M1V 3L3
(416) 298-9200
Hsiang Lin Pai - President
Services:
Manufactures and distributes radon detection products including an active indoor working level surveymeter (patent pending).
Qualifications:
Certified by EPA and NEA in Europe.
Area Served:
Nationwide.
EPA RMP 5:
RP(P).

See Ad Page 206

R.E. WRIGHT ASSOCIATES, INC./RADONIX GROUP
3240 Schoolhouse Rd.
Middletown, PA 17057
(717) 944-5520
Michael Spear - Marketing
Services:
Provides radon detection and mitigation services, distributes charcoal canisters and alpha track radon test kits.
Qualifications:
PA certification pending.
Area Served:
Nationwide.

R.F. SIMON COMPANY, INC.
RD 2, Box 481
Barto, PA 19504
(215) 754-6455
Ron Simon - Project Manager
Peter Maniscalco - Diagnostic Specialist
Services:
Provides mitigation design and installation services, mitigation supply warehouse carrying a full line of radon mitigation and detection products including fans, caulks and sealants, couplings, comprehensive radon services provided to private, commercial, institutional, and government agencies, expert testimony.
Qualifications:
EPA RMP, 13 years of custom building experience, instructor for EPA training courses, consultant to Lawrence Berkeley Laboratory, Princeton University, PA DER.
Area Served:
Nationwide.
EPA RMP 5:
AT, CC, GW(P).

See Ad Page 148

R.I.C. INSPECTION AND ENGINEERING
R.R. 2 P.O. Box 2186-5
Columbia, NJ 07832
(201) 496-4818
E. Richard Galinski
Area Served:
NJ, PA.

R.J. FRIDG COMPANY, INC.
427 Hamill Rd.
Indiana, PA 15701
(412) 463-3225
Ralph J. Fridg, Jr.
Area Served:
PA.

R.J. MOORE AND ASSOCIATES, INC.
1010 West St.
Annapolis, MD 21401
(301) 269-5950
Roderic E. Ordway
Mary T. Bergmann
Area Served:
CO, DC, DE, GA, MD, NJ, PA, VA.
EPA RMP 5:
AT, CC.

R.K. OCCUPATIONAL AND ENVIRONMENTAL ANALYSIS, INC.
19 Burrows Ave.
Bernardsville, NJ 07924
(201) 766-1737
Area Served:
Nationwide.

R.K. VAN SCHAICK AND ASSOCIATES
R.R. 4 Box 4038
Greenwich, NY 12834
(518) 692-2015
Richard Van Schaick - President
Services:
Consulting engineers provide radon detection services.
Area Served:
NY.

R.T.C.
P.O. Box 832
Linbrook, NY 11563
(516) 887-1371
Wayne Cholost - Vice President
Services:
Provides radon detection services.
Area Served:
New York City-Long Island region.

R/K OCCUPATIONAL AND ENVIRONMENTAL ANALYSIS, INC.
616 Warren St.
Alpha, NJ 08865
(215) 250-5948
George D. McGuinness
Area Served:
NJ.

RA CONSULTANTS, INC.
1236 North 7th St.
Grand Junction, CO 81501
(303) 245-5237
Hoyt Mitchell
Area Served:
CO.

RA-TEK/DIVISION OF SHAMROCK, INC.
P.O. Box 496 800 East Main St.
Marshall, MN 56258
(507) 532-7313
Joan Coquyt - Analyst
Services:
Provides radon detection services to residential and commercial markets using continuous radon monitor method, distributes continuous radon monitors.
Area Served:
Nationwide.
EPA RMP 5:
CR(P).

RACHECK, INC.
P.O. Box 548
Southampton, PA 18966
(215) 322-5309
John Barr - President
Area Served:
PA.

RAD ALERT, LTD.
P.O. Box 9238
Springfield, IL 62791
(217) 546-9549
Bob Wilson - Vice President
Services:
Provides radon detection and mitigation services.
Area Served:
IL.

RAD CHEK SERVICES COMPANY
P.O. Box 173
North Granby, CT 06060-0173
(203) 653-4159
Paul Schwartz
Area Served:
CT, MA, NJ, NY.
EPA RMP 5:
CC.

RAD CHEK SERVICES COMPANY
P.O. Box 279
Livingston, NJ 07039
(201) 535-6459
Gary P. Schwartz - President
Paul L. Schwartz - President
Services:
Provides radon detection and consulting services to home owners.
Qualifications:
AARST, EPA RMP, CIH.
Area Served:
CT, MA, NJ, NY.
EPA RMP 5:
CC.

RAD CON, INC.
Box 027
Westminster, MD 21157
(301) 848-6324
Tim Kyle
Area Served:
MD.

RAD ELEC, INC.
5330 J Spectrum Dr.
Frederick, MD 21701
(301) 694-0011
Robert W. Ramsey - President
Paul Kotrappa - Vice President
John C. Dempsey - Vice President
Services:
Manufactures and distributes electret passive environmental radon monitor (E-Perm), performs research and development on new radiometric monitoring devices.
Qualifications:
Inventors of the Electret Perm Measurement System recognized by EPA.
Area Served:
Nationwide.
EPA RMP 5:
AT, CC, EP(P).

See Ad Page 194

RAD ENVIRONMENTAL SERVICES, INC.
2306 Rodeo Dr.
Durham, NC 27704
(919) 688-9226
Marc Y. Menetrez - President
Trey Marchant - Environmental Technician
Mary Menetrez - Physiologist
Steve Boukedes - Environmental Technician
Services:
Provides radon detection and mitigation services using a variety of methods.
Qualifications:
EPA RMP, NC water treatment A license.
Area Served:
Nationwide.
EPA RMP 5:
CC, CR(P), EP(P), GR(P), CW(P).

RAD ENVIRONMENTAL SERVICES, INC.
30 Basswood Ct.
Red Bank, NJ 07701
(201) 758-9324
Bernadette Davis - Vice President
Services:
Provides radon detection services using charcoal canister, E-Perm, continuous working level, and grab sampling methods, provides radon mitigation services.
Area Served:
NC, NJ.

RAD-TEST, INC.
P.O. Box 22576
Louisville, KY 40222
(502) 368-5996
Jeffrey L. Tallent
Area Served:
IN, KY.
EPA RMP 5:
CC.

RADALYTIC LABS, INC.
P.O. Box 202
McConnells, SC 29726
(803) 684-9859
David Culver - President
Services:
Provides laboratory services for radon detection by charcoal canister method.
Area Served:
Nationwide.

RADEX
265 Stone St.
Walpole, MA 02081
(617) 660-1548
Scott Eddy
Area Served:
MA.

RADEX, INC.
1382 Prince Ave. Suite A
Athens, GA 30606
(404) 548-6960
Frederick Tuck - President
Norman Herz - Vice President
Jim Kitchens - Treasurer
Services:
Provides radon detection, mitigation, and training services.
Qualifications:
State certified geologist and university health physicist, experience in commercial and private radon monitoring.
Area Served:
Nationwide.
EPA RMP 5:
AT, CC.

RADIATION CONSULTANTS OF SOUTHWEST VIRGINIA
P.O. Box 4464
Roanoke, VA 24015
(703) 985-5778, (703) 981-7732
Joseph L. Surace
Jeffrey Messi
Area Served:
NC, TN, VA, WV.
EPA RMP 5:
CC.

RADIATION CONSULTANTS, INC.
P.O. Box 34012
Houston, TX 77234
(713) 476-0072
Bruce Bristow
Area Served:
Nationwide.
EPA RMP 5:
AT, CC.

RADIATION DATA
115 Wall St.
Princeton, NJ 08540
(609) 921-7061
Joseph Baicker
Keith Baicker
Jacques Dolphin
Keith Matches
Qualifications:
AARST.
Area Served:
NJ.

RADIATION DATA/DIVISION OF PRINCETON GEOPHYSICAL SERVICES
37 Castleman Dr.
Southington, CT 06489
(203) 621-0678
L. Dudek - Connecticut Manager
J. Dudek - Radiation Safety
Services:
Provides laboratory analysis of charcoal canisters, provides radon detection and mitigation services.
Qualifications:
EPA RMP, AARST, Health Physics Society, certified by NJ and PA, listed in CT.
Area Served:
Nationwide.
EPA RMP 5:
AT, CC, CW(P).

RADIATION DATA/DIVISION OF PRINCETON GEOPHYSICAL SERVICES
1200 State Rd.
Princeton, NJ 08540
(609) 921-7061
Thomas Van Leeuwen
Area Served:
Nationwide.
EPA RMP 5:
AT, CC(P), CW(P).

RADIATION HEALTH
1259 Marlin Dr.
Port Orchard, WA 98366
(206) 871-5454
David M. Grout - Health Physicist
Ralph Alexander - Health Physicist
Services:
Provides radon detection services and maintains a staff of professional health physicists for consultation.
Qualifications:
EPA RMP.
Area Served:
AK, CA, ID, MT, OR, WA.
EPA RMP 5:
CC.

RADIATION MEASUREMENT SERVICES
P.O. Box 3681
Crofton, MD 21114
(301) 261-6966
James P. Allen
Area Served:
DC, MD, ND, VA.
EPA RMP 5:
CC.

RADIATION PROTECTION SERVICES, INC.
P.O. Box 2395
Darien, CT 06820
(203) 324-7967
Stuart Korchin - Engineer
Michael L. Caprio, Jr. - Physicist
Thomas J. LaRocca - Physicist
Services:
Provides radon detction and mitigation services.
Qualifications:
Certified ABHP physicist and professional engineer, NJ license, EPA RMP.
Area Served:
Nationwide.
EPA RMP 5:
CC(P).

RADIATION PROTECTION SERVICES, INC.
646 Rte. 18 Suite 215
East Brunswick, NJ 08816
(201) 390-7360
Michael L. Caprio - Health Physicist
Services:
Provides radon detection and mitigation services, operates a primary laboratory.
Qualifications:
EPA RMP.
Area Served:
CT, NJ, NY.

RADIATION SAFETY ASSOCIATES, INC.
10 Pendleton Dr. P.O. Box 107
Hebron, CT 06248
(203) 228-0487
Paul Steinmeyer
Area Served:
CT, MA, NY, RI.
EPA RMP 5:
CC.

RADIATION SAFETY ENGINEERING
6105 South Ash Suite 1
Tempe, AZ 85283
(602) 897-9459
Robert Metzger
Area Served:
Nationwide.
EPA RMP 5:
CC(P).

RADIATION SAFETY SERVICES, INC.
6312 West Oakton St.
Morton Grove, IL 60053
(312) 965-1999
Eli Port - President
Services:
Manufactures alpha track monitors.
Qualifications:
EPA RMP, certified health physicists on staff.
Area Served:
Nationwide.
EPA RMP 5:
AT(P), GW(P).

RADIATION SERVICE ORGANIZATION, INC.
P.O. Box 1526
Laurel, MD 20707-7444
(301) 953-2482
Kai-Dee Chu
Area Served:
Nationwide.
EPA RMP 5:
CC(P), EP(P).

RADIATION SERVICES, INC.
3412 Doriswood Ct.
Antioch, TN 37013
(615) 361-6492
Michael C. Bradshaw
Area Served:
GA, KY, TN.
EPA RMP 5:
CC.

RADIATION SURVEILLANCE ASSOCIATES, INC.
1230 Princeton Ave.
Salt Lake City, UT 84105
(801) 583-6919, (801) 467-1030
McDonald E. Wrenn - President
Services:
Provides radon detection services using continuous monitor for relocation companies and home owners, sellers and buyers.
Qualifications:
Certified by American Board of Health Physics, health physicist on staff.
Area Served:
CO, UT.

RADIATION TESTING DESIGN
279 County Rd., 84W
Allenspark, CO 80510
(303) 747-2702
Bill Blackman - Chief Executive Officer
Area Served:
CO.

RADIOACTIVITY ALERT DETECTION SERVICES, INC.
P.O. Box L-227
Langhorne, PA 19047
(215) 357-0574
Garry Stklovsky - President
Services:
Provides radon detection, mitigation, and consultation services for home owners,

construction, and real estate industries.
Qualifications:
Master's degree in electrical engineering, PA DER course in reducing radon in structures.
Area Served:
PA.
EPA RMP 5:
AT, CC.

RADIOLOGICAL CONTROL ASSOCIATES
665 Robbins Ave. Unit 2
Dracut, MA 01826
(617) 453-5268
Robert E. Casparius
Area Served:
MA.

RADIOLOGICAL HEALTH SCIENCES, INC.
P.O. Box 4790
Fort Walton Beach, FL 32549
(904) 863-1191
John P. Albright - President
Robert D. Hook
Services:
Provides radon detection services in air and water using charcoal canisters and liquid scintillation techniques.
Qualifications:
EPA RMP.
Area Served:
Nationwide.
EPA RMP 5:
CC(P).

RADIOLOGICAL PHYSICS SERVICES/HEALTH EAST, INC.
P.O. Box 13367
Roanoke, VA 24033
(703) 981-7379
Joseph Surace
Area Served:
VA.

RADIOLOGICAL TESTING SERVICES
HC92, Box2B
Crawford, NE 69339
(308) 665-1587
Rhonda Grantham
Area Served:
NE.

RADION LABORATORY, INC.
70 Walnut St.
Wellesley, MA 02181
(617) 239-8066
Rudolph Hon - Vice President
Kevin A. Doherty - Senior Manager
Services:
Provides radon detection and mitigation services, operates counting instrumentation for radiation detection in air, liquids, and solid samples, distributes charcoal canisters and water testing kits for radon.
Qualifications:
AARST, NRA, listed with MA Department of Public Health, pending certification in PA, 20 years of experience handling and dealing with radioactive material, instrumentation cross-calibrated with DOE labs.
Area Served:
Nationwide.

RADOMET SERVICES
7079 East Briarwood Cr.
Englewood, CO 80112
(303) 694-4519
Robert N. Parkinson
Area Served:
CO.

RADON AIR AND WATER CHECK, INC.
29 Riverside Dr.
Binghamton, NY 13905
(607) 724-5050
S.C. Gottlieb - President
Richard Geachis
Area Served:
NY, PA.
EPA RMP 5:
AT, CC, CW(P).
RADON AIRTEST COMPANY
13800 Bagley Rd.
Middleburg Heights, OH 44130
(216) 845-8050
Thomas R. Lynch
Area Served:
OH.
EPA RMP 5:
CC.

RADON ALERT
P.O. Box 2167
Westminster, MD 21157-9167
(301) 848-2862
Alan M. Lerman
Area Served:
MD.

RADON ALERT SYSTEMS
P.O. Box 1247
Regina, SK S4P 3B8
(306) 949-0445
Services:
Provides radon detection services using charcoal canister method.
Area Served:
SK.

RADON ALERT, INC.
P.O. Box 131378
Tyler, TX 75713
(214) 592-4077
Richard H. Brown - President
Services:
Provides radon detection services using charcoal canisters, alpha track methods, provides water analysis services.
Qualifications:
Laboratory analysis by Texas A&M University.
Area Served:
Nationwide.

RADON ALERT® DETECTION CENTER
P.O. Box 323
Flourtown, PA 19031
(215) 247-9197, (800) 345-6348
Ulrich Hiesinger - President
Leonard Shabason - Radiation
Services:
Distributes charcoal canisters and alpha track radon detectors, provides mitigation diagnosis and design services.
Qualifications:
EPA RMP, certified by NE, NJ, PA.
Area Served:
Nationwide.
EPA RMP 5:
AT, CC.

RADON AMERICA
P.O. Box 13284
Pensacola, FL 32591
Qualifications:
AARST.
Area Served:
FL.

RADON ANALYSIS AND DETECTION OF NEW ENGLAND, INC.
111 Jefferson St.
Laconia, NH 03246
(603) 528-5772
Matthew Raynes - President
Mike Daneau - Vice President
Services:
Provides radon detection services.
Area Served:
CT, MA, ME, NH, OH, PA, RI, VT.
EPA RMP 5:
CC(P).

RADON ANALYSIS, INC.
2520 15th Ave.
Longmont, CO 80501
(303) 651-7314
Jim Bengston

Area Served:
CO.

RADON ANALYSIS, INC.
7 Chamberlain Rd.
Flemington, NJ 08822
(201) 788-0661
David Chippendale - President
Joyce Criss - General Manager
Jeff Parsons - Quality Assurance
Services:
Provides radon detection services, distributes radon test kits.
Qualifications:
EPA RMP, PA DER, NJ DEP.
Area Served:
Nationwide.
EPA RMP 5:
CC, CW(P).

RADON ANALYSIS, INC.
P.O. Box 5224
Toms River, NJ 08754-5224
(201) 929-9575
Fran Johnston
Area Served:
NJ.
EPA RMP 5:
AT, CC.

RADON ANALYSIS, INC.
4635 B South 28 Rd.
Arlington, VA 22206
(703) 379-8614
Eugene Taylor
Area Served:
DC, MD, VA.
EPA RMP 5:
AT, CC.

RADON ANALYSTS, INC.
P.O. Box 509
Livingston Manor, NY 12758
(914) 439-5111, (914) 292-2277
Vincent J. A. Santoro
Area Served:
NY.

RADON ANALYTICAL LABS
9011 North Meridian St.
Indianapolis, IN 46260
(317) 843-0788
Vickie Jordan - Office Manager
Richard Jordan - President
Tom Tillet - Secretary/Treasurer
Services:
Provides laboratory analysis services for the home inspection and relocation markets.
Qualifications:
EPA RMP, ARA, ASHI, various state licenses.
Area Served:
Nationwide.
EPA RMP 5:
CC(P), EP(P).

RADON AND ENVIRONMENTAL PROFESSIONALS
P.O. Box 1045
Independence, KS 67301
(316) 331-5207
Joe Higgins - President
Services:
Provides radon detection services for home owners and businesses.
Qualifications:
EPA RMP, holds radioactive materials license.
Area Served:
AR, AZ, KS, MO, OK.
EPA RMP 5:
AT, CC.

RADON AND ENVIRONMENTAL PROFESSIONALS
P.O. Box 3391
Lawrence, KS 66046
(913) 749-7420
Linda Guinn
Area Served:
IA, KS, LA, MO, ND, NE, OK.
EPA RMP 5:
AT, CC.

RADON AND ENVIRONMENTAL PROFESSIONALS
2768 South Bennett
Wichita, KS 67217-1812
(316) 945-2570
Wayne Gaul - Vice President
Services:
Provides radon detection services using charcoal canisters and alpha track methods for real estate brokers and individuals.
Qualifications:
Master of Science degree in radiation biophysics, professional health physicist.
Area Served:
KS.

RADON AND RADIATION CONSULTANTS
3575 Peninsula Dr.
Akron, OH 44319
(216) 644-7664
Helmar Dollwet
Services:
Provides radon detection and mitigation services, seminars on topics related to radon and radiation.
Qualifications:
AARST, radiation safety officer, licensed by NRC to handle radioactive materials.
Area Served:
OH.

RADON AND WATER TESTING SERVICE
10806 Lafayette-Plain City Rd.
Plain City, OH 43064
(614) 873-8821
Jennifer Mast - Vice President
Services:
Provides radon detection services using charcoal canister method, provides radon in water detection service.
Area Served:
Nationwide.
EPA RMP 5:
CC(P)

RADON ASSOCIATES, INC.
126 Greenlawn Ave.
Clifton, NJ 07013
(201) 773-4955
Lowell C. Thelin
Ellen Thelin
Area Served:
NJ.
EPA RMP 5:
AT, CC.

RADON AWARENESS AND DETECTION
1408 South Rock Hill Rd.
St. Louis, MO 63119
(314) 961-0400
Renee Quackenbush - President
Services:
Provides radon detection services.
Area Served:
Nationwide.
EPA RMP 5:
CC.

RADON AWARENESS, INC.
415 East Galena P.O. Box 566
Hailey, ID 83333
(208) 788-2634
Jeffrey L. Engelhardt - President
Susan M. Engelhardt - Vice President
Services:
Provides radon detection and mitigation services, designs and installs mitigation systems.
Qualifications:
Radon testing specialist course.
Area Served:
ID, MT, NV, OR, UT, WA, WY.
EPA RMP 5:
CC.

RADON BAN
3719 Michelle Way
Baltimore, MD 21208
(301) 484-7120
David H. Nevins - President
Services:
Provides radon detection services and distributes radon test kits.
Qualifications:
EPA RMP.
Area Served:
DC, DE, MD, PA, VA.
EPA RMP 5:
CC.

RADON BUSTERS
1345 Main Ave. P.O. Box 2426
Fargo, ND 58108
(701) 293-5370
Michael Skansgaard - Sales
Services:
Provides radon detection and mitigation services.
Area Served:
IA, MN, MT, ND, SD.

RADON BUSTERS
P.O. Box R247
Radnor, PA 19087
(215) 566-3059
J.B. Turner - President
Services:
Provides radon detection services for home owners and real estate transactions.
Qualifications:
EPA RMP, staff includes a registered professional engineer.
Area Served:
PA.
EPA RMP 5:
CC.

RADON BUSTERS OF CENTRAL NEW YORK
3218 Valley Pl.
Sauquoit, NY 13456
(315) 737-7290
John Stemmer - President
Services:
Provides radon detection services with a gas data acquisition system or canisters with a 12-hour data output.
Area Served:
NY.

RADON CALIBRATIONS
800 Stiles Ct.
Richmond, VA 23235
(804) 320-3761
Glenn P. Morris
John Witt
Area Served:
FL, MD, NC, VA, WV.
EPA RMP 5:
CC.

RADON CHECK
2524 South El Paradiso Suite 50
Mesa, AZ 85202
(602) 730-5559
Hal Dunn - President
Services:
Provides radon detection services.
Qualifications:
AARST, certified geologist.
Area Served:
AZ.

RADON CHECK
P.O. Box 7882
Greensboro, NC 27417
(919) 855-1312
Manfred Krause
Area Served:
Nationwide.
EPA RMP 5:
CC(P).

RADON CHECK
2263 Coppersmith Sq.
Reston, VA 22091
(703) 476-6810
David Klein
Area Served:
Nationwide.
EPA RMP 5:
CC.

RADON CHECK/DIVISION OF HORIZON SYSTEMS
P.O. Box 721
Harvard, MA 01451
(617) 456-8811, (617) 890-2334
Frederick M. Viles
Area Served:
CT, MA, ME, NH, NJ, NY, PA, RI, VT.
EPA RMP 5:
CC.

RADON CONSULTANTS
P.O. Box 5209
Sparks, NV 89432
(800) 327-6826
Russ McCreary - President
Services:
Provides radon detection services and distributes radon test kits.
Qualifications:
EPA RMP.
Area Served:
AK, AL, AZ, CA, CO, ID, MT, ND, NE, NM, NV, OR, SD, UT, WA, WY.
EPA RMP 5:
CC, CR(P).

RADON CONSULTANTS, INC.
P.O. Box 789
Golden, CO 80402
(303) 279-1358
Dennis N. Poffenroth
Area Served:
CO, IL, MT, ND, SD, UT, WY.
EPA RMP 5:
AT, CC.

RADON CONSULTANTS, INC., MARYLAND
4831 Silver Hill Rd.
Suitland, MD 20746
(301) 736-7040
Robert Puskar
Services:
Provides radon detection and mitigation services.
Area Served:
MD.

RADON CONSULTANTS, INC., VIRGINIA
4012 North 35th St.
Arlington, VA 22207
(703) 522-6669, (301) 736-7040
Jacquelyn G. Black
Services:
Provides radon detection and mitigation services.
Area Served:
Nationwide.
EPA RMP 5:
AT, CC.

RADON CONSULTING AND TESTING SERVICE
7449 Highway 44 Box 9
Crystal River, FL 32629
(904) 563-2141, (904) 563-1771
Jon Merkel
Area Served:
Nationwide.
EPA RMP 5:
CC(P), CW(P).

RADON CONTROL PROFESSIONALS, INC.
11920 Fieldthorn Ct.
Reston, VA 22094
(703) 471-9459
See Ad Page 151
Stephen Hall - President
Norman Rodier - Vice President
Services:
Provides full service radon detection and mitigation services, residential, commercial, soil testing, site evaluation, pre-construction

mitigation design.
Qualifications:
County business license, Maryland Home Improvement license, EPA mitigation course, graduate-level scientists.
Area Served:
Nationwide.
EPA RMP 5:
AT.

RADON CONTROL SERVICES CORPORATION
3122 49th St.
Des Moines, IA 50310
(515) 276-9131
Charlie Stephenson - President
Services:
Provides radon detection and mitigation services.
Area Served:
IA.

RADON CONTROL, INC.
337 Route 22 E.
Greenbrook, NJ 08812
(201) 563-4211
Robert Pedersen, Sr.
Area Served:
NJ.

RADON DECON, INC.
P.O. Box 28
Muncy Valley, PA 17758
(717) 482-2076
Calvin Zehring - President
Qualifications:
AARST.
Area Served:
PA.

RADON DETECTION OF MISSISSIPPI
P.O. Box 2404
Columbus, MS 39704
(601) 328-4687
Bill McBryde
Area Served:
Nationwide.
EPA RMP 5:
CC.

RADON DETECTION AGENCY, LTD.
625 5th Ave. Suite 104
New Westminster, BC V3M 1X4
(604) 525-4455
David Keegan - President
David E. McLeran - Vice President
Services:
Provides radon detection services using charcoal canister and alpha track methods, distributes radon mitigation supplies.
Area Served:
Nationwide.

RADON DETECTION AND CONSULTING
705 West Fourth St.
Lawrence, KS 66044
(913) 843-7377
Steven R. Hicks - Director
Services:
Provides full service radon detection, mitigation, and consultation services, information at no charge.
Qualifications:
EPA RMP, operated by medical, engineering, and legal professionals.
Area Served:
Nationwide.
EPA RMP 5:
CC.

RADON DETECTION AND CONTROL
P.O. Box 419 153 Jordan St.
South Hts., PA 15081
(412) 375-1700
John B. Mallon, Jr.
Area Served:
Nationwide.
EPA RMP 5:
CC.

RADON DETECTION AND REDUCTION LABORATORY/DIVISION OF FORMACHEM
10834 Ambassador
St. Louis, MO 63132
(314) 872-9350
Harold Susman - President
Donald Pearline - Biochemist
Services:
Manufactures and distributes the Seal Safe Radon Gas Detector and alpha track detectors, distributes Enkadrain radon mitigation system for new home construction.
Area Served:
Nationwide.

RADON DETECTION AND REMOVAL SERVICES
915 6th Ave., N.
Seattle, WA 98109
(206) 282-9156
Raymond Duffy
James Klosterman
Services:
Provides radon detection and mitigation services.
Area Served:
WA.

RADON DETECTION AND TESTING, INC.
2222 Cumberland Pl.
Smyrna, GA 30080
(404) 434-7299
Terry E. Howell
Area Served:
GA.

RADON DETECTION CENTER OF MICHIGAN
601 South Saginaw St. Suite 514
Flint, MI 48502
(313) 767-8630
Don Potter
Area Served:
MI.

RADON DETECTION CORPORATION OF MICHIGAN
4740 Hall St., SE
Kentwood, MI 49506
(616) 942-0323
Lev Kaganovich - President
Services:
Provides radon detection services and consultation.
Qualifications:
Engineer with 20 years of experience in HVAC area.
Area Served:
MI.
EPA RMP 5:
CC.

RADON DETECTION DEVICES
5 Lauriann Ct.
Parkton, MD 21120
(301) 329-6306
Stephen C. Iachini
Area Served:
MA, MD, ME, NH, NY, PA, VA.
EPA RMP 5:
CC.

RADON DETECTION SERVICE OF KANSAS CITY
8405 Booth
Raytown, MO 64138
(816) 356-9677
Charles Willison
Area Served:
MO.

RADON DETECTION SERVICE OF MONTANA
P.O. Box 1114
Kalispell, MT 59901
(406) 755-9477, (406) 752-8404
Joseph A. Askelson
Area Served:
MT.

RADON DETECTION SERVICES
323 Neusch Way
Saskatoon, SK S7K 7V2
(306) 931-7729
Howard Trischuk - President
Services:
Provides radon detection services .
Area Served:
SK.

RADON DETECTION SERVICES
P.O. Box 434
West Rutland, VT 05777
(802) 775-0250, (802) 438-5603
Leonard V. Knappmiller
Area Served:
NH, NY, VT.
EPA RMP 5:
CW(P).

RADON DETECTION SERVICES OF NEBRASKA
16506 Dorcas St.
Omaha, NE 68130
(402) 330-8968
Jeff Radcliffe - Owner
Area Served:
NE.

RADON DETECTION SERVICES, LTD.
1500 Harbour Oaks Dr.
Tucker, GA 30084
(404) 621-9300
Fred P. Budin
Area Served:
GA.
EPA RMP 5:
CC.

RADON DETECTION SPECIALISTS
7966 South Madison St.
Burr Ridge, IL 60521
(312) 325-4443
Karl Weber - President
Terry Huebner - Vice President
Services:
Provides radon detection services using charcoal canisters, alpha track, and grab sampling methods, mitigation consulting services.
Area Served:
IL, IN, WI.
EPA RMP 5:
CC.

RADON DETECTION SYSTEMS OF VERMONT
P.O. Box 383
Randolph, VT 05060
(802) 728-5190
Shirley Russell - Owner
Don Russell
Services:
Provides radon detection services.
Area Served:
VT.

RADON DETECTION SYSTEMS, INC., OHIO
P.O. Box 473
Aurora, OH 44202
(216) 562-4647
Raymond Sawyer - Owner
Services:
Provides radon detection services.
Qualifications:
ARA.
Area Served:
OH.
EPA RMP 5:
CC, CR(P).

RADON DETECTION SYSTEMS/ALBUQUERQUE, NEW MEXICO BRANCH
4005 Cumino de la Sierra, NE
Albuquerque, NM 87111
(303) 444-5253, (800) 627-2360
Bob Nakiaoka - Geologist
Services:
Broad based company provides radon detection services using charcoal canister, alpha track, E-Perm, and continuous working level methods for corporate relocation and real estate industries.
Qualifications:
EPA RMP, over 20 years of experience in radon detection, qualified geologists on staff.
Area Served:
NM.

RADON DETECTION SYSTEMS/ARIZONA BRANCH
1301 East Dava Dr.
Tempe, AZ 85283
(303) 444-5253, (800) 627-2360
Beth Wrege - Geologist
Services:
Broad based company provides radon detection services using charcoal canister, alpha track, E-Perm, and continuous working level methods for corporate relocation and real estate industries.
Qualifications:
EPA RMP, over 20 years of experience in radon detection, qualified geologists on staff.
Area Served:
AZ, NM.

RADON DETECTION SYSTEMS/ATLANTA, GEORGIA BRANCH
2965 Atterbury Ct.
Decatur, GA 30033
(303) 444-5253, (800) 627-2366
Laura Rose - Scientist
Services:
Broad based company provides radon detection services using charcoal canister, alpha track, E-Perm, and continuous working level methods for corporate relocation and real estate industries.
Qualifications:
EPA RMP, over 20 years of experience in radon detection, qualified geologists on staff.
Area Served:
GA.

RADON DETECTION SYSTEMS/BLOOMINGTON-NORMAL BRANCH
1207 Stephens Rd.
Normal, IL 61761
(303) 444-5253, (800) 627-2366
John Foster - Geologist
Services:
Broad based company provides radon detection services using charcoal canister, alpha track, E-Perm, and continuous working level methods for corporate relocation and real estate industries.
Qualifications:
EPA RMP, over 20 years of experience in radon detection, qualified geologists on staff.
Area Served:
IA, IL, IN, MI, MO, MN, OH, WI.

RADON DETECTION SYSTEMS/BOSTON, MASSACHUSETTS BRANCH
91 Westland Ave. Suite 606
Boston, MA 02115
(303) 444-5253, (800) 627-2366
Hank Tracy - Geologist
Services:
Broad based company provides radon detection services using charcoal canister, alpha track, E-Perm, and continuous working level methods for corporate relocation and real estate industries.
Qualifications:
EPA RMP, over 20 years of experience in radon detection, qualified geologists on staff.
Area Served:
MA.

RADON DETECTION SYSTEMS/BUTTE, MONTANA BRANCH
1123 West Park
Butte, MT 59701
(303) 444-5253, (800) 627-2360
Gerald van Kampen - Geologist

Services:
Broad based company provides radon detection services using charcoal canister, alpha track, E-Perm, and continuous working level methods for corporate relocation and real estate industries.
Qualifications:
EPA RMP, over 20 years of experience in radon detection, qualified geologists on staff.
Area Served:
ID, MT, ND, SD.

RADON DETECTION SYSTEMS/CASPER, WYOMING BRANCH
628 East 14th
Casper, WY 82601
(303) 444-5253, (800) 627-2360
Jerry Nelson - Geologist
Services:
Broad based company provides radon detection services using charcoal canister, alpha track, E-Perm, and continuous working level methods for corporate relocation and real estate industries.
Qualifications:
EPA RMP, over 20 years of experience in radon detection, qualified geologists on staff.
Area Served:
MT, ND, SD, WY.

RADON DETECTION SYSTEMS/CHARLESTON, WEST VIRGINIA BRANCH
2300 McCorkle Ave., SE
Charleston, WV 25304
(303) 444-5253, (800) 627-2366
Zeki Al-Saigh - Chemist
Services:
Broad based company provides radon detection services using charcoal canister, alpha track, E-Perm, and continuous working level methods for corporate relocation and real estate industries.
Qualifications:
EPA RMP, over 20 years of experience in radon detection, qualified geologists on staff.
Area Served:
WV.

RADON DETECTION SYSTEMS/CHICAGO, ILLINOIS BRANCH
12845 Golf View Ln.
Palois Heights, IL 60463
(303) 444-5253, (800) 627-2360
Bill Lyon - Geologist
Services:
Broad based company provides radon detection services using charcoal canister, alpha track, E-Perm, and continuous working level methods for corporate relocation and real estate industries.
Qualifications:
EPA RMP, over 20 years of experience in radon detection, qualified geologists on staff.
Area Served:
IL, IN, WI.

RADON DETECTION SYSTEMS/COLORADO SPRINGS, COLORADO BRANCH
2010 Wooten Ln.
Colorado Springs, CO 80907
(303) 444-5253, (800) 627-2366
Steve Westby - Geologist
Services:
Broad based company provides radon detection services using charcoal canister, alpha track, E-Perm, and continuous working level methods for corporate relocation and real estate industries.
Qualifications:
EPA RMP, over 20 years of experience in radon detection, qualified geologists on staff.
Area Served:
CO.

RADON DETECTION SYSTEMS/COLUMBUS, OHIO BRANCH
3487 Colchester Rd.
Columbus, OH 43221
(303) 444-5253, (800) 627-2366
Russell Utgard - Geologist
Services:
Broad based company provides radon detection services using charcoal canister, alpha track, E-Perm, and continuous working level methods for corporate relocation and real estate industries.
Qualifications:
EPA RMP, over 20 years of experience in radon detection, qualified geologists on staff.
Area Served:
OH.

RADON DETECTION SYSTEMS/CORPORATE HEADQUARTERS
2450 Central Ave. A-1
Boulder, CO 80301
(303) 444-5253, (800) 627-2360
Dan Cutler - General Manager
Mark Williams - Geologist
Bill Lawrence - Geologist
Services:
Provides radon detection services for corporate relocation and real estate transactions. 30 locations nationwide are listed below.

Qualifications:
EPA RMP, 20 years of experience providing radon detection services, qualified geologists and scientists on staff.
Area Served:
Nationwide.
EPA RMP 5:
AT, CC, EP, CW(P).

See Ad Page 74

RADON DETECTION SYSTEMS/DALLAS, TEXAS BRANCH
6607 Northwest Hwy.
Dallas, TX 75225
(303) 444-5253, (800) 627-2360
Gary James - Geologist
Services:
Broad based company provides radon detection services using charcoal canister, alpha track, E-Perm, and continuous working level methods for corporate relocation and real estate industries.
Qualifications:
EPA RMP, over 20 years of experience in radon detection, qualified geologists on staff.
Area Served:
LA, OK, TX.

RADON DETECTION SYSTEMS/DES MOINES, IOWA BRANCH
Box 93 East 57 Rd.
Kelly, IA 50134
(303) 444-5253, (800) 627-2366
Ken Wiggers - Scientist
Services:
Broad based company provides radon detection services using charcoal canister, alpha track, E-Perm, and continuous working level methods for corporate relocation and real estate industries.
Qualifications:
EPA RMP, over 20 years of experience in radon detection, qualified geologists on staff.
Area Served:
IA.

RADON DETECTION SYSTEMS/DETROIT, MICHIGAN BRANCH
6051 Dunn Rd.
Howell, MI 48843
(303) 444-5253, (800) 627-2366
Blair Dudley - Geologist
Services:
Broad based company provides radon detection services using charcoal canister, alpha track, E-Perm, and continuous working level methods for corporate relocation and real estate industries.
Qualifications:
EPA RMP, over 20 years of experience in radon detection, qualified geologists on staff.
Area Served:
MI.

RADON DETECTION SYSTEMS/DURANGO, COLORADO BRANCH
77 Roach Ranch Rd.
Durango, CO 81302
(303) 444-5253, (800) 627-2360
Tony Smith - Technician
Services:
Broad based company provides radon detection services using charcoal canister, alpha track, E-Perm, and continuous working level methods for corporate relocation and real estate industries.
Qualifications:
EPA RMP, over 20 years of experience in radon detection, qualified geologists on staff.
Area Served:
CO.

RADON DETECTION SYSTEMS/FARGO, NORTH DAKOTA BRANCH
1225 Oak St., North
Fargo, ND 58102
(303) 444-5253, (800) 627-2360
Services:
Broad based company provides radon detection services using charcoal canister, alpha track, E-Perm, and continuous working level methods for corporate relocation and real estate industries.
Qualifications:
EPA RMP, over 20 years of experience in radon detection, qualified geologists on staff.
Area Served:
MN, ND.

RADON DETECTION SYSTEMS/GRAND JUNCTION, COLORADO BRANCH
235 Radcliff Dr.
Frulta, CO 81521
(303) 444-5253, (800) 627-2366
Gibson - Geologist
Services:
Broad based company provides radon detection services using charcoal canister, alpha track, E-Perm, and continuous working level methods for corporate relocation and real estate industries.
Qualifications:
EPA RMP, over 20 years of experience in radon detection, qualified geologists on staff.
Area Served:
CO.

RADON DETECTION SYSTEMS/GREENSBORO, NORTH CAROLINA BRANCH
4324 Cooper Rd.
Winston-Salem, NC 27121
(303) 444-5253, (800) 627-2366
Bruce Hageman - Geologist
Services:
Broad based company provides radon detection services using charcoal canister, alpha track, E-Perm, and continuous working level methods for corporate relocation and real estate industries.
Qualifications:
EPA RMP, over 20 years of experience in radon detection, qualified geologists on staff.
Area Served:
NC.

RADON DETECTION SYSTEMS/LOS ANGELES, CALIFORNIA BRANCH
313 8th St. Apt. E
Seal Beach, CA 90740
(213) 596-1948, (800) 627-2366
Tim Smith - Geologist
Services:
Broad based company provides radon detection services using charcoal canister, alpha track, E-Perm, and continuous working level methods for corporate relocation and real estate industries.
Qualifications:
EPA RMP, over 20 years of experience in radon detection, qualified geologists on staff.
Area Served:
AZ, CA, ID, MN, MT, NM, NY, OR, TX, UT, WA, WY.

RADON DETECTION SYSTEMS/MIAMI, FLORIDA BRANCH
17001 SW 109 Ave.
Miami, FL 33157
(303) 444-5253, (800) 627-2366
James DeGrot, Jr. - Geologist
Services:
Broad based company provides radon detection services using charcoal canister, alpha track, E-Perm, and continuous working level methods for corporate relocation and real estate industries.
Qualifications:
EPA RMP, over 20 years of experience in radon detection, qualified geologists on staff.
Area Served:
FL.

RADON DETECTION SYSTEMS/MILWAUKEE BRANCH
2630 North Newhall St.
Milwaukee, WI 53211
(303) 444-5253, (800) 627-2366

John Bruskewitz - Geologist
Services:
Broad based company provides radon detection services using charcoal canister, alpha track, E-Perm, and continuous working level methods for corporate relocation and real estate industries.
Qualifications:
EPA RMP, over 20 years of experience in radon detection, qualified geologists on staff.
Area Served:
WI.

RADON DETECTION SYSTEMS/MINNEAPOLIS-ST. PAUL BRANCH
548 Frontenae
St. Paul, MN 55104
(303) 444-5253, (800) 627-2366
M. Clement - Geologist
Services:
Broad based company provides radon detection services using charcoal canister, alpha track, E-Perm, and continuous working level methods for corporate relocation and real estate industries.
Qualifications:
EPA RMP, over 20 years of experience in radon detection, qualified geologists on staff.
Area Served:
MN.

RADON DETECTION SYSTEMS/NASHVILLE, TENNESSEE BRANCH
3102 Bellwood Ave.
Nashville, TN 37203
(303) 444-5253, (800) 627-2366
Odum Clark - Geologist
Services:
Broad based company provides radon detection services using charcoal canister, alpha track, E-Perm, and continuous working level methods for corporate relocation and real estate industries.
Qualifications:
EPA RMP, over 20 years of experience in radon detection, qualified geologists on staff.
Area Served:
AL, GA, KY, MO, MS, NC, SC, TN, VA, WV.

RADON DETECTION SYSTEMS/NEWARK, NEW JERSEY BRANCH
6 Earl Ct.
Montiville, NJ 07045
(303) 444-5253, (800) 627-2360
Dhiro Mehta - Industrial Hygienist
Services:
Broad based company provides radon detection services using charcoal canister, alpha track, E-Perm, and continuous working level methods for corporate relocation and real estate industries.
Qualifications:
EPA RMP, over 20 years of experience in radon detection, qualified geologists on staff.
Area Served:
CT, DE, MA, NH, NJ, NY, PA, RI, VT.

RADON DETECTION SYSTEMS/NORTHERN CALIFORNIA BRANCH
2567 Greenwich St.
San Francisco, CA 94123
(303) 444-5253, (800) 627-2366
Jeff Herra - Geologist
Services:
Broad based company provides radon detection services using charcoal canister, alpha track, E-Perm, and continuous working level methods for corporate relocation and real estate industries.
Qualifications:
EPA RMP, over 20 years of experience in radon detection, qualified geologists on staff.
Area Served:
CA.

RADON DETECTION SYSTEMS/PITTSBURGH, PENNSYLVANIA BRANCH
180 Indionola Rd.
Pittsburgh, PA 15238
(303) 444-5253, (800) 627-2360
Min Hung - Scientist
Services:
Broad based company provides radon detection services using charcoal canister, alpha track, E-Perm, and continuous working level methods for corporate relocation and real estate industries.
Qualifications:
EPA RMP, over 20 years of experience in radon detection, qualified geologists on staff.
Area Served:
PA.

RADON DETECTION SYSTEMS/PORTLAND, OREGON BRANCH
3710 SE 36th Pl. #27
Portland, OR 97202
(303) 444-5253, (800) 627-2366
Aaron Kance - Geologist
Services:
Broad based company provides radon detection services using charcoal canister, alpha track, E-Perm, and continuous working level methods for corporate relocation and real estate industries.
Qualifications:
EPA RMP, over 20 years of experience in radon detection, qualified geologists on staff.
Area Served:
OR.

RADON DETECTION SYSTEMS/SALT LAKE CITY, UTAH BRANCH
6728 South 1655 East
Salt Lake City, UT 84121
(303) 444-5253, (800) 627-2360
Gloria Kerns - Geologist
Services:
Broad based company provides radon detection services using charcoal canister, alpha track, E-Perm, and continuous working level methods for corporate relocation and real estate industries.
Qualifications:
EPA RMP, over 20 years of experience in radon detection, qualified geologists on staff.
Area Served:
ID, NM, UT, WY.

RADON DETECTION SYSTEMS/SIOUX FALLS, SOUTH DAKOTA BRANCH
2713 South Prairie
Sioux Falls, SD 57105
(303) 444-5253, (800) 627-2360
Peter Hunson - Geologist
Services:
Broad based company provides radon detection services using charcoal canister, alpha track, E-Perm, and continuous working level methods for corporate relocation and real estate industries.
Qualifications:
EPA RMP, over 20 years of experience in radon detection, qualified geologists on staff.
Area Served:
SD.

RADON DETECTION SYSTEMS/SPOKANE, WASHINGTON BRANCH
1115 Lambert Ln.
Coer D'Alene, ID 83814
(303) 444-5253, (800) 627-2360
Jane Fay - Geologist
Services:
Broad based company provides radon detection services using charcoal canister, alpha track, E-Perm, and continuous working level methods for corporate relocation and real estate industries.
Qualifications:
EPA RMP, over 20 years of experience in radon detection, qualified geologists on staff.
Area Served:
ID, WA.

RADON DETECTION, INC.
131 Amherst St. P.O. Box 900
Amherst, NH 03031-0900
(603) 882-7340
Lucian Borduz
Area Served:
CT, MA, ME, NH, VT.
EPA RMP 5:
CR(P).

RADON DETECTORS
RD 1, Box 177
Van Etten, NY 14889
(607) 589-6223
Larry L. Emdee - Owner
Services:
Provides radon detection, mitigation and diagnostic services for commercial and residential buildings.
Qualifications:
EPA RMP, AARST, NY RAS.
Area Served:
CT, DC, DE, MA, ME, NH, NJ, NY, OH, PA, RI, VA, VT, WV.
EPA RMP 5:
AT.

RADON DIAGNOSTIC LABORATORIES
231 Front St. P.O. Box 2890
South Portland, ME 04106
(207) 767-2818
Dave Sankey - President
Don Burrows - Director of Marketing
Jon Dyer - Technical Director
Services:
Provides radon air and water detection and mitigation services, well water and ground water analysis, offers 24 hour turnaround on test results, presents seminars on dealing with radon.
Qualifications:
EPA RMP, AARST, Association of Maine Environmental Laboratories.
Area Served:
Nationwide.
EPA RMP 5:
CC(P).

RADON DIAGNOSTICS
Box 761
Andover, NJ 07821
(201) 729-1251
Charles R. Wayne, Jr. - Owner
Area Served:
Nationwide.
EPA RMP 5:
CC(P), GR(P), CW(P).

RADON DOCTOR, INC., THE
Alpha Building Suite 904
Easton, PA 18042
(215) 252-4545
Ed Roscioli - President
Albert Roscioli - Manager
Services:
Provides radon detection and mitigation services
Qualifications:
President is a nuclear engineer, PA DER, NJ DEP voluntary certification.
Area Served:
PA.

RADON ENGINEERING SERVICES, INC.
3129 Overton Rd.
Birmingham, AL 35223
(205) 967-2665
F. Milton Jessup - President
R. Douglas Macgill - Vice President
Robert H. Hood - Vice President
Services:
Provides radon detection and mitigation services for residential market.
Qualifications:
EPA RMP, staff composed of registered professional engineer, lawyer, and architect.
Area Served:
Nationwide.
EPA RMP 5:
CC.

RADON ENGINEERING/A UNIT OF PSI ENGINEERING, INC.
1 Lethbridge Plaza
Mahwah, NJ 07430
(201) 529-8300
Harvey Greenberg - Manager
William C. Taylor - Quality Assurance
Services:
Provides radon detection services using working level monitors, charcoal canisters, and alpha track for the real estate industry and private residences, provides radon mitigation services.
Qualifications:
EPA RMP, lab certified by NY, listed with CT, NJ.
Area Served:
Nationwide.
EPA RMP 5:
CC(P), EP, CW(P).

RADON ENVIRONMENTAL CONSTRUCTION SERVICES/RECONS, INC., MAIN OFFICE
P.O. Box 3058
Princeton, NJ 08543-3058
(609) 883-1808, (215) 493-0667
Peter J. Ortolano, Jr. - President
Services:
Provides radon detection and mitigation services, retrofit remediation systems installed for new construction, commercial/industrial designs and installations.
Qualifications:
PA DER, AARST, NJ DEP, 22 years of experience in construction industry.
Area Served:
NJ, PA.

RADON ENVIRONMENTAL CONSTRUCTION SERVICES/RECONS, INC., PA OFFICE
979 Mt. Eyre Rd.
Washington Crossing, PA 18977
(215) 321-3274
Peter Ortolano - Owner
Area Served:
PA.

RADON ENVIRONMENTAL MONITORING, INC.
Box 5305
Evanston, IL 60204
(312) 328-6628
Robert Pollock - Vice President
E.V. Benton
Services:
Manufactures and distributes alpha track monitors, primary analysis laboratory.
Area Served:
Nationwide.
EPA RMP 5:
AT(P).

See Ad Page 208

RADON ENVIRONMENTAL SERVICES, INC.
8935 North Meridian St. Suite 110
Indianapolis, IN 46260
(317) 846-1991
Richard Jordan - President
James Mann
Services:
Provides radon detection and mitigation services for corporate relocation companies, offers consulting services.
Qualifications:
EPA RMP.
Area Served:
Nationwide.
EPA RMP 5:
CC, CR(P), EP(P).

RADON ENVIRONMENTAL SERVICES, INC.
901 Mason Headley Rd.
Lexington, KY 40502
(606) 255-6375
Ann Wiley - Office Manager

Services:
Provides radon detection and mitigation services for corporate relocation companies.
Qualifications:
EPA RMP.
Area Served:
AL, AR, FL, GA, IA, IN, KY, MN, MO, NC, NE, OH, SC, TN, WI, WV.
EPA RMP 5:
CC, EP.

RADON ENVIRONMENTAL SERVICES, INC.
4320 Parakeet Rd.
Toledo, OH 43612
(419) 447-0209
James Mann - Office Manager
Services:
Provides radon detection and mitigation services for corporate relocation companies.
Qualifications:
EPA RMP.
Area Served:
OH.

RADON ENVIRONMENTAL TEST EXPERTS, INC.
5271 Holmes Pl.
Boulder, CO 80303
(303) 442-2344
David Oppenheimer
Area Served:
CO.
EPA RMP 5:
CC.

RADON ENVIRONMENTAL TESTING SERVICE INC.
6229 North Dixie Dr.
Dayton, OH 45414
(513) 890-3361
Jeff Schindler
Area Served:
OH.

RADON ENVIRONMENTAL TESTING SERVICES, INC.
P.O. Box 14613
Dayton, OH 45414
(513) 890-3361
M. Arthur Hill - President
Services:
Provides radon detection services using continuous monitor and charcoal canister methods.
Qualifications:
ARA, BBB.
Area Served:
OH.

RADON EVALUATION DIVISION/EVALDO, LTD.
64 Sunset Hill Dr.
Branford, CT 06405
(203) 488-4715
Walter A. Dommers
Area Served:
CT, MA, ME, NH, RI, VT.
EPA RMP 5:
CC.

RADON EVALUATION SERVICES, INC.
P.O. Box 13609
Atlanta, GA 30324
(404) 478-0037
F. Milton Jessup - President
R. Douglas MacGill - Vice President
Robert H. Hood - Vice President
Services:
Provides radon detection and mitigation services for commercial and residential markets.
Qualifications:
EPA RMP, staff composed of registered professional engineer, lawyer, and architect.
Area Served:
AL, FL, GA, KS, KY, LA, MS, SC, TN.
EPA RMP 5:
CC.

RADON FINDERS
1118 Millington Rd.
Schenectady, NY 12309
(518) 372-8395
Douglass Allen
Area Served:
NY.

RADON FREE
5666 Ashbrook
Toledo, OH 43614
(419) 865-0129
Carrie Fleitz - President
Steve Goodall - Vice President
Tim Fleitz - Manager
Services:
Provides radon detection services using charcoal canister, alpha track and continuous radon monitor methods, provides radon mitigation services for new and existing buildings, non-commercial laboratory for research purposes or state agencies' assistance only.
Qualifications:
ARA.
Area Served:
MI, MN, OH.
EPA RMP 5:
AT, CC.

RADON GAS DETECTION SYSTEMS
P.O. Box 15189
Hamilton, OH 45015
(513) 863-6625
Mark D. McCune
Area Served:
ID, KY, OH.
EPA RMP 5:
CC.

RADON GAS TESTING OF CENTRAL OHIO
5548 Autumn Chase
Columbus, OH 43232
(614) 864-0191
J.L. Powell - President
Services:
Provides radon detection services using continuous radon monitor method primarily for the real estate and relocation industry, government agencies, and commercial buildings.
Qualifications:
EPA RMP.
Area Served:
OH.
EPA RMP 5:
CR(P).

RADON GASBUSTERS
892 Green Ln.
Union, NJ 07083
(201) 353-5577
Michael Kellison
Area Served:
Nationwide.
EPA RMP 5:
RP(P), CW(P).

RADON HOME INSPECTION SERVICE
P.O. Box 292
Churchton, MD 20733
(301) 261-5650
Richard K. Jeck - President
Services:
Provides radon detection services for homeowners and real estate transactions.
Qualifications:
EPA listed
Area Served:
DC, MD, VA.

RADON HOME PROTECTION SERVICES
48 Woodruff Ave.
Thomaston, CT 06787
(203) 283-4078
Norman E. Werner - President
Services:
Providesd radon detection services using charcoal canisters, tests well water for

presence of radon.
Qualifications:
EPA RMP.
Area Served:
CT.
EPA RMP 5:
CC.

RADON HOME SEARCH
349 Maxim Dr.
Hopatcong, NJ 07843
(201) 398-4106
James Walsh
Area Served:
NJ.

RADON HUNTER, THE
P.O. Box 6009
McLean, VA 22106
(703) 237-1016
John Hunter - Owner
Services:
Provides radon detection services using Honeywell radon detectors.
Qualifications:
EPA RMP.
Area Served:
DC, MD, VA.

RADON IN-HOUSE DETECTION
Rt. 1, Box 110
Leeds, NY 12451
(518) 622-3509
Walter Prapolsky - Owner
Services:
Provides radon detection services.
Area Served:
NY.

RADON INFORMATION AND DETECTION SERVICE
Rte. 2, Isbill Rd.
Madisonville, TN 37354
(615) 442-5849
John R. Reeves
Area Served:
TN.
EPA RMP 5:
CC.

RADON INFORMATION AND TESTING SERVICE
29 Appleton St.
Manchester, NH 03104
(603) 627-7261
Anthony J. Carita - President
Dave Fenske - Vice President
Services:
Provides radon detection services using continous radon method.
Area Served:
MA, NH.
EPA RMP 5:
CR(P).

RADON INSPECTION SERVICE, INC.
789 East Glen Ave.
Ridgewood, NJ 07450
(201) 670-8821
Robert D. Shumeyko - President
David B. Terry - Vice President
Services:
Provides radon detection and mitigation services for real estate industry and home owners using ten different methods.
Qualifications:
NJ DEP, NY DOH.
Area Served:
CT, NJ, NY.
EPA RMP 5:
AT, CC, CR(P), EP(P), GR(P), GW(P), CW(P).

RADON INSPECTION SERVICES
915 6th Ave., North Unit C
Seattle, WA 98109
(206) 286-0479
James C. Klasterman - Principal
Raymond Duffy - Principal
Services:
Provides radon detection services using charcoal canister, E-Perm, and continuous radon monitor methods.
Qualifications:
Licensed architects, ARA, property appraiser.
Area Served:
WA.

RADON INSTRUMENTAL DETECTION
8000 West 110th Suite 115
Overland Park, KS 66210
(913) 451-9270
Steve Craft
Area Served:
Nationwide.
EPA RMP 5:
CC.

RADON INTERVENTION
P.O. Box 827
Three Rivers, TX 78071
(512) 786-4276
Calvin G. Bunton - Owner
Services:
Provides radon detection services using charcoal adsorption method, primary laboratory.
Qualifications:
EPA RMP.
Area Served:
Nationwide.

EPA RMP 5:
CC(P).

RADON LABORATORIES
P.O. Box 2054 Harvard Sq.
Cambridge, MA 02138
(617) 923-6454
Frank Sedaghati - President
Area Served:
Nationwide.
EPA RMP 5:
CC(P).

RADON LABORATORIES
P.O. Box 829
Middletown, MD 21769
(301) 371-5867
Gregory J. Beal, Jr. - President
Services:
Provides radon detection and consultation services for the public and industry, uses Pico Rad complete radon monitoring system.
Qualifications:
EPA RMP.
Area Served:
DE, MD, PA, VA, WV.
EPA RMP 5:
AT, CC(P), EP.

RADON LABORATORIES, INC.
33 Kensington Pkwy. P.O. Box 867
Abingdon, MD 21009
(301) 679-6494
Robert F. Biser - President
Services:
Provides radon detection services and radon laboratory using the Packard Pico-Rad Analysis System; analysis of air and water uses liquid scintillation analyzers.
Qualifications:
AARST.
Area Served:
Nationwide.

RADON LABORATORIES, INC.
3 Deer Tr.
Tabernacle, NJ 08088
(800) 527-4520, (609) 859-0944
Timothy A. Washburn
Area Served:
Nationwide.
EPA RMP 5:
CC(P).

RADON LABS
P.O. Box 2054 Harvard Square
Cambridge, MA 02238
(617) 923-6454
Frank Sedaghati
Services:
Provides radon detection and mitigation services.
Area Served:
Nationwide.

RADON LABS
P.O. Box 2243
Woburn, MA 01888-0443
(617) 664-0316
Joseph Zanti
John Sumares
Area Served:
CT, MA, ME, NH, NY, RI, VT.
EPA RMP 5:
CC(P).

RADON LABS OF ATLANTA
5400 Hwy. 29
Lilburn, GA 30247
(404) 279-0005, (800) 248-1889
James Stephens - Physicist
Dominic Stephens - Sales Manager
Pat Pettit - Office Manager
Services:
Provides radon test kits and operates laboratory.
Qualifications:
EPA RMP, nuclear physicist on staff.
Area Served:
Nationwide.
EPA RMP 5:
CC(P).

RADON MANAGEMENT COMPANY
4332 Gingham Ct.
Alexandria, VA 22310
(703) 960-4407
Gordon M. Davidson
Area Served:
CO, IL, IN, KY, MD, NC, NJ, NY, OH, OR, PA, SC, VA, WV.
EPA RMP 5:
CC.

RADON MANAGEMENT LABORATORIES
2 Riverdale Rd.
Pompton Lakes, NJ 07442
(201) 835-8721
James Matonti
Area Served:
NJ.

RADON MANAGEMENT LABORATORIES
Overlook Terr.
Chester, NJ 07930
(201) 879-2220 , (201)835-8721
Joseph R. Fastiggi
Area Served:
CT, NJ, NY, PA.

EPA RMP 5:
CC, CR(P), GR(P), RP.

RADON MANAGEMENT OF KENTUCKY
12203 Old Shelbyville Rd.
Middletown, KY 40243
(502) 244-0517
Vance F. Walker, Jr. - Owner
Services:
Provides radon mitigation services for new and existing homes.
Qualifications:
AARST, NAHB.
Area Served:
KY.
EPA RMP 5:
CC.

RADON MANAGEMENT, INC.
474 Easton Rd.
Horsham, PA 19044
(215) 672-8129
John Goldsworthy - Director
Bernard J. Salera - Executive Secretary
Services:
Provides radon detection services to domestic, commercial and industrial markets utilizing charcoal canisters, Honeywell monitors and alpha-track detectors.
Qualifications:
Compliance with Pennsylvania DER interim certification requirement.
Area Served:
PA.

RADON MANAGEMENT/DIVISION OF PHYSICS CONSULTANTS, INC.
P.O. Box 3250
Auburn, ME 04210
(207) 795-2459
Terry Zipper - President
Area Served:
Nationwide.

RADON MEASURE COMPANY, THE
P.O. Box 16243
St. Louis Park, MN 55416
(612) 339-9961, (612) 374-3849
Kenneth B. Sanville - Owner
Area Served:
IA, MN, ND, SD, WI.
EPA RMP 5:
AT, CC.

RADON MEASUREMENT AND CONTROL
7935 Kilbourne Rd.
Sunbury, OH 43074
(614) 866-1197
John R. Hunt

Area Served:
OH.

RADON MEASUREMENT AND CONTROL OF NEW MEXICO
1527 Granite St., NW
Albuquerque, NM 87104
(505) 843-7643
William M. Turner - President
L. David Turner - Vice President
Services:
Provides radon detection services for real estate agents, home owners, government agencies, and relocation firms, manufactures RadonDrain, a ventilation system for removal of indoor radon in on-grade homes, distributes radon "reference" chamber, used by radon detection firms to verify instrument measurement stability over time.
Qualifications:
EPA RMP, registered professional geologist, EPA mitigation course, radon instructor for NM Real Estate Commission.
Area Served:
Nationwide.
EPA RMP 5:
AT, CC, EP.

See Ad Page 238

RADON MEASUREMENT AND SERVICES
13131 West Cedar Dr.
Lakewood, CO 80228
(303) 980-5086, (303) 988-3033
R. Stanley Thompson
Area Served:
Nationwide.
EPA RMP 5:
CC.

RADON MEASUREMENT SERVICE
495 North Maple St.
Ephrata, PA 17522
(717) 738-1219
Thomas E. Woolley
Area Served:
PA.

RADON MEASUREMENTS, INC.
2916 West T.C. Jester
Houston, TX 77018
(713) 683-8227
John H. Cohen
Area Served:
Nationwide.
EPA RMP 5:
AT, CC.

RADON MEASURING SYSTEMS
1129 Wyoming Ave.
Scranton, PA 18509
(717) 961-1666
Jeff Brundage - President
Jack Hill - Technician
Services:
Provides radon detection services using working level monitors, provides radon mitigation services.
Area Served:
NJ, NY, PA.
EPA RMP 5:
CW(P).

RADON MITIGATION SERVICES
4044 Southwest 13th St. Suite 153
Gainesville, FL 32608
(904) 375-5656
Jack W. Revell
Area Served:
Nationwide.
EPA RMP 5:
CC.

RADON MITIGATORS OF ARIZONA
6844 East Thunderbird Rd.
Scottsdale, AZ 85254
(602) 483-7506
Lawrence H. Kaplan - President
Kenneth Gilvary - Vice President
Joan Kaplan - Vice President
Services:
Provides radon detection services using charcoal canister and alpha track methods, provides radon mitigation services, group seminars.
Area Served:
Nationwide.

RADON MITIGATORS, INC.
16 Hillside Dr.
Clinton, NJ 08809
(201) 730-6596
Lawrence Kaplan - President
Joan Kaplan - Vice President
Kenneth Gilvary - Vice President
Services:
Provides radon detection services using charcoal canister and alpha track methods, diagnostics, provides radon mitigation services, group seminars.
Area Served:
Nationwide.
EPA RMP 5:
AT, CC(P).

RADON MONITOR SYSTEMS
P.O. Box 516
Eustis, FL 32727-0516
(904) 735-2929
Harvey Parker - Medical Technologist
Qualifications:
AARST.
Area Served:
FL.

RADON MONITORING AND SERVICES
1350 Butter Lane
Mt. Penn, PA 19606
(215) 779-8314
Ed Gutwein
Area Served:
PA.

RADON MONITORING SERVICES
P.O. Box 9083
Colorado Springs, CO 80932
(719) 577-9662
Charles G. Clute
Area Served:
CO.
EPA RMP 5:
CC.

RADON MONITORING SERVICES, INC.
P.O. Box 3730
Plymouth, MA 02361
(617) 746-8868
Michael O. Moyles - President
Services:
Provides radon detection services for home owners and commercial buildings.
Qualifications:
MA DPH, AARST, NE HP Society, EPA RMP, 20 years of health physics experience.
Area Served:
Nationwide.
EPA RMP 5:
CC.

RADON PROJECT, THE
P.O. Box 90069
Pittsburgh, PA 15224
(412) 687-2886
Bernard L. Cohen - President
Anna F. Cohen - Vice President
Frederick B. Cohen - Director
Services:
Manufactures and distributes diffusion barrier charcoal adsorption canisters
Qualifications:
EPA RMP.
Area Served:
Nationwide.
EPA RMP 5:
CC(P).

RADON PROTECTION SERVICES
P.O. Box 158
Hummelstown, PA 17036
(717) 566-5609
Dan Wilson - Owner

Services:
Provides radon detection and mitigation services, consulting, and training/seminars.
Qualifications:
PA DER.
Area Served:
PA.

RADON PROTECTION, INC.
The Summit Suite 110
Louisville, KY 40207
(502) 893-4563
Roland Hawksley - President
Area Served:
KY.

RADON QC
2857 Nazareth Rd. Palmer Professional Building Suite 6
Palmer, PA 18043
(215) 559-1636
John Rider - Manager
Kelsie Geiger - Assistant Manager

See Ad Page 237

Services:
Provides quality control services for radon measurement companies, known concentrations of radon gas and radon daughters, three chambers of radon gas concentrations, radon chamber services available to radon detection firms.
Area Served:
Nationwide.
EPA RMP 5:
CC(P), CR(P), GR(P), CW(P).

RADON RANGERS
7979 Old Georgetown Rd. Suite 1100
Bethesda, MD 20814
(301) 543-4300
John D. Croley
Area Served:
DC, MD, PA, VA, WV.
EPA RMP 5:
CC.

RADON READINGS
9020 Compton Blvd.
Bellflower, CA 90706
(213) 920-2582, (213) 421-8674
Judy Clark
Joe Livingston
Area Served:
CA.

RADON REDUCTION AND TESTING COMPANY
605 South Sherman Suite 705 J
Richardson, TX 75081
(214) 680-2296
Freeman Stallings - President
Phillip Stallings - General Manager
Services:
Provides radon detection services using charcoal canister method, provides radon mitigation services.
Area Served:
TX.

RADON REDUCTION AND TESTING, INC.
2222 Cumberland Pl.
Smyrna, GA 30080
(404) 434-7296
Terry E. Howell
Dallas Jones
Area Served:
AL, FL, GA, NC, SC, TN.
EPA RMP 5:
CR(P).

RADON REDUCTION ASSOCIATES, INC.
815 Saratoga Rd.
Burnt Hills, NY 12027
(518) 399-6560
L.H. Rutland, Jr. - Professional Engineer
Services:
Provides radon detection services.
Area Served:
NY.

RADON REDUCTION ENGINEERING AND LABORATORY
211 Mill St., NE
Vienna, VA 22180
(703) 281-7865, (703) 281-3730
Carl B. De Azagra - President
Charles H. De Azagra - Mitigation Program
Erika D. De Azagra - Measurement Program
Robin De Azagra - System Installation
Services:
Provides radon detection, mitigation, and monitoring services, and operates a primary laboratory.
Qualifications:
AARST, chemical engineer, architect, geologist, radon professional technician, and licensed contractor on staff.
Area Served:
Nationwide.
EPA RMP 5:
CC(P), CW(P).

RADON REDUCTION SERVICES
5601 East 109th Ter.
Kansas City, MO 64137
(816) 942-7083
Jerry Harper - Partner
Lee Cunningham - Partner
Services:
Provides radon mitigation services, inspects buildings for radon entry routes and designs and installs mitigation systems.
Qualifications:
AARST, training by NY State Energy Office.
Area Served:
IA, KS, MO, NV.

RADON REDUCTION SERVICES, INC.
P.O. Box 627
Mt. Airy, MD 21771
(301) 831-0228
Robert Wolfgang - President
Services:
Provides radon detection and mitigation services.
Area Served:
MD.

RADON REFEREEING/DIVISION OF HOME QUALITY HEADQUARTERS
3 Dogwood Rd.
Tolland, CT 06084
(203) 871-7433
Larry Willette - President
Qualifications:
AARST.
Area Served:
CT.

RADON REGISTRY
819 2nd St. B-178
Manchester, NH 03102
(603) 645-6780
Tom Roller
Area Served:
MA, ME, NH, VT.
EPA RMP 5:
CC.

RADON REMEDIAL SERVICES
P.O. Box 622
Bryn Athyn, PA 19009
(215) 938-7766
Lewis Nelson, IV
Services:
Provides radon mitigation services, mitigation system design and installation, and radon detection services.
Qualifications:
AARST, PA DER.
Area Served:
DE, NJ, PA.

RADON REMEDIES, INC.
2700 West Broad St. P.O. Box 44234
Columbus, OH 43204
(614) 272-7772
Craig S. Wise - President
James M. Farrow - General Manager
Services:
Produces and distributes radon video containing general information, radon

detection and mitigation techniques and products, for homes and businesses.
Qualifications:
EPA RMP, ARA.
Area Served:
Nationwide.
EPA RMP 5:
CR(P).

RADON REMOVAL, INC.
Box 283
Milltown, NJ 08850
(201) 545-2000
Anthony M. Fama - President
Services:
Provides radon detection and mitigation services, distributes a geomatrix, nylon membrane laid over perimeter of gravel foundation to provide an effective radon barrier.
Qualifications:
AARST, engineer on staff.
Area Served:
Nationwide.

RADON REPORTING SYSTEM OF CARROLL COUNTY
P.O. Box 500
Westminster, MD 21157-9167
(301) 857-1807, (301) 795-0775
Alan M. Lerman
Area Served:
MD.
EPA RMP 5:
CC.

RADON RESEARCH
700 Winthrop Pl.
Orlando, FL 32803
(813) 377-3740
Raymond V. Miller - President
Services:
Provides radon detection services using charcoal canister and alpha track methods.
Area Served:
FL.

RADON RESEARCH FOUNDATION, INC.
10922 West 101st St.
Overland Park, KS 66214
(913) 492-1723
John S. Neuberger - President
Services:
Provides radon detection services for homes, schools, businesses, using charcoal canister and alpha track detectors.
Qualifications:
EPA RMP.
Area Served:
KS, MO.
EPA RMP 5:
AT, CC.

RADON RESEARCH GROUP, INC.
P.O. Box 1143
Germantown , MD 20874
(301) 972-3309, (800) 445-8436
Michael J. Myers - Operations Chief
Craig M. Mount - Customer Service
Services:
Distributes radon detection products.
Qualifications:
AARST.
Area Served:
Nationwide.
EPA RMP 5:
AT, CC.

RADON RESEARCH INTERNATIONAL, INC.
7650 South Tamiami Trail , #9
Sarasota, FL 34231
(813) 923-7300
Arthur Lipshutz - Chief Engineer
Stuart Lee - Chief Field Engineer
Services:
Provides radon detection and mitigation services specializing in slab-on-grade construction, consultants to real estate industry.
Qualifications:
AARST, EPA RMP, BBB, three years of radon mitigation experience.
Area Served:
FL.
EPA RMP 5:
AT, CC.

RADON RESOURCE, INC.
49 John St.
Westminster, MD 21157
(301) 857-9330
Neil Edwards - President
Services:
Provides radon detection and mitigation services.
Qualifications:
EPA training courses in testing and mitigation.
Area Served:
MD, PA.
EPA RMP 5:
CC.

RADON RESOURCES
241 Rumson Rd.
Atlanta, GA 30305
(404) 231-4732
Madeleine Jacobsen
Area Served:
Nationwide.
EPA RMP 5:
CC.

RADON RESOURCES, INC.
55 Constitution Dr.
Bedford, NH 03102
(603) 472-5955
Theresa E. Chabot - President
Services:
Provides radon detection services for residential and commercial properties using continuous working level monitor, alpha track and charcoal canister methods.
Qualifications:
EPA RMP.
Area Served:
MA, ME, NH.
EPA RMP 5:
CC, CW(P).

RADON SAFE
P.O. Box 631 Highway 62 W.
Boonville, IN 47601
(812) 897-1700
Carl Westfall
Area Served:
Nationwide.

RADON SAFE
7596 Wheaton Dr.
Canton, MI 48187
(313) 455-6469
Subbiah Harihar
Area Served:
MI.
EPA RMP 5:
CC.

RADON SAFETY AND TECHNOLOGY CORPORATION
4 Charles St.
Pleasant Valley, NY 12569
(914) 635-1138, (800) 552-5257
Ronald J. Gasparro - President
Zia R. Hashmi - Director
Claude J. Noto - Assistant Director
Services:
Radon measurement analytical laboratory, water and soil measurements, surveys for real estate transactions, land sites, and newly constructed and standing buildings, distributes charcoal canisters and alpha track detectors, waterborne radon test kits, radon alarms and auxiliaries offered to public and professionals, provides radon mitigation services.
Qualifications:
EPA RMP, AARST, ASTM.
Area Served:
Nationwide.

EPA RMP 5:
AT, CC(P), EP(P), GR(P), CW(P).

RADON SAFETY SERVICE
2204 Frances Dr.
Loveland, CO 80537
(303) 663-3941
Joseph T. Mamak - Owner
Robert J. Tengler - Owner
Services:
Provides radon detection services using charcoal canisters and alpha-track methods, mitigation consultation or referral.
Area Served:
CO.

RADON SAFETY SERVICES, INC.
P.O. Box 441
New Providence, NJ 07974
(201) 825-8267, (201) 665-1188
Guy J. Agrati
Area Served:
Nationwide.
EPA RMP 5:
CC, EP(P), RP, CW(P).

RADON SAFETY TESTING, INC.
P.O. Box 277
Gales Ferry, CT 06335-0277
(203) 886-8327
Stuart Murray
Area Served:
CT, RI.
EPA RMP 5:
CC.

RADON SCREENING SERVICE, INC.
P.O. Box 37007
Denver, CO 80237
(303) 741-3207
Thomas Staley
Area Served:
Nationwide.
EPA RMP 5:
CC, GW(P), CW(P).

RADON SCREENING, INC.
7 Brookcrest Ct.
Fredericksburg, VA 22401
(703) 786-4320
Robert D. Chapman, Jr.
Qualifications:
AARST.
Area Served:
Nationwide.
EPA RMP 5:
AT, CC.

RADON SEARCH-USA
2701 North Rocky Point Rd. Suite 800
Tampa, FL 33607
(813) 960-1360
Ray Smith - President
Services:
Provides radon detection services using charcoal canister, alpha track, continuous radon monitor, and continuous working level monitor methods.
Area Served:
FL.

RADON SERVICES
1210 Auburn Way N. Suite B-122
Auburn, WA 98002
(206) 939-3023
Bob McCausland
Area Served:
Nationwide.
EPA RMP 5:
CC.

RADON SERVICES ANALYTICAL LABORATORY
111 Benson Rd.
Nashville, TN 37214
(615) 883-4659
John Hughes
Area Served:
Nationwide.
EPA RMP 5:
CC(P).

RADON SERVICES OF MID-AMERICA
935B Southeast 3rd St.
Lee's Summit, MO 64063
(816) 525-7412
Glendora Sparks
Area Served:
MO.

RADON SERVICES OF NORTH AMERICA, INC.
148 East 632 N.
Valparaiso, IN 46383
(219) 462-5866
Linda L. Komp - Vice President
Area Served:
Nationwide.
EPA RMP 5:
CC.

RADON SERVICES OF WYOMING
1714 Rollins
Cheyenne, WY 82001
(307) 634-9505
Ronald J. Sargent - President
Services:
Provides radon detection and mitigation services.
Area Served:
WY.

RADON SERVICES, INC.
2950 Banksville Rd.
Pittsburgh, PA 15216
(412) 563-0553
Andrew R. Schwab
Area Served:
Nationwide.
EPA RMP 5:
CC(P).

RADON SPECIALIST OF AMERICA
12628 Gray Eagle Ct. Suite 12
Germantown, MD 20874
(301) 231-4091
Kevin Sullivan Cavey
Area Served:
Nationwide.
EPA RMP 5:
CC.

RADON SPECIALISTS
1840 Folsom St. Suite 255
Boulder, CO 80302
(303) 673-0494
Sharon A. Durgin
Area Served:
CO.

RADON SURVEY SYSTEMS, INC.
P.O. Box 22288
Cleveland, OH 44122
(216) 464-7997
Darioush T. Ghahremani - President
Services:
Provides radon detection and mitigation services for residential and commercial markets, site surveys for undeveloped land, research and development of radon detection technologies, detectors, and materials, new development on radon mitigation options for various projects.
Qualifications:
AARST, NRA, several national scientific societies.
Area Served:
Nationwide.
EPA RMP 5:
AT, CC.

RADON SURVEYS
40 Pinetree
Seymour, IL 61875
(217) 687-2611
Steven P. Ray - President
Services:
Provides radon detection using progeny grab sample techniques, provides radon mitigation consulting services.
Qualifications:
EPA RMP.
Area Served:

IL, IN, MI, WI.
EPA RMP 5:
GW(P).

RADON SURVEYS/DIVISION OF COPE TECHNICAL SERVICES
3506 87th Pl.
Kenosha, WI 53142
(414) 694-3730
James L. Cope
Area Served:
IA, IL, IN, MI, MN, WI.
EPA RMP 5:
AT, CC, EP.

RADON SYSTEMATICS OF CALIFORNIA
1500 San Remo Dr.
Pacific Palisades, CA 90272
(213) 459-9146, (213) 825-1934
Donald Carlisle
Area Served:
AZ, CA, NV, OR, WA.
EPA RMP 5:
AT, CC, EP.

RADON SYSTEMS OF RHODE ISLAND, INC.
156 Earle St.
Woonsocket, RI 02895
(401) 766-7072
Richard A. Monteiro - President
Services:
Provides radon detection and mitigation services for residential, commercial, and industrial markets.
Qualifications:
EPA RMP, applicant PA licensing.
Area Served:
CT, MA, ME, NH, NJ, NY, PA, RI.
EPA RMP 5:
AT, CC.

RADON SYSTEMS, INC.
P.O. Box 2509
Laurel, MD 20708
(301) 725-9000
Michele Widmayer - President
Patricia Jordan - Vice President
Kenneth Grim - Vice President
Services:
Provides radon detection and mitigation services to home owners, realtors, and builders.
Qualifications:
EPA course on radon testing, NY State Energy Office course on mitigation, AARST.
Area Served:
DC, DE, MD, PA, VA.
EPA RMP 5:
CC.

See Ad Page 157

RADON TECHNICAL SYSTEMS, INC.
37 Tree Brook Dr.
Rochester, NY 14625
(716) 385-4255
Robert J. Salvage
Qualifications:
AARST.
Area Served:
NY.

RADON TECHNOLOGIES, INC.
39 Woodrow Ct.
Troy, NY 12180
(518) 447-5834
Bob Mahoney
Area Served:
CT, MA, NJ, NY, PA, VT.
EPA RMP 5:
AT, CC, GR(P).

RADON TECHNOLOGIES, INC.
698 High St.
Worthington, OH 43085
(614) 441-0550, (614) 431-0322
Thomas E. Warren
Greg Moore
Area Served:
IN, KY, MI, OH, PA, WV.
EPA RMP 5:
EP.

RADON TECHNOLOGIES, INC.
800 Forest Ave.
Norristown, PA 19401
(215) 279-5188
Curtis Red
Area Served:
PA.

RADON TECHNOLOGIES, INC.
12200 Sunrise Valley Dr. Suite 320
Reston, VA 22091
(703) 620-4004, (800) 237-5024
R.A. Brightsen - Chairman
Venus Cain - Secretary
Services:
Distributes No-Rad radon removal system.
Area Served:
Nationwide.
EPA RMP 5:
CC.

RADON TECHNOLOGY
11404A East 24 Hwy.
Sugar Creek, MO 64054
(816) 461-0211
Clay Meyers - President
Services:
Provides radon detection services using continuous radon monitor method, provides radon mitigation services.
Qualifications:
ARA, detection and mitigation workshops.
Area Served:
MO.

RADON TECHNOLOGY AND CONSULTANTS
RFD 2 Louisquisset Pike
North Smithfield, RI 02895
(401) 769-2129, (401) 568-0296
Lucien R. Auger - President
John S. Dolnski - Vice President
Services:
Provides radon detection and mitigation services.
Qualifications:
Radon detection and mitigation training courses, 23 years of experience in waterproofing business.
Area Served:
CT, MA, ME, NH, RI, VT.

RADON TECHNOLOGY, INC.
19595 Merriman Rd.
Romulus, MI 48174
(313) 753-4211
Howard D. Blair - President
Area Served:
MI.

RADON TECHNOLOGY, INC.
7 Tokanel Dr.
Londonderry, NH 03053
(603) 434-4937
Donald A. Moskowitz
Area Served:
MA, ME, NH, VT.
EPA RMP 5:
CC.

RADON TEST
P.O. Box 5210
San Luis Obispo, CA 93403-5210
(805) 481-4069
W. O'Brien - President
Services:
Provides radon detection services using E-Perm, charcoal canister methods.
Area Served:
AZ, CA, CO, ID, MT, NM, NV, OR, UT, WA, WY.
EPA RMP 5:
AT, CC.

RADON TEST AND REDUCTION SPECIALIST
125 Oak Ave. P.O. Box 129
Spruce Pine, NC 28777
(704) 765-1222
Bud Ledford - Consultant
Jerri Ledford - Laboratory Supervisor

Services:
Provides radon detection and mitigation services.
Area Served:
GA, KY, MI, NC, SC, TN, VA, WV.

RADON TEST CENTER
P.O. Box 1542
Ellicott City, MD 21043
(301) 997-2366
Phil Barnacz
Area Served:
MD.

RADON TEST COMPANY
P.O. Box 800
Worcester, MA 01602
(617) 853-7809, (617) 793-7687
Sassan Abdollahzadeh
Area Served:
MA.

RADON TEST SYSTEMS
1418 Upfield
Carlton, TX 75006
(214) 242-1020
Stewart Johnson
Area Served:
TX.

RADON TESTING AND CONSULTING
1771 Post Rd., E. Suite 320
Westport, CT 06880
(203) 454-3886
Lorraine Hodes - Vice President
John Nicholas - Director
Mark Lewis - Senior Geologist
Maria Mody - Communications
Services:
Provides radon detection and mitigation services for home owners, realtors, builders and developers, exclusive New England distributor of Current Indoor Air Model 300 Radon Mitigation System.
Qualifications:
AARST, AGSE.
Area Served:
CO, CT.

RADON TESTING AND CONSULTING SERVICES
34 Jones River Dr.
Kingston, MA 02364
(617) 585-9924
Daniel C. Walker - President
Cathleen Walker - Vice President
Services:
Provides radon detection, mitigation, and consultation services for the real estate industry, construction trade, and private citizens.
Qualifications:
AARST, RTCA professional consultant.
Area Served:
CT, MA, ME, NH, RI, VT.

RADON TESTING AND CONTROL
2910 East Broadway Ave.
Bismarck, ND 58501
(701) 222-1613
Judith A. Atkinson - Owner
Herb Haugen - Owner
Services:
Provides radon detection and mitigation services.
Area Served:
MT, ND, SD.

RADON TESTING AND ENVIRONMENTAL CONTROL
7113 Chippenham Rd.
Louisville, KY 40222
(502) 426-4314
M. Jeana Phelps
Area Served:
Nationwide.
EPA RMP 5:
CC.

RADON TESTING AND MITIGATION OF TEXAS, INC.
P.O. Box 34268
Houston, TX 77234-4268
(800) 920-1663
Jerry Shults - President
Services:
Provides radon detection and mitigation services.
Area Served:
TX.

RADON TESTING AND MITIGATION SERVICES COMPANY
5442 Cary Ave.
Davenport, IA 52807
(319) 386-9005
Tom Gabrilson - President
Services:
Provides radon detection and mitigation services.
Qualifications:
Licensed in IA, certified Air Chek dealer.
Area Served:
IA, IL.

RADON TESTING AND MITIGATION SYSTEMS
P.O. Box 3052
Carmel, IN 46032
(317) 545-5578
Lorand Magyar - Owner
Andrew W. Murrell - Owner
Services:
Provides radon detection and mitigation services for real estate companies and for residential market.
Area Served:
IL, IN, KY, OH.

RADON TESTING AND REDUCTION
P.O. Box 437
Howell, MI 48844
(517) 548-2497
Chris Coffey - General Manager
Craig Coffey - General Manager
Services:
Provides radon detection services using charcoal canister and E-Perm methods, provides radon mitigation services.
Area Served:
MI.

RADON TESTING AND TREATMENT, INC.
723 North Weber Suite 300
Colorado Springs, CO 80903
(719) 520-5693
Douglas L. Kladder - President
Steve Jelinek - Senior Analyst
Jessica Goodwin - Marketing Director
Robert Shideler
Services:
Provides radon detection and mitigation services to the real estate industry, distributes charcoal canisters to the public.
Qualifications:
EPA RMP, licensed contractor.
Area Served:
CO.
EPA RMP 5:
CC, GW(P).

RADON TESTING ASSOCIATION
Rte. 6, 2014 Glenwood Dr.
Martinsburg, WV 25401
(304) 754-7586
Ted Hall
Area Served:
WV.

RADON TESTING COMPANY OF NEW YORK
1790 Meadowbrook Rd.
Merrick, NY 11566
(516) 868-1778
Howard Braff
Area Served:
NY.

RADON TESTING CONSULTANTS, INC.
P.O. Box 25313
Newark, NJ 07101

(201) 290-9217
Mark Jeziorski - President
Services:
Provides radon detection and mitigation consulting services for consumers, businesses, and government agencies, distributes charcoal canisters and alpha track detectors.
Qualifications:
EPA RMP, staff has over 12 years of combined engineering experience.
Area Served:
Nationwide.
EPA RMP 5:
CC.

RADON TESTING CORPORATION
4 Cheyenne Ct.
Bryan's Road, MD 20616
(301) 283-6006
John Moughan
Area Served:
Nationwide.
EPA RMP 5:
AT, CC.

RADON TESTING CORPORATION OF AMERICA
12 West Main St.
Elmsford, NY 10523
(914) 347-5010, (800) 457-2366
Nancy Bredhoff - Vice President
Kurt Dorfi - Sales Manager
Services:
Manufactures and distributes charcoal canister radon test kits, distributes Pro Scan, a product designed to aid detection firms to prevent test tampering.
Qualifications:
EPA RMP, approved by NY DOH.
Area Served:
Nationwide.
EPA RMP 5:
CC(P), EP(P).

RADON TESTING GROUP
P.O. Box 23759
Jacksonville, FL 32241
(904) 260-2752
Dale Shaw - President
Services:
Provides radon detection services using charcoal canister method.
Area Served:
FL.

RADON TESTING LAB
3432 Hollywood
Brookfield, IL 60513
(312) 485-1133
Ron Leslie - President
Services:
Provides radon detection services using charcoal canister and E-Perm methods.
Area Served:
IL.

RADON TESTING LABORATORIES, INC.
10390 Tuscany Ln.
Ellicott City, MD 21043
(301) 465-1093
James L. Boyd
Area Served:
MD.

RADON TESTING LABORATORY OF NEW ENGLAND
P.O. Box 4508
Boston, MA 02101
(617) 426-9224
Raymond Fontana
Area Served:
Nationwide.
EPA RMP 5:
CC(P).

RADON TESTING LABORATORY OF NEW JERSEY
1303 Birch Ave.
Ocean Township, NJ 07712-4514
(201) 502-0727
Daniel D. Cervo - President
Services:
Provides radon detection services for real estate market and general population, professional placement and mail order service.
Area Served:
NJ.
EPA RMP 5:
CC.

RADON TESTING LABS
P.O. Box 1467
Avon, CT 06001
(203) 678-8259
Richard G. Piccolo
Area Served:
CT.

RADON TESTING OF MICHIANA
Colfax Ctr. 914 Lincolnway W.
South Bend, IN 46616
(219) 282-2435
Jim Regelean - Chief Inspector
Robert F. Werner - Vice President
Services:
Provides radon detection services using charcoal canister and alpha track methods.
Area Served:
IL, IN, MI, OH.

RADON TESTING SERVICE, INC.
P.O. Box 1130
Englewood, CO 80110
(303) 762-8598
David L. Egleston, Jr.
Area Served:
Nationwide.
EPA RMP 5:
CC(P), GW(P).

RADON TESTING SERVICE, INC.
Rte. 212
Lake Hill, NY 12488
(914) 679-6835
Tully Slayton - President
Sidney Slayton - Secretary
Services:
Provides radon detection and mitigation services.
Area Served:
NY.

RADON TESTING SERVICE, INC.
545 North 33rd
Milwaukee, WI 53208
(414) 931-0527
Thomas D. Hunter
Area Served:
WI.
EPA RMP 5:
CC.

RADON TESTING SERVICES
P.O. Box 152227
San Diego, CA 92115-0870
(619) 229-9002, (619) 287-5317
Christine Palmieri
Area Served:
Nationwide.
EPA RMP 5:
CC(P).

RADON TESTING SERVICES
4096 Scenic Dr.
Provo, UT 84604
(801) 227-0179, (800) 777-2320
Robert Vivian - Owner
Services:
Provides radon detection services to residences and commercial buildings.
Qualifications:
EPA RMP.
Area Served:
Nationwide.
EPA RMP 5:
CC(P).

RADON TESTING SERVICES OF MISSOURI
P.O. Box 362
Trenton, MO 64683

(816) 359-4442
Norman Baker
Area Served:
Nationwide.
EPA RMP 5:
CC, CR(P).

RADON TESTING SERVICES, BALTIMORE
P.O. Box 7999
Baltimore, MD 21221-0999
(301) 321-8008
Laurence V. Kolego
Area Served:
Nationwide.
EPA RMP 5:
AT, CC.

RADON TESTING SERVICES, CHERRY HILL
P.O. Box 3013
Cherry Hill, NJ 08034
(609) 665-9228
Dorothy Angeloff
Area Served:
NJ.
EPA RMP 5:
CC.

RADON TESTING SERVICES, HARRISBURG
P.O. Box 3013
Cherry Hill, NJ 08034
(717) 232-4616
Hal J. Angeloff
Area Served:
PA.
EPA RMP 5:
CC.

RADON TESTING SERVICES, INC.
4442 18th St.
San Francisco, CA 94114
(415) 552-5496
William E. Hellums - President
Services:
Provides radon detection services, forensic and consultation services, computer dose projections.
Qualifications:
EPA RMP, radioactive materials license, certified health physicist.
Area Served:
AK, AZ, CA, ID, NV, UT, WA.
EPA RMP 5:
CC, EP(P), CW(P).

RADON TESTING SERVICES, INC.
45 Streit Ave.
Poughkeepsie, NY 12603
(914) 485-3060
Jon Behrends - President
Services:
Provides radon detection services.
Area Served:
NY.

RADON TESTING SERVICES, PITTSBURGH
P.O. Box 19425
Pittsburgh, PA 15213
(412) 687-1533
Mark Johnson
Area Served:
Nationwide.
EPA RMP 5:
CC.

RADON TESTING SERVICES, PRINCETON
P.O. Box 3311
Princeton, NJ 08543
(609) 737-8818
Michael Wright
Area Served:
DE, NJ, PA.
EPA RMP 5:
CC.

RADON TESTING SERVICES, SILVER SPRING
1774 Featherwood St.
Silver Spring, MD 20904
(301) 236-9434
Ted Simmons
Area Served:
DC, DE, MD, NJ, PA, VA, WV.
EPA RMP 5:
CC.

RADON TESTING SYSTEMS
P.O. Box 243
Newtown, PA 18940
(215) 639-1933
Thomas D'Arcy
Area Served:
PA.

RADON TESTING, INC.
P.O. Box 7304
Fairfax Station, VA 22039
(703) 591-1111
James P. Schroff - President
Services:
Provides radon detection and mitigation services for the real estate industry.
Qualifications:
Training courses by EPA and NY State Energy Office.
Area Served:
Nationwide.

EPA RMP 5:
CC, CR(P).

RADON TRAC, INC.
P.O. Box 4015
Sarasota, FL 34230
(813) 955-5426, (813) 923-5196
Debora R. Blomster
Qualifications:
AARST.
Area Served:
Nationwide.
EPA RMP 5:
CC, EP(P).

RADON TRACKING SERVICE
1915 Whitney Rd. Suite 1
Clearwater, FL 34230
(904) 260-2752
Services:
Provides radon detection services using charcoal canister method.
Area Served:
FL.

RADON WATCH, INC.
472 North Wooster Way
Lancaster, OH 43130
(614) 837-5748
Tom Daugherty - Owner
Services:
Provides radon detection services for homes and public buildings using charcoal canisters and continuous radon monitors, specializes in real estate industry.
Qualifications:
EPA RMP.
Area Served:
OH.
EPA RMP 5:
CC, CR(P).

RADON X COMPANY, INC.
217 Roxbury St.
Keene, NH 03431
(603) 352-5319
Steven Berger - President
David P. Tkatch - Vice President
Frank Lapinski - Vice President
Services:
Provides radon detection and mitigation services.
Qualifications:
EPA mitigation course, 45 years of radiation experience, professional engineers on staff.
Area Served:
Nationwide.
EPA RMP 5:
GW(P).

RADON, INC.
1637 Downtown West Blvd. Suite 9
Knoxville, TN 37919
(615) 694-4080
Parker Wright
Area Served:
TN.

RADON, LTD.
411 North Weinbach Ave.
Evansville, IN 47711
(812) 479-3110, (800) 634-2901
Ed Shoemaker - President
Services:
Provides radon detection and mitigation services, distributes charcoal canister kit to public and dealers, laboratory facilities for analysis.
Qualifications:
ARA, radon detection and mitigation training courses.
Area Served:
Nationwide.
EPA RMP 5:
CC(P).

RADON-ONE LABORATORY SERVICES
4575 Chandlers Forde
Sarasota, FL 34235
(813) 377-3740
Jim Cook
Services:
Provides radon detection services using charcoal canister method.
Area Served:
FL.

RADON-ONE LABORATORY SERVICES
4700 East Fifth Ave.
Columbus, OH 43219
(614) 239-7171
John R. Bartlett
Area Served:
Nationwide.
EPA RMP 5:
CC(P).

RADON-X CORPORATION
300 Valentine St.
Hackettstown, NJ 07840
(201) 852-6212, (800) 272-3669
John LoPilato - President
Services:
Manufactures air exchange and sub-slab ventilating systems, provides radon mitigation services.
Area Served:
NJ.

RADON-X CORPORATION
3 Edinburgh Ln.
Mt. Laurel, NJ 08054
(609) 778-7771, (800) 272-3669
Paul Gibson
Area Served:
NJ.

RADON/GEOLOGICAL ASSESSMENT SERVICES
727 Crown Ridge Dr.
Colorado Springs, CO 80904
(719) 634-5235, (719) 685-4749
H. Prostka
T. Zacharakis
Area Served:
CO.

RADON/SOILS AND WATER TESTING SERVICE
7080 Price-Hilliard Rd.
Plain City, OH 43064
(614) 873-8821
John Steichen - President
Services:
Manufactures and distributes charcoal canisters, distributes E-Perm detectors, provides radon detection services to the real estate industry.
Qualifications:
AARST.
Area Served:
Nationwide.
EPA RMP 5:
CC(P), EP(P).

RADONAIRE
P.O. Box 3702
Wayne, NJ 07474
(201) 790-3252
H.E. Firschein - President
Area Served:
NJ.

RADONAIRE, INC.
393 Denrose Dr.
Amherst, NY 14150
(716) 837-2366
Karl Koniarczyk - President
Services:
Provides radon detection services using working level meter, charcoal bag, and alpha track methods for home, land, and water.
Qualifications:
AARST, DOE seminar, EPA RMP.
Area Served:
NY.
EPA RMP 5:
AT, CC.

RADONICS
1764 Old Meadow Ln. Suite 125
McLean, VA 22102
(703) 734-0050
Keith S. Fimian
John E. McGreevy
Services:
Provides radon detection services.
Qualifications:
Professional engineers on staff.
Area Served:
DC, MD, VA, WV.
EPA RMP 5:
AT, CC, CR(P), EP(P), CW(P).

RADONICS, INC.
P.O. 489
Pennington, NJ 08534
(609) 737-8888
Theresa R. Marryott - President
Services:
Provides radon testing and mitigation services.
Area Served:
Nationwide.

RADONIX, LTD.
P.O. Box R
Lake Grove, NY 11755
(516) 467-3780
Gene Shafto - President
Services:
Provides radon detection services using charcoal canister method.
Qualifications:
EPA RMP.
Area Served:
Nationwide.
EPA RMP 5:
CC(P).

RADONTECH, INC.
1616 Walnut St. Suite 2200
Philadelphia, PA 19103
(215) 546-7328
Area Served:
PA.

RADONTEST, INC.
4700 East 5th Ave.
Columbia, MO 65201
(314) 875-8045
Karl Hennecke
Area Served:
Nationwide.
EPA RMP 5:
CC(P).

RAF ENGINEERING
Hillside Rd.
Rockaway, NJ 07866
(201) 983-9041
Ralph Favale
Area Served:
Nationwide.
EPA RMP 5:
AT, CC.

RAM EXTERMINATORS
97 Frank St.
Somerset, NJ 08873
(201) 937-6900
James R. Savage
Area Served:
NJ.
EPA RMP 5:
CC.

RAMP INDUSTRIES, INC.
1127 West 46th St.
Denver, CO 80210
John E. Lucas - General Manager
Services:
Provides radon detection and mitigation services, operates a primary laboratory.
Area Served:
CO.

RAMSES II, LTD.
2355 East Dartmouth Ave.
Denver, CO 80210
(303) 758-5209
Robert L. Clark - Vice President
George A. Michael - President
Services:
Manufactures and distributes charcoal canisters, electronic sump well cover kits, electronic PVC ventilation fans, and rotary air-to-air heat exchangers, provides radon mitigation services.
Qualifications:
Licensed professional engineer on staff.
Area Served:
CO.
EPA RMP 5:
CC.

RANDOLPH AND ASSOCIATES, INC.
8901 North Industrial Rd.
Peoria, IL 61615
(309) 692-4160
Bryan W. Arnold
Vahooman Mirkhaef
Area Served:
Nationwide.
EPA RMP 5:
CC(P).

RAY MESSICK RADON TESTING AND MITIGATION
RDI Box 469L
McConnellsburg, PA 17233
(202) 332-1798, (717) 485-5456
Ray Messick
Area Served:
DC, MD, PA, WV.
EPA RMP 5:
CR(P).

RCI SERVICES
11412 Hounds Way
Rockville, MD 20852
(301) 231-5134
Frank M. Clower
Area Served:
MD.

RDS/RADON DETECTION SERVICES, INC.-CORPORATE OFFICE
Rte. 179, Old York Rd. P.O. Box 419
Ringoes, NJ 08551
(201) 788-3080
James G. Davidson - President
Susanne E. Micek - Vice President
Michael S. Terpilak - Health Physicist
Services:
Provides full service radon detection mitigation, and consultation services, offers franchises, offers comprehensive training courses for radon professionals, manufactures and distributes RDS "remediator" fans, locations in four states.
Qualifications:
Has met all licensing requirements in PA, NJ, and KS, courses have been approved by DE, KY, PA, OH.
Area Served:
Nationwide.
EPA RMP 5:
AT, CC, EP(P), GR(P), CW(P).

See Ad Page 153

RDS/RADON DETECTION SERVICES-GEORGIA OFFICE
403 West Ponce de Leon Ave. P.O. Box 1826
Decatur, GA 30031
(404) 377-2366
Barrett P. Walker - President
Peggy Walker - Customer Relations
Judy Reid - Office Manager
Services:
Provides radon detection and mitigation services, informational seminars.
Qualifications:
EPA RMP, BBB, Metro Atlanta Home Builder's Association, AARST.
Area Served:
Nationwide.
EPA RMP 5:
AT, CC, EP(P), CW(P).

RDS/RADON DETECTION SERVICES-MARYLAND OFFICE
10400 Connecticut Ave. Suite 504

Kensington, MD 20895
(301) 933-7900
Milton Iskow - President
David E. Pard - Sales Manager
Services:
Provides radon detection services using continuous working level monitors, E-Perm, alpha track and charcoal canister methods, pre-construction soil testing and consultation to general public, real estate/construction industries, local governments and corporations.
Qualifications:
AARST, EPA training, PA certification, geologist and health physicist on premises.
Area Served:
Nationwide.
EPA RMP 5:
CC, EP(P), CW(P).

RDS/RADON DETECTION SERVICES-PENNSYLVANIA OFFICE
1011 Brookside Rd., Suite 270 P.O. Box 270
Allentown, PA 18106
(215) 481-9555
Bruce Barker - Owner
Donna Garris - Office Manager
Services:
Provides radon detection and mitigation services.
Area Served:
Nationwide.
EPA RMP 5:
AT, CC, EP(P), GR(P), GW(P), CW(P).

REAL ESTATE INSPECTION SERVICE
P.O. Box 3621 CRS
Johnson City, TN 37602
(615) 282-4216
Jerry Buckles
Area Served:
TN, VA.
EPA RMP 5:
CC.

REAL ESTATE INSPECTOR, THE
2195 Stoney Hill Rd.
Boulder, CO 80303
(303) 499-2688
Barry Prentice
Area Served:
CO.
EPA RMP 5:
AT, CC.

REAL ESTATE SUPPORT SERVICES, INC.
7825 Washington Ave., S.
Minneapolis, MN 55435
(612) 941-7558
Kenneth M. Sedor - Vice President
Brian Kane - Director of Marketing
Bruce D. Frimerman - President
Services:
National radon detection firm exclusively serving the needs of the relocation industry.
Qualifications:
EPA RMP, AARST, NRA, ERC, ASHI.
Area Served:
Nationwide.
EPA RMP 5:
AT, CC, EP(P).

REALITY DYNAMICS
4758 Antelope Trail
Elbert, CO 80106
(303) 648-3297
John Micholle
Area Served:
CO.

RECON SYSTEMS, INC.
P.O. Box 460
Three Bridges, NJ 08887
(201) 782-5900
Stephen E. Laney
Peter Rosen
Area Served:
Nationwide.
EPA RMP 5:
AT, CC.

REDIHOME INSPECTIONS, INC.
18 Kennedy Ter.
Westwood, NJ 07675
(201) 358-6888
Reda Fanous - President
Area Served:
NJ.

RELIABLE BASEMENT WATERPROOF COMPANY
2811 Westerville Rd.
Columbus, OH 43224
(614) 267-4477
Matthew Cohen - President
Lloyd Cohen - Vice President
Mike Copley - General Manager
Services:
Provides radon detection and mitigation services for residences, commercial and industrial markets.
Qualifications:
EPA RMP, ARA.
Area Served:
OH.
EPA RMP 5:
CR(P).

REMD/RESIDENTIAL ENVIRONMENTAL MEDICAL DIAGNOSTICS
28 School St. P.O. Box 3285
Stony Creek, CT 06405
(203) 481-8211
Robert McLellan
Area Served:
CT, MA, NY, RI.
EPA RMP 5:
AT, CC.

REMSEARCH, INC.
160 Bleeker St. Suite 5BE
New York, NY 10012
(212) 982-1115
David H. Vories - President
Services:
Provides environmental radioactivity detection services for residences and businesses.
Qualifications:
NY State certification.
Area Served:
CT, NJ, NY.
EPA RMP 5:
AT.

RENSSELAER POLYTECHNICAL INSTITUTE
110 Eighth Street
Troy, NY 12180
(518) 276-8237
Ken Brown - HVAC Engineer
Services:
Provides radon detection and mitigation services or campus buildings, sponsored by state of New York.
Area Served:
NY.

RESIDENCE RADON
28077 West 85th Ter.
Desoto, KS 66018
(913) 383-4610, (913) 585-1094
Richard J. Brazukas - Owner
Services:
Provides radon detection services, distributes charcoal canisters and alpha track detectors to the public.
Area Served:
Nationwide.
EPA RMP 5:
CC.

RESIDENTIAL INSPECTION TECHNOLOGY/RES-I-TEC
145 D Grassy Plain St.
Bethel, CT 06801
(203) 791-1076
Ralph Passaro, Jr. - Vice President

Services:
Provides radon detection services using charcoal canister method.
Qualifications:
NRA.
Area Served:
CT, NY.
EPA RMP 5:
CC.

RESIDENTIAL RADON MEASUREMENTS
P.O. Box 3051
Fairview Heights, IL 62208
(618) 397-0099
Todd A. Jokerst - President
Services:
Provides radon detection services and mitigation consultation services to home owners, realtors, and home builders.
Qualifications:
EPA RMP, ANS, NHPS, Nuclear Engineer on staff.
Area Served:
IL, MO.
EPA RMP 5:
AT, CC.

RESIDENTIAL RADON TEST SERVICE
126 Old Oaken Bucket Rd.
Scituate, MA 02066
(617) 545-9160
Jim Wood
Area Served:
CT, MA, ME, NH, RI, VT.
EPA RMP 5:
CC.

RESTORED ENVIRONMENTS COMPANY, INC.
4940 Nutmeg Knoll
Cincinnati, OH 45244
(513) 831-0743
Bob Smith - President
Services:
Provides radon detection and mitigation services.
Area Served:
IN, KY, OH.

REYNOLDS RADIOLOGICAL SERVICES, INC.
RD #1 , Box 2715 Governor Stable Rd.
Bainbridge, PA 17502
(717) 299-6429, (717) 367-7829
H.W. Reynolds - Health Physicist
Services:
Provides radon detection services for public and private structures using continuous working level monitor method.
Area Served:
PA.
EPA RMP 5:
CW(P).

RIC INSPECTION AND ENGINEERING
R.R. 2 P.O. Box 2186-5
Columbia, NJ 07832
(201) 496-4818
E. Richard Gralinski
Area Served:
NJ, NY.
EPA RMP 5:
CC.

RISE
280 Broadway
Providence, RI 02903
(401) 272-1040
John Masiello - Manager
Services:
Private, non-profit organization which conducts energy audits for residential and commercial buildings, and operates the only laboratory in Rhode Island for radon testing.
Qualifications:
EPA RMP.
Area Served:
CT, MA, ME, NH, RI, VT.
EPA RMP 5:
CC(P).

RMG/PC ENGINEERING, INC.
617 North 17th St. Suite 201
Colorado Springs, CO 80904
(719) 475-9600
John Clark - President
Services:
Provides radon detection services using charcoal canister method.
Area Served:
CO.

ROBERTS ENVIRONMENTAL SERVICES
4110 Old Wake Forest Rd. Suite 109
Raleigh, NC 27609
(919) 790-9378
Glenn D. Osmond
Phillip M. Hooper
Area Served:
NC.
EPA RMP 5:
AT, CC.

ROCKY MOUNTAIN HOME INSPECTIONS
1571 East Mineral Pl.
Littleton, CO 80122
(303) 730-0922
Martin Dommer
Area Served:
CO.

ROCKY MOUNTAIN MEDICAL PHYSICS, INC.
7175 West Jefferson Ave. Suite 2150
Lakewood, CO 80235
(303) 980-7070
John Goddard - President
Services:
Medical physics group provides radon detection and consultation services for industry.
Qualifications:
EPA RMP.
Area Served:
Nationwide.
EPA RMP 5:
CC(P).

ROCKY MOUNTAIN RADON TESTING
6820 South Brook Forest Dr.
Evergreen, CO 80439
(303) 674-7029
Duane A. Riggenbach
Area Served:
CO.
EPA RMP 5:
CC.

ROGERS AND ASSOCIATES ENGINEERING CORPORATION
515 East 4500 S. Suite G-200
Salt Lake City, UT 84107
(801) 263-1600
Arthur Sutherland - Vice President
Kirk K. Nielson - Research
Services:
Provides radon detection services, consulting engineers for radiological and environmental problems, research to develop a unified theory of radon emanation and transport to implement in a computer code for field studies and research programs.
Qualifications:
State licensed for radioactive materials.
Area Served:
Nationwide.
EPA RMP 5:
AT, CC(P), GR(P).

ROMA RADON
P.O. Box 90875
Lakeland, FL 33804-0875
(813) 859-5001
Ron Barrett - President
Services:
Provides radon detection services using charcoal canister method.
Area Served:
FL.

RONALD BARTA AND ASSOCIATES
345 Meadow View Dr.
Evergreen, CO 80439
(303) 674-3025
Ronald Barta - President
Area Served:
CO.

RONALD STEIN CONSTRUCTION, INC.
21 Davis Ave.
Poughkeepsie, NY 12603
(914) 473-0064
Ronald Stein - President
Services:
Provides radon detection and mitigation services.
Qualifications:
EPA RMP.
Area Served:
NY.
EPA RMP 5:
CC.

ROSS RADON TESTING
4200 Chimney Lake Dr.
Roswell, GA 30075
(404) 993-8270
Robert Ross
Area Served:
AL, FL, GA, NC, SC, TN.
EPA RMP 5:
CC.

ROSS SYSTEMS, INC.
955 Connecticut Ave.
Bridgeport, CT 06607
(203) 366-2210
Nancy Adams - Branch Manager
Area Served:
CT, NY.

ROSS SYSTEMS, INC.
21 Main St. Box 670
Blairstown, NJ 07825
(201) 362-5571
Richard Ross - President
Services:
Provides full service radon detection services, distributes charcoal canisters and alpha track detectors, provides radon mitigation consultation services using computer technology, radon mitigation proposals at no charge, offers seminars and workshops.
Qualifications:
AARST, EPA RMP,APCA, NJ DEP.
Area Served:
Nationwide.
EPA RMP 5:
AT, CC, GW(P), CW(P).

ROY F. WESTON, INC.
Building 4-2 1 Weston Way
West Chester, PA 19380
(215) 692-3030
Jean-Claude F. Dehmel
Area Served:
Nationwide.
EPA RMP 5:
GR(P), GW(P).

RPI RADON MEASUREMENT SERVICES
P.O. Box 357
Hershey, PA 17033
(717) 533-8100
William P. Kirk - Manager
Thomas E. Woolley - Radon Consultant
Services:
Provides radon detection services for homes, schools, and industry, distributes charcoal canisters for detection firms or home owner testing.
Qualifications:
EPA RMP, PA certification, AARST.
Area Served:
Nationwide.
EPA RMP 5:
AT, CC(P), CR(P), GR(P), GW(P), CW(P).

RPI RADON MEASUREMENT SERVICES, MARYLAND
10133 Bacon Dr.
Beltsville, MD 20705
(301) 937-4072
Kelly Helweick
Area Served:
MD.

RTK ENGINEERING
2939 Newton St.
Denver, CO 80211
(303) 477-0044
Robert T. Knickerbocker - Engineer
Services:
Provides radon detection services using charcoal canister or alpha track methods for the general public and real estate industry. Non-commercial laboratory for research purposes or state agencies' assistance only.
Qualifications:
Registered professional engineer.
Area Served:
CO, MT, NM, NY, WY.
EPA RMP 5:
AT, CC.

RTS SYSTEMS
P.O. Box 3070
Gaithersburg, MD 20878
(301) 258-7890
Paul R. Ramsey - President
Jay Friedman - Director
Services:
Provides radon detection services to residential and commercial properties using E-Perm detection devices, operates a primary laboratory.
Qualifications:
PA DER training courses.
Area Served:
DC, MD, VA.
EPA RMP 5:
EP(P).

RUECK ASSOCIATES
103 Sage Rd.
Silver Lake, KS 66539-9461
(913) 582-4050
Jon M. Rueck - Chairman
Services:
Provides consultation on radon issues, distributes At-Ease radon monitors, charcoal canisters, and E-Perms.
Qualifications:
Diplomate AAEE, registered professional engineer, degrees in nuclear engineering, mechanical engineering.
Area Served:
Nationwide.
EPA RMP 5:
CC.

RYAN NUCLEAR LABORATORIES
P.O. Box 88
West Jefferson, OH 43162
(614) 848-4414
Wayne Ballantyne
Mary Brenneman
Area Served:
Nationwide.
EPA RMP 5:
CC(P).

S

S & D CUSTOM ESTATES
97 Cuba Ave.
Staten Island, NY 10306
(718) 987-2354, (718) 987-8378
Steven Nunziato
Services:
Non-commercial laboratory for research purposes or state agencies' assistance only.
Area Served:
NJ, NY.
EPA RMP 5:
CC.

S & W ENVIRONMENTAL TESTING
Route 1, Box 185-D
Telephone, TX 75488

(214) 664-3105, (214) 980-4121
Don S. Silagy
Area Served:
TX.

SACCHETTI, R. MICHAEL
103 Intervale St.
Quincy, MA 02169
(617) 479-2111
Mike Sacchetti - President
Services:
Provides radon detection and mitigation services.
Qualifications:
AARST.
Area Served:
MA.

SAF-AIR
P.O. Box 1217
Orange, CT 06477
(203) 799-9609
Thomas F. Meehan - President
Services:
Provides radon detection and mitigation services for residential and commercial properties, seminars and speaking engagements.
Qualifications:
CT mitigation list.
Area Served:
CT.
EPA RMP 5:
CC.

SAFE AIR
P.O. Box 120064
Nashville, TN 37212
(615) 865-5325
Carolyn Tompkins
Area Served:
TN.
EPA RMP 5:
CC.

SAFE AIR COMPANY
9 Eastman Dr.
Derry, NH 03038
(603) 437-1425
Duane L. Howes
Area Served:
NH.

SAFE AND SOUND HOME INSPECTION CONTRACTORS
P.O. Box 195
Cedar Knolls, NJ 07927
(201) 515-0013
Steven W. Freedman
Area Served:
NJ.
EPA RMP 5:
CC.

SAFE SHELTER ENVIRONMENTAL
527 West Lancaster Ave.
Frazer, PA 19355
(215) 640-4839
Rick Haag - President
Steve Cummings - General Manager
Chris Haag - Manager
Joe Batten
Services:
Provides radon detection and mitigation services for real estate transactions and for the general public.
Qualifications:
PA DER course, APCA, EPA RMP.
Area Served:
DE, NJ, PA.
EPA RMP 5:
CW(P).

SAFE-AIRE, INC.
162 East Chestnut St. P.O. Box 160
Canton, IL 61520
(309) 647-0419, (800) 331-2943
Bill Wherley - Sales Manager
Chet Phillips - Products Manager
Harry Ellison - Marketing Director
Don Stees - Vice President
Services:
Provides patent-pending Safe-Aire radon mitigation system to dealers, provides radon detection and mitigation services, distributes extensive line of radon detection and mitigation products, dealership opportunities available.
Area Served:
Nationwide.
EPA RMP 5:
AT, CC, CR(P), EP.

See Ad Page 161

SAFEHOME
P.O. Box 221
Rexford, NY 12148
(518) 399-0135
David Mazzarella
Area Served:
NY.

SAGE RESOURCES,/INDOOR AIR QUALITY DIVISION
P.O. Box 480559
Los Angeles, CA 90048
(213) 825-1934, (213) 651-0155
Donald Carlisle
Area Served:
CA.

SAM STEWART TESTING
P.O. Box 7304
Fairfax Station, VA 22039
(703) 250-9203
Sam Stewart
Area Served:
Nationwide.
EPA RMP 5:
CC.

SANI-DUCT SERVICE OF THE INLAND NORTHWEST
16015 East Sprague Ave.
Spokane, WA 99037
(509) 924-8472
Frank Wagner
Area Served:
ID, MT, NV, OR, UT, WA, WY.
EPA RMP 5:
CC.

SCHWABE, WALTER A.
Rd. 3 Box 533
Branchville, NJ 07826
(201) 948-7500
Walter A. Schwabe
Area Served:
NJ.

SCIENCE AND TECHNOLOGY, INC. (SCAT)
79 Pembroke Crescent
Fredericton, NB E3B 2V1
(506) 457-0232
Stanton T. Friedman - President
Services:
Institute has long history of radon research projects under contract with Canadian government.
Area Served:
NB.

SCIENCE MANAGEMENT SERVICES, LTD.
25 Eshelman Rd.
Lancaster, PA 17601
(717) 392-1425
D.D. Browning - President
D.D. Browning, Jr. - Testing Specialist
L.W. Eckert - Testing Specialist
Services:
Provides radon detection services for residential and commercial markets, offers consultation on mitigation services.
Qualifications:
EPA RMP.
Area Served:
PA.
EPA RMP 5:
AT, CC, CR(P), GR(P), GW(P).

SCIENTIFIC ANALYSIS, INC.
6012 East Shirley Ln.

Montgomery, AL 36117
(205) 271-0643, (800) 345-2575
John Allen Gunn - Laboratory Manager
Rhonda L. Russell - Sales
Services:
Primary radon testing laboratory provides charcoal canister mail order testing services to individuals, radon detecton firms, and commercial groups.
Qualifications:
EPA RMP.
Area Served:
Nationwide.
EPA RMP 5:
CC(P).

SCIENTIFIC TESTING ASSOCIATES, LTD.
Star Route Box 124
Topping, VA 23169
(804) 758-5728
John Reynolds - President
Services:
Manufactures charcoal canister holder for ground radon emission testing, provides radon detection services.
Qualifications:
Doctorate degree in biology.
Area Served:
Nationwide.
EPA RMP 5:
CC.

SCOTT HOME INSPECTION, INC.
87 Beach Ave.
Milford, CT 06460
(203) 874-2414
Walter Scott
Area Served:
CT.
EPA RMP 5:
CC.

SCRUB RIDGE RAD TESTING AND REMEDIATION
RDI Box 469L
McConnellsburg, PA 17233
(202) 332-1798, (717) 485-5456
Ray Messick
Area Served:
PA.

SCS ENGINEERS, INC.
11260 Roger Bacon Dr. Suite 300
Reston, VA 22090
(703) 471-6150
Robert Luce - Scientist
Area Served:
VA.

SEACOAST LABORATORIES
55 Village Ln.
South Hamilton, MA 01982
(508) 468-3136
Stephen Knowlton
Blair Knowlton
Services:
Provides radon detection services using activated charcoal adsorption method.
Qualifications:
AARST.
Area Served:
MA.

SEBRING HOME INSPECTION COMPANY
73 Mulwal Dr.
Painesville, OH 44077
(216) 354-8768
J. Sebring
Area Served:
OH.

SECONDO, THOMAS A.
50 Spruce St.
Hornell, NY 14843
(607) 324-1986
Thomas A. Secondo
Services:
Provides radon detection services.
Area Served:
NY.

SENES CONSULTANTS
499 McNicholl Ave.
Willowdale, ON M2H 2C9
(416) 499-5030
Leo Lowe - Director
Douglas Chambers - Contact
Services:
Environmental and occupational consulting firm analyzes environmental radioactivity, contracts with federal and provincial governments and with industry, analyzes results and sets remedial cleanup levels.
Area Served:
ON.

SEPTECH
7625 North Howard
Spokane, WA 99208
(509) 467-6274
John L. Cox
Area Served:
Nationwide.
EPA RMP 5:
GR(P).

SHANNON HEATING, COOLING AND ELECTRIC
1701 North 74th St.
Omaha, NE 68114
(513) 831-0743
Philip L. Babcock
Area Served:
NE.

SHELBURNE LABORATORIES, INC.
P.O. Box 458 Falls Rd.
Shelburne, VT 05482
(802) 985-3379
Robert J. Emerson - President
Lucie Jean - Laboratory Manager
Services:
Operates an industrial hygiene laboratory providing radon detection services in commercial and residential buildings.
Area Served:
CT, MA, ME, NH, NY, PA, RI, VT.

SHELTERWORKS, INC.
53 New St.
Dover, NJ 07801
(201) 328-2912
Joseph F. Corsetto
Area Served:
NJ.
EPA RMP 5:
CC.

SHERLOCK HOME INSPECTORS
6-01 Fair Lawn Ave.
Fair Lawn, NJ 07410
(201) 796-4942, (800) 458-2202
Raymond LoVecchio
Area Served:
NJ.

SHEVENELL, GALLEN, AND ASSOCIATES
P. O. Box 433 195 Commerce Way
Portsmouth, NH 03801
(603) 436-1490
Tom Shevenell - Principal
Services:
Provides environmental site assessments for real estate transactions, provides analysis of water wells for radon and makes recommendations for mitigation.
Area Served:
NH.

SHIELD HOME INSPECTION SERVICE
R.D. 1, Box 274 Summit Dr.
Branchville, NJ 07826
(201) 875-5922
Leonard A. Philip
Area Served:
NJ.
EPA RMP 5:
CC.

SHOTWELL ASSOCIATES
Box 83
Budd Lake, NJ 07828
(201) 691-9037
Henry P. Shotwell
Area Served:
CT, DE, MD, NJ, NY, PA.
EPA RMP 5:
CC, CR(P), CW(P).

SILVER SPRINGS HOME INSPECTION
P.O. Box 182
Landing, NJ 07850
(201) 770-0026
Glenn A. Bredder
Area Served:
NJ, NY, PA.
EPA RMP 5:
CC.

SIMPSON WEATHER ASSOCIATES
809 East Jefferson St.
Charlottesville, VA 22902
(804) 979-3571
M. Garstang - President
G.D. Emmitt - Vice President
H.G. Goodell
Services:
Provides radon mitigation and environmental consultation services.
Qualifications:
Staff includes persons with doctoral degrees in meteorology, geology, geochemistry, hydrology.
Area Served:
Nationwide.
EPA RMP 5:
CC.

SMALL UNITED LABORATORIES
7606 Patapsco Dr.
Sykesville, MD 21784
(301) 781-6636, (301) 876-7636
Philip Small
Area Served:
MD.
EPA RMP 5:
CC.
SMITH, STANLEY
2901 26th St., W. Suite 702
Bradenton, FL 34205
(813) 756-3635
Stanley Smith
Qualifications:
AARST.
Area Served:
FL.

SOL-AIR, INC.
P.O. Box 458
Plantsville, CT 06479
(203) 621-6211
John Opalenik, Jr.
Area Served:
CT, NY.
EPA RMP 5:
AT, CC.

SOLAR RESOURCE TECHNOLOGY
128 Market St.
South Williamsport, PA 17701
(717) 323-8919
Ernest Kittka - Owner
Area Served:
PA.

SOLAR TESTING LABORATORIES, INC.
9540 Midwest Ave.
Garfield Heights, OH 44125
(216) 663-1800
Carl J. Kranick
Area Served:
Nationwide.
EPA RMP 5:
CC(P).

SOLID STATE RADON DETECTION AND ELIMINATION SERVICES
P.O. Box 107
Glenwood, NJ 07418
(201) 764-8555
Richard A. Davi
Gina L. Davi
Area Served:
Nationwide.
EPA RMP 5:
CC.

SOLID WASTE AND TOXIC TECHNOLOGY
23891 Via Fabricante Suite 616
Mission Viejo, CA 92691
(714) 855-7231
Roland Holmes
Area Served:
CA.
EPA RMP 5:
AT.

SORENSEN ENTERPRISES, INC.
886 West Hunt Rd.
Alcoa, TN 37701
(615) 984-1376
Peter F. Sorensen
Area Served:
Nationwide.
EPA RMP 5:
CC.

SOUTH CAROLINA DEPARTMENT OF HEALTH AND ENVIRONMENTAL CONTROL/BUREAU OF RADIOLOGICAL HEALTH
2600 Bull St.
Columbia, SC 29201
(803) 734-4700
Heyward G. Shealy - Chief
Noland Bivens - Director
Lynn B. Dobbs - Health Physicist
Services:
Non-commercial laboratory for research purposes or state agencies' assistance only.
Area Served:
SC.
EPA RMP 5:
CC(P).

SOUTH CAROLINA ELECTRIC AND GAS
P.O. Box 88
Jenkinsville, SC 29065
(803) 345-1915
Timothy D. Riley - Public Information
Services:
Provides information to state residents on radon-housing related issues.
Area Served:
SC.

SOUTH CENTRAL ENVIRONMENTAL SERVICES
RD 1, Box 265
Pine City, NY 14871
(607) 733-1661
Lawrence P. McLane - Contractor
Robert Lovejoy - Vice President
Area Served:
NY, PA.

SOUTH PARK URBAN RENEWAL CORPORATION
1938 South Park Ave.
Buffalo, NY 14220
(716) 873-4763
Bob Colasanti
Area Served:
NY.

SOUTHEASTERN RADON CONSULTANTS, INC.
161 York Rd.
Nolensville, TN 37135
(615) 776-5569
Otis E. Guinn - President
Services:
Provides radon detection services using Honeywell Professional Radon Monitors for real estate market, provides radon mitigation services.

Area Served:
AL, GA, KY, TN.

SOUTHERN MECHANICAL
1721 Lee St.
Decatur, GA 30035
(404) 289-9637
Myron R. Edelkind
Area Served:
Nationwide.
EPA RMP 5:
CC.

SOUTHERN NEW ENGLAND RADON TESTING AND MITIGATION
P.O. Box 1141
Westbrook, CT 06498
(203) 399-8003
Donald C. Kowalski
George Gavigan
Area Served:
CT.

SOUTHERN OHIO RADON PROTECTION COMPANY
4940 Nutmeg Knoll
Cincinnati, OH 45244
(513) 831-0743
Donna Smith
Bob Smith
Area Served:
ID, KY, OH.
EPA RMP 5:
CC.

SOUTHERN RADON SERVICES, INC.
1000 Johnson Ferry Rd. Suite B-145
Marietta, GA 30068
(404) 565-3886, (800) 537-2366
C. Fred Stafford - President
Services:
Provides radon detection services, primary laboratory for charcoal canister analysis for home owners and radon specialists, provides radon risk assessments for commerical, institutional, and development properties.
Qualifications:
EPA RMP, NJ registration.
Area Served:
Nationwide.
EPA RMP 5:
CC(P), CW(P).

SOUTHERN RESEARCH INSTITUTE
2000 Ninth Ave., S. P.O. Box 55305
Birmingham, AL 35255-5305
(205) 581-2000
B.E. Pyle - Research Physicist
A.D. Williamson - Head, Aerosol Science
C.S. Fowler - Research Physicist
Services:
Fundamental radon mitigation studies and aerosol research. Non-commercial laboratory for research purposes or state agencies assistance only.
Qualifications:
EPA RMP.
Area Served:
Nationwide.
EPA RMP 5:
AT, CC, CR(P), GR(P).

SOUTHWEST ABATEMENT, INC.
4200 Broadway, SE
Albuquerque, NM 87102
(505) 873-2957
Tad West
Area Served:
Nationwide.
EPA RMP 5:
CC.

SPADY, GARY D.
1305 Court St. Box 4 RR #2
Imperial, NE 69033
(308) 882-5198
Gary D. Spady
Area Served:
NE.

SPARTAN HOME INSPECTION
RFD 3, Box 533
Branchville, NJ 07826
(201) 948-7500
Walter A. Schwabe
Area Served:
NJ.

SPARTAN HOME INSPECTION, INC.
54 Woodport Rd.
Sparta, NJ 07871
(201) 729-7972
Robert Haarde - Owner
Area Served:
NJ, PA.

SPECIAL ENVIRONMENTAL SERVICES, INC.
P.O. Box 20058
Atlanta, GA 30325
(404) 355-5600
Kathi Davis
Area Served:
Nationwide.
EPA RMP 5:
CC.

SPECIALIZED SERVICES, INC.
R.R. 1 7 Wall St.
Vernon, NJ 07462
(201) 827-9055
Pete Priestner - President
Services:
Provides radon detection services for homes and larger buildings for owner's personal protection and real estate certifications.
Qualifications:
Certified radon specialist, extensive experience in the building and home inspection industries.
Area Served:
NJ, NY, PA.

SPECTRUM INDUSTRIES
1128 Forrest St.
Dover, DE 19901
(302) 736-0792
Phillip Stubbolo
Area Served:
DE.

SPENCE-GEIGER ASSOCIATES, INC.
1658 Cole Blvd. #260
Golden, CO 80401
(303) 233-8577
Jeffrey R. Geiger - Industrial Hygienist
Services:
Provides radon detection services for large-scale projects.
Qualifications:
Certified industrial hygienist.
Area Served:
AZ, CA, CO, NM, NV, OR, UT, WA.

SPRINGER COMPANY
629 Lancaster Ave.
Malvern, PA 19355
(215) 344-9800
David Springer
Area Served:
PA.

STAFFORD, C. FRED
1000 Johnson Ferry Rd. Suite B145
Marietta, GA 30068
(404) 565-3886
C. Fred Stafford
Area Served:
GA.

STAN A. HUBER CONSULTANTS, INC.
200 North Cedar Rd.
New Lenox, IL 60451
(815) 485-4433
Stan A. Huber - President
Linda Fink - Radon Coordinator
Services:
Provides radon detection services using charcoal canister method.
Qualifications:
EPA RMP, State of IL license for nuclear safety to test for radon.

Area Served:
Nationwide.
EPA RMP 5:
CC(P).

STANDOR RADON DETECTION
598 South Edgewood Ave.
Jacksonville, FL 32205
(904) 388-0284
Stanley J. Szczepanski - President
Services:
Provides radon detection and mitigation services for residential and commercial properties.
Area Served:
Nationwide.
EPA RMP 5:
AT, CC.

STAR LABORATORIES
P.O. Box 1666
Plymouth, MA 02360
(617) 747-5227, (617) 947-4940
Martin Starr
Area Served:
CT, MA, ME, NH, NJ, NY, OH, PA, RI, VT.
EPA RMP 5:
CC.

STERBAL, SYLVESTER
16 Rose Ct.
Staten Island, NY 10301
(718) 273-2422
Sylvester Sterbel
Qualifications:
AARST.
Area Served:
NY.

STONE RIDGE RADON MITIGATORS
424 Mine Rd.
Asbury, NJ 08802
(201) 735-6080, (800) 447-2366
Joanne Hlavinka
Services:
Provides radon mitigation services for residential and commercial buildings.
Qualifications:
NJ DEP remediation program.
Area Served:
NJ, PA.

STRATA ENVIRONMENTAL SERVICE, INC.
401 East Douglas Suite 515
Wichita, KS 67202
(316) 262-0002
Dennis Jones - President
Bruce A. Reed - Vice President
Randall Fornshell - Laboratory Director
Denise Peer - Laboratory Technician
Services:
Provides radon detection services, analytical and geohydrology consultation.
Qualifications:
EPA RMP, certified by KS for drinking water analysis.
Area Served:
CO, IA, KS, MO, NE, OK, TX.
EPA RMP 5:
CC(P).

STUART HUNT AND ASSOCIATES/OCCUPATIONAL AND RADIATION SAFETY SERVICES, LTD.
26 Rayborn Crescent
St. Albert, AB T8N 5D8
(403) 458-0291
Area Served:
AB.

SULLIVAN CONSULTANTS, INC.
302 Forest Dr.
Union, NJ 07083
(201) 964-5316
Eugene Sullivan - President
Services:
Provides radon detection services using charcoal canister method.
Qualifications:
EPA RMP.
Area Served:
CA, FL, NE, NJ, NY, PA.
EPA RMP 5:
CC.

SUN DISTRIBUTING GROUP
813 East Bloomingdale Ave. Suite 313
Brandon, FL 33511
(813) 681-0350
Marc T. Tolley
Services:
Provides radon detection services using charcoal canister method.
Area Served:
FL.

SUN NUCLEAR CORPORATION
415-C Pineda Ct.
Melbourne, FL 32940
(407) 259-6862
Thomas L. Powers - President
William E. Simon - Vice President
Services:
Distributes full line of radon detection instruments, provides radon detection services.
Area Served:
Nationwide.
EPA RMP 5:
CR(P).

SUNDANCE ENTERPRISES
P.O. Box 3607
Dillon, CO 80435
(303) 468-3911, (303) 468-8369
Dan Whalen
Area Served:
CO.
EPA RMP 5:
AT, CC.

SUNSTREAM DEVELOPMENT CORPORATION
300 North St.
Endicott, NY 13760
(607) 754-6200
Thomas Meade, Jr. - President
Peter R. Gotthardt - Production Manager
Ron Andrascik - Technician
Services:
Provides radon detection and mitigation services to home owners, realtors, and commercial properties, distributes VANEE air-to-air heat exchangers.
Qualifications:
NRA, key personnel trained by NY State Energy Office.
Area Served:
NY, PA.

SWALE, INC.
57 Indian Hill Rd.
Collinsville, CT 06022
(203) 693-2146
William Swale
Area Served:
CT.
EPA RMP 5:
CC.

SWAT TERMITE AND PEST CONTROL
1797 Roswell Rd.
Marietta, GA 30062
(404) 565-7928
E. Frank Callawy
Area Served:
Nationwide.
EPA RMP 5:
CC.

SYNERTECH SYSTEMS CORPORATION
472 South Salina St. Suite 800
Syracuse, NY 13202
(315) 422-3828
Laurence Kinney - Energy Research
Gary Klein - Design Engineer
Services:
Energy conservation and research company provides radon detection services.
Area Served:
NY.

SYRACUSE ENERGY SERVICE
100 New Street
Syracuse, NY 13202
(315) 424-0679
Greg Thomas - Energy Consultant
Area Served:
NY.

T

T.R. EDGERTON, INC.
102F Woodwinds Industrial Ct.
Cary, NC 27511
(919) 469-9795
Pete Monica
Area Served:
Nationwide.
EPA RMP 5:
CC.

TARGET RADON SERVICES, INC.
Oakland Ctr. 8940A Rte. 108
Columbia, MD 21045
(301) 992-6622
J. Edward Tillman - President
Jeffrey T. Sims - Vice President
Services:
Provides preconstruction soil gas surveys to building and development industry, radon detection services for commercial, institutional and residential markets.
Qualifications:
AARST, EPA RMP, radon specialist training course.
Area Served:
Nationwide.
EPA RMP 5:
AT(P).

TCS INDUSTRIES, INC.
3805 Paxton St.
Harrisburg, PA 17111
(717) 558-8206
Carl H. Distenfeld - President
Jacquelyn Distenfeld - General Manager
Eric B. Distenfeld - Sales Manager
Services:
Manufactures and distributes charcoal canisters, operates analysis laboratory.
Qualifications:
EPA RMP, certified health physicist, HPS, ANS.
Area Served:
Nationwide.
EPA RMP 5:
CC(P).

TECH OPSOL, INC.
2151 East Dublin-Granville Rd. Suite G522
Columbus, OH 43229
(614) 882-3527
Stacey Hall
Area Served:
OH.

TECHNORAD
P.O. Box 2144
Rockville, MD 20852
(301) 881-8293
C. Tousimis - President
J. Tousimis
C. Mate
J. Mate
Services:
Provides radon detection services for residential and commercial buildings, sells at-home test kits.
Area Served:
Nationwide.
EPA RMP 5:
CC(P).

TECHNORTH
109 Waterloo Ln.
Fairbanks, AK 99712
(907) 456-8378
Philip Loudon
Area Served:
AK.

TECHTRAC
P.O. Box 5050 Suite 303
Lewisville, TX 75067
(214) 221-1047
Michael L. Buvinghausen
Area Served:
Nationwide.
EPA RMP 5:
CC(P).

TECON
2220 Plainfield Pike
Cranston, RI 02920
(401) 946-3720
Phillip B. Renzi
Area Served:
RI.

TECTONIC INSPECTION COMPANY
P.O. Box 278
Denville, NJ 07834
(201) 625-2233
Robert Wisniewski - President
Services:
Provides radon detection services for home owners using charcoal canister method.
Qualifications:
EPA RMP.
Area Served:
NJ.

TEE ENGINEERING COMPANY
320 Cutters Hill Ct.
Lexington, KY 40509
(606) 263-5350
Don F. Blackburn - Professional Engineer
David McFarland - Geologist
Bill Yarnell - Geologist
Services:
Provides radon detection and mitigation services for residential, industrial and commercial properties.
Area Served:
IL, IN, KY, MO, OH, TN, VA, WV.
EPA RMP 5:
CC.

TEK ENGINEERING
7150 East Hampden Ave. #101
Denver, CO 80224
(303) 758-3173
John B. Canady
Area Served:
CO.

TELEDYNE ISOTOPES
50 Van Buren Ave.
Westwood, NJ 07675
(201) 664-7070
D.F. Schutz - President
J.D. Martin - Vice President
D. McKillop - Radon Services Coordinator
Craig Osterhaus - Radon Division
Services:
Manufactures and distributes the Home-Safe Radon Test Kit, performs analysis of results within 24 hours, provides radon detection services.
Qualifications:
NRC licenses, technical and scientific personnel, state licenses.
Area Served:
Nationwide.
EPA RMP 5:
CC(P), EP(P).

TELEDYNE ISOTOPES MIDWEST LABORATORY
700 Landwehr Rd.
Northbrook, IL 60062
(312) 564-0700
L.G. Huebner - General Manager
B. Grob - Laboratory Supervisor
Services:
Operates radiological laboratory used by private citizens, government agencies and the utilities industry.
Qualifications:
EPA RMP, IDNS license, IL drinking water license.
Area Served:
Nationwide.

EPA RMP 5:
CC(P).

TEMPLE UNIVERSITY RADON PROGRAM
Temple University CECSA 084-53
Philadelphia, PA 19122
(215) 787-1665
Robert M. Patterson - Associate Professor
Services:
Provides training in radon detection and mitigation, radon detection services, charcoal canister laboratory analysis, academic courses in radiological health.
Area Served:
Nationwide.
EPA RMP 5:
CC(P).

TEQ CORPORATION
P.O. Box 570
Wilkes-Barre, PA 18703-0570
(717) 825-3939, (800) 877-2366
Russell Hendershot - President
Ed Dructor - Vice President
Brian Redmond - Research Scientist
Services:
Provides radon detection and mitigation services for commercial and residential markets, manufactures and distributes charcoal canisters, maintains calibrated radon chamber, offers radon detection and mitigation training courses.
Qualifications:
PA DER certification.
Area Served:
Nationwide.
See Ad Page 100
EPA RMP 5:
AT, CC(P), CR(P), CW(P).

TERMINIX INTERNATIONAL
855 Ridgelake Blvd.
Memphis, TN 38119
(800) 937-3783
See Ad Page 101
Paul Bert - Vice President
E.D. Gauthreaux - Vice President
Services:
Provides radon detection service for home owners, real estate companies, etc.
Qualifications:
Test results analyzed by licensed and accredited testing laboratories.
Area Served:
Nationwide.

TERRADEX CORPORATION/SUBSIDARY OF TECH/OPS LANDAUER, INC.
2 Science Rd.
Glenwood, IL 60425
(312) 755-7911, (800) 528-8327
Sue Morrow - Customer Service
R. Craig Yoder
Services:
Manufactures and distributes line of radon monitoring products to industry, academic institutions, government agencies, consumers, etc.
Qualifications:
Over 35 years of experience in radiation monitoring services from both occupational and environmental sources, works with ALA, EPA, USAF, universities, national labs, etc.
Area Served:
Nationwide.
See Ad Back Cover
EPA RMP 5:
AT(P).

TERRADON EXPLORATION TECHNOLOGIES, INC.
3931 White Oak Tr. P.O. Box 22288
Cleveland, OH 44122
(216) 464-2002
Darioush T. Ghahremani - President
Services:
Provides radon detection and mitigation services, research and development in radon detection, measurement and abatement technology, development of new alpha track and charcoal canisters for industrial and private use, soil gas radon detection and mapping.
Qualifications:
AARST, NRA, AAPG, OAS, ACS, OG&G, AAGE, other scientific associations.
Area Served:
Nationwide.
EPA RMP 5:
AT, CC.

TERRADYNAMICS CORPORATION
Rte. 1
Aldie, VA 22001
(703) 327-6500, (703) 435-3033
Ned Mamula
Y. Jean Choi
Area Served:
VA.
EPA RMP 5:
GR(P).

TESTING 1.2.3.
50 Freeway Dr.
Cranston, RI 02920
(401) 461-0530
Paul Marangos
Nicholas Kondon
Area Served:
RI.

TESTWELL CRAIG LABORATORIES
11553 U.S. Hwy. 41 South

Gibsonton, FL 33534
(813) 677-0242
Patricia Craig - President
Services:
Provides radon detection services using charcoal canister method.
Qualifications:
NRA.
Area Served:
Nationwide.

TEXAS A & M UNIVERSITY/DEPARTMENT OF NUCLEAR ENGINEERING/RADON ASSAY LABORATORY
129 Zachry Engineering Center
College Station, TX 77843
(409) 845-4107
Milton E. McLain - Professor
Services:
Research centers on measurement techniques. Studying effects of air temperature and humidity on effectiveness of charcoal canisters. Attempting to improve consistency of alpha track detectors. Utilizes scintillation flask technique as way of calibrating measurement devices.
Area Served:
TX.

TFC NUCLEAR ASSOCIATES, INC.
425 Bridgeboro Rd.
Moorestown, NJ 08057
(609) 778-4529
See Ad Page 166
John Chando - President
Eric Trenga - Vice President
Harvey Auger - Radon Testing Division
Edwin Voll - Design Engineer
Services:
Provides comprehensive radon detection and mitigation services for government, schools, commercial and residential structures.
Qualifications:
Recently completed contract with NJ DEP for radioactive waste cleanup of radon contaminated soil, contracts with more than 18 nuclear power plants nationwide, American Society of Mechanical Engineers, AARST, ANS, ARA.
Area Served:
Nationwide.

THERMO ENVIRONMENTAL CORPORATION
101 1st Ave. P.O. Box 9046
Waltham, MA 02254
(617) 622-1103, (505) 345-9931
Jeff Brown - Health Physics
Services:
Manufactures and distributes RAD-SCAN radon screening kits, provides radon detection and mitigation services to homeowners, developers, real estate industry, etc., parent company of TMA Eberline.
Area Served:
Nationwide.

THERMO-SCAN, INC.
1137 3rd Ave., SW P.O. Box 705
Carmel, IN 46032
(317) 846-4655
Christopher L. Maher
Area Served:
IN.

THEROUX ENGINEERING
P.O. Box 4096
Hamden, CT 06514
(203) 932-1313
Dennis R. Theroux - President
Services:
Provides radon detection and mitigation services for schools, industrial and

commercial structures.
Area Served:
CT.

THIRD PARTY INSPECTIONS
P.O. Box 3161
Winchester, VA 22601
(703) 869-1096
Don MacDaniel - President
Services:
Provides radon detection services.
Area Served:
VA.

THOMSON AND NIELSEN ELECTRONICS, LTD.
4019 Carling Ave. Suite 303
Kanata, ON K2K 2A3
(613) 592-3019
Ian Thomson - President
Andrew Singmin - Marketing Director
Line Brabant - Inside Sales
Services:
Manufactures and distributes radon detection instruments, provides radon detection services, conducts research in radiation detection.
See Ad Page 187
Qualifications:
AARST, EPA RMP, HPS.
Area Served:
Nationwide.

THOROUGH CHECK, INC.
2606 Rte. 37 East
Toms River, NJ 08753
(201) 506-0200, (800) 852-0116
William A. DeMarsico
Ronald Ricci
Area Served:
NJ.
EPA RMP 5:
CC, CW(P).

THREE W RADON SERVICES, INC.
1619 Springhill Dr.
Hummelstown, PA 17036
(717) 566-0499
William Weaver
Area Served:
PA.
EPA RMP 5:
CR(P).

TIMBERLINE RADON SERVICE, INC.
P.O. Box 3815
Evergreen, CO 80439
(303) 674-2828
David M. Snyder
Area Served:
CO, ID, MT, UT, WY.

EPA RMP 5:
CC.

TISCO
2615 Skyline Dr.
College Park, GA 30337
(404) 768-2517
K.C. Timmons
Area Served:
GA.

TMA/EBERLINE, HEALTH PHYSICS
5635 Jefferson Blvd., NE
Albuquerque, NM 87109
(505) 345-9931
Stanley Waligora - Sales Manager
Ernie Sanchez
Jeff Brown
Services:
Provides radon detection and mitigation services to commercial and government clients, distributes charcoal canisters, provides consultation services.
Area Served:
Nationwide.
EPA RMP 5:
CC(P).

TMA/EBERLINE, LABORATORY
7021 Pan American Hwy., NE
Albuquerque, NM 87109
(505) 345-3461
Rod Melgard
Area Served:
Nationwide.
EPA RMP 5:
CC(P).

TMB COMPANY
3900 Parkside Dr.
Baltimore, MD 21206
(301) 485-8734
Thomas M. Bills - President
Services:
Provides radon detection services for residential market.
Qualifications:
EPA RMP.
Area Served:
Nationwide.
EPA RMP 5:
CC.

TMW MARKETING
1660 North LaSalle Suite 3301
Chicago, IL 60614
(312) 951-7213
Thomas Ward
Area Served:
IL.

TODD GIDDINGS AND ASSOCIATES
3049 Enterprise Dr.
State College, PA 16801
(814) 238-5927
Todd Giddings
Area Served:
Nationwide.
EPA RMP 5:
EP(P).

TOTAL HOME TECH
808 Sunnyside Ave.
Audubon, PA 19403
(215) 631-9313
Mathew Aresery
Area Served:
NJ, PA.

TRACOR JITCO RADON MEASUREMENT SERVICES
1601 Research Blvd.
Rockville, MD 20850
(301) 984-2800
Ted Junghans - Associate Scientist
Steve Olin - Technical Director
Services:
Distributes charcoal canisters and alpha track detectors to homeowners and provides radon detection services for the real estate industry.
Qualifications:
EPA RMP.
Area Served:
DC, MD, VA.
EPA RMP 5:
CC.

TRADON ENVIRONMENTAL SERVICES
41 Old Pascack Rd.
Woodcliff Lake, NJ 07675
(201) 307-1063, (201) 307-1728
Tracey Schuck
Robert N. Schuck, Jr.
Qualifications:
AARST.
Area Served:
NJ.

TRC-ECI
800 Connecticut Blvd.
East Hartford, CT 06108
(203) 289-8631
Phil J. Hunt
Area Served:
Nationwide.
EPA RMP 5:
CC(P).

TRI-CITIES RADON
30 Clarke St.
Binghamton, NY 13905

(607) 724-4088
Robert G. Smith
Area Served:
NY, PA.
EPA RMP 5:
CC.

TRI-COUNTY RADON, INC.
2796 North Constitution Rd.
Camden, NJ 08104
(609) 963-1533
Susan B. Cusato - Chief Inspector
John Thompson - Mitigation
Services:
Provides radon detection, mitigation, and consulting services for residential and commercial properties.
Area Served:
NJ, PA.
EPA RMP 5:
CC.

TRI-COURT SAFETY/SECURITY
P.O. Box 1257
Sunset Beach, CA 90742
(213) 335-3290, (714) 846-4346
M.L. Coffey - Safety Manager
Leroy Toney - Technician
Services:
Provides radon detection services using charcoal canister method, performs comprehensive safety evaluations.
Qualifications:
Certified safety managers.
Area Served:
Nationwide.
EPA RMP 5:
CC.

TRI-STATE RADON AND DEWATERING
P.O. Box 922
Dayton, NJ 08810
(201) 329-8570
Jack Laurie - President
Area Served:
NJ.

TRI-STATE RADON REDUCTION, INC.
816 Ainsworth Dr.
Linden, NJ 07036
(609) 275-0155, (201) 925-8472
Robert Schalhoub - President
James D. Fox - Vice President
Services:
Provides radon detection and mitigation services.
Qualifications:
AARST.
Area Served:
NJ.

TRI-STATE RADON TESTING AND REDUCTION
530 Eben Ave.
Findlay, OH 45840
(419) 422-0657, (419) 422-6601
Melvin Harvey
Area Served:
OH.

TRIAD BASEMENT DE-WATERING SYSTEM
1443 Spring Hill Cr.
Kernersville, NC 27284
(919) 996-7870, (919) 996-1685
David J. Eckrote
Area Served:
NC.
EPA RMP 5:
CC.

TRIPLE SEAL, INC.
2425 Patterson Rd.
Kettering, OH 45420
(513) 258-8255
A. Brown Dillard
Area Served:
OH.
EPA RMP 5:
CC.

TRU-KIL SERVICE, INC.
P.O. Box 87156
College Park, GA 30337
(404) 768-4533
Gerald O. Vaillancourt, Sr.
Area Served:
GA.

TRUE TEST COMPANY, UTAH
P.O. Box 571
Bountiful, UT 84010
(801) 298-2256
Gary E. Hull
Mark R. Pitts
Area Served:
Nationwide.

TRUE TEST COMPANY, WASHINGTON
1436-C Brandt Rd.
Vancouver, WA 98661
(206) 696-0957
Gary E. Hull - President
Area Served:
WA.
EPA RMP 5:
AT, CC.

TRUESDALE-HAWK INSPECTION CONSULTANTS, INC.
P.O. Box 154
Bedminster, NJ 07921
(201) 234-0323
Walter P. Hawk
Area Served:
NJ.

TSI CONSTRUCTION SERVICES
1137 3rd Ave., SW
Carmel, IN 46032
(317) 846-4655
Christopher L. Maher - Vice President
Services:
Provides radon detection services using continuous radon monitoring using E-Perm and charcoal canister methods, provides radon mitigation services.
Qualifications:
ABC, CSI, NAHB, NACE, eight years of experience in indoor air quality and building diagnostic investigation.
Area Served:
CA, CO, DC, IA, IL, IN, KS, KY, MI, MO, MS, NC, NJ, NY, OH, PA, TN, VA, WI, WV.
EPA RMP 5:
AT, CC.

TUCSON HOME INSPECTIONS, INC.
P.O. Box 50845
Tucson, AZ 85703-1845
(602) 881-5511
Wayne Lindquist
Area Served:
Nationwide.
EPA RMP 5:
CC.

TULANE UNIVERSITY MEDICAL CENTER/OCCUPATIONAL HEALTH AND SAFETY SERVICES
1430 Tulane Ave.
New Orleans, LA 70112-2699
(504) 588-5486
Charles Reindl - Radiation Safety
Services:
Provides radon detection service at medical center using charcoal canister method.
Area Served:
LA.

TWIN CITY TESTING
662 Cromwell
St. Paul, MN 55114
(612) 645-3601
Ann Lori - Industrial Hygienist
Rosemary Breiland - Supervisor
Rennie Smith
Services:
Provides radon detection services.
Qualifications:
EPA RMP.
Area Served:

Nationwide.
EPA RMP 5:
AT, CC(P).

U

U.S. ASSETS RECOVERY INVESTIGATIONS, INC.
P.O. Box 6953
Washington, DC 20032
(202) 362-4764
Steven B. Brenner
Area Served:
DC, DE, MD, NJ, NY, PA, VA.
EPA RMP 5:
CC, GR(P).

U.S. DEPARTMENT OF INTERIOR (DOI)/BUREAU OF MINES(BOM)/DENVER RESEARCH CENTER DRC)/RADIATION LABORATORY
P.O. Box 25086 Denver Federal Ctr. Building 20
Denver, CO 80225
(303) 236-0789
R.F. Holub - Director
Services:
Research to support BOM's Project 2000 program to reduce radiation hazards associated with exposure to radon and radon progeny in mines. One of four labs in world currently setting worldwide standards for calibrating radon daughter measuring and detection equipment.
Area Served:
Nationwide.

U.S. DEPARTMENT OF THE ARMY/ARMY ENVIRONMENTAL HYGIENE AGENCY
HSHB-ML-RR
Aberdeen Proving Ground, MD 21010-5422
(301) 671-2619, (301) 671-2637
Ronald J. Swatski
Services:
Non-commercial laboratory for research purposes or state agencies' assistance only.
Area Served:
Nationwide.
EPA RMP 5:
CC(P).

U.S. DEPARTMENT OF THE ARMY/BELVOIR RESEARCH, DEVELOPMENT AND EDUCATION CENTER
STRBE-VR
Fort Belvoir, VA 22060-5606
(703) 664-5133
Janine Guadagno
Services:
Non-commercial laboratory for research purposes or state agencies' assistance only.
Area Served:
Nationwide.
EPA RMP 5:
CC(P).

U.S. ENVIRONMENTAL PROTECTION AGENCY (EPA)/OFFICE OF AIR AND RADIATION(OAR)/OFFICE OF RADIATION PROGRAMS(ORP)/LAS VEGAS FACILITY(LAF)
Evaluation Branch 4220 South Maryland Pkwy. LaPlaza Building C, Room 532
Las Vegas, NV 89119
(702) 798-2447
Richard D. Hopper - Branch Chief
Services:
Radon chamber at facility enables researchers to evaluate new radon instrumentation and write protocols. Laboratory works almost exclusively on radon research projects. Non-commercial laboratory for research purposes or state agencies' assistance only.
Area Served:
Nationwide
EPA RMP 5:
GR(P).

U.S. HOME INSPECTION SERVICES
2603 Berwyn St.
Richmond, VA 23234
(804) 231-6262, (804) 423-0730
Wallace O. Dorsey, Jr.
Area Served:
VA.
EPA RMP 5:
CR(P).

U.S. RADON TESTING SERVICE
820 S. 24th St.
Arlington, VA 22202
(703) 684-4528
Robert T. Grow
Services:
Performs radon testing services using charcoal canister method.
Area Served:
Nationwide.
EPA RMP 5:
CC.

U.S. RADON TESTING/REDUCTION OF MARYLAND
P.O. Box 3204
Gaithersburg, MD 20878
(301) 330-4187
Stephen A. Webster
Area Served:
MD.

U.S. TOXIC SUBSTANCE TESTING BUREAU
1024 Cottman Ave.
Philadelphia, PA 19111
(215) 745-7778, (800) 535-8378
Harold Stesis - President
Warren Schatz - Director of Marketing
Services:
Provides radon detection services for real estate inspections, distributes and analyzes charcoal canisters, and operates a laboratory.
Qualifications:
AARST, EPA RMP, PA DER, NJ DEP.
Area Served:
Nationwide.
EPA RMP 5:
AT, CC, CW(P).

UCI/ENVIROTEST
834 Old Thorn Run Rd.
Coraopolis, PA 15108
(412) 264-5891
Vincent Gargaro - President
Dennis Geisler - Vice President
Services:
Provides radon detection services and consultation on health physics issues.
Qualifications:
EPA RMP, PA DER certification pending, NRC license, OH radioactive registration.
Area Served:
Nationwide.
EPA RMP 5:
CC.

UNION COUNTY BUILDING INSPECTION SERVICE
502 Twin Oaks Rd.
Union, NJ 07083
(201) 687-6110
Michael L. Jupa
Area Served:
NJ.
EPA RMP 5:
CC.

UNITED SERVICES ALLIANCE, INC.
4900 Leesburg Pike Suite 210
Alexandria, VA 22302
(703) 845-1324
William W. Tu - President
Sheng-jia Leu - Vice President
Services:
Provides radon detection services using E-Perm method, provides radon mitigation services.
Qualifications:
EPA RMP.
Area Served:
Nationwide.

EPA RMP 5:
EP(P).

UNIVERSAL HOME INSPECTION, INC.
5 Boar Ct.
Suffern, NY 10901
(914) 354-1119
Norman Becker
Area Served:
NJ, NY.

UNIVERSITY HOME HEALTH SERVICES, INC.
670 Centre St. Suite 9
Jamaica Plain, MA 02130
(617) 524-2149 (617) 522-2793
Vinod K. Goswami
Area Served:
Nationwide.
EPA RMP 5:
CC.

UNIVERSITY OF COLORADO-COLORADO SPRINGS/PHYSICS DEPARTMENT
1420 Austin Bluffs Pkwy.
Colorado Springs, CO 80933-7150
(719) 593-3214
James F. Burkhart - Associate Professor
Tom Huber - Associate Professor
Services:
Two independent research projects. First is the development of a model for predicting potential for radon based on soil type, permeability, and porosity. Second is a project with EERF providing mathematical analysis for possible modifications of EPA canisters.
Area Served:
Nationwide.
EPA RMP 5:
CC(P), GR(P).

UNIVERSITY OF GEORGIA/CENTER FOR APPLIED ISOTOPE STUDY
120 Riverbend Rd.
Athens, GA 30605
(404) 542-1395
Jim D. Spaulding - Research Associate
Gregory Simones - Research Assistant
Services:
• Studies radon in groundwater. Research conducted at 120 wells in GA. • Studies of radon in air using cc method. Functions as primary lab for radon detection companies. Planning research to determine if identifying radon levels might correlate with other geology processes.
Area Served:
AL, FL, GA, SC, TN.
EPA RMP 5:
CC(P).

UNIVERSITY OF LOWELL/RADIOLOGICAL SCIENCES DEPARTMENT
1 University Ave.
Lowell, MA 01854
(617) 453-1045
Kenneth Skrable - Professor
Services:
Operates primary laboratory and conducts research on radon. Non-commmercial laboratory for research purposes or state agencies' assistance only.
Area Served:
Nationwide.
EPA RMP 5:
CC(P).

UNIVERSITY OF MARYLAND/RADIATION SAFETY OFFICE
Chemical-Nuclear Engineering Building
Room 2124
College Park, MD 20742
(301) 454-5294
Lori Powell - Health Physicist
Services:
Provides radon detection services for campus buildings.
Area Served:
MD.

UNIVERSITY OF NEW MEXICO/DEPARTMENT OF GEOLOGY
Albuquerque, NM 87131
(505) 277-2310, (505) 277-4204
Douglas G. Brookins - Professor
Services:
Soil gas analysis to determine seasonal reactivity of radon levels in Southwest U.S. Research on uranium content in building materials and in soil-radon studies.
Area Served:
Nationwide.
EPA RMP 5:
AT.

UNIVERSITY OF NORTH CAROLINA/SCHOOL OF PUBLIC HEALTH/DEPARTMENT OF ENVIRONMENTAL SCIENCES AND ENGINEERING
Room 201H
Chapel Hill, NC 27599-7400
(919) 966-3840
James E. Watson, Jr. - Professor
Douglas Crawford-Brown - Professor
Services:
• Radiation protection research and radon-water studies over past ten years. Research focused on monitoring radon levels in NC groundwater.
• Analysis of radon/radium in phosphate deposits and their levels in surrounding groundwater.
• Analysis of radon concentrations in geological media to determine methods of predicting radon in groundwater supplies.
• Contributed EPA's criteria document on radon in drinking water.
• Research on transfer coefficient of radon to water to air.
• Participated in statewide survey of radon levels, looking at influence of geologic media on indoor radon. Utilizes radon exposure chamber to study methods of measurement of indoor radon.
• Non-commercial laboratory for research purposes or state agencies' assistance only.
Area Served:
NC.
EPA RMP 5:
CC(P).

UNIVERSITY OF PITTSBURGH/PHYSICS DEPARTMENT
100 Allen Hall
Pittsburgh, PA 15260
(412) 624-9245
Bernard Cohen - Professor
Services:
Research on CC adsorption detection devices analyzing sensitivity to humidity and atmospheric conditions. Studies liquid scintillation alpha counting devices. Participating in EPA program. Developing database on measurement results to detail factors (house characteristics, socioeconomic status) associated with increased radon levels.
Area Served:
Nationwide.
EPA RMP 5:
CC(P), EP.

UNIVERSITY OF TEXAS/SCHOOL OF PUBLIC HEALTH
P.O. Box 20186
Houston, TX 77225
(713) 792-4431
Howard M. Prichard - Associate Professor
Anna Maria Cox - Faculty Associate
Services:
Radon-water research project in 100 homes in Maine. Developed practical method for water sampling and quick technique for using liquid scintillation of radon trapped in

carbon. Design and production of measurement devices. Researching techniques to simplify radon soil-gas measurement. Non-commercial laboratory for research purposes or state agencies' assistance only.
Area Served:
Nationwide.
EPA RMP 5:
CC(P).

UNIVERSITY OF WASHINGTON/DEPARTMENT OF ENVIRONMENTAL HEALTH/RADIATION ECOLOGY LABORATORY
LRE-WH-10
Seattle, WA 98195
(206) 543-4259
Ahmad Ed Nevissi - Research Associate
Services:
Evaluating effects of home weatherization on indoor radon levels. Participating in research on radon and radium in groundwater and hot springs. Analyzing water to air transference and levels of radon in water. Testing levels of radium in sediments. Non-commercial laboratory for research purposes or state agencies' assistance only.
Area Served:
WA.
EPA RMP 5:
CC(P).

UNIVERSITY RADON TESTING ASSOCIATES
145 A Sycamore Ln.
Athens, GA 30605
(404) 549-8057
Kent Sprague
Area Served:
GA, SC.
EPA RMP 5:
CC.

UP-COUNTRY HOME INSPECTORS, INC.
P.O. Box 1568
Scarborough, ME 04074
(207) 883-9876
Stephen O'Connor - President
Catherine O'Connor
Gordon Guest
Services:
Provides radon detection services using charcoal canister, continuous radon monitor, E-Perm, and alpha track methods, provides radon mitigation and consultation services.
Area Served:
Nationwide.
EPA RMP 5:
CC, CR(P), EP(P).

UTILITY ENGINEERS, INC.
37 N. Washington St.
Wilkes Barre, PA 18701
(717) 829-4200
James Palumbo - Vice President
Services:
Provides radon detection services.
Area Served:
PA.

V

VALLEY HOME INSPECTIONS, INC.
P.O. Box 1677
Chandler, AZ 85244-1677
(602) 988-1375
Kirk R. Kraft
Area Served:
Nationwide.
EPA RMP 5:
CC.

VALLEY RADON TESTING, INC.
26 Oak Valley Rd.
Huntington, CT 06484
(203) 929-4109
David Duggan - President
Services:
Provides detection services for air and water.
Area Served:
CT.

VERITECH
47 Carey Ave.
Butler, NJ 07405
(201) 492-8744
Ellen Stein
Area Served:
Nationwide.
EPA RMP 5:
CC.

VERSAR, INC.
6850 Versar Center
Springfield, VA 22151
(703) 750-3000
Michael Markels - President
R.P. Ouellette - Corporate Development
J. Tappan - ARIX
Niren Nagda - Geomet
Services:
Nationwide consulting firm specializes in environmental risk management, provides radon detection services for homes and buildings, construction management for radon-contaminated sites and structures, provides radon mitigation services, recommends appropriate risk mitigation to public and private clients.
Area Served:
Nationwide.

VFC, INC.
1145 Washington St.
Allentown, PA 18102
(215) 437-5010
Michael Perlow - President
Area Served:
PA.

VIKING HOME IMPROVEMENT, INC.
15 South 15th St.
Fargo, ND 58103
(701) 237-6300
Kevin Christianson - President
Services:
Provides radon detection and mitigation services, distributes complete line of radon mitigation products including sealers, membranes, fans, sump equipment, protective equipment.
Area Served:
ND.

VINCE SCHAEFER CONSTRUCTION
3220 Beaver Rd.
Des Moines, IA 50310
(515) 279-4600
Vince Schaefer - Owner
Services:
Provides radon detection services using charcoal canister and alpha track methods, provides mitigation services.
Area Served:
IA.

VIRGINIA DRY-BASE
198 Milnwood Rd. P.O. Box 562
Farmville, VA 23901
(800) 782-5042
J. Wayne Jackson
Area Served:
VA.

VIRGINIA RADON SERVICE, INC.
Box 332-B Route 1
Roseland, VA 22967
(804) 277-5491
Thomas Eick
Area Served:
Nationwide.
EPA RMP 5:
CC.

VISUAL BUILDING INSPECTIONS
15 B Sickletown Rd.
Orangeburg, NY 10962
(914) 358-2282
Virginia Rodi

Area Served:
NY.

VMS AND ASSOCIATES
Route 1, Box 472A
Madison Heights, VA 24572
(804) 595-2413
Vernon Davis
Area Served:
VA.

W

W.R. MARTIN COMPANY
4124-B Clemson Blvd.
Anderson, SC 29621
(803) 225-7010
W.R. Martin - President
Area Served:
SC.

W.S. FLEMING AND ASSOCIATES, INC.
6310 Fly Rd.
East Syracuse, NY 13057
(315) 437-1869
Ian Nitschke - Senior Associate
Wayne Clarke - Technician
Services:
Provides consultation services on radon progeny monitoring, diagnosis, mitigation specifications.
Qualifications:
EPA RMP, past consultation for NY and EPA.
Area Served:
Nationwide.
EPA RMP 5:
AT, CC, CR(P), GR(P).

WALDON ASSOCIATES
80 Hoadley St.
Naugatuck, CT 06770
(203) 729-5045
Wallace Terrill - President
Services:
Provides radon detection services.
Qualifications:
Graduate of Texas A&M school for radon.
Area Served:
CT.

WARREN WOLFE ASSOCIATES
212 South Queen St.
Dover, DE 19901
(302) 734-8275
Gooden T. Warren - Owner
W. Fronk Wolfe - Owner
Area Served:
DE.

WASHINGTON STATE UNIVERSITY/NUCLEAR RADIATION CENTER
Pullman, WA 99164-1300
(509) 335-8917, (509) 335-8641
Dave Barbe - Interim Director
Services:
Radon analysis lab sevices and contracts to provide equipment to monitor radon.
Area Served:
ID, WA.

WASICK, BRUCE
1888 S. Jacoby Rd.
Copley, OH 44321
(216) 666-8078
Bruce Wasick
Area Served:
OH.

WASTE CONTROL INTERNATIONAL, INC.
100 Galleria Pkwy. Suite 400
Atlanta, GA 30339
(404) 443-2061
T. Ruth Huggins
Area Served:
GA.

WATER TEST CORPORATION
33 South Commercial St.
Manchester, NH 03101
(603) 623-7400
Theodore A. Olsson III
Area Served:
Nationwide.
EPA RMP 5:
CC(P).

WATKINS ENVIRONMENTAL SCIENCES
207 McLennan Dr.
Fayetteville, NY 13066
(315) 637-4077
Andrew Watkins - Engineer
Services:
Provides radon detection services for banks, lawyers and realtors.
Qualifications:
Licensed professional engineer.
Area Served:
NY.

WAYNE MUSSER PROPERTY COMPANY, INC.
517 Broad St. P.O. Box 258
Sewickley, PA 15143
(412) 741-9517
Wayne C. Musser - President
Ty W. Musser - Vice President
Services:
Manufactures a tamper proof radon test enclosure that can be electronically activated by attempts to cover or move, provides radon detection services.
Area Served:
Nationwide.

WAYNE, MICHELE
Box 266
Terre Hille, PA 17581
(215) 745-0129
Michele Wayne
Area Served:
PA.

WB CONTRACTORS
3209 Morrison St., NW
Washington, DC 20015
(202) 966-6312
William Brodie
Area Served:
DC.

WEATHER SHIELD, INC.
4071 South Salem Church Rd.
Dover, PA 17315
(717) 292-5613
William Kimmel - President
Services:
Provides radon detection and mitigation services.
Area Served:
PA.

WEBER AND ASSOCIATES
8657 Fallbrook Way
Sacramento, CA 95826
(916) 386-8745
Ronald E. Weber
Area Served:
CA.
EPA RMP 5:
AT.

WEST COAST ENVIRONMENTAL SERVICES, INC.
P.O. Box 18182
Seattle, WA 98118
(206) 324-0920
John Gilmore - President
Services:
Provides radon detection and hazardous material and waste management services to government, homes, and schools, performs inspections.
Qualifications:
EPA certified building inspections and radon certified minority business enterprise.
Area Served:
Nationwide.

EPA RMP 5:
CC.

WEST ENVIRO SYSTEMS, INC.
45 Sycamore Ln.
Skillman, NJ 08558
(609) 924-0680
Fred W. West - President
H.E. West - Vice President
Charles West - Supervisor
Services:
Provides radon detection services using charcoal canister and alpha track methods, provides radon mitigation services.
Qualifications:
PA DER mitigation course, chemical engineer.
Area Served:
NJ.

WESTERN ENVIRONMENTAL, INC.
4320-B Northpark Dr.
Colorado Springs, CO 80907
(719) 528-6132
C.W. Gullickson - Director
Belmont Evans - Field Technician
Services:
Provides radon detection services using charcoal canister and real-time working level methods, operates a laboratory.
Qualifications:
EPA RMP, trained physicists on staff.
Area Served:
Nationwide.
EPA RMP 5:
CC.

WESTERN TECHNOLOGY, INC.
18708 Woodway Dr.
Derwood, MD 20855
(301) 869-2628
Thomas P. Sheahen - President
Services:
Provides radon detection services using alpha track, charcoal canisters, and continuous working level methods.
Qualifications:
EPA RMP.
Area Served:
DC, DE, MD, OH, PA, VA, WV.
EPA RMP 5:
AT.

WILKES COLLEGE/PHYSICS DEPARTMENT
Wilkes-Barre, PA 18766
(717) 824-4651 ext. 481
Roger Maxwell
Services:
Non-commercial laboratory for research purposes or state agencies' assistance only.
Area Served:
Nationwide.
EPA RMP 5:
AT, CC, CR(P), GR(P), CW(P).

WILLIAM G. CHANEY, PE
1708 Park Ave., SW
Albuquerque, NM 87104
(505) 247-2087
William G. Chaney - Owner
Services:
Provides radon detection services using At Ease and charcoal canister method, offers radon education seminars for the real estate industry.
Qualifications:
AARST.
Area Served:
NM.

WILSON JR., WALTER E.
700 Harston Ln.
Erdenheim, PA 19118
(215) 836-7257
Walter E. Wilson, Jr.
Area Served:
PA.

WORLDWIDE RADON, INC.
1941A Nears Pkwy.
Margate, FL 33063
(305) 968-5555
Gordon Marsh - President
Marge Semenuk - Vice President
Services:
Provides radon detection services using charcoal canister and E-Perm methods, provides radon mitigation services.
Qualifications:
General contractor and physicist on staff.
Area Served:
NJ, FL.

WORTH ENGINEERING
650 Palm Ave.
Hialeah, FL 33010
(305) 885-4515
James G. Worth - President
Services:
Provides radon detection services using charcoal canister, alpha track, and continuous radon monitor methods, provides radon mitigation services.
Area Served:
FL.

WPB ENTERPRISES, INC.
RD 1, Box 209
Riegelsville, PA 18077
(215) 346-8004
Bill Brodhead - President
Services:
Provides radon mitigation services, provides training on radon mitigation issues, non-commercial laboratory for research purposes or state agencies' assistance only.
Qualifications:
Developed and instructs EPA three day radon workshops.
Area Served:
NJ, PA.
EPA RMP 5:
CC.

WRIGHT LABORATORY SERVICES, INC.
3240 Schoolhouse Rd.
Middletown, PA 17057
(717) 944-5541
Ian Milnes
Area Served:
Nationwide.
EPA RMP 5:
CC(P).

X

Y

YOUR ENERGY SERVICE, INC.
220 4F Elliston Pl.
Nashville, TN 37203
(615) 329-9747
William B. French
Area Served:
TN.
EPA RMP 5:
CC.

Z

ZANUS CORPORATION
1259 El Camino Real Suite 134
Menlo Park, CA 94025
(415) 591-7783
David G. Sullivan
Area Served:
CA.

1

Radon Detection Companies by States Served

This is an index listing radon detection companies by state served. The index begins with a list of radon detection companies that provide services nationwide. Next, each state is followed by a list of radon detection companies that provide services to that state. The state abbreviation, in parentheses, immediately following each company listing, indicates the state in which the radon detection company is located.

Nationwide

A & A RADON TESTING SERVICES, INC. (CO)
A & W RADON SPECIALISTS, INC. (NH)
A BLACK CROSS SANITATION COMPANY (PA)
A. ARTHUR LIPSHUTZ ASSOCIATES (FL)
A.B.E. RADIATION MEASUREMENTS LABORATORY (PA)
A.S. McCREATH AND SONS, INC. (PA)
A.T.E.I. (IN)
AAA RADON SERVICES, INC. (PA)
AADVANCED AAIR AANALYSIS (OH)
ACADEMY OF IRW, INC. (MD)
ADVANCED RADIATION MONITORING SERVICE, INC. (ARMS) (MI)
AIR CHEK, INC. (NC)
AIR SCIENCES, INC. (MO)
AIR-N-SOL CORPORATION (NJ)
AIRREGS, INC./AIR POLLUTION REGULATORY SERVICE, INC. (GA)
ALABAMA TESTING LABORATORY (AL)
ALLSAFE ENVIRONMENTAL CORPORATION (NJ)
ALLSTATE RADON TESTING SERVICE (GA)
ALPHA CONCEPTS, INC. (NJ)
ALPHA CONTROL, INC. (MD)
ALPHA DETECTORS (UT)
ALPHA ENERGY LABORATORIES, INC. (TX)
ALPHA RADON TESTING, INC. (MN)
ALPHA SPECTRA, INC. (CO)
ALPHANUCLEAR COMPANY (ON)
ALPHATECH, INC. (PA)
AMERICAN COMPLIANCE TECHNOLOGIES, INC. (FL)
AMERICAN RADON CONSULTANTS, INC. (TN)
AMERICAN RADON CORPORATION (AZ)
AMERICAN RADON SERVICES, LTD. (IA)
AMERICAN RADON TESTING LABS, INC. (OH)
AMES ASSOCIATES, INC. (MD)
ANNAPOLIS PROPERTY SERVICES (MD)
APPALACHIAN ENVIRONMENTAL TESTING, INC. (VA)
APPLIED ENERGY COMPANY (NE)
APPLIED HEALTH PHYSICS, INC. (PA)
APPLIED RADON, INC. (OH)
APPLIED TECHNICAL SERVICES, INC. (GA)
AQUATEK LABORATORY/DIVISION OF GRAHAM-MASSEY ANALYTICAL LABORATORY (CT)
ARGONNE NATIONAL LABORATORY (IL)
ARIX CORPORATION (CO)
ARIX CORPORATION/A DIVISION OF VERSAR, INC. (PA)
ASBESTOS AND RADON CONSULTANTS QUALI-PRO-TECH SERVICES, INC. (PA)
ASSAY SERVICES (TX)
ATLANTA TESTING AND ENGINEERING, INC. (GA)
ATOMETRICS (LA)
BARRINGER (CO)
BASEMENT DE-WATERING SYSTEMS, INC. (PA)
BATTELLE MEMORIAL INSTITUTE/NUCLEAR TECHNOLOGY SECTION (OH)
BECKMAN INSTRUMENTS, INC. (CA)
BIOMONITORING SERVICES LABORATORY/TECHNICAL RESOURCES, INC. (MD)
BIONOMIC INDUSTRIES, INC. (NJ)
BOSTON UNIVERSITY MEDICAL CENTER/RADIATION PROTECTION OFFICE (MA)
BRANDEIS UNIVERSITY/ROSENTIEL RESEARCH CENTER (MA)
BRYANT AND BRYANT SERVICE COMPANY (TN)
CAMP DRESSER AND MCKEE, INC. (NJ)
CAMRODEN ASSOCIATES (NY)
CANADIAN INSTITUTE FOR RADIATION SAFETY/ELLIOTT LAKE LABORATORY (ON)
CANBERRA INDUSTRIES, INC. (CT)
CANCERSCREEN, INC. (VA)
CAVIN ANALYTICAL CONSULTANTS, INC. (GA)
CDM FEDERAL PROGRAMS CORPORATION (VA)
CENTARUS CORPORATION (KY)
CERTUS LABORATORIES (GA)
CHEM-SAFE LABORATORIES (WA)
CHENANGO ENVIRONMENTAL LAB, INC. (NY)
CLEAN AIR ENGINEERING, INC. (IL)
CMT, INC. (KS)
COLORADO VENTURE MANAGEMENT, INC. (CO)
COLUMBIA RADIOLOGICAL TESTING LABORATORY (MO)
COMMONWEALTH RADON SPECIALISTS (VA)
COMMUNITY RADON TESTING/ DIVISION OF SANTIAGO ENTERPRISES (VA)
CON-TEST, INC. (MA)
CONTEK/CORPORATE HEADQUARTERS (CA)
CORE LABORATORIES, INC. (WY)
CRIPPEN CONSULTING ASSOCIATES, INC. (DE)
DOUGLAS MARTIN AND ASSOCIATES, INC./DMA ANALYSIS GROUP (PA)
EASTERN ANALYTICAL SERVICES (NY)
EBERLINE INSTRUMENT CORPORATION (NM)
ECI (CO)
ECODEX (MA)
EDA INSTRUMENTS, INC. (CO)
EDINBORO SCIENTIFIC COMPANY (PA)
EG & G ORTEC (TN)
EKS RADTECH, INC. (PA)
ELECTRO ANALYTICAL, INC. (OH)
ENERGY PROFESSIONALS, INC. (PA)
ENERGY RESOURCE CENTER (CO)
ENRAD, INC./RADON TESTING AND ENGINEERING, INC. (MD)
ENTROPY ENVIRONMENTALISTS, INC. (NC)
ENV SERVICES, INC. (PA)
ENVIRO SCIENCES, INC (TX)
ENVIRO TECHNOLOGY CORPORATION (NY)
ENVIROHAZARDS MANAGEMENT, INC. (NJ)
ENVIRONMENTAL AUDIT, INC. (KS)
ENVIRONMENTAL CHEMISTRY, INC. (CO)
ENVIRONMENTAL CONSULTING LABORATORY (CT)
ENVIRONMENTAL DETECTION AGENCY, INC. (TX)
ENVIRONMENTAL DETECTION ASSOCIATES, INC. (NC)
ENVIRONMENTAL DETECTION MANAGEMENT (NC)
ENVIRONMENTAL ENTERPRISES, INC. (GA)
ENVIRONMENTAL HEALTH AND SAFETY (FL)
ENVIRONMENTAL HEALTH CONSULTANTS, INC. (IL)
ENVIRONMENTAL HEALTH INSPECTIONS, INC. (NJ)
ENVIRONMENTAL HEALTH TECHNOLOGIES (CO)
ENVIRONMENTAL PRODUCTS ALLIANCE, LTD. (ND)
ENVIRONMENTAL QUALITY ASSOCIATES, INC. (NY)
ENVIRONMENTAL RADIATION SERVICES, INC. (CO)
ENVIRONMENTAL RADON MEASUREMENTS (FL)
ENVIRONMENTAL RESEARCH ASSOCIATES (PA)
ENVIRONMENTAL SCIENCE AND ENGINEERING, INC. (FL)
ENVIRONMENTAL SCIENCE LABORATORY, INC. (MA)
ENVIRONMENTAL SERVICES AND TECHNOLOGY, INC. (NJ)
ENVIRONMENTAL TESTING ASSOCIATES (GA)
ENVIRONMENTAL TESTING ASSOCIATES (WI)
ENVIRONMENTAL TESTING ASSOCIATES (WI)
ENVIRONMENTAL TESTING ASSOCIATES (WI)
ENVIROSERV/MAIN OFFICE (NJ)
ENVIROSERV/RADON DETECTION OF ALABAMA (AL)
ENVIROTEST (OH)
ERM, INC. (FL)
EVANSHIRE COMPANY, LTD., THE (NY)
FEDERAL RADON, INC. (MO)
FEDERAL RADON, INC. (NJ)
FEMTO-TECH, INC. (OH)
GALLIPOT, INC. (MN)
GARDEN STATE HOME INSPECTION SERVICE (NJ)
GENERAL ENVIRONMENTAL MONITORING (IL)
GENERAL HEALTH PHYSICS, INC. (VA)
GENERAL PHYSICS CORPORATION (MD)
GEO-ENVIRONMENTAL SERVICES, INC. (GA)
GEO-RADON SERVICES (CA)
GEOMET TECHNOLOGIES, INC. (MD)
GEORGIA RADON DETECTION COMPANY (GA)
GLOBAL RADON TESTING SERVICE TRAINING CENTER (FL)
GREAT PLAINS NUCLEAR SERVICES, INC. (NE)
GREAT TRACERS (IL)

GROUNDWATER TECHNOLOGY, INC. (NY)
HALL-KIMBRELL ENVIRONMENTAL SERVICES (OR)
HARVARD ENGINEERING SERVICES COMPANY (MA)
HEALTH AND SAFETY ASSOCIATES (MA)
HEALTH HAZARDS DETECTION (KY)
HEALTH PHYSICS ASSOCIATES, LTD. (IL)
HEALTH PHYSICS NORTHWEST, INC. (OR)
HEALTH SCIENCE SERVICES CORPORATION (NJ)
HELIX ENVIRONMENTAL, INC. (OH)
HERITAGE RADON SERVICES, INC. (PA)
HOME INSPECTION SERVICE, INC. (IL)
HOME RADON DETECTION COMPANY (KS)
HOME RADON DETECTION, INC. (IL)
HOME RADON SURVEYS (VA)
HOME SENTRY, INC. (NJ)
HOME TESTING LABS, INC. (NJ)
HOME TESTING SERVICE (MA)
HOME-TECH ENGINEERING (NJ)
HOMES PLUS RADON SERVICES, INC. (CT)
HORIZON ENGINEERING AND ENVIRONMENTAL SERVICES, INC. (TN)
HOUSE ANALYSIS AND ASSOCIATES (PA)
HOUSE-CHECK, INC. (PA)
HOUSEHOLD LABORATORIES, INC. (IL)
HUDSON VALLEY RADON TESTING CORPORATION (NY)
I-T.E.M., LTD. (MD)
INFILTEC/FACTORY (VA)
INFILTEC/MAIN OFFICE (VA)
INSPEX OF BOULDER, INC. (CO)
INTERNATIONAL RADON TECHNOLOGIES (KY)
J.M. MONTGOMERY ENGINEERS (CA)
JH AND ASSOCIATES, INC. (OK)
JOHN PRIKKEL AND ASSOCIATES (OH)
JOHNSON PLUMBING AND HEATING (PA)
JOHNSON'S ENERGY SERVICE (IA)
JRW INDUSTRIES, INC. (PA)
JUNGBLUT, DAVID W., GEOLOGIST (NJ)
K & J SOLAR SERVICE (CO)
KBC TECHNOLOGY (IL)
KEY TECHNOLOGY, INC. (PA)
KSE, INC. (MA)
LAPTEFF ASSOCIATES (VA)
LEGIS CORPORATION (TX)
LIFELINE ENVIRONMENTAL TESTING (VA)
LOWRY ENGINEERING, INC. (ME)
MAINE DEPARTMENT OF HUMAN SERVICES/BUREAU OF HEALTH/PUBLIC HEALTH LABORATORY (ME)
MARYLAND RADON LABORATORY, INC. (MD)
MASSACHUSETTS INSTITUTE OF TECHNOLOGY/RADON PROJECT (MA)
MBI RADON TECHNOLOGY (MN)
MCCRONE ENVIRONMENTAL SERVICES, INC. (GA)
MEDRAD, INC. (PA)
MENKE CONSULTING, INC. (OH)
METRO RADON MEASUREMENT (MN)
METROPOLITAN RADON SERVICE (VA)
MICROBAC MID-ATLANTIC, INC. (VA)
MID AMERICA RADON TESTING, INC. (KS)
MIDWEST RADON ANALYSIS, INC. (OH)
MONTGOMERY LABORATORIES (CA)
NATIONAL ASSOCIATION OF HOME BUILDERS (NAHB)/NATIONAL RESEARCH CENTER (MD)
NATIONAL PROPERTY INSPECTIONS, INC. (NE)
NATIONAL RADON LABORATORIES, INC. (GA)
NATIONAL RADON LABORATORIES/DIVISION OF ANALYTICAL SERVICES CORPORATION (VA)
NATIONAL RADON TESTING CENTER, INC. (CT)
NATIONWIDE TESTING (NJ)
NATKIN SERVICE COMPANY (IL)
NATLSCO (IL)
NATURAL ENERGY CONSULTANTS (CO)
NATURAL RESOURCE LABORATORY, INC. (CO)
NDL ORGANIZATION, INC., THE (NY)
NERING, INC. (CT)
NITON CORPORATION (MA)
NODAR, INC. (AL)
NORTHEAST RADON TESTING SERVICE, INC. (NY)
NORTHERN ENVIRONMENTAL LABORATORY, INC. (ME)
NORTHERN NECK RADON SERVICE (VA)
NORTON CONSERVATION COMPANY, INC. (NJ)
NUCLEAR SOURCES AND SERVICES, INC. (TX)
NUCLEAR TECHNOLOGY, INC./RADON CHEK (OH)
NUCLEUS, THE (TN)
O.K. REMS CORPORATION (NJ)
OCCUPATIONAL AND ENVIRONMENTAL HEALTH CONSULTING SERVICES (IL)
ORKIN ENVIRONMENTAL SERVICES (GA)
PACKARD INSTRUMENT COMPANY (IL)
PALMER TECHNOLOGY (MD)
PETROLEUM ANALYTICAL LABORATORY, INC. (FL)
PHOENIX LABS (NH)
POTOMAC AIR TESTING (MD)
PRINCETON TESTING LABORATORY (NJ)
PRO CHEK, INC. (NC)
PROFESSIONAL HOME INSPECTIONS, INC. (CO)
PROFESSIONAL HOUSE DOCTORS, INC. (IA)
PYLON ELECTRONIC DEVELOPMENT COMPANY, LTD. (ON)
PYRAMID ENVIRONMENTAL SYSTEMS, INC. (NJ)
QUALITY RADON TESTING CORPORATION (IL)
QUANTUM INFORMATION, LTD. (PA)
R.A.D. SERVICE AND INSTRUMENTS, LTD. (ON)
R.E. WRIGHT ASSOCIATES, INC./RADONIX GROUP (PA)
R.F. SIMON COMPANY, INC. (PA)
R.K. OCCUPATIONAL AND ENVIRONMENTAL ANALYSIS, INC. (NJ)
RA-TEK/DIVISION OF SHAMROCK, INC. (MN)
RAD ELEC, INC. (MD)
RAD ENVIRONMENTAL SERVICES, INC. (NC)
RADALYTIC LABS, INC. (SC)
RADEX, INC. (GA)
RADIATION CONSULTANTS, INC. (TX)
RADIATION DATA/DIVISION OF PRINCETON GEOPHYSICAL SERVICES (CT)
RADIATION DATA/DIVISION OF PRINCETON GEOPHYSICAL SERVICES (NJ)
RADIATION PROTECTION SERVICES, INC. (CT)
RADIATION SAFETY ENGINEERING (AZ)
RADIATION SAFETY SERVICES, INC. (IL)
RADIATION SERVICE ORGANIZATION, INC. (MD)
RADIOLOGICAL HEALTH SCIENCES, INC. (FL)
RADION LABORATORY, INC. (MA)
RADON ALERT, INC. (TX)
RADON ALERT® DETECTION CENTER (PA)
RADON ANALYSIS, INC. (NJ)
RADON ANALYTICAL LABS (IN)
RADON AND WATER TESTING SERVICE (OH)
RADON AWARENESS AND DETECTION (MO)
RADON CHECK (NC)
RADON CHECK (VA)
RADON CONSULTANTS, INC., VIRGINIA (VA)
RADON CONSULTING AND TESTING SERVICE (FL)
RADON CONTROL PROFESSIONALS, INC. (VA)
RADON DETECTION OF MISSISSIPPI (MS)
RADON DETECTION AGENCY, LTD. (BC)
RADON DETECTION AND CONSULTING (KS)
RADON DETECTION AND CONTROL (PA)
RADON DETECTION AND REDUCTION LABORATORY/DIVISION OF FORMACHEM (MO)
RADON DETECTION SYSTEMS/CORPORATE HEADQUARTERS (CO)
RADON DIAGNOSTIC LABORATORIES (ME)
RADON DIAGNOSTICS (NJ)
RADON ENGINEERING SERVICES, INC. (AL)
RADON ENGINEERING/A UNIT OF PSI ENGINEERING, INC. (NJ)
RADON ENVIRONMENTAL MONITORING, INC. (IL)
RADON ENVIRONMENTAL SERVICES, INC. (IN)
RADON GASBUSTERS (NJ)
RADON INSTRUMENTAL DETECTION (KS)
RADON INTERVENTION (TX)
RADON LABORATORIES (MA)
RADON LABORATORIES, INC. (MD)
RADON LABORATORIES, INC. (NJ)
RADON LABS (MA)
RADON LABS OF ATLANTA (GA)
RADON MANAGEMENT/DIVISION OF PHYSICS CONSULTANTS, INC. (ME)
RADON MEASUREMENT AND CONTROL OF NEW MEXICO (NM)
RADON MEASUREMENT AND SERVICES (CO)
RADON MEASUREMENTS, INC. (TX)
RADON MITIGATION SERVICES (FL)
RADON MITIGATORS OF ARIZONA (AZ)
RADON MITIGATORS, INC. (NJ)
RADON MONITORING SERVICES, INC. (MA)
RADON PROJECT, THE (PA)
RADON QC (PA)
RADON REDUCTION ENGINEERING AND LABORATORY (VA)
RADON REMEDIES, INC. (OH)
RADON REMOVAL, INC. (NJ)
RADON RESEARCH GROUP, INC. (MD)
RADON RESOURCES (GA)
RADON SAFE (IN)
RADON SAFETY AND TECHNOLOGY CORPORATION (NY)
RADON SAFETY SERVICES, INC. (NJ)
RADON SCREENING SERVICE, INC. (CO)
RADON SCREENING, INC. (VA)
RADON SERVICES (WA)
RADON SERVICES ANALYTICAL LABORATORY (TN)
RADON SERVICES OF NORTH AMERICA, INC. (IN)
RADON SERVICES, INC. (PA)
RADON SPECIALIST OF AMERICA (MD)
RADON SURVEY SYSTEMS, INC. (OH)
RADON TECHNOLOGIES, INC. (VA)
RADON TESTING AND ENVIRONMENTAL CONTROL (KY)
RADON TESTING CONSULTANTS, INC. (NJ)
RADON TESTING CORPORATION (MD)
RADON TESTING CORPORATION OF AMERICA (NY)
RADON TESTING LABORATORY OF NEW ENGLAND (MA)
RADON TESTING SERVICE, INC. (CO)
RADON TESTING SERVICES (CA)
RADON TESTING SERVICES (UT)
RADON TESTING SERVICES OF MISSOURI (MO)
RADON TESTING SERVICES, BALTIMORE (MD)
RADON TESTING SERVICES, PITTSBURGH (PA)
RADON TESTING, INC. (VA)
RADON TRAC, INC. (FL)
RADON X COMPANY, INC. (NH)
RADON, LTD. (IN)
RADON-ONE LABORATORY SERVICES (OH)
RADON/SOILS AND WATER TESTING SERVICE (OH)
RADONICS, INC. (NJ)
RADONIX, LTD. (NY)
RADONTEST, INC. (MO)
RAF ENGINEERING (NJ)
RANDOLPH AND ASSOCIATES, INC. (IL)
RDS/RADON DETECTION SERVICES, INC.-CORPORATE OFFICE (NJ)
RDS/RADON DETECTION SERVICES-GEORGIA OFFICE (GA)
RDS/RADON DETECTION SERVICES-MARYLAND OFFICE (MD)
RDS/RADON DETECTION SERVICES-PENNSYLVANIA OFFICE (PA)
REAL ESTATE SUPPORT SERVICES, INC. (MN)
RECON SYSTEMS, INC. (NJ)
RESIDENCE RADON (KS)
ROCKY MOUNTAIN MEDICAL PHYSICS, INC. (CO)
ROGERS AND ASSOCIATES ENGINEERING CORPORATION (UT)
ROSS SYSTEMS, INC. (NJ)
ROY F. WESTON, INC. (PA)

RPI RADON MEASUREMENT SERVICES (PA)
RUECK ASSOCIATES (KS)
RYAN NUCLEAR LABORATORIES (OH)
SAFE-AIRE, INC. (IL)
SAM STEWART TESTING (VA)
SCIENTIFIC ANALYSIS, INC. (AL)
SCIENTIFIC TESTING ASSOCIATES, LTD. (VA)
SEPTECH (WA)
SIMPSON WEATHER ASSOCIATES (VA)
SOLAR TESTING LABORATORIES, INC. (OH)
SOLID STATE RADON DETECTION AND ELIMINATION SERVICES (NJ)
SORENSEN ENTERPRISES, INC. (TN)
SOUTHERN MECHANICAL (GA)
SOUTHERN RADON SERVICES, INC. (GA)
SOUTHERN RESEARCH INSTITUTE (AL)
SOUTHWEST ABATEMENT, INC. (NM)
SPECIAL ENVIRONMENTAL SERVICES, INC. (GA)
STAN A. HUBER CONSULTANTS, INC. (IL)
STANDOR RADON DETECTION (FL)
SUN NUCLEAR CORPORATION (FL)
SWAT TERMITE AND PEST CONTROL (GA)
T.R. EDGERTON, INC. (NC)
TARGET RADON SERVICES, INC. (MD)
TCS INDUSTRIES, INC. (PA)
TECHNORAD (MD)
TECHTRAC (TX)
TELEDYNE ISOTOPES (NJ)
TELEDYNE ISOTOPES MIDWEST LABORATORY (IL)
TEMPLE UNIVERSITY RADON PROGRAM (PA)
TEQ CORPORATION (PA)
TERMINIX INTERNATIONAL (TN)
TERRADEX CORPORATION/SUBSIDARY OF TECH/OPS LANDAUER, INC. (IL)
TERRADON EXPLORATION TECHNOLOGIES, INC. (OH)
TESTWELL CRAIG LABORATORIES (FL)
TFC NUCLEAR ASSOCIATES, INC. (NJ)
THERMO ENVIRONMENTAL CORPORATION (MA)
THOMSON AND NIELSEN ELECTRONICS, LTD. (ON)
TMA/EBERLINE, HEALTH PHYSICS (NM)
TMA/EBERLINE, LABORATORY (NM)
TMB COMPANY (MD)
TODD GIDDINGS AND ASSOCIATES (PA)
TRC-ECI (CT)
TRI-COURT SAFETY/SECURITY (CA)
TRUE TEST COMPANY, UTAH (UT)
TUCSON HOME INSPECTIONS, INC. (AZ)
TWIN CITY TESTING (MN)
U.S. DEPARTMENT OF INTERIOR (DOI)/BUREAU OF MINES(BOM)/DENVER RESEARCH CENTER DRC)/RADIATION LABORATORY (CO)
U.S. DEPARTMENT OF THE ARMY/ARMY ENVIRONMENTAL HYGIENE AGENCY (MD)
U.S. DEPARTMENT OF THE ARMY/BELVOIR RESEARCH, DEVELOPMENT AND EDUCATION CENTER (VA)
U.S. ENVIRONMENTAL PROTECTION AGENCY (EPA)/OFFICE OF AIR AND RADIATION(OAR)/OFFICE OF RADIATION PROGRAMS(ORP)/LAS VEGAS FACILITY(LAF) (NV)
U.S. RADON TESTING SERVICE (VA)
U.S. TOXIC SUBSTANCE TESTING BUREAU (PA)
UCI/ENVIROTEST (PA)
UNITED SERVICES ALLIANCE, INC. (VA)
UNIVERSITY HOME HEALTH SERVICES, INC. (MA)
UNIVERSITY OF COLORADO-COLORADO SPRINGS/PHYSICS DEPARTMENT (CO)
UNIVERSITY OF LOWELL/RADIOLOGICAL SCIENCES DEPARTMENT (MA)
UNIVERSITY OF NEW MEXICO/DEPARTMENT OF GEOLOGY (NM)
UNIVERSITY OF PITTSBURGH/PHYSICS DEPARTMENT (PA)
UNIVERSITY OF TEXAS/SCHOOL OF PUBLIC HEALTH (TX)
UP-COUNTRY HOME INSPECTORS, INC. (ME)
VALLEY HOME INSPECTIONS, INC. (AZ)
VERITECH (NJ)
VERSAR, INC. (VA)
VIRGINIA RADON SERVICE, INC. (VA)
W.S. FLEMING AND ASSOCIATES, INC. (NY)
WATER TEST CORPORATION (NH)
WAYNE MUSSER PROPERTY COMPANY, INC. (PA)
WEST COAST ENVIRONMENTAL SERVICES, INC. (WA)
WESTERN ENVIRONMENTAL, INC. (CO)
WILKES COLLEGE/PHYSICS DEPARTMENT (PA)
WRIGHT LABORATORY SERVICES, INC. (PA)

Companies Serving

Alabama

All companies listed under Nationwide provide services for this state.

DAMES AND MOORE (MD)
ENERGY SYSTEMS, INC. (TN)
ENVIRONMENTAL MANAGEMENT GROUP (MD)
HOMEBUYERS INSPECTION SERVICE OF NORTH WEST FLORIDA, INC. (FL)
HUNT AND ASSOCIATES (GA)
LAW ENVIRONMENTAL, INC. (FL)
MACMILLAN RESEARCH, LTD. (GA)
NATURAL FLORIDA RETROFIT, INC. (FL)
RADON CONSULTANTS (NV)
RADON DETECTION SYSTEMS/NASHVILLE, TENNESSEE BRANCH (TN)
RADON ENVIRONMENTAL SERVICES, INC. (KY)
RADON EVALUATION SERVICES, INC. (GA)
RADON REDUCTION AND TESTING, INC. (GA)
ROSS RADON TESTING (GA)
SOUTHEASTERN RADON CONSULTANTS, INC. (TN)
UNIVERSITY OF GEORGIA/CENTER FOR APPLIED ISOTOPE STUDY (GA)

Companies Serving

Alaska

All companies listed under Nationwide provide services for this state.

BCA LABORATORIES, INC. (IL)
GEOCHEM, INC. (AK)
J.A. ASSOCIATES/PUBLIC HEALTH CONSULTANTS (PA)
MERIT ENVIRONMENTAL SERVICES, INC. (KY)
NATURAL PANTRY, THE (AK)
RADIATION HEALTH (WA)
RADON CONSULTANTS (NV)
RADON TESTING SERVICES, INC. (CA)
TECHNORTH (AK)

Companies Serving

Arizona

All companies listed under Nationwide provide services for this state.

AAAIRSCAN ENGINEERING CORPORATION (CT)
ALPHA RADIATION ENGINEERING, INC. (CO)
ARIX CORPORATION (AZ)
ARIZONA RADIATION REGULATORY AGENCY/ENVIRONMENTAL SURVEILLANCE LABORATORY (AZ)
BONDAR-CLEGG, INC. (CO)
ENERGY LABORATORIES, INC. (WY)
ENVIRONMENTAL ENGINEERING AND TESTING (CO)
INDUSTRIAL OPERATIONS, INC./RADON EVALUATION (AZ)
INTEGRAL RESOURCES, INC. (AZ)
PACIFIC RADIATION CORPORATION (CA)
RADON AND ENVIRONMENTAL PROFESSIONALS (KS)
RADON CHECK (AZ)
RADON CONSULTANTS (NV)
RADON DETECTION SYSTEMS/ARIZONA BRANCH (AZ)
RADON DETECTION SYSTEMS/LOS ANGELES, CALIFORNIA BRANCH (CA)
RADON SYSTEMATICS OF CALIFORNIA (CA)
RADON TEST (CA)
RADON TESTING SERVICES, INC. (CA)
SPENCE-GEIGER ASSOCIATES, INC. (CO)

Companies Serving

Arkansas

All companies listed under Nationwide provide services for this state.

ACI CHEMICAL COMPANY, INC. (AR)
ENVIRONMENTAL ENGINEERING LABORATORIES, INC. (CO)
ENVIRONMENTAL SOLUTIONS, INC. (MO)
MERIT ENVIRONMENTAL SERVICES, INC. (KY)
RADON AND ENVIRONMENTAL PROFESSIONALS (KS)
RADON ENVIRONMENTAL SERVICES, INC. (KY)

Companies Serving

California

All companies listed under Nationwide provide services for this state.

AAAIRSCAN ENGINEERING CORPORATION (CT)
ABMS, INC. (CA)
AMERICAN RADON TESTING SERVICES (CA)
BAUBIOLOGIE BY BANTA (CA)
BONDAR-CLEGG, INC. (CO)
CALIFORNIA DEPARTMENT OF HEALTH SERVICES/SANITATION AND RADIATION LABORATORY (CA)
CTL ENVIRONMENTAL SERVICES (CA)
E.C. CHAMBERLAIN, RADIOLOGICAL CONSULTANT (NV)
ENERGY LABORATORIES, INC. (WY)
ENVIRONMENTAL ENGINEERING AND TESTING (CO)
ENVIRONMENTAL ENGINEERING LABORATORIES, INC. (CO)
ENVIRONMENTAL QUALITY CONTROL, INC. (CA)
FAYTEK, INC. (ID)
FORENSIC ANALYTICAL SPECIALTY (CA)
J.A. ASSOCIATES/PUBLIC HEALTH CONSULTANTS (PA)
JMX INDUSTRIES, INC. (CA)
LOS ANGELES URBAN DEVELOPMENT (CA)
LW ENVIRONMENTAL SERVICES (CA)
PACIFIC RADIATION CORPORATION (CA)
R. BERNT AND SONS (CA)
RADIATION HEALTH (WA)
RADON CONSULTANTS (NV)
RADON DETECTION SYSTEMS/LOS ANGELES, CALIFORNIA BRANCH (CA)
RADON DETECTION SYSTEMS/NORTHERN CALIFORNIA BRANCH (CA)
RADON READINGS (CA)
RADON SYSTEMATICS OF CALIFORNIA (CA)
RADON TEST (CA)
RADON TESTING SERVICES, INC. (CA)
SAGE RESOURCES,/INDOOR AIR QUALITY DIVISION (CA)
SOLID WASTE AND TOXIC TECHNOLOGY (CA)
SPENCE-GEIGER ASSOCIATES, INC. (CO)
SULLIVAN CONSULTANTS, INC. (NJ)
TSI CONSTRUCTION SERVICES (IN)

WEBER AND ASSOCIATES (CA)
ZANUS CORPORATION (CA)

Companies Serving
Colorado

All companies listed under Nationwide provide services for this state.

AAA DISASTER FORCE (CO)
AAAIRSCAN ENGINEERING CORPORATION (CT)
ACCENT ENGINEERING, INC. (CO)
ADVANCED RADON SERVICES (CO)
AIR RESOURCE SPECIALISTS, INC. (CO)
AIR TRON MECHANICAL SERVICES (CO)
ALPHA GUARD/TECTROL (CO)
ALPHA RADIATION ENGINEERING, INC. (CO)
ALPHA RADON SCREENING AND HOME INSPECTION, INC. (CO)
AMERICAN RADON CONTROL, INC. (CO)
AMERICAN RADON TESTING LABORATORIES (CO)
ARIX CORPORATION (CO)
ASSOCIATED BUILDERS OF THE ROCKIES, INC. (CO)
BONDAR-CLEGG, INC. (CO)
BOULDER INSPECTION COMPANY (CO)
BRENT, DAVID J. (CO)
BUILDING DIAGNOSTICS, INC. (CO)
C & D HEATING AND CONSTRUCTION (CO)
C. KENT CHAMBERLAIN AND ASSOCIATES (CO)
C.R. ADAMS COMPANY, INC. (CO)
CARPENTER, CHARLES D. (CO)
COLORADO FRONT RANGE BUILDING INSPECTION SERVICE (CO)
COLORADO HOME PROTECTION (CO)
COLORADO RADON DETECTION COMPANY (CO)
COLORADO RADON ENGINEERING, INC. (CO)
COLORADO RADON INSTITUTE (CO)
COLORADO REAL ESTATE INSPECTION SERVICES (CO)
COLORADO SAFE-N-DRY, INC. (CO)
CONCEPT 90, INC. (CO)
CTL/THOMPSON, INC. (CO)
DIAMOND BACK SERVICES, INC. (CO)
E.O. CHURCH, INC. (CO)
ENERGY ALTERNATIVES UNLIMITED, INC. (CO)
ENERGY LABORATORIES, INC. (WY)
ENVIRONMENTAL DATA COMPANY, INC. (CO)
ENVIRONMENTAL DETECTION SERVICES (CO)
ENVIRONMENTAL ENGINEERING AND TESTING (CO)
ENVIRONMENTAL ENGINEERING LABORATORIES, INC. (CO)
ENVIROSCAPE, LTD. (CO)
ERO RESOURCES CORPORATION (CO)
FAYTEK, INC. (ID)
G.A.M. BUILDING INSPECTION AND RADON MEASUREMENTS AND CONTROL, INC. (CO)
GEO-CHEK, INC. (CO)
GLOBAL WHOLESALE (WY)
GREGORY K. MARSH AND ASSOCIATES (CO)
HANDY ANDY (CO)
HARRIS CONSTRUCTION (CO)
HAZEN RESEARCH, INC. (CO)
HOME BUYERS EDGE (CO)
HOME INSPECTION SERVICES (CO)
HOUSE DETECTIVES, INC. (CO)
HOUSE DOCTORS OF LAKEWOOD, INC. (CO)
HOUSE DOCTORS, INC. (CO)
HOUSE MASTER OF AMERICA (CO)
HUTCHISON PEST CONTROL (CO)
INTERMOUNTAIN RADON SERVICE (CO)
J.A. ASSOCIATES/PUBLIC HEALTH CONSULTANTS (PA)
J.J. LAY COMPANY (CO)
LANDMARK RADON LABORATORIES, LTD. (CO)
LINCOLN DEVORE, INC. (CO)
M.V.E., INC. (CO)
METROPOLITAN REAL ESTATE ANALYSIS (CO)
NATIONAL RADON LABORATORIES/DIVISION OF COLORADO SERVICES CORPORATION (CO)
NATURAL RESOURCE TECHNOLOGIES (CO)
NEBRASKA WESLEYAN UNIVERSITY/PHYSICS DEPARTMENT/RADON MEASUREMENTS LABORATORY (NE)
QUALITY ASSURANCE, INC. (CO)
R.J. MOORE AND ASSOCIATES, INC. (MD)
RA CONSULTANTS, INC. (CO)
RADIATION SURVEILLANCE ASSOCIATES, INC. (UT)
RADIATION TESTING DESIGN (CO)
RADOMET SERVICES (CO)
RADON ANALYSIS, INC. (CO)
RADON CONSULTANTS (NV)
RADON CONSULTANTS, INC. (CO)
RADON DETECTION SYSTEMS/COLORADO SPRINGS, COLORADO BRANCH (CO)
RADON DETECTION SYSTEMS/DURANGO, COLORADO BRANCH (CO)
RADON DETECTION SYSTEMS/GRAND JUNCTION, COLORADO BRANCH (CO)
RADON ENVIRONMENTAL TEST EXPERTS, INC. (CO)
RADON MANAGEMENT COMPANY (VA)
RADON MONITORING SERVICES (CO)
RADON SAFETY SERVICE (CO)
RADON SPECIALISTS (CO)
RADON TEST (CA)
RADON TESTING AND CONSULTING (CT)
RADON TESTING AND TREATMENT, INC. (CO)
RADON/GEOLOGICAL ASSESSMENT SERVICES (CO)
RAMP INDUSTRIES, INC. (CO)
RAMSES II, LTD. (CO)
REAL ESTATE INSPECTOR, THE (CO)
REALITY DYNAMICS (CO)
RMG/PC ENGINEERING, INC. (CO)
ROCKY MOUNTAIN HOME INSPECTIONS (CO)
ROCKY MOUNTAIN RADON TESTING (CO)
RONALD BARTA AND ASSOCIATES (CO)
RTK ENGINEERING (CO)
SPENCE-GEIGER ASSOCIATES, INC. (CO)
STRATA ENVIRONMENTAL SERVICE, INC. (KS)
SUNDANCE ENTERPRISES (CO)
TEK ENGINEERING (CO)
TIMBERLINE RADON SERVICE, INC. (CO)
TSI CONSTRUCTION SERVICES (IN)

Companies Serving
Connecticut

All companies listed under Nationwide provide services for this state.

A-1 SHERLOCK HOME INSPECTORS (NJ)
AAA NEW ENGLAND RADON SERVICES (CT)
AAAIRSCAN ENGINEERING CORPORATION (CT)
ABACUS LABS RADON ASSESSMENT AND MITIGATION (ALRAM) (NH)
ADCO (CT)
AIDCO MAINE CORPORATION (NH)
AIR SENSE, INC. (CT)
ALPHA SERVICES/AIR AND WATER RADON TESTING (CT)
APPLIED ANALYTICAL TECHNOLOGIES, INC. (RI)
ASAFE-HOUSE, INC. (CT)
ATESYS, INC. (CT)
ATLANTIC ENVIRONMENTAL LABORATORY (CT)
AUDUBON CONTROLS, INC. (PA)
BARBARA L. SALTHOUSE (CT)
BOXFORD RADON LABORATORY (MA)
BRADLEY HARRISON ASSOCIATES, INC. (NY)
BROOKS LABORATORIES, INC. (CT)
CANNONDALE HEATING AND AIR CONDITIONING (CT)
CERT WATER SERVICES, INC. (CT)
COMMONWEALTH ENGINEERING AND TECHNOLOGY (PA)
CONNECTICUT BASEMENT DE-WATERING AND RADON SYSTEMS, INC. (CT)
CONNECTICUT ENVIRONMENTAL TESTING (CT)
CONTINENTAL ENVIRONMENT COMPANY, INC. (NJ)
COUNTY HOME INSPECTIONS (CT)
DAMES AND MOORE (MD)
DAVID A. MASE REAL ESTATE (CT)
ECOLOGICAL INTERACTIONS, INC. (NJ)
EMPAC, INC. (CT)
EMPIRE INSPECTION, INC. (NJ)
ENVIRONMENTAL ASSOCIATES (CT)
ENVIRONMENTAL CONSULTANTS ASSOCIATES (NJ)
ENVIRONMENTAL FIELD SERVICES, INC. (ME)
ENVIRONMENTAL HAZARD CONTROL (CT)
ENVIRONMENTAL OIL, INC. (NY)
ENVIRONMENTAL RADIOACTIVITY MEASUREMENTS, INC. (NY)
ENVIRONMENTAL SCIENCE SERVICES (RI)
ENVIRONMENTAL SYSTEMS SERVICE (VA)
ENVIRONMENTAL SYSTEMS SERVICE, LTD. (MD)
ENVIROSCIENCE CONSULTANTS, INC. (CT)
FAYTEK, INC. (ID)
FEDERAL NUCLIDE MONITORS (DC)
FMT EQUIPMENT (CT)
HOME RADON ASSOCIATES (CT)
INSPECT-A-HOME, INC. (NJ)
INSURANCE HEALTH SERVICES (CT)
JOHN A. FIXARY, INC. (CT)
MASSACHUSETTS RADON ALERT, INC. (MA)
MICROBAC ENVIRONMENTAL LABORATORY (NC)
MICROBAC LABORATORIES, INC./ERIE TESTING DIVISION-MAIN OFFICE (PA)
MICROBAC LABORATORIES, INC./HANOVER DIVISION (PA)
MICROBAC LABORATORIES, INC./KENTUCKY TESTING DIVISION (KY)
MICROBAC LABORATORIES, INC./MASSACHUSETTS TESTING DIVISION (MA)
MICROBAC LABORATORIES, INC./NEW CASTLE TESTING DIVISION (PA)
MICROBAC LABORATORIES, INC./PITTSBURGH TESTING DIVISION (PA)
MICROBAC LABORATORIES, INC./SCHILLER DIVISION (PA)
MICROBAC LABORATORIES, INC./SEAWAY DIVISION (IN)
MICROBAC LABORATORIES, INC./STRAND TESTING DIVISION (PA)
MIXLEE ENVIRONMENTAL RESEARCH CORPORATION (CT)
MRC HOME INSPECTIONS (NJ)
NEW ENGLAND RADON SERVICES (CT)
NEW ENGLAND RADON, LTD. (NH)
NORDBERG ASSOCIATES (ME)
O.K. REMS CORPORATION (CT)
PELICAN ENVIRONMENTAL CORPORATION (MA)
PNEUMODYNAMICS, INC. (CT)
PROTECT-A-HOME (CT)
QUALITY AIR RADON REDUCTION (CT)
RAD CHEK SERVICES COMPANY (CT)
RAD CHEK SERVICES COMPANY (NJ)
RADIATION PROTECTION SERVICES, INC. (NJ)
RADIATION SAFETY ASSOCIATES, INC. (CT)
RADON ANALYSIS AND DETECTION OF NEW ENGLAND, INC. (NH)
RADON CHECK/DIVISION OF HORIZON SYSTEMS (MA)
RADON DETECTION SYSTEMS/NEWARK, NEW JERSEY BRANCH (NJ)
RADON DETECTION, INC. (NH)
RADON DETECTORS (NY)
RADON EVALUATION DIVISION/EVALDO, LTD. (CT)
RADON HOME PROTECTION SERVICES (CT)

RADON INSPECTION SERVICE, INC. (NJ)
RADON LABS (MA)
RADON MANAGEMENT LABORATORIES (NJ)
RADON REFEREEING/DIVISION OF HOME QUALITY HEADQUARTERS (CT)
RADON SAFETY TESTING, INC. (CT)
RADON SYSTEMS OF RHODE ISLAND, INC. (RI)
RADON TECHNOLOGIES, INC. (NY)
RADON TECHNOLOGY AND CONSULTANTS (RI)
RADON TESTING AND CONSULTING (CT)
RADON TESTING AND CONSULTING SERVICES (MA)
RADON TESTING LABS (CT)
REMD/RESIDENTIAL ENVIRONMENTAL MEDICAL DIAGNOSTICS (CT)
REMSEARCH, INC. (NY)
RESIDENTIAL INSPECTION TECHNOLOGY/RES-I-TEC (CT)
RESIDENTIAL RADON TEST SERVICE (MA)
RISE (RI)
ROSS SYSTEMS, INC. (CT)
SAF-AIR (CT)
SCOTT HOME INSPECTION, INC. (CT)
SHELBURNE LABORATORIES, INC. (VT)
SHOTWELL ASSOCIATES (NJ)
SOL-AIR, INC. (CT)
SOUTHERN NEW ENGLAND RADON TESTING AND MITIGATION (CT)
STAR LABORATORIES (MA)
SWALE, INC. (CT)
THEROUX ENGINEERING (CT)
VALLEY RADON TESTING, INC. (CT)
WALDON ASSOCIATES (CT)

Companies Serving

Delaware

All companies listed under Nationwide provide services for this state.

A-1 SHERLOCK HOME INSPECTORS (NJ)
AAAIRSCAN ENGINEERING CORPORATION (CT)
ABACUS LABS RADON ASSESSMENT AND MITIGATION (ALRAM) (NH)
ACCREDITED WATERPROOFING COMPANY, INC. (NJ)
ACCU-TEST (VA)
ADVANCED AUTOMATIC ALARM COMPANY, INC. (MD)
AFFILIATED BUILDING INSPECTORS, INC. (MD)
AIDCO MAINE CORPORATION (NH)
AIR QUALITY ANALYSTS, INC. (VA)
AIR VENTILATION SYSTEMS, INC. (MD)
ALPHA CHECK (MD)
ALPHA-BAN (MD)
AMERICAN BUILDING INSPECTORS, INC. (PA)
AMERICAN HERITAGE CORPORATION (PA)
AMERICAN HOMECHECK, INC. (PA)
APPLIED ANALYTICAL TECHNOLOGIES, INC. (RI)
ASSOCIATED RADON TESTING SERVICES, INC. (PA)
AUDUBON CONTROLS, INC. (PA)
BAYSURE BUILDING INSPECTION, INC. (PA)
BIOSPHERICS, INC. (MD)
BRANDYWINE ENVIRONMENTAL SERVICES (DE)
CAMBRIDGE ASSOCIATES, LTD. (NJ)
CARL HENRY AND COMPANY, INC. (PA)
CCI-RADON AND HOME INSPECTION/RADON DETECTION (PA)
COMMONWEALTH ENGINEERING AND TECHNOLOGY (PA)
CONTINENTAL ENVIRONMENT COMPANY, INC. (NJ)
DAMES AND MOORE (MD)
DIAGNOSTICS AND DEVICES, INC. (NJ)
ECOLOGICAL INTERACTIONS, INC. (NJ)
ECONOMY HEATING AND AIR CONDITIONING (MD)
ENVIROHEALTH, INC. (PA)
ENVIRONMENTAL MANAGEMENT GROUP (MD)
ENVIRONMENTAL SYSTEMS SERVICE (VA)
ENVIRONMENTAL SYSTEMS SERVICE, LTD. (MD)
FORCHETTI PSI (PA)
FRANKLIN ENVIRONMENTAL ANALYSIS (PA)
GEOCENTRIC ENGINEERING CORPORATION (VA)
HOME CHECK SERVICES (VA)
HOME EXTERMINATING COMPANY (MD)
HOMERAD (MD)
INDUSTRIAL RESOURCE DEVELOPMENT COMPANY (PA)
INSPECT-A-HOME, INC. (NJ)
J.A. ASSOCIATES/PUBLIC HEALTH CONSULTANTS (PA)
L. CLARK AND SONS, INC. (MA)
LYNN PEST MANAGEMENT COMPANY (MD)
MAYA-TEK, INC. (PA)
MEDLAB ENVIRONMENTAL TESTING, INC. (DE)
MICROBAC LABORATORIES, INC./ERIE TESTING DIVISION-MAIN OFFICE (PA)
MICROBAC LABORATORIES, INC./HANOVER DIVISION (PA)
MICROBAC LABORATORIES, INC./KENTUCKY TESTING DIVISION (KY)
MICROBAC LABORATORIES, INC./MASSACHUSETTS TESTING DIVISION (MA)
MICROBAC LABORATORIES, INC./NEW CASTLE TESTING DIVISION (PA)
MICROBAC LABORATORIES, INC./PITTSBURGH TESTING DIVISION (PA)
MICROBAC LABORATORIES, INC./SCHILLER DIVISION (PA)
MICROBAC LABORATORIES, INC./SEAWAY DIVISION (IN)
MICROBAC LABORATORIES, INC./STRAND TESTING DIVISION (PA)
NORDBERG ASSOCIATES (ME)
PENNSYLVANIA BUILDING INSPECTORS, INC. (PA)
PORTER CONSULTANTS, INC. (PA)
PROTECTRON, INC. (MD)
R.J. MOORE AND ASSOCIATES, INC. (MD)
RADON BAN (MD)
RADON DETECTION SYSTEMS/NEWARK, NEW JERSEY BRANCH (NJ)
RADON DETECTORS (NY)
RADON LABORATORIES (MD)
RADON REMEDIAL SERVICES (PA)
RADON SYSTEMS, INC. (MD)
RADON TESTING SERVICES, PRINCETON (NJ)
RADON TESTING SERVICES, SILVER SPRING (MD)
SAFE SHELTER ENVIRONMENTAL (PA)
SHOTWELL ASSOCIATES (NJ)
SPECTRUM INDUSTRIES (DE)
U.S. ASSETS RECOVERY INVESTIGATIONS, INC. (DC)
WARREN WOLFE ASSOCIATES (DE)
WESTERN TECHNOLOGY, INC. (MD)

Companies Serving

District of Columbia

All companies listed under Nationwide provide services for this state.

AAAIRSCAN ENGINEERING CORPORATION (CT)
ABC INSPECTION COMPANY (MD)
ABIONICS, LTD. (VA)
ACCU-TEST (VA)
ADVANCED AUTOMATIC ALARM COMPANY, INC. (MD)
AFFILIATED BUILDING INSPECTORS, INC. (MD)
AIDCO MAINE CORPORATION (NH)
AIR HAZARDS ASSESSMENT, INC. (MD)
AIR QUALITY ANALYSTS, INC. (VA)
AIR QUALITY DETECTION, INC. (MD)
AIR SOLUTIONS (MD)
AIR VENTILATION SYSTEMS, INC. (MD)
AIRMASTERS RADON DIVISION (MD)
ALL SERVICE RADON REDUCERS, INC. (MD)
ALPHA CHECK (MD)
ALPHA-BAN (MD)
APPLIED ANALYTICAL TECHNOLOGIES, INC. (RI)
BENCHMARK INSPECTION SERVICES (VA)
BIOMEDICAL TOXICOLOGY ASSOCIATES (MD)
BIOSPHERICS, INC. (MD)
CAPITAL HOME INSPECTION CORPORATION (VA)
COMMONWEALTH ENGINEERING AND TECHNOLOGY (PA)
CONTINENTAL ENVIRONMENT COMPANY, INC. (NJ)
DAMES AND MOORE (MD)
DREWRY HOME INSPECTIONS, INC. (VA)
ENVIRONMENTAL MANAGEMENT GROUP (MD)
ENVIRONMENTAL SYSTEMS SERVICE (VA)
ENVIRONMENTAL SYSTEMS SERVICE, LTD. (MD)
ETS ENVIRONMENTAL TESTING SERVICE (MD)
FEDERAL NUCLIDE MONITORS (DC)
FORESIGHT ENGINEERING, INC., VIRGINIA (VA)
GEIER CONSTRUCTION COMPANY, INC./DIVISION OF RADON CONTROL (MD)
GEOCENTRIC ENGINEERING CORPORATION (VA)
HILLMANN ENVIRONMENTAL COMPANY (NJ)
HOLIDAY TERMITE AND PEST CONTROL (VA)
HOME ASSURANCE (IL)
HOME EXTERMINATING COMPANY (MD)
HOMERAD (MD)
HOMESURE INSPECTIONS, INC. (MD)
INFILTEC, D.C. (DC)
LYNN PEST MANAGEMENT COMPANY (MD)
MICROBAC LABORATORIES, INC./ERIE TESTING DIVISION-MAIN OFFICE (PA)
MICROBAC LABORATORIES, INC./HANOVER DIVISION (PA)
MICROBAC LABORATORIES, INC./KENTUCKY TESTING DIVISION (KY)
MICROBAC LABORATORIES, INC./MASSACHUSETTS TESTING DIVISION (MA)
MICROBAC LABORATORIES, INC./NEW CASTLE TESTING DIVISION (PA)
MICROBAC LABORATORIES, INC./PITTSBURGH TESTING DIVISION (PA)
MICROBAC LABORATORIES, INC./SCHILLER DIVISION (PA)
MICROBAC LABORATORIES, INC./SEAWAY DIVISION (IN)
MICROBAC LABORATORIES, INC./STRAND TESTING DIVISION (PA)
NATIONAL RADON SERVICE (MD)
PARAMOUNT PEST CONTROL (VA)
PROTECTRON, INC. (MD)
R.J. MOORE AND ASSOCIATES, INC. (MD)
RADIATION MEASUREMENT SERVICES (MD)
RADON ANALYSIS, INC. (VA)
RADON BAN (MD)
RADON DETECTORS (NY)
RADON HOME INSPECTION SERVICE (MD)
RADON HUNTER, THE (VA)
RADON RANGERS (MD)
RADON SYSTEMS, INC. (MD)
RADON TESTING SERVICES, SILVER SPRING (MD)
RADONICS (VA)
RAY MESSICK RADON TESTING AND MITIGATION (PA)
RTS SYSTEMS (MD)
TRACOR JITCO RADON MEASUREMENT SERVICES (MD)
TSI CONSTRUCTION SERVICES (IN)
U.S. ASSETS RECOVERY INVESTIGATIONS, INC. (DC)
WB CONTRACTORS (DC)
WESTERN TECHNOLOGY, INC. (MD)

Companies Serving
Florida

All companies listed under Nationwide provide services for this state.

AAAIRSCAN ENGINEERING CORPORATION (CT)
AARDEN TESTING AND EVALUATION, INC. (FL)
AIDCO MAINE CORPORATION (NH)
ALL AMERICAN RADON SERVICES, INC. (FL)
ALPHA RADIATION ENGINEERING, INC. (CO)
AMERICAN HOME-TECH, INC. (FL)
AMERICAN MEDICAL MARKETING SERVICES (FL)
AMERICAN RADON TECHNICAL SERVICES, INC. (FL)
APPLIED ANALYTICAL TECHNOLOGIES, INC. (RI)
B & W SALES AND MARKETING (FL)
B & W SALES AND MARKETING CORPORATION (GA)
COMMONWEALTH ENGINEERING AND TECHNOLOGY (PA)
CONTINENTAL ENVIRONMENT COMPANY, INC. (NJ)
DAMES AND MOORE (MD)
DAVID E. WOLFE CONSULTANTS, INC. (FL)
ECOLOGICAL INTERACTIONS, INC. (NJ)
ENVIRONMENTAL AIR CONTROL, INC. (FL)
ENVIRONMENTAL ASPECS, INC. (FL)
ENVIRONMENTAL CONSULTANTS ASSOCIATES (NJ)
ENVIRONMENTAL SAFETY PRODUCTS (GA)
ENVIRONMENTAL SYSTEMS SERVICE (VA)
ENVIRONMENTAL SYSTEMS SERVICE, LTD. (MD)
FAYTEK, INC. (ID)
FEDERAL NUCLIDE MONITORS (DC)
FLORIDA DEPARTMENT OF HEALTH AND REHABILITATIVE SERVICES/DEPUTY SECRETARY FOR HEALTH/OFFICE OF RADIATION CONTROL/ENVIRONMENTAL SECTION (FL)
FLORIDA RADON CONSULTANTS, INC. (FL)
GEOHAZARDS, INC. (FL)
HABITABILITY ENGINEERING CONSULTANTS (HECO) (FL)
HILLMANN ENVIRONMENTAL COMPANY (NJ)
HMA HOME INSPECTION AND WARRANTY SERVICE (FL)
HOMEBUYERS INSPECTION SERVICE OF NORTH WEST FLORIDA, INC. (FL)
HOMEBUYERS INSPECTION SERVICE OF SOUTH WEST FLORIDA (FL)
IMPROVED TECHNOLOGY SYSTEMS (OH)
INDOOR POLLUTION CONTROL CENTER (FL)
J.A. ASSOCIATES/PUBLIC HEALTH CONSULTANTS (PA)
LAW ENVIRONMENTAL, INC. (FL)
MACMILLAN RESEARCH, LTD. (GA)
MERIT ENVIRONMENTAL SERVICES, INC. (KY)
MICROBAC LABORATORIES, INC./ERIE TESTING DIVISION-MAIN OFFICE (PA)
MICROBAC LABORATORIES, INC./HANOVER DIVISION (PA)
MICROBAC LABORATORIES, INC./KENTUCKY TESTING DIVISION (KY)
MICROBAC LABORATORIES, INC./MASSACHUSETTS TESTING DIVISION (MA)
MICROBAC LABORATORIES, INC./NEW CASTLE TESTING DIVISION (PA)
MICROBAC LABORATORIES, INC./PITTSBURGH TESTING DIVISION (PA)
MICROBAC LABORATORIES, INC./SCHILLER DIVISION (PA)
MICROBAC LABORATORIES, INC./SEAWAY DIVISION (IN)
MICROBAC LABORATORIES, INC./STRAND TESTING DIVISION (PA)
NATURAL FLORIDA RETROFIT, INC. (FL)
O'RIORDAN INSULATION COMPANY (FL)
POLK COUNTY HEALTH UNIT/RADIOLOGICAL HEALTH SECTION (FL)
PROFESSIONAL HOME ANALYSIS (FL)
RADON AMERICA (FL)
RADON CALIBRATIONS (VA)
RADON DETECTION SYSTEMS/MIAMI, FLORIDA BRANCH (FL)
RADON ENVIRONMENTAL SERVICES, INC. (KY)
RADON EVALUATION SERVICES, INC. (GA)
RADON MONITOR SYSTEMS (FL)
RADON REDUCTION AND TESTING, INC. (GA)
RADON RESEARCH (FL)
RADON RESEARCH INTERNATIONAL, INC. (FL)
RADON SEARCH-USA (FL)
RADON TESTING GROUP (FL)
RADON TRACKING SERVICE (FL)
RADON-ONE LABORATORY SERVICES (FL)
ROMA RADON (FL)
ROSS RADON TESTING (GA)
SMITH, STANLEY (FL)
SULLIVAN CONSULTANTS, INC. (NJ)
SUN DISTRIBUTING GROUP (FL)
UNIVERSITY OF GEORGIA/CENTER FOR APPLIED ISOTOPE STUDY (GA)
WORLDWIDE RADON, INC. (FL)
WORTH ENGINEERING (FL)

Companies Serving
Georgia

All companies listed under Nationwide provide services for this state.

A 'LARY' SERVICE (GA)
AAAIRSCAN ENGINEERING CORPORATION (CT)
ACME EXTERMINATING COMPANY (GA)
ADVANCE TEX CORPORATION (SC)
AGAPE RADON SERVICES, INC. (GA)
AIDCO MAINE CORPORATION (NH)
APPLIED ANTENNA TECHNOLOGIES/RADON MEASUREMENTS DIVISION (GA)
B & W SALES AND MARKETING CORPORATION (GA)
CAROLINA RADON CONSULTANTS (NC)
CAROLINA RADON CONSULTANTS (SC)
CHAPPARAL ENTERPRISES, INC. (GA)
CHEROKEE DE-WATERING, INC. (GA)
COMMONWEALTH ENGINEERING AND TECHNOLOGY (PA)
COURTNEY CONSULTANTS, INC. (GA)
DAMES AND MOORE (MD)
ENVIRONMENTAL POLLUTION CONTROL (GA)
ENVIRONMENTAL SAFETY PRODUCTS (GA)
ENVIRONMENTAL SYSTEMS SERVICE (VA)
ENVIRONMENTAL SYSTEMS SERVICE, LTD. (MD)
GERALD HARGROVE INSPECTION SERVICE (GA)
HUNT AND ASSOCIATES (GA)
J & J QUALITY HOME INSPECTIONS (GA)
LAW ENVIRONMENTAL, INC. (FL)
MACMILLAN RESEARCH, LTD. (GA)
MERIT ENVIRONMENTAL SERVICES, INC. (KY)
MICROBAC ENVIRONMENTAL LABORATORY (NC)
MICROBAC LABORATORIES, INC./ERIE TESTING DIVISION-MAIN OFFICE (PA)
MICROBAC LABORATORIES, INC./HANOVER DIVISION (PA)
MICROBAC LABORATORIES, INC./KENTUCKY TESTING DIVISION (KY)
MICROBAC LABORATORIES, INC./MASSACHUSETTS TESTING DIVISION (MA)
MICROBAC LABORATORIES, INC./NEW CASTLE TESTING DIVISION (PA)
MICROBAC LABORATORIES, INC./PITTSBURGH TESTING DIVISION (PA)
MICROBAC LABORATORIES, INC./SCHILLER DIVISION (PA)
MICROBAC LABORATORIES, INC./SEAWAY DIVISION (IN)
MICROBAC LABORATORIES, INC./STRAND TESTING DIVISION (PA)
NATURAL FLORIDA RETROFIT, INC. (FL)
PEST DOCTOR (GA)
PROFESSIONAL HOME ANALYSIS (FL)
PROTECTION INSPECTION, INC. (TN)
R.J. MOORE AND ASSOCIATES, INC. (MD)
RADIATION SERVICES, INC. (TN)
RADON DETECTION AND TESTING, INC. (GA)
RADON DETECTION SERVICES, LTD. (GA)
RADON DETECTION SYSTEMS/ATLANTA, GEORGIA BRANCH (GA)
RADON DETECTION SYSTEMS/NASHVILLE, TENNESSEE BRANCH (TN)
RADON ENVIRONMENTAL SERVICES, INC. (KY)
RADON EVALUATION SERVICES, INC. (GA)
RADON REDUCTION AND TESTING, INC. (GA)
RADON TEST AND REDUCTION SPECIALIST (NC)
ROSS RADON TESTING (GA)
SOUTHEASTERN RADON CONSULTANTS, INC. (TN)
STAFFORD, C. FRED (GA)
TISCO (GA)
TRU-KIL SERVICE, INC. (GA)
UNIVERSITY OF GEORGIA/CENTER FOR APPLIED ISOTOPE STUDY (GA)
UNIVERSITY RADON TESTING ASSOCIATES (GA)
WASTE CONTROL INTERNATIONAL, INC. (GA)

Companies Serving
Hawaii

All companies listed under Nationwide provide services for this state.

AAAIRSCAN ENGINEERING CORPORATION (CT)
J.A. ASSOCIATES/PUBLIC HEALTH CONSULTANTS (PA)

Companies Serving
Idaho

All companies listed under Nationwide provide services for this state.

AMERICAN INDEPENDENT LABS (ID)
BAR-TEL INFRASCAN (WA)
C.R. ADAMS COMPANY, INC. (CO)
CAVALIER CORPORATION (WA)
E.C. CHAMBERLAIN, RADIOLOGICAL CONSULTANT (NV)
ENERGY LABORATORIES, INC. (WY)
FAYTEK, INC. (ID)
RADIATION HEALTH (WA)
RADON AWARENESS, INC. (ID)
RADON CONSULTANTS (NV)
RADON DETECTION SYSTEMS/BUTTE, MONTANA BRANCH (MT)
RADON DETECTION SYSTEMS/LOS ANGELES, CALIFORNIA BRANCH (CA)
RADON DETECTION SYSTEMS/SALT LAKE CITY, UTAH BRANCH (UT)
RADON DETECTION SYSTEMS/SPOKANE, WASHINGTON BRANCH (ID)
RADON GAS DETECTION SYSTEMS (OH)
RADON TEST (CA)
RADON TESTING SERVICES, INC. (CA)
SANI-DUCT SERVICE OF THE INLAND NORTHWEST (WA)
SOUTHERN OHIO RADON PROTECTION COMPANY (OH)
TIMBERLINE RADON SERVICE, INC. (CO)
WASHINGTON STATE UNIVERSITY/NUCLEAR RADIATION CENTER (WA)

RADON TESTING OF MICHIANA (IN)
RESIDENTIAL RADON MEASUREMENTS (IL)
TEE ENGINEERING COMPANY (KY)
TMW MARKETING (IL)
TSI CONSTRUCTION SERVICES (IN)

Companies Serving
Illinois

All companies listed under Nationwide provide services for this state.

AA RADCO (MI)
ACCURAD TESTING SERVICE (IL)
ACCURATE HOME SERVICE (IL)
AIDCO MAINE CORPORATION (NH)
BCA LABORATORIES, INC. (IL)
CARLSON HEATING, COOLING, AND ELECTRIC COMPANY (IL)
COMMONWEALTH ENGINEERING AND TECHNOLOGY (PA)
DAMES AND MOORE (MD)
ELLIOTT AND ASSOCIATES (IL)
ENCOSYS CORPORATION OF AMERICA (IN)
ENERGY INDEPENDENCE/THE RADON RESPONSE TEAM (MO)
ENERGY MARKETING MIDWEST, INC. (IA)
ENVIRONMENTAL ENGINEERING AND TESTING (CO)
ENVIRONMENTAL RISK CONSULTANTS, INC. (IL)
ENVIRONMENTAL SERVICES OF ILLINOIS, INC. (IL)
ENVIRONMENTAL SOLUTIONS, INC. (MO)
ENVIRONMENTAL SYSTEMS SERVICE (VA)
ENVIRONMENTAL SYSTEMS SERVICE, LTD. (MD)
ENVIROTECH SERVICES (IL)
EXPEDITERS, INC., THE (WI)
HOME ASSURANCE (IL)
HOME ENVIRONMENT COMPANY (IN)
HOMEAIR TESTERS (IN)
HOMEBUYERS INSPECTION SERVICE, INC. (IL)
ILLINOIS DEPARTMENT OF NUCLEAR SAFETY/OFFICE OF ENVIRONMENTAL SAFETY (IL)
ILLINOIS STATE GEOLOGICAL SURVEY (IL)
INSPECT (IL)
MERIT ENVIRONMENTAL SERVICES, INC. (KY)
MICROBAC ENVIRONMENTAL LABORATORY (NC)
MICROBAC LABORATORIES, INC./ERIE TESTING DIVISION-MAIN OFFICE (PA)
MICROBAC LABORATORIES, INC./HANOVER DIVISION (PA)
MICROBAC LABORATORIES, INC./KENTUCKY TESTING DIVISION (KY)
MICROBAC LABORATORIES, INC./MASSACHUSETTS TESTING DIVISION (MA)
MICROBAC LABORATORIES, INC./NEW CASTLE TESTING DIVISION (PA)
MICROBAC LABORATORIES, INC./PITTSBURGH TESTING DIVISION (PA)
MICROBAC LABORATORIES, INC./SCHILLER DIVISION (PA)
MICROBAC LABORATORIES, INC./SEAWAY DIVISION (IN)
MICROBAC LABORATORIES, INC./STRAND TESTING DIVISION (PA)
MIDWEST HEALTH PROGRAMS, INC. (IL)
MOBILE EMISSIONS TESTING CORPORATION (IL)
PROFESSIONAL RADON MEASUREMENTS (IL)
RAD ALERT, LTD. (IL)
RADON CONSULTANTS, INC. (CO)
RADON DETECTION SPECIALISTS (IL)
RADON DETECTION SYSTEMS/BLOOMINGTON-NORMAL BRANCH (IL)
RADON DETECTION SYSTEMS/CHICAGO, ILLINOIS BRANCH (IL)
RADON MANAGEMENT COMPANY (VA)
RADON SURVEYS (IL)
RADON SURVEYS/DIVISION OF COPE TECHNICAL SERVICES (WI)
RADON TESTING AND MITIGATION SERVICES COMPANY (IA)
RADON TESTING AND MITIGATION SYSTEMS (IN)
RADON TESTING LAB (IL)

Companies Serving
Indiana

All companies listed under Nationwide provide services for this state.

AA RADCO (MI)
ACCUTEK LABORATORY (OH)
ACE PEST CONTROL (IN)
AERTEC LABORATORIES, INC. (IN)
BCA LABORATORIES, INC. (IL)
COMMONWEALTH ENGINEERING AND TECHNOLOGY (PA)
DAMES AND MOORE (MD)
ELLIOTT AND ASSOCIATES (IL)
ENCOSYS CORPORATION OF AMERICA (IN)
ENVIRONMENTAL ENGINEERING AND TESTING (CO)
ENVIRONMENTAL SOLUTIONS, INC. (MO)
ENVIRONMENTAL SYSTEMS SERVICE (VA)
ENVIRONMENTAL SYSTEMS SERVICE, LTD. (MD)
FINNEX (RADON) (OH)
HERRICK ENGINEERING, INC. (OH)
HOME ANALYSTS, INC. (OH)
HOME ENVIRONMENT COMPANY (IN)
HOME INSPECTION OF KENTUCKIANA (KY)
HOMEAIR TESTERS (IN)
HOMEBUYERS INSPECTION SERVICE, INC. (IL)
ILLINOIS STATE GEOLOGICAL SURVEY (IL)
INSUL-TECH, INC. (OH)
KALKA MIDWEST CONSTRUCTION, INC. (IN)
MERIT ENVIRONMENTAL SERVICES, INC. (KY)
MICROBAC ENVIRONMENTAL LABORATORY (NC)
MICROBAC LABORATORIES, INC./ERIE TESTING DIVISION-MAIN OFFICE (PA)
MICROBAC LABORATORIES, INC./HANOVER DIVISION (PA)
MICROBAC LABORATORIES, INC./KENTUCKY TESTING DIVISION (KY)
MICROBAC LABORATORIES, INC./MASSACHUSETTS TESTING DIVISION (MA)
MICROBAC LABORATORIES, INC./NEW CASTLE TESTING DIVISION (PA)
MICROBAC LABORATORIES, INC./PITTSBURGH TESTING DIVISION (PA)
MICROBAC LABORATORIES, INC./SCHILLER DIVISION (PA)
MICROBAC LABORATORIES, INC./SEAWAY DIVISION (IN)
MICROBAC LABORATORIES, INC./STRAND TESTING DIVISION (PA)
MOBILE RADON DIAGNOSTICS, INC. (OH)
NUCLEAR MEASUREMENTS CORPORATION (IN)
RAD-TEST, INC. (KY)
RADON DETECTION SPECIALISTS (IL)
RADON DETECTION SYSTEMS/BLOOMINGTON-NORMAL BRANCH (IL)
RADON DETECTION SYSTEMS/CHICAGO, ILLINOIS BRANCH (IL)
RADON ENVIRONMENTAL SERVICES, INC. (KY)
RADON MANAGEMENT COMPANY (VA)
RADON SURVEYS (IL)
RADON SURVEYS/DIVISION OF COPE TECHNICAL SERVICES (WI)
RADON TECHNOLOGIES, INC. (OH)
RADON TESTING AND MITIGATION SYSTEMS (IN)
RADON TESTING OF MICHIANA (IN)
RESTORED ENVIRONMENTS COMPANY, INC. (OH)
TEE ENGINEERING COMPANY (KY)
THERMO-SCAN, INC. (IN)
TSI CONSTRUCTION SERVICES (IN)

Companies Serving
Iowa

All companies listed under Nationwide provide services for this state.

AIDCO MAINE CORPORATION (NH)
BCA LABORATORIES, INC. (IL)
BET, INC. (IA)
BILL EICH CONSTRUCTION COMPANY (IA)
COMMONWEALTH ENGINEERING AND TECHNOLOGY (PA)
CORNELL COLLEGE (IA)
EES OF MINNESOTA, INC. (MN)
EMMI-RADON CHECK LABORATORY (IA)
ENERGY MARKETING MIDWEST, INC. (IA)
ENVIRONMENTAL SOLUTIONS, INC. (MO)
ENVIRONMENTAL SYSTEMS SERVICE (VA)
ENVIRONMENTAL SYSTEMS SERVICE, LTD. (MD)
FRAZIER COMPANY, THE (NE)
GEO-CHEK, INC. (CO)
ILLINOIS STATE GEOLOGICAL SURVEY (IL)
MACK, E.F. (IA)
MAHONEY HEATING AND COOLING (NE)
MECHANICAL EQUIPMENT SERVICES CORPORATION (IA)
MERIT ENVIRONMENTAL SERVICES, INC. (KY)
MICROBAC ENVIRONMENTAL LABORATORY (NC)
MICROBAC LABORATORIES, INC./HANOVER DIVISION (PA)
MICROBAC LABORATORIES, INC./MASSACHUSETTS TESTING DIVISION (MA)
MICROBAC LABORATORIES, INC./NEW CASTLE TESTING DIVISION (PA)
MICROBAC LABORATORIES, INC./PITTSBURGH TESTING DIVISION (PA)
MICROBAC LABORATORIES, INC./SCHILLER DIVISION (PA)
MICROBAC LABORATORIES, INC./SEAWAY DIVISION (IN)
NATIONAL PROPERTY INSPECTIONS (NE)
NEBRASKA WESLEYAN UNIVERSITY/PHYSICS DEPARTMENT/RADON MEASUREMENTS LABORATORY (NE)
NORTHWEST FURNACE COMPANY, INC. (IA)
PROFESSIONAL HOUSE DOCTORS (IA)
RADON AND ENVIRONMENTAL PROFESSIONALS (KS)
RADON BUSTERS (ND)
RADON CONTROL SERVICES CORPORATION (IA)
RADON DETECTION SYSTEMS/BLOOMINGTON-NORMAL BRANCH (IL)
RADON DETECTION SYSTEMS/DES MOINES, IOWA BRANCH (IA)
RADON ENVIRONMENTAL SERVICES, INC. (KY)
RADON MEASURE COMPANY, THE (MN)
RADON REDUCTION SERVICES (MO)
RADON SURVEYS/DIVISION OF COPE TECHNICAL SERVICES (WI)
RADON TESTING AND MITIGATION SERVICES COMPANY (IA)
STRATA ENVIRONMENTAL SERVICE, INC. (KS)
TSI CONSTRUCTION SERVICES (IN)
VINCE SCHAEFER CONSTRUCTION (IA)

Companies Serving
Kansas

All companies listed under Nationwide provide services for this state.

ALPHA RADIATION ENGINEERING, INC. (CO)
B & I SERVICES (KS)
BONDAR-CLEGG, INC. (CO)
C.R. ADAMS COMPANY, INC. (CO)
EGAL, INC. (KS)
ENERGY INDEPENDENCE/THE RADON RESPONSE TEAM (MO)

ENVIRONMENTAL ENGINEERING AND TESTING (CO)
ENVIRONMENTAL SOLUTIONS, INC. (MO)
GUNTER EXTERMINATING COMPANY (MO)
HAWKS INTER-STATE (KS)
HOME CHEK, INC. (KS)
KANSAS DEPARTMENT OF HEALTH AND ENVIRONMENT/BUREAU OF AIR QUALITY AND RADIATION CONTROL/ENVIRONMENTAL SURVEILLANCE AND EMERGENCY PLANNING SECTION (KS)
KANSAS TESTING SERVICES (KS)
LANDMARK RADON LABORATORIES, LTD. (CO)
LINCOLN DEVORE, INC. (CO)
MIDWEST RADON INSPECTION, INC. (KS)
NATIONAL PROPERTY INSPECTIONS (MO)
NEBRASKA WESLEYAN UNIVERSITY/PHYSICS DEPARTMENT/RADON MEASUREMENTS LABORATORY (NE)
RADON AND ENVIRONMENTAL PROFESSIONALS (KS)
RADON AND ENVIRONMENTAL PROFESSIONALS (KS)
RADON AND ENVIRONMENTAL PROFESSIONALS (KS)
RADON EVALUATION SERVICES, INC. (GA)
RADON REDUCTION SERVICES (MO)
RADON RESEARCH FOUNDATION, INC. (KS)
STRATA ENVIRONMENTAL SERVICE, INC. (KS)
TSI CONSTRUCTION SERVICES (IN)

Companies Serving
Kentucky

All companies listed under Nationwide provide services for this state.

ACCUTEK LABORATORY (OH)
AIDCO MAINE CORPORATION (NH)
ALPHA SPECTRA OF KENTUCKY (KY)
ANALYTICAL MANAGEMENT, INC. (KY)
BLUEGRASS RADON SERVICES (KY)
COMMONWEALTH ENGINEERING AND TECHNOLOGY (PA)
DAMES AND MOORE (MD)
ENCOSYS CORPORATION OF AMERICA (IN)
ENERGY SYSTEMS, INC. (TN)
ENVIRONMENTAL MANAGEMENT CONTROL, INC. (KY)
ENVIRONMENTAL MANAGEMENT GROUP (MD)
ENVIRONMENTAL SOLUTIONS, INC. (MO)
ENVIRONMENTAL SYSTEMS SERVICE (VA)
ENVIRONMENTAL SYSTEMS SERVICE, LTD. (MD)
ENVIROTECH (KY)
FINNEX (RADON) (OH)
FRANKS, TERRY M. (KY)
HOME ANALYSTS, INC. (OH)
HOME ENVIRONMENT COMPANY (IN)
HOME INSPECTION OF KENTUCKIANA (KY)
HOMEAIR TESTERS (IN)
HOUSING CONSULTANTS, INC. (KY)
HUNT'S WATER CONTROL, INC. (KY)
INSUL-TECH, INC. (OH)
J.A. ASSOCIATES/PUBLIC HEALTH CONSULTANTS (PA)
J.M.C. ENTERPRISES (TN)
MARTIN MARIETTA ENERGY SYSTEMS (KY)
MERIT ENVIRONMENTAL SERVICES, INC. (KY)
METRO SERVICE LABORATORIES, INC. (KY)
MICROBAC ENVIRONMENTAL LABORATORY (NC)
MICROBAC LABORATORIES, INC./ERIE TESTING DIVISION-MAIN OFFICE (PA)
MICROBAC LABORATORIES, INC./HANOVER DIVISION (PA)
MICROBAC LABORATORIES, INC./KENTUCKY TESTING DIVISION (KY)
MICROBAC LABORATORIES, INC./MASSACHUSETTS TESTING DIVISION (MA)
MICROBAC LABORATORIES, INC./NEW CASTLE TESTING DIVISION (PA)
MICROBAC LABORATORIES, INC./PITTSBURGH TESTING DIVISION (PA)
MICROBAC LABORATORIES, INC./SCHILLER DIVISION (PA)
MICROBAC LABORATORIES, INC./SEAWAY DIVISION (IN)
MICROBAC LABORATORIES, INC./STRAND TESTING DIVISION (PA)
MOBILE RADON DIAGNOSTICS, INC. (OH)
PRE-PURCHASE HOME INSPECTION (KY)
PROTECTION INSPECTION, INC. (TN)
RAD-TEST, INC. (KY)
RADIATION SERVICES, INC. (TN)
RADON DETECTION SYSTEMS/NASHVILLE, TENNESSEE BRANCH (TN)
RADON ENVIRONMENTAL SERVICES, INC. (KY)
RADON EVALUATION SERVICES, INC. (GA)
RADON GAS DETECTION SYSTEMS (OH)
RADON MANAGEMENT COMPANY (VA)
RADON MANAGEMENT OF KENTUCKY (KY)
RADON PROTECTION, INC. (KY)
RADON TECHNOLOGIES, INC. (OH)
RADON TEST AND REDUCTION SPECIALIST (NC)
RADON TESTING AND MITIGATION SYSTEMS (IN)
RESTORED ENVIRONMENTS COMPANY, INC. (OH)
SOUTHEASTERN RADON CONSULTANTS, INC. (TN)
SOUTHERN OHIO RADON PROTECTION COMPANY (OH)
TEE ENGINEERING COMPANY (KY)
TSI CONSTRUCTION SERVICES (IN)

Companies Serving
Louisiana

All companies listed under Nationwide provide services for this state.

ANTI-PEST COMPANY OF SHREVEPORT (LA)
DAMES AND MOORE (MD)
LAW ENVIRONMENTAL, INC. (FL)
MICROBAC LABORATORIES, INC./ERIE TESTING DIVISION-MAIN OFFICE (PA)
MICROBAC LABORATORIES, INC./KENTUCKY TESTING DIVISION (KY)
MICROBAC LABORATORIES, INC./STRAND TESTING DIVISION (PA)
NATURAL FLORIDA RETROFIT, INC. (FL)
R.T.C. (NY)
RADON AND ENVIRONMENTAL PROFESSIONALS (KS)
RADON DETECTION SYSTEMS/DALLAS, TEXAS BRANCH (TX)
RADON EVALUATION SERVICES, INC. (GA)
TULANE UNIVERSITY MEDICAL CENTER/OCCUPATIONAL HEALTH AND SAFETY SERVICES (LA)

Companies Serving
Maine

All companies listed under Nationwide provide services for this state.

AAAIRSCAN ENGINEERING CORPORATION (CT)
ABACUS LABS RADON ASSESSMENT AND MITIGATION (ALRAM) (NH)
AERO-AQUA (ME)
AIDCO MAINE CORPORATION (NH)
AIR SENSE, INC. (CT)
ALPHA RADON ENVIRONMENTAL TESTING (ME)
ALPHA SERVICES/AIR AND WATER RADON TESTING (CT)
APPLIED ANALYTICAL TECHNOLOGIES, INC. (RI)
BOXFORD RADON LABORATORY (MA)
BUILDING INSPECTORS, INC. (ME)
BURNS AND ASSOCIATES (NH)
CAMPTON DEVELOPMENT CORPORATION (NH)
COMMONWEALTH ENGINEERING AND TECHNOLOGY (PA)
DAMES AND MOORE (MD)
EMPAC, INC. (CT)
ENVIRONMENTAL FIELD SERVICES, INC. (ME)
ENVIRONMENTAL SCIENCE SERVICES (RI)
ENVIRONMENTAL SYSTEMS SERVICE (VA)
ENVIRONMENTAL SYSTEMS SERVICE, LTD. (MD)
ENVIROSCIENCE CONSULTANTS, INC. (CT)
FEDERAL NUCLIDE MONITORS (DC)
GRANITE STATE ANALYTICAL, INC. (NH)
GRANITE STATE RADON SERVICES (NH)
HARRIMAN ASSOCIATES (ME)
J.A. ASSOCIATES/PUBLIC HEALTH CONSULTANTS (PA)
JRM ASSOCIATES (NH)
L. CLARK AND SONS, INC. (MA)
MICROBAC ENVIRONMENTAL LABORATORY (NC)
MICROBAC LABORATORIES, INC./ERIE TESTING DIVISION-MAIN OFFICE (PA)
MICROBAC LABORATORIES, INC./HANOVER DIVISION (PA)
MICROBAC LABORATORIES, INC./KENTUCKY TESTING DIVISION (KY)
MICROBAC LABORATORIES, INC./MASSACHUSETTS TESTING DIVISION (MA)
MICROBAC LABORATORIES, INC./NEW CASTLE TESTING DIVISION (PA)
MICROBAC LABORATORIES, INC./PITTSBURGH TESTING DIVISION (PA)
MICROBAC LABORATORIES, INC./SCHILLER DIVISION (PA)
MICROBAC LABORATORIES, INC./SEAWAY DIVISION (IN)
MICROBAC LABORATORIES, INC./STRAND TESTING DIVISION (PA)
MIXLEE ENVIRONMENTAL RESEARCH CORPORATION (CT)
NEW ENGLAND RADON SERVICES (CT)
NEW ENGLAND RADON, LTD. (NH)
NORDBERG ASSOCIATES (ME)
RADON ANALYSIS AND DETECTION OF NEW ENGLAND, INC. (NH)
RADON CHECK/DIVISION OF HORIZON SYSTEMS (MA)
RADON DETECTION DEVICES (MD)
RADON DETECTION, INC. (NH)
RADON DETECTORS (NY)
RADON EVALUATION DIVISION/EVALDO, LTD. (CT)
RADON LABS (MA)
RADON REGISTRY (NH)
RADON RESOURCES, INC. (NH)
RADON SYSTEMS OF RHODE ISLAND, INC. (RI)
RADON TECHNOLOGY AND CONSULTANTS (RI)
RADON TECHNOLOGY, INC. (NH)
RADON TESTING AND CONSULTING SERVICES (MA)
RESIDENTIAL RADON TEST SERVICE (MA)
RISE (RI)
SHELBURNE LABORATORIES, INC. (VT)
STAR LABORATORIES (MA)

Companies Serving
Maryland

All companies listed under Nationwide provide services for this state.

AAAIRSCAN ENGINEERING CORPORATION (CT)
ABC INSPECTION COMPANY (MD)
ABIONICS, LTD. (VA)
ACCU-TEST (VA)
ADVANCED AUTOMATIC ALARM COMPANY, INC. (MD)
AFFILIATED BUILDING INSPECTORS, INC. (MD)

AIDCO MAINE CORPORATION (NH)
AIR HAZARDS ASSESSMENT, INC. (MD)
AIR QUALITY ANALYSTS, INC. (VA)
AIR QUALITY DETECTION, INC. (MD)
AIR SOLUTIONS (MD)
AIR VENTILATION SYSTEMS, INC. (MD)
AIRMASTERS RADON DIVISION (MD)
ALL SERVICE RADON REDUCERS, INC. (MD)
ALPHA CHECK (MD)
ALPHA-BAN (MD)
AMERICAN BUILDING INSPECTORS, INC. (PA)
APPLIED ANALYTICAL TECHNOLOGIES, INC. (RI)
ASSOCIATED RADON TESTING SERVICES, INC. (PA)
BALTIMORE GAS AND ELECTRIC (MD)
BAYSURE BUILDING INSPECTION, INC. (PA)
BENCHMARK INSPECTION SERVICES (VA)
BERG GROUP (MD)
BIOMEDICAL TOXICOLOGY ASSOCIATES (MD)
BIOSPHERICS, INC. (MD)
CAPITAL HOME INSPECTION CORPORATION (VA)
CARROLL COUNTY RADON CONTROL, INC. (MD)
COMMONWEALTH ENGINEERING AND TECHNOLOGY (PA)
CONTINENTAL ENVIRONMENT COMPANY, INC. (NJ)
DAMES AND MOORE (MD)
DAN HIMMELFARB COMPANY (MD)
DIAGNOSTICS AND DEVICES, INC. (NJ)
ECM ENVIRONMENTAL CONSULTANTS (MD)
ECOLOGICAL INTERACTIONS, INC. (NJ)
ECONOMY HEATING AND AIR CONDITIONING (MD)
ENVIRONMENTAL CONTROL SYSTEMS, LTD. (MD)
ENVIRONMENTAL ENGINEERING AND TESTING (CO)
ENVIRONMENTAL HOME SERVICES, INC. (MD)
ENVIRONMENTAL MANAGEMENT GROUP (MD)
ENVIRONMENTAL RADIOACTIVITY MEASUREMENTS, INC. (NY)
ENVIRONMENTAL SYSTEMS SERVICE (VA)
ENVIRONMENTAL SYSTEMS SERVICE, LTD. (MD)
ENVIRONMENTAL TESTING (MD)
ETS ENVIRONMENTAL TESTING SERVICE (MD)
FEDERAL NUCLIDE MONITORS (DC)
FORCHETTI PSI (PA)
FORESIGHT ENGINEERING, INC., VIRGINIA (VA)
GARCO RESEARCH COMPANY (NJ)
GEIER CONSTRUCTION COMPANY, INC./DIVISION OF RADON CONTROL (MD)
GEO-ENVIRONMENTAL SERVICES, INC. (MD)
GEOCENTRIC ENGINEERING CORPORATION (VA)
H & H DECORATING (MD)
HEALTH PHYSICS ASSOCIATES, INC. (PA)
HEALTH PHYSICS SERVICES, INC. (MD)
HILLMANN ENVIRONMENTAL COMPANY (NJ)
HOLIDAY TERMITE AND PEST CONTROL (VA)
HOME ASSURANCE (IL)
HOME EXTERMINATING COMPANY (MD)
HOME-TECH (MD)
HOMERAD (MD)
HOMESURE INSPECTIONS, INC. (MD)
HOUSE CHECK, INC./MARYLAND (MD)
INFILTEC, D.C. (DC)
J. BECK AND ASSOCIATES, INC. (PA)
J.A. ASSOCIATES/PUBLIC HEALTH CONSULTANTS (PA)
JB TESTING COMPANY (MD)
LEONARDS HOME INSPECTIONS (MD)
LYNN PEST MANAGEMENT COMPANY (MD)
MEDLAB ENVIRONMENTAL TESTING, INC. (DE)
MICROBAC ENVIRONMENTAL LABORATORY (NC)
MICROBAC LABORATORIES, INC./ERIE TESTING DIVISION-MAIN OFFICE (PA)
MICROBAC LABORATORIES, INC./HANOVER DIVISION (PA)
MICROBAC LABORATORIES, INC./KENTUCKY TESTING DIVISION (KY)
MICROBAC LABORATORIES, INC./MASSACHUSETTS TESTING DIVISION (MA)
MICROBAC LABORATORIES, INC./NEW CASTLE TESTING DIVISION (PA)
MICROBAC LABORATORIES, INC./PITTSBURGH TESTING DIVISION (PA)
MICROBAC LABORATORIES, INC./SCHILLER DIVISION (PA)
MICROBAC LABORATORIES, INC./SEAWAY DIVISION (IN)
MICROBAC LABORATORIES, INC./STRAND TESTING DIVISION (PA)
MOSS, ROBERT (MD)
NATIONAL RADON SERVICE (MD)
NUCLEON LECTERN ASSOCIATES, INC. (MD)
PARAMOUNT PEST CONTROL (VA)
PENNSYLVANIA BUILDING INSPECTORS, INC. (PA)
PRIGEL RADON TESTING (PA)
PRO-GUARD BUILDING INSPECTIONS, INC. (PA)
PROTECTRON, INC. (MD)
R.J. MOORE AND ASSOCIATES, INC. (MD)
RAD CON, INC. (MD)
RADIATION MEASUREMENT SERVICES (MD)
RADON ALERT (MD)
RADON ANALYSIS, INC. (VA)
RADON BAN (MD)
RADON CALIBRATIONS (VA)
RADON CONSULTANTS, INC., MARYLAND (MD)
RADON DETECTION DEVICES (MD)
RADON HOME INSPECTION SERVICE (MD)
RADON HUNTER, THE (VA)
RADON LABORATORIES (MD)
RADON MANAGEMENT COMPANY (VA)
RADON RANGERS (MD)
RADON REDUCTION SERVICES, INC. (MD)
RADON REPORTING SYSTEM OF CARROLL COUNTY (MD)
RADON RESOURCE, INC. (MD)
RADON SYSTEMS, INC. (MD)
RADON TEST CENTER (MD)
RADON TESTING LABORATORIES, INC. (MD)
RADON TESTING SERVICES, SILVER SPRING (MD)
RADONICS (VA)
RAY MESSICK RADON TESTING AND MITIGATION (PA)
RCI SERVICES (MD)
RPI RADON MEASUREMENT SERVICES, MARYLAND (MD)
RTS SYSTEMS (MD)
SHOTWELL ASSOCIATES (NJ)
SMALL UNITED LABORATORIES (MD)
TRACOR JITCO RADON MEASUREMENT SERVICES (MD)
U.S. ASSETS RECOVERY INVESTIGATIONS, INC. (DC)
U.S. RADON TESTING/REDUCTION OF MARYLAND (MD)
UNIVERSITY OF MARYLAND/RADIATION SAFETY OFFICE (MD)
WESTERN TECHNOLOGY, INC. (MD)

Companies Serving

Massachusetts

All companies listed under Nationwide provide services for this state.

A-1 SHERLOCK HOME INSPECTORS (NJ)
AAAIRSCAN ENGINEERING CORPORATION (CT)
ABACUS LABS RADON ASSESSMENT AND MITIGATION (ALRAM) (NH)
ADVANCED HOME INSPECTION AND TESTING (MA)
ADVANTAGE HOME INSPECTION (MA)
AIDCO MAINE CORPORATION (NH)
AIR SENSE, INC. (CT)
ALPHA RADON ENVIRONMENTAL TESTING (ME)
ALPHA SERVICES/AIR AND WATER RADON TESTING (CT)
AMERICAN HOME INSPECTION SERVICE (MA)
AMERICAN INSULATION, INC. (MA)
APPLIED ANALYTICAL TECHNOLOGIES, INC. (RI)
AREA HOME INSPECTIONS, INC. (MA)
ASAFE-HOUSE, INC. (CT)
ASTLE CORPORATION (MA)
ATLANTIC ENVIRONMENTAL LABORATORY (CT)
BAIN PEST CONTROL SERVICE, INC. (MA)
BIOSOURCES, LTD. (MA)
BIOSOURCES, LTD. (NH)
BOXFORD RADON LABORATORY (MA)
BRADLEY HARRISON ASSOCIATES, INC. (NY)
BUILDING INSPECTOR OF AMERICA, THE (MA)
BURNS AND ASSOCIATES (NH)
C-P LABORATORY (MA)
CAMPTON DEVELOPMENT CORPORATION (NH)
CAPE COD HOME INSPECTIONS (MA)
CARLSON INSPECTION ASSOCIATES, INC. (MA)
CLARK UNIVERSITY/ENVIRONMENT, TECHNOLOGY, AND SOCIETY PROGRAM/DEPARTMENT OF PHYSICS (MA)
CLEAN HARBORS ANALYTICAL SERVICES, INC. (MA)
COMMONWEALTH ENGINEERING AND TECHNOLOGY (PA)
COMMUNITY HEALTH ASSOCIATES, INC. (MA)
CONNECTICUT ENVIRONMENTAL TESTING (CT)
DAMES AND MOORE (MD)
DR. RADON TESTING AND REDUCTION (MA)
EMPAC, INC. (CT)
ENVIRONMENTAL FIELD SERVICES, INC. (ME)
ENVIRONMENTAL OIL, INC. (NY)
ENVIRONMENTAL SCIENCE SERVICES (RI)
ENVIRONMENTAL SYSTEMS SERVICE (VA)
ENVIRONMENTAL SYSTEMS SERVICE, LTD. (MD)
ENVIROSCIENCE CONSULTANTS, INC. (CT)
FEDERAL NUCLIDE MONITORS (DC)
FINN CONSULTANT ENGINEERS, INC. (MA)
FORESIGHT ENGINEERING, INC., VIRGINIA (VA)
GRANITE STATE RADON SERVICES (NH)
HOME INSPECTOR OF NORTH AMERICA, THE (MA)
J.A. ASSOCIATES/PUBLIC HEALTH CONSULTANTS (PA)
J.M. FROHN CORPORATION (MA)
JRM ASSOCIATES (NH)
L. CLARK AND SONS, INC. (MA)
MASSACHUSETTS RADON ALERT, INC. (MA)
MICROBAC ENVIRONMENTAL LABORATORY (NC)
MICROBAC LABORATORIES, INC./ERIE TESTING DIVISION-MAIN OFFICE (PA)
MICROBAC LABORATORIES, INC./HANOVER DIVISION (PA)
MICROBAC LABORATORIES, INC./KENTUCKY TESTING DIVISION (KY)
MICROBAC LABORATORIES, INC./MASSACHUSETTS TESTING DIVISION (MA)
MICROBAC LABORATORIES, INC./NEW CASTLE TESTING DIVISION (PA)
MICROBAC LABORATORIES, INC./PITTSBURGH TESTING DIVISION (PA)
MICROBAC LABORATORIES, INC./SCHILLER DIVISION (PA)
MICROBAC LABORATORIES, INC./SEAWAY DIVISION (IN)
MICROBAC LABORATORIES, INC./STRAND TESTING DIVISION (PA)
MINUTEMAN HOME INSPECTORS, INC. (MA)
MIXLEE ENVIRONMENTAL RESEARCH CORPORATION (CT)
NATIONAL ENVIRONMENTAL ENGINEERING AND DIAGNOSTICS (MA)
NEW ENGLAND RADON SERVICES (CT)
NEW ENGLAND RADON, LTD. (NH)
OLD COLONY PEST CONTROL COMPANY (MA)
PELICAN ENVIRONMENTAL CORPORATION (MA)
PROTECT-A-HOME (CT)
RAD CHEK SERVICES COMPANY (CT)
RAD CHEK SERVICES COMPANY (NJ)
RADEX (MA)

RADIATION SAFETY ASSOCIATES, INC. (CT)
RADIOLOGICAL CONTROL ASSOCIATES (MA)
RADON ANALYSIS AND DETECTION OF NEW ENGLAND, INC. (NH)
RADON CHECK/DIVISION OF HORIZON SYSTEMS (MA)
RADON DETECTION DEVICES (MD)
RADON DETECTION SYSTEMS/BOSTON, MASSACHUSETTS BRANCH (MA)
RADON DETECTION SYSTEMS/NEWARK, NEW JERSEY BRANCH (NJ)
RADON DETECTION, INC. (NH)
RADON DETECTORS (NY)
RADON EVALUATION DIVISION/EVALDO, LTD. (CT)
RADON INFORMATION AND TESTING SERVICE (NH)
RADON LABS (MA)
RADON REGISTRY (NH)
RADON RESOURCES, INC. (NH)
RADON SYSTEMS OF RHODE ISLAND, INC. (RI)
RADON TECHNOLOGIES, INC. (NY)
RADON TECHNOLOGY AND CONSULTANTS (RI)
RADON TECHNOLOGY, INC. (NH)
RADON TEST COMPANY (MA)
RADON TESTING AND CONSULTING SERVICES (MA)
REMD/RESIDENTIAL ENVIRONMENTAL MEDICAL DIAGNOSTICS (CT)
RESIDENTIAL RADON TEST SERVICE (MA)
RISE (RI)
SACCHETTI, R. MICHAEL (MA)
SEACOAST LABORATORIES (MA)
SHELBURNE LABORATORIES, INC. (VT)
STAR LABORATORIES (MA)

Companies Serving
Michigan

All companies listed under Nationwide provide services for this state.

AA RADCO (MI)
ADVANCED AIR ANALYSIS (MI)
AIDCO MAINE CORPORATION (NH)
AIR-TIGHT OF FARMINGTON (MI)
AIR-TIGHT OF GENESSEE (MI)
ALL TIME PEST CONTROL, INC. (MI)
COMMONWEALTH ENGINEERING AND TECHNOLOGY (PA)
DAMES AND MOORE (MD)
DONALD WALKER COMPANY (MI)
ELLIOTT AND ASSOCIATES (IL)
ENCOSYS CORPORATION OF AMERICA (IN)
ENVIRONMENTAL SYSTEMS SERVICE (VA)
ENVIRONMENTAL SYSTEMS SERVICE, LTD. (MD)
EXPEDITERS, INC., THE (WI)
HERRICK ENGINEERING, INC. (OH)
HOME INSPECTIONS OF MICHIGAN (MI)
INSUL-TECH, INC. (OH)
MICROBAC ENVIRONMENTAL LABORATORY (NC)
MICROBAC LABORATORIES, INC./HANOVER DIVISION (PA)
MICROBAC LABORATORIES, INC./MASSACHUSETTS TESTING DIVISION (MA)
MICROBAC LABORATORIES, INC./NEW CASTLE TESTING DIVISION (PA)
MICROBAC LABORATORIES, INC./PITTSBURGH TESTING DIVISION (PA)
MICROBAC LABORATORIES, INC./SCHILLER DIVISION (PA)
MICROBAC LABORATORIES, INC./SEAWAY DIVISION (IN)
ORR-DUVALL AND ASSOCIATES (MI)
RADON DETECTION CENTER OF MICHIGAN (MI)
RADON DETECTION CORPORATION OF MICHIGAN (MI)
RADON DETECTION SYSTEMS/BLOOMINGTON-NORMAL BRANCH (IL)
RADON DETECTION SYSTEMS/DETROIT, MICHIGAN BRANCH (MI)
RADON FREE (OH)
RADON SAFE (MI)
RADON SURVEYS (IL)
RADON SURVEYS/DIVISION OF COPE TECHNICAL SERVICES (WI)
RADON TECHNOLOGIES, INC. (OH)
RADON TECHNOLOGY, INC. (MI)
RADON TEST AND REDUCTION SPECIALIST (NC)
RADON TESTING AND REDUCTION (MI)
RADON TESTING OF MICHIANA (IN)
TSI CONSTRUCTION SERVICES (IN)

Companies Serving
Minnesota

All companies listed under Nationwide provide services for this state.

AA RADCO (MI)
AAAIRSCAN ENGINEERING CORPORATION (CT)
DAMES AND MOORE (MD)
EES OF MINNESOTA, INC. (MN)
ENERGY INDEPENDENCE/THE RADON RESPONSE TEAM (MO)
ENERGY MARKETING MIDWEST, INC. (IA)
ENVIRONMENTAL TECHNOLOGIES, INC. (MN)
EXPEDITERS, INC., THE (WI)
GOY, TERRY (MN)
NEBRASKA WESLEYAN UNIVERSITY/PHYSICS DEPARTMENT/RADON MEASUREMENTS LABORATORY (NE)
PEPPADYNE RADON DETECTION SERVICE (MN)
RADON BUSTERS (ND)
RADON DETECTION SYSTEMS/BLOOMINGTON-NORMAL BRANCH (IL)
RADON DETECTION SYSTEMS/FARGO, NORTH DAKOTA BRANCH (ND)
RADON DETECTION SYSTEMS/LOS ANGELES, CALIFORNIA BRANCH (CA)
RADON DETECTION SYSTEMS/MINNEAPOLIS-ST. PAUL BRANCH (MN)
RADON ENVIRONMENTAL SERVICES, INC. (KY)
RADON FREE (OH)
RADON MEASURE COMPANY, THE (MN)
RADON SURVEYS/DIVISION OF COPE TECHNICAL SERVICES (WI)

Companies Serving
Mississippi

All companies listed under Nationwide provide services for this state.

AA RADCO (MI)
DAMES AND MOORE (MD)
ELLIOTT AND ASSOCIATES (IL)
HOMEBUYERS INSPECTION SERVICE OF NORTH WEST FLORIDA, INC. (FL)
MACMILLAN RESEARCH, LTD. (GA)
PMI CONTRACTORS, INC. (MS)
RADON DETECTION SYSTEMS/NASHVILLE, TENNESSEE BRANCH (TN)
RADON EVALUATION SERVICES, INC. (GA)
TSI CONSTRUCTION SERVICES (IN)

Companies Serving
Missouri

All companies listed under Nationwide provide services for this state.

ACI CHEMICAL COMPANY, INC. (AR)
BCA LABORATORIES, INC. (IL)
CAPE CHEMICAL COMPANY (MO)
EGAL, INC. (KS)
ELLIOTT AND ASSOCIATES (IL)
ENERGY INDEPENDENCE/THE RADON RESPONSE TEAM (MO)
ENVIRONMENTAL ENGINEERING AND TESTING (CO)
ENVIRONMENTAL SERVICES OF ILLINOIS, INC. (IL)
ENVIRONMENTAL SOLUTIONS, INC. (MO)
GEO-CHEK, INC. (CO)
GUNTER EXTERMINATING COMPANY (MO)
MERIT ENVIRONMENTAL SERVICES, INC. (KY)
MIDWEST RADON INSPECTION, INC. (KS)
MISSOURI TESTING LABORATORIES, INC. (MO)
NATIONAL PROPERTY INSPECTIONS (MO)
RADON AND ENVIRONMENTAL PROFESSIONALS (KS)
RADON AND ENVIRONMENTAL PROFESSIONALS (KS)
RADON DETECTION SERVICE OF KANSAS CITY (MO)
RADON DETECTION SYSTEMS/BLOOMINGTON-NORMAL BRANCH (IL)
RADON DETECTION SYSTEMS/NASHVILLE, TENNESSEE BRANCH (TN)
RADON ENVIRONMENTAL SERVICES, INC. (KY)
RADON REDUCTION SERVICES (MO)
RADON RESEARCH FOUNDATION, INC. (KS)
RADON SERVICES OF MID-AMERICA (MO)
RADON TECHNOLOGY (MO)
RESIDENTIAL RADON MEASUREMENTS (IL)
STRATA ENVIRONMENTAL SERVICE, INC. (KS)
TEE ENGINEERING COMPANY (KY)
TSI CONSTRUCTION SERVICES (IN)

Companies Serving
Montana

All companies listed under Nationwide provide services for this state.

ALPHA RADIATION ENGINEERING, INC. (CO)
ARIX CORPORATION (WY)
BAR-TEL INFRASCAN (WA)
C.R. ADAMS COMPANY, INC. (CO)
CAVALIER CORPORATION (WA)
ENERGY LABORATORIES, INC. (WY)
ENVIRONMENTAL ENGINEERING AND TESTING (CO)
ENVIRONMENTAL ENGINEERING LABORATORIES, INC. (CO)
FAYTEK, INC. (ID)
RADIATION HEALTH (WA)
RADON AWARENESS, INC. (ID)
RADON BUSTERS (ND)
RADON CONSULTANTS (NV)
RADON CONSULTANTS, INC. (CO)
RADON DETECTION SERVICE OF MONTANA (MT)
RADON DETECTION SYSTEMS/BUTTE, MONTANA BRANCH (MT)
RADON DETECTION SYSTEMS/CASPER, WYOMING BRANCH (WY)
RADON DETECTION SYSTEMS/LOS ANGELES, CALIFORNIA BRANCH (CA)
RADON TEST (CA)
RADON TESTING AND CONTROL (ND)
RTK ENGINEERING (CO)
SANI-DUCT SERVICE OF THE INLAND NORTHWEST (WA)
TIMBERLINE RADON SERVICE, INC. (CO)

Companies Serving
Nebraska

All companies listed under Nationwide provide services for this state.

AAAIRSCAN ENGINEERING CORPORATION (CT)

ALPHA RADIATION ENGINEERING, INC. (CO)
BONDAR-CLEGG, INC. (CO)
C.R. ADAMS COMPANY, INC. (CO)
ENERGY INDEPENDENCE/THE RADON RESPONSE TEAM (MO)
ENERGY LABORATORIES, INC. (WY)
ENERGY MARKETING MIDWEST, INC. (IA)
ENVIRONMENTAL SOLUTIONS, INC. (MO)
FRAZIER COMPANY, THE (NE)
LANDMARK RADON LABORATORIES, LTD. (CO)
M.V.E., INC. (CO)
MAHONEY HEATING AND COOLING (NE)
NATIONAL PROPERTY INSPECTIONS (NE)
NEBRASKA WESLEYAN UNIVERSITY/PHYSICS DEPARTMENT/RADON MEASUREMENTS LABORATORY (NE)
R.T.C. (NY)
RADIOLOGICAL TESTING SERVICES (NE)
RADON AND ENVIRONMENTAL PROFESSIONALS (KS)
RADON CONSULTANTS (NV)
RADON DETECTION SERVICES OF NEBRASKA (NE)
RADON ENVIRONMENTAL SERVICES, INC. (KY)
SHANNON HEATING, COOLING AND ELECTRIC (NE)
SPADY, GARY D. (NE)
STRATA ENVIRONMENTAL SERVICE, INC. (KS)
SULLIVAN CONSULTANTS, INC. (NJ)

Companies Serving

Nevada

All companies listed under Nationwide provide services for this state.

AAAIRSCAN ENGINEERING CORPORATION (CT)
ALPHA RADIATION ENGINEERING, INC. (CO)
BONDAR-CLEGG, INC. (CO)
C.R. ADAMS COMPANY, INC. (CO)
E.C. CHAMBERLAIN, RADIOLOGICAL CONSULTANT (NV)
ENERGY ALTERNATIVES UNLIMITED, INC. (CO)
ENERGY LABORATORIES, INC. (WY)
ENVIRONMENTAL ENGINEERING AND TESTING (CO)
ENVIRONMENTAL ENGINEERING LABORATORIES, INC. (CO)
FAYTEK, INC. (ID)
HOUSEMASTERS OF AMERICA (NV)
PACIFIC RADIATION CORPORATION (CA)
RADON AWARENESS, INC. (ID)
RADON CONSULTANTS (NV)
RADON REDUCTION SERVICES (MO)
RADON SYSTEMATICS OF CALIFORNIA (CA)
RADON TEST (CA)
RADON TESTING SERVICES, INC. (CA)
SANI-DUCT SERVICE OF THE INLAND NORTHWEST (WA)
SPENCE-GEIGER ASSOCIATES, INC. (CO)

Companies Serving

New Hampshire

All companies listed under Nationwide provide services for this state.

A & N RADON COMPANY (NH)
AAAIRSCAN ENGINEERING CORPORATION (CT)
ABACUS LABS RADON ASSESSMENT AND MITIGATION (ALRAM) (NH)
ADVANCED HOME INSPECTION AND TESTING (MA)
AIDCO MAINE CORPORATION (NH)
AIR SENSE, INC. (CT)
ALPHA RADON ENVIRONMENTAL TESTING (ME)
ALPHA SERVICES/AIR AND WATER RADON TESTING (CT)
APPLIED ANALYTICAL TECHNOLOGIES, INC. (RI)
ATLANTIC ENVIRONMENTAL LABORATORY (CT)
BIOSOURCES, LTD. (NH)
BOXFORD RADON LABORATORY (MA)
BURNS AND ASSOCIATES (NH)
CAMPTON DEVELOPMENT CORPORATION (NH)
COMMONWEALTH ENGINEERING AND TECHNOLOGY (PA)
COMPREHENSIVE RADON SERVICES, INC. (NH)
CONNECTICUT ENVIRONMENTAL TESTING (CT)
DAMES AND MOORE (MD)
EMPAC, INC. (CT)
ENVIRONMENTAL CONTROL TECHNOLOGIES (NH)
ENVIRONMENTAL FIELD SERVICES, INC. (ME)
ENVIRONMENTAL SCIENCE SERVICES (RI)
ENVIRONMENTAL SYSTEMS SERVICE (VA)
ENVIRONMENTAL SYSTEMS SERVICE, LTD. (MD)
ENVIROSCIENCE CONSULTANTS, INC. (CT)
FEDERAL NUCLIDE MONITORS (DC)
GODBEER, GODBEER AND ASSOCIATES (NH)
GRANITE STATE RADON SERVICES (NH)
JRM ASSOCIATES (NH)
L. CLARK AND SONS, INC. (MA)
MASSACHUSETTS RADON ALERT, INC. (MA)
MICROBAC ENVIRONMENTAL LABORATORY (NC)
MICROBAC LABORATORIES, INC./ERIE TESTING DIVISION-MAIN OFFICE (PA)
MICROBAC LABORATORIES, INC./HANOVER DIVISION (PA)
MICROBAC LABORATORIES, INC./KENTUCKY TESTING DIVISION (KY)
MICROBAC LABORATORIES, INC./MASSACHUSETTS TESTING DIVISION (MA)
MICROBAC LABORATORIES, INC./NEW CASTLE TESTING DIVISION (PA)
MICROBAC LABORATORIES, INC./PITTSBURGH TESTING DIVISION (PA)
MICROBAC LABORATORIES, INC./SCHILLER DIVISION (PA)
MICROBAC LABORATORIES, INC./SEAWAY DIVISION (IN)
MICROBAC LABORATORIES, INC./STRAND TESTING DIVISION (PA)
MIXLEE ENVIRONMENTAL RESEARCH CORPORATION (CT)
NEW ENGLAND RADON SERVICES (CT)
NEW ENGLAND RADON, LTD. (NH)
NORDBERG ASSOCIATES (ME)
NORTHEAST RADON CONSULTANTS (NH)
PELICAN ENVIRONMENTAL CORPORATION (MA)
PHOENIX LABS (NH)
RADON ANALYSIS AND DETECTION OF NEW ENGLAND, INC. (NH)
RADON CHECK/DIVISION OF HORIZON SYSTEMS (MA)
RADON DETECTION DEVICES (MD)
RADON DETECTION SERVICES (VT)
RADON DETECTION SYSTEMS/NEWARK, NEW JERSEY BRANCH (NJ)
RADON DETECTION, INC. (NH)
RADON DETECTORS (NY)
RADON EVALUATION DIVISION/EVALDO, LTD. (CT)
RADON INFORMATION AND TESTING SERVICE (NH)
RADON LABS (MA)
RADON REGISTRY (NH)
RADON RESOURCES, INC. (NH)
RADON SYSTEMS OF RHODE ISLAND, INC. (RI)
RADON TECHNOLOGY AND CONSULTANTS (RI)
RADON TECHNOLOGY, INC. (NH)
RADON TESTING AND CONSULTING SERVICES (MA)
RESIDENTIAL RADON TEST SERVICE (MA)
RISE (RI)
SAFE AIR COMPANY (NH)
SHELBURNE LABORATORIES, INC. (VT)
SHEVENELL, GALLEN, AND ASSOCIATES (NH)
STAR LABORATORIES (MA)

Companies Serving

New Jersey

All companies listed under Nationwide provide services for this state.

A-1 SHERLOCK HOME INSPECTORS (NJ)
A.B.E. RADON TESTING, INC. (PA)
A.D.T. INSPECTIONS, INC. (NJ)
A.J. PERRI, INC. (NJ)
A.L. BOUCHARD COMPANY (NJ)
AA HOME INSPECTIONS (NJ)
AAA RADON TESTING (NJ)
AAAIRSCAN ENGINEERING CORPORATION (CT)
ACCREDITED WATERPROOFING COMPANY, INC. (NJ)
ACCU-TEST (VA)
ACCURATE HOME INSPECTION, INC. (NJ)
ACS CONSULTANTS, INC. (NJ)
AFFILIATED BUILDING INSPECTORS, INC. (MD)
AGENTERPRISE, INC. (PA)
AIDCO MAINE CORPORATION (NH)
AIR QUALITY ANALYSTS, INC. (VA)
AIR QUALITY CONTROL, INC. (PA)
AIR RADON SERVICE, INC. (NJ)
AIRMASTER RADON DETECTION, INC. (NJ)
ALL EASTERN HOME INSPECTION COMPANY (NJ)
ALLIED BIOLOGICAL (NJ)
ALPHA CHECK (MD)
ALPHA SERVICES/AIR AND WATER RADON TESTING (CT)
AMBIC BUILDING INSPECTION CONSULTANTS, INC. (NJ)
AMC HOME INSPECTION SERVICES (NJ)
AMERICAN BUILDING INSPECTORS, INC. (PA)
AMERICAN HERITAGE CORPORATION (PA)
AMERICAN HI-TECH (NJ)
AMERICAN HOMECHECK, INC. (PA)
AMERICAN RADON TECHNICAL SERVICES, INC. (FL)
AMERICAN REAL ESTATE INSPECTIONS (NJ)
APPLIED ANALYTICAL TECHNOLOGIES, INC. (RI)
ASSOCIATED RADON TESTING SERVICES, INC. (PA)
ATLANTIC ELECTRIC (NJ)
ATLANTIC ENVIRONMENTAL LABORATORY (CT)
AUDUBON CONTROLS, INC. (PA)
AUGUSTUS W. SMITH, INC. (NJ)
BCT (NJ)
BERGEN HOME INSPECTION SERVICE (NJ)
BERNELL PROFESSIONAL SERVICES (NJ)
BIOSPHERICS, INC. (MD)
BUILDING INSPECTOR OF AMERICA, THE (NJ)
BUILDING INSPECTOR OF AMERICA, THE (NJ)
BULAVA ENVIRONMENTAL, INC. (NJ)
C & D ASSOCIATES HOME INSPECTIONS (NJ)
CALOIERO RADON TECHNOLOGIES (NJ)
CAMBRIDGE ASSOCIATES, LTD. (NJ)
CAPITAL MATERIALS TESTING, INC. (NY)
CARL HENRY AND COMPANY, INC. (PA)
CCI-RADON AND HOME INSPECTION/RADON DETECTION (PA)
CENTER FOR RADON DETECTION (NJ)
CENTER FOR RADON SERVICES (NJ)
CENTRAL HOME INSPECTION, INC. (NJ)
CENTRAL JERSEY RADON, INC. (NJ)
CENTRAL RADON TESTING (NJ)
CHIEF HOME AND BUILDING INSPECTION SERVICES (NJ)
CHILTON ASSOCIATES (RADON GROUP) (NJ)
COMMONWEALTH ENGINEERING AND TECHNOLOGY (PA)
CONTINENTAL ENVIRONMENT COMPANY, INC. (NJ)
COUNTY COLLEGE OF MORRIS/CHEMISTRY DEPARTMENT (NJ)

DAMES AND MOORE (MD)
DEAN ARRINGTON ASSOCIATES (NJ)
DELAWARE VALLEY RADON INSPECTIONS/ARSENAL BUSINESS CENTER (PA)
DEVELOPMENT SERVICES, LTD. (PA)
DIAGNOSTICS AND DEVICES, INC. (NJ)
DITTMAR INSPECTION COMPANY, INC. (NJ)
DK ROCKWELL (NJ)
DMC ASSOCIATES (NJ)
DPR BUILDERS (NJ)
EASTERN INTEGRATED SERVICES CORPORATION (NJ)
ECOLOGICAL INTERACTIONS, INC. (NJ)
EMERGENCY RADON TESTING, INC. (NJ)
EMPAC, INC. (CT)
EMPIRE INSPECTION, INC. (NJ)
ENVIRO TESTING (NY)
ENVIROHEALTH, INC. (PA)
ENVIRONMENTAL CONSULTANTS ASSOCIATES (NJ)
ENVIRONMENTAL ENGINEERING AND TESTING (CO)
ENVIRONMENTAL FIELD SERVICES, INC. (ME)
ENVIRONMENTAL MANAGEMENT GROUP (MD)
ENVIRONMENTAL OIL, INC. (NY)
ENVIRONMENTAL PRO-TECH SERVICES, INC. (NJ)
ENVIRONMENTAL RADIOACTIVITY MEASUREMENTS, INC. (NY)
ENVIRONMENTAL SYSTEMS SERVICE (VA)
ENVIRONMENTAL SYSTEMS SERVICE, LTD. (MD)
ENVIRONMENTAL TESTING AND CONSULTING (NJ)
ENVIROSCIENCE CONSULTANTS, INC. (CT)
ERNEST W. BLOCHINGER, INC. (NJ)
EXCEL INSPECTION SERVICE, INC. (NJ)
EXECUTIVE HOME CONSULTANTS (NJ)
FAYTEK, INC. (ID)
FEDERAL NUCLIDE MONITORS (DC)
FORCHETTI PSI (PA)
FORESIGHT ENGINEERING (NJ)
FORESIGHT ENGINEERING, INC. (NJ)
FORESIGHT ENGINEERING, INC. (NJ)
FORESIGHT ENGINEERING, INC. (NJ)
FRANKLIN ENVIRONMENTAL ANALYSIS (PA)
G.S. TECHNOLOGIES, INC. (NJ)
GARCO RESEARCH COMPANY (NJ)
GARDEN STATE RADON DEWATERING SERVICES, INC. (NJ)
GARNER CONSULTANTS INSPECTION SERVICE (NJ)
GASOMETRY ASSOCIATES (NJ)
GENUS CORPORATION, THE (NJ)
GEOCENTRIC ENGINEERING CORPORATION (VA)
GLOBAL BUILDING INSPECTORS (NJ)
GS ANALYTICS, INC. (NJ)
H & J FREILE HOME INSPECTIONS, INC. (NJ)
H.P.J. INSPECTION SERVICE, INC. (NJ)
HEIMER INSPECTIONS OF NEW JERSEY, INC. (NJ)
HILLMANN ENVIRONMENTAL COMPANY (NJ)
HOLMES INSPECTION SERVICE (NJ)
HOME INSPECTION ONE (NJ)
HOME INSPECTIONS BY CHANEL, INC. (NJ)
HOMERAD (MD)
HOMESAFE RADON ENGINEERING (NJ)
HOMESTEAD INSPECTION SERVICE (NJ)
HOUSE CHEK (NJ)
INDEPENDENT HOME INSPECTION SERVICE, INC. (NJ)
INDUSTRIAL RESOURCE DEVELOPMENT COMPANY (PA)
INFOTECH HOME INSPECTION SERVICES (NJ)
INSPECT-A-HOME, INC. (NJ)
INTEGRATED PEST MANAGEMENT (NJ)
INTERNATIONAL PEST CONTROL, INC. (NJ)
J. BECK AND ASSOCIATES, INC. (PA)
J.A. ASSOCIATES/PUBLIC HEALTH CONSULTANTS (PA)
JBK INSPECTION SERVICE, INC. (NJ)
JP HOME INSPECTION, INC. (NJ)
L. CLARK AND SONS, INC. (MA)
LAMSON HOME INSPECTIONS (NJ)
LANIDEX CORPORATION (NJ)
LEAFE, EDWARD (NJ)
LEWIS AND KAHLSON RADON CONSULTANTS (NJ)
LEWIS HOME INSPECTION (NJ)
MARTIN J. EGAN ASSOCIATES (NJ)
MAYA-TEK, INC. (PA)
MEDLAB ENVIRONMENTAL TESTING, INC. (DE)
MICROBAC ENVIRONMENTAL LABORATORY (NC)
MICROBAC LABORATORIES, INC./ERIE TESTING DIVISION-MAIN OFFICE (PA)
MICROBAC LABORATORIES, INC./HANOVER DIVISION (PA)
MICROBAC LABORATORIES, INC./MASSACHUSETTS TESTING DIVISION (MA)
MICROBAC LABORATORIES, INC./NEW CASTLE TESTING DIVISION (PA)
MICROBAC LABORATORIES, INC./PITTSBURGH TESTING DIVISION (PA)
MICROBAC LABORATORIES, INC./SCHILLER DIVISION (PA)
MICROBAC LABORATORIES, INC./SEAWAY DIVISION (IN)
MICROBAC LABORATORIES, INC./STRAND TESTING DIVISION (PA)
MRC HOME INSPECTIONS (NJ)
NATIONAL RADON CONTROL, INC. (NJ)
NEW JERSEY DEPARTMENT OF ENVIRONMENTAL PROTECTION/DIVISION OF ENVIRONMENTAL QUALITY/BUREAU OF ENVIRONMENTAL RADIATION (NJ)
NEW JERSEY DEPARTMENT OF ENVIRONMENTAL PROTECTION/DIVISION OF ENVIRONMENTAL QUALITY/BUREAU OF RADIATION AND INORGANIC ANALYTICAL SERVICES (NJ)
NEW JERSEY DEPARTMENT OF ENVIRONMENTAL PROTECTION/DIVISION OF SCIENCE AND RESEARCH (NJ)
NEW JERSEY HOME SERVICE, INC. (NJ)
NORTH EAST ENVIRONMENTAL SERVICES (NJ)
NORTHEAST VECTOR MANAGEMENT, INC. (NJ)
NORTHEASTERN RADON SERVICE (NJ)
O.K. REMS COMPANY (NJ)
PAGANO ASSSOCIATES, INC. (NJ)
PENNSYLVANIA BUILDING INSPECTORS, INC. (PA)
PORTER CONSULTANTS, INC. (PA)
PRINCETON HOME INSPECTION SERVICE (NJ)
PRINCETON PLASMA PHYSICS LABORATORY (NJ)
PRO-TEC HOME INSPECTORS, INC. (NJ)
PROFESSIONAL HOME INSPECTIONS, INC. (NJ)
PROTECT-A-HOME (CT)
QUALITY HOME INSPECTIONS, INC. (NJ)
QUANTEC (NJ)
R.I.C. INSPECTION AND ENGINEERING (NJ)
R.J. MOORE AND ASSOCIATES, INC. (MD)
R/K OCCUPATIONAL AND ENVIRONMENTAL ANALYSIS, INC. (NJ)
RAD CHEK SERVICES COMPANY (CT)
RAD CHEK SERVICES COMPANY (NJ)
RAD ENVIRONMENTAL SERVICES, INC. (NJ)
RADIATION DATA (NJ)
RADIATION PROTECTION SERVICES, INC. (NJ)
RADON ANALYSIS, INC. (NJ)
RADON ASSOCIATES, INC. (NJ)
RADON CHECK/DIVISION OF HORIZON SYSTEMS (MA)
RADON CONTROL, INC. (NJ)
RADON DETECTION SYSTEMS/NEWARK, NEW JERSEY BRANCH (NJ)
RADON DETECTORS (NY)
RADON ENVIRONMENTAL CONSTRUCTION SERVICES/RECONS, INC., MAIN OFFICE (NJ)
RADON HOME SEARCH (NJ)
RADON INSPECTION SERVICE, INC. (NJ)
RADON MANAGEMENT COMPANY (VA)
RADON MANAGEMENT LABORATORIES (NJ)
RADON MANAGEMENT LABORATORIES (NJ)
RADON MEASURING SYSTEMS (PA)
RADON REMEDIAL SERVICES (PA)
RADON SYSTEMS OF RHODE ISLAND, INC. (RI)
RADON TECHNOLOGIES, INC. (NY)
RADON TESTING LABORATORY OF NEW JERSEY (NJ)
RADON TESTING SERVICES, CHERRY HILL (NJ)
RADON TESTING SERVICES, PRINCETON (NJ)
RADON TESTING SERVICES, SILVER SPRING (MD)
RADON-X CORPORATION (NJ)
RADON-X CORPORATION (NJ)
RADONAIRE (NJ)
RAM EXTERMINATORS (NJ)
REDIHOME INSPECTIONS, INC. (NJ)
REMSEARCH, INC. (NY)
RIC INSPECTION AND ENGINEERING (NJ)
S & D CUSTOM ESTATES (NY)
SAFE AND SOUND HOME INSPECTION CONTRACTORS (NJ)
SAFE SHELTER ENVIRONMENTAL (PA)
SCHWABE, WALTER A. (NJ)
SHELTERWORKS, INC. (NJ)
SHERLOCK HOME INSPECTORS (NJ)
SHIELD HOME INSPECTION SERVICE (NJ)
SHOTWELL ASSOCIATES (NJ)
SILVER SPRINGS HOME INSPECTION (NJ)
SPARTAN HOME INSPECTION (NJ)
SPARTAN HOME INSPECTION, INC. (NJ)
SPECIALIZED SERVICES, INC. (NJ)
STAR LABORATORIES (MA)
STONE RIDGE RADON MITIGATORS (NJ)
SULLIVAN CONSULTANTS, INC. (NJ)
TECTONIC INSPECTION COMPANY (NJ)
THOROUGH CHECK, INC. (NJ)
TOTAL HOME TECH (PA)
TRADON ENVIRONMENTAL SERVICES (NJ)
TRI-COUNTY RADON, INC. (NJ)
TRI-STATE RADON AND DEWATERING (NJ)
TRI-STATE RADON REDUCTION, INC. (NJ)
TRUESDALE-HAWK INSPECTION CONSULTANTS, INC. (NJ)
TSI CONSTRUCTION SERVICES (IN)
U.S. ASSETS RECOVERY INVESTIGATIONS, INC. (DC)
UNION COUNTY BUILDING INSPECTION SERVICE (NJ)
UNIVERSAL HOME INSPECTION, INC. (NY)
WEST ENVIRO SYSTEMS, INC. (NJ)
WORLDWIDE RADON, INC. (FL)
WPB ENTERPRISES, INC. (PA)

Companies Serving

New Mexico

All companies listed under Nationwide provide services for this state.

ALPHA RADIATION ENGINEERING, INC. (CO)
BONDAR-CLEGG, INC. (CO)
C.R. ADAMS COMPANY, INC. (CO)
COLORADO FRONT RANGE BUILDING INSPECTION SERVICE (CO)
ENERGY LABORATORIES, INC. (WY)
ENVIRONMENTAL ENGINEERING LABORATORIES, INC. (CO)
FINNEX (RADON) (OH)
LANDMARK RADON LABORATORIES, LTD. (CO)
LINCOLN DEVORE, INC. (CO)
NATURAL RESOURCE TECHNOLOGIES (CO)
NEW MEXICO DEPARTMENT OF HEALTH AND ENVIRONMENT/COMMUNITY SERVICES BUREAU (NM)
RADON CONSULTANTS (NV)
RADON DETECTION SYSTEMS/ALBUQUERQUE, NEW MEXICO BRANCH (NM)
RADON DETECTION SYSTEMS/ARIZONA BRANCH (AZ)

RADON DETECTION SYSTEMS/LOS ANGELES, CALIFORNIA BRANCH (CA)
RADON DETECTION SYSTEMS/SALT LAKE CITY, UTAH BRANCH (UT)
RADON TEST (CA)
RTK ENGINEERING (CO)
SPENCE-GEIGER ASSOCIATES, INC. (CO)
WILLIAM G. CHANEY, PE (NM)

Companies Serving

New York

All companies listed under Nationwide provide services for this state.

A & A CONSULTING AND INSPECTION SERVICES, INC. (NY)
A-1 SHERLOCK HOME INSPECTORS (NJ)
AAAIRSCAN ENGINEERING CORPORATION (CT)
ACCORD CORPORATION (NY)
ACRES INTERNATIONAL CORPORATION (NY)
ACS CONSULTANTS, INC. (NJ)
ADIRONDACK ENVIRONMENTAL INVESTIGATIONS (NY)
AFFTREX, LTD. (PA)
AFI ENVIRONMENTAL/DIVISION OF BUFFALO CONSULTANTS, INC. (NY)
AIDCO MAINE CORPORATION (NH)
AIR ANALYSIS ASSOCIATES (NY)
AIR ANALYSIS ASSOCIATES (NY)
ALL EASTERN HOME INSPECTION COMPANY (NJ)
ALPHA RADON ENVIRONMENTAL TESTING (ME)
ALPHA SERVICES/AIR AND WATER RADON TESTING (CT)
AMERICAN HERITAGE CORPORATION (PA)
AMERICAN HOME SERVICE COMPANY (NY)
APPLIED ANALYTICAL TECHNOLOGIES, INC. (RI)
ASAFE-HOUSE, INC. (CT)
ASSOCIATION FOR IMPROVEMENT IN MOHEGAN, INC. (NY)
ASTECO, INC. (NY)
ATLANTIC ENVIRONMENTAL LABORATORY (CT)
AUDUBON CONTROLS, INC. (PA)
BIOHAZARDOUS MATERIALS CONSULTING (NY)
BIOSPHERICS, INC. (MD)
BOXFORD RADON LABORATORY (MA)
BRADLEY HARRISON ASSOCIATES, INC. (NY)
BRAMA CONSTRUCTION CORPORATION (NY)
CAPITAL CONSTRUCTION SERVICES (NY)
CAPITAL MATERIALS TESTING, INC. (NY)
CHIEF HOME AND BUILDING INSPECTION SERVICES (NJ)
CHILTON ASSOCIATES (RADON GROUP) (NJ)
COMMONWEALTH ENGINEERING AND TECHNOLOGY (PA)
CONNECTICUT ENVIRONMENTAL TESTING (CT)
CONSTRUCTION EVALUATION, INC. (NY)
CONTINENTAL ENVIRONMENT COMPANY, INC. (NJ)
CORNERSTONES (NY)
DAMES AND MOORE (MD)
DEGENSHEIN DENKER (NY)
DIAGNOSTICS AND DEVICES, INC. (NJ)
DPR BUILDERS (NJ)
DUDLEY, PAUL (NY)
E-THREE, INC. (NY)
ECOLOGICAL INTERACTIONS, INC. (NJ)
EMPAC, INC. (CT)
EMPIRE INSPECTION, INC. (NJ)
EN-THERM, INC. (NY)
ENERGY CONSERVATION SERVICES (NY)
ENGINEERING ASSOCIATES (NY)
ENSA, INC. (NY)
ENVIRO TESTING (NY)
ENVIRO-CARE RADON SERVICES (NY)
ENVIRONMENTAL CONSULTANTS ASSOCIATES (NJ)
ENVIRONMENTAL CONTROL SYSTEMS, LTD. (MD)
ENVIRONMENTAL FIELD SERVICES, INC. (ME)
ENVIRONMENTAL MANAGEMENT GROUP (MD)
ENVIRONMENTAL OIL, INC. (NY)
ENVIRONMENTAL RADIOACTIVITY MEASUREMENTS, INC. (NY)
ENVIRONMENTAL SYSTEMS SERVICE (VA)
ENVIRONMENTAL SYSTEMS SERVICE, LTD. (MD)
ENVIROSCIENCE CONSULTANTS, INC. (CT)
FAYTEK, INC. (ID)
FEDERAL NUCLIDE MONITORS (DC)
FORCHETTI PSI (PA)
GALSON TECHNICAL SERVICES (NY)
GARCO RESEARCH COMPANY (NJ)
GEOCENTRIC ENGINEERING CORPORATION (VA)
GOLDEN CONSTRUCTION SERVICE (NY)
GREEN, GEORGE (NY)
HOLMES INSPECTION SERVICE (NJ)
HOME DETOX SPECIALISTS (NY)
HOME INSPECTION SERVICE, INC. (NY)
HOMERAD (MD)
INSPECT-A-HOME, INC. (NJ)
J.A. ASSOCIATES/PUBLIC HEALTH CONSULTANTS (PA)
L. CLARK AND SONS, INC. (MA)
LMS ENGINEERS (NY)
LYNN PEST MANAGEMENT COMPANY (MD)
MASTER BUILDING SERVICE (NY)
MATTER RESEARCH, INC. (NY)
MICROBAC ENVIRONMENTAL LABORATORY (NC)
MICROBAC LABORATORIES, INC./ERIE TESTING DIVISION-MAIN OFFICE (PA)
MICROBAC LABORATORIES, INC./HANOVER DIVISION (PA)
MICROBAC LABORATORIES, INC./KENTUCKY TESTING DIVISION (KY)
MICROBAC LABORATORIES, INC./MASSACHUSETTS TESTING DIVISION (MA)
MICROBAC LABORATORIES, INC./NEW CASTLE TESTING DIVISION (PA)
MICROBAC LABORATORIES, INC./PITTSBURGH TESTING DIVISION (PA)
MICROBAC LABORATORIES, INC./SCHILLER DIVISION (PA)
MICROBAC LABORATORIES, INC./SEAWAY DIVISION (IN)
MICROBAC LABORATORIES, INC./STRAND TESTING DIVISION (PA)
MIXLEE ENVIRONMENTAL RESEARCH CORPORATION (CT)
MR. RENOVATOR (NY)
MRC HOME INSPECTIONS (NJ)
NALEPA, DAN (NY)
NATIONAL FUEL (NY)
NEW ENGLAND RADON SERVICES (CT)
NEW YORK STATE HEALTH DEPARTMENT/WADSWORTH CENTER FOR LABORATORIES AND RESEARCH/LABORATORY FOR INORGANIC AND NUCLEAR CHEMISTRY (NY)
NIGRO CONSTRUCTION COMPANY, INC. (NY)
NITE HAWK ENTERPRISES, INC. (NY)
NUWS LABORATORIES, INC. (NY)
PENNSYLVANIA BUILDING INSPECTORS, INC. (PA)
PROTECT-A-HOME (CT)
R.K. VAN SCHAICK AND ASSOCIATES (NY)
RAD CHEK SERVICES COMPANY (CT)
RAD CHEK SERVICES COMPANY (NJ)
RADIATION PROTECTION SERVICES, INC. (NJ)
RADIATION SAFETY ASSOCIATES, INC. (CT)
RADON AIR AND WATER CHECK, INC. (NY)
RADON ANALYSTS, INC. (NY)
RADON BUSTERS OF CENTRAL NEW YORK (NY)
RADON CHECK/DIVISION OF HORIZON SYSTEMS (MA)
RADON DETECTION DEVICES (MD)
RADON DETECTION SERVICES (VT)
RADON DETECTION SYSTEMS/LOS ANGELES, CALIFORNIA BRANCH (CA)
RADON DETECTION SYSTEMS/NEWARK, NEW JERSEY BRANCH (NJ)
RADON DETECTORS (NY)
RADON FINDERS (NY)
RADON IN-HOUSE DETECTION (NY)
RADON INSPECTION SERVICE, INC. (NJ)
RADON LABS (MA)
RADON MANAGEMENT COMPANY (VA)
RADON MANAGEMENT LABORATORIES (NJ)
RADON MEASURING SYSTEMS (PA)
RADON REDUCTION ASSOCIATES, INC. (NY)
RADON SYSTEMS OF RHODE ISLAND, INC. (RI)
RADON TECHNICAL SYSTEMS, INC. (NY)
RADON TECHNOLOGIES, INC. (NY)
RADON TESTING COMPANY OF NEW YORK (NY)
RADON TESTING SERVICE, INC. (NY)
RADON TESTING SERVICES, INC. (NY)
RADONAIRE, INC. (NY)
REMD/RESIDENTIAL ENVIRONMENTAL MEDICAL DIAGNOSTICS (CT)
REMSEARCH, INC. (NY)
RENSSELAER POLYTECHNICAL INSTITUTE (NY)
RESIDENTIAL INSPECTION TECHNOLOGY/RES-I-TEC (CT)
RIC INSPECTION AND ENGINEERING (NJ)
RONALD STEIN CONSTRUCTION, INC. (NY)
ROSS SYSTEMS, INC. (CT)
RTK ENGINEERING (CO)
S & D CUSTOM ESTATES (NY)
SAFEHOME (NY)
SECONDO, THOMAS A. (NY)
SHELBURNE LABORATORIES, INC. (VT)
SHOTWELL ASSOCIATES (NJ)
SILVER SPRINGS HOME INSPECTION (NJ)
SOL-AIR, INC. (CT)
SOUTH CENTRAL ENVIRONMENTAL SERVICES (NY)
SOUTH PARK URBAN RENEWAL CORPORATION (NY)
SPECIALIZED SERVICES, INC. (NJ)
STAR LABORATORIES (MA)
STERBAL, SYLVESTER (NY)
SULLIVAN CONSULTANTS, INC. (NJ)
SUNSTREAM DEVELOPMENT CORPORATION (NY)
SYNERTECH SYSTEMS CORPORATION (NY)
SYRACUSE ENERGY SERVICE (NY)
TRI-CITIES RADON (NY)
TSI CONSTRUCTION SERVICES (IN)
U.S. ASSETS RECOVERY INVESTIGATIONS, INC. (DC)
UNIVERSAL HOME INSPECTION, INC. (NY)
VISUAL BUILDING INSPECTIONS (NY)
WATKINS ENVIRONMENTAL SCIENCES (NY)

Companies Serving

North Carolina

All companies listed under Nationwide provide services for this state.

A-1 TERMITE AND PEST CONTROL, INC. (NC)
AAAIRSCAN ENGINEERING CORPORATION (CT)
ABACUS LABS RADON ASSESSMENT AND MITIGATION (ALRAM) (NH)
ADVANCE PEST CONTROL (NC)
ADVANCE TEX CORPORATION (SC)
AIDCO MAINE CORPORATION (NH)
AIR QUALITY ANALYSTS, INC. (VA)
BIOSPHERICS, INC. (MD)
C SENTINEL ENTERPRISES (VA)
CAROLINA RADON CONSULTANTS (NC)
CAROLINA RADON CONSULTANTS (SC)
COMMONWEALTH ENGINEERING AND TECHNOLOGY (PA)
CONSULTING CHEMISTS (TN)
DAMES AND MOORE (MD)
DJ'S RADIATION PROTECTION (NC)
ENVIRADON, INC. (NC)
ENVIRADON, INC. (NC)

ENVIRONMENTAL MANAGEMENT GROUP (MD)
ENVIRONMENTAL SYSTEMS SERVICE (VA)
ENVIRONMENTAL SYSTEMS SERVICE, LTD. (MD)
GEOCENTRIC ENGINEERING CORPORATION (VA)
HARTSELL HOME INSPECTION SERVICE (NC)
J.M.C. ENTERPRISES (TN)
JOYCE EXTERMINATING COMPANY (NC)
LAW ENVIRONMENTAL, INC. (FL)
MECKLENBURG PURIFIER (NC)
MERIT ENVIRONMENTAL SERVICES, INC. (KY)
MICROBAC ENVIRONMENTAL LABORATORY (NC)
MICROBAC LABORATORIES, INC./ERIE TESTING DIVISION-MAIN OFFICE (PA)
MICROBAC LABORATORIES, INC./HANOVER DIVISION (PA)
MICROBAC LABORATORIES, INC./KENTUCKY TESTING DIVISION (KY)
MICROBAC LABORATORIES, INC./MASSACHUSETTS TESTING DIVISION (MA)
MICROBAC LABORATORIES, INC./NEW CASTLE TESTING DIVISION (PA)
MICROBAC LABORATORIES, INC./PITTSBURGH TESTING DIVISION (PA)
MICROBAC LABORATORIES, INC./SCHILLER DIVISION (PA)
MICROBAC LABORATORIES, INC./SEAWAY DIVISION (IN)
MICROBAC LABORATORIES, INC./STRAND TESTING DIVISION (PA)
NATIONAL PROPERTY INSPECTIONS (NC)
NORTH CAROLINA DEPARTMENT OF HUMAN RESOURCES/DIVISION OF FACILITY SERVICES/RADIATION PROTECTION SECTION (NC)
PROTECTION INSPECTION, INC. (TN)
QUALITY ENVIRONMENTS, INC. (NC)
RAD ENVIRONMENTAL SERVICES, INC. (NJ)
RADIATION CONSULTANTS OF SOUTHWEST VIRGINIA (VA)
RADON CALIBRATIONS (VA)
RADON DETECTION SYSTEMS/GREENSBORO, NORTH CAROLINA BRANCH (NC)
RADON DETECTION SYSTEMS/NASHVILLE, TENNESSEE BRANCH (TN)
RADON ENVIRONMENTAL SERVICES, INC. (KY)
RADON MANAGEMENT COMPANY (VA)
RADON REDUCTION AND TESTING, INC. (GA)
RADON TEST AND REDUCTION SPECIALIST (NC)
ROBERTS ENVIRONMENTAL SERVICES (NC)
ROSS RADON TESTING (GA)
TRIAD BASEMENT DE-WATERING SYSTEM (NC)
TSI CONSTRUCTION SERVICES (IN)
UNIVERSITY OF NORTH CAROLINA/SCHOOL OF PUBLIC HEALTH/DEPARTMENT OF ENVIRONMENTAL SCIENCES AND ENGINEERING (NC)

Companies Serving

North Dakota

All companies listed under Nationwide provide services for this state.

ALPHA RADIATION ENGINEERING, INC. (CO)
C.R. ADAMS COMPANY, INC. (CO)
EES OF MINNESOTA, INC. (MN)
ENERGY LABORATORIES, INC. (WY)
NEBRASKA WESLEYAN UNIVERSITY/PHYSICS DEPARTMENT/RADON MEASUREMENTS LABORATORY (NE)
R.T.C. (NY)
RADIATION MEASUREMENT SERVICES (MD)
RADON AND ENVIRONMENTAL PROFESSIONALS (KS)
RADON BUSTERS (ND)
RADON CONSULTANTS (NV)
RADON CONSULTANTS, INC. (CO)
RADON DETECTION SYSTEMS/BUTTE, MONTANA BRANCH (MT)
RADON DETECTION SYSTEMS/CASPER, WYOMING BRANCH (WY)
RADON DETECTION SYSTEMS/FARGO, NORTH DAKOTA BRANCH (ND)
RADON MEASURE COMPANY, THE (MN)
RADON TESTING AND CONTROL (ND)
VIKING HOME IMPROVEMENT, INC. (ND)

Companies Serving

Ohio

All companies listed under Nationwide provide services for this state.

A-1 RELIABLE BASEMENT WATERPROOFING (OH)
AA RADCO (MI)
ACCUTEK LABORATORY (OH)
AIDCO MAINE CORPORATION (NH)
AMERICAN MONITORING AND ENGINEERING SERVICES (PA)
AMERICAN RADON TESTING SERVICES, INC. (OH)
BBC & M ENGINEERING (OH)
CLASSIC RENOVATIONS AND BUILDERS (OH)
COMMONWEALTH ENGINEERING AND TECHNOLOGY (PA)
CONSUMER ENERGY SERVICES, INC. (OH)
DAMES AND MOORE (MD)
ENCOSYS CORPORATION OF AMERICA (IN)
ENERGY UNLIMITED (PA)
ENVIRONMENTAL HEALTH WATCH OF GREATER CLEVELAND (OH)
ENVIRONMENTAL MANAGEMENT GROUP (MD)
ENVIRONMENTAL OIL, INC. (NY)
ENVIRONMENTAL SOLUTIONS, INC. (MO)
ENVIRONMENTAL SYSTEMS SERVICE (VA)
ENVIRONMENTAL SYSTEMS SERVICE, LTD. (MD)
ENVIROSERV/RADON DETECTION OF OHIO (OH)
FINNEX (RADON) (OH)
FTLC SERVICES, INC. (OH)
FUTURA MEDICAL SYSTEMS, INC. (OH)
GEOCENTRIC ENGINEERING CORPORATION (VA)
GITERMAN ASSOCIATES (MN)
GLOGORE ENGINEERING COMPANY (OH)
HERRICK ENGINEERING, INC. (OH)
HOME ANALYSTS, INC. (OH)
HOME ASSURANCE (IL)
HOME CHEK, INC. (KS)
HOME ENVIRONMENT COMPANY (IN)
HOMEAIR TESTERS (IN)
IMPROVED TECHNOLOGY SYSTEMS (OH)
INSUL-TECH, INC. (OH)
INSULATING SYSTEMS (OH)
J. BECK AND ASSOCIATES, INC. (PA)
MERIT ENVIRONMENTAL SERVICES, INC. (KY)
MICROBAC ENVIRONMENTAL LABORATORY (NC)
MICROBAC LABORATORIES, INC./ERIE TESTING DIVISION-MAIN OFFICE (PA)
MICROBAC LABORATORIES, INC./HANOVER DIVISION (PA)
MICROBAC LABORATORIES, INC./KENTUCKY TESTING DIVISION (KY)
MICROBAC LABORATORIES, INC./MASSACHUSETTS TESTING DIVISION (MA)
MICROBAC LABORATORIES, INC./NEW CASTLE TESTING DIVISION (PA)
MICROBAC LABORATORIES, INC./PITTSBURGH TESTING DIVISION (PA)
MICROBAC LABORATORIES, INC./SCHILLER DIVISION (PA)
MICROBAC LABORATORIES, INC./SEAWAY DIVISION (IN)
MICROBAC LABORATORIES, INC./STRAND TESTING DIVISION (PA)
MOBILE RADON DIAGNOSTICS, INC. (OH)
MONARCH ANALYTICAL LABORATORIES, INC. (OH)
OAK HILL RADON TESTING AND REPAIR (OH)
PRO-TECH RADON (OH)
RADON AIRTEST COMPANY (OH)
RADON ANALYSIS AND DETECTION OF NEW ENGLAND, INC. (NH)
RADON AND RADIATION CONSULTANTS (OH)
RADON DETECTION SYSTEMS, INC., OHIO (OH)
RADON DETECTION SYSTEMS/BLOOMINGTON-NORMAL BRANCH (IL)
RADON DETECTION SYSTEMS/COLUMBUS, OHIO BRANCH (OH)
RADON DETECTORS (NY)
RADON ENVIRONMENTAL SERVICES, INC. (KY)
RADON ENVIRONMENTAL SERVICES, INC. (OH)
RADON ENVIRONMENTAL TESTING SERVICE INC. (OH)
RADON ENVIRONMENTAL TESTING SERVICES, INC. (OH)
RADON FREE (OH)
RADON GAS DETECTION SYSTEMS (OH)
RADON GAS TESTING OF CENTRAL OHIO (OH)
RADON MANAGEMENT COMPANY (VA)
RADON MEASUREMENT AND CONTROL (OH)
RADON TECHNOLOGIES, INC. (OH)
RADON TESTING AND MITIGATION SYSTEMS (IN)
RADON TESTING OF MICHIANA (IN)
RADON WATCH, INC. (OH)
RELIABLE BASEMENT WATERPROOF COMPANY (OH)
RESTORED ENVIRONMENTS COMPANY, INC. (OH)
SEBRING HOME INSPECTION COMPANY (OH)
SOUTHERN OHIO RADON PROTECTION COMPANY (OH)
STAR LABORATORIES (MA)
TECH OPSOL, INC. (OH)
TEE ENGINEERING COMPANY (KY)
TRI-STATE RADON TESTING AND REDUCTION (OH)
TRIPLE SEAL, INC. (OH)
TSI CONSTRUCTION SERVICES (IN)
WASICK, BRUCE (OH)
WESTERN TECHNOLOGY, INC. (MD)

Companies Serving

Oklahoma

All companies listed under Nationwide provide services for this state.

ALPHA RADIATION ENGINEERING, INC. (CO)
ENVIRONMENTAL SOLUTIONS, INC. (MO)
RADON AND ENVIRONMENTAL PROFESSIONALS (KS)
RADON AND ENVIRONMENTAL PROFESSIONALS (KS)
RADON DETECTION SYSTEMS/DALLAS, TEXAS BRANCH (TX)
STRATA ENVIRONMENTAL SERVICE, INC. (KS)

Companies Serving

Oregon

All companies listed under Nationwide provide services for this state.

A & M ASSOCIATES (OR)
BAR-TEL INFRASCAN (WA)
CAVALIER CORPORATION (WA)
E.C. CHAMBERLAIN, RADIOLOGICAL CONSULTANT (NV)
ENERGY LABORATORIES, INC. (WY)
ENVIRONMENTAL ENGINEERING AND TESTING (CO)
ENVIRONMENTAL ENGINEERING LABORATORIES, INC. (CO)
FAYTEK, INC. (ID)
NEA, INC. (OR)
R.T.C. (NY)
RADIATION HEALTH (WA)

RADON AWARENESS, INC. (ID)
RADON CONSULTANTS (NV)
RADON DETECTION SYSTEMS/LOS ANGELES, CALIFORNIA BRANCH (CA)
RADON DETECTION SYSTEMS/PORTLAND, OREGON BRANCH (OR)
RADON MANAGEMENT COMPANY (VA)
RADON SYSTEMATICS OF CALIFORNIA (CA)
RADON TEST (CA)
SANI-DUCT SERVICE OF THE INLAND NORTHWEST (WA)
SPENCE-GEIGER ASSOCIATES, INC. (CO)

Companies Serving

Pennsylvania

All companies listed under Nationwide provide services for this state.

A-1 SHERLOCK HOME INSPECTORS (NJ)
A.B.E. RADON TESTING, INC. (PA)
AAA RADON TESTING (NJ)
AAAIRSCAN ENGINEERING CORPORATION (CT)
ACCREDITED WATERPROOFING COMPANY, INC. (NJ)
ACCU-TEST (VA)
AFFILIATED BUILDING INSPECTORS, INC. (MD)
AFFTREX, LTD. (PA)
AGENTERPRISE, INC. (PA)
AIDCO MAINE CORPORATION (NH)
AIR HAZARDS ASSESSMENT, INC. (MD)
AIR QUALITY ANALYSTS, INC. (VA)
AIR QUALITY CONTROL, INC. (PA)
AIR QUALITY CONTROL, INC. (PA)
AIR VENTILATION SYSTEMS, INC. (MD)
AIRCARE SYSTEMS (PA)
ALL EASTERN HOME INSPECTION COMPANY (NJ)
ALPHA BETA GAMMA TECHNOLOGIES, INC. (PA)
ALPHA CHECK (MD)
ALPHA RADON ENVIRONMENTAL TESTING (ME)
ALPHA SERVICES/AIR AND WATER RADON TESTING (CT)
ALPHA-BAN (MD)
AMBIC BUILDING INSPECTION CONSULTANTS, INC. (NJ)
AMERICAN BUILDING INSPECTORS, INC. (PA)
AMERICAN HERITAGE CORPORATION (PA)
AMERICAN HOMECHECK, INC. (PA)
AMERICAN INSPECTION AGENCY (PA)
AMERICAN INSPECTIONS SERVICES (PA)
AMERICAN MONITORING AND ENGINEERING SERVICES (PA)
AMERICAN RADON DETECTION COMPANY, INC. (PA)
AMERICAN RADON TESTING, INC. (PA)
ANGELO, MARC (PA)
APPLIED ANALYTICAL TECHNOLOGIES, INC. (RI)
ASSOCIATED RADON TESTING SERVICES, INC. (PA)
ATLAS WATERPROOFING COMPANY, INC. (PA)
AUDUBON CONTROLS, INC. (PA)
B & H RADON SERVICES (PA)
B-DRY SYSTEM OF DELAWARE AND CHESTER COUNTY, INC. (PA)
BAYSURE BUILDING INSPECTION, INC. (PA)
BERGEN HOME INSPECTION SERVICE (NJ)
BIOSPHERICS, INC. (MD)
BORO RADON TESTING SERVICE, INC. (PA)
BRUN-REUSS ASSOCIATES (PA)
BUILDING INSPECTION SERVICES, INC. (PA)
BUILDING INSPECTOR OF AMERICA, THE (PA)
BURENS ENVIRONMENTAL SERVICES CORPORATION (PA)
CAMBRIDGE ASSOCIATES, LTD. (NJ)
CAPITAL MATERIALS TESTING, INC. (NY)
CARL HENRY AND COMPANY, INC. (PA)
CARROLL COUNTY RADON CONTROL, INC. (MD)
CCI-RADON AND HOME INSPECTION/RADON DETECTION (PA)
CERTIFIED TESTING SERVICES, INC. (PA)
CHESTER COUNTY DEVELOPMENT SERVICE (PA)
CHIEF HOME AND BUILDING INSPECTION SERVICES (NJ)
CHILTON ASSOCIATES (RADON GROUP) (NJ)
COMMONWEALTH ENGINEERING AND TECHNOLOGY (PA)
COMMONWEALTH ENTERPRISES, INC. (PA)
COMPREHENSIVE MANAGEMENT SERVICES (PA)
CONSHOHOCKEN RADON TESTING SERVICES (PA)
CONTINENTAL ENVIRONMENT COMPANY, INC. (NJ)
CUMBERLAND ANALYTICAL LABS, INC. (PA)
D-TECH, INC. (PA)
DAMES AND MOORE (MD)
DELAWARE VALLEY GEOTECHNICAL ASSOCIATES (PA)
DELAWARE VALLEY RADON INSPECTIONS/ARSENAL BUSINESS CENTER (PA)
DEMAR PEST CONTROL COMPANY (PA)
DESERT-DRY WATERPROOFING (PA)
DEVELOPMENT SERVICES, LTD. (PA)
DIAGNOSTICS AND DEVICES, INC. (NJ)
DIAMOND REALTY INSPECTIONS, INC. (PA)
DK ROCKWELL (NJ)
DPR BUILDERS (NJ)
ECOLOGICAL INTERACTIONS, INC. (NJ)
ECONOMY HEATING AND AIR CONDITIONING (MD)
ELECTRO MECHANICAL CONCEPTS (PA)
EMPIRE INSPECTION, INC. (NJ)
ENERGY UNLIMITED (PA)
ENVIRO TECH SYSTEMS (PA)
ENVIRO TESTING (NY)
ENVIROHEALTH, INC. (PA)
ENVIRONMENTAL CONSULTANTS ASSOCIATES (NJ)
ENVIRONMENTAL FIELD SERVICES, INC. (ME)
ENVIRONMENTAL INDOOR AIR QUALITY CONSULTANTS, INC. (EIC) (PA)
ENVIRONMENTAL MANAGEMENT GROUP (MD)
ENVIRONMENTAL OIL, INC. (NY)
ENVIRONMENTAL SYSTEMS SERVICE (VA)
ENVIRONMENTAL SYSTEMS SERVICE, LTD. (MD)
ENVIRONMENTAL TESTING ASSOCIATES, INC. (PA)
FAYTEK, INC. (ID)
FORCHETTI PSI (PA)
FORESIGHT ENGINEERING, INC. (NJ)
FRANKLIN ENVIRONMENTAL ANALYSIS (PA)
GALLAGHER ENGINEERING AND INSPECTION (PA)
GARCO RESEARCH COMPANY (NJ)
GARDNER, PHILIP (PA)
GAVIN, GERARD M. (PA)
GEIER CONSTRUCTION COMPANY, INC./DIVISION OF RADON CONTROL (MD)
GEOCENTRIC ENGINEERING CORPORATION (VA)
GIDDINGS AND ASSOCIATES, INC. (PA)
GRAVES RADON CONTRACTING (PA)
HEALTH PHYSICS ASSOCIATES, INC. (PA)
HOLMES INSPECTION SERVICE (NJ)
HOME TESTING LABS, INC. (PA)
HOMERAD (MD)
HOUSEHOLD RADON TESTERS, INC. (PA)
HOUSEMASTER (PA)
HOUSING INSPECTION SERVICES, INC. (PA)
HUGHES, RONALD L. (PA)
INDUSTRIAL RESOURCE DEVELOPMENT COMPANY (PA)
INSPECT-A-HOME, INC. (NJ)
INSPECTIONS UNLIMITED (PA)
INSUL-TECH, INC. (OH)
INTERNATIONAL EXPLORATIONS (PA)
J. BECK AND ASSOCIATES, INC. (PA)
J.A. ASSOCIATES/PUBLIC HEALTH CONSULTANTS (PA)
KELLOGG ENVIRONMENTAL SERVICES, INC. (PA)
L. CLARK AND SONS, INC. (MA)
L.G. PAUL, SR., INC. (PA)
LEHIGH VALLEY HOUSE INSPECTION SERVICE (PA)
LONCHAR, NIKOLA (PA)
LV SAFETY AND HEALTH CONSULTANTS (PA)
LYNN PEST MANAGEMENT COMPANY (MD)
M.A.D. EXTERMINATORS, INC. (PA)
MAYA-TEK, INC. (PA)
MCGUIRE BUILDERS, INC. (PA)
MCHILL, A.D. AND SONS, INC. (PA)
MCWILLIAMS PEST CONTROL (PA)
MEDLAB ENVIRONMENTAL TESTING, INC. (DE)
MICROBAC ENVIRONMENTAL LABORATORY (NC)
MICROBAC LABORATORIES, INC./ERIE TESTING DIVISION-MAIN OFFICE (PA)
MICROBAC LABORATORIES, INC./HANOVER DIVISION (PA)
MICROBAC LABORATORIES, INC./KENTUCKY TESTING DIVISION (KY)
MICROBAC LABORATORIES, INC./MASSACHUSETTS TESTING DIVISION (MA)
MICROBAC LABORATORIES, INC./NEW CASTLE TESTING DIVISION (PA)
MICROBAC LABORATORIES, INC./PITTSBURGH TESTING DIVISION (PA)
MICROBAC LABORATORIES, INC./SCHILLER DIVISION (PA)
MICROBAC LABORATORIES, INC./SEAWAY DIVISION (IN)
MICROBAC LABORATORIES, INC./STRAND TESTING DIVISION (PA)
MISCAVAGE AND COMPANY (PA)
NATIONAL FUEL (PA)
NATIONAL RADON SERVICE (MD)
NATIONAL RADON TESTING SERVICE (PA)
NELSON, C. STUART (PA)
NUCHEM CORPORATION (PA)
NUCLEAR RESEARCH CORPORATION (PA)
PENNSYLVANIA BUILDING INSPECTORS, INC. (PA)
PENNSYLVANIA RADON CENTER (PA)
PENNSYLVANIA STATE UNIVERSITY, THE/BREAZEALE NUCLEAR REACTOR/LOW LEVEL RADIATION MONITORING LABORATORY (PA)
PENRAD, INC. (PA)
PERISON, JANICE B. (PA)
PEST CONTROL TECHNICIANS, INC. (PA)
PORTER CONSULTANTS, INC. (PA)
PRECISION RADON TESTING (PA)
PRIGEL RADON TESTING (PA)
PRINCETON HOME INSPECTION SERVICE (NJ)
PRO-GUARD BUILDING INSPECTIONS, INC. (PA)
PROTECTRON, INC. (MD)
PUKAL CONTRACTING COMPANY (PA)
R.I.C. INSPECTION AND ENGINEERING (NJ)
R.J. FRIDG COMPANY, INC. (PA)
R.J. MOORE AND ASSOCIATES, INC. (MD)
RACHECK, INC. (PA)
RADIOACTIVITY ALERT DETECTION SERVICES, INC. (PA)
RADON AIR AND WATER CHECK, INC. (NY)
RADON ANALYSIS AND DETECTION OF NEW ENGLAND, INC. (NH)
RADON BAN (MD)
RADON BUSTERS (PA)
RADON CHECK/DIVISION OF HORIZON SYSTEMS (MA)
RADON DECON, INC. (PA)
RADON DETECTION DEVICES (MD)
RADON DETECTION SYSTEMS/NEWARK, NEW JERSEY BRANCH (NJ)
RADON DETECTION SYSTEMS/PITTSBURGH, PENNSYLVANIA BRANCH (PA)
RADON DETECTORS (NY)
RADON DOCTOR, INC., THE (PA)
RADON ENVIRONMENTAL CONSTRUCTION SERVICES/RECONS, INC., MAIN OFFICE (NJ)

RADON ENVIRONMENTAL CONSTRUCTION SERVICES/RECONS, INC., PA OFFICE (PA)
RADON LABORATORIES (MD)
RADON MANAGEMENT COMPANY (VA)
RADON MANAGEMENT LABORATORIES (NJ)
RADON MANAGEMENT, INC. (PA)
RADON MEASUREMENT SERVICE (PA)
RADON MEASURING SYSTEMS (PA)
RADON MONITORING AND SERVICES (PA)
RADON PROTECTION SERVICES (PA)
RADON RANGERS (MD)
RADON REMEDIAL SERVICES (PA)
RADON RESOURCE, INC. (MD)
RADON SYSTEMS OF RHODE ISLAND, INC. (RI)
RADON SYSTEMS, INC. (MD)
RADON TECHNOLOGIES, INC. (NY)
RADON TECHNOLOGIES, INC. (OH)
RADON TECHNOLOGIES, INC. (PA)
RADON TESTING SERVICES, HARRISBURG (NJ)
RADON TESTING SERVICES, PRINCETON (NJ)
RADON TESTING SERVICES, SILVER SPRING (MD)
RADON TESTING SYSTEMS (PA)
RADONTECH, INC. (PA)
RAY MESSICK RADON TESTING AND MITIGATION (PA)
REYNOLDS RADIOLOGICAL SERVICES, INC. (PA)
SAFE SHELTER ENVIRONMENTAL (PA)
SCIENCE MANAGEMENT SERVICES, LTD. (PA)
SCRUB RIDGE RAD TESTING AND REMEDIATION (PA)
SHELBURNE LABORATORIES, INC. (VT)
SHOTWELL ASSOCIATES (NJ)
SILVER SPRINGS HOME INSPECTION (NJ)
SOLAR RESOURCE TECHNOLOGY (PA)
SOUTH CENTRAL ENVIRONMENTAL SERVICES (NY)
SPARTAN HOME INSPECTION, INC. (NJ)
SPECIALIZED SERVICES, INC. (NJ)
SPRINGER COMPANY (PA)
STAR LABORATORIES (MA)
STONE RIDGE RADON MITIGATORS (NJ)
SULLIVAN CONSULTANTS, INC. (NJ)
SUNSTREAM DEVELOPMENT CORPORATION (NY)
THREE W RADON SERVICES, INC. (PA)
TOTAL HOME TECH (PA)
TRI-CITIES RADON (NY)
TRI-COUNTY RADON, INC. (NJ)
TSI CONSTRUCTION SERVICES (IN)
U.S. ASSETS RECOVERY INVESTIGATIONS, INC. (DC)
UTILITY ENGINEERS, INC. (PA)
VFC, INC. (PA)
WAYNE, MICHELE (PA)
WEATHER SHIELD, INC. (PA)
WESTERN TECHNOLOGY, INC. (MD)
WILSON JR., WALTER E. (PA)
WPB ENTERPRISES, INC. (PA)

Companies Serving

Puerto Rico

All companies listed under Nationwide provide services for this state.

DAMES AND MOORE (MD)
NEW ENGLAND RADON, LTD. (NH)

Companies Serving

Rhode Island

All companies listed under Nationwide provide services for this state.

AAAIRSCAN ENGINEERING CORPORATION (CT)
ABACUS LABS RADON ASSESSMENT AND MITIGATION (ALRAM) (NH)
AIDCO MAINE CORPORATION (NH)
AIR SENSE, INC. (CT)
ALPHA RADON ENVIRONMENTAL TESTING (ME)
ALPHA SERVICES/AIR AND WATER RADON TESTING (CT)
APPLIED ANALYTICAL TECHNOLOGIES, INC. (RI)
ASAFE-HOUSE, INC. (CT)
ATLANTIC ENVIRONMENTAL LABORATORY (CT)
BOXFORD RADON LABORATORY (MA)
BRADLEY HARRISON ASSOCIATES, INC. (NY)
CAMPTON DEVELOPMENT CORPORATION (NH)
COMMONWEALTH ENGINEERING AND TECHNOLOGY (PA)
CONNECTICUT ENVIRONMENTAL TESTING (CT)
DAMES AND MOORE (MD)
EMPAC, INC. (CT)
ENVIRONMENTAL SCIENCE SERVICES (RI)
ENVIRONMENTAL SYSTEMS SERVICE (VA)
ENVIRONMENTAL SYSTEMS SERVICE, LTD. (MD)
ENVIROSCIENCE CONSULTANTS, INC. (CT)
FORESIGHT ENGINEERING, INC., VIRGINIA (VA)
L. CLARK AND SONS, INC. (MA)
MASSACHUSETTS RADON ALERT, INC. (MA)
MICROBAC ENVIRONMENTAL LABORATORY (NC)
MICROBAC LABORATORIES, INC./ERIE TESTING DIVISION-MAIN OFFICE (PA)
MICROBAC LABORATORIES, INC./HANOVER DIVISION (PA)
MICROBAC LABORATORIES, INC./KENTUCKY TESTING DIVISION (KY)
MICROBAC LABORATORIES, INC./MASSACHUSETTS TESTING DIVISION (MA)
MICROBAC LABORATORIES, INC./NEW CASTLE TESTING DIVISION (PA)
MICROBAC LABORATORIES, INC./PITTSBURGH TESTING DIVISION (PA)
MICROBAC LABORATORIES, INC./SCHILLER DIVISION (PA)
MICROBAC LABORATORIES, INC./SEAWAY DIVISION (IN)
MICROBAC LABORATORIES, INC./STRAND TESTING DIVISION (PA)
MIXLEE ENVIRONMENTAL RESEARCH CORPORATION (CT)
NEW ENGLAND RADON SERVICES (CT)
NEW ENGLAND RADON, LTD. (NH)
PELICAN ENVIRONMENTAL CORPORATION (MA)
PROTECT-A-HOME (CT)
RADIATION SAFETY ASSOCIATES, INC. (CT)
RADON ANALYSIS AND DETECTION OF NEW ENGLAND, INC. (NH)
RADON CHECK/DIVISION OF HORIZON SYSTEMS (MA)
RADON DETECTION SYSTEMS/NEWARK, NEW JERSEY BRANCH (NJ)
RADON DETECTORS (NY)
RADON EVALUATION DIVISION/EVALDO, LTD. (CT)
RADON LABS (MA)
RADON SAFETY TESTING, INC. (CT)
RADON SYSTEMS OF RHODE ISLAND, INC. (RI)
RADON TECHNOLOGY AND CONSULTANTS (RI)
RADON TESTING AND CONSULTING SERVICES (MA)
REMD/RESIDENTIAL ENVIRONMENTAL MEDICAL DIAGNOSTICS (CT)
RESIDENTIAL RADON TEST SERVICE (MA)
RISE (RI)
SHELBURNE LABORATORIES, INC. (VT)
STAR LABORATORIES (MA)
TECON (RI)
TESTING 1.2.3. (RI)

Companies Serving

South Carolina

All companies listed under Nationwide provide services for this state.

AAAIRSCAN ENGINEERING CORPORATION (CT)
ADDY'S RADON DETECTION SERVICE (SC)
ADVANCE TEX CORPORATION (SC)
CAROLINA RADON CONSULTANTS (NC)
CAROLINA RADON CONSULTANTS (SC)
CAROLINA RADON DETECTION (SC)
COMMONWEALTH ENGINEERING AND TECHNOLOGY (PA)
DAMES AND MOORE (MD)
ENTESCO (SC)
ENVIRADON, INC. (NC)
ENVIRADON, INC. (NC)
ENVIRONMENTAL MANAGEMENT GROUP (MD)
ENVIRONMENTAL SYSTEMS SERVICE (VA)
ENVIRONMENTAL SYSTEMS SERVICE, LTD. (MD)
GEOCENTRIC ENGINEERING CORPORATION (VA)
HARTSELL HOME INSPECTION SERVICE (NC)
LAW ENVIRONMENTAL, INC. (FL)
MECKLENBURG PURIFIER (NC)
MICROBAC ENVIRONMENTAL LABORATORY (NC)
MICROBAC LABORATORIES, INC./ERIE TESTING DIVISION-MAIN OFFICE (PA)
MICROBAC LABORATORIES, INC./HANOVER DIVISION (PA)
MICROBAC LABORATORIES, INC./KENTUCKY TESTING DIVISION (KY)
MICROBAC LABORATORIES, INC./MASSACHUSETTS TESTING DIVISION (MA)
MICROBAC LABORATORIES, INC./NEW CASTLE TESTING DIVISION (PA)
MICROBAC LABORATORIES, INC./PITTSBURGH TESTING DIVISION (PA)
MICROBAC LABORATORIES, INC./SCHILLER DIVISION (PA)
MICROBAC LABORATORIES, INC./SEAWAY DIVISION (IN)
MICROBAC LABORATORIES, INC./STRAND TESTING DIVISION (PA)
QUALITY ENVIRONMENTS, INC. (NC)
RADON DETECTION SYSTEMS/NASHVILLE, TENNESSEE BRANCH (TN)
RADON ENVIRONMENTAL SERVICES, INC. (KY)
RADON EVALUATION SERVICES, INC. (GA)
RADON MANAGEMENT COMPANY (VA)
RADON REDUCTION AND TESTING, INC. (GA)
RADON TEST AND REDUCTION SPECIALIST (NC)
ROSS RADON TESTING (GA)
SOUTH CAROLINA DEPARTMENT OF HEALTH AND ENVIRONMENTAL CONTROL/BUREAU OF RADIOLOGICAL HEALTH (SC)
SOUTH CAROLINA ELECTRIC AND GAS (SC)
UNIVERSITY OF GEORGIA/CENTER FOR APPLIED ISOTOPE STUDY (GA)
UNIVERSITY RADON TESTING ASSOCIATES (GA)
W.R. MARTIN COMPANY (SC)

Companies Serving

South Dakota

All companies listed under Nationwide provide services for this state.

C.R. ADAMS COMPANY, INC. (CO)
EES OF MINNESOTA, INC. (MN)
ENERGY LABORATORIES, INC. (WY)
ENERGY MARKETING MIDWEST, INC. (IA)
NEBRASKA WESLEYAN UNIVERSITY/PHYSICS DEPARTMENT/RADON MEASUREMENTS LABORATORY (NE)
RADON BUSTERS (ND)
RADON CONSULTANTS (NV)
RADON CONSULTANTS, INC. (CO)
RADON DETECTION SYSTEMS/BUTTE, MONTANA BRANCH (MT)
RADON DETECTION SYSTEMS/CASPER, WYOMING BRANCH (WY)
RADON DETECTION SYSTEMS/SIOUX FALLS, SOUTH DAKOTA BRANCH (SD)
RADON MEASURE COMPANY, THE (MN)

RADON TESTING AND CONTROL (ND)

Companies Serving

Tennessee

All companies listed under Nationwide provide services for this state.

ACI CHEMICAL COMPANY, INC. (AR)
ADVANCED MATERIALS ENTERPRISES, INC. (TN)
ADVANCED RADON DETECTION SERVICES, INC. (TN)
ASSOCIATED PROTECTIVE SYSTEMS (TN)
COMFORT SYSTEMS (TN)
COMMONWEALTH ENGINEERING AND TECHNOLOGY (PA)
CONSULTING CHEMISTS (TN)
DAMES AND MOORE (MD)
ENERGY SYSTEMS, INC. (TN)
ENVIRONMENTAL MANAGEMENT GROUP (MD)
ENVIRONMENTAL SOLUTIONS, INC. (MO)
ENVIRONMENTAL SYSTEMS SERVICE (VA)
ENVIRONMENTAL SYSTEMS SERVICE, LTD. (MD)
ETE CONSULTING ENGINEERING, INC. (TN)
HAYNES REALTY (TN)
HUNT AND ASSOCIATES (GA)
J.A. ASSOCIATES/PUBLIC HEALTH CONSULTANTS (PA)
J.M.C. ENTERPRISES (TN)
LAW ENVIRONMENTAL, INC. (FL)
MERIT ENVIRONMENTAL SERVICES, INC. (KY)
MICROBAC ENVIRONMENTAL LABORATORY (NC)
MICROBAC LABORATORIES, INC./ERIE TESTING DIVISION-MAIN OFFICE (PA)
MICROBAC LABORATORIES, INC./HANOVER DIVISION (PA)
MICROBAC LABORATORIES, INC./KENTUCKY TESTING DIVISION (KY)
MICROBAC LABORATORIES, INC./MASSACHUSETTS TESTING DIVISION (MA)
MICROBAC LABORATORIES, INC./NEW CASTLE TESTING DIVISION (PA)
MICROBAC LABORATORIES, INC./PITTSBURGH TESTING DIVISION (PA)
MICROBAC LABORATORIES, INC./SCHILLER DIVISION (PA)
MICROBAC LABORATORIES, INC./SEAWAY DIVISION (IN)
MICROBAC LABORATORIES, INC./STRAND TESTING DIVISION (PA)
PROFESSIONAL RADON ASSESSMENT (TN)
PROTECTION INSPECTION, INC. (TN)
QUALITY ENVIRONMENTS, INC. (NC)
RADIATION CONSULTANTS OF SOUTHWEST VIRGINIA (VA)
RADIATION SERVICES, INC. (TN)
RADON DETECTION SYSTEMS/NASHVILLE, TENNESSEE BRANCH (TN)
RADON ENVIRONMENTAL SERVICES, INC. (KY)
RADON EVALUATION SERVICES, INC. (GA)
RADON INFORMATION AND DETECTION SERVICE (TN)
RADON REDUCTION AND TESTING, INC. (GA)
RADON TEST AND REDUCTION SPECIALIST (NC)
RADON, INC. (TN)
REAL ESTATE INSPECTION SERVICE (TN)
ROSS RADON TESTING (GA)
SAFE AIR (TN)
SOUTHEASTERN RADON CONSULTANTS, INC. (TN)
TEE ENGINEERING COMPANY (KY)
TSI CONSTRUCTION SERVICES (IN)
UNIVERSITY OF GEORGIA/CENTER FOR APPLIED ISOTOPE STUDY (GA)
YOUR ENERGY SERVICE, INC. (TN)

Companies Serving

Texas

All companies listed under Nationwide provide services for this state.

ACCURATE AIR TESTING (TX)
AIR CONCERNS (TX)
ALL PRO PEST CONTROL (TX)
ALPHA RADIATION ENGINEERING, INC. (CO)
ANTI-PEST COMPANY (TX)
CONTINENTAL ENVIRONMENT COMPANY, INC. (NJ)
E.D.A., INC. (TX)
ENERGY LABORATORIES, INC. (WY)
ENVIRONMENTAL ENGINEERING AND TESTING (CO)
FINNEX (RADON) (OH)
GENE CALVERT AND ASSOCIATES (TX)
INTERNATIONAL EXTERMINATOR CORPORATION (TX)
NATIONAL RADON TEST COMPANY (TX)
PARADISE COMPANY (TX)
RADON DETECTION SYSTEMS/DALLAS, TEXAS BRANCH (TX)
RADON DETECTION SYSTEMS/LOS ANGELES, CALIFORNIA BRANCH (CA)
RADON REDUCTION AND TESTING COMPANY (TX)
RADON TEST SYSTEMS (TX)
RADON TESTING AND MITIGATION OF TEXAS, INC. (TX)
S & W ENVIRONMENTAL TESTING (TX)
STRATA ENVIRONMENTAL SERVICE, INC. (KS)
TEXAS A & M UNIVERSITY/DEPARTMENT OF NUCLEAR ENGINEERING/RADON ASSAY LABORATORY (TX)

Companies Serving

Utah

All companies listed under Nationwide provide services for this state.

ALPHA RADIATION ENGINEERING, INC. (CO)
ARIX CORPORATION (UT)
BONDAR-CLEGG, INC. (CO)
C.R. ADAMS COMPANY, INC. (CO)
E.C. CHAMBERLAIN, RADIOLOGICAL CONSULTANT (NV)
ENERGY ALTERNATIVES UNLIMITED, INC. (CO)
ENERGY LABORATORIES, INC. (WY)
ENVIRONMENTAL ENGINEERING AND TESTING (CO)
ENVIRONMENTAL ENGINEERING LABORATORIES, INC. (CO)
FAYTEK, INC. (ID)
INSPECT-A-HOME OF UTAH (UT)
LANDMARK RADON LABORATORIES, LTD. (CO)
LINCOLN DEVORE, INC. (CO)
RADIATION SURVEILLANCE ASSOCIATES, INC. (UT)
RADON AWARENESS, INC. (ID)
RADON CONSULTANTS (NV)
RADON CONSULTANTS, INC. (CO)
RADON DETECTION SYSTEMS/LOS ANGELES, CALIFORNIA BRANCH (CA)
RADON DETECTION SYSTEMS/SALT LAKE CITY, UTAH BRANCH (UT)
RADON TEST (CA)
RADON TESTING SERVICES, INC. (CA)
SANI-DUCT SERVICE OF THE INLAND NORTHWEST (WA)
SPENCE-GEIGER ASSOCIATES, INC. (CO)
TIMBERLINE RADON SERVICE, INC. (CO)

Companies Serving

Vermont

All companies listed under Nationwide provide services for this state.

AAAIRSCAN ENGINEERING CORPORATION (CT)
ABACUS LABS RADON ASSESSMENT AND MITIGATION (ALRAM) (NH)
AIDCO MAINE CORPORATION (NH)
AIR SENSE, INC. (CT)
ALPHA RADON ENVIRONMENTAL TESTING (ME)
ALPHA SERVICES/AIR AND WATER RADON TESTING (CT)
APPLIED ANALYTICAL TECHNOLOGIES, INC. (RI)
ATLANTIC ENVIRONMENTAL LABORATORY (CT)
BOXFORD RADON LABORATORY (MA)
BURNS AND ASSOCIATES (NH)
CAMPTON DEVELOPMENT CORPORATION (NH)
COMMONWEALTH ENGINEERING AND TECHNOLOGY (PA)
CONNECTICUT ENVIRONMENTAL TESTING (CT)
DAMES AND MOORE (MD)
EMPAC, INC. (CT)
ENVIRONMENTAL FIELD SERVICES, INC. (ME)
ENVIRONMENTAL OIL, INC. (NY)
ENVIRONMENTAL SCIENCE SERVICES (RI)
ENVIRONMENTAL SYSTEMS SERVICE (VA)
ENVIRONMENTAL SYSTEMS SERVICE, LTD. (MD)
ENVIROSCIENCE CONSULTANTS, INC. (CT)
L. CLARK AND SONS, INC. (MA)
MICROBAC ENVIRONMENTAL LABORATORY (NC)
MICROBAC LABORATORIES, INC./ERIE TESTING DIVISION-MAIN OFFICE (PA)
MICROBAC LABORATORIES, INC./HANOVER DIVISION (PA)
MICROBAC LABORATORIES, INC./KENTUCKY TESTING DIVISION (KY)
MICROBAC LABORATORIES, INC./MASSACHUSETTS TESTING DIVISION (MA)
MICROBAC LABORATORIES, INC./NEW CASTLE TESTING DIVISION (PA)
MICROBAC LABORATORIES, INC./PITTSBURGH TESTING DIVISION (PA)
MICROBAC LABORATORIES, INC./SCHILLER DIVISION (PA)
MICROBAC LABORATORIES, INC./SEAWAY DIVISION (IN)
MICROBAC LABORATORIES, INC./STRAND TESTING DIVISION (PA)
MIXLEE ENVIRONMENTAL RESEARCH CORPORATION (CT)
NEW ENGLAND RADON SERVICES (CT)
NEW ENGLAND RADON, LTD. (NH)
NORDBERG ASSOCIATES (ME)
RADON ANALYSIS AND DETECTION OF NEW ENGLAND, INC. (NH)
RADON CHECK/DIVISION OF HORIZON SYSTEMS (MA)
RADON DETECTION SERVICES (VT)
RADON DETECTION SYSTEMS OF VERMONT (VT)
RADON DETECTION SYSTEMS/NEWARK, NEW JERSEY BRANCH (NJ)
RADON DETECTION, INC. (NH)
RADON DETECTORS (NY)
RADON EVALUATION DIVISION/EVALDO, LTD. (CT)
RADON LABS (MA)
RADON REGISTRY (NH)
RADON TECHNOLOGIES, INC. (NY)
RADON TECHNOLOGY AND CONSULTANTS (RI)
RADON TECHNOLOGY, INC. (NH)
RADON TESTING AND CONSULTING SERVICES (MA)
RESIDENTIAL RADON TEST SERVICE (MA)
RISE (RI)
SHELBURNE LABORATORIES, INC. (VT)
STAR LABORATORIES (MA)

Companies Serving
Virgin Islands

All companies listed under Nationwide provide services for this state.

AAAIRSCAN ENGINEERING CORPORATION (CT)
DAMES AND MOORE (MD)

Companies Serving
Virginia

All companies listed under Nationwide provide services for this state.

AAAIRSCAN ENGINEERING CORPORATION (CT)
ABC INSPECTION COMPANY (MD)
ABIONICS, LTD. (VA)
ACCU-TEST (VA)
ADVANCED AUTOMATIC ALARM COMPANY, INC. (MD)
AFFILIATED BUILDING INSPECTORS, INC. (MD)
AIDCO MAINE CORPORATION (NH)
AIR HAZARDS ASSESSMENT, INC. (MD)
AIR QUALITY ANALYSTS, INC. (VA)
AIR QUALITY DETECTION, INC. (MD)
AIR SOLUTIONS (MD)
AIR VENTILATION SYSTEMS, INC. (MD)
AIRMASTERS RADON DIVISION (MD)
ALL SERVICE RADON REDUCERS, INC. (MD)
ALPHA CHECK (MD)
AMERICAN RADON TESTERS (VA)
APPLIED ANALYTICAL TECHNOLOGIES, INC. (RI)
BENCHMARK INSPECTION SERVICES (VA)
BIOMEDICAL TOXICOLOGY ASSOCIATES (MD)
BIOSPHERICS, INC. (MD)
C SENTINEL ENTERPRISES (VA)
CAPITAL HOME INSPECTION CORPORATION (VA)
CARROLL COUNTY RADON CONTROL, INC. (MD)
COMMONWEALTH ENGINEERING AND TECHNOLOGY (PA)
CONSULTING CHEMISTS (TN)
CONTINENTAL ENVIRONMENT COMPANY, INC. (NJ)
DAMES AND MOORE (MD)
DPR BUILDERS (NJ)
DREWRY HOME INSPECTIONS, INC. (VA)
ECOLOGICAL INTERACTIONS, INC. (NJ)
ECONOMY HEATING AND AIR CONDITIONING (MD)
ENVIRONMENTAL ENGINEERING AND TESTING (CO)
ENVIRONMENTAL MANAGEMENT GROUP (MD)
ENVIRONMENTAL SYSTEMS SERVICE (VA)
ENVIRONMENTAL SYSTEMS SERVICE, LTD. (MD)
ETS ENVIRONMENTAL TESTING SERVICE (MD)
FEDERAL NUCLIDE MONITORS (DC)
FORESIGHT ENGINEERING, INC. (NJ)
FORESIGHT ENGINEERING, INC., VIRGINIA (VA)
GARCO RESEARCH COMPANY (NJ)
GEIER CONSTRUCTION COMPANY, INC./DIVISION OF RADON CONTROL (MD)
GEOCENTRIC ENGINEERING CORPORATION (VA)
H.M. COOK, JR., PROFESSIONAL ENGINEER (VA)
HILLMANN ENVIRONMENTAL COMPANY (NJ)
HOLIDAY TERMITE AND PEST CONTROL (VA)
HOME CHECK SERVICES (VA)
HOME EXTERMINATING COMPANY (MD)
HOMEPRO SYSTEMS, INC. (VA)
HOMERAD (MD)
HOMESURE INSPECTIONS, INC. (MD)
HURT AND PROFFITT, PROFESSIONAL ENGINEERS (VA)
INFILTEC, D.C. (DC)
J.A. ASSOCIATES/PUBLIC HEALTH CONSULTANTS (PA)
J.M.C. ENTERPRISES (TN)
LYNN PEST MANAGEMENT COMPANY (MD)
MERIT ENVIRONMENTAL SERVICES, INC. (KY)
MICROBAC ENVIRONMENTAL LABORATORY (NC)
MICROBAC LABORATORIES, INC./ERIE TESTING DIVISION-MAIN OFFICE (PA)
MICROBAC LABORATORIES, INC./HANOVER DIVISION (PA)
MICROBAC LABORATORIES, INC./KENTUCKY TESTING DIVISION (KY)
MICROBAC LABORATORIES, INC./MASSACHUSETTS TESTING DIVISION (MA)
MICROBAC LABORATORIES, INC./NEW CASTLE TESTING DIVISION (PA)
MICROBAC LABORATORIES, INC./PITTSBURGH TESTING DIVISION (PA)
MICROBAC LABORATORIES, INC./SCHILLER DIVISION (PA)
MICROBAC LABORATORIES, INC./SEAWAY DIVISION (IN)
MICROBAC LABORATORIES, INC./STRAND TESTING DIVISION (PA)
MORRISON WATERPROOFING COMPANY, INC. (VA)
NATIONAL RADON SERVICE (MD)
NORTH EAST ENVIRONMENTAL SERVICES (NJ)
PARAMOUNT PEST CONTROL (VA)
PERMA TREAT, INC. (VA)
PLUM GROVE CORPORATION (VA)
PROTECTION INSPECTION, INC. (TN)
PROTECTRON, INC. (MD)
QUALITY ENVIRONMENTS, INC. (NC)
R.J. MOORE AND ASSOCIATES, INC. (MD)
RADIATION CONSULTANTS OF SOUTHWEST VIRGINIA (VA)
RADIATION MEASUREMENT SERVICES (MD)
RADIOLOGICAL PHYSICS SERVICES/HEALTH EAST, INC. (VA)
RADON ANALYSIS, INC. (VA)
RADON BAN (MD)
RADON CALIBRATIONS (VA)
RADON DETECTION DEVICES (MD)
RADON DETECTION SYSTEMS/NASHVILLE, TENNESSEE BRANCH (TN)
RADON DETECTORS (NY)
RADON HOME INSPECTION SERVICE (MD)
RADON HUNTER, THE (VA)
RADON LABORATORIES (MD)
RADON MANAGEMENT COMPANY (VA)
RADON RANGERS (MD)
RADON SYSTEMS, INC. (MD)
RADON TEST AND REDUCTION SPECIALIST (NC)
RADON TESTING SERVICES, SILVER SPRING (MD)
RADONICS (VA)
REAL ESTATE INSPECTION SERVICE (TN)
RTS SYSTEMS (MD)
SCS ENGINEERS, INC. (VA)
TEE ENGINEERING COMPANY (KY)
TERRADYNAMICS CORPORATION (VA)
THIRD PARTY INSPECTIONS (VA)
TRACOR JITCO RADON MEASUREMENT SERVICES (MD)
TSI CONSTRUCTION SERVICES (IN)
U.S. ASSETS RECOVERY INVESTIGATIONS, INC. (DC)
U.S. HOME INSPECTION SERVICES (VA)
VIRGINIA DRY-BASE (VA)
VMS AND ASSOCIATES (VA)
WESTERN TECHNOLOGY, INC. (MD)

Companies Serving
Washington

All companies listed under Nationwide provide services for this state.

A & M ASSOCIATES (OR)
AIR HAZARDS ASSESSMENT, INC. (MD)
AMERICAN PEST MANAGEMENT (WA)
BAR-TEL INFRASCAN (WA)
C SENTINEL ENTERPRISES (VA)
CAVALIER CORPORATION (WA)
ENERGY LABORATORIES, INC. (WY)
ENVIRONMENTAL ENGINEERING LABORATORIES, INC. (CO)
FAYTEK, INC. (ID)
INSUL-TECH, INC. (OH)
MADRONA DIVERSIFIED, INC. (WA)
RADIATION HEALTH (WA)
RADON AWARENESS, INC. (ID)
RADON CONSULTANTS (NV)
RADON DETECTION AND REMOVAL SERVICES (WA)
RADON DETECTION SYSTEMS/LOS ANGELES, CALIFORNIA BRANCH (CA)
RADON DETECTION SYSTEMS/SPOKANE, WASHINGTON BRANCH (ID)
RADON INSPECTION SERVICES (WA)
RADON SYSTEMATICS OF CALIFORNIA (CA)
RADON TEST (CA)
RADON TESTING SERVICES, INC. (CA)
SANI-DUCT SERVICE OF THE INLAND NORTHWEST (WA)
SPENCE-GEIGER ASSOCIATES, INC. (CO)
TRUE TEST COMPANY, WASHINGTON (WA)
UNIVERSITY OF WASHINGTON/DEPARTMENT OF ENVIRONMENTAL HEALTH/RADIATION ECOLOGY LABORATORY (WA)
WASHINGTON STATE UNIVERSITY/NUCLEAR RADIATION CENTER (WA)

Companies Serving
West Virginia

All companies listed under Nationwide provide services for this state.

AAAIRSCAN ENGINEERING CORPORATION (CT)
ADVANCED AUTOMATIC ALARM COMPANY, INC. (MD)
AIDCO MAINE CORPORATION (NH)
AIR QUALITY ANALYSTS, INC. (VA)
BIOMEDICAL TOXICOLOGY ASSOCIATES (MD)
BIOSPHERICS, INC. (MD)
COMMONWEALTH ENGINEERING AND TECHNOLOGY (PA)
DAMES AND MOORE (MD)
ECONOMY HEATING AND AIR CONDITIONING (MD)
ENVIRONMENTAL MANAGEMENT GROUP (MD)
ENVIRONMENTAL SYSTEMS SERVICE (VA)
ENVIRONMENTAL SYSTEMS SERVICE, LTD. (MD)
FORESIGHT ENGINEERING, INC., VIRGINIA (VA)
GEOCENTRIC ENGINEERING CORPORATION (VA)
H.M. COOK, JR., PROFESSIONAL ENGINEER (VA)
HOMERAD (MD)
INSUL-TECH, INC. (OH)
J. BECK AND ASSOCIATES, INC. (PA)
J.A. ASSOCIATES/PUBLIC HEALTH CONSULTANTS (PA)
MERIT ENVIRONMENTAL SERVICES, INC. (KY)
MICROBAC ENVIRONMENTAL LABORATORY (NC)
MICROBAC LABORATORIES, INC./ERIE TESTING DIVISION-MAIN OFFICE (PA)
MICROBAC LABORATORIES, INC./HANOVER DIVISION (PA)
MICROBAC LABORATORIES, INC./KENTUCKY TESTING DIVISION (KY)
MICROBAC LABORATORIES, INC./MASSACHUSETTS TESTING DIVISION (MA)
MICROBAC LABORATORIES, INC./NEW CASTLE TESTING DIVISION (PA)
MICROBAC LABORATORIES, INC./PITTSBURGH TESTING DIVISION (PA)
MICROBAC LABORATORIES, INC./SCHILLER DIVISION (PA)
MICROBAC LABORATORIES, INC./SEAWAY DIVISION (IN)
NATIONAL RADON SERVICE (MD)

PROTECTRON, INC. (MD)
RADIATION CONSULTANTS OF SOUTHWEST VIRGINIA (VA)
RADON CALIBRATIONS (VA)
RADON DETECTION SYSTEMS/CHARLESTON, WEST VIRGINIA BRANCH (WV)
RADON DETECTION SYSTEMS/NASHVILLE, TENNESSEE BRANCH (TN)
RADON DETECTORS (NY)
RADON ENVIRONMENTAL SERVICES, INC. (KY)
RADON LABORATORIES (MD)
RADON MANAGEMENT COMPANY (VA)
RADON RANGERS (MD)
RADON TECHNOLOGIES, INC. (OH)
RADON TEST AND REDUCTION SPECIALIST (NC)
RADON TESTING ASSOCIATION (WV)
RADON TESTING SERVICES, SILVER SPRING (MD)
RADONICS (VA)
RAY MESSICK RADON TESTING AND MITIGATION (PA)
TEE ENGINEERING COMPANY (KY)
TSI CONSTRUCTION SERVICES (IN)
WESTERN TECHNOLOGY, INC. (MD)

Companies Serving

Wisconsin

All companies listed under Nationwide provide services for this state.

BCA LABORATORIES, INC. (IL)
DAMES AND MOORE (MD)
EES OF MINNESOTA, INC. (MN)
ELLIOTT AND ASSOCIATES (IL)
ENERGY INDEPENDENCE/THE RADON RESPONSE TEAM (MO)
EXPEDITERS, INC., THE (WI)
HOMEBUYERS INSPECTION SERVICE, INC. (IL)
ILLINOIS STATE GEOLOGICAL SURVEY (IL)
NATIONAL PROPERTY INSPECTIONS (WI)
NIMCO/NESSI (WI)
RADON DETECTION SPECIALISTS (IL)
RADON DETECTION SYSTEMS/BLOOMINGTON-NORMAL BRANCH (IL)
RADON DETECTION SYSTEMS/CHICAGO, ILLINOIS BRANCH (IL)
RADON DETECTION SYSTEMS/MILWAUKEE BRANCH (WI)
RADON ENVIRONMENTAL SERVICES, INC. (KY)
RADON MEASURE COMPANY, THE (MN)
RADON SURVEYS (IL)
RADON SURVEYS/DIVISION OF COPE TECHNICAL SERVICES (WI)
RADON TESTING SERVICE, INC. (WI)
TSI CONSTRUCTION SERVICES (IN)

Companies Serving

Wyoming

All companies listed under Nationwide provide services for this state.

ALPHA RADIATION ENGINEERING, INC. (CO)
ARIX CORPORATION (WY)
BONDAR-CLEGG, INC. (CO)
COLORADO FRONT RANGE BUILDING INSPECTION SERVICE (CO)
COLORADO RADON ENGINEERING, INC. (CO)
ENERGY ALTERNATIVES UNLIMITED, INC. (CO)
ENERGY LABORATORIES, INC. (WY)
ENVIRONMENTAL ENGINEERING AND TESTING (CO)
ENVIRONMENTAL ENGINEERING LABORATORIES, INC. (CO)
FAYTEK, INC. (ID)
GLOBAL WHOLESALE (WY)
LANDMARK RADON LABORATORIES, LTD. (CO)
LINCOLN DEVORE, INC. (CO)
M.V.E., INC. (CO)
NATURAL RESOURCE TECHNOLOGIES (CO)
RADON AWARENESS, INC. (ID)
RADON CONSULTANTS (NV)
RADON CONSULTANTS, INC. (CO)
RADON DETECTION SYSTEMS/CASPER, WYOMING BRANCH (WY)
RADON DETECTION SYSTEMS/LOS ANGELES, CALIFORNIA BRANCH (CA)
RADON DETECTION SYSTEMS/SALT LAKE CITY, UTAH BRANCH (UT)
RADON SERVICES OF WYOMING (WY)
RADON TEST (CA)
RTK ENGINEERING (CO)
SANI-DUCT SERVICE OF THE INLAND NORTHWEST (WA)
TIMBERLINE RADON SERVICE, INC. (CO)

Companies Serving

Canada

All companies listed under Nationwide provide services to Canada.

BARRINGER MAGENTA, LTD. (ON)
GLOBAL WHOLESALE (WY)
MACLAREN ENGINEERS, INC. (ON)
ONTARIO RESEARCH FOUNDATION (ORF) (ON)
R.T.C. (NY)
RADON ALERT SYSTEMS (SK)
RADON DETECTION SERVICES (SK)
SCIENCE AND TECHNOLOGY, INC. (SCAT) (NB)
SENES CONSULTANTS (ON)
STUART HUNT AND ASSOCIATES/OCCUPATIONAL AND RADIATION SAFETY SERVICES, LTD. (AB)

Radon Mitigation Companies

This section lists companies in the U.S. and Canada that have indicated capabilities for the design, installation, and evaluation of radon mitigation systems. Included here are building contractors, engineers, architects, health physicists, industrial hygienists, and research scientists that use various techniques to reduce radon levels in existing structures and new construction. Many of the companies listed here also provide other radon-related services and can be found in additional sections of the Directory.

Because radon mitigation companies frequently use radon measurement methods to aid in the selection, design, and evaluation of a radon mitigation system, a significant number of companies listed here have participated in The National Radon Measurement Proficiency (RMP) Program. The designation "EPA RMP 5", when found at the end of a listing, designates the company's successful participation in this program. For a full explanation of the EPA RMP program, refer to Section 1.

Please note that the design and installation of radon mitigation systems is still a developing field. When seeking a radon mitigation company, it is advised that proper evaluation include direct contact and careful assessment of the company's qualifications and ability to meet the necessary requirements.

An index listing radon mitigation companies by state served can be found immediately following this section.

A

A & A ENVIRONMENTAL ENGINEERING COMPANY
1805 South Bellaire St.
Denver, CO 80222
(303) 753-1034
Darrell C. Rosen - Senior Partner
Services:
Provides radon detection and mitigation services.
Area Served:
Nationwide.
EPA RMP 5:
AT, CC.

A & P WATERPROOFING, INC.
1635 Spring Ave.
Jenkintown, PA 19046
(215) 886-3398, (609) 772-1112
John B. Ambler - President
Services:
Provides radon mitigation services for local real estate companies and general public.
Qualifications:
Affiliate member of Montgomery Co., PA Board of Realtors.
Area Served:
NJ, PA.

A. ARTHUR LIPSHUTZ ASSOCIATES
7650 South Tamiami Trail Suite 9
Sarasota, FL 34231
(813) 923-7300
A. Arthur Lipshutz - Manager
Services:
Provides radon detection and mitigation services for residential and commercial markets, specializes in slab-on-grade construction, consultants to builders and real estate professionals.
Qualifications:
AARST, BBB, Executive Network, three years of radon mitigation experience.
Area Served:
Nationwide.
EPA RMP 5:
CC(P), CR(P).

A.L. BOUCHARD COMPANY
56 Franklin Pl.
Totowa, NJ 07512
(201) 790-6259
A.L. Bouchard - President
Services:
Provides radon detection and mitigation services.
Qualifications:
AARST.
Area Served:
NJ.

A.T.E.I.
809 East 11th St.
Michigan City, IN 46360
(800) 833-0999
George Perry - President
Charlie Geren - Regional Representative
Services:
Provides radon detection and mitigation services for residential and commercial structures.
Area Served:
Nationwide.

AAA DISASTER FORCE
18219 West 3rd Ave. Suite 3
Golden, CO 80401
(303) 399-4203
Mary Knudson

Area Served:
CO.
EPA RMP 5:
CC.

AAA RADON SERVICES, INC.
P.O. Box F
Hellertown, PA 18055
(215) 868-6025, (215) 821-2938
Robert Grosset - President
Services:
Provides radon detection and mitigation services.
Qualifications:
PA DER approved course.
Area Served:
Nationwide.
EPA RMP 5:
CC.

AAA RADON TESTING
26 Flynn Pl.
Wayne, NJ 07470
(201) 772-7166
Nicholas Verhagen - President
Services:
Provides radon detection and mitigation services.
Qualifications:
NJ, PA certification.
Area Served:
NJ, PA.
EPA RMP 5:
CC.

AADVANCED AAIR AANALYSIS
2670 Billingsley Rd.
Columbus, OH 43235
(614) 792-3400
James Cutcher - President
Services:
Provides radon detection, mitigation, and reduction repair services for residential, commercial, and industrial sites.
Qualifications:
EPA RMP, ARA, BBB, 18 years of engineering and building experience.
Area Served:
Nationwide.
EPA RMP 5:
CR(P).

ABACUS LABS RADON ASSESSMENT AND MITIGATION (ALRAM)
P.O. Box 5490
Portsmouth, NH 03801
(603) 433-3332, (800) 228-3332
Gary W. Seesman
Area Served:
CT, DE, MA, ME, NC, NH, RI, VT.
EPA RMP 5:
CC(P).

ACCREDITED WATERPROOFING COMPANY, INC.
2504 Church Rd.
Cherry Hill, NJ 08002
(609) 482-2630, (215) 265-2888
Allan Konkus - Vice President
Services:
Provides radon mitigation services for residential market.
Area Served:
DE, NJ, PA.

ACS HOVAL
935 Lively Blvd.
Wood Dale, IL 60191
(312) 860-6800
William Fitzpatrick
Area Served:
MD.

ADEPT REMEDIATION CORPORATION
12710 Jesse Smith Rd.
Mt. Airy, MD 21771
(301) 831-9607
Ernest Brightwell
Services:
Provides radon mitigation services.
Area Served:
MD.

ADVANCED BASEMENT DE-WATERING AND RADON CONTROL SYSTEMS
20 Giblin Dr.
West Seneca, NY 14224
(716) 675-5898, (716) 674-6070
John Mayer
L.T. Kempski
Area Served:
NY.

ADVANCED RADIATION MONITORING SERVICE, INC. (ARMS)
2219 Prosperity
Portage, MI 49081
(616) 323-9483
Philip B. Noack - President
Mary K. Noack - Secretary
Services:
Manufactures and distributes activated charcoal test monitors and radon laboratory equipment, provides radon mitigation services.
Qualifications:
Nuclear medicine technologist, physicists, engineers on staff.
Area Served:
Nationwide.
EPA RMP 5:
CC(P).

AERO-AQUA
54 Miller St.
Belfast, ME 04915
(207) 338-2809
Tom Reuwer - Field Office Manager
Area Served:
ME.

AFFTREX, LTD.
3530 Washington Ave.
Finleyville, PA 15332
(412) 348-9211
William H. Button - President
Douglas J. Button - Vice President
Services:
Provides radon detection services using charcoal canister, alpha track, and grab sample methods, provides mitigation services.
Qualifications:
PA DER interim radon testing and mitigation certification.
Area Served:
NY, PA.

AGENTERPRISE, INC.
227 North 7th St.
Allentown, PA 18102
(215) 776-0777
John Murphy
Area Served:
NJ, PA.
EPA RMP 5:
CC.

AIDCO MAINE CORPORATION
Grassy Pond House
Rindge, NH 03461
(603) 899-5166
R. Multer - President
Services:
Provides radon detection, mitigation, and on-site consultation services.
Area Served:
CT, DC, DE, FL, GA, IA, IL, KY, MA, MD, ME, MI, NC, NH, NJ, NY, OH, PA, RI, VT, VA, WV.
EPA RMP 5:
CC.

AIR QUALITY CONTROL, INC.
4 Terry Dr. Suite 3
Newtown, PA 18940
(215) 860-6198
Dan Smith - Branch Manager

Services:
Provides radon detection and mitigation services.
Area Served:
NJ, PA.

AIR QUALITY CONTROL, INC.
585 Phoenixville Pike
Malvern, PA 19335
(215) 889-3100, (800) 654-2144
Dave Smith - Owner
Buzz Clark - Owner
Services:
Provides radon detection and mitigation services.
Area Served:
PA.

AIR SCIENCES, INC.
508 North Truman Blvd. P.O. Box 320
Crystal City, MO 63019
(314) 933-3915, (800) 242-8886
Kerry Van House - Chief Executive Officer
Ralph Overman - Vice President
Services:
Provides radon detection and mitigation services, distributes charcoal canister home detection kits.
Qualifications:
Staff includes nuclear scientist with over 40 years of radiation work and physicist skilled in radiation measurement.
Area Served:
Nationwide.
EPA RMP 5:
CC(P).

AIR SENSE, INC.
25 William St.
Pawcatuck, CT 06379-2108
(203) 599-5599
Douglas C. Hoagland - President
Services:
Provides radon detection and mitigation services.
Qualifications:
AARST.
Area Served:
CT, MA, ME, NH, RI, VT.

AIR SOLUTIONS
1004 Dennis Ave.
Silver Spring, MD 20901
(301) 681-9242
Kevin St. Clair - President
Services:
Provides radon mitigation services and consultation for residential market.
Qualifications:
Registered architect, participant in Maryland indoor radon evaluation methods course.
Area Served:
DC, MD, VA.
EPA RMP 5:
CC.

AIR TRON MECHANICAL SERVICES
3178 South Halifax St.
Aurora, CO 80013
(303) 699-1836
Rob Brooks - Owner
Area Served:
CO.

AIRCARE SYSTEMS
P.O. Box 4942
Lancaster, PA 17604
(717) 293-9223
John E. Henderson
Area Served:
PA.

AIRMASTER RADON DETECTION, INC.
P.O. Box 3177
Wayne, NJ 07474-3177
(201) 794-3733
William Mirabile - Owner
Services:
Provides radon detection services using charcoal canister method, provides radon mitigation services.
Area Served:
NJ.

AIRMASTERS RADON DIVISION
1746 Boggs Rd.
Forest Hill, MD 21050
(301) 893-1993
Kelly F. McGill - President
Services:
Provides radon detection and mitigation services.
Area Served:
DC, MD, VA.

AIRREGS, INC./AIR POLLUTION REGULATORY SERVICE, INC.
1640 Powers Ferry Rd. Building 17, Suite 210
Atlanta, GA 30067
(404) 984-1919
Robert Geddis - Hygienist
Richard Shutt - Engineer
Services:
Provides radon detection and mitigation services, training to homeowners and realtors.
Qualifications:
EPA RMP, 38 years of combined experience, both professionals have master's degrees.
Area Served:
Nationwide.
EPA RMP 5:
AT, CC.

ALL AMERICAN RADON SERVICES, INC.
P.O. Box 1079
Orange Park, FL 32067
(904) 642-1963
Paul Bliton
J.E. Bliton
Qualifications:
AARST.
Area Served:
FL.

ALL-DRY BASEMENTS AND SAFE AIR, INC.
P.O. Box 205
Collinsville, CT 06022
(203) 693-0245
Gary Drapeau - President
Area Served:
CT.

ALLISON CONSTRUCTION
2821 Remington
Fort Collins, CO 80525
(303) 226-1740
Chris Allison
Area Served:
CO.

ALPHA BETA GAMMA TECHNOLOGIES, INC.
1829 Cawley Ave.
Bethlehem, PA 18017
(215) 866-8895
Glenn E. Aranyos - Owner
Area Served:
PA.

ALPHA CONCEPTS, INC.
P.O. Box 119
Hopatcong, NJ 07843
(201) 770-3046
Thomas Hatton
Area Served:
Nationwide.
EPA RMP 5:
AT, CC, CR(P).

ALPHA GUARD/TECTROL
5613 Yarrow St.
Arvada, CO 80002
(303) 421-7955
Greg Piette - Owner
Area Served:
CO.

ALPHA RADIATION ENGINEERING, INC.
P.O. Box 90
Littleton, CO 80160-0090
(303) 721-0068, (303) 721-0872
Howard E. Dixon
Area Served:
AZ, CO, FL, KS, MT, ND, NE, NM, NV, OK, TX, UT, WY.
EPA RMP 5:
AT, CC.

ALPINE RADIATION MITIGATION
35 Kris Ln.
Manitou Springs, CO 80829
(719) 685-1024
Virgil McCallson - Owner
Area Served:
CO.

AMERICAN COMPLIANCE TECHNOLOGIES, INC.
4404 South Florida Ave. Suite 12
Lakeland, FL 33813
(813) 647-1200
Robert O. Kincart - Hazardous Materials
Services:
Comprehensive environmental services programs including mobile laboratory, radon detection and mitigation, real estate inspections and assessments.
Qualifications:
Certified hazardous materials manager, radon chemistry and detection courses.
Area Served:
Nationwide.

AMERICAN EXCELSIOR COMPANY
1 Lovellee Ln.
Thornton, PA 19373
(215) 459-1146
Abe Pollackov - Sales Manager
Services:
Distributes venting mat used in pre-construction to build radon-resistant homes, provides radon mitigation services.
Area Served:
PA.

AMERICAN HERITAGE CORPORATION
P.O. Box 1145
Paoli, PA 19301
(717) 275-0300, (800) 247-0300
Kenneth G. Enochs - Director
Services:
Provides radon detection and mitigation services, fully equipped vehicles and teams (200 mile radius).
Qualifications:
PA DER, NY DOE.
Area Served:
DE, NJ, NY, PA.

AMERICAN IMPROVEMENT COMPANY
2101 South Branch Ave.
Altoona, PA 16602
(814) 946-1664, (800) 922-5857
James M. Werner - President
Services:
Provides radon mitigation services, distributes charcoal canisters.
Qualifications:
Attended DER seminar, Kutztown, PA.
Area Served:
PA.

AMERICAN INSULATION, INC.
15 Cherry St. P.O. Box 266
Danvers, MA 01923
(617) 777-0617
David O'Heir - President
Services:
Provides radon mitigation services using insulation, distributes radon test kits.
Area Served:
MA.

AMERICAN MEDICAL MARKETING SERVICES
2429 West Davie Blvd. Suite 17
Ft. Lauderdale, FL 33312
(305) 462-2575
Jim Phillips - Vice President
Services:
Provides radon detection and mitigation services.
Area Served:
FL.

AMERICAN RADON CORPORATION
8300 North Hayden Rd. Suite 100
Scottsdale, AZ 85258
(602) 967-8029
James Krueger - President
Susan Bollin - Senior Director
Services:
Provides radon detection and mitigation services, provides educational services for real estate and legal professions.
Qualifications:
President is member of ASTM-E6 task group subcommittee and director of AARST.
Area Served:
Nationwide.
EPA RMP 5:
AT, CC, EP(P).

AMERICAN RADON SERVICES, LTD.
303 I.S.I.S. Center Iowa State University
Ames, IA 50011-2130
(515) 296-9926
Kenneth D. Wiggers - President
Services:
Provides radon detection services using E-Perm and charcoal canister methods, provides radon mitigation services.
Area Served:
Nationwide.
EPA RMP 5:
CC(P), EP(P).

AMERICAN RADON TECHNICAL SERVICES, INC.
1941A Nears Pkwy.
Margate, FL 33063
(305) 968-5555
Gordon Marsh - President
Marge Semenuk - Vice President
Services:
Provides radon detection services using charcoal canister and E-Perm methods, provides radon mitigation services.
Qualifications:
General contractor and physicist on staff.
Area Served:
NJ, FL.

APPLIED CONSTRUCTION, INC.
The Friary, Winchester-on-Severn 1600 Winchester Rd.
Annapolis, MD 21401
(301) 757-7887
Leroy Phillips
Area Served:
DC, MD, PA, VA.

APPLIED HEALTH PHYSICS, INC.
2986 Industrial Blvd.
Bethel Park, PA 15102
(412) 563-2242, (518) 477-7974
R.G. Gallagher - President
John Douglas - Instrument Manager
John Harshman - Radon Specialist
Services:
Professional radiation safety firm distributes charcoal canisters, provides instrument calibration, repair, and leasing.
Qualifications:
EPA RMP, twenty-five years of experience, NRC and state licensed, all services by board certified registered professional engineers.
Area Served:
Nationwide.
EPA RMP 5:
AT, CC, EP(P).

APPLIED RADON, INC.
52205 Warehime Rd.
Cumberland, OH 43732-9804
(614) 638-2078, (800) 638-2078
Frank J. Donia - Vice President

Nancy A. Donia - President
Services:
Provides radon detection and mitigation services, distributes At-Ease radon monitors, presents seminars to real estate industry.
Qualifications:
ARA, EPA RMP, health physicist on staff.
Area Served:
Nationwide.
EPA RMP 5:
AT, CC, CR(P).

ARID BASEMENT COMPANY
237 East Utica
Buffalo, NY 14208
(716) 885-2162
Harrison R. Woolworth
Area Served:
NY.

ARIX CORPORATION
325 East Southern Ave. Suite 18
Tempe, AZ 85282
(602) 968-0052
Lee Compton
Area Served:
AZ.

ARIX CORPORATION
760 Horizon Dr.
Grand Junction, CO 81506
(303) 242-6203
J. Tell Tappan - President
David Reinersten
Area Served:
Nationwide.
EPA RMP 5:
AT, CC(P), GR(P).

ARIX CORPORATION
P.O. Box 2021
Greeley, CO 80632-2021
(303) 356-3930
William F. McKelvey
Patti Terry
Area Served:
CO.
EPA RMP 5:
CC.

ARIX CORPORATION
2155 North 200 West Suite 100
Provo, UT 84604
(801) 373-2749
Toni Sullivan
Area Served:
UT.
EPA RMP 5:
CC.

ARIX CORPORATION
877 North Eighth, W. Suite 8
Riverton, WY 82501
(307) 858-6506
Melody Reeb
Patty Gartrell
Area Served:
MT, WY.
EPA RMP 5:
CC.

ARIX CORPORATION/A DIVISION OF VERSAR, INC.
1200 East High St.
Pottstown, PA 19464
(215) 327-1140
W. McKelvey - Operations Manager
R. Meier - Radiologic Technician
L. Schlegel - Business Development
Services:
Radon consultants provide mitigation, design, and testing for residential and commercial buildings, and schools in the public and private sectors, provides education and training courses.
Qualifications:
EPA RMP, PA DER testing and mitigation courses, listed by NJ DEP.
Area Served:
Nationwide.
EPA RMP 5:
AT, CC, GR(P).

ARTHUR SCOTT AND ASSOCIATES/AMERICAN ATCON
2150 Burnhamthorpe Rd., W. Box 164
Mississauga, ON L5L 3A2
(416) 828-2389
Arthur G. Scott - President
Andrew Robertson - Research
Services:
Operates under the name American Atcon in the U.S., presently managing EPA project in Ohio that deals with radon mitigation techniques in buildings, has contracted with FIPR for mitigation projects in Florida, radon projects deal with experimental radon mitigation procedures and construction techniques.
Area Served:
Nationwide.

ARTHUR W. MATTES, INC.
134 Washington Ave.
Bethlehem, PA 18018
(215) 867-5268
Area Served:
PA.

ASSOCIATED DESERT DRY WATERPROOFING COMPANY, INC.
504 South Van Dorn St.
Alexandria, VA 22304
(703) 360-7100
H. Edward Reiff - President
Services:
Provides radon mitigation services using underdrains, sump pits, fans, etc.
Qualifications:
Listed by state of VA as mitigation firm.
Area Served:
MD, NJ, PA, VA.

ASTLE CORPORATION
146 Old Farm Rd.
Leominister, MA 01453-2029
(508) 534-0979
William Astle - President
Services:
Provides radon detection and mitigation services, manufactures a patented air-to-air energy recovery device that reclaims 95% of heat lost up flue.
Area Served:
MA.

ATLAS WATERPROOFING COMPANY, INC.
P.O. Box 14953
Reading, PA 19612-4953
(215) 374-6454
Leo Runge - Owner
Area Served:
PA.

AURORA INDUSTRIES
P.O. Box 936
Concordville, PA 19331
(215) 459-1146
Abe Pollackov - Owner
Services:
Provides radon mitigation services and construction consultation.
Area Served:
NJ, PA.

B

B & H RADON SERVICES
P.O. Box 247
Silver Spring, PA 17575-0247
(717) 684-2612, (717) 426-3723
James Bradley - Owner
Alfred Henne - Owner
Area Served:
PA.

BASEMENT DE-WATERING SYSTEMS, INC.
156 Curry Hill Rd.

Canonsburg, PA 15317
(412) 745-5959
Robert Johnston - Owner
Services:
Provides radon detection and mitigation services.
Area Served:
Nationwide.

BATEMAN HOME IMPROVEMENTS
17024 Sioux Ln.
Gaithersburg, MD 20878
(301) 330-1528
Michael R. Bateman - Owner
Services:
Provides radon mitigation services for residential and commercial buildings.
Qualifications:
AARST.
Area Served:
MD.

BAYSTATE ENVIRONMENTAL
53 Cushing St.
Stoughton, MA 02072
(617) 341-0002
Frederick J. Hagerty
Qualifications:
AARST.
Area Served:
MA.

BERG GROUP
1401 West Hamburg
Baltimore, MD 21230
(301) 837-5525
David Berg - Vice President
Services:
Provides radon detection services using charcoal canister method, provides mitigation services.
Qualifications:
NRA.
Area Served:
MD.

BIEDERER CONSTRUCTION SERVICE
421 Hillandale
Round Lake, IL 60073
(312) 546-3890
Ray Biederer
Area Served:
IL.

BIONOMIC INDUSTRIES, INC.
479 Route 17 N.
Mahwah, NJ 07430
(201) 529-1094, (800) 338-3210
John Enhoffer - Vice President
Steve Augulis - Design Engineer
Barbara Purcel - Customer Service
Services:
Provides complete engineered radon mitigation systems for residential, commercial, and industrial buildings, designs for both new and existing construction.
Qualifications:
ARRST, registered with EPA, NJ DEP, and other state organizations.
Area Served:
Nationwide.
EPA RMP 5:
AT, CC, CW(P).

BOHNSACK GROUP, THE
4329 Tamalga
South Euclid, OH 44121
(216) 291-8720
Robert Bohnsack - Owner

BONDAR-CLEGG, INC.
12980 West Cedar Dr.
Lakewood, CO 80228
(303) 989-1404
W.G. Sherwood
Services:
Provides radon detection and mitigation services for residential and commercial markets.
Qualifications:
AARST, EPA RMP.
Area Served:
AZ, CA, CO, KS, NE, NM, NV, UT, WY.
EPA RMP 5:
CC.

BRANDYWINE ENVIRONMENTAL SERVICES
P.O. Box 3343 206-B East Ayre St.
Newport, DE 19804
(302) 995-7565
Thomas S. Anthony - President
Area Served:
DE.

BROOKS LABORATORIES, INC.
44 Codfish Ln.
Weston, CT 06883
(203) 226-6384
Pramod Patel
Services:
Provides radon detection, mitigation, and consultation services.
Qualifications:
Industrial hygiene consultants.
Area Served:
CT.

BRUN-REUSS ASSOCIATES
P.O. Box 0842
Lionville, PA 19353-0482
(800) 522-0072
J.F. Brown - President
Area Served:
PA.

BUILDING DIAGNOSTICS, INC.
380 27th St.
Boulder, CO 80303
(303) 494-2431
Mike Holecek
Area Served:
CO.

BURNS AND ASSOCIATES
265 Kearsarge Way
Portsmouth, NH 03801
(603) 431-3400
Robert J. Burns - Principal
Virginia Steckowych - Q/A Manager
Services:
Provides radon air and water detection services, diagnostic work ,and design of mitigation techniques.
Qualifications:
Registered professional engineer, completed AARST Radon Specialist course.
Area Served:
MA, ME, NH, VT.
EPA RMP 5:
AT, CC, EP(P).

BUZZANO CONTRACTING
62 Mt. Pleasant Rd.
Newtown, CT 06470
(203) 426-7196
Steve J. Nicolosi - Owner
Chris Meehan
Services:
Provides radon reduction services to public and government, water mitigation by aeration, radon diagnostics using EPA methods and techniques.
Qualifications:
NY workshop and CT, MA and NH state lists.
Area Served:
CT, MA, NJ, NY.

C

C & D HEATING AND CONSTRUCTION
1111 South Sheridan Blvd.
Lakewood, CO 80226
(303) 935-6461
Charles Smith
Sami Todd
Services:
Provides radon detection and mitigation services.
Area Served:
CO.

C. KENT CHAMBERLAIN AND ASSOCIATES
6579 West David Dr.
Littleton, CO 80123
(303) 979-3916
C. Kent Chamberlain - President
Services:
Provides radon detection services using charcoal canister and alpha track methods, provides radon mitigation services.
Qualifications:
Principal holds doctorate in geology, certified by two professional societies.
Area Served:
CO.

CALOIERO RADON TECHNOLOGIES
RD5, Box 84 North Main St.
Williamstown, NJ 08094
(609) 728-9257
Joseph Caloiero
Services:
Provides radon detection and mitigation services.
Qualifications:
AARST.
Area Served:
NJ.

CAMPAGNA BUILDERS
398B Union Blvd.
Totowa, NJ 07512
(201) 942-9661
Anthony Campagna
Qualifications:
AARST.
Area Served:
NJ.

CAMRODEN ASSOCIATES
RFD 4, Box 62
Rome, NY 13440
(315) 865-4269
Terry Brennan - President
Services:
Provides radon mitigation services, indoor air quality investigation, and training programs for federal, state, and private sector groups.
Qualifications:
APCA, ASTM, NAHB.
Area Served:
Nationwide.
EPA RMP 5:
AT, CC.

CAPITAL MATERIALS TESTING, INC.
P.O. Box 81 5 Hemlock St.
Latham, NY 12110
(518) 783-3150, (800) 541-4109
James Swartz - Regional Manager
Bob Field - Technician
Services:
Provides radon detection and mitigation services, distributes radon test kits.
Qualifications:
NY state certified lab, EPA RMP.
Area Served:
NJ, NY, PA.

CARE SYSTEMS
P.O. Box 122
Brentwood, TN 37027
(615) 373-4275
Bob Ingham
Area Served:
TN.

CAROLINA RADON CONSULTANTS
P.O. Box 2000
Etowah, NC 28729
(704) 891-9559
Ned Doyle - President
Area Served:
GA, NC, SC.

CAROLINA RADON CONSULTANTS
P.O. Box 6683
Columbia, SC 29260
(803) 779-2043
Ned Doyle - President
Area Served:
GA, NC, SC.
EPA RMP 5:
CC.

CAROLINA RADON DETECTION
209 Tebblewood Dr.
Simpsonville, SC 29681
(803) 282-0494
Alan Huggins
Services:
Provides radon detection and mitigation services.
Qualifications:
AARST.
Area Served:
SC.

CAVALIER CORPORATION
11516 East Sprague
Spokane, WA 99206
(509) 926-6217
Warren Riddle
Sylvia Riddle
Services:
Provides radon detection and mitigation services specializing in real estate, commercial, and public structures, distributes Fantech products.
Qualifications:
EPA training courses.
Area Served:
ID, MT, OR, WA.
EPA RMP 5:
AT, CC.

CCI-RADON AND HOME INSPECTION/RADON DETECTION
P.O. Box 1045
Malvern, PA 19355
(215) 889-1554, (215) 363-9015
Charles E. Weber - President
Bruce Jordan - Vice President
Norman Warner - Vice President
Services:
Provides radon detection and mitigation services for home owners and buyers, distributes charcoal canisters and alpha track detectors to retail outlets.
Qualifications:
AARST, EPA RMP PA DER certification.
Area Served:
DE, NJ, PA.
EPA RMP 5:
CC.

CENTRAL JERSEY RADON, INC.
P.O. Box 600
Lincroft, NJ 07738
(201) 741-0400
John A. Murray - President
Services:
Provides radon detection and mitigation services and consultations.
Area Served:
NJ.

CERTIFIED TESTING SERVICES, INC.
474 Easton Rd.
Horsham, PA 19044
(215) 672-8100
James Bucciarelli - President
David Delavan - Mitigation Specialist
Bernard J. Salera - Executive Secretary
Richard Salera
Services:
Provides radon detection and mitigation services, specializes in real estate transactions.
Qualifications:
Full compliance with PA DER interim requirements for certification.
Area Served:
PA.

COLORADO SAFE-N-DRY, INC.
7950 East Prentice Ave. Suite 201
Englewood, CO 80111
(303) 771-2212
Charles Austin - President
Services:
Provides radon detection and mitigation services.

Area Served:
CO.

COLORADO VENTURE MANAGEMENT, INC.
2995 Wilderness Pl.
Boulder, CO 80301
(303) 440-4055
Kyle Lefkoff - General Manager
Services:
Provides radon detection services for residential and commercial properties using the charcoal canister method, provides radon mitigation services.
Area Served:
Nationwide.

COMMONWEALTH ENTERPRISES, INC.
P.O. Box S187
St. Davids, PA 19087
(215) 688-7097
William DiAddezio - President
Doris Natale
Services:
Provides radon detection and mitigation services to property owners and buyers privately and through realtors.
Qualifications:
EPA detection and mitigation courses.
Area Served:
PA.

COMMUNITY HEALTH ASSOCIATES, INC.
316 Sherman St.
Canton, MA 02021
(617) 828-3123
Thomas Gecewizcz - President
William Thibeault - Vice President
Services:
Provides radon detection and mitigation services.
Area Served:
MA.

COMPREHENSIVE RADON SERVICES, INC.
P.O. Box 548
Manchester, NH 03105
(603) 644-7301
Mark Malbon - President
Services:
Provides radon detection services using charcoal canister, alpha track, continuous radon, grab sampling, working level methods, water analysis, and open land testing, provides radon mitigation services, distributes radon mitigation supplies.
Area Served:
NH.
EPA RMP 5:
CC, CR(P), GR(P).

CONNECTICUT BASEMENT DE-WATERING AND RADON SYSTEMS, INC.
1 Clayton St.
Milford, CT 06460
(203) 874-3491
Larry Janeski - President
Services:
Provides radon detection and mitigation services.
Area Served:
CT.
EPA RMP 5:
CC.

CONSTRUCTION SERVICES OF INDIANA, INC.
9341 Sargeant Rd.
Indianapolis, IN 46256
(317) 849-2686
Barbara Carrow - Vice President
Services:
Provides radon mitigation services.
Area Served:
IN.

CONSUMER ENERGY SERVICES, INC.
12 Westerville Sq. Suite 200
Westerville, OH 43081
(614) 861-0600
Stephen R. Helmuth - President
Services:
Provides radon detection services using continuous monitor method, provides radon mitigation services for the real estate industry and residential markets.
Area Served:
OH.
EPA RMP 5:
CC, CR(P).

CONTAMINATION MANAGEMENT ENGINEERING
P.O. Box 871
Greenbrook, NJ 08812
(201) 752-0078
Rande Funkhauser - President
Area Served:
NJ.

CONTINENTAL ENVIRONMENT COMPANY, INC.
34 Maple St.
Summit, NJ 07901
(201) 277-2255
Alvin M. Natkin - President
Services:
Provides radon detection and consultation services using charcoal canister, alpha-track detectors, and E-Perm for private and public sectors, real estate transactions, provides radon mitigation services.
Qualifications:
EPA RMP, AARST, ACS, APCA, listed with NJ DEP for testing and mitigation.
Area Served:
CT, DC, DE, FL, MD, NJ, NY, PA, TX, VA.
EPA RMP 5:
AT, CC, EP(P).

CRIPPEN CONSULTING ASSOCIATES, INC.
242 North James St. Suite 202
Wilmington, DE 19804
(302) 994-8000
Lee W. Thomas - Vice President
William D. Garlington - Vice President
Services:
Provides radon detection, mitigation and consultation services for residential, commercial, and public buildings.
Qualifications:
Licensed professional engineers and radon certified consultants.
Area Served:
Nationwide.

D

D. PAUL BUILDING ENTERPRISE
P.O. Box 1265
Waldorf, MD 20601
(301) 843-5751
Don Paul - President
Joyce Paul - Vice President
Services:
Provides radon mitigation services.
Area Served:
MD.

DELAWARE VALLEY GEOTECHNICAL ASSOCIATES
P.O. Box 98
Washington Crossing, PA 18977
(215) 357-0907
Rocky Richards - President
Area Served:
PA.

DEMPSEY, JOSEPH H.
P.O. Box 247
Eldorado Springs, CO 80025
(303) 499-7578
Joseph H. Dempsey
Area Served:
CO.

DIAMOND BACK SERVICES, INC.
7396 Reynolds Dr.

Sedalia, CO 80135
(303) 791-1984
Richard Rocquemore
Larry Wujcikowski
Area Served:
CO.

DINOFF ASSOCIATES, INC.
112 South Decatur
Strasburg, PA 17579
(717) 687-0938
Lawrence Dinoff - President
Area Served:
PA.

DOUGLAS MARTIN AND ASSOCIATES, INC./MARTIN CONSULTING
See Ad Page 199
520 Chestnut St.
Emmaus, PA 18049
(215) 965-2914
Richard D. Martin - President
Ken Smith - Health and Safety Services
Jim Sherlock - Mitigation Supervisor
Services:
Provides radon mitigation services, mitigation design, engineering procedure development, nuclear consulting, manufactures Professional Series water filtration system, provides radon mitigation training program for state of New Jersey.
Qualifications:
Certified in NJ, PA.
Area Served:
Nationwide.

DPR BUILDERS
21 Stacy Dr.
Belle Mead, NJ 08502
(201) 874-3419
Ralph Viola - President
Services:
Provides radon diagnostic, consultation, mitigation (pre- or post-construction), and detection services using charcoal canister method.
Qualifications:
AARST, NJ DEP, licensed electrical contractor, graduate construction engineer.
Area Served:
NJ, NY, PA, VA.

DREWRY HOME INSPECTIONS, INC.
2625 Oakton Glen Dr.
Vienna, VA 22180
(703) 938-1044
David R. Drewry - President
Services:
Provides radon detection, mitigation, and consultation services.
Qualifications:
AARST, EPA training courses.
Area Served:
DC, VA.

E

ECM ENVIRONMENTAL CONSULTANTS
7425 Hawkins Dr.
Hanover, MD 21076
(301) 761-1136
Ernest C. Michaelson
Services:
Provides radon detection and mitigation services.
Qualifications:
AARST.
Area Served:
MD.

ECODEX
1 Aaron Rd.
Lexington, MA 02173
(617) 862-3440
Steven Wightman - Operations Chief
Margaret Wacks - Quality Control
Services:
Provides radon mitigation services, distributes radon home test kits.
Qualifications:
DOD certified in radiation monitoring, medical technology, AIHA.
Area Served:
Nationwide.
EPA RMP 5:
CC(P).

ECOLOGICAL INTERACTIONS, INC.
96 East Main St.
Rockaway, NJ 07866
(201) 625-7700, (800) 543-2116
F. Michael Fitzpatrick - Vice President
Bernadette E. Secallus - Director
Services:
Provides radon detection and mitigation services, distributes radon test kits.
Qualifications:
AARST board of directors, EPA RMP, NJ DEP approved testing and mitigation course.
Area Served:
CT, DE, FL, MD, NJ, NY, PA, VA.
EPA RMP 5:
AT, CC.

ELLIOTT AND ASSOCIATES
146 North Drexel Ave.
La Grange, IL 60525
(312) 354-1481
Elliott Wall - President
E. Hughes
Area Served:
IL, IN, MI, MO, MS, WI.
EPA RMP 5:
AT.

EN-THERM, INC.
509 West Fayette St.
Syracuse, NY 13204
(315) 474-6549
Richard Kornbluth - President
Services:
Provides radon detection and mitigation services.
Area Served:
NY.

ENERGY ALTERNATIVES UNLIMITED, INC.
8384 Middlecrest Rd.
Boulder, CO 80302
(303) 444-6063
David Nyquist
Services:
Provides radon detection, mitigation and consultation services for disposal of toxic and hazardous materials.
Qualifications:
Charter member AARST.
Area Served:
CO, NV, UT, WY.

ENERGY CONSERVATION SERVICES
6290 North Vine Valley Rd.
Rushville, NY 14544
(716) 554-3994
Peter Duffy - Principal
Services:
Provides radon detection, mitigation, and energy consultation services.
Area Served:
NY.

ENERGY INDEPENDENCE/THE RADON RESPONSE TEAM
9940 East 36th St.
Independence, MO 64052
(816) 353-3755
Kenneth C. Riead - President
Services:
Provides radon detection and mitigation services for residential and commercial markets, distributes Vanee line of heat recovery ventilation systems.
Qualifications:
EPA RMP, 12 years of experience in the field.
Area Served:
IL, KS, MN, MO, NE, WI.

ENERGY OUTFITTERS, LTD.
245 North Aldrich Ave.

Minneapolis, MN 55405
(612) 377-6502
Michael Noble - President
Services:
Provides radon mitigation services.
Area Served:
MN, WI.

ENERGY PROFESSIONALS, INC.
387 Irwin St.
Lock Haven, PA 17745
(717) 748-5910, (717) 726-3825
Jay Crust - President
Mike Miller - Vice President
Jim Donley - Officer
Services:
Provides radon detection and mitigation services.
Qualifications:
PA state certified in testing and mitigation.
Area Served:
Nationwide.

ENERGY RESOURCE CENTER
P.O. Box 267
Colorado Springs, CO 80901
(719) 636-2447
Lee Salmon - Radon Specialist
Services:
Provides radon detection and mitigation services, project management, independent evaluation, presents seminars, distributes charcoal canisters and alpha tracks to homeowners.
Qualifications:
Graduate, Colorado School of Mines course on radon and indoor environment, 15 years of experience in energy management.
Area Served:
Nationwide.

ENERGY RESOURCE TECHNOLOGY
1494 North Charlotte St.
Pottstown, PA 19464
(215) 327-4641
John Mauger
Area Served:
PA.

ENERGY SAVINGS WAREHOUSE, INC.
3490 Hwy. 22 W.
Somerville , NJ 08876
(201) 534-6922
Ivan A. Franceus - President
Paul E. Franceus - Vice President
Services:
Provides radon mitigation services, manufactures ventilation equipment for radon mitigation.
Area Served:
Nationwide.

ENERGY UNLIMITED
RFD 1, Box 1090
Mercer, PA 16137
(412) 662-3591
Mitchell Moseley - Partner
Services:
Provides radon detection and mitigation services.
Area Served:
OH, PA.

ENERGY VENT, INC.
915 Valley St. P.O. Box 14145
Dayton, OH 45414
(513) 222-7749
John Prikkel - President
Services:
Provides radon mitigation services for residential, office, factory, and commercial buildings.
Area Served:
KY, OH.

ENGINEERING ASSOCIATES
4682 Watch Hill Rd.
Manlius, NY 13104
(315) 682-4389
David Miller - Principal
Services:
Provides radon detection and mitigation services.
Qualifications:
Licensed professional engineer, NYSERDA trained in radon detection and mitigation.
Area Served:
NY.

ENVIRO TECH SYSTEMS
430 Fernhill Ave.
Pittsburgh, PA 15226
(412) 561-2100
Jack Stanizzo - Owner
Services:
Provides radon detection and mitigation services, distributes radon mitigation products.
Qualifications:
PA DER, 450+ mitigations completed.
Area Served:
PA.

ENVIRO-CARE RADON SERVICES
661 Wyoming St.
Buffalo, NY 14215
(716) 834-1788
Charles Mertz - President
Services:
Provides radon detection and mitigation services.
Area Served:
NY.

ENVIRONMENTAL AIR CONTROL, INC.
2321 Northwest 30th Ct.
Fort Lauderdale, FL 33311
(305) 739-5159
Ray Narushko - President
Services:
Provides radon detection and mitigation services.
Area Served:
FL.

ENVIRONMENTAL ENTERPRISES, INC.
P.O. Box 16847
Atlanta, GA 30321
(404) 766-5551
Chris Esworthy - President
Services:
Provides radon detection and mitigation services, distributes full line of radon mitigation products including fans, sealers, etc.
Area Served:
Nationwide.
EPA RMP 5:
CC.

ENVIRONMENTAL HOME SERVICES, INC.
12354 Howard Lodge Dr.
Sykesville, MD 21784
(301) 442-5861
Charles K. Clopein - President
Frank Checkes - Vice President
Services:
Provides radon detection and mitigation services for realtors, builders and homeowners.
Qualifications:
EPA RMP, EPA training courses in radon mitigation methods.
Area Served:
MD.
EPA RMP 5:
CC.

ENVIRONMENTAL INDOOR AIR QUALITY CONSULTANTS, INC. (EIC)
58 Keller Ave.
Lancaster, PA 17601
(717) 393-5774, (800) 367-7223
William H. Miller - Radon Manager
H. Edward Carr, Jr.
Services:
Provides radon detection and mitigation services using ventilation equipment.
Area Served:
PA.

ENVIRONMENTAL MAINTENANCE SPECIALISTS COMPANY, INC.
1429 Danby Rd.
Ithaca, NY 14850
(607) 273-5938
John Kaminsky - President
Services:
Provides radon mitigation services in residences and commercial buildings.
Qualifications:
NY State Energy Office workshop on reducing indoor radon.
Area Served:
NY.

ENVIRONMENTAL POLLUTION CONTROL
195 Cliffwood Dr.
Atlanta, GA 30328
(400) 425-6454
Eugene A. Bryson, Sr.
Area Served:
GA.

ENVIRONMENTAL RADON MEASUREMENTS
49 North Orange Ave.
Orlando, FL 32801
(407) 648-1969, (800) 237-3065
Terrie Keene
Paul Pagello
Mark Riffe
Area Served:
Nationwide.
EPA RMP 5:
CC(P).

ENVIRONMENTAL SERVICES AND TECHNOLOGY, INC.
22 Oakland Ave.
Atlantic Highlands, NJ 07716
(201) 291-5525
Allan Dean - Radiation Physicist
Services:
Provides radon detection and mitigation services serving residential, commercial, industrial, and government clients.
Qualifications:
On-staff radiation physicist, EPA RMP.
Area Served:
Nationwide.
EPA RMP 5:
AT, CC.

ENVIRONMENTAL SERVICES OF ILLINOIS, INC.
302 West State St.
O'Fallon, IL 62269
(618) 624-8079, (314) 421-4374
Stephen J. Willig - Air Quality Division
Douglas R. Wall - Vice President
Mark Warchol - Project Manager
G.R. Wall - Project Manager
Services:
Provides radon detection and mitigation services.
Qualifications:
IL Department of Safety voluntary certification program.
Area Served:
IL, MO.
EPA RMP 5:
CC.

ENVIRONMENTAL TESTING ASSOCIATES
9 White Pine Trail
Madison, WI 53717
(608) 833-4073
Don Steiger - President
Services:
Provides radon detection and mitigation services.
Area Served:
Nationwide.
EPA RMP 5:
CC.

ERM, INC.
49 North Orange Ave.
Orlando, FL 32801
(407) 648-1969, (800) 237-3065
Wayne Dean - President
Alan Baker - Marketing Director
Paul Pagello - Laboratory Director
Services:
Provides radon detection and mitigation services to real estate and mortgage lending institutions.
Qualifications:
EPA RMP.
Area Served:
Nationwide.

EVANSHIRE COMPANY, LTD., THE
Rte. 4, Box 57
Rome, NY 13440
(315) 865-5941, (800) 962-5941
Edwin Evans - President
Wade Evans - Vice President
Services:
Provides radon detection services using charcoal canister, alpha track detector, and grab sampling methods, provides radon mitigation services, distributes full line of radon detection and mitigation products, including sealers, sump covers, fans.
Qualifications:
Experience in state and federally funded projects.
Area Served:
Nationwide.

F

FCA DEVELOPMENT CORPORATION
P.O. Box 1268
Exton, PA 19341
(215) 363-5460
Ash Sudakar - President
Services:
Provides radon mitigation services.
Area Served:
PA.

FINN CONSULTANT ENGINEERS, INC.
227 North St.
Dalton, MA 01226
(413) 684-2019
George Finn - President
Services:
Provides radon detection and mitigation services, consultation in mitigation design.
Area Served:
MA.

FORCHETTI PSI
P.O. Box 1234
Morrisville, PA 19067
(215) 493-1646, (609) 695-1229
Vince Forchetti - President
Area Served:
DE, MD, NJ, NY, PA.
EPA RMP 5:
CC.

FORESIGHT ENGINEERING
P.O. Box 349
Pittstown, NJ 08867
(201) 996-3090
Charles J. Yanulevich - Manager
Services:
Provides radon detection using charcoal canisters and on-site working level monitors, provides radon mitigation services.
Qualifications:
AARST.
Area Served:
NJ.

FORESIGHT ENGINEERING, INC.
P.O. Box 3111
Wayne, NJ 07474-3111
(201) 942-3061
Ernie Borsellino - Engineer
Services:
Provides radon detection services using charcoal canisters and on-site working level monitors, provides radon mitigation services.
Area Served:
NJ.

FORESIGHT ENGINEERING, INC.
P.O. Box 362

Monasquan, NJ 08736
(201) 223-3316
Ken Lucas - Engineer
Services:
Provides radon detection and mitigation services for residential and commercial properties.
Area Served:
NJ.

FORESIGHT ENGINEERING, INC.
17 Cook Ave. P.O. Box 621
Madison, NJ 07940
(201) 377-0600
Christine Hammitt - President
F.B. Kaufmann - Vice President
T.D. Kaufmann - Treasurer
F.G. Hammitt - Secretary
Services:
Provides radon detection and mitigation services.
Qualifications:
AARST, EPA RMP, NJ professional engineers.
Area Served:
NJ, PA, VA.
EPA RMP 5:
CC.

FULL HOUSE CONTRACTING
1900 South Washington
Casper, WY 82601
(307) 266-0895
Michael E. Hedberg - Owner
Services:
Provides radon mitigation services.
Qualifications:
Authorized Safe-Aire distributor, completed EPA training course in mitigation.
Area Served:
WY.

G

G.A.M. BUILDING INSPECTION AND RADON MEASUREMENTS AND CONTROL, INC.
17223 East Purdue Pl.
Aurora, CO 80013
(303) 699-2827
George A. Michael - President
Services:
Provides radon detection and mitigation services for the real estate industry, home buyers and sellers, and the general public.
Qualifications:
Licensed professional engineer.
Area Served:
CO.
EPA RMP 5:
AT, CC, GW(P).

G.M.D. CONSTRUCTION, INC.
P.O. Box 204
Lisbon, MD 21765
(301) 854-6186
Mike Deitz - President
Services:
Provides radon mitigation services using sub-slab or block wall depressurization, sealing, ventilation, air exchangers and reducing negative pressure.
Qualifications:
EPA trained, licensed MHIC, AARST, BBB, NHBA.
Area Served:
MD, PA, VA.

GARCO RESEARCH COMPANY
144A Sutton Rd.
Lebanon, NJ 08833
(201) 832-7895
Gary J. Gilbert - Technical Director
Laslo Veradi - Safety Specialist
Garry M. Annibal - Env. Specialist
Services:
Provides radon detection and mitigation services for homeowners and businesses.
Qualifications:
Certified industrial hygienist on staff.
Area Served:
MD, NJ, NY, PA, VA.
EPA RMP 5:
CC.

GARDEN STATE RADON DEWATERING SERVICES, INC.
2 Edgewater Pl. Suite 1A
New Brunswick, NJ 08901
(201) 846-1948
Ron Weshnak - Owner
Area Served:
NJ.

GEO-RADON SERVICES
P.O. Box 5985
Berkeley, CA 94705
(415) 843-8844
Barbara Moed - Owner
Services:
Provides radon detection and mitigation services, consultation and research on influence of geology and soil properties on indoor radon.
Qualifications:
Eight years of indoor radon research funded by DOE, EPA, BPA, masters' degree in geology, consultant to California radon survey.
Area Served:
Nationwide.

GEOCHEM, INC.
3200 New Seward Hwy. Suite 201
Anchorage, AK 99503
(907) 272-2121
Joseph Neubauer - President
David Neubauer - Vice President
Donald Gibson - Marketing
Services:
Provides radon detection and mitigation services.
Area Served:
AK.

GEOHAZARDS, INC.
P.O. Box 14956
Gainesville, FL 32604
(904) 371-7243
Douglas L. Smith - President
Services:
Provides radon detection services and consultation on radon mitigation options.
Area Served:
FL.

GEOMET TECHNOLOGIES, INC.
20251 Century Blvd.
Germantown, MD 20874
(301) 428-9898
Niren L. Nagda - Vice President
Bill Mountel - Marketing
Services:
Research on radon detection and mitigation technologies. Presently studying two homes in Maryland to analyze particular mitigation strategies. Large scale surveys in Florida and Massachusetts on charcoal canister and alpha track detection methods.
Area Served:
Nationwide.
EPA RMP 5:
AT.

GIDDINGS AND ASSOCIATES, INC.
3049 Enterprise Dr.
State College, PA 16801
(814) 238-5927
Todd Giddings
Area Served:
PA.

GILMONT CONSTRUCTION COMPANY
P.O. Box 1268
Exton, PA 19341
(215) 363-5460
Bob Fremont
Area Served:
DE, NJ, PA.

GMAC HOME IMPROVEMENT
111 East Broadway St.

Campbellsville, KY 42718
(800) 222-5008
Alan Chandler
Area Served:
KY.

GRAVES RADON CONTRACTING
P.O. Box 266
Terre Hill, PA 17581
(215) 445-7886
Thomas Graves
Area Served:
PA.

GROUNDWATER TECHNOLOGY, INC.
12 Walker Way
Albany, NY 12205
(518) 456-2444
William Welling - Geologist
Services:
Provides radon detection and mitigation services.
Area Served:
Nationwide.

GURNEY KERR CONTRACTORS, INC.
Box 753
Bryn Athyn, PA 19009
(215) 947-7796
Glenn Gurney - Vice President
Area Served:
PA.

H

HALL-KIMBRELL ENVIRONMENTAL SERVICES
5319 Southwest Westgate
Portland, OR 97221
(503) 292-9406
David Butts - Technical Director
Services:
Provides radon detection, mitigation, and consultation services.
Qualifications:
NRA.
Area Served:
Nationwide.

HARAD WATERPROOFING
3301 North Second St.
Harrisburg, PA 17110
(717) 233-6143
Stanley Harad - President
Services:
Provides radon mitigation services.
Qualifications:
EPA courses and seminars.
Area Served:
PA.

HARDICO RADON MANAGEMENT, CENTRAL OFFICE
P.O. Box 40
Stevens, PA 17578
(215) 267-6006
Jerry Harding - Owner
Services:
Provides radon mitigation services.
Area Served:
PA.

HARDICO RADON MANAGEMENT, COLUMBUS OFFICE
95 Cherry St.
Groveport, OH 43215
(614) 836-2788
Bob Mitchell - Manager
Area Served:
OH.

HARDICO RADON MANAGEMENT, D.C. OFFICE
10720 Main St. Suite 205
Fairfax, VA 22030
(703) 385-4200
Cheryl Kreider - Manager
Area Served:
VA.

HARDICO RADON MANAGEMENT, GREENVILLE OFFICE
744 Isaqueena Trail Ridgecrest Apt. #21
Central, SC 29630
(803) 654-6734
Tom Harding - Manager
Area Served:
SC.

HARDICO RADON MANAGEMENT, HARRISBURG OFFICE
730 Tower Rd.
Enola, PA 17025
(717) 732-9469
Fred Peters - Branch Manager
Services:
Provides radon mitigation services.
Area Served:
PA.

HARDICO RADON MANAGEMENT, PITTSBURGH OFFICE
216 West Sycamore St.
Pittsburgh, PA 15211
(412) 381-6155
Ray Rowe, Sr. - Manager
Area Served:
PA.

HARDICO RADON MANAGEMENT, ROANOKE OFFICE
P.O. Box 10494
Blacksburg, VA 24062-0494
(703) 951-0808
Steve Price - Manager
Area Served:
VA.

HARRIS CONSTRUCTION
P.O. Box 1203
Breckenridge, CO 80424
(303) 453-2045
Tony Harris - Owner
Services:
Provides radon detection and mitigation services.
Area Served:
CO.

HARVARD ENGINEERING SERVICES COMPANY
28 Shaker Rd.
Harvard, MA 01451
(617) 456-3716
James Galeski - President
Lorelei Galeski - Vice President
Services:
Provides radon detection services, design and installation of radon reduction methods.
Qualifications:
Bachelor's degree in mechanical engineering, construction supervisor's license.
Area Served:
Nationwide.

HERITAGE RADON SERVICES, INC.
6968 Birchwood Rd.
Harrisburg, PA 17112
(717) 599-5060
Richard C. Smith - President
Glenn Livetsberger - Vice President
Services:
Provides radon detection and mitigation services.
Area Served:
Nationwide.
EPA RMP 5:
CC.

HOME CHECK SERVICES
P.O. Box 98
Front Royal, VA 22630
(703) 635-1274
William H. McSpadden - President
Glen Hickerson - Geologist
Mary D. Sitlon - Geologist
Services:
Provides radon detection, mitigation, and consulting services.
Qualifications:
AARST, EPA contractor's course.
Area Served:
DE, VA.

HOME DETOX SPECIALISTS
81 Candy Ln.
Hamburg, NY 14075
(716) 648-3321
Thomas Deere - President
Guy Gessner - Vice President
Services:
Provides radon detection and mitigation services.
Qualifications:
Trained by NY State Health Department.
Area Served:
NY.

HOME ENVIRONMENT COMPANY
P.O. Box 24744
Speedway, IN 46224
(317) 247-1881
Gil Ward - Owner
Services:
Manufactures Radon Sentry Alarm System, alarm signaling radon mitigation system failure, provides radon mitigation services.
Qualifications:
Completed certified course by Radon Detection Services, Inc.
Area Served:
IL, IN, KY, OH.

HOME INSPECTION SERVICE, INC.
1916 Southland Ave.
Highland Park, IL 60035
(312) 831-4545
Jerry T. Comer - President
Services:
Provides radon detection and mitigation services.
Qualifications:
EPA RMP.
Area Served:
Nationwide.
EPA RMP 5:
AT, CC.

HUGHES, RONALD L.
68 Hanshue Rd.
Hummelstown, PA 17036
(717) 534-1168
Ronald L. Hughes
Area Served:
PA.

HUNT'S WATER CONTROL, INC.
607 Tower View Dr.
Taylor Mill, KY 41015
(606) 292-0333
Jim Hunt - President
Services:
Provides radon detection services using charcoal canister and alpha track methods, provides mitigation services.
Area Served:
KY.

HUNTER-DON PEST CONTROL
777 Mountain View Rd.
Asbury , NJ 08802
(201) 638-4024
William DeBald
Area Served:
NJ.

I

I & M RADON MITIGATION
529 Browar
Eaton, OH 45320
(513) 456-6784
Earl Moore - Associate
Mark Ingrams - Associate
Services:
Provides radon detection services using charcoal canister method, provides mitigation services for rural and residential areas and the real estate industry.
Area Served:
NJ.

INDOOR POLLUTION CONTROL CENTER
3921 Sawyer Rd.
Sarasota, FL 34233
(813) 923-3641, (800) 226-2636
Ronald V. Miller - President
Robert L. Brogan - Manager
David V. Miller - Vice President
Services:
Provides radon detection and mitigation services.
Qualifications:
EPA RMP, certified by Honeywell, Inc.
Area Served:
FL.
EPA RMP 5:
AT, CC.

INDUSTRIAL RESOURCE DEVELOPMENT COMPANY
2490 General Armistead Ave. Suite 303
Norristown, PA 19403
(215) 630-9150
Maurice Silvestris - Manager
William Minnick - Vice President
Services:
Provides radon detection and mitigation services.
Area Served:
DE, NJ, PA.

INEX
8704 Gutierrez, NE
Albuquerque, NM 87111
(505) 292-0805
William L. Mansker
Area Served:
NM.

INFILTEC/MAIN OFFICE
P.O. Box 8007
Falls Church, VA 22041
(703) 820-7696
David Saum - President
Marc Messing - Vice President
Services:
Provides radon mitigation, research and development, nationwide training, consulting services, specializes in large buildings.
Qualifications:
EPA RMP, EPA subcontractor, EPA trainer.
Area Served:
Nationwide.
EPA RMP 5:
AT, CC, EP(P), GR(P).

INSUL-TECH, INC.
5559 Westerville Rd.
Westerville, OH 43081
(614) 891-9555
Dan West - President
Services:
Provides radon detection services using charcoal canister, alpha track, and continuous radon monitor methods, provides radon mitigation services.
Qualifications:
ARA, ICAA, NRA, contractor's license.
Area Served:
IN, KY, MI, OH, PA, WA, WV.
EPA RMP 5:
AT, CC, CR(P).

INSULATING SYSTEMS
340 Gargrave Rd.
West Carrollton, OH 45449
(513) 435-0079
James Dunn - President
Services:
Provides radon detection and mitigation services.
Qualifications:
EPA workshops.
Area Served:
OH.
EPA RMP 5:
CC, CR(P).

INTERMOUNTAIN RADON SERVICE
P.O. Box 3
Salida, CO 81201
(719) 539-6726
Monty Holmes - President
Services:
Provides radon detection and mitigation

services, conducts training seminars for real estate personnel, is currently producing a radon video for Coldwell-Banker.
Area Served:
CO.

INTERNATIONAL EXPLORATIONS
577 Sackettsford Rd.
Warminster, PA 18974
(215) 598-7137
Jacob Jacobsen - President
Area Served:
PA.

J

J. BECK AND ASSOCIATES, INC.
35 Meetinghouse Ln.
Bradford Woods, PA 15015
(412) 935-6711
Joseph R. Beck - President
Services:
Provides radon detection and mitigation services to home owners, real estate companies, and relocation companies.
Area Served:
MD, NJ, OH, PA, WV.
EPA RMP 5:
CR(P).

JANCE ASSOCIATES, INC.
P.O. Box 234
East Texas, PA 18046
(215) 398-0434
John Helwig
Area Served:
PA.

JOHN L. RENSHAW, INC.
10738 Tucker St.
Beltsville, MD 20705
(301) 937-4440
Douglas Stephenson - Vice President
Services:
Provides radon mitigation services.
Qualifications:
NRA.
Area Served:
DC, MD, VA.

JOHNSON PLUMBING AND HEATING
232 South Main St.
Chambersburg, PA 17201
(717) 264-5028
Kirk Johnson - Partner
Services:
Provides radon detection and mitigation services.
Area Served:
Nationwide.

JOHNSON'S ENERGY SERVICE
104 Central Dr.
Forest City, IA 50436
(515) 582-2998
Harvey Johnson - Owner
Services:
Provides radon detection and mitigation services.
Qualifications:
Master's degree in radiation biology.
Area Served:
Nationwide.

K

KELLOGG ENVIRONMENTAL SERVICES, INC.
816 Anthony Dr.
Mechanicsburg, PA 17055
(717) 732-3364, (717) 732-9263
Ed Kellogg - President
Services:
Provides radon detection and mitigation services.
Area Served:
PA.

KITTREDGE CONSTRUCTION COMPANY
2758 South Langley Ct.
Denver, CO 80210
(303) 825-4940
David Brockway - President
Services:
Provides radon mitigation services.
Qualifications:
AARST, Class A construction license, professional engineer on staff.
Area Served:
CO.

KOVACS CONSULTANTS
4515 Willard Ave. Suite 1921S
Chevy Chase, MD 20815
(301) 652-7133
J. Kovacs - President
Services:
Provides radon mitigation services to commercial accounts.

KUNKEL, HAROLD E.
6 Radnor Rd.
Newtown Square, PA 19073
(215) 356-5096
Harold E. Kunkel
Qualifications:
AARST.
Area Served:
PA.

L

L. CLARK AND SONS, INC.
986 Fisher Rd.
Fitchburg, MA 01420
(508) 343-7849
LeRoy E. Clark - President
Services:
Provides radon mitigation services, installation of radon control systems, monitoring, and consultation services to the home owner/buyer.
Area Served:
DE, NH, NJ, NY, MA, ME, PA, RI, VT.

L.J. QUIGLIANO, II, ENTERPRISES
2173 Seneca Ave.
Niagara Falls, NY 14305
(716) 284-4074
Lawrence J. Quigliano - Owner
Services:
Provides radon mitigation services.
Area Served:
Nationwide.

LAKE BUILDERS, INC.
Star Rte. 2, Box 39 Rte. 6
Hawley, PA 18428
(717) 226-1850
David Stear - Vice President
Area Served:
PA.

LARRY LIBERTORE, INC.
230 West Highland Dr.
Lakeland, FL 33803
(813) 644-2828
Larry Libertore, Jr. - President
Services:
Provides radon mitigation services.
Area Served:
FL.

LDH WATERPROOFING AND RADON CONTROL
286 Old Delp Rd.
Lancaster, PA 17601
(717) 569-2062
Lenny Hornberger - Owner
Services:
Provides radon mitigation services.
Area Served:
PA.

LEHIGH VALLEY HOUSE INSPECTION SERVICE
200 Delaware Ln.
Whitehall, PA 18052
(215) 435-0208
Pete Potochney - Director
Services:
Provides radon detection and mitigation services.
Qualifications:
BOCA, CSI, NAHRO.
Area Served:
PA.

LININGER BROTHERS, INC.
12600 Penn Mar Rd.
Waynesboro, PA 17268
(717) 762-4024
Rick Lininger
Area Served:
PA.

LOWRY ENGINEERING, INC.
P.O. Box 536 Back Troy Rd.
Unity, ME 04988
(207) 948-3790
See Ad Page 211
Sylvia Lowry - President
Jerry Lowry - Resarch Director
Services:
Engineering research and consulting firm specializes in radionuclide analysis and removal from water supplies, and removal equipment for homes and public water supplies, manufactures and distributes the Stripper Aeration System, removes 99.9% of radon from water.
Qualifications:
Director is a leading expert on radon removal from water supplies, teaches EPA courses on mitigation.
Area Served:
Nationwide.

LUKENBUILT CONTRACTORS
965 Dearborn St.
Aurora, CO 80011
(303) 364-9526
Paul D. Lukenbill
Area Served:
CO.

LV SAFETY AND HEALTH CONSULTANTS
446 Timothy Dr.
Bethlehem, PA 18017
(215) 758-9072
Laslo Varadi - President
Area Served:
PA.

LW ENVIRONMENTAL SERVICES
2111 Jennings St.
San Francisco, CA 94124
(415) 822-4555
Richard S. DellaValle
Qualifications:
AARST.
Area Served:
CA.

M

MACK, E.F.
11 South 8th Ave.
Marshalltown, IA 50158
(515) 753-7328
Thomas E. Mack - Partner
Area Served:
IA.

MACLAREN ENGINEERS, INC.
33 Yonge St.
Toronto, ON M5E 1E7
(416) 756-4919
E. Chart
Area Served:
ON.

MAHONEY HEATING AND COOLING
13620 C St.
Omaha, NE 68144
(402) 895-3333
Pat Mahoney - President
Bob Mahoney - General Manager
Services:
Provides radon detection and mitigation services.
Qualifications:
Licensed radon mitigation contractor for NE.
Area Served:
IA, NE.
EPA RMP 5:
CC.

MARTIN BUILDERS COMPANY, INC., THE
7114 Rising Sun Ave.
Philadelphia, PA 19111
(215) 745-5398
Saul J. Golub
Area Served:
PA.

MASTER BUILDING SERVICE
245 Huguenot St. Suite 107
New Rochelle, NY 10801
(914) 235-8282
Marvin Kapus
Area Served:
NY.

MAYA-TEK, INC.
P.O. Box 1092
Chadds Ford, PA 19317
(215) 358-5453
Mark D. Daniels - President
Len W. Daniels - Specialist
Services:

Provides radon detection and mitigation services.
Area Served:
DE, NJ, PA.

MCGUIRE BUILDERS, INC.
Rte. 2, Box 1840
Warfordsburg, PA 17267
(717) 294-3108
Robert McGuire - Vice President
Area Served:
PA.

MCHILL, A.D. AND SONS, INC.
1121 Second St. Pike
Richboro, PA 18954
(215) 364-2784
A.D. McHill - General Contractor
Area Served:
PA.

MCINERNY, T.M.
1214 Church St.
Elmira, NY 14905
(607) 733-1994
T.M. McInerny - General Contractor
Area Served:
NY.

MECHANICAL EQUIPMENT SERVICES CORPORATION
3122 49th St.
Des Moines, IA 50310
(515) 276-9131
Charlie Stephenson - Owner
Services:
Provides radon detection and mitigation services.
Area Served:
IA.

MERIT ENVIRONMENTAL SERVICES, INC.
P.O. Box 3083
Berea, KY 40403
(606) 986-7070
David McFarland - President
Services:
Provides radon detection and mitigation services.
Qualifications:
ARA, professional geologist and certified soil scientist.
Area Served:
AK, AR, FL, GA, IA, IL, IN, KY, MO, NC, OH, TN, VA, WV.
EPA RMP 5:
CC.

METROPOLITAN RADON SERVICE
4011 Newbys Bridge Rd.
Chesterfield, VA 23832
(804) 745-4702
Linda Alt - President
Wayne Alt - Vice President
Services:
Provides radon detection and mitigation services.
Qualifications:
EPA RMP.
Area Served:
Nationwide.
EPA RMP 5:
CC.

MIDWEST RADON INSPECTION, INC.
25005 West 129 Ter.
Olathe, KS 66061
(913) 780-2000
Gary Hodgden - President
Steve Hodgden - Vice President
Services:
Provides radon detection services and mitigation consultation.
Area Served:
KS, MO.

MITIGATING CIRCUMSTANCES
153 Forest Park
Durham, NH 03824
(603) 868-1557
Clayton Ogilvie - Owner
Services:
Provides radon mitigation services.
Area Served:
MA, ME, NH.

MUSSO, FRANCIS P.
P.O. Box 651
Bath, NY 14810
(607) 776-6245
Francis P. Musso
Area Served:
NY.

N

NATURAL ENERGY CONSULTANTS
P.O. Box 586
Conifer, CO 80433
(303) 838-5291
Rob Haines - Consultant
Services:
Provides radon detection, mitigation, and consultation services.
Area Served:
Nationwide.

NEW ENGLAND DE-WATERING, INC.
70 Heritage Ave. Suite 14 Heritage Pl.
Portsmouth, NH 03801
(800) 535-1211
Wayne Bowers - President
Ron Morse - Regional Supervisor
Services:
Provides radon mitigation services, manufactures ventilated flooring used in radon mitigation.
Area Served:
CT, MA, MS, NH, NY, RI, VT.

NEW ENGLAND RADON, LTD.
373 Main St.
Salem, NH 03079
(603) 893-4260, (800) 637-2366
Daniel Espinal - Vice President
Services:
Provides radon detection and mitigation services, distributes radon test kits, laboratory analysis.
Qualifications:
EPA RMP.
Area Served:
CT, MA, ME, NH, PR, RI, VT.
EPA RMP 5:
CC(P).

NEW WAVE ENERGY SYSTEMS, INC.
3645 Pearl St.
Boulder, CO 80301
(303) 449-8628
Douglas Neithercut
Area Served:
CO.

NITE HAWK ENTERPRISES, INC.
17 Lincoln Ave.
White Plains, NY 10606
(914) 949-3207
Jerry Shepherd
Area Served:
NY.

NORTH EAST ENVIRONMENTAL PRODUCTS, INC.
See Ad Page 432
85 Mechanic St.
Lebanon, NH 03766
(603) 448-3944
Bruce L. Lamarre - President
Jack Nelson - Chief Executive Officer
Kevin Lang - Vice President
Services:
Manufactures and distributes Clearadon, an aeration system for removal of radon or VOCs from residential water supplies, provides consultation services for removal of waterborne radon.
Qualifications:
Water Quality Association.
Area Served:
Nationwide.

NORTHEASTERN RADON SERVICE
21 Rte. 202
Montville, NJ 07045
(201) 402-1030
Bogdan Choruza - President
Services:
Provides radon detection and mitigation services.
Area Served:
NJ.

NORTHERN NECK RADON SERVICE
Rte. 1, Box 259-E
Hague, VA 22469
(804) 472-4252
Milford L. Brann - Owner
Jean E. Brann - Manager
Services:
Provides radon detection services using EPA-approved charcoal canister method, provides radon mitigation services.
Area Served:
Nationwide.
EPA RMP 5:
CC.

NORTON CONSERVATION COMPANY, INC.
P.O. Box 583
Hackettstown, NJ 07840
(201) 852-6046
Joseph Norton - President
Bryan Tafuri - Pennsylvania Manager
Services:
Provides radon mitigation and energy management services for residential and commercial markets.
Qualifications:
NJ DOE, NJ DEP, PA DER.
Area Served:
Nationwide.
EPA RMP 5:
CC.

NUCHEM CORPORATION
P.O. Box 3369 2900 William Penn Hwy.
Palmer, PA 18043
(215) 252-9050
Leroi R. Yaffey - Owner
Services:
Provides radon detection and mitigation services, operates a primary laboratory.
Area Served:
PA.

O

O'RIORDAN INSULATION COMPANY
8901 North Nebraska Ave.
Tampa, FL 33604
(813) 932-0008
Joe O'Riordan - Vice President
Services:
Provides radon detection and mitigation services.
Area Served:
FL.

OAK HILL RADON TESTING AND REPAIR
15 1/2 North State St.
Westerville, OH 43081
(614) 895-8365
Jim Lubinsky - President
Services:
Provides radon detection, mitigation and consultation services.
Qualifications:
EPA RMP, ARA.
Area Served:
OH.
EPA RMP 5:
AT, CC, CR(P), EP.

OBAR SYSTEMS, INC. See Ad Page 215
RR 3, Box 759
Highland Lakes, NJ 07422
(201) 728-7679, (201) 697-0112
Gunner Bar - Owner
David O'Brien - Owner
Services:
Distributes complete range of radon mitigation products, including fans, heat exchangers, sealers and caulks, pumps, filters, water systems, couplings, sumps, manometers, expansion joint materials, and other accessories, provides radon mitigation services.
Area Served:
Nationwide.

OCCUPATIONAL AND ENVIRONMENTAL HEALTH CONSULTING SERVICES
635 Harding Rd.
Hinsdale, IL 60521
(312) 325-2083
Robert C. Brandys - President
Services:
Provides radon detection and mitigation services to industry, architects, and homeowners.
Qualifications:
Certified industrial hygienist, registered environmental assessor.
Area Served:
Nationwide.

OXYTHERM, INC.
P.O. Box 3467
Greenville, SC 29602
(803) 246-2055
Harry Link - President
Services:
Provides radon mitigation services.
Qualifications:
NRA.
Area Served:
AL, GA, LA, NC, OK, SC, TX, VA.

P

P.C. ENGINEERING
Rte. 1, Box 347-B
Lexington, VA 24450
(703) 463-9116
Kurt Chittum
Area Served:
VA.

PARADISE COMPANY
109 Greenleaf
Lumberton, TX 77656
(409) 755-3524
Linda McKinney - President
Ed McKinney - Vice President
Dale Vandine - Vice President
Liz McKinney - Secretary
Services:
Provides radon detection, mitigation, and consulting services.
Area Served:
TX.

PAYNE, LARRY D.
P.O. Box 1292
Purcellville, VA 22132
Larry D. Payne, Jr.
Qualifications:
AARST.
Area Served:
VA.

PELICAN ENVIRONMENTAL CORPORATION
42 Lehigh Rd.
Wellesley, MA 02181
(617) 235-5997
Don Crawshaw - President
Jeff Crawshaw - Engineer
Randy Foster - Mitigator
Services:
Provides radon detection, mitigation and consultation services.
Qualifications:
Master's degree in civil engineering, accredited in MA for testing.
Area Served:
CT, MA, NH, RI.

PENN NUCLEAR SERVICES
3190 Industrial Blvd.
Bethel Park, PA 15102

(412) 833-5300
Alan Zeller - Vice President
Services:
Provides radon mitigation services.
Area Served:
PA.

PENNSYLVANIA RADON CENTER
21 Hillbrook Cr.
Malvern, PA 19355
(215) 647-9999
Area Served:
PA.

PENRAD, INC.
54 South Market St.
Elizabethtown, PA 17022
(717) 367-8410, (717) 367-3995
Bud Longenecker
Area Served:
PA.

PHOENIX SAFETY SYSTEMS
7917 Carnation Dr.
Louisville, KY 40258
(502) 935-8317, (502) 964-0865
Dave Morrison
Services:
Provides radon mitigation services, designs surveys.
Area Served:
KY.

PLUM GROVE CORPORATION
33 West Queens Way Mall Old Towne
Hampton, VA 23669
(804) 722-9841
Garland H. Harrington - President
Services:
Provides radon detection and mitigation services.
Area Served:
VA.

PRINCETON HOME INSPECTION SERVICE
197 Wall St.
Princeton, NJ 08540
(609) 921-3775
Clive Usiskin - President
Services:
Provides radon detection and mitigation services.
Area Served:
NJ, PA.
EPA RMP 5:
CC.

PRO-GUARD BUILDING INSPECTIONS, INC.
RD 1, Box 193B
Auburn, PA 17922
(717) 754-7673
William X. Cairnes
Services:
Provides radon mitigation services.
Area Served:
MD, PA.

PROFESSIONAL HOME ANALYSIS
1025 Northeast Sanchez Ave.
Ocala, FL 32670
(904) 694-3982
G. Douglas Foreman - Owner
Keith Brandies - Analysis
Jeff Foreman - Analysis
Services:
Provides radon detection services.
Area Served:
FL, GA.
EPA RMP 5:
CR(P).

PROFESSIONAL HOUSE DOCTORS
599 Short Ave.
Marengo, IA 52301
(319) 642-7133
Tom Voightman
Area Served:
IA.

PROFESSIONAL HOUSE DOCTORS, INC.
1406 East 14th St.
Des Moines, IA 50316
(515) 265-6667
Dane J. Shearer - President
Services:
Provides radon detection and mitigation services, manufactures blower doors with 25-6,000 cfm capacity, manufactures structural analyzers providing complete and detailed indoor air and energy analysis, performs mitigation training nationwide.
Area Served:
Nationwide.

PROFESSIONAL RESTORATION AND REPAIR
P.O. Box 38932
Colorado Springs, CO 80937
(719) 632-3996
Craig L. Nelson - Owner
Services:
Provides radon mitigation services and consultation, radon information slide show available for groups.
Qualifications:
AARST, BS in mechanical engineering.
Area Served:
CO.

PROJECT ENVIRONMENTAL GROUP
10 Duff Rd.
Pittsburgh, PA 15235
(412) 243-3100
Larry Horvat - Corporate Secretary
Services:
Provides radon mitigation services.
Qualifications:
NRA.
Area Served:
CT, NJ, NY, OH, PA.

PROTECTION INSPECTION, INC.
Rte. 11 Hill Rd.
Knoxville, TN 37938
(615) 688-9779
Norman R. Kurtz - President
Thomas A. Walkiewicz - Consultant
Services:
Provides radon detection and mitigation services to residential and commercial markets, provides "watchdog" service to prevent tampering during radon testing.
Qualifications:
ASHI, EPA RMP.
Area Served:
GA, KY, NC, TN, VA.
EPA RMP 5:
CC.

PUKAL CONTRACTING COMPANY
390 Moorehead Rd.
Cabot, PA 16023
(412) 352-4030
Richard Pukal - Owner
Area Served:
PA.

PYRAMID ENVIRONMENTAL SYSTEMS, INC.
P.O. Box 548
Sparta, NJ 07871
(201) 729-9375
Jeffrey R. DeChacon
Area Served:
Nationwide.
EPA RMP 5:
CC, CW(P).

Q

QUADREX
1700 Dell Ave.
Campbell, CA 95008
(408) 866-4510
Stan Springel - Financial Officer
Services:
Provides radon mitigation services.
Area Served:
Nationwide.

QUOIN PARTNERSHIP, ARCHITECTS AND ENGINEERS
P.O. Box 147
St. Peters, PA 19470
(215) 469-9595
Norman Grant - Principal
Area Served:
PA.

R

R.E. WRIGHT ASSOCIATES, INC./RADONIX GROUP
3240 Schoolhouse Rd.
Middletown, PA 17057
(717) 944-5520
Michael Spear - Marketing
Services:
Provides radon detection and mitigation services, distributes charcoal canisters and alpha track radon test kits.
Qualifications:
PA certification pending.
Area Served:
Nationwide.

R.F. SIMON COMPANY, INC.
RD 2, Box 481
Barto, PA 19504
(215) 754-6455
Ron Simon - Project Manager
Peter Maniscalco - Diagnostic Specialist
Services:
Provides mitigation design and installation services, mitigation supply warehouse carrying a full line of radon mitigation and detection products including fans, caulks and sealants, couplings, comprehensive radon services provided to private, commercial, institutional, and government agencies, expert testimony.
Qualifications:
EPA RMP, 13 years of custom building experience, instructor for EPA training courses, consultant to Lawrence Berkeley Laboratory, Princeton University, PA DER.
Area Served:
Nationwide.
EPA RMP 5:
AT, CC, GW(P).

See Ad Page 148

RA CONSULTANTS, INC.
1236 North 7th St.
Grand Junction, CO 81501
(303) 245-5237
Hoyt Mitchell
Area Served:
CO.

RA-MED
P.O. Box 4413
Alexandria, VA 22303
(703) 960-0514
Scott Owens - President
Area Served:
DC, MD, VA.

RAD ALERT, LTD.
P.O. Box 9238
Springfield, IL 62791
(217) 546-9549
Bob Wilson - Vice President
Services:
Provides radon detection and mitigation services.
Area Served:
IL.

RAD ENVIRONMENTAL SERVICES, INC.
2306 Rodeo Dr.
Durham, NC 27704
(919) 688-9226

Marc Y. Menetrez - President
Trey Marchant - Environmental Technician
Mary Menetrez - Physiologist
Steve Boukedes - Environmental Technician
Services:
Provides radon detection and mitigation services using a variety of methods.
Qualifications:
EPA RMP, NC water treatment A license.
Area Served:
Nationwide.
EPA RMP 5:
CC, CR(P), EP(P), GR(P), CW(P).

RAD ENVIRONMENTAL SERVICES, INC.
30 Basswood Ct.
Red Bank, NJ 07701
(201) 758-9324
Bernadette Davis - Vice President
Services:
Provides radon detection services using charcoal canister, E-Perm, continuous working level, and grab sampling methods, provides radon mitigation services.
Area Served:
NC, NJ.

RADALYTIC LABS, INC.
P.O. Box 202
McConnells, SC 29726
(803) 684-9859
David Culver - President
Services:
Provides laboratory services for radon detection by charcoal canister method.
Area Served:
Nationwide.

RADEX, INC.
1382 Prince Ave. Suite A
Athens, GA 30606
(404) 548-6960
Frederick Tuck - President
Norman Herz - Vice President
Jim Kitchens - Treasurer
Services:
Provides radon detection, mitigation, and training services.
Qualifications:
State certified geologist and university health physicist, experience in commercial and private radon monitoring.
Area Served:
Nationwide.
EPA RMP 5:
AT, CC.

RADIATION DATA
115 Wall St.
Princeton, NJ 08540
(609) 921-7061
Joseph Baicker
Keith Baicker
Jacques Dolphin
Keith Matches
Qualifications:
AARST.
Area Served:
NJ.

RADIATION DATA/DIVISION OF PRINCETON GEOPHYSICAL SERVICES
37 Castleman Dr.
Southington, CT 06489
(203) 621-0678
L. Dudek - Connecticut Manager
J. Dudek - Radiation Safety
Services:
Provides laboratory analysis of charcoal canisters, provides radon detection and mitigation services.
Qualifications:
EPA RMP, AARST, Health Physics Society, certified by NJ and PA, listed in CT.
Area Served:
Nationwide.
EPA RMP 5:
AT, CC, CW(P).

RADIATION DATA/DIVISION OF PRINCETON GEOPHYSICAL SERVICES
1200 State Rd.
Princeton, NJ 08540
(609) 921-7061
Thomas Van Leeuwen
Area Served:
Nationwide.
EPA RMP 5:
AT, CC(P), CW(P).

RADIATION PROTECTION SERVICES, INC.
P.O. Box 2395
Darien, CT 06820
(203) 324-7967
Stuart Korchin - Engineer
Michael L. Caprio, Jr. - Physicist
Thomas J. LaRocca - Physicist
Services:
Provides radon detction and mitigation services.
Qualifications:
Certified ABHP physicist and professional engineer, NJ license, EPA RMP.
Area Served:
Nationwide.
EPA RMP 5:
CC(P).

RADIATION PROTECTION SERVICES, INC.
646 Rte. 18 Suite 215
East Brunswick, NJ 08816
(201) 390-7360
Michael L. Caprio - Health Physicist
Services:
Provides radon detection and mitigation services, operates a primary laboratory.
Qualifications:
EPA RMP.
Area Served:
CT, NJ, NY.

RADIOACTIVITY ALERT DETECTION SERVICES, INC.
P.O. Box L-227
Langhorne, PA 19047
(215) 357-0574
Garry Stklovsky - President
Services:
Provides radon detection, mitigation, and consultation services for home owners, construction, and real estate industries.
Qualifications:
Master's degree in electrical engineering, PA DER course in reducing radon in structures.
Area Served:
PA.
EPA RMP 5:
AT, CC.

RADIOLOGICAL CONTROL ASSOCIATES
665 Robbins Ave. Unit 2
Dracut, MA 01826
(617) 453-5268
Robert E. Casparius
Area Served:
MA.

RADION LABORATORY, INC.
70 Walnut St.
Wellesley, MA 02181
(617) 239-8066
Rudolph Hon - Vice President
Kevin A. Doherty - Senior Manager
Services:
Provides radon detection and mitigation services, operates counting instrumentation for radiation detection in air, liquids, and solid samples, distributes charcoal canisters and water testing kits for radon.
Qualifications:
AARST, NRA, listed with MA Department of Public Health, pending certification in PA, 20 years of experience handling and dealing with radioactive material, instrumentation cross-calibrated with DOE labs.
Area Served:
Nationwide.

RADON ALERT® DETECTION CENTER
P.O. Box 323
Flourtown, PA 19031
(215) 247-9197, (800) 345-6348
Ulrich Hiesinger - President
Leonard Shabason - Radiation
Services:
Distributes charcoal canisters and alpha track radon detectors, provides mitigation diagnosis and design services.
Qualifications:
EPA RMP, certified by NE, NJ, PA.
Area Served:
Nationwide.
EPA RMP 5:
AT, CC.

RADON AWARENESS, INC.
415 East Galena P.O. Box 566
Hailey, ID 83333
(208) 788-2634
Jeffrey L. Engelhardt - President
Susan M. Engelhardt - Vice President
Services:
Provides radon detection and mitigation services, designs and installs mitigation systems.
Qualifications:
Radon testing specialist course.
Area Served:
ID, MT, NV, OR, UT, WA, WY.
EPA RMP 5:
CC.

RADON BUSTERS
1345 Main Ave. P.O. Box 2426
Fargo, ND 58108
(701) 293-5370
Michael Skansgaard - Sales
Services:
Provides radon detection and mitigation services.
Area Served:
IA, MN, MT, ND, SD.

RADON CONSULTANTS, INC., MARYLAND
4831 Silver Hill Rd.
Suitland, MD 20746
(301) 736-7040
Robert Puskar
Services:
Provides radon detection and mitigation services.
Area Served:
MD.

RADON CONSULTANTS, INC., VIRGINIA
4012 North 35th St.
Arlington, VA 22207
(703) 522-6669, (301) 736-7040
Jacquelyn G. Black
Services:
Provides radon detection and mitigation services.
Area Served:
Nationwide.
EPA RMP 5:
AT, CC.

RADON CONTROL PROFESSIONALS, INC.
11920 Fieldthorn Ct.
Reston, VA 22094
(703) 471-9459
See Ad Page 151
Stephen Hall - President
Norman Rodier - Vice President
Services:
Provides full service radon detection and mitigation services, residential, commercial, soil testing, site evaluation, pre-construction mitigation design.
Qualifications:
County business license, Maryland Home Improvement license, EPA mitigation course, graduate-level scientists.
Area Served:
Nationwide.
EPA RMP 5:
AT.

RADON CONTROL SERVICES CORPORATION
3122 49th St.
Des Moines, IA 50310
(515) 276-9131
Charlie Stephenson - President
Services:
Provides radon detection and mitigation services.
Area Served:
IA.

RADON CONTROL SYSTEMS
138 Riverview Plantation Dr.
Williamsburg, VA 23185
(804) 564-1893
Gary A. Hall - President
Tom Girard - Vice President
Services:
Provides radon mitigation services using sub-slab suction, drain-tile suction, block-wall ventilation, and sealing entry points, etc.
Qualifications:
AARST, EPA training courses.
Area Served:
VA.

RADON CORRECTIONS INSTITUTE, INC.
P.O. Box 314
Longmont, CO 80502
(303) 651-6190, (800) 648-9281
Wayne Barcewski - President
Douglas Wallace - Vice President
Services:
Provides radon mitigation and consultation services for existing housing and new construction.
Qualifications:
AARST, Colorado School of Mines course on indoor radon.
Area Served:
CO.

RADON DECON, INC.
P.O. Box 28
Muncy Valley, PA 17758
(717) 482-2076
Calvin Zehring - President
Qualifications:
AARST.
Area Served:
PA.

RADON DETECTION AND CONSULTING
705 West Fourth St.
Lawrence, KS 66044
(913) 843-7377
Steven R. Hicks - Director
Services:
Provides full service radon detection, mitigation, and consultation services, information at no charge.
Qualifications:
EPA RMP, operated by medical, engineering, and legal professionals.
Area Served:
Nationwide.
EPA RMP 5:
CC.

RADON DETECTION AND REDUCTION LABORATORY/DIVISION OF FORMACHEM
10834 Ambassador
St. Louis, MO 63132
(314) 872-9350
Harold Susman - President
Donald Pearline - Biochemist
Services:
Manufactures and distributes the Seal Safe Radon Gas Detector and alpha track detectors, distributes Enkadrain radon mitigation system for new home construction.
Area Served:
Nationwide.

RADON DETECTION AND REMOVAL SERVICES
915 6th Ave., N.
Seattle, WA 98109
(206) 282-9156
Raymond Duffy
James Klosterman
Services:
Provides radon detection and mitigation services.
Area Served:
WA.

RADON DETECTORS
RD 1, Box 177
Van Etten, NY 14889
(607) 589-6223
Larry L. Emdee - Owner
Services:
Provides radon detection, mitigation and diagnostic services for commercial and residential buildings.
Qualifications:
EPA RMP, AARST, NY RAS.
Area Served:
CT, DC, DE, MA, ME, NH, NJ, NY, OH, PA, RI, VA, VT, WV.
EPA RMP 5:
AT.

RADON DIAGNOSTIC LABORATORIES
231 Front St. P.O. Box 2890
South Portland, ME 04106
(207) 767-2818
Dave Sankey - President
Don Burrows - Director of Marketing
Jon Dyer - Technical Director
Services:
Provides radon air and water detection and mitigation services, well water and ground water analysis, offers 24 hour turnaround on test results, presents seminars on dealing with radon.
Qualifications:
EPA RMP, AARST, Association of Maine Environmental Laboratories.
Area Served:
Nationwide.
EPA RMP 5:
CC(P).

RADON DIAGNOSTICS
Box 761
Andover, NJ 07821
(201) 729-1251
Charles R. Wayne, Jr. - Owner
Area Served:
Nationwide.
EPA RMP 5:
CC(P), GR(P), CW(P).

RADON DOCTOR, INC., THE
Alpha Building Suite 904
Easton, PA 18042
(215) 252-4545
Ed Roscioli - President
Albert Roscioli - Manager
Services:
Provides radon detection and mitigation services
Qualifications:
President is a nuclear engineer, PA DER, NJ DEP voluntary certification.
Area Served:
PA.

RADON ENGINEERING SERVICES, INC.
3129 Overton Rd.
Birmingham, AL 35223
(205) 967-2665
F. Milton Jessup - President
R. Douglas Macgill - Vice President
Robert H. Hood - Vice President
Services:
Provides radon detection and mitigation services for residential market.
Qualifications:
EPA RMP, staff composed of registered professional engineer, lawyer, and architect.
Area Served:
Nationwide.
EPA RMP 5:
CC.

RADON ENGINEERING/A UNIT OF PSI ENGINEERING, INC.
1 Lethbridge Plaza
Mahwah, NJ 07430
(201) 529-8300
Harvey Greenberg - Manager
William C. Taylor - Quality Assurance
Services:
Provides radon detection services using working level monitors, charcoal canisters, and alpha track for the real estate industry and private residences, provides radon mitigation services.
Qualifications:
EPA RMP, lab certified by NY, listed with CT, NJ.
Area Served:
Nationwide.
EPA RMP 5:
CC(P), EP, CW(P).

RADON ENVIRONMENTAL CONSTRUCTION SERVICES/RECONS, INC., MAIN OFFICE
P.O. Box 3058
Princeton, NJ 08543-3058
(609) 883-1808, (215) 493-0667
Peter J. Ortolano, Jr. - President
Services:
Provides radon detection and mitigation services, retrofit remediation systems installed for new construction, commercial/industrial designs and installations.
Qualifications:
PA DER, AARST, NJ DEP, 22 years of experience in construction industry.
Area Served:
NJ, PA.

RADON ENVIRONMENTAL CONSTRUCTION SERVICES/RECONS, INC., PA OFFICE
979 Mt. Eyre Rd.
Washington Crossing, PA 18977
(215) 321-3274
Peter Ortolano - Owner
Area Served:
PA.

RADON ENVIRONMENTAL SERVICES, INC.
8935 North Meridian St. Suite 110
Indianapolis, IN 46260
(317) 846-1991
Richard Jordan - President
James Mann
Services:
Provides radon detection and mitigation services for corporate relocation companies, offers consulting services.
Qualifications:
EPA RMP.
Area Served:
Nationwide.
EPA RMP 5:
CC, CR(P), EP(P).

RADON ENVIRONMENTAL SERVICES, INC.
901 Mason Headley Rd.
Lexington, KY 40502
(606) 255-6375
Ann Wiley - Office Manager
Services:
Provides radon detection and mitigation services for corporate relocation companies.
Qualifications:
EPA RMP.
Area Served:
AL, AR, FL, GA, IA, IN, KY, MN, MO, NC, NE, OH, SC, TN, WI, WV.
EPA RMP 5:
CC, EP.

RADON ENVIRONMENTAL SERVICES, INC.
4320 Parakeet Rd.
Toledo, OH 43612
(419) 447-0209
James Mann - Office Manager
Services:
Provides radon detection and mitigation services for corporate relocation companies.
Qualifications:
EPA RMP.
Area Served:
OH.

RADON EVALUATION SERVICES, INC.
P.O. Box 13609
Atlanta, GA 30324
(404) 478-0037
F. Milton Jessup - President
R. Douglas MacGill - Vice President
Robert H. Hood - Vice President
Services:
Provides radon detection and mitigation services for commercial and residential markets.
Qualifications:
EPA RMP, staff composed of registered professional engineer, lawyer, and architect.
Area Served:
AL, FL, GA, KS, KY, LA, MS, SC, TN.
EPA RMP 5:
CC.

RADON FREE
5666 Ashbrook
Toledo, OH 43614
(419) 865-0129
Carrie Fleitz - President
Steve Goodall - Vice President
Tim Fleitz - Manager
Services:
Provides radon detection services using charcoal canister, alpha track and continuous radon monitor methods, provides radon mitigation services for new and existing buildings, non-commercial laboratory for research purposes or state agencies' assistance only.
Qualifications:
ARA.
Area Served:
MI, MN, OH.
EPA RMP 5:
AT, CC.

RADON HOME SEARCH
349 Maxim Dr.
Hopatcong, NJ 07843
(201) 398-4106
James Walsh
Area Served:
NJ.

RADON INSPECTION SERVICE, INC.
789 East Glen Ave.
Ridgewood, NJ 07450
(201) 670-8821
Robert D. Shumeyko - President
David B. Terry - Vice President
Services:
Provides radon detection and mitigation services for real estate industry and home owners using ten different methods.
Qualifications:
NJ DEP, NY DOH.
Area Served:
CT, NJ, NY.
EPA RMP 5:
AT, CC, CR(P), EP(P), GR(P), GW(P), CW(P).

RADON LABS
P.O. Box 2054 Harvard Square
Cambridge, MA 02238
(617) 923-6454
Frank Sedaghati
Services:
Provides radon detection and mitigation services.
Area Served:
Nationwide.

RADON MANAGEMENT OF KENTUCKY
12203 Old Shelbyville Rd.
Middletown, KY 40243
(502) 244-0517
Vance F. Walker, Jr. - Owner
Services:
Provides radon mitigation services for new and existing homes.
Qualifications:
AARST, NAHB.
Area Served:
KY.
EPA RMP 5:
CC.

RADON MEASUREMENT AND SERVICES
13131 West Cedar Dr.
Lakewood, CO 80228
(303) 980-5086, (303) 988-3033
R. Stanley Thompson
Area Served:
Nationwide.
EPA RMP 5:
CC.

RADON MEASURING SYSTEMS
1129 Wyoming Ave.
Scranton, PA 18509
(717) 961-1666
Jeff Brundage - President
Jack Hill - Technician
Services:
Provides radon detection services using working level monitors, provides radon mitigation services.
Area Served:
NJ, NY, PA.
EPA RMP 5:
CW(P).

RADON MITIGATION SERVICE
421 Hillandale
Round Lake, IL 60073
(312) 546-3890
Ray Biederer - President
Area Served:
IL, WI.

RADON MITIGATORS OF ARIZONA
6844 East Thunderbird Rd.
Scottsdale, AZ 85254
(602) 483-7506
Lawrence H. Kaplan - President
Kenneth Gilvary - Vice President
Joan Kaplan - Vice President
Services:
Provides radon detection services using charcoal canister and alpha track methods, provides radon mitigation services, group seminars.
Area Served:
Nationwide.

RADON MITIGATORS, INC.
16 Hillside Dr.
Clinton, NJ 08809
(201) 730-6596
Lawrence Kaplan - President
Joan Kaplan - Vice President
Kenneth Gilvary - Vice President
Services:
Provides radon detection services using charcoal canister and alpha track methods, diagnostics, provides radon mitigation services, group seminars.
Area Served:
Nationwide.
EPA RMP 5:
AT, CC(P).

RADON PROTECTION SERVICES
P.O. Box 158
Hummelstown, PA 17036
(717) 566-5609
Dan Wilson - Owner
Services:
Provides radon detection and mitigation services, consulting, and training/seminars.
Qualifications:
PA DER.
Area Served:
PA.

RADON PROTECTION, INC.
The Summit Suite 110
Louisville, KY 40207
(502) 893-4563
Roland Hawksley - President
Area Served:
KY.

RADON REDUCTION AND TESTING COMPANY
605 South Sherman Suite 705 J
Richardson, TX 75081
(214) 680-2296
Freeman Stallings - President
Phillip Stallings - General Manager
Services:
Provides radon detection services using charcoal canister method, provides radon mitigation services.
Area Served:
TX.

RADON REDUCTION ENGINEERING AND LABORATORY
211 Mill St., NE
Vienna, VA 22180
(703) 281-7865, (703) 281-3730
Carl B. De Azagra - President
Charles H. De Azagra - Mitigation Program
Erika D. De Azagra - Measurement Program
Robin De Azagra - System Installation
Services:
Provides radon detection, mitigation, and monitoring services, and operates a primary laboratory.
Qualifications:
AARST, chemical engineer, architect, geologist, radon professional technician, and licensed contractor on staff.
Area Served:
Nationwide.
EPA RMP 5:
CC(P), CW(P).

RADON REDUCTION SERVICES
5601 East 109th Ter.
Kansas City, MO 64137
(816) 942-7083
Jerry Harper - Partner
Lee Cunningham - Partner
Services:
Provides radon mitigation services, inspects buildings for radon entry routes and designs and installs mitigation systems.
Qualifications:
AARST, training by NY State Energy Office.
Area Served:
IA, KS, MO, NV.

RADON REDUCTION SERVICES, INC.
P.O. Box 627
Mt. Airy, MD 21771
(301) 831-0228
Robert Wolfgang - President
Services:
Provides radon detection and mitigation services.
Area Served:
MD.

RADON REDUCTION SYSTEMS
169 Old Gage Hill Rd.
Pelham, NH 03076
(603) 893-5549
Joseph Maraldo - President
Area Served:
NH.

RADON REDUCTION, INC.
1125 Mechanic St.
Jeffersonville, IN 47130
(812) 284-4466
Jim Ward - President
Carolyn Ward - Vice President
Services:
Provides radon mitigation services for homeowners, builders, and realtors.
Qualifications:
ARA, EPA seminars.
Area Served:
IN, KY.

RADON REDUCTION, INC.
363 Massachusetts Ave.
Lexington, MA 02173
(617) 862-8462
John A. Chapin - President
Scott Chapin - Vice President
Services:
Provides radon mitigation services.
Area Served:
MA, ME, NH, RI.

RADON REMEDIAL SERVICES
P.O. Box 622
Bryn Athyn, PA 19009
(215) 938-7766
Lewis Nelson, IV
Services:
Provides radon mitigation services, mitigation system design and installation, and radon detection services.
Qualifications:
AARST, PA DER.

Area Served:
DE, NJ, PA.

RADON REMEDIATION COMPANY, INC.
38 Whippoorwill Way
Belle Mead, NJ 08502
(201) 359-0050
Area Served:
NJ.

RADON REMEDIATION SERVICES
10400 Connecticut Ave. Suite 504
Kensington, MD 20895
(301) 933-7900
Milton Iskow - President
David E. Pard - Sales Manager
Services:
Provides radon mitigation services, fully guaranteed, for residential, comercial, government and industrial buildings.
Qualifications:
AARST, EPA training courses.
Area Served:
DC, DE, MD, VA.

RADON REMOVAL, INC.
Box 283
Milltown, NJ 08850
(201) 545-2000
Anthony M. Fama - President
Services:
Provides radon detection and mitigation services, distributes a geomatrix, nylon membrane laid over perimeter of gravel foundation to provide an effective radon barrier.
Qualifications:
AARST, engineer on staff.
Area Served:
Nationwide.

RADON RESEARCH INTERNATIONAL, INC.
7650 South Tamiami Trail , #9
Sarasota, FL 34231
(813) 923-7300
Arthur Lipshutz - Chief Engineer
Stuart Lee - Chief Field Engineer
Services:
Provides radon detection and mitigation services specializing in slab-on-grade construction, consultants to real estate industry.
Qualifications:
AARST, EPA RMP, BBB, three years of radon mitigation experience.
Area Served:
FL.
EPA RMP 5:
AT, CC.

RADON RESIDENTIAL SYSTEMS, INC.
1474 Rte. 23 N. Suite 3
Wayne, NJ 07470
(201) 633-3976
Mike Agrati - President
Services:
Provides radon mitigation services.
Area Served:
NJ.

RADON RESOURCE, INC.
49 John St.
Westminster, MD 21157
(301) 857-9330
Neil Edwards - President
Services:
Provides radon detection and mitigation services.
Qualifications:
EPA training courses in testing and mitigation.
Area Served:
MD, PA.
EPA RMP 5:
CC.

RADON SAFETY AND TECHNOLOGY CORPORATION
4 Charles St.
Pleasant Valley, NY 12569
(914) 635-1138, (800) 552-5257
Ronald J. Gasparro - President
Zia R. Hashmi - Director
Claude J. Noto - Assistant Director
Services:
Radon measurement analytical laboratory, water and soil measurements, surveys for real estate transactions, land sites, and newly constructed and standing buildings, distributes charcoal canisters and alpha track detectors, waterborne radon test kits, radon alarms and auxiliaries offered to public and professionals, provides radon mitigation services.
Qualifications:
EPA RMP, AARST, ASTM.
Area Served:
Nationwide.
EPA RMP 5:
AT, CC(P), EP(P), GR(P), CW(P).

RADON SCREENING SERVICE, INC.
P.O. Box 37007
Denver, CO 80237
(303) 741-3207
Thomas Staley
Area Served:
Nationwide.
EPA RMP 5:
CC, GW(P), CW(P).

RADON SERVICES OF WYOMING
1714 Rollins
Cheyenne, WY 82001
(307) 634-9505
Ronald J. Sargent - President
Services:
Provides radon detection and mitigation services.
Area Served:
WY.

RADON SURVEY SYSTEMS, INC.
P.O. Box 22288
Cleveland, OH 44122
(216) 464-7997
Darioush T. Ghahremani - President
Services:
Provides radon detection and mitigation services for residential and commercial markets, site surveys for undeveloped land, research and development of radon detection technologies, detectors, and materials, new development on radon mitigation options for various projects.
Qualifications:
AARST, NRA, several national scientific societies.
Area Served:
Nationwide.
EPA RMP 5:
AT, CC.

RADON SYSTEMS OF RHODE ISLAND, INC.
156 Earle St.
Woonsocket, RI 02895
(401) 766-7072
Richard A. Monteiro - President
Services:
Provides radon detection and mitigation services for residential, commercial, and industrial markets.
Qualifications:
EPA RMP, applicant PA licensing.
Area Served:
CT, MA, ME, NH, NJ, NY, PA, RI.
EPA RMP 5:
AT, CC.

RADON SYSTEMS, INC.
P.O. Box 2509
Laurel, MD 20708
(301) 725-9000
See Ad Page 213
Michele Widmayer - President
Patricia Jordan - Vice President
Kenneth Grim - Vice President
Services:
Provides radon detection and mitigation services to home owners, realtors, and builders.

Qualifications:
EPA course on radon testing, NY State Energy Office course on mitigation, AARST.
Area Served:
DC, DE, MD, PA, VA.
EPA RMP 5:
CC.

RADON TECHNICIANS, INC
27 Boren Shore Ln.
Elkton, MD 21921
(301) 854-6186
Bob Meliodon
Area Served:
MD.

RADON TECHNOLOGY
11404A East 24 Hwy.
Sugar Creek, MO 64054
(816) 461-0211
Clay Meyers - President
Services:
Provides radon detection services using continuous radon monitor method, provides radon mitigation services.
Qualifications:
ARA, detection and mitigation workshops.
Area Served:
MO.

RADON TECHNOLOGY AND CONSULTANTS
RFD 2 Louisquisset Pike
North Smithfield, RI 02895
(401) 769-2129, (401) 568-0296
Lucien R. Auger - President
John S. Dolnski - Vice President
Services:
Provides radon detection and mitigation services.
Qualifications:
Radon detection and mitigation training courses, 23 years of experience in waterproofing business.
Area Served:
CT, MA, ME, NH, RI, VT.

RADON TEST AND REDUCTION SPECIALIST
125 Oak Ave. P.O. Box 129
Spruce Pine, NC 28777
(704) 765-1222
Bud Ledford - Consultant
Jerri Ledford - Laboratory Supervisor
Services:
Provides radon detection and mitigation services.
Area Served:
GA, KY, MI, NC, SC, TN, VA, WV.

RADON TESTING AND CONSULTING
1771 Post Rd., E. Suite 320
Westport, CT 06880
(203) 454-3886
Lorraine Hodes - Vice President
John Nicholas - Director
Mark Lewis - Senior Geologist
Maria Mody - Communications
Services:
Provides radon detection and mitigation services for home owners, realtors, builders and developers, exclusive New England distributor of Current Indoor Air Model 300 Radon Mitigation System.
Qualifications:
AARST, AGSE.
Area Served:
CO, CT.

RADON TESTING AND CONSULTING SERVICES
34 Jones River Dr.
Kingston, MA 02364
(617) 585-9924
Daniel C. Walker - President
Cathleen Walker - Vice President
Services:
Provides radon detection, mitigation, and consultation services for the real estate industry, construction trade, and private citizens.
Qualifications:
AARST, RTCA professional consultant.
Area Served:
CT, MA, ME, NH, RI, VT.

RADON TESTING AND CONTROL
2910 East Broadway Ave.
Bismarck, ND 58501
(701) 222-1613
Judith A. Atkinson - Owner
Herb Haugen - Owner
Services:
Provides radon detection and mitigation services.
Area Served:
MT, ND, SD.

RADON TESTING AND MITIGATION OF TEXAS, INC.
P.O. Box 34268
Houston, TX 77234-4268
(800) 920-1663
Jerry Shults - President
Services:
Provides radon detection and mitigation services.
Area Served:
TX.

RADON TESTING AND MITIGATION SERVICES COMPANY
5442 Cary Ave.
Davenport, IA 52807
(319) 386-9005
Tom Gabrilson - President
Services:
Provides radon detection and mitigation services.
Qualifications:
Licensed in IA, certified Air Chek dealer.
Area Served:
IA, IL.

RADON TESTING AND MITIGATION SYSTEMS
P.O. Box 3052
Carmel, IN 46032
(317) 545-5578
Lorand Magyar - Owner
Andrew W. Murrell - Owner
Services:
Provides radon detection and mitigation services for real estate companies and for residential market.
Area Served:
IL, IN, KY, OH.

RADON TESTING AND REDUCTION
P.O. Box 437
Howell, MI 48844
(517) 548-2497
Chris Coffey - General Manager
Craig Coffey - General Manager
Services:
Provides radon detection services using charcoal canister and E-Perm methods, provides radon mitigation services.
Area Served:
MI.

RADON TESTING AND TREATMENT, INC.
723 North Weber Suite 300
Colorado Springs, CO 80903
(719) 520-5693
Douglas L. Kladder - President
Steve Jelinek - Senior Analyst
Jessica Goodwin - Marketing Director
Robert Shideler
Services:
Provides radon detection and mitigation services to the real estate industry, distributes charcoal canisters to the public.
Qualifications:
EPA RMP, licensed contractor.
Area Served:
CO.
EPA RMP 5:
CC, GW(P).

RADON TESTING CONSULTANTS, INC.
P.O. Box 25313
Newark, NJ 07101
(201) 290-9217
Mark Jeziorski - President
Services:
Provides radon detection and mitigation consulting services for consumers, businesses, and government agencies, distributes charcoal canisters and alpha track detectors.
Qualifications:
EPA RMP, staff has over 12 years of combined engineering experience.
Area Served:
Nationwide.
EPA RMP 5:
CC.

RADON TESTING SERVICE, INC.
Rte. 212
Lake Hill, NY 12488
(914) 679-6835
Tully Slayton - President
Sidney Slayton - Secretary
Services:
Provides radon detection and mitigation services.
Area Served:
NY.

RADON TESTING, INC.
P.O. Box 7304
Fairfax Station, VA 22039
(703) 591-1111
James P. Schroff - President
Services:
Provides radon detection and mitigation services for the real estate industry.
Qualifications:
Training courses by EPA and NY State Energy Office.
Area Served:
Nationwide.
EPA RMP 5:
CC, CR(P).

RADON X COMPANY, INC.
217 Roxbury St.
Keene, NH 03431
(603) 352-5319
Steven Berger - President
David P. Tkatch - Vice President
Frank Lapinski - Vice President
Services:
Provides radon detection and mitigation services.
Qualifications:
EPA mitigation course, 45 years of radiation experience, professional engineers on staff.
Area Served:
Nationwide.
EPA RMP 5:
GW(P).

RADON, LTD.
411 North Weinbach Ave.
Evansville, IN 47711
(812) 479-3110, (800) 634-2901
Ed Shoemaker - President
Services:
Provides radon detection and mitigation services, distributes charcoal canister kit to public and dealers, laboratory facilities for analysis.
Qualifications:
ARA, radon detection and mitigation training courses.
Area Served:
Nationwide.
EPA RMP 5:
CC(P).

RADON-X CORPORATION
300 Valentine St.
Hackettstown, NJ 07840
(201) 852-6212, (800) 272-3669
John LoPilato - President
Services:
Manufactures air exchange and sub-slab ventilating systems, provides radon mitigation services.
Area Served:
NJ.

RADON-X CORPORATION
3 Edinburgh Ln.
Mt. Laurel, NJ 08054
(609) 778-7771, (800) 272-3669
Paul Gibson
Area Served:
NJ.

RADONAIRE
P.O. Box 3702
Wayne, NJ 07474
(201) 790-3252
H.E. Firschein - President
Area Served:
NJ.

RADONICS, INC.
P.O. 489
Pennington, NJ 08534
(609) 737-8888
Theresa R. Marryott - President
Services:
Provides radon testing and mitigation services.
Area Served:
Nationwide.

RAMP INDUSTRIES, INC.
1127 West 46th St.
Denver, CO 80210
John E. Lucas - General Manager
Services:
Provides radon detection and mitigation services, operates a primary laboratory.
Area Served:
CO.

RAMSES II, LTD.
2355 East Dartmouth Ave.
Denver, CO 80210
(303) 758-5209
Robert L. Clark - Vice President
George A. Michael - President
Services:
Manufactures and distributes charcoal canisters, electronic sump well cover kits, electronic PVC ventilation fans, and rotary air-to-air heat exchangers, provides radon mitigation services.
Qualifications:
Licensed professional engineer on staff.
Area Served:
CO.
EPA RMP 5:
CC.

RANDOLPH AND ASSOCIATES, INC.
8901 North Industrial Rd.
Peoria, IL 61615
(309) 692-4160
Bryan W. Arnold
Vahooman Mirkhaef
Area Served:
Nationwide.
EPA RMP 5:
CC(P).

RCI SERVICES
11412 Hounds Way
Rockville, MD 20852
(301) 231-5134
Frank M. Clower
Area Served:
MD.

RDS/RADON DETECTION SERVICES, INC.-CORPORATE OFFICE
Rte. 179, Old York Rd. P.O. Box 419
Ringoes, NJ 08551
(201) 788-3080
James G. Davidson - President
Susanne E. Micek - Vice President
Michael S. Terpilak - Health Physicist
Services:
Provides full service radon detection mitigation, and consultation services, offers franchises, offers comprehensive training courses for radon professionals, manufactures and distributes RDS "remediator" fans, locations in four states.
Qualifications:
Has met all licensing requirements in PA, NJ, and KS, courses have been approved by DE, KY, PA, OH.
Area Served:
Nationwide.
See Ad Page 153
EPA RMP 5:
AT, CC, EP(P), GR(P), CW(P).

RDS/RADON DETECTION SERVICES-GEORGIA OFFICE
403 West Ponce de Leon Ave. P.O. Box 1826
Decatur, GA 30031
(404) 377-2366
Barrett P. Walker - President
Peggy Walker - Customer Relations
Judy Reid - Office Manager
Services:
Provides radon detection and mitigation services, informational seminars.
Qualifications:
EPA RMP, BBB, Metro Atlanta Home Builder's Association, AARST.
Area Served:
Nationwide.
EPA RMP 5:
AT, CC, EP(P), CW(P).

RDS/RADON DETECTION SERVICES-PENNSYLVANIA OFFICE
1011 Brookside Rd., Suite 270 P.O. Box 270
Allentown, PA 18106
(215) 481-9555
Bruce Barker - Owner
Donna Garris - Office Manager
Services:
Provides radon detection and mitigation services.
Area Served:
Nationwide.
EPA RMP 5:
AT, CC, EP(P), GR(P), GW(P), CW(P).

REALTY REPAIR AND DEVELOPMENT
8 Mount Hope Rd.
Rockaway, NJ 07866
(201) 625-0030
Keith Batley
Area Served:
NJ.

RELIABLE BASEMENT WATERPROOF COMPANY
2811 Westerville Rd.
Columbus, OH 43224
(614) 267-4477
Matthew Cohen - President
Lloyd Cohen - Vice President

Mike Copley - General Manager
Services:
Provides radon detection and mitigation services for residences, commercial and industrial markets.
Qualifications:
EPA RMP, ARA.
Area Served:
OH.
EPA RMP 5:
CR(P).

REMEDIAL TECHNOLOGIES, INC.
10110 Kile Rd.
Chardon, OH 44024
(216) 765-2442
Richard Crombie - President
Services:
Provides radon mitigation services, distributes radon detection and mitigation products including sealers, fans, Honeywell, Pylon, EDA radon monitoring equipment.
Area Served:
Nationwide.

RENSSELAER POLYTECHNICAL INSTITUTE
110 Eighth Street
Troy, NY 12180
(518) 276-8237
Ken Brown - HVAC Engineer
Services:
Provides radon detection and mitigation services or campus buildings, sponsored by state of New York.
Area Served:
NY.

RESOURCE CENTER FOR ENVIRONMENTAL PROTECTION
P.O. Box 25
Schuylerville, NY 12871
(518) 695-3218
See Ad Page 160
Peter Lane - President
Fred Staing - Business Manager
Services:
Provides radon mitigation services using ventilation, barrier, and water remediation systems, consultation with contractors to develop new radon mitigation products and designs.
Area Served:
NY.

RESTORED ENVIRONMENTS COMPANY, INC.
4940 Nutmeg Knoll
Cincinnati, OH 45244
(513) 831-0743
Bob Smith - President
Services:
Provides radon detection and mitigation services.
Area Served:
IN, KY, OH.

RIEHS AND RIEHS
501 George St.
New Bern, NC 28560
(919) 637-9752
H. Erich Riehs - President
Services:
Provides radon mitigation services, distributes air-to-air heat exchangers for residential applications.
Qualifications:
Ten years of experience in heat recovery and ventilation industry.
Area Served:
Nationwide.

RITE WAY WATERPROOFING
7337 Lake Dr.
Lino Lakes, MN 55014
(612) 786-0550
Sharon Thompson - Vice President
Larry Russ - Sales Representative
Services:
Provides radon mitigation services.
Area Served:
MN.

ROBERT A. SWOYER ASSOCIATES
241 South Fifth St.
Reading, PA 19602
(215) 376-2580
Robert A. Swoyer - Consulting Engineer
Olin R. Marth, II - Project Engineer
Services:
Provides radon mitigation services, provides consulting engineering services.
Qualifications:
PA DER course.
Area Served:
PA.

ROCKY MOUNTAIN HOME INSPECTIONS
1571 East Mineral Pl.
Littleton, CO 80122
(303) 730-0922
Martin Dommer
Area Served:
CO.

RONALD STEIN CONSTRUCTION, INC.
21 Davis Ave.
Poughkeepsie, NY 12603
(914) 473-0064
Ronald Stein - President

Services:
Provides radon detection and mitigation services.
Qualifications:
EPA RMP.
Area Served:
NY.
EPA RMP 5:
CC.

ROSS SYSTEMS, INC.
955 Connecticut Ave.
Bridgeport, CT 06607
(203) 366-2210
Nancy Adams - Branch Manager
Area Served:
CT, NY.

ROSS SYSTEMS, INC.
21 Main St. Box 670
Blairstown, NJ 07825
(201) 362-5571
Richard Ross - President
Services:
Provides full service radon detection services, distributes charcoal canisters and alpha track detectors, provides radon mitigation consultation services using computer technology, radon mitigation proposals at no charge, offers seminars and workshops.
Qualifications:
AARST, EPA RMP,APCA, NJ DEP.
Area Served:
Nationwide.
EPA RMP 5:
AT, CC, GW(P), CW(P).

S

SACCHETTI, R. MICHAEL
103 Intervale St.
Quincy, MA 02169
(617) 479-2111
Mike Sacchetti - President
Services:
Provides radon detection and mitigation services.
Qualifications:
AARST.
Area Served:
MA.

SAF-AIR
P.O. Box 1217
Orange, CT 06477
(203) 799-9609
Thomas F. Meehan - President
Services:
Provides radon detection and mitigation services for residential and commercial properties, seminars and speaking engagements.
Qualifications:
CT mitigation list.
Area Served:
CT.
EPA RMP 5:
CC.

SAFE SHELTER ENVIRONMENTAL
527 West Lancaster Ave.
Frazer, PA 19355
(215) 640-4839
Rick Haag - President
Steve Cummings - General Manager
Chris Haag - Manager
Joe Batten
Services:
Provides radon detection and mitigation services for real estate transactions and for the general public.
Qualifications:
PA DER course, APCA, EPA RMP.

Area Served:
DE, NJ, PA.
EPA RMP 5:
CW(P).

SAFE-AIR
2599 East Main St. Suite 525
Bexley, OH 43209
(614) 447-8300, (800) 553-3117
James Diamond - President
Services:
Provides radon mitigation services using roof venting systems.
Qualifications:
ARA.
Area Served:
OH, TN.

SAFE-AIRE, INC.
162 East Chestnut St. P.O. Box 160
Canton, IL 61520
(309) 647-0419, (800) 331-2943
Bill Wherley - Sales Manager
Chet Phillips - Products Manager
Harry Ellison - Marketing Director
Don Stees - Vice President
Services:
Provides patent-pending Safe-Aire radon mitigation system to dealers, provides radon detection and mitigation services, distributes extensive line of radon detection and mitigation products, dealership opportunities available.
Area Served:
Nationwide.
See Ad Page 161
EPA RMP 5:
AT, CC, CR(P), EP.

SAFEHOUSE, INC.
2335 Sigman Industrial Ct.
Conyers, GA 30207
(404) 388-0481
Stanley Shaw - President
Raymond McQueen - Vice President
Qualifications:
NRA.
Area Served:
GA.

SALAMON, PAUL
2621 North Knoxville Ave.
Peoria, IL 61604
(309) 686-9392, (309) 686-0046
Paul Salamon
Area Served:
IL.

SCHUCK HOME SERVICES
P.O. Box 15581
Ft. Wayne, IN 46885
(219) 486-7945
Gregory R. Schuck - President
Services:
Provides radon mitigation services for residential and commercial properties.
Qualifications:
20 years of experience in construction management, AARST, attended two workshops.
Area Served:
IN, MI, OH.

SHAPIRO, MARK
7709 Morgan Ln.
Laverock, PA 19118
(215) 886-3060
Mark Shapiro
Area Served:
PA.

SIGMA ENVIRONMENTAL SYSTEMS, INC.
P.O. Box 229
Wappingers Falls, NY 12590
(914) 297-4000
Jack Armstrong - Chief Engineer
Area Served:
NY.

SIMPSON WEATHER ASSOCIATES
809 East Jefferson St.
Charlottesville, VA 22902
(804) 979-3571
M. Garstang - President
G.D. Emmitt - Vice President
H.G. Goodell
Services:
Provides radon mitigation and environmental consultation services.
Qualifications:
Staff includes persons with doctoral degrees in meteorology, geology, geochemistry, hydrology.
Area Served:
Nationwide.
EPA RMP 5:
CC.

SMITH, STANLEY
2901 26th St., W. Suite 702
Bradenton, FL 34205
(813) 756-3635
Stanley Smith
Qualifications:
AARST.
Area Served:
FL.

SOLAR RESOURCE TECHNOLOGY
128 Market St.
South Williamsport, PA 17701
(717) 323-8919
Ernest Kittka - Owner
Area Served:
PA.

SOOS RADON SYSTEMS, INC.
25 Pheasant Run Dr.
Skillman, NJ 08558
(201) 359-0356
John Soos - President
Services:
Provides radon mitigation services for commercial and residential buildings.
Qualifications:
NJ DEP list of approved companies.
Area Served:
NJ.

SOUTH CENTRAL ENVIRONMENTAL SERVICES
RD 1, Box 265
Pine City, NY 14871
(607) 733-1661
Lawrence P. McLane - Contractor
Robert Lovejoy - Vice President
Area Served:
NY, PA.

SOUTHEASTERN RADON CONSULTANTS, INC.
161 York Rd.
Nolensville, TN 37135
(615) 776-5569
Otis E. Guinn - President
Services:
Provides radon detection services using Honeywell Professional Radon Monitors for real estate market, provides radon mitigation services.
Area Served:
AL, GA, KY, TN.

SOUTHERN NEW ENGLAND RADON TESTING AND MITIGATION
P.O. Box 1141
Westbrook, CT 06498
(203) 399-8003
Donald C. Kowalski
George Gavigan
Area Served:
CT.

SPECTRUM INDUSTRIES
1128 Forrest St.
Dover, DE 19901
(302) 736-0792
Phillip Stubbolo
Area Served:
DE.

SPRINGER COMPANY
629 Lancaster Ave.

Malvern, PA 19355
(215) 344-9800
David Springer
Area Served:
PA.

STANDOR RADON DETECTION
598 South Edgewood Ave.
Jacksonville, FL 32205
(904) 388-0284
Stanley J. Szczepanski - President
Services:
Provides radon detection and mitigation services for residential and commercial properties.
Area Served:
Nationwide.
EPA RMP 5:
AT, CC.

STONE RIDGE RADON MITIGATORS
424 Mine Rd.
Asbury, NJ 08802
(201) 735-6080, (800) 447-2366
Joanne Hlavinka
Services:
Provides radon mitigation services for residential and commercial buildings.
Qualifications:
NJ DEP remediation program.
Area Served:
NJ, PA.

SUBURBAN JERSEY WATERPROOFING, INC.
30-32 Abbett Ave.
Morristown, NJ 07960
(201) 539-1538
Henry Hill - Manager
Services:
Provides radon mitigation services using all recognized industry methods.
Area Served:
NJ.

SULLIVAN, ROBERT
Rte. 17, Box 11
Lowman, NY 14861
(607) 732-8617
Robert Sullivan - Contractor
Area Served:
NY.

SUNN CORPORATION
Box M, Dept. 213 3520 Forbes Ave.
Pittsburgh, PA 15213
(412) 828-9100
Nick Gromicko - Chief Executive Officer
Bob Dunkle
Mike Russell
Services:
Provides radon mitigation services, manufactures and distributes charcoal canisters and tamper-resistant cages.
Qualifications:
BBB, NSBA, several state mitigation lists.
Area Served:
Nationwide.

SUNSTREAM DEVELOPMENT CORPORATION
300 North St.
Endicott, NY 13760
(607) 754-6200
Thomas Meade, Jr. - President
Peter R. Gotthardt - Production Manager
Ron Andrascik - Technician
Services:
Provides radon detection and mitigation services to home owners, realtors, and commercial properties, distributes VANEE air-to-air heat exchangers.
Qualifications:
NRA, key personnel trained by NY State Energy Office.
Area Served:
NY, PA.

SURE-WAY SYSTEMS/DIVISION OF E.C.S., INC.
107 Pickard Dr.
Syracuse, NY 13211
(315) 455-2431, (800) 622-1002
Ronald G. Newvine - President
Thomas J. Vitale - Vice President
Services:
Provides radon mitigation services, interior drainage systems to correct wet basement problems, sub-slab suction and foundation sealing to reduce radon levels.
Area Served:
NY.

T

TAPPAN ZEE DEVELOPMENT
25 Webb Ln.
Highland Falls, NY 10928
(914) 446-7704
Roger Heimann - President
Area Served:
NY.

TEE ENGINEERING COMPANY
320 Cutters Hill Ct.
Lexington, KY 40509
(606) 263-5350
Don F. Blackburn - Professional Engineer
David McFarland - Geologist
Bill Yarnell - Geologist
Services:
Provides radon detection and mitigation

services for residential, industrial and commercial properties.
Area Served:
IL, IN, KY, MO, OH, TN, VA, WV.
EPA RMP 5:
CC.

TEQ CORPORATION
P.O. Box 570
Wilkes-Barre, PA 18703-0570
(717) 825-3939, (800) 877-2366
Russell Hendershot - President
Ed Dructor - Vice President
Brian Redmond - Research Scientist
Services:
Provides radon detection and mitigation services for commercial and residential markets, manufactures and distributes charcoal canisters, maintains calibrated radon chamber, offers radon detection and mitigation training courses.
Qualifications:
PA DER certification.
Area Served:
Nationwide.
EPA RMP 5:
AT, CC(P), CR(P), CW(P).

See Ad Page 100

TERRADON EXPLORATION TECHNOLOGIES, INC.
3931 White Oak Tr. P.O. Box 22288
Cleveland, OH 44122
(216) 464-2002
Darioush T. Ghahremani - President
Services:
Provides radon detection and mitigation services, research and development in radon detection, measurement and abatement technology, development of new alpha track and charcoal canisters for industrial and private use, soil gas radon detection and mapping.
Qualifications:
AARST, NRA, AAPG, OAS, ACS, OG&G, AAGE, other scientific associations.
Area Served:
Nationwide.
EPA RMP 5:
AT, CC.

TFC NUCLEAR ASSOCIATES, INC.
425 Bridgeboro Rd.
Moorestown, NJ 08057
(609) 778-4529
John Chando - President
Eric Trenga - Vice President
Harvey Auger - Radon Testing Division
Edwin Voll - Design Engineer
Services:
Provides comprehensive radon detection and mitigation services for government, schools, commercial and residential structures.
Qualifications:
Recently completed contract with NJ DEP for radioactive waste cleanup of radon contaminated soil, contracts with more than 18 nuclear power plants nationwide, American Society of Mechanical Engineers, AARST, ANS, ARA.
Area Served:
Nationwide.

See Ad Page 166

THERMO ENVIRONMENTAL CORPORATION
101 1st Ave. P.O. Box 9046
Waltham, MA 02254
(617) 622-1103, (505) 345-9931
Jeff Brown - Health Physics
Services:
Manufactures and distributes RAD-SCAN radon screening kits, provides radon detection and mitigation services to homeowners, developers, real estate industry, etc., parent company of TMA Eberline.
Area Served:
Nationwide.

THEROUX ENGINEERING
P.O. Box 4096
Hamden, CT 06514
(203) 932-1313
Dennis R. Theroux - President
Services:
Provides radon detection and mitigation services for schools, industrial and commercial structures.
Area Served:
CT.

THOMAS L. HOFFMAN, COMPANY, INC.
RD 3, Box 173M
Coopersburg, PA 18036
(215) 433-5887
Barry Hoffman - Manager
Services:
Provides radon mitigation services.
Area Served:
PA.

TMA/EBERLINE, HEALTH PHYSICS
5635 Jefferson Blvd., NE
Albuquerque, NM 87109
(505) 345-9931
Stanley Waligora - Sales Manager
Ernie Sanchez
Jeff Brown
Services:
Provides radon detection and mitigation services to commercial and government clients, distributes charcoal canisters, provides consultation services.
Area Served:
Nationwide.
EPA RMP 5:
CC(P).

TRI-COUNTY RADON, INC.
2796 North Constitution Rd.
Camden, NJ 08104
(609) 963-1533
Susan B. Cusato - Chief Inspector
John Thompson - Mitigation
Services:
Provides radon detection, mitigation, and consulting services for residential and commercial properties.
Area Served:
NJ, PA.
EPA RMP 5:
CC.

TRI-STATE RADON AND DEWATERING
P.O. Box 922
Dayton, NJ 08810
(201) 329-8570
Jack Laurie - President
Area Served:
NJ.

TRI-STATE RADON REDUCTION, INC.
816 Ainsworth Dr.
Linden, NJ 07036
(609) 275-0155, (201) 925-8472
Robert Schalhoub - President
James D. Fox - Vice President
Services:
Provides radon detection and mitigation services.
Qualifications:
AARST.
Area Served:
NJ.

TRIANGLE WATERPROOFING AND REPAIR SERVICE
P.O. Box 741 206 North Arendell Ave.
Zebulon, NC 27597
(919) 269-8021
John Walk - President
Van Hogan - Vice President
Services:
Provides radon mitigation services.
Area Served:
NC.

TSI CONSTRUCTION SERVICES
1137 3rd Ave., SW
Carmel, IN 46032
(317) 846-4655
Christopher L. Maher - Vice President

Services:
Provides radon detection services using continuous radon monitoring using E-Perm and charcoal canister methods, provides radon mitigation services.
Qualifications:
ABC, CSI, NAHB, NACE, eight years of experience in indoor air quality and building diagnostic investigation.
Area Served:
CA, CO, DC, IA, IL, IN, KS, KY, MI, MO, MS, NC, NJ, NY, OH, PA, TN, VA, WI, WV.
EPA RMP 5:
AT, CC.

U

UNITED SERVICES ALLIANCE, INC.
4900 Leesburg Pike Suite 210
Alexandria, VA 22302
(703) 845-1324
William W. Tu - President
Sheng-jia Leu - Vice President
Services:
Provides radon detection services using E-Perm method, provides radon mitigation services.
Qualifications:
EPA RMP.
Area Served:
Nationwide.
EPA RMP 5:
EP(P).

UP-COUNTRY HOME INSPECTORS, INC.
P.O. Box 1568
Scarborough, ME 04074
(207) 883-9876
Stephen O'Connor - President
Catherine O'Connor
Gordon Guest
Services:
Provides radon detection services using charcoal canister, continuous radon monitor, E-Perm, and alpha track methods, provides radon mitigation and consultation services.
Area Served:
Nationwide.
EPA RMP 5:
CC, CR(P), EP(P).

V

VALLEY DE-WATERING
RD 2, Box 265 A-1
Sugarloaf, PA 18249
(717) 384-3315
Bob Plitnick
Area Served:
PA.

VER-GON BUILDERS
193 Christie Ave.
Clifton, NJ 07011
(201) 772-4040
Nicholas Verhagen
Area Served:
NJ.

VERSAR, INC.
6850 Versar Center
Springfield, VA 22151
(703) 750-3000
Michael Markels - President
R.P. Ouellette - Corporate Development
J. Tappan - ARIX
Niren Nagda - Geomet
Services:
Nationwide consulting firm specializes in environmental risk management, provides radon detection services for homes and buildings, construction management for radon-contaminated sites and structures, provides radon mitigation services, recommends appropriate risk mitigation to public and private clients.
Area Served:
Nationwide.

VICK ENTERPRISES
2060 East Easter Ave.
Littleton, CO 80122
(303) 798-7918
Robert I. Vick
Area Served:
CO.

VIKING HOME IMPROVEMENT, INC.
15 South 15th St.
Fargo, ND 58103
(701) 237-6300
Kevin Christianson - President
Services:
Provides radon detection and mitigation services, distributes complete line of radon mitigation products including sealers, membranes, fans, sump equipment, protective equipment.
Area Served:
ND.

VINCE SCHAEFER CONSTRUCTION
3220 Beaver Rd.
Des Moines, IA 50310
(515) 279-4600
Vince Schaefer - Owner
Services:
Provides radon detection services using charcoal canister and alpha track methods, provides mitigation services.
Area Served:
IA.

W

WEATHER SHIELD, INC.
4071 South Salem Church Rd.
Dover, PA 17315
(717) 292-5613
William Kimmel - President
Services:
Provides radon detection and mitigation services.
Area Served:
PA.

WEST ENVIRO SYSTEMS, INC.
45 Sycamore Ln.
Skillman, NJ 08558
(609) 924-0680
Fred W. West - President
H.E. West - Vice President
Charles West - Supervisor
Services:
Provides radon detection services using charcoal canister and alpha track methods, provides radon mitigation services.
Qualifications:
PA DER mitigation course, chemical engineer.
Area Served:
NJ.

WILSON JR., WALTER E.
700 Harston Ln.
Erdenheim, PA 19118
(215) 836-7257
Walter E. Wilson, Jr.
Area Served:
PA.

WOODLINE, INC.
Box 1170
Frisco, CO 80443
(303) 668-5603
Garry Neff - President
Services:
Provides radon mitigation services for new construction.
Area Served:
CO.

WORLDWIDE RADON, INC.
1941A Nears Pkwy.
Margate, FL 33063
(305) 968-5555
Gordon Marsh - President
Marge Semenuk - Vice President
Services:
Provides radon detection services using charcoal canister and E-Perm methods, provides radon mitigation services.
Qualifications:
General contractor and physicist on staff.

Area Served:
NJ, FL.

WORTH ENGINEERING
650 Palm Ave.
Hialeah, FL 33010
(305) 885-4515
James G. Worth - President
Services:
Provides radon detection services using charcoal canister, alpha track, and continuous radon monitor methods, provides radon mitigation services.
Area Served:
FL.

WPB ENTERPRISES, INC.
RD 1, Box 209
Riegelsville, PA 18077
(215) 346-8004
Bill Brodhead - President
Services:
Provides radon mitigation services, provides training on radon mitigation issues, non-commercial laboratory for research purposes or state agencies' assistance only.
Qualifications:
Developed and instructs EPA three day radon workshops.
Area Served:
NJ, PA.
EPA RMP 5:
CC.

X Y Z

Radon Mitigation Companies by States Served

This is an index listing radon mitigation companies by state served. The index begins with a list of radon mitigation companies that provide services nationwide. Next, each state is followed by a list of radon mitigation companies that provide services to that state. The state abbreviation, in parentheses, immediately following each company listing, indicates the state in which the radon mitigation company is located.

Nationwide

A & A ENVIRONMENTAL ENGINEERING COMPANY (CO)
A. ARTHUR LIPSHUTZ ASSOCIATES (FL)
A.T.E.I. (IN)
AAA RADON SERVICES, INC. (PA)
AADVANCED AAIR AANALYSIS (OH)
ADVANCED RADIATION MONITORING SERVICE, INC. (ARMS) (MI)
AIR SCIENCES, INC. (MO)
AIRREGS, INC./AIR POLLUTION REGULATORY SERVICE, INC. (GA)
ALPHA CONCEPTS, INC. (NJ)
AMERICAN COMPLIANCE TECHNOLOGIES, INC. (FL)
AMERICAN RADON CORPORATION (AZ)
AMERICAN RADON SERVICES, LTD. (IA)
APPLIED HEALTH PHYSICS, INC. (PA)
APPLIED RADON, INC. (OH)
ARIX CORPORATION (CO)
ARIX CORPORATION/A DIVISION OF VERSAR, INC. (PA)
ARTHUR SCOTT AND ASSOCIATES/AMERICAN ATCON (ON)
BASEMENT DE-WATERING SYSTEMS, INC. (PA)
BIONOMIC INDUSTRIES, INC. (NJ)
CAMRODEN ASSOCIATES (NY)
COLORADO VENTURE MANAGEMENT, INC. (CO)
CRIPPEN CONSULTING ASSOCIATES, INC. (DE)
DOUGLAS MARTIN AND ASSOCIATES, INC./MARTIN CONSULTING (PA)
ECODEX (MA)
ENERGY PROFESSIONALS, INC. (PA)
ENERGY RESOURCE CENTER (CO)
ENERGY SAVINGS WAREHOUSE, INC. (NJ)
ENVIRONMENTAL ENTERPRISES, INC. (GA)
ENVIRONMENTAL RADON MEASUREMENTS (FL)
ENVIRONMENTAL SERVICES AND TECHNOLOGY, INC. (NJ)
ENVIRONMENTAL TESTING ASSOCIATES (WI)
ERM, INC. (FL)
EVANSHIRE COMPANY, LTD., THE (NY)
GEO-RADON SERVICES (CA)
GEOMET TECHNOLOGIES, INC. (MD)
GROUNDWATER TECHNOLOGY, INC. (NY)
HALL-KIMBRELL ENVIRONMENTAL SERVICES (OR)
HARVARD ENGINEERING SERVICES COMPANY (MA)
HERITAGE RADON SERVICES, INC. (PA)
HOME INSPECTION SERVICE, INC. (IL)
INFILTEC/MAIN OFFICE (VA)
JOHNSON PLUMBING AND HEATING (PA)
JOHNSON'S ENERGY SERVICE (IA)
L.J. QUIGLIANO, II, ENTERPRISES (NY)
LOWRY ENGINEERING, INC. (ME)
METROPOLITAN RADON SERVICE (VA)
NATURAL ENERGY CONSULTANTS (CO)
NORTH EAST ENVIRONMENTAL PRODUCTS, INC. (NH)
NORTHERN NECK RADON SERVICE (VA)
NORTON CONSERVATION COMPANY, INC. (NJ)
OBAR SYSTEMS, INC. (NJ)
OCCUPATIONAL AND ENVIRONMENTAL HEALTH CONSULTING SERVICES (IL)
PROFESSIONAL HOUSE DOCTORS, INC. (IA)
PYRAMID ENVIRONMENTAL SYSTEMS, INC. (NJ)
QUADREX (CA)
R.E. WRIGHT ASSOCIATES, INC./RADONIX GROUP (PA)
R.F. SIMON COMPANY, INC. (PA)
RAD ENVIRONMENTAL SERVICES, INC. (NC)
RADALYTIC LABS, INC. (SC)
RADEX, INC. (GA)
RADIATION DATA/DIVISION OF PRINCETON GEOPHYSICAL SERVICES (CT)
RADIATION DATA/DIVISION OF PRINCETON GEOPHYSICAL SERVICES (NJ)
RADIATION PROTECTION SERVICES, INC. (CT)
RADION LABORATORY, INC. (MA)
RADON ALERT® DETECTION CENTER (PA)
RADON CONSULTANTS, INC., VIRGINIA (VA)
RADON CONTROL PROFESSIONALS, INC. (VA)
RADON DETECTION AND CONSULTING (KS)
RADON DETECTION AND REDUCTION LABORATORY/DIVISION OF FORMACHEM (MO)
RADON DIAGNOSTIC LABORATORIES (ME)
RADON DIAGNOSTICS (NJ)
RADON ENGINEERING SERVICES, INC. (AL)
RADON ENGINEERING/A UNIT OF PSI ENGINEERING, INC. (NJ)
RADON ENVIRONMENTAL SERVICES, INC. (IN)
RADON LABS (MA)
RADON MEASUREMENT AND SERVICES (CO)
RADON MITIGATORS OF ARIZONA (AZ)
RADON MITIGATORS, INC. (NJ)
RADON REDUCTION ENGINEERING AND LABORATORY (VA)
RADON REMOVAL, INC. (NJ)
RADON SAFETY AND TECHNOLOGY CORPORATION (NY)
RADON SCREENING SERVICE, INC. (CO)
RADON SURVEY SYSTEMS, INC. (OH)
RADON TESTING CONSULTANTS, INC. (NJ)
RADON TESTING, INC. (VA)
RADON X COMPANY, INC. (NH)
RADON, LTD. (IN)
RADONICS, INC. (NJ)
RANDOLPH AND ASSOCIATES, INC. (IL)
RDS/RADON DETECTION SERVICES, INC.-CORPORATE OFFICE (NJ)
RDS/RADON DETECTION SERVICES-GEORGIA OFFICE (GA)
RDS/RADON DETECTION SERVICES-PENNSYLVANIA OFFICE (PA)
REMEDIAL TECHNOLOGIES, INC. (OH)
RIEHS AND RIEHS (NC)
ROSS SYSTEMS, INC. (NJ)
SAFE-AIRE, INC. (IL)
SIMPSON WEATHER ASSOCIATES (VA)
STANDOR RADON DETECTION (FL)
SUNN CORPORATION (PA)
TEQ CORPORATION (PA)
TERRADON EXPLORATION TECHNOLOGIES, INC. (OH)
TFC NUCLEAR ASSOCIATES, INC. (NJ)
THERMO ENVIRONMENTAL CORPORATION (MA)
TMA/EBERLINE, HEALTH PHYSICS (NM)
UNITED SERVICES ALLIANCE, INC. (VA)
UP-COUNTRY HOME INSPECTORS, INC. (ME)
VERSAR, INC. (VA)

Companies Serving

Alabama

All companies listed under Nationwide provide services for this state.

OXYTHERM, INC. (SC)
RADON ENVIRONMENTAL SERVICES, INC. (KY)
RADON EVALUATION SERVICES, INC. (GA)
SOUTHEASTERN RADON CONSULTANTS, INC. (TN)

Companies Serving

Alaska

All companies listed under Nationwide provide services for this state.

GEOCHEM, INC. (AK)
MERIT ENVIRONMENTAL SERVICES, INC. (KY)

Companies Serving

Arizona

All companies listed under Nationwide provide services for this state.

ALPHA RADIATION ENGINEERING, INC. (CO)
ARIX CORPORATION (AZ)
BONDAR-CLEGG, INC. (CO)

Companies Serving
Arkansas

All companies listed under Nationwide provide services for this state.

MERIT ENVIRONMENTAL SERVICES, INC. (KY)
RADON ENVIRONMENTAL SERVICES, INC. (KY)

Companies Serving
California

All companies listed under Nationwide provide services for this state.

BONDAR-CLEGG, INC. (CO)
LW ENVIRONMENTAL SERVICES (CA)
TSI CONSTRUCTION SERVICES (IN)

Companies Serving
Colorado

All companies listed under Nationwide provide services for this state.

AAA DISASTER FORCE (CO)
AIR TRON MECHANICAL SERVICES (CO)
ALLISON CONSTRUCTION (CO)
ALPHA GUARD/TECTROL (CO)
ALPHA RADIATION ENGINEERING, INC. (CO)
ALPINE RADIATION MITIGATION (CO)
ARIX CORPORATION (CO)
BONDAR-CLEGG, INC. (CO)
BUILDING DIAGNOSTICS, INC. (CO)
C & D HEATING AND CONSTRUCTION (CO)
C. KENT CHAMBERLAIN AND ASSOCIATES (CO)
COLORADO SAFE-N-DRY, INC. (CO)
DEMPSEY, JOSEPH H. (CO)
DIAMOND BACK SERVICES, INC. (CO)
ENERGY ALTERNATIVES UNLIMITED, INC. (CO)
G.A.M. BUILDING INSPECTION AND RADON MEASUREMENTS AND CONTROL, INC. (CO)
HARRIS CONSTRUCTION (CO)
INTERMOUNTAIN RADON SERVICE (CO)
KITTREDGE CONSTRUCTION COMPANY (CO)
LUKENBUILT CONTRACTORS (CO)
NEW WAVE ENERGY SYSTEMS, INC. (CO)
PROFESSIONAL RESTORATION AND REPAIR (CO)
RA CONSULTANTS, INC. (CO)
RADON CORRECTIONS INSTITUTE, INC. (CO)
RADON TESTING AND CONSULTING (CT)
RADON TESTING AND TREATMENT, INC. (CO)
RAMP INDUSTRIES, INC. (CO)
RAMSES II, LTD. (CO)
ROCKY MOUNTAIN HOME INSPECTIONS (CO)
TSI CONSTRUCTION SERVICES (IN)
VICK ENTERPRISES (CO)
WOODLINE, INC. (CO)

Companies Serving
Connecticut

All companies listed under Nationwide provide services for this state.

ABACUS LABS RADON ASSESSMENT AND MITIGATION (ALRAM) (NH)
AIDCO MAINE CORPORATION (NH)
AIR SENSE, INC. (CT)
ALL-DRY BASEMENTS AND SAFE AIR, INC. (CT)
BROOKS LABORATORIES, INC. (CT)
BUZZANO CONTRACTING (CT)
CONNECTICUT BASEMENT DE-WATERING AND RADON SYSTEMS, INC. (CT)
CONTINENTAL ENVIRONMENT COMPANY, INC. (NJ)
ECOLOGICAL INTERACTIONS, INC. (NJ)
NEW ENGLAND DE-WATERING, INC. (NH)
NEW ENGLAND RADON, LTD. (NH)
PELICAN ENVIRONMENTAL CORPORATION (MA)
PROJECT ENVIRONMENTAL GROUP (PA)
RADIATION PROTECTION SERVICES, INC. (NJ)
RADON DETECTORS (NY)
RADON INSPECTION SERVICE, INC. (NJ)
RADON SYSTEMS OF RHODE ISLAND, INC. (RI)
RADON TECHNOLOGY AND CONSULTANTS (RI)
RADON TESTING AND CONSULTING (CT)
RADON TESTING AND CONSULTING SERVICES (MA)
ROSS SYSTEMS, INC. (CT)
SAF-AIR (CT)
SOUTHERN NEW ENGLAND RADON TESTING AND MITIGATION (CT)
THEROUX ENGINEERING (CT)

Companies Serving
Delaware

All companies listed under Nationwide provide services for this state.

ABACUS LABS RADON ASSESSMENT AND MITIGATION (ALRAM) (NH)
ACCREDITED WATERPROOFING COMPANY, INC. (NJ)
AIDCO MAINE CORPORATION (NH)
AMERICAN HERITAGE CORPORATION (PA)
BRANDYWINE ENVIRONMENTAL SERVICES (DE)
CCI-RADON AND HOME INSPECTION/RADON DETECTION (PA)
CONTINENTAL ENVIRONMENT COMPANY, INC. (NJ)
ECOLOGICAL INTERACTIONS, INC. (NJ)
FORCHETTI PSI (PA)
GILMONT CONSTRUCTION COMPANY (PA)
HOME CHECK SERVICES (VA)
INDUSTRIAL RESOURCE DEVELOPMENT COMPANY (PA)
L. CLARK AND SONS, INC. (MA)
MAYA-TEK, INC. (PA)
RADON DETECTORS (NY)
RADON REMEDIAL SERVICES (PA)
RADON REMEDIATION SERVICES (MD)
RADON SYSTEMS, INC. (MD)
SAFE SHELTER ENVIRONMENTAL (PA)
SPECTRUM INDUSTRIES (DE)

Companies Serving
District of Columbia

All companies listed under Nationwide provide services for this state.

AIDCO MAINE CORPORATION (NH)
AIR SOLUTIONS (MD)
AIRMASTERS RADON DIVISION (MD)
APPLIED CONSTRUCTION, INC. (MD)
CONTINENTAL ENVIRONMENT COMPANY, INC. (NJ)
DREWRY HOME INSPECTIONS, INC. (VA)
JOHN L. RENSHAW, INC. (MD)
RA-MED (VA)
RADON DETECTORS (NY)
RADON REMEDIATION SERVICES (MD)
RADON SYSTEMS, INC. (MD)
TSI CONSTRUCTION SERVICES (IN)

Companies Serving
Florida

All companies listed under Nationwide provide services for this state.

AIDCO MAINE CORPORATION (NH)
ALL AMERICAN RADON SERVICES, INC. (FL)
ALPHA RADIATION ENGINEERING, INC. (CO)
AMERICAN MEDICAL MARKETING SERVICES (FL)
AMERICAN RADON TECHNICAL SERVICES, INC. (FL)
CONTINENTAL ENVIRONMENT COMPANY, INC. (NJ)
ECOLOGICAL INTERACTIONS, INC. (NJ)
ENVIRONMENTAL AIR CONTROL, INC. (FL)
GEOHAZARDS, INC. (FL)
INDOOR POLLUTION CONTROL CENTER (FL)
LARRY LIBERTORE, INC. (FL)
MERIT ENVIRONMENTAL SERVICES, INC. (KY)
O'RIORDAN INSULATION COMPANY (FL)
PROFESSIONAL HOME ANALYSIS (FL)
RADON ENVIRONMENTAL SERVICES, INC. (KY)
RADON EVALUATION SERVICES, INC. (GA)
RADON RESEARCH INTERNATIONAL, INC. (FL)
SMITH, STANLEY (FL)
WORLDWIDE RADON, INC. (FL)
WORTH ENGINEERING (FL)

Companies Serving
Georgia

All companies listed under Nationwide provide services for this state.

AIDCO MAINE CORPORATION (NH)
CAROLINA RADON CONSULTANTS (NC)
CAROLINA RADON CONSULTANTS (SC)
ENVIRONMENTAL POLLUTION CONTROL (GA)
MERIT ENVIRONMENTAL SERVICES, INC. (KY)
OXYTHERM, INC. (SC)
PROFESSIONAL HOME ANALYSIS (FL)
PROTECTION INSPECTION, INC. (TN)
RADON ENVIRONMENTAL SERVICES, INC. (KY)
RADON EVALUATION SERVICES, INC. (GA)
RADON TEST AND REDUCTION SPECIALIST (NC)
SAFEHOUSE, INC. (GA)
SOUTHEASTERN RADON CONSULTANTS, INC. (TN)

Companies Serving
Hawaii

All companies listed under Nationwide provide services for this state.

Companies Serving
Idaho

All companies listed under Nationwide provide services for this state.

CAVALIER CORPORATION (WA)
RADON AWARENESS, INC. (ID)

Companies Serving
Illinois

All companies listed under Nationwide provide services for this state.

AIDCO MAINE CORPORATION (NH)
BIEDERER CONSTRUCTION SERVICE (IL)
ELLIOTT AND ASSOCIATES (IL)
ENERGY INDEPENDENCE/THE RADON RESPONSE TEAM (MO)

ENVIRONMENTAL SERVICES OF ILLINOIS, INC. (IL)
HOME ENVIRONMENT COMPANY (IN)
MERIT ENVIRONMENTAL SERVICES, INC. (KY)
RAD ALERT, LTD. (IL)
RADON MITIGATION SERVICE (IL)
RADON TESTING AND MITIGATION SERVICES COMPANY (IA)
RADON TESTING AND MITIGATION SYSTEMS (IN)
SALAMON, PAUL (IL)
TEE ENGINEERING COMPANY (KY)
TSI CONSTRUCTION SERVICES (IN)

Companies Serving
Indiana

All companies listed under Nationwide provide services for this state.

CONSTRUCTION SERVICES OF INDIANA, INC. (IN)
ELLIOTT AND ASSOCIATES (IL)
HOME ENVIRONMENT COMPANY (IN)
INSUL-TECH, INC. (OH)
MERIT ENVIRONMENTAL SERVICES, INC. (KY)
RADON ENVIRONMENTAL SERVICES, INC. (KY)
RADON REDUCTION, INC. (IN)
RADON TESTING AND MITIGATION SYSTEMS (IN)
RESTORED ENVIRONMENTS COMPANY, INC. (OH)
SCHUCK HOME SERVICES (IN)
TEE ENGINEERING COMPANY (KY)
TSI CONSTRUCTION SERVICES (IN)

Companies Serving
Iowa

All companies listed under Nationwide provide services for this state.

AIDCO MAINE CORPORATION (NH)
MACK, E.F. (IA)
MAHONEY HEATING AND COOLING (NE)
MECHANICAL EQUIPMENT SERVICES CORPORATION (IA)
MERIT ENVIRONMENTAL SERVICES, INC. (KY)
PROFESSIONAL HOUSE DOCTORS (IA)
RADON BUSTERS (ND)
RADON CONTROL SERVICES CORPORATION (IA)
RADON ENVIRONMENTAL SERVICES, INC. (KY)
RADON REDUCTION SERVICES (MO)
RADON TESTING AND MITIGATION SERVICES COMPANY (IA)
TSI CONSTRUCTION SERVICES (IN)
VINCE SCHAEFER CONSTRUCTION (IA)

Companies Serving
Kansas

All companies listed under Nationwide provide services for this state.

ALPHA RADIATION ENGINEERING, INC. (CO)
BONDAR-CLEGG, INC. (CO)
ENERGY INDEPENDENCE/THE RADON RESPONSE TEAM (MO)
MIDWEST RADON INSPECTION, INC. (KS)
RADON EVALUATION SERVICES, INC. (GA)
RADON REDUCTION SERVICES (MO)
TSI CONSTRUCTION SERVICES (IN)

Companies Serving
Kentucky

All companies listed under Nationwide provide services for this state.

AIDCO MAINE CORPORATION (NH)
ENERGY VENT, INC. (OH)
GMAC HOME IMPROVEMENT (KY)
HOME ENVIRONMENT COMPANY (IN)
HUNT'S WATER CONTROL, INC. (KY)
INSUL-TECH, INC. (OH)
MERIT ENVIRONMENTAL SERVICES, INC. (KY)
PHOENIX SAFETY SYSTEMS (KY)
PROTECTION INSPECTION, INC. (TN)
RADON ENVIRONMENTAL SERVICES, INC. (KY)
RADON EVALUATION SERVICES, INC. (GA)
RADON MANAGEMENT OF KENTUCKY (KY)
RADON PROTECTION, INC. (KY)
RADON REDUCTION, INC. (IN)
RADON TEST AND REDUCTION SPECIALIST (NC)
RADON TESTING AND MITIGATION SYSTEMS (IN)
RESTORED ENVIRONMENTS COMPANY, INC. (OH)
SOUTHEASTERN RADON CONSULTANTS, INC. (TN)
TEE ENGINEERING COMPANY (KY)
TSI CONSTRUCTION SERVICES (IN)

Companies Serving
Louisiana

All companies listed under Nationwide provide services for this state.

OXYTHERM, INC. (SC)
RADON EVALUATION SERVICES, INC. (GA)

Companies Serving
Maine

All companies listed under Nationwide provide services for this state.

ABACUS LABS RADON ASSESSMENT AND MITIGATION (ALRAM) (NH)
AERO-AQUA (ME)
AIDCO MAINE CORPORATION (NH)
AIR SENSE, INC. (CT)
BURNS AND ASSOCIATES (NH)
L. CLARK AND SONS, INC. (MA)
MITIGATING CIRCUMSTANCES (NH)
NEW ENGLAND RADON, LTD. (NH)
RADON DETECTORS (NY)
RADON REDUCTION, INC. (MA)
RADON SYSTEMS OF RHODE ISLAND, INC. (RI)
RADON TECHNOLOGY AND CONSULTANTS (RI)
RADON TESTING AND CONSULTING SERVICES (MA)

Companies Serving
Maryland

All companies listed under Nationwide provide services for this state.

ACS HOVAL (IL)
ADEPT REMEDIATION CORPORATION (MD)
AIDCO MAINE CORPORATION (NH)
AIR SOLUTIONS (MD)
AIRMASTERS RADON DIVISION (MD)
APPLIED CONSTRUCTION, INC. (MD)
ASSOCIATED DESERT DRY WATERPROOFING COMPANY, INC. (VA)
BATEMAN HOME IMPROVEMENTS (MD)
BERG GROUP (MD)
CONTINENTAL ENVIRONMENT COMPANY, INC. (NJ)
D. PAUL BUILDING ENTERPRISE (MD)
ECM ENVIRONMENTAL CONSULTANTS (MD)
ECOLOGICAL INTERACTIONS, INC. (NJ)
ENVIRONMENTAL HOME SERVICES, INC. (MD)
FORCHETTI PSI (PA)
G.M.D. CONSTRUCTION, INC. (MD)
GARCO RESEARCH COMPANY (NJ)
J. BECK AND ASSOCIATES, INC. (PA)
JOHN L. RENSHAW, INC. (MD)
PRO-GUARD BUILDING INSPECTIONS, INC. (PA)
RA-MED (VA)
RADON CONSULTANTS, INC., MARYLAND (MD)
RADON REDUCTION SERVICES, INC. (MD)
RADON REMEDIATION SERVICES (MD)
RADON RESOURCE, INC. (MD)
RADON SYSTEMS, INC. (MD)
RADON TECHNICIANS, INC (MD)
RCI SERVICES (MD)

Companies Serving
Massachusetts

All companies listed under Nationwide provide services for this state.

ABACUS LABS RADON ASSESSMENT AND MITIGATION (ALRAM) (NH)
AIDCO MAINE CORPORATION (NH)
AIR SENSE, INC. (CT)
AMERICAN INSULATION, INC. (MA)
ASTLE CORPORATION (MA)
BAYSTATE ENVIRONMENTAL (MA)
BURNS AND ASSOCIATES (NH)
BUZZANO CONTRACTING (CT)
COMMUNITY HEALTH ASSOCIATES, INC. (MA)
FINN CONSULTANT ENGINEERS, INC. (MA)
L. CLARK AND SONS, INC. (MA)
MITIGATING CIRCUMSTANCES (NH)
NEW ENGLAND DE-WATERING, INC. (NH)
NEW ENGLAND RADON, LTD. (NH)
PELICAN ENVIRONMENTAL CORPORATION (MA)
RADIOLOGICAL CONTROL ASSOCIATES (MA)
RADON DETECTORS (NY)
RADON REDUCTION, INC. (MA)
RADON SYSTEMS OF RHODE ISLAND, INC. (RI)
RADON TECHNOLOGY AND CONSULTANTS (RI)
RADON TESTING AND CONSULTING SERVICES (MA)
SACCHETTI, R. MICHAEL (MA)

Companies Serving
Michigan

All companies listed under Nationwide provide services for this state.

AIDCO MAINE CORPORATION (NH)
ELLIOTT AND ASSOCIATES (IL)
INSUL-TECH, INC. (OH)
RADON FREE (OH)
RADON TEST AND REDUCTION SPECIALIST (NC)
RADON TESTING AND REDUCTION (MI)
SCHUCK HOME SERVICES (IN)
TSI CONSTRUCTION SERVICES (IN)

Companies Serving
Minnesota

All companies listed under Nationwide provide services for this state.

ENERGY INDEPENDENCE/THE RADON RESPONSE TEAM (MO)
ENERGY OUTFITTERS, LTD. (MN)
RADON BUSTERS (ND)
RADON ENVIRONMENTAL SERVICES, INC. (KY)
RADON FREE (OH)
RITE WAY WATERPROOFING (MN)

Companies Serving
Mississippi

All companies listed under Nationwide provide services for this state.

ELLIOTT AND ASSOCIATES (IL)
NEW ENGLAND DE-WATERING, INC. (NH)
RADON EVALUATION SERVICES, INC. (GA)
TSI CONSTRUCTION SERVICES (IN)

Companies Serving
Missouri

All companies listed under Nationwide provide services for this state.

ELLIOTT AND ASSOCIATES (IL)
ENERGY INDEPENDENCE/THE RADON RESPONSE TEAM (MO)
ENVIRONMENTAL SERVICES OF ILLINOIS, INC. (IL)
MERIT ENVIRONMENTAL SERVICES, INC. (KY)
MIDWEST RADON INSPECTION, INC. (KS)
RADON ENVIRONMENTAL SERVICES, INC. (KY)
RADON REDUCTION SERVICES (MO)
RADON TECHNOLOGY (MO)
TEE ENGINEERING COMPANY (KY)
TSI CONSTRUCTION SERVICES (IN)

Companies Serving
Montana

All companies listed under Nationwide provide services for this state.

ALPHA RADIATION ENGINEERING, INC. (CO)
ARIX CORPORATION (WY)
CAVALIER CORPORATION (WA)
RADON AWARENESS, INC. (ID)
RADON BUSTERS (ND)
RADON TESTING AND CONTROL (ND)

Companies Serving
Nebraska

All companies listed under Nationwide provide services for this state.

ALPHA RADIATION ENGINEERING, INC. (CO)
BONDAR-CLEGG, INC. (CO)
ENERGY INDEPENDENCE/THE RADON RESPONSE TEAM (MO)
MAHONEY HEATING AND COOLING (NE)
RADON ENVIRONMENTAL SERVICES, INC. (KY)

Companies Serving
Nevada

All companies listed under Nationwide provide services for this state.

ALPHA RADIATION ENGINEERING, INC. (CO)
BONDAR-CLEGG, INC. (CO)
ENERGY ALTERNATIVES UNLIMITED, INC. (CO)
RADON AWARENESS, INC. (ID)
RADON REDUCTION SERVICES (MO)

Companies Serving
New Hampshire

All companies listed under Nationwide provide services for this state.

ABACUS LABS RADON ASSESSMENT AND MITIGATION (ALRAM) (NH)
AIDCO MAINE CORPORATION (NH)
AIR SENSE, INC. (CT)
BURNS AND ASSOCIATES (NH)
COMPREHENSIVE RADON SERVICES, INC. (NH)
L. CLARK AND SONS, INC. (MA)
MITIGATING CIRCUMSTANCES (NH)
NEW ENGLAND DE-WATERING, INC. (NH)
NEW ENGLAND RADON, LTD. (NH)
PELICAN ENVIRONMENTAL CORPORATION (MA)
RADON DETECTORS (NY)
RADON REDUCTION SYSTEMS (NH)
RADON REDUCTION, INC. (MA)
RADON SYSTEMS OF RHODE ISLAND, INC. (RI)
RADON TECHNOLOGY AND CONSULTANTS (RI)
RADON TESTING AND CONSULTING SERVICES (MA)

Companies Serving
New Jersey

All companies listed under Nationwide provide services for this state.

A & P WATERPROOFING, INC. (PA)
A.L. BOUCHARD COMPANY (NJ)
AAA RADON TESTING (NJ)
ACCREDITED WATERPROOFING COMPANY, INC. (NJ)
AGENTERPRISE, INC. (PA)
AIDCO MAINE CORPORATION (NH)
AIR QUALITY CONTROL, INC. (PA)
AIRMASTER RADON DETECTION, INC. (NJ)
AMERICAN HERITAGE CORPORATION (PA)
AMERICAN RADON TECHNICAL SERVICES, INC. (FL)
ASSOCIATED DESERT DRY WATERPROOFING COMPANY, INC. (VA)
AURORA INDUSTRIES (PA)
BUZZANO CONTRACTING (CT)
CALOIERO RADON TECHNOLOGIES (NJ)
CAMPAGNA BUILDERS (NJ)
CAPITAL MATERIALS TESTING, INC. (NY)
CCI-RADON AND HOME INSPECTION/RADON DETECTION (PA)
CENTRAL JERSEY RADON, INC. (NJ)
CONTAMINATION MANAGEMENT ENGINEERING (NJ)
CONTINENTAL ENVIRONMENT COMPANY, INC. (NJ)
DPR BUILDERS (NJ)
ECOLOGICAL INTERACTIONS, INC. (NJ)
FORCHETTI PSI (PA)
FORESIGHT ENGINEERING (NJ)
FORESIGHT ENGINEERING, INC. (NJ)
FORESIGHT ENGINEERING, INC. (NJ)
FORESIGHT ENGINEERING, INC. (NJ)
GARCO RESEARCH COMPANY (NJ)
GARDEN STATE RADON DEWATERING SERVICES, INC. (NJ)
GILMONT CONSTRUCTION COMPANY (PA)
HUNTER-DON PEST CONTROL (NJ)
I & M RADON MITIGATION (OH)
INDUSTRIAL RESOURCE DEVELOPMENT COMPANY (PA)
J. BECK AND ASSOCIATES, INC. (PA)
L. CLARK AND SONS, INC. (MA)
MAYA-TEK, INC. (PA)
NORTHEASTERN RADON SERVICE (NJ)
PRINCETON HOME INSPECTION SERVICE (NJ)
PROJECT ENVIRONMENTAL GROUP (PA)
RAD ENVIRONMENTAL SERVICES, INC. (NJ)
RADIATION DATA (NJ)
RADIATION PROTECTION SERVICES, INC. (NJ)
RADON DETECTORS (NY)
RADON ENVIRONMENTAL CONSTRUCTION SERVICES/RECONS, INC., MAIN OFFICE (NJ)
RADON HOME SEARCH (NJ)
RADON INSPECTION SERVICE, INC. (NJ)
RADON MEASURING SYSTEMS (PA)
RADON REMEDIAL SERVICES (PA)
RADON REMEDIATION COMPANY, INC. (NJ)
RADON RESIDENTIAL SYSTEMS, INC. (NJ)
RADON SYSTEMS OF RHODE ISLAND, INC. (RI)
RADON-X CORPORATION (NJ)
RADON-X CORPORATION (NJ)
RADONAIRE (NJ)
REALTY REPAIR AND DEVELOPMENT (NJ)
SAFE SHELTER ENVIRONMENTAL (PA)
SOOS RADON SYSTEMS, INC. (NJ)
STONE RIDGE RADON MITIGATORS (NJ)
SUBURBAN JERSEY WATERPROOFING, INC. (NJ)
TRI-COUNTY RADON, INC. (NJ)
TRI-STATE RADON AND DEWATERING (NJ)
TRI-STATE RADON REDUCTION, INC. (NJ)
TSI CONSTRUCTION SERVICES (IN)
VER-GON BUILDERS (NJ)
WEST ENVIRO SYSTEMS, INC. (NJ)
WORLDWIDE RADON, INC. (FL)
WPB ENTERPRISES, INC. (PA)

Companies Serving
New Mexico

All companies listed under Nationwide provide services for this state.

ALPHA RADIATION ENGINEERING, INC. (CO)
BONDAR-CLEGG, INC. (CO)
INEX (NM)

Companies Serving
New York

All companies listed under Nationwide provide services for this state.

ADVANCED BASEMENT DE-WATERING AND RADON CONTROL SYSTEMS (NY)
AFFTREX, LTD. (PA)
AIDCO MAINE CORPORATION (NH)
AMERICAN HERITAGE CORPORATION (PA)
ARID BASEMENT COMPANY (NY)
BUZZANO CONTRACTING (CT)
CAPITAL MATERIALS TESTING, INC. (NY)
CONTINENTAL ENVIRONMENT COMPANY, INC. (NJ)
DPR BUILDERS (NJ)
ECOLOGICAL INTERACTIONS, INC. (NJ)
EN-THERM, INC. (NY)
ENERGY CONSERVATION SERVICES (NY)
ENGINEERING ASSOCIATES (NY)
ENVIRO-CARE RADON SERVICES (NY)
ENVIRONMENTAL MAINTENANCE SPECIALISTS COMPANY, INC. (NY)
FORCHETTI PSI (PA)
GARCO RESEARCH COMPANY (NJ)
HOME DETOX SPECIALISTS (NY)
L. CLARK AND SONS, INC. (MA)
MASTER BUILDING SERVICE (NY)
MCINERNY, T.M. (NY)
MUSSO, FRANCIS P. (NY)
NEW ENGLAND DE-WATERING, INC. (NH)
NITE HAWK ENTERPRISES, INC. (NY)
PROJECT ENVIRONMENTAL GROUP (PA)
RADIATION PROTECTION SERVICES, INC. (NJ)
RADON DETECTORS (NY)
RADON INSPECTION SERVICE, INC. (NJ)
RADON MEASURING SYSTEMS (PA)
RADON SYSTEMS OF RHODE ISLAND, INC. (RI)
RADON TESTING SERVICE, INC. (NY)
RENSSELAER POLYTECHNICAL INSTITUTE (NY)
RESOURCE CENTER FOR ENVIRONMENTAL PROTECTION (NY)
RONALD STEIN CONSTRUCTION, INC. (NY)
ROSS SYSTEMS, INC. (CT)

SIGMA ENVIRONMENTAL SYSTEMS, INC. (NY)
SOUTH CENTRAL ENVIRONMENTAL SERVICES (NY)
SULLIVAN, ROBERT (NY)
SUNSTREAM DEVELOPMENT CORPORATION (NY)
SURE-WAY SYSTEMS/DIVISION OF E.C.S., INC. (NY)
TAPPAN ZEE DEVELOPMENT (NY)
TSI CONSTRUCTION SERVICES (IN)

Companies Serving
North Carolina

All companies listed under Nationwide provide services for this state.

ABACUS LABS RADON ASSESSMENT AND MITIGATION (ALRAM) (NH)
AIDCO MAINE CORPORATION (NH)
CAROLINA RADON CONSULTANTS (NC)
CAROLINA RADON CONSULTANTS (SC)
MERIT ENVIRONMENTAL SERVICES, INC. (KY)
OXYTHERM, INC. (SC)
PROTECTION INSPECTION, INC. (TN)
RAD ENVIRONMENTAL SERVICES, INC. (NJ)
RADON ENVIRONMENTAL SERVICES, INC. (KY)
RADON TEST AND REDUCTION SPECIALIST (NC)
TRIANGLE WATERPROOFING AND REPAIR SERVICE (NC)
TSI CONSTRUCTION SERVICES (IN)

Companies Serving
North Dakota

All companies listed under Nationwide provide services for this state.

AIDCO MAINE CORPORATION (NH)
CONSUMER ENERGY SERVICES, INC. (OH)
ENERGY UNLIMITED (PA)
ENERGY VENT, INC. (OH)
HARDICO RADON MANAGEMENT, COLUMBUS OFFICE (OH)
HOME ENVIRONMENT COMPANY (IN)
INSUL-TECH, INC. (OH)
INSULATING SYSTEMS (OH)
J. BECK AND ASSOCIATES, INC. (PA)
MERIT ENVIRONMENTAL SERVICES, INC. (KY)
OAK HILL RADON TESTING AND REPAIR (OH)
PROJECT ENVIRONMENTAL GROUP (PA)
RADON DETECTORS (NY)
RADON ENVIRONMENTAL SERVICES, INC. (KY)
RADON ENVIRONMENTAL SERVICES, INC. (OH)
RADON FREE (OH)
RADON TESTING AND MITIGATION SYSTEMS (IN)
RELIABLE BASEMENT WATERPROOF COMPANY (OH)
RESTORED ENVIRONMENTS COMPANY, INC. (OH)
SAFE-AIR (OH)
SCHUCK HOME SERVICES (IN)
TEE ENGINEERING COMPANY (KY)
TSI CONSTRUCTION SERVICES (IN)

Companies Serving
Ohio

All companies listed under Nationwide provide services for this state.

ALPHA RADIATION ENGINEERING, INC. (CO)
OXYTHERM, INC. (SC)

Companies Serving
Oregon

All companies listed under Nationwide provide services for this state.

CAVALIER CORPORATION (WA)
RADON AWARENESS, INC. (ID)

Companies Serving
Pennsylvania

All companies listed under Nationwide provide services for this state.

A & P WATERPROOFING, INC. (PA)
AAA RADON TESTING (NJ)
ACCREDITED WATERPROOFING COMPANY, INC. (NJ)
AFFTREX, LTD. (PA)
AGENTERPRISE, INC. (PA)
AIDCO MAINE CORPORATION (NH)
AIR QUALITY CONTROL, INC. (PA)
AIR QUALITY CONTROL, INC. (PA)
AIRCARE SYSTEMS (PA)
ALPHA BETA GAMMA TECHNOLOGIES, INC. (PA)
AMERICAN EXCELSIOR COMPANY (PA)
AMERICAN HERITAGE CORPORATION (PA)
AMERICAN IMPROVEMENT COMPANY (PA)
APPLIED CONSTRUCTION, INC. (MD)
ARTHUR W. MATTES, INC. (PA)
ASSOCIATED DESERT DRY WATERPROOFING COMPANY, INC. (VA)
ATLAS WATERPROOFING COMPANY, INC. (PA)
AURORA INDUSTRIES (PA)
B & H RADON SERVICES (PA)
BRUN-REUSS ASSOCIATES (PA)
CAPITAL MATERIALS TESTING, INC. (NY)
CCI-RADON AND HOME INSPECTION/RADON DETECTION (PA)
CERTIFIED TESTING SERVICES, INC. (PA)
COMMONWEALTH ENTERPRISES, INC. (PA)
CONTINENTAL ENVIRONMENT COMPANY, INC. (NJ)
DELAWARE VALLEY GEOTECHNICAL ASSOCIATES (PA)
DINOFF ASSOCIATES, INC. (PA)
DPR BUILDERS (NJ)
ECOLOGICAL INTERACTIONS, INC. (NJ)
ENERGY RESOURCE TECHNOLOGY (PA)
ENERGY UNLIMITED (PA)
ENVIRO TECH SYSTEMS (PA)
ENVIRONMENTAL INDOOR AIR QUALITY CONSULTANTS, INC. (EIC) (PA)
FCA DEVELOPMENT CORPORATION (PA)
FORCHETTI PSI (PA)
FORESIGHT ENGINEERING, INC. (NJ)
G.M.D. CONSTRUCTION, INC. (MD)
GARCO RESEARCH COMPANY (NJ)
GIDDINGS AND ASSOCIATES, INC. (PA)
GILMONT CONSTRUCTION COMPANY (PA)
GRAVES RADON CONTRACTING (PA)
GURNEY KERR CONTRACTORS, INC. (PA)
HARAD WATERPROOFING (PA)
HARDICO RADON MANAGEMENT, CENTRAL OFFICE (PA)
HARDICO RADON MANAGEMENT, HARRISBURG OFFICE (PA)
HARDICO RADON MANAGEMENT, PITTSBURGH OFFICE (PA)
HUGHES, RONALD L. (PA)
INDUSTRIAL RESOURCE DEVELOPMENT COMPANY (PA)
INSUL-TECH, INC. (OH)
INTERNATIONAL EXPLORATIONS (PA)
J. BECK AND ASSOCIATES, INC. (PA)
JANCE ASSOCIATES, INC. (PA)
KELLOGG ENVIRONMENTAL SERVICES, INC. (PA)
KUNKEL, HAROLD E. (PA)
L. CLARK AND SONS, INC. (MA)
LAKE BUILDERS, INC. (PA)
LDH WATERPROOFING AND RADON CONTROL (PA)
LEHIGH VALLEY HOUSE INSPECTION SERVICE (PA)
LININGER BROTHERS, INC. (PA)
LV SAFETY AND HEALTH CONSULTANTS (PA)
MARTIN BUILDERS COMPANY, INC., THE (PA)
MAYA-TEK, INC. (PA)
MCGUIRE BUILDERS, INC. (PA)
MCHILL, A.D. AND SONS, INC. (PA)
NUCHEM CORPORATION (PA)
PENN NUCLEAR SERVICES (PA)
PENNSYLVANIA RADON CENTER (PA)
PENRAD, INC. (PA)
PRINCETON HOME INSPECTION SERVICE (NJ)
PRO-GUARD BUILDING INSPECTIONS, INC. (PA)
PROJECT ENVIRONMENTAL GROUP (PA)
PUKAL CONTRACTING COMPANY (PA)
QUOIN PARTNERSHIP, ARCHITECTS AND ENGINEERS (PA)
RADIOACTIVITY ALERT DETECTION SERVICES, INC. (PA)
RADON DECON, INC. (PA)
RADON DETECTORS (NY)
RADON DOCTOR, INC., THE (PA)
RADON ENVIRONMENTAL CONSTRUCTION SERVICES/RECONS, INC., MAIN OFFICE (NJ)
RADON ENVIRONMENTAL CONSTRUCTION SERVICES/RECONS, INC., PA OFFICE (PA)
RADON MEASURING SYSTEMS (PA)
RADON PROTECTION SERVICES (PA)
RADON REMEDIAL SERVICES (PA)
RADON RESOURCE, INC. (MD)
RADON SYSTEMS OF RHODE ISLAND, INC. (RI)
RADON SYSTEMS, INC. (MD)
ROBERT A. SWOYER ASSOCIATES (PA)
SAFE SHELTER ENVIRONMENTAL (PA)
SHAPIRO, MARK (PA)
SOLAR RESOURCE TECHNOLOGY (PA)
SOUTH CENTRAL ENVIRONMENTAL SERVICES (NY)
SPRINGER COMPANY (PA)
STONE RIDGE RADON MITIGATORS (NJ)
SUNSTREAM DEVELOPMENT CORPORATION (NY)
THOMAS L. HOFFMAN, COMPANY, INC. (PA)
TRI-COUNTY RADON, INC. (NJ)
TSI CONSTRUCTION SERVICES (IN)
VALLEY DE-WATERING (PA)
WEATHER SHIELD, INC. (PA)
WILSON JR., WALTER E. (PA)
WPB ENTERPRISES, INC. (PA)

Companies Serving
Puerto Rico

All companies listed under Nationwide provide services for this state.

NEW ENGLAND RADON, LTD. (NH)

Companies Serving
Rhode Island

All companies listed under Nationwide provide services for this state.

ABACUS LABS RADON ASSESSMENT AND MITIGATION (ALRAM) (NH)
AIDCO MAINE CORPORATION (NH)
AIR SENSE, INC. (CT)
L. CLARK AND SONS, INC. (MA)
NEW ENGLAND DE-WATERING, INC. (NH)
NEW ENGLAND RADON, LTD. (NH)
PELICAN ENVIRONMENTAL CORPORATION (MA)
RADON DETECTORS (NY)

RADON REDUCTION, INC. (MA)
RADON SYSTEMS OF RHODE ISLAND, INC. (RI)
RADON TECHNOLOGY AND CONSULTANTS (RI)
RADON TESTING AND CONSULTING SERVICES (MA)

Companies Serving
South Carolina

All companies listed under Nationwide provide services for this state.

CAROLINA RADON CONSULTANTS (NC)
CAROLINA RADON CONSULTANTS (SC)
CAROLINA RADON DETECTION (SC)
HARDICO RADON MANAGEMENT, GREENVILLE OFFICE (SC)
OXYTHERM, INC. (SC)
RADON ENVIRONMENTAL SERVICES, INC. (KY)
RADON EVALUATION SERVICES, INC. (GA)
RADON TEST AND REDUCTION SPECIALIST (NC)

Companies Serving
South Dakota

All companies listed under Nationwide provide services for this state.

RADON BUSTERS (ND)
RADON TESTING AND CONTROL (ND)

Companies Serving
Tennessee

All companies listed under Nationwide provide services for this state.

CARE SYSTEMS (TN)
MERIT ENVIRONMENTAL SERVICES, INC. (KY)
PROTECTION INSPECTION, INC. (TN)
RADON ENVIRONMENTAL SERVICES, INC. (KY)
RADON EVALUATION SERVICES, INC. (GA)
RADON TEST AND REDUCTION SPECIALIST (NC)
SAFE-AIR (OH)
SOUTHEASTERN RADON CONSULTANTS, INC. (TN)
TEE ENGINEERING COMPANY (KY)
TSI CONSTRUCTION SERVICES (IN)

Companies Serving
Texas

All companies listed under Nationwide provide services for this state.

ALPHA RADIATION ENGINEERING, INC. (CO)
CONTINENTAL ENVIRONMENT COMPANY, INC. (NJ)
OXYTHERM, INC. (SC)
PARADISE COMPANY (TX)
RADON REDUCTION AND TESTING COMPANY (TX)
RADON TESTING AND MITIGATION OF TEXAS, INC. (TX)

Companies Serving
Utah

All companies listed under Nationwide provide services for this state.

ALPHA RADIATION ENGINEERING, INC. (CO)
ARIX CORPORATION (UT)
BONDAR-CLEGG, INC. (CO)
ENERGY ALTERNATIVES UNLIMITED, INC. (CO)
RADON AWARENESS, INC. (ID)

Companies Serving
Vermont

All companies listed under Nationwide provide services for this state.

CAVALIER CORPORATION (WA)
INSUL-TECH, INC. (OH)
RADON AWARENESS, INC. (ID)
RADON DETECTION AND REMOVAL SERVICES (WA)

Companies Serving
Virgin Islands

All companies listed under Nationwide provide services for this state.

Companies Serving
Virginia

All companies listed under Nationwide provide services for this state.

AIDCO MAINE CORPORATION (NH)
AIR SOLUTIONS (MD)
AIRMASTERS RADON DIVISION (MD)
APPLIED CONSTRUCTION, INC. (MD)
ASSOCIATED DESERT DRY WATERPROOFING COMPANY, INC. (VA)
CONTINENTAL ENVIRONMENT COMPANY, INC. (NJ)
DPR BUILDERS (NJ)
DREWRY HOME INSPECTIONS, INC. (VA)
ECOLOGICAL INTERACTIONS, INC. (NJ)
FORESIGHT ENGINEERING, INC. (NJ)
G.M.D. CONSTRUCTION, INC. (MD)
GARCO RESEARCH COMPANY (NJ)
HARDICO RADON MANAGEMENT, D.C. OFFICE (VA)
HARDICO RADON MANAGEMENT, ROANOKE OFFICE (VA)
HOME CHECK SERVICES (VA)
JOHN L. RENSHAW, INC. (MD)
MERIT ENVIRONMENTAL SERVICES, INC. (KY)
OXYTHERM, INC. (SC)
P.C. ENGINEERING (VA)
PAYNE, LARRY D. (VA)
PLUM GROVE CORPORATION (VA)
PROTECTION INSPECTION, INC. (TN)
RA-MED (VA)
RADON CONTROL SYSTEMS (VA)
RADON DETECTORS (NY)
RADON REMEDIATION SERVICES (MD)
RADON SYSTEMS, INC. (MD)
RADON TEST AND REDUCTION SPECIALIST (NC)
TEE ENGINEERING COMPANY (KY)
TSI CONSTRUCTION SERVICES (IN)

Companies Serving
Washington

All companies listed under Nationwide provide services for this state.

CAVALIER CORPORATION (WA)
INSUL-TECH, INC. (OH)
RADON AWARENESS, INC. (ID)
RADON DETECTION AND REMOVAL SERVICES (WA)

Companies Serving
West Virginia

All companies listed under Nationwide provide services for this state.

AIDCO MAINE CORPORATION (NH)
INSUL-TECH, INC. (OH)
J. BECK AND ASSOCIATES, INC. (PA)
MERIT ENVIRONMENTAL SERVICES, INC. (KY)
RADON DETECTORS (NY)
RADON ENVIRONMENTAL SERVICES, INC. (KY)
RADON TEST AND REDUCTION SPECIALIST (NC)
TEE ENGINEERING COMPANY (KY)
TSI CONSTRUCTION SERVICES (IN)

Companies Serving
Wisconsin

All companies listed under Nationwide provide services for this state.

ELLIOTT AND ASSOCIATES (IL)
ENERGY INDEPENDENCE/THE RADON RESPONSE TEAM (MO)
ENERGY OUTFITTERS, LTD. (MN)
RADON ENVIRONMENTAL SERVICES, INC. (KY)
RADON MITIGATION SERVICE (IL)
TSI CONSTRUCTION SERVICES (IN)

Companies Serving
Wyoming

All companies listed under Nationwide provide services for this state.

ALPHA RADIATION ENGINEERING, INC. (CO)
ARIX CORPORATION (WY)
BONDAR-CLEGG, INC. (CO)
ENERGY ALTERNATIVES UNLIMITED, INC. (CO)
FULL HOUSE CONTRACTING (WY)
RADON AWARENESS, INC. (ID)
RADON SERVICES OF WYOMING (WY)

Companies Serving
Canada

All companies listed under Nationwide provide services to Canada.

BARRINGER MAGENTA, LTD. (ON)
GLOBAL WHOLESALE (WY)
MACLAREN ENGINEERS, INC. (ON)
ONTARIO RESEARCH FOUNDATION (ORF) (ON)
R.T.C. (NY)
RADON ALERT SYSTEMS (SK)
RADON DETECTION SERVICES (SK)
RADON DETECTION SYSTEMS/CORPORATE HEADQUARTERS (CO)
SCIENCE AND TECHNOLOGY, INC. (SCAT) (NB)
SENES CONSULTANTS (ON)
STUART HUNT AND ASSOCIATES/OCCUPATIONAL AND RADIATION SAFETY SERVICES, LTD. (AB)

Other Professionals and Consultants

This section lists organizations and individuals who offer services and consultation on a broad range of radon-related issues and activities. Among the areas of expertise included here are radon mitigation design, site evaluation, risk assessment, waste management, instrument calibration and repair, industrial safety and health, legal counsel, insurance, dealership opportunities, marketing, and information management.

A

A & A CONSULTING AND INSPECTION SERVICES, INC.
349 Chenango St.
Binghamton, NY 13901
(607) 723-7968
Larry Richards - President
John Meade - Vice President
Peter Connell - Environmental Consulting
John Link - Industrial Hygienist
Services:
Provides radon detection services, radon screening, house characterization for residential and commercial structures.
Qualifications:
NRA, AARST, staff includes professional engineer, geologist, NYS DOE workshops.

AA HOME INSPECTIONS
P.O. Box 248
Springfield, NJ 07081
(201) 635-6060
Paul Uslan - President
Services:
Provides radon detection services for residential and commercial properties, serves as a consultant and expert witness in litigation matters.

AIR-N-SOL CORPORATION
3 Cedar St. P.O. Box 437
Frenchtown, NJ 08825
(201) 996-2028
Dale L. Johnson - President
Services:
Provides radon detection services for residential and commercial buildings, diagnostic testing and preparation of specifications for radon mitigation.
Qualifications:
NJ DEP, EPA RMP, EPA courses, AARST.

ARTHUR SCOTT AND ASSOCIATES/AMERICAN ATCON
2150 Burnhamthorpe Rd., W. Box 164
Mississauga, ON L5L 3A2
(416) 828-2389
Arthur G. Scott - President
Andrew Robertson - Research
Services:
Operates under the name American Atcon in the U.S., presently managing EPA project in Ohio that deals with radon mitigation techniques in buildings, has contracted with FIPR for mitigation projects in Florida, radon projects deal with experimental radon mitigation procedures and construction techniques.

B

BARNES AND CONE, INC.
5894 Court St.
Syracuse, NY 13206
(315) 437-0305
Rick Roach - President
Services:
Provides consultation for building foundation design.

BENCHMARK INSPECTION SERVICES
939 South 26th St.
Arlington, VA 22202
(703) 548-5690
A.R. Dahms - Field Inspector
K.F. Dahms - Administrative Asst.
Services:
Provides radon detection services, pre- and post-mitigation measurement, and

consultation.
Qualifications:
Certified radon technologist.

BERNELL PROFESSIONAL SERVICES
361 12th St.
Cresskill, NJ 07626
(201) 568-7083
Bernard Kasmir - President
Services:
Provides radon detection services using alpha track and charcoal canister methods, provides consultation on radon mitigation methods.

BIOHAZARDOUS MATERIALS CONSULTING
P.O. Box 4376
Ithaca, NY 14852
(607) 257-5010, (315) 478-0653
Mark Schenerman - Radon Specialist
Kirk Leister - Radon Specialist
Services:
Provides radon detection services using charcoal canister method, provides consultation services for associated contractors to perform radon mitigation procedures.
Qualifications:
NY certified for radon measurement and mitigation.

BIOMEDICAL TOXICOLOGY ASSOCIATES
P.O. Box 3568
Frederick, MD 21701
(301) 662-0783
W.G. Palmer - Director
Services:
Provides radon detection and consultation services in toxicology and health sciences.

BRADLEY HARRISON ASSOCIATES, INC.
55 Virginia Ave.
Freeport, NY 11520
(516) 546-2283, (516) 546-7760
Bradley Harrison - President
Services:
Consulting firm specializing in environmental matters, waste management, and industrial safety and health.

BRIGGS LONGBOTHUM THERMINSPECTION
1440 Prospect Ave.
Plainfield, NJ 07060
(201) 755-8116
G. Briggs Longbothum - President
Services:
Provides analytical consulting services on radon mitigation issues, design services, installation of mitigation systems for indoor air quality, offers seminars and training for real estate community on air and water quality issues.

BROOKS LABORATORIES, INC.
44 Codfish Ln.
Weston, CT 06883
(203) 226-6384
Pramod Patel
Services:
Provides radon detection, mitigation, and consultation services.
Qualifications:
Industrial hygiene consultants.

C

CADWALADER, WICKERSHAM AND TAFT
1333 New Hampshire Ave., NW
Washington, DC 20036
(202) 862-2317
Laurence S. Kirsch - Attorney
Services:
Legal counseling, administrative proceedings, and litigation for building-related companies, real estate brokers, and radon testing companies.

CAMRODEN ASSOCIATES
RFD 4, Box 62
Rome, NY 13440
(315) 865-4269
Terry Brennan - President
Services:
Provides radon mitigation services, indoor air quality investigation, and training programs for federal, state, and private sector groups.
Qualifications:
APCA, ASTM, NAHB.

CANADIAN INSTITUTE FOR RADIATION SAFETY/RADON IN HOMES COURSE
595 Bay St. Suite 1050
Toronto, ON M5G 2C2
(416) 596-1617
Fergal Nolan - President
Ganice Kawka - Administrative Assistant
Services:
Offers radon training programs for Canadian federal and provincial radiation policy developers.

COGITO TECHNICAL SERVICES, INC.
1799 Rte. 13 P.O. Box 86
Sheds, NY 13151
(315) 662-7529
Susan Galbraith - Principal
Services:
Researcher and technical writer has prepared reports and instructional materials on radon for the New York State Energy Office and EPA.
Qualifications:
Subcontractor for Camroden Associates, co-author of the EPA three-day workshop manual.

COMMUNICATION SCIENCES INSTITUTE
16440 Emory Ln.
Rockville, MD 20853
(301) 570-0984
Raymond Johnson
Services:
Provides health communication and consultation to homeowners regarding radon fundamentals, behavior, detection, mitigation and risks.

COTHERN, C. RICHARD
4732 Merivale Rd.
Chevy Chase, MD 20815
(301) 652-2863
C. Richard Cothern
Services:
Provides consultation on waterborne radon issues.

CRIPPEN CONSULTING ASSOCIATES, INC.
242 North James St. Suite 202
Wilmington, DE 19804
(302) 994-8000
Lee W. Thomas - Vice President
William D. Garlington - Vice President
Services:
Provides radon detection, mitigation and consultation services for residential, commercial, and public buildings.
Qualifications:
Licensed professional engineers and radon certified consultants.

CUTTER INFORMATION CORPORATION
1100 Massachusetts Ave.
Arlington, MA 02174
(617) 648-8700
Karen Coburn - Publisher
Ed Coburn - Bulk Sales
Kim Gay - Customer Service
Services:
Publishers of information in the field of advanced building technology, including radon mitigation and indoor air quality.

D

DECISION RESEARCH
1201 Oak
Eugene, OR 97401
(503) 485-2400
Paul Slovic - Research Associate
Services:
Research on human judgment and risk assessment. Analyzes failure of individuals to have homes tested for radon. All publications available to public.

DOUGLAS MARTIN AND ASSOCIATES, INC./MARTIN CONSULTING
520 Chestnut St.
Emmaus, PA 18049
See Ad Page 199
(215) 965-2914
Richard D. Martin - President
Ken Smith - Health and Safety Services
Jim Sherlock - Mitigation Supervisor
Services:
Provides radon mitigation services, mitigation design, engineering procedure development, nuclear consulting, manufactures Professional Series water filtration system, provides radon mitigation training program for state of New Jersey.
Qualifications:
Certified in NJ, PA.

DUDLEY, PAUL
140 Railroad Mills Rd.
Pittsford, NY 14534
(716) 385-2368
Paul Dudley - Environmental Consultant
Services:
Provides radon detection and consultation services for residential markets.

E

EASON AND EASON CHARTERED/ATTORNEYS AT LAW
4423 Montgomery Ave.
Bethesda, MD 20814-4491
(301) 652-2666
See Ad Page 175
Charles Eason - Partner
Services:
Provides legal consultation to firms and individuals involved in radon related services, drafting or review of applications for state licenses or certification, preparation of customer agreements and reporting forms, defense of claims arising from radon-related activities, general corporate representation, evaluation of insurance requirements and coverage.
Qualifications:
Over 10 years of experience in nuclear and environmental law.

EDINBORO SCIENTIFIC COMPANY
P.O. Box 589 Wetsell Ridge Rd.
Edinboro, PA 16412
(814) 734-4648
Thomas A. Walkiewicz - Vice President
Services:
Provides radon detection services, laboratory analysis, consultation in experimental nuclear/radiation physics, distributes charcoal canisters.
Qualifications:
EPA RMP.

ENERGY ALTERNATIVES UNLIMITED, INC.
8384 Middlecrest Rd.
Boulder, CO 80302
(303) 444-6063

David Nyquist
Services:
Provides radon detection, mitigation and consultation services for disposal of toxic and hazardous materials.
Qualifications:
Charter member AARST.

ENERGY CONSERVATION SERVICES
6290 North Vine Valley Rd.
Rushville, NY 14544
(716) 554-3994
Peter Duffy - Principal
Services:
Provides radon detection, mitigation, and energy consultation services.

ENERGY RESOURCE CENTER
P.O. Box 267
Colorado Springs, CO 80901
(719) 636-2447
Lee Salmon - Radon Specialist
Services:
Provides radon detection and mitigation services, project management, independent evaluation, presents seminars, distributes charcoal canisters and alpha tracks to homeowners.
Qualifications:
Graduate, Colorado School of Mines course on radon and indoor environment, 15 years of experience in energy management.

ENVIROHEALTH, INC.
1090 General Greene Rd.
Wayne, PA 19087
(215) 964-0668
R.M. Patterson - President
Services:
Provides consultation on radon detection, industrial hygiene, and air pollution technical and management services.
Qualifications:
Certified industrial hygienist, PA certified.

ENVIRONMENTAL CONTROL SYSTEMS, LTD.
15325 Jones Ln.
Gaithersburg, MD 20878
(301) 421-5150
Moe Vakili - Environmental Specialist
Jasmin Karim - Virologist
Services:
Provides radon detection services, environmental research on indoor air quality, radon detection and mitigation.
Qualifications:
ASHRAE, ASME.

ENVIRONMENTAL DETECTION ASSOCIATES, INC.
P.O. Box 1006 21 Middlemount Rd.
Pisgah Forest, NC 28768-1006
(704) 877-4300
Wolfgang Selle - President
Harry Hill - Vice President
Services:
Provides radon detection services, distributes radon detection devices, consultation services to architects. Non-commercial laboratory for research purposes or state agencies' assistance only.
Qualifications:
EPA RMP.

ENVIRONMENTAL HEALTH INSPECTIONS, INC.
1200 Rte. 9 Cross River Mill Suite 5E
Lakewood, NJ 08701
(201) 905-8888
Robert Gogats - President
Mark Guarino - Vice President
Services:
Specializes in mortgage commitment inspections and certifications, radon detection and consultation, well water sampling, inspection and certification.
Qualifications:
Partners in firm are NJ licensed health officers and sanitarians.

ENVIRONMENTAL RADIATION GROUP
4900 Northwest 19th Pl.
Gainesville, FL 32605
(904) 378-5581
W. Emmett Bolch - Consultant
Services:
Provides land evaluation prior to development, information used in materials management during excavations or fills, in selection of home sites versus commercial sites or for parks and golf courses, and in predicting indoor radon levels and the associated statistics.

ENVIRONMENTAL RESEARCH ASSOCIATES
P.O. Box 219 490 Darby-Paoli Rd.
Villanova, PA 19085
(215) 687-5629
Michael H. Levin - Director
Services:
Provides radon detection services, research and consulting to real estate, trade, and legal professionals.
Qualifications:
Services since 1970.

ENVIRONMENTAL SOLUTIONS, INC.
P.O. Box 11323
St. Louis, MO 63105
(314) 863-9801
William M. Vaughan - Senior Scientist
Trudy Y. Smith - Technical Assistant
Services:
Provides consultation services on air quality and air toxins, distributes radon test kits.
Qualifications:
EPA RMP.

ENVIROSCIENCE CONSULTANTS, INC.
66 Cedar St.
Newington, CT 06111
(203) 666-7167
Neal B. Freudan - President
Scott D. Herzog - Vice President
Gregg Stevens - Technician
Services:
Provides radon detection services, full service environmental engineering and industrial hygiene firm.
Qualifications:
Certified industrial hygienist, chemical engineer.

ENVIROSPHERE
10900 Northeast Eighth St.
Bellevue, WA 98004
(206) 451-4600
David Waite - Owner
Services:
Engineering firm primarily working with building foundations.

FAST
929 Gloucester Ave.
Bricktown, NJ 08723
(201) 920-2369
Antonio D. Pare
Services:
Provides radon mitigation consultation services.

FGHB ASSOCIATES
1141 Hohlfelder Rd.
Glencoe, IL 60022
(312) 835-4433
Theodore Fields - President
Services:
Provides consultation for the medical community safety issues and radiation quality control.

FINN CONSULTANT ENGINEERS, INC.
227 North St.
Dalton, MA 01226
(413) 684-2019
George Finn - President

Services:
Provides radon detection and mitigation services, consultation in mitigation design.

G

GEO-RADON SERVICES
P.O. Box 5985
Berkeley, CA 94705
(415) 843-8844
Barbara Moed - Owner
Services:
Provides radon detection and mitigation services, consultation and research on influence of geology and soil properties on indoor radon.
Qualifications:
Eight years of indoor radon research funded by DOE, EPA, BPA, masters' degree in geology, consultant to California radon survey.

GEOHAZARDS, INC.
P.O. Box 14956
Gainesville, FL 32604
(904) 371-7243
Douglas L. Smith - President
Services:
Provides radon detection services and consultation on radon mitigation options.

GEOMET TECHNOLOGIES, INC.
20251 Century Blvd.
Germantown, MD 20874
(301) 428-9898
Niren L. Nagda - Vice President
Bill Mountel - Marketing
Services:
Research on radon detection and mitigation technologies. Presently studying two homes in Maryland to analyze particular mitigation strategies. Large scale surveys in Florida and Massachusetts on charcoal canister and alpha track detection methods.

GORDON, SAMUEL
146 S. Little Tor Rd.
New City, NY 10956
(914) 634-9331
Samuel Gordon
Services:
Provides consultations for radon detection and mitigation processes at no charge to public.

GROUNDWATER TECHNOLOGY, INC.
Meadowbrook Industrial Park
Milford, NH 03055
(603) 672-5303
Paul Andrew
Services:
Provides consultation on removal of waterborne radon.

GTS CONSULTANTS
5599 North Maplewood Pl.
Littleton, CO 80123
(303) 794-1153
Nicholas F. Davis - Consultant
Services:
Provides consultation services on radon issues.

H

HALL-KIMBRELL ENVIRONMENTAL SERVICES
5319 Southwest Westgate
Portland, OR 97221
(503) 292-9406
David Butts - Technical Director
Services:
Provides radon detection, mitigation, and consultation services.
Qualifications:
NRA.

HEALTH AND SAFETY ASSOCIATES
35 Grove St.
Medford, MA 02155
(617) 488-7081
Neil A. Gaeta - Health Physicist
Services:
Provides radon detection services, training in radon for realtors and corporate firms and consultation services for the industry.
Qualifications:
Certified health physicists and safety professionals.

HEALTH PHYSICS ASSOCIATES, LTD.
3312 Commercial Ave.
Northbook, IL 60062-1909
(312) 480-1900
Ted Fields - President
Locke Bailey - General Manager
William Rivkin - Vice President
Services:
Radiation service company provides radon detection services using charcoal canister, alpha track, and liquid scintillation methods, instrument calibration using radon chamber, and consultation on radiation safety.
Qualifications:
Certified health physicists, technicians registered for radon.

HEALTH PHYSICS NORTHWEST, INC.
11535 Southwest 67th Ave.
Tigard, OR 97223
(503) 620-6617, (800) 762-8444
Ross L. Mercer - President
Corwin M. Hymes - Sales Manager
Services:
Radiation service company provides radiation consultation, instrumentation, materials and calibration and repair services to medical, industrial and commercial clients.

HELIX ENVIRONMENTAL, INC.
416 Triangle Ave.
Dayton, OH 45419
(513) 298-2990
Ralph A. Froehlich - President
R.C. Froehlich - Asbestos Planner
E.E. Rothwell - Industrial Hygienist
Services:
Occupational and environmental health consultants specializing in the recognition, evaluation, and control of indoor air contaminants.
Qualifications:
CIH, Ohio Asbestos Building Inspectors, Ohio Abatement Supervisors.

HENIGAR AND RAY ENGINEERING ASSOCIATES, INC.
640 East Hwy. 44
Crystal River, FL 32629
(904) 795-6551, (904) 447-2889
Michael G. Czerwinski - Geologist
Services:
Provides consultation services for the real estate industry.
Qualifications:
Currently studying the entire town of Dunnellon, FL for land ordinances. This model study may be used for other Florida towns.

HERRICK ENGINEERING, INC.
2004 Belvedere Dr.
Toledo, OH 43614
(419) 385-5655
Robert Herrick - Professional Engineer
Services:
Provides radon detection and consultation services.
Qualifications:
NRA.

HOME TESTING LABS, INC.
299 West Fort Lee Rd.
Bogota, NJ 07603
(201) 343-1199
Thomas Parisi - Director
Services:
Provides consultation for radon mitigation design.

HOME TESTING LABS, INC.
44 Crestview Rd.

Phoenixville, PA 19460
(215) 768-7960
Thomas Parisi - Director
Services:
Provides consultation for radon mitigation design.

HOMEPRO SYSTEMS, INC.
6715 Lowell Ave. Suite 2
McLean, VA 22101
(703) 761-1400
Mike Lennon
Services:
Provides consultation for analysis of structures, feasibility studies, specifications, radon detection and mitigation training.

I

INFILTEC/MAIN OFFICE
P.O. Box 8007
Falls Church, VA 22041
(703) 820-7696
David Saum - President
Marc Messing - Vice President
Services:
Provides radon mitigation, research and development, nationwide training, consulting services, specializes in large buildings.
Qualifications:
EPA RMP, EPA subcontractor, EPA trainer.

IQBAL CONSULTING ENGINEERS
5 Ironstone Dr.
Reading , PA 19606
(215) 779-8345
M.A. Iqbal - President
Services:
Geotechnical firm provides radon consultation services.

J

K

K & S ASSOCIATES
1854 Airlane Dr.
Nashville, TN 37210
(615) 883-9760
Bob Dillard - Product Manager
Services:
Dosimeter calibration laboratory.

KINSLOW, ROY H.
9935 Coppedge Ln.
Dallas, TX 75220
(214) 351-1446
Roy H. Kinslow - Professional Engineer
Services:
Provides consultation services on radon detection devices.

L

L.H. FRIEND AND COMPANY
1875 Century Park, E.
Los Angeles, CA 90210
(213) 556-5488, (800) 351-9923
Stephen Weinress - Research Director
Services:
Investment analysis for the radon industry.
Qualifications:
NASD securities registered.

LOS ANGELES URBAN DEVELOPMENT
1725 Radcliff St.
Los Angeles, CA 90026-1131
(213) 668-2655
W.R. Dowling - CEO
Randy Morceau - Vice President
Services:
Provides radon detection services, computer Bulletin Board includes radon information exchange, accessable 24 hours a day at (213) 666-6639.
Qualifications:
AARST.

LOWRY ENGINEERING, INC.
P.O. Box 536 Back Troy Rd.
Unity, ME 04988
(207) 948-3790
See Ad Page 211
Sylvia Lowry - President
Jerry Lowry - Resarch Director
Services:
Engineering research and consulting firm specializes in radionuclide analysis and removal from water supplies, and removal equipment for homes and public water supplies, manufactures and distributes the Stripper Aeration System, removes 99.9% of radon from water.
Qualifications:
Director is a leading expert on radon removal from water supplies, teaches EPA courses on mitigation.

LRM, INC.
632 Spooner Rd.
Virginia Beach, VA 23462
(804) 497-6505
Larry R. Martin
Services:
Provides consultation services for radon detection, mitigation, and surveys.

LUDGATE ENGINEERING CORPORATION
9A North Fifth St.
Reading, PA 19522
(215) 374-2239
Robert Ludgate - President
Services:
Civil engineering firm provides consultations on design for new home builders.

M

M.R. HOLLAND ASSOCIATES
70 Woodburn Dr.
Methuen, MA 01844
(508) 681-0821
Mark R. Holland - Principal
Services:
Consulting architectural firm provides design specifications and solutions for radon problems in structures.
Qualifications:
Registered architects in MA, NRA.

MANSFIELD, PONTICELLI AND LIBBY
2100 M St., NW
Washington, DC 20036
(202) 296-1330
Richard Mansfield - General Counsel
Services:
Employment Relocation Council general counsel provides radon information for ERC. Developing proposals for establishment of a radon policy for Council.

MEDRAD, INC.
271 Kappa Dr.
Pittsburgh, PA 15238
(412) 782-4600
John P. Friel - Vice President
A. Kenneth Grob - Vice President
Services:
Interested in the license of radon technology to strategic partners.

MICRO-TRENDS
25 Caton Terr.
Caldwell, NJ 07006
(201) 228-5907
Charles Trapani
Services:
Publishes and distributes radon mitigation manual.

N

N.A.K. PRODUCTION ASSOCIATES
1317 Apple Ave.
Silver Spring, MD 20910
(301) 565-0355
Norm Klotz - President
Larry Petcovic - Director
Fred Rehbein - Director

Services:
Produced and distributes videotape: "Living With Radon: Protecting Your Health and Protecting Your Investment," a nontechnical radon overview intended for consumers, real estate industry, libraries, etc.
Qualifications:
Several award-winning health series videotapes.

NATURAL ENERGY CONSULTANTS
P.O. Box 586
Conifer, CO 80433
(303) 838-5291
Rob Haines - Consultant
Services:
Provides radon detection, mitigation, and consultation services.

NEILSON ASSOCIATES, INC.
100 Dickinson Dr. Suites 109-110
Chadds Ford, PA 19317
(215) 793-1389
Arthur Neilson - President
David Heckman - Industrial Hygienist
Brenda Hansen - Industrial Hygienist
Services:
Provides consultation services on occupational health and safety issues to industry, government, and private sector.
Qualifications:
CSP, CHP, CIH.

NEW ENGLAND RADON SERVICES
95 Vauxhall St.
New London, CT 06320
(203) 442-6675
Martin Grisham - Owner
Services:
Provides radon detection and mitigation consulting services for the do-it-yourself homeowner.

NORTH EAST ENVIRONMENTAL PRODUCTS, INC.
85 Mechanic St.
Lebanon, NH 03766
(603) 448-3944
See Ad Page 432
Bruce L. Lamarre - President
Jack Nelson - Chief Executive Officer
Kevin Lang - Vice President
Services:
Manufactures and distributes Clearadon, an aeration system for removal of radon or VOCs from residential water supplies, provides consultation services for removal of waterborne radon.
Qualifications:
Water Quality Association.

O

P

PELICAN ENVIRONMENTAL CORPORATION
42 Lehigh Rd.
Wellesley, MA 02181
(617) 235-5997
Don Crawshaw - President
Jeff Crawshaw - Engineer
Randy Foster - Mitigator
Services:
Provides radon detection, mitigation and consultation services.
Qualifications:
Master's degree in civil engineering, accredited in MA for testing.

PENNSYLVANIA STATE UNIVERSITY, THE/RADON DETECTION AND MITIGATION CERTIFICATION TRAINING PROGRAM
231 Sackett Building Energy Technology Projects
University Park, PA 16802
(814) 863-2133
James K. Shillen - Professor
Services:
Offers radon detection and mitigation training programs approved by PA DER.

PHOENIX ENGINEERS
10041 El Monte
Overland Park, KS 66207
(301) 975-5906
Preston McNall
Services:
Provides consultation on radon mitigation design.

PHOENIX SAFETY SYSTEMS
7917 Carnation Dr.
Louisville, KY 40258
(502) 935-8317, (502) 964-0865
Dave Morrison
Services:
Provides radon mitigation services, designs surveys.

PORTER CONSULTANTS, INC.
125 Argyle Rd.
Ardmore, PA 19003
(215) 896-5353
S.W. Porter, Jr. - Health Physicist
M.C. Roberts - Health Physicist
J. Sykes, III - Health Physicist
Services:
Offers training courses on radon detection and mitigation, provides radon detection services using charcoal canister, grab sample, alpha track methods, water and soil.
Qualifications:
ABH certified health physicists on staff, certified radon testing specialists, EMP RMP, Vice President of AARST.

PROFESSIONAL HOME INSPECTIONS, INC.
28790 Pine Dr.
Evergreen, CO 80439
(303) 674-4620
G. Thomas Martin - President
Services:
Provides radon detection services and consultation on mitigation design and building materials.
Qualifications:
AARST, ASHI, Colorado School of Mines course on radon and the indoor environment.

PROTECTION INSPECTION, INC.
Rte. 11 Hill Rd.
Knoxville, TN 37938
(615) 688-9779
Norman R. Kurtz - President
Thomas A. Walkiewicz - Consultant
Services:
Provides radon detection and mitigation services to residential and commercial markets, provides "watchdog" service to prevent tampering during radon testing.
Qualifications:
ASHI, EPA RMP.

Q

R

R.F. SIMON COMPANY, INC.
RD 2, Box 481
Barto, PA 19504
(215) 754-6455
See Ad Page 148
Ron Simon - Project Manager
Peter Maniscalco - Diagnostic Specialist
Services:
Provides mitigation design and installation services, mitigation supply warehouse carrying a full line of radon mitigation and detection products including fans, caulks and sealants, couplings, comprehensive radon services provided to private, commercial, institutional, and government agencies, expert testimony.
Qualifications:
EPA RMP, 13 years of custom building experience, instructor for EPA training courses, consultant to Lawrence Berkeley Laboratory, Princeton University, PA DER.

RADEX, INC.
1382 Prince Ave. Suite A

Athens, GA 30606
(404) 548-6960
Frederick Tuck - President
Norman Herz - Vice President
Jim Kitchens - Treasurer
Services:
Provides radon detection, mitigation, and training services.
Qualifications:
State certified geologist and university health physicist, experience in commercial and private radon monitoring.

RADION LABORATORY, INC.
70 Walnut St.
Wellesley, MA 02181
(617) 239-8066
Rudolph Hon - Vice President
Kevin A. Doherty - Senior Manager
Services:
Provides radon detection and mitigation services, operates counting instrumentation for radiation detection in air, liquids, and solid samples, distributes charcoal canisters and water testing kits for radon.
Qualifications:
AARST, NRA, listed with MA Department of Public Health, pending certification in PA, 20 years of experience handling and dealing with radioactive material, instrumentation cross-calibrated with DOE labs.

RADON ASSOCIATES MARKETING COMPANY
1099 18th St. Suite 1830
Denver, CO 80202
(303) 293-3141
Steven V. Harrison - President
Services:
Collecting information for development of comprehensive insurance program for radon professionals, currently offers a radon exposure bodily injury liability policy to corporate clients.

RADON CONTROL PROFESSIONALS, INC.
11920 Fieldthorn Ct.
Reston, VA 22094
(703) 471-9459
See Ad Page 151
Stephen Hall - President
Norman Rodier - Vice President
Services:
Provides full service radon detection and mitigation services, residential, commercial, soil testing, site evaluation, pre-construction mitigation design.
Qualifications:
County business license, Maryland Home Improvement license, EPA mitigation course, graduate-level scientists.

RADON CORRECTIONS INSTITUTE, INC.
P.O. Box 314
Longmont, CO 80502
(303) 651-6190, (800) 648-9281
Wayne Barcewski - President
Douglas Wallace - Vice President
Services:
Provides radon mitigation and consultation services for existing housing and new construction.
Qualifications:
AARST, Colorado School of Mines course on indoor radon.

RADON DETECTION AND CONSULTING
705 West Fourth St.
Lawrence, KS 66044
(913) 843-7377
Steven R. Hicks - Director
Services:
Provides full service radon detection, mitigation, and consultation services, information at no charge.
Qualifications:
EPA RMP, operated by medical, engineering, and legal professionals.

RADON LABORATORIES
P.O. Box 829
Middletown, MD 21769
(301) 371-5867
Gregory J. Beal, Jr. - President
Services:
Provides radon detection and consultation services for the public and industry, uses Pico Rad complete radon monitoring system.
Qualifications:
EPA RMP.

RADON PRESS, INC.
500 North Washington St.
Alexandria, VA 22314
(703) 548-2756, (800) 548-1567
Larry Siegelman - Publisher
Robert Morrow - Editor
Marc O' Brien - Associate Editor
Services:
Radon information management and publishing firm. Publishes The Radon Industry Directory and other radon related media including newsletters, catalogs, list rentals, etc. Monitors and projects radon industry trends.
Qualifications:
AARST, ARA, extensive experience in data collection, marketing, and publishing.

RADON QC
2857 Nazareth Rd. Palmer Professional Building Suite 6
Palmer, PA 18043
(215) 559-1636
See Ad Page 237
John Rider - Manager
Kelsie Geiger - Assistant Manager
Services:
Provides quality control services for radon measurement companies, known concentrations of radon gas and radon daughters, three chambers of radon gas concentrations, radon chamber services available to radon detection firms.

RADON REMEDIES, INC.
2700 West Broad St. P.O. Box 44234
Columbus, OH 43204
(614) 272-7772
Craig S. Wise - President
James M. Farrow - General Manager
Services:
Produces and distributes radon video containing general information, radon detection and mitigation techniques and products, for homes and businesses.
Qualifications:
EPA RMP, ARA.

RADON SAFETY AND TECHNOLOGY CORPORATION
4 Charles St.
Pleasant Valley, NY 12569
(914) 635-1138, (800) 552-5257
Ronald J. Gasparro - President
Zia R. Hashmi - Director
Claude J. Noto - Assistant Director
Services:
Radon measurement analytical laboratory, water and soil measurements, surveys for real estate transactions, land sites, and newly constructed and standing buildings, distributes charcoal canisters and alpha track detectors, waterborne radon test kits, radon alarms and auxiliaries offered to public and professionals, provides radon mitigation services.
Qualifications:
EPA RMP, AARST, ASTM.

RADON SURVEY SYSTEMS, INC.
P.O. Box 22288
Cleveland, OH 44122
(216) 464-7997
Darioush T. Ghahremani - President
Services:
Provides radon detection and mitigation services for residential and commercial markets, site surveys for undeveloped land, research and development of radon detection technologies, detectors, and materials, new development on radon mitigation options for various projects.

Qualifications:
AARST, NRA, several national scientific societies.

RADON TESTING AND CONSULTING SERVICES
34 Jones River Dr.
Kingston, MA 02364
(617) 585-9924
Daniel C. Walker - President
Cathleen Walker - Vice President
Services:
Provides radon detection, mitigation, and consultation services for the real estate industry, construction trade, and private citizens.
Qualifications:
AARST, RTCA professional consultant.

RADON TESTING SERVICES, INC.
4442 18th St.
San Francisco, CA 94114
(415) 552-5496
William E. Hellums - President
Services:
Provides radon detection services, forensic and consultation services, computer dose projections.
Qualifications:
EPA RMP, radioactive materials license, certified health physicist.

RAMAPO COLLEGE OF NEW JERSEY/INSTITUTE FOR ENVIRONMENTAL STUDIES
505 Ramapo Valley Rd.
Mahwah, NJ 07430
(201) 529-7742
William Makofske - Director
Michael Edelstein - Associate Professor
Services:
Provides radon consultation, education, and training workshops to public and private sector.

RDS/RADON DETECTION SERVICES, INC.-CORPORATE OFFICE
Rte. 179, Old York Rd. P.O. Box 419
Ringoes, NJ 08551
(201) 788-3080
See Ad Page 153
James G. Davidson - President
Susanne E. Micek - Vice President
Michael S. Terpilak - Health Physicist
Services:
Provides full service radon detection mitigation, and consultation services, offers franchises, offers comprehensive training courses for radon professionals, manufactures and distributes RDS "remediator" fans, locations in four states.
Qualifications:
Has met all licensing requirements in PA, NJ, and KS, courses have been approved by DE, KY, PA, OH.

RDS/RADON DETECTION SERVICES-MARYLAND OFFICE
10400 Connecticut Ave. Suite 504
Kensington, MD 20895
(301) 933-7900
Milton Iskow - President
David E. Pard - Sales Manager
Services:
Provides radon detection services using continuous working level monitors, E-Perm, alpha track and charcoal canister methods, pre-construction soil testing and consultation to general public, real estate/construction industries, local governments and corporations.
Qualifications:
AARST, EPA training, PA certification, geologist and health physicist on premises.

RE FORMATION
P.O. Box 328
Pine Island, NY 10969
(914) 258-4711
William McDonnell - Owner
Services:
Provides second opinions on radon analysis results, policy evaluations of radon programs.
Qualifications:
EPA, NY State training seminars.

RESOURCE CENTER FOR ENVIRONMENTAL PROTECTION
P.O. Box 25
Schuylerville, NY 12871
(518) 695-3218
See Ad Page 160
Peter Lane - President
Fred Staing - Business Manager
Services:
Provides radon mitigation services using ventilation, barrier, and water remediation systems, consultation with contractors to develop new radon mitigation products and designs.

RICHARD P. ARBER ASSOCIATES
44 Cook St., Suite 601
Denver, CO 80206
(303) 393-6363
Robert Jelinek - Project Manager
Services:
Consulting engineering systems design firm works with EPA on drinking water study. Designed ion exchange treatment system for removal of radionuclides from drinking water in school system in Jefferson County, CO.

RMC ENVIRONMENTAL SERVICES
919 Edgewood Rd.
Havertown, PA 19083
(215) 841-5268
Douglas F. Wahl
Services:
Provides environmental consultation services.

RMC ENVIRONMENTAL SERVICES/ROYERSFORD DIVISION
323 Church St.
Royersford, PA 19468
(215) 948-4949
John V. Kostige
Services:
Provides environmental consultation services.

ROCKY MOUNTAIN MEDICAL PHYSICS, INC.
7175 West Jefferson Ave. Suite 2150
Lakewood, CO 80235
(303) 980-7070
John Goddard - President
Services:
Medical physics group provides radon detection and consultation services for industry.
Qualifications:
EPA RMP.

ROGERS AND ASSOCIATES ENGINEERING CORPORATION
515 East 4500 S. Suite G-200
Salt Lake City, UT 84107
(801) 263-1600
Arthur Sutherland - Vice President
Kirk K. Nielson - Research
Services:
Provides radon detection services, consulting engineers for radiological and environmental problems, research to develop a unified theory of radon emanation and transport to implement in a computer code for field studies and research programs.
Qualifications:
State licensed for radioactive materials.

ROSS SYSTEMS, INC.
21 Main St. Box 670
Blairstown, NJ 07825
(201) 362-5571
Richard Ross - President
Services:
Provides full service radon detection services, distributes charcoal canisters and alpha track detectors, provides radon mitigation consultation services using computer technology, radon mitigation proposals at no charge, offers seminars and workshops.
Qualifications:
AARST, EPA RMP,APCA, NJ DEP.

RPI INTERNATIONAL ENVIRONMENTAL TECHNOLOGY DIVISION
P.O. Box 328
Columbia, SC 29202
(803) 256-7322
Jacqueline Michel - Director
Services:
Scientific consulting firm provides expertise on radon sources, behavior, and aquifer characterization for identifying radon in ground water and homes.

RSA CORPORATION/RADON SERVICE AGREEMENT
918 Woodley Rd.
Dayton, OH 45403
(513) 254-4102
Mike Kimmell - President
David B. Frawley - Service Representative
Services:
Radon service agreements offered to homeowners through relocation or RSA certified radon detection and mitigation firms, company underwrites radon testing and mitigation protocols to be used by RSA-certified firms.
Qualifications:
Endorsed by ARA.

RUECK ASSOCIATES
103 Sage Rd.
Silver Lake, KS 66539-9461
(913) 582-4050
Jon M. Rueck - Chairman
Services:
Provides consultation on radon issues, distributes At-Ease radon monitors, charcoal canisters, and E-Perms.
Qualifications:
Diplomate AAEE, registered professional engineer, degrees in nuclear engineering, mechanical engineering.

S

S. COHEN AND ASSOCIATES, INC.
8200 Riding Ridge Pl.
McLean, VA 22102
(703) 893-6592
Sanford Cohen - President
John Mauro - Vice President
Services:
Performs research and development, offers training in radon detection and mitigation.
Qualifications:
EPA Office of Radiation Programs prime contractor.

SABAL INVESTMENTS
915 Oakfield Drive Suite A
Brandon, FL 33511
(813) 689-2548
James Lee - President
Services:
Educational programs and seminars provide radon information.

SAFE-AIRE, INC.
162 East Chestnut St. P.O. Box 160
Canton, IL 61520
(309) 647-0419, (800) 331-2943 See Ad Page 161
Bill Wherley - Sales Manager
Chet Phillips - Products Manager
Harry Ellison - Marketing Director
Don Stees - Vice President
Services:
Provides patent-pending Safe-Aire radon mitigation system to dealers, provides radon detection and mitigation services, distributes extensive line of radon detection and mitigation products, dealership opportunities available.

SAFETY AND HEALTH SERVICES
3307 North Second St.
Harrisburg, PA 17110
(717) 238-2122
Jimmie Masich - Consultant
Jack Bregman - Consultant
Bill Burdick - Consultant
Services:
Provides residential and commercial safety and health consultation services, provides radon detection, contractor monitoring and performance evaluation services.
Qualifications:
All consultants are certified industrial hygienists, have doctorate degrees, or are certified safety professionals.

SCHMELTZER, APTAKER AND SHEPPARD
1800 Massachusetts Ave., NW Suite 500
Washington, DC 20036
(202) 828-1000
Mike Brown - Attorney
Services:
Practices environmental, health and safety law, consultation on legal policies and strategies for EPA rulemaking and enforcement.

SCHNABEL ENGINEERING ASSOCIATES/GEOLOGICAL SERVICES
4909 Cordell Ave.
Bethesda, MD 20814
(301) 652-8922
Bryan Beard - Senior Associate
Services:
Provides geological consultation and analysis.

SCIENCE AND TECHNOLOGY, INC. (SCAT)
79 Pembroke Crescent
Fredericton, NB E3B 2V1
(506) 457-0232
Stanton T. Friedman - President
Services:
Institute has long history of radon research projects under contract with Canadian government.

SCIENCE APPLICATION INTERNATIONAL CORPORATION
1710 Goodridge Dr.
McLean, VA 22102
(703) 821-4300
Reginald Gotchy - Senior Scientist
Services:
Provides consultation services on radon measurement technology.

SCIENTIFIC RESEARCH AND DEVELOPMENT COMPANY
P.O. Box 5590
Lakeland, FL 33807
(813) 644-0404
Peter Orlowsky - President
Services:
Research and development of an electronic unit that can mitigate radon (not yet on the market), provides consultation services.

SCINTA, INC.
2421 Homestead Dr.
Silver Spring, MD 20902
(301) 593-9478
Bernard Shleien - Owner
Services:
Provides consultation services on radiation protection and risk.

SELSER, ROBERT C.
442 Broadway
Hillsdale, NJ 07642
(201) 666-4060
Robert C. Selser

SENES CONSULTANTS
499 McNicholl Ave.
Willowdale, ON M2H 2C9
(416) 499-5030
Leo Lowe - Director
Douglas Chambers - Contact
Services:
Environmental and occupational consulting firm analyzes environmental radioactivity, contracts with federal and provincial governments and with industry, analyzes results and sets remedial cleanup levels.

SHEVENELL, GALLEN, AND ASSOCIATES
P. O. Box 433 195 Commerce Way
Portsmouth, NH 03801
(603) 436-1490
Tom Shevenell - Principal
Services:
Provides environmental site assessments for real estate transactions, provides analysis of water wells for radon and makes recommendations for mitigation.

SIMPSON WEATHER ASSOCIATES
809 East Jefferson St.
Charlottesville, VA 22902
(804) 979-3571
M. Garstang - President
G.D. Emmitt - Vice President
H.G. Goodell
Services:
Provides radon mitigation and environmental consultation services.
Qualifications:
Staff includes persons with doctoral degrees in meteorology, geology, geochemistry, hydrology.

SOLAPLEXUS/PRODUCT FILE DIVISION
21 Tamarack Rd.
Fishkill, NY 12524
(914) 896-4796, (914) 384-6798
Milt Silver
Services:
Publishes The Solaplexus Radon Guide, a catalogue of manufacturer's literature on radon-related products and services.

SOUTHERN RADON SERVICES, INC.
1000 Johnson Ferry Rd. Suite B-145
Marietta, GA 30068
(404) 565-3886, (800) 537-2366
C. Fred Stafford - President
Services:
Provides radon detection services, primary laboratory for charcoal canister analysis for home owners and radon specialists, provides radon risk assessments for commerical, institutional, and development properties.
Qualifications:
EPA RMP, NJ registration.

SPENCE-GEIGER ASSOCIATES, INC.
1658 Cole Blvd. #260
Golden, CO 80401
(303) 233-8577
Jeffrey R. Geiger - Industrial Hygienist
Services:
Provides radon detection services for large-scale projects.
Qualifications:
Certified industrial hygienist.

STRATA ENVIRONMENTAL SERVICE, INC.
401 East Douglas Suite 515
Wichita, KS 67202
(316) 262-0002
Dennis Jones - President
Bruce A. Reed - Vice President
Randall Fornshell - Laboratory Director
Denise Peer - Laboratory Technician
Services:
Provides radon detection services, analytical and geohydrology consultation.
Qualifications:
EPA RMP, certified by KS for drinking water analysis.

T

TARGET RADON SERVICES, INC.
Oakland Ctr. 8940A Rte. 108
Columbia, MD 21045
(301) 992-6622
J. Edward Tillman - President
Jeffrey T. Sims - Vice President
Services:
Provides preconstruction soil gas surveys to building and development industry, radon detection services for commercial, institutional and residential markets.
Qualifications:
AARST, EPA RMP, radon specialist training course.

TECHNICAL ANALYSIS CORPORATION
6723 Whittier Ave. Suite 202
McLean, VA 22101
(703) 883-3700
David Rosenbaum - President
Services:
Former director of EPA Office of Radiation Programs provides radon consultation.

TMA/EBERLINE, HEALTH PHYSICS
5635 Jefferson Blvd., NE
Albuquerque, NM 87109
(505) 345-9931

Stanley Waligora - Sales Manager
Ernie Sanchez
Jeff Brown
Services:
Provides radon detection and mitigation services to commercial and government clients, distributes charcoal canisters, provides consultation services.

U

U.S. DEPARTMENT OF INTERIOR (DOI)/BUREAU OF MINES(BOM)/DENVER RESEARCH CENTER DRC)/RADIATION LABORATORY
P.O. Box 25086 Denver Federal Ctr. Building 20
Denver, CO 80225
(303) 236-0789
R.F. Holub - Director
Services:
Research to support BOM's Project 2000 program to reduce radiation hazards associated with exposure to radon and radon progeny in mines. One of four labs in world currently setting worldwide standards for calibrating radon daughter measuring and detection equipment.

UCI/ENVIROTEST
34100 Center Ridge Rd.
North Ridgeville, OH 44039
(216) 327-9432
Vincent Gargaro - President
Dennis Geisler - Vice President
Services:
Provides radon detection services and consultation on health physics issues.
Qualifications:
EPA RMP, PA DER certification pending, NRC license, OH radioactive registration.

UCI/ENVIROTEST
834 Old Thorn Run Rd.
Coraopolis, PA 15108
(412) 264-5891
Vincent Gargaro - President
Dennis Geisler - Vice President
Services:
Provides radon detection services and consultation on health physics issues.
Qualifications:
EPA RMP, PA DER certification pending, NRC license, OH radioactive registration.

UNIVERSITY OF MICHIGAN/SCHOOL OF PUBLIC HEALTH/DEPARTMENT OF ENVIRONMENTAL AND INDUSTRIAL HEALTH/RADON RESOURCE AND TRAINING CENTER
109 Observatory
Ann Arbor, MI 48109
(313) 764-0523
A.P. Jacobson - Director
Services:
Training program authorized by the Michigan Department of Public Health includes • Radon mitigation training as prerequisite to certification in the state of Michigan (includes a three-day contractors course). Training considers building, soil, and meteorlogical characteristics unique to Michigan. • Radon detection training addressing instrument operation and laboratory analysis.

V

VENABLE, BAETJER AND HOWARD
2 Hopkins Plaza
Baltimore, MD 21201-2978
(301) 244-7772
Kevin L. Shepherd - Attorney
Services:
Provides legal counsel on radon-related issues.

VERSAR, INC.
6850 Versar Center
Springfield, VA 22151
(703) 750-3000
Michael Markels - President
R.P. Ouellette - Corporate Development
J. Tappan - ARIX
Niren Nagda - Geomet
Services:
Nationwide consulting firm specializes in environmental risk management, provides radon detection services for homes and buildings, construction management for radon-contaminated sites and structures, provides radon mitigation services, recommends appropriate risk mitigation to public and private clients.

W

W.S. FLEMING AND ASSOCIATES, INC.
6310 Fly Rd.
East Syracuse, NY 13057
(315) 437-1869
Ian Nitschke - Senior Associate
Wayne Clarke - Technician
Services:
Provides consultation services on radon

progeny monitoring, diagnosis, mitigation specifications.
Qualifications:
EPA RMP, past consultation for NY and EPA.

WASTREN, INC.
2 Waverly Dr.
Hummelstown, PA 17036
(717) 566-6780
Philip J. Grant
Services:
Provides consultation on hazardous waste removal.

WATER STORE, THE
P.O. Box 176
Bath, PA 18014
(215) 837-9660
Scott Moore
Services:
Certified water treatment specialist provides consultation services on waterborne radon.

WESLEY R. VAN PELT ASSOCIATES, INC.
773 Paramus Rd.
Paramus, NJ 07652
(201) 445-5124
Wesley R. Van Pelt - President
Services:
Conducts environmental and industrial health and safety audits, evaluation and consultation on radiation safety, industrial hygiene and indoor air pollution.
Qualifications:
Certified health physicist, certified industrial hygienist.

WILLIAM G. CHANEY, PE
1708 Park Ave., SW
Albuquerque, NM 87104
(505) 247-2087
William G. Chaney - Owner
Services:
Provides radon detection services using At Ease and charcoal canister method, offers radon education seminars for the real estate industry.
Qualifications:
AARST.

WJLA TV
4461 Connecticut Ave.
Washington, DC 20008
(212) 364-7777
Roberta Baskin - Reporter
Services:
Broadcast journalist has focused series of reports on radon issues.

X

Y

Z

Radon WL Meter · Radon WL Meter · Radon WL Meter

Radon Meter
Quality and Precision

From a Name You can Trust

USED INTERNATIONALLY BY GOVERNMENT AND INDUSTRY RADON EXPERTS

Features

- **Silicon-chip/CMOS technology**
- **Direct reading**
- **6 pre-set sample periods or continuous sampling**
- **AC power or battery power**
- **Full data-logging capability**

Field-Proven

When radon is a concern, the Thomson and Nielsen Radon WL Meter is your best choice for measuring radon working level (WL) inside buildings. Professionals world-wide depend on it for analysis and routine monitoring in a variety of laboratory, residential, and industrial environments.

Whether you need an instrument for rapid screening or long-term averaging, the Radon WL Meter offers reliability, convenience and accurate results.

Versatile

The Radon WL Meter has what professionals look for in a monitoring device.

It is flexible, with a choice of six preselected sample periods or continuous sampling. It is sensitive, reading as low as 0.001 working level. And it's tough — designed to operate continuously and reliably at temperatures from 0 to 60 degrees Celsius.

Our standard kit includes the Radon WL Meter, box of 25 Millipore filters, tweezers, AC power adapter, air-flow meter and carrying case.

For more advanced applications, such as mitigation, a PC-compatible computerized data-recording unit is available to produce graphic, time-registered data.

- **Fast, easy to use, cost-effective**
- **Uses established EPA methodology**
- **Many applications: real estate, mitigation, etc.**
- **Duty-paid courier delivery**

Tomorrow's Technology...Today

THOMSON & NIELSEN ELECTRONICS LTD.

4019 Carling Avenue Kanata, Ontario, Canada, K2K 2A3 **Tel.: (613) 592-3019 Telex: 053-4741**

Radon Product Manufacturers and Distributors

This section lists companies that manufacture or distribute products used in radon detection and mitigation. The section is organized into 11 product categories as follows:

1. Radon Measurement Product Manufacturers - Manufacturers of charcoal canisters, alpha track detectors, grab samplers, E-Perms, continuous monitors, time integrating units, and laboratory analysis systems.

2. Radon Measurement Product Distributors - Distributors of charcoal canisters, alpha track detectors, grab samplers, E-Perms, continuous monitors, time integrating units, and laboratory analysis systems.

3. Radon Mitigation System Manufacturers - Manufacturers of single unit systems used for indoor air and waterborne radon mitigation.

4. Radon Mitigation Product Distributors - Distributors of a broad inventory of products used in the installation of radon mitigation systems.

5. Sealants - Manufacturers and distributors of sealers, caulks, fillers, foams, paints, and coatings used to seal cracks, pores, and other radon entry routes.

6. Radon Barriers - Manufacturers and distributors of membranes and films used as gas-tight radon barriers during the installation of radon mitigation systems.

7. Ventilation Equipment - Manufacturers and distributors of fans, air-to-air heat exchangers, recovery systems, and other ventilation equipment.

8. Air Flow Measurement Equipment - Manufacturers and distributors of smoke sticks, air samplers, flow meters, manometers, blower doors, and other products used to measure air movement.

9. Sump Equipment - Manufacturers and distributors of sumps, pumps, covers, water traps, and drains.

10. Protective Equipment - Suppliers of respirators, safety glasses, hearing protectors, and protective clothing.

11. Miscellaneous - Manufacturers and distributors of a broad range of products used in radon detection and mitigation including alarms, test stands, tamper-proof cages, couplings, vent hoods, louvres, tooling systems, and radon chambers.

1. Radon Measurement Product Manufacturers

ADVANCED RADIATION MONITORING SERVICE, INC. (ARMS)
2219 Prosperity
Portage, MI 49081
(616) 323-9483
Philip B. Noack - President
Mary K. Noack - Secretary
Products & Services:
Manufactures and distributes activated charcoal test monitors and radon laboratory equipment, provides radon mitigation services.
Qualifications:
Nuclear medicine technologist, physicists, engineers on staff.

AIR CHEK, INC.
180 Glenn Bridge Rd. Box 2000
Arden, NC 28704
See Ad Page 188
(704) 684-0893, (800) 257-2366
B.V. Alvarez - President
C.S. Amos - Vice President
J.L. Alvarez - Director of Research
Ned Doyle
Products & Services:
Manufactures, distributes, and provides analysis for radon detection kits nationwide, direct to consumers as well as through dealership network.

ALPHANUCLEAR COMPANY
1125 Derry Rd., E.
Mississauga, ON L5T 1P3
(416) 676-1364
G.M. Denham - President
See Ad Page 197
Products & Services:
Develops, manufactures, and distributes radon/radon progeny measurement instrumentation, provides charcoal canister radon test kits and analysis in Canada.
Qualifications:
Manufacturer of radon detection instrumentation since 1975, supplier to EPA, DOE, and many state agencies.

ATLANTIC NUCLEAR
6 Village Gate Rd.
Canton, MA 02021
(617) 828-9118
See Ad Page 193
John Anderson - Sales Manager
Products & Services:
Manufactures and distributes turnkey radon counting and analysis system, including shielding, detectors, automatic analysis, report generation, etc.

BECKMAN INSTRUMENTS, INC.
2500 Harbor Blvd.
Fullerton, CA 92634
(714) 871-4848
See Ad Page 431
Robert J. Jarnutowski - Marketing
Sheila Brann - Communications
Products & Services:
Manufactures and distributes laboratory instrumentation and supplies, radon detection systems, including samplers and software for report generation.

BICRON CORPORATION
12345 Kinsman Rd.
Newbury, OH 44065
(216) 564-8000

J.G. Bellian - Vice President
R.D. Terry - Sales Manager
J. Price - Inside Sales
Dan Love
Products & Services:
Manufactures scintillation detectors and radiation detection instruments, distributes Thomson and Nielsen Radon Working Level Meter.

CANBERRA INDUSTRIES, INC.
1 State St.
Meriden, CT 06450
(203) 238-2351
See Ad Page 186
Tom Perkins - Marketing Specialist
Michela Hirnak - Marketing Specialist
Bruce M. Gillespie
Products & Services:
Manufactures and distributes turnkey systems for radon detection using the charcoal canister method.
Qualifications:
EPA RMP.

CHEM-SAFE LABORATORIES
North 131 Grand Ave. P.O. Box 546
Pullman, WA 99163-0546
(509) 334-0922
Cliff Cooper - General Manager
See Ad Page 21
Products & Services:
Manufactures and distributes charcoal canisters and alpha track detectors for the indoor environment, operates a full service laboratory.
Qualifications:
EPA RMP, Institute for Environmental Sciences.

DOSIMETER CORPORATION
Box 897
Fayetteville, GA 30214
(404) 461-5639
Robert Rich - Technical Sales
Products & Services:
Manufactures and distributes instrumentation for radiation monitoring, including radon.
Qualifications:
NRA.

DOUGLAS MARTIN AND ASSOCIATES, INC./DMA ANALYSIS GROUP
520 Chestnut St.
Emmaus, PA 18049
(215) 965-2914
See Ad Page 199
JoAnne D. Martin - Senior Scientist
Ken Smith - Health and Safety Services
Richard Martin - President
Products & Services:
Nuclear analysis laboratory provides radon detection services, manufactures and provides gamma spectroscopy analysis of four inch EPA-specification canisters, water analysis.
Qualifications:
AARST, EPA RMP, licensed in NJ, PA, interim/voluntary paticipant in NJ programs.

EBERLINE INSTRUMENT CORPORATION
P.O. Box 2108
Santa Fe, NM 87504-2108
See Ad Inside Back Cover
(505) 471-3232
Eric L. Geiger - Marketing Manager
Wayne Gemmill - Sales Manager
Michael D. Shepherd - Sales Manager
Products & Services:
Manufactures and distributes measuring instruments for radon concentrations and working levels.

EDA INSTRUMENTS, INC.
9200 East Mineral Ave. Suite 370
Englewood, CO 80112
(303) 790-2541
Jerry McJunkin - General Manager
Products & Services:
Manufactures and distributes radon detection instruments including grab radon, grab working level, continuous radon, continuous working level, and gamma scintillometers.

EG & G ORTEC
See Ad Page 200
100 Midland Rd.
Oak Ridge, TN 37831
(615) 482-4411, (800) 251-9750
Ken Knight
Products & Services:
Manufactures and distributes radiation detection, measurement and analysis instruments and systems.

EKS RADTECH, INC.
See Ad Page 201
1000 Herald Sq.
Ashton, PA 19014
(215) 358-9425, (800) 682-5378
John Rooks - General Manager
Products & Services:
Manufactures and distributes humidity independent charcoal canister detectors.
Qualifications:
NJ, PA certified, EPA RMP.

F & J SPECIALTY PRODUCTS, INC.
P.O. Box 660065
Miami Springs, FL 33166
(305) 888-0383, (800) 832-5037
Frank M. Gavila - President
Products & Services:
Manufactures charcoal canisters of all sizes (4", 2 3/4", 2 1/4", 1 3/4"), all materials (plastic, metal), canisters available with diffusion barriers, open face or flowthrough geometry, active radon sampling systems.
Qualifications:
Numerous customers have passed EPA RMP using F & J products.

FEMTO-TECH, INC.
150 C Industry Dr.
Carlisle, OH 45005
(513) 746-4427
E.I. Burgess - President
C.J. Kershner - Vice President
R. Manning - Director of Marketing
Products & Services:
Manufactures and distributes continuous radon monitors.
Qualifications:
ARA.

GALLIPOT, INC.
2401 Pilot Knob Rd.
St. Paul, MN 55120
(612) 681-9517, (800) 423-6967
Michael J. Jones - President
Products & Services:
Manufactures and distributes charcoal canisters, distributes alpha track detectors, provides radon detection services.

GAMMADATA/MATTEKNIK AB
Klostergatan 10 Box 1815
Uppsala S-75148, Sweden
46-18155870
Leif Glantz
Products & Services:
Manufactures complete radioactivity measurement system.

HONEYWELL, INC./RESIDENTIAL CONTROLS DIVISION
MN-10-2522 1985 Douglas Dr., N
Golden Valley, MN 55422
(612) 542-3356
Mike Schell - Radon Manager
See Ad Page 191
Brian C. Krafthefer
Products & Services:
Manufactures and provides distributorships for Honeywell Home and Professional Radon Monitors, microprocessor based monitors providing a direct reading of ambient radon gas.

ILMASTI ELECTRONICS, LTD.
P.B. 50
Vantaa SF-01711, Finland
011-358-0-840711
Products & Services:
Manufactures IlmaRadon meter providing instant measuring and continuous monitoring function.

Available March 1989

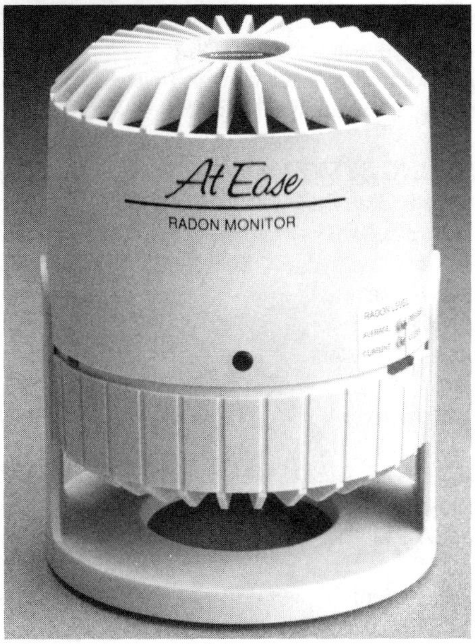

Get "On-The-Spot" Radon Measurements.

Honeywell Introduces New Continuous Radon Monitors.

The Professional.

Useful for researchers, mitigators, home inspectors, and professional radon testers wanting quick, accurate results in premitigation and postmitigation data collection.

Honeywell's new *Professional* Radon Monitor gives you precise radon gas monitoring capability. And does it at a price/performance point that used to cost thousands more.

This highly accurate, yet low cost, monitor was originally developed with assistance from the U. S. Environmental Protection Agency (EPA).

When you want to measure radon gas concentrations —and you want results in hours, and not days later after lab tests — you need the Honeywell Professional Radon Monitor.

Digital readout for instant indication of current and average radon levels.
Continuously monitors radon gas concentrations. Measures actual Alpha decay from radon/radon progeny.
Switch selectable monitoring period. Choose 4, 8, 12 or 24-hour periods. Memory stores up to 96 reading periods.
Optional printer. For hardcopy reports in tabular or graph form. Output either hourly or daily readings.
Tamper-sensing electronics indicate if radon monitor was moved or if power interruption occurred during testing.
Self-diagnostic routines assure unit is in proper working order.
Battery backup system.
U.S. EPA evaluated and accepted.

The *At Ease*®

Gives homeowners accurate radon/radon progeny readings at the push of a button. Useful for homeowners wanting to monitor radon levels on an ongoing basis.

Honeywell's new *At Ease*® Home Radon Monitor gives homeowners peace of mind with its precise radon/radon progeny monitoring capability. And does it at a price that won't break the family budget.

Digital readout for instant indication of current and average radon levels.
Continuously monitors radon gas concentrations. Measures actual Alpha decay from radon/radon progeny.
Displays current and average radon level information. Readings of *current* radon gas concentration, or *average* (all readings to date) level are always available at the push of a button.
Self-diagnostic routines assure unit is in proper working order.
U.S. EPA evaluated and accepted.

For further information on Honeywell Radon Products call 1-800-328-5111, ext. 7102. Or Write: Inquiries, Honeywell Plaza, Minneapolis, MN 55408.

Honeywell

Helping you control your world

Qualifications:
Proven reliability through test by UNC, University Laboratories, and extensive field tests in Finland, accepted in EPA RMP.

INK FILTERING SYSTEMS
3402 Habersham Club Dr.
Cumming, GA 30130
(404) 889-1061
A.T. Cole - President
Products & Services:
Manufactures 4" canisters.
Qualifications:
Canisters used by EPA.

INTERNATIONAL MEDCOM
7497 Kennedy Rd.
Sebastopol, CA 95472
(707) 823-0336
Dan Sythe - President
Joe Jaffe - Chairman
Products & Services:
Manufactures digital geiger counter used by radon testing personnel when in high radon concentration areas, research centering on other radon detection equipment.

INTERNATIONAL RADON TECHNOLOGIES
2417 Regency Rd.
Lexington, KY 40503
(606) 278-0104, (800) 537-5066
Ed Davis - President
William G. Buchman - Inventor
Products & Services:
Distributes radon-related products including a water reducible, two-part polyamide epoxy coating to block radon passage, manufactures continuous radon progeny monitor and electret cylindrical airborne radon monitor, operates a primary laboratory.

JRW INDUSTRIES, INC.
2312 North 2nd St.
Harrisburg, PA 17110
(717) 236-1418
See Ad Page 236
James R. Williams - President
Kevin J. Williams - General Manager
Valeria Hollern - Secretary
Products & Services:
Provides radon detection services, manufactures and distributes radon test kits, charcoal canisters, test stands, warning signs, certification kits, and other accessories.

Qualifications:
EPA RMP, PA DER, AARST.

KEY TECHNOLOGY, INC.
P.O. Box 562
Jonestown, PA 17038
(717) 274-8310, (800) 523-4964
See Ad Inside Front Cover
Claude Wiblin - President
Robert Garland - Marketing Director
Robby Marquette - Secretary/Treasurer
Edward Suter - Training Director
Products & Services:
Manufactures and distributes a complete line of radon detection products and distributes radon mitigation products.
Qualifications:
EPA RMP, PA DER testing and mitigation courses.

LUDLUM MEASUREMENTS, INC.
501 Oak St. P.O. Box 810
Sweetwater, TX 79556
(915) 235-5494
Dwayne Stevens - Sales Manager
Bill Norred
Chris Maxwell

Products & Services:
Manufactures radiation detection equipment.

NITON CORPORATION
P.O. Box 9153
Lexington, MA 02173
(617) 229-4929, (800) 451-6467
See Ad Page 202
Lee Grodzins - President
Ethel Romm - Vice President
Diana B. Doran - Operations Manager
Ann McGuineas - Marketing
Products & Services:
Manufactures and distributes radon detection equipment including ultra-sensitive charcoal canisters and liquid scintillation counting method licensed by Packard Instrument Pico-Rad Sysyem.

NITRONICS CORPORATION
P.O. Box 9153
Lexington, MA 02173
(617) 229-4929, (800) 451-6467
See Ad Page 202
Lee Grodzins - President
Ethel Romm - Vice President
Diana B. Doran - Operations Manager
Ann McGuineas - Marketing
Products & Services:
Manufactures and distributes digital electronic survey meters, monitors, and screen testers for radon detection.

NORLAND CORPORATION
Norland Dr.
Ft. Atkinson, WI 53538
(414) 563-8456, (800) 333-8456
See Ad Page 193
Eric Lysne - Nuclear Products
C.J. Havlik - Nuclear Products
Products & Services:
Manufactures and distributes turnkey radon counting and analysis system, including shielding, detectors, automatic analysis, report generation, etc.

NUCLEAR TECHNOLOGY, INC./RADON CHEK
20 H. South State St.
Westerville, OH 43081
(614) 899-2435
James M. Copp - President
Daniel L. Bowman - Secretary
Products & Services:
Manufactures and distributes charcoal adsorption canisters.
Qualifications:
ARA, EPA RMP.

NUCLEUS, THE
P.O. Box 2561
Oak Ridge, TN 37830-2561
(615) 482-4041
See Ad Page 192
Larry Webb - Marketing Manager
Products & Services:
Manufactures and distributes a complete radon monitoring system for charcoal canisters including data analysis software and calibration sources. Non-commercial laboratory for research purposes or state agencies' assistance only.

PACKARD INSTRUMENT COMPANY
2200 Warrenville Rd.
Downers Grove, IL 60515
(312) 969-6000, (800) 323-1891
See Ad Page 205
Jim Floeckher - Product Manager
Products & Services:
Manufactures and distributes the Pico-Rad complete radon system for air and water, including liquid scintillation counters, software, canisters, etc. Non-commercial laboratory for research purposes or state agencies assistance only.
Qualifications:
EPA RMP.

POLLENEX/ASSOCIATED MILLS, INC.
165 North Canal St.
Chicago, IL 60606
(312) 454-5400
Richard J. Stern - President
John Jiambalvo - Director of Research
Ira Pruzan - Vice President of Marketing
Doug Moss - Vice President for Sales
Products & Services:
Pollenex markets charcoal canister radon test kits under the names Pollenex Radon Analyst and Pollenex Radon Alarm.
Qualifications:
Leading manufacturer of consumer health care products.

PYLON ELECTRONIC DEVELOPMENT COMPANY, LTD.
147 Colonnade Rd.
Nepean, ON K2E 7L9
(613) 226-7920
See Ad Page 209
Dave Hanneson - Sales Manager
Chuck Palmer - Marketing
Aurel Lebel
Products & Services:
Manufactures and distributes radon gas monitors, working level monitors, radon in water analyzers, radon gas calibration sources, scintillation cells.
Qualifications:
EPA RMP.

QUANTUM TECHNOLOGY, INC.
830 Franklin Ct.
Marietta, GA 30067
(404) 429-8324
Mahmoud Ghavi - President
Products & Services:

Manufactures and distributes RadonOne-PC, a complete radon monitoring system including electronics, software, pc package.

R.A.D. SERVICE AND INSTRUMENTS, LTD.
50 Silver Star Blvd. Unit 208
Scarborough, ON M1V 3L3
(416) 298-9200
Hsiang Lin Pai - President
See Ad Page 206
Products & Services:
Manufactures and distributes radon detection products including an active indoor working level surveymeter (patent pending).
Qualifications:
Certified by EPA and NEA in Europe.

RAD ELEC, INC.
5330 J Spectrum Dr.
Frederick, MD 21701
(301) 694-0011
Robert W. Ramsey - President
Paul Kotrappa - Vice President
John C. Dempsey - Vice President
See Ad Page 194
Products & Services:
Manufactures and distributes electret passive environmental radon monitor (E-Perm), performs research and development on new radiometric monitoring devices.
Qualifications:
Inventors of the Electret Perm Measurement System recognized by EPA.

RADIATION SAFETY SERVICES, INC.
6312 West Oakton St.
Morton Grove, IL 60053
(312) 965-1999
Eli Port - President
Products & Services:
Manufactures alpha track monitors.
Qualifications:
EPA RMP, certified health physicists on staff.

RADON DETECTION AND REDUCTION LABORATORY/DIVISION OF FORMACHEM
10834 Ambassador
St. Louis, MO 63132
(314) 872-9350
Harold Susman - President
Donald Pearline - Biochemist
Products & Services:
Manufactures and distributes the Seal Safe Radon Gas Detector and alpha track detectors, distributes Enkadrain radon mitigation system for new home construction.

RADON ENVIRONMENTAL MONITORING, INC.
Box 5305
Evanston, IL 60204
(312) 328-6628
Robert Pollock - Vice President
E.V. Benton
See Ad Page 208
Products & Services:
Manufactures and distributes alpha track monitors, primary analysis laboratory.

RADON PROJECT, THE
P.O. Box 90069
Pittsburgh, PA 15224
(412) 687-2886
Bernard L. Cohen - President
Anna F. Cohen - Vice President
Frederick B. Cohen - Director
Products & Services:
Manufactures and distributes diffusion barrier charcoal adsorption canisters
Qualifications:
EPA RMP.

RADON SURVEY SYSTEMS, INC.
P.O. Box 22288
Cleveland, OH 44122
(216) 464-7997
Darioush T. Ghahremani - President
Products & Services:
Provides radon detection and mitigation services for residential and commercial markets, site surveys for undeveloped land, research and development of radon detection technologies, detectors, and materials, new development on radon mitigation options for various projects.
Qualifications:
AARST, NRA, several national scientific societies.

RADON TESTING CORPORATION OF AMERICA
12 West Main St.
Elmsford, NY 10523
(914) 347-5010, (800) 457-2366
Nancy Bredhoff - Vice President
Kurt Dorfi - Sales Manager
Products & Services:
Manufactures and distributes charcoal canister radon test kits, distributes Pro Scan, a product designed to aid detection firms to prevent test tampering.
Qualifications:
EPA RMP, approved by NY DOH.

RADON/SOILS AND WATER TESTING SERVICE
7080 Price-Hilliard Rd.
Plain City, OH 43064
(614) 873-8821
John Steichen - President
Products & Services:
Manufactures and distributes charcoal canisters, distributes E-Perm detectors, provides radon detection services to the real estate industry.
Qualifications:
AARST.

RAMSES II, LTD.
2355 East Dartmouth Ave.
Denver, CO 80210
(303) 758-5209
Robert L. Clark - Vice President
George A. Michael - President
Products & Services:
Manufactures and distributes charcoal canisters, electronic sump well cover kits, electronic PVC ventilation fans, and rotary air-to-air heat exchangers, provides radon mitigation services.
Qualifications:
Licensed professional engineer on staff.

SUNN CORPORATION
Box M, Dept. 213 3520 Forbes Ave.
Pittsburgh, PA 15213
(412) 828-9100
Nick Gromicko - Chief Executive Officer
Bob Dunkle
Mike Russell
Products & Services:
Provides radon mitigation services, manufactures and distributes charcoal canisters and tamper-resistant cages.
Qualifications:
BBB, NSBA, several state mitigation lists.

TCS INDUSTRIES, INC.
3805 Paxton St.
Harrisburg, PA 17111
(717) 558-8206
Carl H. Distenfeld - President
Jacquelyn Distenfeld - General Manager
Eric B. Distenfeld - Sales Manager
Products & Services:
Manufactures and distributes charcoal canisters, operates analysis laboratory.
Qualifications:
EPA RMP, certified health physicist, HPS, ANS.

TECH/OPS LANDAUER, INC.
2 Science Rd.
Glenwood, IL 60425-1586
(312) 755-7000
See Ad Back Cover
Robert Wheeler - Radon Product
C. Yoder
Products & Services:
Parent company of Terradex Corporation, world leader in radon monitoring services for indoor air, water, and soil.

TELEDYNE ISOTOPES
50 Van Buren Ave.
Westwood, NJ 07675
(201) 664-7070
D.F. Schutz - President
J.D. Martin - Vice President
D. McKillop - Radon Services Coordinator
Craig Osterhaus - Radon Division
Products & Services:
Manufactures and distributes the Home-Safe Radon Test Kit, performs analysis of results within 24 hours, provides radon detection services.
Qualifications:
NRC licenses, technical and scientific personnel, state licenses.

TEQ CORPORATION
P.O. Box 570
Wilkes-Barre, PA 18703-0570
(717) 825-3939, (800) 877-2366
See Ad Page 100
Russell Hendershot - President
Ed Dructor - Vice President
Brian Redmond - Research Scientist
Products & Services:
Provides radon detection and mitigation services for commercial and residential markets, manufactures and distributes charcoal canisters, maintains calibrated radon chamber, offers radon detection and mitigation training courses.
Qualifications:
PA DER certification.

TERRADEX CORPORATION/SUBSIDARY OF TECH/OPS LANDAUER, INC.
2 Science Rd.
Glenwood, IL 60425
(312) 755-7911, (800) 528-8327
See Ad Back Cover
Sue Morrow - Customer Service
R. Craig Yoder
Products & Services:
Manufactures and distributes line of radon monitoring products to industry, academic institutions, government agencies, consumers, etc.
Qualifications:
Over 35 years of experience in radiation monitoring services from both occupational and environmental sources, works with ALA, EPA, USAF, universities, national labs, etc.

TERRADON EXPLORATION TECHNOLOGIES, INC.
3931 White Oak Tr. P.O. Box 22288
Cleveland, OH 44122
(216) 464-2002
Darioush T. Ghahremani - President
Products & Services:
Provides radon detection and mitigation services, research and development in radon detection, measurement and abatement technology, development of new alpha track and charcoal canisters for industrial and private use, soil gas radon detection and mapping.
Qualifications:
AARST, NRA, AAPG, OAS, ACS, OG&G, AAGE, other scientific associations.

THERMO ENVIRONMENTAL CORPORATION
101 1st Ave. P.O. Box 9046
Waltham, MA 02254
(617) 622-1103, (505) 345-9931
Jeff Brown - Health Physics
Products & Services:
Manufactures and distributes RAD-SCAN radon screening kits, provides radon detection and mitigation services to homeowners, developers, real estate industry, etc., parent company of TMA Eberline.

THOMSON AND NIELSEN ELECTRONICS, LTD.
4019 Carling Ave. Suite 303
Kanata, ON K2K 2A3
(613) 592-3019
Ian Thomson - President
Andrew Singmin - Marketing Director
Line Brabant - Inside Sales

See Ad Page 187

Products & Services:
Manufactures and distributes radon detection instruments, provides radon detection services, conducts research in radiation detection.
Qualifications:
AARST, EPA RMP, HPS.

THRESHOLD TECHNICAL PRODUCTS
7225 Edington Dr.
Cincinnati, OH 45249
(800) 458-4931
John Grether - President
Ken Marguet - Vice President
Products & Services:
Manufactures and distributes Survivor II continuous radon monitor directly to consumer.

TOM SCURRY ASSOCIATES
1976 Linden Dr.
Boulder, CO 80302
(303) 444-4802
Tom Scurry - President
Robert Droullard - Vice President

See Ad Page 210

Products & Services:
Maintains a radon chamber to test and calibrate detection instruments, manufactures and distributes continuous and grab working level monitors, provides radon detection services.

VICTOREEN, INC.
6000 Cochran Rd.
Cleveland, OH 44139
(216) 248-9300
Dan Davis - Product Manager
Products & Services:
Manufactures radiation measurement devices.

WILLIAM B. JOHNSON AND ASSOCIATES, INC.
P.O. Box 98 Research Park
Montville, NJ 07045
(201) 334-9222
Rudy Schober - Vice President
Products & Services:
Manufactures and distributes an alpha scintillation probe connected to a digital scaler/timer with its own high voltage source.

2. Radon Measurement Product Distributors

A.J. ABRAMS COMPANY
P.O. Box 5171
Westport, CT 06881
(203) 226-3010
Arthur Abrams - President
Jetta Reynolds - Office Manager
Products & Services:
Distributes radon monitoring instruments.

ADCO SERVICES, INC.
17650 DuVan Dr.
Tinley Park, IL 60477
(312) 429-1660
Stan Huber - General Manager
Products & Services:
Distributes "Radon Snooper" charcoal canister detection kit.
Qualifications:
NRA.

ADVANCED RADIATION MONITORING SERVICE, INC. (ARMS)
2219 Prosperity
Portage, MI 49081
(616) 323-9483
Philip B. Noack - President
Mary K. Noack - Secretary
Products & Services:
Manufactures and distributes activated charcoal test monitors and radon laboratory equipment, provides radon mitigation services.
Qualifications:
Nuclear medicine technologist, physicists, engineers on staff.

AFFILIATED BUILDING INSPECTORS, INC.
P.O. Box 15770
Chevy Chase, MD 20815
(301) 986-8866
Larry Wasson - President
Products & Services:
Provides radon detection services using charcoal canister, alpha track, E-Perm and continuous radon monitor methods, distributes E-Perms, charcoal canisters, and alpha track detectors.
Qualifications:
EPA RMP.

AIR CHEK, INC.
180 Glenn Bridge Rd. Box 2000
Arden, NC 28704

See Ad Page 188

(704) 684-0893, (800) 257-2366
B.V. Alvarez - President
C.S. Amos - Vice President
J.L. Alvarez - Director of Research
Ned Doyle
Products & Services:
Manufactures, distributes, and provides analysis for radon detection kits nationwide, direct to consumers as well as through dealership network.

AIR QUALITY ANALYSTS, INC.
5400 Eisenhower Ave.
Alexandria, VA 22304
(703) 823-7567
Bruce C. Phillips - President
Products & Services:
Distributes residential radon detection devices.

AIR SCIENCES, INC.
508 North Truman Blvd. P.O. Box 320
Crystal City, MO 63019
(314) 933-3915, (800) 242-8886
Kerry Van House - Chief Executive Officer
Ralph Overman - Vice President
Products & Services:
Provides radon detection and mitigation services, distributes charcoal canister home detection kits.
Qualifications:
Staff includes nuclear scientist with over 40 years of radiation work and physicist skilled in radiation measurement.

ALPHANUCLEAR COMPANY
1125 Derry Rd., E.
Mississauga, ON L5T 1P3
(416) 676-1364
G.M. Denham - President

See Ad Page 197

Products & Services:
Develops, manufactures, and distributes radon/radon progeny measurement instrumentation, provides charcoal canister radon test kits and analysis in Canada.
Qualifications:
Manufacturer of radon detection instrumentation since 1975, supplier to EPA, DOE, and many state agencies.

AMERICAN PEST MANAGEMENT
3929 Martin Way Suite D
Olympia, WA 98506
(206) 491-1809
David Scott - President
Products & Services:
Provides radon detection services for relocation companies and distributes charcoal canisters and alpha track detectors.

APPLIED HEALTH PHYSICS, INC.
2986 Industrial Blvd.
Bethel Park, PA 15102
(412) 563-2242, (518) 477-7974
R.G. Gallagher - President
John Douglas - Instrument Manager
John Harshman - Radon Specialist
Products & Services:
Professional radiation safety firm distributes charcoal canisters, provides instrument calibration, repair, and leasing.
Qualifications:
EPA RMP, twenty-five years of experience, NRC and state licensed, all services by board certified registered professional engineers.

APPLIED RADON, INC.
52205 Warehime Rd.
Cumberland, OH 43732-9804
(614) 638-2078, (800) 638-2078
Frank J. Donia - Vice President
Nancy A. Donia - President
Products & Services:
Provides radon detection and mitigation services, distributes At-Ease radon monitors, presents seminars to real estate industry.
Qualifications:
ARA, EPA RMP, health physicist on staff.

APPLIED TECHNICAL SERVICES, INC.
1190 Atlanta Industrial Dr.
Marietta, GA 30066
(404) 423-1400, (800) 451-3405
Donnissa Duvic - Environmental Science
Products & Services:
Distributes radon test kits including canister, instructions, pre-paid return postage, analysis and report, retail packaging available.

ATLANTIC NUCLEAR
6 Village Gate Rd.
Canton, MA 02021
(617) 828-9118
See Ad Page 193
John Anderson - Sales Manager
Products & Services:
Manufactures and distributes turnkey radon counting and analysis system, including shielding, detectors, automatic analysis, report generation, etc.

BARRINGER
15000 West 6th Ave. Suite 300
Golden, CO 80401
(303) 277-1687
B.E. Sabels - Radon Manager
David Lasher
Products & Services:
Distributes charcoal canisters and alpha track detectors, mail order services, mitigation, and consultation services.
Qualifications:
AARST, EPA RMP, general environmental laboratory listed with EPA, Health Department, etc.

BCA LABORATORIES, INC.
1102 South Main St.
Bloomington, IL 61701
(309) 828-7772
Kurt Benckendorf - Lab Manager
Dave Martin - Business Manager
Don Curry
Products & Services:
Provides radon detection services by grab sampling, charcoal canisters and other methods, distributes charcoal canisters and alpha track detectors to the public.
Qualifications:
EPA RMP.

BECKMAN INSTRUMENTS, INC.
2500 Harbor Blvd.
Fullerton, CA 92634
(714) 871-4848
See Ad Page 431
Robert J. Jarnutowski - Marketing
Sheila Brann - Communications
Products & Services:
Manufactures and distributes laboratory instrumentation and supplies, radon detection systems, including samplers and software for report generation.

BICRON CORPORATION
12345 Kinsman Rd.
Newbury, OH 44065
(216) 564-8000
J.G. Bellian - Vice President
R.D. Terry - Sales Manager
J. Price - Inside Sales
Dan Love
Products & Services:
Manufactures scintillation detectors and radiation detection instruments, distributes Thomson and Nielsen Radon Working Level Meter.

BOULDER INSPECTION COMPANY
P.O. Box 3369
Boulder, CO 80307
(303) 444-6657
Ian Rosebrook - Engineer
John Gilbert - Engineer
Don Drake - Engineer
Products & Services:
Provides radon detection services, distributes radon detection products to the public.
Qualifications:
EPA courses in radon detection.

CANBERRA INDUSTRIES, INC.
1 State St.
Meriden, CT 06450
(203) 238-2351
See Ad Page 186
Tom Perkins - Marketing Specialist
Michela Hirnak - Marketing Specialist
Bruce M. Gillespie
Products & Services:
Manufactures and distributes turnkey systems for radon detection using the charcoal canister method.
Qualifications:
EPA RMP.

CAPITAL MATERIALS TESTING, INC.
P.O. Box 81 5 Hemlock St.
Latham, NY 12110
(518) 783-3150, (800) 541-4109
James Swartz - Regional Manager
Bob Field - Technician
Products & Services:
Provides radon detection and mitigation services, distributes radon test kits.
Qualifications:
NY state certified lab, EPA RMP.

CCI-RADON AND HOME INSPECTION/RADON DETECTION
P.O. Box 1045
Malvern, PA 19355
(215) 889-1554, (215) 363-9015
Charles E. Weber - President
Bruce Jordan - Vice President
Norman Warner - Vice President
Products & Services:
Provides radon detection and mitigation services for home owners and buyers, distributes charcoal canisters and alpha track detectors to retail outlets.
Qualifications:
AARST, EPA RMP PA DER certification.

CERT WATER SERVICES, INC.
25 Duval Ln.
Plainville, CT 06062
(203) 793-9142
David Gworek - President
Products & Services:
Provides radon detection services using charcoal canister, alpha track , and water testing methods for residential, commercial and industrial structures, distributes continuous radon monitors.
Qualifications:
President is a registered professional engineer.

CHEM-SAFE LABORATORIES
North 131 Grand Ave. P.O. Box 546
Pullman, WA 99163-0546
(509) 334-0922
See Ad Page 21

Cliff Cooper - General Manager
Products & Services:
Manufactures and distributes charcoal canisters and alpha track detectors for the indoor environment, operates a full service laboratory.
Qualifications:
EPA RMP, Institute for Environmental Sciences.

DIAGNOSTICS AND DEVICES, INC.
1201 Mt. Kemble Ave.
Morristown, NJ 07960
(201) 766-7111
J.G. Murnick - President
Products & Services:
Distributes radon test kits for radon in air or water.

DOSIMETER CORPORATION
Box 897
Fayetteville, GA 30214
(404) 461-5639
Robert Rich - Technical Sales
Products & Services:
Manufactures and distributes instrumentation for radiation monitoring, including radon.
Qualifications:
NRA.

DOUGLAS MARTIN AND ASSOCIATES, INC./DMA ANALYSIS GROUP
See Ad Page 199
520 Chestnut St.
Emmaus, PA 18049
(215) 965-2914
JoAnne D. Martin - Senior Scientist
Ken Smith - Health and Safety Services
Richard Martin - President
Products & Services:
Nuclear analysis laboratory provides radon detection services, manufactures and provides gamma spectroscopy analysis of four inch EPA-specification canisters, water analysis.
Qualifications:
AARST, EPA RMP, licensed in NJ, PA, interim/voluntary paticipant in NJ programs.

EBERLINE INSTRUMENT CORPORATION
See Ad Inside Back Cover
P.O. Box 2108
Santa Fe, NM 87504-2108
(505) 471-3232
Eric L. Geiger - Marketing Manager
Wayne Gemmill - Sales Manager
Michael D. Shepherd - Sales Manager
Products & Services:
Manufactures and distributes measuring instruments for radon concentrations and working levels.

ECOLOGICAL INTERACTIONS, INC.
96 East Main St.
Rockaway, NJ 07866
(201) 625-7700, (800) 543-2116
F. Michael Fitzpatrick - Vice President
Bernadette E. Secallus - Director
Products & Services:
Provides radon detection and mitigation services, distributes radon test kits.
Qualifications:
AARST board of directors, EPA RMP, NJ DEP approved testing and mitigation course.

EDA INSTRUMENTS, INC.
9200 East Mineral Ave. Suite 370
Englewood, CO 80112
(303) 790-2541
Jerry McJunkin - General Manager
Products & Services:
Manufactures and distributes radon detection instruments including grab radon, grab working level, continuous radon, continuous working level, and gamma scintillometers.

EDINBORO SCIENTIFIC COMPANY
P.O. Box 589 Wetsell Ridge Rd.
Edinboro, PA 16412
(814) 734-4648
Thomas A. Walkiewicz - Vice President
Products & Services:
Provides radon detection services, laboratory analysis, consultation in experimental nuclear/radiation physics, distributes charcoal canisters.
Qualifications:
EPA RMP.

EG & G ORTEC
See Ad Page 200
100 Midland Rd.
Oak Ridge, TN 37831
(615) 482-4411, (800) 251-9750
Ken Knight
Products & Services:
Manufactures and distributes radiation detection, measurement and analysis instruments and systems.

EKS RADTECH, INC.
See Ad Page 201
1000 Herald Sq.
Ashton, PA 19014
(215) 358-9425, (800) 682-5378
John Rooks - General Manager
Products & Services:
Manufactures and distributes humidity independent charcoal canister detectors.
Qualifications:
NJ, PA certified, EPA RMP.

EMPAC, INC.
25 Science Pk.
New Haven, CT 06511

(203) 786-5420, (800) 872-3662
Albert W. Doolittle, Jr.
Products & Services:
Provides radon detection services, distributes radon detection products to commercial, municipal, and residential markets.
Qualifications:
Listed with CT DOH-Toxic Hazards.

EMPIRE INSPECTION, INC.
1020 Springfield Ave.
Mountainside, NJ 07092
(201) 654-3111
Thomas J. Sanford - President
Products & Services:
Provides radon detection services, 24-hour test for radon progeny, radon gas sampling, distributes charcoal canisters.

ENERGY RESOURCE CENTER
P.O. Box 267
Colorado Springs, CO 80901
(719) 636-2447
Lee Salmon - Radon Specialist
Products & Services:
Provides radon detection and mitigation services, project management, independent evaluation, presents seminars, distributes charcoal canisters and alpha tracks to homeowners.
Qualifications:
Graduate, Colorado School of Mines course on radon and indoor environment, 15 years of experience in energy management.

ENVIRONMENTAL CHEMISTRY, INC.
1233 Ogden Suite 504
Denver, CO 80218
(303) 831-0437
John Taylor - President
Susan Romeo - Vice President
Edward Shirley - Secretary
Products & Services:
Provides radon detection services, distributes radon test kits.
Qualifications:
EPA RMP.

ENVIRONMENTAL DETECTION AGENCY, INC.
P.O. Box 4854
Austin, TX 78765
(512) 322-9013
Lorrie Brown - President
Products & Services:
Provides radon detection services for home inspectors, realtors, etc., distributes radon test kits, services in conjunction with the Texas A&M University radon program.

ENVIRONMENTAL DETECTION ASSOCIATES, INC.
P.O. Box 1006 21 Middlemount Rd.
Pisgah Forest, NC 28768-1006
(704) 877-4300
Wolfgang Selle - President
Harry Hill - Vice President
Products & Services:
Provides radon detection services, distributes radon detection devices, consultation services to architects. Non-commercial laboratory for research purposes or state agencies' assistance only.
Qualifications:
EPA RMP.

ENVIRONMENTAL PRODUCTS ALLIANCE, LTD.
218 North 4th St., Suite 21 P.O. Box 2156
Bismarck, ND 58502
(701) 258-5050
Dan Schmaltz - Director of Marketing
Products & Services:
Distributes charcoal canister radon test kits.

ENVIRONMENTAL SCIENCE LABORATORY, INC.
26 Pearl St. Suite 110
Bellingham, MA 02019
(508) 966-4344, (800) 343-0707
Bob Juliano - President
Products & Services:
Provides radon detection services using charcoal canister method, distributes radon test kits.
Qualifications:
EPA RMP.

ENVIRONMENTAL SOLUTIONS, INC.
P.O. Box 11323
St. Louis, MO 63105
(314) 863-9801
William M. Vaughan - Senior Scientist
Trudy Y. Smith - Technical Assistant
Products & Services:
Provides consultation services on air quality and air toxins, distributes radon test kits.
Qualifications:
EPA RMP.

FEDERAL RADON, INC.
206 Third St. P.O. Box 374
Cameron, MO 64429
(816) 632-1318
R.M. Yetter - President
Products & Services:
Provides radon detection services, distributes charcoal canisters to public.

FEMTO-TECH, INC.
150 C Industry Dr.

Carlisle, OH 45005
(513) 746-4427
E.I. Burgess - President
C.J. Kershner - Vice President
R. Manning - Director of Marketing
Products & Services:
Manufactures and distributes continuous radon monitors.
Qualifications:
ARA

GALLIPOT, INC.
2401 Pilot Knob Rd.
St. Paul, MN 55120
(612) 681-9517, (800) 423-6967
Michael J. Jones - President
Products & Services:
Manufactures and distributes charcoal canisters, distributes alpha track detectors, provides radon detection services

GEOPHYSICAL INSTRUMENT AND SUPPLY COMPANY, INC./GISCO
4665 Joliet St.
Denver, CO 80239
(303) 371-1940
Heinz Siegel - General Manager
Products & Services:
Distributes full product line of radon detection instruments, scintillation counters, and survey meters.

GEOTHERMAL ENERGY COMPANY, LTD.
P.O. Box 59
Wolfeboro, NH 03894
(603) 569-2519
Arthur L. Robinson - President
Products & Services:
Distributes radon detection and mitigation products and supplies.
Qualifications:
ARA, NAHB, EPA and NAHB training.

GRANITE STATE ANALYTICAL, INC.
16 Manning St. Unit 107
Derry, NH 03038
(603) 432-3044
Philip West - President
Products & Services:
Distributor of Niton radon monitors.

GREAT TRACERS
3 Schoenbeck Rd.
Prospect Heights, IL 60070-1435
(312) 255-0436
David Blustein - Industrial Hygienist
Josh Blustein - President
Teryln Blustein - Quality Assurance
Products & Services:
Provides radon detection services using charcoal canister method, distributes charcoal canisters.
Qualifications:
EPA RMP.

HENRY INTERNATIONAL, INC.
14742 Plaza Dr.
Tustin, CA 92680
(714) 730-8451
See Ad Page 193
Y. David LaTouche - President
Products & Services:
Distributes Alpha Nuclear and Norland turnkey radon monitor systems, including shielding, detectors, automatic analysis, report generation, etc.

HOMES PLUS RADON SERVICES, INC.
490 Main St. P.O. Box 654
Norwich, CT 06360
(203) 887-5128
Scott Harrington - President
Tony Benoit - Scientist
Martin Grisham - Marketing
Susan Axline - Assistant Manager
Products & Services:
Provides radon detection services, distributes charcoal canister test kits.
Qualifications:
AARST.

INFILTEC/FACTORY
108 South Delphine Ave. P.O. Box 1125
Waynesboro, VA 22980
(703) 943-2776
See Ad Page 214
David Brubaker - Factory Manager
Frankie Patterson - Warehouse Manager
Arthur Saum - Research Manager
Products & Services:
Manufactures and distributes a full line of radon detection and mitigation products including fans, couplings, pressure gauges, alarms, sumps and drain covers, radon test kits, continuous radon monitors, caulks and sealants, manometers, blower doors, etc.
Qualifications:
EPA RMP.

JRW INDUSTRIES, INC.
2312 North 2nd St.
Harrisburg, PA 17110
(717) 236-1418
See Ad Page 236
James R. Williams - President
Kevin J. Williams - General Manager
Valeria Hollern - Secretary
Products & Services:
Provides radon detection services, manufactures and distributes radon test kits, charcoal canisters, test stands, warning signs, certification kits, and other accessories.
Qualifications:
EPA RMP, PA DER, AARST.

KEY TECHNOLOGY, INC. See Ad Inside Front Cover
P.O. Box 562
Jonestown, PA 17038
(717) 274-8310, (800) 523-4964
Claude Wiblin - President
Robert Garland - Marketing Director
Robby Marquette - Secretary/Treasurer
Edward Suter - Training Director
Products & Services:
Manufactures and distributes a complete line of radon detection products and distributes radon mitigation products.
Qualifications:
EPA RMP, PA DER testing and mitigation courses.

LACO
P.O. Box 666
Chesterland, OH 44026
(216) 729-3034
Carl Lantz - President
Products & Services:
Distributes Pylon, Sun Nuclear and Norland radon detection and analysis equipment.

MEDLAB ENVIRONMENTAL TESTING, INC.
P.O. Box 2045
Wilmington, DE 19899
(302) 655-5227, (800) 633-5221
James Boylan - Vice President
Sev Sinasian - Laboratory Director
Products & Services:
Provides radon detection services, distributes charcoal canisters to public.
Qualifications:
NRA.

METRO RADON MEASUREMENT
1565 Cliff Rd. Suite 3-101
Eagan, MN 55122
(612) 681-1829
Steven Joyce - President
Sandra Joyce - Owner
Products & Services:
Distributes charcoal canisters and alpha track detectors.
Qualifications:
EPA RMP.

MICROBAC LABORATORIES, INC./ERIE TESTING DIVISION-MAIN OFFICE See Ad Page 55
2411 West 26th St.
Erie, PA 16506
(814) 833-4790
James K. Willmot - Operations Manager
Mark Bannister - Project Supervisor
Don Tyler - Radon Analyst
Jeff Porte
Products & Services:
Environmental testing laboratory for radon detection companies, primary laboratory facility, distributes charcoal canisters with 24 hour results.
Qualifications:
EPA RMP.

NEW ENGLAND RADON, LTD.
373 Main St.
Salem, NH 03079
(603) 893-4260, (800) 637-2366
Daniel Espinal - Vice President
Products & Services:
Provides radon detection and mitigation services, distributes radon test kits, laboratory analysis.
Qualifications:
EPA RMP.

NITON CORPORATION See Ad Page 202
P.O. Box 9153
Lexington, MA 02173
(617) 229-4929, (800) 451-6467
Lee Grodzins - President
Ethel Romm - Vice President
Diana B. Doran - Operations Manager
Ann McGuineas - Marketing
Products & Services:
Manufactures and distributes radon detection equipment including ultra-sensitive charcoal canisters and liquid scintillation counting method licensed by Packard Instrument Pico-Rad Sysyem.

NITRONICS CORPORATION See Ad Page 202
P.O. Box 9153
Lexington, MA 02173
(617) 229-4929, (800) 451-6467
Lee Grodzins - President
Ethel Romm - Vice President
Diana B. Doran - Operations Manager
Ann McGuineas - Marketing
Products & Services:
Manufactures and distributes digital electronic survey meters, monitors, and screen testers for radon detection.

NORLAND CORPORATION See Ad Page 193
Norland Dr.
Ft. Atkinson, WI 53538
(414) 563-8456, (800) 333-8456
Eric Lysne - Nuclear Products
C.J. Havlik - Nuclear Products
Products & Services:
Manufactures and distributes turnkey radon counting and analysis system, including shielding, detectors, automatic analysis, report generation, etc.

NUCLEAR ASSOCIATES/DIVISION OF VICTOREEN
100 Voice Rd.
Carle Place, NY 11514
(516) 741-6360
Marty Ratner - Marketing Director
Al Zirkes - Product Manager
Products & Services:
Distributes Honeywell Professional Radon Monitor and analytical detection equipment.

NUCLEAR TECHNOLOGY, INC./RADON CHEK
20 H. South State St.
Westerville, OH 43081
(614) 899-2435
James M. Copp - President
Daniel L. Bowman - Secretary
Products & Services:
Manufactures and distributes charcoal adsorption canisters.
Qualifications:
ARA, EPA RMP.

NUCLEUS, THE
See Ad Page 192
P.O. Box 2561
Oak Ridge, TN 37830-2561
(615) 482-4041
Larry Webb - Marketing Manager
Products & Services:
Manufactures and distributes a complete radon monitoring system for charcoal canisters including data analysis software and calibration sources. Non-commercial laboratory for research purposes or state agencies' assistance only.

PACKARD INSTRUMENT COMPANY
See Ad Page 205
2200 Warrenville Rd.
Downers Grove, IL 60515
(312) 969-6000, (800) 323-1891
Jim Floeckher - Product Manager
Products & Services:
Manufactures and distributes the Pico-Rad complete radon system for air and water, including liquid scintillation counters, software, canisters, etc. Non-commercial laboratory for research purposes or state agencies assistance only.
Qualifications:
EPA RMP.

PACKARD INSTRUMENT COMPANY
1640 Worcester Rd. Unit 61D
Framingham, MA 01701
(617) 872-5415, (800) 638-4058
Ronald Bedell
Products & Services:
Distributes Pico-Rad complete radon monitoring system.

PYLON ELECTRONIC DEVELOPMENT COMPANY, LTD.
See Ad Page 209
147 Colonnade Rd.
Nepean, ON K2E 7L9
(613) 226-7920
Dave Hanneson - Sales Manager
Chuck Palmer - Marketing
Aurel Lebel
Products & Services:
Manufactures and distributes radon gas monitors, working level monitors, radon in water analyzers, radon gas calibration sources, scintillation cells.
Qualifications:
EPA RMP.

QUANTUM TECHNOLOGY, INC.
830 Franklin Ct.
Marietta, GA 30067
(404) 429-8324
Mahmoud Ghavi - President
Products & Services:
Manufactures and distributes RadonOne-PC, a complete radon monitoring system including electronics, software, pc package.

R.A.D. SERVICE AND INSTRUMENTS, LTD.
See Ad Page 206
50 Silver Star Blvd. Unit 208
Scarborough, ON M1V 3L3
(416) 298-9200
Hsiang Lin Pai - President
Products & Services:
Manufactures and distributes radon detection products including an active indoor working level surveymeter (patent pending).
Qualifications:
Certified by EPA and NEA in Europe.

R.E. WRIGHT ASSOCIATES, INC./RADONIX GROUP
3240 Schoolhouse Rd.
Middletown, PA 17057
(717) 944-5520
Michael Spear - Marketing
Products & Services:
Provides radon detection and mitigation services, distributes charcoal canisters and alpha track radon test kits.
Qualifications:
PA certification pending.

R.F. SIMON COMPANY, INC.
See Ad Page 148
RD 2, Box 481
Barto, PA 19504
(215) 754-6455
Ron Simon - Project Manager
Peter Maniscalco - Diagnostic Specialist
Products & Services:
Provides mitigation design and installation services, mitigation supply warehouse carrying a full line of radon mitigation and detection products including fans, caulks and sealants, couplings, comprehensive radon services provided to private, commercial, institutional, and government agencies, expert testimony.
Qualifications:
EPA RMP, 13 years of custom building experience, instructor for EPA training courses, consultant to Lawrence Berkeley Laboratory, Princeton University, PA DER.

RA-TEK/DIVISION OF SHAMROCK, INC.
P.O. Box 496 800 East Main St.
Marshall, MN 56258
(507) 532-7313
Joan Coquyt - Analyst
Products & Services:
Provides radon detection services to residential and commercial markets using continuous radon monitor method, distributes continuous radon monitors.

RAD ELEC, INC.
See Ad Page 194
5330 J Spectrum Dr.
Frederick, MD 21701
(301) 694-0011
Robert W. Ramsey - President
Paul Kotrappa - Vice President
John C. Dempsey - Vice President
Products & Services:
Manufactures and distributes electret passive environmental radon monitor (E-Perm), performs research and development on new radiometric monitoring devices.
Qualifications:
Inventors of the Electret Perm Measurement System recognized by EPA.

RAD-LAB SERVICES, INC.
1311 Washington Blvd.
Williamsport, PA 17701
(717) 326-2628
Stephen L. Rosenfarb - President
Donald G. Gorasi - Vice President
Jeff C. Wheeland - Secretary
Products & Services:
Wholesale supplier distributes radon detection kits directly to consumers.

RADIATION INSTRUMENT COMPANY (RICO)
See Ad Page 203
9165 Helaine Hamlet Way
Columbia, MD 21045
(301) 997-7426
Paul V. Jennemann - Owner
Products & Services:
Specializes in radon instrumentation, distributes full line of radon detection instruments for Pylon and Ludlum, including

radon monitors, grab sample and charcoal canisters, etc.

RADION LABORATORY, INC.
70 Walnut St.
Wellesley, MA 02181
(617) 239-8066
Rudolph Hon - Vice President
Kevin A. Doherty - Senior Manager
Products & Services:
Provides radon detection and mitigation services, operates counting instrumentation for radiation detection in air, liquids, and solid samples, distributes charcoal canisters and water testing kits for radon.
Qualifications:
AARST, NRA, listed with MA Department of Public Health, pending certification in PA, 20 years of experience handling and dealing with radioactive material, instrumentation cross-calibrated with DOE labs.

RADON ALERT® DETECTION CENTER
P.O. Box 323
Flourtown, PA 19031
(215) 247-9197, (800) 345-6348
Ulrich Hiesinger - President
Leonard Shabason - Radiation
Products & Services:
Distributes charcoal canisters and alpha track radon detectors, provides mitigation diagnosis and design services.
Qualifications:
EPA RMP, certified by NE, NJ, PA.

RADON ANALYSIS, INC.
7 Chamberlain Rd.
Flemington, NJ 08822
(201) 788-0661
David Chippendale - President
Joyce Criss - General Manager
Jeff Parsons - Quality Assurance
Products & Services:
Provides radon detection services, distributes radon test kits.
Qualifications:
EPA RMP, PA DER, NJ DEP.

RADON BAN
3719 Michelle Way
Baltimore, MD 21208
(301) 484-7120
David H. Nevins - President
Products & Services:
Provides radon detection services and distributes radon test kits.
Qualifications:
EPA RMP.

RADON CONSULTANTS
P.O. Box 5209
Sparks, NV 89432
(800) 327-6826
Russ McCreary - President
Products & Services:
Provides radon detection services and distributes radon test kits.
Qualifications:
EPA RMP.

RADON DETECTION AND REDUCTION LABORATORY/DIVISION OF FORMACHEM
10834 Ambassador
St. Louis, MO 63132
(314) 872-9350
Harold Susman - President
Donald Pearline - Biochemist
Products & Services:
Manufactures and distributes the Seal Safe Radon Gas Detector and alpha track detectors, distributes Enkadrain radon mitigation system for new home construction.

RADON ENVIRONMENTAL MONITORING, INC.
Box 5305
Evanston, IL 60204
See Ad Page 208
(312) 328-6628
Robert Pollock - Vice President
E.V. Benton
Products & Services:
Manufactures and distributes alpha track monitors, primary analysis laboratory.

RADON LABORATORIES
P.O. Box 829
Middletown, MD 21769
(301) 371-5867
Gregory J. Beal, Jr. - President
Products & Services:
Provides radon detection and consultation services for the public and industry, uses Pico Rad complete radon monitoring system.
Qualifications:
EPA RMP.

RADON PRODUCTS, INC.
3215 West Hampden Ave.
Englewood, CO 80110
(303) 761-4705
William Hall - President
Products & Services:
Distributes caulking compounds, foams, plastic, and radon test canisters.

RADON PROJECT, THE
P.O. Box 90069
Pittsburgh, PA 15224
(412) 687-2886
Bernard L. Cohen - President
Anna F. Cohen - Vice President
Frederick B. Cohen - Director
Products & Services:
Manufactures and distributes diffusion barrier charcoal adsorption canisters
Qualifications:
EPA RMP.

RADON RESEARCH GROUP, INC.
P.O. Box 1143
Germantown , MD 20874
(301) 972-3309, (800) 445-8436
Michael J. Myers - Operations Chief
Craig M. Mount - Customer Service
Products & Services:
Distributes radon detection products.
Qualifications:
AARST.

RADON SAFETY AND TECHNOLOGY CORPORATION
4 Charles St.
Pleasant Valley, NY 12569
(914) 635-1138, (800) 552-5257
Ronald J. Gasparro - President
Zia R. Hashmi - Director
Claude J. Noto - Assistant Director
Products & Services:
Radon measurement analytical laboratory, water and soil measurements, surveys for real estate transactions, land sites, and newly constructed and standing buildings, distributes charcoal canisters and alpha track detectors, waterborne radon test kits, radon alarms and auxiliaries offered to public and professionals, provides radon mitigation services.
Qualifications:
EPA RMP, AARST, ASTM.

RADON TEST COMPANY
P.O. Box 345
Eagle Grove, IA 50533
(515) 448-4600
Harlan L. Tofteberg - President
Robert Possehl - Vice President
Products & Services:
Wholesalers of charcoal canister and alpha track radon detection kits.

RADON TESTING AND TREATMENT, INC.
723 North Weber Suite 300
Colorado Springs, CO 80903
(719) 520-5693
Douglas L. Kladder - President
Steve Jelinek - Senior Analyst
Jessica Goodwin - Marketing Director
Robert Shideler

Products & Services:
Provides radon detection and mitigation services to the real estate industry, distributes charcoal canisters to the public.
Qualifications:
EPA RMP, licensed contractor.

RADON TESTING CONSULTANTS, INC.
P.O. Box 25313
Newark, NJ 07101
(201) 290-9217
Mark Jeziorski - President
Products & Services:
Provides radon detection and mitigation consulting services for consumers, businesses, and government agencies, distributes charcoal canisters and alpha track detectors.
Qualifications:
EPA RMP, staff has over 12 years of combined engineering experience.

RADON TESTING CORPORATION OF AMERICA
12 West Main St.
Elmsford, NY 10523
(914) 347-5010, (800) 457-2366
Nancy Bredhoff - Vice President
Kurt Dorfi - Sales Manager
Products & Services:
Manufactures and distributes charcoal canister radon test kits, distributes Pro Scan, a product designed to aid detection firms to prevent test tampering.
Qualifications:
EPA RMP, approved by NY DOH.

RADON TESTING SERVICES, INC.
4442 18th St.
San Francisco, CA 94114
(415) 552-5496
William E. Hellums - President
Products & Services:
Provides radon detection services, forensic and consultation services, computer dose projections.
Qualifications:
EPA RMP, radioactive materials license, certified health physicist.

RADON, LTD.
411 North Weinbach Ave.
Evansville, IN 47711
(812) 479-3110, (800) 634-2901
Ed Shoemaker - President
Products & Services:
Provides radon detection and mitigation services, distributes charcoal canister kit to public and dealers, laboratory facilities for analysis.
Qualifications:
ARA, radon detection and mitigation training courses.

RADON/SOILS AND WATER TESTING SERVICE
7080 Price-Hilliard Rd.
Plain City, OH 43064
(614) 873-8821
John Steichen - President
Products & Services:
Manufactures and distributes charcoal canisters, distributes E-Perm detectors, provides radon detection services to the real estate industry.
Qualifications:
AARST.

RAMSES II, LTD.
2355 East Dartmouth Ave.
Denver, CO 80210
(303) 758-5209
Robert L. Clark - Vice President
George A. Michael - President
Products & Services:
Manufactures and distributes charcoal canisters, electronic sump well cover kits, electronic PVC ventilation fans, and rotary air-to-air heat exchangers, provides radon mitigation services.
Qualifications:
Licensed professional engineer on staff.

REMEDIAL TECHNOLOGIES, INC.
10110 Kile Rd.
Chardon, OH 44024
(216) 765-2442
Richard Crombie - President
Products & Services:
Provides radon mitigation services, distributes radon detection and mitigation products including sealers, fans, Honeywell, Pylon, EDA radon monitoring equipment.

RESIDENCE RADON
28077 West 85th Ter.
Desoto, KS 66018
(913) 383-4610, (913) 585-1094
Richard J. Brazukas - Owner
Products & Services:
Provides radon detection services, distributes charcoal canisters and alpha track detectors to the public.

RESIDENTIAL SECURITY ADVISORS
265 Kenny Brook Ln.
Elyria, OH 44035
(216) 329-2365
James Steadman - Consultant
Bernadette Steadman - Financial Director
Products & Services:

Provides mail order radon detection and mitigation supplies.
Qualifications:
Licensed in PA, OH.

ROSS SYSTEMS, INC.
21 Main St. Box 670
Blairstown, NJ 07825
(201) 362-5571
Richard Ross - President
Products & Services:
Provides full service radon detection services, distributes charcoal canisters and alpha track detectors, provides radon mitigation consultation services using computer technology, radon mitigation proposals at no charge, offers seminars and workshops.
Qualifications:
AARST, EPA RMP,APCA, NJ DEP.

RPI RADON MEASUREMENT SERVICES
P.O. Box 357
Hershey, PA 17033
(717) 533-8100
William P. Kirk - Manager
Thomas E. Woolley - Radon Consultant
Products & Services:
Provides radon detection services for homes, schools, and industry, distributes charcoal canisters for detection firms or home owner testing.
Qualifications:
EPA RMP, PA certification, AARST.

RSO, INC.
P.O. Box 1526 711 Gorman Ave.
Laurel , MD 20707-0953
(301) 953-2482
Steven McDaniel - Marketing
Products & Services:
Provides radon detection services, distributes charcoal canisters and water samplers.
Qualifications:
EPA RMP.

RUECK ASSOCIATES
103 Sage Rd.
Silver Lake, KS 66539-9461
(913) 582-4050
Jon M. Rueck - Chairman
Products & Services:
Provides consultation on radon issues, distributes At-Ease radon monitors, charcoal canisters, and E-Perms.
Qualifications:
Diplomate AAEE, registered professional engineer, degrees in nuclear engineering, mechanical engineering.

SAFE-AIRE, INC.
162 East Chestnut St. P.O. Box 160
Canton, IL 61520
(309) 647-0419, (800) 331-2943 See Ad Page 161
Bill Wherley - Sales Manager
Chet Phillips - Products Manager
Harry Ellison - Marketing Director
Don Stees - Vice President
Products & Services:
Provides patent-pending Safe-Aire radon mitigation system to dealers, provides radon detection and mitigation services, distributes extensive line of radon detection and mitigation products, dealership opportunities available.

SCIENTIFIC ANALYSIS, INC.
6012 East Shirley Ln.
Montgomery, AL 36117
(205) 271-0643, (800) 345-2575

John Allen Gunn - Laboratory Manager
Rhonda L. Russell - Sales
Products & Services:
Primary radon testing laboratory provides charcoal canister mail order testing services to individuals, radon detecton firms, and commercial groups.
Qualifications:
EPA RMP.

SECURE-ALL COMPANY, INC.
4300 W. 38th Ave.
Denver, CO 80212
(303) 458-5577
Lou Nowakowski - President
Products & Services:
Distributes retail home radon test kits.

SUN NUCLEAR CORPORATION
415-C Pineda Ct.
Melbourne, FL 32940
(407) 259-6862
Thomas L. Powers - President
William E. Simon - Vice President
Products & Services:
Distributes full line of radon detection instruments, provides radon detection services.

SUNN CORPORATION
Box M, Dept. 213 3520 Forbes Ave.
Pittsburgh, PA 15213
(412) 828-9100
Nick Gromicko - Chief Executive Officer
Bob Dunkle
Mike Russell
Products & Services:
Provides radon mitigation services, manufactures and distributes charcoal canisters and tamper-resistant cages.
Qualifications:
BBB, NSBA, several state mitigation lists.

TCS INDUSTRIES, INC.
3805 Paxton St.
Harrisburg, PA 17111
(717) 558-8206
Carl H. Distenfeld - President
Jacquelyn Distenfeld - General Manager
Eric B. Distenfeld - Sales Manager
Products & Services:
Manufactures and distributes charcoal canisters, operates analysis laboratory.
Qualifications:
EPA RMP, certified health physicist, HPS, ANS.

TELEDYNE ISOTOPES
50 Van Buren Ave.
Westwood, NJ 07675
(201) 664-7070
D.F. Schutz - President
J.D. Martin - Vice President
D. McKillop - Radon Services Coordinator
Craig Osterhaus - Radon Division
Products & Services:
Manufactures and distributes the Home-Safe Radon Test Kit, performs analysis of results within 24 hours, provides radon detection services.
Qualifications:
NRC licenses, technical and scientific personnel, state licenses.

TELEDYNE WATER PIK
1730 East Prospect
Fort Collins, CO 80525
(303) 484-1352
John Spencer - Product Manager
Products & Services:
Distributes Home-Safe radon test kit.

TEQ CORPORATION
P.O. Box 570
Wilkes-Barre, PA 18703-0570
(717) 825-3939, (800) 877-2366
See Ad Page 100
Russell Hendershot - President
Ed Dructor - Vice President
Brian Redmond - Research Scientist
Products & Services:
Provides radon detection and mitigation services for commercial and residential markets, manufactures and distributes charcoal canisters, maintains calibrated radon chamber, offers radon detection and mitigation training courses.
Qualifications:
PA DER certification.

TERRADEX CORPORATION
460 North Widget Lane
Walnut Creek, CA 94598
(415) 938-2545
See Ad Back Cover
H. Ward Alter - Senior Consultant
Richard Oswald - Regional Sales Manager
Products & Services:
West coast branch sales office.

TERRADEX CORPORATION/SUBSIDARY OF TECH/OPS LANDAUER, INC.
2 Science Rd.
Glenwood, IL 60425
(312) 755-7911, (800) 528-8327
See Ad Back Cover
Sue Morrow - Customer Service
R. Craig Yoder
Products & Services:
Manufactures and distributes line of radon monitoring products to industry, academic institutions, government agencies, consumers, etc.

Qualifications:
Over 35 years of experience in radiation monitoring services from both occupational and environmental sources, works with ALA, EPA, USAF, universities, national labs, etc.

TERRADON EXPLORATION TECHNOLOGIES, INC.
3931 White Oak Tr. P.O. Box 22288
Cleveland, OH 44122
(216) 464-2002
Darioush T. Ghahremani - President
Products & Services:
Provides radon detection and mitigation services, research and development in radon detection, measurement and abatement technology, development of new alpha track and charcoal canisters for industrial and private use, soil gas radon detection and mapping.
Qualifications:
AARST, NRA, AAPG, OAS, ACS, OG&G, AAGE, other scientific associations.

THERMO ENVIRONMENTAL CORPORATION
101 1st Ave. P.O. Box 9046
Waltham, MA 02254
(617) 622-1103, (505) 345-9931
Jeff Brown - Health Physics
Products & Services:
Manufactures and distributes RAD-SCAN radon screening kits, provides radon detection and mitigation services to homeowners, developers, real estate industry, etc., parent company of TMA Eberline.

THOMSON AND NIELSEN ELECTRONICS, LTD.
4019 Carling Ave. Suite 303
Kanata, ON K2K 2A3
(613) 592-3019
See Ad Page 187
Ian Thomson - President
Andrew Singmin - Marketing Director
Line Brabant - Inside Sales
Products & Services:
Manufactures and distributes radon detection instruments, provides radon detection services, conducts research in radiation detection.
Qualifications:
AARST, EPA RMP, HPS.

THRESHOLD TECHNICAL PRODUCTS
7225 Edington Dr.
Cincinnati, OH 45249
(800) 458-4931
John Grether - President
Ken Marguet - Vice President
Products & Services:
Manufactures and distributes Survivor II continuous radon monitor directly to consumer.

TMA/EBERLINE, HEALTH PHYSICS
5635 Jefferson Blvd., NE
Albuquerque, NM 87109
(505) 345-9931
Stanley Waligora - Sales Manager
Ernie Sanchez
Jeff Brown
Products & Services:
Provides radon detection and mitigation services to commercial and government clients, distributes charcoal canisters, provides consultation services.

TOM SCURRY ASSOCIATES
1976 Linden Dr.
Boulder, CO 80302
(303) 444-4802
See Ad Page 210
Tom Scurry - President
Robert Droullard - Vice President

Products & Services:
Maintains a radon chamber to test and calibrate detection instruments, manufactures and distributes continuous and grab working level monitors, provides radon detection services.

TRACOR JITCO RADON MEASUREMENT SERVICES
1601 Research Blvd.
Rockville, MD 20850
(301) 984-2800
Ted Junghans - Associate Scientist
Steve Olin - Technical Director
Products & Services:
Distributes charcoal canisters and alpha track detectors to homeowners and provides radon detection services for the real estate industry.
Qualifications:
EPA RMP.

U.S. TOXIC SUBSTANCE TESTING BUREAU
1024 Cottman Ave.
Philadelphia, PA 19111
(215) 745-7778, (800) 535-8378
Harold Stesis - President
Warren Schatz - Director of Marketing
Products & Services:
Provides radon detection services for real estate inspections, distributes and analyzes charcoal canisters, and operates a laboratory.
Qualifications:
AARST, EPA RMP, PA DER, NJ DEP.

WILLIAM B. JOHNSON AND ASSOCIATES, INC.
P.O. Box 98 Research Park
Montville, NJ 07045
(201) 334-9222
Rudy Schober - Vice President
Products & Services:
Manufactures and distributes an alpha scintillation probe connected to a digital scaler/timer with its own high voltage source.

3. Radon Mitigation System Manufacturers

AMERICAN ENVIRONMENTAL SYSTEMS, INC.
4699 Nautilus Ct., S.
Boulder, CO 80301
(303) 530-7077
James Yehl - President
Tom Archbold - Marketing Manager
Products & Services:
Manufactures and distributes electrophoresis effect system which mitigates radon via unipolar electrical ionization.

BIO-PURE COMPANY
P.O. Box 20593
Tampa, FL 33622-0593
(813) 839-8617
Don Saceman - President
Y.A. Demirjian - Consultant
Donah Parent - Vice President
Products & Services:
Manufactures radon mitigation system using plants with a specially formulated soil medium, process involves use of carbon rich soil media that filters out particulate matter with some adsorption of radon gas.
Qualifications:
Two years of field testing and development.

CONTROL RESOURCE SYSTEMS, INC.
670 Mariner Dr.
Michigan City, IN 46360
(219) 872-5591, (800) 272-3786
See Ad Page 234
Ronald W. Marsh - Vice President
Don Musilli - National Sales Manager
Products & Services:
Manufactures and distributes air purification systems for residential, commercial and industrial facilities, manufactures the CRSI 600H Radon Removal System.
Qualifications:
Registered medical device manufacturer.

CURRENT INDOOR AIR SYSTEMS, INC.
P.O. Box 18075
Boulder, CO 80308
(303) 440-8555
See Ad Page 212
Kyle Lefkoff - General Manager
Kim Kenney - Marketing Manager
Products & Services:
Manufactures The Model 300 Radon Mitigation System, a pressurization/dilution mitigation device for crawlspace and basement pressurizations or where sub-slab alternatives are impractical, incorporates a passive air-to-air heat exchange in its design.

DOUGLAS MARTIN AND ASSOCIATES, INC./MARTIN CONSULTING
520 Chestnut St.
Emmaus, PA 18049
(215) 965-2914
See Ad Page 199
Richard D. Martin - President
Ken Smith - Health and Safety Services
Jim Sherlock - Mitigation Supervisor
Products & Services:
Provides radon mitigation services, mitigation design, engineering procedure

development, nuclear consulting, manufactures Professional Series water filtration system, provides radon mitigation training program for state of New Jersey.
Qualifications:
Certified in NJ, PA.

ION SYSTEMS, INC.
2546 10th St.
Berkeley, CA 94710
(415) 548-3640
Scott Gehlke - President
Dave Taylor - Vice President
Bob Stewart - Sales Manager
Products & Services:
Manufactures and distributes the No-Rad radon removal system, removes attached and free floating radon decay products.
Qualifications:
Product was developed, patented and tested by Drs. Moeller, Maher, and Rudnick of the School of Public Health, Harvard University.

LOWRY ENGINEERING, INC.
P.O. Box 536 Back Troy Rd.
Unity, ME 04988
See Ad Page 211
(207) 948-3790
Sylvia Lowry - President
Jerry Lowry - Resarch Director
Products & Services:
Engineering research and consulting firm specializes in radionuclide analysis and removal from water supplies, and removal equipment for homes and public water supplies, manufactures and distributes the Stripper Aeration System, removes 99.9% of radon from water.
Qualifications:
Director is a leading expert on radon removal from water supplies, teaches EPA courses on mitigation.

NORTH EAST ENVIRONMENTAL PRODUCTS, INC.
85 Mechanic St.
Lebanon, NH 03766
See Ad Page 432
(603) 448-3944
Bruce L. Lamarre - President
Jack Nelson - Chief Executive Officer
Kevin Lang - Vice President
Products & Services:
Manufactures and distributes Clearadon, an aeration system for removal of radon or VOCs from residential water supplies, provides consultation services for removal of waterborne radon.
Qualifications:
Water Quality Association.

RAD SYSTEMS, INC.
21 East Main St.
Westborough, MA 01581
See Ad Page 213
(508) 366-5051
Richard F. Abrams - President
Products & Services:
Manufactures and distributes radon gas removal unit based on unique charcoal adsorption process for use in homes to treat air or water.
Qualifications:
AARST.

RADON MEASUREMENT AND CONTROL OF NEW MEXICO
1527 Granite St., NW
Albuquerque, NM 87104
See Ad Page 238
(505) 843-7643
William M. Turner - President
L. David Turner - Vice President
Products & Services:
Provides radon detection services for real estate agents, home owners, government agencies, and relocation firms, manufactures RadonDrain, a ventilation system for removal

of indoor radon in on-grade homes, distributes radon "reference" chamber, used by radon detection firms to verify instrument measurement stability over time.
Qualifications:
EPA RMP, registered professional geologist, EPA mitigation course, radon instructor for NM Real Estate Commission.

SCIENTIFIC RESEARCH AND DEVELOPMENT COMPANY
P.O. Box 5590
Lakeland, FL 33807
(813) 644-0404
Peter Orlowsky - President
Products & Services:
Research and development of an electronic unit that can mitigate radon (not yet on the market), provides consultation services.

SYBRON CHEMICAL
P.O. Box 66 Birmingham Rd.
Birmingham, NJ 08011
(609) 893-1100
Frank McGarvey - Technical Manager
Products & Services:
Developing ion exchange system for removal of uranium and radium in water.

WESTERN WATER INTERNATIONAL, INC.
7715 Penn Belt Dr.
Forestville, MD 20747
(301) 568-0200
Mike Pedersen - President
Products & Services:
Manufactures and distributes a filter with charged, modified composite, blend media achieving a higher rate of organic chemical removal than ordinary carbon, effective in mitigation of waterborne radon.

4. Radon Mitigation Product Distributors

AMERICAN ENVIRONMENTAL SYSTEMS, INC.
4699 Nautilus Ct., S.
Boulder, CO 80301
(303) 530-7077
James Yehl - President
Tom Archbold - Marketing Manager
Products & Services:
Manufactures and distributes electrophoresis effect system which mitigates radon via unipolar electrical ionization.

COMPREHENSIVE RADON SERVICES, INC.
P.O. Box 548
Manchester, NH 03105
(603) 644-7301
Mark Malbon - President
Products & Services:
Provides radon detection services using charcoal canister, alpha track, continuous radon, grab sampling, working level methods, water analysis, and open land testing, provides radon mitigation services, distributes radon mitigation supplies.

CONCRETE EPOXY TECHNICAL SYSTEMS, INC.
P.O. Box F-25 41 Terry Dr.
Trevose, PA 19047
(215) 322-7310
David W. Markloff - President
Products & Services:
Distributes radon approved sealants, coatings, etc.

CONTROL RESOURCE SYSTEMS, INC.
670 Mariner Dr.
Michigan City, IN 46360 See Ad Page 234
(219) 872-5591, (800) 272-3786
Ronald W. Marsh - Vice President
Don Musilli - National Sales Manager
Products & Services:
Manufactures and distributes air purification systems for residential, commercial and industrial facilities, manufactures the CRSI 600H Radon Removal System.
Qualifications:
Registered medical device manufacturer.

CURRENT INDOOR AIR SYSTEMS, INC.
P.O. Box 18075 See Ad Page 212
Boulder, CO 80308
(303) 440-8555
Kyle Lefkoff - General Manager
Kim Kenney - Marketing Manager
Products & Services:
Manufactures The Model 300 Radon Mitigation System, a pressurization/dilution mitigation device for crawlspace and basement pressurizations or where sub-slab alternatives are impractical, incorporates a passive air-to-air heat exchange in its design.

ENVIRO TECH SYSTEMS
430 Fernhill Ave.
Pittsburgh, PA 15226
(412) 561-2100
Jack Stanizzo - Owner
Products & Services:
Provides radon detection and mitigation

services, distributes radon mitigation products.
Qualifications:
PA DER, 450+ mitigations completed.

ENVIRONMENTAL ENTERPRISES, INC.
P.O. Box 16847
Atlanta, GA 30321
(404) 766-5551
Chris Esworthy - President
Products & Services:
Provides radon detection and mitigation services, distributes full line of radon mitigation products including fans, sealers, etc.

EVANSHIRE COMPANY, LTD., THE
Rte. 4, Box 57
Rome, NY 13440
(315) 865-5941, (800) 962-5941
See Ad Page 215
Edwin Evans - President
Wade Evans - Vice President
Products & Services:
Provides radon detection services using charcoal canister, alpha track detector, and grab sampling methods, provides radon mitigation services, distributes full line of radon detection and mitigation products, including sealers, sump covers, fans.
Qualifications:
Experience in state and federally funded projects.

GEOTHERMAL ENERGY COMPANY, LTD.
P.O. Box 59
Wolfeboro, NH 03894
(603) 569-2519
Arthur L. Robinson - President
Products & Services:
Distributes radon detection and mitigation products and supplies.
Qualifications:
ARA, NAHB, EPA and NAHB training.

INFILTEC/FACTORY
108 South Delphine Ave. P.O. Box 1125
Waynesboro, VA 22980
(703) 943-2776
See Ad Page 214
David Brubaker - Factory Manager
Frankie Patterson - Warehouse Manager
Arthur Saum - Research Manager
Products & Services:
Manufactures and distributes a full line of radon detection and mitigation products including fans, couplings, pressure gauges, alarms, sumps and drain covers, radon test kits, continuous radon monitors, caulks and sealants, manometers, blower doors, etc.
Qualifications:
EPA RMP.

ION SYSTEMS, INC.
2546 10th St.
Berkeley, CA 94710
(415) 548-3640
Scott Gehlke - President
Dave Taylor - Vice President
Bob Stewart - Sales Manager
Products & Services:
Manufactures and distributes the No-Rad radon removal system, removes attached and free floating radon decay products.
Qualifications:
Product was developed, patented and tested by Drs. Moeller, Maher, and Rudnick of the School of Public Health, Harvard University.

KEY TECHNOLOGY, INC.
P.O. Box 562
Jonestown, PA 17038
(717) 274-8310, (800) 523-4964
See Ad Inside Front Cover
Claude Wiblin - President
Robert Garland - Marketing Director
Robby Marquette - Secretary/Treasurer
Edward Suter - Training Director
Products & Services:
Manufactures and distributes a complete line of radon detection products and distributes radon mitigation products.
Qualifications:
EPA RMP, PA DER testing and mitigation courses.

LOWRY ENGINEERING, INC.
P.O. Box 536 Back Troy Rd.
Unity, ME 04988
(207) 948-3790
See Ad Page 211
Sylvia Lowry - President
Jerry Lowry - Resarch Director
Products & Services:
Engineering research and consulting firm specializes in radionuclide analysis and removal from water supplies, and removal equipment for homes and public water supplies, manufactures and distributes the Stripper Aeration System, removes 99.9% of radon from water.
Qualifications:
Director is a leading expert on radon removal from water supplies, teaches EPA courses on mitigation.

NORTH EAST ENVIRONMENTAL PRODUCTS, INC.
85 Mechanic St.
Lebanon, NH 03766
(603) 448-3944
See Ad Page 432
Bruce L. Lamarre - President
Jack Nelson - Chief Executive Officer
Kevin Lang - Vice President
Products & Services:
Manufactures and distributes Clearadon, an aeration system for removal of radon or VOCs from residential water supplies, provides consultation services for removal of waterborne radon.
Qualifications:
Water Quality Association.

OBAR SYSTEMS, INC.
RR 3, Box 759
Highland Lakes, NJ 07422
(201) 728-7679, (201) 697-0112
See Ad Page 215
Gunner Bar - Owner
David O'Brien - Owner
Products & Services:
Distributes complete range of radon mitigation products, including fans, heat exchangers, sealers and caulks, pumps, filters, water systems, couplings, sumps, manometers, expansion joint materials, and other accessories, provides radon mitigation services.

R.F. SIMON COMPANY, INC.
RD 2, Box 481
Barto, PA 19504
(215) 754-6455
See Ad Page 148
Ron Simon - Project Manager
Peter Maniscalco - Diagnostic Specialist
Products & Services:
Provides mitigation design and installation services, mitigation supply warehouse carrying a full line of radon mitigation and detection products including fans, caulks and sealants, couplings, comprehensive radon services provided to private, commercial, institutional, and government agencies, expert testimony.
Qualifications:
EPA RMP, 13 years of custom building experience, instructor for EPA training courses, consultant to Lawrence Berkeley Laboratory, Princeton University, PA DER.

RAD SYSTEMS, INC.
21 East Main St.
Westborough, MA 01581
(508) 366-5051
See Ad Page 213
Richard F. Abrams - President
Products & Services:
Manufactures and distributes radon gas removal unit based on unique charcoal adsorption process for use in homes to treat air or water.
Qualifications:
AARST.

RADON AND DEWATERING SYSTEMS, INC. OF MARYLAND
Box 1814
Silver Spring, MD 20902
(301) 593-1861
Joel Wolensky - President
Products & Services:
Distributes full line of radon mitigation supplies including Fantech products.

See Ad Page 215

RADON CONTROL PRODUCTS
511 Industrial Dr.
Carmel, IN 46032
(317) 846-7442, (800) 523-2082
Steven Weaver - President
Roger Ice - Sales Director
Products & Services:
Distributes radon mitigation products including fan products, couplings, sealers, smoke guns, etc., manufactures the Weaver Werks retrofit sump pump cover and related retrofit products, distributes the Radon Control Products Catalog.

See Ad Page 217

RADON DETECTION AGENCY, LTD.
625 5th Ave. Suite 104
New Westminster, BC V3M 1X4
(604) 525-4455
David Keegan - President
David E. McLeran - Vice President
Products & Services:
Provides radon detection services using charcoal canister and alpha track methods, distributes radon mitigation supplies.

RADON REDUCTION EQUIPMENT OF COLORADO
921 Main Suite C
Louisville, CO 80027
(303) 665-1776
Kevin Lightsey - President
Pete Werner - Manager
Lou Spier - Secretary
Products & Services:
Distributes Kanalflact fans, Current Indoor Air System, and sealing and venting products.

RADON TECHNOLOGIES, INC.
12200 Sunrise Valley Dr. Suite 320
Reston, VA 22091
(703) 620-4004, (800) 237-5024
R.A. Brightsen - Chairman
Venus Cain - Secretary
Products & Services:
Distributes No-Rad radon removal system.

RADON TESTING AND CONSULTING
1771 Post Rd., E. Suite 320
Westport, CT 06880
(203) 454-3886
Lorraine Hodes - Vice President
John Nicholas - Director
Mark Lewis - Senior Geologist
Maria Mody - Communications
Products & Services:
Provides radon detection and mitigation services for home owners, realtors, builders and developers, exclusive New England distributor of Current Indoor Air Model 300 Radon Mitigation System.
Qualifications:
AARST, AGSE.

RAMSES II, LTD.
2355 East Dartmouth Ave.
Denver, CO 80210
(303) 758-5209
Robert L. Clark - Vice President
George A. Michael - President
Products & Services:
Manufactures and distributes charcoal canisters, electronic sump well cover kits, electronic PVC ventilation fans, and rotary air-to-air heat exchangers, provides radon mitigation services.
Qualifications:
Licensed professional engineer on staff.

REMEDIAL TECHNOLOGIES, INC.
10110 Kile Rd.
Chardon, OH 44024
(216) 765-2442
Richard Crombie - President
Products & Services:
Provides radon mitigation services, distributes radon detection and mitigation products including sealers, fans, Honeywell, Pylon, EDA radon monitoring equipment.

RESIDENTIAL SECURITY ADVISORS
265 Kenny Brook Ln.
Elyria, OH 44035
(216) 329-2365
James Steadman - Consultant
Bernadette Steadman - Financial Director
Products & Services:
Provides mail order radon detection and mitigation supplies.
Qualifications:
Licensed in PA, OH.

RESOURCE CONSERVATION TECHNOLOGY, INC.
2633 North Calvert St.
Baltimore, MD 21218
(301) 366-1146
Lee Jaslow - President
Products & Services:
Manufactures and distributes full line of radon mitigation products including membranes, film barriers, sealants, radon

sumps, fans, etc.

SAFE-AIRE, INC.
162 East Chestnut St. P.O. Box 160
Canton, IL 61520
(309) 647-0419, (800) 331-2943 See Ad Page 161
Bill Wherley - Sales Manager
Chet Phillips - Products Manager
Harry Ellison - Marketing Director
Don Stees - Vice President
Products & Services:
Provides patent-pending Safe-Aire radon mitigation system to dealers, provides radon detection and mitigation services, distributes extensive line of radon detection and mitigation products, dealership opportunities available.

SCWEST, INC.
4950 East Asbury Ave. Suite 5
Denver, CO 80222
(303) 759-1286
Jeffery Kramer - President
Products & Services:
Distributes radon mitigation equipment, cross-laminated polyethylene, urethane caulks, heat recovery ventilation systems, and Current Indoor Air Systems.

SMALLEY AND COMPANY
861 South Jason
Denver, CO 80223
(303) 777-3435
John Coy - Sales Representative
Products & Services:
Distributes radon mitigation products and one component polyurethane sealant.

VIKING HOME IMPROVEMENT, INC.
15 South 15th St.
Fargo, ND 58103
(701) 237-6300
Kevin Christianson - President
Products & Services:
Provides radon detection and mitigation services, distributes complete line of radon mitigation products including sealers, membranes, fans, sump equipment, protective equipment.

W.W. GRAINGER, INC. DISTRIBUTION GROUP
5959 West Howard St.
Chicago, IL 60648
(312) 647-8900
David Mills - Product Manager
Products & Services:
Broad line distributor of ventilation and related building equipment, components and supplies to the commercial, industrial and contractor markets.
Qualifications:
265 stocking locations in major cities.

5. Sealants

ACME CHEMICALS AND INSULATION COMPANY
P.O. Box 1404
New Haven, CT 06505
(203) 562-2171
Marvin Leonard - Vice President
Products & Services:
Manufactures two-component, water-based epoxy.

ALBION ENGINEERING COMPANY
2080-A Wheat Sheaf Ln.
Philadelphia, PA 19124
(215) 535-3476
Franz Schneider - President
Products & Services:
Manufactures caulking guns.

AMERICAN CYANAMID COMPANY
1 Cyanamid Plaza
Wayne, NJ 07470
(201) 831-2000
Kevin Lynch - Marketing Manager
Products & Services:
Manufactures pore sealants, one-component acetate latex.

APPLIED EXTRUSION TECHNOLOGIES, INC.
P.O. Box 582
Middletown, DE 19709
(302) 378-8888, (800) 521-6713
J.B. Havewala - Vice President of Marketing
M. J. Schroeder - Marketing Manager
Products & Services:
Closed and open cell backer rod and strip-off XL expansion joint forming system.

ARDEX, INC.
630 Stoops Ferry Rd.
Corapolis, PA 15108
Lewis Rea - National Sales Manager
Products & Services:
Manufactures self leveling toppings and underlayments.

BONDEX INTERNATIONAL, INC.
3616 Scarlet Oak Blvd.
St. Louis, MO 63122
(314) 225-5001
Charles G. Miller - President
David T. Vass - Vice President
John Sutter - Vice President of Sales
Products & Services:
Manufactures wall and ceiling patching and mitigation paints useful in radon mitigation.

BOSTIK
Boston St.
Middleton, MA 01949
(508) 777-0100, (800) 343-8178
C. Moss - Marketing Services
Products & Services:
Manufactures sealants and adhesives for industry.

BUILDERS EQUIPMENT AND SUPPLY
1500 South Hanover St.
Pottstown, PA 19464
(215) 323-3320
Harry Wynn - General Manager
Products & Services:
Distributes Vulkem and Sika line of sealants and non-shrink grout.

C.R. LAURENCE
2503 East Vernon
Los Angeles, CA 90058
(800) 421-6144
Jim Webb - Technical Services
Judy Muller - Sales
Products & Services:
Distributes adhesives, silicone products, and urethane sealants.

CALBAR, INC.
2626 North Martha
Philadelphia, PA 19125
(215) 739-9141
Bill Reynolds - Sales Manager
Products & Services:
Manufactures caulking compounds.

CHEMICAL SYSTEMS, INC.
P.O. Box 68
Dunbridge, OH 43414
(419) 833-4492
Ron Lane - President
Products & Services:
Distributes hydropoxy, waterproofing material for use as a moisture barrier.

CONCRETE EPOXY TECHNICAL SYSTEMS, INC.
P.O. Box F-25 41 Terry Dr.
Trevose, PA 19047
(215) 322-7310
David W. Markloff - President
Products & Services:
Distributes radon approved sealants, coatings, etc.

CONVENIENCE PRODUCTS, INC.
4205 Forest Park Blvd.
St. Louis, MO 63108-2892
(314) 535-6226
Byron Lapin - President

Steve Sprehe - Vice President of Sales
Sherry Lawrence - Customer Relations
Products & Services:
Supplies one component, polyurethane expandable foam.

DAP, INC.
P.O. Box 277
Dayton, OH 45401
(513) 667-4461
Bill Sherwin - Marketing
Products & Services:
Manufactures full line of sealants, caulks and adhesives.

DOW CORNING CORPORATION
Mail Stop CO2306
Midland, MI 48686-0994
(517) 496-4578
Ed Hutchinson - Product Manager
Products & Services:
Manufactures silicone sealants.

ENVIRONMENTAL ENTERPRISES, INC.
P.O. Box 16847
Atlanta, GA 30321
(404) 766-5551
Chris Esworthy - President
Products & Services:
Provides radon detection and mitigation services, distributes full line of radon mitigation products including fans, sealers, etc.

ESSEX CHEMICAL CORPORATION
1333 Broad St.
Clifton, NJ 07015
(201) 773-6300
Steven Drozdow - Director of Sales
Linda LaMonico - Marketing Supervisor
Products & Services:
Manufactures brush-on urethane.

EVANSHIRE COMPANY, LTD., THE
Rte. 4, Box 57
Rome, NY 13440
(315) 865-5941, (800) 962-5941
See Ad Page 215
Edwin Evans - President
Wade Evans - Vice President
Products & Services:
Provides radon detection services using charcoal canister, alpha track detector, and grab sampling methods, provides radon mitigation services, distributes full line of radon detection and mitigation products, including sealers, sump covers, fans.
Qualifications:
Experience in state and federally funded projects.

FOAM ENTERPRISES, INC.
13630 Watertower Cir.
Minneapolis, MN 55441
(612) 559-3266, (800) 888-3342
James L. Anderson - Vice President
Products & Services:
Manufactures spray-applied polyurethane foam for seamless insulation to reduce radon migration.
Qualifications:
Joint project with Westinghouse Materials Company.

FOMO PRODUCTS, INC.
2775 Barber Rd.
Norton, OH 44203
(216) 753-4585, (800) 321-5585
W. Steding - President
Gene Testa - Marketing Manager
Products & Services:
Manufactures disposable one- and two-component polyurethane foam systems used to fill gaps and cavities.

GALLAGHER AND ASSOCIATES, INC.
950 Sussex Blvd.
Broomall, PA 19008
(215) 328-2200
John Gallagher - Vice President
Products & Services:
Distributes a two-component epoxy product.

GARDWELL MANUFACTURING CORPORATION
602 John F. Kennedy St. P.O. Box 7
Wampum, PA 16157
(412) 535-4779
Charles Schrankel - President
Products & Services:
Manufactures Radon Gard, a two component polyamide epoxy.

GARON PRODUCTS, INC.
1974 Hwy. 35 CN20
Wall, NJ 07719
(201) 499-1776, (800) 631-5380
Art Crowley - President
Christy Karl - Marketing Manager
Lee Miller - Product Consultant
Products & Services:
Manufactures full line of sealers and protective equipment.
Qualifications:
NJ CC, National Safety Council, BBB, Direct Marketing Association, OSHA compliance.

GE SILICONES
260 Hudson River Rd.
Waterford, NY 12188
(518) 237-3330, (800) 255-8886
Greg Bronk - Marketing Manager
Products & Services:
Manufactures silicone sealants, caulks, adhesives, and coatings.

GEOCEL CORPORATION
P.O. Box 398
Elkhart, IN 46515
(219) 264-0645, (800) 348-7615
Harry Funk - Technical Service
Doug Mattix - Sales Manager
Products & Services:
Manufactures caulk and sealants for radon mitigation.
Qualifications:
Adhesive and Sealant Council.

H.B. FULLER
315 South Hicks
Palatine, IL 60067
(800) 323-7407
Jim Conaty - Sales Manager
Products & Services:
Manufactures adhesives used as radon barriers.

HALLTECH, INC.
465 Coronation Dr.
West Hill, ONT M1E 2K2
(416) 284-6111
Art Brooker - Sales Manager
Products & Services:
Manufactures adhesives.

HI-TEK POLYMERS
9808 Bluegrass Pkwy.
Louisville, KY 40299
(502) 499-4300
Doug Abbott - Eastern Sales
Jim Melloan - Central Sales
Products & Services:
Manufactures and sells specialty resins and water soluble polymers used in coatings, structural applications and oil/gas extraction.

HOECHST-CELANESE, INC.
1250 West Mockingbird Ln.
Dallas, TX 75247
(214) 689-4000
C. Joe Schaefer - Product Manager
Products & Services:
Manufactures a water dispersed, polyamine-cured epoxy.

INFILTEC/FACTORY
108 South Delphine Ave. P.O. Box 1125
Waynesboro, VA 22980
(703) 943-2776
See Ad Page 214
David Brubaker - Factory Manager
Frankie Patterson - Warehouse Manager
Arthur Saum - Research Manager
Products & Services:

Manufactures and distributes a full line of radon detection and mitigation products including fans, couplings, pressure gauges, alarms, sumps and drain covers, radon test kits, continuous radon monitors, caulks and sealants, manometers, blower doors, etc.
Qualifications:
EPA RMP.

INSTA-FOAM PRODUCTS, INC.
1500 Cedarwood Dr.
Joliet, IL 60435
(800) 435-9359
Sharon McFarland - Sales Manager
Products & Services:
Manufactures caulks and urethane foams.

INTERNATIONAL RADON TECHNOLOGIES
2417 Regency Rd.
Lexington, KY 40503
(606) 278-0104, (800) 537-5066
Ed Davis - President
William G. Buchman - Inventor
Products & Services:
Distributes radon-related products including a water reducible, two-part polyamide epoxy coating to block radon passage, manufactures continuous radon progeny monitor and electret cylindrical airborne radon monitor, operates a primary laboratory.

MAMECO INTERNATIONAL, INC.
4475 East 175th St.
Cleveland, OH 44128
(216) 752-4400, (800) 321-6412
John Horvath - Vice President of Sales
Wes Senseman - Technical Services
Jeff Kovach - President
Products & Services:
Manufactures polyurethane sealants, membranes and coatings for construction and restorations; manufactures Vulkem 45 and Vulkem 116 weatherproofing sealants.
Qualifications:
Meets federal specifications, USDA and EPA approvals.

MATESON CHEMICAL COMPANY
1025 East Montgomery Ave.
Philadelphia, PA 19125
(215) 358-5453
Jean F. Mateson - President
Herman Rossi - Vice President of Sales
See Ad Page 220
Products & Services:
Manufactures Radox, a water-based elastomer coating used in radon mitigation.

OBAR SYSTEMS, INC.
RR 3, Box 759
Highland Lakes, NJ 07422
(201) 728-7679, (201) 697-0112
Gunner Bar - Owner
David O'Brien - Owner
See Ad Page 215
Products & Services:
Distributes complete range of radon mitigation products, including fans, heat exchangers, sealers and caulks, pumps, filters, water systems, couplings, sumps, manometers, expansion joint materials, and other accessories, provides radon mitigation services.

PECORA CORPORATION
165 Wambold Rd.
Harleysville, PA 19438
Al Kilchesty - Technical Service
Products & Services:
Manufactures one- and two-part polyurethane sealants.

PH SALES COMPANY
P.O. Box 372
Edison, NJ 08818
(201) 287-5300
Frank Lancellotti
Products & Services:
Distributes environmental sealing products.

PLASTICRAFTS
600 West Bayaud Ave.
Denver, CO 80223
(303) 744-3700
Paul Trujillo - Marketing Manager
Products & Services:
Distributes and fabricates line of sealants.

POLYMERIC SYSTEMS, INC.
Wheatland and Mason Sts.
Phoenixville, PA 19460
(215) 935-1170, (800) 228-5548
Jack Walsh - Sales Manager
Products & Services:
Manufacturers one- and two-part urethane sealants, silicone sealants, acrylic and butyl caulks.

PROGRESS UNLIMITED, INC.
200 Madison Ave. Suite 1806
New York, NY 10016
(212) 689-7030
Gustav Szabo - President
Tom Szabo - Manager
Products & Services:
Joint fillers, compression seals, building gaskets, and vapor barriers.

R.F. SIMON COMPANY, INC.
RD 2, Box 481
Barto, PA 19504
(215) 754-6455
See Ad Page 148
Ron Simon - Project Manager
Peter Maniscalco - Diagnostic Specialist
Products & Services:
Provides mitigation design and installation services, mitigation supply warehouse carrying a full line of radon mitigation and detection products including fans, caulks and sealants, couplings, comprehensive radon services provided to private, commercial, institutional, and government agencies, expert testimony.
Qualifications:
EPA RMP, 13 years of custom building experience, instructor for EPA training courses, consultant to Lawrence Berkeley Laboratory, Princeton University, PA DER.

RADON CONTROL PRODUCTS
511 Industrial Dr.
Carmel, IN 46032
(317) 846-7442, (800) 523-2082
See Ad Page 217
Steven Weaver - President
Roger Ice - Sales Director
Products & Services:
Distributes radon mitigation products including fan products, couplings, sealers, smoke guns, etc., manufactures the Weaver Werks retrofit sump pump cover and related retrofit products, distributes the Radon Control Products Catalog.

RADON PRODUCTS, INC.
3215 West Hampden Ave.
Englewood, CO 80110
(303) 761-4705
William Hall - President
Products & Services:
Distributes caulking compounds, foams, plastic, and radon test canisters.

RADON REDUCTION EQUIPMENT OF COLORADO
921 Main Suite C
Louisville, CO 80027
(303) 665-1776
Kevin Lightsey - President
Pete Werner - Manager
Lou Spier - Secretary
Products & Services:
Distributes Kanalflact fans, Current Indoor Air System, and sealing and venting products.

RESOURCE CONSERVATION TECHNOLOGY, INC.
2633 North Calvert St.
Baltimore, MD 21218
(301) 366-1146
Lee Jaslow - President
Products & Services:
Manufactures and distributes full line of radon mitigation products including membranes, film barriers, sealants, radon sumps, fans, etc.

RHH FOAM SYSTEMS, INC.
P.O. Box 752 6001 S. Pennsylvania Ave.
Cudahy, WI 53110-0752
(414) 744-6066
See Ad Page 222
Richard Heitzer - President
Dick Christenson - Vice President
Dan Soby - Sales Manager
Products & Services:
Manufactures and distributes Versi-foam, a polyurethane and foam dispensing system to contractors and others for radon mitigation.

SAFE-AIRE, INC.
162 East Chestnut St. P.O. Box 160
Canton, IL 61520
(309) 647-0419, (800) 331-2943
See Ad Page 161
Bill Wherley - Sales Manager
Chet Phillips - Products Manager
Harry Ellison - Marketing Director
Don Stees - Vice President
Products & Services:
Provides patent-pending Safe-Aire radon mitigation system to dealers, provides radon detection and mitigation services, distributes extensive line of radon detection and mitigation products, dealership opportunities available.

SCWEST, INC.
4950 East Asbury Ave. Suite 5
Denver, CO 80222
(303) 759-1286
Jeffery Kramer - President
Products & Services:
Distributes radon mitigation equipment, cross-laminated polyethylene, urethane caulks, heat recovery ventilation systems, and Current Indoor Air Systems.

SIKA CORPORATION
201 Polito Ave.
Lyndhurst, NJ 07071
(201) 933-8800, (800) 631-7270
Jim Schwartz - National Sales Manager
Products & Services:
Manufactures polyurethane sealants, epoxies, cementitious mortar, grouts and admixture products.
Qualifications:
Company is over 50 years old, several products submitted for approval to EPA.

SMALLEY AND COMPANY
861 South Jason
Denver, CO 80223
(303) 777-3435
John Coy - Sales Representative
Products & Services:
Distributes radon mitigation products and one component polyurethane sealant.

SMOOTH-ON, INC.
1000 Valley Rd.
Gillette, NJ 07933
(201) 647-5800
David Hoadley - Vice President of Sales
Products & Services:
Epoxy resins, polyurethanes, and other polymers.

THORO SYSTEM PRODUCTS
7800 Northwest 38th St.
Miami, FL 33166
(305) 592-2081
Gerald J. Knudsen - Sales Director
Products & Services:
Manufactures Thoroseal, a heavy duty waterproof coating for basements and

foundations that also suppresses the infiltration of radon gas.

TODOL PRODUCTS, INC.
P.O. Box 398
Natick, MA 01760
(508) 879-7741
Tom Conlon - President
Pete Conlon - Sales Manager
Products & Services:
Exclusive national distributor of Multi-Foam 1-G and NBS dispensing gun.

TRA-CON, INC.
55 North St.
Medford, MA 02155
(617) 391-5550, (800) 872-2661
Dan Romanow - Sales Manager
Products & Services:
Epoxies, specialty packaging of polymer materials and sealants.

TREMCO, INC.
3735 Green Rd.
Beachwood, OH 44122
(216) 292-5000
Dave Larkin - Communications Manager
Products & Services:
Manufactures flowable, urethane sealants.

UNITED GILSONITE LABORATORIES
P.O. Box 70
Scranton, PA 18501
(717) 344-1202
Frank John - National Sales Manager
Products & Services:
Non-flowable, silicone caulk and other sealants.

URETECH, INC.
154 Potomac Ave.
Tallmadge, OH 44278
(216) 633-3586
George Karayianes - Vice President
Products & Services:
Manufactures one- and two- component polyurethane foam systems.

VIMCO, INC.
300 Hansen Access Rd.
King of Prussia, PA 19406
(215) 768-0500
David Morrison - Sales
Products & Services:
Distributes products for restoration and repair of basement slabs and walls for contractors, radon sumps for contractors.
Qualifications:
28 years experience supplying concrete and masonry related materials.

W.L. GORE AND ASSOCIATES
100 Airport Rd. Box 1010
Elkton, MD 21921
(301) 392-3200
John McMillan - Sales
Products & Services:
Manufactures joint/gap fillers, backer rods, and sealers.

W.R. GRACE/CONSTRUCTION PRODUCTS DIVISION
62 Whittemore Ave.
Cambridge, MA 02140
(617) 876-1400
Greg Ciampa - Product Manager/Polycel
Paul Korenberg - Product Manager/Zonolite
Products & Services:
Manufactures spray and expanding foams.

W.R. GRACE/ORGANIC CHEMICALS DIVISION
55 Hayden Ave.
Lexington, MA 02173
(617) 861-6600
Marc Mulready - Product Manager
Products & Services:
Manufacures Saran latex.

W.R. MEADOWS, INC.
P.O. Box 543
Elgin, IL 60121
(312) 683-4500
Brent Weber - Sales Director
Products & Services:
Manufactures non-ferrous, non-shrink grout.

WEATHER BARRIER SYSTEMS, INC./CAM GROUP
111 West Second St.
Jamestown, NY 14701
(716) 484-7183, (800) 624-6434
Charles A. Morando - President
Products & Services:
Manufactures water-borne safe, rubberized, fluid, radon mitigation waterproof membranes, sealants and caulkings.

See Ad Page 223

WESTINGHOUSE ELECTRIC CORPORATION
P.O. Box 355
Pittsburgh, PA 15230
(412) 374-5716
Jim Flaherty - Senior Engineer
Products & Services:
Distributes sprayable and flowable, two component polyurethane foam used in radon mitigation.

6. Radon Barriers

AKZO INDUSTRIAL SYSTEMS
Box 7249 1 North Pack Sq.
Asheville, NC 28704
(704) 258-5074
George Dodson - Project Engineer
Products & Services:
Manufactures and distributes Enkadrain, a new construction and retrofit radon venting conduit material for the subsurface.
Qualifications:
AARST, BOCA certified.

APPLIED ULTRASONICS, INC.
145 C Grassy Plains St.
Bethel, CT 06801
(203) 790-1102
Ray Sullivan - Vice President of Marketing
Products & Services:
Manufactures ultrascanner that welds membranes for use as a radon barrier.

BRENTWOOD INDUSTRIES, INC.
P.O. Box 605
Reading, PA 19603
(215) 374-5109
David Fry - Sales Manager
Products & Services:
Distributes Aqua Flex, a PVC membrane that can be used as a passive barrier system along with sub-slab ventilation, in both retro-fit and new construction.

ENERGY SAVER IMPORTS, INC.
P.O. Box 387 2150 West 6th Ave. Suite E
Broomfield, CO 80020
(303) 469-1787, (800) 288-3647
Bowen Hyma - Vice President of Sales
Products & Services:
Manufactures and distributes Foil Ray products, reflective insulating radon barriers for crawlspace and sub-slab radon ventilation systems and some exterior perimeter wall ventilation systems.
Qualifications:
Certified product testing available.

GLOBAL WHOLESALE
2410 East Lincolnway
Cheyenne, WY 82001
(307) 634-7318
Ron Langley - President
Products & Services:
Provides radon detection services using charcoal canister, alpha track and Honeywell Professional Radon Monitors, distributes Foil Ray products, useful as radon barrier.

GUNDLE LINING SYSTEMS, INC.
19103 Gundle Rd.
Houston, TX 77073
(713) 443-8564, (800) 435-2008
Hal Pastner - Vice President of Sales
Products & Services:
Manufactures high density polyethylene liners.

JPS ELASTOMERICS CORPORATION
395 Pleasant St.
Northampton, MA 01060
(413) 586-8750
Richard Taylor - Marketing Manager
Products & Services:
Manufactures impermeable rubber membranes.

NATIONAL SEAL COMPANY
1255 Monmouth Blvd.
Galesburg, IL 61401
(309) 343-3418, (800) 323-3820
John Hardison - President
Products & Services:
Manufactures and distributes geomembrane liner systems.

PALCO LININGS, INC.
2500-B Hamilton Blvd.
South Plainfield, NJ 07080
(201) 753-6262, (800) 457-2526
Frank Taylor - Sales Engineer
Wayne Behrsin - Sales Engineer
Products & Services:
Manufactures and custom fabricates extra-wide, impermeable membrane liners and covers to serve as gas migration barriers.

POLY-AMERICA, INC.
2000 West Marshall Dr.
Grand Prairie, TX 75051
(214) 647-4374, (800) 527-3322
Paul Kramer - National Sales Manager
Products & Services:
Manufactures polyethylene film products used as radon barriers.

RADON DETECTION AND REDUCTION LABORATORY/DIVISION OF FORMACHEM
10834 Ambassador
St. Louis, MO 63132
(314) 872-9350
Harold Susman - President
Donald Pearline - Biochemist
Products & Services:
Manufactures and distributes the Seal Safe Radon Gas Detector and alpha track detectors, distributes Enkadrain radon mitigation system for new home construction.

RADON REMOVAL, INC.
Box 283
Milltown, NJ 08850
(201) 545-2000
Anthony M. Fama - President
Products & Services:
Provides radon detection and mitigation services, distributes a geomatrix, nylon membrane laid over perimeter of gravel foundation to provide an effective radon barrier.
Qualifications:
AARST, engineer on staff.

RAVEN INDUSTRIES, INC.
P.O. Box 1007
Sioux Falls, SD 57117
(605) 336-2750, (800) 227-2836
Tom Denevan - Marketing Representative
Tom Stoebner - Marketing Representative
Hilda Schryver - Customer Service
Products & Services:
Provides under slab polyethylene moisture/vapor barrier for radon mitigation.
Qualifications:
Over 30 years of experience in plastics.

REDICORP PROTECTIVE MATERIALS
212 Dominion Rd., NE
Vienna, VA 22180
(703) 938-2928
J.R. Suda - President
G. Shea - Vice President
Products & Services:
Manufactures Radtex 8400; co-extruded, stabilized film laminate of PET, is non-woven and paintable.

RESOURCE CONSERVATION TECHNOLOGY, INC.
2633 North Calvert St.
Baltimore, MD 21218
(301) 366-1146
Lee Jaslow - President
Products & Services:
Manufactures and distributes full line of radon mitigation products including membranes, film barriers, sealants, radon sumps, fans, etc.

STO-COTE, INC.
P.O. Drawer 310
Richmond, IL 60071
(815) 675-2358
Jim Stoller - Sales Manager
Products & Services:
Manufactures a cross-laminated, high density polyethylene sheeting product.

TROCAL
10 Link Drive
Rockleigh, NJ 07647
(201) 767-1660
Paul Riesebieter - Technical Services
Products & Services:
Supplies polymer, PVC sheeting.

WEATHER BARRIER SYSTEMS, INC./CAM GROUP
See Ad Page 223
111 West Second St.
Jamestown, NY 14701
(716) 484-7183, (800) 624-6434
Charles A. Morando - President
Products & Services:
Manufactures water-borne safe, rubberized, fluid, radon mitigation waterproof membranes, sealants and caulkings.

7. Ventilation Equipment

AIRXCHANGE, INC.
401 VFW Dr.
Rockland, MA 02370
(617) 871-4816
Bede Wolford - Product Manager
Ed Leptich - Vice President of Sales
D.F. Steele - President
Products & Services:
Manufactures heat recovery ventilators for homes and commercial buildings.
Qualifications:
Products certified by HUD, R2000, UL listed, CSA approved.

ALTERNATE HEATING
621 Rothesay Ave.
Saint John, NB E2H 2G9
(506) 633-1500
Don Hemmings - Owner
Products & Services:
Distributes Vanee line of heat recovery ventilation systems and accessories.

AMERICAN ALDES VENTILATION CORPORATION
4539 Northgate Ct.
Sarasota, FL 34234-4864
(813) 351-3441, (800) 255-7749
Dwight Shackelford - Sales Manager
Products & Services:
Residential heat recovery ventilators, in-line duct fans, and other ventilation products.

ASTLE CORPORATION
146 Old Farm Rd.
Leominister, MA 01453-2029
(508) 534-0979
William Astle - President
Products & Services:
Provides radon detection and mitigation services, manufactures a patented air-to-air energy recovery device that reclaims 95% of heat lost up flue.

BERNER INTERNATIONAL CORPORATION
P.O. Box 5205
New Castle, PA 16105
(412) 658-1530, (800) 852-5015
Ken Lapinski - Regional Sales
Mike Coscarelli - Regional Sales
Products & Services:
Manufactures an indoor air quality appliance containing a heat exchanger for homes with 20 pCi/l or less of radon.

BLACKHAWK INDUSTRIES, INC.
607 Park St.
Regina, SK S4N 5N1
(306) 924-1551
Michael Horlick - President
Products & Services:
Manufactures and distributes heat recovery ventilation systems.

BOSSAIRE, INC.
1321 Tyler St., NE
Minneapolis, MN 55413
(612) 781-0179
Norma Swanson - President
Walter Patrek - Sales Manager
Larry Boeser - Vice President
Products & Services:
Manufactures and distributes air-to-air heat exchangers.

CONSERVATION ENERGY SYSTEMS
P.O. Box 10146
Minneapolis, MN 55458
(306) 242-3663, (800) 667-3717
Silvia L. Martini - Director of Sales
Doug Wicks - Technical Applications Specialist
Products & Services:
Manufactures Vanee product line of heat recovery ventilation systems and accessories.
Qualifications:
CSA certified, UL approved components.

CONTROL RESOURCE SYSTEMS, INC.
See Ad Page 234
670 Mariner Dr.
Michigan City, IN 46360
(219) 872-5591, (800) 272-3786
Ronald W. Marsh - Vice President
Don Musilli - National Sales Manager
Products & Services:
Manufactures and distributes air purification systems for residential, commercial and industrial facilities, manufactures the CRSI 600H Radon Removal System.
Qualifications:
Registered medical device manufacturer.

CURRENT INDOOR AIR SYSTEMS, INC.
See Ad Page 212
P.O. Box 18075
Boulder, CO 80308
(303) 440-8555
Kyle Lefkoff - General Manager
Kim Kenney - Marketing Manager
Products & Services:
Manufactures The Model 300 Radon Mitigation System, a pressurization/dilution mitigation device for crawlspace and basement pressurizations or where sub-slab alternatives are impractical, incorporates a passive air-to-air heat exchange in its design.

DES CHAMPS LABORATORIES, INC.
P.O. Box 440 17 Farinella Dr.
East Hanover, NJ 07936
(201) 994-4660
Nicholas Des Champs - President
Richard Schick - Residential Sales
Christina Del Guercio - Marketing Director
Products & Services:
Manufactures air-to-air heat recovery ventilators for use in radon mitigation.
Qualifications:
ETL listed.

DUOVENT INDUSTRIES, LTD.
1375 Bl. Charest Ouest Suite 6
Quebec, PQ G1N 2E7
(418) 688-8805
Pierre Chaput - President
Products & Services:
Manufactures residential air-to-air heat exchangers
Qualifications:
Licensed by Canadian Standards Association, Heating and Refrigeration Association Institute.

EER PRODUCTS, INC.
3536 East 28th St.
Minneapolis, MN 55406
(612) 721-4231
Gerry Martin - Owner
Steve Everson - Sales Engineer
Jim Cazett - Sales Engineer
Products & Services:
Distributor of air-to-air heat exchangers.

ENERGY INDEPENDENCE/THE RADON RESPONSE TEAM
9940 East 36th St.
Independence, MO 64052
(816) 353-3755
Kenneth C. Riead - President

Products & Services:
Provides radon detection and mitigation services for residential and commercial markets, distributes Vanee line of heat recovery ventilation systems.
Qualifications:
EPA RMP, 12 years of experience in the field.

ENERGY SAVINGS WAREHOUSE, INC.
3490 Hwy. 22 W.
Somerville , NJ 08876
(201) 534-6922
Ivan A. Franceus - President
Paul E. Franceus - Vice President
Products & Services:
Provides radon mitigation services, manufactures ventilation equipment for radon mitigation.

ENERMATRIX, INC.
P.O. Box 466
Fargo, ND 58107
(701) 232-3330
John Muhleman - General Manager
Products & Services:
Manufactures the EMX 20 air exchanger/humidity controller.

ENGINEERING DEVELOPMENT, INC.
4850 Northpark
Colorado Springs, CO 80918
(717) 599-9080
Scott Ford - President
Products & Services:
Distributes Vent-Aire air-to-air exchangers.

ENVIRONMENT AIR, LTD.
P.O. Box 459
Bouctouche, NB E0A 1G0
(506) 743-8901
Alvin Brun - President
See Ad Page 226
Products & Services:
Manufactures air-to-air heat recovery ventilators and exhaust equipment.

ENVIRONMENTAL ENTERPRISES, INC.
P.O. Box 16847
Atlanta, GA 30321
(404) 766-5551
Chris Esworthy - President
Products & Services:
Provides radon detection and mitigation services, distributes full line of radon mitigation products including fans, sealers, etc.

EVANSHIRE COMPANY, LTD., THE
Rte. 4, Box 57
Rome, NY 13440
(315) 865-5941, (800) 962-5941
Edwin Evans - President
Wade Evans - Vice President
See Ad Page 215
Products & Services:
Provides radon detection services using charcoal canister, alpha track detector, and grab sampling methods, provides radon mitigation services, distributes full line of radon detection and mitigation products, including sealers, sump covers, fans.
Qualifications:
Experience in state and federally funded projects.

F.H. TIMBLIN COMPANY
P.O. Box 254
Industry, PA 15052
(412) 643-1755
Frank Timblin - Owner
Products & Services:
Distributes RB Kanalflakt fans used in radon mitigation.

FANTECH, INC.
2225 Industrial Blvd.
Sarasota, FL 34234
(813) 351-2947
See Ad Page 228
Robert W. Robinson - General Manager
Products & Services:
Manufactures ABS plastic fans for use in radon mitigation, sells to distributors, manufacturer's representatives, and provides referrals for contractors.

HEIL COMPANY
P.O. Box 25995
Colorado Springs, CO 80918
(719) 593-8407
James Heil - Owner
Dave Strong - Sales Agent
Products & Services:
Provides ventilation filtration services.

INFILTEC/FACTORY
108 South Delphine Ave. P.O. Box 1125
Waynesboro, VA 22980
(703) 943-2776
David Brubaker - Factory Manager
See Ad Page 214
Frankie Patterson - Warehouse Manager
Arthur Saum - Research Manager
Products & Services:
Manufactures and distributes a full line of radon detection and mitigation products including fans, couplings, pressure gauges, alarms, sumps and drain covers, radon test kits, continuous radon monitors, caulks and sealants, manometers, blower doors, etc.
Qualifications:
EPA RMP.

MEMPHREMAGOG HEAT EXCHANGERS, INC.
P.O. Box 456
Newport, VT 05855
(802) 334-5412
Products & Services:
Supplies air-to-air heat exchangers.

MOUNTAIN ENERGY AND RESOURCES, INC.
15800 W. 6th Ave.
Golden, CO 80401
(303) 279-4971
Brian Becker - President
Dale Rowan - General Manager
Products & Services:
Air-to-air heat exchangers.

NEW ENGLAND DE-WATERING, INC.
70 Heritage Ave. Suite 14 Heritage Pl.
Portsmouth, NH 03801
(800) 535-1211
Wayne Bowers - President
Ron Morse - Regional Supervisor
Products & Services:
Provides radon mitigation services, manufactures ventilated flooring used in radon mitigation.

NUTECH ENERGY SYSTEMS, INC.
124 Newbold Ct.
London, ON N6E 1Z7
(519) 686-0797
Bob Henderson - President
Jerry Fenwick - General Manager
Products & Services:
Manufactures a full line of air and heat recovery systems.

OBAR SYSTEMS, INC.
RR 3, Box 759
Highland Lakes, NJ 07422
(201) 728-7679, (201) 697-0112
Gunner Bar - Owner
David O'Brien - Owner
See Ad Page 215
Products & Services:
Distributes complete range of radon mitigation products, including fans, heat exchangers, sealers and caulks, pumps, filters, water systems, couplings, sumps, manometers, expansion joint materials, and other accessories, provides radon mitigation services.

P.M. WRIGHT, LTD.
1300 Jules Poitras Blvd.
St. Lawrence, PQ H4N 1X8
(514) 337-3331
Marcel Patenaude - President
Harry Topikian - National Sales Manager
Products & Services:
Air-to-air heat exchangers.

QDT, LTD.
1000 Singleton Blvd.
Dallas, TX 75212
(214) 741-1993
John Survil - Regional Sales
Products & Services:
Manufactures heat recovery systems.

R.B. KANALFLAKT, INC.
1121 Lewis Ave.
Sarasota, FL 34237
(813) 366-7505, (813) 365-4118
See Ad Page 229
Rainer Blomster - President
John Crockett - Sales Manager
Products & Services:
Manufactures exhaust fans used for radon mitigation, products include in-line, wall type, roof exhausters, etc.
Qualifications:
Over 40,000 fans installed worldwide.

R.F. SIMON COMPANY, INC.
RD 2, Box 481
Barto, PA 19504
(215) 754-6455
See Ad Page 148
Ron Simon - Project Manager
Peter Maniscalco - Diagnostic Specialist
Products & Services:
Provides mitigation design and installation services, mitigation supply warehouse carrying a full line of radon mitigation and detection products including fans, caulks and sealants, couplings, comprehensive radon services provided to private, commercial, institutional, and government agencies, expert testimony.
Qualifications:
EPA RMP, 13 years of custom building experience, instructor for EPA training courses, consultant to Lawrence Berkeley Laboratory, Princeton University, PA DER.

RADON AND DEWATERING SYSTEMS, INC. OF MARYLAND
Box 1814
Silver Spring, MD 20902
(301) 593-1861
See Ad Page 215
Joel Wolensky - President
Products & Services:
Distributes full line of radon mitigation supplies including Fantech products.

RADON APPLIANCES, INC.
P.O. Box 1461
Glen Burnie, MD 21061
(301) 766-5711
Douglas W. Townsend - President
Duane O. Terry - Vice President

Products & Services:
Manufactures custom PVC sump covers, in-line fan, and water traps for home owners and contractors.
Qualifications:
AARST, PA course on reducing radon in structures.

RADON CONTROL PRODUCTS
511 Industrial Dr.
Carmel, IN 46032
(317) 846-7442, (800) 523-2082
Steven Weaver - President
Roger Ice - Sales Director
See Ad Page 217
Products & Services:
Distributes radon mitigation products including fan products, couplings, sealers, smoke guns, etc., manufactures the Weaver Werks retrofit sump pump cover and related retrofit products, distributes the Radon Control Products Catalog.

RADON MEASUREMENT AND CONTROL OF NEW MEXICO
1527 Granite St., NW
Albuquerque, NM 87104
(505) 843-7643
See Ad Page 238
William M. Turner - President
L. David Turner - Vice President
Products & Services:
Provides radon detection services for real estate agents, home owners, government agencies, and relocation firms, manufactures RadonDrain, a ventilation system for removal of indoor radon in on-grade homes, distributes radon "reference" chamber, used by radon detection firms to verify instrument measurement stability over time.
Qualifications:
EPA RMP, registered professional geologist, EPA mitigation course, radon instructor for NM Real Estate Commission.

RADON REDUCTION EQUIPMENT OF COLORADO
921 Main Suite C
Louisville, CO 80027
(303) 665-1776
Kevin Lightsey - President
Pete Werner - Manager
Lou Spier - Secretary
Products & Services:
Distributes Kanalflact fans, Current Indoor Air System, and sealing and venting products.

RADON-X CORPORATION
300 Valentine St.
Hackettstown, NJ 07840
(201) 852-6212, (800) 272-3669
John LoPilato - President
Products & Services:
Manufactures air exchange and sub-slab ventilating systems, provides radon mitigation services.

RAMSES II, LTD.
2355 East Dartmouth Ave.
Denver, CO 80210
(303) 758-5209
Robert L. Clark - Vice President
George A. Michael - President
Products & Services:
Manufactures and distributes charcoal canisters, electronic sump well cover kits, electronic PVC ventilation fans, and rotary air-to-air heat exchangers, provides radon mitigation services.
Qualifications:
Licensed professional engineer on staff.

RAYDOT, INC.
145 Jackson Ave.
Cokato, MN 55321
(612) 286-2103, (800) 328-3813
Niilo Raisanen - President
David Raisanen - Vice President
Products & Services:
Manufactures and distributes home air-to-air heat exchangers for industrial and residential use.

RDS/RADON DETECTION SERVICES, INC.-CORPORATE OFFICE
Rte. 179, Old York Rd. P.O. Box 419
Ringoes, NJ 08551
(201) 788-3080
James G. Davidson - President
See Ad Page 153
Susanne E. Micek - Vice President
Michael S. Terpilak - Health Physicist
Products & Services:
Provides full service radon detection mitigation, and consultation services, offers franchises, offers comprehensive training courses for radon professionals, manufactures and distributes RDS "remediator" fans, locations in four states.
Qualifications:
Has met all licensing requirements in PA, NJ, and KS, courses have been approved by DE, KY, PA, OH.

RESOURCE CONSERVATION TECHNOLOGY, INC.
2633 North Calvert St.
Baltimore, MD 21218
(301) 366-1146
Lee Jaslow - President
Products & Services:
Manufactures and distributes full line of radon mitigation products including membranes, film barriers, sealants, radon sumps, fans, etc.

RIEHS AND RIEHS
501 George St.
New Bern, NC 28560
(919) 637-9752
H. Erich Riehs - President
Products & Services:
Provides radon mitigation services, distributes air-to-air heat exchangers for residential applications.
Qualifications:
Ten years of experience in heat recovery and ventilation industry.

SAFE-AIRE, INC.
162 East Chestnut St. P.O. Box 160
Canton, IL 61520
(309) 647-0419, (800) 331-2943
Bill Wherley - Sales Manager
Chet Phillips - Products Manager
See Ad Page 161
Harry Ellison - Marketing Director
Don Stees - Vice President
Products & Services:
Provides patent-pending Safe-Aire radon mitigation system to dealers, provides radon detection and mitigation services, distributes extensive line of radon detection and mitigation products, dealership opportunities available.

SCWEST, INC.
4950 East Asbury Ave. Suite 5
Denver, CO 80222
(303) 759-1286
Jeffery Kramer - President
Products & Services:
Distributes radon mitigation equipment, cross-laminated polyethylene, urethane caulks, heat recovery ventilation systems, and Current Indoor Air Systems.

SKUTTLE MANUFACTURING COMPANY
Rte. 1, Box 16A
Marietta, OH 45750
(614) 373-9169
David B. Powers - President
Marty Hansell - Vice President
Earl K. Lewis - Product Engineer
Products & Services:
Manufactures Skuttle Make-Up Air Controls which provide controlled amounts of fresh air into the home.

SUNSTREAM DEVELOPMENT CORPORATION
300 North St.
Endicott, NY 13760
(607) 754-6200

Thomas Meade, Jr. - President
Peter R. Gotthardt - Production Manager
Ron Andrascik - Technician
Products & Services:
Provides radon detection and mitigation services to home owners, realtors, and commercial properties, distributes VANEE air-to-air heat exchangers.
Qualifications:
NRA, key personnel trained by NY State Energy Office.

VACU-RAY CORPORATION
P.O. Box 291
Whitehouse Station, NJ 08889
(201) 534-5471
See Ad Page 227
Ivan Franceus - President
Paul Franceus - Vice President
Ron Feasprack
Products & Services:
Manufactures and distributes the DC-1 low voltage DC radon fan specifically designed to be installed on the outside of a building as opposed to conventional in-line fans.

W.W. GRAINGER, INC. DISTRIBUTION GROUP
5959 West Howard St.
Chicago, IL 60648
(312) 647-8900
David Mills - Product Manager
Products & Services:
Broad line distributor of ventilation and related building equipment, components and supplies to the commercial, industrial and contractor markets.
Qualifications:
265 stocking locations in major cities.

8. Air Flow Measurement Equipment

AEROVIRONMENT, INC.
825 Myrtle Ave.
Monrovia, CA 91016-3424
(813) 357-9983
Mike Chan - Atmospherics Sciences Group
Tom Tombach - Environmental Programs
Products & Services:
Manufactures air sampling bags.

ALNOR INSTRUMENT COMPANY
7555 N. Linder Ave.
Skokie, IL 60076
(312) 677-3500
Fran Roche - Sales Office Manager
Frank Spevak - Product Manager
Products & Services:
Manufactures electromanometers, thermoanemometers and micromanometers to monitor air flow and for ventilation surveys.

BGI, INC.
58 Guinan St.
Waltham, MA 02154
(617) 891-9380
Products & Services:
Gas sampling bags.

BRAILSFORD AND COMPANY, INC.
670 Milton Rd.
Rye, NY 10580
(914) 967-1820
Bob Drummond - President
Products & Services:
Manufactures miniature brushless, DC, motor-driven diaphragm pumps and blowers for air sampling.

COLE-PARMER INSTRUMENT COMPANY
7425 N. Oak Park Ave.
Chicago, IL 60648
(312) 647-7600, (800) 323-4340
Products & Services:
Manufactures smoke candles, flowmeters, anemometers, and assorted scientific supplies.

CONTROL POWER SYSTEMS, INC.
1010 E. 31 St.
Hialeah, FL 33013
(305) 691-2260
Mark Grandholm - Vice President of Sales
Products & Services:
Manufactures tracer gas equipment.

DWYER INSTRUMENTS, INC.
P.O. Box 373
Michigan City, IN 46360
(219) 872-9141
Al Robertson - Vice President of Sales
Products & Services:
Manufactures low pressure gauges and switches, flowmeters, manometers.

E. VERNON HILL, INC.
P.O. Box 7053
Corte Madera, CA 94925
(415) 924-6837
Kirk Bocek - Vice President
Products & Services:
Manufactures air test equipment, smoke guns, smoke sticks and white and colored smoke candles for air movement testing.

GILIAN INSTRUMENT CORPORATION
8 Dawes Hwy.
Wayne, NJ 07470
(201) 831-0440
Kathy Denauex - Sales Manager
Phil Heiney - Technical Advisor
Products & Services:
Air pumps, calibrators, flow and sampling equipment.

INFILTEC/FACTORY
108 South Delphine Ave. P.O. Box 1125
Waynesboro, VA 22980
(703) 943-2776
See Ad Page 214
David Brubaker - Factory Manager
Frankie Patterson - Warehouse Manager
Arthur Saum - Research Manager
Products & Services:
Manufactures and distributes a full line of radon detection and mitigation products including fans, couplings, pressure gauges, alarms, sumps and drain covers, radon test kits, continuous radon monitors, caulks and sealants, manometers, blower doors, etc.
Qualifications:
EPA RMP.

MINE SAFETY APPLIANCES COMPANY
P.O. Box 426
Pittsburgh, PA 15230
(412) 967-3000
Jack Joy
Leonard Short
Products & Services:
Manufactures smoke tubes, air samplers, and protective equipment.

MINNEAPOLIS BLOWER DOOR
920 W. 53rd St.
Minneapolis, MN 55419
(612) 827-1117
Gary Nelson - Partner
Gary Anderson - Partner
Products & Services:
Manufactures Minneapolis Blower Door, a device used for locating leaks that allow radon entry, measures flow requirements for pressurizing space, determines leakage rates of buildings.

NATIONAL DRAEGER, INC.
101 Technology Dr.
Pittsburgh, PA 15275
(412) 787-8383
Wesley J. Kenneweg - Marketing Manager
Products & Services:
Provides detection and protection equipment.

NEOTRONICS
P.O. Box 370
Gainesville, GA 30501
(404) 535-0600

Shannon Simpson - Sales Manager
Products & Services:
Manufactures energy and air flow instrumentation.

OBAR SYSTEMS, INC.
RR 3, Box 759
Highland Lakes, NJ 07422
(201) 728-7679, (201) 697-0112
Gunner Bar - Owner
David O'Brien - Owner
See Ad Page 215
Products & Services:
Distributes complete range of radon mitigation products, including fans, heat exchangers, sealers and caulks, pumps, filters, water systems, couplings, sumps, manometers, expansion joint materials, and other accessories, provides radon mitigation services.

PROFESSIONAL HOUSE DOCTORS, INC.
1406 East 14th St.
Des Moines, IA 50316
(515) 265-6667
Dane J. Shearer - President
Products & Services:
Provides radon detection and mitigation services, manufactures blower doors with 25-6,000 cfm capacity, manufactures structural analyzers providing complete and detailed indoor air and energy analysis, performs mitigation training nationwide.

RADON CONTROL PRODUCTS
511 Industrial Dr.
Carmel, IN 46032
(317) 846-7442, (800) 523-2082
Steven Weaver - President
Roger Ice - Sales Director
See Ad Page 217
Products & Services:
Distributes radon mitigation products including fan products, couplings, sealers, smoke guns, etc., manufactures the Weaver Werks retrofit sump pump cover and related retrofit products, distributes the Radon Control Products Catalog.

RETROTEC
P.O. Box 939
Ogdensburg, NY 13669
(613) 723-2453
Brendan Reid - President
Products & Services:
Blower door and blower door accessories.

SAFE-AIRE, INC.
162 East Chestnut St. P.O. Box 160
Canton, IL 61520
(309) 647-0419, (800) 331-2943
Bill Wherley - Sales Manager
See Ad Page 161
Chet Phillips - Products Manager
Harry Ellison - Marketing Director
Don Stees - Vice President
Products & Services:
Provides patent-pending Safe-Aire radon mitigation system to dealers, provides radon detection and mitigation services, distributes extensive line of radon detection and mitigation products, dealership opportunities available.

SENSIDYNE, INC.
12345 Starkey Rd. Suite E
Largo, FL 34643
(813) 530-3602, (800) 451-9444
Bryan Truex - President
Tom Hanway - National Sales Manager
Products & Services:
Manufactures gas detection products including smoke tube kits for air flow determination.

SETRA SYSTEMS, INC.
45 Nagog Pk.
Acton, MA 01720
(617) 263-1400
Ed Cope - National Sales Manager
Richard Pansire - Application Engineer
Products & Services:
Manufactures digital pressure measurement systems.

SHORTRIDGE INSTRUMENTS, INC.
7855 E. Redfield
Scottsdale, AZ 85260
(602) 991-6744
Ernest Shortridge - President
Frankie Dowell - Sales Manager
Products & Services:
Air balancing systems, flow heads, micromanometers, air velocity, temperature and flow instruments.

SKC, INC.
334 Valley View Rd.
Eighty Four, PA 15330
(412) 941-9701, (800)752-8472
Carol Knouse - Sales Manager
Products & Services:
Pump calibrators, air flow, and sampling equipment.

SUPERIOR SIGNAL, INC.
P.O. Box 96
Spotswood, NJ 08884
(201) 251-0800
Frank Nichols, Jr. - President
Barbara Marques - Product Coordinator
Products & Services:
Manufactures full product line of leak detectors.

9. Sump Equipment

BLUE ANGEL PUMP COMPANY
801 Glasgow Ave. P.O. Box 2257
Fort Wayne, IN 46801
(219) 426-4000
Richard J. Ostroski - President
Lloyd A. Shaffer - Sales
Products & Services:
Manufactures submersible sump pumps, sewage pumps, and effluent pumps.
Qualifications:
Water Systems Council, Water Quality Association, National Water Well Association, Sump and Sewage Pump Manufacturers Association.

DRANJER CORPORATION
1441 Pembina Hwy.
Winnipeg, MB R3T 2C4
(204) 474-0451
See Ad Page 233
Harold Westdal - President
Products & Services:
Manufactures and distributes floor drain with U-shaped water trap and cup-type water trap designed to fit pipes as small as two inches in diameter.

EVANSHIRE COMPANY, LTD., THE
Rte. 4, Box 57
Rome, NY 13440
(315) 865-5941, (800) 962-5941
See Ad Page 215
Edwin Evans - President
Wade Evans - Vice President
Products & Services:
Provides radon detection services using charcoal canister, alpha track detector, and grab sampling methods, provides radon mitigation services, distributes full line of radon detection and mitigation products, including sealers, sump covers, fans.
Qualifications:
Experience in state and federally funded projects.

HANCOR, INC.
401 Olive St. P.O. Box 1047
Findlay, OH 45839
(800) 537-9520
Joe McNall - Sales Application Engineer
Products & Services:
Manufactures polyurethane, radon-sealed, dual-purpose sump for distribution to radon mitigation product distributors and radon mitigation contractors.

INFILTEC/FACTORY
108 South Delphine Ave. P.O. Box 1125
Waynesboro, VA 22980

(703) 943-2776
David Brubaker - Factory Manager
Frankie Patterson - Warehouse Manager
Arthur Saum - Research Manager
Products & Services:
Manufactures and distributes a full line of radon detection and mitigation products including fans, couplings, pressure gauges, alarms, sumps and drain covers, radon test kits, continuous radon monitors, caulks and sealants, manometers, blower doors, etc.
Qualifications:
EPA RMP.

See Ad Page 214

OBAR SYSTEMS, INC.
RR 3, Box 759
Highland Lakes, NJ 07422
(201) 728-7679, (201) 697-0112
Gunner Bar - Owner
David O'Brien - Owner

See Ad Page 215

Products & Services:
Distributes complete range of radon mitigation products, including fans, heat exchangers, sealers and caulks, pumps, filters, water systems, couplings, sumps, manometers, expansion joint materials, and other accessories, provides radon mitigation services.

RADON APPLIANCES, INC.
P.O. Box 1461
Glen Burnie, MD 21061
(301) 766-5711
Douglas W. Townsend - President
Duane O. Terry - Vice President
Products & Services:
Manufactures custom PVC sump covers, in-line fan, and water traps for home owners and contractors.
Qualifications:
AARST, PA course on reducing radon in structures.

RAMSES II, LTD.
2355 East Dartmouth Ave.
Denver, CO 80210
(303) 758-5209
Robert L. Clark - Vice President
George A. Michael - President
Products & Services:
Manufactures and distributes charcoal canisters, electronic sump well cover kits, electronic PVC ventilation fans, and rotary air-to-air heat exchangers, provides radon mitigation services.
Qualifications:
Licensed professional engineer on staff.

RESOURCE CONSERVATION TECHNOLOGY, INC.
2633 North Calvert St.
Baltimore, MD 21218
(301) 366-1146
Lee Jaslow - President
Products & Services:
Manufactures and distributes full line of radon mitigation products including membranes, film barriers, sealants, radon sumps, fans, etc.

SAFE-AIRE, INC.
162 East Chestnut St. P.O. Box 160
Canton, IL 61520
(309) 647-0419, (800) 331-2943
Bill Wherley - Sales Manager
Chet Phillips - Products Manager
Harry Ellison - Marketing Director
Don Stees - Vice President

See Ad Page 161

Products & Services:
Provides patent-pending Safe-Aire radon mitigation system to dealers, provides radon detection and mitigation services, distributes extensive line of radon detection and mitigation products, dealership opportunities available.

TRUSTY WARNS, INC.
246 East Irving Park Rd.
Wood Dale, IL 60191
(312) 766-9015
Carl Niedermeyer - President
Products & Services:
Manufactures a battery-operated back-up sump pump system that signals excessive water levels and removes water through discharge pipe above ground.

VIMCO, INC.
300 Hansen Access Rd.
King of Prussia, PA 19406
(215) 768-0500
David Morrison - Sales
Products & Services:
Distributes products for restoration and repair of basement slabs and walls for contractors, radon sumps for contractors.
Qualifications:
28 years experience supplying concrete and masonry related materials.

WEAVER WERKS

See Ad Page 151

511 Industrial Dr.
Carmel, IN 46032
(317) 846-7442, (800) 523-2082
Steven Weaver - President
Products & Services:
Distributes radon mitigation products including fans, couplings, sealers, smoke guns, etc., manufactures retrofit sump pump covers.
Qualifications:
EPA RMP.

10. Protective Equipment

GARON PRODUCTS, INC.
1974 Hwy. 35 CN20
Wall, NJ 07719
(201) 499-1776, (800) 631-5380
Art Crowley - President
Christy Karl - Marketing Manager
Lee Miller - Product Consultant
Products & Services:
Manufactures full line of sealers and protective equipment.
Qualifications:
NJ CC, National Safety Council, BBB, Direct Marketing Association, OSHA compliance.

LYONS SAFETY
North 92 W. 14350 Anthony Ave.
Menomonee Falls, WI 53051
(800) 558-8900
Gail Talabac - Marketing Coordinator
Joe Steffen - Purchasing
Products & Services:
Distributes safety products for industry.

MINE SAFETY APPLIANCES COMPANY
P.O. Box 426
Pittsburgh, PA 15230
(412) 967-3000
Jack Joy
Leonard Short
Products & Services:
Manufactures smoke tubes, air samplers, and protective equipment.

NATIONAL DRAEGER, INC.
101 Technology Dr.
Pittsburgh, PA 15275
(412) 787-8383
Wesley J. Kenneweg - Marketing Manager
Products & Services:
Provides detection and protection equipment.

VALLEN SAFETY SUPPLY COMPANY
13333 Northwest Fwy.
Houston, TX 77040
(713) 462-8700
Bob Sisterson - Marketing Services
Products & Services:
Distributes industrial safety equipment.

11. Miscellaneous

ADIRONDACK ALTERNATE ENERGY
County Rte. 4
Edinburg, NY 12134
(518) 863-4338
Bruce Brownell - President
Tom Doherty - Engineer
Products & Services:
Designs, engineers, and manufactures post and beam pre-cut structures, fully resistant to radon infiltration.

ALBION ENGINEERING COMPANY
2080-A Wheat Sheaf Ln.
Philadelphia, PA 19124
(215) 535-3476
Franz Schneider - President
Products & Services:
Manufactures caulking guns.

AMERICAN EXCELSIOR COMPANY
1 Lovellee Ln.
Thornton, PA 19373
(215) 459-1146
Abe Pollackov - Sales Manager
Products & Services:
Distributes venting mat used in pre-construction to build radon-resistant homes, provides radon mitigation services.

APPLIED HEALTH PHYSICS, INC.
2986 Industrial Blvd.
Bethel Park, PA 15102
(412) 563-2242, (518) 477-7974
R.G. Gallagher - President
John Douglas - Instrument Manager
John Harshman - Radon Specialist
Products & Services:
Professional radiation safety firm distributes charcoal canisters, provides instrument calibration, repair, and leasing.
Qualifications:
EPA RMP, twenty-five years of experience, NRC and state licensed, all services by board certified registered professional engineers.

BEAVER INDUSTRIES
890 Hersey St.
St. Paul, MN 55114
(612) 644-9933
Jim Cotten - Vice President of Sales
Products & Services:
Manufactures Beaver Basement Water Control System, a hollow baseboard that collects water where floor and wall join and directs to sump pump.

BIO-PURE COMPANY
P.O. Box 20593
Tampa, FL 33622-0593
(813) 839-8617
Don Saceman - President
Y.A. Demirjian - Consultant
Donah Parent - Vice President
Products & Services:
Manufactures radon mitigation system using plants with a specially formulated soil medium, process involves use of carbon rich soil media that filters out particulate matter with some adsorption of radon gas.
Qualifications:
Two years of field testing and development.

CONSTRUCTION SPECIALTIES, INC.
55 Winans Ave.
Cranford, NJ 07016
(201) 272-5200
Charlie Gross - Product Sales Manager
Bob Phillips - General Manager
Products & Services:
Manufactures full line of louvres.

DEFLECTO CORPORATION
7035 East 86th St.
Indianapolis, IN 46250
(317) 849-9555
Dick Fleck - Marketing
Products & Services:
Manufactures vent hoods.

DRANJER CORPORATION
1441 Pembina Hwy.
Winnipeg, MB R3T 2C4
(204) 474-0451
Harold Westdal - President

See Ad Page 233

Products & Services:
Manufactures and distributes floor drain with U-shaped water trap and cup-type water trap designed to fit pipes as small as two inches in diameter.

EASTERN FASTENERS, INC.
2 Commerce Dr.
Moorestown, NJ 08057
(800) 257-5830
Alan Wood
Products & Services:
Construction fasteners, anchors, drilling and tooling systems.

FERNCO, INC.
300 S. Dayton
Davison, MI 48423
(313) 653-9626, (800) 521-1283
Tom Galstezer - National Sales Manager
Products & Services:
Manufactures flexible couplings.

GENOVA PRODUCTS, INC.
7034 East Court St.
Davison, MI 48423

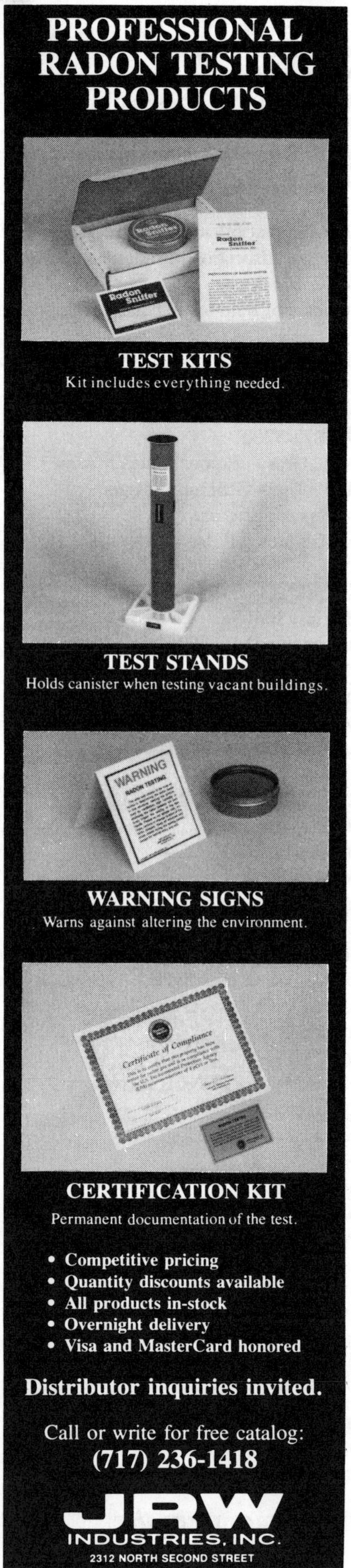

(313) 744-4500, (800) 521-7488
R.M. Williams - President
Dwight Van Steenkiste - Vice President
Jeanette Kellogg - Marketing Director
Products & Services:
Manufactures Raingo solid vinyl gutter and downspouts and Genova plastic plumbing pipe and fittings.

HEALTH PHYSICS ASSOCIATES, LTD.
3312 Commercial Ave.
Northbook, IL 60062-1909
(312) 480-1900
Ted Fields - President
Locke Bailey - General Manager
William Rivkin - Vice President
Products & Services:
Radiation service company provides radon detection services using charcoal canister, alpha track, and liquid scintillation methods, instrument calibration using radon chamber, and consultation on radiation safety.
Qualifications:
Certified health physicists, technicians registered for radon.

HEALTH PHYSICS NORTHWEST, INC.
11535 Southwest 67th Ave.
Tigard, OR 97223
(503) 620-6617, (800) 762-8444
Ross L. Mercer - President
Corwin M. Hymes - Sales Manager
Products & Services:
Radiation service company provides radiation consultation, instrumentation, materials and calibration and repair services to medical, industrial and commercial clients.

HOME ENVIRONMENT COMPANY
P.O. Box 24744
Speedway, IN 46224
(317) 247-1881
Gil Ward - Owner
Products & Services:
Manufactures Radon Sentry Alarm System, alarm signaling radon mitigation system failure, provides radon mitigation services.
Qualifications:
Completed certified course by Radon Detection Services, Inc.

INDIANA SEAL COMPANY
12000 Mitchell Rd.
Carmel, IN 46032
(317) 844-8700, (800) 428-5309
Peter D. Palasini - National Sales Manager
Brian R. McConnell - Vice President
Products & Services:
Manufactures flexible couplings.
Qualifications:
BOCA, IAPMO.

INFILTEC/FACTORY
108 South Delphine Ave. P.O. Box 1125
Waynesboro, VA 22980
(703) 943-2776
See Ad Page 214
David Brubaker - Factory Manager
Frankie Patterson - Warehouse Manager
Arthur Saum - Research Manager
Products & Services:
Manufactures and distributes a full line of radon detection and mitigation products including fans, couplings, pressure gauges, alarms, sumps and drain covers, radon test kits, continuous radon monitors, caulks and sealants, manometers, blower doors, etc.
Qualifications:
EPA RMP.

JRW INDUSTRIES, INC.
2312 North 2nd St.
Harrisburg, PA 17110
See Ad Page 236
(717) 236-1418
James R. Williams - President
Kevin J. Williams - General Manager
Valeria Hollern - Secretary
Products & Services:
Provides radon detection services, manufactures and distributes radon test kits, charcoal canisters, test stands, warning signs, certification kits, and other accessories.
Qualifications:
EPA RMP, PA DER, AARST.

K & S ASSOCIATES
1854 Airlane Dr.
Nashville, TN 37210
(615) 883-9760
Bob Dillard - Product Manager
Products & Services:
Dosimeter calibration laboratory.

KENMAR, INC.
2626 North Martha St.
Philadelphia, PA 19125-1493
(215) 739-1245
Bill Reynolds, III - Sales Manager
Products & Services:
Distributes caulking guns and accessories.

MILWAUKEE ELECTRIC TOOL
13135 West Lisbon
Brookfield, WI 53005
(414) 781-3600
Daniel Perry - General Sales Manager
Products & Services:
Full line of rotary percussion hammers and hammer drills and diamond drilling equipment.

N.A.K. PRODUCTION ASSOCIATES
1317 Apple Ave.
Silver Spring, MD 20910

(301) 565-0355
Norm Klotz - President
Larry Petcovic - Director
Fred Rehbein - Director
Products & Services:
Produced and distributes videotape: "Living With Radon: Protecting Your Health and Protecting Your Investment," a nontechnical radon overview intended for consumers, real estate industry, libraries, etc.
Qualifications:
Several award-winning health series videotapes.

NEW ENGLAND DE-WATERING, INC.
70 Heritage Ave. Suite 14 Heritage Pl.
Portsmouth, NH 03801
(800) 535-1211
Wayne Bowers - President
Ron Morse - Regional Supervisor
Products & Services:
Provides radon mitigation services, manufactures ventilated flooring used in radon mitigation.

OBAR SYSTEMS, INC.
RR 3, Box 759
Highland Lakes, NJ 07422
(201) 728-7679, (201) 697-0112
Gunner Bar - Owner
David O'Brien - Owner
See Ad Page 215
Products & Services:
Distributes complete range of radon mitigation products, including fans, heat exchangers, sealers and caulks, pumps, filters, water systems, couplings, sumps, manometers, expansion joint materials, and other accessories, provides radon mitigation services.

R.F. SIMON COMPANY, INC.
RD 2, Box 481
Barto, PA 19504
(215) 754-6455
See Ad Page 148
Ron Simon - Project Manager
Peter Maniscalco - Diagnostic Specialist
Products & Services:
Provides mitigation design and installation services, mitigation supply warehouse carrying a full line of radon mitigation and detection products including fans, caulks and sealants, couplings, comprehensive radon services provided to private, commercial, institutional, and government agencies, expert testimony.
Qualifications:
EPA RMP, 13 years of custom building experience, instructor for EPA training courses, consultant to Lawrence Berkeley Laboratory, Princeton University, PA DER.

RADON APPLIANCES, INC.
P.O. Box 1461
Glen Burnie, MD 21061
(301) 766-5711
Douglas W. Townsend - President
Duane O. Terry - Vice President
Products & Services:
Manufactures custom PVC sump covers, in-line fan, and water traps for home owners and contractors.
Qualifications:
AARST, PA course on reducing radon in structures.

RADON CONTROL PRODUCTS
511 Industrial Dr.
Carmel, IN 46032
(317) 846-7442, (800) 523-2082
See Ad Page 151

Steven Weaver - President
Roger Ice - Sales Director
Products & Services:
Distributes radon mitigation products including fan products, couplings, sealers, smoke guns, etc., manufactures the Weaver Werks retrofit sump pump cover and related retrofit products, distributes the Radon Control Products Catalog.

RADON MEASUREMENT AND CONTROL OF NEW MEXICO
1527 Granite St., NW
Albuquerque, NM 87104
(505) 843-7643
See Ad Page 238
William M. Turner - President
L. David Turner - Vice President
Products & Services:
Provides radon detection services for real estate agents, home owners, government agencies, and relocation firms, manufactures RadonDrain, a ventilation system for removal of indoor radon in on-grade homes, distributes radon "reference" chamber, used by radon detection firms to verify instrument measurement stability over time.
Qualifications:
EPA RMP, registered professional geologist, EPA mitigation course, radon instructor for NM Real Estate Commission.

RADON QC
2857 Nazareth Rd. Palmer Professional Building Suite 6
Palmer, PA 18043
(215) 559-1636
See Ad Page 237
John Rider - Manager
Kelsie Geiger - Assistant Manager
Products & Services:
Provides quality control services for radon measurement companies, known concentrations of radon gas and radon daughters, three chambers of radon gas concentrations, radon chamber services available to radon detection firms.

RADON REMEDIES, INC.
2700 West Broad St. P.O. Box 44234
Columbus, OH 43204
(614) 272-7772
Craig S. Wise - President
James M. Farrow - General Manager
Products & Services:
Produces and distributes radon video containing general information, radon detection and mitigation techniques and products, for homes and businesses.
Qualifications:
EPA RMP, ARA.

RADON SAFETY AND TECHNOLOGY CORPORATION
4 Charles St.
Pleasant Valley, NY 12569
(914) 635-1138, (800) 552-5257
Ronald J. Gasparro - President
Zia R. Hashmi - Director
Claude J. Noto - Assistant Director
Products & Services:
Radon measurement analytical laboratory, water and soil measurements, surveys for real estate transactions, land sites, and newly constructed and standing buildings, distributes charcoal canisters and alpha track detectors, waterborne radon test kits, radon alarms and auxiliaries offered to public and professionals, provides radon mitigation services.
Qualifications:
EPA RMP, AARST, ASTM.

RADON TESTING CORPORATION OF AMERICA
12 West Main St.
Elmsford, NY 10523
(914) 347-5010, (800) 457-2366
Nancy Bredhoff - Vice President
Kurt Dorfi - Sales Manager
Products & Services:
Manufactures and distributes charcoal canister radon test kits, distributes Pro Scan, a product designed to aid detection firms to prevent test tampering.
Qualifications:
EPA RMP, approved by NY DOH.

ROMAES CORPORATION
6395 Gunpark Dr. Suite C
Boulder, CO 80301
(303) 530-1637
Reuben Maes - President
Products & Services:
Repairs and calibrates radiation detection instruments including counting systems and air sampler pumps.
Qualifications:
Radioactive materials license.

SAFE-AIRE, INC.
162 East Chestnut St. P.O. Box 160
Canton, IL 61520
(309) 647-0419, (800) 331-2943
See Ad Page 161
Bill Wherley - Sales Manager
Chet Phillips - Products Manager
Harry Ellison - Marketing Director
Don Stees - Vice President
Products & Services:
Provides patent-pending Safe-Aire radon mitigation system to dealers, provides radon detection and mitigation services, distributes extensive line of radon detection and

mitigation products, dealership opportunities available.

SCIENTIFIC TESTING ASSOCIATES, LTD.
Star Route Box 124
Topping, VA 23169
(804) 758-5728
John Reynolds - President
Products & Services:
Manufactures charcoal canister holder for ground radon emission testing, provides radon detection services.
Qualifications:
Doctorate degree in biology.

SUNN CORPORATION
Box M, Dept. 213 3520 Forbes Ave.
Pittsburgh, PA 15213
(412) 828-9100
Nick Gromicko - Chief Executive Officer
Bob Dunkle
Mike Russell
Products & Services:
Provides radon mitigation services, manufactures and distributes charcoal canisters and tamper-resistant cages.
Qualifications:
BBB, NSBA, several state mitigation lists.

TEQ CORPORATION
P.O. Box 570
Wilkes-Barre, PA 18703-0570
(717) 825-3939, (800) 877-2366
See Ad Page 100
Russell Hendershot - President
Ed Dructor - Vice President
Brian Redmond - Research Scientist
Products & Services:
Provides radon detection and mitigation services for commercial and residential markets, manufactures and distributes charcoal canisters, maintains calibrated radon chamber, offers radon detection and mitigation training courses.
Qualifications:
PA DER certification.

TOM SCURRY ASSOCIATES
1976 Linden Dr.
Boulder, CO 80302
(303) 444-4802
See Ad Page 210
Tom Scurry - President
Robert Droullard - Vice President
Products & Services:
Maintains a radon chamber to test and calibrate detection instruments, manufactures and distributes continuous and grab working level monitors, provides radon detection services.

VIMCO, INC.
300 Hansen Access Rd.
King of Prussia, PA 19406
(215) 768-0500
David Morrison - Sales
Products & Services:
Distributes products for restoration and repair of basement slabs and walls for contractors, radon sumps for contractors.
Qualifications:
28 years experience supplying concrete and masonry related materials.

WAYNE MUSSER PROPERTY COMPANY, INC.
517 Broad St. P.O. Box 258
Sewickley, PA 15143
(412) 741-9517
Wayne C. Musser - President
Ty W. Musser - Vice President
Products & Services:
Manufactures a tamper proof radon test enclosure that can be electronically activated by attempts to cover or move, provides radon detection services.

5

Research Facilities

This section lists research facilities that have recently sponsored basic and applied radon-related research. These projects involve research in a number of areas of interest including epidemiology, geology, hydrology, radiology, environmental engineering, building science, and public policy. Please note that studies funded and conducted at federally owned and operated research facilities are not listed in this section and can be found in Section 8.

ALPHANUCLEAR COMPANY
1125 Derry Rd. East
Mississauga, ON L5T 1P3
(416) 676-1364
G.M. Denham - President
Activities:
Develops and manufactures radon/radon daughter measurement instrumentation, provides charcoal canister radon test kits/analysis in Canada.

AMERICAN WATER WORKS ASSOCIATION RESEARCH FOUNDATION (AWWARF)
6666 West Quincy Ave.
Denver, CO 80235
(303) 794-7711
Rick Corland - Director, Research Management
Activities:
Assesses research needs of members concerning radon-in-water issues.

ARIZONA STATE UNIVERSITY/COLLEGE OF ENGINEERING AND APPLIED SCIENCES/DEPARTMENT OF ELECTRICAL AND COMPUTER ENGINEERING
Department Of Electrical And Computer Engineering, Radiological Imaging Facility
Tempe, AZ 85287-5706
(602) 965-2727
Kimberly Kearfott - Director
Activities:
Investigation of radon gas contamination in Arizona.

ARTHUR SCOTT AND ASSOCIATES/AMERICAN ATCON
2150 Burnhamthorpe Rd., West
Box 164
Mississauga, ON L5L 3A2
(416) 828-2389
Arthur G. Scott - President
Andrew Robertson - Research
Activities:
Operates under the name American Atcon in the U.S. Presently managing EPA project in Ohio that deals with mitigation techniques in buildings. Has contracted with FIPR for mitigation projects in Florida. Radon projects deal with experimental mitigation procedures and construction techniques.

BATTELLE MEMORIAL INSTITUTE/NUCLEAR TECHNOLOGY SECTION
505 King Ave.
Columbus, OH 43201-2693
(614) 879-5192
V. Pasupathi - Section Manager
Mike Stenhouse - Principal Research Scientist

Activities:
Contracts with federal government and other oraganizations for radon-related research projects. Radon analysis performed at laboratory.

BOSTON COLLEGE/DEPARTMENT OF GEOLOGY AND GEOPHYSICS
Devlin Hall
Chestnut Hill, MA 02167
(617) 552-3656
Rudolph Hon - Associate Professor
Activities:
Analysis of emanation rates of radon from solid materials (granites and shales) into surrounding environments (air, water) to determine what controls mechanisims of escape. Initial stages of research indicate rates are higher than expected due to microcracks and channeling. Goal is to establish region of predictability to determine potential for radon in homes in an area. Second phase of research will be to analyze migration rates through soils.

BRANDEIS UNIVERSITY/ROSENTIEL RESEARCH CENTER
415 South St.
Waltham, MA 02154
(617) 736-2461
Daniel Perlman - Senior Scientist
Activities:
Developing new passive and active radon detector based on adsorption. Developer of liquid scintillation counting technology in radon detection. Developed first 100 percent humidity independent cc based detector.

CAMRODEN ASSOCIATES
R.F.D. 4
Box 62
Rome, NY 13440
(315) 865-4269
Terry Brennan - President
Activities:
Indoor air quality investigation and training programs for federal, state, and private sectors groups.

CANADIAN INSTITUTE FOR RADIATION SAFETY/ELLIOTT LAKE LABORATORY
7 Timmins Rd.
Suite 7-15
Elliott Lake, ON P5A 2R7
(705) 848-3687
Ernest Becker - Senior Scientist, Laboratory Manager
Activities:
- Non-profit agency operates as a registered charity. Has enacted radon-daughter assessments in uranium mines (550 miners).
- Provides mitigation and assessment training and measures radon levels in private homes.
- Presently developing detection devices.
- Environmental monitoring and applied research.

CASE WESTERN RESERVE UNIVERSITY/SCHOOL OF MEDICINE
2048 Adelbert Rd..
Cleveland, OH 44186
(216) 844-3530
Helen W. Evans - Professor
Activities:
DOE funded animal research project analyzing mutagenic aspects of low dose rate exposure to radon and radon progeny.

CATHOLIC UNIVERSITY OF AMERICA
Pangborn Hall
Radiation Safety Dept.
Washington, DC 20065
(202) 635-5206
Warren Keene - Professor

CLARK UNIVERSITY/ENVIRONMENT, TECHNOLOGY, AND SOCIETY PROGRAM/DEPARTMENT OF PHYSICS
950 Main St.
Worcester, MA 01610
(617)793-7687
Robert L. Goble - Research Associate Professor
Cristoph Hohenemser - Chair, Environment, Society,
Halina Brown - Associate Professor
Activities:
Analyzes distribution of indoor radon exposure and radon epidemiology (possibly expanding to environmental epidemiology). Performs some radon measurements and experiments with measurement devices utilizing physics laboratory.

CLEMSON UNIVERSITY/ENVIRONMENTAL SYSTEMS ENGINEERING
501 Rhodes Engineering Research Center
Clemson, SC 29634-0919
(803) 656-5569
Robert A. Fjeld - Associate Professor
Activities:
Performed string survey for state of South Carolina. Ongoing work with South Carolina Department of Health and Environmental Control. Performs radon research on a continuing basis.

COLORADO STATE UNIVERSITY/DEPARTMENT OF RADIOLOGY AND RADIATION BIOLOGY
Ft. Collins, CO 80523
(303) 491-5222
F. W. Whicker - Professor
Thomas B. Borak - Professor
Activities:
Studies of fundamental processes in soils and transport of radon into basements. Two experimental structures in open field serve as test sites. Monitors radon in structures on a continuous basis for analysis of variations of levels over time. Sensors register soil moisture, soil temperature, and pressure differentials. Barometric pressure and temperature are measured on surface. Also outdoor concentrations of radon are measured to determine what best correlates with radon concentrations in basements. Attempt is being made to determine whether events behave identically to facilitate a range of experiments to identify methods to obstruct radon.

CORNELL UNIVERSITY/INSTITUTE FOR COMPARATIVE AND ENVIRONMENTAL TOXICOLOGY
Fernow Hall
Ithaca, NY 14850
(607) 255-9107
Jeanne Appling - Assistant Director for Extension
Activities:
Research on private drinking water sources. Pilot study assessing radon exposure from dug and drilled wells in 150-200 homes in central Finger Lakes region of upstate NY.

DREXEL UNIVERSITY/DEPARTMENT OF CHEMISTRY/ATMOSPHERIC CHEMISTRY PROGRAM
32nd and Chestnut Sts.
3rd Floor
Philadelphia, PA 19104
(215) 895-2638
Jack Kay - Professor
Activities:
NSF grant awarded for survey of radon-222 and lead-210 in tropical and subtropical regions of Pacific Ocean.

DUKE UNIVERSITY MEDICAL CENTER/DEPARTMENT OF MEDICINE/DIVISION OF PULMONARY MEDICINE
353 Bell Building
Box 3177
Durham, NC 27710
(919) 684-2513
Robert R. Mercer - Assistant Professor of Medical Research
Activities:
DOE funded research project to develop data important for the calculation of radiation dose to cells after inhalation of radon progeny.

EAST STROUDSBURG UNIVERSITY/ PHYSICS DEPARTMENT/RADON PROJECT
East Stroudsburg, PA 18301
(717) 424-3350
Paul N. Houle - Physics Department Chairman
Activities:
Research on radon concentration distribution (geographic) based on statistics gathered from radon testing analysis performed by the department. Also researching diffusion coefficients of radon.

ELECTRICAL POWER RESEARCH INSTITUTE (EPRI)/ENERGY MANAGEMENT AND UTILIZATION DIVISION/RESIDENTIAL AND COMMERCIAL PROGRAMS
3412 Hillview Ave.
Palo Alto, CA 94304
(415) 855-2000
Arvo Lannus - Project Manager
Activities:
Assists electric utilities in helping homeowners, builders, and new home buyers understand broad range of issues related to indoor air quality. Involved in research on indoor air quality as it affects the energy industry.

ELECTRICAL POWER RESEARCH INSTITUTE (EPRI)/LAWRENCE BERKELEY LABORATORY(LBL)/APPLIED SCIENCE DIVISION
University of California
One Cyclotran Rd.
Berkeley, CA 94720
(415) 486-4000
David Grimsrud - Building Ventilation and Indoor Air
Richard Diamond - Energy Performance of Buildings
Activities:
LBL contracts with EPRI to research topics on building ventilation and indoor air quality including radon. Produced two major indoor air quality reports.

FEMTO-TECH, INC.
150 C Industry Dr.
Carlisle, OH 45005
(513) 746-4427
E.I. Burgess - President
C.J. Kershner - Vice President
R. Manning - Director of Marketing
Activities:
Manufactures radon detection instrument and radon testing laboratory.

FLORIDA STATE UNIVERSITY/DEPARTMENT OF GEOLOGY
305 Carroway Bldg.
Tallahassee, FL 32306
(904) 644-5784
James B. Cowart - Associate Professor
William C. Burnette - Professor of Oceanography
Activities:
FIPR funded research analyzing the vertical and horizontal distribution of radon in specific locations. Analysis of radon in shallow groundwater to determine whether or not it is a predictor for amount of radon in home.

FORUM FOR SCIENTIFIC EXCELLENCE, INC., THE
P.O. Box 682
Sparta, NJ 07871
(201) 729-4823
George Thompson - Director
Activities:
Private company dealing with hazardous chemical information and services. Has released informational overview on radon and real estate concerns. Provides impact analysis on radon-related topics. Future plans include providing full range of radon sampling techniques. Publishes books and booklets based on research.

FRED HUTCHINSON CANCER RESEARCH CENTER/PUBLIC HEALTH SCIENCES DIVISION
1124 Columbia St.
Seattle, WA 98104
(206) 467-4267
Suresh H. Moolgavkar - Professor
Activities:
DOE funded research project analyzing the interaction between radon inhalation and cigarette smoke in the development of lung cancer in rats.

G.K. YUILL AND ASSOCIATES, LTD.
1441 Pembina Hwy.
Winnipeg, MB R3T 2C4
(204) 474-2461
Gren Yuill - President
Gord Comeau - Professional Engineer
Activities:
Mechanical engineering and consulting firm develops and evaluates mitigation measures. Presently designing a model for construction techniques best suited for predicting radon levels in housing.

GAS RESEARCH INSTITUTE (GRI)/INDOOR AIR QUALITY RESEARCH PROGRAM
8600 West Bryn Mawr Ave.
Chicago, IL 60631
(312) 399-8100
Irwin Billick - Director
Activities:
Involved in research project with NAHB analyzing effects of home combustion sources on radon.

GEORGE MASON UNIVERSITY/CENTER FOR BASIC AND APPLIED SCIENCE
4400 University Dr.
Fairfax, VA 22030
(703) 323-4318
Douglas Mose - Chairman, Geology Department
George Mushrush - Chairman, Chemistry Department
Stephen Kline - Associate Professor
Activities:
• Analysis of seasonal nature of indoor radon concentration (two year study) in Virginia and Maryland.
• Studying geological units, home construction and home use to determine how these factors affect radon exposure in the home.
• Radon-water analysis to determine what geologic units will produce radon enriched levels of groundwater.
• Soil gas and soil permeability research.
• Analyzing sub-slab ventilation method to determine efficiency in home remediation programs.
• Planning cancer-indoor radon exposure study.

GEORGE WASHINGTON UNIVERSITY/DEPARTMENT OF GEOLOGY
2029 G St., NW
Washington, DC 20052
(202) 994-6194
Frederic R. Siegel - Professor of Geochemistry
Activities:
Radiation measurement near ground surface using thermoluminescent dosimetry. Less reactive than radon measurement devices to grade changes seasonal and atmospheric variations (temperature, barometric pressure).

GEORGIA INSTITUTE OF TECHNOLOGY/NUCLEAR ENGINEERING AND HEALTH PHYSICS PROGRAM
ESM Building
Atlanta, GA 30332
(404) 894-3720
Geoffrey Eichholz - Regents Professor
Activities:
Analysis of radon evolution from building materials. Research to develop coatings (paints, wall papers) to reduce evolution. Utilizes radon chamber to analyze radon evolution from watersprays.

HARVARD UNIVERSITY/SCHOOL OF PUBLIC HEALTH
677 Huntington Ave.
Boston, MA 02115
(617) 732-1000
Dade W. Moeller - Professor of Environmental Engineering

INFILTEC/MAIN OFFICE
P.O. Box 8007
Falls Church, VA 22041
(703) 820-7696
David Saum - President
Marc Messing - Vice President
Activities:
Provides radon mitigation, research and development, nationwide training, consulting services; specializes in large buildings.

IOWA STATE UNIVERSITY/BUSINESS AND ENGINEERING EXTENSION/IOWA RADON PROJECT
EES-Building, Room 101
Haber Road
Ames, IA 50011-0588
(515) 294-0588
Laurent Hodges - Professor of Physics and Research Coordinator
Activities:
• Statewide sample (random and selected) to determine radon levels in all 99 counties in state.
• Survey identifies characteristics of homes (construction methods, design) sample is drawn from.
• Data are being analyzed to determine correlation among factors such as geography and construction methods.
Iowa Radon Project provides public information and sells CC and ATD.

JOHNS HOPKINS UNIVERSITY/DEPARTMENT OF ENVIRONMENTAL HEALTH SCIENCES/DIVISION OF ENVIRONMENTAL HEALTH ENGINEERING
615 Wolfe St.
Baltimore, MD 21205
(301) 955-3720
David L. Swift - Professor/Environmental Health Engineering
Activities:
DOE funded epidemiological research to analyze factors that influence deposition of aerosols in human airways above the larynx. Age is hypothesized to be a critical factor in the deposition process and information is important for the development of dosimetric models for radon progeny.

KUTZTOWN UNIVERSITY/DEPARTMENT OF PHYSICAL SCIENCES/RADON PROJECT
110 Grim Science Building
Kutztown, PA 19530
(215) 683-4443
Edward Vitz - Director

Activities:
Three-year nationwide study on radon in private water supplies (18.3 percent of U.S. population gets water from private supplies).

LOWRY ENGINEERING, INC.
P.O. Box 536
Unity, ME 04988
(207) 948-3790
Jerry Lowry - Director of Research
Sylvia B. Lowry - President
Activities:
Engineering research and consulting firm specializes in radionuclide analysis and removal from water supplies, and removal equipment for homes and public water supplies.

MALCOLM PIRNIE, INC.
100 Eisenhower Ave.
Paramus, NJ 07652
(201) 845-0400
David Hiltebrand - Water Quality and Treatment
Activities:
Environmental engineering consulting firm participating in EPA-funded water analysis project. Developing cost/technology report for removal of radon from drinking water supplies.

MASSACHUSETTS INSTITUTE OF TECHNOLOGY/RADON PROJECT
Building 26
Room 421
Cambridge, MA 02139
(617) 253-4151
Lee Grodzins - Professor

MCMASTER UNIVERSITY/HEALTH SCIENCES CENTER/DEPARTMENT OF EPIDEMIOLOGY
1200 Main St., West
Hamilton, ON L8N 3Z5
(416) 525-3892
Harry Shannon - Professor
Activities:
Epidemiological studies of mortality rates among Canadian uranium miners.

MCMASTER UNIVERSITY/HEALTH SCIENCES CENTER/OCCUPATIONAL HEALTH PROGRAM
1200 Main St., West
Room 3H50
Hamilton, ON L8N 3Z5
(416) 525-9140
David Muir - Program Director
Activities:
Participant in the NAS research program that produced the BEIR IV report analyzing the health effects of radon progeny exposure.

NATIONAL ACADEMY OF SCIENCES (NAS)/NATIONAL RESEARCH COUNCIL (NRC)/FEDERAL CONSTRUCTION COUNCIL
2101 Constitution Ave., NW
Washington, DC 20418
(202) 334-3378
Henry Borger - Executive Secretary
Activities:
Private non-profit organization researches topics such as radon in buildings and structures on an ad hoc basis.

NATIONAL ASSOCIATION OF HOME BUILDERS (NAHB)/NATIONAL RESEARCH CENTER
400 Prince Georges Center Blvd.
Upper Marlboro, MD 20772
(301) 249-2000
John Spears - Program Manager
Activities:
• Research arm of NAHB. Conducts training, provides technical assistance and conducts field research. Involved in research projects focusing on radon as it affects building industry.
• Works with EPA to provide information on radon-related building and construction issues. Plans are to develop a draft building code to support building code development.

NATIONAL CENTER FOR ATMOSPHERIC RESEARCH/ATMOSPHERIC CHEMISTRY DIVISION
Box 3000
Boulder, CO 80307
(303) 497-1000
John Winchester - Research Scientist
Edward Martell - Research Scientist
Activities:
National Science Foundation and member organizations provide grants and funding for this private institution that has been involved in a number of radon research projects.

NATIONAL COUNCIL ON RADIATION PROTECTION AND MEASUREMENTS (NCRP)
7910 Woodmont Ave.
Suite 800
Bethesda, MD 20814
(301) 657-2652
Warren Sinclair - President
W. Rosen Ney - Executive Director
William Beckner - Staff Scientist
Activities:
Organization of scientists employing cooperative effort for advances in radiation research and measurement. Has released two major reports on radon hazards and is active in radon-related research.

NATIONAL INSTITUTE OF BUILDING SCIENCES (NIBS)
1015 15th St., NW
Suite 700
Washington, DC 20005
(202) 347-5710
David Harris - Acting President
Leslie Levine - Manager of Communications
Activities:
Non-profit organization brings together government, labor, consumer, and private interests to focus on construction of safe structures in U.S. Actively involved in radon issues that affect the building industry.

NATIONAL INSTITUTE OF BUILDING SCIENCES (NIBS)/INDOOR AIR QUALITY COMMITTEE
C/O Armstrong World Industries
333 Liberty St.
Lancaster, PA 17603
(717) 397-0611
Nat E. Hager, Jr. - Senior Research Associate
Activities:
Develops strategies for indoor air quality research. Directs research on radon as it affects the building industry.

NEW JERSEY INSTITUTE OF TECHNOLOGY/DEPARTMENT OF CIVIL AND ENVIRONMENTAL ENGINEERING
323 King Blvd.
Newark, NJ 07102
(201) 761-0977
G. Fred Lee - Distinguished Professor
R. Anne Jones - Associate Professor
Activities:
Evaluation of leaching of radon-226 from uranium mill tailings and radium watchdial painting wastes. Risk assessment of contamination of ground/surface by radium or radon.

NEW MEXICO INSTITUTE OF MINING AND TECHNOLOGY/DEPARTMENT OF PHYSICS/RADON LABORATORY
Weir 104
Socorro, NM 87801
(505) 835-5328
Stephen Schery - Associate Professor
Marvin Wilkening - Professor of Physics Emeritus
Activities:
• Radon research began in the early 1950's when NM Tech used natural radon as a tracer in atmospheric studies.
• Investigated origin of radon in soil and rocks in the crust. Methods were developed to measure radon movement from soil to air. Developed instrumentation for measuring radon flux density as related to measuring indoor radon.
• Investigated physics of radon in air-electrical effects on radon and how that influences radon levels. Studies suggested radon daughters might be more damaging in ion form.
• Studies on radon in natural environments indicate unusually high radon levels at Carlsbad Caverns, and identified air exchange as a critical factor determining radon levels.
• DOE funded project analyzing the process of sorption and whether or not the phenomenon is significant enough in porous media to explain releases of radon that have been reported.

NEW MEXICO INSTITUTE OF MINING AND TECHNOLOGY/GEOSCIENCE DEPARTMENT
Socorro, NM 87801
(505) 835-5308
John Wilson - Professor of Hydrology
Chia S. Chen - Associate Professor
Activities:
Mathematical modeling of radon movement in vadose zone (diffusion). Analysis of radon movement through water (movement to wells). Estimates rate and how to operate to minimize risk.

NEW YORK UNIVERSITY MEDICAL CENTER/INSTITUTE OF ENVIRONMENTAL MEDICINE
550 1st Ave.
New York, NY 10016
(212) 340-5280
Arthur Upton - Professor and Chairman
Gerald Laurer - Research Professor
Frederick Burns - Professor
Naomi Harley - Research Professor
Activities:
DOE funded research projects providing:
• Research on health effects of cumulative radon exposure; radon-induced lung tumors; and development of radon detecting devices.
• Analysis of carcinogenic risk of radon inhaled alone or in combination with other indoor air pollutants.
• Analysis of mechanisms for radon entry into different types of houses and level of radon attributable to different types of sources.

NEW YORK UNIVERSITY MEDICAL CENTER/INSTITUTE OF ENVIRONMENTAL MEDICINE
550 1st Ave.
New York, NY 10016
(212) 340-5280
Beverly S. Cohen - Research Associate Professor
Edith S. Robbins - Research Professor
Activities:
DOE funded projects to:
• Assess the effects of particle size, breathing patterns and respitory tract morphology on the alpha radiation dose from radon progeny to cells of the bronchial epithelium.
• Perform comparative animal/human research to increase knowledge of characteristics of cells in the bronchi. This will enhance researchers ability to make more accurate radon daughter risk calculations.

OHIO UNIVERSITY/GEOLOGICAL SCIENCES DEPARTMENT
Athens, OH 45701
(614) 593-1109
Geoffrey Smith - Professor
Royal Mapes - Associate Professor
Rich Darr - Research Assistant
Jon Hinkel - Research Assistant
Activities:
Research funded by Ohio Air Quality Development Authority. Performing baseline analysis of radon distribution in Ohio looking at two contributors: black shales and glacial deposits. Attempting to identify correlation between glacial deposits and radon levels. Other research centers on types of geologic materials at home sites and indoor radon levels.

PENNSYLVANIA STATE UNIVERSITY, THE/BREAZEALE NUCLEAR REACTOR/LOW LEVEL RADIATION MONITORING LABORATORY
101 Academic Projects Building
University Park, PA 16802
(814) 863-2647

William A. Jester - Professor of Nuclear Engineering
Bonnie C. Ford - Laboratory Supervisor
Activities:
Laboratory established for monitoring research and service including airborne radon and radon and radium in water.

PENNSYLVANIA STATE UNIVERSITY, THE/COLLEGE OF EARTH AND MINERAL SCIENCE/GEOSCIENCES DEPARTMENT
332 Deike Building
University Park, PA 16802
(814) 865-7261
Arthur W. Rose - Professor of Geochemistry
Activities:
Analyses of radon in soils and controls for radon concentration in types of soil. Analyzes properties that control soil gas movement. DOE funding research that is being expanded to include much of Eastern U.S.

PENNSYLVANIA STATE UNIVERSITY, THE/DEPARTMENT OF CHEMISTRY
152 Davey Laboratory
University Park, PA 16802
(814) 865-6553
A. Welford Castleman - Professor of Chemistry
Activities:
DOE funded project to analyze thermochemical properties and rates of formation of radon progeny cluster ions.

PRINCETON UNIVERSITY/SCHOOL OF ENGINEERING AND APPLIED SCIENCES/CENTER FOR ENGINEERING AND APPLIED STUDIES
Engineering Quadrangle
Princeton, NJ 08544
(609) 452-5445
R.H. Socolow - Director
Activities:
DOE funded research to analyze movement of radon/radon progeny within a structure and the attachment of progeny to indoor aerosols.

R.A.D. SERVICE AND INSTRUMENTS, LTD.
50 Silver Star Blvd.
Unit 208
Scarborough, ON M1V 3L3
(416) 298-9200
Hsiang Lin Pai - President
Activities:
Manufactures an active indoor working level surveyometer (patent pending).

RAD ELEC, INC.
5330 J Spectrum Dr.
Frederick, MD 21701
(301) 694-0011
Robert W. Ramsey - President
Paul Kotrappa - Vice President
John C. Dempsey - Vice President
Activities:
Provides electret radon monitoring system and working leve measurement instrument; performs R&D on new radiometric monitoring devices.

RADIAN CORPORATION/ENGINEERING DIVISION
P.O. Box 1300
Research Triangle Park, NC 27709
(919) 541-9100
James C. Dickerman - Vice President
Clint Berkland - Engineer
Activities:
Contracts with EPA for measuring radon exposure levels in the home. Organizers of EPA's radon symposium in Colorado in October, 1988.

RADIATION RESEARCH SOCIETY (RRS)
1101 Market St.
14th Floor
Philadelphia, PA 19107
(215) 574-3153
Meg Keiser - Administrative Director
Activities:
Professional society studying radiation and its effects. Affiliated with International Association for Radiation Research and National Academy of Sciences.

RAMAPO COLLEGE OF NEW JERSEY/INSTITUTE FOR ENVIRONMENTAL STUDIES
505 Ramapo Valley Rd.
Mahwah, NJ 07430
(201) 529-7742
William Makofske - Director
Michael Edelstein
Activities:
Provides consulting, education, training workshops on radon to public and private sector.

RICHARD P. ARBER ASSOCIATES
44 Cook St., Suite 601
Denver, CO 80206
(303) 393-6363
Robert Jelinek - Project Manager
Activities:
Consulting engineering systems design firm works with EPA on drinking water study. Designed ion exchange treatment system for removal of radionuclides from drinking water in school system in Jefferson County, CO.

ROGERS AND ASSOCIATES ENGINEERING CORPORATION
515 East 4500 South
Suite G-200
Salt Lake City, UT 84107
(801) 263-1600
Arthur Sutherland - Vice President
Kirk K. Nielson - Research
Activities:
DOE is funding this research project where attempts are being made to develop a unified theory of radon emanation and transport and to implement it in a computer code for field studies and research programs.

RPI INTERNATIONAL/ENVIRONMENTAL TECHNOLOGY DIVISION
P.O. Box 328
Columbia, SC 29202
(803) 256-7322

Jacqueline Michel - Director
Activities:
Scientific consulting firm provides expertise on radon sources, behavior, and aquifer characterization for identifying radon in ground water and homes.

RUTGERS UNIVERSITY/DEPARTMENT OF COMMUNICATIONS/ENVIRONMENTAL COMMUNICATIONS RESEARCH PROGRAM
Cook College
Ryders Lane
New Brunswick, NJ 08903
(201) 932-8795
Peter Sandman - Professor
Activities:
Conducts research, provides consulting services, and holds training workshops concerning how to communicate effectively with the public about environmental health issues.

SCIENCE AND TECHNOLOGY, INC. (SCAT)
79 Pembroke Crescent
Fredericton, NB E3B 2V1
(506) 457-0232
Stanton T. Friedman - President
Activities:
Institute has long history of radon research projects under contract with Canadian government.

SCIENTIFIC AND COMMERCIAL SYSTEMS CORPORATION (SCSC)
4651 King St.
Suite 200
Alexandria, VA 22302
(703) 824-8240
Susan Hills - Project Manager
Activities:
Develops radon-related education and training materials. Under current EPA contract to develop various educational tools through the use of different media such as audio-visual aids, information packets and instructional materials.

SOUTHERN RESEARCH INSTITUTE
2000 Ninth Ave. South
P.O. Box 55305
Birmingham, AL 35255-5305
(205) 581-2000
B.E. Pyle - Research Physicist
A.D. Williamson - Head, Aerosol Science
C.S. Fowler - Research Physicist
Activities:
Fundamental radon mitigation studies and aerosol research.

STANFORD UNIVERSITY/DEPARTMENT OF CIVIL ENGINEERING
Stanford, CA 94306
(415) 723-0861
Lewis Semprini - Research Associate
Activities:
Radon measurements in fluids produced from geothermal reservoirs. Analysis of radon levels in thermodynamic systems.

TEMPLE UNIVERSITY RADON PROGRAM
Temple University
CECSA 084-53
Philadelphia, PA 19122
(215) 787-1665
Robert M. Patterson - Associate Professor
Activities:
Provides training in radon detection and mitigation, radon detection services, charcoal canister laboratory analysis, academic courses in radiological health.

TENNESSEE TECH UNIVERSITY/CENTER FOR THE MANAGEMENT, UTILIZATION AND PROTECTION OF WATER RESOURCES
P.O. Box 5082
Cookeville, TN 38505
(615) 372-3507
Albert Ogden - Associate Professor Of Hydrogeology
Activities:
Measuring radon in homes as part of radon project (using CC and ATD detectors). Assesses radioactivity of groundwater in wells. Analyzes soil types and corresponding levels of indoor radon. Evaluates home construction techniques to determine effect on indoor radon levels (research is to determine possible techniques of homebuilding along the Chattanooga Shale.

TERRADON EXPLORATION TECHNOLOGIES, INC.
3931 White Oak Tr.
P.O. Box 22288
Cleveland, OH 44122
(216) 464-2002
Darioush T. Ghahremani - President
Activities:
Research and development on radon detection, measurement and abatement; development of new alpha-track and charcoal canisters for industrial and private use; state-of-the-art soil and gas radon detection.

TEXAS A & M UNIVERSITY/DEPARTMENT OF NUCLEAR ENGINEERING
129 Zachry Engineering Center
College Station, TX 77843-3133
(409) 845-2271
William Marlow - Associate Professor
Activities:
Primary involvement is in basic research. One aspect--a theoretical explanation of immediate value--diffusivity fraction of radon daughters. Goal is to lessen inconvenience of having to recalibrate equipment. Research centers on computing rates of processes of radon daughter atom formation.

TEXAS A & M UNIVERSITY/DEPARTMENT OF NUCLEAR ENGINEERING/RADON ASSAY LABORATORY
129 Zachry Engineering Center
College Station, TX 77843
(409) 845-4107
Milton E. McLain - Professor
Activities:
Research centers on measurement techniques. Studying effects of air temperature and humidity on effectiveness of

CCs. Attempting to improve consistency of ATDs. Utilizes scintillation flask technique as way of calibrating CC and ATD devices.

THOMSON AND NIELSEN ELECTRONICS, LTD.
4019 Carling Ave.
Suite 303
Kanata, ON K2K 2A3
(613) 592-3019
Ian Thomson - President
Andrew Singmin - Director of Marketing
Line Brabant - Inside Sales
Activities:
Manufactures radon detection instruments, provides radon detection services, conducts research in radiation detection.

TOM SCURRY ASSOCIATES
1976 Linden Dr.
Boulder, CO 80302
(303) 444-4802
Tom Scurry - President
Robert Droullard - Vice President
Activities:
Manufactures continuous working level monitor and grab sampler.

UNC GEOTECH
2594 B 314 Road
P.O. Box 14000
Grand Junction, CO 81502-5504
(303) 248-6702
John R. Duray - Project Manager
Steve Colony - Architectural Engineer
Nick Abramiuk
Activities:
• Prime contractor with DOE is involved with work on the UMTRA project, the Grand Junction Remediation Project and with remediation strategies for structures near uranium mill tailing sites.
• Assists radon analysis firms by providing research/information on radon and radon daughter exposure in certain environments (temperature, pressure). Service is nonprofit and cannot interfere with DOE obligations.

UNIVERSITY OF ALASKA/COOPERATIVE EXTENSION SERVICE
207D Eielson Building
Fairbanks, AK 99775-5200
(907) 474-7201
Rick Seifert - Energy and Building Specialist
Activities:
Research indicating high radon levels in certain areas of Fairbanks.

UNIVERSITY OF CALIFORNIA-SAN DIEGO/DEPARTMENT OF RADIOLOGY/RADIATION BIOLOGY
M-010
La Jolla, CA 92093
(619) 534-4916
John F. Ward - Professor
Activities:
DOE funded research project analyzing types of DNA damage caused by alpha particles emitted by radon progeny.

UNIVERSITY OF CALIFORNIA-SAN FRANCISCO/DEPARTMENT OF RADIATION ONCOLOGY/PHYSICS DEPARTMENT
L-75, Box 0226
San Francisco, CA 94143-0226
(415) 476-9000
C. Clifton Ling - Vice Chairman
Activities:
DOE funded research project detailing carcinogenic effect of X rays and alpha particles that simulate radon progeny at the molecular level. Researchers will be able to examine the LET dependence of the radiation activation of oncogenes.

UNIVERSITY OF CALIFORNIA-SAN FRANCISCO/LABORATORY OF RADIOBIOLOGY AND ENVIRONMENTAL HEALTH
San Francisco, CA 94143-0750
(415) 476-1636
James E. Cleaver - Associate Director
Sheldon Wolff - Professor
Activities:
DOE funded research projects detailing:
• DNA damage caused by radon exposure by attempting to define types of genetic mutations brought on by radon exposure.
• Whether cell repair mechanisms can be activated by low doses of densly ionizing alpha particles and how the radiation induced repair mecanism affects alpha particle induced cell damage.

UNIVERSITY OF CALIFORNIA/LAWRENCE BERKELEY LABORATORY/DONNER LABORATORY BIOMEDICAL SCIENCES
One Cyclotran Rd./Donner Pavilion
Cowell Hospital, Room 420
Berkeley, CA 94720
(415) 486-4000
Jacob I. Fabrikant - Professor of Radiology
Activities:
Radon-related epidemiological research. LBL also conducts research for DOE.

UNIVERSITY OF CHICAGO MEDICAL CENTER/DEPARTMENT OF RADIATION ONCOLOGY
5841 South Maryland Ave.
Chicago, IL 60637
(312) 702-6860
Jeffrey L. Schwartz - Research Professor
Activities:
DOE funded comparative animal/human research project to examine DNA damage due to exposure to radon progeny to better understand the variety of mutations induced by exposure.

UNIVERSITY OF COLORADO-COLORADO SPRINGS/PHYSICS DEPARTMENT
1420 Austin Bluffs Pkwy.
Colorado Springs, CO 80933-7150

(719) 593-3214
James F. Burkhart - Department Chair and Associate Professor
Tom Huber - Associate Professor of Geography
Activities:
Two independent research projects. First is the development of a model for predicting potential for radon based on soil type, permeability, and porosity. Second is a project with EERF providing mathematical analysis for possible modifications of EPA canisters.

UNIVERSITY OF FLORIDA INDOOR RADON TASK FORCE
A.P. Black Hall
Gainesville, FL 32611
(904) 392-0841
Charles E. Roessler - Task Force Chairman
Activities:
Task Force is an interdisciplinary group of investigators who have worked together on radon research in various combinations since 1975 and who have been formally joinec as a task force since 1987. Group includes researchers from UF's College of Engineering, College of Architecture, College of Liberal Arts and Sciences and College of Communications and Journalism. Task Force meets weekly to prepare project proposals or cooperate on projects and reports. Research includes:
• "House Identification Project" to develop a list of houses to be considered for an EPA funded North Florida mitigation demonstration project.
• "Mitigation Demonstration Project" to demonstrate mitigation techniques in homes, typical of Florida construction, with high radon levels.
• "Mitigation Techniques Project" to investigate specific radon diagnostic and mitigation techniques utilized in North Central Florida structures.
• "Dunnellon Project" to classify specific areas thought to have the potential for elevated indoor radon.
• "Mobil Oil Project" working with officials from Mobil Oil Corp. to evaluate radon emanating properties of phosphogypsum aggregates in Texas.
• "Emanation Study" to determine potential for simultaneous determination of radium and radon coefficient using gamma spectroscopy and time frame counting.

UNIVERSITY OF FLORIDA/COLLEGE OF ARCHITECTURE/SCHOOL OF BUILDING CONSTRUCTION
AFC-101
Gainesville, FL 32611
(904) 392-5965
Weilin P. Chang - Associate Professor
Gary D. Cook - Assistant Professor
Richard A. Furman - Assistant Professor
Activities:
Members of UF Radon Task Force. Responsibilities include construction aspects of the research (Chang). Ventilation and energy performance aspects of radon infiltration and mitigation (Cook). Supervision of installation of mitigation design for Mitigation Demonstration Project (Furman).

UNIVERSITY OF FLORIDA/COLLEGE OF ENGINEERING/ENVIRONMENTAL ENGINEERING DEPARTMENT
217 Black Hall
Gainesville, FL 32611
(904) 392-0841
Charles E. Roessler - Professor
W. Emmett Bolch - Professor
Activities:
Task force members. Research includes Mitigation Demonstration Project (Roessler) and interpretation of radioactivity data, evaluation of lands for radon potentials, and laboratory radioactivity analysis (Bolch).

UNIVERSITY OF FLORIDA/COLLEGE OF ENGINEERING/NUCLEAR ENGINEERING DEPARTMENT
202 Nuclear Science Center
Gainesville, FL 32611
(904) 392-1401
Genevieve Roessler - Associate Professor
David E. Hintenlang - Assistant Professor
Activities:
Project Director for Housing Identification Project (Roessler). Maintains ongoing data review and interpretations. Co-principal investigator for radon measurement aspects of research (Hintenlang). Details of measurement, instrumentation, computer programming and data analysis.

UNIVERSITY OF FLORIDA/COLLEGE OF JOURNALISM AND COMMUNICATION/COMMUNICATION RESEARCH CENTER
2042 Weimer
Gainesville, FL 32611
(904) 392-6660
JoAnn Myer Valenti - Assistant Professor
Activities:
Member of UF Radon Task Force. Research on homeowner attitudes toward radon and the communication of radon information to the public.

UNIVERSITY OF FLORIDA/COLLEGE OF LIBERAL ARTS AND SCIENCES/DEPARTMENT OF GEOLOGY
1112 Turlington Hall
Gainesville, FL 32611
(904) 392-2231
Douglas L. Smith - Professor
Activities:
Member of UF Radon Task Force. Provides geological mapping of regions with high radon levels, interpretation of substrate characteristics. Description of the microgeology of study house sites.

UNIVERSITY OF GEORGIA/CENTER FOR APPLIED ISOTOPE STUDY
120 Riverbend Rd.
Athens, GA 30605
(404) 542-1395
Jim D. Spaulding - Research Associate
Gregory Simones - Research Assistant

Activities:
• Studies radon in groundwater. Research conducted at 120 wells in GA.
• Studies of radon in air using cc method. Functions as primary lab for radon detection companies.
• Planning research to determine if identifying radon levels might correlate with other geology processes.

UNIVERSITY OF HAWAII-MANOA/HAWAII INSTITUTE OF GEOPHYSICS/DIVISION OF EARTH SCIENCES
2525 Correa
Honolulu, HI 96822
(808) 948-8855
Donald Thomas - Professor
Activities:
• NSF grants for investigation of radon release by seismic and volcanic phenomena at Kilauea Volcano.
• DOE funded project to develop mathematical model of gas transport processes occurring in subsurface environment that can be applied to radon transport into structures.

UNIVERSITY OF HOUSTON/DEPARTMENT OF CIVIL AND ENVIRONMENTAL ENGINEERING
4800 Calhoun
Houston, TX 77204-4791
(713) 749-4481
Dennis Clifford - Director of Environmental Engineering
Activities:
Research on the development of techniques for removal of radium from water.

UNIVERSITY OF ILLINOIS/ENVIRONMENTAL RESEARCH LABORATORY
1005 West Western Ave.
Urbana, IL 61801
(217) 333-6230
Phil Hopke - Professor
Activities:
DOE funded research project includes two aspects:
• Analysis of the physics and chemistry of radon decay products relative to the nature of indoor air. This research details how the composition of indoor air changes the behavior of radon decay products.
• Development of instruments to measure the size and concentration of radon progeny on a semicontinuous basis.

UNIVERSITY OF IOWA/UNIVERSITY HYGIENIC LABORATORY
Oakdale Campus
Iowa City, IA 52242
(319) 335-4500
Ivan Schwabbauer - Chief of Environmental Monitoring
Activities:
Research centers on radon in water analyses.

UNIVERSITY OF LOWELL/RADIOLOIGICAL SCIENCES DEPARTMENT
1 University Ave.
Lowell, MA 01854
(617) 453-1045
Kenneth Skrable - Professor
Activities:
Operates primary laboratory and conducts research on radon.

UNIVERSITY OF MAINE-ORONO/DEPARTMENT OF CIVIL ENGINEERING
103 Boardman
Orono, ME 04469
(207) 581-2171
Willem Brutsaert - Professor
Activities:
Research on hydrological components of radon in groundwater analysis.

UNIVERSITY OF MAINE-ORONO/DEPARTMENT OF ENGINEERING
455 Aubert Hall
Orono, ME 04469
(207) 581-2216
Jerry Lowry - Professor
Activities:
Research on radon levels in drinking water. Studies have included investigation of radon levels in private supplies; point of entry removal of radon from drinking water; removal of radon from water using granular-activated carbon adsorption.

UNIVERSITY OF MAINE-ORONO/DEPARTMENT OF GEOLOGY
110 Boardman Hall
Orono, ME 04469
(207) 581-2156
Stephen Norton - Professor
Activities:
Research on distribution of radon in groundwater, providing a delineation of the geologic and hydrologic controls. Emphasis has recently shifted to studies of radon in air. Provided educational and instructional information to Norway on radon-groundwater studies.

UNIVERSITY OF MAINE-ORONO/DEPARTMENT OF PHYSICS
120 Bennett Hall
Orono, ME 04469
(207) 581-1016
Charles T. Hess - Professor
Activities:
Involved in radon research since 1972. Analyzes ventilation techniques and radon in structures. Involved in radon in water and transference to air. Project is collaborative effort with University of Texas and University of North Carolina. Twenty-plus projects since 1972.

UNIVERSITY OF MAINE-ORONO/LAND AND WATER RESOURCES CENTER
11 Coburn Hall
Orono, ME 04469
(207) 581-1490
Gregory White - Director
Activities:
Sponsors research through faculty departments at UMO. Brings interdisciplinary groups together and manages research programs.

UNIVERSITY OF NEW HAMPSHIRE/DEPARTMENT OF EARTH SCIENCES
James Hall
Durham, NH 03824
(603) 862-1234
William Olszewski - Research Scientist
Frances Hall - Hydrologist
Activities:
Research on rocks and soils in the natural environment. Mapping of structures and their uranium content. Identifies potential sites for radon problems. Studies water flow entry of radon into water sources.

UNIVERSITY OF NEW MEXICO SCHOOL OF MEDICINE/CANCER RESEARCH AND TREATMENT CENTER/NEW MEXICO TUMOR REGISTRY
900 Camino Dey Salud, NE
Albuquerque, NM 87131
(505) 277-5541
Jonathan Samet - Professor of Medicine
Activities:
DOE funded longitudinal study (in fifth year) assesses lung cancer mortality among New Mexico uranium miners. Research on relationship between lung cancer rates, radon progeny exposure, and cigarette smoking.

UNIVERSITY OF NEW MEXICO/DEPARTMENT OF GEOLOGY
Albuquerque, NM 87131
(505) 277-2310, (505) 277-4204
Douglas G. Brookins - Professor
Activities:
Soil gas analysis to determine seasonal reactivity of radon levels in Southwest U.S. Research on uranium content in building materials and in soil-radon studies.

UNIVERSITY OF NORTH CAROLINA/SCHOOL OF PUBLIC HEALTH/DEPARTMENT OF ENVIRONMENTAL SCIENCES AND ENGINEERING
Room 201H
Chapel Hill, NC 27599-7400
(919) 966-3840
James E. Watson, Jr. - Professor
Douglas Crawford-Brown - Professor
Activities:
- Radiation protection research and radon-water studies over past ten years. Research focused on monitoring radon levels in NC groundwater.
- Analysis of radon/radium in phosphate deposits and their levels in surrounding groundwater.
- Analysis of radon concentrations in geological media to determine methods of predicting radon in groundwater supplies.
- Contributed EPA's criteria document on radon in drinking water.
- Research on transfer coefficient of radon to water to air.
- Participated in statewide survey of radon levels, looking at influence of geologic media on indoor radon. Utilizes radon exposure chamber to study methods of measurement of indoor radon.

UNIVERSITY OF PENNSYLVANIA/DEPARTMENT OF GEOLOGY
138 West Rose Valley Rd.
Wallingford, PA 19086
(215) 898-5191
Robert Giegengack - Professor
Activities:
Analysis of radon concentrations and soil composition. Investigating process where radon is liberated from rock. Involved with research on decay processes of parent isotopes to progeny isotopes.

UNIVERSITY OF PITTSBURGH/PHYSICS DEPARTMENT
100 Allen Hall
Pittsburgh, PA 15260
(412) 624-9245
Bernard Cohen - Professor
Activities:
Research on CC adsorption detection devices analyzing sensitivity to humidity and atmospheric conditions. Studies liquid scintillation alpha counting devices. Participating in EPA program. Developing database on measurement results to detail factors (house characteristics, socioeconomic status) associated with increased radon levels.

UNIVERSITY OF SOUTHERN CALIFORNIA/GEOLOGICAL SCIENCE DEPARTMENT
University Park
Los Angeles, CA 90089-0740
(213) 743-6973
Douglas Hammond - Professor of Geochemistry
William Berelson - Research Professor
Activities:
Analysis of radon in deep ocean and in water columns to determine flux. Also measuring in estuaries between sediments and water column.

UNIVERSITY OF TEXAS/SCHOOL OF PUBLIC HEALTH
P.O. Box 20186
Houston, TX 77225
(713) 792-4431
Howard M. Prichard - Associate Professor
Anna Maria Cox - Faculty Associate
Activities:
Radon-water research project in 100 homes in Maine. Developed practical method for water sampling and quick technique for using liquid scintillation of radon trapped in carbon. Design and production of measurement devices. Researching techniques to simplify radon soil-gas measurement.

UNIVERSITY OF TORONTO/NATIONAL CANCER INSTITUTE OF CANADA/EPIDEMIOLOGICAL UNIT
12 Queens Park Crescent West
Toronto, ON M5S 1A8
(416) 978-4793
Geoffrey Howe - Professor
Activities:
Epidemiological studies of relationship between radon

daughter exposure and lung cancer rates of Canadian uranium miners. Examines cigarette smoking to determine combined effects. Preliminary data on the risk of radon daughter exposure to the general population.

UNIVERSITY OF UTAH/SCHOOL OF MEDICINE/RADIOBIOLOGICAL LABORATORY
Building 351
Salt Lake City, UT 84112
(801) 581-7201
McDonald Wrenn

UNIVERSITY OF VIRGINIA/DEPARTMENT OF ENVIRONMENTAL SCIENCES
Clark Hall
Charlottesville, VA 22903
(804) 924-0559
Grant Goodell - Professor
Activities:
Analyzes natural levels of radon in groundwater and soils in VA to develop statewide radon profile.

UNIVERSITY OF WASHINGTON/DEPARTMENT OF ENVIRONMENTAL HEALTH
LRE-WH-10
Seattle, WA 98195
(206) 543-4259
Ahmad Ed Nevissi - Research Associate Professor
Activities:
Evaluating effects of home weatherization on indoor radon levels. Participating in research on radon and radium in groundwater and hot springs. Analyzing water to air transference and levels of radon in water. Testing levels of radium in sediments.

UNIVERSITY OF WISCONSON-GREEN BAY/COLLEGE OF ENVIRONMENTAL SCIENCES
Natural and Applied Sciences
Green Bay, WI 54301-7001
(414) 465-2371
Jack C. Norman - Professor of Applied and Natural Sciences
Activities:
Research on building technologies and effects of weatherization on indoor radon levels. Studies efficiency of air-to-air heat exchangers in radon remediation. Studies relationships between geologic features and indoor radon levels.

WESTERN CAROLINA UNIVERSITY/DEPARTMENT OF CHEMISTRY AND PHYSICS
Natural Sciences Building
Cullowhee, NC 28723
(704) 227-7260
Glenn Liming - Associate Professor of Physics
Activities:
Studies correlation between Pb-210 levels in floors of crawlspaces and levels of detected radon.

WESTERN KENTUCKY UNIVERSITY/DEPARTMENT OF GEOGRAPHY AND GEOLOGY
WKU EST-304
Bowling Green, KY 42101
(502) 745-5989
Nick Crawford - Professor, Cave and Carst Studies

WESTINGHOUSE MATERIAL COMPANY OF OHIO/FEED MATERIALS PRODUCTION CENTER
P.O. Box 398704
Cincinnati, OH 45239-8704
(513) 738-6200, x6280
Joseph Grumski, Jr. - Manager of Waste Operations
Activities:
Facility is a uranium processing plant run by Westinghouse for DOE. Research has centered on the prevention of the diffusion of radon into homes around mill tailing sites. Experiments were performed on two concrete silos that were emitting radon due to the materials used to construct them. A rigid polyurethane foam was applied and a factor of ten reduction was achieved. Researchers have applied for a patent application because of the applicability of the product for use in private homes. It can be used as a continuous shell in basements.

6

Training/Workshops/Courses

This section lists organizations that sponsor radon detection and mitigation training, workshops, courses, and seminars. Training is scheduled in several areas of the country throughout the year. Additional information on the content and schedule of these activities can be obtained by directly contacting the sponsoring organization.

AIR CHEK, INC./NATIONAL TRAINING SEMINAR FOR RADON TESTING PROFESSIONALS
180 Glenn Bridge Rd.
Box 2000
Arden, NC 28704
(704) 684-0893
Carol S. Amos - Vice President
Programs:
First of series of training programs is tentatively scheduled for March, 1989. Program includes sessions by top radon professionals in fields of geology, construction, hydrogeologic analysis, ethics, product manufacturing, testing techniques, etc.

AIR MOVEMENT AND CONTROL ASSOCIATION (AMCA)/HOME VENTILATING INSTITUTE(HVI)
30 West University Dr.
Arlington Heights, IL 60004
(312) 394-0150
Dale Rammien - Project Manager
Programs:
Under contract with Bonneville Power Administration (DOE), HVI trains approximately 400 installers of heat recovery ventilators in the Pacific Northwest. Ventilators maintain indoor air quality, including the control of radon exposure.

AIRREGS, INC./RADON AND REAL ESTATE TRAINING PROGRAM
1640 Powers Ferry Rd.
Building 17, Suite 210
Atlanta, GA 30067
(404) 984-1919
Robert Geddis - Hygienist
Programs:
Series of three-hour courses to realtors providing a general overview of radon and how the real estate industry is affected by radon-related regulations and legislation.

AMERICAN ASSOCIATION OF RADON SCIENTISTS AND TECHNOLOGISTS (AARST)
P.O. Box 70
Park Ridge, NJ 07656
(201) 391-6445
James Craffey - Executive Secretary
Programs:
Comprehensive series of radon detection and mitigation training programs are scheduled throughout 1989. Contact AARST for further information.

AMERICAN RADON ASSOCIATION (ARA)/RADON TESTING AND MITIGATION TRAINING SEMINAR
4949 South 25-A
Tipp City, OH 45371
(513) 667-2464
Janice Lee - Executive Director
Programs:
Series of radon detection and mitigation training programs are offered in 1989. Contact ARA for additional details.

AMERICAN RADON CORPORATION
8300 N. Hayden Rd.
Suite 100
Scottsdale, AZ 85258
(602) 967-8029
James Krueger - President
Programs:
Provides a series of courses designed to

educate and inform the real estate industry on topics concerning radon and real estate issues.

ARIX CORPORATION/DIVISION OF VERSAR, INC./RADON TESTING AND MITIGATION TRAINING PROGRAM
1200 E. High St.
Pottstown, PA 19464
(215) 327-1140
L. Schlegal - Business Development Manager
Programs:
Comprehensive radon training programs are developed and designed to meet the specific needs of individual clients.

CAMRODEN ASSOCIATES
R.F.D. 4
Box 62
Rome, NY 13440
(315) 865-4269
Terry Brennan - President
Programs:
Indoor air quality investigation and training programs developed for federal, state, and private sector groups including EPA, DOE, several state government offices, 3M Company, etc.

CANADIAN INSTITUTE FOR RADIATION SAFETY/RADON IN HOMES COURSE
595 Bay St.
Suite 1050
Toronto, ON M5G 2C2
(416) 596-1617
Fergal Nolan - President
Ganice Kawka - Administrative Assistant
Programs:
Radon training programs are generally offered to Canadian federal and provincial radiation policy developers.

CERTUS LABORATORIES/RADON TRAINING AND EDUCATION SEMINARS
P.O. Box 105633
Atlanta, GA 30348
(404) 762-5411
Thomas R. Stephenson - President
Programs:
Sponsors one-day seminars for executives, policy makers and managers investigating the radon industry. Topics include the fundamentals of radon, testing and mitigation methods, state-by-state legislative update, and legal considerations.

COLORADO SCHOOL OF MINES/DEPARTMENT OF CONTINUING EDUCATION/INDOOR RADON WORKSHOP
Golden, CO 80401
(303) 273-3329
Jerome G. Morse - Professor of Physics
Phillip Nyberg - EPA Region 8
Robert Beckman - MSHA
Programs:
Three-day training program covers principles of radiation physics, hands-on operation of instruments and current aspects of the radon mitigation process. A panel discussion covers broad aspects of radon and legal concerns, real estate, health issues, geology, etc. Certificate of completion awarded upon successful completion of course and units earned apply toward professional society requirements. Information available on technical refresher course.

DOUGLAS, MARTIN, AND ASSOCIATES, INC.
P.O. Box 288
Coopersburg, PA 18049
(215) 965-2914
Ken Smith - Health and Safety Services
Programs:
Authorized by NJ DEP to administer mitigation program for certification. Provides three-day workshop for radon testing and mitigation.

EG & G ORTEC/RADON TRAINING PROGRAMS
100 Midland Rd.
Oak Ridge, TN 37831
(615) 482-4411
Lester Sideropoulos - Sales Manager
Programs:
Provides series of radon detection training programs. Call for additional information.

ENERGY RESOURCE CENTER/RADON TESTING AND MITIGATION SEMINARS
P.O. Box 267
Colorado Springs, CO 80901
(719) 636-2447
Lee Salmon - Radon Specialist
Programs:
Two series of one- or three-day seminars offered in 1989 are designed for radon detection and mitigation specialties and are cosponsored by EPA or Colorado Office of Energy Conservation.

FLORIDA DEPARTMENT OF HEALTH AND REHABILITATIVE SERVICES/DEPUTY SECRETARY FOR HEALTH/OFFICE OF RADIATION CONTROL
1317 Winewood Blvd.
Tallahassee, FL 32399-0700
(904) 487-1004
Walter Klein - Public Health Physicist
Clark Eldredge - Chemist
Programs:
Three-day course and examination for radon detection and/or mitigation fims as prerequisite for certification in the state of Florida.

GLOBAL RADON TESTING SERVICE TRAINING CENTER
163 Padgett Dr.
Lakeland, FL 33809
(813) 859-5891
Alexander Z. Kobryn - Director of Training
Programs:
Regional training facility conducts seminars and training programs addressing radon business issues, radon detection and mitigation. Facility includes 3 bedroom/2 bath demonstration structure. Programs offered in other regions of state as well as Lakeland facility.

HARVARD SCHOOL OF PUBLIC HEALTH/OFFICE OF CONTINUING EDUCATION/RADON IN BUILDINGS: SOURCES, BIOLOGICAL EFFECTS, MONITORING AND CONTROL
677 Huntington Ave.
Boston, MA 02115
(617) 732-1171
Dade W. Moeller - Course Director
Programs:
Course addresses the nature and sources of indoor radon, methods for monitoring radon, biological effects of breathing air contaminated with radon, methods for control of radon. A review of building modifications and construction techniques to reduce indoor radon levels is also presented. Credits earned during the course may be put toward certification and maintenance units with the American Board of Industrial Hygiene or for Council of Continuing Education Unit.

INFILTEC/RADON MITIGATION PRESENTATIONS
P.O. Box 8007
Falls Church, VA 22041

(703) 820-7696
David Saum - President
Marc Messing - Vice President
Programs:
Presentations nationwide addressing radon mitigation techniques and issues.

INTERMOUNTAIN RADON SERVICE/RADON AND REAL ESTATE TRAINING PROGRAM
P.O. Box 3
Salida, CO 81201
(719) 539-6726
Monty Holmes - President
Programs:
Conducts training seminars on radon-related topics for real estate industry.

IOWA DEPARTMENT OF PUBLIC HEALTH/DIVISION OF DISEASE PREVENTION/BUREAU OF RADIOLOGICAL HEALTH
Lucas State Office Building
Des Moines, IA 50319
(515) 281-6549
Rick Welke - State Radon Contact
Programs:
Developing certification program for radon measurement "specialists" and "technicians" as mandated by Iowa state government.

KEY TECHNOLOGY, INC./RADON TESTING AND MITIGATION SPECIALIST TRAINING PROGRAM
P.O. Box 562
Jonestown, PA 17038
(717) 274-8310
Edward Suter - Training Manager
Programs:
Four-day radon testing and radon mitigation courses satisfy Pennsylvania DER's training requirements toward certification as Radon Testing Specialist and Radon Mitigation Specialist.

NATIONAL ASSOCIATION OF HOME BUILDERS (NAHB)/NATIONAL RESEARCH CENTER
400 Prince Georges Center Blvd.
Upper Marlboro, MD 20772
(301) 249-2000
John Spears - Program Manager
Programs:
Conducts comprehensive training programs on radon mitigation and construction techniques, provides technical assistance, and conducts field research.

NATIONAL RADON ASSOCIATION (NRA)
2786 North Decatur Rd.
Suite 240
Decatur, GA 30033
(404) 292-1113
Stewart Huey - Director
Programs:
Series of radon training programs are scheduled for 1989. Contact NRA for further details.

NATIONAL RADON TESTING CORPORATION
1099 18th St.
Suite 1830
Denver, CO 80202
(303) 293-3141
Steve Harrison - Executive Vice President
Programs:
Company provides training in radon detection and mitigation for corporate clients, specializing in franchise networks.

NEW JERSEY DEPARTMENT OF ENVIRONMENTAL PROTECTION/DIVISION OF ENVIRONMENTAL QUALITY/BUREAU OF ENVIRONMENTAL RADIATION/RADON PROGRAM
CN-415
Trenton, NJ 08625-0415
(609) 987-6378
Harvey Worrell - Testing Certification
Nick DiPierro - Mitigation Certification
Programs:
Replacing state's voluntary certification procedures with mandatory certification programs for radon detection and mitigation companies.

NEW YORK STATE ENERGY OFFICE/BUREAU OF RESIDENTIAL CONSERVATION
2 Rockefeller Plaza
8th Floor
Albany, NY 12223
(518) 473-7243
John Reese - Director, Residential Assistance
Craig Kneeland - Radon Specialist
Mark Watson - Radon Specialist
Al Adams - Radon Specialist
Programs:
Authorized by state of New York to conduct radon training and education assistance. Radon detection and mitigation workshops presented on an ongoing basis throughout the state. Maintains list of detection and mitigation companies in New York, derived from training program participants.

PENNSYLVANIA DEPARTMENT OF ENVIRONMENTAL RESOURCES/BUREAU OF RADIATION PROTECTION/DIVISION OF SPECIAL PROJECTS
Fulton Building, 16th Floor
Third and Locust Sts.
Harrisburg, PA 17120
(717) 787-2480
Marcee Myers - Certification Program
Programs:
Oversees state of Pennsylvania radon detection and mitigation training programs.

PENNSYLVANIA STATE UNIVERSITY/RADON DETECTION AND MITIGATION CERTIFICATION TRAINING PROGRAM
231 Sackett Building
Penn State University
Energy Technology Projects
University Park, PA 16802
(814) 863-2133
James K. Shillen
Programs:
PA DER approved radon detection and mitigation training programs.

PORTER CONSULTANTS, INC./CERTIFICATION REVIEW COURSE FOR RADON TESTING SPECIALISTS
125 Argyle Rd.
Ardmore, PA 19003
(215) 896-5353
S.W. Porter, Jr. - Health Physicist
Programs:
Series of two-day radon detection and radon mitigation workshops are approved for New Jersey and Pennsylvania certification programs.

PROFESSIONAL HOUSE DOCTORS, INC./RADON MITIGATION AND REDUCTION TECHNIQUES TRAINING PROGRAM
1406 East 14th St.
Jonestown, PA 17038
(717) 274-8310
Dane J. Shearer - President
Programs:
One- and three-day training programs on radon mitigation methods. Topics include new construction techniques and mitigation procedures in homes with elevated radon levels.

R.F. SIMON COMPANY, INC./MITIGATION TRAINING PROGRAMS
R.D. 2
Box 481
Barto, PA 19504
(215) 754-6455
Ronald Simon - Project Manager
Programs:
Radon mitigation training through EPA and Penn State University.

RADON ALERT/RADON TRAINING SEMINAR
P.O. Box 131378
Tyler, TX 75713
(214) 592-3407
Rick Brown - President
Programs:
Three-day radon detection and mitigation training program offered in conjunction with Texas A & M University.

RADON DETECTION SERVICES, INC./1989 RADON TESTING SPECIALIST AND RADON MITIGATION SPECIALIST TRAINING COURSES
P.O. Box 419
Old York Rd.
Ringoes, NJ 08551
(201) 788-3080
Deborah Fugazzotto - Director of Training
Programs:
Comprehensive training scheduled throughout the year is designed to offer information on specific topics of radon detection and mitigation. Courses provide an overview on radiation fundamentals, radon sources and transport, health risk assessment, current federal/state programs, media and public perceptions of radon issues, implications surrounding radon testing and mitigation, etc.

ROSS SYSTEMS, INC./RADON TESTING AND MITIGATION TRAINING PROGRAMS
21 Main St.
Box 670
Blairstown, NJ 07825
(201) 362-5571
Richard Ross - President
Programs:
Series of two- or three-day radon detection and mitigation training programs.

RTS SYSTEMS
P.O. Box 3070
Gaithersburg, MD 20870
(301) 258-7890
Paul Ramsey - President
Programs:
Radon detection training on the use, placement, and interpretation of E-Perm detection devices.

SAFE-AIRE, INC./RADON DETECTION AND MITIGATION TRAINING PROGRAM
162 E. Chestnut St.
P.O. Box 160
Canton, IL 61520
(309) 647-0419, (800) 331-2943
Don Stees - Vice President
Programs:
Offers three-day detection/mitigation training programs throughout the year. Dealership training programs are scheduled weekly.

SANFORD COHEN AND ASSOCIATES, INC.
8200 Riding Ridge Pl.
McLean, VA 22102
(703) 893-6592
Sanford Cohen - President
John Mauro - Vice President
Programs:
Provides three-day EPA sanctioned radon mitigation workshops using EPA trained instructors.

SCIENTIFIC AND COMMERCIAL SYSTEMS CORPORATION
4651 King St.
Suite 200
Alexandria, VA 22302
(703) 824-8240
Susan Hills - Project Manager
Programs:
Develops full range of training and educational programs related to radon. Programs are tailored to the needs of federal and private clients. Corporation has worked with experts in radon research to develop training programs.

SOLAPLEXUS
R.D. 3, Box 507
New Paultz, NY 12561
(914) 384-6798
Chuck Silver - President
Programs:
Develops curriculum for radon training workshops offered by EPA , NY State Energy Office, and several other clients.

TEMPLE UNIVERSITY/RADON PROGRAM
CECSA 084-53
Philadelphia, PA 19122
(215) 787-1665
Robert M. Patterson - Associate Professor
Programs:
Provides radon detection and mitigation training and academic courses in radiological health.

TEQ CORPORATION RADON SERVICES/TRAINING PROGRAMS
P.O. Box 570
Wilkes-Barre, PA 18703-0570
(717) 457-0233
Ed Dructor - Vice President
Programs:
Three-day radon detection and mitigation training programs offered jointly with Wilkes College and approved by PA DER.

TEXAS A & M RESEARCH FOUNDATION/RADON TRAINING SEMINAR
Radon Assay Laboratory
129 Zachary Engineering Ctr.
College Station, TX 77843
(409) 845-4107
Milton E. McLain - Professor of Nuclear
Programs:
Three-day radon detection and mitigation training program offered in conjunction with Radon Alert.

U.S. ENVIRONMENTAL PROTECTION AGENCY/OFFICE OF AIR AND RADIATION/RADON DIVISION/1989 REDUCING RADON IN STRUCTURES TRAINING PROGRAMS
401 M St., SW
ANR-464
Washington, DC 20460
(202) 475-9605
Jed Harrison
Programs:
Ongoing, comprehensive radon detection and mitigation training programs scheduled for various locations nationwide.

U.S. ENVIRONMENTAL PROTECTION AGENCY/OFFICE OF AIR AND RADIATION/RADON DIVISION/REGIONAL RADON TRAINING CENTERS
401 M St., SW
ANR-464
Washington, DC 20460
(202) 475-9605
Jed Harrison
Programs:
Three EPA regional radon training

centers will be operating at partial capacity during 1989. Contact EPA for additional information.

UNIVERSITY OF MICHIGAN/SCHOOL OF PUBLIC HEALTH/DEPARTMENT OF ENVIRONMENTAL AND INDUSTRIAL HEALTH/RADON RESOURCE AND TRAINING CENTER
109 Observatory
Ann Arbor, MI 48109
(313) 764-0523
A.P. Jacobson - Director
Programs:
Training program authorized by the Michigan Department of Public Health includes
• Radon mitigation training as prerequisite to certification in the state of Michigan (includes a three-day contractors course). Training considers building, soil, and meteorlogical characteristics unique to Michigan.
• Radon detection training addressing instrument operation and laboratory analysis.

WASHINGTON DEPARTMENT OF SOCIAL AND HEALTH SERVICES/OFFICE OF RADIATION PROTECTION
Mail Stop LE-13
Olympia, WA 98504
(206) 753-5962
Robert Mooney - Supervisor (State Contact)
Bruce Pickett - Information
Programs:
Offers ongoing, comprehensive three-day radon detection, radon mitigation, and radon diagnosis workshops.

WEST VIRGINIA DEPARTMENT OF HEALTH/INDUSTRIAL HYGIENE DIVISION
151 11th Ave.
South Charleston, WV 25303
(304) 348-3526
William H. Aaroe - Director (State Contact)
Beattie L. DeBord - Assistant Director
Programs:
Offers three-day radon detection and mitigation workshops.

WPB ENTERPRISES, INC.
Box 209, RD 1
Riegelsville, PA 18077
(215) 346-8004
Bill Brodhead - President
Programs:
Offers radon mitigation training. Past clients include New York State Energy Office, EPA, etc.

Radon Associations

This section lists the national professional associations that represent the interests of the radon industry.

AMERICAN ASSOCIATION OF RADON SCIENTISTS AND TECHNOLOGISTS, INC., THE (AARST)
P.O. Box 70
Park Ridge, NJ 07656
(201) 391-6445

Jim Craffey - Executive Secretary
Activities:
AARST is a national, nonprofit, professional and trade association serving those engaged in the research, testing, measurement and remediation of radon who are committed to the organization's objectives:
• To formulate and promulgate measurement and remediation standards so that the public maintains confidence in those on whom they will rely to identify and correct the problem of radon in the home or workplace.
• To inform and educate the public about radon, and the benefits and limitations of analytical and remediation standards.
• To upgrade the technical capabilities and know-how of those engaged in providing radon related services.
• To promote improved technology to more clearly identify and enhance the remediation of radon.
• To develop high professional and ethical business standards.
AARST was organized in 1986 by scientists and professionals experienced in radon and its potentially life-threatening impact. Since its founding, the association has attracted members throughout the U.S. and Canada. AARST offers a variety of services to its members. Continuing education and professional development is offered through training programs, technical meetings and seminars as part of a carefully designed program to serve the public by expanding the knowledge and skills of those in the radon industry. The association also works with federal and state officials, real estate brokers, property developers and building contractors in an effort to influence actions of benefit to the public and the membership.

AMERICAN RADON ASSOCIATION (ARA)
4949 South 25-A
Tipp City, OH 45371
(513) 667-2464

Janice Lee - Executive Director
Activities:
Formed in early 1988, the ARA is a nonprofit national foundation based in Ohio. The ARA brings together manufacturers of radon equipment, laboratories, testing and mitigation contractors, real estate associates, insurance companies, homeowners, researchers and other concerned individuals. These diverse groups and individuals have in common a desire to learn more about radon and its effects, to make the public aware of the problem and its solutions and to deal with radon in an ethical manner. The ARA is actively involved in encouraging states to locate the areas within their borders where radon levels are high, establishing standards on a state-by-state basis for radon detection, working with states to develop certification programs for radon testing and mitigation contractors, helping the real estate industry set testing and remediation guidelines, publicizing the results of important radon research, and sponsoring educational programs for radon professionals and homeowners. The

association depends on the involvement and expertise of its members in all of these areas.

ARA programs include:
- Information Dissemination.
- Identifying Problems.
- Establishing Standards.
- Certification.
- Real Estate Guidelines.

NATIONAL RADON ASSOCIATION, INC., THE (NRA)
2786 North Decatur Rd.
Suite 220
Decatur, GA 30033
(404) 292-1113

Stewart Huey - Executive Director
Activities:
The NRA was formed as a nonprofit organization to enhance professionalism and quality of services to the membership and the public in all aspects of radon. NRA's objectives are:
- To provide a mechanism for identifying and improving radon analytical methods and instrumentation, mitigation procedures, quality assurance and protocols.
- To enhance the professional actions and quality of services of members through a code of ethics and a source of international recognition.
- To provide a unified voice of radon professionals in the formation of policies in regulatory negotiations and other issues in the radon field.
- To provide a means of continuing education and training.
- To provide a source for state guidelines in the establishment of radon standards.

Federal Government

This section lists federal government departments and agencies involved in radon-related programs. These activities include administration, policy, research, and disseminating information. A federal government listing begins with the primary administrative department or agency and proceeds hierarchically to the end office to which all information in the listing pertains. Please note that subheads are included within the listings to aid in locating departments or agencies with four or more listings.

CONSUMER PRODUCT SAFETY COMMISSION (CPSC)
5401 Westbard Ave.
Room 420
Bethesda, MD 20207
(301) 492-6554
Sandra Eberle - CIAQ Co-Chair
Activities:
Jurisdiction over indoor air pollutants resulting from consumer products. Oversees complaints about passive monitors and has responsibility to inform consumers of any dangerous malfunction of products. Representative from this office officially serves as one of the four cochairpersons of the CIAQ.

Executive Office Of The President

EXECUTIVE OFFICE OF THE PRESIDENT (EOP)/COUNCIL ON ENVIRONMENTAL QUALITY(CEQ)
722 Jackson Pl., NW
Washington, DC 20503
(202) 395-5080
A. Alan Hill - Chairman
Activities:
Advises President on environmental issues. Prepares annual report on environmental quality for Congress. Provides information on environmental affairs.

EXECUTIVE OFFICE OF THE PRESIDENT (EOP)/DOMESTIC POLICY COUNCIL(DPC)
Old Executive Office Building
Washington, DC 20500
(202) 456-6640
Ralph C. Bledsoe - Executive Secretary
Activities:
Members include: Attorney General, Administrator of EPA, Director of OMB, Secretaries of HHS, HUD, Energy and Education. Advises President on domestic and social policy, including policies relating to environmental concerns.

EXECUTIVE OFFICE OF THE PRESIDENT (EOP)/FEDERAL COORDINATING COUNCIL FOR SCIENCE, ENGINEERING AND TECHNOLOGY POLICY(FCSETP)
New Executive Office Building
Washington, DC 20506
(202) 456-6272
Jonathan Thompson - Executive Assistant
Activities:
Provides for coordination among agencies with R&D responsibilities or programs. Committees include radiation policy.

EXECUTIVE OFFICE OF THE PRESIDENT (EOP)/OFFICE OF SCIENCE AND TECHNOLOGY POLICY(OSTP)
New Executive Office Building
Room 5005
Washington, DC 20500
(202) 395-4697
Richard Johnson - Director
Activities:
Analysis of policies and technical programs of federal government. Requires Committee on Interagency Radiation Programs and Policy Coordination to provide information on radon-related issues to President of the U.S.

EXECUTIVE OFFICE OF THE PRESIDENT (EOP)/OFFICE OF SCIENCE AND TECHNOLOGY POLICY(OSTP)/OAK RIDGE ASSOCIATED UNIVERSITIES(ORAU)/COMMITTEE ON INTERAGENCY RADIATION RESEARCH AND POLICY COORDINATION(CIRRPC)
1019 19th St., NW
Washington, DC 20036
(202) 653-5505
Anthony Ewing - ORAU/CIRRPC Program Director
William Mills - Senior Technical Advisor to CIRRPC
Activities:
ORAU provides personnel and facilities in support of technical, administrative and logistical functions for executive office's CIRRPC group.

(End EOP Listings)

FEDERAL EMERGENCY MANAGEMENT AGENCY (FEMA)/STATE AND LOCAL PROGRAMS AND SUPPORT(SLPS)
500 C St., SW
Washington, DC 20472
(202) 646-3692
Grant Peterson - Associate Director
Activities:
Administers disaster relief programs including disaster research, preparedness, temporary housing, and hazard mitigation. Provides financial and technical assistance. Coordinates other federal agency disaster assistance.

LIBRARY OF CONGRESS (LOC)/CONGRESSIONAL RESEARCH SERVICE(CRS)
101 Independence Ave., SE
LM 413
Washington, DC 20540
(202) 287-5000
Christopher Dodge
Activities:
LOC liaison to CIAQ.

NATIONAL AERONAUTICS AND SPACE ADMINISTRATION (NASA)/NATIONAL SPACE TECHNOLOGY LABORATORY(NSTL)
Building T2423
NSTL, MS 39529
(601) 688-3155
Billy Wolverton - Research Scientist
Activities:
Member of Council on Indoor Air Quality addressing radon-related issues. Developer of closed-air system with applications to radon mitigation.

NATIONAL SCIENCE FOUNDATION (NSF)/DIRECTORATE FOR GEOSCIENCES/COMMITTEE ON ENVIRONMENTAL MATTERS
1800 G St., NW
Washington, DC 20550
(202) 357-9752
Julian Shedlovsky - Chairman
Activities:
Budget and Environmental Policy group provides federal funds for several radon research projects in U.S.

NATIONAL SCIENCE FOUNDATION (NSF)/LOWER ATMOSPHERIC RESEARCH DIVISION
1800 G St., NW
Washington, DC 20550
(202) 357-8645
Richard S. Greenfield - Division Head
Activities:
Provides grant funding for research in atmospheric science.

NATIONAL SCIENCE FOUNDATION (NSF)/LOWER EARTH SCIENCES DIVISION
1800 G St., NW
Washington, DC 20550
(202) 357-7958
Ian MacGregor - Division Head
Activities:
Provides grant funding for research in geology, geophysics, and related fields.

TENNESSEE VALLEY AUTHORITY (TVA)/DIVISION OF OCCUPATIONAL HEALTH AND SAFETY(DOHS)
Radiation Safety
324 MPB
Mussel Shoals, AL 35660
(205) 386-2767
Ron Maxwell - Supervisor
Jesse Coleman - Health Physicist
Activities:
Monitors TVA structures and workplace environments to determine hazards posed by radon. Participated in research on structures built with phosphate slag and by-products to determine exposure hazards.

U.S. Department Of Agriculture

U.S. DEPARTMENT OF AGRICULTURE (USDA)/AGRICULTURE RESEARCH SERVICE(ARS)/RADIATION SAFETY(RS)
Building 001
Room 226
BARC-West
Beltsville, MD 20705
(301) 344-4492

Robert Jarrett - Director
Peter Stang - Health Physicist
Barbara A. Flook - Health Physicist
Activities:
USDA's research wing analyzes environmental safety issues and radiation exposure.

U.S. DEPARTMENT OF AGRICULTURE (USDA)/FARMERS HOME ADMINISTRATION (FMHA)
Room 6309 South Building
Washington, DC 20250
(202) 382-9653
Sam Hodges - Architect
Activities:
Provides radon-related information to those seeking loans from FmHA.

U.S. DEPARTMENT OF AGRICULTURE (USDA)/FARMERS HOME ADMINISTRATION (FMHA)/LEGISLATIVE AFFAIRS AND PUBLIC INFORMATION STAFF(LAPIS)
14th St. and Independence Ave., SW
Room 5037
Washington, DC 20250
(202) 447-6903
Ron Emce - Director
Activities:
Provides information concerning FmHA's plans to develop an indoor air pollution guideline for housing financed through the agency.

U.S. DEPARTMENT OF AGRICULTURE (USDA)/SPECIAL SERVICES
14th St. and Independence Ave., SW
Washington, DC 20250
(202) 447-7173
George S. Dunlop - Assistant Secretary
Activities:
Develops and executes policy related to environmental activities and natural resources. Administers National Environmental Policy Act.

U.S. Department Of Commerce

U.S. DEPARTMENT OF COMMERCE (DOC)/NATIONAL INSTITUTE OF STANDARDS AND TECHNOLOGY(NIST)/CENTER FOR BUILDING TECHNOLOGY(CBT)
Building 226
Room B306
Washington, DC 20899
(202) 975-5900
Richard Wright - Director
Activities:
Administers CBT's research programs. Performs in-depth research on building technology. Produces performance criteria and evaluation, testing, and measurement criteria.

U.S. DEPARTMENT OF COMMERCE (DOC)/NATIONAL INSTITUTE OF STANDARDS AND TECHNOLOGY(NIST)/CENTER FOR BUILDING TECHNOLOGY(CBT)/BUILDING ENVIRONMENT DIVISION
Building 226
Room B306
Gaithersburg, MD 20899
(301) 975-5851
James Hill - Director
Activities:
Research on indoor environmental quality issues related to building technology.

U.S. DEPARTMENT OF COMMERCE (DOC)/NATIONAL INSTITUTE OF STANDARDS AND TECHNOLOGY(NIST)/CENTER FOR BUILDING TECHNOLOGY(CBT)/BUILDING ENVIRONMENT DIVISION/INDOOR AIR QUALITY GROUP
Building 226
Room A313
Gaithersburg, MD 20899
(301) 975-6431
Richard Grot - Group Leader
Activities:
Research involving radon measurement in commercial buildings.

U.S. DEPARTMENT OF COMMERCE (DOC)/NATIONAL INSTITUTE OF STANDARDS AND TECHNOLOGY(NIST)/CENTER FOR RADIATION RESEARCH(CRR)
Radiation Physics Building
Room C235
Gaithersburg, MD 20899
(301) 975-6093
Chris E. Kuyatt - Director
Activities:
Administers CRR's research projects and programs.

U.S. DEPARTMENT OF COMMERCE (DOC)/NATIONAL INSTITUTE OF STANDARDS AND TECHNOLOGY(NIST)/CENTER FOR RADIATION RESEARCH(CRR)
Building 245
Room C114
Gaithersburg, MD 20899
(301) 975-5532
Dale D. Hoppes - Leader, Radioactivity Group
J.M. Robin Hutchinson - Research Scientist
Activities:
Involved in radon standards research project that sets national standards, ensures measurement quality assurance, and analyzes calibration techniques. Works with EPA and DOE for comparison of measurement scales.

U.S. DEPARTMENT OF COMMERCE (DOC)/NATIONAL INSTITUTE OF STANDARDS AND TECHNOLOGY(NIST)/CENTER FOR RADIATION RESEARCH(CRR)/IONIZING RADIATION DIVISION
Building 245

Room C210
Gaithersburg, MD 20899
(301) 975-5524
Randall S. Caswell - Chief
Activities:
Oversees research project: Energy Deposition and Radiation Quality of Radon and Radon Daughters (co-funded with DOE).

U.S. DEPARTMENT OF COMMERCE (DOC)/NATIONAL INSTITUTE OF STANDARDS AND TECHNOLOGY(NIST)/CENTER FOR RADIATION RESEARCH(CRR)/NEUTRON INTERACTIONS AND DOSIMETRY GROUP
Building 245
Room C308
Gaithersburg, MD 20899
(301) 975-5525
James A Grundle - Group Leader
J. Joseph Coyne - Physicist
Activities:
Participates in research project involving analysis of high linear energy transfer (LET) radiation (e.g., alpha particles emitted by radon) and how radiation acts when it penetrates matter.

U.S. DEPARTMENT OF COMMERCE (DOC)/NATIONAL INSTITUTE OF STANDARDS AND TECHNOLOGY(NIST)/CENTER FOR RADIATION RESEARCH(CRR)/OFFICE OF RADIATION MEASUREMENTS
Building 245
Room C-225
Gaithersburg, MD 20899
(301) 975-6042
Elmer Eisenhower - Chief
Ronald Colle - Research Scientist
Activities:
Participates in radon standards research project involving standards development, measurement quality assurance, and calibration refinement. Compares results with information gathered by DOE and EPA.

U.S. DEPARTMENT OF COMMERCE (DOC)/NATIONAL INSTITUTE OF STANDARDS AND TECHNOLOGY(NIST)/NATIONAL ENGINEERING LABORATORY(NEL)
Technology Building
Room B119
Gaithersburg, MD 20899
(301) 975-2300
John W. Lyon - Director
Activities:
Oversees CBT's research programs on building structures and environments.

U.S. DEPARTMENT OF COMMERCE (DOC)/NATIONAL INSTITUTE OF STANDARDS AND TECHNOLOGY(NIST)/NATIONAL MEASUREMENTS LABORATORY(NML)
Physics Building
Room A363
Gaithersburg, MD 20899
(301) 975-4500
Donald Johnson - Director
Activities:
Oversees NIST's radiation research programs and sets and certifies national standards for radiation measurement (including radon).

U.S. DEPARTMENT OF COMMERCE (DOC)/NATIONAL TECHNICAL INFORMATION SERVICE(NTIS)
5248 Port Royal Rd.
Springfield, VA 22161
(703) 487-4650
Joseph F. Caponio - Director
Activities:
Principal office in U.S. for public sale of research, development, and other federally funded reports prepared by federal agencies and their contractors.

U.S. Department Of Defense

U.S. DEPARTMENT OF DEFENSE (DOD)/ARMED FORCES RADIOBIOLOGY RESEARCH INSTITUTE(AFRRI)
8901 Wisconsin Ave.
WH-550
Bethesda, MD 20814
(301) 295-1210
George W. Irving - Director
Activities:
Principal ionizing radiation radiobiology research laboratory under auspices of DOD.

U.S. DEPARTMENT OF DEFENSE (DOD)/ARMY CORPS OF ENGINEERS(ACE)/ENVIRONMENTAL DIVISION(ED)/CONSTRUCTION ENGINEERING RESEARCH LABORATORY(CERL)
P.O. Box 4005
Champaign, IL 61280-1305
(217) 373-7227
Walter Mikucki - Environmental Engineer
Activities:
Researches techniques for radon mitigation in large structures.

U.S. DEPARTMENT OF DEFENSE (DOD)/DEPARTMENT OF THE AIR FORCE(USAF)/OCCUPATIONAL AND ENVIRONMENTAL HEALTH LABORATORY(OEHL)/RADIATION SERVICES DIVISION
USAFOEHL/RZA
Brooks AFB, TX 78235-5501
(512) 536-2061
Ed Maher - Chief
Activities:
Testing of structures at military bases for mitigation program.

U.S. DEPARTMENT OF DEFENSE (DOD)/DEPARTMENT OF THE ARMY(DOA)/BELVOIR RESEARCH, DEVELOPMENT AND ENGINEERING CENTER/MATERIALS, FUELS AND LUBRICANTS LABORATORY
Radiation Research Group
Ft. Belvoir, VA 22060
(703) 664-5813, (703) 664-4175
Ramachandra Bhat - Chief
Michael Funkhauser - Radiation Protection Officer
Janine Gudagno - Environmental Specialist
Lindsey Bender - Health Physicist
Activities:
Identifies unknown radionuclides used in Army commodities. Group participates in EPA's RMP program. RRG has patent application for design of radon-water detection system. Applied research in radon-water studies.

U.S. DEPARTMENT OF DEFENSE (DOD)/DEPARTMENT OF THE NAVY(DON)/RADIOLOGICAL CONTROL HEALTH BRANCH(RCHB)/CHIEF OF NAVAL OPERATIONS OFFICE(CNO)
Crystal Plaza 5
(Code 455)
Washington, DC 20350
(202) 692-5604
Carl Mendenhall - Chief
Activities:
Office of Chief of Naval Operations. Policy-setting group, initiating comprehensive program for radon throughout the country. Navy's lead group for radon policy.

U.S. Department Of Energy

U.S. DEPARTMENT OF ENERGY (DOE)
1000 Independence Ave., SW
Washington, DC 20585
(202) 586-6476
Joseph Selgado - Undersecretary
Activities:
Manages DOE programs requiring major funding such as energy research, development and application, and associate environmental programs. EPA and DOE have established a Memorandum of Understanding on radon where EPA has responsibility for applied research and DOE is responsible for basic research.

U.S. DEPARTMENT OF ENERGY (DOE)/ALBUQUERQUE OPERATIONS OFFICE(AOO)/LOS ALAMOS NATIONAL LABORATORY(LANL)
P.O. Box 1663
M 888
Los Alamos, NM 87545
(505) 667-5061
Mudindi Raju - Research Scientist
Michael Cornforth - Research Scientist
David Chen - Research Scientist
Activities:
Research on chromosome damage in the 1 rad region, molecular mechanisms of radiation induced mutation, dosimetry, and cell killing.

U.S. DEPARTMENT OF ENERGY (DOE)/ALBUQUERQUE OPERATIONS OFFICE(AOO)/URANIUM MILL TAILINGS REMEDIAL ACTION PROJECT(UMTRA)
P.O. Box 5400
Albuquerque, NM 87115
(505) 844-3941
W. John Arthur III - Project Manager
Activities:
Project established by "Uranium Mill Tailings Radiation Control Act of 1978", Public Law 95-604. Gives AOO responsibility of managing program to minimize health hazards posed by uranium decay products (radon, radon daughters) at 24 inactive, private uranium mill sites and an estimated 4,800 vicinities in eleven states.

U.S. DEPARTMENT OF ENERGY (DOE)/ALBUQUERQUE OPERATIONS OFFICE(AOO)LOVELACE INHALATION TOXICOLOGY INSTITUTE(LITI)
P. O. Box 5890
Albuquerque, NM 87185
(505) 844-2203
Richard Cuddihy - Associate Scientist
Neil Johnson - Associate Scientist
Ronald Wolff - Research Scientist
David Thomassen - Research Scientist
Activities:
Researching radiation dose and injury to critical cells of respiratory system. Deposition of radon and radon-progeny in the respiratory tract. Radon induced DNA damage and cell transformation.

U.S. DEPARTMENT OF ENERGY (DOE)/ASSISTANT SECRETARY FOR NUCLEAR ENERGY(ASNE)/OFFICE OF HAZARDOUS WASTE AND REMEDIAL ACTION
DP-14
Germantown, MD 20874
(301) 353-4971
Tony Kluk - Health Physicist
Activities:
DOE liaison to CIAQ's radon subcommittee.

U.S. DEPARTMENT OF ENERGY (DOE)/BONNEVILLE POWER ADMINISTRATION (BPA)/OFFICE OF ENERGY RESOURCES(OE)/DIVISION OF RESOURCE MANAGEMENT/RESIDENTIAL TECHNOLOGY SECTION
905 Northeast 11th St.
Portland, OR 97208
(503) 230-5488
Phillip Thor - Chief
Charles Alton - Environmental Coordinator
Chuck Eastwood - Electrical Engineer
Activities:
• Offers, through member utilities, weatherization incentive program that monitors indoor radon levels for participants. BPA subsidizes proven mitigation methods for those homes showing an increase above 5.0 pCi/l.
• Funds research at LBL--"Pacific Northwest Existing Homes Indoor Air Quality Survey and Weatherization Sensitivity Study." Preliminary reports indicate

weatherization can reduce indoor radon levels for homes with crawl spaces by up to 33%.
• Works with USGS to analyze correlation between EPA research and NURE data on airborne radioactive particles to determine if low risk sites can be determined from NURE data.
• Participates in preconstruction radon assessment with engineering firm to review literature on movement of radon through soils. Looks at soil permeability, evaluates available mitigation equipment and develops mitigation prototypes. Final step involves testing 100 new home sites previously measured.
• Proposing to work with EPA to provide training to Pacific Northwest contractors on proven mitigation systems.

U.S. DEPARTMENT OF ENERGY (DOE)/CHICAGO OPERATIONS OFFICE(COO)/ARGONNE NATIONAL LABORATORY(ANL)/BIOLOGY DEPARTMENT
9700 South Cass Ave.
Argonne, IL 60439
(312) 972-3865
Thomas Seed - Research Scientist
Activities:
Research on cellular mechanisms of radon.

U.S. DEPARTMENT OF ENERGY (DOE)/CHICAGO OPERATIONS OFFICE(COO)/ARGONNE NATIONAL LABORATORY(ANL)/DIVISION OF BIOLOGICAL, ENVIRONMENTAL AND MEDICAL RESEARCH/ENVIRONMENTAL HEALTH SECTION
900 South Cass Ave.
Argonne, IL 60439
(312) 972-7678
J.H. Stebbings - Acting Section Head
Activities:
Contracts with federal government and other organizations for radon-related research projects. Currently qualifying for EPA radon measurement programs.

U.S. DEPARTMENT OF ENERGY (DOE)/CHICAGO OPERATIONS OFFICE(COO)/BROOKHAVEN NATIONAL LABORATORY(BNL)
Building 510-A
Upton, NY 11973
(516) 282-2123
Peter Bond - Chairman, Physics Department
Garman Harbottle - Chemistry Department
Michael Bender - Senior Scientist, Medical Department
Activities:
An HSEF for chromosome aberrations (Bond).
Fundamental studies of radon release from soil (Harbottle).

U.S. DEPARTMENT OF ENERGY (DOE)/CHICAGO OPERATIONS OFFICE(COO)/ENVIRONMENTAL MEASUREMENTS LABORATORY (EML)
376 Hudson St.
New York, NY 10014-3621
(212) 620-3609
Andreas George - Director
Activities:
DOE's premier environmental measurements research laboratory with long-term involvement in radon research.

U.S. DEPARTMENT OF ENERGY (DOE)/CHICAGO OPERATIONS OFFICE(COO)/ENVIRONMENTAL SAFETY AND HEALTH(ESH)
9800 South Cass Ave.
Argonne, IL 60439
(312) 972-4800
Ed Jazewsky - Health Physicist
Activities:
Participates in measurement characterization study of sites that include radon measurement. Passes on recommendations to fulfill DOE requirements.

U.S. DEPARTMENT OF ENERGY (DOE)/IDAHO OPERATIONS OFFICE(IOO)/ENVIRONMENTAL SAFETY AND HEALTH PROGRAM
785 DOE Pl.
Idaho Falls, ID 83402
(208) 526-1318
John Barry - Assistant Manager
Activities:
Ensures that DOE facilities meet department regulations for radon levels.

U.S. DEPARTMENT OF ENERGY (DOE)/OAK RIDGE ASSOCIATED UNIVERSITIES(ORAU)
P.O. Box 117
Oak Ridge, TN 37831-0117
(615) 576-3152
Susan Barnes - Information Officer
Activities:
Consortium of 49 colleges and universities; management and operating contractor for DOE, other government agencies, and private sector companies.

U.S. DEPARTMENT OF ENERGY (DOE)/OAK RIDGE OPERATIONS OFFICE(OROO)/OAK RIDGE NATIONAL LABORATORY(ORNL)/DIVISION OF ENVIRONMENTAL RESEARCH(DER)
P.O. Box 2008
Oak Ridge, TN 37831-6113
(615) 574-5898
Marvin Payne - Research Scientist
Mitra Sankar - Research Scientist
Activities:
Analysis of sticking times of radon atoms to metal surfaces; unfolds alpha spectra to determine location of daughter products in an experimental cell; develops new techniques of studying diffusion, mobility, and cluster formations of ions of radon.

U.S. DEPARTMENT OF ENERGY (DOE)/OAK RIDGE OPERATIONS OFFICE(OROO)/OAK RIDGE NATIONAL LABORATORY(ORNL)/HEALTH AND SAFETY RESEARCH DIVISION/MEASUREMENTS AND APPLICATIONS SECTION
P.O. Box 2008
Building 7509 South
MS-6383
Oak Ridge, TN 37831
(615) 576-2100
Steve Kaye - Division Director
Barry Bervin - Group Leader

Activities:
Studies basic mechanisms of radon entry into structures. Researches geologic formations and radon movement through those strata. Research to determine best methods of radon mitigation. Involved in radon surveys and the DOE's UMTRA project (through offices in Grand Junction).

U.S. DEPARTMENT OF ENERGY (DOE)/OAK RIDGE OPERATIONS OFFICE(OROO)/OAK RIDGE NATIONAL LABORATORY(ORNL)/HEALTH STUDIES DIVISION(HSD)/METABOLISM AND DOSIMETRY DIVISION
P.O. Box 2008
Building 7509
MS-6383
Oak Ridge, TN 37831
(615) 574-6251
Alan Hawthorne - Division Director
Keith Eckerman - Group Leader
Activities:
Researches radon dosimetry and fundamental interaction of alpha particles with lining of respiratory system at cellular level. Co-author of ICRP-30 document.

U.S. DEPARTMENT OF ENERGY (DOE)/OFFICE OF CONSERVATION AND RENEWABLE ENERGY(OCRE)/OFFICE OF BUILDINGS AND COMMUNITY SYSTEMS
1000 Independence Ave.
Room 5E-080
Washington, DC 20585
(202) 586-1510
John Milhone - Director
Activities:
Administers DOE's research and programs in building technology, including indoor air research.

U.S. DEPARTMENT OF ENERGY (DOE)/OFFICE OF CONSERVATION AND RENEWABLE ENERGY(OCRE)/OFFICE OF BUILDINGS AND COMMUNITY SYSTEMS(OBCS)/BUILDING SYSTEMS DIVISION(BSD)
CE-131
Washington, DC 20585
(202) 586-9445
James A. Smith - Director
John Talbot - Program Manager
Activities:
Indoor air quality research directly related to energy conservation measures and indoor air quality implications programs.

U.S. DEPARTMENT OF ENERGY (DOE)/OFFICE OF ENERGY RESEARCH(OER)/HEALTH AND ENVIRONMENTAL RESEARCH(OHER)
ER-70
Washington, DC 20545
(202) 353-3251
Robert W. Wood - Acting Associate Director
Activities:
Evaluation of health and environmental effects associated with energy technologies and conservation methods. Evaluates indoor pollutants such as radon and identifies health effects. Extensive research history on radon, radon daughters and inhalation toxicology not specifically tied to technological developments.

U.S. DEPARTMENT OF ENERGY (DOE)/OFFICE OF ENERGY RESEARCH(OER)/HEALTH AND ENVIRONMENTAL RESEARCH(OHER)
ER-75, GTN
Washington, DC 20545
(202) 353-5329
William Forster - Technical Advisor
Activities:
Provides interagency assistance to members of radon research projects being cosponsored with DOE/OHER.

U.S. DEPARTMENT OF ENERGY (DOE)/OFFICE OF ENERGY RESEARCH(OER)/HEALTH AND ENVIRONMENTAL RESEARCH(OHER)/HUMAN HEALTH AND ASSESSMENTS DIVISION(HHAD)
ER-73, GTN
Washington, DC 20545
(202) 353-5355
James S. Robinson - Director, HHAD
Susan Rose - Director, Radon Project
Activities:
OHER has lead responsibility for conducting basic research related to indoor radon. Office sponsors an extensive and comprehensive research program carried out at universities and national laboratories throughout the country. Over 60 radon research projects funded in 1988.

U.S. DEPARTMENT OF ENERGY (DOE)/OFFICE OF ENVIRONMENT, SAFETY AND HEALTH (ESH)
1000 Independence Ave., SW
Washington, DC 20585
(202) 586-6151
Ernest Baynard III - Assistant Secretary
Activities:
Oversees DOE programs to ensure compliance with federal policies and standards designed to protect environment and government property. Oversees health and safety conditions at DOE facilities.

U.S. DEPARTMENT OF ENERGY (DOE)/OFFICE OF ENVIRONMENT, SAFETY AND HEALTH(ESH)/ENVIRONMENTAL ANALYSIS(EA)
1000 Independence Ave, SW
Room 4G-036
MS-EH22
Washington, DC 20585
(202) 586-2061
David Moses - Deputy Director
Activities:
Responsibility to ensure that DOE's regulations for radiatior exposure at its facilities are enforced.

U.S. DEPARTMENT OF ENERGY (DOE)/OFFICE OF INTERGOVERNMENTAL AFFAIRS(OIA)
CP-23 Forrestal Building
Room 8E044
1000 Independence Ave., SW

Washington, DC 20585
(202) 586-5544
J. Michael Wisniewski - Intergovernmental Affairs Officer
Activities:
DOE liaison to Council of Radiation Control Program Directors.

U.S. DEPARTMENT OF ENERGY (DOE)/OFFICE OF NUCLEAR ENERGY(ONE)
1000 Independence Ave., SW
Room SA-115
Washington , DC 20585
(202) 586-8728
Dan Butler - Director
Activities:
Administers programs that ensure that ONE's facilities meet DOE requirements for indoor air quality.

U.S. DEPARTMENT OF ENERGY (DOE)/RICHLANDS OPERATIONS OFFICE(ROO)/PACIFIC NORTHWEST LABORATORY(PNL)/CELL BIOLOGY DIVISION
M/S K416
Richland, WA 99352
(509) 375-2121
F. T. Cross - Senior Staff Scientist
Activities:
Research on mechanisms of radon injury involves conducting molecular-cellular analyses of the inhalation toxicology of radon-radon progeny exposure. The study "Inhalation Hazards to Uranium Miners," (a longitudinal study going on over two decades) utilizes animal research and simulates mine environments to determine toxicology of radon. One of two major labs in the world (the other is Cogeman Laboratory in southwest France) studying long-term health effects of radon exposure. Research indicates lung cancer is most sensitive indicator of radon exposure. Animal and human studies show close correlation for health effects.

U.S. DEPARTMENT OF ENERGY (DOE)/RICHLANDS OPERATIONS OFFICE(ROO)/PACIFIC NORTHWEST LABORATORY(PNL)/CELL BIOLOGY DIVISION
M/S K-416
Richland, WA 99352
(509) 375-2169
F.C. Leung - Research Scientist
Activities:
Researching the role of growth factors and receptors in radon carcinogenesis in laboratory animals. Searching for potential molecular markers of radon exposure.

U.S. DEPARTMENT OF ENERGY (DOE)/RICHLANDS OPERATIONS OFFICE(ROO)/PACIFIC NORTHWEST LABORATORY(PNL)/CHEMICAL SCIENCE DEPARTMENT
Atomic And Molecular Chemistry Section
M/S P7-07
Richland, WA 99352
(509) 376-3294
T.J. Whitaker - Manager
Activities:
Researching a laser determination of radon exposure. Analyzes the health effects of long-term storage of radon daughters. By determining amount of Pb 210 in bone, determines cumulative radon daughter exposure.

U.S. DEPARTMENT OF ENERGY (DOE)/RICHLANDS OPERATIONS OFFICE(ROO)/PACIFIC NORTHWEST LABORATORY(PNL)/DEPARTMENT OF ATMOSPHERIC SCIENCES
M/S-K607
Richland, WA 99352
(509) 376-1701
P.C. Owczarski - Senior Research Engineer
Activities:
Research involves the modeling of radon transport in soils to atmosphere and indoor environments.

U.S. DEPARTMENT OF ENERGY (DOE)/RICHLANDS OPERATIONS OFFICE(ROO)/PACIFIC NORTHWEST LABORATORY(PNL)/DIVISION OF BIOLOGY AND CHEMISTRY
Radiological Physics Section
Richland, WA 99352
(509) 376-5916
L.A. Braby - Staff Scientist
Activities:
Researching the biological effectiveness of radon alpha particles by employing alpha particle microbeams. Determines the inverse exposure rate effect with alpha particles. Involves hitting single cells with beam of discreet alpha particles to determine the dose rate effects.

U.S. DEPARTMENT OF ENERGY (DOE)/RICHLANDS OPERATIONS OFFICE(ROO)/PACIFIC NORTHWEST LABORATORY(PNL)/HEALTH PHYSICS DIVISION
Dosimetry Technology Section
M/S K3-57
Richland, WA 99352
(509) 375-6852
D.P. Fisher - Senior Research Scientist
Activities:
Microdosimetry of radon daughters. Research determines cells at risk from radon daughter exposure.

U.S. DEPARTMENT OF ENERGY (DOE)/SAN FRANCISCO OPERATIONS OFFICE(SFOO)/LAWRENCE BERKELEY LABORATORY(LBL)
1 Cyclotran Rd.
Building 90
Room 3058
Berkeley, CA 94720
(415) 486-6377
Anthony Nero - Applied Science/ Indoor Environment Program
Activities:
Researching radon characterization. Co-leader of program on building ventilation and indoor air quality. Produces major reports on radon and radon daughters.

U.S. Department Of Health And Human Services

U.S. DEPARTMENT OF HEALTH AND HUMAN SERVICES (HHS)/PUBLIC HEALTH SERVICE PHS)/FOOD AND DRUG ADMINISTRATION FDA)/CENTER FOR DEVICES AND RADIOLOGICAL HEALTH(CDRH)
Office Of Health Affairs(OHA)
8757 Georgia Ave.
Silver Spring, MD 20910
(301) 427-7004
Gordon C. Johnson - Acting Director
Activities:
Research on health effects of radiation exposure.

U.S. DEPARTMENT OF HEALTH AND HUMAN SERVICES (HHS)/PUBLIC HEALTH SERVICE(PHS)/CENTER FOR HEALTH STATISTICS(CHS)
3700 East-West Hwy.
Hyattsville, MD 20872
(301) 436-8500
Manning Feinlab - Director
Activities:
Compiles, analyzes and disseminates U.S. health statistics on health hazards.

U.S. DEPARTMENT OF HEALTH AND HUMAN SERVICES (HHS)/PUBLIC HEALTH SERVICE(PHS)/CENTERS FOR DISEASE CONTROL (CDC)/ENVIRONMENTAL HAZARDS(EH)
1600 Clifton Rd., NE
Koger Center S-27
Atlanta, GA 30333
(404) 639-3311
Christie Eheman - Health Physicist
Activities:
CDC contact for radon exposure research as related to public health.

U.S. DEPARTMENT OF HEALTH AND HUMAN SERVICES (HHS)/PUBLIC HEALTH SERVICE(PHS)/CENTERS FOR DISEASE CONTROL(CDC)/NATIONAL INSTITUTE FOR OCCUPATIONAL SAFETY AND HEALTH(NIOSH)
Division Of Standards Development And Technology Transfer (DSOTT)
4676 Columbia Pkwy., MS-C14
Cincinnati, OH 45226
(513) 533-8236
Richard Lemen - Director
Richard Neimeier - Deputy Director
William D. Wagner - Chief, Document Development Branch
Activities:
Developed NIOSH document "A Recommended Standard for Occupational Exposure to Radon Progeny in Underground Mines." Report provides analysis of exposure hazards faced by underground miners and provides suggestions for remedial efforts and areas for further research.

U.S. DEPARTMENT OF HEALTH AND HUMAN SERVICES (HHS)/PUBLIC HEALTH SERVICE(PHS)/CENTERS FOR DISEASE CONTROL(CDC)/NATIONAL INSTITUTE FOR OCCUPATIONAL SAFETY AND HEALTH(NIOSH)
Ionizing And Nonionizing Radiation Division(INRD)/Hazard And Evaluation Branch
4676 Columbia Pkwy
Cincinnati, OH 45226
(513) 841-4374
Eugene Moss - Chief
Activities:
NIOSH facility for research of potential health hazards of exposure to ionizing radiation in the workplace.

U.S. DEPARTMENT OF HEALTH AND HUMAN SERVICES (HHS)/PUBLIC HEALTH SERVICE(PHS)/CENTERS FOR DISEASE CONTROL(CDC)/NATIONAL INSTITUTES FOR OCCUPATIONAL SAFETY AND HEALTH(NIOSH)
Division Of Respiratory Disease Studies
944 Chestnut Ridge Rd.
Morgantown, WV 26505
(304) 291-4474
Robert Glenn - Director
Activities:
Analysis of long-term relationship between radon exposure and respiratory disorders.

U.S. DEPARTMENT OF HEALTH AND HUMAN SERVICES (HHS)/PUBLIC HEALTH SERVICE(PHS)/FOOD AND DRUG ADMINISTRATION(FDA)
5600 Fishers Ln.
HFD-733
Rockville, MD 20857
(301) 443-3380
Frank Lundin - Medical Epidemiologist
Activities:
Liaison to executive office's CIRRPC group.

U.S. DEPARTMENT OF HEALTH AND HUMAN SERVICES (HHS)/PUBLIC HEALTH SERVICE(PHS)/FOOD AND DRUG ADMINISTRATION(FDA)/CENTER FOR DEVICE AND RADIOLOGICAL HEALTH
Office Of Device Evaluation (ODE)
8757 Georgia Ave.
Silver Spring, MD 20910
(301) 427-7004
Kshitij J. Mohan - Director
Activities:
Conducts research and develops standards for radiation-emitting products.

U.S. DEPARTMENT OF HEALTH AND HUMAN SERVICES (HHS)/PUBLIC HEALTH SERVICE(PHS)/FOOD AND DRUG ADMINISTRATION(FDA)/CENTER FOR DEVICES AND RADIOLOGICAL HEALTH(CDRH)
5600 Fishers Ln.
Rockville, MD 20857
(301) 443-4690

John C. Villforth - Director
Activities:
Administers national program to limit exposure to radiation. Conducts research on health effects of radiation exposure. Offers educational programs. Library open to public.

U.S. DEPARTMENT OF HEALTH AND HUMAN SERVICES (HHS)/PUBLIC HEALTH SERVICE(PHS)/FOOD AND DRUG ADMINISTRATION(FDA)/CENTER FOR DEVICES AND RADIOLOGICAL HEALTH(CDRH)
Office Of Health Physics
5600 Fishers Ln.
Rockville, MD 20857
(301) 443-2850
Marvin Rosenstein - Director
Activities:
Information on health hazards of radiation exposure.

U.S. DEPARTMENT OF HEALTH AND HUMAN SERVICES (HHS)/PUBLIC HEALTH SERVICE(PHS)/HEALTH RESOURCES AND SERVICES ADMINISTRATION(HRSA)/INDIAN HEALTH SERVICE(IHS)
Room 5A55
Parklawn Building
5600 Fishers Ln.
Rockville, MD 20857
(301) 443-1083
Everett M. Rhoades - Director
Activities:
Director of agency cooperating with EPA for radon survey on a number of Indian lands in Minnesota, Michigan and Wisconsin.

U.S. DEPARTMENT OF HEALTH AND HUMAN SERVICES (HHS)/PUBLIC HEALTH SERVICE(PHS)/NATIONAL CANCER INSTITUTE(NCI)/OFFICE OF CANCER COMMUNICATIONS/CANCER INFORMATION CLEARINGHOUSE
Building 31
Room 101A18
Bethesda, MD 20892
(301) 496-4070
Kate Ruddon - Public Information
Activities:
Collects and disseminates information concerning existing communications programs for patient and public education. Information includes screening, detection, causes, and prevention of cancer.

U.S. DEPARTMENT OF HEALTH AND HUMAN SERVICES (HHS)/PUBLIC HEALTH SERVICE(PHS)/NATIONAL CANCER INSTITUTE(NCI)/RADIATION EPIDEMIOLOGY BRANCH
Landow Building
7910 Woodmont Ave.
Room 3A22
Bethesda, MD 20205
(301) 496-6600
John Boice - Chief
Activities:
Information concerning four major radon exposure-cancer risk studies funded by NCI.

U.S. DEPARTMENT OF HEALTH AND HUMAN SERVICES (HHS)/PUBLIC HEALTH SERVICE(PHS)/NATIONAL INSTITUTE OF ENVIRONMENTAL HEALTH SCIENCES(NIEHS)
P.O. Box 12233
Research Triangle Park, NC 27711
(919) 541-3441
David Hoel - Director of Diometry and Risk Assessment
Activities:
Funds radon research as it relates to environmental health.

U.S. DEPARTMENT OF HEALTH AND HUMAN SERVICES (HHS)/PUBLIC HEALTH SERVICE(PHS)/NATIONAL INSTITUTES OF HEALTH(NIH)/DIVISION OF SAFETY
Radiation Safety Branch
Building 21, Room 231
9000 Rockville Pike
Bethesda, MD 20892
(301)496-5774
Peter Doob - Radiation Safety Officer
Activities:
Radon mitigation program office at HHS.

U.S. DEPARTMENT OF HEALTH AND HUMAN SERVICES (HHS)/PUBLIC HEALTH SERVICE(PHS)/OFFICE OF THE SURGEON GENERAL(OSG)
200 Independence Ave., SW
Washington, DC 20201
(202) 245-6467
C. Everett Koop - Surgeon General of the United States
Vernon J. Houk - Assistant Surgeon General
Activities:
Surgeon General's Office heads the U.S. Public Health Service. During joint press conference with officials from EPA September 13, 1988, PHS suggested severity of the threat to public health posed by radon is great enough to warrant the testing of every home in the U.S.

U.S. DEPARTMENT OF HEALTH AND HUMAN SERVICES (HHS)/PUBLIC HEALTH SERVICE(PHS)/OFFICE OF THE SURGEON GENERAL(OSG)
5600 Fisher Ln.
Parklawn Building
Room 1867
Washington, DC 20857
(301) 443-6496
Kay Strawder - Deputy Staff Director
Activities:
OSG's contact for all radon inquiries and information.

U.S. Department Of Housing And Urban Development

U.S. DEPARTMENT OF HOUSING AND URBAN DEVELOPMENT (HUD)/COMMUNITY PLANNING AND DEVELOPMENT(CPD)/OFFICE OF ENVIRONMENT AND ENERGY(OEE)
451 7th St., SW
Room 7154
Washington, DC 20410
(202) 755-7849
Richard Broun - Director
Activities:
Issues policies and standards for environmental planning and management practices.

U.S. DEPARTMENT OF HOUSING AND URBAN DEVELOPMENT (HUD)/COMMUNITY PLANNING AND DEVELOPMENT(CPD)/OFFICE OF ENVIRONMENT AND ENERGY(OEE)/ENVIRONMENTAL PLANNING DIVISION(EPD)
451 7th St., SW
Room 7212
Washington, DC 20410
(202) 755-7225
James Miller - Director
Activities:
Provides technical radon information to other offices and bureaus at HUD.

U.S. DEPARTMENT OF HOUSING AND URBAN DEVELOPMENT (HUD)/DEPUTY ASSISTANT SECRETARY FOR PROGRAM DEVELOPMENT
415 7th St., SW
Room 7212
Washington, DC 20410
(202) 755-5566
Janice S. Golack - Deputy Assistant Secretary
Barbara Dorf - Special Assistant
Activities:
Responsible for comprehensive development of HUD radon policy.

U.S. DEPARTMENT OF HOUSING AND URBAN DEVELOPMENT (HUD)/OFFICE OF POLICY DEVELOPMENT AND RESEARCH(OPDR)
451 7th St. SW
Room 8320
Washington, DC 20410
(202) 755-5600
Henry C. Fekder - Deputy Assistant Secretary
Activities:
Advises HUD secretary in policy, program evaluation and research. Evaluates existing programs and analyzes radon resaerch programs at HUD.

U.S. DEPARTMENT OF HOUSING AND URBAN DEVELOPMENT (HUD)/OFFICE OF POLICY DEVELOPMENT AND RESEARCH(OPDR)/DIVISION OF INNOVATIVE TECHNOLOGY AND SPECIAL PROJECTS(DITSP)
451 7th St., SW
Room 8232
Washington, DC 20410
(202) 755-0640
Ron Morony - Director
Conrad Arnold - Architect
William Freeborne - Mechanical Engineer
Activities:
Develops indoor air policy for HUD-assisted programs. Provides demonstrations on innovative construction techniques; reviews and assesses changes in building codes and standards.

U.S. Department Of Interior

U.S. DEPARTMENT OF INTERIOR (DOI)/BUREAU OF MINES(BOM)/DENVER RESEARCH CENTER DRC)/RADIATION LABORATORY
P.O. Box 25086
Denver Federal Ctr.
Building 20
Denver, CO 80225
(303) 236-0789
R.F. Holub - Director
Activities:
Research to support BOM's Project 2000 program to reduce radiation hazards associated withexposure to radon and radon progeny in mines. One of four labs in world currently setting worldwide standards for calibrating radon daughter measuring and detection equipment.

U.S. DEPARTMENT OF INTERIOR (DOI)/BUREAU OF MINES(BOM)/RESEARCH DIVISION
2401 E St., NW
Washington, DC 20241
(202) 634-1210
David R. Forshey - Associate Director
Activities:
Develops new procedures for mining technology, studies health and safety in mining and mine reclaimation programs.

U.S. DEPARTMENT OF INTERIOR (DOI)/NATIONAL PARK SERVICE(NPS)/ENVIRONMENTAL SANITATION PROGRAM
Box 37127
Washington, DC 20013
(202) 343-7090
Joseph Schock - Director
Allan Kingbury - Public Health Officer
Activities:
Organizes NPS Radon Gas Detection Program to test and mitigate federal park buildings and homes by FY 1990. Participates in EPA's House Evaluation Program which provides training for radon diagnosis and mitigation. Has surveyed over 3,000 housing units and conducts remediation programs.

U.S. DEPARTMENT OF INTERIOR (DOI)/NATIONAL PARK SERVICE(NPS)/MID ATLANTIC REGIONAL OFFICE/INDOOR RADON PROGRAM
143 South Third St.

Philadelphia, PA 19106
(215) 597-3679
Tom Bartholomew - Public Health Service Consultant
Ted Hillmer - Chief of Maintenance
Activities:
Radon Gas Detection Program covers federal parks in Pennsylvania, Delaware, West Virginia, Maryland, Virginia, except parks assigned to North Capital Region.

U.S. DEPARTMENT OF INTERIOR (DOI)/NATIONAL PARK SERVICE(NPS)/MIDWEST REGIONAL OFFICE/INDOOR RADON PROGRAM
1709 Jackson St.
Omaha, NE 68102
(402) 221-3448
Bert Mitchell - Public Health Service Consultant
Activities:
Radon Gas Detection Program covers federal parks in Ohio, Indiana, Illinois, Iowa, Michigan, Wisconsin, Minnesota, Kansas, Nebraska, Missouri.

U.S. DEPARTMENT OF INTERIOR (DOI)/NATIONAL PARK SERVICE(NPS)/NATIONAL CAPITAL REGION/INDOOR RADON PROGRAM
1100 Ohio Dr., SW
Washington, DC 20242
(202) 426-7700
Tom Bartholomew - Public Health Service Consultant
Activities:
Radon Gas Detection Program covers federal parks in District of Columbia and some parks in Maryland, Virginia and West Virginia.

U.S. DEPARTMENT OF INTERIOR (DOI)/NATIONAL PARK SERVICE(NPS)/NORTH ATLANTIC REGIONAL OFFICE/INDOOR RADON PROGRAM
15 State St.
Boston, MA 02109
(617) 565-8888
Tom Bartholomew - Public Health Service Consultant
Vern Hurt - Safety Officer
Activities:
Radon Gas Detection Program covers federal parks in Rhode Island, New York, New Jersey, Maine, New Hampshire, Vermont, Massachusetts, Connecticut.

U.S. DEPARTMENT OF INTERIOR (DOI)/NATIONAL PARK SERVICE(NPS)/PACIFIC NORTHWEST REGION/INDOOR RADON PROGRAM
2001 6th Ave.
Seattle, WA 98121
(206) 442-4830
Wayne Mohler - Public Health Service Consultant
Activities:
Radon Gas Detection Program covers federal parks in Idaho, Oregon, Washington.

U.S. DEPARTMENT OF INTERIOR (DOI)/NATIONAL PARK SERVICE(NPS)/ROCKY MOUNTAIN REGIONAL OFFICE/INDOOR RADON PROGRAM
P.O. Box 25287
Denver, CO 80225
(303) 969-2503
Mike Jensen - Environmental Sanitation Consultant
Activities:
Radon Gas Detection Program covers federal parks in Montana, North Dakota, South Dakota, Wyoming, Utah, Colorado.

U.S. DEPARTMENT OF INTERIOR (DOI)/NATIONAL PARK SERVICE(NPS)/SOUTHEAST REGIONAL OFFICE/INDOOR RADON PROGRAM
Richard B. Russell Federal Building And U.S. Courthouse
75 Spring St., SW
Atlanta, GA 30303
(404) 331-5187
Ron Speedy - Public Health Service Consultant
Activities:
Radon Gas Detection Program covers federal parks in Kentucky, Tennessee, North Carolina, South Carolina, Mississippi, Alabama, Georgia, Florida, Puerto Rico, Virgin Islands.

U.S. DEPARTMENT OF INTERIOR (DOI)/NATIONAL PARK SERVICE(NPS)/SOUTHWEST REGIONAL OFFICE/INDOOR RADON PROGRAM
P.O. Box 728
Santa Fe, NM 87501
(505) 988-6340
Chuck Bowman - Environmental Sanitation Consultant
Activities:
Radon Gas Detection Program covers federal parks in Arkansas, Louisiana, Texas, Oklahoma, New Mexico and NE Arizona.

U.S. DEPARTMENT OF INTERIOR (DOI)/NATIONAL PARK SERVICE(NPS)/WESTERN REGIONAL OFFICE/INDOOR RADON PROGRAM
450 Golden Gate Ave.
P.O. Box 36063
San Francisco, CA 94102
(415) 556-5186
Morris Publis - Environmental Sanitation Consultant
Activities:
Radon Gas Detection Program covers federal parks in California, Nevada, Hawaii, most of Arizona.

U.S. DEPARTMENT OF INTERIOR (DOI)/NATIONAL PARK SERVICE(NPS)REGIONAL OFFICE FOR ALASKA/INDOOR RADON PROJECT
2525 Gambell St.
Anchorage, AK 99503
(907) 257-2690
Wayne Mohler - Public Health Service Consultant
Activities:
Radon Gas Detection program covering federal parks in Alaska.

U.S. DEPARTMENT OF INTERIOR (DOI)/U.S. GEOLOGICAL SURVEY(USGS)
7638 Federal Blvd.
300 North Los Angeles St.
Los Angeles, CA 80012
(213) 894-2850
Dave Compas - Information
Activities:
Public Inquiries Office for information disseminated by USGS.

U.S. DEPARTMENT OF INTERIOR (DOI)/U.S. GEOLOGICAL SURVEY(USGS)
Denver Federal Center
MS-963
Denver, CO 80225
(303) 236-5400
G. Michael Reimer - Research Geophysicist
R. Randall Schuman - Research Geologist
Douglass E. Owen - Geologist
Segrid Asher-Bolinder - Geologist
Activities:
Researches geologic controls on distribution, concentration and migration of radon in the natural environment.

U.S. DEPARTMENT OF INTERIOR (DOI)/U.S. GEOLOGICAL SURVEY(USGS)
Box 25425
Denver Federal Ctr.
Denver, CO 80225
(303) 236-7476
Richard Davidson - Chief, Distribution Branch
Activities:
NURE information and open file reports on radon available to public.

U.S. DEPARTMENT OF INTERIOR (DOI)/U.S. GEOLOGICAL SURVEY(USGS)/ENERGY AND MARINE GEOLOGY(EMG)
12201 Sunrise Valley Dr.
Reston, VA 22092
(703) 648-6472
Gary Hill - Chief
Activities:
Develops research programs in certain geographic regions to determine radon levels present.

U.S. DEPARTMENT OF INTERIOR (DOI)/U.S. GEOLOGICAL SURVEY(USGS)/GEOLOGIC DIVISION(GD)
MS-990, National Ctr.
Reston, VA 22092
(703) 648-6996
Allan Tanner - Geophysicist
G. Michael Reimer - Research Geologist
Activities:
Cosponsored project with DOE analyzing radon movement in soil. Investigates soil as a homogenous porous medium. Assesses vertical and horizontal rates of migration.

U.S. DEPARTMENT OF INTERIOR (DOI)/U.S. GEOLOGICAL SURVEY(USGS)/GEOLOGIC DIVISION(GD)
MS-955, National Ctr.
Reston, VA 22092
(703) 648-4000
James Otten - Coordinator, Radon Projects
Alan Tanner - Geophysicist
Linda Gunderson - Research Geologist
Joseph S. Duval - Research Geophysicist
Activities:
• Conducts research on geology of uranium, radium and radon and potential application for predicting radon hazards.
• Maps radon potential in rocks and soils using NURE data, other radiometric data and gamma spectrometers.
• Contact for BPA's research program with USGS
• Research projects with DOE on geology and geochemistry in shear zones and gas transport in soils and relationship to radon levels.

U.S. DEPARTMENT OF INTERIOR (DOI)/U.S. GEOLOGICAL SURVEY(USGS)/GEOLOGIC DIVISION(GD)
MS-926, National Ctr.
Reston, VA 22092
(703) 648-7470
Joseph Duval - Research Geophysicist
G. Michael Reimer - Research Geologist
Activities:
Project with DOE utilizing aeroradiometric data to discern its application in predicting radon levels for particular geologic regions.

U.S. DEPARTMENT OF INTERIOR (DOI)/U.S. GEOLOGICAL SURVEY(USGS)/GEOLOGIC DIVISION(GD)
MS-990, National Ctr.
Reston, VA 22092
(703) 648-4000
Linda Gunderson - Research Geologist
G. Michael Reimer - Research Geologist
Activities:
Project with DOE analyzing distribution of radon in different geologic environments, specifically as it relates to points of stress in strata. Information could be missed on a regional survey level.

U.S. DEPARTMENT OF INTERIOR (DOI)/U.S. GEOLOGICAL SURVEY(USGS)/LIBRARY
12201 Sunrise Valley Dr.
Reston, VA 22092
(703) 648-4302
Elizabeth I. Yeates - Chief Librarian, Reference Section
Activities:
USGS collection of books, periodicals, maps and technical reports geology, environmental sciences, water resources and related areas. Open to public.

U.S. DEPARTMENT OF INTERIOR (DOI)/U.S. GEOLOGICAL SURVEY(USGS)/NATIONAL CARTOGRAPHIC INFORMATION CENTER HEADQUARTERS
MS-907, National Ctr.
Reston, VA 22092
(703) 860-6045
John Wood - Chief
Activities:
Geologic maps detailing radon information for certain regions of the U.S. available.

U.S. DEPARTMENT OF INTERIOR (DOI)/U.S. GEOLOGICAL SURVEY(USGS)/NATIONAL CARTOGRAPHIC INFORMATION CENTER-ALASKA(NCIC)
4230 University Dr.
Anchorage, AK 99508-4664
(907) 271-4159
Ellen Hallam - Chief
Activities:
Geologic maps pertinent to radon occurrence in selected regions of US available.

U.S. DEPARTMENT OF INTERIOR (DOI)/U.S. GEOLOGICAL SURVEY(USGS)/NATIONAL CARTOGRAPHIC INFORMATION CENTER-MID-CONTINENT(NCIC)
1400 Independence Rd.
Rolla, MO 65401
(314) 341-0851
Eddie Schwertz - Chief
Activities:
Geologic maps detailing radon information for certain regions of the US available.

U.S. DEPARTMENT OF INTERIOR (DOI)/U.S. GEOLOGICAL SURVEY(USGS)/NATIONAL CARTOGRAPHIC INFORMATION CENTER-NSTL STATION(NCIC)
Building 3101
NSTL Station, MS 39529
(601)688-3544
Hank Svehlak - Director
Activities:
Geologic maps detailing radon information for certain regions of the U.S. available.

U.S. DEPARTMENT OF INTERIOR (DOI)/U.S. GEOLOGICAL SURVEY(USGS)/NATIONAL CARTOGRAPHIC INFORMATION CENTER-ROCKY MOUNTAIN(NCIC)
Box 25046,
Denver Federal Ctr.
MS-504
Denver, CO 80225
(303) 236-5829
William F. Graser - Chief
Activities:
Geologic maps detailing radon information for certain regions of the U.S. available.

U.S. DEPARTMENT OF INTERIOR (DOI)/U.S. GEOLOGICAL SURVEY(USGS)/NATIONAL CARTOGRAPHIC INFORMATION CENTER-WEST(NCIC)
345 Middlefield Rd
Menlo Park, CA 94025
(415) 329-4309
Richard Zorker - Chief
Activities:
Geologic maps and materials detailing radon information for certain regions of the U.S. available.

U.S. DEPARTMENT OF INTERIOR (DOI)/U.S. GEOLOGICAL SURVEY(USGS)/NATIONAL WATER DATA EXCHANGE(NWDE)
417 National Ctr.
Reston, VA 22092
(703) 648-5677
Owen Williams - Program Manager
Activities:
Maintains informational database on radiochemical levels in water resources. Information available to other government agencies.

U.S. DEPARTMENT OF INTERIOR (DOI)/U.S. GEOLOGICAL SURVEY(USGS)/OFFICE OF WATER DATA COORDINATION(OWDC)
417 National Ctr.
Reston, VA 22092
(703) 648-5016
Nancy Lopez - Chief
Activities:
Coordinates dissemination an updating of information concerning radiochemicals (including radon) and water resources.

U.S. DEPARTMENT OF INTERIOR (DOI)/U.S. GEOLOGICAL SURVEY(USGS)/WATER RESOURCES DIVISION(WRD)/WATER RESOURCES SCIENTIFIC INFORMATION CENTER(WRSIC)
425 National Ctr.
Reston, VA 22092
(703) 648-6821
Raymond Jansen - Chief
Activities:
Provides information on radon-water research and analyses.

(End DOI Listings)

U.S. DEPARTMENT OF JUSTICE (DOJ)/LAND AND NATURAL RESOURCES DIVISION/POLICY, LEGISLATION AND SPECIAL LITIGATION
10 Constitution Ave., NW
Room 2615
Washington, DC 20530
(202) 633-1442
Amelia Salzman - Attorney
Activities:
DOJ liaison to CIAQ's radon subcommittee.

U.S. Department Of Labor

U.S. DEPARTMENT OF LABOR (DOL)/MINE SAFETY AND HEALTH ADMINISTRATION(MSHA)/HEALTH AND SAFETY TECHNOLOGY(HST)
2401 E St., NW
Washington, DC 20241
(202) 634-1251
Alex Bachco - Chief
Activities:
Research an responsibility to improve health and safety of miners including uranium miners at risk to radon exposure.

U.S. DEPARTMENT OF LABOR (DOL)/MINE SAFETY AND HEALTH ADMINISTRATION(MSHA)/OAK RIDGE ASSOCIATED UNIVERSITIES(ORAU)/CENTER FOR EPIDEMILOLOGICAL RESEARCH
P.O. Box 117
210 Badger Rd.
Oak Ridge, TN 37831
(615) 576-3532
Peter Groer - Research Scientist
Activities:
Research on risk estimates and radon daughter exposure.

U.S. DEPARTMENT OF LABOR (DOL)/MINE SAFETY AND HEALTH ADMINISTRATION(MSHA)/OFFICE OF STANDARDS AND REGULATIONS(OSR)
MS-112
Arlington, VA 22203
(703) 235-1910
Patricia Silvey - Director
Activities:
Information concerning MSHA's proposed rule to revise existing standards for ionizing radiation at underground metal and nonmetal mines.

U.S. DEPARTMENT OF LABOR (DOL)/MINE SAFETY AND HEALTH ADMINISTRATION(MSHA)/SAFETY AND TECHNOLGY CENTER(STC)/PHYSICAL AGENTS DIVISION
P.O. Box 25367
Denver, CO 80225
(303) 236-2650
Robert Beckman - Chief
Activities:
Provides technological assistance to the mining industry on radon and other radiation health and safety problems.

U.S. DEPARTMENT OF LABOR (DOL)/OCCUPATIONAL SAFETY AND HEALTH ADMINISTRATION(OSHA)/DIRECTORATE OF TECHNICAL SUPPORT
200 Constitution Ave., NW
Room N-3651
Washington, DC 20210
(202) 523-7031
Ed Baier - Director
Activities:
OSHA liaison to the CIAQ radon subcommittee.

U.S. DEPARTMENT OF LABOR (DOL)/OCCUPATIONAL SAFETY AND HEALTH ADMINISTRATION(OSHA)/OFFICE OF CONSULTATION PROGRAMS
200 Constitution Ave., NW
Room N-3700
Washington, DC 20210
(202) 523-7102
Joseph Collier - Director
Activities:
Health and safety issues governing environmental hazards in the workplace regulations.

(End DOL Listings)

U.S. DEPARTMENT OF STATE (DOS)/OFFICE OF MEDICAL SERVICES
2210 C St., NW
Room 4535 A
Washington, DC 20520
(202) 647-6923
Rudolph Morrazzo - Environmental Health Engineer
Activities:
Provides guidance/support for DOS safety directors's development of departmental radon policy.

U.S. DEPARTMENT OF STATE (DOS)/SAFETY OFFICE
2210 C St., NW
Room 2728
Washington, DC 20520
(202) 647-4302
Steve Urman - Safety Director
Activities:
Develops radon detection/mitigation policy and procedural guidelines for department's overseas facilities and housing.

U.S. DEPARTMENT OF TRANSPORTATION (DOT)/ENVIRONMENT DIVISION (TED)
400 7th St., SW
Nassif Building
Room 9217
Washington, DC 20590
(202) 366-4860
Eugene L. Lehr - Chief
Activities:
Maintains liaison to CIAQ on indoor air quality issues important to DOT, including tunnels and enclosed areas.

U.S. Environmental Protection Agency

U.S. ENVIRONMENTAL PROTECTION AGENCY (EPA)
401 M St., SW
Washington, DC 20460
(202) 382-4700
Lee M. Thomas - Administrator
Eileen Claussen - Acting Deputy Assistant Administrator

Activities:
Administers federal environmental policies, research and regulations. Oversees programs that provide information on radon exposure, radon-water issues, radon-resistant buildings/structures.

U.S. ENVIRONMENTAL PROTECTION AGENCY (EPA)/NATIONAL DRINKING WATER ADVISORY COUNCIL(NDWAC)
401 M St., SW
WH-550A
Washington, DC 20460
(202) 382-5700
Charlene Shaw - Executive Secretary
Activities:
Advises administrator on issues related to Safe Drinking Water Act.

U.S. ENVIRONMENTAL PROTECTION AGENCY (EPA)/OFFICE OF AIR AND RADIATION(OAR)
401 M St., SW
ANR-443
Washington, DC 20460
(202) 382-7400
Donald Clay - Assistant Administrator
Activities:
Oversees EPA's Office of Radiation Programs and Radon Division. Has cooperated with a number of federal agencies on a variety of radon projects. Administers air quality standard and planning programs.

U.S. ENVIRONMENTAL PROTECTION AGENCY (EPA)/OFFICE OF AIR AND RADIATION(OAR)/OFFICE OF RADIATION PROGRAMS(ORP)
401 M St., SW
ANR-458
Washington, DC 20460
(202) 475-9600
Richard J. Guimond - Director
Activities:
Administers radiation programs, major radiation research facilities in Las Vegas and Montgomery (AL). Establishes standards to regulate amount of radiation discharged from uranium mining and milling projects and other activities that result in radioactive emissions.

U.S. ENVIRONMENTAL PROTECTION AGENCY (EPA)/OFFICE OF AIR AND RADIATION(OAR)/OFFICE OF RADIATION PROGRAMS(ORP)
401 M St., SW
ANR-461
Washington, DC 20460
(202) 475-9640
Neal S. Nelson - Radiobiologist
Activities:
EPA liaison to executive office's CIRRPC group.

U.S. ENVIRONMENTAL PROTECTION AGENCY (EPA)/OFFICE OF AIR AND RADIATION(OAR)/OFFICE OF RADIATION PROGRAMS(ORP)/EASTERN ENVIRONMENTAL RADIATION FACILITY (EERF)
Environmental Studies Branch
1890 Federal Dr.
Montgomery, AL 36109
(205) 272-3403
Sam Windham - Chief
Edwin Sensintaffer - Health Physicist
Mike Smith - Nuclear Physicist
Activities:
Maintains two radon chambers and has been actively involved in RMP program. Performs calibrations for state survey charcoal canisters program and maintains calibration of chambers at facility.

U.S. ENVIRONMENTAL PROTECTION AGENCY (EPA)/OFFICE OF AIR AND RADIATION(OAR)/OFFICE OF RADIATION PROGRAMS(ORP)/EASTERN ENVIRONMENTAL RADIATION FACILITY(EERF)
1890 Federal Dr.
Montgomery, AL 36109
(205) 272-3402
Charles R. Porter - Director
Activities:
Administers facility's programs including radon research projects.

U.S. ENVIRONMENTAL PROTECTION AGENCY (EPA)/OFFICE OF AIR AND RADIATION(OAR)/OFFICE OF RADIATION PROGRAMS(ORP)/LAS VEGAS FACILITY(LAF)
Evaluation Branch
4220 South Maryland Pkwy.
LaPlaza Building C, Room 532
Las Vegas, NV 89119
(702) 798-2447
Richard D. Hopper - Branch Chief
Activities:
Radon chamber at facility enables researchers to evaluate new radon instrumentation and write protocols. Laboratory works almost exclusively on radon research projects.

U.S. ENVIRONMENTAL PROTECTION AGENCY (EPA)/OFFICE OF AIR AND RADIATION(OAR)/OFFICE OF RADIATION PROGRAMS(ORP)/LAS VEGAS FACILITY(LAF)
P.O. Box 98517
Las Vegas, NV 89193-8517
(702) 798-2476
Wayne A. Bliss - Director
Activities:
Administers facility's programs including comprehensive instrumentation evaluation program at laboratory.

U.S. ENVIRONMENTAL PROTECTION AGENCY (EPA)/OFFICE OF AIR AND RADIATION(OAR)/OFFICE OF RADIATION PROGRAMS(ORP)/RADON DIVISION(RD)
Radon Action Program

401 M Street SW
ANR-464
Washington, DC 20460
(202) 475-9605
Margo T. Oge - Director
Activities:
Administers national radon survey and provides states with assistance and training for radon detection/mitigation. Develops measurement standards for radon; conducts soil and gas studies. Provides technical and public information (see bibliography section). Issues model building codes for state/local governments.

U.S. ENVIRONMENTAL PROTECTION AGENCY (EPA)/OFFICE OF AIR AND RADIATION(OAR)/OFFICE OF RADIATION PROGRAMS(ORP)/RADON DIVISION (RD)
Mitigation Prevention And Quality Assurance Branch
ANR-460
401 M St., SW
Washington, DC 20460
(202) 475-9605
Michael Mardis - Branch Chief
Dave Murane - Project Director
Eugene Fisher - Program Analyst
John Hoornbeck - Program Analyst
Joseph Gearo - National Coordinator-RMP
Activities:
• Develops new construction techniques that limit possibility of radon intrusion in houses (Murane).
• Directing House (US House of Representatives) Evaluation Program (includes aspects of federally mandated Superfund support programs (Fisher).
• Policy coordination contact in branch . Program evaluation and policy development. Developing mitigation/prevention survey of private firms in US (Hoornbeck).
• Coordinates National RMP program (Gearo).

U.S. ENVIRONMENTAL PROTECTION AGENCY (EPA)/OFFICE OF AIR AND RADIATION(OAR)/OFFICE OF RADIATION PROGRAMS(ORP)/RADON DIVISION(RD)
Problem Assessment Branch
401 M St. SW
ANR-464
Washington, D.C. 20460
(202) 475-9605
Kirk Maconaughey - Branch Chief
Melinda Ronca-Battista - Health Physicist
Jacolyn Dzubian - Biologist
Thomas Peak - Geologist
Activities:
• Oversees development of uniform measurement protocols and provides technical assistance for state surveys.
• Manages EPA's National Radon Residential Survey (Ronca-Battista) and provides technical expertise for other branches on RMP program and testing in school projects.
• Functions as staff geologist (Peak) and coordinates work with USGS and state geologists. Provides information on radon in soils for particular locales.
• Manages agency's State Radon Survey (Dzubian).

U.S. ENVIRONMENTAL PROTECTION AGENCY (EPA)/OFFICE OF AIR AND RADIATION(OAR)/OFFICE OF RADIATION PROGRAMS(ORP)/RADON DIVISION(RD)
Policy And Public Information Branch
401 M St., SW
ANR-464
Washington, DC 20460
(202) 475-9605
Stephen D. Page - Branch Chief
Bonnie Chiles
Jed Harrison - Environmental Protection Specialist
Activities:
Develops comprehensive radon information program for public and initiates cooperative activites with states and pertinent national organizations.

U.S. ENVIRONMENTAL PROTECTION AGENCY (EPA)/OFFICE OF ENFORCEMENT AND COMPLIANCE MONITORING(OECM)
401 M St., SW
LE-133
Washington, DC 20460
(202) 382-4700
Thomas L. Adams - Assistant Administrator
Activities:
Principal advisor to Administrator on enforcement of standards for radiation, air and water.

U.S. ENVIRONMENTAL PROTECTION AGENCY (EPA)/OFFICE OF EXTERNAL AFFAIRS(OEA)/OFFICE OF FEDERAL ACTIVITIES(OFA)
401 M St., SW
A-104
Washington, DC 20460
(202) 382-5053
Richard E. Sanderson - Director
Activities:
Liaison office for all federal agencies. Enacts environmental review of other agency projects and activities.

U.S. ENVIRONMENTAL PROTECTION AGENCY (EPA)/OFFICE OF HEALTH RESEARCH(OHR)/HEALTH EFFECTS RESEARCH LABORATORY(HERL)
MD-51
Research Triangle Park, NC 27711
(919) 382-7400
Ted Martonen - Research Scientist
Activities:
Research on physiological implications of deposition of radon progeny in the lungs.

U.S. ENVIRONMENTAL PROTECTION AGENCY (EPA)/OFFICE OF POLICY, PLANNING AND EVALUATION(OPPE)
401 M St., SW
PM-219
Washington, DC 20460
(202) 382-4332
Linda J. Fisher - Assistant Administrator

Activities:
Coordinates agency planning, standard-setting and policy development actions.

U.S. ENVIRONMENTAL PROTECTION AGENCY (EPA)/OFFICE OF RESEARCH AND DEVELOPMENT(ORD)
401 M St., SW
RD-672
Washington, DC 20460
(202) 382-7676
Vaun A. Newell - Assistant Administrator
Activities:
Records research data and develops methods to support EPA standards and regulations. Conducts exposure risk assessment and researches procedures to reduce risk exposure.

U.S. ENVIRONMENTAL PROTECTION AGENCY (EPA)/OFFICE OF RESEARCH AND DEVELOPMENT(ORD)/AIR AND ENERGY ENGINEERING RESEARCH LABORATORY (AEERL)
MD-49
Research Triangle Park, NC 27711
(919) 541-2905
Jack Greene - Supervisor, Technical Support Staff
Activities:
Organized EPA's Radon Symposium in Denver, CO in October, 1988

U.S. ENVIRONMENTAL PROTECTION AGENCY (EPA)/OFFICE OF RESEARCH AND DEVELOPMENT(ORD)/AIR AND ENERGY ENGINEERING RESEARCH LABORATORY(AEERL)/RADON MITIGATION PROJECT
MD-60
Research Triangle Park, NC 27711
(919) 541-2745
Bruce Henschel - Project Officer
David C. Sanchez - Project Officer
Michael C. Osborne - Project Officer
Reynold Mosley - Project Officer
Kelly Witter Leovic - Project Officer
John Ruppersberger - Project Officer
Bruce Harris - Project Officer
Activities:
Develops and demonstrates cost effective techniques (soil gas flow diversion, soil-structure barriers, indoor mitigation technologies, treatment of other sources such as water) for radon reduction in private homes and large structures.

U.S. ENVIRONMENTAL PROTECTION AGENCY (EPA)/OFFICE OF RESEARCH AND DEVELOPMENT(ORD)/AIR AND ENGINEERING RESEARCH LABORATORY(AEERL)/INDOOR AIR BRANCH
MD-54
Research Triangle Park, NC 27711
(919) 541-2746
W. Gene Tucker - Chief
Activities:
Administers ORD's Radon Mitigation Project.

U.S. ENVIRONMENTAL PROTECTION AGENCY (EPA)/OFFICE OF RESEARCH AND DEVELOPMENT(ORD)/OFFICE OF ENVIRONMENTAL ENGINEERING AND TECHNOLOGY DEMONSTRATION(OEETD)
Radon Mitigation Development And Demonstration Program
401 M St., NW
RD 691
Washington, DC 20460
(202) 382-2583
John H. Skinner - Director
Lynn Ann Hitchins - Chemical Engineer (Contact)
Activities:
Program to develop and demonstrate cost-effective mitigation techniques for new and existing structures.

U.S. ENVIRONMENTAL PROTECTION AGENCY (EPA)/OFFICE OF RESEARCH AND DEVELOPMENT(ORD)/OFFICE OF TECHNOLOGY TRANSFER AND REGULATORY SUPPORT (OTTRS)/AIR TEAM
401 M St., SW
RD-672
Washington, DC 20460
(202) 382-7669
Kevin Teichman - Acting Chief
Activities:
Program management of indoor air quality research and review of EPA program documents for scientific and technical quality.

U.S. ENVIRONMENTAL PROTECTION AGENCY (EPA)/OFFICE OF WATER(OW)
401 M St., SW
WH-556
Washington, DC 20460
(202) 382-5700
Rebecca Hanmer - Assistant Administrator
Activities:
Oversees EPA's water responsibilities including drinking water, ground water, water pollution and water research.

U.S. ENVIRONMENTAL PROTECTION AGENCY (EPA)/OFFICE OF WATER(OW)/OFFICE OF DRINKING WATER(ODW)
401 M St., SW
WH-550
Washington, D.C. 20460
(202) 475-9605
Michael B. Cook - Director
Activities:
Implements standards for quality in drinking water supply systems. Provides information on public water drinking systems.

U.S. ENVIRONMENTAL PROTECTION AGENCY (EPA)/OFFICE OF WATER(OW)/OFFICE OF DRINKING WATER(ODW)/CRITERIA AND STANDARDS DIVISION(CSD)
Science And Technology Branch
401 M St. SW
WH-550D
Washington, DC 20460
(202) 382-3022
Arthur Perler - Chief
Activities:
Analyzes treatment technologies for radon-water remediation, detailing costs and feasability of systems. Conducts exposure analyses to determine level of radon exposure from drinking water sources.

U.S. ENVIRONMENTAL PROTECTION AGENCY (EPA)/OFFICE OF WATER(OW)/OFFICE OF DRINKING WATER(ODW)/CRITERIA AND STANDARDS DIVISION(CSD)
Health Effects Branch
401 M St., SW
WH-550D
Washington, DC 20460
(202) 382-7571
Ed Ohanian - Chief
Activities:
Develops standards for radionuclides (uranium, radium, radon) in drinking water. Analyzes local drinking water systems across the US to establish health threat posed by radon in water. Establishing new radon-drinking water standards by 1989.

U.S. ENVIRONMENTAL PROTECTION AGENCY (EPA)/OFFICE OF WATER(OW)/OFFICE OF DRINKING WATER(ODW)/CRITERIA AND STANDARDS DIVISION(CSD)
401 M St., SW
WH-550D
Washington, DC 20460
(202) 382-7575
Joseph Cotruro - Director
Activities:
Directs ODW's radon projects dealing with setting of standards for drinking water, health issues and technologica information.

U.S. ENVIRONMENTAL PROTECTION AGENCY (EPA)/OFFICE OF WATER(OW)/OFFICE OF DRINKING WATER(ODW)/TECHNICAL SUPPORT DIVISION(TSD)
Water Supply Technology Branch
26 West Martin Luther King Dr.
Cincinnati, OH 45268
(513) 569-7919
James Westrick - Chief
Jon P. Longtin - Project Engineer
Activities:
Conducts National Inorganics and Radionuclides Survey (NIRS) to obtain information on uranium, radium and radon in drinking water from ground water sources. Information utilized for establishment of revised drinking water regulations.

U.S. ENVIRONMENTAL PROTECTION AGENCY (EPA)/OFFICE OF WATER(OW)/OFFICE OF DRINKING WATER(ODW)/WATER ENGINEERING RESEARCH LABORATORY(WERL)
Drinking Water Research Division (DWRD)
26 West Martin Luther King Dr.
Cincinnati, OH 45268
(513) 569-7370
Tom Sorg - Chief, Inorganics/Particulates Control Branch
Activities:
Research in radon removal from drinking water. WERL scientists provide technical support to AEERL radon-water research.

U.S. ENVIRONMENTAL PROTECTION AGENCY (EPA)/RESEARCH TRIANGLE INSTITUTE(RTI)/RADON TECHNICAL INFORMATION SERVICE(RTIS)/NATIONAL RADON MEASUREMENT PROFICIENCY PROGRAM(RMP)
3040 Cornwallis Road, Building 7
Research Triangle Park, NC 27709
(919) 541-7131
Henry Singletary - Director
Activities:
Contracts with EPA to perform specific research projects on radon-related topics. Offers technical assistance and information to laboratories and companies who want to participate in the National Radon Measurement Proficiency (RMP) programs. The national RMP program was established to assist states and public in selecting companies to measure radon/radon progeny. The program is designed to promote standars measurement and quality assurance procedures for all indoor radon measurement companies.

U.S. ENVIRONMENTAL PROTECTION AGENCY (EPA)/SCIENCE ADVISORY BOARD(SAB)
401 M St., SW
A-101
Washington, DC 20460
(202) 382-2552
Donald G. Barnes - Director
Activities:
Advises administrator on scientific/technical aspects of environmental issues. Evaluates agency research projects and policy statements and guidelines.

U.S. ENVIRONMENTAL PROTECTION AGENCY (EPA)/SCIENCE ADVISORY BOARD(SAB)/ENVIRONMENTAL HEALTH COMMITTEE/DRINKING WATER SUBCOMMITTEE
401 M St., SW
A-101F
Washington, DC 20460
(202) 382-2552
C. Richard Cothern - Executive Secretary
Activities:
Advises Administrator on issues concerning the quality of drinking water.

U.S. ENVIRONMENTAL PROTECTION AGENCY (EPA)/SCIENCE ADVISORY BOARD(SAB)/INDOOR AIR AND TOTAL HUMAN EXPOSURE SUBCOMMITTEE
401 M St., SW
A-101F
Washington, DC 20460
(202) 382-2552
Robert Flaak - Executive Secretary
Activities:
Advises Administrator on indoor air quality issues including radon exposure and its potentially dangerous effects.

U.S. ENVIRONMENTAL PROTECTION AGENCY (EPA)/SCIENCE ADVISORY BOARD(SAB)/RADIATION ADVISORY COMMITTEE
401 M St. SW
A-101F
Washington , DC 20460
(202) 382-2552
Kathleen Conway - Executive Secretary
Activities:
Advises Administrator on issues concerning radiation and its effects.

U.S. ENVIRONMENTAL PROTECTION AGENCY REGION 1 (EPA1)/AIR AND RADIATION DIVISION(OAR)
JFK Federal Building
Room 2203
Boston , MA 02203
(617) 565-3234
Byron Keene - Radon Chief
Activities:
Provides radon information(technical or general) for residents of Connecticut, Maine, Massachusetts, New Hampshire, Rhode Island and Vermont.

U.S. ENVIRONMENTAL PROTECTION AGENCY REGION 2 (EPA2)/RADIATION DIVISION(ORP)
26 Federal Plaza
New York, NY 10278
(212) 264-4418
Paul Giardina - Chief
Activities:
Provides general radon information for residents of New York, New Jersey, Puerto Rico,Virgin Islands.

U.S. ENVIRONMENTAL PROTECTION AGENCY REGION 3 (EPA3)/RADIATION DIVISION(ORP)/RADON PROJECT
841 Chestnut St.
MS-3AM12
Philadelphia, PA 19107
(215) 597-1256
Thomas Maslany - Director, Radiation Division
Lou Felleisen - Section Chief, Radon Project
Activities:
Provides detailed information on radon, directs mitigation inquiries and informs residents of state officials offering services in the states of Delaware, District of Columbia, Maryland, Pennsylvania, West Virginia and Virginia.

U.S. ENVIRONMENTAL PROTECTION AGENCY REGION 4 (EPA4)/RADIATION DIVISION(ORP)
345 Courtland St. NE
Atlanta, GA 30365
(404) 347-3907
Chuck Wakamo - Director
Activities:
Provides general radon information to residents of Alabama, Florida, Georgia, Kentucky, Mississippi, North Carolina, South Carolina, Tennessee.

U.S. ENVIRONMENTAL PROTECTION AGENCY REGION 5 (EPA5)/RADIATION DIVISION(ORP)
230 S.Dearborn St.
Chicago, IL 60604
(312) 886-6175
Larry Jensen - Director
Activities:
Provides general and technical EPA information to residents of Illinois, Indiana, Michigan, Minnesota, Ohio and Wisconson.

U.S. ENVIRONMENTAL PROTECTION AGENCY REGION 6 (EPA6)/OFFICE OF AIR AND RADIATION(OAR)
1201 Elm St.
Dallas, TX 75270
(214) 655-7208
Mike Miller - Radon Coordinator
Activities:
Provides limited radon information to residents of: Louisiana, New Mexico, Oklahoma and Texas.

U.S. ENVIRONMENTAL PROTECTION AGENCY REGION 7 (EPA7)/RADIATION TECHNOLOGY(ORP)
726 Minnesota Ave.
Kansas City, KS 66101
(913) 236-2893
Bill Brinck - Radiation Representative
Activities:
Refers inquiries to proper state authorities in Iowa, Kansas, Missouri, and Nebraska.

U.S. ENVIRONMENTAL PROTECTION AGENCY REGION 8 (EPA8)/RADIATION DIVISION
One Denver Pl.
999 18th St.
8HWM-RP
Denver, CO 80202
(303) 293-1603
Milton Lammering - Director, Radon Section
Activities:
Provides comprehensive information concerning region's testing and mitigation companies and radon product manufacturers to residents of: Colorado, Montana, North Dakota, South Dakota, Utah and Wyoming.

U.S. ENVIRONMENTAL PROTECTION AGENCY REGION 9 (EPA9)/RADIATION DIVISION(ORP)
215 Fremont St.
San Francisco, CA 94105
(415) 974-8378
David Duncan - Radiation Specialist
Activities:
General radon information for residents of Arizona, California, Hawaii, Nevada, Guam and American Samoa.

U.S. ENVIRONMENTAL PROTECTION AGENCY REGION 10 (EPA10)/RADIATION DIVISION(ORP)
1200 Sixth Ave.
Seattle, WA 98101
(206) 442-7299
Jerry Leach - Director
Misha Vackoc - Environmental Protection Specialist-Radon
Activities:
Provides radon information (technical or general) to residents of Alaska, Idaho, Oregon, and Washington.

(End EPA Listings)

U.S. GENERAL ACCOUNTING OFFICE (GAO)/ENVIRONMENTAL PROGRAMS(EP)
441 G St. NW
Washington, DC 20548
(202) 275-5489
Hugh J. Wessinger - Senior Associate Director
Activities:
Independent agency in legislative branch. Audits, evaluates and analyzes EPA programs and makes information available to public.

U.S. GENERAL ACCOUNTING OFFICE (GAO)/RESOURCES, COMMUNITY AND ECONOMIC DEVELOPMENT DIVISION(RCED)
441 G St. NW
Washington, DC 20548
(202) 426-1780
John H. Luke - Associate Director
Frank Sabalusky - Group Director
Activities:
Independent agency in legislative branch reports on federal government's effectiveness in development of radon policy.

U.S. GENERAL SERVICES ADMINISTRATION REGIONAL HEADQUARTERS(GSAR)
7th and D Sts.
Room 1126
Washington, DC 20407
(202) 453-5236
Fred Sisson - Branch Chief
Mary Beth Stewart - Section Chief
Cherylynn Williams - Industrial Hygienist
Activities:
Evaluation of testing procedures for radon detection and mitigation in U.S. government-owned buildings.

U.S. GENERAL SERVICES ADMINSTRATION (GSA)/SAFETY AND ENVIRONMENTAL MANAGEMENT DIVISION(SEMD)
18th and F St., NW
Room 4320
Washington, DC 20405
(202) 472-1082
Wing T. Chan - Industrial Hygienist
Jeffrey M. Bauer - Environmental Engineer
Activities:
Developing an indoor air program where GSA sites are used in test program for development of EPA measurement protocols for workplaces.

VETERANS ADMINISTRATION CENTRAL OFFICE (VA)
810 Vermont Ave.
Washington, DC 20420
(202) 233-2726
Carl Howe - Government Trial Attorney
Activities:
Maintains CIAQ Radon Subcommittee liaison through this office. Keeps agency informed on radon issues.

Federal Legislation

9

This section lists radon-related legislation either proposed or passed by members of the U.S. Senate and the U.S. House of Representatives during the 100th Congress which convened January 21, 1987 to January 20, 1989. Consideration of proposed legislation listed in this section will resume during the 101st Congress. Please note that The Superfund Amendments and Reauthorization Act (SARA) of 1986 is also listed here.

An index of congressional contacts can be found immediately following this section. This index lists all sponsors and co-sponsors of legislation listed in this section.

Bills, Sponsors, And Co-Sponsors

THE SUPERFUND AMENDMENTS AND REAUTHORIZATION ACT OF 1986
U.S. SENATE/S.2840 (99TH CONGRESS)

Amends the "Comprehensive Environmental Response, Compensation and Liability Act of 1980" to reauthorize Superfund. Contains two provisions related to indoor radon:

• Requires EPA to assess radon levels throughout the US (in structures where people "normally live and work"). The EPA is also required to develop a radon mitigation demonstration program.
• Requires EPA to establish a program to assess indoor air and radon problems, coordinate federal, state, local, and private efforts and address appropriate federal actions to mitigate risk of indoor radon.

Sponsor

Senator Robert Stafford (R-VT)

Co-Sponsors

Sen. Chafee
Sen Durenburger
Sen. Mitchell
Sen. Burdick
Sen. Moynihan
Sen. Humphrey
Sen. Lautenberg
Sen. Broyhill
Sen. Roth, W.
Sen. Baucus
Sen. Abnor
Sen. Simpson
Sen. Domenici
Sen. Riegle
Sen. Bentson
Sen. Kennedy
Sen. Hart

U.S. HOUSE OF REPRESENTATIVES/H.R.1108

• Bill introduced February, 1987 proposes to amend the Internal Revenue Code to allow tax credits for expenses incurred for radon corrective measures within a principal residence.
• Provides for coverage of up to 40 percent of costs up to $2000.00 in residences where radon levels exceed 2 WLMs per year.
• Bill referred to House Committee on Ways and Means.

Sponsor

Representative Donald Ritter (R-PA)

Co-Sponsors

Rep. Yatron
Rep. Courter
Rep. Kostmayer
Rep. A. Murphy
Rep. Murtha
Rep. Ridge
Rep. Towns
Rep. Gilman
Rep. Byron
Rep. Coughlin
Rep. Kolter
Rep. R. Davis

U.S. HOUSE OF REPRESENTATIVES/H.R.1610

• Proposed legislation would require HUD to provide state and local officials with technical assistance to encourage incorporating radon testing requirements into building codes.

• Testing would be performed by companies EPA labels proficient.
• Funds for FY 1988, 1989, 1990 would be appropriated "as necessary" and activities would be covered in HUD's yearly report.
• Referred to Subcommittee on Housing and Community Development.

Sponsor

Representative Gus Yatron (D-PA)

Co-Sponsors

No Co-Sponsors

U.S. HOUSE OF REPRESENTATIVES/H.R.1697

• Proposal to amend the Toxic Substances Control Act.
• Bill would provide for assistance to states in responding to threat to human health posed by radon.
• Assistance includes technical aid and grants.
• Directs EPA to study the extent of radon in school buildings.
• Referred to Subcommittee on Transportation, Tourism, and Hazardous Materials.

Sponsor

Representative Thomas A. Luken (D-OH)

Co-Sponsors

Rep. Sikorski
Rep. T. Hall
Rep. Obey
Rep. Yatron
Rep. Kolter

U.S. HOUSE OF REPRESENTATIVES/H.R.2187

• Proposed legislation to direct the EPA to continue the voluntary Radon Mitigation Proficiency Program and quality assurance programs to inform the public of reliable companies and individuals that measure indoor radon.
• Referred to Subcommittee on Health and the Environment.

Sponsor

Representative James H. Scheuer (D-NY)

Co-Sponsors

No Co-Sponsors

U.S. HOUSE OF REPRESENTATIVES/H.R.2355

• Bill authorizes monies for EPA research, development, and demonstration for FY 1988 and 1989.
• Radon provisions addressed by specified environmental laws.
• Measure passed House amended and has been read twice in Senate and referred to Senate Committee on Environment and Public Works.

Sponsor

Representative James H. Scheuer (D-NY)

Co-Sponsors

No Co-Sponsors

U.S. HOUSE OF REPRESENTATIVES/H.R.2837

• Companion bill to S.744.
• Requires EPA to update materials informing the public on the development of model construction techniques and standards.
• Authorizes EPA to provide technical assistance to states for remediation in non-residential day care facilities.
• Authorizes EPA to make grants to establish regional radon training centers.
• Bill passed by the House of Representatives October 5, 1988 and has been sent before a joint House-Senate committee for final adjustments.

Sponsor

Representative Thomas Luken (D-OH)

Co-Sponsors

Rep. Whittaker
Rep. Sikorski
Rep. Ritter
Rep. Hall, T.
Rep. Traficant
Rep. Obey
Rep. Yatron
Rep. Roukema
Rep. Gilman
Rep. Brennan
Rep. Gordon
Rep. Kleczka
Rep. Davis, R.
Rep. Gallo
Rep. Lowry
Rep. Vento

U.S. HOUSE OF REPRESENTATIVES/H.R.3110

• Proposal for amendments to the Clean Air Act that would require EPA to establish a radon exposure standard for indoor environments.
• Referred to House Subcommittee on Health and Environment.

Sponsor

Representative James J. Florio (D-NJ)

Co-Sponsors

No Co-Sponsors

U.S. HOUSE OF REPRESENTATIVES/H.R.3915

• Proposed bill to amend the Internal Revenue Code of 1986.
• Bill would provide tax relief for developers, builders, and individuals who take steps to prevent radon penetration of their structures.
• Includes proposal to amend federal law to make funds available to reimburse federal employees for costs incurred when radon reduction measures must be taken due to job relocation.
• Referred to House Committee on Ways and Means.

Sponsor

Representative Patricia Schroeder (D-CO)

Co-Sponsors

Rep. Campbell
Rep. Yatron
Rep. Fuster
Rep. Collins, C.
Rep. Parris
Rep. Byron
Rep. Rodino
Rep. Howard
Rep. Neal
Rep. Martinez
Rep. Murphy, A.
Rep. Kolter
Rep. Davis, R.
Rep. Gilman
Rep. Towns
Rep. Gordon
Rep. Garcia
Rep. Dymally
Rep. Hawkins
Rep. Foglietta
Rep. Lloyd
Del. Fauntroy
Rep. Sensenbrenner
Rep. Dellums
Rep. Wolfe
Rep. Roe
Rep. Rangel
Rep. Kaptur
Rep. Pelosi
Rep. Feighan
Rep. Owens, M.
Rep. Traficant
Rep. Lewis, John
Rep. Espy

U.S. SENATE/S.1067

• Directs EPA to develop and implement the voluntary radon RMP program to inform the public of reliable firms and individuals that provide radon testing.
• Proposal has been merged into S.744 to become part of the Radon Program Development Act of 1987.

Sponsor
Senator Arlen Specter (R-PA)
Co-Sponsors
No Co-Sponsors

U.S. SENATE/S.1629
• Proposed amendment to S.1629, "The Indoor Air Quality Act of 1987."
• Legislation would require EPA to develop protocols for measuring radon in child care facilities, provide measuring devices for those facilities, and to develop programs for remediation of child care centers where necessary.
• Proposes HUD and EPA enter into a memorandum of understanding on their respective roles concerning radon.
• Amendment establishes funds for proposed regional radon training centers.
• Bill referred to Senate Finance Committee.
Sponsor
Sen. Frank Lautenberg (D-NJ)
Co-Sponsors
No Co-Sponsors

U.S. SENATE/S.743
• Legislation proposes EPA conduct a study to determine the extent to which radon exposure in US schools poses a threat tc students and employees.
• Bill was merged into S.744 and became part of the Radon Program Development Act of 1987.
Sponsor
Senator John Chafee (R-RI)
Co-Sponsors
Sen. Mitchell
Sen. Stafford
Sen. Heinz
Sen. Reid
Sen. Mikulski
Sen. Lautenberg
Sen. Specter
Sen. Durenberger
Sen. Gore
Sen. Warner

U.S. SENATE/S.744-THE RADON PROGRAM DEVELOPMENT ACT OF 1987
S.744 was approved by the U.S. Senate July 8, 1987. The bill directs the Administrator of the EPA to develop and initiate means to assist state radon programs. It includes:
• $10 million annually for fiscal years 1988, 1989, and 1990 fo: grants to help states establish radon reduction programs, conduct surveys, and conduct demonstration and mitigation projects.
• $1.5 million for EPA training seminars so EPA may evaluate and report on proficiency of private radon control companies. This program to be funded through user fee provision.
• $1 million for EPA to begin analysis of radon contamination in nation's schools. $500,000 for demonstrations of radon reduction techniques in schools.
• Baucas Amendment to S.744 authorizes study of radon infiltration of buildings in high radon risk areas owned by GSA, VA, DOI, USDA, and DOD.
Sponsor
Senator George Mitchell (D-ME)
Co-Sponsors
Sen. Chaffee
Sen. Stafford
Sen. Durenberger
Sen. Lautenberg
Sen. Moynihan
Sen. Mikulski
Sen. Reid
Sen. Dodd
Sen. Kerry
Sen. Heinz
Sen. Gore
Sen. Daschle
Sen. Specter
Sen. Graham
Sen. Wirth
Sen. Bradley
Sen. Baucas
Sen. Warner
Sen. Sarbanes
Sen. DeConcini
Sen. Glenn

U.S. SENATE/S.756
• Legislation to ensure when persons pay for home improvements to mitigate harmful levels of radon, their expenses shall be treated as medical expenses and as such qualify for the medical expense tax deduction.
• Read twice and referred to Senate Committee on Finance.
Sponsor
Senator Frank Lautenberg (D-NJ)
Co-Sponsors
Sen. Bradley
Sen. Moynihan
Sen. Heinz
Sen. Specter
Sen. Burdick
Sen. Heflin
Sen. Melcher
Sen. Gore
Sen. Kennedy
Sen. Mikulski
Sen. Riegle
Sen. Daschle
Sen. Durenberger
Sen. Chafee
Sen. Shelby
Sen. Pell
Sen. Ford, W.H.
Sen. Reid
Sen. Cranston
Sen. Boren
Sen. Bumpers
Sen. Dole
Sen. Rockefeller
Sen. Armstrong
Sen. Wirth

Congressional Contacts

Sen. William L. Armstrong (R) CO
528 HSOB
Washington, DC 20515
(202) 224-5941

Sen. Max Baucus (D) MT
706 HSOB
Washington, DC 20515
(202) 224-2651

Sen. David Lyle Boren (D) OK
453 RSOB
Washington, DC 20515
(202) 224-4721

Sen. Bill Bradley (D) NJ
731 HSOB
Washington, DC 20515
(202) 224-3224

Rep. Joseph E. Brennan (D) ME
1428 LHOB
Washington, DC 20515
(202) 225-6116

Sen. Dale Bumpers (D) AR
229 DSOB
Washington, DC 20515
(202) 224-4843

Sen. Quentin Burdick (D) ND
511 HSOB
Washington, DC 20515
(202) 224-2551

Rep. Beverly Byron (D) MD
1216 LHOB
Washington, DC 20515
(202) 225-2721

Rep. Ben Nighthorse Campbell (D) CO
1724 LHOB
Washington, DC 20515
(202) 225-4761

Sen. John H. Chafee (R) RI
567 DSOB
Washington, DC 20515
(202) 224-2921

Rep. Cardiss Collins (D) IL
2264 RHOB
Washington, DC 20515
(202) 225-5006

Rep. Lawrence Coughlin (R) PA
2467 RHOB
Washington, DC 20515
(202) 225-6111

Rep. James A. Courter (R) NJ
2422 RHOB
Washington, DC 20515
(202) 225-5801

Sen. Alan Cranston (D) CA
112 HSOB
Washington, DC 20515
(202) 224-3553

Sen. Thomas A. Daschle (D) SD
317 HSOB
Washington, DC 20515
(202) 224-2321

Rep. Robert W. Davis (R) MI
1124 LHOB
Washington, DC 20515
(202) 225-4735

Sen. Dennis DeConcini (D) AZ
328 HSOB
Washington, DC 20515
(202) 224-4521

Rep. Ronald V. Dellums (D) CA
2136 RHOB
Washington, DC 20515
(202) 225-2661

Sen. Christopher J. Dodd (D) CT
444 RSOB
Washington, DC 20515
(202) 224-2823

Sen. Robert Dole (R) KS
141 HSOB
Washington, DC 20515
(202) 224-6521

Sen. David Durenberger (R) MN
154 RSOB
Washington, DC 20515
(202) 224-3244

Rep. Mervyn M. Dymally (D) CA
1717 LHOB
Washington, DC 20515
(202) 225-5425

Rep. Mike Espy (D) MS
216 CHOB
Washington, DC 20515
(202) 225-5876

Del. Walter E. Fauntroy (D) DC
2135 RHOB
Washington, DC 20515
(202) 225-8050

Rep. Edward F. Feighan (D) OH
1124 LHOB
Washington, DC 20515
(202) 225-5731

Rep. James J. Florio (D) NJ
2162 RHOB
Washington, DC 20515
(202) 225-6501

Rep. Thomas M. Foglietta (D) PA
1217 LHOB
Washington, DC 20515
(202) 225-4731

Sen. Wendell H. Ford (D) KY
173A RSOB
Washington, DC 20515
(202) 224-4343

Del. Jaime Fuster (D) PR
4427 CHOB
Washington, DC 20515
(202) 225-2615

Rep. Dean A. Gallo (R) NJ
1318 LHOB
Washington, DC 20515
(202) 225-5035

Rep. Robert Garcia (D) NY
2338 RHOB
Washington, DC 20515
(202) 225-4361

Rep. Benjamin A. Gilman (R) NY
2160 RHOB
Washington, DC 20515
(202) 225-3776

Sen. John H. Glenn (D) OH
503 HSOB
Washington, DC 20515
(202) 224-3353

Rep. Bert Gordon (D) TN
1517 LHOB
Washington, DC 20515
(202) 225-4231

Sen. Albert Gore, Jr. (D) TN
393 RSOB
Washington, DC 20515
(202) 224-4944

Sen. Robert Graham (D) FL
241 DSOB
Washington, DC 20515
(202) 224-3041

Rep. Tony P. Hall (D) OH
2448 RHOB
Washington, DC 20515
(202) 225-6465

Rep. Augustus F. Hawkins (D) CA
2371 RHOB
Washington, DC 20515
(202) 225-2201

Sen. Howell Heflin (D) AL
728 HSOB
Washington, DC 20515
(202) 224-4124

Sen. H. John Heinz III (R) PA
227 RSOB
Washington, DC 20515
(202) 224-6324

Rep. James J. Howard (D) NJ
2188 RHOB
Washington, DC 20515
(202) 225-4671

Rep. Marcy Kaptur (D) OH
1228 LHOB
Washington, DC 20515
(202) 225-4146

Sen. Edward M. Kennedy (D) MA
315 RSOB
Washington, DC 20515
(202) 224-4543

Sen. John F. Kerry (D) MA
166 RSOB
Washington, DC 20515
(202) 224-2742

Rep. Gerald D. Kleczka (D) WI
226 CHOB
Washington, DC 20515
(202) 225-4572

Rep. Jospeh P. Kolter (D) PA
212 CHOB
Washington, DC 20515
(202) 225-2565

Rep. Peter H. Kostmayer (D) PA
123 CHOB
Washington, DC 20515
(202) 225-4276

Sen. Frank R. Lautenberg (D) NJ
717 HSOB
Washington, DC 20515
(202) 224-4744

Rep. John Lewis (D) GA
501 CHOB
Washington, DC 20515
(202) 225-3801

Rep. Marilyn Lloyd (D) TN
2266 RHOB
Washington, DC 20515
(202) 225-3271

Rep. Michael E. Lowry (D) WA
2454 RHOB
Washington, DC 20515
(202) 225-3106

Rep. Thomas A. Luken (D) OH
2368 RHOB
Washington, DC 20515
(202) 225-2216

Rep. Matthew G. Martinez (D) CA
109 CHOB
Washington, DC 20515
(202) 225-5464

Sen. John Melcher (D) MT
730 HSOB
Washington, DC 20515
(202) 224-2644

Sen. Barbara A. Mikulski (D) MD
320 HSOB
Washington, DC 20515
(202) 224-4654

Sen. George J. Mitchell (D) ME
176 RSOB
Washington, DC 20515
(202) 224-5344

Sen. Patrick Moynihan (D) NY
464 RSOB
Washington, DC 20515
(202) 224-4451

Rep. Austin J. Murphy (D) PA
2210 RHOB
Washington, DC 20515
(202) 225-4665

Rep. John P. Murtha (D) PA
2423 RHOB
Washington, DC 20515
(202) 225-2065

Rep. Stephen L. Neal (D) NC
2463 RHOB
Washington, DC 20515
(202) 225-2071

Rep. David R. Obey (D) WI
2217 RHOB
Washington, DC 20515
(202) 225-3365

Rep. Major R. Owens (D) NY
114 CHOB
Washington, DC 20515
(202) 225-6231

Rep. Stanford Parris (R) VA
1526 LHOB
Washington, DC 20515
(202) 225-4376

Sen. Claiborne Pell (D) RI
335 RSOB
Washington, DC 20515
(202) 224-4642

Rep. Nancy Pelosi (D) CA
1517 LHOB
Washington, DC 20515
(202) 225-4965

Rep. Charles B. Rangel (D) NY
2330 RHOB
Washington, DC 20515
(202) 225-4365

Sen. Harry Reid (D) NV
702 HSOB
Washington, DC 20515
(202) 224-3542

Rep. Thomas J. Ridge (R) PA
1714 LHOB
Washington, DC 20515
(202) 225-5406

Sen. Donald W. Riegle (D) MI
105 DSOB
Washington, DC 20515
(202) 224-4822

Rep. Donald L. Ritter (R) PA
2447 RHOB
Washington, DC 20515
(202) 225-5801

Sen. John D. Rockefeller IV (D) WV
241 DSOB
Washington, DC 20515
(202) 224-6472

Rep. Peter W. Rodino, Jr. (D) NJ
2462 RHOB
Washington, DC 20515
(202) 225-3436

Rep. Robert A. Roe (D) NJ
2243 RHOB
Washington, DC 20515
(202) 225-5751

Rep. Margaret S. Roukema (R) NJ
303 CHOB
Washington, DC 20515
(202) 225-4465

Sen. Paul S. Sarbanes (D) MD
332 DSOB
Washington, DC 20515
(202) 224-4524

Rep. James H. Scheuer (D) NY
2466 RHOB
Washington, DC 20515
(202) 225-5471

Rep. F. James Sensenbrenner (R) WI
2444 RHOB
Washington, DC 20515
(202) 225-5101

Sen. Richard C. Shelby (D) AL
313 HSOB
Washington, DC 20515
(202) 224-5744

Rep. Gerry Sikorski (D) MN
414 CHOB
Washington, DC 20515
(202) 225-2271

Sen. Arlen Specter (R) PA
331 HSOB
Washington, DC 20515
(202) 224-4254

Sen. Robert T. Stafford (R) VT
133 HSOB
Washington, DC 20515
(202) 224-5141

Rep. Edolphus Towns (D) NY
1726 LHOB
Washington, DC 20515
(202) 225-5936

Rep. James A. Traficant (D) OH
128 CHOB
Washington, DC 20515
(202) 225-5261

Rep. Bruce F. Vento (D) MN
2304 RHOB
Washington, DC 20515
(202) 225-6631

Sen. John Warner (R) VA
421 RSOB
Washington, DC 20515
(202) 224-2023

Rep. Robert Whittaker (R) KS
2436 RHOB
Washington, DC 20515
(202) 225-3911

Sen. Timothy E. Wirth (D) CO
380 RSOB
Washington, DC 20515
(202) 224-5852

Rep. Howard E. Wolpe (D) MI
1535 LHOB
Washington, DC 20515
(202) 225-5011

Rep. Gus Yatron (D) PA
2267 RHOB
Washington, DC 20515
(202) 225-5546

10

State and Local Government

This section lists lead and support state government agencies involved in radon-related programs. These activities include administration, policy, research, and disseminating information. Also included here are listings of municipal governments nationwide that have indicated involvement in radon-related activities.

Alabama

ALABAMA DEPARTMENT OF PUBLIC HEALTH/RADIOACTIVE MATERIALS BRANCH/COMPLIANCE SECTION
434 Monroe St.
Room 510
Montgomery, AL 36130
(205) 261-5315
James McNees - Director (State Contact)
Activities:
Lead agency for radon information. Provides EPA information to public on per request basis. Participated with EPA and Alabama Geological Survey on statewide radon survey in 1987.

ALABAMA GEOLOGICAL SURVEY
P.O. Box O
420 Hackberry Ln.
Tuscaloosa, AL 35486-9780
(205) 349-2852
Ernest A. Mancini - State Geologist
Activities:
Mapping capabilities pertinent to radon studies in state. Provides information for state's radon survey.

ALABAMA OCCUPATIONAL SAFETY AND HEALTH CONSULTATION PROGRAM
P.O. Box 6005
University, AL 35486
(205) 348-7136
Bill Weems - Director, Environmental/industrial Programs
Betty Schney - Associate Director
Robert Thompson - Assistant Director Of Safety
Activities:
Lead agency for occupational health and safety issues.

ALABAMA REAL ESTATE COMMISSION
State Capitol
Montgomery, AL 36130
(205) 261-5544
Joseph D. Brindley - Executive Director
Patricia Anderson - Education Advisor
Activities:
Lead agency for real estate regulatory and enforcement issues.

Local Government Agencies

HUNTSVILLE/AIR POLLUTION CONTROL DEPARTMENT
2033 City Airport Rd.
Huntsville, AL 35802
(205) 881-7803
I.N. Vaughan - Air Pollution Control Officer

JEFFERSON COUNTY DEPARTMENT OF HEALTH
1400 Sixth Avenue South
Birmingham, AL 35233
(205) 933-9110
Carole W. Samuelson - Health Officer

Alaska

ALASKA DEPARTMENT OF HEALTH AND SOCIAL SERVICES/DIVISION OF PUBLIC HEALTH/RADIOLOGICAL HEALTH
Box H
Juneau, AK 99811-0613
(907) 465-3019
Sidney Heidersdorf - Chief (State Contact)
Activities:
Disseminates information to public on radon-related issues. Provides detectors, for a fee, for home use in state detection program. State's lead agency for radon information.

ALASKA DEPARTMENT OF ENVIRONMENTAL CONSERVATION/AIR AND HAZARDOUS WASTE MANAGEMENT SECTION
Box O
Juneau, AK 99811-1800
(907) 465-2666
Stan Hungerford - Chief
Activities:
Provides information on radon-related issues to citizens of Alaska.

ALASKA DEPARTMENT OF LABOR/ OCCUPATIONAL SAFETY AND HEALTH
3301 Eagle St.
Suite 303
Pouch 7-022
Anchorage, AK 99510
(907) 276-5013
Tom Stuart - Director
Activities:
Lead agency for occupational safety and health issues.

ALASKA DEPARTMENT OF NATURAL RESOURCES/DIVISION OF GEOLOGICAL AND GEOPHYSICAL SURVEYS
794 University Ave.
Fairbanks, AK 99701
(907) 474-7147
Dr. Robert Forbes - Director
Activities:
Geologic mapping capabilities important for analysis and evaluation of extent of radon threat in Alaska.

ALASKA DIVISION OF OCCUPATIONAL LICENSING
3601 C St.
Suite 722
Anchorage, AK 99503
(907) 563-2169
Grayce A. Oakley - Executive Secretary
Terry McGillivary - Education Director
Activities:
Lead agency for real estate regulatory and enforcement issues.

Local Government Agencies

ANCHORAGE AIR POLLUTION CONTROL AGENCY
825 L St.
Anchorage, AK 99503
(907) 343-4713
W. Randall Poteet - Air Pollution Control Supervisor

FAIRBANKS NORTH STAR BOROUGH/AIR POLLUTION CONTROL DIVISION
P.O. Box 1267
Fairbanks, AK 99701
(907) 452-4761
Richard Joy - Health And Safety Director

Arizona

ARIZONA RADIATION REGULATORY AGENCY
4814 South 40th St.
Phoenix, AZ 85040
(602) 255-4845
John T. O'Neill - (State Contact)
Charles Tedford - Director
Activities:
Provides information to public on radon issues. Participating with EPA on statewide radon study. State's lead radon information agency.

ARIZONA BUREAU OF GEOLOGY AND MINERAL TECHNOLOGY
845 North Park Ave.
Tucson, AZ 85719
(602) 621-7906
Larry D. Fellows - Director And State Geologist
Activities:
Mapping service pertinent to radon surveys in state. Useful for radon in soil or water studies.

ARIZONA DEPARTMENT OF REAL ESTATE
202 East Earll Dr.
Phoenix, AZ 85012
(602) 255-3481
Joe Sotelo - Commissioner
Donna Williams - Education Director
Activities:
Lead agency for real estate regulatory and enforcement issues.

ARIZONA DIVISION OF OCCUPATIONAL SAFETY AND HEALTH/CONSULTATION AND TRAINING
P.O. Box 19070
1624 West Adams
Phoenix, AZ 85005
(602) 255-5795
Tim Arbogast - Director
Activities:
Lead agency for occupational health and safety.

ARIZONA RADIATION REGULATORY AGENCY/ENVIRONMENTAL SURVEILLANCE LABORATORY
4814 South 40th St.
Phoenix, AZ 85040
(602) 255-4845
Paul J. Weeden - Program Manager (State Contact)
Activities:
Provides recommendations to state legislature concerning results from statewide radon study.

Local Government Agencies

COCONINO COUNTY AIR POLLUTION CONTROL DISTRICT/ENVIRONMENTAL HEALTH DIVISION
2500 North Fort Valley Rd.
Flagstaff, AZ 86001
(602) 779-5164
Dan Smith - Director

MARICOPA COUNTY DEPARTMENT OF HEALTH SERVICES/ENVIRONMENTAL HEALTH SERVICES DIVISION
1825 East Roosevelt St.
Phoenix, AZ 85006
(602) 258-6381
John A. Power - Assistant Director

PIMA COUNTY HEALTH DEPARTMENT/AIR QUALITY CONTROL DISTRICT
150 West Congress St.
Tucson, AZ 85701
(602) 792-8803
C. Lee Fox - Director

PINAL-GILA COUNTIES AIR QUALITY CONTROL DISTRICT
P.O. Box 1076
Florence, AZ 85232
(602) 685-5801 x486
Dorothy M. Rankin - Director and Air Pollution Control Officer

Arkansas

ARKANSAS DEPARTMENT OF HEALTH/RADIATION CONTROL AND EMERGENCY MANAGEMENT/NUCLEAR AND ENVIRONMENTAL SAFETY SECTION
4815 West Markham St.
Little Rock , AR 72201
(501) 661-2301
Bernie Bevill - Supervisor (State Contact)
Greta Dicus - Division Director (State Contact)
Activities:
Arkansas lead radon information agency. Distributes information to the public upon request. Developing proposals for statewide radon program.

ARKANSAS DEPARTMENT OF LABOR/OSHA CONSULTATION
1022 High St.
Little Rock, AR 72202
(501) 375-8442
Bud Daven - Supervisor
Activities:
Lead agency for occupational health and safety issues.

ARKANSAS GEOLOGICAL COMMISSION
3815 West Roosevelt Rd.
Little Rock, AR 72204
(501) 371-1488
Norman F. Williams - Director And State Geologist
Activities:
Mapping capabilities useful in event of statewide or regional radon survey.

ARKANSAS REAL ESTATE COMMISSION
One Riverfront Pl.
Suite 660
North Little Rock, AR 72114
(501) 371-1247
Roy L. Bilheimer - Executive Secretary
Activities:
Lead agency for real estate regulatory and enforcement issues.

California

CALIFORNIA DEPARTMENT OF HEALTH SERVICES/AIR AND INDUSTRIAL HYGIENE LABORATORY
2151 Berkeley Way
Room 334
Berkeley, CA 94704
(415) 540-2134
Steve Hayward - Manager (State Contact)
Activities:
Develops state's radon policy guidelines and provides public with information on radon issues.

CALIFORNIA AIR RESOURCES BOARD/AIR QUALITY DIVISION
1102 Q St.
Sacramento, CA 95814
(916) 323-5043
Peggy Jenkins - Director

CALIFORNIA DEPARTMENT OF CONSERVATION/DIVISION OF MINES
1416 Ninth St.
Room 1341
Sacramento, CA 95814
(916) 445-1923
Brian Tucker - Acting State Geologist
Activities:
Provides mapping service pertinent to radon surveys in state. Information useful for radon in soil or water studies.

CALIFORNIA DEPARTMENT OF HEALTH SERVICES/RADIOLOGICAL HEALTH BRANCH
714 P St.
Building #8
Sacramento, CA 95814
(916) 322-2040
Paul Szalinski - Chief

Activities:
Radioactive materials licensing bureau contributing to development of statewide radon policy.

CALIFORNIA DEPARTMENT OF HEALTH SERVICES/SANITATION AND RADIATION LABORATORY
2151 Berkeley Way
Room 465
Berkeley, CA 94704
(415) 540-2201
George S. Uyesugi - Director
Activities:
Performs radon-water testing. Provides information on radon-related issues.

CALIFORNIA DEPARTMENT OF INDUSTRIAL RELATIONS/DIVISION OF OCCUPATIONAL SAFETY AND HEALTH
P.O. Box 603
San Francisco, CA 94101-0603
(415) 557-2870
R.W. Stranberg - Chief
F.R. Ciosalo - Deputy Chief
Activities:
Lead agency for occupational health and safety issues.

CALIFORNIA DEPARTMENT OF REAL ESTATE
2201 Broadway
P.O. Box 187000
Sacramento, CA 95818-7000
(916) 739-3721
James A. Edmonds, Jr. - Commissioner
Thomas L. Mabry - Manager, Education and Research
Activities:
Lead agency for real estate regulatory and enforcement issues.

Local Government Agencies

AMADOR COUNTY AIR POLLUTION CONTROL DISTRICT
108 Court St.
Jackson, CA 95642-2379
(209) 223-6406
James B. McClenahan - Air Pollution Control Officer

BAY AREA AIR QUALITY MANAGEMENT DISTRICT
939 Eillis St.
San Francisco, CA 94109
(415) 771-6000
Milton Feldstein - Air Pollution Control Officer

BUTTE COUNTY AIR POLLUTION CONTROL DISTRICT
316 Nelson Ave.
P.O. Box 1229
Oroville, CA 95965
(916) 534-4383
Joe E. Bandy - Air Pollution Control Officer

CALAVERAS COUNTY HEALTH DEPARTMENT/AIR POLLUTION DIVISION
Government Center
San Andreas, CA 95249
(209) 754-3742
Robert E. Marshall - Air Pollution Control Officer

COLUSA COUNTY AIR POLLUTION CONTROL DISTRICT
P.O. Box 1029
Colusa, CA 95932
(916) 458-5891
Harry A. Krug - Air Pollution Control Officer

EL DORADO COUNTY AIR POLLUTION CONTROL DISTRICT
7553 Green Valley Rd.
Placerville, CA 95667
(916) 626-2407
Vernon C. Peterson - Air Pollution Control Officer

FRESNO COUNTY AIR POLLUTION CONTROL DISTRICT
1221 Fulton Mall
Fresno, CA 93710
(209) 445-3239
William J. Roddy - Air Quality Control Chief

GLENN COUNTY AIR POLLUTION CONTROL DISTRICT
720 North Colusa St.
Willows, CA 95988
(916) 934-4651
Ed Romano - Air Pollution Control Chief

GREAT BASIN UNIFIED AIR POLLUTION CONTROL DISTRICT
157 Short St.
Suite 6
Bishop, CA 93514
(619) 872-8211
Ed Romano - Air Pollution Control Officer

KERN COUNTY AIR POLLUTION CONTROL DISTRICT
2700 M St.
Bakersfield, CA 93305-4198
(805) 861-3682
Stephen L. Birdsall - Air Pollution Control Officer

KINGS COUNTY AIR POLLUTION CONTROL DISTRICT
330 Campus Dr.
Hanford, CA 93230
(209) 584-1411
Sheldon R. Minkin - Air Pollution Control Officer

LAKE COUNTY AIR QUALITY MANAGEMENT DISTRICT
883 Lakeport Blvd.

Lakeport, CA 95453
(707) 263-7000
Robert L. Reynolds - Air Pollution Control Officer

LASSEN COUNTY AIR POLLUTION CONTROL DISTRICT
175 Russell Ave.
Susanville, CA 96130
(916) 257-8311
Robert L. Reynolds - Air Pollution Control Officer

LOS ANGELES COUNTY DEPARTMENT OF HEALTH SERVICES/RADIATION MANAGEMENT
2615 South Grand Ave.
Los Angeles, CA 90007
(213) 744-3244
A. Ferguson - County Radon Contact

MADERA COUNTY AIR POLLUTION CONTROL DISTRICT
135 West Yosemite Ave.
Madera, CA 93637
(209) 675-7823
W.E. Sturk - Acting Air Pollution Control Officer

MARIPOSA COUNTY HEALTH DEPARTMENT
4988 11th St.
P.O. Box 5
Mariposa, CA 95338
(209) 966-3689
Avery E. Sturm - Air Pollution Control Officer

MENDOCINO COUNTY AIR POLLUTION CONTROL DISTRICT
Courthouse Sq.
Ukiah, CA 95482
(707) 463-4354
Craig M. McMillian - Air Pollution Control Officer

MERCED COUNTY AIR POLLUTION CONTROL DISTRICT
385 East13th St.
P.O. Box 471
Merced, CA 95341
(209) 385-7391
Richard Welch - Air Pollution Control Officer

MODOC COUNTY AIR POLLUTION CONTROL DISTRICT
202 West Fourth St.
Alturas, CA 96101
(916) 233-3939 x 401
Clinton B. Greenbank - Air Pollution Control Officer

MONTEREY BAY UNIFIED AIR POLLUTION CONTROL DISTRICT
1164 Monroe St.
Suite 10
Salinas, CA 93906
(408) 443-1135
Lawrence D. Odle - Air Pollution Control Officer

NORTHERN SIERRA AIR QUALITY MANAGEMENT DISTRICT
P.O. Box 5
Sierra City, CA 96125
(916) 862-1233
Earl Withycombe - Air Pollution Control Officer

NORTHERN SIERRA AIR QUALITY MANAGEMENT DISTRICT
10433 Willow Valley Rd.
Nevada City, CA 95959
(916) 265-1398
Wayne Morgan - Air Pollution Control Officer

NORTHERN SONOMA COUNTY AIR POLLUTION CONTROL DISTRICT
109 North St.
Healdsburg, CA 95448
(707) 433-5911 (707) 433-5742
Michael W. Tolmasoff - Air Pollution Control Officer

PLACER COUNTY AIR POLLUTION CONTROL DISTRICT
11484 B Ave.
Auburn, CA 95603
(916) 823-4443
Noel A. Bonderson - Air Pollution Control Officer

PLUMAS COUNTY DEPARTMENT OF ENVIRONMENTAL HEALTH/AIR POLLUTION CONTROL DIVISION
P.O. Box 480
Quincy, CA 95971
(916) 283-1255
William F. Crigler - Director

SACRAMENTO COUNTY AIR POLLUTION CONTROL DISTRICT
9323 Tech Center Dr.
Suite 800
Sacramento, CA 95826
(916) 366-2107
Norman Covell - Air Pollution Control Officer

SAN BERNARDINO COUNTY AIR POLLUTION CONTROL DISTRICT
15505 Civic Dr.
Victorville, CA 92392
(619) 243-8147
Walter R. Mook - Executive Officer

SAN DIEGO COUNTY AIR POLLUTION CONTROL DISTRICT
9150 Chesapeake Dr.
San Diego, CA 92123-1095
(619) 694-3307
R.J. Sommerville - Air Pollution Control Officer

SAN JOAQUIN COUNTY AIR POLLUTION CONTROL DISTRICT
1601 East Hazelton Ave.

P.O. Box 2009
Stockton, CA 95201
(209) 468-3470
Jogi Khanna - Air Pollution Control Officer

SAN LUIS OBISPO COUNTY AIR POLLUTION CONTROL DISTRICT
2156 Sierra Way
Suite B
San Luis Obispo, CA 93401
(805) 549-5912
George B. Rowland - Air Pollution Control Officer

SANTA BARBARA COUNTY AIR POLLUTION CONTROL DISTRICT
5540 Ekwill St.
Suite B
Santa Barbara, CA 93111
(805) 964-8111
James M. Ryerson - Air Pollution Control Officer

SHASTA COUNTY AIR QUALITY MANAGEMENT DISTRICT
1855 Placer St.
Redding, CA 96001
(916) 225-5674
Richard B. Booth - Air Pollution Control Officer

SISKIYOU COUNTY AIR POLLUTION CONTROL DISTRICT
525 South Foothill Dr.
Yreka, CA 96097
(916) 842-3906
Edmond W. Hale - Air Pollution Control Officer

SOUTH COAST AIR QUALITY MANAGEMENT DISTRICT/AIR POLLUTION CONTROL DIVISION
9150 Flair Dr.
El Monte, CA 91731
(818) 572-6200
James Lents - Executive Officer

STANISLAUS COUNTY AIR POLLUTION CONTROL DISTRICT
1716 Morgan Rd.
Modesto, CA 95351
(209) 571-6908
Gordon Dewers - Air Pollution Control Officer

SUTTER COUNTY AIR POLLUTION CONTROL DISTRICT
142 Garden Way
Yuba City, CA 95991
(916) 674-2851
Alfred C Perrin, Jr. - Air Pollution Control Officer

TEHAMA COUNTY AIR POLLUTION CONTROL DISTRICT
1760 Walnut St.
P.O. Box 38
Red Bluff, CA 96080
(916) 527-4504
Donald R. Hill - Air Pollution Control Officer

TULARE COUNTY AIR POLLUTION CONTROL DISTRICT/ENVIRONMENTAL HEALTH DIVISION
Health Building
County Civic Ctr.
Visalia, CA 93291
(209) 733-6441
Martin Keast - Environmental Health Engineer

TUOLUMNE COUNTY AIR POLLUTION CONTROL DISTRICT
2 South Green St.
Sonora, CA 95730
(209) 533-5698
Gerald A. Benincasa - Air Pollution Control Officer

VENTURA COUNTY AIR POLLUTION CONTROL DISTRICT
800 South Victoria Ave.
Ventura, CA 93009
(805) 654-2806
Richard H. Baldwin - Air Pollution Control Officer

YOLO-SOLANO AIR POLLUTION CONTROL DISTRICT
323 First St.
Suite 5
P.O. Box 1006
Woodland, CA 95695
(916) 666-8146
James A. Koslow - Air Pollution Control Officer

YUBA COUNTY AIR POLLUTION CONTROL DISTRICT
938 14th St.
Marysville, CA 95901
(916) 741-6484
D.B. Engle - Air Pollution Control Officer

Colorado

COLORADO DEPARTMENT OF HEALTH/RADIATION CONTROL DIVISION
4210 East 11th Ave.
Denver, CO 80220
(303) 331-4812
Richard Gamewell - State Contact
Warren E. Jacobi - Section Chief
A.J. Hazle - Division Director
Activities:
One of original 10 states to be surveyed for radon by EPA. Provides EPA pamphlets and list of detection companies upon request. Information from EPA/state survey indicates more than 25% of houses in Colorado might have screening levels greater than 4 pCi/l.

COLORADO GEOLOGICAL SURVEY
1313 Sherman St.
Room 715
Denver, CO 80203

(303) 866-2611
John W. Rold - Director And State Geologist
Mark Davis - Mineral Resources Chief (State Contact)
Ann Scanlon - Mineral Fuels Geologist
Activities:
Developed program for original Colorado state radon survey. Presently offers radon information concerning building and development at particular sites.

COLORADO INSTITUTE OF ENVIRONMENTAL HEALTH/OCCUPATIONAL SAFETY AND HEALTH SECTION
110 Veterinary Science Bldg.
Colorado State University
Fort Collins, CO 80523
(303) 491-6151
Bill Sandfort - Project Supervisor
Activities:
Lead agency for occupational health and safety issues.

COLORADO REAL ESTATE COMMISSION
1776 Logan St.
4th Floor
Denver, CO 80203
(303) 894-2166
Michael B. Gorham - Director
Activities:
Lead agency for real estate regulatory and enforcement issues.

Local Government Agencies

BOULDER COUNTY HEALTH DEPARTMENT/AIR POLLUTION CONTROL DIVISION
3450 Broadway
Boulder, CO 80302
(303) 441-1180
Susan Martino - Environmental Health Sanitarian

DENVER/ CITY-COUNTY DEPARTMENT OF HEALTH AND HOSPITALS/ENVIRONMENTAL HEALTH SERVICE
605 Bannock St.
Denver, CO 80204
(303) 893-6243
Steven Foute - Director, Air Pollution Control

EL PASO COUNTY HEALTH DEPARTMENT/AIR QUALITY CONTROL SECTION
501 North Foote Ave.
Colorado Springs, CO 80909-4598
(303) 578-3137
John B. Muth - Director-Medical
Donna Schneiderman-Scheer - Air Quality Aide

INSTITUTE OF ENVIRONMENTAL HEALTH/OCCUPATIONAL SAFETY AND HEALTH SECTION
Colorado State University
110 Veterinary Science Building
Fort Collins, CO 80523
(303) 491-6151

JEFFERSON COUNTY HEALTH DEPARTMENT
260 South Kipling
Lakewood, CO 80226
(303) 232-6301
Richard A. Bell - Air Pollution Control Officer

LARIMER COUNTY HEALTH DEPARTMENT/ENVIRONMENTAL HEALTH DIVISION
363 Jefferson St.
Fort Collins, CO 80524
(303) 221-7497
Laurence Johnston - Director

MESA COUNTY HEALTH DEPARTMENT/AIR POLLUTION CONTROL DIVISION
515 Patterson Rd.
Grand Junction, CO 81501
(303) 244-1750
Kenneth J. Lampert - Director

PUEBLO CITY-COUNTY HEALTH DEPARTMENT
151 Central Main
Pueblo, CO 81003
(303) 544-8376
Gerald Hollandsworth - Program Manager

WELD COUNTY HEALTH DEPARTMENT/AIR POLLUTION CONTROL DIVISION
1516 Hospital Rd.
Greeley, CO 80631
(303) 353-0635
Jeff Stoll - Environmental Specialist

Connecticut

CONNECTICUT DEPARTMENT OF HEALTH SERVICES/TOXIC HAZARDS SECTION
150 Washington St.
Hartford, CT 06106
(203) 566-3122
Alan Siniscalchi - Radon Program Coordinator (State Contact)
Lynn Rothney - Health Program Assistant
Maria Van der Werss - Environmental Sanitarian
Activities:
Lead agency for radon programs in Connecticut. Information package includes list of companies providing testing, detection and mitigation services. Participated in radon surveys. Department, with assistance from EPA, Connecticut Department of Environmental Protection, and CONNSAVE, has surveyed state to determine distribution of indoor radon levels throughout state.

CONNECTICUT DEPARTMENT OF CONSUMER PROTECTION/REAL ESTATE DIVISION
State Office Building
Room G-8
165 Capitol Ave.
Hartford, CT 06106
(203) 566-5130
Laurence L. Hannafin - Director

Activities:
Lead agency for real estate regulatory and enforcement issues.

CONNECTICUT DEPARTMENT OF ENVIRONMENTAL PROTECTION/RADIATION CONTROL UNIT
State Office Building
165 Capital Ave.
Hartford, CT 06106
(203) 566-5668
Kevin McCarthy - Director
Donald Karn - Radiation Control Physicist
Activities:
Oversees radiation control programs in Connecticut. Provides public with information on radon and maintains level of involvement in state radon programs.

CONNECTICUT DEPARTMENT OF LABOR/DIVISION OF OCCUPATIONAL SAFETY AND HEALTH
200 Folly Brook Rd.
Wethersfield, CT 06109
(203) 566-4550
Emil Caruso - Director
Activities:
Lead agency for occupational health and safety issues.

CONNECTICUT GEOLOGICAL AND NATURAL HISTORY SURVEY/NATURAL RESOURCES CENTER
165 Capitol Ave.
Room 553
Hartford, CT 06106
(203) 566-3540
Hugo Thomas - Director
Activities:
Gamma readings and mapping data points. Reviews of NURE maps covering Connecticut.

Local Government Agencies

BRIDGEPORT/DEPARTMENT OF AIR POLLUTION CONTROL/OFFICE OF HUMANE AFFAIRS
752 East Main St.
Bridgeport, CT 06608
(203) 576-7451
Henry Gross - Director, Environmental Health

BRISTOL-BURLINGTON HEALTH DEPARTMENT
240 Stafford Ave.
Bristol, CT 06010-4617
(203) 584-7682
William E. Furniss - Director Of Health

GREENWICH HEALTH DEPARTMENT/DIVISON OF ENVIRONMENTAL HEALTH AND LABORATORY
Town Hall
Field Point Rd.
Greenwich, CT 06830
(203) 622-7843
Robert C. Brown - Director

MERIDEN DEPARTMENT OF HEALTH/HUMAN SERVICES
165 Miller St.
Meriden, CT 06450
(203) 634-0003
Raymond E. West - Chief, Environmental Health

NEW HAVEN HEALTH DEPARTMENT/ENVIRONMENTAL HEALTH DIVISION
One State St.
New Haven, CT 06511
(203) 787-8173
Paul A. Kowalski - Director

NORWALK DEPARTMENT OF HEALTH
137 East Ave.
Norwalk, CT 06851
(203) 854-7776
Louise S. Leary - Director of Health

STAMFORD HEALTH DEPARTMENT
229 North St.
Stamford, CT 06902
(203) 358-4396
Vincent J. Radke - Director of Environmental Health

STRATFORD DEPARTMENT OF HEALTH
2730 Main St.
Stratford, CT 06497
(203) 385-4090
William J. Watson - Director Of Health

Delaware

DELAWARE DEPARTMENT OF HEALTH AND SOCIAL SERVICES/BUREAU OF ENVIRONMENTAL HEALTH/OFFICE OF RADIATION CONTROL/DIVISION OF PUBLIC HEALTH
802 Silver Lake Blvd. and Walker Rd.
P.O. Box 637
Dover, DE 19903
(302) 736-4731
Allan C. Tapert - Program Administrator (State Contact)
Gary K. Lynn - Radiation Control Specialist
John J. Hedden - Radon Control Inspection
Glen Roberts - Inspector
Activities:
State's lead radon information agency. Provides radon inspection test kit for screening purposes to all citizens upon request. Informs homeowners of radon levels based on test results. Requires certification of indoor measurement companies.

DELAWARE DEPARTMENT OF LABOR/DIVISION OF INDUSTRIAL AFFAIRS
820 North French St.
6th Fl.
Wilmington, DE 19801
(302) 571-3908
William Whigham - Project Administrator

Activities:
Lead agency for occupational health and safety issues.

DELAWARE DIVISION OF PROFESSIONAL REGULATION
Margaret O'Neill Building
P.O. Box 1401
Dover, DE 19903
(302) 736-4522
Dorothy F. Baker - Administrative Assistant
Activities:
Lead agency for real estate regulatory and enforcement issues.

DELAWARE GEOLOGICAL SURVEY
University Of Delaware
Newark, DE 19716
(302) 452-2833
Robert Jordan - Director And State Geologist
Activities:
Review of NURE maps to analyze potential radon "hot spots" in state.

District Of Columbia

DISTRICT OF COLUMBIA DEPARTMENT OF CONSUMER AND REGULATORY AFFAIRS/SERVICE FACILITY REGULATION ADMINISTRATION/PHARMACEUTICAL AND MEDICAL DEVICES CONTROL DIVISION
614 H St.
Room 1014
Washington, DC 20004
(202) 727-7190
Norma Stewart - Program Manager (State Contact)
Veronica Singh - Information Specialist
Frances A. Bowie - Administrator
Activities:
Provides radon information on request.

DISTRICT OF COLUMBIA DEPARTMENT OF CONSUMER AND REGULATORY AFFAIRS/REAL ESTATE COMMISSION
614 H St., NW
Room 923
Washington, DC 20013
(202) 727-7468, (202) 727-7469
Barbara Anthon - Board/Commission Support Officer
Holly Jones Kincaide - Realty Education Program Specialist
Activities:
Lead agecy for real estate regulatory and enforcement issues.

DISTRICT OF COLUMBIA DEPARTMENT OF EMPLOYMENT SERVICES/OCCUPATIONAL SAFETY AND HEALTH DIVISION
950 Upshur St., NW
Washington, DC 20011
(202) 576-6651
Lorenzo White - Associate Director
Activities:
Lead agency for occupational health and safety issues.

DISTRICT OF COLUMBIA/ENERGY OFFICE
613 G St., NW
Suite 500
Washington, DC 20001
(202) 724-2100
Jack Werner - Special Assistant
Activities:
Public information hotline on issues concerning potential for rising radon levels in weatherized homes.

Florida

FLORIDA DEPARTMENT OF HEALTH AND REHABILITATIVE SERVICES/DEPUTY SECRETARY FOR HEALTH/OFFICE OF RADIATION CONTROL
1317 Winewood Blvd.
Tallahassee, FL 32399-0700
(904) 487-1004
Lyle E. Jerrett - Director (State Contact)
Mary Clark - Public Health Physicist Consultant
Activities:
State's lead agency for radon information. Provides radon information upon request and offers free alpha track test kits to some counties. Funds research projects that study radon-related issues. Sets policy guidelines and standards for Florida's radon program.

FLORIDA DEPARTMENT OF NATURAL RESOURCES/BUREAU OF MINE RECLAMATION
Innovation Park
2051 East Dirac Dr.
Tallahassee, FL 32304-3760
(904) 488-8217
Steve Windham - Environmental Administrator
Activities:
Ensures that environmental regulations concerning radon at phosphate mines are followed.

FLORIDA DEPARTMENT OF COMMUNITY AFFAIRS/DIVISION OF HOUSING AND COMMUNITY DEVELOPMENT/CODES AND STANDARDS SECTION
2740 Centerview Dr.
The Rhyne Building
Tallahassee, FL 32399
(904) 487-1824
Mary K. Smith - Program Director
Activities:
Oversees radon detection and mitigation standards in large structures and homes.

FLORIDA DEPARTMENT OF HEALTH AND REHABILITATIVE SERVICES/DEPUTY SECRETARY FOR HEALTH/OFFICE OF RADIATION CONTROL
1317 Winewood Blvd.
Tallahassee, FL 32399-0700
(904) 487-1004
Walter Klein - Public Health Physicist Manager
Clark Eldredge - Chemist
Activities:
3-day course and examination for detection and/or mitigation fims to gain certification in Florida. Analysis

labs are certifitied on basis of EPA's RMP program. Labs that provide any service other than analysis must participate in certification program.

FLORIDA DEPARTMENT OF HEALTH AND REHABILITATIVE SERVICES/DEPUTY SECRETARY FOR HEALTH/OFFICE OF RADIATION CONTROL/ENVIRONMENTAL SECTION
P.O. Box 15490
Orlando, FL 32858
(407) 628-2274 (407) 297-2097
Harlan Keaton - Public Health Physicist Manager
Benjamin Prewitt
Activities:
Provides research on radon-related issues for DHR's radon programs.

FLORIDA DEPARTMENT OF HEALTH AND REHABILITATIVE SERVICES/DEPUTY SECRETARY FOR HEALTH/OFFICE OF TOXICOLOGY AND HAZARDS ASSESSMENT
1317 Winewood Blvd.
Building 6
Room 466
Tallahassee, FL 32399-0700
(904) 488-2905
Richard Freeman - Toxicologist
Activities:
Research on epidemiological implications of radon exposure.

FLORIDA DEPARTMENT OF LABOR AND EMPLOYMENT SECURITY/BUREAU OF INDUSTRIAL SAFETY AND HEALTH
Lafayette Building
Room 204
2551 Executive Center Circle West
Talahassee, FL 32301
(904) 488-3044
Gary Strobel - Bureau Chief
Activities:
Lead agency for occupational health and safety issues.

FLORIDA DEPARTMENT OF NATURAL RESOURCES
Cedars Executive Center
2639 North Monroe
Box 20, Building B
Tallahassee, FL 32399
(904) 488-3177
Charles Hendry - State Geologist
Activities:
Measuring radon flux and radium in soils for pre- and post-mining levels.

FLORIDA DEPARTMENT OF PROFESSIONAL REGULATION/DIVISION OF REAL ESTATE
P.O. Box 1900
Orlando, FL 32802-1900
(407) 423-6048
Darlene F. Keller - Director
Charles Hoeck - Education and Examination Supervisor
Activities:
Lead agency for real estate regulatory and enforcement issues.

FLORIDA INSTITUTE OF PHOSPHATE RESEARCH
1855 West Main St.
Bartow, FL 33830
(813) 533-0983
Gordon Nifong, Ph.D. - Research Director
David Borris - Executive Director
Activities:
State-funded agency produced state-wide radon study bringing about enactment of new radon legislation. FIPR released 1987 study demonstrating proper construction of radon-resistant homes on lands elevated in radon potential.

Local Government Agencies

AIR POLLUTION CONTROL PROGRAM OF THE ENVIRONMENTAL PROTECTION COMMISSION OF HILLSBOROUGH COUNTY
1410 North 21st St.
Tampa, FL 33605
(813) 272-5530
Gary Carlson - Chief, Air Section

ALACHUA COUNTY DEPARTMENT OF ENVIRONMENTAL SERVICES
825 NW 23rd Ave.
Suite F-10
Gainesville, FL 32609
(904) 373-8509
John Schert - Director

BROWARD COUNTY ENVIRONMENTAL QUALITY CONTROL BOARD
500 SW 14th Ct.
Fort Lauderdale, FL 33315
(305) 765-5881
Gary Carlson - Chief, Air Section

DADE COUNTY CONSUMER ADVOCATES OFFICE
140 West Flagler St.
Room 1605
Miami, FL 33130
(305) 375-4206
Leonard Elias - Director
Activities:
Provides information on radon-related issues to consumers.

IMPERIAL POLK COUNTY/BUILDING DIVISION
P.O. Box 726
Bartow, FL 33830
(813) 534-6573
Richard L. Iden - Building Director
Activities:
Issues building permits for unincorporated areas of Polk County.

JACKSONVILLE/DUVAL COUNTY BIO-ENVIRONMENTAL SERVICES DIVISION
515 West 6th St.
Jacksonville, FL 32206

(904) 630-3210
Khurshid Mehta - Air Pollution Engineer

MANATEE COUNTY PUBLIC HEALTH UNIT/POLLUTION CONTROL DIVISION
202 Sixth Ave. East
Bradenton, FL 34208
(813) 748-0666
William L. Priesmeyer - Pollution Control Director

METROPOLITAN DADE COUNTY DEPARTMENT OF ENVIRONMENTAL RESOURCES MANAGEMENT
Metro Dade Ctr.
111 NW First St.
Suite 1310
Miami, FL 33128
(305) 375-3326
Carl Pfaffenberger - Chief, Environmental Monitoring Division

PALM BEACH COUNTY HEALTH DEPARTMENT/AIR POLLUTION CONTROL PROGRAM
901 Evernia St.
West Palm Beach, FL 33401
(305) 820-3070
E.J. Sacco - Administrator

PINELLAS COUNTY DEPARTMENT OF ENVIRONMENTAL MANAGEMENT/AIR QUALITY DIVISION
16100 Fairchild Dr.
Building V102
Clearwater, FL 34622
(813) 530-6522
Peter A. Hessling - Administrator

REEDY CREEK IMPROVEMENT DISTRICT/POLLUTION CONTROL DEPARTMENT
P.O. Box 36
Lake Buena Vista, FL 32830
(305) 824-3460
Gary Gornto - Director

SARASOTA COUNTY AIR POLLUTION CONTROL/SARASOTA COUNTY ENVIRONMENTAL SERVICES DEPARTMENT
1301 Cattlemen Rd.
Sarasota, FL 34232
(813) 378-6137
Jose M. Guira - Air Program Manager

Georgia

GEORGIA DEPARTMENT OF NATURAL RESOURCES/PROGRAM COORDINATION BRANCH
205 Butler St., SE
Floyd Towers East
Atlanta, Ga 30334
(404) 656-4713
James Setser - Branch Chief
James C. Hardeman - (State Contact)
Richard Schreiber - Environmental Specialist
Activities:
Collects radon data and makes it available to the general public. Also provides EPA information upon request. Georgia's lead agency for radon information.

GEORGIA GEOLOGICAL SURVEY
19 Martin Luther King, Jr., Dr., SW
Room 400
Atlanta, GA 30334
(404) 656-3214
William McLemore - State Geologist
Activities:
Geologic mapping service could be used for radon surveys in state.

GEORGIA ONSITE CONSULTATION PROGRAM
Georgia Institute Of Technology
EES/EDL/OSHG
Atlanta, GA 30332
(404) 894-3806
Kenneth A. Smith - Program Director
Activities:
Lead agency for occupational health and safety issues.

GEORGIA REAL ESTATE COMMISSION
40 Pryor St., SW
Atlanta, GA 30303-3184
(404) 656-3916
Charles Clark - Commissioner
Tripp Anderson - Education Advisor
Activities:
Lead agency for real estate regulatory and enforcement issues.

Hawaii

HAWAII DEPARTMENT OF HEALTH/ENVIRONMENTAL PROTECTION AND HEALTH SERVICES DIVISION/NOISE AND RADIATION BRANCH
591 Ala Moana Blvd.
Honolulu, HI 96813
(808) 548-4383
Thomas Anamizu - Chief (State Contact)
Activities:
State's lead radon information agency provides information upon request. Limited resources.

HAWAII DEPARTMENT OF COMMERCE AND CONSUMER AFFAIRS/PROFESSIONAL AND VOCATIONAL LICENSING DIVISION
P.O. Box 3469
Honolulu, HI 96801
(808) 548-7579
Calvin Kimura - Executive Secretary
Cyntia M.L. Yee - Information Officer
Activities:
Lead agency for real estate regulatory and enforcement issues.

HAWAII DEPARTMENT OF LABOR AND INDUSTRIAL RELATIONS/DIVISION OF OCCUPATIONAL SAFETY AND HEALTH
677 Ala Moana
Suite 910
Honolulu, HI 96813
(808) 548-2511
Harold W. Barks - Administrator
Activities:
Lead agency for occupational health and safety issues.

HAWAII DIVISION OF WATER AND LAND DEVELOPMENT
P. O. Box 373
Honolulu, HI 96809
(808) 548-7533
Manabu Tagomori - Manager/Chief Engineer and State Geologist
Activities:
Geologic maps could be used for statewide radon survey.

Idaho

IDAHO DEPARTMENT OF HEALTH AND WELFARE/DIVISION OF ENVIRONMENTAL QUALITY/BUREAU OF HAZARDOUS MATERIALS/RADIATION CONTROL DIVISION
450 West State St.
Boise, ID 83720
(208) 334-5879
Ernie Ranieri - Supervisor (State Contact)
Activities:
Lead agency for radon information. Provides EPA information to public upon request.

IDAHO GEOLOGICAL SURVEY
University Of Idaho
Morrill Hall, Room 332
Moscow, ID 83843
(208) 885-7991
Earl H. Bennett - Director and State Geologist
Activities:
Geologic mapping capabilities to identify radon threat in regions of state.

IDAHO OSHA ONSITE CONSULTATION PROGRAM
Boise State University
Community And Environmental Health
1910 University Drive
Boise, ID 83725
(208) 385-3929
Lee Stokes - Director
Activities:
Lead agency for occupational health and safety issues.

IDAHO REAL ESTATE COMMISSION
Statehouse Mail
Boise, ID 83720
(208) 334-3285
Michael P. Gray - Executive Director
Jeri Pyeatt - Director of Education, Examination and Licensure
Activities:
Lead agency for real estate regulatory and enforcement issues.

Illinois

ILLINOIS DEPARTMENT OF NUCLEAR SAFETY/OFFICE OF ENVIRONMENTAL SAFETY
1035 Outer Park Dr.
Springfield, IL 62704
(217) 786-6399
Terry Lash - Department Director (State Contact)
John Cooper - Office Manager
Melanie Anne Hamel - Health Physicist
Richard Allen - Chief, Radioecology Division, Radon Program
Activities:
Department has compiled 2,300 radon measurements throughout state in ongoing radon survey. Will sponsor legislation for certification of mitigation companies. State's lead agency for radon information.

ILLINOIS DEPARTMENT OF COMMERCE AND COMMUNITY AFFAIRS/DIVISION OF INDUSTRIAL SERVICES
State Of Illinois Center
100 West Randolph St.
Suite 3-400
Chicago, IL 60601
(800) 972-4140, (800) 972-4216
Stan Fryzel - Division Manager
Activities:
Lead agency for occupational health and safety issues.

ILLINOIS DEPARTMENT OF PROFESSIONAL REGULATION
320 West Washington St.
3rd Floor
Springfield , IL 62786
(217) 785-0800
Ronald L. See - Commissioner of Real Estate
Activities:
Lead agency for real estate regulatory and enforcement issues.

ILLINOIS ENVIRONMENTAL PROTECTION AGENCY/DIVISON OF AIR POLLUTION CONTROL
2200 Churchill Rd.
P.O. Box 19276
Springfield, IL 62794-9276
(217) 782-7326
Michael J. Hayes - Division Manager

ILLINOIS STATE GEOLOGICAL SURVEY
Natural Resources Building
615 East Peabody Dr.
Room 121
Champaign, IL 61820
(217) 333-5111
Morris W. Leighton - Chief and State Geologist
Richard A. Cahill
Activities:
Provides geologic mapping service pertinent to radon soil and water analyses in state.

Local Government Agencies

BENSENVILLE AIR POLLUTION CONTROL DISTRICT
1253 Roosevelt Ave.
Glenview, IL 60025
(312) 498-9840
John Evanoff - Director and Chief Pollution Control Officer

CHICAGO/DEPARTMENT OF CONSUMER SERVICES/ENERGY AND ENVIRONMENTAL PROGRAMS DIVISION
121 North LaSalle St.
Room 810
Chicago, IL 60602
(312) 744-4034
John Evanoff - Director

COOK COUNTY DEPARTMENT OF ENVIRONMENTAL CONTROL
Maybrook Civic Center
1500 Maybrook Dr.
Maywood, IL 60153
(312) 865-6165
Mario G. Tonelli - Director

DUPAGE COUNTY HEALTH DEPARTMENT/ENVIRONMENTAL HEALTH DIVISION
111 North County Farm Rd.
Wheaton, IL 60187
(312) 682-7400
Bruce Schroer - Air Pollution Sanitarian

EAST ST. LOUIS/AIR POLLUTION CONTROL
7 Collinsville Ave.
East St. Louis, IL 62201
(618) 874-2115
Carl E. Officer - Chairperson, City Commission

EVANSTON/DEPARTMENT OF BUILDING AND ZONING
2100 Ridge Ave.
Evanston, IL 60204
(312) 328-2100
Reed Carlson - Director, Buildling And Zoning

VILLAGE OF BEDFORD PARK/AIR POLLUTION CONTROL
6701 South Archer Rd.
Bedford Park, IL 60501
(312) 458-2975
Karl Mixson - Chairperson

Indiana

INDIANA STATE BOARD OF HEALTH/RADIOLOGICAL HEALTH SECTION
1330 West Michigan St.
Box 1964
Indianapolis, IN 46206-1964
(317) 633-0153
Hal S. Stocks - Chief (State Contact)
David E. Nauth - Coordinator, Radon In Structures
Marcia L. Howard - Coordinator, Contract And Special Projects
Activities:
Evaluating radon exposure of residents through state survey. Provides EPA information guidelines upon request.

INDIANA DEPARTMENT OF LABOR/BUREAU OF SAFETY, EDUCATION AND TRAINING
1013 State Office Building
Indianapolis, IN 46204
(317) 232-2688
Robert Malcolm - Director
Dave Knessy - Department Commissioner/IOSHA Compliance
Activities:
Lead agency for occupational health and safety issues.

INDIANA GEOLOGICAL SURVEY
611 North Walnut Grove
Bloomington, IN 47405
(812) 335-1950
Norman C. Hester - Director And State Geologist
Activities:
Mapping service important for radon surveys in state. Information can be used for radon in soil or water studies.

INDIANA REAL ESTATE COMMISSION
100 North Senate Ave.
1021 State Office Building
Indianapolis, IN 46204
(317) 232-2980
Alan K. Updike - Deputy Director
Denise Shockley - Education Director
Activities:
Lead agency for real estate regulatory and enforcement issues.

Local Government Agencies

ANDERSON AIR POLLUTION CONTROL DEPARTMENT
P.O. Box 2100
Anderson, IN 46018
(317) 646-9835
William A. Dorff - Administrator

EAST CHICAGO/DEPARTMENT OF AIR QUALITY CONTROL
4525 Indianapolis Blvd.
East Chicago, IN 46312
(219) 391-8297
M.M. Ali Khan - Administrator

EVANSVILLE/AIR POLLUTION CONTROL OFFICER
Administration Building
Room 207
Evansville, IN 47708
(812) 426-5597
Joan D. Shelton - Director

GARY/AIR POLLUTION CONTROL OFFICE
846 Broadway
2nd Floor

Gary, IN 46402
(219) 882-3000
Dennis J. McGuire - Chief

HAMMOND/AIR POLLUTION CONTROL
5925 Calumet Ave.
Hammond, IN 46320
(219) 853-6306
Ronald L. Novak - Chief

INDIANAPOLIS/AIR POLLUTION CONTROL DIVISION
2700 South Belmont Ave.
Indianpolis, IN 46221
(317) 633-5496
David R. Jordan - Adminstrator

LAKE COUNTY HEALTH DEPARTMENT
2293 North Main St.
Crown Point, IN 46307
(219) 738-2020
Peter Stecy - Health Commissioner

ST. JOSEPH COUNTY AIR POLLUTION CONTROL
County-City Building
8th & 9th Floors
South Bend, IN 46601
(219) 284-9721
Paul E. Trost - Pollution Control Officer

VIGO COUNTY AIR POLLUTION CONTROL
201 Cherry St.
Terre Haute, IN 47807
(812) 238-8433
C. Larry Beddow - Director

Iowa

IOWA DEPARTMENT OF PUBLIC HEALTH/DIVISION OF DISEASE PREVENTION
Lucas State Office Building
Des Moines, IA 50319
(515) 281-6549
John A. Eure - Assistant To Division Director
Activities:
Iowa's lead agency for radon policy and public information.

IOWA BUREAU OF LABOR/IOWA OCCUPATIONAL SAFETY AND HEALTH ADMINISTRATION
307 East Seventh St.
Des Moines, IA 50319
(515) 281-5352
Mary L. Bryant - Commissioner
Activities:
Lead agency for occupational health and safety issues.

IOWA DEPARTMENT OF COMMERCE/PROFESSIONAL LICENSING AND REGULATION DIVISION
1918 SE Hulsizer
Ankeny, IA 50021
(515) 281-3183
Kenneth L. Smith - Executive Secretary
Activities:
Lead agency for real estate regulatory and enforcement issues.

IOWA DEPARTMENT OF NATURAL RESOURCES/ENVIRONMENTAL PROTECTION DIVISION
Henry A. Wallace Building
Des Moines, IA 50319-0034
(515) 281-5145
Allan Stokes - Administrator

IOWA DEPARTMENT OF PUBLIC HEALTH/DIVISION OF DISEASE PREVENTION/BUREAU OF RADIOLOGICAL HEALTH
Lucas State Office Building
Des Moines, IA 50319
(515) 281-7781
Rick Welke - (State Contact)
Donald A. Flater - Chief
Activities:
Provides information to public on radon-related issues.

IOWA GEOLOGICAL SURVEY BUREAU
123 North Capitol St.
Iowa City, IA 52242
(319) 335-1575
Donald A. Koch - Director And State Geologist
Activities:
Mapping capabilities pertinent to radon surveys in state.

Local Government Agencies

LINN COUNTY HEALTH DEPARTMENT/AIR POLLUTION CONTROL DIVISION
751 Center Point Rd., NE
Cedar Rapids, IA 52402-4692
(318) 398-3551
Gregory Slager - Air Pollution Control Officer

POLK COUNTY PHYSICAL PLANNING
5895 NE 14th St.
Des Moines, IA 50313
(515) 286-3351
Gary L. Pryor - Director, Physical Planning and Health

Kansas

KANSAS DEPARTMENT OF HEALTH AND ENVIRONMENT/BUREAU OF AIR QUALITY AND RADIATION CONTROL/ENVIRONMENTAL SURVEILLANCE AND EMERGENCY PLANNING SECTION
Forbes Field Building
Suite 740
Topeka, KS 66620
(913) 296-1567
Harold Spiker - Chief (State Contact)
Dominic To - Radiation Laboratory Director
John Irwin - Manager, Bureau Of Air Quality

Activities:
Provides public information packet with EPA recommendations and guidelines and new home construction information. Consults with state's legislatively-mandated radon advisory committee. State's lead agency for radon information.

KANSAS DEPARTMENT OF HUMAN RESOURCES/DIVISION OF INDUSTRIAL SAFETY AND HEALTH
512 West Sixth St.
Topeka, KS 66603
(913) 296-4386
David F. Willsie - Director
Activities:
Lead agency for occupational health and safety issues.

KANSAS GEOLOGICAL SURVEY
The University Of Kansas
1930 Constant Ave.
Campus West
Lawrence, KS 66046
(913) 864-3965
Lee C. Gerhard - Director and State Geologist
Activities:
Studying radon-bearing rock formations in Kansas.

KANSAS REAL ESTATE COMMISSION
Landon State Office Building
900 Jackson St.
Room 501
Topeka, KS 66612-1220
(913) 296-3411
Eugene W. Yockers - Director
Shirley Wheeler - Education Specialist
Activities:
Lead agency for real estate regulatory and enforcement issues.

Local Government Agencies

KANSAS CITY,KANSAS-WYANDOTTE COUNTY/ HEALTH DEPARTMENT/AIR POLLUTION CONTROL DIVISION
619 Ann Ave.
Kansas City, KS 66101
(913) 321-4803
Richard S. Michael - Director

TOPEKA-SHAWNEE COUNTY HEALTH AGENCY
1615 West Eighth St.
Topeka, KS 66606
(913) 233-8961
Randall R. Walker - Environmental Engineer

WICHITA-SEDGWICK COUNTY DEPARTMENT OF COMMUNITY HEALTH
1900 East 9th St.
Wichita, KS 67214
(316) 268-8351
Richard G. Hunter - Environmental Health Director

Kentucky

KENTUCKY DEPARTMENT OF HUMAN RESOURCES/DIVISION OF RADIATION AND PRODUCT SAFETY/RADIATION CONTROL BRANCH
275 East Main St.
Frankfort, KY 40621
(502) 564-3700
Don Nodler - Section Supervisor (State Contact)
Donald R. Hughes - Branch Manager
Activities:
Provides public with information concerning radon detection, testing, and mitigation services. Participated in EPA's pilot radon survey of ten states. State's lead agency for radon information.

KENTUCKY DEPARTMENT OF LABOR/EDUCATION AND TRAINING, OCCUPATIONAL SAFETY AND HEALTH
U.S. Highway 127 South
Frankfort, KY 40601
(502) 564-6895

KENTUCKY GEOLOGICAL SURVEY
University Of Kentucky
311 Breckinridge Hall
Lexington, KY 40506
(606) 257-5863
Donald Haney - Director and State Geologist
Activities:
Provides geologic mapping service pertinent to radon surveys in state.

KENTUCKY REAL ESTATE COMMISSION
222 South First St.
Suite 300
Louisville, KY 40202
(502) 588-4462
Susan G. Stopher - Executive Director
Activities:
Lead agency for real estate regulatory and enforcement issues.

Local Government Agencies

JEFFERSON COUNTY/AIR POLLUTION CONTROL
914 East Broadway
Louisville, KY 40204
(502) 584-8151
Harvel M. Rogers, Jr. - Air Pollution Control Officer

Louisiana

LOUISIANA DEPARTMENT OF ENVIRONMENTAL QUALITY/OFFICE OF AIR QUALITY AND NUCLEAR ENERGY/NUCLEAR ENERGY DIVISION
P.O. Box 14690
Baton Rouge, LA 70898-4690
(504) 925-4518
William H. Spell - Administrator (State Contact)
Jay Mason - Health Physicist
Activities:
State's lead radon information agency. Provides test kits for pilot survey. Provides specific information to written

requests or general information upon request. Pilot survey indicates minimal radon risk in state.

LOUISIANA GEOLOGICAL SURVEY/DEPARTMENT OF NATURAL RESOURCES
Box G
University Station
Baton Rouge, LA 70803
(504) 342-6754
Charles Groat - Director And State Geologist
Activities:
Provides topographical mapping service pertinent to radon surveys in state. Information can be used for radon-soil or water studies.

LOUISIANA REAL ESTATE COMMISSION
P.O. Box 14785
Baton Rouge, LA 70898
(504) 925-4771
Anna-Kathryn Williams - Executive Director
Belle Dunaway - Education Coordinator
Activities:
Lead agency for real estate regulatory and enforcement issues.

Maine

MAINE DEPARTMENT OF HUMAN SERVICES/BUREAU OF HEALTH/DIVISION OF HEALTH ENGINEERING/OCCUPATIONAL AND RESIDENTIAL HEALTH PROGRAM
State House
Station 10
Augusta , ME 04333
(207) 289-3826
Eugene Moreau - Program Manager (State Contact)
Eleanor King - Sanitarian
Activities:
Lead agency in Maine to provide public with information concerning radon. Collects information on radon levels throughout state. Funds cancer-radon study with University of Maine and Maine Medical Center.

MAINE DEPARTMENT OF ENVIRONMENTAL PROTECTION/BUREAU OF AIR QUALITY CONTROL
State House, Station 17
Augusta, ME 04333
(207) 289-2437
Dean Marriott - Commissioner

MAINE DEPARTMENT OF HUMAN SERVICES/BUREAU OF HEALTH/DIVISION OF HEALTH ENGINEERING
State House
Station 10
Augusta, ME 04333
(207) 289-3826
Don Hoxie - Division Director
Activities:
Administers Maine's radon program.

MAINE DEPARTMENT OF HUMAN SERVICES/BUREAU OF HEALTH/DIVISION OF HEALTH ENGINEERING/RADIOLOGICAL HEALTH PROGRAM
State House
Station 10
Augusta, ME 04333
(207) 289-5676
W. Clough Toppan - P.E. Manager
Activities:
Provides technical information to Maine's Occupational and Residential Health Program.

MAINE DEPARTMENT OF HUMAN SERVICES/BUREAU OF HEALTH/PUBLIC HEALTH LABORATORY
State House
Station 12
Augusta, ME 04333
(207) 289-2727
Philip Haines - Director
Cheryl L. Baker - Chemist
Activities:
Laboratory provides testing services to determine radon levels in homes and buildings.

MAINE DEPARTMENT OF LABOR/DIVISION OF INDUSTRIAL SAFETY
Labor Station 82
283 State St.
Augusta, ME 04333
(207) 289-2591
Lester Wood - Director
Activities:
lead agency for occupational health and safety issues.

MAINE GEOLOGICAL SURVEY/DEPARTMENT OF CONSERVATION
State House
Station 22
Augusta, ME 04333
(207) 289-2801
Walter Anderson - Director and State Geologist
Activities:
Detailed bedrock map pertinent to radon studies available.

MAINE REAL ESTATE COMMISSION
State House Station # 35
Augusta, ME 04333
(207) 783-9303
Paul A. Sawyer - Director
Terry Hayes - Education Director
Activities:
Lead agency for real estate regulatory and enforcement issues.

Maryland

MARYLAND DEPARTMENT OF THE ENVIRONMENT/OFFICE OF TOXICS, ENVIRONMENTAL SAFETY AND HEALTH
201 West Preston St.
7th Floor Mail Room
Baltimore, MD 21201

(301) 333-3130, (800) 872-3666
Roland G. Fletcher - Administrator (State Contact)
Richard Brisson - Section Chief, Radon
Leon Rochuba - Assistant Radon Chief
Activities:
Central office for development of state radon policy and dissemination of radon information to public.

MARYLAND DEPARTMENT OF THE ENVIRONMENT/AIR MANAGEMENT ADMINISTRATION
201 West Preston St.
Baltimore, MD 21201
(301) 225-5235
George P. Ferreri - Director

MARYLAND DIVISION OF LABOR AND INDUSTRY/CONSULTATION PROGRAM
501 St. Paul Pl.
3rd Fl.
Baltimore, MD 21202
(301) 659-4218
Milton Saul - Assistant Commissioner
Activities:
Lead agency for occupational health and safety issues.

MARYLAND GEOLOGICAL SURVEY
2300 St. Paul St.
Baltimore, MD 21218
(301) 554-5500
Kenneth Weaver - Director
Activities:
Provides mapping and geologic services pertinent to radon studies.

MARYLAND REAL ESTATE COMMISSION
501 St. Paul Pl.
Suite 804
Baltimore, MD 21202-2272
(301) 333-6230
Charles G. Chambers - Executive Director
Activities:
Lead agency for real estate regulatory and enforcement issues.

Local Government Agencies

ALLEGHENY COUNTY HEALTH DEPARTMENT/ENVIRONMENTAL HEALTH DIVISION
Willowbrook Rd.
Cumberland, MD 21502
(301) 777-5653
Conrad B. Zimmermann - Assistant Director

ANNE ARUNDEL COUNTY DEPARTMENT OF HEALTH/DIVISON OF ENVIRONMENTAL HEALTH/AIR QUALITY CONTROL SECTION
3 Harry S. Truman Pkwy.
Annapolis, MD 21401
(301) 244-7360
Joseph G. Abey - Chief

BALTIMORE COUNTY DEPARTMENT OF ENVIRONMENTAL PROTECTION AND RESOURCE MANAGEMENT
401 Bosley Ave.
Towson, MD 21204
(301) 494-3775
David Filbert - Chief, Bureau Of Air Quality Management

BALTIMORE COUNTY HEALTH DEPARTMENT/BUREAU OF INDUSTRIAL HYGIENE/AIR POLLUTION CONTROL
303 East Fayette St.
Room 400
Baltimore, MD 21202
(301) 396-4427
Elkins W. Dahle, Jr. - Director

FREDERICK COUNTY HEALTH DEPARTMENT
12 East Church St.
Frederick, MD 21701
(301) 694-1700
Charles L. Gillis, Jr. - Air Quality Sanitarian

HARFORD COUNTY HEALTH DEPARTMENT
P.O. Box 191
Bel Air, MD 21014-0191
(301) 838-1500 x246
Thomas M. Thomas - Health Officer

HOWARD COUNTY HEALTH DEPARTMENT/BUREAU OF ENVIRONMENTAL HEALTH
P.O. Box 476
Ellicott City, MD 21043
(301) 461-9955
Frank A. Skinner - Director

MONTGOMERY COUNTY GOVERNMENT
101 Monroe St.
Rockville, MD 20850
(301) 217-2500
Sidney Kramer - County Executive
Activities:
Organized radon task force to study significance of health problems in county caused by radon exposure.

MONTGOMERY COUNTY GOVERNMENT/DIVISION OF ENVIRONMENTAL PLANNING AND MONITORING
101 Monroe St.
6th Fl.
Rockville, MD 20850
(301) 217-2380
Eric Mendelsohn - Air Resources Engineer
Activities:
Maintains ongoing record of reports of radon levels in county's communities.

MONTGOMERY COUNTY GOVERNMENT/MONTGOMERY COUNTY RADON TASK FORCE
100 Maryland Ave.

6th Fl.
Rockville, MD 20850
(301) 217-7900
John L. Mencke - Chairman
Stewart McKenzie - Legislative Analyst for the Environment
Activities:
Developing radon policy recommendation for county government.

PRINCE GEORGE'S COUNTY HEALTH DEPARTMENT/DIVISION OF AIR QUALITY CONTROL
10210 Greenbelt Rd.
Seabrook, MD 20706
(301) 794-6800
Donald A. Nork - Chief

Massachusetts

MASSACHUSETTS DEPARTMENT OF PUBLIC HEALTH/OFFICE OF HUMAN SERVICES/RADIATION CONTROL PROGRAM
150 Tremont St.
Boston, MA 02111
(617) 727-6214
Robert Hallisey - Director (State Contact)
George Swible - Radiation Scientist
Robert T. Watkins - Radiation Scientist
William J. Bell - Radiation Scientist
Activities:
State's lead agency for radon information. Provides EPA guidelines and information upon request.

MASSACHUSETTS BOARD OF REGISTRATION OF REAL ESTATE BROKERS AND SALESMEN
100 Cambridge St.
Room 1518
Boston, MA 02202
(617) 727-7376
Joseph R. Autilio - Executive Secretary
Activities:
Lead agency for real estate regulatory and enforcement issues.

MASSACHUSETTS DEPARTMENT OF LABOR AND INDUSTRIES/DIVISION OF INDUSTRIAL SAFETY
100 Cambridge St.
Boston, MA 02202
(617) 727-3463
Vacant - Director
Richard J. Collins - Consultation Program Manager
Activities:
Lead agency for occupational safety and health issues.

MASSACHUSETTS DIVISION OF AIR QUALITY CONTROL
One Winter St.
Boston, MA 02108
(617) 292-5593
Bruce K. Maillet - Director

MASSACHUSETTS STATE GEOLOGIST/COASTAL ZONE MANAGEMENT
100 Cambridge St.
20th Floor
Boston, MA 02202
(617) 292-5690
Joseph Sinnott - State Geologist
Activities:
Provides geological mapping service pertinent to radon surveys in state. Information can be used for radon-soil or water studies.

Local Government Agencies

BERKSHIRE AND PIONEER VALLEY AIR POLLUTION CONTROL DISTRICTS
State House West
436 Dwight St., 4th Fl.
Springfield, MA 01103
(413) 785-5327
David Howland - Air Quality Chief

BOSTON AIR POLLUTION CONTROL COMMISSION
Boston City Hall
Room 805
Boston, MA 02201
(617) 725-4416
Denise K. Breiteneicher - Executive Director

FITCHBURG/BOARD OF HEALTH
718 Main St.
Fitchburg, MA 01420
(617) 342-9843
Jeffrey P. Jerszyk - Director, Public Health

MERRIMACK VALLEY AND METROPOLITAN BOSTON AIR POLLUTION CONTROL DISTRICTS
5 Commonwealth Ave.
Woburn, MA 01801
(617) 935-2160
James E. Beisky - Air Quality Section Chief

SOUTHEASTERN MASSACHUSETTS AIR POLLUTION CONTROL DISTRICT
Lakeville Hospital
Lakeville, MA 02346
(617) 947-1231 x680
Vaughan M. Steeves - Regional Air Quality Chief

Michigan

MICHIGAN DEPARTMENT OF PUBLIC HEALTH/BUREAU OF ENVIRONMENTAL AND OCCUPATIONAL HEALTH/DIVISION OF RADIOLOGICAL HEALTH
3423 North Logan St.
P.O. Box 30195
Lansing, MI 48909
(517) 335-8200
Robert DeHaan - Chief, (State Contact)
George Bruchman - Chief, Division Of Radiological Health
Dennis Hahn - Chief, Nuclear Facilities

Activities:
Lead agency in state for radon information. Completed statewide random statistical survey to identify radon threat to state residents. Contracting with University of Michigan for training of diagnosticians. Provides information to public on request.

MICHIGAN DEPARTMENT OF LICENSING AND REGULATION
P.O. Box 30018
Lansing, MI 48909
(517) 373-0490
Ann Millben - Administrative Secretary
Activities:
Lead agency for real estate regulatory and enforcement issues.

MICHIGAN DEPARTMENT OF NATURAL RESOURCES
P.O. Box 30028
Lansing, MI 48909
(517) 334-6923
Thomas Segall - State Geologist
Activities:
Mapping service capabilities pertinent to radon surveys in state. Information utilized for a number of radon surveys.

MICHIGAN DEPARTMENT OF PUBLIC HEALTH/DIVISION OF OCCUPATIONAL HEALTH/SPECIAL PROGRAMS SECTION
3500 North Logan
Lansing, MI 48909
(517) 322-1809
Flint C. Watt - Chief
Activities:
Office handles occupational health issues for state of Michigan.

Local Government Agencies

GRAND RAPIDS/ENVIRONMENTAL SERVICES DEPARTMENT/AIR POLLUTION CONTROL DIVISION
1300 Market, SW
Grand Rapids, MI 49503
(616) 456-3158
Richard Olson - Air Pollution Control Officer

WAYNE COUNTY HEALTH DEPARTMENT/AIR POLLUTION CONTROL
2211 East Jefferson
Detroit, MI 48207
(313) 567-4100
Alan M. Greenberg - Director

Minnesota

MINNESOTA DEPARTMENT OF HEALTH/ENVIRONMENTAL HEALTH DIVISION/SECTION OF RADIATION CONTROL
717 Delaware St., SE
P.O. Box 9441
Minneapolis, MN 55440
(612) 623-5351
Alice T. Dolezal Hennigan - Chief (State Contact)
Earnest E. Tate - Health Physicist
Activities:
State's lead radon information agency, works with EPA. Developed information booklet to inform public and list testing, detection, and mitigation companies servicing state. Booklet available on request. Participant in EPA's 1988 State Radon Survey Assistance Program where results showed 47 percent of homes tested above 4.0 pCi/l.

MINNESOTA COMMERCE DEPARTMENT
500 Metro Square
St. Paul, MN 55101
(612) 297-3379
Melbourne R. Boynton - Director of Licensing
Cherl McLean - Education Coordinator
Activities:
Lead agency for real estate regulatory and enforcement issues.

MINNESOTA DEPARTMENT OF LABOR AND INDUSTY
444 Lafayette Rd.
5th Floor
St. Paul, MN 55101
(612) 297-1953
Timothy Tierny - Director, Consultation Unit
Ivan Russell - Director, Compliance Unit
Activities:
Lead agency for occupational health and safety issues in state.

MINNESOTA DEPARTMENT OF PUBLIC SERVICE/ENERGY INFORMATION CENTER
900 American Center Building
150 East Kellogg Blvd.
St. Paul, MN 55101
(612) 297-2291, (800) 652-9747
Chris Gilchrist - Manager
Activities:
Provides information on how to maintain low radon levels in energy efficient housing.

MINNESOTA GEOLOGICAL SURVEY
2642 University Ave.
St. Paul, MN 55114-1057
(612) 627-4780
Priscilla C. Grew - Director and State Geologist
Activities:
Radon in water analyses have been important in identifying radon threat in state.

MINNESOTA POLLUTION CONTROL AGENCY/DIVISION OF AIR QUALITY
520 Lafayette Rd.
St. Paul, MN 55155
(612) 296-7802
Gary Eckhardt - Supervisor, Air Monitoring
Activities:
Information dissemination concerning how to monitor for radon. EPA requires office to study indoor air quality.

Local Government Agencies

BLOOMINGTON/ENVIRONMENTAL POLLUTION SECTION
2215 West Old Shakopee Rd.
Bloomington, MN 55431
(612) 881-5811
Robert A. Mood - Manager, Building And Environmental

MINNEAPOLIS ENERGY OFFICE
City Hall
Room 330
Minneapolis, MN 55415
(612) 348-4830
David Bohac - Energy Engineer
Activities:
City government office provides information on radon mitigation and effects of home weatherization on indoor radon levels.

MINNEAPOLIS POLLUTION CONTROL DIVISION/DEPARTMENT OF INSPECTIONS
250 South 4th St.
Public Health Ctr.
Room 300
Minneapolis, MN 55415
(612) 348-7897
Sol J. Jacobs - Director

RICHFIELD/SUPPORT SERVICES DIVISION
6700 Portland Ave.
Richfield, MN 55423
(612) 869-7521
Betsy Christensen - Health Administrator

ST. LOUIS PARK INSPECTIONAL SERVICES
5005 Minnetonka Blvd.
St. Louis Park, MN 55416
(612) 924-2588
Harvey J. McPhee - Director

Mississippi

MISSISSIPPI STATE DEPARTMENT OF HEALTH/DIVISION OF RADIOLOGICAL HEALTH
3150 Lawson St.
P.O. Box 1700
Jackson, MS 39215-1700
(601) 354-6657
Eddie S. Fuente - Director (State Contact)
Renee Echols - Senior Health Physicist
Activities:
State regulatory agency dealing with all radiation matters in Mississippi.

MISSISSIPPI REAL ESTATE COMMISSION
1920 Dunbarton Dr.
Jackson, MS 39216-5087
(601) 987-3969
John W. Neelley - Administrator
Activities:
Lead agency for real estate regulatory and enforcement issues.

MISSISSIPPI STATE DEPARTMENT OF HEALTH/DIVISION OF OCCUPATIONAL SAFETY AND HEALTH
P.O. Box 1700
Jackson, MS 39205
(601) 987-3981
Henry Laird - Director
Activities:
Lead agency for occupational health and safety issues.

MISSISSIPPI STATE DEPARTMENT OF NATURAL RESOURCES/BUREAU OF GEOLOGY
P.O. Box 5348
Jackson, MS 39216
(601) 354-6228
Conrad Gazzier - Director and State Geologist
Activities:
Mapping capabilities pertinent to radon surveys in state. Information can be used for radon-soil or water studies.

Missouri

MISSOURI DEPARTMENT OF HEALTH/DIVISION OF ENVIRONMENTAL HEALTH AND EPIDEMIOLOGY SERVICES
1730 East Elm Plaza
P.O. Box 570
Jefferson City, MO 65102
(314) 751-6083
Kennith V. Miller - Chief (State Contact)
Gary W. McNutt - Radiological Health Analyst
Activities:
State's lead radon information agency. Participated in 1988 EPA radon survey of state. Provides EPA information to public upon request.

MISSOURI DEPARTMENT OF LABOR AND INDUSTRIAL RELATIONS
1001-D Southwest Blvd.
P.O. Box 449
Jefferson City, MO 65102
(314) 751-3403
Fred Seaman - Assistant Director
Activities:
Lead agency for occupational health and safety issues.

MISSOURI DEPARTMENT OF NATURAL RESOURCES/DIVISION OF GEOLOGY AND LAND SURVEY
111 Fairgrounds Rd.
P.O. Box 250
Rolla, MO 65401
(314) 364-1752
James H. Williams - Director and State Geologist
Activities:
Provides mapping service pertinent to radon surveys.

MISSOURI REAL ESTATE COMMISSION
P.O. Box 1339
Jefferson City, MO 65102
(314) 751-2334
Marie B. Campbell - Executive Director
Gail Eldridge - Education Coordinator

Activities:
Lead agency for real estate regulatory and enforcement issues.

Local Government Agencies

GREENE COUNTY-CITY OF SPRINGFIELD/AIR POLLUTION CONTROL AUTHORITY
227 East Chestnut Expressway
Springfield, MO 65802
(417) 864-1662
Ronald Boyer - Chief, Air Pollution Control

INDEPENDENCE PLANNING DEPARTMENT/AIR POLLUTION CONTROL DIVISION
111 East Maple Ave.
Independence, MO 64050
(816) 836-8300 x363
Kathryn Akins - Air Pollution Control Officer

KANSAS CITY, MISSOURI/AIR QUALITY SECTION
21st Floor, City Hall
414 East 12th St.
Kansas City, MO 64106
(816) 274-2501
Paul F. Stablein - Acting Program Manager

ST. LOUIS COUNTY AIR POLLUTION CONTROL BRANCH
801 South Brentwood Blvd.
Clayton, MO 63105
(314) 854-6921
Blaine J. Rhoades - Program Manager

ST. LOUIS/DIVISION OF AIR POLLUTION CONTROL
City Hall
Room 419
St. Louis, MO 63103
(314) 622-3334
Earl D. Ray - Commissioner

Montana

MONTANA DEPARTMENT OF HEALTH AND ENVIRONMENTAL SERVICES/OCCUPATIONAL HEALTH BUREAU/ENVIRONMENTAL SCIENCES DIVISION
Cogswell Building
Helena, MT 59620
(406) 444-3671
Adrian C. Howe - Chief (State Contact)
Activities:
State's lead agency for radon information. Montana does not maintain a funded radon program.

MONTANA BOARD OF REALTY REGULATION
1424 9th Ave.
Helena, MT 59620
(406) 444-2961
Grace A. Berger - Administrative Assistant
Activities:
Lead agency for real estate regulatory and enforcement issues.

MONTANA BUREAU OF MINES AND GEOLOGY
Montana College of Mineral Science and Technology
Butte, MT 59701
(406) 496-4180
Edward Ruppel - Director and State Geologist
Susan Foster - Geologist
Activities:
Provide mapping service pertinent to radon surveys in state.

MONTANA DEPARTMENT OF LABOR AND INDUSTRY/BUREAU OF SAFETY AND HEALTH/DIVISION OF WORKERS' COMPENSATION
5 South Last Chance Gulch
Helena, MT 59601
(406) 444-6401
Ed Gatzemeier - Director
Lyle Chapman - Loss Control Supervisor
Activities:
Lead agency for occupational health and safety issues.

Local Government Agencies

CASCADE/CITY-COUNTY AIR POLLUTION CONTROL PROGRAM
1130 17 Ave. South
Great Falls, MT 59405
(406) 761-1190
Donald E. Pizzini - Health Officer

MISSOULA CITY-COUNTY HEALTH DEPARTMENT/ENVIRONMENTAL HEALTH DIVISION
301 West Adler St.
Missoula, MT 59802
(406) 721-5700
Linda Hedstrom - Environmental Health Director

Nebraska

NEBRASKA DEPARTMENT OF HEALTH/DIVISION OF RADIOLOGICAL HEALTH
301 Centennial Mall South
P.O. Box 93007
Lincoln, NE 68509
(402) 471-2168
Harold Borchert - Director (State Contact)
James DeFrain - Health Physicist
H. Ellis Simmons - Radiological Health Coordinator
Activities:
Provides radon information to citizens on request. Maintains state's certification program for radon testing, detection, and mitigation companies. Lead agency for radon information.

NEBRASKA DEPARTMENT OF ENVIRONMENTAL CONTROL/AIR POLLUTION CONTROL DIVISION
301 Centennial Mall South
P.O. Box 94877
Lincoln, NE 68509
(402) 471-2186
Gene Robinson - Chief

NEBRASKA DEPARTMENT OF LABOR
State Office Building
301 Centennial Mall, South
Lincoln, NE 68509
(402) 471-2239
Charlie Calcaterra - Program Manager
Activities:
Lead agency for occupational health and safety issues.

NEBRASKA INSTITUTE OF AGRICULTURE AND NATURAL RESOURCES/CONSERVATION AND SURVEY DIVISION
113 Nebraska Hall
The University Of Nebraska
Lincoln, NE 68588-0517
(402) 472-3471
Perry Wigley - Director And State Geologist
Activities:
Provides mapping service pertinent to radon surveys in state. Information can be used for radon-soil or water studies.

NEBRASKA REAL ESTATE COMMISSION
301 Centennial Mall South
P.O. Box 94667
Lincoln, NE 68508
(402) 471-2004
Paul Quinlan - Director
Teresa Hoffman - Education Director
Activities:
Lead agency for real estate regulatory and enforcement issues.

Local Government Agencies

LINCOLN-LANCASTER COUNTY/AIR POLLUTION CONTROL AGENCY
2200 St. Mary's Ave.
Lincoln, NE 68502
(402) 471-8021
M. Jane Ford - Air Pollution Control Officer

OMAHA/AIR QUALITY CONTROL
1819 Farnam
Omaha, NE 68183
(402) 444-5379
R.L. Timmerman - Air Pollution Control Officer

Nevada

NEVADA DEPARTMENT OF HUMAN RESOURCES/HEALTH DIVISION/RADIOLOGICAL HEALTH SECTION
505 East King St.
Room 202
Carson City, NV 89710
(702) 885-5394
Stanley R. Marshall - Supervisor (State Contact)
Larry Franks - Radiation Control Specialist
Paul Simpson - Radiation Control Specialist
Mary Thompson - Radiation Control Specialist, Las Vegas Office
Activities:
Provides EPA radon information pamphlets upon request. Maintains list of testing companies providing services to Nevadans.

NEVADA BUREAU OF MINES AND GEOLOGY
University Of Nevada-Reno
Reno, NV 89557-0088
(702) 784-6691
James Taranik - Acting Director And State Geologist
Activities:
Provides mapping service pertinent to radon surveys in state. Information can be used for radon in soil or water studies.

NEVADA DEPARTMENT OF COMMERCE/ REAL ESTATE DIVISION
201 South Fall St.
Carson City, NV 89710
(702) 885-4280
R. Lynn Lumon - Administrator
Jan Windsor - Education Director
Activities:
Lead agency for real estate regulatory and enforcement issues.

NEVADA DIVISION OF OCCUPATIONAL SAFETY AND HEALTH
Capitol Complex
1370 South Curry St.
Carson City, NV 89710
(702) 885-5240
James I. Barnes - Director of Industrial Relations
Michael Tyler - Administrator
Activities:
Lead Agency for occupational health and safety issues.

NEVADA STATE AIR QUALITY DEPARTMENT/DIVISION OF ENVIRONMENTAL PROTECTION
201 South Fall St.
Carson City, NV 89710
(702) 885-5065
Lowell H. Shifley, Jr. - State Air Quality Officer

Local Government Agencies

CLARK COUNTY HEALTH DISTRICT/AIR POLLUTION CONTROL DIVISION
625 Shadow Ln.
Las Vegas, NV 89127
(702) 383-1276
Otto Ravenhold - Chief Health Officer

WASHOE COUNTY DISTRICT HEALTH DEPARTMENT
1001 East Ninth
Reno, NV 89520
(702) 328-2400
Michael Ford - Health Officer

New Hampshire

NEW HAMPSHIRE DEPARTMENT OF HEALTH AND HUMAN SERVICES/DIVISION OF PUBLIC HEALTH SERVICES/BUREAU OF RADIOLOGICAL HEALTH
P.O. Box 148
Concord, NH 03301
(603) 271-4588

Diane Tefft - Bureau Director (State Contact)
Activities:
State's lead agency in development of radon program. Coordinates activity with Department of Environmental Services. Enacting random survey test of 2200 homes with Governor's Energy Office. Provides general information to public.

NEW HAMPSHIRE DEPARTMENT OF ENVIRONMENTAL SERVICES/WATER SUPPLY AND POLLUTION CONTROL DIVISION/BUREAU OF WATER SUPPLY AND ENGINEERING
6 Hazen Dr.
P.O. Box 95
Concord, NH 03301-6527
(603) 271-3139
Richard Thayer - Sanitary Engineer
Activities:
Oversees state water supply. Covers issues on radon in water Joint study with University of New Hampshire to determine radon-water mitigation techniques.

NEW HAMPSHIRE DEPARTMENT OF LABOR/ONSITE CONSULTATION PROGRAM
19 Pillsbury St.
Concord, NH 03301
(603) 271-3170
Steven Byer - Chief of Consulting Services
Activities:
Lead agency for occupational health and safety issues.

NEW HAMPSHIRE DEPARTMENT OF RESOURCES AND ECONOMIC DEVELOPMENT/ENVIRONMENTAL SERVICES/GEOLOGY UNIT
University Of New Hampshire
117 James Hall
Durham, NH 03824
(603) 862-3160
Eugene Boudette - State Geologist
Activities:
Released geologic map detailing radon potential in New Hampshire.

NEW HAMPSHIRE GOVERNOR'S ENERGY OFFICE
2 1/2 Beacon St.
Concord, NH 03301-4498
(603) 271-4498
Scott Maltzie
Activities:
Participating in random survey test of 2200 homes to investigate radon levels.

NEW HAMPSHIRE REAL ESTATE COMMISSION
State Office Park South
Johnson Hall, 3rd Fl.
107 Pleasant St.
Concord, NH 03301
(603) 271-2701
Valerie B. Lanigan - Executive Director
Activities:
Lead agency for real estate regulatory and enforcement issues.

New Jersey

NEW JERSEY DEPARTMENT OF ENVIRONMENTAL PROTECTION/DIVISION OF ENVIRONMENTAL QUALITY/BUREAU OF ENVIRONMENTAL RADIATION
729 Alexander Rd.
CN-415
Trenton, NJ 08625
(609) 987-6396, (800) 648 0394
Bob Stern - Chief (State Contact)
Gerald Nicholls - Director, Radiation Programs
Bill Hoyle - Supervisor, Radon Programs
Elaine Makatura - Community Relations Coordinator
Activities:
New Jersey's designated lead agency for radon-related issues. Provides information on radon testing and mitigation to residents of state. Researches radon issues, particularly mitigation techniques and transfers information to potential users. Develops policy guidelines on radon measurement/mitigation techniques.

NEW JERSEY DEPARTMENT OF COMMUNITY AFFAIRS/DIVISION OF HOUSING AND DEVELOPMENT/OFFICE OF PLANNING AND OPERATIONS
101 South Broad St.
CN-802
Trenton, NJ 08625
(609) 633-6060
Mitch Malec - Supervisor, Technical Policy
Activities:
Develops and enforces building code standards for construction of radon-free structures in New Jersey.

NEW JERSEY DEPARTMENT OF ENVIRONMENTAL PROTECTION/DIVISION OF ENVIRONMENTAL PROTECTION/BUREAU OF ENVIRONMENTAL RADIATION/RADON PROGRAM
380 Scotch Rd.
CN-411
Trenton, NJ 08625
(201) 879-2612
Harvey Worrell - Director, Radon Testing Certification Prc
Nick DiPierro - Director, Radon Mitigation Certification Pr
Activities:
Replacing state's voluntary certification procedures with mandatory certification programs for detection and mitigation companies.

NEW JERSEY DEPARTMENT OF ENVIRONMENTAL PROTECTION/DIVISION OF ENVIRONMENTAL QUALITY
401 East State St.
2nd Floor, East Wing
Trenton, NJ 08625
(609) 292-5383
Jorge H. Berkowitz - Director
Activities:
Oversees Division of Environmental Quality's radon program.

NEW JERSEY DEPARTMENT OF ENVIRONMENTAL PROTECTION/DIVISION OF ENVIRONMENTAL QUALITY/BUREAU OF ENVIRONMENTAL RADIATION/RADIOLOGICAL ENVIRONMENTAL ASSESSMENT
380 Scotch Rd.
CN-411
Trenton, NJ 08625
(609) 530-4051
Jeanette Eng - Supervisor
Activities:
Analysis of technologically enhanced radon. Environmental assessment at radiologically contaminated sites.

NEW JERSEY DEPARTMENT OF ENVIRONMENTAL PROTECTION/DIVISION OF ENVIRONMENTAL QUALITY/BUREAU OF RADIATION AND INORGANIC ANALYTICAL SERVICES
380 Scotch Rd.
CN-411
Trenton, NJ 08628
(609) 530-4116
Patricia Gardner - Chief
Activities:
Laboratory provides carbon canister radon detection kit analysis.

NEW JERSEY DEPARTMENT OF ENVIRONMENTAL PROTECTION/DIVISION OF SCIENCE AND RESEARCH
401 East State St.
7th Floor, West Wing
Trenton, NJ 08625
(609) 984-6070
Robert Tucker - Director
Barbara Litt - Program Manager, Radon
Activities:
Funding development of personal radon monitor to compare human exposure to radon with levels in the home. Testing 150 wells in state to compare level of radon contamination in different strata.

NEW JERSEY DEPARTMENT OF HEALTH/DIVISION OF ENVIRONMENTAL AND OCCUPATIONAL HEALTH/ENVIRONMENTAL HEALTH SERVICES/AIR AND RADIATION DIVISION
John Fitch Plaza
CN-360
Trenton, NJ 08625
(609) 633-2043
Judy Klotz - Program Manager
Laurie Pyrch - Health Educator
Activities:
Provides public education programs on radon. Coordinates public health offices throughout state concerning radon policy. Contributes to development of comprehensive state radon policy. Maintains radon registry, recording health status of persons exposed to levels of radon.

NEW JERSEY DEPARTMENT OF HEALTH/DIVISION OF ENVIRONMENTAL AND OCCUPATIONAL HEALTH/PUBLIC EMPLOYEE OCCUPATIONAL SAFETY AND HEALTH PROGRAM
CN-360
Trenton, NJ 08625
(609) 984-1863
Alice Freund - Program Director
Gary Ludwig - Research Scientist
Activities:
Radon exposure hazards to public employees is covered by New Jersey legislation. Program investigates and processes state employee complaints relating to radon.

NEW JERSEY DEPARTMENT OF HEALTH/DIVISION OF EPIDEMIOLOGY AND DISEASE CONTROL/CANCER EPIDEMIOLOGICAL PROGRAM
CN-369
Trenton, NJ 08625
(609) 588-3500
Janet Schoenberg - Chief
Activities:
Case control study of lung cancer rates and radon exposure.

NEW JERSEY DEPARTMENT OF LABOR AND INDUSTRY/DIVISION OF WORKPLACE STANDARDS
36 West State
Room 315
Trenton, NJ 08625
(609) 984-3507
Activities:
Lead agency for occupational health and safety issues.

NEW JERSEY GEOLOGICAL SURVEY/DIVISION OF WATER RESOURCES
CN-029
Trenton, NJ 08625
(609) 292-1185
Haig Kasabach - State Geologist
Activities:
Involved in New Jersey's comprehensive program to assess radon exposure threat.

NEW JERSEY REAL ESTATE COMMISSION
Commerce Building
20 West State St.
CN 325
Trenton, NJ 08625
(609) 987-2010
Daryl B. Bell - Executive Director
Activities:
Lead agency for real estate regulatory and enforcement issues.

Local Government Agencies

ELIZABETH/DEPARTMENT OF HEALTH, WELFARE AND HOUSING
60 Winfield Scott Plaza
Elizabeth, NJ 07201

(201) 820-4068
Kenneth J. Sandor - Health Officer

HUDSON REGIONAL HEALTH COMMISSION
215 Harrison Ave.
Harrison, NJ 07029
(201) 485-7001
Robert Ferraiuolo - Executive Director

MIDDLESEX COUNTY AIR POLLUTION CONTROL PROGRAM
280 Hobart St.
Room 518
Perth Amboy, NJ 08861
(201) 826-3100
Richard J. Hills - Program Coordinator

SUBURBAN REGIONAL HEALTH COMMISSION
141 South Harrison St.
Suite 11
East Orange, NJ 07018
(201) 675-1774
John J. Farley - Executive Director

New Mexico

NEW MEXICO DEPARTMENT OF HEALTH AND ENVIRONMENT/ENVIRONMENTAL IMPROVEMENT DIVISION/COMMUNITY SERVICES BUREAU
P.O. Box 968
Santa Fe, NM 87504-0968
(505) 827-2948
J. Margo Metzger-Keele - Radon Program Manager (State Contact)
Activities:
State's lead radon information agency. Provides testing and mitigation services information. Has tested public schools. Will participate in 1989 EPA State Radon Survey Assistance Program.

NEW MEXICO BUREAU OF MINES AND MINERAL RESOURCES
Campus Station
Socorro, NM 87801
(505) 835-5420
Frank Kottlowski - Director and State Geologist
Activities:
Provides mapping service pertinent to radon surveys in state. Information can be used for radon-soil or water studies.

NEW MEXICO DEPARTMENT OF HEALTH AND ENVIRONMENT/OCCUPATIONAL HEALTH AND SAFETY BUREAU
1476 St. Francis Dr.
Santa Fe, NM 87504
(505) 827-2877
Sam Rogers - Bureau Chief
Activities:
Lead agency for occupational health and safety issues.

NEW MEXICO REAL ESTATE COMMISSION
4125 Carlisle Blvd., NE
Albuquerque, NM 87107
(505) 841-6524
Randy Ernest Lovato - Executive Secretary
Activities:
Lead agency for real estate regulatory and enforcement issues.

Local Government Agencies

ALBUQUERQUE/ENVIRONMENTAL HEALTH DEPARTMENT/AIR POLLUTON CONTROL DIVISION
P.O. Box 1293
Albuquerque, NM 87103
(505) 768-2600
Steven W. Walker - Manager

New York

NEW YORK STATE ENERGY OFFICE/BUREAU OF RESIDENTIAL CONSERVATION
2 Rockefeller Plaza
8th Floor
Albany, NY 12223
(518) 473-7243
John Reese - Director, Residential Assistance/Radon
Craig Kneeland - Radon Specialist
Mark Watson - Radon Specialist
Al Adams - Radon Specialist
Activities:
Authorized to conduct radon training and education assistance. Maintains list of testing, detection and mitigatior companies in New York, derived from training program participants.

NEW YORK DEPARTMENT OF STATE/DIVISION OF LICENSING SERVICES
162 Washington Ave.
Albany, NY 12231-0001
(518) 473-2419
Gail A. Bates - Director
Activities:
Lead agency for real estate regulatory and enforcement issues.

NEW YORK STATE DEPARTMENT OF ENVIRONMENTAL CONSERVATION/BUREAU OF AIR RESEARCH
50 Wolf Rd.
Albany, NY 12223-3250
(518) 457-3200
S. Rao - Information Officer

NEW YORK STATE DEPARTMENT OF ENVIRONMENTAL CONSERVATION/BUREAU OF AIR TOXICS
50 Wolf Rd.
Albany, NY 12223-3250
(518) 457-7688
B. Majewski - Information Officer

NEW YORK STATE DEPARTMENT OF ENVIRONMENTAL CONSERVATION/BUREAU OF TOXICS AIR SAMPLING
50 Wolf Rd.
Albany, NY 12223-3250
(518) 457-7454
A. Fossa - Information Officer

NEW YORK STATE DEPARTMENT OF ENVIRONMENTAL CONSERVATION/DIVISION OF AIR RESOURCES
50 Wolf Rd.
Albany, NY 12223-3250
(518) 457-7230
Thomas M. Allen - Associate Director

NEW YORK STATE DEPARTMENT OF LABOR/DIVISION OF SAFETY AND HEALTH
Building 12
New York State Office Building Campus
Albany, NY 12240
(518) 457-2574
Robert Goldnick - Director
Activities:
Lead agency for occupational health and safety issues.

NEW YORK STATE ENERGY RESEARCH AND DEVELOPMENT AUTHORITY/ENERGY EFFICIENCY AND ECONOMIC DEVELOPMENT/BUILDING ENERGY SYSTEMS PROGRAM
2 Rockefeller Plaza
14th Floor
Albany, NY 12223
(518) 465-6251
Joe Rizzuto - Program Manager
Activities:
Conducts research on radon in structures. Develops mitigation techniques.

NEW YORK STATE GEOLOGICAL SURVEY
3136 Cultural Education Center
Empire State Plaza
Albany, NY 12230
(518) 474-5816
Robert Fakundiny - Chief and State Geologist
Activities:
Researching soil characteristics (porosity, permeability, composition) in four regions of state.

NEW YORK STATE HEALTH DEPARTMENT/BUREAU OF ENVIRONMENTAL AND OCCUPATIONAL EPIDEMIOLOGY/ENVIRONMENTAL HEALTH SECTION
Radon Registry
Albany, NY 12203
(518) 458-6212
Carolyn Ju - Director
Activities:
Longitudinal study of health consequences of exposure to radon.

NEW YORK STATE HEALTH DEPARTMENT/BUREAU OF ENVIRONMENTAL RADIATION PROTECTION
2 University Place
Albany, NY 12203
(518) 458-6461
Karim Rimawi - Director
Activities:
Bureau oversees all radiation programs, including radon, in Health Department.

NEW YORK STATE HEALTH DEPARTMENT/BUREAU OF ENVIRONMENTAL RADIATION PROTECTION/ENVIRONMENTAL RADIATION SECTION
2 University Place
Albany, NY 12203
(800) 458-1158
Lawrence Keefe - Chief
Activities:
Radon detector distribution program. Detectors sold at cost upon request. Free detectors for eligible persons. Conducts statewide radon information and education campaign. Maintains confidential database on radon testing results. 9,000 homes tested to date.

NEW YORK STATE HEALTH DEPARTMENT/WADSWORTH CENTER FOR LABORATORIES AND RESEARCH/LABORATORY FOR INORGANIC AND NUCLEAR CHEMISTRY
Empire State Plaza
Room D366-B
Albany, NY 12201
(518) 474-6649
Charles Kunz - Research Scientist
Activities:
Provides radon analysis services; geological studies. Responsible for listing analysis and testing companies in New York. Using data, laboratory attempts to pinpoint areas where highest radon levels can be expected.

Local Government Agencies

ALBANY COUNTY DEPARTMENT OF HEALTH/DIVISION OF ENVIRONMENTAL HEALTH SERVICES
South Ferry and Green Sts.
Albany, NY 12201
(518) 447-4580
Stephen S. Lukowski - Director

ALBANY-SCHENECTADY-SCHOHARIE BOARD OF COOPERATIVE EDUCATIONAL SERVICES (BOCES)
47 Cornell Rd.
Latham, NY 12110
(518) 786-3211
Brian Backus - Risk Management
Activities:
BOCES coordinates radon detection program for public schools in Albany, Schoharie, and Schenectady area.

CATTARAUGUS COUNTY HEALTH DEPARTMENT
301 Laurens St.
Olean, NY 14760
(716) 375-4121
Tom Power - Contact Person
Activities:
Distributes radon detectors as part of New York State's radon program. Installs and removes detectors.

CHATAUQUA COUNTY HEALTH DEPARTMENT
Health and Social Services Building
Mayville, NY 14757
(716) 753-4481
Steve Johnson - Contact Person
Jim Metzger - Contact Person
Activities:
Distributes radon detectors as part of New York State's radon program.

DUTCHESS COUNTY HEALTH DEPARTMENT
County Office Building
22 Market St.
Poughkeepsie, NY 12601
(914) 294-7961
Steve Capowski - Contact Person
Activities:
Distributes radon detectors as part of New York State's radon program.

ERIE BOARD OF COOPERATIVE EDUCATIONAL SERVICES (BOCES)
2 Pleasant Ave.
West Lancaster, NY 14086
(716) 686-2094
Jack Weisenbaum - Risk Manager
Activities:
BOCES coordinates radon detection program for public school districts in Buffalo region.

LIVINGSTON-STEUBEN-WYOMING BOARD OF COOPERATIVE EDUCATIONAL SERVICES (BOCES)
Lackawanna Ave.
Mt. Morris, NY 14510
(716) 658-2291
Carol Garey - Safety Coordinator
Activities:
BOCES coordinates radon detection program for 90-plus public school districts in region.

NASSAU COUNTY DEPARTMENT OF HEALTH/DIVISION OF ENVIRONMENTAL HEALTH SERVICES/BUREAU OF AIR QUALITY MANAGEMENT
240 Old County Rd.
Mineola, NY 11501
(516) 535-3232
Robert V. Close - Director

NEW YORK CITY BUREAU OF AIR RESOURCES
295 Lafayette St.
New York, NY 10012
(212) 323-8580
Samuel Stempler - Assistant Commissioner And Director
Martha M. Holstein - Deputy Commissioner

NEW YORK METROPOLITAN AREA INTERSTATE SANITATION COMMISSION
311 West 43 St.
New York, NY 10036
(212) 582-0380
Alan I. Mytelka - Director and Chief Engineer

NIAGARA COUNTY HEALTH DEPARTMENT/AIR RESOURCES BUREAU
10th and East Falls Sts.
Niagara Falls, NY 14302
(716) 284-3128
Maurice Vaughan - Associate Sanitarian

ONEIDA COUNTY BOARD OF COOPERATIVE EDUCATIONAL SERVICES (BOCES)
Middle Settlement Rd.
Box 70
New Hartford, NY 13413
(315) 768-4583
Cassandra Ingham - Safety Coordinator
Craig Dean - Safety Coordinator
Activities:
BOCES coordinates radon detection program for public schools in Oneida, Herkimer and Madison counties.

ORANGE COUNTY HEALTH DEPARTMENT
124 Main St.
Goshen, NY 10924
(914) 294-7961
Dave Bechtle - Contact Person
Activities:
Distributes radon detectors as part of New York State's radon program.

ORLEANS-NIAGRA BOARD OF COOPERATIVE EDUCATIONAL SERVICES (BOCES)
4232 Shelby Basin Rd.
Medina, NY 14103
(716) 798-4800
Bruce Potter - Safety Coordinator
Activities:
BOCES coordinates radon detection program for public schools in region.

PUTNAM COUNTY HEALTH DEPARTMENT
Putnam County Office Building
110 Old Route Six Center
Building #3
Carmel, NY 10512
(914) 225-2115
John E. Karell Jr. - Contact Person
Larry Werper - Contact Person
Activities:
Distributes radon detectors as part of New York State's radon program.

RENSSELAER COUNTY DEPARTMENT OF HEALTH/DIVISION OF ENVIRONMENTAL HEALTH
County Office Building
1600 Seventh Ave.
Troy , NY 12180
(518) 270-2664
Kenneth Van Praag - Public Health Director

ROCKLAND COUNTY HEALTH DEPARTMENT
Sanatorium Rd.
Building D
Pomona, NY 10970
(914) 354-0250
George Brozowski - Contact Person
Activities:
Distributes radon detectors as part of New York State's radon program.

SCHENECTADY CITY HEALTH DEPARTMENT
City Hall
Schenectady, NY 12305
(518) 382-5046
Jack Parisi - Contact Person
Activities:
Distributes radon detectors as part of New York State's radon program.

SCHOHARIE COUNTY HEALTH DEPARTMENT
342 Main St.
Schoharie, NY 12157
(518) 295-8182
Michael Cole - Contact Person
Activities:
Distributes radon detectors as part of New York State's radon program.

SENECA NATION HEALTH DEPARTMENT
Box 231
Salamanca, NY 14779
(716) 532-5582
Lisa Maybee - Environmental Health Technician

Activities:
Participating in radon action program cosponsored with EPA.

SUFFOLK COUNTY DEPARTMENT OF HEALTH SERVICES/AIR POLLUTION CONTROL BUREAU
15 Horse Block Pl.
Farmingville, NY 11738
(516) 451-4637
James C. Maloney - Chief

SUFFOLK II BOARD OF COOPERATIVE EDUCATIONAL SERVICES (BOCES)
201 Sunrise Hwy.
Patchogue, NY 11772
(516) 289-2200
John Thomas - Safety Coordinator
Activities:
BOCES coordinates radon detection program for public school districts in the Long Island region.

TOWN OF HAMBURG
S-6100 South Park Ave.
Hamburg, NY 14075
(716) 649-6111
Jack Lauchert - Information
Mary Jane Kruszka - Information
Thomas F. Gilmartin - Information
Activities:
Provides radon information to local citizens.

WESTCHESTER COUNTY HEALTH DEPARTMENT
County Office Building #2
White Plains, NY 10601
(914) 285-1500
Neil DiBenedetto - Contact Person
Activities:
Distributes radon detectors as part of New York State's radon program.

North Carolina

NORTH CAROLINA DEPARTMENT OF HUMAN RESOURCES/DIVISION OF FACILITY SERVICES/RADIATION PROTECTION SECTION
701 Barbour Dr.
Raleigh, NC 27603-2008
(919) 733-4283
Dayne H. Brown - Chief (State Contact)
Richard M. Fry - Deputy Chief
Felix Fong - Head Environmental Radiation Surveillance
Harry J. Newman - Environment Radiation Specialist
Activities:
Information dissemination. Provides technical assistance based on requests. State's lead radon information agency.

NORTH CAROLINA DEPARTMENT OF LABOR/OCCUPATIONAL SAFETY AND HEALTH
Shore Memorial Building
214 West Jones St.
Raleigh, NC 27603
(919) 733-4880
James Oppold - Director
Activities:
Lead agency for occupational health and safety issues.

NORTH CAROLINA DEPARTMENT OF NATURAL RESOURCES AND COMMUNITY DEVELOPMENT/DIVISION OF LAND RESOURCES
P.O. Box 27687
Raleigh, NC 27611
(919) 733-3833
Stephen Conrad - Director and State Geologist
Activities:
Provides mapping service pertinent to radon surveys in state. Information can be used for radon in soil or water studies.

NORTH CAROLINA REAL ESTATE COMMISSION
1313 Navaho Dr.
Raleigh, NC 27619
(919) 733-9580
Phillip T. Fisher - Executive Director
Larry A. Outlaw - Director of Education and Licensing

Activities:
Lead agency for real estate regulatory and enforcement issues.

Local Government Agencies

CLEVELAND COUNTY HEALTH DEPARTMENT/ENVIRONMENTAL HEALTH DIVISION
315 Grover St.
Shelby, NC 28150
(704) 487-1138
Sparty M. Webb - Environmental Health Coordinator

CUMBERLAND COUNTY/E. NEWTON SMITH PUBLIC HEALTH CENTER/ENVIRONMENTAL HEALTH DIVISION
227 Fountainhead Lane
Fayetteville, NC 28301
(919) 483-9046
Lacy Williams - Director, Environmental Health

FORSYTH COUNTY ENVIRONMENTAL AFFAIRS DEPARTMENT
537 North Spruce St.
Winston-Salem, NC 27101
(919) 727-8060
Robert R. Fulp - Director

GUILFORD COUNTY DEPARTMENT HEALTH/DIVISON OF ENVIRONMENTAL HEALTH/AIR QUALITY PROGRAM
301 North Eugene St.
P.O. Box 3508
Greensboro, NC 27401
(919) 373-3771
R. W. Prentice, Jr. - Supervisor
Scott Reading - Toxic and Health Hazard Specialist

MACKLENBURG COUNTY DEPARTMENT OF ENVIRONMENTAL PROTECTION/AIR QUALITY DIVISION
1200 Blythe Blvd.
Charlotte, NC 28203
(704) 376-4603
Bill Harister - Manager

WESTERN NORTH CAROLINA REGIONAL AIR POLLUTION CONTROL BOARD
Buncombe County Courthouse
Asheville, NC 28802-3569
(704) 255-5655
Ronald Boone - Director

North Dakota

NORTH DAKOTA STATE DEPARTMENT OF HEALTH AND CONSOLIDATED LABORATORIES/DIVISION OF ENVIRONMENTAL ENGINEERING
1200 Missouri Ave.
P.O. Box 5520
Bismarck, ND 58502-5520
(701) 224-2348
Dana K. Mount - Director (State Contact)
Terry Lindsey - Manager, Radiation Control Project
James Killingbeck - Scientist
Tom Morth - Environmental Scientist
Activities:
Participants in EPA mapping project. State's lead agency provides general information to public concerning radon, its health effects, its measurement, and mitigation. Results from 1988 North Dakota/EPA Indoor Radon Survey showec 64 percent of homes tested there had radon levels exceeding 4.0 pCi/l.

NORTH DAKOTA GEOLOGICAL SURVEY
University Station
Grand Forks, ND 58202-8156
(701) 777-5771
Sidney Anderson - Acting State Geologist
Activities:
Developing maps indicating radon potential in state.

NORTH DAKOTA STATE DEPARTMENT OF HEALTH/DIVISION OF ENVIRONMENTAL ENGINEERING/OCCUPATIONAL HEALTH PROGRAM
1200 Missouri Ave.
Room 304
Bismarck, ND 58501
(701) 224-2348
Lee Huber - Manager
Activities:
Lead agency for occupational health and safety issues.

NORTH DAKOTA STATE REAL ESTATE COMMISSION
314 East Thayer Ave.
Bismarck, ND 58502
(701) 224-2749, (701) 224-2737
Dennis D. Schulz - Secretary-Treasurer
Activities:
Lead agency for real estate regulatory and enforcement issues.

Ohio

OHIO DEPARTMENT OF HEALTH/RADIOLOGICAL HEALTH PROGRAM/OHIO RADON PROJECT
1224 Kinnear Rd.
Columbus, OH 43212
(614) 644-2727
Robert M. Quillin - Administrator (State Contact)
Dave Farnsworth - Health Physicist
Deborah Steva - Health Physicist
Activities:
State's lead radon information agency provides general or technical information to public upon request.

OHIO DEPARTMENT OF COMMERCE/DIVISION OF REAL ESTATE
Two Nationwide Plaza
5th Floor
Columbus, OH 43266-0547
(614) 466-4100
Margaret J. Ritenour - Executive Officer and Superintendent

Nelson Snider - Supervisor, Education and Testing
Activities:
Lead agency for real estate regulatory and enforcement issues.

OHIO DEPARTMENT OF INDUSTRIAL RELATIONS/OCCUPATIONAL SAFETY AND HEALTH/DIVISION OF ONSITE CONSULTATION
P.O. Box 825
2323 Fifth Ave.
Columbus, OH 43216
(614) 644-2223, (800) 282-1425
John Sahayda - Chief, Occupational Safety and Health
Activities:
Lead agency for occupational health and safety issues.

OHIO DEPARTMENT OF NATURAL RESOURCES/DIVISION OF GEOLOGICAL SURVEY
Fountain Square
Building B
Columbus, OH 43224
(614) 265-6605
Horace Collins - Division Chief and State Geologist
Activities:
Topographical mapping capabilities pertinent to radon surveys in state. Information utilized for radon in soil or water studies.

OHIO ENVIRONMENTAL PROTECTION AGENCY/DIVISION OF AIR POLLUTION CONTROL
1800 Watermark Dr.
Columbus, OH 43266-0149
(614) 644-3020
Patricia P. Walling - Chief

Local Government Agencies

AKRON REGIONAL AIR QUALITY MANAGEMENT DISTRICT/AIR POLLUTION CONTROL DIVISION
177 South Broadway
Akron, OH 44308
(216) 375-2480
Norman F. Keckler - Administrator, Air Pollution Control

CANTON CITY HEALTH DEPARTMENT/AIR POLLUTION CONTROL DIVISION
City Hall
Canton, OH 44702
(216) 489-3385
Bruce E. Blankenship - Assistant Director

CLEVELAND DEPARTMENT OF PUBLIC HEALTH AND WELFARE/DIVISION OF AIR POLLUTION CONTROL
2735 Broadway Ave.
Cleveland, OH 44115
(216) 664-3500
Gary J. Nied - Commissioner

COLUMBUS HEALTH DEPARTMENT
1815 Washington Blvd.
Columbus, OH 43215
(614) 222-6992
Harry E. Grafton
Activities:
Distributes radon detectors as part of New York State's radon program.

DAYTON REGIONAL AIR POLLUTION CONTROL AGENCY
451 West Third St.
P.O. Box 972
Dayton, OH 45422
(513) 225-4435
John A. Paul - Supervisor

HAMILTON COUNTY-SOUTHWESTERN OHIO/AIR POLLUTION CONTROL AGENCY
2400 Beekman St.
Cincinnati, OH 45214
(513) 251-8777
Thomas W. Wenz - Administrator

LAKE COUNTY GENERAL HEALTH DISTRICT/AIR POLLUTION CONTROL DIVISION
105 Main St.
P.O. Box 490
Painesville, OH 44077
(216) 357-2543
Leon A. Weitzel - Supervisor

MAHONING-TRUMBULL AIR POLLUTION CONTROL AGENCY
9 West Front St.
Youngstown, OH 44503
(216) 744-1928
Larry C. Himes - Administrator

NORTH OHIO VALLEY AIR AUTHORITY
814 Adams St.
Steubenville, OH 43952
(614) 282-3908
P. J. Deluca - Director

PORTSMOUTH LOCAL AIR AGENCY/AIR POLLUTION CONTROL DIVISION
740 Second St.
Portsmouth, OH 45662
(614) 353-5156
Donald R. Cavote - Director-Commissioner

TOLEDO/ENVIRONMENTAL SERVICES DIVISION
26 Main St.
Toledo, OH 43605
(419) 693-0350
Donald M. Moline - Commissioner

Oklahoma

OKLAHOMA DEPARTMENT OF HEALTH/RADIATION AND SPECIAL HAZARDS SERVICE
1000 NW Tenth St.

P.O. Box 53551
Oklahoma City, OK 73152
(405) 271-5221
Dale McHard - Chief (State Contact)
Paul Brown - Director, Radiation Protection Division
Activities:
Lead agency for radon information in Oklahoma.

OKLAHOMA GEOLOGICAL SURVEY
The University Of Oklahoma
830 Van Vleet Oval
Room 163
Norman, OK 73019
(405) 325-3031
Charles Mankin - Director (State Geologist)
Activities:
Mapping capabilities pertinent to radon-potential survey of state.

OKLAHOMA REAL ESTATE COMMISSION
4040 North Lincoln Blvd.
Suite 100
Oklahoma City, OK 73105
(405) 521-3387
Norris Price - Executive Director
Robert Gray - Educational Program Director
Activities:
Lead agency for real estate regulatory and enforcement issues.

Local Government Agencies

OKLAHOMA CITY-COUNTY HEALTH DEPARTMENT/AIR QUALITY CONTROL SECTION
921 NE 23rd St.
Oklahoma City, OK 73105
(405) 427-8651
Curt Goeller - Environmental Specialist Supervisor

TULSA CITY-COUNTY HEALTH DEPARTMENT
4616 East 15th St.
Tulsa, OK 74112
(918) 744-1000
Roger D. Randolph - Supervisor

Oregon

OREGON STATE DEPARTMENT OF HUMAN RESOURCES/HEALTH DIVISION/RADIATION CONTROL SECTION
1400 SW 5th Ave.
P.O. Box 231
Portland, OR 97201
(503) 229-5797
Ray D. Paris - Manager (State Contact)
George L. Toombs - Health Physicist
Activities:
Initiating public educational campaign and offers residents radon detectors at reduced fees.Vendor reports results to Health Division in statewide voluntary program. State's lead radon information agency.

OREGON REAL ESTATE AGENCY
158 12th St, NE
Salem, OR 97310-0240
(503) 378-8414
Morella Larsen - Real Estate Commissioner
Jerry Cripe - Education Director
Activities:
Lead agency for real estate regulatory and enforcement issues.

OREGON STATE DEPARTMENT OF ENVIRONMENTAL QUALITY
811 SW Sixth Ave.
Portland, OR 97204
(503) 229-5696
Ron Householder - Acting Administrator

OREGON STATE DEPARTMENT OF GEOLOGY AND MINERAL INDUSTRIES
910 State Office Building
Portland, OR 97201-5528
(503) 229-5580
Donald A. Hull - Director and State Geologist
Activities:
Geologic maps pertinent to radon surveys in state. Information can be used for geologic regions radon analysis.

OREGON STATE DEPARTMENT OF WORKERS' COMPENSATION/ACCIDENT PREVENTION DIVISION/CONSULTATIVE SECTION
Labor And Industries Building
Room 115
Salem, OR 97310
(503) 378-3272
Jack Pompei - Administrator
Activities:
Lead agency for occupational health and safety issues.

Local Government Agencies

LANE REGIONAL AIR POLLUTION AUTHORITY
225 North 5th St.
Suite 501
Springfield, OR 97477
(503) 726-2514
Donald A. Arkell - Director

Pennsylvania

PENNSYLVANIA DEPARTMENT OF ENVIRONMENTAL RESOURCES/BUREAU OF RADIATION PROTECTION/DIVISION OF SPECIAL PROJECTS
Fulton Building, 16th Floor
Third and Locust Sts.
Harrisburg, PA 17120
(717) 787-2480
Carl Granlund - Acting Chief (State Contact)
Michael Pyles - Technical Supervisor
Jason Gatner - Public Relations
Marcee Meyers - Certification Program
Activities:
Policy development for state of Pennsylvania's response to radon problems. Involved in research on distribution and health threat posed by radon. Certification of testing, detection, and mitigation companies as well as product manufacturers.

PENNSYLVANIA DEPARTMENT OF ENVIRONMENTAL RESOURCES/BUREAU OF RADIATION PROTECTION
P.O. Box 2063
Harrisburg, PA 17120
(717) 787-2480
Thomas M. Gerusky - Director
Activities:
Oversees radon information programs in Pennsylvania.

PENNSYLVANIA DEPARTMENT OF ENVIRONMENTAL RESOURCES/BUREAU OF RADIATION PROTECTION/DIVISION OF ENVIRONMENTAL RADIATION
Fulton Building, 16th Fl.
Third and Locust Sts.
Harrisburg, PA 17120
(717) 787-3479
Margaret A. Reilly - Chief
Activities:
Provides support information and technical services to Division of Special Projects.

PENNSYLVANIA DEPARTMENT OF ENVIRONMENTAL RESOURCES/BUREAU OF TOPOGRAPHIC AND GEOLOGIC SURVEY
P.O. Box 2357
Harrisburg, PA 17120
(717) 787-2169
Donald Hoskins - State Geologist
Activities:
Performs gamma radiation road surveys.

PENNSYLVANIA DEPARTMENT OF STATE/BUREAU OF PROFESSIONAL AND OCCUPATIONAL AFFAIRS/REAL ESTATE AND VEHICLE DIVISION
P.O. Box 2649
Harrisburg, PA 17105-3658
(717) 783-3658
Pierce L. Clouser, Jr. - Chief

Local Government Agencies

ALLEGHENY COUNTY HEALTH DEPARTMENT/BUREAU OF AIR POLLUTION CONTROL
301 39th St.
Pittsburgh, PA 15201
(412) 578-8111
Ronald J. Chleboski - Deputy Director-Air Pollution Control

PHILADELPHIA DEPARTMENT OF PUBLIC HEALTH/AIR MANAGEMENT SERVICES
500 South Broad St.
Philadelphia, PA 19146
(215) 875-5623
William Reilly - Assistant Health Commissioner

Rhode Island

RHODE ISLAND DEPARTMENT OF HEALTH/DIVISION OF OCCUPATIONAL HEALTH AND RADIATION CONTROL
Cannon Building
75 Davis St.
Providence, RI 02908
(401) 277-2438
James E. Hickey - Division Chief (State Contact)
Activities:
Will participate in EPA mitigation demonstration program. State's lead agency for radon information. Also lead agency for occupational health and safety issues.

RHODE ISLAND DEPARTMENT OF BUSINESS REGULATION, LICENSING AND CONSUMER PROTECTION
100 North Main St.
Providence, RI 02903
(401) 277-2255
Valerie Voccio - License Examiner
Activities:
Lead agency for real estate regulatory and enforcement issues

RHODE ISLAND DEPARTMENT OF GEOLOGY
Green Hall
The University of Rhode Island
Kingston, RI 02881
(401) 792-2265
Allan Cain - State Geologist and Chairman
Activities:
Will participate in 1989 EPA/Rhode Island statewide radon survey.

South Carolina

SOUTH CAROLINA DEPARTMENT OF HEALTH AND ENVIRONMENTAL CONTROL/BUREAU OF RADIOLOGICAL HEALTH
2600 Bull St.
Columbia, SC 29201
(803) 734-4700
Heyward G. Shealy - Chief (State Contact)
Noland Bivens - Director
Lynn B. Dobbs - Health Physicist
Activities:
State's lead agency provides radon analysis services, general information to public. Has enacted surveys on health effects of radon.

SOUTH CAROLINA DEPARTMENT OF LABOR/OCCUPATIONAL SAFETY AND HEALTH/CONSULTATION AND MONITORING
P.O. Box 11329
Columbia, SC 29211
(803) 734-9599
Robert Peck - Director, Education, Training and Consultation
Activities:
Lead agency for occupational health and safety issues.

SOUTH CAROLINA GEOLOGICAL SURVEY
Harbison Forest Rd.
Columbia, SC 29210
(803) 737-9440
Norman Olson - State Geologist
Activities:
Provides geologic mapping service pertinent to radon surveys in state.

SOUTH CAROLINA REAL ESTATE COMMISSION
1201 Main St.
Suite 1500
Columbia, SC 29201
(803) 737-0700
Henry L. Jolly - Commissioner
Eleanor Lightsey Pruet - Education Director
Activities:
Lead agency for real estate regulatory and enforcement issues.

Local Government Agencies

COLUMBIA/DEPARTMENT OF HEALTH/AIR POLLUTION CONTROL DIVISION
1225 Laurel St.
Columbia, SC 29217
(803) 733-8320
Darlene Boyd - Pollution Control Officer

South Dakota

SOUTH DAKOTA STATE DEPARTMENT OF HEALTH/OFFICE OF AIR QUALITY AND SOLID WASTE
Joe Foss Office Building
523 East Capital
Pierre, SD 57501
(605) 773-3329
James P. Platt - Director (State Contact)
Joel Smith - Administrator
Mike Pochop - Natural Resources Scientist
Tammy LaBeau - Natural Resources Analyst
Activities:
Lead agency for radon information in South Dakota.

SOUTH DAKOTA DEPARTMENT OF WATER AND NATURAL RESOURCES/SOUTH DAKOTA GEOLOGICAL SURVEY
Science Center
University of South Dakota
Vermillion, SD 57069
(605) 677-5227
Merlin Tipton - State Geologist
Activities:
Mapping capabilities pertinent to radon surveys in state.

SOUTH DAKOTA REAL ESTATE COMMISSION
P.O. Box 490
Pierre, SD 57501
(605) 773-3600
Larry G. Lyngstad - Executive Secretary
Activities:
Lead agency for real estate regulatory and enforcement issues.

SOUTH DAKOTA STATE UNIVERSITY/S.T.A.T.E. ENGINEERING DIVISION
Box 2218
Brookings, SD 57007
(605) 688-4101
Jim Ceglian - Program Director
Activities:
Lead agency for occupational health and safety issues.

Tennessee

TENNESSEE DEPARTMENT OF HEALTH AND ENVIRONMENT/DIVISION OF AIR POLLUTION CONTROL
701 Broadway
4th Floor
Nashville, TN 37219-5403
(615) 741-4634, (615) 741-3931
Jackie Waynick - (State Contact)
Susie Shimek - Environmental Specialist
Activities:
State's lead radon information agency. Provides general information from EPA upon request. Any of Division's 16 regional offices can provide radon information.

TENNESSEE DEPARTMENT OF CONSERVATION/DIVISION OF GEOLOGY
Customs House
701 Broadway
Nashville, TN 37219-5237
(615) 742-6691
William Hill - State Geologist
Activities:
Geologic maps with potential utilization in radon soil or water studies.

TENNESSEE DEPARTMENT OF LABOR/TENNESSEE OCCUPATIONAL SAFETY AND HEALTH
501 Union Building
3rd Floor
Nashville, TN 37219
(615) 741-2793
Robert B. Taylor - Director
Activities:
Lead agency for occupational health and safety issues.

TENNESSEE REAL ESTATE COMMISSION
1808 West End Building
11th Floor
Nashville, TN 37219-5322
(615) 741-2273
Ann Litton Tuck - Executive Director
Sharon Langford - Education Director
Activities:
Lead agency for real estate regulatory and enforcement issues.

Local Government Agencies

CHATTANOOGA-HAMILTON COUNTY/AIR POLLUTION CONTROL BUREAU
3511 Rossville Blvd.
Chattanooga, TN 37407
(615) 867-4321
J. Wayne Cropp - Director

KNOX COUNTY DEPARTMENT OF AIR POLLUTION CONTROL
400 Main Ave.
City-County Building, Room 459
Knoxville, TN 37902
(615) 521-2488
Terry C. Harris - Director

MEMPHIS AND SHELBY COUNTY HEALTH DEPARTMENT/AIR POLLUTION CONTROL BRANCH
814 Jefferson Ave.
Memphis, TN 38105
(901) 576-7775
Denis Fritchie - Supervisor

NASHVILLE AND DAVIDSON COUNTYMETROPOLITAN HEALTH DEPARTMENT/POLLUTION CONTROL DIVISION
311 23rd Ave. North
Nashville, TN 37203
(615) 327-9313
Paul J. Bontrager - Director

Texas

TEXAS DEPARTMENT OF HEALTH/BUREAU OF RADIATION CONTROL
1100 West 49th St.
Austin, TX 78756-3189
(512) 835-7000
David K. Lacker - Chief (State Contact)
Gary Smith - Chief, Radiological Assessment Program
Activities:
State's lead radon information agency provides general information upon request.

TEXAS BUREAU OF ECONOMIC GEOLOGY
The University Of Texas, Austin
Balcones Research Center
University Station, Box X
Austin, TX 78713-7508
(512) 471-1534
William Fisher - Director And State Geologist
Activities:
Geologic information compiled can be used for radon-soil or water studies.

TEXAS DEPARTMENT OF HEALTH/DIVISION OF OCCUPATIONAL SAFETY AND STATE SAFETY ENGINEER
1100 West 49th St.
Austin, TX 78756
(512) 458-7287
Jerry F. Waterdale - Director
Activities:
Lead agency for occupational health and safety issues.

TEXAS REAL ESTATE COMMISSION
P.O. Box 12188
Capitol Station
Austin, TX 78711
(512) 465-3980
Wallace Collins - Administrator
Donald C. Roose - Education Officer
Activities:
Lead agency for real estate regulatory and enforcement issues.

Local Government Agencies

CORPUS CHRISTI-NUECES COUNTY DEPARTMENT OF PUBLIC HEALTH/ENVIRONMENTAL CONTROL DIVISION
1702 Horne Rd.
Corpus Christi, TX 78416
(512) 851-7226
R. Leon Decker - Director

DALLAS/HEALTH AND HUMAN SERVICES/ENVIRONMENTAL HEALTH DIVISION/AIR POLLUTION CONTROL SECTION
400 South Zang Blvd.
Dallas, TX 75208
(214) 670-7520
H. Gary Burlbaw - Manager, Air Pollution Control

ECTOR COUNTY HEALTH DEPARTMENT/AIR POLLUTION CONTROL DIVISION
221 North Texas
Odessa, TX 79761
(214) 670-7520
Johnny O. Johnson - Health Educator

EL PASO COUNTY HEALTH UNIT/ENVIRONMENTAL HEALTH SERVICES DIVISION
222 South Campbell St.
El Paso, TX 79901
(915) 541-4262
Raul V. Munoz, Jr. - Chief

FORT WORTH AIR POLLUTION CONTROL/DEPARTMENT OF PUBLIC HEALTH
1800 University Dr.
Fort Worth, TX 76107
(817) 870-7280
Gerald Bearden - Supervisor

GALVESTON COUNTY HEALTH DISTRICT/AIR POLLUTION CONTROL DIVISION
1207 Oak St.
P.O. Box 939
La Marque, TX 77568
(409) 938-7221
Edward R. Ibert - Air Control Director

HARRIS COUNTY POLLUTION CONTROL DEPARTMENT
107 North Munger
Pasadena, TX 77506
(713) 920-2831
A.R. Pierce - Director

HOUSTON/BUREAU OF AIR QUALITY CONTROL
7411 Park Place Blvd.
Houston, TX 77087
(713) 640-4200
Dallas E. Evans - Chief

LUBBOCK CITY HEALTH DEPARTMENT/AIR POLLUTION CONTROL DIVISION
1902 Texas Ave.
P.O. Box 2548
Lubbock, TX 79408
(806) 762-6411 x2918
Anthony Way - Health Director

Utah

UTAH DEPARTMENT OF HEALTH/BUREAU OF RADIATION CONTROL
288 North 1460 West
P.O. Box 16690
Salt Lake City, UT 84116-0690
(801) 538-6734
Larry Anderson - Director (State Contact)
Diane Finerfrock - Section Manager
Activities:
Statewide study to determine exposure levels. Provide at-cost radon detection devices. Consulting services provided for citizens who need detection and mitigation services. State's lead radon information agency.

UTAH DEPARTMENT OF BUSINESS REGULATIONS/DIVISION OF REAL ESTATE
P.O. Box 45802
160 East 300 South
Salt Lake City, UT 84145
(801) 530-6747
Richard M. Mercer - Director
Karen Post - Education Coordinator
Activities:
Lead agency for real estate regulatory and enforcement issues.

UTAH GEOLOGICAL AND MINERAL SURVEY
606 Black Hawk Way
Salt Lake City, UT 84108-1280
(801) 244-5164
Genevieve Atwood - Director and State Geologist
Activities:
Provides geologic mapping and research services.

Vermont

VERMONT DEPARTMENT OF HEALTH/DIVISION OF OCCUPATIONAL AND RADIOLOGICAL HEALTH
Administration Building
10 Baldwin St.
Montpelier, VT 05602
(802) 828-2886
Raymond N. McCandless - Director
Paul E. Clemons - Radiological Health Specialist (State Contact)
Activities:
EPA information pamphlets provided upon request.

VERMONT DEPARTMENT OF LABOR AND INDUSTRY/DIVISION OF OCCUPATIONAL SAFETY AND HEALTH
120 State St.
Montpelier, VT 05602
(802) 828-2765
Rober McCloud - Manager
Activities:
Lead agency for occupational safety and health issues.

VERMONT GEOLOGICAL SURVEY
103 South Main St.
Center Building
Waterbury, VT 05676
(802) 244-5164
Charles Ratte - State Geologist
Activities:
Geologic mapping and research capabilities.

VERMONT OFFICE OF THE SECRETARY OF STATE/REAL ESTATE COMMISSION
Pavilion Office Building
Montpelier, VT 05602
(802) 828-3228
Ellen H. Graham - Executive Director
Activities:
Lead agency for real estate regulatory and enforcement issues.

Virginia

VIRGINIA DEPARTMENT OF HEALTH/DIVISION OF HEALTH HAZARDS CONTROL/BUREAU OF RADIOLOGICAL HEALTH
109 Governor St.
Richmond, VA 23219
(804) 786-6029
Phillip Shaheen - Acting Director (State Contact)
Jim DeKroff - Radiation Safety Specialist
Activities:
Provides information to public on radon issues. Keeps list of EPA-approved testing and detection firms in state.

VIRGINIA DEPARTMENT OF AGRICULTURE AND CONSUMER SERVICES/OFFICE OF CONSUMER AFFAIRS
Washington Building
1100 Bank St.
Richmond, VA 23219
(800) 552-9963
Betty Blakemore - Director
Activities:
Mediation board. Fields complaints on quality of radon detection, testing, and mitigation services. Can request investigation by state attorney general.

VIRGINIA DEPARTMENT OF COMMERCE/STATE REAL ESTATE BOARD
3600 West Broad St.
5th Floor
Richmond, VA 23230
(804) 367-8552
Joan Langford White - Assistant Director
Activities:
Lead agency for real estate regulatory and enforcement issues.

VIRGINIA DEPARTMENT OF HEALTH/DIVISION OF HEALTH HAZARDS CONTROL
109 Governor St.

Room 700
Richmond, VA 23219
(804) 786-6029
Carl Armstrong - Director
Activities:
Oversees Virginia's radon programs at state level.

VIRGINIA DEPARTMENT OF HEALTH/DIVISION OF HEALTH HAZARDS CONTROL/BUREAU OF RADIOLOGICAL HEALTH
Northern Regional Office
9504-A Lee Highway
Fairfax, VA 22031
(703) 430-8609
Fred Goriach - Regional Director
Activities:
Provides radon information to citizens of state's heavily populated northern region.

VIRGINIA DIVISION OF MINERAL RESOURCES
P.O. Box 3667
Charlottesville, VA 22903
(804) 293-5121
Robert Milici - State Geologist
Activities:
Provides geologic mapping service and research capabilities pertinent to radon surveys in state. Information can be used for radon-soil or water studies.

Local Government Agencies

ALEXANDRIA HEALTH DEPARTMENT/OFFICE OF ENVIRONMENTAL QUALITY
517 North St. Asaph St.
Alexandria, VA 22314
(703) 838-4850, 838-4865
Vacant - Chief, Office Of Environmental Quality

FAIRFAX CITY DEPARTMENT.OF HEALTH
10777 Main St.
Suite 100
Fairfax, VA 22030
(703) 246-2411
Edgar Chase - Information

FAIRFAX COUNTY AIR POLLUTION CONROL
10777 Main St.
Suite 100A
Fairfax, VA 22030
(703) 246-2541
Edgar M. Chase - Director, Air Pollution Control

ROANOKE CITY ENGINEERING DEPARTMENT/AIR POLLUTION CONTROL DIVISION
215 West Campbell Ave.
Roanoke, VA 240111
(703) 981-2731
Richard V. Hamilton - Air Pollution Control Officer

ROANOKE COUNTY AND CITY OF SALEM/AIR POLLUTION CONTROL
P.O. Box 1074
Salem, VA 24153
(703) 387-6224
W. Carter Bradley - Air Pollution Control Director

Washington

WASHINGTON STATE DEPARTMENT OF SOCIAL AND HEALTH SERVICES/OFFICE OF RADIATION PROTECTION
Mail Stop LE-13
Olympia, WA 98504
(206) 753-5962, (800) 323-9727
Robert Mooney - Supervisor (State Contact)
Bruce Pickett - Information
Activities:
State's lead agency provides information to public and maintains list of detection and mitigation companies servicing state. Provides technical training and health assessment.

WASHINGTON STATE DEPARTMENT OF ENERGY/ENERGY EXTENSION SERVICE
914 East Jefferson
Suite 300
Seattle, WA 98122
(509) 624-4180
Stan Price - Manager
Activities:
Program of state energy office provides education services through seminars, phone consultations and fact sheets on building construction issues related to control and mitigation of radon in residential structures. Office is currently developing a model radon control ordinance for low rise residential construction under contract with BPA.

WASHINGTON STATE DEPARTMENT OF LABOR AND INDUSTY
814 East 4th
P.O. Box 207
Olympia, WA 98504
(206) 753-6500
David Hutchins - Assistant Director

WASHINGTON STATE DEPARTMENT OF LICENSING/REAL ESTATE AND ESCROW/PROFESSIONAL PROGRAMS MANAGEMENT DIVISION
P.O. Box 9012
Olympia, WA 98504
(206) 753-0775
Syd Beckett - Executive Secretary
Robert Nocoloff - Program Manager, Education/Exam Administration
Activities:
Lead agency for real estate regulatory and enforcement issues.

WASHINGTON STATE DEPARTMENT OF NATURAL RESOURCES/DIVISION OF GEOLOGY AND EARTH RESOURCES DIVISION
MS-PY-12
Olympia, WA 98504
(206) 459-6372
Raymond Lasmanis - State Geologist

Activities:
Extensive uranium survey developed a number of years ago that is now useful in determining radon potential in state's regions.

Local Government Agencies

BENTON-FRANKLIN-WALLA WALLA COUNTIES AIR POLLUTION AUTHORITY/AIR POLLUTION CONTROL DIVISION
650 George Washington Way
Richland, WA 99352
(509) 545-2354
J. Philip Cooke - Control Officer

NORTHWEST AIR POLLUTION AUTHORITY
207 Pioneer Building
Mount Vernon, WA 98273
(206) 336-5705
Terry L. Nyman - Air Pollution Control Officer

OLYMPIA AIR POLLUTION CONTROL AUTHORITY
120 East State Ave.
Olympia, WA 98501
(206) 352-4881
Charles E. Peace - Control Officer

PUGET SOUND AIR POLLUTION CONTROL AGENCY
200 West Mercer St.
Room 205
P.O. Box 9863
Seattle, WA 98119-3958
(206) 344-7330
Arthur R. Dammkoehler - Air Pollution Control Officer

SOUTHWEST AIR POLLUTION CONTROL AUTHORITY
1306 NE 134th St.
Suite D
Vancouver, WA 98685
(206) 574-3058
Dick Serdos - Executive Director

SPOKANE COUNTY AIR POLLUTION CONTROL AUTHORITY
1101 West College Ave.
Room 230
Spokane, WA 99201
(509) 456-4727
Fred O. Gray - Director

YAKIMA COUNTY CLEAN AIR AUTHORITY
County Courthouse
Yakima, WA 98901
(509) 575-4116
Robert L. Crossland - Director

West Virginia

WEST VIRGINIA DEPARTMENT OF HEALTH/INDUSTRIAL HYGIENE DIVISION
151 11th Ave.
South Charleston, WV 25303
(304) 348-3526
William H. Aaroe - Director (State Contact)
Beattie L. DeBord - Assistant Director
Activities:
Assists residents with guidance on testing for radon and evaluation of test results and mitigation measures. State's lead agency for radon information.

WEST VIRGINIA DEPARTMENT OF LABOR
State Capitol
Room 415
1900 Washington St.
Charleston, WV 25305
(304) 348-7890
David Hutchins - Assistant Director

WEST VIRGINIA GEOLOGICAL SURVEY
Mt. Chateau Research Center
P.O. Box 879
Morgantown, WV 26507-0879
(304) 594-2331
Robert Erwin - Director and State Geologist
Activities:
Mapping services pertinent to radon surveys in state.

WEST VIRGINIA REAL ESTATE COMMISSION
1033 Quarrier St.
Suite 400
Charleston, WV 25301
(304) 348-3555
Donald E. Portis - Executive Secretary
Activities:
Lead agency for real estate regulatory and enforcement issues.

Wisconsin

WISCONSIN DEPARTMENT OF HEALTH AND SOCIAL SERVICES/DIVISION OF HEALTH/RADIATION PROTECTION SECTION
5708 Odana Rd.
Madison, WI 53719
(608) 273-5180
Lawrence J. McDonnell - Chief (State Contact)
John Micka - Radiation Consultant
Activities:
State's lead agency on radon. Provides information to public. Limited funding for radon measurement program.

WISCONSIN DEPARTMENT OF HEALTH AND SOCIAL SERVICES/DIVISION OF HEALTH/BUREAU OF ENVIRONMENTAL HEALTH/SECTION OF OCCUPATIONAL HEALTH
P.O. Box 309
Madison, WI 53701
(608) 266-0417
Terry Moen - Section Chief

Activities:
Lead agency for occupational health and safety issues.

WISCONSIN DEPARTMENT OF REGULATION AND LICENSING/BUREAU OF DIRECT LICENSING AND REAL ESTATE
1400 East Washington
P.O. Box 8935
Madison, WI 53708
(608) 267-7132
Cletus J. Hansen - Bureau Director
Michael Daly - Education
Activities:
Lead agency for real estate regulatory and enforcement issues.

WISCONSIN GEOLOGICAL AND NATURAL HISTORY SURVEY
University of Wisconsin
3817 Mineral Point Rd.
Madison, WI 53705
(608) 262-1705
Meredith Ostrom - Director and State Geologist
Activities:
Geologic information can be used for radon-soil or water studies.

Local Government Agencies

EAU CLAIRE/CITY-COUNTY HEALTH DEPARTMENT/ENVIRONMENTAL HEALTH DIVISION
720 2nd Ave.
Eau Claire, WI 54701
(715) 839-4718
Darryll D. Farmer - Director

MADISON DEPARTMENT OF PUBLIC HEALTH/ENVIRONMENTAL HEALTH AND LABORATORY
210 Martin Luther King, Jr. Blvd.
Madison, WI 53710
(608) 266-4821
Jill Schmidt - Director

MILWAUKEE COUNTY DEPARTMENT OF PUBLIC WORKS/PROFESSIONAL SERVICES DIVISION/ENGINEERS ENVIRONMENTAL AND ENERGY SERVICES SECTION
901 North 9th St.
Milwaukee, WI 53233
(414) 278-4874
Carl W. Birks - Director

Wyoming

WYOMING DEPARTMENT OF HEALTH AND SOCIAL SERVICES/DIVISION OF HEALTH AND MEDICAL SERVICES/RADIOLOGICAL HEALTH SERVICES
Hathaway Building
Cheyenne, WY 82002-0710
(307) 777-6015
Julius E. Haes - Director (State Contact)
Activities:
Information service for general public. No legislative authority for radon programs at present. Lead information agency in state. In late 1988 Wyoming's legislature voted to cut funding for radon programs.

WYOMING DEPARTMENT OF OCCUPATIONAL HEALTH AND SAFETY/COMMISSIONER OF LABOR
604 East 25th St.
Cheyenne, WY 82002
(307) 777-7786
Mike Sullivan - Commissioner
Activities:
Lead agency for occupational health and safety issues.

WYOMING GEOLOGICAL SURVEY
P.O. Box 3008
University Station
Laramie, WY 82071
(307) 742-2054
Gary Glass - State Geologist
Activities:
Maps hot spots, provides EPA/state survey map.

WYOMING REAL ESATE COMMISSION
4301 Herschler Building
Cheyenne, WY 82002
(307) 777-7141
Constance K. Anderson - Executive Director
Lawrence D. Billis - Investigator and Education Coordinator
Activities:
Lead agency for real estate regulatory and enforcement issues.

Territories

AMERICAN SAMOA/ENVIRONMENTAL QUALITY COMMISSION/GOVERNOR'S OFFICE
Pago Pago, American Samoa, AS 96799
(684) 633-4116, (684) 633-2304
Lyle Richmond - Chairperson (Territorial Contact)

GOVERNMENT OF GUAM/DEPARTMENT OF PUBLIC HEALTH AND SOCIAL SERVICES/DIVISION OF ENVIRONMENTAL HEALTH
P.O. Box 2816
Agana, GU 96910
(671) 734-2671
Vicente D. Quitoriano - Administrator (Territorial Contact)
Activities:
Provides general radon information to public upon request.

GOVERNMENT OF GUAM/DEPARTMENT OF LABOR
P.O. Box 23548
Guam Main Facility
Agana, GU 96921
(671) 477-9821
Activities:
Lead agency for occupational health and safety issues.

GOVERNMENT OF GUAM/DEPARTMENT OF REVENUE AND TAXATION/REAL ESTATE AGENCY
855 West Marine Dr.
Agana, GU 96910
(671) 477-1040
Rafael C. Pangelinan - Administrator Coordinator
Activities:
Lead agency for real estate regulatory and enforcement issues.

GOVERNMENT OF GUAM/ENVIRONMENTAL PROTECTION AGENCY
P.O. Box 2999
Agana, GU 96910
(671) 646-8863
D. Dawn Cahoon - Program Coordinator IV
Activities:
Provides citizens of Guam with radon information upon request. Limited information program.

PUERTO RICO/DEPARTMENT OF HEALTH/ RADIOLOGICAL HEALTH DIVISION
G.P.O. Call Box 70184
Rio Piedras, PR 00936
(809) 767-3563
David Saldana - Director (Commonwealth Contact)
Jose Perez - Radiological Health Specialist
Activities:
Provides EPA radon information pamphlets upon request. Commonwealth's lead agency for radon information.

PUERTO RICO DEPARTMENT OF LABOR AND HUMAN RESOURCES/OCCUPATIONAL SAFETY AND HEALTH
505 Munoz Rivera Ave.
21st Fl.
Hato Rey, PR 00918
(809) 754-2134, (809) 754-2171
Filiberto Cruz - Auxilary Secretary for OSHA
Activities:
Lead agency for occupational safety and health issues.

PUERTO RICO GEOLOGICAL SURVEY/DEPARTMENT DE RECURSOS NATURALES
Apartado 5887
Puerta de Tierra
San Juan, PR 00906
(809) 723-2716
Ramon Alonzo - Director
Activities:
Geologic mapping capabilities pertinent to radon studies.

PUERTO RICO/ENVIRONMENTAL QUALITY BOARD
Del Parque St. #204
Corner Pumarada St.
Santurce, PR 00910
(809) 725-5140, (809) 722-0077
Juan Merced Mateo - Director

U.S. VIRGIN ISLANDS DEPARTMENT OF PLANNING AND NATURAL RESOURCES/ENVIRONMENTAL PROTECTION DIVISION
P.O. Box 3594
St. Thomas, VI 00801
(809) 776-4844
Francine Lange - Director (Territorial Contact)
Activities:
Territory's lead agency for radon information.

U.S. VIRGIN ISLANDS DEPARTMENT OF CONSERVATION AND CULTURAL AFFAIRS/DIVISION OF NATURAL RESOURCES MANAGEMENT/AIR POLLUTION PROGRAM
P.O. Box 4399
St. Thomas, VI 00801-4399
(809) 774-3320
Vernon Richards - Director

U.S. VIRGIN ISLANDS DEPARTMENT OF LABOR/DIVISION OF OCCUPATIONAL SAFETY AND HEALTH
Lagoon St.
Frederiksted, VI 00840
(809) 772-1315
Vaughn Hendricks - Director
Activities:
Lead agency for occupational health and safety issues.

U.S. VIRGIN ISLANDS DEPARTMENT OF LICENSING AND CONSUMER SERVICES
Sub Base #1
Property and Procurement Building
St. Thomas, VI 00802
(809) 774-3130
Prudence Wesselhoft - Director
Activities:
Lead agency for real estate regulatory and enforcement issues.

U.S. VIRGIN ISLANDS/COLLEGE OF THE VIRGIN ISLANDS/CARIBBEAN RESEARCH INSTITUTE

St. Thomas, VI 809774-9200
(809) 774-9200
Norwell Harrigan - Director and Territorial Geologist
Activities:
Chief office for geologic mapping services in territory.

Indian Nations

NAVAJO NATION/OCCUPATIONAL SAFETY AND HEALTH
P.O. Box 1690
Window Rock, 86515
(614) 644-2223, (800) 282-1425
Verne Curtis - Director
Activities:
Lead agency for occupational health and safety issues.

NAVAJO NATION/OCCUPATIONAL HEALTH AND SAFETY
P.O. Box 1690
Window Rock, 86515
(602) 871-6335
Verne Curtis - Director
Activities:
Lead agency for occupational health and safety issues.

SENECA NATION HEALTH DEPARTMENT
Box 231
Salamanca, NY 14779
(716) 532-5582
Lisa Maybee - Environmental Health Technician

Activities:
Participating in radon action program cosponsored with EPA.

Canadian Governments and Associations

This section lists Canadian federal and provincial government agencies and associations involved in radon-related programs. These activities include administration, policy, research, and disseminating information. Canadian radon detection and mitigation companies, as well as radon product manufacturers can be found in Sections 1, 2, and 4 respectively.

Canada Federal

ATOMIC ENERGY CONTROL BOARD (AECB)/TECHNICAL INFORMATION DIVISION/OFFICE OF PUBLIC INFORMATION
P.O. Box 1046
Ottawa, ON K1P 559
(613) 992-9206
Hugh Spence - Public Information Officer
Activities:
Distributes informational packet to Canadians that addresses radon exposure and potential health effects. Sponsors research on radon-related issues.

ATOMIC ENERGY OF CANADA, LTD. (AECL)/ADMINISTRATIVE DIVISION/RADIATION INDUSTRIAL SAFETY BRANCH
Tinawa, MB R0E 1L0
(204) 753-2720
Colin Brown - Safety Officer
Activities:
Crown Corporation (funded with both public and private monies) operating under auspices of Ministry of Energy, Mines and Resources. Industrial hygiene program offers indoor air quality assessment program for businesses or private citizens. Radon detection program part of package. Also performs commercial monitoring for radon.

ATOMIC ENERGY OF CANADA, LTD. (AECL)/WASTE MANAGEMENT/APPLIED GEOSCIENCE/HYDROGEOCHEMISTRY SECTION
Tinawa, MB R0E 1L0
(204) 753-2311
Melvyn Gascoyne - Section Head
Activities:
Research centers on radon levels in groundwater. Administers program measuring radon in some structures.

CANADA MORTGAGE AND HOUSING CORPORATION (CMHC)/RESEARCH DIVISION
Montreal Rd.
Ottawa, ON K1A 0P7
(613) 748-2306
Peter Russell - Consultant on Remedial Action
Activities:
Federal housing authority in Canada informs public and industry on radon issues pertaining to housing industry. Three radon informational publications in preparation: public information, construction techniques, and renovation suggestions.

HEALTH AND WELFARE CANADA/HEALTH PROTECTION BRANCH/BUREAU OF RADIATION AND MEDICAL DEVICES
775 Brookfield Rd.
Ottawa, ON K1A 1C1

(613) 954-6677
E.G. Letourneau - Director
R.G. McGregor - Contact
Activities:
Canada's lead federal agency for radon health policy, research, and information. Involved in radon studies for past ten years and has produced a number of reports on the topic.

Canada Provincial

ALBERTA COMMUNITY AND OCCUPATIONAL HEALTH DEPARTMENT/OCCUPATIONAL HEALTH AND SAFETY DIVISION/RADIATION HEALTH SECTION/OCCUPATIONAL HEALTH BRANCH
Industrial and Special Applications
4th Floor, Donsdale Pl.
10709 Jasper Ave.
Edmonton, AB T5J 3N3
(403) 427-2691
Frederick J. Anderson - Radiation Health Officer
Activities:
Provides comprehensive radon information package upon request, including information from AECB and a list of radon detection companies serving the province. Citizens are also provided information concerning results of nationwide radon survey and circumstances that may lead to indoor radon problems.

ALBERTA DEPARTMENT OF CONSUMER AND CORPORATE AFFAIRS/REAL ESTATE STANDARDS
10025 Jasper Ave.
19th Floor
Edmonton, AB T5J 3Z5
(403) 422-1588
Rudolph J. Palovcik - Superintendent of Real Estate
Activities:
Lead agency for real estate regulatory and enforcement issues.

BRITISH COLUMBIA MINISTRY OF HEALTH/RADIATION PROTECTION SERVICE
307 West Broadway
Suite 200
Vancouver, BC V5Y 1P9
(604) 660-6630
Brian Phillips - Director
Activities:
Provides documentation on a case-by-case basis. Information provided on research and radon's effects on health. Developing radon monitoring information.

BRITISH COLUMBIA REAL ESTATE COUNCIL
626 West Pender St.
5th Floor
Vancouver, BC V6B 1V9
(604) 683-9664
P. Dermot Murphy - Secretary
Activities:
Lead agency for real estate regulatory and enforcement issues.

GOVERNMENT OF NEWFOUNDLAND AND LABRADOR/DEPARTMENT OF HEALTH/PUBLIC HEALTH INSPECTION DIVISION
P.O. Box 4750
St. John's, NF A1C 5T7
(709) 576-2679
Warrick Swyers - Contact
Activities:
Provides radon information packet to provincial citizens upon request.

GOVERNMENT OF NEWFOUNDLAND AND LABRADOR/DEPARTMENT OF LABOUR/OCCUPATIONAL HEALTH AND SAFETY DIVISION/MEDICAL AND HYGIENE SERVICES
Beothuck Building
Crosbie Pl.
St. John's, NF A1C 5T7
(709) 576-2645
Evertt Price - Manager
Activities:
Office handles questions concerning association between radon and uranium mining and uranium mill tailings on a case-by-case basis.

MANITOBA DEPARTMENT OF ENERGY AND MINES/INFORMATION CENTER/HOME CHECK-UP PROGRAM
330 Graham Ave.
Suite 555
Winnipeg, MB R3C 4E3
(204) 945-4154, (800) 282-8069
Ken Klassen - Program Manager
Activities:
Provides information booklet on radon-resistant construction techniques developed by the Canadian Home Builders Association.

MINISTERE DE L'ENVIRONMENT DU QUEBEC/DIVISION DE RADIOPROTECTION
5199 Est, rue Sherbrooke
Suite 3860
Montreal, PQ H1T 3X9
(514) 873-1978
J. M. Legare - Chief
Activities:
Provides citizens with radon research publications from APCA which include studies in Quebec. Provides information by phone and refers citizens to detection companies and analysis labs.

MINISTERE DE LA JUSTICE DU QUEBEC/SERVICE DU COURTAGE IMMOBILIER DU QUEBEC
220, Grande Allee Est
Suite 910
Quebec City, PQ G1R 2J1
(418) 643-4597
Real Martel - Superintendent
Activities:
Lead agency for real estate regulatory and enforcement issues.

NEW BRUNSWICK DEPARTMENT OF HEALTH AND COMMUNITY SERVICES/RADIATION PROTECTION SERVICES
P.O. Box 5100
Frederickton, NB E3B 5G8
(506) 453-2067
Keith Davies - Radiation Protection Officer
Activities:
Provides general radon information upon request.

NEW BRUNSWICK DEPARTMENT OF JUSTICE/CONSUMER AFFAIRS BRANCH/REAL ESTATE COUNCIL
P.O. Box 785
Fredericton, NB E3B 5B4
(506) 455-9733
Robert H. Kerr - Licensing Supervisor
Activities:
Lead agency for real estate regulatory and enforcement issues.

NOVA SCOTIA DEPARTMENT OF CONSUMER AFFAIRS/CONSUMER SERVICES DIVISION
P.O. Box 998
Halifax, NS B3J 2X3
(902) 424-4690
Robert G. Martin - Superintendent
Activities:
Lead agency for real estate regulatory and enforcement issues.

NOVA SCOTIA DEPARTMENT OF HEALTH
P.O. Box 488
Halifax, NS B3H 2R8
(902) 424-4077
Ted Dalgleish - Contact
Activities:
Department has purchased detection equipment to monitor for radon and is surveying certain areas to determine the extent of the problem in the province. Provides radon information packet upon request.

ONTARIO MINISTRY OF CONSUMER AND COMMERCIAL RELATIONS/REAL ESTATE AND BUSINESS BROKERS ACT
Toronto, ON M7A 2H6
(416) 965-0406
Cora M. de la Cruz - Registrar
Activities:
Lead agency for real estate regulatory and enforcement issues.

ONTARIO MINISTRY OF HEALTH/DIVISION OF COMMUNITY HEALTH/PUBLIC HEALTH BRANCH/DISEASE CONTROL AND EPIDEMIOLOGICAL SERVICES
15 Orelea Blvd.
Toronto, ON M4H 1A9
(416) 963-2238
Les Smith - Provincial Contact
Activities:
Provides scientific information on radon as related to public health.

ONTARIO MINISTRY OF LABOUR/RADIATION PROTECTION SERVICES
81 Resources Rd.
Weston, ON M9P 3T1
(416) 235-5922
John Tai-Pow - Director
Activities:
- List of testing companies and their fees for services are provided upon request.
- Department provides WLM detection devices for occupational health testing.
- Local medical offices of health can request that the Ministry of Labour perform radon testing.

PRINCE EDWARD ISLAND/QUEEN ELIZABETH HOSPITAL
Charlottetown, PE C1A 8T5
(902) 566-6277
Gayle Murphy - Radiology Manager
Activities:
Provides radon information to provincial citizens upon request.

SASKATCHEWAN CONSUMER AND COMMERCIAL AFFAIRS
1871 Smith St.
Regina, SK S4P 3V7
(306) 787-2958
O.A. MacGillivray - Superintendent of Insurance
Activities:
Lead agency for real estate regulatory and enforcement issues.

SASKATCHEWAN DEPARTMENT OF HUMAN RESOURCES, LABOUR AND EMPLOYMENT
Saskatchewan Place
1870 Albert St.
Regina, SK S4P 3V4
(306) 787-4486
Dennis Brown - Head
Stephen Webster - Contact
Activities:
Provides radon measurement information upon request and answers inquiries in writing. Is proposing public building study to analyze radon problems in the province.

Canada Organizations And Associations

CANADIAN HOMEBUILDERS ASSOCIATION (CHA)
200 Elgin St.
Suite 502
Ottawa, ON K2P 1L5
(613) 230-3060
Robert Sloat - Director Of Technical Services
Activities:
Research on housing construction techniques to prevent intrusion of radon. Information booklet available.

CANADIAN INSTITUTE FOR RADIATION SAFETY/ELLIOTT LAKE LABORATORY
7 Timmins Rd.

Suite 7-15
Elliott Lake, ON P5A 2R7
(705) 848-3687
Ernest Becker - Senior Scientist, Laboratory Manager
Activities:
• Non-profit agency operates as a registered charity. Has enacted radon-daughter assessments in uranium mines (550 miners).
• Provides mitigation and assessment training and measures radon levels in private homes.
• Presently developing detection devices.
• Environmental monitoring and applied research.

CANADIAN LUNG ASSOCIATION (CLA)
75 Albert St.
Suite 908
Ottawa, ON K1P 5E7
(613) 237-1208
Robert Mech - Executive Director
Activities:
Canadian affiliate of ALA.

MINING ASSOCIATION OF CANADA (MAC)
350 Sparks St.
Suite 705
Ottawa, ON K1R 7S8
(613) 233-9391
C. George Miller - President
Activities:
Confederation of provincial mining associations protects interest of mining companies throughout Canada. Companies include uranium mines subject to federal and provincial regulations for radon exposure.

NO POSTAGE
NECESSARY
IF MAILED
IN THE
UNITED STATES

BUSINESS REPLY MAIL

FIRST-CLASS MAIL PERMIT NO. 8873 ALEXANDRIA, VIRGINIA

POSTAGE WILL BE PAID BY ADDRESSEE

RADON PRESS, INC.

P.O. BOX 25551
ALEXANDRIA, VA. 22313

NO POSTAGE
NECESSARY
IF MAILED
IN THE
UNITED STATES

BUSINESS REPLY MAIL

FIRST-CLASS MAIL PERMIT NO. 8873 ALEXANDRIA, VIRGINIA

POSTAGE WILL BE PAID BY ADDRESSEE

RADON PRESS, INC.

P.O. BOX 25551
ALEXANDRIA, VA. 22313

NO POSTAGE
NECESSARY
IF MAILED
IN THE
UNITED STATES

BUSINESS REPLY MAIL

FIRST-CLASS MAIL PERMIT NO. 8873 ALEXANDRIA, VIRGINIA

POSTAGE WILL BE PAID BY ADDRESSEE

RADON PRESS, INC.

P.O. BOX 25551
ALEXANDRIA, VA. 22313

International Associations

This section lists associations, international in scope, that are involved in radon-related activities with respect to radiation measurement, radiation protection, public health, and disseminating information.

COMMISSION OF THE EUROPEAN COMMUNITIES (CEC)/RADIATION PROTECTION PROGRAM
rue de la Loi, 200
B-1049 Brussels, Belgium
(011) 322-235-5211, (011) 322-235-1111
Jaak Sinnaeve - Director
Activities:
Government-appointed representatives from 12 nations that make up the European Economic Community (EEC). Ensures that rules and regulations of the EEC are correctly applied. Promotes scientific research and acts as informational resource for Europeans with questions concerning radon-related issues. Drafts proposals for programs of research, policy development and dissemination of information on radon. Official languages are Danish, Dutch, English, French, German, Greek, and Italian.

ENVIRONMENT LIAISON CENTRE (ELC)
P.O. Box 72461
Nairobi, Kenya
(011) 254-224-770
David Bull - Executive Officer
Activities:
Nongovernmental environment and development organizations. Purpose is to develop ties among nongovernmental groups working on environmental and related issues. Provides information, holds conferences and seminars and maintains a library. Official languages are English, French and Spanish.

EUROPEAN ENVIRONMENTAL BUREAU (EEB)
20, rue du Luxembourg
B-1040 Brussels, Belgium
No Phone Available
Ernst R. Klatt - Secretary-General
Activities:
Environmental organizations within the EEC. Provides information on environmental concerns and act as liaison between institutions and nongovernmental environmental organizations. Official languages are English and French.

EUROPEAN INSTITUTE FOR WATER (EIW)
2, rue Boussingault
F-67000 Strasbourg, France
(011) 33-88-611631
Jean J. Fried - President
Activities:
Evaluates adequacy of directives of CEC. Evaluates research needs that could benefit from collaboration with other countries. English, French, and Italian are official languages.

EUROPEAN UNION OF GEOSCIENCES (EUG)/EUROPEAN SCIENCE FOUNDATION
1, quai Lezay-Marnesia
F-67000 Strasbourg, France
(011) 33-88-353036
E.R. Oxburgh - President
Activities:
Encourages cooperation among scientists involved in geology, geophysics, and geochemistry research. Sponsors conferences and symposia. English is official language.

INSTITUTE FOR EUROPEAN ENVIRONMENTAL POLICY (IEEP)
Aloys-Schulte-Strasse 6
D-5300 Bonn 1, West Germany
(011) 49-228-213810
Ernst von Weizsacker - Director
Activities:
Environmental improvement through contact with government officials, researchers, and representatives of industry and agriculture. Works to strengthen impact of environmental policies. Maintains 300-volume library. English, French, and German are official languages.

INTERNATIONAL ASSOCIATION OF HYDROGEOLOGISTS (IAH)
Postbus 9090
NL-6800 Arnhem, Netherlands
(011) 3185-340396
Erik Romijn - Secretary General
Activities:
Disseminates geological information through meetings and publications. Develops hydrogeological maps and conducts research. English and French are official languages.

INTERNATIONAL ASSOCIATION OF HYDROLOGICAL SCIENCES (IAHS)/INSTITUTE OF HYDROLOGY
Maclean Building
Crowmarsh Gifford
Wallingford
Oxfordshire, UK 0X10 8BB
(011) 44-491-8800
John C. Rodda - Secretary General
Activities:
Promotes study of hydrology. Coordinates hydrological research that requires international cooperation. Holds symposia (6-8 yearly). English and French are official languages.

INTERNATIONAL ASSOCIATION OF RADIATION RESEARCH (IARR)/RADIOBIOLOGICAL INSTITUTE TNO
Lange Kleiweg 151
NL-2288 GJ Rijswijk, Netherlands
(011) 3115-136940
J.J. Broerse - Secretary-Treasurer
Activities:
Societies that promote radiation research. Fosters scientific cooperation and encourages scientifc conferences and symposia. English is the official language.

INTERNATIONAL COMMISSION ON RADIATION UNITS AND MEASUREMENTS (ICRU)
7910 Woodmont Ave.
Suite 1016
Bethesda, MD 20814
(301) 657-2652
Harold O. Wyckoff - Chairman
Activities:
Develops internationally acceptable recommendations regarding quantities and units of radiation and radioactivity, suitable measurement procedures and physical data needed in applying these procedures. Formulates recommendations on radiation quantities, units and measurement for radiation protection. Members: USA, Canada, Japan, Austria, Belgium, Finland, France, West Germany, Italy, Netherlands, Sweden, Switzerland, United Kingdom.

INTERNATIONAL COMMISSION ON RADIOLOGICAL PROTECTION (ICRP)
Clifton Ave.
Sutton (Surrey), UK SM2 5PU
(011) 44-642-4680
F.D. Sowby - Scientific Secretary
Activities:
Ensures progress in field of radiation protection and publishes recommendations on radiation safety standards. Recommendations are kept under review to cover number and scope of potential radiation hazards and to take into account new information on effects of ionizing radiation. Members: South Africa, Argentina, Canada, USA, China, India, Israel, Japan, Czechoslovakia, France, West Germany, Italy, Netherlands, Poland, Sweden, USSR, United Kingdom.

INTERNATIONAL GEOGRAPHICAL UNION (IGU)
University of Alberta
Department of Geology
Edmonton, AB T6G 2H4
(403) 432-3329
L.A. Kosinski - Secretary General and Treasurer
Activities:
Promotes study of geographical problems; initiates and coordinates research that requires international cooperation. Facilitates communication of geographical data among members. English and French are official languages.

INTERNATIONAL GEOLOGICAL CORRELATION PROGRAM (IGCP)
C/O UNESCO
Maison de l' UNESCO
7 Place de Fontenoy
F-75700 Paris, France
(011) 331-45684123
Endre Dudich - Director
Activities:
Joint enterprise of UNESCO and IUGS to facilitate information exchange. Encourages cooperation among governments, institutions, and scientists in earth science research. Geological processes and their environmental implication is important area of research. English and French are official languages.

INTERNATIONAL RADIATION PROTECTION ASSOCIATION (IRPA)
BP N Q 33
F-92260 Fontenay-aux-Roses, France
(011) 331-746547139
Gilbert Bresson - Executive Officer
Activities:
Provides channels for international contacts among scientists engaged in radiation protection. Emphasizes protection of man in his environment from hazards caused by ionizing radiation. General assembly meets every 3-4 years to elect executive counsel (meets 1-2 times yearly). Members: South Africa, Argentina, Canada, Mexico, USA, India, Israel, Japan, South Korea, Philippines, Australia, Austria, Belgium, Czechoslovakia, France, East Germany, West Germany, Hungary, Italy, Luxembourg, Netherlands, Poland, Spain, Switzerland, USSR, United Kingdom, Yugoslavia, and regional associate grouping of five Nordic countries (Nordic Radiation Protection Institute).

INTERNATIONAL UNION AGAINST TUBERCULOSIS AND LUNG DISEASE (IUATLD)
68, Boulevard Saint-Michel
75006 Paris, France
No Phone Available
Annik Rouillon - Executive Director
Activities:
International agency associated with the ALA.

INTERNATIONAL UNION OF GEOLOGICAL SCIENCES (IUGS)
Postboks 3006
N-7002 Tronheim, Norway
Activities:
Encourages study of problems pertaining to geology and related disciplines. Members include representatives of national committees for geological sciences.

SCIENTIFIC COMMITTEE ON PROBLEMS OF THE ENVIRONMENT (SCOPE)
51 Boulevard De Montmorency
F-75016 Paris, France
(011) 331-45250498
J.W.M. LaRiviere - President
Activities:
Association of groups working to analyze effects of environmental changes on people and their health and welfare. Serves as consultant to a variety of agencies (governmental and nongovernmental) on environmental concerns. English and French are official languages.

UNITED NATIONS SCIENTIFIC COMMITTEE ON THE EFFECTS OF ATOMIC RADIATION (UNSCEAR)
P.O. Box 500
Vienna, Austria A1440
(011) 43-222-26310
Burton Bennett - Executive Secretary
Activities:
Publishing summary of international radon studies with general assessment of status of problem. Comprehensive report released November, 1988. (Last report 1982)

WORLD FEDERATION OF PUBLIC HEALTH ASSOCIATIONS (WFPHA)
C/O American Public Health Association
1015 15th St., NW
Washington, DC 20005
(202) 789-5691
William H. McBeath - Contact
Activities:
Affiliated with World Health Organization. National public health associations dedicated to strengthening public health professions and improving community health.

WORLD HEALTH ORGANIZATION (WHO)/DIVISION OF ENVIRONMENTAL HEALTH/ENVIRONMENTAL HAZARDS AND FOOD PROTECTION UNIT
1211 Geneva 27, Switzerland
(011) 4122-34-6061
Wilfred Krisel - Director
Activities:
Responsible for the identification, assessment of control and surveillance of environmental health hazards.

WORLD HEALTH ORGANIZATION (WHO)/WASHINGTON OFFICE ENVIRONMENTAL HEALTH PROGRAM
525 23rd St., NW
Washington, DC 20037
(202) 861-3311
Guillermo Davila - Coordinator
Activities:
Provides information on environmental health concerns and directs inquirer to proper WHO agency or affiliate to answer questions on issues such as radon exposure.

WORLD INFORMATION SERVICE ON ENERGY (WISE)/WASHINGTON OFFICE
P.O. Box 53373
Washington, DC 20009
(202) 745-7467
Elaine Douglas - National Coordinator
Activities:
Citizens action group with offices located in 12 countries. Conducts research on uranium mining around the world.

Building and Construction

This section lists associations, organizations, and companies involved in radon-related activities as they pertain to the building and construction industry. Among the areas of interest listed here are construction design and standards, materials, energy, heating, and ventilation.

ADHESIVE AND SEALANT COUNCIL (ASC)
1627 K St., NW
Suite 1000
Arlington, VA 22209
(202) 452-1500
Kerry Lake - Deputy Executive Vice President
Activities:
Members are manufacturers and suppliers of adhesives and sealants.

ADIRONDACK ALTERNATE ENERGY
County Rt. 4
Edinburg, NY 12134
(518) 863-4338
Bruce Brownell - President
Tom Doherty - Engineer
Activities:
Designs and engineers post and beam pre-cut structures, fully resistant to radon infiltration.

AIR CONDITIONING CONTRACTORS OF AMERICA (ACCA)
1513 16th St., NW
Washington, DC 20036
(202) 296-7610
Ann Warner - Director, Government Relations
Activities:
Contractors involved in installation/service of heating and cooling systems. Represents interests and monitors legislation, such as indoor air quality legislation, that affects the industry. Maintains small library for members.

AIR MOVEMENT AND CONTROL ASSOCIATION (AMCA)/HOME VENTILATING INSTITUTE(HVI)
30 West University Dr.
Arlington Heights, IL 60004
(312) 394-0150
Dale Rammien - Director
Activities:
Manufacturers of centrifugal fans, power roof ventilators, propeller fans, air curtains, louvres, dampers and shutters. Administers Certified Ratings Program to ensure products bearing AMCA seal meet established regulations and standards.

AMERICAN CONCRETE INSTITUTE (ACI)
P.O. Box 19150
22400 West Seven Mile Rd.
Detroit, MI 48219
(313) 532-2600
Arthur Mullkoff - Staff Engineer
Activities:
Provides radon information packet to members upon request. Promotes use and improvement of concrete structures.

AMERICAN COUNCIL FOR AN ENERGY EFFICIENT ECONOMY (ACEEE)
1001 Connecticut Ave., NW
Suite 535
Washington, DC 20036
(202) 429-8873
Howard Geller - Associate Director
Activities:
Private, non-profit organization gathers, evaluates, and disseminates information to stimulate energy efficiency. Provides information on indoor air quality issues associated with energy efficient dwellings.

AMERICAN FAN ASSOCIATION (AFA)
P.O. Box 888
East Lansing, MI 48826
No Phone Available
D. Gravatt Huber - President
Activities:
Members are manufacturers of fans. Associate members include manufacturers, distributors and suppliers of component parts. Functions as legislative liaison for fan industry. Supports Air Movement Institute.

AMERICAN INSTITUTE OF ARCHITECTS (AIA)
1735 New York Ave., NW
Washington, DC 20006
(202) 626-7452
Bruce Kriviskey - Director
Activities:
Professional society of architects. Works to improve capability of nation's architects and be of service for safety and welfare of public. Actively involved in questions concerning indoor air quality and associated risk of radon exposure. Maintains library open to public. Several state AIA affiliates have been invited to participate in development of state-level action programs to resolve radon problems.

AMERICAN SOCIETY OF HEATING, REFRIGERATING AND AIR CONDITIONING ENGINEERS (ASHRAE)
1791 Tullie Circle Rd., NE
Atlanta, GA 30329
(404) 636-8400
Lou Flagg - President-Elect
Jim Norman - Technical Services
Activities:
Provides radon information to members based on special committee findings. Has actively pursued questions on indoor air quality and ventilation systems. Recommends radon remedial action for levels above 2pCi/l.

ASHRAE/ENVIRONMENTAL HEALTH COMMITTEE/INDOOR AIR QUALITY SUBCOMMITTEE
Atlanta, GA 30329
(404) 636-8400
Stanley A. Mumma - Penn State University
Preston McNall - Consultant
Michael J. Hodgson - U. of Pittsburgh/Occupational Medicine Center
Activities:
Outlines association's interests in and responsibilities for indoor air quality.

ASHRAE/ENVIRONMENTAL HEALTH COMMITTEE/INDOOR AIR QUALITY SUBCOMMITTEE
1791 Tullie Cr., NE
Atlanta, GA 30329
(404) 636-8400
Susan Rose - IAQ Chairman/DOE
Jan A.J. Stolwijk - Vice Chairman/Yale University
Ralph Goldman - Multitech Corporation
Nancy Balter - Georgetown University
Activities:
Outlines association's interests in and responsibilities for indoor air quality.

ASSOCIATION OF MAJOR CITY BUILDING OFFICIALS (AMCBO)
481 Carlisle
Herndon, VA 22070
(703) 437-0100
Lee Jensen - Chairman
Activities:
Provides information on building and construction issues in major U.S. metropolitan areas.

BALTIMORE GAS & ELECTRIC
P.O. Box 1475
Baltimore, MD 21203
(301) 281-3781
Lawrence Bartal - Environmental Health Department
Activities:
Provides local citizens with information on effect of home weatherization on indoor air quality (including radon).

BUILDING OFFICIALS AND CODE ADMINISTRATORS INTERNATIONAL (BOCA)
4051 West Flossmore Rd.
Country Club Hills, IL 60477
(312) 799-2300
Clarence R. Bechtel - Executive Director
Activities:
Government officials or agencies responsible for administering or formulating zoning, building and housing regulations. Promotes minimum code standards and works to ensure their timeliness.

BUILDING OWNERS AND MANAGERS ASSOCIATION INTERNATIONAL (BOMA)
1250 I St., NW
Suite 200
Washington, DC 20005
(202) 289-7000
Douglas Greenaway - Government Affairs
Activities:
Managers and owners of office buildings and developers of commercial properties. Represents interests before Congress on issues such as radon legislation affecting existing structures. Primarily North American membership.

CLANCY & COMPANY
405 West Rockrimmon
Colorado Springs, CO 80919
(303) 599-4660
Al Clancy - President

Activities:
Housing construction company participated in EPA volunteer program for housing construction methods to reduce possibility of radon intrusion.

CLASSIC CONSTRUCTION
7001 Orchard Lake Rd.
Suite 130
West Bloomfield, MI 48322
(313) 737-8800
Donald Marhofer - Project Manager
Activities:
Building company worked with EPA on development of construction specifications to reduce indoor radon. Uses guidelines on EPA standards in condominium development.

CONSTRUCTION SPECIFICATION INSTITUTE (CSI)
601 Madison St.
Alexandria, VA 22314
(703) 684-0300
Penny Goodman - Information Officer
Activities:
Individuals involved with documents and specifications used in connection with design, construction, maintenance and equipment for construction projects.

COUNCIL OF AMERICAN BUILDING OFFICIALS (CABO)
5203 Leesburg Pike
Falls Church, VA 22041
(703) 931-4533
Richard P. Kuchnicki - President
Bob Spangler - Information Director
Activities:
National organization oversees the three regional model building codes. Maintains one- and two-family dwelling code, model energy code, and manufactured home and safety construction standards. Oversees building certification program. Has responsibility to provide public forum on issues if changes are proposed to develop standard radon policy.

EDISON ELECTRIC INSTITUTE (EEI)
1111 19th St., NW
Washington, DC 20036
(202) 778-6400
William McCollam - President
Activities:
Investor-owned electric power companies and electric utilities holding companies. Concerns include indoor air quality in energy efficient homes.

ELECTRICAL POWER RESEARCH INSTITUTE (EPRI)/ENERGY MANAGEMENT AND UTILIZATION DIVISION/RESIDENTIAL AND COMMERCIAL PROGRAMS
3412 Hillview Ave.
Palo Alto, CA 94304
(415) 855-2000
Arvo Lannus - Project Manager
Activities:
Assists electric utilities in helping homeowners, builders, and new home buyers understand broad range of issues related to indoor air quality. Involved in research on indoor air quality as it affects the energy industry.

ELECTRICAL POWER RESEARCH INSTITUTE (EPRI)/LAWRENCE BERKELEY LABORATORY(LBL)/APPLIED SCIENCE DIVISION
University Of California
One Cyclotran Rd.
Berkeley, CA 94720
(415) 486-4000
David Grimsrud - Building Ventilation and Indoor Air
Richard Diamond - Energy Performance of Buildings
Activities:
LBL contracts with EPRI to research topics on building ventilation and indoor air quality including radon. Produced two major indoor air quality reports.

EXPANDED SHALE, CLAY AND SLATE INSTITUTE (ESCSI)
6218 Montrose Rd.
Rockville, MD 20852
(301) 231-9497
Harry Robinson - Managing Director
Activities:
Producers of shale, clay and slate aggregates used in lightweight structural concrete and concrete masonry. Provides radon information as it affects those industries.

FLORIDA POWER AND LIGHT COMPANY/ENVIRONMENTAL AFFAIRS DEPARTMENT
P.O. Box 14000
Juno Beach, FL 33408
(305) 694-3651
Martin Smith - Information
Activities:
Public information programs for radon mitigation in energy efficient homes.

GARNET HOMES, INC.
3900 Germantown Rd.
Suite 400
Fairfax, VA 22030
(703) 591-4663
Regis Skeehan - President
Activities:
Construction company installs radon abatement package in new homes. Ongoing research with NAHB to ensure sites remain below 1.0 pCi/l.

GAS RESEARCH INSTITUTE (GRI)/INDOOR AIR QUALITY RESEARCH PROGRAM
8600 West Bryn Mawr Ave.
Chicago, IL 60631
(312) 399-8100
Irwin Billick - Director
Activities:
Involved in research project with NAHB analyzing effects of home combustion sources on radon.

MELODY HOMES, INC.
11031 Sheridan Sq.
Westminster, CO 80020
(303) 466-1831
Jim Henry - Vice President
Activities:
Home builder worked with EPA on development of housing construction techniques to reduce likelihood of radon intrusion.

NATIONAL ACADEMY OF SCIENCES (NAS)/NATIONAL RESEARCH COUNCIL (NRC)/FEDERAL CONSTRUCTION COUNCIL
2101 Constitution Ave., NW
Washington, DC 20418
(202) 334-3378
Henry Borger - Executive Secretary
Activities:
Private non-profit organization researches topics such as radon in buildings and structures on an ad hoc basis.

NATIONAL APARTMENT ASSOCIATION (NAA)
1101 14th St., NW
Suite 804
Washington, DC 20005
(202) 842-4050
Colleen M. Fisher - Vice President, Gov. Affairs
Julie Desmond - Assistant Director, Gov. Affairs
Activities:
Provides information to membership on proposed or enacted radon legislation that affects the industry. Represents industry's views before federal government.

NATIONAL ASSOCIATION OF HOME BUILDERS OF THE UNITED STATES (NAHB)
15th and M Sts., NW
Washington, DC 20005
(202) 822-0200
Robert Schenck - Government Affairs Department
Danny George - Technology And Codes Department
Activities:
Supports Homeowners Warranty Corporation and NAHB Research Foundation. Provides radon information packet to members upon request.

NATIONAL ASSOCIATION OF HOME BUILDERS (NAHB)/NATIONAL RESEARCH CENTER
400 Prince Georges Center Blvd.
Upper Marlboro, MD 20772
(301) 249-2000
John Spears - Program Manager
Activities:
Research arm of NAHB. Conducts training, provides technical assistance and conducts field research. Involved in research projects focusing on radon as it affects building industry.

NATIONAL ASSOCIATION OF THE REMODELING INDUSTRY (NARI)
1901 North Moore St.
Suite 808
Arlington, VA 22209
(703) 276-7600
Patti Knoff - Director, Government Affairs
Activities:
Updates and disseminates information on federal legislative actions concerning remodeling and restructuring buildings.

NATIONAL CONCRETE MASONRY ASSOCIATION (NCMA)
P.O. Box 781
2302 Horse Pen Rd.
Herndon, VA 22070
(703) 435-4900
Bob Van Laningham - Engineer
Activities:
Manufacturers of concrete blocks and suppliers. Radon information pertinent to those industries available.

NATIONAL CONFERENCE OF STATES ON BUILDING CODES AND STANDARDS (NCSBCS)
481 Carlisle Dr.
Herndon, VA 22070
(703) 437-0100
Robert C. Wible - Executive Director
Marla McIntyre - Communications Director
Activities:
Problem solving forum for resolution of issues on uniform state building codes and regulations. Provides information on federal radon legislation to state and regional officials.

NATIONAL CONFERENCE ON WEIGHTS AND MEASURES (NCWM)
P.O. Box 3137
Gaithersburg, MD 20870
(301) 975-4004
Ann Turner - Conference Coordinator
Activities:
Sponsored by Commerce Department's National Institute of Standards and Technology. Members are weights and measures enforcement officials from state and local governments; associate members from industry.

NATIONAL INSTITUTE OF BUILDING SCIENCES (NIBS)
1015 15th St., NW
Suite 700
Washington, DC 20005
(202) 347-5710
David Harris - Acting President
Leslie Levine - Manager of Communications
Activities:
Non-profit organization brings together government, labor, consumer, and private interests to focus on construction of safe structures in U.S. Actively involved in radon issues that affect the building industry.

NATIONAL INSTITUTE OF BUILDING SCIENCES (NIBS)/INDOOR AIR QUALITY COMMITTEE
C/O Armstrong World Industries
333 Liberty St.
Lancaster, PA 17603
(717) 397-0611
Nat E. Hager, Jr. - Senior Research Associate
Activities:
Develops strategies for indoor air quality research. Directs research on radon as it affects the building industry.

NATIONAL READY MIXED CONCRETE ASSOCIATION (NRMCA)
900 Spring St.
Silver Spring, MD 20910
(301) 587-1400
Steve Lane - Staff Engineer
Activities:
Provides radon information packet to members upon request.

NATIONAL SOCIETY OF PROFESSIONAL ENGINEERS (NSPE)
1420 King St.
Alexandria, VA 22314
(703) 684-2800
Donald G. Weinert - Executive Director
Jean Robertson - Information Director
Activities:
Monitors legislation that affects the engineering profession including proposed guidelines concerning radon. Addresses social, ethical, economic and professional aspects of engineering.

NATIONAL STONE ASSOCIATION (NSA)
1415 Elliot Pl., NW
Washington, DC 20007
(202) 342-1100
Bruce Remick - Director
Activities:
Quarry owners and operators, equipment manufacturers and service organizations associated with crushed stone industry. Provides information on radon to members upon request.

NORTHEAST UTILITIES
P.O. Box 270
Hartford, CT 06141-0270
(203) 665-5000
Russell K. Johnson - Market Research Engineer
Activities:
Research involves home weatherization and implications for indoor air quality. Extensive analysis of radon. Informational guide available.

NORTHWEST POWER PLANNING COUNCIL (NPPC)/POWER PLANNING DIVISION
851 SW Sixth Ave.
Suite 1100
Portland, OR 97204-1337
(800) 222-3355
Margaret Gardner - Conservation Analyst
Activities:
Interstate compact agency (representing Washington, Oregon, Idaho and Montana) provides information to its customers on questions concerning radon related issues and housing ventilation.

PORTLAND CEMENT ASSOCIATION (PCA)
5420 Old Orchard Rd.
Skokie, IL 60077-4321
(312) 966-6200
George Barney - Divisional Director
Activities:
Provides members with radon information relevant to industry.

RYAN HOMES/PRODUCTS DIVISION
100 Ryan Court
Pittsburgh, PA 15205
(412) 276-8000
Dick Tracey - Systems Design Standard Manager
Activities:
Homebuilder has worked with EPA for over two years to incorporate techniques to reduce entry of radon into new and existing homes.

SEALANT AND WATERPROOFERS INSTITUTE (SWI)
3101 Broadway
Suite 300
Kansas City, MO 6411
(816) 561-8230
Kenneth A. Bowman - Executive Vice President
Activities:
Sealant contractors,suppliers and and related personnel. Promotes development of industry wide standards and specifications. Works to improve conditions affecting sealant and waterproofing industries.

SHEET METAL AND AIR CONDITIONING CONTRACTORS NATIONAL ASSOCIATION (SMACCNA)
P.O. Box 70
Merrifield, VA 22116
(703) 790-9890
John Sroka - Executive Vice President
Rosalynn Raymond - Director Of Public Relations
Activities:
Committee on Energy Management provides information on indoor air quality including ventilation systems for radon remediation.

SHINN COMPANY
145 Wolf Creek Trail
Broomfield, CO 80020
(303) 465-1331
Chuck Shinn - President
Activities:
Construction company has worked with EPA to develop a basement design to decrease likelihood of radon infiltration.

SOUTH CAROLINA ELECTRIC AND GAS
P.O. Box 88
Jenkinsville, SC 29065
(803) 345-1915 x5003
Timothy D. Riley - Public Information
Activities:
Provides information to state residents on radon-housing related issues.

SOUTHERN BUILDING CODE CONGRESS INTERNATIONAL (SBCCI)
900 Montclair Rd.
Birmingham, AL 35213

(205) 591-1853
William J. Tangye - Executive Vice President
Activities:
Members work to develop and maintain standard building, gas, fire prevention, plumbing, mechanical, and housing codes. Stresses standardization for codes and their applications and enforcement.

SUMP AND SEWAGE PUMP MANUFACTURERS ASSOCIATION (SSPMA)
560 West Washington St.
Suite 301
Chicago, IL 60606
(312) 332-4146
Pam Francis - Information Officer
Activities:
Manufacturers of residential sump pumps (cellar drainers) with application to radon remediation procedures. Develops and promulgates quality standards and has a certification and labelling program for products meeting those standards.

THERMAL INSULATION MANUFACTURERS ASSOCIATION (TIMA)
8341 Sangre De Cristo Rd.
Littleton, CO 80127
(303) 933-9774
William B. Reitze - Executive Director
Activities:
Conducts research, experiments and testing. Involved in issues concerning energy conservation and air handling systems and implications for indoor air quality.

14

Environment

This section lists agencies, associations, and organizations concerned with environmental hazards resulting from radon. Among the areas of interest listed here are air and water pollution, geology, radiation control, testing, and measurement.

AIR POLLUTION CONTROL ASSOCIATION (APCA)
P.O. Box 2861
Pittsburgh, PA 15230
(412) 232-3444
Paul Kueser - Radon Specialist
Activities:
Researchers, educators, manufacturers, government officials, industrial officials, and meterologists seeking long-term environmental solutions to air pollution/air quality problems. Takes active role in analysis of indoor air concerns. Report of proceedings from EPA/APCA co-sponsored symposium dealing with indoor air toxicants available for fee.

AMERICAN ACADEMY OF ENVIRONMENTAL ENGINEERS (AAEE)
132 Holiday Ct.
Annapolis, MD 21401
(301) 266-3311
William C. Anderson - Executive Director
Activities:
State-licensed environmental engineers who have passed exams given by Academy in engineering specialties such as air pollutio water supply, and radiation protection.

AMERICAN ACADEMY OF ENVIRONMENTAL MEDICINE (AAEM)
P.O. Box 16106
Denver, CO 80216
(303) 622-9755
Judith A. Howard - Executive Administrator
Activities:
Concerned with adverse reactions in relationship between man an environment. Calls for environmental manipulation to decrease likelihood of radiation exposure.

AMERICAN ASSOCIATION FOR THE ADVANCEMENT OF SCIENCE (AAAS)
1333 H St., NW
Washington , DC 20005
(202) 326-6400
Alvin Trivelpiece - Executive Director
Activities:
Largest general scientific organization. Increasing effectiveness o science to further promote human welfare is major concern. Publication "Science" has featured information concerning radon.

AMERICAN CHEMICAL SOCIETY (ACS)
1155 16th St., NW
Washington, DC 20036

(202) 872-6062
John Crum - Executive Director
Roger Minear - Chairman, Environmental Chemistry
Richard W. Hoff - Chairman, Nuclear Technology
Activities:
Maintains complete collection of ACS publications including reports on radon, radon progeny, radon occurrance and health effects from radon. Provides information through library; makes interlibrary loans and permits onsite inspections.

AMERICAN GEOLOGICAL INSTITUTE (AGI)
4220 King St.
Alexandria, VA 22310
(703) 379-2480
Marvin E. Kaufman - Executive Director
Activities:
Federation of earth sciences societies. Seeks to increase public awareness of geological science. Provides information on geology of regions in U.S.

AMERICAN GEOPHYSICAL UNION (AGU)
2000 Florida Ave.
Washington, DC 20009
(202) 462-6903
Fred Spilhaus - Executive Director
Activities:
Promotes research concerned with physics and chemistry of earth Coordinates research projects that depend upon national and international cooperation.

AMERICAN NATIONAL STANDARDS INSTITUTE (ANSI)
1430 Broadway
New York, NY 10018
(212) 354-3300
Dorothy Hogan - Vice President/Communications
Activities:
Approves measurement standards (proposed radon standards) submitted by associations, agencies, and boards. Participated in early radon exposure studies.

AMERICAN NUCLEAR SOCIETY (ANS)
555 North Kensington
La Grange Park, IL 60525
(312) 352-6611
Octave DuTemple - Executive Director
Kim Flesher - Public Communication Assistant
Activities:
Disseminates information, promotes research, conducts meetings devoted to scientific and technical research; works with government agencies and educational institutions dealing with nuclear issues. Maintains library of more than 3000 volumes. Issued major technical information brief in late 1988 entitled "Radon in Homes."

American Nuclear Society Local Sections

AMERICAN NUCLEAR SOCIETY/ALABAMA-BIRMINGHAM SECTION
C/O Southern Company Services
P.O. Box 2625
Inverness 40
Room 330
Birmingham, AL 35202
(205) 868-5832
J.R. Crone - ANS Local Section Chair

AMERICAN NUCLEAR SOCIETY/ALABAMA-NORTH ALABAMA SECTION
C/O Tennessee Valley Authority
Browns Ferry Nuclear Plant/Administrative Bldg.
P.O. Box 2000
Decatur, AL 35602
(205) 259-5677
Michael J. May - ANS Local Section Chair

AMERICAN NUCLEAR SOCIETY/ARIZONA-ARIZONA SECTION
C/O Combustion Engineering
10010 North 25th Dr.
Corporate Center, Suite 310
Phoenix, AZ 85021
(602) 944-4664
Jerry Dilk - ANS Local Section Chair

AMERICAN NUCLEAR SOCIETY/ARKANSAS-ARKANSAS SECTION
400 West 7th St.
Suite 201
Little Rock, AR 72201
(501) 375-2060
Garry Young - ANS Local Section Chair

AMERICAN NUCLEAR SOCIETY/CALIFORNIA-LOS ANGELES SECTION
C/O ICN Biomedicals, Inc.
2727 Campus Dr.
Irvine, CA 92715
(714) 833-2500
Albert Baietti - ANS Local Section Chair

AMERICAN NUCLEAR SOCIETY/CALIFORNIA-NORTHERN CALIFORNIA SECTION
C/O James Engineering
6329 Fairmount Ave.
El Cerrito, CA 94530
(415) 525-7350
J.Z. James - ANS Local Section Chair

AMERICAN NUCLEAR SOCIETY/CALIFORNIA-SAN DIEGO SECTION
C/O GA Technologies, Inc.
P.O. Box 85608
San Diego, CA 92138
(619) 455-4555
Larry Parme - ANS Local Section Chair

AMERICAN NUCLEAR SOCIETY/COLORADO-COLORADO SECTION
C/O S.M. Stoller Corporation
Nuclear Power Experience
1919 14th St.
Suite 500
Boulder, CO 80302
(303) 449-7220
James F. Franks - ANS Local Section Chair

AMERICAN NUCLEAR SOCIETY/CONNECTICUT-CONNECTICUT SECTION
C/O Combustion Engineering 9422-2405
1000 Prospect Hill Rd.
Windsor, CT 06095
(203) 285-3881
David Moelling - ANS Local Section Chair

AMERICAN NUCLEAR SOCIETY/DISTRICT OF COLUMBIA-WASHINGTON SECTION
C/O Baltimore Electric And Gas
Calvert Cliffs Nuclear Plant
P.O. Box 1475
Baltimore, MD 21203
(301) 254-8529
Malcolm D. Patterson - ANS Local Section Chair

AMERICAN NUCLEAR SOCIETY/FLORIDA-FLORIDA SECTION
C/O University of Florida
Nuclear Engineering
Room 202 NSC
Gainesville, FL 32611
(904) 392-1408
William G. Vernetson - ANS Local Section Chair

AMERICAN NUCLEAR SOCIETY/GEORGIA-ATLANTA SECTION
C/O Georgia Power Company
P.O. Box 4545
Atlanta, GA 30302
(404) 526-7952
Emsley F. Cobb - ANS Local Section Chair

AMERICAN NUCLEAR SOCIETY/IDAHO-IDAHO SECTION
C/O Intermountain Technology, Inc.
Science Engineering and Analysis Division
P.O. Box 1604
Idaho Falls, ID 83403-1604
(208) 523-7255
Gail Cordes - ANS Local Section Chair

AMERICAN NUCLEAR SOCIETY/ILLINOIS-CENTRAL ILLINOIS SECTION
C/O Illinois Power Company
500 South 27th St.
Decatur, IL 62525
(217) 424-8301
John S. Spencer - ANS Local Section Chair

AMERICAN NUCLEAR SOCIETY/ILLINOIS-CHICAGO SECTION
C/O Argonne National Laboratory
9700 South Cass Ave.
Argonne, IL 60439
(312) 972-8352
Roger Tilbrook - ANS Local Section Chair

AMERICAN NUCLEAR SOCIETY/IOWA-IOWA SECTION
C/O Iowa Electric Light And Power
P.O. Box 351
Cedar Rapids, IA 52406
(319) 851-5611
Kenneth Peveler - ANS Local Section Chair

AMERICAN NUCLEAR SOCIETY/KANSAS-KANSAS SECTION
MS6-02
P.O. Box 2908
Wichita, KS 67201
(316) 261-6799
Wayne Gaul - ANS Local Section Chair

AMERICAN NUCLEAR SOCIETY/LOUISIANA-LOUISIANA SECTION
C/O GSU-River Bend
P.O. Box 220
St. Francisville, LA 70775
(504) 381-4622
Thomas C. Crouse - ANS Local Section Chair

AMERICAN NUCLEAR SOCIETY/MAINE, MASSACHUSETTS, NEW HAMPSHIRE, RHODE ISLAND, VERMONT-NORTHEASTERN SECTION
32 Knight Rd.
Framingham, MA 01701
(617) 872-8100
Michael W. Scott - ANS Local Section Chair

AMERICAN NUCLEAR SOCIETY/MICHIGAN-MICHIGAN SECTION
5340 Deerridge Rd.

Jackson, MI 49201
George H. Reicks - ANS Local Section Chair

AMERICAN NUCLEAR SOCIETY/MINNESOTA-UPPER MIDWEST SECTION
C/O Northern States Power
414 Nicollet Mall
Minneapolis, MN 55401
(612) 330-6341
Laura McCarten - ANS Local Section Chair

AMERICAN NUCLEAR SOCIETY/MISSISSIPPI-MISSISSIPPI SECTION
110 Bottino Dr.
Vicksburg, MS 39180-9721
(601) 437-6214
Phillip Giles, Jr. - ANS Local Section Chair

AMERICAN NUCLEAR SOCIETY/MISSOURI-CENTRAL/EASTERN MISSOURI SECTION
C/O University Of Missouri
1026 Engineering Complex
Rolla, MO 65211
(314) 882-3550
Jay F. Kunze - ANS Local Section Chair

AMERICAN NUCLEAR SOCIETY/MISSOURI-MISSOURI/KANSAS SECTION
C/O Kansas City Power And Light
P.O. Box 670
Kansas City, MO 64141
(816) 556-2942
J. Lee Sippel - ANS Local Section Chair

AMERICAN NUCLEAR SOCIETY/NEBRASKA-NEBRASKA SECTION
C/O Omaha Public Power District
1623 Harney St.
Omaha, NE 68102
(402) 536-4468
Richard L. Jaworski - ANS Local Section Chair

AMERICAN NUCLEAR SOCIETY/NEVADA-NEVADA SECTION
7164 South Abbeyville Dr.
Las Vegas, NV 89119
(702) 794-7639
Joy Fiore - ANS Local Section Chair

AMERICAN NUCLEAR SOCIETY/NEW JERSEY-NORTHERN NEW JERSEY SECTION
C/O William B. Johnson And Associates
P.O. Box 98
Research Park
Montville, NJ 07045
(201) 334-9222
William B. Johnson - ANS Local Section Chair

AMERICAN NUCLEAR SOCIETY/NEW MEXICO-TRINITY SECTION
C/O Public Service Company Of New Mexico
Alvarado Square
Albuquerque, NM 87158
(505) 848-2574
Burt Koske - ANS Local Section Chair

AMERICAN NUCLEAR SOCIETY/NEW YORK-METROPOLITAN NEW YORK SECTION
C/O Con-Edison Company
4 Irving Pl.
Room 1014
New York, NY 10003
(212) 460-6179
Surendra N. Purohit - ANS Local Section Chair

AMERICAN NUCLEAR SOCIETY/NEW YORK-LONG ISLAND SECTION
C/O Brookhaven National Laboratory
Building 475 B
Upton, NY 11973
(516) 282-2146
Dimitrios Cokinos - ANS Local Section Chair

AMERICAN NUCLEAR SOCIETY/NEW YORK-NIAGARA FINGER LAKES SECTION
C/O New York Power Authority
P.O. Box 41
Lycoming, NY 13093
(416) 449-6706
George J. Vargo - ANS Local Section Chair

AMERICAN NUCLEAR SOCIETY/NEW YORK-NORTHEASTERN NEW YORK SECTION
C/O Knolls Atomic Power Laboratory
P.O. Box 1072
Building E6
Schenectady, NY 12301
(518) 393-6611
Edward J. Brown - ANS Local Section Chair

AMERICAN NUCLEAR SOCIETY/NORTH CAROLINA-EASTERN CAROLINAS SECTION
C/O Carolina Power And Light
P.O. Box 1551
Raleigh, NC 27602
(919) 836-6628
Douglas L. Bensinger - ANS Local Section Chair

AMERICAN NUCLEAR SOCIETY/NORTH CAROLINA-PIEDMONT CAROLINAS SECTION
C/O Duke Power Comapny
Nuclear Production
P.O. Box 33189
Charlotte, NC 28242
(704) 373-4504
Mary Birch - ANS Local Section Chair

AMERICAN NUCLEAR SOCIETY/OHIO-NORTHERN OHIO SECTION
C/O Harshaw/Filtrol
14450 Summerfield Rd.
University Heights, OH 44118
(216) 349-6590
Harry Koppell - ANS Local Section Chair

AMERICAN NUCLEAR SOCIETY/OHIO-SOUTHWESTERN OHIO SECTION
2520 Westmont Blvd.
Columbus, OH 43221
(614) 422-0445
Richard N. Christensen - ANS Local Section Chair

AMERICAN NUCLEAR SOCIETY/OREGON-OREGON SECTION
C/O Portland General Electric
121 Southwest Salmon St.
Portland, OR 97204
(503) 226-8126
Alvin R. Ankrum - ANS Local Section Chair

AMERICAN NUCLEAR SOCIETY/PENNSYLVANIA- DELAWARE VALLEY SECTION
6327 Fairfield Dr.
Flourtown, PA 19031
(215) 327-1200
Thomas G. Szonntagh - ANS Local Section Chair

AMERICAN NUCLEAR SOCIETY/PENNSYLVANIA-CENTRAL PENNSYLVANIA SECTION
447 Lovell Ct.
Hummelstown, PA 17036
(717) 948-1043
John R. Saunders - ANS Local Section Chair

AMERICAN NUCLEAR SOCIETY/PENNSYLVANIA-NORTHERN PENNSYLVANIA SECTION
1326 Brassie St.
Wescosville, PA 18106
(215) 770-7893
David W. Miller - ANS Local Section Chair

AMERICAN NUCLEAR SOCIETY/PENNSYLVANIA-PITTSBURGH SECTION
C/O Westinghouse Electric Corporation
P.O. Box 355
Pittsburgh, PA 15230-0355
(412) 374-3632
John M. Senko - ANS Local Section Chair

AMERICAN NUCLEAR SOCIETY/SOUTH CAROLINA-COLUMBIA SECTION
335 Tram Rd.
Columbia, SC 29210
(803) 345-5209
Sam H. Bailey, Jr. - ANS Local Section Chair

AMERICAN NUCLEAR SOCIETY/SOUTH CAROLINA-SAVANNAH RIVER SECTION
C/O Georgia Power
Route 2
Box 299A
Waynesboro, GA 30830
(404) 554-9961
Larry Russell - ANS Local Section Chair

AMERICAN NUCLEAR SOCIETY/TENNESSEE-CHATTANOOGA SECTION
7408 Twinbrook Dr.
Chattanooga, TN 37421
(615) 751-2067
William H. Hannum - ANS Local Section Chair

AMERICAN NUCLEAR SOCIETY/TENNESSEE-OAK RIDGE/KNOXVILLE SECTION
C/O Oak Ridge National Laboratory
P.O. Box X
Oak Ridge, TN 37831-6089
(615) 574-6852
Ted Besmann - ANS Local Section Chair

AMERICAN NUCLEAR SOCIETY/TEXAS-NORTH TEXAS SECTION
C/O TU Electric
Box 2300
Glen Rose, TX 76043
(817) 897-5688
Matthew Sunseri - ANS Local Section Chair

AMERICAN NUCLEAR SOCIETY/TEXAS-SOUTH TEXAS SECTION
C/O Houston Lighting And Power Company
STP-5400
P.O. Box 1700
Houston, TX 77001
(713) 993-1335
Jack N. Bailey - ANS Local Section Chair

AMERICAN NUCLEAR SOCIETY/VIRGINIA-VIRGINIA SECTION
10108 Heritage Ln.
Glen Arlen, VA 23060
(804) 771-4595
Charles T. Snow - ANS Local Section Chair

AMERICAN NUCLEAR SOCIETY/WASHINGTON-EASTERN WASHINGTON SECTION
C/O Westinghouse Hanford Company
P.O. Box 1970
Richland, WA 99352
(509) 376-4953
Wanda Munn - ANS Local Section Chair

AMERICAN NUCLEAR SOCIETY/WASHINGTON-PUGET SOUND SECTION
University Of Washington/Nuclear Engineering Department
Seattle, WA 98105
(206) 543-2754
Robert W. Albrecht - ANS Local Section Chair

AMERICAN NUCLEAR SOCIETY/WISCONSIN-WISCONSIN SECTION
C/O Wisconsin Power And Light
P.O. Box 192
Madison, WI 53701-0192
(608) 252-3345
Stephan Bohlman - ANS Local Section Chair

(End ANS Listings)

AMERICAN SOCIETY FOR TESTING AND MATERIALS (ASTM)
1916 Race St.
Philadelphia, PA 19103
(215) 299-5571
George Luciw - Senior Staff Manager/Group Manager
David Bradley - Senior Staff Manager/Group Manager
Activities:
World's largest source for voluntary consensus standards for products services, systems, and materials. ASTM has set up a number of working committees to analyze radon issues.

AMERICAN WATER WORKS ASSOCIATION (AWWA)
6666 West Quincy Ave.
Denver, CO 80235
(303) 794-7711
Kurt M. Keeley - Director/Informational Services
Activities:
Provides radon in water information through library service and publications.

AMERICAN WATER WORKS ASSOCIATION RESEARCH FOUNDATION (AWWARF)
6666 West Quincy Ave.
Denver, CO 80235
(303) 794-7711
Rick Corland - Director, Research Management
Activities:
Assesses research needs of members concerning radon-in-water issues.

ASSOCIATION OF AMERICAN STATE GEOLOGISTS (AASG)
C/O Louisiana Geological Survey
P.O. Box G
Baton Rouge, LA 70893
(504) 342-6754
Charles Groot - President
Activities:
State geologists nationwide. Directs exchange of accumulated research and technical information, including radon studies by sta geological surveys.

ASSOCIATION OF GROUND WATER SCIENTISTS AND ENGINEERS (AGWSE)
6375 Riverside Dr.
Dublin, OH 43017
(614) 761-1711
Jay H. Lehr - Executive Director
Activities:
Division of National Water Well Association. Operates library o groundwater science and water well technology (including radon i groundwater studies).

ASSOCIATION OF STATE AND INTERSTATE WATER POLLUTION CONTROL ADMINISTRATORS (ASIWPCA)
444 North Capitol St.
Suite 330
Washington, DC 20001
(202) 624-7782
Robbi Savage - Executive Director
Activities:
Establishes policies and standards for state water pollution contro activities. Promotes state-to-state exchanges on important issues such as radon in water.

ASSOCIATION OF STATE DRINKING WATER ADMINISTRATORS (ASDWA)
1911 North Fort Myer Dr.
Arlington, VA 22209
(703) 524-2428
G. Wade Miller - Executive Director
Activities:
Represents state drinking water interests before Congress. Involved with radon issues in response to EPA's proposed standards for radon in drinking water.

CENTER FOR CLEAN AIR POLICY (CCAP)
444 North Capitol St., NW
Washington, DC 20001
(202) 624-7709
Ned Helme - Executive Director
Activities:
Analyzes economic and environmental effects of air pollution. Provides mediation role on environmental issues among government, corporate community, and environmental groups.

CENTER FOR ENVIRONMENTAL INFORMATION, INC. (CEI)
33 South Washington St.
Rochester, NY 14608
(716) 546-3796
Frederick W. Stoss - Director of Information Services
Activities:
Provides radon information through publications, computerized information sources, educational programs, conferences, and library.

CONFERENCE OF RADIATION CONTROL PROGRAM DIRECTORS (CRCPD)
71 Fountain Pl.
Frankfort, KY 40601
(502) 227-4543
Charles Hardin - Executive Secretary
Activities:
National non-profit professional association made up of state radiation experts and federal government contacts from pertinent agencies. Involved in comprehensive radon policy development program.

CONFERENCE OF RADIATION CONTROL PROGRAM DIRECTORS (CRCPD)/COMMITTEE ON RADON POLICY
71 Fountain Pl.
Frankfort, KY 40601
(502) 227-4543
Harlan Keaton - Chairman
Albert J. Hazle - Committee Member
Mary K. Cahill - Committee Member
Paul Weeden - Committee Member
Activities:
Recommends actions by CRCPD to assist states on radon issues Helps states develop criteria to assess potential for radon problem and responses to those problems. Standardizes recommended guidelines for dealing with public on radon issues. Communicate states' concerns to federal agencies and assesses federal government's long- and short-term goals on radon research and policy development.

CONFERENCE OF RADIATION CONTROL PROGRAM DIRECTORS (CRCPD)/TASK FORCE FOR ADEQUATE RADIATION CONTROL PROGRAM-RADON
71 Fountain Pl.
Frankfort, KY 40601
(502) 227-4543
Don Hoxie - Chairperson
Vladimir Dvorak - Task Force Member
Robert Mooney - Task Force Member
Paul Weeden - Task Force Member
Activities:
Developing document outlining adequate radiation control program for radon.

ENVIRONMENTAL POLICY INSTITUTE (EPI)/RADIATION HEALTH INFORMATION PROJECT
218 D St., SE
Washington, DC 20003
(202) 544-2600
Bob Alvarez - Director
Activities:
Supports reductions in ionizing radiation (such as radon) exposures. Provides policy options for state/federal occupational or public health radiation protection guidelines. Library open to public.

FLORIDA CONSERVATION FOUNDATION/ENVIRONMENTAL INFORMATION CENTER
1191 Orange Ave.
Winter Park, FL 32789
(305) 644-5377
William M. Partington - Director
Activities:
Provides radon information to Florida citizens.

GEOLOGICAL SOCIETY OF AMERICA (GSA)
P.O. Box 9140
3300 Penrose Pl.
Boulder, CO 80301
(303) 447-2020
C. Edward Buchwald - Chairman,Public Policy Committee
John M. Sharp - Chairman, Hydrogeology Division
Activities:
Members of select committees address topical issues (radon) to provide timely reports.

INTERSTATE CONFERENCE ON WATER PROBLEMS (ICWP)
2300 M St., NW
Suite 800
Washington, DC 20037
(202) 466-7287
Dirk Hofman - First Vice Chairman
Activities:
Represents state positions on radon in water issues to Congress.

MALCOLM PIRNIE, INC.
100 Eisenhower Ave.
Paramus, NJ 07652
(201) 845-0400
David Hiltebrand - Water Quality and Treatment

Activities:
Environmental engineering consulting firm participating in EPA-funded water analysis project. Developing cost/technology report for removal of radon from drinking water supplies.

METROPOLITAN WASHINGTON COUNCIL OF GOVERNMENTS/DEPARTMENT OF ENVIRONMENTAL PROGRAMS
1875 Eye St., NW
Suite 200
Washington, DC 20006
(202) 223-6800
George Nicholls - Senior Environmental Planner
Activities:
Reconvening radon contact group to develop a proposal to create metropolitan Washington "Radon Measurement Data Bank." The objective would be to respond to needs of regional academic programs, aid radon policy personnel and provide general information concerning local radon problems.

NATIONAL ASSOCIATION OF ENVIRONMENTAL PROFESSIONALS (NAEP)
P.O. Box 9400
Washington, DC 20016
(202) 683-3746
Joan Schroeder - Executive Secretary
Activities:
Environmental administrators, educators and persons from government and industry interested in environmental issues and policies.

NATIONAL ASSOCIATION OF WATER COMPANIES (NAWC)
1725 K St., NW
Suite 1212
Washington, DC 20006
(202) 833-8383
James Groff - Executive Director
Activities:
Trade association represents members in states greatly affected by new EPA radon-water regulations.

NATIONAL CONFERENCE OF STATE LEGISLATURES (NCSL)
1050 17 St.
#2100
Denver, CO 80265
(303) 623-7800
Larry Morandi - Program Director
Paul Doyle - Principal Research Analyst
Activities:
Gathers information for state lawmakers on scientific, environmental and technological aspects of public policy issues (radon).

NATIONAL COUNCIL ON RADIATION PROTECTION AND MEASUREMENTS (NCRP)
7910 Woodmont Ave.
Suite 800
Bethesda, MD 20814
(301) 657-2652
Warren Sinclair - President
W. Rosen Ney - Executive Director
William Beckner - Staff Scientist
Activities:
Organization of scientists employing cooperative effort for advances in radiation research and measurement. Has released two major reports on radon hazards and is active in radon-related research.

NATIONAL GOVERNORS ASSOCIATION (NGA)/COMMITTEE ON ENERGY AND ENVIRONMENT
Hall of States, Suite 250
444 North Capitol St., NW
Washington, DC 20001
(202) 624-5300
Tom Curtis - Staff Director
Activities:
Governors of 50 states and 5 U.S. territories. Informs members of federal legislation impacting upon states. Policy statements voted upon by members.

NATIONAL WATER WELL ASSOCIATION (NWWA)
6375 Riverside Dr.
Dublin, OH 43017
(614) 761-1711
Kevin McCray - Director
Activities:
Water well contractors, makers, and suppliers of well-drilling equipment and researchers interested in techniques for locating and utilizing underground water. Funds major research on radon-water issues and provides detailed information on the topic.

NORTH AMERICAN ASSOCIATION FOR ENVIRONMENTAL EDUCATION (NAAEE)
P.O. Box 400
Troy, OH 45373
(513) 698-6493
Joan C. Heidelberg - Executive Vice President
Activities:
Educators, students, and organizations devoted to environmental education and studies. Promotes and disseminates environmental education and information and fosters sharing of information about environmental programs. Radiation research is a major concern.

PRECISION MEASUREMENTS ASSOCIATION (PMA)
9681 Business Center Dr.
Rancho Cucomong, CA 91730
(714) 980-6166
Phillip A. Painchard - Executive Director
Activities:
Engineers, scientists, and technicians in measurement science.

RADIAN CORPORATION/ENGINEERING DIVISION
P.O. Box 1300
Research Triangle Park, NC 27709
(919) 541-9100
James C. Dickerman - Vice President
Clint Berkland - Engineer
Activities:
Contracts with EPA for measuring radon exposure levels in the home. Organizers of EPA's radon symposium in Colorado in October, 1988.

RADIATION RESEARCH SOCIETY (RRS)
1101 Market St.
14th Floor
Philadelphia, PA 19107
(215) 574-3153
Meg Keiser - Administrative Director
Activities:
Professional society studying radiation and its effects. Affiliated with International Association for Radiation Research and National Academy of Sciences.

STATE AND TERRITORIAL AIR POLLUTION PROGRAM ADMINISTRATORS (STAPPA)/ASSOCIATION OF LOCAL AIR POLLUTION CONTROL OFFICIALS (ALAPCO)
444 North Capitol St., NW
Washington, DC 20001
(202) 624-7864
William Becker - Executive Director
Activities:
Joint headquarters strive to improve effectiveness of state and loca air pollution control programs and coordinate activities of state ar local officials. Serves as liaison to EPA, Congress, industrial associations, and others involved in questions concerning air resources. Many STAPPA and ALAPCO officials are the lead personnel in their states and locales for radon information and all can direct inquirer to proper sources for information.

UNIVERSITY OF MAINE-ORONO/COOPERATIVE EXTENSION SERVICE
100 Winslow Hall
Orono, ME 04469
(207) 581-3188
Louis Fourman - Director
Activities:
Information and educational service provides information and results from UM's extensive radon research programs to public. Any of the Service's 16 local offices can be contacted for information about radon in Maine.

WATER POLLUTION CONTROL FEDERATION (WPCF)
601 Wyeth St.
Alexandria, VA 22314-1994
(703) 684-2400
Berinda Ross - Assistant Director
Gerald Schnorr - Co-Director of Radon Programs
Francis DiGiano - Co-Director of Radon Programs
Jay DeMichael
Activities:
Provides research and information on radon levels in water. Members are autonomous regional organizations that are concerned with disposal of domestic and industrial waste water.

WATER QUALITY ASSOCIATION (WQA)
4151 Naperville Rd.
Lisle, IL 60532
(312) 369-1600
Mary Beth Robb - Director of Communications
Activities:
Provides library service and research information concerning wate treatment equipment and services.

Health

This section lists associations and organizations involved in radon-related activities as they pertain to human health. Among the areas of interest listed here are epidemiology, health physics, public health management, and policy.

AMERICAN ASSOCIATION OF PUBLIC HEALTH PHYSICIANS (AAPHP)
C/O DeKalb County Health Department
P.O. Box 987
Decatur, GA 30030
(404) 294-3789
Gunar Bohan - Secretary/Treasurer
Activities:
Physicians actively engaged in public health. Speaks for membership on proposed health legislation at various levels.

AMERICAN CANCER SOCIETY (ACS)/NATIONAL HEADQUARTERS
3340 Peachtree Rd.
Atlanta, GA 30326
(404) 320-3333
Gerald D. Dodd - Medical and Scientific Committee
Activities:
Developing proposal for radon exposure information. General information available from local society chapters listed below.

AMERICAN CANCER SOCIETY (ACS)/NATIONAL PUBLIC AFFAIRS OFFICE
224 East Capitol St.
Washington, DC 20003
(202) 546-4011
John Madigan - Director of Government Relations
Activities:
Passed a resolution to urging U.S. Congress to approve legislation which reduces the potential health threat posed by radon. The Society's National Committee on Cancer Prevention and Detection, in conjunction with the National Research Department is reviewing the scientific evidence concerning the extent of the health threat posed by radon gas, and will present its report to Board of Directors.

American Cancer Society Charter Divisions

AMERICAN CANCER SOCIETY ALABAMA DIVISION, INC.
402 Office Park Dr.
Suite 300
Birmingham, AL 35223
(205) 879-2242

AMERICAN CANCER SOCIETY ALASKA DIVISION, INC.
406 West Fireweed Ln.
Suite 101
Anchorage, AK 99503
(907) 277-8696

AMERICAN CANCER SOCIETY
ARIZONA DIVISION, INC.
2929 East Thomas Rd.
Phoenix, AZ 85016
(602) 224-0524

AMERICAN CANCER SOCIETY
ARKANSAS DIVISION, INC.
5520 West Markham St.
P.O. Box 3822
Little Rock, AR 72203
(501) 664-3480, (501) 664-3481

AMERICAN CANCER SOCIETY
CALIFORNIA DIVISION, INC.
1710 Webster St.
P.O. Box 2061
Oakland, CA 94604
(415) 893-7900

AMERICAN CANCER SOCIETY
COLORADO DIVISION, INC.
2255 South Oneida
P.O. Box 24669
Denver, CO 80224
(303) 758-2030

AMERICAN CANCER SOCIETY
CONNECTICUT DIVISION, INC.
Barnes Park South
14 Village Ln.
P.O. Box 410
Wallingford, CT 06492
(203) 265-7161

AMERICAN CANCER SOCIETY
DELAWARE DIVISION, INC.
1708 Lovering Ave.
Suite 202
Wilmington, DE 19806
(302) 654-6267

AMERICAN CANCER SOCIETY
DISTRICT OF COLUMBIA DIVISION, INC.
Universal Building South
1825 Connecticut Ave., NW
Washington, DC 20009
(202) 483-2600

AMERICAN CANCER SOCIETY
FLORIDA DIVISION, INC.
1001 South MacDill Ave.
Tampa, FL 33629
(813) 253-0541

AMERICAN CANCER SOCIETY
GEORGIA DIVISION, INC.
46 Fifth St., NE
Atlanta, GA 30308
(404) 892-0026

AMERICAN CANCER SOCIETY
HAWAII PACIFIC DIVISION, INC.
Community Services Center Building
200 North Vineyard Blvd.
Honolulu, HI 96817
(808) 531-1662, (808) 531-1663

AMERICAN CANCER SOCIETY
IDAHO DIVISION, INC.
1609 Abbs St.
P.O. Box 5386
Boise, ID 83705
(208) 343-4609

AMERICAN CANCER SOCIETY
ILLINOIS DIVISION, INC.
37 South Wabash Ave.
Chicago, IL 60603
(312) 372-0472

AMERICAN CANCER SOCIETY
INDIANA DIVISION, INC.
9575 North Valparaiso Ct.
Indianapolis, IN 46268
(317) 872-4432

AMERICAN CANCER SOCIETY
IOWA DIVISION, INC.
8364 Hickman Rd.
Suite D
Des Moines, IA 50322
(515) 253-0147

AMERICAN CANCER SOCIETY
KANSAS DIVISION, INC.
3003 Van Buren St.
Topeka, KS 66611
(913) 267-0131

AMERICAN CANCER SOCIETY
KENTUCKY DIVISION, INC.
Medical Arts Building
1169 Eastern Pkwy.
Louisville, KY 40217
(502) 459-1867

AMERICAN CANCER SOCIETY
LONG ISLAND DIVISION, INC.
145 Pidgeon Hill Rd.

Huntington Station, NY 11746
(516) 385-9100

AMERICAN CANCER SOCIETY
LOUISIANA DIVISION, INC.
Masonic Temple Building
Suite 700
333 St. Charles Ave.
New Orleans, LA 70130-3118
(504) 523-2029

AMERICAN CANCER SOCIETY
MAINE DIVISION, INC.
Federal And Green Streets
Brunswick, ME 04011
(207) 729-3339

AMERICAN CANCER SOCIETY
MARYLAND DIVISION, INC.
1840 York Rd.
Suite K-M
P.O. Box 544
Timonium, MD 21093
(301) 561-4790

AMERICAN CANCER SOCIETY
MASSACHUSETTS DIVISION, INC.
247 Commonwealth Ave.
Boston, MA 02116
(617) 267-6250

AMERICAN CANCER SOCIETY
MICHIGAN DIVISION, INC.
1205 East Saginaw St.
Lansing, MI 48906
(517) 371-2920

AMERICAN CANCER SOCIETY
MINNESOTA DIVISION, INC.
3316 West 66th St.
Minneapolis, MN 55435
(612) 925-2772

AMERICAN CANCER SOCIETY
MISSISSIPPI DIVISION, INC.
1380 Livingston Ln.
Lakeover Office Park
Jackson, MS 39213
(601) 362-8874

AMERICAN CANCER SOCIETY
MISSOURI DIVISION, INC.
3322 American Ave.
P.O. Box 1066
Jefferson City, MO 65102
(314) 893-4800

AMERICAN CANCER SOCIETY
MONTANA DIVISION, INC.
313 North 32nd St.
Suite 1
Billings, MT 59101
(406) 252-7111

AMERICAN CANCER SOCIETY
NEBRASKA DIVISION, INC.
8502 West Center Rd.
P.O. Box 241255
Omaha, NE 68124-5255
(402) 393-5800

AMERICAN CANCER SOCIETY
NEVADA DIVISION, INC.
1325 East Harmon
Las Vegas, NV 89109
(702) 798-6877

AMERICAN CANCER SOCIETY
NEW HAMPSHIRE DIVISION, INC.
686 Mast Rd.
Manchester, NH 03102
(603) 669-3270

AMERICAN CANCER SOCIETY
NEW JERSEY DIVISION, INC.
CN2201
2600 Route 1
North Brunswick, NJ 08902
(201) 297-8000

AMERICAN CANCER SOCIETY
NEW MEXICO DIVISION, INC.
5800 Lomas Blvd., NE
Albuquerque, NM 87110
(505) 262-2336

AMERICAN CANCER SOCIETY
NEW YORK CITY DIVISION, INC.
19 West 56th St.
New York, NY 10019
(212) 586-8700

AMERICAN CANCER SOCIETY
NEW YORK STATE DIVISION, INC.
6725 Lyons St.
P.O. Box 7
East Syracuse, NY 13057
(315) 437-7025

AMERICAN CANCER SOCIETY
NORTH CAROLINA DIVISION, INC.
11 South Boylan Ave.
Suite 221

Raleigh, NC 27603
(919) 834-8463

AMERICAN CANCER SOCIETY NORTH DAKOTA DIVISION, INC.
Hotel Graver Annex Building
115 Roberts St.
P.O. Box 426
Fargo, ND 58102
(701) 232-1385

AMERICAN CANCER SOCIETY OHIO DIVISION, INC.
5555 Frantz Rd.
Dublin, OH 43017
(614) 889-9565

AMERICAN CANCER SOCIETY OKLAHOMA DIVISION, INC.
3800 North Cromwell
Oklahoma City, OK 73112
(405) 946-5000

AMERICAN CANCER SOCIETY OREGON DIVISION, INC.
9330 Southwest Curry
Portland , OR 97201
(503) 295-6422

AMERICAN CANCER SOCIETY PENNSYLVANIA DIVISION, INC.
Route 422 And Sipe Ave.
P.O. Box 897
Hershey, PA 17033-0897
(717) 533-6144

AMERICAN CANCER SOCIETY PHILADELPHIA DIVISION, INC.
1422 Chestnut St.
Philadelphia, PA 19102
(215) 665-2900

AMERICAN CANCER SOCIETY PUERTO RICO DIVISION, INC.
GPO Box 6004
San Juan, PR 00936
(809) 764-2295

AMERICAN CANCER SOCIETY QUEENS DIVISION, INC.
112-25 Queens Blvd.
Forest Hills, NY 11375
(718) 263-2224

AMERICAN CANCER SOCIETY RHODE ISLAND DIVISION, INC.
400 Main St.
Pawtucket, RI 02860
(401) 722-8480

AMERICAN CANCER SOCIETY SOUTH CAROLINA DIVISION, INC.
2442 Devine St.
Columbia, SC 29205
(803) 256-0245

AMERICAN CANCER SOCIETY SOUTH DAKOTA DIVISION, INC.
1025 North Minnesota Ave.
Hillcrest Plaza
Sioux Falls, SD 57104
(605) 336-0897

AMERICAN CANCER SOCIETY TENNESSEE DIVISION, INC.
1315 Eighth Ave. South
Nashville, TN 37203
(615) 255-1227

AMERICAN CANCER SOCIETY TEXAS DIVISION, INC.
2433 Ridgepoint Dr.
Austin, TX 78754
(512) 928-2262

AMERICAN CANCER SOCIETY UTAH DIVISION, INC.
610 East South Temple
Salt Lake City, UT 84102
(801) 322-0431

AMERICAN CANCER SOCIETY VERMONT DIVISION, INC.
13 Loomis St.
Drawer C
Montpelier, VT 05602
(802) 223-2348

AMERICAN CANCER SOCIETY VIRGINIA DIVISION, INC.
4240 Park Place Ct.
P.O. Box 1547
Glen Allen, VA 23060
(804) 270-0142

AMERICAN CANCER SOCIETY WASHINGTON DIVISION, INC.
2120 First Ave. North
Seattle, WA 98019
(206) 283-1152

AMERICAN CANCER SOCIETY WEST VIRGINIA DIVISION, INC.
2428 Kanawha Blvd. East
Charleston, WV 25311
(304) 344-3611

AMERICAN CANCER SOCIETY WESTCHESTER DIVISION, INC.
30 Glenn St.
White Plains, NY 10603
(914) 949-4800

AMERICAN CANCER SOCIETY WISCONSIN DIVISION, INC.
615 North Sherman Ave.
P.O. Box 8370
Madison, WI 53708
(608) 249-0487

AMERICAN CANCER SOCIETY WYOMING DIVISION, INC.
3109 Boxelder Dr.
Cheyenne, WY 82001
(307) 638-3331

(End ACS Listings)

AMERICAN COUNCIL ON SCIENCE AND HEALTH (ACSH)
1995 Broadway
18th Floor
New York, NY 10023-5860
(212) 362-7044
Edward G. Remmers - Associate Director
Activities:
Produces media seminars to educate media on relevant topics sucl as radon.

AMERICAN LUNG ASSOCIATION (ALA)/NATIONAL HEADQUARTERS
1740 Broadway
New York, NY 10019
(212) 315-8700
Ron White - Information Officer (Radon)
Activities:
Non-profit organization provides educational materials on environmental and occupational health. Lung diseases, smoking, and indoor air quality are all major concerns. Information on specific topics can be found from state offices listed below (many of these state associations have active radon information programs.)

American Lung Association Constituents

ALA OF ALABAMA
P.O. Box 55209
Birmingham, AL 35255
(205) 933-8821
James L. Hughes - Executive Director

ALA OF ALASKA
P.O. Box 103056
Anchorage, AK 99510-3056
(907) 276-5864
Deborah Lee Williams - Executive Director

ALA OF ARKANSAS
P.O. Box 3857
Little Rock, AR 72203
(501) 374-3726
David P. Cook - Executive Director

ALA OF CALIFORNIA
424 Pendelton Way
Oakland, CA 94621-2189
(415) 638-5864
George R. Williams - Executive Director

ALA OF COLORADO
1600 Race St.
Denver, CO 802067-1198
(303) 388-4327
Lea Ann Purvis - Executive Director

ALA OF CONNECTICUT
45 Ash St.
East Hartford, CT 06108
(203) 289-5401
John E. Zinn - Executive Director

ALA OF DELAWARE
1021 Gilpin Ave.
Suite 202
Wilmington, DE 19806
(302) 655-7258
Martha C. Bogdan - Executive Director

ALA OF DISTRICT OF COLUMBIA
475 H St., NW
Washington, DC 20001
(202) 682-5864
Robert Day - Environmental Health Project
Janet M. Widmer - Executive Director

ALA OF EASTERN MISSOURI
1118 Hampton Ave.
St. Louis, MO 63139-3147
(314) 645-5505
Richard D. Stevenson - Executive Director

ALA OF FLORIDA
P.O. Box 8127
Jacksonville, FL 32239-8127
(904) 743-2933
William F. White - Executive Director

ALA OF FLORIDA GOVERNMENTAL AFFAIRS OFFICE
345 West Madison St.
Suite 2
Tallahassee, FL 32301-1625
(904) 224-7222
Carolyn Christensen-Spore - Director

ALA OF GEORGIA
2452 Spring Rd.
Smyrna, GA 30080
(404) 434-5864
Ronald C. Miller - Managing Director

ALA OF HAWAII
245 North Kukui St.
Honolulu, HI 96817
(808) 537-5966
Donald R. Ford - Executive Director

ALA OF IDAHO
2621 Camas St.
Boise, ID 83705-2447
(208) 344-6567
Elizabeth C. Miller - Executive Director

ALA OF INDIANA
8777 Purdue Rd.
Suite 310
Indianapolis, IN 46268-3109
(317) 872-9685
Larry A. McAllister - Managing Director

ALA OF IOWA
1321 Walnut St.
Des Moines, IA 50309
(515) 243-1225
Craig P. Redshaw - Executive Director

ALA OF KANSAS
P.O. Box 4426
Topeka, KS 66604-2419
(913) 272-9290
Steve W. Berndsen - Executive Director

ALA OF KENTUCKY
P.O. Box 969
Louisville, KY 40201-0969
(502) 363-2652
Barry Gottschalk - Executive Director

ALA OF LOUISIANA
333 St. Charles Ave.
Suite 500
New Orleans, LA 70130-3180
(504) 523-5864
W. Findlay Raymond - Executive Director

ALA OF MAINE
128 Sewall St.
Augusta, ME 04330
(207) 622-6394
Edward F. Miller - Executive Director

ALA OF MARYLAND
Heaver Plaza
1301 York Rd.
Suite 705
Lutherville, MD 21093-6010
(301) 494-1100
Joyce C. Waite - Executive Director

ALA OF MASSACHUSETTS
803 Summer St.
South Boston, MA 02127-1609
(617) 269-9720
Terry R. Lehman - Executive Director

ALA OF MICHIGAN
403 Seymour Ave
Lansing, MI 48933-1179
(517) 484-4541
Robert G. Smith - Executive Director

ALA OF MINNESOTA
614 Portland Ave.
St. Paul, MN 55102
(612) 227-8014
Dorothy Berg - Managing Director

ALA OF MISSISSIPPI
P.O. Box 9865
Jackson, MS 39206-9865
(601) 362-5453
Judson M. Allard - Executive Director

ALA OF MONTANA
Christmas Seal Building
825 Helena Ave.
Helena, MT 59601
(406) 442-6556
Earl W. Thomas - Executive Director

ALA OF NEBRASKA
8901 Indian Hills Dr.
Suite 107
Omaha, NE 68114-4057

(402) 393-2222
Alan R. Wass - Executive Director

ALA OF NEVADA
P.O. Box 7056
Reno, NV 89510-7056
(702) 323-5864
John L. Holck - Executive Director

ALA OF NEW HAMPSHIRE
P.O. Box 1014
Manchester, NH 03105
(603) 669-2411
Steven P. Smith - Executive Director

ALA OF NEW JERSEY
1600 Route 22 East
Union, NJ 07083
(201) 687-9340
Robert F. Corso - Managing Director

ALA OF NEW MEXICO
216 Truman, NE
Albuquerque, NM 87108
(505) 265-0732
Billie Dytzel - Executive Director

ALA OF NEW YORK CITY-MANHATTAN/BRONX/STATEN ISLAND
432 Park Ave. South
8th Fl.
New York, NY 10016
(212) 889-3370
Edith Ewenstein - General Director

ALA OF NEW YORK CITY/BROOKLYN
165 Cadman Plaza East
Room 210
Brooklyn, NY 11201
(718) 624-8531
Carole A. Snyder - Executive Director

ALA OF NEW YORK CITY/QUEENSBORO
112-25 Queens Blvd.
Forest Hills, NY 11375
(718) 263-5656
Charlotte Schwartz - Executive Director

ALA OF NEW YORK STATE
8 Mountain View Ave.
Albany, NY 11205-2899
(518) 459-4197
Philip W. Woodrow - Executive Director

ALA OF NORTH CAROLINA
P.O. Box 27985
Raleigh, NC 27611-7985
(919) 832-8326
C. Scott Venable - Executive Director

ALA OF NORTH DAKOTA
P.O. Box 5004
Bismarck, ND 58502-5014
(701) 223-5613
Duane K. Flick - Executive Director

ALA OF OHIO
P.O. Box 16677
Columbus, OH 43216-6677
(614) 279-1700
Dale C. Hollern - Executive Director

ALA OF OKLAHOMA
P.O. Box 53303
Oklahoma City, OK 73152
(405) 524-8471
John G. Rogers - Executive Vice President

ALA OF OREGON
P.O. Box 115
Portland, OR 97207
(503) 224-5145
George B. Wright - Executive Director

ALA OF PENNSYLVANIA
P.O. Box 4213
Harrisburg, PA 17111-0213
(717) 564-4850
Charles J. White - Executive Director

ALA OF PUERTO RICO
G.P.O. 3468
San Juan, PR 00936
(809) 756-5664
Annie Miller Dieppa - Executive Director

ALA OF RHODE ISLAND
10 Abbott Park Pl.
Providence, RI 02903-3703
(401) 421-6487
Margaret E. Kane - Executive Director

ALA OF SOUTH CAROLINA
1817 Gadsden St.
Columbia, SC 29201
(803) 765-9066
Theodore W. Beiderwieden - Executive Director

ALA OF SOUTH DAKOTA
208 East 13th St.
Sioux Falls, SD 57102

(605) 336-7222
Kathleen Wiebers - Executive Director

ALA OF SOUTHEAST MICHIGAN
18860 West Ten Mile Rd.
Southfield, MI 48933-1179
(313) 559-5100
Thomas P. Banas - Executive Director

ALA OF TENNESSEE
P.O. Box 399
Nashville, TN 37202-0399
(615) 329-1151
William F. Busse - Executive Director

ALA OF TEXAS
3520 Executive Center Dr.
Suite G-100
Austin, TX 78731-1606
(512) 343-0502
Edward Carter - Managing Director

ALA OF THE VIRGIN ISLANDS
P.O. Box 974
St. Thomas, VI 00801
(809) 774-2077
Myriam James - Chairman

ALA OF UTAH
1930 South 1100 East
Salt Lake City, UT 84106-2317
(801) 484-4456
Katharin Gay Saunders - Executive Director

ALA OF VERMONT
30 Farrell St.
South Burlington, VT 05401
(802) 863-6817
Robert P. Jones - Executive Director

ALA OF VIRGINIA
P.O. Box 7065
Richmond, VA 23221-0065
(804) 355-3295
Carl C. Booberg - Executive Director

ALA OF WASHINGTON
2625 3rd Ave.
Seattle, WA 98121
(206) 441-5100
Hubert R. Slack - Executive Director

ALA OF WEST VIRGINIA
P.O. Box 3980
Charleston, WV 25339
(304) 342-6600
John R. Carlson - Executive Director

ALA OF WESTERN MISSOURI
2007 Broadway
Kansas City, MO 64108
(816) 842-5242
William J. Pfeifer - Executive Director

ALA OF WISCONSIN
10001 West Lisbon Ave.
Milwaukee, WI 53222
(414) 463-3232
James E. Pender - Executive Director

ALA OF WYOMING
P.O. Box 1128
Cheyenne, WY 82003-1128
(307) 636-5342
Dee Arndt - Executive Director

(End ALA Listings)

AMERICAN MEDICAL ASSOCIATION (AMA)/COUNCIL ON SCIENTIFIC AFFAIRS
535 North Dearborn St.
Chicago, IL 60610
(312) 645-5000
William R. Hendee - Executive Secretary
Mary Beth Tuhoy - EPA Contact/Washington, DC
Activities:
Publishes reports on health effects of radon. Working with EPA to develop program for local health officials to educate them on their responsibilities to public. Producing radon information videotape for consumer groups. Preparing six regional conference on radon for 1989.

AMERICAN PUBLIC HEALTH ASSOCIATION (APHA)
1015 15th St., NW
Washington, DC 20005
(202) 789-5663
Katherine MacCarter - Associate Executive Director
Activities:
Represents state public health officials. Protects and promotes environmental health. Promulgates standards and enacts public health research.

ARIZONA LUNG ASSOCIATION
102 West McDowell Rd.
Phoenix, AZ 85003
(602) 258-7505
Ben Chaiken - Executive Director

ASSOCIATION OF STATE AND TERRITORIAL HEALTH OFFICIALS (ASTHO)
1311 Dolly Madison Blvd.
McLean, VA 22101
(703) 556-9222
George Dagnon - Executive Director
Theresa Benincasa - Assistant Director
Activities:
Addresses needs of states before Congress. Lobbies for appropriations for mandated programs on radon. Publication "Prevention Notes" features radon related issues.

CENTER FOR ENVIRONMENTAL MEDICINE
60 Washington Ave.
Hamden, CT 06518
(203) 789-1911
Robert McLellan - Director
Activities:
Analyzes residential health hazards, particularly water issues. Researched relationship between household use of water and radon exposure and well-operator radon exposure. Researches aeration techniques and what to do with gases, uranium and radium in water.

CONFERENCE OF PUBLIC HEALTH LABORATORIANS (COPHL)
P.O. Box 9083
Austin, TX 78766
(512) 458-7318
Charles E. Sweet - Secretary/Treasurer
Activities:
Develops and maintains adequate standards for professional training of laboratory workers. Membership includes directors/assistants in directing public health, military, hospital and research laboratories.

CONFERENCE OF STATE HEALTH AND ENVIRONMENTAL MANAGERS (CSHEM)
P.O. Box 5254
Hudson, FL 34674
Meredith Thompson - Executive Officer
Activities:
Sanitary engineers and environmental officials at departments of health/environment agencies in U.S. Conducts research and surveys. Provides policy information and promotes exchange of ideas among members.

HEALTH AND ENERGY INSTITUTE
236 Massachusetts Ave., NE
Suite 506
Washington, DC 20002
(202) 543-1070
Kathleen M. Tucker - President
Activities:
Addresses health and environmental effects of radiation on women.

HEALTH PHYSICS SOCIETY (HPS)
800 Westpark Dr.
Suite 400
McLean, VA 22102
(703) 790-1745
Richard Burke - Executive Secretary
Activities:
Protection of humanity and environment from radiation. Radon exposure is a major issue concerning society. HPS has sponsored seminars and journal reports providing information on radon.

NATIONAL ACADEMY OF SCIENCES (NAS)/NATIONAL RESEARCH COUNCIL (NRC)/MEDICAL FOLLOW-UP AGENCY
2101 Constitution Ave., NW
Washington, DC 20418
(202) 334-2000
William Ellett - Director, BEIR IV
Raymond Cooper - Senior Program Officer
Activities:
Research and reports on biological effects of ionizing radiation--including cancer risks from radon exposure. Latest publication is known as the "BEIR IV Report."

NATIONAL ASSOCIATION OF COUNTY HEALTH OFFICIALS (NACHO)
440 First St., NW
Washington, DC 20001
(202) 393-6226
Matthew Namoviev - Publications Associate
Activities:
Represents all U.S. county health officials before Congress. Sponsors quarterly conferences to address issues such as environmental health.

NATIONAL CONFERENCE OF LOCAL ENVIRONMENTAL HEALTH ADMINISTRATORS (NCLEHA)
3415 28th St.
Port Huron, MI 48060
(313) 987-5306
John O. Tironi - Chairman
Activities:
City, county, and district environmental health personnel. Information on practice of environmental health in community.

NATIONAL ENVIRONMENTAL HEALTH ASSOCIATION (NEHA)
720 South Colorado Blvd.
South Tower, Suite 970
Denver, CO 80222
(303) 756-9090
Nelson E. Fabian - Executive Director
Terry Johnson - Environmental Specialist
Activities:
Provides health information on indoor air quality research. Radon information available.

NATIONAL SANITATION FOUNDATION (NSF)
P.O. Box 1468
Ann Arbor, MI 48016
(313) 769-8010
Tom S. Gable - Executive Vice President
Activities:
Representatives of public health, business, and industry. Develop standards in field of environmental sanitation and health.

PUBLIC HEALTH FOUNDATION (PHF)
1220 L St., NW
Suite 350
Washington, DC 20005
(202) 898-5600
Ellen Shapiro - Program Director
Activities:
Non-profit organization established by Association of State and Territorial Health Organizations, but separately incorporated. Supports public health organizations and general public health community. Released indoor air quality guide in September, 1988.

Occupational and Workplace

This section lists associations and organizations concerned with radon as it affects the workplace. Among the areas of interest listed here are industrial hygiene, mining, occupational medicine, and health.

AMERICAN COUNCIL OF GOVERNMENTAL INDUSTRIAL HYGIENISTS (ACGIH)
6500 Glenway Ave.
Building D-7
Cincinnati, OH 45211
(513) 661-7881
William Kelley - Executive Secretary
Activities:
Government and university employees engaged in full-time industrial hygiene. Worker health protection issues are priority.

AMERICAN INDUSTRIAL HYGIENE ASSOCIATION (AIHA)
475 Wolf Ledges Pkwy.
Akron, OH 44311
(216) 762-7294
Daniel Agopsowicz - Ionizing Radiation Committee
Activities:
Promotes study of environmental factors affecting health and well being of workers exposed to ionizing radiation.

AMERICAN MINING CONGRESS (AMC)
1920 N St., NW
Suite 300
Washington, DC 20036
(202) 861-2879
James Gilchrist - Senior Counsel
Activities:
Mining companies, manufacturers, engineers, and leasing companies. Long term involvement in radon research as it relates to the mining industry. AMC has produced two major studies on radon exposure in the mining profession.

AMERICAN OCCUPATIONAL MEDICAL ASSOCIATION (AOMA)
2340 South Arlington Heights Rd.
Arlington Heights, IL 60005
(312) 228-6850
Donald L. Hoops - Executive Director
Activities:
Promotes understanding of importance of high-quality medical care for workers. Encourages awareness of potential health hazards in the workplace.

AMERICAN SOCIETY OF SAFETY ENGINEERS (ASSE)
1800 East Oakton
Des Plaines, IL 60018
(312) 692-4121
Ray Boylston - President
Judy Weel - Executive Director
Jack Dobson - Director, Technical Services
Activities:
Issues guidelines to ensure adherence to pending radon standards.

IMC DEVELOPMENT CORPORATION
P.O. Box 7047
Lakeland, FL 33807
(813) 646-5061
Robert S. Hearon - Environmental Manager
Activities:
Phosphate mining company has worked with EPA since 1975 when ORP evaluated radon threat to phosphate miners in Florida. IMC funded epidemiological study of its employees. Corporate members have served on state's radon task force.

INDUSTRIAL SAFETY EQUIPMENT ASSOCIATION (ISEA)
1901 North Moore St.
Arlington, VA 22209
(703) 525-1695
Thomas A. Augherton - Executive Director
Activities:
Members are makers of clothing, supplies, and equipment used for protection of industrial workers.

INTERNATIONAL CHEMICAL WORKERS UNION (ICWU)
1655 West Market
Akron, OH 44313
(216) 867-2444
David Ortleib - Director, Health Safety
Activities:
Represents worker occupational safety concerns, including those of phosphate mine workers.

MINE INSPECTORS INSTITUTE OF AMERICA (MIIA)
1900 Grant Building
Pittsburgh, PA 15219
(412) 281-2620
Raymond G. Helfrich - President
Activities:
Includes members of state mine inspection agencies and federal mine inspectors. Provides information concerning radon levels in uranium mines.

NATIONAL SOCIETY OF PROFESSIONAL SANITARIANS (NSPS)
Missouri Health Department
Route 4, Hoffman Dr.
Jefferson, MO 65101
(314) 751-6095
John G. Norris - Executive Director
Activities:
Provides information on services important to public health and welfare, including radon exposure and indoor air quality.

OIL, CHEMICAL AND ATOMIC WORKERS INTERNATIONAL UNION (OCAW)
P.O. Box 2812
Denver, CO 80201
(303) 987-2229
Sylvia Kreckel - Occupational Health Specialist
Activities:
Represents occupational health concerns of uranium miners who belong to union.

ORGANIZATION OF RESOURCES COUNSELORS (ORC)
1910 Sunderland Pl., NW
Washington, DC 20036
(202) 737-6300
Darrel Matthias - Consultant
Activities:
Disseminates information on occupational safety and health issues (including radon) to member corporations.

SCHMELTZER, APTAKER AND SHEPPARD
1800 Massachusetts Ave., NW
Suite 500
Washington, DC 20036
(202) 828-1000
Mike Brown - Counselor-At-Law
Activities:
Practices environmental, health and safety law. Responsible for legal policies and strategies concerning EPA rulemaking and enforcement.

WASHINGTON OCCUPATIONAL HEALTH ASSOCIATES, INC.
1120 19th St., NW
Suite 410
Washington, DC 20036
(202) 463-6698
Kenneth Chase - President
Robert Swotinsky - Senior Clinical Associate
Activities:
Developer of risk assessment plan for radon testing. Center has also developed a program for interpretation (clinical examination) of radon exposures in the workplace or environment.

Public Interest

This section lists associations and organizations involved in radon-related activities as they pertain to the public interest. Among the activities listed here are lobbying, litigation, and disseminating information.

ASSOCIATION FOR IMPROVEMENT IN MOHEGAN, INC.
East Main St.
P.O. Box 103
Mohegan Lake, NY 10547
(914) 528-3300
William Stevens - Rehabilitation Specialist
Activities:
Housing preservation company funded by New York State Energy Office to assist residents with referrals for radon detection, mitigation, and education concerning radon. Helps coordinate state radon effort at the local level.

BUYERS UP RADON PROJECT
Box 53005
Washington, DC 20036
(202) 328-3800
Chris Dyson - Project Director
Activities:
Buyers Up is affiliated with Public Citizen, Inc., founded by Ralph Nader in 1971. Radon project aims to inform public on radon and the radon industry.

CITIZEN'S CLEARINGHOUSE FOR HAZARDOUS WASTES (CCHW)
P. O. Box 926
Arlington, VA 22216
(703) 276-7070
Steven Lester - Science Adviser
Brian Lipsett - Science Associate
Activities:
Provides information to citizens/groups throughout U.S. on issues concerning exposure to toxic or hazardous materials including radon.

CITIZENS FOR A BETTER ENVIRONMENT (CBE)
506 South Wabash
Chicago, IL 60605
(312) 939-1530
Joan D'Argo - Research Assistant
Activities:
Distributes information to increase public awareness of radon issues.

CONSUMER FEDERATION OF AMERICA (CFA)
1424 16th St., NW
Suite 604
Washington, DC 20036
(202) 387-6121
Steve Brobeck - Executive Director
David Swankin - IAQ Program Coordinator
Activities:
Promotes rights of consumers. Coordinates consumer programs in fair pricing of products and services and acts as clearinghouse for exchange of ideas. Indoor air quality program developed to provide consumers with information on threat of toxicants and services available.

CONSUMERS UNION OF UNITED STATES (CU)
256 Washington St.
Mt. Vernon, NY 06608
(914) 667-9400
Edward Groth - Associate Technical Director
Activities:
Testing, rating, and reporting organization provides information on competing brands of products. Represents consumer interests in legislature, courts, and administrative agencies. Publisher of "Consumer Reports."

DECISION RESEARCH
1201 Oak
Eugene, OR 97401
(503) 485-2400
Paul Slovic - Research Associate
Activities:
Research on human judgment and risk assessment. Has analyzed issue of failure of individuals to have homes tested for radon even though the cost is low or non-existent. All publications available to public.

ENVIRONMENTAL DEFENSE FUND (EDF)/NATIONAL HEADQUARTERS
257 Park Avenue South
New York, NY 10010
(212) 505-2100
Janice Caswell - Public Information
Ellen K. Silbergeld - Chair, Toxic Chemicals Program
Robert Yuhnke - EDF Radon Specialist
Activities:
Citizens' interest group staffed by lawyers, economists, and scientists. Takes legal action on environmental issues. Provides public with information on radon-related issues. Radon information booklet available.

ENVIRONMENTAL HEALTH WATCH OF GREATER CLEVELAND
4115 Bridge Ave.
Cleveland, OH 44113
(216) 961-4646
Bob Staib - Education Director
Stuart Greenberg - Research Director
Activities:
Previously known as Council on Hazardous Materials. Provides general information on radon hazards, products and services. Provides radon testing services (charcoal and continuous monitor) for the general public. Sponsors "Blueprint for a Healthy House" conference annually and is releasing "Healthy House Catalog" for information on indoor air pollutants.

FLORIDA DEFENDERS OF THE ENVIRONMENT (FDE)
1523 Northwest 4th St.
Gainesville, FL 32608
(904) 372-6965
Marjorie Carr - President
Activities:
FDE has called for comprehensive research on radon issues and provides citizens with information on radon.

LEAGUE OF WOMEN VOTERS EDUCATIONAL FUND/NATURAL RESOURCES
1730 M St., NW
Washington, DC 20036
(202) 429-1965
David Gray Loveland - Director
Activities:
Conducts national project on community drinking water issues. Sponsors research and develops educational material on drinking water.

MANASOTA 88
5314 Bay State Rd.
Palmetto, FL 34221
(813) 722-7413
Gloria Rains - Director
Activities:
Nonprofit organization proposing tighter federal, state, and local regulations concerning radon exposure and health issues.

MINISTRY OF CONCERN FOR PUBLIC HEALTH (MCPH)
5495 Main St.
Suite 147
Buffalo, NY 14221
No Phone Available
Joyce Troy - Executive Director
Activities:
Promotes understanding of pollution-health-environment systems. Initiates research projects and administers grant monies. Maintains Citizen Health Register recording human health effects of environmental pollutants. Publishes "Handbook for Estimating Health Effects from Exposure to Ionizing Radiation."

NATIONAL CLEAN AIR COALITION (NCAC)
801 Pennsylvania Ave., SE
Washington, DC 20003
(202) 543-8200
Richard Ayers - Chairman
Activities:
Civic, health, labor, and environmental groups that support clean air standards and enforcement. Interests include toxic air pollutants.

NATIONAL HOME BUYERS AND HOME OWNERS ASSOCIATION (NHBHOA)
1050 17th St., NW
Washington, DC 20036
(202) 659-6500
Benny L. Kass - Executive Director
Activities:
Promotes consumer interests in housing. Publishes "Home Buyers Checklist."

ORANGE ENVIRONMENT, INC.
RD 2
Maple Ave.
Goshen, NY 10924
(914) 294-5023
Michael R. Edelstein - President
Liana Hoodes - Vice President

William Makofske - Radon Committee
Activities:
Nonprofit community environmental organization provides information, conducts research, holds public programs, and lobbies for government action on radon-related issues.

PUBLIC CITIZEN LITIGATION GROUP (PCLG)/HEALTH RESEARCH GROUP
2000 P St., NW
Suite 700
Washington, DC 20036
(202) 872-0320
David Valdec - Attorney
Activities:
Founded by Ralph Nader. Files consumer cases including those involving health and safety issues.

SOUTHFACE ENERGY INSTITUTE
P.O. Box 5506
Atlanta, GA 30307
(404) 525-7657
Dennis Creech - Information Officer
Activities:
Private, non-profit, energy education facility that sponsors seminars and public education campaigns. Presently offering public education programs on radon including information on construction techniques.

SOUTHWEST RESEARCH AND INFORMATION CENTER
P.O. Box 4524
Albuquerque, NM 87106
(505) 262-1862
Don Hancock - Administrator
Activities:
Consultant to community organizations, conducts training workshops and seminars. Provides information on uranium mining issues (including radon exposure).

U.S. PUBLIC INTEREST RESEARCH GROUP (US PIRG)
215 Pennsylvania Ave., SE
Washington, DC 20003
(202) 546-9707
Gene Karpinski - Director
Activities:
Grass roots organizer of efforts to advance environmental protection laws. Conducts research on environmental issues (several related to radon). Provides public information and testifies before Congress on proposed environmental legislation.

UNION OF CONCERNED SCIENTISTS (UCS)
26 Church St.
Cambridge, MA 02238
(617) 547-5552
Howard C. Ris - Executive Director
Activities:
Advocate organization concerned with impact of advanced technology on society. Has issued recommended radon exposure action level at less than 5 pCi/l.

Real Estate

This section lists associations and organizations involved in radon-related activities as they pertain to the real estate industry. Among the areas of interest listed here are home inspection, property appraisal, lending, real estate brokerage, and relocation.

AMERICAN INSTITUTE OF REAL ESTATE APPRAISERS (AIREA)
430 North Michigan Ave.
Chicago, IL 60611
(312) 329-8437
W. Lee Minnerly - Research Manager
Activities:
Tracks radon issues as they relate to the real estate appraisa industry. Has published a policy paper on radon for appraisers. Maintains reference library that includes radon-related topics. Publication "Environmental Watch" covers current events involving radon issues.

AMERICAN SOCIETY OF HOME INSPECTORS (ASHI)
3299 K St., NW
Suite 700
Washington, DC 20007
(202) 842-3096
Bob Dolibois - Executive Director
Activities:
Sets home inspection standards and provides members with information and techniques.

AMERICAN SOCIETY OF REAL ESTATE COUNSELORS (ASREC)
430 North Michigan Ave.
Chicago, IL 60611
(312) 329-8429
Susan Grossman - Information
Activities:
Individuals with comprehensive training and experience of all aspects of real estate. Members use the professional title CRE (Counselor of Real Estate). These professionals provide counseling services and receive compensation on a fee, per diem, or retainer basis and not on a commission basis. Committees include education and government affairs.

EMPLOYEE RELOCATION COUNCIL (ERC)
1720 N St., NW
Washington, DC 20036
(202) 857-0857
Anita Brienza - Director, Public Relations
Activities:
Representatives of large companies that transfer personnel; representatives from the relocation industry. Analyzes and evaluates issues associated with relocation. Works to provide members with latest information on federal/state/local regulations concerning radon. Bimonthly publication "Mobility" has examined radon and real estate issues.

FEDERAL HOME LOAN MORTGAGE CORPORATION (FREDDIE MAC)
1776 G St., NW
Washington, DC 20006
(703) 759-8000
John Hemschoot - Director, Home Mortgage Standards

Activities:
Congressionally chartered corporation is leading seller of conventional mortgage securities in U.S. Corporation buys mortgages, packages them as securities and sells the securities to investors. Presently drafting guidelines for lenders to use to determine when to test homes for environmental hazards such as radon. Already requires title insurance policies to include an environmental endorsement policy before Freddie Mac will buy a mortgage.

FEDERAL NATIONAL MORTGAGE ASSOCIATION (FANNIE MAE)
3900 Wisconson Ave., NW
Washington, DC 20016
(202) 537-7000
Robert Engelstad - Vice President, Mortgage Standards
Activities:
Congressionally chartered shareholder owned corporation is the largest single investor in home mortgage funds in the U.S. Under guidelines drafted in 1988, title insurance policies must include an environmental endorsement before Fannie Mae will buy a mortgage. New guidelines are being drafted to determine when to test for environmental hazards such as radon.

HOME OWNERS WARRANTY CORPORATION
2000 L St., NW
Suite 400
Washington, DC 20036
(202) 463-4600
Tom Hallenbach - Architect and Engineering Supervisor
Activities:
Administers Home Owners Warranty Program that provides for 1ten year insured warranty on new homes through builder's registration with corporation. Original member of Radon Information Council is developing radon policy for its programs.

INSTITUTE OF REAL ESTATE MANAGEMENT
430 North Michigan Ave.
Chicago, IL 60611
(312) 661-1930
Ronald Vukas - Executive Vice President
Activities:
Organization of real property managers. Offers management courses, gathers statistics and maintains speakers bureau. Committees include education, legislation and research. Provides information on timely real estate issues.

MANSFIELD, PONTICELLI AND LIBBY
2100 M St., NW
Washington, DC 20036
(202) 296-1330
Richard Mansfield - General Counsel
Activities:
Employment Relocation Council general counsel provides radon information for ERC. Developing proposals for establishment of a radon policy for Council.

MORTGAGE BANKERS ASSOCIATION OF AMERICA (MBAA)
1125 15th St., NW
Washington, DC 20005
(202) 861-6500
Michael Taliefero - Government Agency Relations
Burton Wood - Legislation
Ann Vom Eigen - Associate Legislative Counsel
Carol Wagner - Single Family Division
Activities:
Provides radon education programs for members. Lobbies Congress on radon legislation affecting mortgage banking industry.

NATIONAL ASSOCIATION OF REAL ESTATE BROKERS (REALTIST)
P.O. Box 56483
Washington, DC 20041
(202) 289-6655
Albert A. Johnson - President
Activities:
Persons working in the real estate industry. Association provides research, educational and certification programs such as the National Society of Real Estate Appraisers, Real Estate Brokerage Institute and United Developers Council. Works to ensure public is protected from fraudulent or unethical practices within the real estate industry. Conducts research on developments within the real estate profession.

NATIONAL ASSOCIATION OF REAL ESTATE LICENSE LAW OFFICIALS (NARELLO)
P.O. Box 129
Centerville, UT 84010-0129
(801) 531-8202
Les Tyrrell - Executive Vice President
Activities:
Membership is made up of state officials administering real estate licensing laws. NARELLO members are those state real estate officials who maintain enforcement of regulations and ensure standards within the real estate industry.

NATIONAL ASSOCIATION OF REALTORS (NAR)
Education Division
430 North Michigan Ave.
Chicago, IL 60611
(312) 329-8388
Charles Morrison - Vice President, Education
Activities:
Organization of state real estate associations and local real estate boards. Chicago office sanctions all activities that involve NAR. Washington, DC office deals with legislative and public policy issues. Educational programs are developed by Chicago office which maintains 17,000 volume library.

NATIONAL ASSOCIATION OF REALTORS (NAR)/GOVERNMENT AFFAIRS OFFICE
777 14th St., NW
Washington, DC 20005
(202) 383-1087
David Weiss - Energy, Environment and Development
Activities:
Oversees realtor's interests before the federal government.

Has issued NAR policy statement on radon which supports:
• Federal legislation for further funding of indoor air quality research.
• Public education to better understand the need to test and mitigate indoor air quality problems.
• Education of homeowners as to their responsibilities to identify and disclose to agents and brokers any indoor environmental hazards that could pose a health threat.
• Efforts by state NAR associations to determine desirability of disclosure language in sales contracts.
• Development of products and construction techniques that would reduce the threat of radon problems in new structures.
• Education of building owners on proper operation and maintenance of HVAC systems to ensure acceptable indoor air quality.
• Public education concerning relative risk of indoor air contamination.
• Development by EPA of certification programs for testing and mitigation companies.
• Further investigation of needs of homeowners concerning the radon issue.
NRA has pledged to continue its cooperation with government to further understanding of the radon problem and to work with state and local offices on the issue.

NATIONAL ASSOCIATION OF REALTORS (NAR)/GOVERNMENT AFFAIRS OFFICE
777 14th St., NW
Washington, DC 20005
(202) 383-1087
David Weiss - Energy, Environment and Development
Activities:
Oversees realtor's interests before the federal government.
Has issued NAR policy statement on radon which supports:
• Federal legislation for further funding of indoor air quality research.
• Public education to better understand the need to test and mitigate indoor air quality problems.
• Education of homeowners as to their responsibilities to identify and disclose to agents and brokers any indoor environmental hazards that could pose a health threat.
• Efforts by state NAR associations to determine desirability of disclosure language in sales contracts.
• Development of products and construction techniques that would reduce the threat of radon problems in new structures.
• Education of building owners on proper operation and maintenance of HVAC systems to ensure acceptable indoor air quality.
• Public education concerning relative risk of indoor air contamination.
• Development by EPA of certification programs for testing and mitigation companies.
• Further investigation of needs of homeowners concerning the radon issue.

NATIONAL ASSOCIATION OF REALTORS (NAR)/PUBLIC AFFAIRS OFFICE
777 14th St., NW
Washington, DC 20005
(202) 383-1177
Walt Maloney - Public Affairs Spokesman
Activities:
Represents NAR to national media and provides news releases detailing position on topics such as radon legislation and policy.

State NAR Affiliates

ALABAMA ASSOCIATION OF REALTORS
P.O. Box 4070
Montgomery, AL 36103
(205) 262-3808
Thomas Walker - Executive Vice President

ALASKA ASSOCIATION OF REALTORS
1836 West Northern Lights Blvd.
Anchorage, AK 99517
(907) 276-0655
Dea Turner - Executive Vice President

ARIZONA ASSOCIATION OF REALTORS
4414 North 19th Ave.
Suite R
Phoenix, AZ 85015
(602) 248-7785
Ty Strout - CAE

ARKANSAS REALTORS ASSOCIATION
Realtors Building
201 Executive Ct., Suite 100
Little Rock, AR 72205
(602) 248-7785
Robert J. Balhorn - Executive Vice President

CALIFORNIA ASSOCIATION OF REALTORS
P.O. Box 76917
Los Angeles, CA 90067
(213) 739-8201
Donald A. Wiedmann - Executive Vice President

COLORADO ASSOCIATION OF REALTORS
6558 South Yosemite Cr.
Englewood, CO 80111
(303) 770-6550
Arnold Van Zandt - Executive Vice President

CONNECTICUT ASSOCIATION OF REALTORS
316 Farmington Ave.
Hartford, CT 06105
(203) 522-7255
Harry W. Wenz - Executive Vice President

DELAWARE ASSOCIATION OF REALTORS
P.O. Box 1364
Dover, DE 19903
(302) 734-4444
Milman Prettyman - Executive Vice President

FLORIDA ASSOCIATION OF REALTORS
P.O. Box 1231
Orlando, FL 32802
(407) 849-1400
Gerald Matthews - Executive President
Larry Schwartz - Legal Department

GEORGIA ASSOCIATION OF REALTORS
3200 Presidential Dr.
Atlanta, GA 30340
(404) 451-1831
John W. Cox - Executive Vice President

HAWAII ASSOCIATION OF REALTORS
505 Ward Ave.
Honolulu, HI 96814
(808) 538-3641
Peggy Comeau - Executive Vice President

IDAHO ASSOCIATION OF REALTORS
1450 West Bannock
Boise, ID 83702
(208) 342-3585
Mark Dunham - Executive Vice President

ILLINOIS ASSOCIATION OF REALTORS
Illinois Realtors Pl.
Suite 400
Springfield, IL 62703
(217) 529-2600
Gary Clayton - Executive Vice President

INDIANA ASSOCIATION OF REALTORS
7301 North Shadeland
Indianapolis, IN 46250
(317) 842-0890
Al Unser - Executive Vice President

IOWA ASSOCIATION OF REALTORS
999 Oakridge Dr.
Des Moines, IA 50308
(515) 244-2294
Bud Ewell - Executive Vice President

KANSAS ASSOCIATION OF REALTORS
3644 Southwest Burlingamr Road
Topeka, KS 66611
(913) 267-3610
James Mayer - Executive Vice President

KENTUCKY ASSOCIATION OF REALTORS
2224 Young Dr.
Lexington, KY 40505
(606) 266-2185
Richard A. Rushing - Executive Vice President

LOUISIANA REALTORS ASSOCIATION
4639 Bennington
Baton Rouge, LA 70808
(504) 923-2210
Malcolm Young - Executive Vice President

MAINE ASSOCIATION OF REALTORS
21 Bangor St.
Augusta, ME 04330
(207) 622-7501
Donna Snethen - Executive Vice President

MARYLAND ASSOCIATION OF REALTORS
2083 West St.
Suite 4-A
Annapolis, MD 21401
(305) 841-6080
Edgar C. Hilley - Executive Vice President

MASSACHUSETTS ASSOCIATION OF REALTORS
P.O. Box 650
Waltham, MA 02254-0650
(617) 890-3700
Robert Authier - Executive Vice President

MICHIGAN ASSOCIATION OF REALTORS
P.O. Box 40725
Lansing, MI 48901-7925
(517) 372-8890
Don Freels - Executive Vice President

MINNESOTA ASSOCIATION OF REALTORS
5750 Lincoln Dr.
Edina, MN 55436
(612) 935-8313
James Manke - Executive Vice President

MISSISSIPPI ASSOCIATION OF REALTORS
P.O. Box 2454
Jackson, MS 39205
(601) 355-0216
Thad I. Vann - Executive Vice President

MISSOURI ASSOCIATION OF REALTORS
P.O. Box 1327
Columbia, MO 65205
(314) 445-8400
Dennis McDermott - Executive Vice President

MONTANA ASSOCIATION OF REALTORS
910 Helena Pk.
Helena, MT 59601
(406) 443-4032
Sharon Cleary - Executive Vice President

NEBRASKA REALTORS ASSOCIATION
1115 K St.
Suite 102
Lincoln, NE 68508
(402) 475-6334
Vacant - Executive Vice President
Karen Gehle - Administrative Coordinator

NEVADA ASSOCIATION OF REALTORS
P.O. Box 7338
Reno, NV 89510
(702) 329-6648
Patsy Redmond - Executive Vice President

NEW HAMPSHIRE ASSOCIATION
115 A. Airport Rd.
Box 515A

Concord, NH 03301
(603) 225-5549
Ashton Welch - Executive Vice President

NEW JERSEY ASSOCIATION OF REALTORS
P.O. Box 2098
Edison, NJ 08818
(201) 494-5616
Robert Ferguson - Executive Vice President

NEW MEXICO, REALTORS ASSOCIATION OF
P.O. Box 1569
Santa Fe, NM 87504
(505) 258-2361
William A. Neal - Executive Vice President

NEW YORK STATE ASSOCIATION
107 Washington Ave.
Box 122
Albany, NY 12260
(518) 462-9563
Charles M. Staro - Executive Vice President

NORTH CAROLINA ASSOCIATION OF REALTORS
P.O. Box 7918
Greensboro, NC 27417
(919) 294-1415
James L. Bichsel - Executive Vice President

NORTH DAKOTA ASSOCIATION OF REALTORS
1910 North 11th St.
Bismark, ND 58501
(701) 258-2361
Claus H. Lembke - Executive Vice President

OHIO ASSOCIATION OF REALTORS
200 East Town St.
Columbus, OH 43215
(614) 228-6675
Almon R. Smith - Executive Vice President

OKLAHOMA ASSOCIATION OF REALTORS
9807 North Broadway Extension
Oklahoma City, OK 73114
(405) 848-9944
Stan Orr - Executive Vice President

OREGON ASSOCIATION OF REALTORS
P.O. Box 351
Salem, OR 97308
(503) 362-3645
Bill Cozart - Executive Vice President

PENNSYLVANIA ASSOCIATION OF REALTORS
4501 Chambers Hill Rd.
Harrisburg, PA 17111
(717) 561-1303
John J. Bernardo - Executive Vice President
Beth Winters - Director of Member Services

RHODE ISLAND REALTORS ASSOCIATION
120 Lavan St.
Warwick, RI 02888
(401) 785-3650
Thomas E. Mulhearn - Executive Vice President

SOUTH CAROLINA ASSOCIATION OF REALTORS
3780 Fernandina Rd.
Box 21827
Columbia, SC 29221
(803) 772-5206
Richard Davis - Executive Vice President

SOUTH DAKOTA ASSOCIATION OF REALTORS
411 East Capitol St.
Box 386
Pierre, SD 57501
(605) 224-9211
Wayne Olson - Executive Vice President

TENNESSEE ASSOCIATION OF REALTORS
1910 Adelicia St.
Nashville TN 37212
(615) 321-0515
Stephen D. Harding - Executive Vice President

TEXAS ASSOCIATION OF REALTORS
P.O. Box 14488
Austin, TX 78761
(512) 836-8630
James T. Sawyer - Executive Vice President

UTAH ASSOCIATION OF REALTORS
5710 South Green St.
Murray, UT 84123
(801) 268-4747
Pat Iannone - Executive Vice President

VERMONT ASSOCIATION OF REALTORS
P.O. Box 1074
Montpelier, VT 05602
(802) 229-0513
Gar Anderson - Executive Vice President

VIRGINIA ASSOCIATION OF REALTORS
P.O. Box 15719
Richmond, VA 23227
(804) 264-5033
Ralph K. Anderson - Executive Vice President

WASHINGTON ASSOCIATION OF REALTORS
P.O. Box 719
Olympia, WA 98507
(206) 943-3100
Steve A. Hyer - Executive Vice President

WASHINGTON DC ASSOCIATION
777 14th St., NW
Suite 200
Washington DC 20005

(202) 628-4494
David Strachen - Executive Vice President

WEST VIRGINIA ASSOCIATION OF REALTORS
814 Virginia St.
Charleston, WV 25301
(304) 342-7600
J. Jackson Pauley - Executive Vice President

WISCONSIN ASSOCIATION OF REALTORS
4801 Hayes Rd.
Madison, WI 53704
(608) 241-2047
William E. Malkasian - Executive Vice President

WYOMING ASSOCIATION OF REALTORS
P.O. Box 2312
Casper, WY 82602-2312
(307) 237-4085
Rosemary Cash Tucker - Executive Vice President

(End NAR Listings)

NATIONAL HOMEOWNERS ASSOCIATION (NHA)
1906 Sunderland Pl., NW
Washington, DC 20036
(202) 223-1453
John L. Weidlein - President
Activities:
Advocacy organization for home owners. Provides education and consumer programs. Monitors legislation and regulations affecting homeowners.

NATIONAL MULTI HOUSING COUNCIL (NMHC)
1250 Connecticut Ave., NW
Suite 620
Washington, DC 20036
(202) 659-3381
Jonathan L. Kemner - Executive Officer
Activities:
Builders, developers, owners, and managers of condominium developments and multi-family housing units. Represents industry on federal legislation and regulatory activity.

REAL ESTATE EDUCATORS ASSOCIATION (REEA)
230 North Michigan Ave.
Suite 1200
Chicago, IL 60601
(312) 372-9800
George Otto - Executive Director
Activities:
Members provide real estate education and training. Ensures consistent training programs in real estate education and standardization of content in training programs.

REAL ESTATE SECURITIES AND SYNDICATION INSTITUTE
430 North Michigan Ave.
Chicago, IL 60611
(312) 670-6760
Cindy Martin
Activities:
Members of the real estate industry involved in real estate syndication or other types of real estate securities. Represents members interests concerning legislation and regulations of the industry.

REALTORS LAND INSTITUTE (RLI)
430 North Michigan Ave.
Chicago, IL 60611
(312) 329-8440
Linda J. Thompson - Executive Vice President
Activities:
Brokers and salespersons specializing in management, development, appraisal, or selling of land. Maintains education program for real estate brokers.

REALTORS NATIONAL MARKETING INSTITUTE (RNMI)
430 North Michigan Ave.
Chicago, IL 60611
(312) 670-3780
Nancy Gentile - Vice President, Administrative Affairs
Activities:
Facilitates cooperation among brokers engaged in activities such as buying, selling, trading, leasing, or renting real estate. Members must be Certified Residential Specialists, Certified Real Estate Brokerage Managers, or Certified Commercial Investment Members. Offers over 200 courses annually on real estate issues.

SOCIETY OF INDUSTRIAL AND OFFICE REALTORS (SIRO)
777 14th St., NW
Washington, DC 20005
(202) 383-1150
Nancy Bryant - Executive Vice President
Activities:
Brokers specializing in industrial/office properties and representatives of corporations, utilities, and industrial park developments. Develops and oversees surveys of plants or site locations and analyzes special problems facing industria development.

Conferences and Events

This section lists conferences and events sponsored by organizations within fields of relevance to the radon industry. These areas of interest include building and construction, environment, health, industry, and real estate. Additional information on the content and schedule of these events can be obtained by directly contacting the sponsoring organization.

ADHESIVE AND SEALANT COUNCIL (ASC)/SEMI-ANNUAL CONVENTIONS 1989
1627 K St., NW
Suite 1000
Washington DC 20006
(202) 452-1500
Contact: Kathy Oates, Conference Information
Conference Activities:
Forum for presentation and discussion of issues developing within the industry. Includes exhibits.
Location:
Airport Marriott, Los Angeles, CA (March).
O'Hare Hyatt, Chicago, IL (October).
Dates:
March 19-25, 1989.
October 18-22, 1989.

AIR CONDITIONING CONTRACTORS OF AMERICA (ACCA)/ANNUAL CONVENTION 1989
1513 16th St., NW
Washington, DC 20036
(202) 483-9370
Contact: Christie Higgins, Meetings Coordinator
Conference Activities:
Seminars and workshops include topics such as new designs for improved indoor air quality. Includes exhibits.
Location:
Charleston, SC.
Dates:
February, 1989.

AIR MOVEMENT AND CONTROL ASSOCIATION (AMCA)/HOME VENTILIATING INSTITUTE (HVI)/1989 ANNUAL CONVENTION
30 West University Dr.
Arlington Heights, IL 60004
(312) 394-0150
Contact: Peter Hanly, Conference Information
Conference Activities:
Presentations cover latest technological advancements in home ventilation systems. Indoor air quality issues are also addressed. Includes exhibits.
Location:
La Paloma Resort, Phoenix, AZ.
Dates:
October 13-17, 1989.

AIR POLLUTION CONTROL ASSOCIATION (APCA)/1989 ANNUAL CONVENTION
Three Gateway Center, Four West
Pittsburgh, PA 15222
(412) 232-3444

Contact: Sharon Deandra, Convention Information
Conference Activities:
Presentations address major factors contributing to indoor air pollution and technological developments affecting indoor air issues. Includes exhibits.
Location:
Anaheim, CA.
Dates:
June 25-30, 1989.

AIR POLLUTION CONTROL ASSOCIATOIN (APCA)/SYMPOSIA AND SEMINARS
Three Gateway Center, Four West
Pittsburgh, PA 15222
(412) 232-3444
Contact: Steve Stasko, Manager of Technical Programs
Conference Activities:
Programs include:
• EPA/APCA Hazardous Waste Symposium
February 21-22, 1989, Cincinnati, OH.
• US Government Affairs Seminar
March 14-15, 1989, Washington, DC.
• EPA/APCA International Symposium on "Measurement of Toxic and Related Air Pollutants"
May 2-4, 1989, Raleigh, NC.

AMERICAN ACADEMY OF ENVIRONMENTAL MEDICINE (AAEM)/23RD SCIENTIFIC SESSION-24TH ANNUAL MEETING
P.O. Box 16106
Denver, CO 80216
(303) 622-9755
Contact: Judy Howard, Convention Information
Conference Activities:
Presentations on latest developments in environmental medicine, environmental hazards and indoor environments.
Location:
Atlanta, GA.
Dates:
October 29-November 2, 1989.

AMERICAN ASSOCIATION FOR THE ADVANCEMENT OF SCIENCE (AAAS)/1989 ANNUAL CONFERENCE
1333 H St., NW
Washington, DC 20005
(202) 326-6400
Contact: Rosita Johnson, Meetings Planner
Conference Activities:
Complete schedule of workshops, seminars and events available from AAAS.
Location:
San Francisco Hilton, San Francisco, CA.
Dates:
January 14-19, 1989.

AMERICAN ASSOCIATOIN OF RADON SCIENTISTS AND TECHNOLOGISTS (AARST)/1989 ANNUAL CONFERENCE
P.O. Box 70
Park Ridge, NJ 07656
(201) 391-6445
Contact: Jim Craffey, Executive Secretary
Conference Activities:
Presentations on radon/radon progeny, measurement and mitigation techniques, government legislation and regulation, geologic and hydrogeologic developments and indications, real estate, legal issues, and general trends within the radon industry. Includes exhibits.

AMERICAN COLLEGE OF OCCUPATIONAL MEDICINE (ACOM)/SEMI-ANNUAL CONFERENCES
2340 South Arlington Heights Rd.
Arlington Heights, IL 60005
(312) 228-6850
Contact: Nancy Kay Olsson, Conference Director
Conference Activities:
Presentations on research and major issues concerning occupational medicine.
Location:
Boston, MA (Spring).
Baltimore, MD (Autumn).
Dates:
April 29-May 5, 1989.
October 30-November 3, 1989.

AMERICAN CONCRETE INSTITUTE (ACI)/SEMI-ANNUAL MEETINGS 1989
P.O. Box 19150
22400 West Seven Mile Rd.
Detroit, MI 48219
(313) 352-2600
Contact: Meetings Information
Conference Activities:
Presentations on issues and developments within the concrete industry.
Location:
Atlanta, GA (February).
San Diego, CA (November).
Dates:
February 19-24, 1989.
October 29-November 3, 1989.

AMERICAN COUNCIL OF GOVERNMENT INDUSTRIAL HYGIENISTS (ACGIH)-AMERICAN INDUSTRIAL HYGIENE ASSOCIATION (AIHA)/1989 COMBINED CONVENTION
475 Wolf Ledges Hwy.
Akron, OH 44311
(216) 762-7294
Contact: Kim Milush, Meetings Coordinator
Conference Activities:
Co-sponsored annual convention covers technological

developments as well as timely issues concerning industrial hygiene. Includes exhibits.
Location:
Cervantes Convention Center, St. Louis, MO.
Dates:
May 21-26, 1989.

AMERICAN GEOLOGICAL INSTITUTE (AGI)/1989 ANNUAL CONVENTION
4220 King St.
Alexandria, VA 22310
(703) 379-2480
Contact: Patricia Okaya, Meetings Information
Conference Activities:
Information on the 1989 convention may be obtained by contacting AGI headquarters.

AMERICAN GEOPHYSICAL UNION (AGU)/1989 SEMI-ANNUAL CONFERENCES
2000 Florida Ave.
Washington, DC 20009
(202) 462-6903
Contact: Meetings Information
Conference Activities:
Presentations on broad range of geologic, physical, geochemical, and environmental topics.
Location:
Baltimore, MD (Spring).
San Francisco, CA (Autumn).
Dates:
May 8-12, 1989.
December 4-8, 1989.

AMERICAN INSTITUTE OF ARCHITECTS (AIA)/1989 ANNUAL CONVENTION
1735 New York Ave.
Washington, DC 20006
(202) 626-7452
Contact: John Gaillard, Senior Manager-Conventions
Conference Activities:
Addresses major architecural issues and provides seminars and workshops on issues such as design priorities, construction techniques, and indoor air quality.
Location:
St. Louis, MO.
Dates:
May 7-10, 1989.

AMERICAN INSTITUTE OF REAL ESTATE APPRAISERS (AIRA)/1989 ANNUAL CONVENTION AND TRADE SHOW
430 North Michigan Ave.
Chicago, IL 60611
(312) 329-8437
Contact: Convention Information
Conference Activities:
Includes topics on effect of radon issue upon the real estate appraisal industry. Includes exhibits.
Location:
Dallas, TX.
Dates:
November 9-14, 1989.

AMERICAN LUNG ASSOCIATION (ALA)-AMERICAN THORACIC SOCIETY (ATS)/ANNUAL MEETING
1740 Broadway
New York, NY 10019
(212) 315-8700
Contact: Meetings and Conventions Division
Conference Activities:
Presentations cover issues on health threat posed by radon and other indoor air quality concerns.
Location:
Cincinnati, OH.
Dates:
May 14-17, 1989.

AMERICAN MEDICAL ASSOCIATION (AMA)/1989 RADON EDUCATION PROGRAM
535 North Dearborn St.
Chicago IL 60610
(312) 645-5000
Contact: William R. Hendee, Science and Technology
Conference Activities:
Presentations to health professionals and community leaders concerned about radon and its health effects.
Location:
Various Locations.

AMERICAN NUCLEAR SOCIETY (ANS)/1989 ANNUAL CONFERENCE
555 North Kensington
La Grange Park, IL 60525
(312) 352-6611
Contact: Mary Keenan, Meetings Director
Conference Activities:
Presentations address nuclear research, energy, and radiation. Includes exhibits.
Location:
San Francisco Hilton, San Francisco, CA.
Dates:
May 8-12, 1989.

AMERICAN PUBLIC HEALTH ASSOCIATION (APHA)/1989 ANNUAL CONFERENCE
1015 15th St., NW
Washington, DC 20005
(202) 789-5663
Contact: Bob Johnson, Convention Manager
Conference Activities:
Convention addresses important issues facing public health officials, including their responsibilities concerning radon-related issues.

Location:
Chicago, IL.
Dates:
October 22-26, 1989.

AMERICAN RADON ASSOCIATON/1989 ANNUAL TRADE CONFERENCE
4949 South 25-A
Tipp City, OH 45371
(513) 667-2464
Contact: Joan Kaup, Administrative Assistant
Conference Activities:
Covers developments in testing methods, new instrumentation, and remediation techniques. Contact ARA headquarters for further details.
Dates:
February, 1988.

AMERICAN SOCIETY FOR TESTING AND MATERIALS (ASTM)/1989 COMMITTEE ON ATMOSPHERE CONFERENCE
1916 Race St.
Philadelphia, PA 19103
(215) 299-5400
Contact: George Luciw, Committee Chairman
Conference Activities:
Meetings address issues concerning radon, its measurement, and general indoor air quality questions.
Location:
Walt Disney World Village, Orlando, FL.
Dates:
December 4-6, 1989.

AMERICAN SOCIETY OF HEATING, REFRIGERATING AND AIR CONDITIONING ENGINEERS (ASHRAE)/1989 ANNUAL CONVENTION
1791 Tullie Circle Rd., NE
Atlanta, GA 30329
(404) 636-8400
Contact: Judy Marshall, Convention Coordinator
Conference Activities:
Series of seminars, workshops, and presentations covering important issues on ventilation, indoor air quality, and radon.
Location:
Hyatt Four Seasons, Vancouver, BC.
Dates:
June 25-28, 1989.

AMERICAN SOCIETY OF HOME INSPECTORS (ASHI)/ASHI ANNUAL CONVENTION
3299 K St., NW
7th Floor
Washington, DC 20007
(202) 842-3096
Contact: Marsha Lyons, Conference Coordinator
Conference Activities:
Issues concerning all aspects of home inspection are addressed at annual convention. Includes exhibits.
Location:
Chicago, IL.
Dates:
January 12-15, 1989.

AMERICAN SOCIETY OF REAL ESTATE COUNSELORS (ASREC)/1989 ANNUAL CONVENTION AND TRADE EXPOSITION
430 North Michigan Ave.
Chicago, IL 60611
(312) 329-8437
Contact: Convention Information
Conference Activities:
Presentations address major issues concerning education and government regulatory activity.
Location:
Dallas, TX.
Dates:
November 9-14, 1989.

AMERICAN SOCIETY OF SAFETY ENGINEERS (ASSE)/1989 ANNUAL CONVENTION
1800 East Oakton
Des Plains, IL 60018
(312) 692-4121
Contact: Pat DeLaney, Convention Coordinator
Conference Activities:
Presentations include topics covering workplace and industrial safety issues.
Location:
Cobol Hall, Detroit, MI.
Dates:
June 11-14, 1989.

AMERICAN WATER WORKS ASSOCIATION (AWWA)/1989 ANNUAL CONFERENCE
6666 West Quincy Ave.
Denver, CO 80235
(303) 794-7711
Contact: Kurt M. Keeley, Director of Information Services
Conference Activities:
Conference addresses topics such as radionuclides in water and federal/state legislation affecting water quality issues.
Location:
San Diego Convention Center, San Diego, CA.
Dates:
June 8-12, 1989.

ASSOCIATION OF STATE AND INTERSTATE WATER POLLUTION CONTROL ADMINISTRATORS (ASIWPCA)/1989 ANNUAL MEETING
444 North Capitol St.
Washington, DC 20001

(202) 624-7782
Contact: Robbi Savage, Executive Director
Conference Activities:
Meeting addresses major issues facing drinking water officials including government regulations and coordination of state effort to ensure safe drinking water supplies.
Location:
Nashville, TN.
Dates:
August 13-16, 1989.

ASSOCIATION OF STATE AND TERRITORIAL HEALTH OFFICIALS (ASTHO)/1989 ANNUAL MEETING
1311 Dolly Madison Blvd.
McLean, VA 22101
(703) 556-9222
Contact: Maxine Mantell, Associate Director
Conference Activities:
Presentations on health regulations, major health issues, and health costs.
Location:
Vail, CO.
Dates:
April 4-7, 1989.

ASSOCIATION OF STATE DRINKING WATER ADMINISTRATORS (ASDWA)/FOURTH ANNUAL CONFERENCE
1911 North Ft. Myer Dr.
Arlington, VA 22209
(703) 524-2428
Contact: Janet McGoldrick, Assistant to Director
Conference Activities:
Program includes panel discussions on:
• Review of the Safe Drinking Water Act Amendments of 1986, including a discussion of radionuclides regulation.
• State Initiatives and Issues of Importance to State Programs, including presentation on radon.
Location:
Hilton East, Tucson, AZ.
Dates:
February 20-23, 1989.

BUILDING OFFICIALS AND CODE ADMINISTRATORS INTERNATIONAL (BOCA)/1989 ANNUAL CONVENTION
4051 West Flossmore Rd.
Country Club Hills, IL 60477
(312) 799-2300
Contact: Karen Biggs, Assistant to the Director
Conference Activities:
Seminars and workshops covering legislation, codes, education, and new developments. Includes exhibits.
Location:
Schaumberg, IL.
Dates:
June 18-23, 1989.

BUILDING OWNERS AND MANAGERS ASSOCIATION INTERNATIONAL (BOMA)/82ND ANNUAL CONVENTION AND OFFICE BUILDING SHOW
1250 Eye St., NW
Suite 200
Washington, DC 20005
(202) 289-7000
Contact: Chuck Glazer, Staff Vice President
Conference Activities:
Seminars and workshops covering building operations, specialized buildings, government relations, and statistics and research. Includes exhibits.
Location:
Merchandise Mart, Atlanta, GA.
Dates:
June 18-21, 1989.

CANADIAN HOMEBUILDERS ASSOCIATION/46TH ANNUAL CONFERENCE AND EXPOSITION
5218 Young St.
Willowdale, ON M2H 5P6
(416) 224-1436
Contact: Linda Barrett, Conference Coordinator
Conference Activities:
Presentations cover major trends in Canadian homebuilding industry, including construction issues, radon intrusion, and ventilation systems. Includes exhibits.
Location:
Hamilton, ON.
Dates:
February 5-8, 1989.

CONFERENCE OF RADIATION CONTROL PROGRAM DIRECTORS (CRCPD)/ANNUAL MEETING
71 Fountain Pl.
Frankfort, KY 40601
(502) 227-4543
Contact: Charles Hardin, Director
Conference Activities:
Major current radiation protection issues are addressed by participants from state and local radiation protection program, state agencies responsible for radiation protection, professional associations, industry, and the general public.
Location:
Baton Rouge Hilton, Baton Rouge, LA.
Dates:
May 21-25, 1989.

EMPLOYEE RELOCATION COUNCIL (ERC)/SPRING CONFERENCE
1720 N St., NW
Washington, DC 20036
(202) 857-0857
Contact: Cici Thompson, Conference Director

Conference Activities:
Addresses major issues facing relocation industry such as real estate concerns and questions concerning development of a relocation industry radon policy.
Location:
San Antonio, TX.
Dates:
May 17-20, 1989.

ENERGY EFFICIENT BUILDING ASSOCIATION (EEBA)/SEVENTH ANNUAL INTERNATIONAL ENERGY EFFICIENT BUILDING CONFERENCE AND EXPOSITION: EXCELLENCE IN HOUSING '89
University of Southern Maine
EEBA/Technology Center
Gorham, ME 04038
(207) 780-5143
Contact: Howard Faulkner, Executive Director
Conference Activities:
Presenters introduce the latest developments in building for health, durability, affordability, and energy efficiency. A special series of one-day workshops will be offered the day before the general conference. Topics include radon mitigation techniques and ventilation systems. Includes exhibits.
Location:
Ft. Garry Place, Winnipeg, MB.
Dates:
March 2-4, 1989.

ENVIRONMENTAL HEALTH WATCH OF GREATER CLEVELAND/BLUEPRINT FOR A HEALTHY HOUSE ANNUAL SEMINAR
4115 Bridge Ave.
Suite 104
Cleveland, OH 44113
(216) 961-4646
Contact: Robert Staib, Education Director
Conference Activities:
Comprehensive program on indoor air quality issues facing homebuilders. Includes sections on radon, testing, abatement, building/design alternatives and strategies, and health issues.
Location:
Cleveland, OH.
Dates:
October, 1989.

EXECUTIVE ENTERPRISES, INC./ENVIRONMENTAL REGULATION COURSE
22 West 21st St.
New York, NY 10010-6904
(800) 831-8333
Contact: Susan MacNamara, Program Director
Conference Activities:
Policy and regulatory issues addressed include:
• The process of federal environmental regulation.
• The National Environmental Policy Act.
• The Clean Air Act.
• The Clean Water Act.
• Superfund/CERCLA.
• The Toxic Substances Control Act.
• Enforcement of rules/regulations.
Faculty includes EPA personnel and industry representatives.
Location:
The Peabody, Orlando, FL.
Dates:
January 17-19, 1989.

EXPANDED SHALE, SLATE AND CLAY INSTITUTE (ESSCI)/ANNUAL CONVENTION
6218 Montrose Rd.
Rockville, MD 20852
(301) 231-9497
Contact: Harry Robinson, Managing Director
Conference Activities:
Seminars and workshops covering research and new development within the industry.
Location:
Seattle, WA.
Dates:
October 1-4, 1989.

GEOLOGICAL SOCIETY OF AMERICA (GSA)/1989 ANNUAL CONFERENCE
P.O. Box 9140
3300 Penrose Pl.
Boulder, CO 80301
(303) 447-2020
Contact: Sue Beggs, Meetings Chairperson
Conference Activities:
Covers important issues concerning geophysics, geochemistry, and hydrogeology.
Location:
St. Louis, MO.
Dates:
November 6-9, 1988.

HARVARD SCHOOL OF PUBLIC HEALTH/RADON IN BUILDINGS: SOURCES, BIOLOGICAL EFFECTS, MONITORING AND CONTROL
Office of Continuing Education
677 Huntington Ave.
Boston, MA 02115
(617) 732-1171
Contact: Beth Ellis, Marketing Coordinator
Conference Activities:
Seminar designed to educate contractors, real estate agents, homeowners, and federal, state, and local government officials involved with radon issues.
Location:
Boston, MA.
Dates:
Late 1989.

HEALTH PHYSICS SOCIETY/1989 ANNUAL CONVENTION
800 Westpark Dr.
Suite 400
McLean, VA 22102
(703) 790-1745
Contact: Sue Burke, Meetings Manager
Conference Activities:
Presentations cover radiation protection, radiation safety, radon exposure, and radon hazards.
Location:
Albuquerque, NM.
Dates:
June 25-29, 1989.

MINING ASSOCIATION OF CANADA/ANNUAL GENERAL MEETING
350 Sparks St.
Suite 705
Ottawa, ON K1R 7S8
(613) 233-9391
Contact: Geraldine Higgs, Information
Conference Activities:
Presentations covering mining issues including uranium mine concerns and regulations.
Location:
Westin Hotel, Ottawa, ON.
Dates:
May 24, 1989.

NATIONAL ASSOCIATION OF COUNTY HEALTH OFFICIALS (NACHO)/1989 ANNUAL CONVENTION
440 1st St., NW
Washington, DC 20001
(202) 783-5550
Contact: Kathy Dobbins, Information.
Conference Activities:
Seminars and workshops covering issues of concern to local health officials. Presentations on new regulations and requirements.
Location:
Cincinnati, OH.
Dates:
July 15-18, 1989.

NATIONAL ASSOCIATION OF ENVIRONMENTAL PROFESSIONALS (NAEP)/1989 ANNUAL CONFERENCE
P.O. Box 9400
Washington, DC 20016
(202) 660-2364, (505) 884-0950
Contact: Donald Van Buskirk
Conference Activities:
Covers major developments within areas of environmental health education, training, and public policy.
Location:
Hyatt Hotel, Lake Tahoe, CA.
Dates:
June 19-21, 1989.

NATIONAL ASSOCIATION OF HOMEBUILDERS (NAHB)/1989 ANNUAL CONVENTION
15th and M Sts., NW
Washington, DC 20005
(202) 822-0200, (800) 368-5242
Contact: Donald Reynolds, Convention Information
Conference Activities:
Conference presentations include series of reports on radon issues and the home building industry.
Location:
Georgia World Congress Center, Atlanta, GA.
Dates:
January 20-23, 1989.

NATIONAL ASSOCIATION OF REAL ESTATE LICENSE LAW OFFICIALS (NARELLO)/1989 ANNUAL CONVENTION
P.O. Box 129
Centerville, UT 84010-0129
(801) 531-8202
Contact: Linda Newman, Administrative Assistant
Conference Activities:
Conference covers regulatory issues and new developments related to the real estate industry.
Location:
San Diego, CA.
Dates:
October 10-13, 1989.

NATIONAL ASSOCIATION OF REALTORS (NAR)/1989 ANNUAL CONVENTION AND TRADE EXPOSITION
430 North Michigan Ave.
Chicago, IL 60611
(312) 329-8200
Contact: Carol Hart, Convention Information
Conference Activities:
Conference includes workshops, presentations and seminars that cover all aspects of the real estate industry. Includes topics on radon and real estate values, regulatory activities, legal issues, and realtor responsibilities.
Location:
Dallas, TX.
Dates:
November 9-14, 1989.

NATIONAL ASSOCIATION OF WATER COMPANIES (NAWC)/1989 ANNUAL CONVENTION
1725 K St., NW
Suite 1212
Washington, DC 20006
(202) 833-8383
Contact: Marilyn Miller, Director of Administration

Conference Activities:
Addresses issues of concern to privately owned and operated water companies.
Location:
The Broadmoor, Colorado Springs, CO.
Dates:
October 8-12, 1989.

NATIONAL CONFERENCE OF STATE LEGISLATURES (NCSL)/1989 ANNUAL CONFERENCE
1050 17th St.
#2100
Denver, CO 80265
(303) 623-7800
Contact: Linda Worrell, Information
Conference Activities:
Addresses legislative issues concerning states, including environmental and technology policy.
Location:
Tulsa, OK.
Dates:
August 6-11, 1989.

NATIONAL CONFERENCE OF STATES ON BUILDING CODES AND STANDARDS (NCSBCS)/22ND ANNUAL CONVENTION
481 Carlisle Dr.
Herndon, VA 22070
(703) 437-0100
Contact: Debbie Baillargeon, Assistant to the Director
Conference Activities:
Presentations center on topical issues concerning development and modification of building codes and standards.
Location:
Marriott Hotel, Newport, RI.
Dates:
September 17-20, 1989.

NATIONAL COUNCIL ON RADIATION PROTECTION AND MEASUREMENTS (NCRP)/1989 ANNUAL SCIENCE MEETING: RADIATION PROTECTION TODAY-NCRP AT 60
7910 Woodmont Ave.
Suite 800
Bethesda, MD 20814
(301) 657-2652
Contact: Cindy O'Bryan, Conference Director
Conference Activities:
Covers radiation research topics and developments in field.
Location:
Washington, DC.
Dates:
April 5-6, 1989.

NATIONAL ENVIRONMENTAL HEALTH ASSOCIATION (NEHA)/1989 ANNUAL CONVENTION
720 South Colorado Blvd.
South Tower, Suite 970
Denver, CO 80222
(303) 756-9090
Contact: Mel Monkelis, Convention Information
Conference Activities:
Presentations cover major environmental health issues including radiation exposure topics.
Location:
Westin Hotel, Seattle, WA.
Dates:
June 24-28 1989.

NATIONAL GOVERNORS ASSOCIATION (NGA)/1989 ANNUAL CONVENTION
Hall of States, Suite 250
444 North Capitol St., NW
Washington, DC 20001
(202) 624-5300
Contact: Raymond C. Scheppach
Conference Activities:
Forum for discussion on states' role in major issues including environment, housing, and education.
Location:
Hyatt Regency, Chicago, IL.
Dates:
July 30-August 1, 1989.

NATIONAL INSTITUTE OF BUILDING SCIENCES (NIBS)/1989 ANNUAL CONVENTION
1015 15th St., NW
Suite 700
Washington, DC 20005
(202) 347-5710
Contact: Dean Phelus
Conference Activities:
Major issues concerning building industry are addressed. Topics include radon concerns in the building/construction industry. New construction techniques are highlighted.

NATIONAL RADON ASSOCIATION (NRA)/NATIONAL RADON CONFERENCE
2786 North Decatur Rd.
Suite 240
Decatur, GA 30033
(404) 292-1113
Contact: Stewart Huey, Executive Secretary
Conference Activities:
Issues such as radon fundamentals, radon testing/evaluation and remediation, impact upon the real estate industry, government legislation and regulation, and latest research developments are addressed at a series of meetings by leading persons involved in radon issues. Workshops designed for members of the real estate, construction, and radon industries. Includes exhibits.

Location:
Hyatt Regency, Cincinnati, OH.
Dates:
March 21-23, 1989.

NATIONAL SOCIETY OF PROFESSIONAL ENGINEERS (NSPE)/1989 ANNUAL CONVENTION
1420 King St.
Alexandria, VA 22314
(703) 684-2800
Contact: Wanda Parten, Convention Information
Conference Activities:
Presentations include issues concerning mining engineering, environmental engineering, construction engineering, and technical developments.
Location:
Minneapolis, MN.
Dates:
July 23-29, 1989.

NATIONAL WATER WELL ASSOCIATION (NWWA)/1989 ANNUAL CONFERENCE
6375 Riverside Drive
Dublin, OH 43017
(614) 761-1711
Contact: Kathy Butcher, Director of Conferences
Conference Activities:
Presentations address a variety of water well issues including environmental concerns, pollutants, and technological developments.
Location:
Astrodome, Houston, TX.
Dates:
October 13-November 2, 1989.

RADIATION RESEARCH SOCIETY (RRS)/1989 ANNUAL CONFERENCE
1101 Market St.
14th Floor
Philadelphia, PA 19107
(215) 574-3153
Contact: Meg Kaiser, Information
Conference Activities:
Reports, workshops, and seminars addressing important radiation issues.
Location:
Seattle, WA.
Dates:
March 19-23, 1989.

RAMAPO COLLEGE OF NEW JERSEY/INSTITUTE OF ENVIRONMENTAL STUDIES/TIGHT HOUSES INFORMANTION PROGRAM
505 Ramapo Valley Rd.
Mahwah, NJ 07430
(201) 529-7749
Contact: Michael Edelstein, Professor
Conference Activities:
Program is designed to explore relationships between energy efficiency and radon and other indoor air contaminants.
Location:
Mahwah, NJ.
Dates:
April 28, 1989.

REAL ESTATE EDUCATORS ASSOCIATION (REEA)/ANNUAL CONVENTION
230 North Michigan Ave.
Suite 1200
Chicago, IL 60611
(312) 372-9800
Contact: Convention Information
Conference Activities:
Addresses major issues concerning real estate education and training programs.
Location:
Chicago, IL.
Dates:
June, 1989.

REALTORS LAND INSTITUTE (RLI)/1989 ANNUAL CONVENTION AND TRADE EXPOSITION
430 North Michigan Ave.
Chicago, IL 60611
(312) 329-8440
Contact: Convention Information.
Conference Activities:
Addresses topics related to the management, development, appraisal, and selling of land. Includes exhibits.
Location:
Dallas, TX.
Dates:
November 9-14, 1989.

REALTORS NATIONAL MARKETING INSTITUTE (RNMI)/1989 ANNUAL CONVENTION AND TRADE EXPOSITION
430 North Michigan Ave.
Chicago, IL 60611
(312) 670-3780
Contact: Convention Information
Conference Activities:
Addresses issues concerning brokers and their interests in the real estate market. Includes exhibits.
Location:
Dallas, TX.
Dates:
November 9-14, 1989.

SHEET METAL AND AIR CONDITIONING CONTRACTORS NATIONAL ASSOCIATION (SMACCNA)/1989 ANNUAL CONVENTION
P.O. Box 70
Merrifield, VA 22116
(703) 790-9890
Contact: Mary Lou Taylor, Director of Conferences
Conference Activities:
Conference addresses major issues in ventilation systems and indoor air quality issues. Includes exhibits.
Location:
Honolulu, HI.
Dates:
October 2-5, 1989.

SOUTHERN BUILDING CODE CONGRESS INTERNATIONAL (SBCCI)/1989 ANNUAL CONVENTION
900 Montclair Rd.
Birmingham, AL 35213
(205) 591-1853
Contact: Bill Tangye, Convention Information
Conference Activities:
Conference workshops and seminars cover major issues concerning building code regulations and new developments. Includes exhibits.
Location:
Charlotte, NC.
Dates:
October 23-26, 1989.

Radon Publications

This section lists recent books, periodicals, articles, and other publications representative of the broad range of radon-related information. The section is organized into seven categories as follows:

Building and Construction
General Overviews
Health Effects
Radon In Water
Radon Measurement
Radon Mitigation
Radon Newsletters

Building And Construction

Airborne Radionuclides And Radiation In Buildings: A Review.
Nero, A.V.
Health Physics 45:2:303-322, AUG 1983.

Control Of Formaldehyde And Radon In The Built Environment: A Survey Of The Literature.
Alder, K., Et Al.
Lawrence Berkeley Laboratory, Berkeley, CA, SEP 1982.
Order #: LBID-625

Handbook Of Radon In Buildings: Detection, Safety, And Control.
Science Information Resource Center.
Hemisphere Publishing Corporation, Philadelphia, PA, 1988.

Indoor Air Quality In Tight Houses: A Literature Review.
Coon, D.
Ontario Ministry Of Municipal Affairs And Housing, Toronto, ON, 1984.
Order #: NTIS PB85-162022

Method For Measuring The Exhalation Of Radon From Building Materials.
Ingersoll, J.G., Stitt, B.D., and Zapalac, G.M.
Lawrence Berkeley Laboratory, Berkely, CA, 1982.
Order #: NTIS DE82013877

Radon Dosimetry: A Review Of Radon And Radon Daughter Exposure Conditions In Dwellings And Other Structures.
Ryan, M.T., et al.
Oak Ridge National Laboratory, Oak Ridge, TN, 1983.
Order #: NTIS DE83-015033

Radon In Buildings: Proceedings Of A Roundtable Discussion Of Radon In Buildings.
Colle, R. and McNall, P.E., Jr.
National Bureau Of Standards, Washington, DC, 1980.
Order #: NTIS PB80-195712

Radon Reduction In New Construction.
U.S. Environmental Protection Agency, Washington,

DC, 1987.
Order #: OPA-87-009

Radon Transport Through And Exhalation From Building Materials: A Review And Assessment.
Colle, R., Et Al.
National Bureau Of Standards, Washington, DC, 1981.
Order #: NTIS PB82-112384

Report Of The Task Force On Radon In Structures.
Radiation Policy Council, Washington, DC, AUG 1980.
Order #: RPC-80-002

Survey Of Radionuclide Contents And Radon Emanation Rates In Building Materials Used In The U.S., A.
Ingersoll, J.G.
Health Physics, AUG 1983.

General Overviews

Air Pollution: Hazards Of Indoor Radon Could Pose A National Health Problem.
Report To Pennsylvania Congressional Delegation, House Of Representatives
U.S. General Accounting Office, Washington, DC, JUN 1986.
Order #: GAO/RCED-86-170

Background Atmospheric 222-Rn Concentrations Outdoors And Indoors: A Review.
Gesell, Thomas F.
Health Physics 45:2:289-302, AUG 1983.

Best's Safety Directory, 28th Edition
A. M. Best Company
A. M. Best Company, Inc., Oldwick, NJ, 1988

Citizen's Guide To Radon, A: What It Is And What To Do About It.
U.S. Environmental Protection Agency
Government Printing Office, Washington, DC, 1986.
Order #: OPA 86-004

Current Federal Indoor Air Quality Activities.
U.S. Environmental Protection Agency, Office Of Program Development, Office Of Air And Radiation, Washington, DC, MAR 1988.
Order #: EPA 520/1-87-19-2

Distribution Of Airborne Radon-222 Concentrations In U.S. Homes.
Nero, A.V., Schwehr, M.B., Nazaroff, W.W., and Revzan, K.L.
Lawrence Berkeley Laboratory, Berkeley, CA, JUL 1985.
Order #: LBL-18274

Elevated Radon Levels In Today's "Air Tight" Homes.
Steva, D.P.
Proceedings Of The Ninth National Passive Solar Conference.
American Solar Energy Society, Boulder, CO, 1984.

Environmental Radon.
Cothern, C.R. and Smith, J.E., Jr., Eds.
Plenum Press, New York, NY, 1987.
See Ad Page 394

Factors Controlling Indoor Radon Levels, Annual Report, June, 1983-May, 1984.
Harley, N.H.
New York University, Department Of Environmental Medicine, New York, NY, 1984.
Order #: NTIS DE84012712

Federal Radon Activities Inventory.
Radon Workgroup, Interagency Committee On Indoor Air Quality
U.S. Environmental Protection Agency/Department Of Energy, Washington, DC, SEP, 1988.

Florida Statewide Radiation Study.
Geomet Technologies
Florida Institute Of Phosphate Research, DEC 1987.
Order #: 05-029-057

Homeowner's Guide To Radon, A.
Dynamic Health Education, Newton, NJ, 1986.

Indoor Air And Human Health.
Gammage, R.B. and Kaye, S.V., Eds.
Lewis Publishers, Inc., Chelsea, MI, 1985.

Indoor Air Pollution.
Hileman, B.
Environmental Science And Technology 17:10:469a-472a, OCT 1983.

Indoor Air Pollution.
Wadden, R.A. and Scheff, P.A.
John Wiley And Sons, Inc., New York, NY, 1983.

Indoor Air Quality And Human Health.
Turiel, I.
Stanford University Press, Stanford, CA, 1985.

Indoor Air Quality Environmental Information Handbook: Radon.
U.S. Department Of Energy, Washington, DC, JAN

1986.
Order #: DOE/PE/72013-2

Indoor Air Quality.
Meyer, B.
Addison-Wesley Publishing Co., Inc., Reading, MA, 1983.

Indoor Concentrations Of Radon-222 And Its Daughters: Sources, Ranges And Environmental Influences.
Nero, A.V.
Indoor Air And Human Health, Proceedings Of The 7th Life Sciences Symposium, Knoxville, TN, Oct. 29-31, 1984.
Lewis, Chelsea, MS, 1985.

Indoor Pollutants.
National Research Council, Committee On Indoor Pollutants.
National Academy Press, Washington, DC, 1986.

Indoor Radiation Exposures From Radon And Its Daughters: A View Of The Issue.
Lawrence Berkeley Laboratory, Berkeley, CA, 1981.
Order #: NTIS DE82005144

Indoor Radon (Special Issue).
Nero, A.V. and Lowder, W.M., Eds.
Health Physics, 45:2, AUG 1983.

Indoor Radon And Its Hazards.
Bodansky, D., Robkin, M.A, and Stadler, D.R., Eds.
University Of Washington Press, Seattle, WA, 1987.

Indoor Radon I: Proceedings Of An APCA International Specialty Conference.
Air Pollution Control Association, Philadelphia, PA, 1986.

Indoor Radon II.
Air Pollution Control Association, Pittsburgh, PA, 1987.
Order #: SP-60

Indoor Radon Pollution: Bibliographic Series.
U.S. Environmental Protection Agency, Information Services And Library, Washington, DC, MAY 1986.
Order #: EPA IMSD/86-002

Indoor Radon: A Selected And Annotated Bibliography.
Youngen, G.
Council Of Planning Librarians, Chicago, IL, FEB 1987.
Order #: CPL Bibliography 184

Indoor Radon: Limited Federal Response To Reduce Contamination In Housing; Report To Honorable Frank R. Lautenberg, U.S. Senate. April 6, 1988.
U.S. General Accounting Office, Washington, DC, APRIL 1988.
Order #: GAO/RCED-88-103

Indoor Radon: The Deadliest Pollutant.
Kerr, R.
Science, 240, APR 29, 1988.

Inventory Of Current Indoor Air Quality-Related Research.
Meyer, B. and Hartley, R.
Industrial Environmental Research Laboratory, Cincinnati, OH, 1981.
Order #: NTIS PB82-127952

Ionizing Radiation Exposure Of The Population Of The United States.
National Council On Radiation Protection, SEP 1987.
Order #: NCRP Report No. 93

Manual On Indoor Air Quality: EPRI Manual.
Diamond, R.C. and Grimsrud, D.T.
Lawrence Berkely Laboratory, Berkeley, CA, FEB 1984.
Order #: EM-3469 Project 2034-3

National Radon Measurement Proficiency Program Cumulative Proficiency Report: EPA Test Round Five.
Research Triangle Institute/Radon Technical Information Service
U.S. Environmental Protection Agency, Washington, DC, 1988.
Order #: EPA 520/1-88-024

Radon Action Program.
U.S. Environmental Protection Agency, Washington, DC, 1986.

Radon And Indoor Air Pollution.
U.S. Congress, Committee On Science And Technology
The Committee (GPO), Washington, DC, 1985.

Radon And Its Decay Products In Indoor Air.
Nazaroff, W.W. and Nero, A.V.
J. Wiley, New York City, NY, 1988.

Radon And Its Decay Products: Occurrence, Properties And Health Effects.
Hopke, P.K., Ed.
American Chemical Society, Washington, DC, 1987.

Radon And The Environment.
Makofske, W.J. and Edelstein, M.R., Eds.
Noyes Publications, Park Ridge, NJ, 1988.

Radon Criteria For Indoor Air Quality Guidelines.
Health And Welfare Canada, Bureau Of Radiation And Medical Devices, Ottawa, ON, JUL 1986.

Radon Facts: Sources And Characteristics Of Radon.
U.S. Environmental Protection Agency, Office Of Air And Radiation, Washington, DC, 1987.

Radon In Homes.
American Medical Association, Council On Scientific Affairs
Journal Of The American Medical Assocation, 258:668-672, AUG 7, 1987.

Radon In Water And Air: Health Risks And Control Measures.
University Of Maine, Land And Water Resources Center, Orono, ME, 1983.

Radon Industry Directory, The, 1989 Edition.
Radon Press, Inc.
Radon Press, Alexandria, Va, 1989.

Radon Reference Manual.
U.S. Environmental Protection Agency, Washington, DC, 1987.
Order #: EPA-520/1-87-020

Radon Research Program.
U.S. Department Of Energy, Office Of Energy Research, Office Of Health And Environment, Washington, DC, NOV 1987.
Order #: DOE/EV/01013-HA

Radon Transform And Some Of Its Applications, The.
Deans, S.
Wiley, New York, NY, 1983.

Radon Transform, The.
Helgason, S.
Birkhauser, Boston, MA, 1980.

Radon, Water, And Air Pollution: Risks And Control.
Wieffenbach, C.V.
University Of Maine, Orono, ME, 1982.
Order #: NTIS PB93-206789

Radon: A Bibliography.
Lepman, S.R., Boegel, M.L., and Hollowell, C.D.
Lawrence Berkeley Laboratory, Berkeley, CA, 1981.

Radon: A Homeowner's Guide To Detection And Control.
Cohen, B.
Consumer Reports Books, Bridgeport, CT, 1987.

Radon: Basic Guide For Appraisers.
American Institute Of Real Estate Appraisers, Chicago, IL, JULY 1988.`

Radon: New Jersey And Florida.
Nicholls, G.P., Deieso, D.A., and Nash, J.D.
Environment, 29:12-14, 34-38, MAR 1987.

Radon: The Citizen's Guide.
Yuhnke, R., Silbergeld, E., and Caswell, J.
Environmental Defense Fund, 1987.

Radon: The Quiet Killer.
LaFavore, M.
Rodale Press, Emmaus, PA, 1987.

See Ad Page 395

Residential Indoor Air Quality: Problems And Solutions.
Nissan, J.N.
Cutter Information Corporation, Boston, MA, 1989.

Selected Radionuclides.
Environmental Health Criteria 25
World Health Organization, Geneva, Switzerland, 1983.

Sources Of Indoor Radon In Houses: A Review.
Bruno, R.C.
Journal Of Air Pollution Control Association, 33:2:105-109, 1983.

State Radon Programs: The Role Of Legislation.
Doyle, P. and Morandi, L.
State Legislative Report 13:10.
National Conference Of State Legislatures, Denver CO, 1988.

Summary Of State Radon Programs.
U.S. Environmental Protection Agency, Office Of Radiation Programs, Washington, DC, AUG 1987.
Order #: EPA 520/1-87-19-1

Health Effects

Effects On Populations Of Exposure To Low Levels Of Ionizing Radiation, The.
National Academy Of Sciences, Committee On The Biological Effects Of Ionizing Radiations (BEIR).
National Academy Press, Washington, DC, 1987.

Evaluation Of Occupational And Environmental Exposures To Radon And Radon Daughters In The United States.
National Council On Radiation Protection And Measurement, Bethesda, MD, 1984.
Order #: NCRP Report No. 78

Exposures From The Uranium Series With Emphasis On Radon And Its Daughters.
National Council On Radiation Protection And Measurements, Bethesda, MD, 1984.
Order #: NCRP Report No. 77

Health Risks Of Radon And Other Internally Deposited Alpha-Emitters: BEIR IV.
National Research Council, Commission On Life Sciences, Board On Radiation Effects Research, Committee On The Biological Effects Of Ionizing Radiations
National Academy Press, Washington, DC, 1988.

Human Disease From Radon Exposures: The Impact Of Energy Conservation In Buildings.
Budnitz, R.J., Et Al.
Energy And Environmental Division, Lawrence Radiation Laboratory, Berkeley, CA, AUG 1978.
Order #: LBL-7809

Indoor Air Pollution: An Emerging Health Problem.
U.S. General Accounting Office, GPO, Washington, DC, 1980.

Radon Daughter Exposure And Respiratory Cancer: Quantitative And Temporal Aspects.
Lundin, F.E., Jr., Wagoner, J.K., and Archer, V.E.
U.S. Department Of Health, Education And Welfare, National Institute For Occupational Safety And Health, and National Institute Of Environmental Health Sciences, Joint Monograph No. 1, 1971.

Radon In Water

Contribution Of Radon In Tap Water To Indoor Radon Concentrations, The.
Gesell, T.F. and Prichard, H.M.
Natural Radiation In The Environment III, 2:1347.
Oak Ridge National Laboratory, Oak Ridge, TN, 1980.

Estimating The Health Risks Of Radon In Drinking Water.
Cothern, C.R.
Journal Of The American Water Works Association, 79:153-158, APR 1987.

Evaluation Of Waterborne Radon Impact On Indoor Air Quality And Assessment Of Control Options.
Becker, A.P., Et Al.
U.S. Environmental Protection Agency, Washington, DC, SEP 1984.
Order #: PB84246404

Health Effects And Risks From 222Rn In Drinking Water.
Cross, F.T., Harley, N.H., And Hofmann, W.
Health Physics, 48:649-670, 1985.

Methods And Results Of EPA's Study Of Radon In Drinking Water.
U.S. Environmental Protection Agency, Washington, DC, 1986.
Order #: EPA/625/5-86/019

Nationwide Occurrence Of Radon And Other Natural Radioactivity In Public Water Supplies.
Horton, T.R.
U.S. Environmental Protection Agency, Eastern Environmental Radiation Facility, Montgomery, AL, 1985.

New Potable Water Radium-Radon Removal System, A. Proceedings From 1983 Annual Conference.
American Water Works Association, Las Vegas, NV, 1983.

Occurrence Of Radioactivity In Public Water Supplies In The United States, The.
Hess, C.T., Michel, J., Horton, T.R., Prichard, H.M., and Coniglio, W.A.
Health Physics 48:5:553-586, 1985.

Potable Water As A Source Of Airborne Radon-222 In U.S. Dwellings: A Review And Assessment.
Nazaroff, W.W., Doyle, S.M., and Nero, A.V.
Lawrence Berkeley Laboratory, Berkeley, CA, DEC 1985.
Order #: LBL-18154

Radionuclides In Drinking Water.
Lowry, J.D. and Lowry, S.B.
Journal Of The American Water Works Association, 80:50-64 JUL 1988.

Radionuclides In Drinking Water: An Overview.
Aieta, E.M., Singley, J.E., Trussell, A.R., Thorbjarnarson, K.W., and McGuire, M.J.
Journal Of The American Water Works Association, 79:4:144-152, 1987.

Radon In Groundwater (Proceedings From NWWA Annual Conference).
Graves, B., Ed.
National Well Water Association, APR 1987.

Radon In Water And Air: Health Risks And Control Measures.
University Of Maine, Land And Water Resources Center, Orono, ME, 1983.

Radon In Water Sampling Program.
U.S. Environmental Protection Agency, Washington, DC, 1978.
Order #: EPA/EERF-Manual 78-1

Radon Measurement And Control In Reducing Radon In Structures.
Lowry, J.D.
U.S. Environmental Protection Agency, Washington, DC, JAN 1987.

Removal Of Radon From Household Water.
U.S. Environmental Protection Agency
U.S. Environmental Protection Agency, Washington, DC, SEP 1987.
Order #: OPA-87-011

Removal Of Radon Fron Drinking Water.
Journal Of The American Water Works Association, 79:162-173, 176, APR 1987.

Study Of Radon-222 Released From Water During Typical Household Activities, A.
Partridge, J.E., Horton, T.R., and Sensinstaffer, E.L.
U.S. Environmental Protection Agency, Montgomery, AL, FEB 1979.
Order #: ORP/EERF-79-1

Transfer Of Radon From Domestic Water To Indoor Air, The.
Prichard, H.M.
Journal Of The American Water Works Association, 79:159-161, APR 1987.

Radon Measurement

Implementaion Strategy For The Radon/Radon Progeny Measurement Proficiency Evaluation And Quality Assurance Program.
Singletary, H.M., Starner, K., and Howard, C.E.
U.S. Environmental Protection Agency, Washington, DC, FEB 1986.
Order #: EPA-520 1-86-03

Indoor Radon Testing Specialists Training Manual, A Comprehensive Training Series For Radon Professionals.
Fugazzato, D., Terpilak, M., and Micek, S.
Radon Detection Services, Ringoes, NJ, 1988.

Indoor Radon Workshop.
George, A.C., et al, Eds.
Environmental Measurements Laboratory, New York, NY, 1982.
Order #: NTIS DE83017341

Instruments And Methods For Measuring Indoor Radon And Radon Progeny Concentrations. Proceedings Of The Conference: Radon And The Environment.
George, A.
Ramapo College, Mahwah, NJ, MAY 1986.

Interim Indoor Radon And Radon Decay Product Measurement Protocols.
Ronca-Battista, M., Magno, P., Windham, S. and Sensintaffar, E.
U.S. Environmental Protection Agency, Office Of Radiation Programs, Washington, DC, 1986.
Order #: EPA-520/1-86-04

Interim Protocols For Screening And Follow-up Radon And Radon Decay Product Measurements.
Ronca-Battista, M., Magno, P., Nyberg, P., and Sensintaffar, E.
U.S. Environmental Protection Agency, Office Of Radiation Programs, Washington, DC, 1988.
Order #: EPA 520/1-88-014

International Meeting On Radon-Radon Progeny Measurments.
U.S. Environmental Protection Agency, Washington, DC, 1983.

Measuring Radon Source Magnitude In Residential Buildings.
Nazaroff, W.W., Boegel, M.L., and Nero, A.V.
Lawrence Berkeley Laboratory, Berkely, CA, 1981.
Order #: NTIS DE82000774

National Radon Measurement Proficiency Program: Application And Participation Manual.
Research Triangle Institute
U.S. Environmental Protection Agency, Washington, DC, FEB 1988.
Order #: EPA-520/1-87-022

Nationwide Distribution Of Indoor Radon Measurements: A Preliminary Database.
Alter, H.W. and Oswald, R.A.
Journal Of Air Pollution Control Association, 37:3:227-231, 1987.

Passive Integrating Radon Monitor For Environmental Monitoring.
Alter, H.W. and Fleischer, R.L.
Health Physics, 40:5:693-702, MAY 1981.

Passive Radon Daughter Dosimeters.
McElroy, R.G.C. and Johnson, J.R.
Atomic Energy Control Board, Ottawa, ON, MAR 1986.

Procedure Manual For The Estimation Of Average Indoor Radon Daughter Concentrations Using The Radon Grab-Sampling Method.
George, J.L.
Bendix Field Engineering, Grand Junction, CO, 1986.
Order #: NTIS DE86012078

Progress In Indoor Radon Measurement: Review Of Previous Research.
University Of Pittsburgh, Pittsburgh, PA, 1985.
Order #: NTIS DE8600448

Review Of Existing Instrumentation And Evaluation Of Possibilities For Research And Development Of Instrumentation To Determine Future Levels Of Radon At A Proposed Building Site.
Canada Atomic Energy Control Board, Ottawa, ON, 1983.

State/EPA Indoor Radon Survey Results, Winter 1986-1987.
U.S. Environmental Protection Agency, Washington, DC, 1987.

Workshop On Dosimetry For Radon And Radon Daughters, Oak Ridge National Laboratory, April 12-13, 1977.
Turner, J.E., Holoway, C.F., and Loebl, A.S., Eds.
U.S. Department Of Energy, Oak Ridge National Laboratory, Oak Ridge, TN, 1978.
Order #: ORNL-4348

Radon Mitigation

Control Of Indoor Radon Decay Products By Air Treatment Devices.
Hinds, W.C., Et Al.
Journal Of The Air Pollution Control Association, 33:2:134-136, FEB 1983.

Control Of Respirable Particles And Radon Progeny With Portable Air Cleaners.
Offermann, F.J., Et Al.
Lawrence Berkeley Laboratory, Berkeley, CA, 1984.
Order #: LBL-16659

Cost Evaluation Of Control Measures For Indoor Radon Progeny.
Moeller, D.W. and Fujimoto, K.
Health Physics, 46:1181-1193, JUN 1984.

Cost Survey For Radon Daughter Control By Ventilation And Other Techniques, Volume I.
Bloomster, C.H., Enderlin, W.I., Young, J.K., and Dirks, J.A.
Batelle Memorial Institute, Pacific Northwest Laboratories, Richland, WA, 1984.
Order #: NTIS PB85-152932

Cost Survey For Radon Daughter Control By Ventilation And Other Techniques, Volume II.
Bloomster, C.H., Jackson, P.O., Dirks, J.A., and Reis, J.W.
Batelle Memorial Institute, Pacific Northwest Laboratories, Richland, WA, 1984.
Order #: NTIS PB85-152940

Demonstration Of Remedial Techniques Against Radon In Florida Phosphate Lands.
Scott, A.G. and Findlay, W.O.
American Atcon, Inc., Columbia, MD, 1983.
Order #: EPA-520/5-83-009

Effects Of Home Ventilation Systems On Indoor Radon-Radon Daughter Levels, The.
Windham, S.T., Savage, E.D., and Philips, C.R.
Eastern Environmental Radiation Facility, Montgomery, AL, 1978.
Order #: NTIS PB-291 925/6

EPA Program To Demonstrate Mitigation Measures For Indoor Radon: Initial Results.
Henschel, D.B. and Scott, A.G.
U.S. Environmental Protection Agency, Air And Energy Engineering Research Laboratory, Research Triangle Park, NC, 1986.
Order #: NTIS PB86-176039

Evaluation Of Indoor Aerosol Control Devices And Their Effects On Radon Progeny Concentrations.
Sextro, R.G., Et Al.
Lawrence Berkeley Laboratory, Berkeley, CA, FEB 1984.
Order #: DE840160001

Four Common Diagnostic Problems That Inhibit Radon Mitigation.
Osborne, M.C.
Journal Of Air Pollution Control Association 37:604-606 MAY 1987.

General Remedial Action Details For Radon Gas Mitigation.
State Of Pennsylvania, Department Of Environmental Resources, Bureau Of Radiation Protection, Harrisburg, PA, 1985.

General Remedial Action Details For Radon Gas Mitigation.
Pennsylvania Department Of Environmental Resources, Bureau Of Radiation Protection, MAY 1985.

How To Operate A Heat Recovery Ventilator.
Canada Department Of Energy, Mines And Resources, Ottawa, ON, 1985.

Indoor Air Quality Control Techniques: A Critical Review.
Fisk, W.J., Spencer, R.K., Grimsrud, D.T., Offermann, F.J., Pedersen, B., and Sextro, R.G.
Lawrence Berkeley Laboratory, Berkeley, CA, MAR 1984.
Order #: LBL-16493

Indoor Air Quality Handbook: For Designers, Builders, And Users Of Energy Efficient Residences.
Sandia National Laboratory, Albuquerque, NM, 1982.
Order #: NTIS DE83002315

Indoor Radon Mitigation Specialists Training Manual, "A Comprehensive Training Series For Radon Professionals".
Fugazzato, D., Terpilak, M., and Davidson, J.
Radon Detection Services, Ringoes, NJ, 1988.

Infiltec Catalog Of Hard To Find Radon Control Supplies.
Saum, D.
Saum Enterprises, Waynesboro, VA, 1988.

Interim Guide To Radon Reduction In New Construction.
U.S. Environmental Protection Agency, Washington, DC, AUG 1987.
Order #: OPA-87-009

Low Cost Reduction Of Indoor Radon: Industrial Foam Study.
American ATCON, Inc., APR 1986.

Mitigation Quality Control Checklist In Reducing Radon In Structures.
Brodhead, W.
U.S. Environmental Protection Agency, Washington, DC, JAN 1987.

Practical Radon Control For Homes.
Brennan, T. and Galbraith, S.
Cutter Information Corporation, Boston, MA, 1988.

Preliminary Evaluation Of The Control Of Indoor Radon Daughter Levels In New Structures, A.
Fitzgerald, J.E., Jr., Guimond, R.J., and Shaw, R.A.
U.S. Environmental Protection Agency, Office Of Radiation Programs, Washington, DC, 1976.
Order #: EPA-520/4-76-018

R.F. Simon Catalog Of Radon Products.
Simon, R.F.
R.F. Simon Company, Barto, PA, 1988.

Radon Mitigation Products Catalog.
Phillips, C.
Safe-Aire, Inc., Canton, IL, 1988.

Radon Mitigation Remedial Action Demonstration At The Watras Residence, Report To Philadelphia Electric Co.
Tappan, J.T.
Arix Corporation, JUN 1985.

Radon Reduction Methods: A Homeowner's Guide.
U.S. Environmental Protection Agency, Washington, DC, 1986.
Order #: OPA-86-005

Radon Reduction Methods: A Homeowner's Guide: 2nd Edition.
U.S. Environmental Protection Agency, Washington, DC, 1987.
Order #: OPA-87-010

Radon Reduction Techniques For Detached Houses: Technical Guidance.
Sanchez, D.C. and Henschel, D.B.
U.S. Environmental Protection Agency, Washington, DC, JUN 1986.
Order #: EPA/625/5-86/019

Radon Reduction Techniques For Detached Houses: Technical Guidance: 2nd Edition.
Henschel, D.B.
U.S. Environmental Protection Agency, Washington, DC, JAN 1988.
Order #: EPA/625/5-87/019

Survey Of Radionuclide Contents And Radon Emanation Rates In Building Materials Used In The U.S.A.
Ingersoll, J.G.
Health Physics, AUG 1983.

Radon Newsletters

Energy Design Update.
Nissan, J.N.
Cutter Information Corporation, Boston, MA.

Indoor Air Quality Update.
Levin, H.
Cutter Information Corporation, Boston, MA.

Indoor Pollution Law Report.
Laucheimer, M.
Leader Publications, New York, NY.

Indoor Pollution News.
Lesser, G.H.
Buraff Publications, Washington, D.C.

National Radon Association Newsletter.
Huey, S.
National Radon Association, Decatur, GA.

Radon Hotline Letter, The.
Siegelman, L.
Radon Press Inc., Alexandria, VA

Radon News Digest. See Ad Page 401
Jordan, R.
Hoosier Environmental Publishing, Carmel, IN.

Radon Reporter, The.
Craffey, J.
AARST, Park Ridge, NJ.

Radon Review, The.
Lee, J.
American Radon Association, Tipp City, OH.

It's Like Having The Entire Radon Industry Delivered To Your Door!

Everything You Need To Know. Everyone You Want To Reach.

Finally there's a reference that contains all the information on the radon issue — everything you need to know — everyone you want to reach!

The Radon Industry Directory is an indispensable sourcebook for anyone concerned about radon.

Edited for a wide and varied audience, it serves as a frequently consulted reference to the rapidly emerging radon industry.

Thoroughly up-to-date, with over 3,000 listings, **The Radon Industry Directory** gives you instant access to thousands of organizations and radon professionals — radon detection and mitigation firms, product manufacturers, federal, state, and local government agencies, special interest groups, workshops, conferences, indexes, and much, much more. In all, 30 vital categories are concisely listed, described, and indexed. There's even a section providing the reader a complete overview of radon — its origins, testing for radon, and radon mitigation.

The Radon Industry Directory puts the names, numbers, and addresses of key radon people and organizations right in your hands — everything you need to know, everyone you want to reach! Published annually and available on standing order.

$75.^{oo} ea.

ISBN: 0-929840-00-3
ISSN: 0899-9023

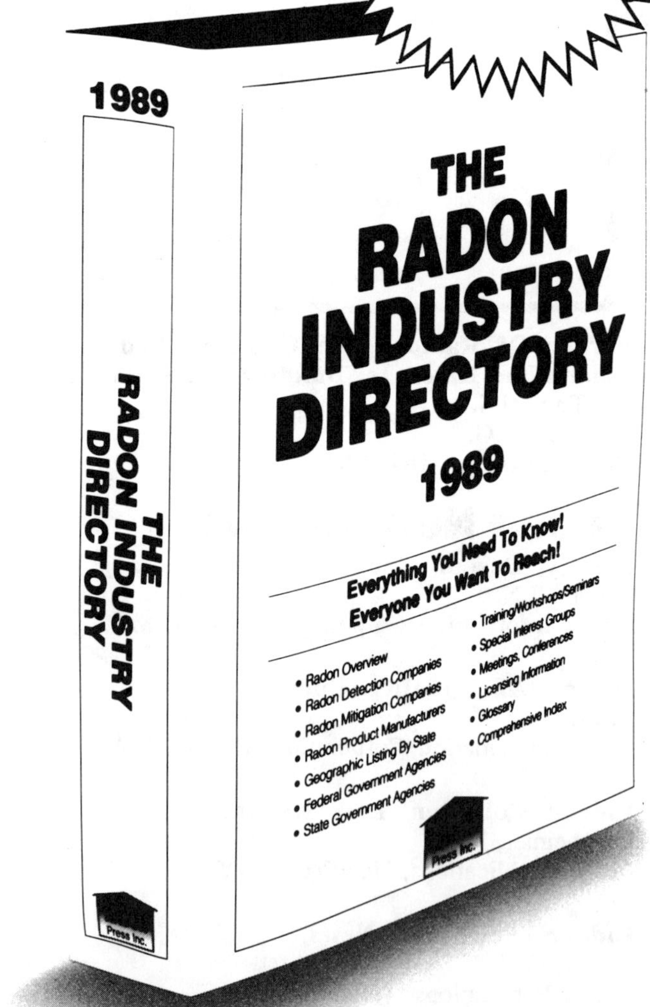

To Order
Call 1-800-548-1567 or fill out and return the order form at the back of the directory.

Satisfaction Guaranteed

You must be completely satisfied with The 1989 Radon Industry Directory or simply return it within 10 days of receipt for a prompt refund.

RADON
Press Inc.

Glossary of Terms

The terms listed in this section range from basic terms commonly used within the radon industry to more technical terms used within the related fields of biology, chemistry, dosimetry, ecology, epidemiology, geology, health, physics, and research theory.

Absolute Risk Projection Model
A model which estimates the risk of radon exposure beyond the years of observation of a studied population by projecting the average observed number of excess cancers per unit dose into the future years at risk.

Absorbed Dose
The amount of energy per unit mass transferred to human or animal tissue from radiation. Measured in rads or grays standard international (SI unit). One gray equals one jouie per kilogram. One rad equals 0.01 gray.

Acceptable Air Quality
Air in which there are no known contaminants at harmful concentrations and with which a substantial majority (usually 80%) of the people exposed do not express dissatisfaction.

Actinium Series
The naturally radioactive series starting with U235.

Actinon
The isotope of radon in the actinium series.

Activity
The rate of atomic disintegration measured in curies.

Activity Concentration
Activity per unit volume. Commonly measured in picocuries per liter in air or picocuries per gram in solids.

Additive Relative Risk Model
The relative risk from the combined exposure to radon progeny and smoking equals the sum of risks from each exposure considered separately.

Adsorption
The condensation of gases, liquids, or dissolved substances on the surfaces of solids.

Aeration
Exposing to the action of air. For example, a liquid can be aerated by agitation or by means of a fine spray.

Aerodynamic Diameter
The diameter of a unit density particle that has the same settling velocity as the particle described.

Aerosols
Atomized particles of a substance suspended in air.

Air Age
A concept which describes the condition of non-equilibrium of radon and its progeny.

Air Changes Per Hour (ACH)
A measure of the movement of a volume of air in a given period of time (i.e., of the Air Exchange Rate). One air exchange per hour is equivalent to replacing all of the air in a house in a one-hour period. Air changes also may be expressed in cubic feet per minute.

Air Cleaner
A device used to remove airborne impurities such as dust, gases, vapors, fumes, and smoke.

Air Exchange Rate
The number of times the volume of air in a space is exchanged with outdoor air in a given period of time (usually expressed as "air changes per hour" or "ach").

Air Filtration
A radon reduction technique in which air is passed through high efficiency filters or electronic devices. The devices collect dust and other airborne materials to which radon particles attach themselves.

Air-To-Air Heat Exchanger
A ventilation device used to reduce radon levels in homes by retaining indoor heat while exchanging indoor air and accompanying pollutants for fresh, outdoor air.

Airflow Bypass
Any opening through the floors between stories of a house (or through the ceiling between the living area and the attic) which facilitates the upward movement of house air under the influence of the stack effect. By facilitating the upward movement, airflow bypasses also, in effect, facilitate exfiltration at the upper levels, which in turn will increase infiltration of outdoor air and soil gas.

Alpha Activity
The number of atoms of a particular radionuclide disintegrating by radioactive decay per second resulting in the emission of an alpha particle.

Alpha Emission Density
The number of alphas per unit space emerging from a source element. For radon progeny in the lung, the emission source density is the number of alphas per unit surface area in each bronchial tube.

Alpha Energy
The energy released when an alpha particle emitted during radioactive decay is halted by collision with a substance (e.g., lung tissue). The amount of energy depends on the velocity of the alpha particle, which in turn depends on the source of radioactive decay (e.g., decay of U-238 versus Ra-226).

Alpha Particle
An energized particle made up of two protons and two neutrons that is ejected from a radioactive atom. It is indistinguishable from a helium atom nucleus.

Alpha Track Detector
A type of time-integrated sampling detector that records alpha particles from radon decay products by looking at radiation damage on plastic film. After exposure times of from 30 days to one year, the film is etched with sodium hydroxide solution revealing the tracks which are then counted manually or automatically under a microscope.

Alum Shale Brick
Brick building material made with a type of shale rock that has been found to contain naturally elevated levels of uranium.

Alveola
Tiny air sacs in the lung.

Ambient Air
That portion of the atmosphere, external to buildings, to which the general public has access.

Analytical Model
A model developed from theoretical principles.

Anatomical Dead Space
The region of the lung given by the tracheobronchial and nasopharyngeal regions, which normally constitute the part of the lung not responsible for gas exchange with blood.

Aneuploidy
Change in the normal chromsome number of cells.

Artificial Radioactivity
Man-made radioactivity produced by fission, fusion, particle bombardment, or electromagnetic irradiation.

Association
Two variables are associated if one is more (or less) common in the presence of the second.

Atomic Mass Number
The total number of nucleons in the nucleus or the total number of protons and neutrons in the nucleus.

Atomic Number
The number of protons in the nucleus of a neutral atom of a nuclide.

Attached Fraction
Fraction of airborne radon progeny attached to aerosol particles.

Attached Radon Decay Product
A radon decay product that is attached to a particle of dust or other material in the air.

Back-drafting
A condition where the normal movement of combustion products up a flue is reversed, so that the combustion products can enter the house. Back-drafting of combustion appliances (such as fireplaces and furnaces) can occur when depressurization in the house overwhelms the buoyant force on the hot gases. Back-drafting can also be caused by high air pressures at the chimney or flue termination.

Backer Rod
A rope of compressible plastic foam. Backer rod can be force-fit into wide cracks and similar openings to serve as support for caulking material.

Backfilling
The placement of soil in and around a building foundation following construction of the walls and slab.

Background Radiation
The average radiation from all sources that is contained in outdoor air.

Band Joist
A board (typically 2"x 8" or 2"x 10" in.) that rests (on its 2" dimension) on top of the sill plate around the perimeter of the house. The ends of the floor joists are nailed into the header joist that maintains spacing between the floor joists. Also called header joist or rim joist.

Barrier Coating
A layer of material that obstructs or prevents passage of a particular substance through a surface that is to be protected. More specifically, grout, caulk, or various sealing compounds, perhaps used with polyurethane membranes to prevent soil-gasborne radon from moving through walls, cracks, or joints in a house.

Basal Cells
Dividing cells in the bronchial epithelium that produce replacement cells for those normally lost.

Basaltic Lavas
A dense, dark igneous rock formed by cooling volcanic material that can contain uranium minerals at low levels. Uranium minerals are finely disseminated through the rock and concentrations are typically of low grade.

Baseboard Duct
A continuous system of sheet metal or plastic channel ducting that is sealed over the joint between the wall and floor around the entire perimeter of the basement. Holes drilled into hollow blocks in the wall allow suction to be drawn on the walls and joint to remove radon through the ducts to a release point outside the house.

Basement
A type of house construction where the bottom livable level has a slab (or earthen floor) which averages three feet or more below grade level on one or more sides of the house.

Basement Foundation
Underlying structure of a building or dwelling. Usually constructed of cinder block or cement. May have a dirt floor.

Basement Pressurization
A radon reduction technique in which air pressure in a basement is increased in order to raise the basement's air pressure above that of the soil beneath and around a home.

Becquerel (Bq)
The SI unit for the rate of radioactive decay in a sample of material. One Becquerel equals one disintegration per second. One picocurie equals 0.037 Bequerel.

Beta Energy
The energy released when a beta particle is halted by collision with a substance. The amount of energy depends on the source of radioactive decay.

Beta Particle
A negatively charged subatomic particle (electron) emitted from a nucleus during some types of radioactive decay.

Bias
An error in the measure of the association between two variables.

Bifurcation
The point at which a tube in the lung divides to form the next generation.

Black Shales
A laminar type of sedimentary rock found in parts of the central United States characteristically rich in organic matter, such as marine black shale and carbonaceous shale. Although the uranium content is normally less than 0.005 percent U3O8, the uranium level tends to be higher than in other sedimentary rock.

Blower Door
A device consisting of an instrumented fan which can be mounted in an existing doorway of a house. By determining the air flows through this fan required to achieve different degrees of house depressurization, the blower door permits determination of the tightness of the house shell, and an estimation of the natural infiltration rate.

Breathing Rate
The rate of oxygen/air intake by the lungs.

Brick Veneer
An ornamental layer of brick that is sometimes applied over cinder block walls.

British Thermal Unit (BTU)
The quantity of heat required to raise the temperature of one pound of water by one degree Fahrenheit.

Bronchi
The two branches leading from the trachea to the lungs.

Bronchial Epithelium
The cellular lining of the bronchi.

Bronchial Generation
The branches of airways from the trachea to the gas exchange region of the lung. The branches are numbered sequentially from the trachea.

Bronchialoalveolar Carcinomas
Cancers of the respiratory tract airways that are epithelial in origin.

Building Envelope
The exterior surfaces of a building--specifically, walls, roof, and floor (if over unheated basement, slab, or crawlspace).

Bypassing
The tendency of certain areas of a structure to be less well ventilated than others.

Carbonaceous
Relating to, containing, or composed of the element carbon.

Carbonate Rock
Rock composed principally of carbonates, especially if at least 50 percent by weight. Carbonate is an ester or salt of carbonic acid. Calcium carbonate and calcium magnesium carbonate are principal constituents of limestone and dolomite rock.

Carcinogen
Cancer-causing agent. Any agent that incites development of a carcinoma or any other form of malignancy.

Carcinoma
A malignant tumor which can effect almost any organ; tends to metastasize.

Case-Control Study
Selection of study groups to be compared based on presence or absence of disease.

Caulk
To stop up and make airtight against leakage by forcing in a sealing substance.

Charcoal Canister (CC)
Passive device allows continuous adsorption and desorption of radon which undergoes radioactive decay during the measuring process. Canister is analyzed in a laboratoy, after three days to a week, by placing it directly on a gamma detector that counts the gamma rays emitted by radon decay products on the charcoal.

Chromosome
The cell organelle that contains the DNA and is responsible for segmenting DNA during cell division.

Chromosome Aberrations
Rearrangements of chromosomes in the cell nucleus either in number or structure.

Cilia
The hairlike structures on some bronchial surface cells which propel the mucus and deposited material upward toward the pharynx.

Cinder Block
Molded concrete building blocks, usually rectangular with two hollow chambers in the center. They are no longer made with cinders, but the name remains.

Cohort
A large homogenous group of people tested in epidemiological or socioeconomic studies. EPA's lung cancer estimates are based on calculations for a cohort of 100,000 people.

Cold Air Return
The registers and ducting which withdraw house air from various parts of the house and direct it to a central forced-air furnace or heat pump. The return ducting is at low pressure relative to the house.

Combustion Appliance
A fuel-burning (oil, gas, coal, or wood) device such as a range, furnace, or water heater.

Concentration
The quantity of one constituent dispersed in a defined amount of another; for example, 1b per cubic foot, kg/kg, or parts per million (ppm).

Concentration Gradient
The change in concentration (mass per unit volume) per unit distance.

Confidence
The degree of trust that a method will achieve the radon reduction estimated.

Confounding Bias
A potential attribute of data. In measuring an association between an exposure and a disease, a confounding factor is one that is associated with the exposure and independently is a cause of the disease. Confounding bias can be controlled if information on the confounding factor is present.

Constant-Relative-Risk Model
A risk model which assumes that, after a certain time, the ratio of the risk at a specific dose to the risk in the absence of the dose does not change with time.

Contaminant
Any airborne constituent that reduces the quality of the air.

Continuous Radon Monitor (CRM)
Samples air by pumping it through a scintillation cell after passing it through a special filter that removes dust and radon decay products. As the radon decays the ionized radon decay products plate out on the interior surface of the scintillation cell. The radon decay products emit alpha particles, which strike the coatings on the inside of the cell, causing scintillations to occur. The scintillations are detected by a photo tube in the detector, which generates electrical signals. The signals are processed and the results are either stored in the memory of the CRM or printed on paper tape. The CRM must be calibrated in a known radon environment to obtain the conversion factor used to convert count rate to radon concentration.

Continuous Working Level Monitor (CWLM)
Samples ambient air by filtering airborne particles as the air is drawn through a filter at a low flow rate. An alpha detector counts the alpha particles produced by the radon decay products as they decay on the filter. The unit contains a microprocessor that stores the number of counts and the elapsed time. It can be set to record the total counts registered over a specified time period and must be calibrated to to convert count rate to WL values.

Contractor
A building trades professional who works for profit to correct radon problems; a remediation expert.

Convective Movement
The bulk flow of radon-containing soil gas into the house as the result of pressure differences between the house and soil. Distinguished from diffusive movement.

Coulomb
The charge flowing past a point of a circuit in one second, when there is a current of one ampere in the circuit; also, the charge by 6x10 to the eighteenth power electrons.

Crawl-Space
An area beneath some types of houses which are constructed so that the floor is raised slightly above the ground, leaving a space between the floor and ground to allow access to utilities and other services.

Critical Cells
Those cells deemed most likely to act as essential targets for producing an effect such as cancer.

Cubic Feet Per Minute
A measure of the volume of matter flowing within a fixed period of time.

Cumulative Working Level Month (CWLM)
A unit of cumulative radon exposure. The sum of lifetime exposure to radon working levels expressed in total working level months.

Curie (Ci)
A unit used to describe the rate of radioactive decay in a sample of material. One curie equals 37 billion disintegrations per second. One gram of radium-226 has an activity of one curie.

Cytogenic Analysis
Evaluation of chromosome numbers and structure.

Daughters
The new radionuclides formed by the radioactive decay of a parent radionuclide.

De-gassing
The release of dissolved radon gas into the house air when radon-containing well water is used in the house.

Decay Chain
A sequence of radioactive decays in the same nucleus.

Decay Series
The consecutive members of a family of radioactive isotopes formed by sequential radioactive decay. A complete series commences with a long-lived parent such as U-238 and ends with a stable element such as Pb-206.

Deoxyribonucleic Acid (DNA)
The basic instruction code which a living cell uses to direct its life processes.

Depressurization
A phenomenon that occurs when the air pressure outside of a building is higher than the pressure inside. Normally, buildings have higher pressure than outdoor air, but buildings may be depressurized when wood stoves, fireplaces, or furnaces draw indoor air for combustion. The slight vacuum that is created by depressurization can draw radon-laden air into a basement through cracks and openings.

Depth-Dose Curve
A curve describing the absorbed dose at points located at varying depths into a material such as the walls of a passageway in the human body.

Detached Houses
Single family dwellings as opposed to apartments, duplexes, townhouses, or condominiums. Those dwellings which are typically occupied by one family unit and which do not share foundations and/or walls with other family dwellings.

Diagnostic Measurements
Measurements typically done after screening and follow-up tests to help determine the dynamics of radon gas entry and to help decide on a mitigation or remedial approach.

Diffusion Barrier
Any material placed over a detector which delays, through increased diffusion time, the rate at which radon or progeny in air reach the detector.

Diffusive Flux
The weight of a material diffusing across unit area per unit of time in response to a pressure or concentration gradient. Measured in grams per square centimeter per second. Used to characterize the rate of radon movement into a home.

Diffusive Movement
The random movement of individual atoms or molecules, such as radon atoms, in the absence of (or independent of) bulk (convective) gas flow. Atoms of radon can diffuse through tiny openings, or even through unbroken concrete slabs. Distinguished from mass airflow.

Diffusivity
The constant by which diffusive flux and concentration gradient are related at constant pressure under Fick's First Law of Diffusion.

Direct-Recoil Fraction
Those atoms which terminate their recoil path in a pore space.

Distal Lung
The region of the lung farthest from the nose.

Dolomite
A brittle calcium magnesium carbonate rock occurring abundantly in white to pale pink rhombohedral crystals.

Dose
The amount of radiation that a person, or specific organ, is exposed to.

Dose Conversion Factor
The absorbed dose in rad per working level month.

Dose Equivalent
The product of the absorbed dose and the quality factor (units: rem or sievert).

Dose-Response Curve
The functional relationship between dose and effect.

Dosimetry
The measurement or calculation of the energy absorbed by matter.

Drain Tile
A perforated pipe used to drain water away from the foundation of a house.

Drain Tile Ventilation
A radon reduction technique in which radon is drawn away from a home via an underground drainage system surrounding the home at the basement floor level. Drain tile ventilation is one of several soil ventilation techniques.

Duct
A round, oval, or rectangular pipe-like passageway for moving air. A duct can be made of plastic, metal, wood, or fiberglass.

Dust
An air suspension of particles of any solid material, usually less than 100 micrometers in size.

Earth-Based Building Materials
Materials such as cinder block and brick which are formed using naturally-occurring minerals and rocks.

Effective Dose Equivalent
Dose equivalent weighted by a factor which measures the relative sensitivity of the tissue to a radiation-induced cancer.

Effective Leakage Area (ELA)
A parameter determined from blower door testing, giving a measure of the tightness of the house shell. Conceptually, this leakage area reflects the square inches of open area through the house shell through which air can infiltrate or exfiltrate.

Electron Volt
A unit of energy commonly used to measure energy releases during radioactive decay.

Electrons
Negatively charged particles that orbit around the nucleus of an atom.

Electrostatic Precipitator
An air cleaner that removes dust particles from the air by electrically charging them so they'll be attracted to a plate with the opposite charge within the unit.

Emanation
Product of radioactive disintegration.

Emanation Rate
The rate of flow or flux of radon across a surface.

Empirical
A mathematical constant or relationship derived from observed data.

Entry Routes
Pathways by which soil gas can flow into a house. Openings through the flooring and walls where the house contacts the soil.

Environmental Exposure
Exposure to radiation in nonoccupational situations.

Epidemiology
A science that deals with the incidence, distribution, and control of diseases in a population.

Epithelial Lining
Cellular tissue covering surfaces, forming glands, and lining most cavities of the body.

Epithelium
Surface layers of cells that line respiratory tract airways.

Equilibrium
A state of balance or rest; radioactive equilibrium is the state in which the rate of formation of atoms is equal to the rate of their disintegration by radioactive decay, so that the amount of the element or isotope is constant.

Equilibrium Factor
An adjustment used in converting from picocuries per liter to working level concentration which takes into account the possible absence of radioactive equilibrium between radon and its decay products.

Equilibrium Ratio
Equilibrium is achieved when the activity of all the short-lived radon decay products is equal to the parent radon activity. At equilibrium (i.e., at an equilibrium ratio of 1.0), 1 WL of decay products would be present when the radon concentration was 100 pCi/L. The ratio is never 1.0 in a house; that is, the decay products never reach equilibrium in a house environment due to ventilation and plate-out. In practice, equilibrium ratios of 0.3 to 0.7 are commonly observed.

Equivalent Dose
The absorbed dose weighted to account for its relative biological effectiveness by use of quality and modifying factors. Measured in rem or sievert (SI unit).

Eukaryotic Cells
Cells that contain a nucleus.

Exfiltration
The movement of indoor air out of the house.

Exhalation Rate
The rate at which radon is released from a solid surface into the adjacent air. It is a function of the radon production rate and the diffusion coefficient of the solid.

Exhaust Fan
A fan oriented so that it blows indoor air out of the house. Exhaust fans cause outdoor air (and soil gas) to infiltrate at other locations in the house, to compensate for the exhausted air.

Exposure
The amount of radiation present in an environment, not necessarily indicative of absorbed energy, but representative of potential health damage to the individual present. Measured in roentgen.

Far-Wall Contribution
The component of dose to cells in tube walls that arises from radiation which must first pass through the air tube to reach tissue on the "far side" from the point of decay.

Fibroblasts
Cells of connective tissue that produce tissue stroma.

Flux
Amount of radiation crossing a specified area per unit time. The unit of flux is the number of particles, energy, etc., per square centimeter per second.

Flux Measurement
Measurements made to determine the rate at which radon is emanating from a surface or source into some part of the house. These diagnostic measurements allow the radon source strengths to be estimated and help in planning the most effective remediation procedures.

Follow-Up Measurement
A measurement taken after an initial screening measurement to determine annual average exposure to home occupants.

Follow-Up Period
The length of time between a person entering an epidemiological study cohort and the present report (or the end of the study).

Footing
A concrete or stone base which supports a foundation wall and which is used to distribute the weight of the house over the soil or subgrade underlying the house.

Footing Drains
A perforated perimeter pipe installed alongside the footings (interior or exterior) to collect water and drain it away.

Forced Or Mechanical Ventilation
Ventilation induced by means of a fan or other mechanical device.

Forced-air Furnace
A central furnace or heat pump that functions by recirculating the house air through a heat exchanger in the furnace. A forced-air furnace is distinguished from a central hot-water space heating system, or electric resistance heating.

French Drain
A water drainage technique installed in basements of some houses during initial construction. Typically consists of a 1 or 2 inch gap between the basement block wall and the concrete floor slab around the entire perimeter of the basement. Also known as a perimeter drain or channel drain.

Gamma Ray Or Photon Radiation
Emission of a high-energy photon, especially as emitted by a nucleus in a transition between two energy levels.

Gamma Rays
Short-wave electromagnetic radiation.

Gaseous Diffusion
The natural tendency for a gas to flow from a region of relatively high concentration to a region of relatively low concentration.

General Sealing
Radon reduction techniques that involve placing barriers between the radon source and the living space in a home. Techniques include sealing cracks in concrete floors and walls and filling concrete block walls.

Generation
A loosely defined set of passageways in the lung characterized by roughly equal anatomical dimensions and depth into the lung.

Generational Volume
The sum of the volumes contained in all of the passages in a generation.

Genome
The total genetic information in an organism.

Gneiss
A coarse-grained, high-grade metamorphic rock that has a distinct banding due to segregation of light and dark minerals.

Grab Radon Progeny
Testing method that refers to very short term--about five minutes--sampling. The radon progeny grab sample is collected by drawing air through a filter upon which the progeny are collected. Operators can perform analysis on location in less than an hour.

Grab Radon Sample
Testing method that refers to very short term--about five minutes--sampling. The radon grab sample is collected in a special flask holding 100 to 2,000 cubic centimeters of air. Flask must be returned to a laboratory for analysis which can usually be accomplished in about four hours.

Grade
The surface of the ground surrounding the house. Below grade means that part of the house is lower than the surface level of the soil.

Grain Size
The size of a microscopic particle of regular crystalline

structure. Grain size affects a number of important macroscopic physical properties, such as yield point and porosity.

Granular Activate Carbon
A unique material made of wood, coal, or other carbon base that has been carbonized (to charcoal) and activated to develop a vast internal pore structure and adsorption capacity.

Gray
A standard international unit of absorbed dose.

Growth Factor
Chemical substance, often a peptide that stimulates cellular proliferation.

Half-life (biologic)
Time required for the body to eliminate half of an administered dose of any substance by regular processes of elimination; it is approximately the same for both stable and radioactive isotopes of a particular element.

Half-life (radioactive)
The time it takes for a radioactive substance to decay. If the half-life of a substance is two years, half its radiation will be gone after two years.

Healthy-Worker Effect
Lower mortality rate among workers than among the general population if their health has not been adversely affected by work.

Heat Exchanger
A device that transfers heat from one medium to another, such as an air-to-air heat exchanger used for heat-recovery ventilation.

Heat Recovery Ventilators (HRV)
Also known as air-to-air exchangers or heat exchangers.

Henry's Constant
The ratio of the air concentration to the water concentration of a gas or volatile compound.

Hollow Block Walls
Walls made of concrete blocks that have hollow spaces in the middle. The spaces can be connected either horizontally or vertically.

Home Ventilation
A radon reduction technique in which a home's existing indoor air supply is replaced by fresh, outdoor air. Structural ventilation may be done naturally (by opening windows or vents) or mechanically (by using fans).

House Air
Synonymous with indoor air. The air that occupies the space within the interior of a house.

HVAC System
The heating, ventilating, and air conditioning system for a house. Generally refers to a central furnace and air conditioner.

Hybridization
Positioning of complementary DNA strands to give a duplex structure.

Igneous Rock
Rock formed by the cooling of liquid material; for example, basalt and granite.

In-Vivo Counting
The measurement of the amount of radioactivity in the human body through counting of the number of gammas which emerge from that body.

Incidence
A number of cases of disease or other characteristics in a given period of time (monthly or annually).

Indirect-Recoil Fraction
Those atoms which terminate their recoil path in a solid grain but diffuse out of the solid through the recoil path, while the solid is still molten.

Indoor Air
That air that occupies the space within the interior of a house or other building.

Indoor Radiation Concentration
Concentration of radon or radon decay products in a house. Concentration is dependent on the geologic formation under the house and the structural conditions of the house, as well as other factors.

Infiltration
Air leakage through cracks and gaps, and through ceilings, floors, and walls of a building.

Interaction
The association of one factor with disease modified by the effect of another factor. The measure of association can be the rate or odds ratio. This follows the nonmultiplicative model (may be additive).

International System
System of units commonly used in most of the world outside the United States.

Inversion Condition
A meteorologic condition in which temperature increases with altitude, generally caused by a warm air mass overlying a colder one.

Ion
A charged atom or molecularly bound group of atoms; sometimes also a free electron or other charged subatomic particle.

Ionization Chamber
An instrument which detects radiation through collection of ionization products produced by the radiation in a gas-filled cavity.

Ionizing Radiation
Radiation capable of exerting enough energy to disrupt atoms by knocking electrons loose. This is the most dangerous type of radiation to human health.

Isotope
One of two or more types of atoms having the same atomic number but different mass number.

Joist
Any of the parallel horizontal beams set from wall to wall to support the boards of a floor or ceiling.

Lagging Exposures
Lagging of the cumulative exposure assigned to a miner. Some consider that radon progeny exposures are "redundant" if they occur after lung cancer is induced. Some believe that cumulative exposures should be lagged by a certain number of years (5 or 10), to exclude redundant exposures occurring during these years.

Latency Period
The time during which disease exists without manifesting itself; a period of incubation.

Latent Heat
Heat that is associated with the change in physical form of a substance (e.g., with vaporization of liquid water).

Leukemia
A form of cancer in the blood. Any of several diseases of the hemopoietic system characterized by uncontrolled leukocyte proliferation.

Lifetime Risk
Exposure-induced risks reported as a function of the distribution of age in a population.

Linear Dose Mode
This model postulates that the excess risk is linearly proportional to the dose.

Linear Energy Transfer (LET)
The energy lost by a charged particle passing through a substance per unit length of path.

Load Bearing
A term referring to walls or other structures in a house that contribute to supporting the weight of the house.

Lognormal Distribution
The distribution of a random variable whose logarithm obeys the normal or Gaussian law of probability.

Makeup Air
Outdoor air supplied into the house to compensate for house air which is exhausted by combustion appliances or other devices such as exhaust fans. Provision of makeup air can reduce the house depressurization that might otherwise result from the use of these appliances.

Manometer
Instrument used for measuring pressure of gases and vapors both above and below atmospheric pressure.

Mass Airflow
The bulk flow of radon-containing soil gas into the house as the result of pressure differences between the house and the soil. Distinguished from diffusive movement.

Matching
A procedure to reduce the biasing effect of a confounding variable. A feature of selection to study groups.

Maximum Contaminant Level Goals (MCLG)
A non-enforceable health goal under the Safe Drinking Water Act, set at levels which would result in no known or anticipated adverse health effects and allow an adequate margin of safety.

Maximum Contaminant Levels
An enforceable standard under the Safe Drinking Water Act, set as close to the MCLG as feasible, taking into consideration cost, availability of treatment technologies, and other practical considerations.

Metamorphic Rocks
Rocks that have been changed in their composition or texture by heat and pressure. The degree of change can be very small, such as that which occurs when shale changes into slate, or large enough to completely recrystallize the rock, as occurs in the formation of schist or gneiss.

Micron
A unit of length equal to one millionth of a meter or one thousandth of a millimeter.

Microrem
A unit of measure of dose equivalence which reflects health risk resulting from a given absorbed dose of radiation. A microrem is one millionth of a rem.

Microrem Per Hour
A unit of measure of the rate at which health risk is being incurred as a result of exposure to radiation.

Million Electric Volts (MeV)
A unit of energy defined as the energy gained by an electron accelerated through a potential difference of one million electron volts.

Minute Volume
The amount of air moving through the lung per minute; the product of the breathing rate times the volume of air per breath.

Mitigation (Long-Term)
Mitigation by a homeowner to permanently reduce elevated radon levels in a house.

Mitigation (Short-Term)
Temporary measures to reduce elevated radon levels in a house. Usually performed during the time between screening and follow-up measurement results.

Molecular Diffusion
The transfer of mass between adjacent layers of fluid in laminar flow.

Monoclonal Antibodies
Antibodies produced in culture from cell hybrids that exhibit extreme binding specificity.

Monodisperse
Particles falling within a very narrow range of sizes so as to behave identically as far as deposition in the lung.

Mucociliary Blanket
The layer of mucous lining the generations of a lung above the pulmonary region and responsible for transporting deposited materials out of the lung and into the gastrointestinal tract.

Natural Radioactivity
The property of radioactivity exhibited by more than 50 naturally occurring radionuclides.

Natural Ventilation
The unaided movement of air into and out of an enclosed space through intentionally provided openings, such as open windows and doors.

Naturally-Occurring
Occurring naturally in the soil. Not caused by industrial or other human activity.

Near-Wall Contribution
The component of dose to cells in tube walls which arises from radiation passing only through tissue (and not air).

Neoplasm
Any new and abnormal growth, such as a tumor.

Neoplastic Cells
Cells capable of forming cancers.

Neoplastic Disease
Any disease that forms tumors, whether malignant or benign.

Neutral Plane
A roughly horizontal plane through a house defining the level at which the pressure indoor equals the pressure outdoors. During cold weather, when the thermal stack effect is occuring, indoor pressures below the neutral plane will be lower than outdoors, so that outdoor air and soil gas will infiltrate. Above the neutral plane, indoor pressures will be higher than outdoors, so that house air will exfiltrate.

Neutrons
Electrically neutral particles in the nucleus of an atom.

Nuclide
A species of atom characterized by the constitution of its nucleus.

Occupational Exposure
Exposure of a worker during a period of work. For radon, it usually refers to miners, but may also refer to workers involved in radon mitigation.

Off-Gassing
The exhalation of gas by materials.

Oncogenes
Genes whose products have the ability to transform eukaryotic cells so they grow in a manner similar to tumor cells.

Parent
A radionuclide which, upon disintegration, yields a specified nuclide--either directly or as a later member of a radioactive series.

Particles
Matter in which solid or liquid substances exist in the form of aggregated molecules. Airborne particulate matter typically range in size from 0.01 to 20 micrometers.

Pascal
A unit of pressure equal to 248 inches of water column (WC).

Pegmatite
Igneous rock type that is usually found as dikes associated with granites. It can have high uranium count.

Permeability
The capacity of a porous rock, solid, or sediment to transmit fluid without damage to the structure of the medium.

Person-Years (PY)
A standard technique for handling variable follow-up periods; multiply the number of persons by the number of years of follow-up.

Phosphate Slag
Waste materials from phosphate mining, sometimes used as foundation material in construction work. Phosphate is often laced with radon-producing uranium.

Photomultiplier
A device that produces electrical pulses when struck by light pulses (such as from a scintillator).

Photons
Units of energy of electromagnetic radiations.

Picocurie
A unit of measurement of radioactivity. A curie is the amount of any radionuclide that undergoes exactly 3.7 x 10 to the tenth power radioactive disintegrations per second. A picocurie is one trillionth of a curie, or 0.037 disintegrations per second.

PicoCuries Per Liter (pCi/l)
A unit of measure used for expressing levels of radon gas.

Plate Out
The tendency of radon daughter products to attach themselves to indoor surfaces such as furniture, walls, and the insides of heating ducts. Radioactive particles that plate out are essentially rendered harmless because they won't be inhaled.

Polyurethane
Any of a class of synthetic resinous, fibrous, or elastomeric compounds belonging to the family of organic polymers made by the reaction of diisocyantes (organic

compounds containg two functional groups of structure-NCO) with other difunctional compounds such as glycols. The best known polyurethanes are flexible foams.

Polyvinylchloride (PVC)
Synthetic resin belonging to the family of polymeric organic compounds, manufactured by treating vinyl chloride with a peroxide catalyst usually in aqueous suspension or emulsion.

Porosity
Property of a solid which contains many minute channels or open spaces that are capable of absorbing liquids.

Potential Alpha Activity
The total amount of alpha-particle radiation to be expected from a unit volume of air containing a mixture of radon and its progeny.

Potential Alpha Energy
Total energy emitted during alpha decays if a quantity of radon progeny is allowed to decay through the entirety of the decay chain.

Preneoplastic Cells
One of the stages through which cells progress in changing from normal to neoplastic.

Pressure Driven Air Flow
Air flow in a home that is caused by differences in pressure between the indoor and outdoor air.

Progeny
The decay products of a radioactive series decay.

Proportional Mortality Ratio (PMR)
The ratio of two mortality proportions, expressed as a percentage, often adjusted for chronological or temporal differences between the two groups being compared.

Prospective
A study characteristic. Disease has not occurred in study groups at the start of a study.

Protons
Positively charged particles in the atomic nucleus. The number of protons in an atom is the atomic number of the element.

Quality Factor
The factor by which absorbed dose is to be multiplied to obtain a quantity that expresses, on a common scale, for all ionizing radiations, the irradiation incurred by exposed persons.

Rad
The standard unit of absorbed dose, equal to energy absorption of 100 ergs per gram.

Radiation Dose
The total amount of ionizing radiation absorbed by material or tissues, in the sense of absorbed dose (expressed in rads), exposure (expressed in roentgens), or dose equivalent (expressed in rems).

Radioactive Decay
The spontaneous transformation of a nuclide into one or more different nuclides accompanied by either the emission of energy and/or particles from the nucleus, nuclear capture or ejection of orbital elements, or fission.

Radioactive Equilibrium
A state in which the rate of formation of atoms by decay of a parent radioactive isotope is equal to its rate of disintegration by radioactive decay, so that the activity of the parent and the decay product assume a constant proportion (equal to one to one, if the parent has only one mode of radioactive decay). Because radon decay products tend to attach readily to surfaces (plate out), equilibrium between radon and its decay products is seldom reached.

Radioactive Particles
Products of radioactive decay.

Radioactivity
The spontaneous release of energy by a nucleus which always results in a change in mass and often results in a different element being formed.

Radionuclide
An unstable nucleus of an element that decays or disintegrates spontaneously, emitting radiation.

Radium (Ra)
A naturally-occurring radioactive element found in rock and soil. Atomic number 88.

Radon (Rn)
A naturally occuring radioactive gas that's colorless, odorless, and tasteless. It is formed by the decay of radium, which is itself a by-product of the radioactive disintegration of uranium. Radon gas has a half-life of 3.8 days. Atomic number 86.

Radon Bearing
Containing or emitting radon.

Radon Daughter
One of the radioactive decay products of radon gas--polonium, bismuth, and lead. Each decay product is the ultra-fine radioactive particle that decays into other radioactive products until, finally, a stable nonradioactive lead atom is formed and no further radioactivity is produced. The Rn-222 decay chain has 12 decay products, including the stable lead isotope, Pb-206. Also referred to as radon progeny.

Radon Progeny
The short-lived radionuclides formed as a result of decay of radon-222. They consist of polonium-218, polonium-214, lead-214 and bismuth-214. The effective combined half-life is about 30 minutes.

Radon Progeny Integrating Sampling Unit (RPISU)
An accoustically shielded, low-volume air sampling device that collects radon decay products on a membrane filter located near a thermoluminescence detector (TLD).

Glossary of Terms

Radon Source Strength
The intensity, power, or concentration of radon action from its point of origin. Refers to the general intensity of radon evolution from a specific soil or rock type beneath a house.

Rate
The number of cases per unit of population.

Rate Ratio
One rate divided by another rate with the same dimensions. A measure of association without a unit.

Reading Prong
A geological formation extending through New Jersey, New York, and Pennsylvania where high levels of naturally-occuring radon have been found.

Receptors
A protein that binds extracellular signals involved in cell-cell communication (usually found on a cell surface).

Recoil
The disattachment of radon decay products from solid surfaces as a result of alpha decay, causing the atom to re-enter the air.

Relative Biological Effect (RBE)
An evaluation of the impact of a given type of radiation on tissue based on the LET of the radiation type. RBE equals the product of the quality factor and the modifying factor.

Relative Risk
The ratio of the rate of disease in exposed to unexposed populations.

Relative Risk Projection Model
A model which estimates the risk of exposure beyond the years of observation of a studied population by projecting the currently observed percentage increase in cancer risk per unit dose into future years.

Relaxation Length
The depth of the soil which delivers a surface flux of radon that is 1/e of that delivered by soil at the surface.

Rem
A unit of dose equivalent.

Remediation
Work undertaken to lower radon levels in a structure.

Respiratory System
The organ system concerned with the exchange of oxygen and carbon dioxide. The respiratory system consists of the air passages through the nose, mouth, pharynx, the conducting airways (trachea, bronchi, bronchioles) and the gas exchange surfaces (alveoli) in the pulmonary parenchyma.

Retrospective
A study characteristic. Disease has already occurred in study groups at the start of a study.

Risk
The probability that something will happen, usually used to refer to something bad.

Roentgen Equivalent Man (REM)
A unit of equivalent dose equal to the amount that produces the same damage to humans as one roentgen of high voltage x-rays. Equal to 0.01 sievert. Related to absorbed dose by the RBE: Equivalent Dose (REM) = RBE * Absorbed Dose (RAD).

Sarcoma
Cancer arising from underlying tissues (muscle, bone, connective tissue).

Schist
A medium-to-coarse-grained, high-grade metamorphic rock with a texture of thin, irregular plates due to abundant mica.

Scintillator
A material which produces light pulses when radiation interacts with it.

Screening Measurement
A measurement taken under closed-house conditions in the lowest livable area of a house. This initial measurement is recommended to determine whether a follow-up measurement is necessary.

Sealant
Any organic substance soft enough to pour or extrude and is capable of subsequent hardening to form a permanent bond with the substrate. Most sealants are synthetic polymers (silicones, urethanes, acrylics) which are semisolid before application and later become elastomeric.

Sealer
A coat applied to keep subsequent coats from permeating a surface.

Secretory Cells
Cells in the respiratory tract epithelium that are capable of producing mucoid substances and dividing to replace cells normally lost.

Sedimentation
The process by which a particle deposits on the wall of a tube in the lung due to gravitational settling.

Sensible Heat
Heat which is associated with the change in temperature of a substance.

Sensitivity
The ability of a measurement system to determine the presence or quantity of radiation above background signals.

Sievert (Sv)
A standard international unit for dose equivalent which equals one gray (Gy), times a quality factor, times a modifying factor; 1Sv = 100 rem.

Silicone
Any of a diverse class of chemical polymers manufactured in the form of fluids, resins, or elastomers. Although partially organic in composition, molecules of silicones differ from most polymers because the backbone of their structure does not contain carbon, the characteristic element of organic compounds. Silicone resins are employed in protective coatings.

Sill Or Sill Plate
The horizontal wooden band that rests on top of a cinder block or poured foundation and extends around the entire perimeter of the house. The floor joists, which support the floor above, rest on the sill plate.

Slab
A level layer of concrete, usually 3 to 6 inches thick, on which some houses are built. Slab-on-grade houses, as they are called, usually don't have any basement or crawl space.

Slab Below Grade
A type of house construction where the bottom floor is a slab which averages between 1 and 3 feet below grade level on one or more sides.

Slab-on-Grade Foundation
A substructure type in which the foundation is at approximately the same level ("on-grade") as the surrounding ground surface. Generally, a poured concrete slab.

Smoke Stick
A small tube, several inches long, which releases a small stream of inert smoke when a rubber bulb at one end of the tube is compressed. Can be used to visually define bulk air movement in a small area, such as the direction of air flow through small openings in slabs and foundation walls.

Soil Gas
Those gaseous elements and compounds that occur in the small spaces between particles of earth or soil. Rock can contain gas also. Such gases can move through or leave the soil or rock depending on changes in pressure. Radon is a gas which forms in the soil wherever radioactive decay of radium occurs.

Soil Ventilation
A radon reduction technique in which radon is drawn away, or ventilated, from a home before it can enter. Several types of soil ventilation exist.

Solubility Coefficient
A measure of gas solubility in a liquid. Radon solubility coefficient is defined as the ratio of radon concentration in water to that of air (Co86).

Spectrum
A mathematical description of the relative frequencies for radiations of different energies.

Stack Effect
In houses and other buildings, the tendency toward displacement (due to the difference in temperature) of internal heated air by unheated outside air which is caused by the difference in density of outside and inside air. Similar to the air and gas in a duct, flue, or chimney rising when heated due to its lower density compared with that of surrounding air or gas.

Standardization
A procedure to reduce the biasing effect of a confounding variable. A feature of data analysis.

Standardization Mortality Ratio (SMR)
The ratio of mortality rates, expressed as percentage, usually adjusted for age or time differences between the two groups being compared.

Steady State
A condition in which a system exhibits no variability over time.

Stochastic Process
A process in which the probability of an effect occurring rather than its severity is regarded as a function of dose without threshold.

Stopping Power
The rate at which radiation loses energy in traversing a small distance in a material, given as the ratio of the energy loss to the distance traveled.

Sub-Slab Ventilation
A radon reduction technique in which pipes are placed into the stone aggregate beneath a home's basement floor to ventilate radon. Sub-slab ventilation is one of several soil ventilation techniques.

Subatomic
Smaller than an atom. In other words, the parts of an atom: electrons, protons, and neutrons.

Subatomic Particles
The electrons, protons, and neutrons comprising atoms.

Submicron
Below one micron in size or diameter. Less than one-millionth of a meter.

Substructure/Understructure
The underlying structure of a building or house. May be a basement, slab-on-grade foundation, crawl-space, etc.

Sump
A hole in the basement floor designed to collect water. The water may drain out by itself or be removed by a sump pump.

Sump Pump
A pump to move collected water out of the sump pit to an above-grade discharge remote from the house.

Surficial Material
Surface or near surface soil deposits.

Symmetric Branching
An anatomical model which assumes that each lung passageway divides into two identical subpassages. Contrasts with asymmetric branching.

Synergism
Harmoniously combined action of two or more factors such that the resulting effect is greater than the sum of the effects produced by the actions individually.

Synergistic Model
A form of cancer causality whereby two or more carcinogens act synergistically to cause cancer with a greater probability than if each were acting alone.

Thermal Bypass
Any opening through the floor between stories of a house (or through the ceiling between the living areas and the attic) which facilitates the upward movement of house air under the influence of the stack effect. By facilitating the upward movement, thermal bypasses increase exfiltration at the upper levels, which in turn will increase infiltration of outdoor air and soil gas.

Thermoluminescent Dosimeter (TLD)
A material which stores the energy absorbed from radiation passing through it, and then releases this energy as light pulses upon subsequent heating.

Thorium-232
A radioactive isotope of the element Thorium in the actinium series. Symbol Th, atomic number 90, atomic weight 232.

Thoron
Conventional name for the isotope of radon with an atomic weight of 220 (radon-220). Sometimes symbolized as Tn.

Threshold
The lowest level of a stimulus above which symptoms appear, or conversely, below which no effect is observed.

Tidal Volume
The amount of air taken into the lung during each breath.

Tight House
A house with a very low air exchange rate. If 0.5 to 0.9 air changes per hour is typical of modern housing, a tight house would be one with an exchange rate well below 0.5 ACH.

Time Integrated
A type of sampling instrument that measures a single average concentration value over an extended time interval.

Top Voids
Air space(s) within masonry walls. Specifically refers to the openings in the top course of such walls; the course of block to which the sill plate is attached and on which the walls of the house rest. Also known as block voids or voids.

Trachea
The windpipe. The duct by which air passes from the larynx (in the throat) to the bronchi and the lungs.

Tracheobronchial Tree
The portion of the respiratory system which conducts air to the alveoli.

Transfection
Acquisition of new genetic markers by incorporating added DNA.

Transformation
Conversion of eukaryotic cells to a state of unrestricted growth in culture, resembling or identical to the tumor condition.

Transient Equilibrium
Relatively short-term equilibrium within a portion of a radioactive series where the parent has a long half-life only relative to successive series members. An example is the radon progeny series.

Turbulent Deposition
The process by which air streams fail to follow the contour of passageways thereby transporting particles to the walls.

Unattached Fraction
The fraction of airborne radon progeny unassociated with aerosol particles.

Unattached Radon Decay Product
A radon decay product that is not electrostatically attached to dust or particles in the air. Capable of attaching to lung tissue if inhaled.

Unattached Radon Progeny
The fraction of radon progeny that is not attached to the ambient soil.

Uraniferous
Of or containing uranium.

Uranium Ore
A mineral deposit that can be mined to recover uranium.

Uranium-238
A highly toxic, radioactive metallic isotope of the element uranium in the actinide series. Undergoes radioactive decay, eventually to radium-226. Symbol U, atomic number 92, atomic weight 238.

Veneer
A single layer or tier of masonry or similar materials securely attached to a wall for the purposes of providing ornamentation, protection, or insulation, but not bonded or attached to intentionally exert common action under load.

Ventilation
The process of supplying and removing air by natural and/or mechanical means to and from any space.

Voids
The air spaces created within cinder block walls by the interconnecting of the holes in the middles of the blocks.

Wall Ventilation
A radon reduction technique in which the network of openings within a home's concrete block walls is ventilated. Wall ventilation is a type of soil ventilation.

Warm Air Supply
The ducting and registers which direct heated house air from the forced-air furnace to the various parts of the house. The supply ducting is elevated pressure relative to the house because the central furnace fan is blowing air through its ducting.

Water Column (WC)
A unit of pressure equal to 1/248 of a pascal.

Weighting Factor
A numerical factor used to modify calculation of the equivalent dose to account for variation of LET and radiation effects with tissue type and exposure.

Working Level (WL)
A unit of measure for documenting exposure to radon decay products. One WL is equal to approximately 200 pCi/l.

Working Level Month (WLM)
A unit of measure used for cumulative exposure to radon. One WLM equals exposure to one WL for 173 hours.

Abbreviations and Acronyms

This section lists abbreviations and acronyms found throughout the Directory.

AAAS
American Association for the Advancement of Science

AAEE
American Academy of Environmental Engineers

AAEM
American Academy of Environmental Medicine

AAPHP
American Association of Public Health Physicians

AARST
American Association of Radon Scientists and Technologists

AASG
Association of American State Geologists

AB
Alberta

ACCA
Air Conditioning Contractors of America

ACE
Army Corps of Engineers

ACEEE
American Council for an Energy Efficient Economy

ACGIH
American Council of Govermental Industrial Hygienists

ACI
American Concrete Institute

ACS
American Cancer Society

ACSH
American Council on Science and Health

AECB
Atomic Energy Control Board

AECL
Atomic Energy of Canada, Ltd.

AEERL
Air and Energy Engineering Research Laboratory

AEHRC
Army Engineering and Housing Research Center

AEO
Army Environment Office

AFA
American Fan Association

AFRP
Air Force Radon Project

AGI
American Geological Institute

AGU
American Geophysical Union

AGWSE
Association of Ground Water Scientists and Engineers

AIA
American Institute of Architects

AIHA
American Industrial Hygiene Association

AIREA
American Institute of Real Estate Appraisers

AK
Alaska

AL
Alabama

ALA
American Lung Association

ALAPCO
Association of Local Air Pollution Control Officials

AMA
American Medical Association

AMC
American Mining Congress

AMCA
Air Movement and Control Association

AMCBO
Association of Major City Building Officials

ANL
Argonne National Laboratory

ANS
American Nuclear Society

ANSI
American National Standards Institute

AOMA
American Occupational Medical Association

AOO
Albuquerque Operations Office

APCA
Air Pollution Control Association

APHA
American Public Health Association

ARP
Army Radon Project

ARS
Agriculture Research Service

ASC
Adhesive and Sealant Council

ASDWA
Association of State Drinking Water Administrators

ASHI
American Society of Home Inspectors

ASHRAE
American Society of Heating, Refrigerating and Air Conditioning Engineers

ASIWPCA
Association of State and Interstate Water Pollution Control Administrators

ASREC
American Society of Real Estate Counselors

ASSE
American Society of Safety Engineers

ASTHO
Association of State and Territorial Health Officials

ASTM
American Society for Testing and Materials

AT
Alpha-track Radon Measurement Method

ATD
Alpha-Track Detector

AWWA
American Water Works Association

AWWARF
American Water Works Association Research Foundation

AZ
Arizona

BBB
Better Business Bureau

BC
British Columbia

BED
Building Environment Division

BEIR IV
Biological Effects of Ionizing Radiation Study

BLA
Brooklyn Lung Association

BNL
Brookhaven National Laboratory

BOCA
Building Officials and Code Administrators, International

BOM
Bureau of Mines

BOMA
Building Owners and Managers Association, International

BPA
Bonneville Power Administration

Bq/m3
Becquerel per cubic meter

BTU
British Thermal Unit

CA
California

CABO
Council of American Building Officials

CBE
Citizen's for a Better Environment

CBT
Center for Building Technology

CC
Activated Charcoal Adsorption Radon Measurement Method

CCAP
Center for Clean Air Policy

CCHW
Citizen's Clearinghouse for Hazardous Wastes

CDC
Centers for Disease Control

CDRH
Center for Devices and Radiological Health

CEC
Commission of the European Communities

CEI
Center for Environmental Information

CEQ
Council on Environmental Quality

CER
Center for Epidemiological Research

CERL
Construction Engineering Research Laboratory

CFA
Consumer Federation of America

CHA
Canadian Homebuilders Association

CHOB
Cannon House Office Building

CHS
Center for Health Statistics

Ci
Curies

CIAQ
Council on Indoor Air Quality

CIRRPC
Committee on Interagency Radiation Research and Policy Coordination

CLA
Canadian Lung Association

CMHC
Canada Mortgage and Housing Corporation

CO
Colorado

COO
Chicago Operations Office

COPHL
Conference of Public Health Laboratorians

CPSC
Consumer Product Safety Commission

CR
Continuous Radon Measurement Method

CRCPD
Conference of Radiation Control Program Directors

CRE
Counselor of Real Estate

CRI
Center for Research Information

CRM
Continuous Radon Monitor

CRR
Center for Radiation Research

CSD
Criteria and Standards Division

CSHEM
Conference of State Health and Environmental Managers

CSI
Construction Specification Institute

CT
Connecticut

CU
Consumers Union of United States

CW
Continuous Working-Level Radon Measurement Method

CWLM
Continuous Working Level Monitor

D-
Democrat

DASR
Deputy Assistant Secretary for Research

DBS
Division of Building Systems

DC
District of Columbia

DE
Delaware

DEE
Division of Environmental Engineering

Del.
Delegate

DER
Division of Environmental Research

DITSP
Division of Innovative Technology and Special Projects

DNA
Deoxyribonucleic Acid

DOA
Department of the Army

DOC
Department of Commerce

DOD
Department of Defense

DOE
Department of Energy

DOHS
Division of Occupational Health and Safety

DOI
Department of Interior

DOJ
Department of Justice

DOL
Department of Labor

DON
Department of Navy

DOS
Department of State

DOS
Division of Safety

DOT
Department of Transportation

DPC
Domestic Policy Council

DRC
Denver Research Center

DRCRL
Denver Research Center Research Laboratory

DRP
Division of Residential Programs

DSDTT
Division of Standards Development and Technology Transfer

DSOB
Dirkson Senate Office Building

DTS
Directorate of Technical Support

E-PERM
Electret-PERM Radon Measurement Method

EA
Environmental Analysis Section

ED
Energy Division

EDF
Environmental Defense Fund

EEB
European Environmental Bureau

EEI
Edison Electric Institute

EERF
Eastern Environmental Radiation Facility

EH
Environmental Health

EHA
Environmental Hygiene Agency

EIW
European Institute for Water

ELA
Effective Leakage Area

ELC
Environment Liaison Centre

EMD
Environmental Management Division

EMG
Energy and Marine Geology

EML
Environmental Measurements Laboratory

EOP
Executive Office of the President

EP
Electret-PERM Radon Measurement Method

EPA
Environmental Protection Agency

EPA 1
Environmental Protection Agency Region 1

EPA 2
Environmental Protection Agency Region 2

EPA 3
Environmental Protection Agency Region 3

EPA 4
Environmental Protection Agency Region 4

EPA 5
Environmental Protection Agency Region 5

EPA 6
Environmental Protection Agency Region 6

EPA 7
Environmental Protection Agency Region 7

EPA 8
Environmental Protection Agency Region 8

EPA 9
Environmental Protection Agency Region 9

EPA 10
Environmental Protection Agency Region 10

EPD
Environmental Planning Division

EPD
Environmental Policy Division

EPI
Environmental Policy Institute

EPRI
Electrical Power Research Institute

EQ
Environmental Quality Staff

ERC
Employee Relocation Council

ERD
Environmental Review Division

ESCSI
Expanded Shale, Clay and Slate Institute

ESH
Environmental Safety and Health

EUG
European Union of Geosciences

Fannie Mae
Federal National Mortgage Association

FCSETP
Federal Coordinating Council for Science, Engineering and Technology Policy

FDA
Food and Drug Administration

FDE
Florida Defenders of the Environment

FEMA
Federal Emergency Management Agency

FIPR
Florida Institute of Phosphate Research

FL
Florida

FmHA
Farmers Home Administration

Freddie Mac
Federal Home Loan Mortgage Corporation

FY
Fiscal Year

GAO
General Accounting Office

GD
Geologic Division

GJAO
Grand Junction Area Office

GPO
Government Printing Office

GR
Grab Radon Measurement Method

GRI
Gas Research Institute

GSA
General Services Administration

GSA
Geological Society of America

GSAR
General Services Administration Regional Office

GU
Guam

GW
Grab Radon Progeny Measurement Method

Gy
Gray

HEB
Health Effects Branch

HETAB
Hazards Evaluation and Technical Assistance Branch

HHAD
Human Health Assessment Division

HHS
U.S. Department of Health and Human Services

HI
Hawaii

HOR
US House of Representatives

HPS
Health Physics Society

HRSA
Health Resources and Services Administration

HRV
Heat Recovery Ventilator

HSOB
Hart Senate Office Building

HST
Health and Safety Technology

HUD
Department of Housing and Urban Development

HVAC
Heating, Ventilating, Air Conditioning

HVI
Home Ventilating Institute

IAH
International Association of Hydrogeologists

IAHS
International Association of Hydrological Sciences

IAQIVP
Indoor Air Quality, Infiltration and Ventilation Program

IAQVG
Indoor Air Quality/Ventilation Group

IARR
International Association of Radiation Research

IAS
Indoor Air Staff

ICRP
International Commission on Radiological Protection

ICRU
International Commission on Radiation Units and Measurements

ICWP
Interstate Conference on Water Problems

ICWU
International Chemical Workers Union

ID
Idaho

IEEP
Institute for European Environmental Policy

IGCP
International Geological Correlation Program

IGU
International Geographical Union

IHB
Industrial Hygene Branch

IHS
Indian Health Service

INRD
Ionizing/Nonionizing Radiation Division

IOO
Idaho Operations Office

IRD
Ionizing Radiation Division

IREM
Institute of Real Estate Management

IRPA
International Radiation Protection Association

ISEA
Industrial Safety Equipment Association

IUATLD
International Union Against Tuberculosis and Lung Disease

IUGS
International Union of Geological Sciences

KS
Kansas

KY
Kentucky

LA
Louisiana

LANL
Los Alamos National Laboratory

LAOO
Los Alamos Operations Office

LAPIS
Legislative Affairs and Public Information Staff

LBL
Lawrence Berkeley Laboratory

LET
Linear Energy Transfer

LHOB
Longworth House Office Building

LLNL
Lawrence Livermore National Laboratory

LNRD
Land and Natural Resources Division

LOC
Library of Congress

LVF
Las Vegas Facility

MA
Massachusetts

MAC
Mining Association of Canada

MB
Manitoba

MBAA
Mortgage Bankers Association of America

MCPH
Ministry of Concern for Public Health

MD
Maryland

ME
Maine

MeV
Million Electric Volts

MI
Michigan

MIIA
Mine Inspectors Institute of America

MN
Minnesota

MNMSH
Metal and Non Metal Mine Safety and Health

MO
Missouri

MS
Mississippi

MSHA
Mine Safety and Health Administration

MT
Montana

NAA
National Apartment Association

NAAEE
North American Association for Environmental Education

NAAEM
North American Association of Environmental Medicine

NACHO
National Association of County Health Officials

NAEP
National Association of Environmental Professionals

NAHB
National Association of Home Builders of the United States

NAR
National Association of Realtors

NARI
National Association of the Remodeling Industry

NAS
National Academy of Sciences

NASA
National Aeronautics and Space Administration

NAWC
National Association of Water Companies

NB
New Brunswick

NC
North Carolina

NCAC
National Clean Air Coalition

NCI
National Cancer Institute

NCIC
National Cartographic Information Center

NCLEHA
National Conference of Local Environmental Health Administrators

NCMA
National Concrete Masonry Association

NCRP
National Council on Radiation Protection and Measurements

NCSBCS
National Conference of States on Building Codes and Standards

NCSL
National Conference of State Legislatures

NCWM
National Conference on Weights and Measures

ND
North Dakota

NE
Nebraska

NEHA
National Environmental Health Association

NF
Newfoundland

NFES
Naval Facilities Engineering Staff

NGA
National Governors Association

NH
New Hampshire

NIBS
National Institute of Building Sciences

NIEHS
National Institute of Enviromental Health Science

NIH
National Institutes of Health

NIOSH
National Institute for Occupational Safety and Health

NIRS
National Inorganics and Radionuclide Survey

NIST
National Institute for Standards and Technology

NJ
New Jersey

NJ DEP
New Jersey Department of Environmental Protection

NM
New Mexico

NML
National Measurements Laboratory

NOO
Nevada Operations Office

NPPC
Northwest Power Planning Council

NPS
National Park Service

NRC
Nuclear Regulatory Commission

NRC
National Research Council

NRMCA
National Ready Mixed Concrete Association

NRP
Navy Radon Project

NSA
National Stone Association

NSF
National Science Foundation

NSF
National Sanitation Foundation

NSPE
National Society of Professional Engineers

NSPS
National Society of Professional Sanitarians

NSTL
National Space Technology Laboratory

NURE
National Uranium Resource Evaluation

NV
Nevada

NWDE
National Water Data Exchange

NWWA
National Water Well Association

NY
New York

NYSDOH
New York State Department of Health

NYSEO
New York State Energy Office

NYSERDA
New York State Energy Research and Development Agency

OAR
Office of Air and Radiation

OBCS
Office of Building and Community Systems

OC
Office of Conservation

OCAW
Oil, Chemical and Atomic Workers International Union

OCPD
Office of Community Planning and Development

OCRE
Office of Conservation and Renewable Energy

ODE
Office of Device Evaluation

ODP
Office of Defense Programs

ODW
Office of Drinking Water

OEA
Office of Environmental Analysis

OEE
Office of Environment and Energy

OEETD
Office of Environmental Engineering and Technical Demonstation

OER
Office of Energy Research

OH
Ohio

OHER
Office of Health and Environmental Research

OHP
Office of Health Physics

OHWRA
Office of Hazardous Waste and Remedial Action

OIR
Office of Information and Regulation

OK
Oklahoma

OMB
Office of Management and Budget

OMS
Office of Medical Services

ON
Ontario

ONE
Office of Nuclear Energy

OPD
Office of Policy Development

OPDR
Office of Policy Development and Research

OPP
Office of Policy and Program Development

OR
Oregon

ORC
Organization of Resources Counselors

ORDE
Office of Research and Development

ORF
Ontario Research Foundation

ORM
Office of Radiation Measurements

ORNL
Oak Ridge National Laboratory

OROO
Oak Ridge Operations Office

ORP
Office of Radiation Programs

ORS
Office of Regulatory Support

OSG
Office of Surgeon General

OSHA
Occupational Safety and Health Administration

OSHTIB
Occupational Safety and Health Technical Information Branch

OSTI
Office of Scientific and Technical Information

OSTP
Office of Science and Technology Policy

OTTPD
Office of Technology Transfer and Program Development

OW
Office of Water

OWDC
Office of Water Data Coordination

PA
Pennsylvania

PA DER
Pennsylvania Department of Environmental Resources

PAB
Problem Assessment Branch

PAD
Physical Agents Division

Pb 206
Lead 206

Pb 214
Lead 214

PCA
Portland Cement Association

PCi/l
Picocuries per liter

PCLG
Public Citizen Litigation Group

PE
Prince Edward Island

PHF
Public Health Foundation

PHS
Public Health Service

PIC
Public Information Center

PMA
Precision Measurements Association

PMR
Proportion Mortality Ratio

PNL
Pacific Northwest Laboratory

PPIB
Policy and Public Information Branch

ppm
Parts per Million

PQ
Quebec

PR
Puerto Rico

PY
Person Years

PYR
Person Years At Risk

R-
Republican

Ra 226
Radium 226

RAC
Radiation Advisory Committee

RCEDD
Resources, Community and Economic Development Division

RCHB
Radiation Control Health Branch

RD
Radon Division

REALTIST
National Association of Real Estate Brokers

Rep.
US Representative

RESSI
Real Estate Securities and Syndication Institute

RG
Radioactive Group

RHD
Radiation Health Division

RHOB
Rayburn House Office Building

RI
Rhode Island

RICD
Radiological and Inorganic Chemicals Division

RIDG
Radiation Interaction And Dosimetry Group

RILA
Rhode Island Lung Association

RIT
Rochester Institute of Technology

RLI
Realtors Land Institute

RMP
Radon Measurement Proficiency Program

Rn 222
Radon 222

RNMI
Realtors National Marketing Institute

ROO
Richland Operations Office

RP
Radon Progeny Integrating Sampling Unit

RPD
Radiaiton Physics Division

RPISU
Radon Progeny Integrated Sampling Unit

RPO
Radon Project Office

RRP
Radon Research Project

RRS
Radiation Research Society

RS
Radiation Staff

RS
Radiation Safety

RSB
Radiation Safety Branch

RSI
Radiation Source and Instrumentation

RSOB
Russell Senate Office Building

RTI
Research Triangle Institute

RTS
Residential Technology Section

SAB
Science Advisory Board

SARA
Superfund Amendments and Reauthorization Act (of 1986)

SBCCI
Southern Building Code Congress International

SC
South Carolina

SCOPE
Scientific Committee on Problems of the Environment

SD
South Dakota

SDLA
South Dakota Lung Association

SEMD
Safety and Environmental Measurements Division

Sen
US Senator

SENATE
US Senate

SFOO
San Francisco Operations Office

SHTA
Safety and Health Technology Center

SIOR
Society of Industrial and Office Realtors

SK
Saskatchewan

SLPS
State and Local Programs and Support

SMACCNA
Sheet Metal and Air Conditioning Contractors National Association

SMR
Standardized Mortality Ratio

SNL
Sandia National Laboratory

SNLL
Sandia National Laboratory Livermore

SO
Safety Office

SROO
Savannah River Operations Office

SSPMA
Sump and Sewage Pump Manufacturers Association

STAPPA
State and Territorial Air Pollution Program Administrators

STC
Safety and Technology Center

Sv
Sievert

SWI
Sealant and Waterproofers Institute

SWI
Sealant and Waterproofing Institute

TB
Torts Branch

TBD
Technical Support Division

TED
Environmental Division

TIMA
Thermal Insulation Manufacturers Association

TLD
Thermoluminescence Dosimeter

Tn
Thoron

TN
Tennessee

TVA
Tennessee Valley Authority

TX
Texas

U 235
Uranium 235

U 238
Uranium 238

UCS
Union of Concerned Scientists

UL
Underwriters Laboratory

UMTRA
Uranium Mine Tailings Remediation Act

UNESCO
United Nations Educational, Scientific, and Cultural Organization

UNSCEAR
United Nations Scientific Committee on the Effects of Atomic Radiation

USAF
Department of Air Force

USDA
Department of Agriculture

USGS
US Geological Survey

USPIRG
U.S. Public Interest Reseach Group

UT
Utah

VA
Virginia

VA
Veterans Administration

VI
Virgin Islands

VLA
Vermont Lung Association

VT
Vermont

WA
Washington

WC
Water Column

WERL
Water Engineering Research Laboratory

WFPHA
World Federation of Public Health Associations

WHO
World Health Organization

WI
Wisconsin

WISE
World Information Service on Energy

WL
Working Level

WLM
Working Level Month

WPCF
Water Pollution Control Federation

WQA
Water Quality Association

WRSIC
Water Resources Scientific Information Center

WV
West Virginia

WY
Wyoming

Index of Advertisers

This section lists display advertisers in the Directory. Additional information on products and services can be obtained by contacting the company directly.

Index of Names

This section lists all key personnel referenced in the Directory. Each individual entry is followed by the affiliated organization, address, telephone number, and the sections where the listing can be located.

Please note that individuals affiliated with more than one organization are accorded additional entries.

A

Bonamo, Paul T., COUNTY HOME INSPECTIONS, P.O. Box 391, Wilton, CT, 06897, (203) 762-2608, (1).

Bond, Peter, U.S. DEPARTMENT OF ENERGY (DOE)/CHICAGO OPERATIONS OFFICE(COO)/BROOKHAVEN NATIONAL LABORATORY(BNL), Building 510-A, Upton, NY, 11973, (516) 282-2123, (8).

Bonderson, Noel A., PLACER COUNTY AIR POLLUTION CONTROL DISTRICT, 11484 B Ave., Auburn, CA, 95603, (916) 823-4443, (10).

Boneham, Roger, HOMEAIR TESTERS, 1234 Old Town South Dr., Indianapolis, IN, 46260, (317) 846-5711, (1).

Boneham, Tom, HOMEAIR TESTERS, 1234 Old Town South Dr., Indianapolis, IN, 46260, (317) 846-5711, (1).

Bontrager, Paul J., NASHVILLE AND DAVIDSON COUNTYMETROPOLITAN HEALTH DEPARTMENT/POLLUTION CONTROL DIVISION, 311 23rd Ave. North, Nashville, TN, 37203, (615) 327-9313, (10).

Booberg, Carl C., ALA OF VIRGINIA, P.O. Box 7065, Richmond, VA, 23221-0065, (804) 355-3295, (15).

Boone, Ronald, WESTERN NORTH CAROLINA REGIONAL AIR POLLUTION CONTROL BOARD, Buncombe County Courthouse, Asheville, NC, 28802-3569, (704) 255-5655, (10).

Booth, Richard B., SHASTA COUNTY AIR QUALITY MANAGEMENT DISTRICT, 1855 Placer St., Redding, CA, 96001, (916) 225-5674, (10).

Borak, Thomas B., COLORADO STATE UNIVERSITY/DEPARTMENT OF RADIOLOGY AND RADIATION BIOLOGY, Ft. Collins, CO, 80523, (303) 491-5222, (5).

Borchert, Harold, NEBRASKA DEPARTMENT OF HEALTH/DIVISION OF RADIOLOGICAL HEALTH, 301 Centennial Mall South P.O. Box 93007, Lincoln, NE, 68509, (402) 471-2168, (10).

Borden, Tim, BUILDING INSPECTOR OF AMERICA, THE, 243 Reed Rd., Westport, MA, 02790, (508) 771-2477, (508) 636-6767, (1).

Borden, Timothy E., CAPE COD HOME INSPECTIONS, 243 Reed Rd., Westport, MA, 02790, (508) 771-2477, (800) 828-1919, (1).

Borduz, Lucian, RADON DETECTION, INC., 131 Amherst St. P.O. Box 900, Amherst, NH, 03031-0900, (603) 882-7340, (1).

Borger, Henry, NATIONAL ACADEMY OF SCIENCES (NAS)/NATIONAL RESEARCH COUNCIL (NRC)/FEDERAL CONSTRUCTION COUNCIL, 2101 Constitution Ave., NW, Washington, DC, 20418, (202) 334-3378, (5)(13).

Borris, David, FLORIDA INSTITUTE OF PHOSPHATE RESEARCH, 1855 West Main St., Bartow, FL, 33830, (813) 533-0983, (10).

Borsellino, Ernie, FORESIGHT ENGINEERING, INC., P.O. Box 3111, Wayne, NJ, 07474-3111, (201) 942-3061, (1)(2).

Boswell, Edward, NATIONAL RADON SERVICE, P.O. Box 3596, Gaithersburg, MD, 20878-0596, (301) 953-4598, (1).

Bou-Matar, Walid, ENVIRONMENTAL ENGINEERING LABORATORIES, INC., P.O. Box 55304, Grand Junction, CO, 81505, (303) 245-2500, (1).

Bouchard, A.L., A.L. BOUCHARD COMPANY, 56 Franklin Pl., Totowa, NJ, 07512, (201) 790-6259, (1)(2).

Boudette, Eugene, NEW HAMPSHIRE DEPARTMENT OF RESOURCES AND ECONOMIC DEVELOPMENT/ENVIRONMENTAL SERVICES/GEOLOGY UNIT, University Of New Hampshire 117 James Hall, Durham, NH, 03824, (603) 862-3160, (10).

Boukedes, Steve, RAD ENVIRONMENTAL SERVICES, INC., 2306 Rodeo Dr., Durham, NC, 27704, (919) 688-9226, (1)(2).

Bowers, Wayne, NEW ENGLAND DE-WATERING, INC., 70 Heritage Ave. Suite 14 Heritage Pl., Portsmouth, NH, 03801, (800) 535-1211, (2)(4).

Bowers, Wendy, MIXLEE ENVIRONMENTAL RESEARCH CORPORATION, 684-B Commerce Dr., Fairfield, CT, 06430, (203) 336-1211, (800) 545-1211, (1).

Bowie, Frances A., DISTRICT OF COLUMBIA DEPARTMENT OF CONSUMER AND REGULATORY AFFAIRS/SERVICE FACILITY REGULATION ADMINISTRATION/PHARMACEUTICAL AND MEDICAL DEVICES CONTROL DIVISION, 614 H St. Room 1014, Washington, DC, 20004, (202) 727-7190, (10).

Bowman, Chuck, U.S. DEPARTMENT OF INTERIOR (DOI)/NATIONAL PARK SERVICE(NPS)/SOUTHWEST REGIONAL OFFICE/INDOOR RADON PROGRAM, P.O. Box 728, Santa Fe, NM, 87501, (505) 988-6340, (8).

Bowman, Daniel L., NUCLEAR TECHNOLOGY, INC./RADON CHEK, 20 H. South State St., Westerville, OH, 43081, (614) 899-2435, (1)(4).

Bowman, Kenneth A., SEALANT AND WATERPROOFERS INSTITUTE (SWI), 3101 Broadway Suite 300, Kansas City, MO, 6411, (816) 561-8230, (13).

Boyce, J. Trevor, MICROBAC MID-ATLANTIC, INC., 5808 Jefferson Ave., Newport News, VA, 23605, (804) 245-8613, (1).

Boyd, Darlene, COLUMBIA/DEPARTMENT OF HEALTH/AIR POLLUTION CONTROL DIVISION, 1225 Laurel St., Columbia, SC, 29217, (803) 733-8320, (10).

Boyd, James L., RADON TESTING LABORATORIES, INC., 10390 Tuscany Ln., Ellicott City, MD, 21043, (301) 465-1093, (1).

Boyer, Ronald, GREENE COUNTY-CITY OF SPRINGFIELD/AIR POLLUTION CONTROL AUTHORITY, 227 East Chestnut Expressway, Springfield, MO, 65802, (417) 864-1662, (10).

Boylan, James, MEDLAB ENVIRONMENTAL TESTING, INC., P.O. Box 2045, Wilmington, DE, 19899, (302) 655-5227, (800) 633-5221, (1)(4).

Boylston, Ray, AMERICAN SOCIETY OF SAFETY ENGINEERS (ASSE), 1800 East Oakton, Des Plaines, IL, 60018, (312) 692-4121, (16).

Boynton, Melbourne R., MINNESOTA COMMERCE DEPARTMENT, 500 Metro Square, St. Paul, MN, 55101, (612) 297-3379, (10).

Brabant, Line, THOMSON AND NIELSEN ELECTRONICS, LTD., 4019 Carling Ave. Suite 303, Kanata, ON, K2K 2A3, (613) 592-3019, (1)(4)(5).

Braby, L.A., U.S. DEPARTMENT OF ENERGY (DOE)/RICHLANDS OPERATIONS OFFICE(ROO)/PACIFIC NORTHWEST LABORATORY(PNL)/DIVISION OF BIOLOGY AND CHEMISTRY, Radiological Physics Section, Richland, WA, 99352, (509) 376-5916, (8).

Bradley, David, AMERICAN SOCIETY FOR TESTING AND MATERIALS (ASTM), 1916 Race St., Philadelphia, PA, 19103, (215) 299-5571, (14).

Bradley, James, B & H RADON SERVICES, P.O. Box 247, Silver Spring, PA, 17575-0247, (717) 684-2612, (717) 426-3723, (1)(2).

Bradley, Roy, HURT AND PROFFITT, PROFESSIONAL ENGINEERS, 2524 Langhorne Rd., Lynchburg, VA, 24501, (804) 847-7796, (1).

Bradley, Valerie, HOUSEMASTER, 601 N. Olive St., Media, PA, 19063, (302) 658-2880, (1).

Bradley, W. Carter, ROANOKE COUNTY AND CITY OF SALEM/AIR POLLUTION CONTROL, P.O. Box 1074, Salem, VA, 24153, (703) 387-6224, (10).

Bradshaw, Michael C., RADIATION SERVICES, INC., 3412 Doriswood Ct., Antioch, TN, 37013, (615) 361-6492, (1).

Braff, Howard, RADON TESTING COMPANY OF NEW YORK, 1790 Meadowbrook Rd., Merrick, NY, 11566, (516) 868-1778, (1).

Brak, Tracey, DAMES AND MOORE, 7101 Wisconsin Ave. Suite 700, Bethesda, MD, 20814, (301) 652-2215, (1).

Brandies, Keith, PROFESSIONAL HOME ANALYSIS, 1025 Northeast Sanchez Ave., Ocala, FL, 32670, (904) 694-3982, (1)(2).

Brandys, Robert C., OCCUPATIONAL AND ENVIRONMENTAL HEALTH CONSULTING SERVICES, 635 Harding Rd., Hinsdale, IL, 60521, (312) 325-2083, (1)(2).

Brann, Jean E., NORTHERN NECK RADON SERVICE, Rte. 1, Box 259-E, Hague, VA, 22469, (804) 472-4252, (1)(2).

Brann, Milford L., NORTHERN NECK RADON SERVICE, Rte. 1, Box 259-E, Hague, VA, 22469, (804) 472-4252, (1)(2).

Brann, Sheila, BECKMAN INSTRUMENTS, INC., 2500 Harbor Blvd., Fullerton, CA, 92634, (714) 871-4848, (1)(4).

Brasof, Arthur, PENNSYLVANIA BUILDING INSPECTORS, INC., 1273 Wilton Crescent, Yardley, PA, 19067, (215) 493-9426, (1).

Brazukas, Richard J., RESIDENCE RADON, 28077 West 85th Ter., Desoto, KS, 66018, (913) 383-4610, (913) 585-1094, (1)(4).

Bredder, Glenn A., SILVER SPRINGS HOME INSPECTION, P.O. Box 182, Landing, NJ, 07850, (201) 770-0026, (1).

Bredhoff, Nancy, RADON TESTING CORPORATION OF AMERICA, 12 West Main St., Elmsford, NY, 10523, (914) 347-5010, (800) 457-2366, (1)(4)(6).

Bregman, Jack, SAFETY AND HEALTH SERVICES, 3307 North Second St., Harrisburg, PA, 17110, (717) 238-2122, (3).

Breiland, Rosemary, TWIN CITY TESTING, 662 Cromwell, St. Paul, MN, 55114, (612) 645-3601, (1).

Breiteneicher, Denise K., BOSTON AIR POLLUTION CONTROL COMMISSION, Boston City Hall Room 805, Boston, MA, 02201, (617) 725-4416, (10).

Brennan, Jack, HOME EXTERMINATING COMPANY, 4608 Bucks Schoolhouse Rd., Baltimore, MD, 21237, (301) 661-7911, (800) 492-5544, (1).

Brennan, John W., JB TESTING COMPANY, P.O. Box 1303, Frederick, MD, 21701, (301) 662-7043, (1).

Brennan, Terry, CAMRODEN ASSOCIATES, RFD 4, Box 62, Rome, NY, 13440, (315) 865-4269, (1)(2)(3)(5)(6).

Brenneman, Mary, RYAN NUCLEAR LABORATORIES, P.O. Box 88, West Jefferson, OH, 43162, (614) 848-4414, (1).

Brenner, Steven B., U.S. ASSETS RECOVERY INVESTIGATIONS, INC., P.O. Box 6953, Washington, DC, 20032, (202) 362-4764, (1).

Brent, David J., BRENT, DAVID J., 3000 South Jamaica Ct. Suite 130, Aurora, CO, 80014, (303) 755-7012, (1).

Bresson, Gilbert, INTERNATIONAL RADIATION PROTECTION ASSOCIATION (IRPA), BP N Q 33, F-92260 Fontenay-aux-Roses, France, (011) 331-746547139, (12).

Brienza, Anita, EMPLOYEE RELOCATION COUNCIL (ERC), 1720 N St., NW, Washington, DC, 20036, (202) 857-0857, (18).

Brightsen, R.A., RADON TECHNOLOGIES, INC., 12200 Sunrise Valley Dr. Suite 320, Reston, VA, 22091, (703) 620-4004, (800) 237-5024, (1)(4).

Brightwell, Ernest, ADEPT REMEDIATION CORPORATION, 12710 Jesse Smith Rd., Mt. Airy, MD, 21771, (301) 831-9607, (2).

Brinck, Bill, U.S. ENVIRONMENTAL PROTECTION AGENCY REGION 7 (EPA7)/RADIATION TECHNOLOGY(ORP), 726 Minnesota Ave., Kansas City, KS, 66101, (913) 236-2893, (8).

Brindley, Joseph D., ALABAMA REAL ESTATE COMMISSION, State Capitol, Montgomery, AL, 36130, (205) 261-5544, (10).

Brisson, Richard, MARYLAND DEPARTMENT OF THE ENVIRONMENT/OFFICE OF TOXICS, ENVIRONMENTAL SAFETY AND HEALTH, 201 West Preston St. 7th Floor Mail Room, Baltimore, MD, 21201, (301) 333-3130, (800) 872-3666, (10).

Bristow, Bruce, RADIATION CONSULTANTS, INC., P.O. Box 34012, Houston, TX, 77234, (713) 476-0072, (1).

Brobeck, Steve, CONSUMER FEDERATION OF AMERICA (CFA), 1424 16th St., NW Suite 604, Washington, DC, 20036, (202) 387-6121, (17).

Brockway, Dave, AMERICAN RADON CONTROL, INC., 2758 S. Langley Ct., Denver, CO, 80210, (303) 758-6580, (1).

Brockway, David, KITTREDGE CONSTRUCTION COMPANY, 2758 South Langley Ct., Denver, CO, 80210, (303) 825-4940, (2).

Brodeur, Robert, PHOENIX LABS, P.O. Box 472, Hollis, NH, 03049, (603) 465-7868, (1).

Brodhead, Bill, WPB ENTERPRISES, INC., RD 1, Box 209, Riegelsville, PA, 18077, (215) 346-8004, (1)(2)(6).

Brodie, William, WB CONTRACTORS, 3209 Morrison St., NW, Washington, DC, 20015, (202) 966-6312, (1).

Broerse, J.J., INTERNATIONAL ASSOCIATION OF RADIATION RESEARCH (IARR)/RADIOBIOLOGICAL INSTITUTE TNO, Lange Kleiweg 151, NL-2288 GJ Rijswijk, Netherlands, (011) 3115-136940, (12).

Brogan, Robert L., INDOOR POLLUTION CONTROL CENTER, 3921 Sawyer Rd., Sarasota, FL, 34233, (813) 923-3641, (800) 226-2636, (1)(2).

Bronk, Greg, GE SILICONES, 260 Hudson River Rd., Waterford, NY, 12188, (518) 237-3330, (800) 255-8886, (4).

Brooker, Art, HALLTECH, INC., 465 Coronation Dr., West Hill, ONT, M1E 2K2, (416) 284-6111, (4).

Brookins, Douglas G., UNIVERSITY OF NEW MEXICO/DEPARTMENT OF GEOLOGY, Albuquerque, NM, 87131, (505) 277-2310, (505) 277-4204, (1)(5).

Brooks, Rob, AIR TRON MECHANICAL SERVICES, 3178 South Halifax St., Aurora, CO, 80013, (303) 699-1836, (1)(2).

Broun, Richard, U.S. DEPARTMENT OF HOUSING AND URBAN DEVELOPMENT (HUD)/COMMUNITY PLANNING AND DEVELOPMENT(CPD)/OFFICE OF ENVIRONMENT AND ENERGY(OEE), 451 7th St., SW Room 7154, Washington, DC, 20410, (202) 755-7849, (8).

Brown, Colin, ATOMIC ENERGY OF CANADA, LTD. (AECL)/ADMINISTRATIVE DIVISION/RADIATION INDUSTRIAL SAFETY BRANCH, Tinawa, MB, R0E 1L0, (204) 753-2720, (11).

Brown, Dayne H., NORTH CAROLINA DEPARTMENT OF HUMAN RESOURCES/DIVISION OF FACILITY SERVICES/RADIATION PROTECTION SECTION, 701 Barbour Dr., Raleigh, NC, 27603-2008, (919) 733-4283, (1)(10).

Brown, Dennis, SASKATCHEWAN DEPARTMENT OF HUMAN RESOURCES, LABOUR AND EMPLOYMENT, Saskatchewan Place 1870 Albert St., Regina, SK, S4P 3V4, (306) 787-4486, (11).

Brown, Edward J., AMERICAN NUCLEAR SOCIETY/NEW YORK-NORTHEASTERN NEW YORK SECTION, C/O Knolls Atomic Power Laboratory P.O. Box 1072 Building E6, Schenectady, NY, 12301, (518) 393-6611, (14).

Brown, Halina, CLARK UNIVERSITY/ENVIRONMENT, TECHNOLOGY, AND SOCIETY PROGRAM/DEPARTMENT OF PHYSICS, 950 Main St., Worcester, MA, 01610, (617)793-7687, (1)(5).

Brown, J. Keith, COLORADO HOME PROTECTION, 6318 South Gallop Ct., Littleton, CO, 80120, (303) 797-2007, (1).

Brown, J.F., BRUN-REUSS ASSOCIATES, P.O. Box 0842, Lionville, PA, 19353-0482, (800) 522-0072, (1)(2).

Brown, Jeff, THERMO ENVIRONMENTAL CORPORATION, 101 1st Ave. P.O. Box 9046, Waltham, MA, 02254, (617) 622-1103, (505) 345-9931, (1)(2)(4).

Brown, Jeff, TMA/EBERLINE, HEALTH PHYSICS, 5635 Jefferson Blvd., NE, Albuquerque, NM, 87109, (505) 345-9931, (1)(2)(3)(4).

Brown, Ken, RENSSELAER POLYTECHNICAL INSTITUTE, 110 Eighth Street, Troy, NY, 12180, (518) 276-8237, (1)(2).

Brown, Lorrie, ENVIRONMENTAL DETECTION AGENCY, INC., P.O. Box 4854, Austin, TX, 78765, (512) 322-9013, (1)(4).

Brown, Mike, SCHMELTZER, APTAKER AND SHEPPARD, 1800 Massachusetts Ave., NW Suite 500, Washington, DC, 20036, (202) 828-1000, (3)(16).

Brown, Munroe, ELECTRO ANALYTICAL, INC., 7118 Industrial Park Blvd., Mentor, OH, 44060, (216) 951-3514, (1).

Brown, Paul, OKLAHOMA DEPARTMENT OF HEALTH/RADIATION AND SPECIAL HAZARDS SERVICE, 1000 NW Tenth St. P.O. Box 53551, Oklahoma City, OK, 73152, (405) 271-5221, (10).

Brown, Richard, BBC & M ENGINEERING, 195 Chittenden Ave., Columbus, OH, 43201, (614) 294-3745, (1).

Brown, Robert C., GREENWICH HEALTH DEPARTMENT/DIVISON OF ENVIRONMENTAL HEALTH AND LABORATORY, Town Hall Field Point Rd., Greenwich, CT, 06830, (203) 622-7843, (10).

Brown, Richard H., RADON ALERT, INC., P.O. Box 131378, Tyler, TX, 75713, (214) 592-4077, (1)(6).

Brownell, Bruce, ADIRONDACK ALTERNATE ENERGY, County Rte. 4, Edinburg, NY, 12134, (518) 863-4338, (4)(13).

Browning, D.D., SCIENCE MANAGEMENT SERVICES, LTD., 25 Eshelman Rd., Lancaster, PA, 17601, (717) 392-1425, (1).

Browning, Jr., D.D., SCIENCE MANAGEMENT SERVICES, LTD., 25 Eshelman Rd., Lancaster, PA, 17601, (717) 392-1425, (1).

Brozowski, George, ROCKLAND COUNTY HEALTH DEPARTMENT, Sanatorium Rd. Building D, Pomona, NY, 10970, (914) 354-0250, (10).

Brubaker, David, INFILTEC/FACTORY, 108 South Delphine Ave. P.O. Box 1125, Waynesboro, VA, 22980, (703) 943-2776, (1)(4).

Bruchman, George, MICHIGAN DEPARTMENT OF PUBLIC HEALTH/BUREAU OF ENVIRONMENTAL AND OCCUPATIONAL HEALTH/DIVISION OF RADIOLOGICAL HEALTH, 3423 North Logan St. P.O. Box 30195, Lansing, MI, 48909, (517) 335-8200, (6)(10).

Brun, Alvin, ENVIRONMENT AIR, LTD., P.O. Box 459, Bouctouche, NB, E0A 1G0, (506) 743-8901, (4).

Brundage, Jeff, RADON MEASURING SYSTEMS, 1129 Wyoming Ave., Scranton, PA, 18509, (717) 961-1666, (1)(2).

Bruno, Len, BORO RADON TESTING SERVICE, INC., 501 Sharp Ave., Glenolden, PA, 19036, (215) 532-8378, (215) 586-5640, (1).

Bruno, Rob, BORO RADON TESTING SERVICE, INC., 501 Sharp Ave., Glenolden, PA, 19036, (215) 532-8378, (215) 586-5640, (1).

Bruno, Ron, BORO RADON TESTING SERVICE, INC., 501 Sharp Ave., Glenolden, PA, 19036, (215) 532-8378, (215) 586-5640, (1).

Bruskewitz, John, RADON DETECTION SYSTEMS/MILWAUKEE BRANCH, 2630 North Newhall St., Milwaukee, WI, 53211, (303) 444-5253, (800) 627-2366, (1).

Brutsaert, Willem, UNIVERSITY OF MAINE-ORONO/DEPARTMENT OF CIVIL ENGINEERING, 103 Boardman, Orono, ME, 04469, (207) 581-2171, (5).

Bryant, Harold, BRYANT AND BRYANT SERVICE COMPANY, 992 1/2 Davidson Dr., Nashville, TN, 37205, (615) 352-2570, (1).

Bryant, Mary L., IOWA BUREAU OF LABOR/IOWA OCCUPATIONAL SAFETY AND HEALTH ADMINISTRATION, 307 East Seventh St., Des Moines, IA, 50319, (515) 281-5352, (10).

Bryant, Nancy, SOCIETY OF INDUSTRIAL AND OFFICE REALTORS (SIRO), 777 14th St., NW, Washington, DC, 20005, (202) 383-1150, (18).

Bryant, Robert, HOME INSPECTOR OF NORTH AMERICA, THE, 301 North Main St., Randolph, MA, 02368, (617) 963-0343, (1).

Bryson, Sr., Eugene A., ENVIRONMENTAL POLLUTION CONTROL, 195 Cliffwood Dr., Atlanta, GA, 30328, (400) 425-6454, (1)(2).

Bucciarelli, James, CERTIFIED TESTING SERVICES, INC., 474 Easton Rd., Horsham, PA, 19044, (215) 672-8100, (1)(2).

Buchman, William G., INTERNATIONAL RADON TECHNOLOGIES, 2417 Regency Rd., Lexington, KY, 40503, (606) 278-0104, (800) 537-5066, (1)(4).

Buchwald, C. Edward, GEOLOGICAL SOCIETY OF AMERICA (GSA), P.O. Box 9140 3300 Penrose Pl., Boulder, CO, 80301, (303) 447-2020, (14).

Buckles, Jerry, REAL ESTATE INSPECTION SERVICE, P.O. Box 3621 CRS, Johnson City, TN, 37602, (615) 282-4216, (1).

Buddenbaum, Jack, AMERICAN RADON CONSULTANTS, INC., 109 Ladbrook Rd., Knoxville, TN, 37922, (615) 891-9581, (1).

Budin, Fred P., RADON DETECTION SERVICES, LTD., 1500 Harbour Oaks Dr., Tucker, GA, 30084, (404)

621-9300, (1).

Bulava, Edward J., BULAVA ENVIRONMENTAL, INC., 13 Hunt Club Rd., Belle Mead, NJ, 08502, (201) 874-6207, (1).

Bull, David, ENVIRONMENT LIAISON CENTRE (ELC), P.O. Box 72461, Nairobi, Kenya, (011) 254-224-770, (12).

Bunton, Calvin G., RADON INTERVENTION, P.O. Box 827, Three Rivers, TX, 78071, (512) 786-4276, (1).

Burden, Jr., Nathaniel, HOUSE ANALYSIS AND ASSOCIATES, 747 Hallowell Dr., Huntingdon Valley, PA, 19006, (215) 947-1169, (215) 947-9209, (1).

Burdick, Bill, SAFETY AND HEALTH SERVICES, 3307 North Second St., Harrisburg, PA, 17110, (717) 238-2122, (3).

Burgess, E.I., FEMTO-TECH, INC., 150 C Industry Dr., Carlisle, OH, 45005, (513) 746-4427, (1)(4)(5).

Burke, Richard, HEALTH PHYSICS SOCIETY (HPS), 800 Westpark Dr. Suite 400, McLean, VA, 22102, (703) 790-1745, (15).

Burkert, Chris, CAPITAL HOME INSPECTION CORPORATION, P.O. Box 2541, Manassas, VA, 22110, (703) 369-4206, (1).

Burkhart, James F., UNIVERSITY OF COLORADO-COLORADO SPRINGS/PHYSICS DEPARTMENT, 1420 Austin Bluffs Pkwy., Colorado Springs, CO, 80933-7150, (719) 593-3214, (1)(5).

Burlbaw, H. Gary, DALLAS/HEALTH AND HUMAN SERVICES/ENVIRONMENTAL HEALTH DIVISION/AIR POLLUTION CONTROL SECTION, 400 South Zang Blvd., Dallas, TX, 75208, (214) 670-7520, (10).

Burner, Doris A., PRECISION RADON TESTING, 302 Leidy Rd., Souderton, PA, 18964, (215) 721-1958, (800) 545-8378, (1).

Burnette, William C., FLORIDA STATE UNIVERSITY/DEPARTMENT OF GEOLOGY, 305 Carroway Bldg., Tallahassee, FL, 32306, (904) 644-5784, (5).

Burns, Frederick, NEW YORK UNIVERSITY MEDICAL CENTER/INSTITUTE OF ENVIRONMENTAL MEDICINE, 550 1st Ave., New York, NY, 10016, (212) 340-5280, (5).

Burns, Robert J., BURNS AND ASSOCIATES, 265 Kearsarge Way, Portsmouth, NH, 03801, (603) 431-3400, (1)(2).

Buro, Fred A., GENUS CORPORATION, THE, 2400 Morris Ave., Union, NJ, 07083, (201) 687-5000, (1).

Burrows, Don, RADON DIAGNOSTIC LABORATORIES, 231 Front St. P.O. Box 2890, South Portland, ME, 04106, (207) 767-2818, (1)(2).

Burry, Craig, CONNECTICUT ENVIRONMENTAL TESTING, 6 East Main St., Westport, CT, 06880, (203) 454-1544, (1).

Busse, William F., ALA OF TENNESSEE, P.O. Box 399, Nashville, TN, 37202-0399, (615) 329-1151, (15).

Butler, Dan, U.S. DEPARTMENT OF ENERGY (DOE)/OFFICE OF NUCLEAR ENERGY(ONE), 1000 Independence Ave., SW Room SA-115, Washington, DC, 20585, (202) 586-8728, (8).

Butler, Steve, HOMEBUYERS INSPECTION SERVICE OF NORTH WEST FLORIDA, INC., P.O. Box 9433, Pensacola, FL, 32513-9433, (904) 479-3276, (1).

Button, Douglas J., AFFTREX, LTD., 3530 Washington Ave., Finleyville, PA, 15332, (412) 348-9211, (1)(2).

Button, William H., AFFTREX, LTD., 3530 Washington Ave., Finleyville, PA, 15332, (412) 348-9211, (1)(2).

Butts, David, HALL-KIMBRELL ENVIRONMENTAL SERVICES, 5319 Southwest Westgate, Portland, OR, 97221, (503) 292-9406, (1)(2)(3).

Buvinghausen, Michael L., ENVIRO SCIENCES, INC, 1548 Valwood Pkwy. Suite 118, Carrollton, TX, 75066, (214) 406-8100 (817) 792-3434, (1).

Buvinghausen, Michael L., TECHTRAC, P.O. Box 5050 Suite 303, Lewisville, TX, 75067, (214) 221-1047, (1).

Buvinghausen, Mike, ALPHA ENERGY LABORATORIES, INC., 2100 Road to Six Flags, Arlington, TX, 76011, (817) 792-3434, (1).

Byer, Steven, NEW HAMPSHIRE DEPARTMENT OF LABOR/ONSITE CONSULTATION PROGRAM, 19 Pillsbury St., Concord, NH, 03301, (603) 271-3170, (10).

Byrd, Rickie, MARTIN MARIETTA ENERGY SYSTEMS, P.O. Box 1410 Hobbs Rd., Paducah, KY, 42001, (502) 444-6311, (1).

Bytz, Myron M., HOME INSPECTION ONE, P.O. Box 1177 2 Valley Rd., Denville, NJ, 07834, (201) 586-2730, (1).

C

Cahill, Mary K., CONFERENCE OF RADIATION CONTROL PROGRAM DIRECTORS (CRCPD)/COMMITTEE ON RADON POLICY, 71 Fountain Pl., Frankfort, KY, 40601, (502) 227-4543, (14).

Cahill, Richard A., ILLINOIS STATE GEOLOGICAL SURVEY, Natural Resources Building 615 East Peabody Dr. Room 121, Champaign, IL, 61820, (217) 333-5111, (1)(10).

Cahoon, D. Dawn, GOVERNMENT OF GUAM/ENVIRONMENTAL PROTECTION AGENCY, P.O. Box 2999, Agana, GU, 96910, (671) 646-8863, (10).

Cain, Allan, RHODE ISLAND DEPARTMENT OF GEOLOGY, Green Hall The University of Rhode Island, Kingston, RI, 02881, (401) 792-2265, (5)(10).

Cain, Richard F., ALLSTATE RADON TESTING SERVICE, 2231 Pinecliff Dr., Atlanta, GA, 30345, (404) 938-2238, (1).

Cain, Venus, RADON TECHNOLOGIES, INC., 12200 Sunrise Valley Dr. Suite 320, Reston, VA, 22091, (703) 620-4004, (800) 237-5024, (1)(4).

Cairnes, William X., PRO-GUARD BUILDING INSPECTIONS, INC., RD 1, Box 193B, Auburn, PA, 17922, (717) 754-7673, (1)(2).

Calcaterra, Charlie, NEBRASKA DEPARTMENT OF LABOR, State Office Building 301 Centennial Mall, South, Lincoln, NE, 68509, (402) 471-2239, (10).

Callaway, Kevin, NATIONAL RADON TEST COMPANY, P.O. Box 50468, Austin, TX, 78763, (512) 495-9330, (1).

Callawy, E. Frank, SWAT TERMITE AND PEST CONTROL, 1797 Roswell Rd., Marietta, GA, 30062, (404) 565-7928, (1).

Caloiero, Joseph, CALOIERO RADON TECHNOLOGIES, RD5, Box 84 North Main St., Williamstown, NJ, 08094, (609) 728-9257, (1)(2).

Calvert, G. Wayne, CENTARUS CORPORATION, P.O. Box 513, Glasgow, KY, 42141, (502) 651-2396, (1).

Calvert, Gene, GENE CALVERT AND ASSOCIATES, P.O. Box 217, Waxahachie, TX, 75165, (214) 937-2223, (1).

Campagna, Anthony, CAMPAGNA BUILDERS, 398B Union Blvd., Totowa, NJ, 07512, (201) 942-9661, (2).

Campbell, Marie B., MISSOURI REAL ESTATE COMMISSION, P.O. Box 1339, Jefferson City, MO, 65102, (314) 751-2334, (10).

Campbell, R.A., COLORADO REAL ESTATE INSPECTION SERVICES, P.O. Box 24962, Denver, CO, 80224-0962, (303) 753-0543, (1).

Canady, John B., TEK ENGINEERING, 7150 East Hampden Ave. #101, Denver, CO, 80224, (303) 758-3173, (1).

Cannon, Edmund M., NATIONAL ENVIRONMENTAL ENGINEERING AND DIAGNOSTICS, Box 519, Rockland, MA, 02370, (617) 871-4426, (1).

Caponio, Joseph F., U.S. DEPARTMENT OF COMMERCE (DOC)/NATIONAL TECHNICAL INFORMATION SERVICE(NTIS), 5248 Port Royal Rd., Springfield, VA, 22161, (703) 487-4650, (8).

Capowski, Steve, DUTCHESS COUNTY HEALTH DEPARTMENT, County Office Building 22 Market St., Poughkeepsie, NY, 12601, (914) 294-7961, (10).

Caprio, Michael L., RADIATION PROTECTION SERVICES, INC., 646 Rte. 18 Suite 215, East Brunswick, NJ, 08816, (201) 390-7360, (1)(2)(6).

Caprio, Jr., Michael L., RADIATION PROTECTION SERVICES, INC., P.O. Box 2395, Darien, CT, 06820, (203) 324-7967, (1)(2).

Carapetian, Ara, HOMERAD, 908 Wild Forest Dr., Gaithersburg, MD, 20879, (301) 889-5769, (1).

Carita, Anthony J., RADON INFORMATION AND TESTING SERVICE, 29 Appleton St., Manchester, NH, 03104, (603) 627-7261, (1).

Carlisle, Donald, SAGE RESOURCES/INDOOR AIR QUALITY DIVISION, P.O. Box 480559, Los Angeles, CA, 90048, (213) 825-1934, (213) 651-0155, (1).

Carlisle, Donald, RADON SYSTEMATICS OF CALIFORNIA, 1500 San Remo Dr., Pacific Palisades, CA, 90272, (213) 459-9146, (213) 825-1934, (1).

Carlson, Gary, AIR POLLUTION CONTROL PROGRAM OF THE ENVIRONMENTAL PROTECTION COMMISSION OF HILLSBOROUGH COUNTY, 1410 North 21st St., Tampa, FL, 33605, (813) 272-5530, (10).

Carlson, Gary, BROWARD COUNTY ENVIRONMENTAL QUALITY CONTROL BOARD, 500 SW 14th Ct., Fort Lauderdale, FL, 33315, (305) 765-5881, (10).

Carlson, John R., ALA OF WEST VIRGINIA, P.O. Box 3980, Charleston, WV, 25339, (304) 342-6600, (15).

Carlson, Ken, CARLSON HEATING, COOLING, AND ELECTRIC COMPANY, 3340 West Lake Ave., Glenview, IL, 60025-1297, (312) 729-0123, (1).

Carlson, Reed, EVANSTON/DEPARTMENT OF BUILDING AND ZONING, 2100 Ridge Ave., Evanston, IL, 60204, (312) 328-2100, (10).

Carlson, Warner, CARLSON INSPECTION ASSOCIATES, INC., 51 R Dudley St., Arlington, MA, 02174, (617) 641-0600, (1).

Caron, Valerie A., ENVIRONMENTAL FIELD SERVICES, INC., 63 Elm St., Topsham, ME, 04086, (207) 725-4112, (1).

Carpenter, Charles D., CARPENTER, CHARLES D., 13462 West Florida Dr., Lakewood, CO, 80228, (303) 988-6818, (1).

Carr, Marjorie, FLORIDA DEFENDERS OF THE ENVIRONMENT (FDE), 1523 Northwest 4th St., Gainesville, FL, 32608, (904) 372-6965, (17).

Carr, Jr., H. Edward, ENVIRONMENTAL INDOOR AIR QUALITY CONSULTANTS, INC. (EIC), 58 Keller Ave., Lancaster, PA, 17601, (717) 393-5774, (800) 367-7223, (1)(2).

Carrow, Barbara, CONSTRUCTION SERVICES OF INDIANA, INC., 9341 Sargeant Rd., Indianapolis, IN, 46256, (317) 849-2686, (2).

Carter, Edward, ALA OF TEXAS, 3520 Executive Center Dr. Suite G-100, Austin, TX, 78731-1606, (512) 343-0502, (15).

Caruso, Emil, CONNECTICUT DEPARTMENT OF LABOR/DIVISION OF OCCUPATIONAL SAFETY AND HEALTH, 200 Folly Brook Rd., Wethersfield, CT, 06109, (203) 566-4550, (10).

Casparius, Robert E., RADIOLOGICAL CONTROL ASSOCIATES, 665 Robbins Ave. Unit 2, Dracut, MA, 01826, (617) 453-5268, (1)(2).

Castleman, A. Welford, PENNSYLVANIA STATE UNIVERSITY, THE/DEPARTMENT OF CHEMISTRY, 152 Davey Laboratory, University Park, PA, 16802, (814) 865-6553, (5).

Caswell, Janice, ENVIRONMENTAL DEFENSE FUND (EDF)/NATIONAL HEADQUARTERS, 257 Park Avenue South, New York, NY, 10010, (212) 505-2100, (17).

Caswell, Randall S., U.S. DEPARTMENT OF COMMERCE (DOC)/NATIONAL INSTITUTE OF STANDARDS AND TECHNOLOGY(NIST)/CENTER FOR RADIATION RESEARCH(CRR)/IONIZING RADIATION DIVISION, Building 245 Room C210, Gaithersburg, MD, 20899, (301) 975-5524, (8).

Catizone, Pietro A., ALPHA SERVICES/AIR AND WATER RADON TESTING, 296 Grindle Brook Rd., South Glastonberry, CT, 06073, (203) 659-8099, (1)(4).

Cavey, Kevin Sullivan, RADON SPECIALIST OF AMERICA, 12628 Gray Eagle Ct. Suite 12, Germantown, MD, 20874, (301) 231-4091, (1).

Cavin, Donald K., CAVIN ANALYTICAL CONSULTANTS, INC., 2165K West Park Ct., Stone Mountain, GA, 30087, (404) 498-8295, (1).

Cavote, Donald R., PORTSMOUTH LOCAL AIR AGENCY/AIR POLLUTION CONTROL DIVISION, 740 Second St., Portsmouth, OH, 45662, (614) 353-5156, (10).

Cazett, Jim, EER PRODUCTS, INC., 3536 East 28th St., Minneapolis, MN, 55406, (612) 721-4231, (4).

Ceglian, Jim, SOUTH DAKOTA STATE UNIVERSITY/S.T.A.T.E. ENGINEERING DIVISION, Box 2218, Brookings, SD, 57007, (605) 688-4101, (10).

Cervo, Daniel D., RADON TESTING LABORATORY OF NEW JERSEY, 1303 Birch Ave., Ocean Township, NJ, 07712-4514, (201) 502-0727, (1).

Chabot, Theresa E., RADON RESOURCES, INC., 55 Constitution Dr., Bedford, NH, 03102, (603) 472-5955, (1).

Chaiken, Ben, ARIZONA LUNG ASSOCIATION, 102 West McDowell Rd., Phoenix, AZ, 85003, (602) 258-7505, (15).

Chamberlain, C. Kent, C. KENT CHAMBERLAIN AND ASSOCIATES, 6579 West David Dr., Littleton, CO, 80123, (303) 979-3916, (1)(2).

Chamberlain, E. Carl, E.C. CHAMBERLAIN, RADIOLOGICAL CONSULTANT, 3304 Susileen Dr., Reno, NV, 89509, (702) 329-5638, (702) 323-8341, (1).

Chambers, Charles G., MARYLAND REAL ESTATE COMMISSION, 501 St. Paul Pl. Suite 804, Baltimore, MD, 21202-2272, (301) 333-6230, (10).

Chambers, Douglas, SENES CONSULTANTS, 499 McNicholl Ave., Willowdale, ON, M2H 2C9, (416) 499-5030, (1)(3).

Champlin, Paul B., APPLIED ANALYTICAL TECHNOLOGIES, INC., P.O. Box 668, Narragansett, RI, 02882, (401) 782-8188, (1).

Chan, Desmond, GENERAL PHYSICS CORPORATION, 10650 Hickory Ridge Rd., Columbia, MD, 21044, (301) 290-2300, (301) 997-6241, (1).

Chan, Mike, AEROVIRONMENT, INC., 825 Myrtle Ave., Monrovia, CA, 91016-3424, (813) 357-9983, (4).

Chan, Wing T., U.S. GENERAL SERVICES ADMINSTRATION (GSA)/SAFETY AND ENVIRONMENTAL MANAGEMENT DIVISION(SEMD), 18th and F St., NW Room 4320, Washington, DC, 20405, (202) 472-1082, (8).

Chandler, Alan, GMAC HOME IMPROVEMENT, 111 East Broadway St., Campbellsville, KY, 42718, (800) 222-5008, (2).

Chando, John, TFC NUCLEAR ASSOCIATES, INC., 425 Bridgeboro Rd., Moorestown, NJ, 08057, (609) 778-4529, (1)(2).

Chaney, William G., WILLIAM G. CHANEY, PE, 1708 Park Ave., SW, Albuquerque, NM, 87104, (505) 247-2087, (1)(3).

Chang, Weilin P., UNIVERSITY OF FLORIDA/COLLEGE OF ARCHITECTURE/SCHOOL OF BUILDING CONSTRUCTION, AFC-101, Gainesville, FL, 32611, (904) 392-5965, (5).

Chapin, John A., RADON REDUCTION, INC., 363 Massachusetts Ave., Lexington, MA, 02173, (617) 862-8462, (2).

Chapin, Scott, RADON REDUCTION, INC., 363 Massachusetts Ave., Lexington, MA, 02173, (617) 862-8462, (2).

Chapman, Lyle, MONTANA DEPARTMENT OF LABOR AND INDUSTRY/BUREAU OF SAFETY AND HEALTH/DIVISION OF WORKERS' COMPENSATION, 5 South Last Chance Gulch, Helena, MT, 59601, (406) 444-6401, (10).

Chapman, Jr., Robert D., RADON SCREENING, INC., 7 Brookcrest Ct., Fredericksburg, VA, 22401, (703) 786-4320, (1).

Chaput, Pierre, DUOVENT INDUSTRIES, LTD., 1375 Bl. Charest Ouest Suite 6, Quebec, PQ, G1N 2E7, (418) 688-8805, (4).

Chart, E., MACLAREN ENGINEERS, INC., 33 Yonge St., Toronto, ON, M5E 1E7, (416) 756-4919, (1)(2).

Chase, Edgar, FAIRFAX CITY DEPARTMENT.OF HEALTH, 10777 Main St. Suite 100, Fairfax, VA, 22030, (703) 246-2411, (10).

Chase, Edgar M., FAIRFAX COUNTY AIR POLLUTION CONROL, 10777 Main St. Suite 100A, Fairfax, VA, 22030, (703) 246-2541, (10).

Chase, Kenneth, WASHINGTON OCCUPATIONAL HEALTH ASSOCIATES, INC., 1120 19th St., NW Suite 410, Washington, DC, 20036, (202) 463-6698, (16).

Checkes, Frank, ENVIRONMENTAL HOME SERVICES, INC., 12354 Howard Lodge Dr., Sykesville, MD, 21784, (301) 442-5861, (1)(2).

Chen, Chia S., NEW MEXICO INSTITUTE OF MINING AND TECHNOLOGY/GEOSCIENCE DEPARTMENT, Socorro, NM, 87801, (505) 835-5308, (5).

Chen, David, U.S. DEPARTMENT OF ENERGY (DOE)/ALBUQUERQUE OPERATIONS OFFICE(AOO)/LOS ALAMOS NATIONAL LABORATORY(LANL), P.O. Box 1663 M 888, Los Alamos, NM, 87545, (505) 667-5061, (8).

Chiles, Bonnie, U.S. ENVIRONMENTAL PROTECTION AGENCY (EPA)/OFFICE OF AIR AND RADIATION(OAR)/OFFICE OF RADIATION PROGRAMS(ORP)/RADON DIVISION(RD), Policy And Public Information Branch 401 M St., SW ANR-464, Washington, DC, 20460, (202) 475-9605, (8).

Chilton, George E., CHILTON ASSOCIATES (RADON GROUP), 144 Valley Rd., Haworth, NJ, 07641, (201) 385-1660, (1).

Chippendale, David, RADON ANALYSIS, INC., 7 Chamberlain Rd., Flemington, NJ, 08822, (201) 788-0661, (1)(4).

Chirby, A.E., HUTCHISON PEST CONTROL, 2662 North Freeway, Pueblo, CO, 81003, (719) 544-7378, (719) 633-7378, (1).

Chittum, Kurt, P.C. ENGINEERING, Rte. 1, Box 347-B, Lexington, VA, 24450, (703) 463-9116, (2).

Chleboski, Ronald J., ALLEGHENY COUNTY HEALTH DEPARTMENT/BUREAU OF AIR POLLUTION CONTROL, 301 39th St., Pittsburgh, PA, 15201, (412) 578-8111, (10).

Choi, Y. Jean, TERRADYNAMICS CORPORATION, Rte. 1, Aldie, VA, 22001, (703) 327-6500, (703) 435-3033, (1).

Cholost, Wayne, R.T.C., P.O. Box 832, Linbrook, NY, 11563, (516) 887-1371, (1).

Choruza, Bogdan, NORTHEASTERN RADON SERVICE, 21 Rte. 202, Montville, NJ, 07045, (201) 402-1030, (1)(2).

Chow, William K.H., NEW JERSEY HOME SERVICE, INC., 16 Westwood Dr., N., West Orange, NJ, 07052, (201) 669-8166, (1).

Christensen, Betsy, RICHFIELD/SUPPORT SERVICES DIVISION, 6700 Portland Ave., Richfield, MN, 55423, (612) 869-7521, (10).

Christensen, Michael H., INSPECT-A-HOME OF UTAH, 455 East 400 S., Salt Lake City, UT, 84111, (801) 532-4663, (1).

Christensen, Richard N., AMERICAN NUCLEAR SOCIETY/OHIO-SOUTHWESTERN OHIO SECTION, 2520 Westmont Blvd., Columbus, OH, 43221, (614) 422-0445, (14).

Christensen-Spore, Carolyn, ALA OF FLORIDA GOVERNMENTAL AFFAIRS OFFICE, 345 West Madison St. Suite 2, Tallahassee, FL, 32301-1625, (904) 224-7222, (15).

Christenson, Dick, RHH FOAM SYSTEMS, INC., P.O. Box 752 6001 S. Pennsylvania Ave., Cudahy, WI, 53110-0752, (414) 744-6066, (4).

Christianson, Kevin, VIKING HOME IMPROVEMENT, INC., 15 South 15th St., Fargo, ND, 58103, (701) 237-6300, (1)(2)(4).

Chu, Kai-Dee, RADIATION SERVICE ORGANIZATION, INC., P.O. Box 1526, Laurel, MD, 20707-7444, (301) 953-2482, (1).

Church, E.O., E.O. CHURCH, INC., 925 East 17th Ave., Denver, CO, 80218, (303) 832-9692, (1).

Ciampa, Greg, W.R. GRACE/CONSTRUCTION PRODUCTS DIVISION, 62 Whittemore Ave., Cambridge, MA, 02140, (617) 876-1400, (4).

Ciosalo, F.R., CALIFORNIA DEPARTMENT OF INDUSTRIAL RELATIONS/DIVISION OF OCCUPATIONAL SAFETY AND HEALTH, P.O. Box 603, San Francisco, CA, 94101-0603, (415) 557-2870, (10).

Clancy, Al, CLANCY & COMPANY, 405 West Rockrimmon, Colorado Springs, CO, 80919, (303) 599-4660, (13).

Clark, Buzz, AIR QUALITY CONTROL, INC., 585 Phoenixville Pike, Malvern, PA, 19335, (215) 889-3100, (800) 654-2144, (1)(2).

Clark, Charles, GEORGIA REAL ESTATE COMMISSION, 40 Pryor St., SW, Atlanta, GA, 30303-3184, (404) 656-3916, (10).

Clark, John, RMG/PC ENGINEERING, INC., 617 North

17th St. Suite 201, Colorado Springs, CO, 80904, (719) 475-9600, (1).

Clark, Judy, RADON READINGS, 9020 Compton Blvd., Bellflower, CA, 90706, (213) 920-2582, (213) 421-8674, (1).

Clark, LeRoy E., L. CLARK AND SONS, INC., 986 Fisher Rd., Fitchburg, MA, 01420, (508) 343-7849, (1)(2).

Clark, Mary, FLORIDA DEPARTMENT OF HEALTH AND REHABILITATIVE SERVICES/DEPUTY SECRETARY FOR HEALTH/OFFICE OF RADIATION CONTROL, 1317 Winewood Blvd., Tallahassee, FL, 32399-0700, (904) 487-1004, (10).

Clark, Odum, RADON DETECTION SYSTEMS/NASHVILLE, TENNESSEE BRANCH, 3102 Bellwood Ave., Nashville, TN, 37203, (303) 444-5253, (800) 627-2366, (1).

Clark, Robert L., RAMSES II, LTD., 2355 East Dartmouth Ave., Denver, CO, 80210, (303) 758-5209, (1)(2)(4).

Clarke, Wayne, W.S. FLEMING AND ASSOCIATES, INC., 6310 Fly Rd., East Syracuse, NY, 13057, (315) 437-1869, (1)(3).

Claussen, Eileen, U.S. ENVIRONMENTAL PROTECTION AGENCY (EPA), 401 M St., SW, Washington, DC, 20460, (202) 382-4700, (8).

Clay, Donald, U.S. ENVIRONMENTAL PROTECTION AGENCY (EPA)/OFFICE OF AIR AND RADIATION(OAR), 401 M St., SW ANR-443, Washington, DC, 20460, (202) 382-7400, (8).

Clayton, Gary, ILLINOIS ASSOCIATION OF REALTORS, Illinois Realtors Pl. Suite 400, Springfield, IL, 62703, (217) 529-2600, (18).

Cleary, Sharon, MONTANA ASSOCIATION OF REALTORS, 910 Helena Pk., Helena, MT, 59601, (406) 443-4032, (18).

Cleaver, James E., UNIVERSITY OF CALIFORNIA-SAN FRANCISCO/LABORATORY OF RADIOBIOLOGY AND ENVIRONMENTAL HEALTH, San Francisco, CA, 94143-0750, (415) 476-1636, (5).

Clement, M., RADON DETECTION SYSTEMS/MINNEAPOLIS-ST. PAUL BRANCH, 548 Frontenae, St. Paul, MN, 55104, (303) 444-5253, (800) 627-2366, (1).

Clemons, Paul E., VERMONT DEPARTMENT OF HEALTH/DIVISION OF OCCUPATIONAL AND RADIOLOGICAL HEALTH, Administration Building 10 Baldwin St., Montpelier, VT, 05602, (802) 828-2886, (10).

Clifford, Dennis, UNIVERSITY OF HOUSTON/DEPARTMENT OF CIVIL AND ENVIRONMENTAL ENGINEERING, 4800 Calhoun, Houston, TX, 77204-4791, (713) 749-4481, (5).

Cline, Joseph H., ORKIN ENVIRONMENTAL SERVICES, P.O. Box 647, Atlanta, GA, 30301, (404) 888-2801, (1).

Clopein, Charles K., ENVIRONMENTAL HOME SERVICES, INC., 12354 Howard Lodge Dr., Sykesville, MD, 21784, (301) 442-5861, (1)(2).

Close, Robert V., NASSAU COUNTY DEPARTMENT OF HEALTH/DIVISION OF ENVIRONMENTAL HEALTH SERVICES/BUREAU OF AIR QUALITY MANAGEMENT, 240 Old County Rd., Mineola, NY, 11501, (516) 535-3232, (10).

Clouser, Jr., Pierce L., PENNSYLVANIA DEPARTMENT OF STATE/BUREAU OF PROFESSIONAL AND OCCUPATIONAL AFFAIRS/REAL ESTATE AND VEHICLE DIVISION, P.O. Box 2649, Harrisburg, PA, 17105-3658, (717) 783-3658, (10).

Clower, Frank M., RCI SERVICES, 11412 Hounds Way, Rockville, MD, 20852, (301) 231-5134, (1)(2).

Clute, Charles G., RADON MONITORING SERVICES, P.O. Box 9083, Colorado Springs, CO, 80932, (719) 577-9662, (1).

Cobb, Emsley F., AMERICAN NUCLEAR SOCIETY/GEORGIA-ATLANTA SECTION, C/O Georgia Power Company P.O. Box 4545, Atlanta, GA, 30302, (404) 526-7952, (14).

Coburn, Ed, CUTTER INFORMATION CORPORATION, 1100 Massachusetts Ave., Arlington, MA, 02174, (617) 648-8700, (3).

Coburn, Karen, CUTTER INFORMATION CORPORATION, 1100 Massachusetts Ave., Arlington, MA, 02174, (617) 648-8700, (3).

Coffey, Chris, RADON TESTING AND REDUCTION, P.O. Box 437, Howell, MI, 48844, (517) 548-2497, (1)(2).

Coffey, Craig, RADON TESTING AND REDUCTION, P.O. Box 437, Howell, MI, 48844, (517) 548-2497, (1)(2).

Coffey, M.L., TRI-COURT SAFETY/SECURITY, P.O. Box 1257, Sunset Beach, CA, 90742, (213) 335-3290, (714) 846-4346, (1).

Cohen, Anna F., RADON PROJECT, THE, P.O. Box 90069, Pittsburgh, PA, 15224, (412) 687-2886, (1)(4).

Cohen, Bernard, UNIVERSITY OF PITTSBURGH/PHYSICS DEPARTMENT, 100 Allen Hall, Pittsburgh, PA, 15260, (412) 624-9245, (1)(5).

Cohen, Bernard L., RADON PROJECT, THE, P.O. Box 90069, Pittsburgh, PA, 15224, (412) 687-2886, (1)(4).

Cohen, Beverly S., NEW YORK UNIVERSITY MEDICAL CENTER/INSTITUTE OF ENVIRONMENTAL MEDICINE, 550 1st Ave., New York, NY, 10016, (212) 340-5280, (5).

Cohen, Frederick B., RADON PROJECT, THE, P.O. Box 90069, Pittsburgh, PA, 15224, (412) 687-2886, (1)(4).

Cohen, John H., RADON MEASUREMENTS, INC., 2916 West T.C. Jester, Houston, TX, 77018, (713) 683-8227, (1).

Cohen, Lloyd, RELIABLE BASEMENT WATERPROOF COMPANY, 2811 Westerville Rd., Columbus, OH, 43224, (614) 267-4477, (1)(2).

Cohen, Lloyd D., A-1 RELIABLE BASEMENT WATERPROOFING, 3008 Gracemore St., Dayton, OH, 45420, (513) 293-6699, (1).

Cohen, Matthew, RELIABLE BASEMENT WATERPROOF COMPANY, 2811 Westerville Rd., Columbus, OH, 43224, (614) 267-4477, (1)(2).

Cohen, Sanford, S. COHEN AND ASSOCIATES, INC., 8200 Riding Ridge Pl., McLean, VA, 22102, (703) 893-6592, (3)(6).

Cokinos, Dimitrios, AMERICAN NUCLEAR SOCIETY/NEW YORK-LONG ISLAND SECTION, C/O Brookhaven National Laboratory Building 475 B, Upton, NY, 11973, (516) 282-2146, (14).

Colasanti, Bob, SOUTH PARK URBAN RENEWAL CORPORATION, 1938 South Park Ave., Buffalo, NY, 14220, (716) 873-4763, (1).

Cole, A.T., INK FILTERING SYSTEMS, 3402 Habersham Club Dr., Cumming, GA, 30130, (404) 889-1061, (4).

Cole, Michael, SCHOHARIE COUNTY HEALTH DEPARTMENT, 342 Main St., Schoharie, NY, 12157, (518) 295-8182, (10).

Coleman, Jesse, TENNESSEE VALLEY AUTHORITY (TVA)/DIVISION OF OCCUPATIONAL HEALTH AND SAFETY(DOHS), Radiation Safety 324 MPB, Mussel Shoals, AL, 35660, (205) 386-2767, (8).

Colle, Ronald, U.S. DEPARTMENT OF COMMERCE (DOC)/NATIONAL INSTITUTE OF STANDARDS AND TECHNOLOGY(NIST)/CENTER FOR RADIATION RESEARCH(CRR)/OFFICE OF RADIATION MEASUREMENTS, Building 245 Room C-225, Gaithersburg, MD, 20899, (301) 975-6042, (8).

Collier, Joseph, U.S. DEPARTMENT OF LABOR (DOL)/OCCUPATIONAL SAFETY AND HEALTH ADMINISTRATION(OSHA)/OFFICE OF CONSULTATION PROGRAMS, 200 Constitution Ave., NW Room N-3700, Washington, DC, 20210, (202) 523-7102, (8)

Collins, Horace, OHIO DEPARTMENT OF NATURAL RESOURCES/DIVISION OF GEOLOGICAL SURVEY, Fountain Square Building B, Columbus, OH, 43224, (614) 265-6605, (10).

Collins, Mary Ellen, DMC ASSOCIATES, P.O. Box 11, Bergenfield, NJ, 07621, (201) 384-6060, (1).

Collins, Richard J., MASSACHUSETTS DEPARTMENT OF LABOR AND INDUSTRIES/DIVISION OF INDUSTRIAL SAFETY, 100 Cambridge St., Boston, MA, 02202, (617) 727-3463, (10).

Collins, Wallace, TEXAS REAL ESTATE COMMISSION, P.O. Box 12188 Capitol Station, Austin, TX, 78711, (512) 465-3980, (10).

Colony, Steve, UNC GEOTECH, 2594 B 314 Road P.O. Box 14000, Grand Junction, CO, 81502-5504, (303) 248-6702, (5).

Comeau, Gord, G.K. YUILL AND ASSOCIATES, LTD., 1441 Pembina Hwy., Winnipeg, MB, R3T 2C4, (204) 474-2461, (5).

Comeau, Peggy, HAWAII ASSOCIATION OF REALTORS, 505 Ward Ave., Honolulu, HI, 96814, (808) 538-3641, (18).

Comer, Jerry T., HOME INSPECTION SERVICE, INC., 1916 Southland Ave., Highland Park, IL, 60035, (312) 831-4545, (1)(2).

Compas, Dave, U.S. DEPARTMENT OF INTERIOR (DOI)/U.S. GEOLOGICAL SURVEY(USGS), 7638 Federal Blvd. 300 North Los Angeles St., Los Angeles, CA, 80012, (213) 894-2850, (8).

Compton, Lee, ARIX CORPORATION, 325 East Southern Ave. Suite 18, Tempe, AZ, 85282, (602) 968-0052, (1)(2).

Comtois, R.R., NORTHEAST RADON CONSULTANTS, 180 Zachary Rd., Manchester, NH, 03103, (603) 622-9391, (1).

Conaty, Jim, H.B. FULLER, 315 South Hicks, Palatine, IL, 60067, (800) 323-7407, (4).

Conklin, Richard D., CHIEF HOME AND BUILDING INSPECTION SERVICES, Box 284, Towaco, NJ, 07082, (201) 299-9087, (1).

Conley, John M., J.M.C. ENTERPRISES, 700 Ferndale Rd., Johnson City, TN, 37604, (615) 282-2136, (1).

Conlon, Pete, TODOL PRODUCTS, INC., P.O. Box 398, Natick, MA, 01760, (508) 879-7741, (4).

Conlon, Tom, TODOL PRODUCTS, INC., P.O. Box 398, Natick, MA, 01760, (508) 879-7741, (4).

Connell, Peter, A & A CONSULTING AND INSPECTION SERVICES, INC., 349 Chenango St., Binghamton, NY, 13901, (607) 723-7968, (1)(3).

Conrad, Stephen, NORTH CAROLINA DEPARTMENT OF NATURAL RESOURCES AND COMMUNITY DEVELOPMENT/DIVISION OF LAND RESOURCES, P.O. Box 27687, Raleigh, NC, 27611, (919) 733-3833, (10).

Conway, Kathleen, U.S. ENVIRONMENTAL PROTECTION AGENCY (EPA)/SCIENCE ADVISORY BOARD(SAB)/RADIATION ADVISORY COMMITTEE, 401 M St. SW A-101F, Washington, DC, 20460, (202) 382-2552, (8).

Cook, David P., ALA OF ARKANSAS, P.O. Box 3857, Little Rock, AR, 72203, (501) 374-3726, (15).

Cook, Gary D., UNIVERSITY OF FLORIDA/COLLEGE OF ARCHITECTURE/SCHOOL OF BUILDING CONSTRUCTION, AFC-101, Gainesville, FL, 32611, (904) 392-5965, (5).

Cook, Jim, RADON-ONE LABORATORY SERVICES, 4575 Chandlers Forde, Sarasota, FL, 34235, (813) 377-3740, (1).

Cook, Michael B., U.S. ENVIRONMENTAL PROTECTION AGENCY (EPA)/OFFICE OF WATER(OW)/OFFICE OF DRINKING WATER(ODW), 401 M St., SW WH-550, Washington, D.C., 20460, (202) 475-9605, (8).

Cook, Jr., H.M., H.M. COOK, JR., PROFESSIONAL ENGINEER, 14632 Lufthansa Cr., Chantilly, VA, 22021, (703) 378-0114, (703) 248-3809, (1).

Cooke, J. Philip, BENTON-FRANKLIN-WALLA WALLA COUNTIES AIR POLLUTION AUTHORITY/AIR POLLUTION CONTROL DIVISION, 650 George Washington Way, Richland, WA, 99352, (509) 545-2354, (10).

Cooper, Ben, B & I SERVICES, P.O. Box 291, Edmond, KS, 67636, (913) 622-4341, (1).

Cooper, Bernard Wilson, B & W SALES AND MARKETING, P.O. Box 130, Chiefland, FL, 32626, (904) 528-3600, (1).

Cooper, Cliff, CHEM-SAFE LABORATORIES, North 131 Grand Ave. P.O. Box 546, Pullman, WA, 99163-0546, (509) 334-0922, (1)(4).

Cooper, John, ILLINOIS DEPARTMENT OF NUCLEAR SAFETY/OFFICE OF ENVIRONMENTAL SAFETY, 1035 Outer Park Dr., Springfield, IL, 62704, (217) 786-6399, (1)(10).

Cooper, John A., NEA, INC., 10950 Southwest 5th St. Suite 380, Beaverton, OR, 97005, (503) 643-4661, (1).

Cooper, Raymond, NATIONAL ACADEMY OF SCIENCES (NAS)/NATIONAL RESEARCH COUNCIL (NRC)/MEDICAL FOLLOW-UP AGENCY, 2101 Constitution Ave., NW, Washington, DC, 20418, (202) 334-2000, (15).

Cope, Ed, SETRA SYSTEMS, INC., 45 Nagog Pk., Acton, MA, 01720, (617) 263-1400, (4).

Cope, James L., RADON SURVEYS/DIVISION OF COPE TECHNICAL SERVICES, 3506 87th Pl., Kenosha, WI, 53142, (414) 694-3730, (1).

Copley, Mike, RELIABLE BASEMENT WATERPROOF COMPANY, 2811 Westerville Rd., Columbus, OH, 43224, (614) 267-4477, (1)(2).

Copp, James M., NUCLEAR TECHNOLOGY, INC./RADON CHEK, 20 H. South State St., Westerville, OH, 43081, (614) 899-2435, (1)(4).

Coquyt, Joan, RA-TEK/DIVISION OF SHAMROCK, INC., P.O. Box 496 800 East Main St., Marshall, MN, 56258, (507) 532-7313, (1)(4).

Cordes, Gail, AMERICAN NUCLEAR SOCIETY/IDAHO-IDAHO SECTION, C/O Intermountain Technology, Inc. Science Engineering and Analysis Division P.O. Box 1604, Idaho Falls, ID, 83403-1604, (208) 523-7255, (14).

Corland, Rick, AMERICAN WATER WORKS ASSOCIATION RESEARCH FOUNDATION (AWWARF), 6666 West Quincy Ave., Denver, CO, 80235, (303) 794-7711, (5)(14).

Cornforth, Michael, U.S. DEPARTMENT OF ENERGY (DOE)/ALBUQUERQUE OPERATIONS OFFICE(AOO)/LOS ALAMOS NATIONAL LABORATORY(LANL), P.O. Box 1663 M 888, Los Alamos, NM, 87545, (505) 667-5061, (8).

Corsetto, Joseph F., SHELTERWORKS, INC., 53 New St., Dover, NJ, 07801, (201) 328-2912, (1).

Corso, Robert F., ALA OF NEW JERSEY, 1600 Rte. 22 East, Union, NJ, 07083, (201) 687-9340, (15).

Coscarelli, Mike, BERNER INTERNATIONAL CORPORATION, P.O. Box 5205, New Castle, PA, 16105, (412) 658-1530, (800) 852-5015, (4).

Costello, J.O., ATLANTA TESTING AND ENGINEERING, INC., 11420 Johns Creek Pkwy., Duluth, GA, 30136, (404) 476-3555, (1).

Costello, Paul, HUDSON VALLEY RADON TESTING CORPORATION, P.O. Box 790, Poughkeepsie, NY, 12601, (914) 431-6424, (914) 485-5561, (1).

Cothern, C. Richard, COTHERN, C. RICHARD, 4732 Merivale Rd., Chevy Chase, MD, 20815, (301) 652-2863, (3).

Cothern, C. Richard, U.S. ENVIRONMENTAL PROTECTION AGENCY (EPA)/SCIENCE ADVISORY BOARD(SAB)/ENVIRONMENTAL HEALTH COMMITTEE/DRINKING WATER SUBCOMMITTEE, 401 M St., SW A-101F, Washington, DC, 20460, (202) 382-2552, (8).

Cotruro, Joseph, U.S. ENVIRONMENTAL PROTECTION AGENCY (EPA)/OFFICE OF WATER(OW)/OFFICE OF DRINKING WATER(ODW)/CRITERIA AND STANDARDS DIVISION(CSD), 401 M St., SW WH-550D, Washington, DC, 20460, (202) 382-7575, (8).

Cotten, Jim, BEAVER INDUSTRIES, 890 Hersey St., St. Paul, MN, 55114, (612) 644-9933, (4).

Cotton, Don, ALABAMA TESTING LABORATORY, P.O. Box 2337, Gasden, AL, 35903, (205) 547-5311, (1).

Cotton, James, HOME EXTERMINATING COMPANY, 4608 Bucks Schoolhouse Rd., Baltimore, MD, 21237, (301) 661-7911, (800) 492-5544, (1).

Courtney, F.E., COURTNEY CONSULTANTS, INC., 520 Carriage Dr., Atlanta, GA, 30328, (404) 256-2487, (1).

Courtney, Peter, COURTNEY CONSULTANTS, INC., 520 Carriage Dr., Atlanta, GA, 30328, (404) 256-2487, (1).

Covell, Norman, SACRAMENTO COUNTY AIR POLLUTION CONTROL DISTRICT, 9323 Tech Center Dr. Suite 800, Sacramento, CA, 95826, (916) 366-2107, (10).

Cowart, James B., FLORIDA STATE UNIVERSITY/DEPARTMENT OF GEOLOGY, 305 Carroway Bldg., Tallahassee, FL, 32306, (904) 644-5784, (5).

Cox, Anna Maria, UNIVERSITY OF TEXAS/SCHOOL OF PUBLIC HEALTH, P.O. Box 20186, Houston, TX, 77225, (713) 792-4431, (1)(5).

Cox, John L., SEPTECH, 7625 North Howard, Spokane, WA, 99208, (509) 467-6274, (1).

Cox, John W., GEORGIA ASSOCIATION OF REALTORS, 3200 Presidential Dr., Atlanta, GA, 30340, (404) 451-1831, (18).

Coy, John, SMALLEY AND COMPANY, 861 South Jason, Denver, CO, 80223, (303) 777-3435, (4).

Coyne, J. Joseph, U.S. DEPARTMENT OF COMMERCE (DOC)/NATIONAL INSTITUTE OF STANDARDS AND TECHNOLOGY(NIST)/CENTER FOR RADIATION RESEARCH(CRR)/NEUTRON INTERACTIONS AND DOSIMETRY GROUP, Building 245 Room C308, Gaithersburg, MD, 20899, (301) 975-5525, (8).

Cozart, Bill, OREGON ASSOCIATION OF REALTORS, P.O. Box 351, Salem, OR, 97308, (503) 362-3645, (18).

Craffey, Jim, AMERICAN ASSOCIATION OF RADON SCIENTISTS AND TECHNOLOGISTS, INC., THE (AARST), P.O. Box 70, Park Ridge, NJ, 07656, (201) 391-6445, (7).

Craft, Steve, RADON INSTRUMENTAL DETECTION, 8000 West 110th Suite 115, Overland Park, KS, 66210, (913) 451-9270, (1).

Craig, Patricia, TESTWELL CRAIG LABORATORIES, 11553 U.S. Hwy. 41 South, Gibsonton, FL, 33534, (813) 677-0242, (1).

Crawford, Nick, WESTERN KENTUCKY UNIVERSITY/DEPARTMENT OF GEOGRAPHY AND GEOLOGY, WKU EST-304, Bowling Green, KY, 42101, (502) 745-5989, (5).

Crawford-Brown, Douglas, UNIVERSITY OF NORTH CAROLINA/SCHOOL OF PUBLIC HEALTH/DEPARTMENT OF ENVIRONMENTAL SCIENCES AND ENGINEERING, Room 201H, Chapel Hill, NC, 27599-7400, (919) 966-3840, (1)(5).

Crawshaw, Don, PELICAN ENVIRONMENTAL CORPORATION, 42 Lehigh Rd., Wellesley, MA, 02181, (617) 235-5997, (1)(2)(3).

Crawshaw, Jeff, PELICAN ENVIRONMENTAL CORPORATION, 42 Lehigh Rd., Wellesley, MA, 02181, (617) 235-5997, (1)(2)(3).

Creech, Dennis, SOUTHFACE ENERGY INSTITUTE, P.O. Box 5506, Atlanta, GA, 30307, (404) 525-7657, (17).

Crigler, William F., PLUMAS COUNTY DEPARTMENT OF ENVIRONMENTAL HEALTH/AIR POLLUTION CONTROL DIVISION, P.O. Box 480, Quincy, CA, 95971, (916) 283-1255, (10).

Cripe, Jerry, OREGON REAL ESTATE AGENCY, 158 12th St, NE, Salem, OR, 97310-0240, (503) 378-8414, (10).

Criss, Joyce, RADON ANALYSIS, INC., 7 Chamberlain Rd., Flemington, NJ, 08822, (201) 788-0661, (1)(4).

Crockett, John, R.B. KANALFLAKT, INC., 1121 Lewis Ave., Sarasota, FL, 34237, (813) 366-7505, (813) 365-4118, (4).

Croley, John D., RADON RANGERS, 7979 Old Georgetown Rd. Suite 1100, Bethesda, MD, 20814, (301) 543-4300, (1).

Crombie, Richard, REMEDIAL TECHNOLOGIES, INC., 10110 Kile Rd., Chardon, OH, 44024, (216) 765-2442, (2)(4).

Crone, J.R., AMERICAN NUCLEAR SOCIETY/ALABAMA-BIRMINGHAM SECTION, C/O Southern Company Services P.O. Box 2625 Inverness 40 Room 330, Birmingham, AL, 35202, (205) 868-5832, (14).

Cronin, Kevin, AFI ENVIRONMENTAL/DIVISION OF BUFFALO CONSULTANTS, INC., 6103 Robinson Rd., Lockport, NY, 14094, (716) 625-8434, (1).

Cropp, J. Wayne, CHATTANOOGA-HAMILTON COUNTY/AIR POLLUTION CONTROL BUREAU, 3511 Rossville Blvd., Chattanooga, TN, 37407, (615) 867-4321, (10).

Cross, F. T., U.S. DEPARTMENT OF ENERGY (DOE)/RICHLANDS OPERATIONS OFFICE(ROO)/PACIFIC NORTHWEST LABORATORY(PNL)/CELL BIOLOGY DIVISION, M/S K416, Richland, WA, 99352, (509) 375-2121, (8).

Crossland, Robert L., YAKIMA COUNTY CLEAN AIR AUTHORITY, County Courthouse, Yakima, WA, 98901, (509) 575-4116, (10).

Crouse, Thomas C., AMERICAN NUCLEAR SOCIETY/LOUISIANA-LOUISIANA SECTION, C/O GSU-River Bend P.O. Box 220, St. Francisville, LA, 70775, (504) 381-4622, (14).

Crowley, Art, GARON PRODUCTS, INC., 1974 Hwy. 35 CN20, Wall, NJ, 07719, (201) 499-1776, (800) 631-

5380, (4).

Crum, John, AMERICAN CHEMICAL SOCIETY (ACS), 1155 16th St., NW, Washington, DC, 20036, (202) 872-6062, (14).

Crust, Jay, ENERGY PROFESSIONALS, INC., 387 Irwin St., Lock Haven, PA, 17745, (717) 748-5910, (717) 726-3825, (1)(2).

Cruz, Filiberto, PUERTO RICO DEPARTMENT OF LABOR AND HUMAN RESOURCES/OCCUPATIONAL SAFETY AND HEALTH, 505 Munoz Rivera Ave. 21st Fl., Hato Rey, PR, 00918, (809) 754-2134, (809) 754-2171, (10).

Cuddihy, Richard, U.S. DEPARTMENT OF ENERGY (DOE)/ALBUQUERQUE OPERATIONS OFFICE(AOO)LOVELACE INHALATION TOXICOLOGY INSTITUTE(LITI), P. O. Box 5890, Albuquerque, NM, 87185, (505) 844-2203, (8).

Culbertson, Mark, NATIONAL PROPERTY INSPECTIONS, 11212 Davenport St. Suite 15, Omaha, NE, 68154, (402) 330-1074, (1).

Culver, David, RADALYTIC LABS, INC., P.O. Box 202, McConnells, SC, 29726, (803) 684-9859, (1)(2).

Culwell, Bill, ALL SERVICE RADON REDUCERS, INC., 166 Southeast Crain Hwy., Upper Marlboro, MD, 20772, (301) 390-7440, (1).

Cummings, Steve, SAFE SHELTER ENVIRONMENTAL, 527 West Lancaster Ave., Frazer, PA, 19355, (215) 640-4839, (1)(2).

Cunning, Ray, INDUSTRIAL OPERATIONS, INC./RADON EVALUATION, 3736 South 7th Ave., Phoenix, AZ, 85041, (602) 268-8772, (1).

Cunningham, Lee, RADON REDUCTION SERVICES, 5601 East 109th Ter., Kansas City, MO, 64137, (816) 942-7083, (1)(2).

Curry, Don, BCA LABORATORIES, INC., 1102 South Main St., Bloomington, IL, 61701, (309) 828-7772, (1)(4).

Curry, George, ENVIRONMENTAL HAZARD CONTROL, 590 Whitney Ave., New Haven, CT, 06511, (203) 787-1871, (1).

Curtis, Tom, NATIONAL GOVERNORS ASSOCIATION (NGA)/COMMITTEE ON ENERGY AND ENVIRONMENT, Hall of States, Suite 250 444 North Capitol St., NW, Washington, DC, 20001, (202) 624-5300, (14).

Curtis, Verne, NAHAJO NATION/OCCUPATIONAL SAFETY AND HEALTH, P.O. Box 1690, Window Rock, 86515, (614) 644-2223, (800) 282-1425, (10).

Curtis, Verne, NAVAJO NATION/OCCUPATIONAL HEALTH AND SAFETY, P.O. Box 1690, Window Rock, AZ, 86515, (602) 871-6335, (10).

Cusato, Susan B., TRI-COUNTY RADON, INC., 2796 North Constitution Rd., Camden, NJ, 08104, (609) 963-1533, (1)(2).

Cutcher, James, AADVANCED AAIR AANALYSIS, 2670 Billingsley Rd., Columbus, OH, 43235, (614) 792-3400, (1)(2).

Cutler, Dan, RADON DETECTION SYSTEMS/CORPORATE HEADQUARTERS, 2450 Central Ave. A-1, Boulder, CO, 80301, (303) 444-5253, (800) 627-2360, (1).

Czerwinski, Michael G., HENIGAR AND RAY ENGINEERING ASSOCIATES, INC., 640 East Hwy. 44, Crystal River, FL, 32629, (904) 795-6551, (904) 447-2889, (3).

D

D'Angelo, Ralph, AMERICAN BUILDING INSPECTORS, INC., P.O. Box 1105, Haverford, PA, 19083, (215) 449-9361, (1).

D'Angelo, Ralph, AMERICAN HOMECHECK, INC., P.O. Box 565, Wayne, PA, 19087, (215) 259-8707, (1).

D'Arcy, Thomas, RADON TESTING SYSTEMS, P.O. Box 243, Newtown, PA, 18940, (215) 639-1933, (1).

D'Argo, Joan, CITIZENS FOR A BETTER ENVIRONMENT (CBE), 506 South Wabash, Chicago, IL, 60605, (312) 939-1530, (17).

Dagnon, George, ASSOCIATION OF STATE AND TERRITORIAL HEALTH OFFICIALS (ASTHO), 1311 Dolly Madison Blvd., McLean, VA, 22101, (703) 556-9222, (15).

Dahle, Jr., Elkins W., BALTIMORE COUNTY HEALTH DEPARTMENT/BUREAU OF INDUSTRIAL HYGIENE/AIR POLLUTION CONTROL, 303 East Fayette St. Room 400, Baltimore, MD, 21202, (301) 396-4427, (10).

Dahms, A.R., BENCHMARK INSPECTION SERVICES, 939 South 26th St., Arlington, VA, 22202, (703) 548-5690, (1)(3).

Dahms, K.F., BENCHMARK INSPECTION SERVICES, 939 South 26th St., Arlington, VA, 22202, (703) 548-5690, (1)(3).

Dalgleish, Ted, NOVA SCOTIA DEPARTMENT OF HEALTH, P.O. Box 488, Halifax, NS, B3H 2R8, (902) 424-4077, (11).

Dalphonse, David, ALPHA RADON ENVIRONMENTAL TESTING, Government Center P.O. Box 737, Portland, ME, 04104, (207) 773-8004, (1).

Daly, Michael, WISCONSIN DEPARTMENT OF REGULATION AND LICENSING/BUREAU OF DIRECT LICENSING AND REAL ESTATE, 1400 East Washington P.O. Box 8935, Madison, WI, 53708, (608) 267-7132, (10).

Damato, Anthony, PRINCETON TESTING LABORATORY, Princeton Service Center 3490 U.S. Rte. 1, Princeton, NJ, 08540, (609) 452-2037, (1).

Damiano, David, INTEGRATED PEST MANAGEMENT, Box 93, RD #2 White Horse Pike, Hammonton, NJ, 08037, (609) 567-0100, (1).

Dammkoehler, Arthur R., PUGET SOUND AIR POLLUTION CONTROL AGENCY, 200 West Mercer St. Room 205 P.O. Box 9863, Seattle, WA, 98119-3958, (206) 344-7330, (10).

Danckwerth, Paul A., FEDERAL NUCLIDE MONITORS, 2023 Park Rd., NW, Washington, DC, 20010, (202) 265-0429, (1).

Daneau, Mike, RADON ANALYSIS AND DETECTION OF NEW ENGLAND, INC., 111 Jefferson St., Laconia, NH, 03246, (603) 528-5772, (1).

Daniels, Len W., MAYA-TEK, INC., P.O. Box 1092, Chadds Ford, PA, 19317, (215) 358-5453, (1)(2).

Daniels, Mark D., MAYA-TEK, INC., P.O. Box 1092, Chadds Ford, PA, 19317, (215) 358-5453, (1)(2).

Danis, David J., MICROBAC LABORATORIES, INC./SCHILLER DIVISION, 449 Rochester Rd., Pittsburgh, PA, 15237-1733, (412) 369-4830, (1).

Darr, Rich, OHIO UNIVERSITY/GEOLOGICAL SCIENCES DEPARTMENT, Athens, OH, 45701, (614) 593-1109, (5).

Daugherty, Tom, RADON WATCH, INC., 472 North Wooster Way, Lancaster, OH, 43130, (614) 837-5748, (1).

Daven, Bud, ARKANSAS DEPARTMENT OF LABOR/OSHA CONSULTATION, 1022 High St., Little Rock, AR, 72202, (501) 375-8442, (10).

Davi, Gina L., SOLID STATE RADON DETECTION AND ELIMINATION SERVICES, P.O. Box 107, Glenwood, NJ, 07418, (201) 764-8555, (1).

Davi, Richard A., SOLID STATE RADON DETECTION AND ELIMINATION SERVICES, P.O. Box 107, Glenwood, NJ, 07418, (201) 764-8555, (1).

Davidson, Gordon M., RADON MANAGEMENT COMPANY, 4332 Gingham Ct., Alexandria, VA, 22310, (703) 960-4407, (1).

Davidson, James G., RDS/RADON DETECTION SERVICES, INC.-CORPORATE OFFICE, Rte. 179, Old York Rd. P.O. Box 419, Ringoes, NJ, 08551, (201) 788-3080, (1)(2)(3)(4)(6).

Davidson, Richard, U.S. DEPARTMENT OF INTERIOR (DOI)/U.S. GEOLOGICAL SURVEY(USGS), Box 25425 Denver Federal Ctr., Denver, CO, 80225, (303) 236-7476, (8).

Davies, Keith, NEW BRUNSWICK DEPARTMENT OF HEALTH AND COMMUNITY SERVICES/RADIATION PROTECTION SERVICES, P.O. Box 5100, Frederickton, NB, E3B 5G8, (506) 453-2067, (11).

Davila, Guillermo, WORLD HEALTH ORGANIZATION (WHO)/WASHINGTON OFFICE ENVIRONMENTAL HEALTH PROGRAM, 525 23rd St., NW, Washington, DC, 20037, (202) 861-3311, (12).

Davis, Bernadette, RAD ENVIRONMENTAL SERVICES, INC., 30 Basswood Ct., Red Bank, NJ, 07701, (201) 758-9324, (1)(2).

Davis, Dan, VICTOREEN, INC., 6000 Cochran Rd., Cleveland, OH, 44139, (216) 248-9300, (4).

Davis, Ed, INTERNATIONAL RADON TECHNOLOGIES, 2417 Regency Rd., Lexington, KY, 40503, (606) 278-0104, (800) 537-5066, (1)(4).

Davis, Jay W., CAMP DRESSER AND MCKEE, INC., Plaza 1 Raritan Ctr., Raritan, NJ, 08818, (201) 225-7000, (1).

Davis, John B., GENERAL HEALTH PHYSICS, INC., 7217 Lockport Pl., Lorton, VA, 22079, (703) 550-7525, (1).

Davis, John B., HOUSE CHECK, INC./MARYLAND, 2025 Maryland Ave., Baltimore, MD, 21218, (301) 727-5553, (703) 550-7525, (1).

Davis, Kathi, SPECIAL ENVIRONMENTAL SERVICES, INC., P.O. Box 20058, Atlanta, GA, 30325, (404) 355-5600, (1).

Davis, Mark, COLORADO GEOLOGICAL SURVEY, 1313 Sherman St. Room 715, Denver, CO, 80203, (303) 866-2611, (5)(10).

Davis, Nicholas F., GTS CONSULTANTS, 5599 North Maplewood Pl., Littleton, CO, 80123, (303) 794-1153, (3).

Davis, Randolph S., HOME RADON DETECTION COMPANY, 5824 El Monte, Fairway, KS, 66205, (913) 262-1555, (1).

Davis, Richard, SOUTH CAROLINA ASSOCIATION OF REALTORS, 3780 Fernandina Rd. Box 21827, Columbia, SC, 29221, (803) 772-5206, (18).

Davis, Vernon, VMS AND ASSOCIATES, Route 1, Box 472A, Madison Heights, VA, 24572, (804) 595-2413, (1).

Davis, Wade A., ENVIRONMENTAL DETECTION MANAGEMENT, P.O. Box 12316, Research Triangle Park, NC, 27709, (919) 469-8005, (1).

Day, Robert, ALA OF DISTRICT OF COLUMBIA, 475 H St., NW, Washington, DC, 20001, (202) 682-5864, (15).

De Azagra, Carl B., RADON REDUCTION ENGINEERING AND LABORATORY, 211 Mill St., NE, Vienna, VA, 22180, (703) 281-7865, (703) 281-

3730, (1)(2).

De Azagra, Charles H., RADON REDUCTION ENGINEERING AND LABORATORY, 211 Mill St., NE, Vienna, VA, 22180, (703) 281-7865, (703) 281-3730, (1)(2).

De Azagra, Erika D., RADON REDUCTION ENGINEERING AND LABORATORY, 211 Mill St., NE, Vienna, VA, 22180, (703) 281-7865, (703) 281-3730, (1)(2).

De Azagra, Robin, RADON REDUCTION ENGINEERING AND LABORATORY, 211 Mill St., NE, Vienna, VA, 22180, (703) 281-7865, (703) 281-3730, (1)(2).

De La Cruz, Cora M., ONTARIO MINISTRY OF CONSUMER AND COMMERCIAL RELATIONS/REAL ESTATE AND BUSINESS BROKERS ACT, Toronto, ON, M7A 2H6, (416) 965-0406, (11).

Dean, Allan, ENVIRONMENTAL SERVICES AND TECHNOLOGY, INC., 22 Oakland Ave., Atlantic Highlands, NJ, 07716, (201) 291-5525, (1)(2).

Dean, Craig, ONEIDA COUNTY BOARD OF COOPERATIVE EDUCATIONAL SERVICES (BOCES), Middle Settlement Rd. Box 70, New Hartford, NY, 13413, (315) 768-4583, (10).

Dean, Wayne, ERM, INC., 49 North Orange Ave., Orlando, FL, 32801, (407) 648-1969, (800) 237-3065, (1)(2).

DeAngelis, Robert, HEALTH SCIENCE SERVICES CORPORATION, P.O. Box 102, Marlboro, NJ, 07746, (201) 536-2066, (1).

DeBald, William, HUNTER-DON PEST CONTROL, 777 Mountain View Rd., Asbury, NJ, 08802, (201) 638-4024, (2).

DeBord, Beattie L., WEST VIRGINIA DEPARTMENT OF HEALTH/INDUSTRIAL HYGIENE DIVISION, 151 11th Ave., South Charleston, WV, 25303, (304) 348-3526, (6)(10).

DeChacon, Jeffrey R., PYRAMID ENVIRONMENTAL SYSTEMS, INC., P.O. Box 548, Sparta, NJ, 07871, (201) 729-9375, (1)(2).

DeCinque, Mark, M.A.D. EXTERMINATORS, INC., 3116 Methacton Ave., Norristown, PA, 19403, (215) 539-7647, (1).

Decker, R. Leon, CORPUS CHRISTI-NUECES COUNTY DEPARTMENT OF PUBLIC HEALTH/ENVIRONMENTAL CONTROL DIVISION, 1702 Horne Rd., Corpus Christi, TX, 78416, (512) 851-7226, (10).

Deere, Thomas, HOME DETOX SPECIALISTS, 81 Candy Ln., Hamburg, NY, 14075, (716) 648-3321, (1)(2).

Deerwater, Ronald, COLORADO RADON DETECTION COMPANY, 8824 West Ontario Ave., Littleton, CO, 80123, (303) 973-3261, (1).

DeForest, Jim, HEALTH HAZARDS DETECTION, Route 3, Box 372, Albany, KY, 42602, (606) 387-7021, (1).

DeFrain, James, NEBRASKA DEPARTMENT OF HEALTH/DIVISION OF RADIOLOGICAL HEALTH, 301 Centennial Mall South P.O. Box 93007, Lincoln, NE, 68509, (402) 471-2168, (10).

DeGrot, Jr., James, RADON DETECTION SYSTEMS/MIAMI, FLORIDA BRANCH, 17001 SW 109 Ave., Miami, FL, 33157, (303) 444-5253, (800) 627-2366, (1).

DeHaan, Robert, MICHIGAN DEPARTMENT OF PUBLIC HEALTH/BUREAU OF ENVIRONMENTAL AND OCCUPATIONAL HEALTH/DIVISION OF RADIOLOGICAL HEALTH, 3423 North Logan St. P.O. Box 30195, Lansing, MI, 48909, (517) 335-8200, (6)(10).

DeHaven, Mark, ERO RESOURCES CORPORATION, 2801 Youngfield Suite 121, Golden, CO, 80401, (303) 231-9022, (1).

Dehgan, M., ENVIRONMENTAL HEALTH CONSULTANTS, INC., 1505 North Prospect, Champaign, IL, 61820, (217) 356-4797, (1).

Dehmel, Jean-Claude F., ROY F. WESTON, INC., Building 4-2 1 Weston Way, West Chester, PA, 19380, (215) 692-3030, (1).

Deitz, Mike, G.M.D. CONSTRUCTION, INC., P.O. Box 204, Lisbon, MD, 21765, (301) 854-6186, (2).

DeKroff, Jim, VIRGINIA DEPARTMENT OF HEALTH/DIVISION OF HEALTH HAZARDS CONTROL/BUREAU OF RADIOLOGICAL HEALTH, 109 Governor St., Richmond, VA, 23219, (804) 786-6029, (10).

Del Guercio, Christina, DES CHAMPS LABORATORIES, INC., P.O. Box 440 17 Farinella Dr., East Hanover, NJ, 07936, (201) 994-4660, (4).

Delavan, David, CERTIFIED TESTING SERVICES, INC., 474 Easton Rd., Horsham, PA, 19044, (215) 672-8100, (1)(2).

DeLizza, Philip, C-P LABORATORY, 783 State Rd. Suite 174, Plymouth, MA, 02360, (617) 477-8128, (1).

DellaValle, Richard S., LW ENVIRONMENTAL SERVICES, 2111 Jennings St., San Francisco, CA, 94124, (415) 822-4555, (1)(2).

Deluca, P. J., NORTH OHIO VALLEY AIR AUTHORITY, 814 Adams St., Steubenville, OH, 43952, (614) 282-3908, (10).

DeLuca, Vincent, HOLMES INSPECTION SERVICE, 159 Liberty St., Little Ferry, NJ, 07643, (201) 641-8377, (800) 642-4603, (1).

DeMar, E. David, DEMAR PEST CONTROL COMPANY, 3522 Kensington Ave., Philadelphia, PA, 19134, (215) 744-0336, (1).

DeMarsico, William A., THOROUGH CHECK, INC., 2606 Rte. 37 East, Toms River, NJ, 08753, (201) 506-0200, (800) 852-0116, (1).

Demerest, David L., CORE LABORATORIES, INC., 420 West First St., Casper, WY, 82601, (307) 235-5741, (800) 666-0306, (1).

DeMichael, Jay, WATER POLLUTION CONTROL FEDERATION (WPCF), 601 Wyeth St., Alexandria, VA, 22314-1994, (703) 684-2400, (14).

Demirjian, Y.A., BIO-PURE COMPANY, P.O. Box 20593, Tampa, FL, 33622-0593, (813) 839-8617, (4).

Dempsey, John C., RAD ELEC, INC., 5330 J Spectrum Dr., Frederick, MD, 21701, (301) 694-0011, (1)(4)(5).

Dempsey, Joseph H., DEMPSEY, JOSEPH H., P.O. Box 247, Eldorado Springs, CO, 80025, (303) 499-7578, (2).

Denauex, Kathy, GILIAN INSTRUMENT CORPORATION, 8 Dawes Hwy., Wayne, NJ, 07470, (201) 831-0440, (4).

DeNeufville, David T., NATIONAL RADON CONTROL, INC., P.O. Box 5342, North Branch, NJ, 08876, (201) 231-0844, (1).

Denevan, Tom, RAVEN INDUSTRIES, INC., P.O. Box 1007, Sioux Falls, SD, 57117, (605) 336-2750, (800) 227-2836, (4).

Denham, G.M., ALPHANUCLEAR COMPANY, 1125 Derry Rd., E., Mississauga, ON, L5T 1P3, (416) 676-1364, (1)(4)(5).

Depascale, Aniello, ADCO, P.O. Box 1267, Burlington, CT, 06013, (203) 589-3333, (1).

Des Champs, Nicholas, DES CHAMPS LABORATORIES, INC., P.O. Box 440 17 Farinella Dr., East Hanover, NJ, 07936, (201) 994-4660, (4).

Desmond, Julie, NATIONAL APARTMENT ASSOCIATION (NAA), 1101 14th St., NW Suite 804, Washington, DC, 20005, (202) 842-4050, (13).

Dewers, Gordon, STANISLAUS COUNTY AIR POLLUTION CONTROL DISTRICT, 1716 Morgan Rd., Modesto, CA, 95351, (209) 571-6908, (10).

DiAddezio, William, COMMONWEALTH ENTERPRISES, INC., P.O. Box S187, St. Davids, PA, 19087, (215) 688-7097, (1)(2).

Diamond, Arthur M., MR. RENOVATOR, 395 South End Ave. Apt. 27P, New York, NY, 10280, (212) 321-9129, (1).

Diamond, Barry, DIAMOND REALTY INSPECTIONS, INC., 211 Monocacy Creek Rd., Birdsboro, PA, 19508, (215) 582-2070, (1).

Diamond, Glenn, AA RADCO, 1447 Old Sauk Trail, Niles, MI, 49120, (616) 684-8115, (1).

Diamond, James, SAFE-AIR, 2599 East Main St. Suite 525, Bexley, OH, 43209, (614) 447-8300, (800) 553-3117, (2).

Diamond, Richard, ELECTRICAL POWER RESEARCH INSTITUTE (EPRI)/LAWRENCE BERKELEY LABORATORY(LBL)/APPLIED SCIENCE DIVISION, University of California One Cyclotran Rd., Berkeley, CA, 94720, (415) 486-4000, (5)(13).

DiBenedetto, Neil, WESTCHESTER COUNTY HEALTH DEPARTMENT, County Office Building #2, White Plains, NY, 10601, (914) 285-1500, (10).

Dickerman, James C., RADIAN CORPORATION/ENGINEERING DIVISION, P.O. Box 1300, Research Triangle Park, NC, 27709, (919) 541-9100, (5)(14).

Dicus, Greta, ARKANSAS DEPARTMENT OF HEALTH/RADIATION CONTROL AND EMERGENCY MANAGEMENT/NUCLEAR AND ENVIRONMENTAL SAFETY SECTION, 4815 West Markham St., Little Rock, AR, 72201, (501) 661-2301, (10).

Diebolt, Timothy J., ADVANCED AIR ANALYSIS, 61836 30th St., Lawton, MI, 49065, (616) 624-2741, (1).

Dietrich, David L., AIR RESOURCE SPECIALISTS, INC., 1901 Sharp Point Dr. Suite E, Ft. Collins, CO, 80525, (303) 484-7941, (1).

DiGiano, Francis, WATER POLLUTION CONTROL FEDERATION (WPCF), 601 Wyeth St., Alexandria, VA, 22314-1994, (703) 684-2400, (14).

Dilk, Jerry, AMERICAN NUCLEAR SOCIETY/ARIZONA-ARIZONA SECTION, C/O Combustion Engineering 10010 North 25th Dr. Corporate Center, Suite 310, Phoenix, AZ, 85021, (602) 944-4664, (14).

Dillard, Bob, K & S ASSOCIATES, 1854 Airlane Dr., Nashville, TN, 37210, (615) 883-9760, (3)(4).

Dillard, A. Brown, TRIPLE SEAL, INC., 2425 Patterson Rd., Kettering, OH, 45420, (513) 258-8255, (1).

Dinoff, Lawrence, DINOFF ASSOCIATES, INC., 112 South Decatur, Strasburg, PA, 17579, (717) 687-0938, (2).

Diodati, Al, BRAMA CONSTRUCTION CORPORATION, 47 Buena Vista Ave., Yonkers, NY, 10701, (914) 968-4700, (1).

DiPierro, Nick, NEW JERSEY DEPARTMENT OF ENVIRONMENTAL PROTECTION/DIVISION OF ENVIRONMENTAL PROTECTION/BUREAU OF ENVIRONMENTAL RADIATION/RADON PROGRAM, 380 Scotch Rd. CN-411, Trenton, NJ, 08625, (201) 879-2612, (6)(10).

Distenfeld, Carl H., TCS INDUSTRIES, INC., 3805 Paxton St., Harrisburg, PA, 17111, (717) 558-8206, (1)(4).

Distenfeld, Eric B., TCS INDUSTRIES, INC., 3805 Paxton St., Harrisburg, PA, 17111, (717) 558-8206, (1)(4).

Distenfeld, Jacquelyn, TCS INDUSTRIES, INC., 3805 Paxton St., Harrisburg, PA, 17111, (717) 558-8206, (1)(4).

Distenthal, Carl, A.B.E. RADON TESTING, INC., P.O. Box 508, Trexlertown, PA, 18087, (215) 481-9916, (1)(6).

Dittenhauser, Michael J., ENSA, INC., 4550 Bailey Ave., Buffalo, NY, 14226, (716) 834-3111, (1).

Dittmar, Willy C., DITTMAR INSPECTION COMPANY, INC., 20 Redding Pl., Towaco, NJ, 07082, (201) 263-4745, (1).

Dittmar, Willy C., HOME INSPECTIONS BY CHANEL, INC., 20 Redding Pl., Towaco, NJ, 07082, (201) 263-9163, (1).

Dixon, Howard E., ALPHA RADIATION ENGINEERING, INC., P.O. Box 90, Littleton, CO, 80160-0090, (303) 721-0068, (303) 721-0872, (1)(2).

Dobbs, Lynn B., SOUTH CAROLINA DEPARTMENT OF HEALTH AND ENVIRONMENTAL CONTROL/BUREAU OF RADIOLOGICAL HEALTH, 2600 Bull St., Columbia, SC, 29201, (803) 734-4700, (1)(10).

Dobson, Jack, AMERICAN SOCIETY OF SAFETY ENGINEERS (ASSE), 1800 East Oakton, Des Plaines, IL, 60018, (312) 692-4121, (16).

Dockery, Stan, PEST DOCTOR, P.O. Box 32, Lithia Springs, GA, 30057, (404) 920-2288, (1).

Dodd, Gerald D., AMERICAN CANCER SOCIETY (ACS)/NATIONAL HEADQUARTERS, 3340 Peachtree Rd., Atlanta, GA, 30326, (404) 320-3333, (15).

Dodge, Christopher, LIBRARY OF CONGRESS (LOC)/CONGRESSIONAL RESEARCH SERVICE(CRS), 101 Independence Ave., SE LM 413, Washington, DC, 20540, (202) 287-5000, (8).

Dodson, George, AKZO INDUSTRIAL SYSTEMS, Box 7249 1 North Pack Sq., Asheville, NC, 28704, (704) 258-5074, (4).

Doherty, Kevin A., RADION LABORATORY, INC., 70 Walnut St., Wellesley, MA, 02181, (617) 239-8066, (1)(2)(3)(4)(5)(6)(13).

Doherty, Tom, ADIRONDACK ALTERNATE ENERGY, County Rte. 4, Edinburg, NY, 12134, (518) 863-4338, (4)(13).

Dolezal Hennigan, Alice T., MINNESOTA DEPARTMENT OF HEALTH/ENVIRONMENTAL HEALTH DIVISION/SECTION OF RADIATION CONTROL, 717 Delaware St., SE P.O. Box 9441, Minneapolis, MN, 55440, (612) 623-5351, (10).

Dolibois, Bob, AMERICAN SOCIETY OF HOME INSPECTORS (ASHI), 3299 K St., NW Suite 700, Washington, DC, 20007, (202) 842-3096, (18).

Dollwet, Helmar, RADON AND RADIATION CONSULTANTS, 3575 Peninsula Dr., Akron, OH, 44319, (216) 644-7664, (1).

Dolnski, John S., RADON TECHNOLOGY AND CONSULTANTS, RFD 2 Louisquisset Pike, North Smithfield, RI, 02895, (401) 769-2129, (401) 568-0296, (1)(2).

Dolphin, Jacques, RADIATION DATA, 115 Wall St., Princeton, NJ, 08540, (609) 921-7061, (1)(2).

Dominski, Peter, ENVIRONMENTAL PRO-TECH SERVICES, INC., 13 Edgar Dr. P.O. Box 599, West Milford, NJ, 07480, (201) 728-2666, (1).

Dommer, Martin, ROCKY MOUNTAIN HOME INSPECTIONS, 1571 East Mineral Pl., Littleton, CO, 80122, (303) 730-0922, (1)(2).

Dommers, Walter A., RADON EVALUATION DIVISION/EVALDO, LTD., 64 Sunset Hill Dr., Branford, CT, 06405, (203) 488-4715, (1).

Donia, Frank J., APPLIED RADON, INC., 52205 Warehime Rd., Cumberland, OH, 43732-9804, (614) 638-2078, (800) 638-2078, (1)(2)(4).

Donia, Nancy A., APPLIED RADON, INC., 52205 Warehime Rd., Cumberland, OH, 43732-9804, (614) 638-2078, (800) 638-2078, (1)(2)(4).

Doob, Peter, U.S. DEPARTMENT OF HEALTH AND HUMAN SERVICES (HHS)/PUBLIC HEALTH SERVICE(PHS)/NATIONAL INSTITUTES OF HEALTH(NIH)/DIVISION OF SAFETY, Radiation Safety Branch Building 21, Room 231 9000 Rockville Pike, Bethesda, MD, 20892, (301)496-5774, (8).

Doolittle, Jr., Albert W., EMPAC, INC., 25 Science Pk., New Haven, CT, 06511, (203) 786-5420, (800) 872-3662, (1)(4).

Doran, Diana B., NITON CORPORATION, P.O. Box 9153, Lexington, MA, 02173, (617) 229-4929, (800) 451-6467, (1)(4).

Doran, Diana B., NITONICS CORPORATION, P.O. Box 9153, Lexington, MA, 02173, (617) 229-4929, (800) 451-6467, (4).

Dorf, Barbara, U.S. DEPARTMENT OF HOUSING AND URBAN DEVELOPMENT (HUD)/DEPUTY ASSISTANT SECRETARY FOR PROGRAM DEVELOPMENT, 415 7th St., SW Room 7212, Washington, DC, 20410, (202) 755-5566, (8).

Dorff, William A., ANDERSON AIR POLLUTION CONTROL DEPARTMENT, P.O. Box 2100, Anderson, IN, 46018, (317) 646-9835, (10).

Dorfi, Kurt, RADON TESTING CORPORATION OF AMERICA, 12 West Main St., Elmsford, NY, 10523, (914) 347-5010, (800) 457-2366, (1)(4)(6).

Doron, Henry H., MRC HOME INSPECTIONS, P.O. Box 157, Wyckoff, NJ, 07481, (201) 891-7057, (1).

Dorsey, Jr., Wallace O., U.S. HOME INSPECTION SERVICES, 2603 Berwyn St., Richmond, VA, 23234, (804) 231-6262, (804) 423-0730, (1).

Doruff, Mark A., ACCURAD TESTING SERVICE, P.O. Box 8766, Rolling Meadows, IL, 60008, (312) 529-6633, (1).

Douglas, Elaine, WORLD INFORMATION SERVICE ON ENERGY (WISE)/WASHINGTON OFFICE, P.O. Box 53373, Washington, DC, 20009, (202) 745-7467, (12).

Douglas, John, APPLIED HEALTH PHYSICS, INC., 2986 Industrial Blvd., Bethel Park, PA, 15102, (412) 563-2242, (518) 477-7974, (1)(2)(4)(6).

Dowell, Frankie, SHORTRIDGE INSTRUMENTS, INC., 7855 E. Redfield, Scottsdale, AZ, 85260, (602) 991-6744, (4).

Dowling, W.R., LOS ANGELES URBAN DEVELOPMENT, 1725 Radcliff St., Los Angeles, CA, 90026-1131, (213) 668-2655, (1)(3).

Downing, Lance, NATIONAL RADON LABORATORIES/DIVISION OF COLORADO SERVICES CORPORATION, 14677 East Easter Ave. Suite H, Englewood, CO, 80112, (303) 699-7378, (1).

Downing, Mary Lou, M.L. DOWNING AND ASSOCIATES, 2806 Farm Rd., Alexandria, VA, 22314, (703) 548-8607, (1).

Doyle, Ned, AIR CHEK, INC., 180 Glenn Bridge Rd. Box 2000, Arden, NC, 28704, (704) 684-0893, (800) 257-2366, (1)(4)(6).

Doyle, Ned, CAROLINA RADON CONSULTANTS, P.O. Box 2000, Etowah, NC, 28729, (704) 891-9559, (1)(2).

Doyle, Ned, CAROLINA RADON CONSULTANTS, P.O. Box 6683, Columbia, SC, 29260, (803) 779-2043, (1)(2).

Doyle, Paul, NATIONAL CONFERENCE OF STATE LEGISLATURES (NCSL), 1050 17 St. #2100, Denver, CO, 80265, (303) 623-7800, (14).

Drake, Don, BOULDER INSPECTION COMPANY, P.O. Box 3369, Boulder, CO, 80307, (303) 444-6657, (1)(4).

Drapeau, Gary, ALL-DRY BASEMENTS AND SAFE AIR, INC., P.O. Box 205, Collinsville, CT, 06022, (203) 693-0245, (2)(4).

Drewry, David R., DREWRY HOME INSPECTIONS, INC., 2625 Oakton Glen Dr., Vienna, VA, 22180, (703) 938-1044, (1)(2).

Droullard, Robert, TOM SCURRY ASSOCIATES, 1976 Linden Dr., Boulder, CO, 80302, (303) 444-4802, (4).

Drozdow, Steven, ESSEX CHEMICAL CORPORATION, 1333 Broad St., Clifton, NJ, 07015, (201) 773-6300, (4).

Dructor, Ed, TEQ CORPORATION, P.O. Box 570, Wilkes-Barre, PA, 18703-0570, (717) 825-3939, (800) 877-2366, (1)(2)(4).

Drummond, Bob, BRAILSFORD AND COMPANY, INC., 670 Milton Rd., Rye, NY, 10580, (914) 967-1820, (4).

Duarte, Steven, FORESIGHT ENGINEERING, INC., VIRGINIA, P.O. Box 7333, Arlington, VA, 22207-0333, (703) 525-4790, (1).

Dudek, J., RADIATION DATA/DIVISION OF PRINCETON GEOPHYSICAL SERVICES, 37 Castleman Dr., Southington, CT, 06489, (203) 621-0678, (1)(2).

Dudek, L., RADIATION DATA/DIVISION OF PRINCETON GEOPHYSICAL SERVICES, 37 Castleman Dr., Southington, CT, 06489, (203) 621-0678, (1)(2).

Dudich, Endre, INTERNATIONAL GEOLOGICAL CORRELATION PROGRAM (IGCP), C/O UNESCO Maison de l' UNESCO 7 Place de Fontenoy, F-75700 Paris, France, (011) 331-45684123, (12).

Dudley, Blair, RADON DETECTION SYSTEMS/DETROIT, MICHIGAN BRANCH, 6051 Dunn Rd., Howell, MI, 48843, (303) 444-5253, (800) 627-2366, (1).

Dudley, Paul, DUDLEY, PAUL, 140 Railroad Mills Rd., Pittsford, NY, 14534, (716) 385-2368, (1)(3).

Duffy, Peter, ENERGY CONSERVATION SERVICES, 6290 North Vine Valley Rd., Rushville, NY, 14544, (716) 554-3994, (1)(2)(3).

Duffy, Raymond, RADON INSPECTION SERVICES, 915 6th Ave., North Unit C, Seattle, WA, 98109, (206) 286-0479, (1).

Duffy, Raymond, RADON DETECTION AND REMOVAL SERVICES, 915 6th Ave., N., Seattle, WA, 98109, (206) 282-9156, (1)(2).

Duggan, David, VALLEY RADON TESTING, INC., 26 Oak Valley Rd., Huntington, CT, 06484, (203) 929-4109, (1).

Dunaway, Belle, LOUISIANA REAL ESTATE COMMISSION, P.O. Box 14785, Baton Rouge, LA, 70898, (504) 925-4771, (10).

Duncan, David, U.S. ENVIRONMENTAL PROTECTION AGENCY REGION 9 (EPA9)/RADIATION DIVISION(ORP), 215 Fremont St., San Francisco, CA, 94105, (415) 974-8378, (8).

Dunham, Mark, IDAHO ASSOCIATION OF REALTORS, 1450 West Bannock, Boise, ID, 83702, (208) 342-3585, (18).

Dunkle, Bob, SUNN CORPORATION, Box M, Dept. 213 3520 Forbes Ave., Pittsburgh, PA, 15213, (412) 828-9100, (2)(4).

Dunlop, George S., U.S. DEPARTMENT OF AGRICULTURE (USDA)/SPECIAL SERVICES, 14th St. and Independence Ave., SW, Washington, DC, 20250, (202) 447-7173, (8).

Dunn, Hal, RADON CHECK, 2524 South El Paradiso Suite 50, Mesa, AZ, 85202, (602) 730-5559, (1).

Dunn, James, INSULATING SYSTEMS, 340 Gargrave Rd., West Carrollton, OH, 45449, (513) 435-0079, (1)(2).

Duray, John R., UNC GEOTECH, 2594 B 314 Road P.O. Box 14000, Grand Junction, CO, 81502-5504, (303) 248-6702, (5).

Durgin, Sharon A., RADON SPECIALISTS, 1840 Folsom St. Suite 255, Boulder, CO, 80302, (303) 673-0494, (1).

DuTemple, Octave, AMERICAN NUCLEAR SOCIETY (ANS), 555 North Kensington, La Grange Park, IL, 60525, (312) 352-6611, (14).

Duval, Joseph S., U.S. DEPARTMENT OF INTERIOR (DOI)/U.S. GEOLOGICAL SURVEY(USGS)/GEOLOGIC DIVISION(GD), MS-955, National Ctr., Reston, VA, 22092, (703) 648-4000, (8).

DuVall, Owen, ORR-DUVALL AND ASSOCIATES, 22346 Metamora, Birmingham, MI, 48010, (313) 644-0706, (1).

Duvic, Donnissa, APPLIED TECHNICAL SERVICES, INC., 1190 Atlanta Industrial Dr., Marietta, GA, 30066, (404) 423-1400, (800) 451-3405, (1)(4).

Dvorak, Vladimir, CONFERENCE OF RADIATION CONTROL PROGRAM DIRECTORS (CRCPD)/TASK FORCE FOR ADEQUATE RADIATION CONTROL PROGRAM-RADON, 71 Fountain Pl., Frankfort, KY, 40601, (502) 227-4543, (14).

Dyer, Jon, RADON DIAGNOSTIC LABORATORIES, 231 Front St. P.O. Box 2890, South Portland, ME, 04106, (207) 767-2818, (1)(2).

Dyroff, George A., DR. RADON TESTING AND REDUCTION, 687 North Cary St., Brockton, MA, 02402, (508) 583-2950, (1).

Dyson, Chris, BUYERS UP RADON PROJECT, Box 53005, Washington, DC, 20036, (202) 328-3800, (17).

Dytzel, Billie, ALA OF NEW MEXICO, 216 Truman, NE, Albuquerque, NM, 87108, (505) 265-0732, (15).

Dzubian, Jacolyn, U.S. ENVIRONMENTAL PROTECTION AGENCY (EPA)/OFFICE OF AIR AND RADIATION(OAR)/OFFICE OF RADIATION PROGRAMS(ORP)/RADON DIVISION(RD), Problem Assessment Branch 401 M St. SW ANR-464, Washington, D.C., 20460, (202) 475-9605, (8).

E

Eason, Charles, EASON AND EASON CHARTERED/ATTORNEYS AT LAW, 4423 Montgomery Ave., Bethesda, MD, 20814-4491, (301) 652-2666, (3).

Eastaugh, David, FTLC SERVICES, INC., P.O. Box 563, Amherst, OH, 44001, (216) 960-2550, (1).

Eastaugh, Nancy, FTLC SERVICES, INC., P.O. Box 563, Amherst, OH, 44001, (216) 960-2550, (1).

Eastwood, Chuck, U.S. DEPARTMENT OF ENERGY (DOE)/BONNEVILLE POWER ADMINISTRATION (BPA)/OFFICE OF ENERGY RESOURCES(OE)/DIVISION OF RESOURCE MANAGEMENT/RESIDENTIAL TECHNOLOGY SECTION, 905 Northeast 11th St., Portland, OR, 97208, (503) 230-5488, (8).

Eaton, Andrew, J.M. MONTGOMERY ENGINEERS, 555 East Walnut St., Pasadena, CA, 91101, (818) 796-9141, (1).

Eaton, Andrew D., MONTGOMERY LABORATORIES, 555 East Walnut, Pasadena, CA, 91109, (818) 796-9141, (1).

Eberle, Sandra, CONSUMER PRODUCT SAFETY COMMISSION (CPSC), 5401 Westbard Ave. Room 420, Bethesda, MD, 20207, (301) 492-6554, (8).

Echols, Renee, MISSISSIPPI STATE DEPARTMENT OF HEALTH/DIVISION OF RADIOLOGICAL HEALTH, 3150 Lawson St. P.O. Box 1700, Jackson, MS, 39215-1700, (601) 354-6657, (10).

Eckerman, Keith, U.S. DEPARTMENT OF ENERGY (DOE)/OAK RIDGE OPERATIONS OFFICE(OROO)/OAK RIDGE NATIONAL LABORATORY(ORNL)/HEALTH STUDIES DIVISION(HSD)/METABOLISM AND DOSIMETRY DIVISION, P.O. Box 2008 Building 7509 MS-6383, Oak Ridge, TN, 37831, (615) 574-6251, (8).

Eckert, L.W., SCIENCE MANAGEMENT SERVICES, LTD., 25 Eshelman Rd., Lancaster, PA, 17601, (717) 392-1425, (1).

Eckhardt, Gary, MINNESOTA POLLUTION CONTROL AGENCY/DIVISION OF AIR QUALITY, 520 Lafayette Rd., St. Paul, MN, 55155, (612) 296-7802, (10).

Eckrote, David J., TRIAD BASEMENT DE-WATERING SYSTEM, 1443 Spring Hill Cr., Kernersville, NC, 27284, (919) 996-7870, (919) 996-1685, (1).

Ecksel, Scott, A BLACK CROSS SANITATION COMPANY, 1024 Cottman Ave., Philadelphia, PA, 19111, (215) 342-5100, (1).

Ed Nevissi, Ahmad, UNIVERSITY OF WASHINGTON/DEPARTMENT OF ENVIRONMENTAL HEALTH/RADIATION ECOLOGY LABORATORY, LRE-WH-10, Seattle, WA, 98195, (206) 543-4259, (1)(5).

Eddy, Scott, RADEX, 265 Stone St., Walpole, MA, 02081, (617) 660-1548, (1).

Edelkind, Myron R., SOUTHERN MECHANICAL, 1721 Lee St., Decatur, GA, 30035, (404) 289-9637, (1).

Edelstein, Michael, RAMAPO COLLEGE OF NEW JERSEY/INSTITUTE FOR ENVIRONMENTAL STUDIES, 505 Ramapo Valley Rd., Mahwah, NJ, 07430, (201) 529-7742, (3)(5)(6).

Edelstein, Michael R., ORANGE ENVIRONMENT, INC., RD 2 Maple Ave., Goshen, NY, 10924, (914) 294-5023, (17).

Edmonds, Jr., James A., CALIFORNIA DEPARTMENT OF REAL ESTATE, 2201 Broadway P.O. Box 187000, Sacramento, CA, 95818-7000, (916) 739-3721, (10).

Edwards, Neil, RADON RESOURCE, INC., 49 John St., Westminster, MD, 21157, (301) 857-9330, (1)(2).

Egan, Martin J., MARTIN J. EGAN ASSOCIATES, 37 Waldron Ave., Summit, NJ, 07901, (201) 522-0931, (1).

Egitto, Tony, ENVIRONMENTAL ASPECS, INC., 1527 North Dale Mabry Suite 103, Lutz, FL, 33549, (813) 948-1387, (1).

Egleston, Jr., David L., RADON TESTING SERVICE, INC., P.O. Box 1130, Englewood, CO, 80110, (303) 762-8598, (1).

Eheman, Christie, U.S. DEPARTMENT OF HEALTH AND HUMAN SERVICES (HHS)/PUBLIC HEALTH SERVICE(PHS)/CENTERS FOR DISEASE CONTROL (CDC)/ENVIRONMENTAL HAZARDS(EH), 1600 Clifton Rd., NE Koger Center S-27, Atlanta, GA, 30333, (404) 639-3311, (8).

Eichholz, Geoffrey, GEORGIA INSTITUTE OF TECHNOLOGY/NUCLEAR ENGINEERING AND HEALTH PHYSICS PROGRAM, ESM Building, Atlanta, GA, 30332, (404) 894-3720, (5).

Eick, Thomas, VIRGINIA RADON SERVICE, INC., Box 332-B Route 1, Roseland, VA, 22967, (804) 277-5491, (1).

Eisenhower, Elmer, U.S. DEPARTMENT OF COMMERCE (DOC)/NATIONAL INSTITUTE OF STANDARDS AND TECHNOLOGY(NIST)/CENTER FOR RADIATION RESEARCH(CRR)/OFFICE OF RADIATION MEASUREMENTS, Building 245 Room C-225, Gaithersburg, MD, 20899, (301) 975-6042, (8).

Eldredge, Clark, FLORIDA DEPARTMENT OF HEALTH AND REHABILITATIVE SERVICES/DEPUTY SECRETARY FOR HEALTH/OFFICE OF RADIATION CONTROL, 1317 Winewood Blvd., Tallahassee, FL, 32399-0700, (904) 487-1004, (6)(10).

Eldridge, Gail, MISSOURI REAL ESTATE COMMISSION, P.O. Box 1339, Jefferson City, MO, 65102, (314) 751-2334, (10).

Elias, Leonard, DADE COUNTY CONSUMER ADVOCATES OFFICE, 140 West Flagler St. Room 1605, Miami, FL, 33130, (305) 375-4206, (10).

Ellett, William, NATIONAL ACADEMY OF SCIENCES (NAS)/NATIONAL RESEARCH COUNCIL (NRC)/MEDICAL FOLLOW-UP AGENCY, 2101 Constitution Ave., NW, Washington, DC, 20418, (202) 334-2000, (15).

Ellins, Katherine, ENVIRONMENTAL RADIOACTIVITY MEASUREMENTS, INC., P.O. Box 226, Piermont, NY, 10968, (914) 534-8679, (201) 768-7651, (1).

Ellis, Bruce O., ALLIED BIOLOGICAL, 240 Main St., Gladstone, NJ, 07934, (201) 234-2476, (1).

Ellis, Pamela A., MICROBAC LABORATORIES, INC./KENTUCKY TESTING DIVISION, 1121 West Broadway, Louisville, KY, 40203, (502) 583-5256, (502) 589-6019, (1).

Ellison, Harry, SAFE-AIRE, INC., 162 East Chestnut St. P.O. Box 160, Canton, IL, 61520, (309) 647-0419, (800) 331-2943, (1)(2)(3)(4)(6).

Emce, Ron, U.S. DEPARTMENT OF AGRICULTURE (USDA)/FARMERS HOME ADMINISTRATION (FMHA)/LEGISLATIVE AFFAIRS AND PUBLIC INFORMATION STAFF(LAPIS), 14th St. and Independence Ave., SW Room 5037, Washington, DC, 20250, (202) 447-6903, (8).

Emdee, Larry L., RADON DETECTORS, RD 1, Box 177, Van Etten, NY, 14889, (607) 589-6223, (1)(2).

Emerson, Robert J., SHELBURNE LABORATORIES, INC., P.O. Box 458 Falls Rd., Shelburne, VT, 05482, (802) 985-3379, (1).

Emmitt, G.D., SIMPSON WEATHER ASSOCIATES, 809 East Jefferson St., Charlottesville, VA, 22902, (804) 979-3571, (1)(2)(3).

Eng, Jeanette, NEW JERSEY DEPARTMENT OF ENVIRONMENTAL PROTECTION/DIVISION OF ENVIRONMENTAL QUALITY/BUREAU OF ENVIRONMENTAL RADIATION/RADIOLOGICAL ENVIRONMENTAL ASSESSMENT, 380 Scotch Rd. CN-411, Trenton, NJ, 08625, (609) 530-4051, (10).

Engelhardt, Jeffrey L., RADON AWARENESS, INC., 415 East Galena P.O. Box 566, Hailey, ID, 83333, (208) 788-2634, (1)(2).

Engelhardt, Susan M., RADON AWARENESS, INC., 415 East Galena P.O. Box 566, Hailey, ID, 83333, (208)

788-2634, (1)(2).

Engelstad, Robert, FEDERAL NATIONAL MORTGAGE ASSOCIATION (FANNIE MAE), 3900 Wisconsin Ave., NW, Washington, DC, 20016, (202) 537-7000, (18).

Engle, D.B., YUBA COUNTY AIR POLLUTION CONTROL DISTRICT, 938 14th St., Marysville, CA, 95901, (916) 741-6484, (10).

Enhoffer, John, BIONOMIC INDUSTRIES, INC., 479 Route 17 N., Mahwah, NJ, 07430, (201) 529-1094, (800) 338-3210, (1)(2).

Enochs, Kenneth G., AMERICAN HERITAGE CORPORATION, P.O. Box 1145, Paoli, PA, 19301, (717) 275-0300, (800) 247-0300, (1)(2).

Eriksen, Harold W., ENVIRONMENTAL CONTROL TECHNOLOGIES, 400 Amherst St., Nashua, NH, 03063, (603) 881-7133, (1).

Erny, Karen A., MICROBAC LABORATORIES, INC./SEAWAY DIVISION, 542-544 Conkey St., Hammond, IN, 46324, (219) 932-1770, (1).

Erwin, Robert, WEST VIRGINIA GEOLOGICAL SURVEY, Mt. Chateau Research Center P.O. Box 879, Morgantown, WV, 26507-0879, (304) 594-2331, (5)(10).

Esperseth, James, HOUSING INSPECTION SERVICES, INC., P.O. Box 373, Murrysville, PA, 15668, (412) 374-1900, (1).

Espinal, Daniel, NEW ENGLAND RADON, LTD., 373 Main St., Salem, NH, 03079, (603) 893-4260, (800) 637-2366, (1)(2)(4).

Esworthy, Chris, ENVIRONMENTAL ENTERPRISES, INC., P.O. Box 16847, Atlanta, GA, 30321, (404) 766-5551, (1)(2)(4).

Eure, John A., IOWA DEPARTMENT OF PUBLIC HEALTH/DIVISION OF DISEASE PREVENTION, Lucas State Office Building, Des Moines, IA, 50319, (515) 281-6549, (10).

Evanoff, John, BENSENVILLE AIR POLLUTION CONTROL DISTRICT, 1253 Roosevelt Ave., Glenview, IL, 60025, (312) 498-9840, (10).

Evanoff, John, CHICAGO/DEPARTMENT OF CONSUMER SERVICES/ENERGY AND ENVIRONMENTAL PROGRAMS DIVISION, 121 North LaSalle St. Room 810, Chicago, IL, 60602, (312) 744-4034, (10).

Evans, Belmont, WESTERN ENVIRONMENTAL, INC., 4320-B Northpark Dr., Colorado Springs, CO, 80907, (719) 528-6132, (1).

Evans, Dallas E., HOUSTON/BUREAU OF AIR QUALITY CONTROL, 7411 Park Place Blvd., Houston, TX, 77087, (713) 640-4200, (10).

Evans, Edwin, EVANSHIRE COMPANY, LTD., THE, Rte. 4, Box 57, Rome, NY, 13440, (315) 865-5941, (800) 962-5941, (1)(2)(4).

Evans, Helen W., CASE WESTERN RESERVE UNIVERSITY/SCHOOL OF MEDICINE, 2048 Adelbert Rd.., Cleveland, OH, 44186, (216) 844-3530, (5).

Evans, Wade, EVANSHIRE COMPANY, LTD., THE, Rte. 4, Box 57, Rome, NY, 13440, (315) 865-5941, (800) 962-5941, (1)(2)(4).

Evdokimoff, Victor, BOSTON UNIVERSITY MEDICAL CENTER/RADIATION PROTECTION OFFICE, 80 East Concord, Boston, MA, 02118, (617) 638-7052, (1).

Everson, Steve, EER PRODUCTS, INC., 3536 East 28th St., Minneapolis, MN, 55406, (612) 721-4231, (4).

Ewell, Bud, IOWA ASSOCIATION OF REALTORS, 999 Oakridge Dr., Des Moines, IA, 50308, (515) 244-2294, (18).

Ewenstein, Edith, ALA OF NEW YORK CITY-MANHATTAN/BRONX/STATEN ISLAND, 432 Park Ave., S. 8th Fl., New York, NY, 10016, (212) 889-3370, (15).

Ewing, Anthony, EXECUTIVE OFFICE OF THE PRESIDENT (EOP)/OFFICE OF SCIENCE AND TECHNOLOGY POLICY(OSTP)/OAK RIDGE ASSOCIATED UNIVERSITIES(ORAU)/COMMITTEE ON INTERAGENCY RADIATION RESEARCH AND POLICY COORDINATION(CIRRPC), 1019 19th St., NW, Washington, DC, 20036, (202) 653-5505, (8).

F

Fabian, Nelson E., NATIONAL ENVIRONMENTAL HEALTH ASSOCIATION (NEHA), 720 South Colorado Blvd. South Tower, Suite 970, Denver, CO, 80222, (303) 756-9090, (15).

Fabrikant, Jacob I., UNIVERSITY OF CALIFORNIA/LAWRENCE BERKELEY LABORATORY/DONNER LABORATORY BIOMEDICAL SCIENCES, One Cyclotran Rd./Donner Pavilion Cowell Hospital, Room 420, Berkeley, CA, 94720, (415) 486-4000, (5).

Fakundiny, Robert, NEW YORK STATE GEOLOGICAL SURVEY, 3136 Cultural Education Center Empire State Plaza, Albany, NY, 12230, (518) 474-5816, (10).

Fama, Anthony M., RADON REMOVAL, INC., Box 283, Milltown, NJ, 08850, (201) 545-2000, (1)(2)(4).

Fanous, Reda, REDIHOME INSPECTIONS, INC., 18 Kennedy Ter., Westwood, NJ, 07675, (201) 358-6888, (1).

Farley, John J., SUBURBAN REGIONAL HEALTH COMMISSION, 141 South Harrison St. Suite 11, East Orange, NJ, 07018, (201) 675-1774, (10).

Farmer, Darryll D., EAU CLAIRE/CITY-COUNTY HEALTH DEPARTMENT/ENVIRONMENTAL HEALTH DIVISION, 720 2nd Ave., Eau Claire, WI, 54701, (715) 839-4718, (10).

Farnsworth, Barbara, INSPEX OF BOULDER, INC., 2955 Valmont Suite 300, Boulder, CO, 80301, (303) 442-8995, (1).

Farnsworth, Dave, OHIO DEPARTMENT OF HEALTH/RADIOLOGICAL HEALTH PROGRAM/OHIO RADON PROJECT, 1224 Kinnear Rd., Columbus, OH, 43212, (614) 644-2727, (10).

Farrow, James M., RADON REMEDIES, INC., 2700 West Broad St. P.O. Box 44234, Columbus, OH, 43204, (614) 272-7772, (1)(3)(4).

Fastiggi, Joseph R., RADON MANAGEMENT LABORATORIES, Overlook Terr., Chester, NJ, 07930, (201) 879-2220, (201)835-8721, (1).

Faust, John, AIR RESOURCE SPECIALISTS, INC., 1901 Sharp Point Dr. Suite E, Ft. Collins, CO, 80525, (303) 484-7941, (1).

Favale, Ralph, RAF ENGINEERING, Hillside Rd., Rockaway, NJ, 07866, (201) 983-9041, (1).

Fay, Jan Elizabeth, FAYTEK, INC., P.O. Box 2229, Coeur D'Alene, ID, 83814, (208) 667-3263, (1).

Fay, Jane, RADON DETECTION SYSTEMS/SPOKANE, WASHINGTON BRANCH, 1115 Lambert Ln., Coer D'Alene, ID, 83814, (303) 444-5253, (800) 627-2360, (1).

Feasprack, Ron, VACU-RAY CORPORATION, P.O. Box 291, Whitehouse Station, NJ, 08889, (201) 534-5471, (4).

Feinlab, Manning, U.S. DEPARTMENT OF HEALTH AND HUMAN SERVICES (HHS)/PUBLIC HEALTH SERVICE(PHS)/CENTER FOR HEALTH STATISTICS(CHS), 3700 East-West Hwy., Hyattsville, MD, 20872, (301) 436-8500, (8).

Feist, Ernest W., INFOTECH HOME INSPECTION SERVICES, 984 Route 9, Parlin, NJ, 08859, (201) 727-5335, (1).

Fekder, Henry C., U.S. DEPARTMENT OF HOUSING AND URBAN DEVELOPMENT (HUD)/OFFICE OF POLICY DEVELOPMENT AND RESEARCH(OPDR), 451 7th St. SW Room 8320, Washington, DC, 20410, (202) 755-5600, (8).

Feldman, Carl, CARL HENRY AND COMPANY, INC., Benson East Suite 207B, Jenkintown, PA, 19046, (215) 576-7766, (1).

Feldman, Dorothy, CARL HENRY AND COMPANY, INC., Benson East Suite 207B, Jenkintown, PA, 19046, (215) 576-7766, (1).

Feldman, Jack, CARL HENRY AND COMPANY, INC., Benson East Suite 207B, Jenkintown, PA, 19046, (215) 576-7766, (1).

Feldstein, Milton, BAY AREA AIR QUALITY MANAGEMENT DISTRICT, 939 Eillis St., San Francisco, CA, 94109, (415) 771-6000, (10).

Felleisen, Lou, U.S. ENVIRONMENTAL PROTECTION AGENCY REGION 3 (EPA3)/RADIATION DIVISION(ORP)/RADON PROJECT, 841 Chestnut St. MS-3AM12, Philadelphia, PA, 19107, (215) 597-1256, (8).

Fellows, Larry D., ARIZONA BUREAU OF GEOLOGY AND MINERAL TECHNOLOGY, 845 North Park Ave., Tucson, AZ, 85719, (602) 621-7906, (10).

Fenske, Dave, RADON INFORMATION AND TESTING SERVICE, 29 Appleton St., Manchester, NH, 03104, (603) 627-7261, (1).

Fenwick, Jerry, NUTECH ENERGY SYSTEMS, INC., 124 Newbold Ct., London, ON, N6E 1Z7, (519) 686-0797, (4).

Feola, Frank, ATLANTIC ELECTRIC, 65 W. Highland Ave., Atlantic Highlands, NJ, 07716, (201) 291-3468, (1).

Ferguson, A., LOS ANGELES COUNTY DEPARTMENT OF HEALTH SERVICES/RADIATION MANAGEMENT, 2615 South Grand Ave., Los Angeles, CA, 90007, (213) 744-3244, (10).

Ferguson, Robert, NEW JERSEY ASSOCIATION OF REALTORS, P.O. Box 2098, Edison, NJ, 08818, (201) 494-5616, (18).

Ferraiuolo, Robert, HUDSON REGIONAL HEALTH COMMISSION, 215 Harrison Ave., Harrison, NJ, 07029, (201) 485-7001, (10).

Ferreri, George P., MARYLAND DEPARTMENT OF THE ENVIRONMENT/AIR MANAGEMENT ADMINISTRATION, 201 West Preston St., Baltimore, MD, 21201, (301) 225-5235, (10).

Field, Bob, CAPITAL MATERIALS TESTING, INC., P.O. Box 81 5 Hemlock St., Latham, NY, 12110, (518) 783-3150, (800) 541-4109, (1)(2)(4).

Fields, Ted, HEALTH PHYSICS ASSOCIATES, LTD., 3312 Commercial Ave., Northbook, IL, 60062-1909, (312) 480-1900, (1)(3)(4).

Fields, Theodore, FGHB ASSOCIATES, 1141 Hohlfelder Rd., Glencoe, IL, 60022, (312) 835-4433, (3).

Fields, Theodore, GENERAL ENVIRONMENTAL MONITORING, 3312 Commercial Ave., Northbrook, IL, 60062-1909, (312) 291-0772, (1).

Filbert, David, BALTIMORE COUNTY DEPARTMENT OF ENVIRONMENTAL PROTECTION AND RESOURCE MANAGEMENT, 401 Bosley Ave., Towson, MD, 21204, (301) 494-3775, (10).

Filley, Norman E., NATIONAL FUEL, 800 State St., Erie, PA, 16501, (814) 871-8000, (1).

Fimian, Keith S., RADONICS, 1764 Old Meadow Ln. Suite 125, McLean, VA, 22102, (703) 734-0050, (1).

Finerfrock, Diane, UTAH DEPARTMENT OF HEALTH/BUREAU OF RADIATION CONTROL, 288 North 1460 West P.O. Box 16690, Salt Lake City, UT, 84116-0690, (801) 538-6734, (10).

Fink, Linda, STAN A. HUBER CONSULTANTS, INC., 200 North Cedar Rd., New Lenox, IL, 60451, (815) 485-4433, (1).

Finn, George, FINN CONSULTANT ENGINEERS, INC., 227 North St., Dalton, MA, 01226, (413) 684-2019, (1)(2)(3).

Fiore, Joy, AMERICAN NUCLEAR SOCIETY/NEVADA-NEVADA SECTION, 7164 South Abbeyville Dr., Las Vegas, NV, 89119, (702) 794-7639, (14).

Firschein, H.E., RADONAIRE, P.O. Box 3702, Wayne, NJ, 07474, (201) 790-3252, (1)(2).

Fisher, Bryan, AMERICAN MONITORING AND ENGINEERING SERVICES, 200 High Tower Blvd. Suite 205, Pittsburgh, PA, 15205, (215) 547-5460, (609) 392-1422, (1).

Fisher, Cathy, BOXFORD RADON LABORATORY, 52 Wildmeadow Rd., Boxford, MA, 01921, (617) 887-9641, (1).

Fisher, Colleen M., NATIONAL APARTMENT ASSOCIATION (NAA), 1101 14th St., NW Suite 804, Washington, DC, 20005, (202) 842-4050, (13).

Fisher, D.P., U.S. DEPARTMENT OF ENERGY (DOE)/RICHLANDS OPERATIONS OFFICE(ROO)/PACIFIC NORTHWEST LABORATORY(PNL)/HEALTH PHYSICS DIVISION, Dosimetry Technology Section M/S K3-57, Richland, WA, 99352, (509) 375-6852, (8).

Fisher, Eugene, U.S. ENVIRONMENTAL PROTECTION AGENCY (EPA)/OFFICE OF AIR AND RADIATION(OAR)/OFFICE OF RADIATION PROGRAMS(ORP)/RADON DIVISION (RD), Mitigation Prevention And Quality Assurance Branch ANR-460 401 M St., SW, Washington, DC, 20460, (202) 475-9605, (8).

Fisher, John, BOXFORD RADON LABORATORY, 52 Wildmeadow Rd., Boxford, MA, 01921, (617) 887-9641, (1).

Fisher, Linda J., U.S. ENVIRONMENTAL PROTECTION AGENCY (EPA)/OFFICE OF POLICY, PLANNING AND EVALUATION(OPPE), 401 M St., SW PM-219, Washington, DC, 20460, (202) 382-4332, (8).

Fisher, Phillip T., NORTH CAROLINA REAL ESTATE COMMISSION, 1313 Navaho Dr., Raleigh, NC, 27619, (919) 733-9580, (10).

Fisher, Robert, HOME BUYERS EDGE, 880 Benton St., Lakewood, CO, 80214, (303) 238-6275, (1).

Fisher, William, TEXAS BUREAU OF ECONOMIC GEOLOGY, The University Of Texas, Austin Balcones Research Center University Station, Box X, Austin, TX, 78713-7508, (512) 471-1534, (10).

Fitzpatrick, F. Michael, ECOLOGICAL INTERACTIONS, INC., 96 East Main St., Rockaway, NJ, 07866, (201) 625-7700, (800) 543-2116, (1)(2)(4).

Fitzpatrick, Frank, BIOSPHERICS, INC., 4928 Wyaconda Rd., Rockville, MD, 20852, (301) 770-7700, (1).

Fitzpatrick, William, ACS HOVAL, 935 Lively Blvd., Wood Dale, IL, 60191, (312) 860-6800, (2).

Fix, Richard, CLEAN HARBORS ANALYTICAL SERVICES, INC., P.O. Box 327 325 Wood Rd., Braintree, MA, 02184, (617) 849-6070, (1).

Fixary, John A., JOHN A. FIXARY, INC., P.O. Box 490, Riverside, CT, 06878, (203) 622-6661, (1).

Fjeld, Robert A., CLEMSON UNIVERSITY/ENVIRONMENTAL SYSTEMS ENGINEERING, 501 Rhodes Engineering Research Center, Clemson, SC, 29634-0919, (803) 656-5569, (5).

Flaak, Robert, U.S. ENVIRONMENTAL PROTECTION AGENCY (EPA)/SCIENCE ADVISORY BOARD(SAB)/INDOOR AIR AND TOTAL HUMAN EXPOSURE SUBCOMMITTEE, 401 M St., SW A-101F, Washington, DC, 20460, (202) 382-2552, (8).

Flagg, Lou, AMERICAN SOCIETY OF HEATING, REFRIGERATING AND AIR CONDITIONING ENGINEERS (ASHRAE), 1791 Tullie Circle Rd., NE, Atlanta, GA, 30329, (404) 636-8400, (13).

Flaherty, Jim, WESTINGHOUSE ELECTRIC CORPORATION, P.O. Box 355, Pittsburgh, PA, 15230, (412) 374-5716, (4).

Flater, Donald A., IOWA DEPARTMENT OF PUBLIC HEALTH/DIVISION OF DISEASE PREVENTION/BUREAU OF RADIOLOGICAL HEALTH, Lucas State Office Building, Des Moines, IA, 50319, (515) 281-7781, (10).

Fleck, Dick, DEFLECTO CORPORATION, 7035 East 86th St., Indianapolis, IN, 46250, (317) 849-9555, (4).

Fleitz, Carrie, RADON FREE, 5666 Ashbrook, Toledo, OH, 43614, (419) 865-0129, (1)(2).

Fleitz, Tim, RADON FREE, 5666 Ashbrook, Toledo, OH, 43614, (419) 865-0129, (1)(2).

Fleming, G.D., ENVIRONMENTAL SYSTEMS SERVICE, LTD., P.O. Box 220, College Park, MD, 20740, (301) 779-0606, (1).

Flesher, Kim, AMERICAN NUCLEAR SOCIETY (ANS), 555 North Kensington, La Grange Park, IL, 60525, (312) 352-6611, (14).

Fletcher, Roland G., MARYLAND DEPARTMENT OF THE ENVIRONMENT/OFFICE OF TOXICS, ENVIRONMENTAL SAFETY AND HEALTH, 201 West Preston St. 7th Floor Mail Room, Baltimore, MD, 21201, (301) 333-3130, (800) 872-3666, (10).

Flick, Duane K., ALA OF NORTH DAKOTA, P.O. Box 5004, Bismarck, ND, 58502-5014, (701) 223-5613, (15).

Floeckher, Jim, PACKARD INSTRUMENT COMPANY, 2200 Warrenville Rd., Downers Grove, IL, 60515, (312) 969-6000, (800) 323-1891, (1)(4).

Flook, Barbara A., U.S. DEPARTMENT OF AGRICULTURE (USDA)/AGRICULTURE RESEARCH SERVICE(ARS)/RADIATION SAFETY(RS), Building 001 Room 226 BARC-West, Beltsville, MD, 20705, (301) 344-4492, (8).

Fong, Felix, NORTH CAROLINA DEPARTMENT OF HUMAN RESOURCES/DIVISION OF FACILITY SERVICES/RADIATION PROTECTION SECTION, 701 Barbour Dr., Raleigh, NC, 27603-2008, (919) 733-4283, (1)(10).

Fontana, Raymond, RADON TESTING LABORATORY OF NEW ENGLAND, P.O. Box 4508, Boston, MA, 02101, (617) 426-9224, (1).

Forbes, Dr. Robert, ALASKA DEPARTMENT OF NATURAL RESOURCES/DIVISION OF GEOLOGICAL AND GEOPHYSICAL SURVEYS, 794 University Ave., Fairbanks, AK, 99701, (907) 474-7147, (10).

Forchetti, Vince, FORCHETTI PSI, P.O. Box 1234, Morrisville, PA, 19067, (215) 493-1646, (609) 695-1229, (1)(2).

Ford, Bonnie C., PENNSYLVANIA STATE UNIVERSITY, THE/BREAZEALE NUCLEAR REACTOR/LOW LEVEL RADIATION MONITORING LABORATORY, 101 Academic Projects Building, University Park, PA, 16802, (814) 863-2647, (1)(5).

Ford, Donald R., ALA OF HAWAII, 245 North Kukui St., Honolulu, HI, 96817, (808) 537-5966, (15).

Ford, M. Jane, LINCOLN-LANCASTER COUNTY/AIR POLLUTION CONTROL AGENCY, 2200 St. Mary's Ave., Lincoln, NE, 68502, (402) 471-8021, (10).

Ford, Michael, WASHOE COUNTY DISTRICT HEALTH DEPARTMENT, 1001 East Ninth, Reno, NV, 89520, (702) 328-2400, (10).

Ford, Scott, ENGINEERING DEVELOPMENT, INC., 4850 Northpark, Colorado Springs, CO, 80918, (717) 599-9080, (4).

Foreman, G. Douglas, PROFESSIONAL HOME ANALYSIS, 1025 Northeast Sanchez Ave., Ocala, FL, 32670, (904) 694-3982, (1)(2).

Foreman, Jeff, PROFESSIONAL HOME ANALYSIS, 1025 Northeast Sanchez Ave., Ocala, FL, 32670, (904) 694-3982, (1)(2).

Fornshell, Randall, STRATA ENVIRONMENTAL SERVICE, INC., 401 East Douglas Suite 515, Wichita, KS, 67202, (316) 262-0002, (1)(3).

Forshey, David R., U.S. DEPARTMENT OF INTERIOR (DOI)/BUREAU OF MINES(BOM)/RESEARCH DIVISION, 2401 E St., NW, Washington, DC, 20241, (202) 634-1210, (8).

Forster, William, U.S. DEPARTMENT OF ENERGY (DOE)/OFFICE OF ENERGY RESEARCH(OER)/HEALTH AND ENVIRONMENTAL RESEARCH(OHER), ER-75, GTN, Washington, DC, 20545, (202) 353-5329, (8).

Fossa, A., NEW YORK STATE DEPARTMENT OF ENVIRONMENTAL CONSERVATION/BUREAU OF TOXICS AIR SAMPLING, 50 Wolf Rd., Albany, NY, 12223-3250, (518) 457-7454, (10).

Foster, John, RADON DETECTION SYSTEMS/BLOOMINGTON-NORMAL BRANCH, 1207 Stephens Rd., Normal, IL, 61761, (303) 444-5253, (800) 627-2366, (1).

Foster, John W., PROFESSIONAL RADON MEASUREMENTS, P.O. Box 791, Normal, IL, 61761, (309) 827-3377, (309)452-9904, (1).

Foster, Randy, PELICAN ENVIRONMENTAL CORPORATION, 42 Lehigh Rd., Wellesley, MA, 02181, (617) 235-5997, (1)(2)(3).

Foster, Susan, MONTANA BUREAU OF MINES AND GEOLOGY, Montana College of Mineral Science and Technology, Butte, MT, 59701, (406) 496-4180, (10).

Fourman, Louis, UNIVERSITY OF MAINE-ORONO/COOPERATIVE EXTENSION SERVICE, 100 Winslow Hall, Orono, ME, 04469, (207) 581-3188, (14).

Foute, Steven, DENVER/ CITY-COUNTY DEPARTMENT OF HEALTH AND HOSPITALS/ENVIRONMENTAL HEALTH SERVICE, 605 Bannock St., Denver, CO, 80204, (303) 893-6243, (10).

Fowler, C.S., SOUTHERN RESEARCH INSTITUTE, 2000 Ninth Ave., S. P.O. Box 55305, Birmingham, AL, 35255-5305, (205) 581-2000, (1)(5).

Fowler, Carl C., HOMEBUYERS INSPECTION SERVICE OF SOUTH WEST FLORIDA, 2062 Stella St., Ft. Myers, FL, 33901, (813) 334-6116, (1).

Fox, C. Lee, PIMA COUNTY HEALTH DEPARTMENT/AIR QUALITY CONTROL DISTRICT, 150 West Congress St., Tucson, AZ, 85701, (602) 792-8803, (10).

Fox, James D., TRI-STATE RADON REDUCTION, INC., 816 Ainsworth Dr., Linden, NJ, 07036, (609) 275-0155, (201) 925-8472, (1)(2).

Franceus, Ivan, VACU-RAY CORPORATION, P.O. Box

291, Whitehouse Station, NJ, 08889, (201) 534-5471, (4).

Franceus, Ivan A., ENERGY SAVINGS WAREHOUSE, INC., 3490 Hwy. 22 W., Somerville, NJ, 08876, (201) 534-6922, (2)(4).

Franceus, Paul, VACU-RAY CORPORATION, P.O. Box 291, Whitehouse Station, NJ, 08889, (201) 534-5471, (4).

Franceus, Paul E., ENERGY SAVINGS WAREHOUSE, INC., 3490 Hwy. 22 W., Somerville, NJ, 08876, (201) 534-6922, (2)(4).

Francis, Pam, SUMP AND SEWAGE PUMP MANUFACTURERS ASSOCIATION (SSPMA), 560 West Washington St. Suite 301, Chicago, IL, 60606, (312) 332-4146, (13).

Franks, James F., AMERICAN NUCLEAR SOCIETY/COLORADO-COLORADO SECTION, C/O S.M. Stoller Corporation Nuclear Power Experience 1919 14th St. Suite 500, Boulder, CO, 80302, (303) 449-7220, (14).

Franks, Larry, NEVADA DEPARTMENT OF HUMAN RESOURCES/HEALTH DIVISION/RADIOLOGICAL HEALTH SECTION, 505 East King St. Room 202, Carson City, NV, 89710, (702) 885-5394, (10).

Franks, Terry M., FRANKS, TERRY M., P.O. Box 5224, Louisville, KY, 40205, (603) 437-1425, (1).

Frawley, David B., RSA CORPORATION/RADON SERVICE AGREEMENT, 918 Woodley Rd., Dayton, OH, 45403, (513) 254-4102, (3).

Frazier, Dean, FRAZIER COMPANY, THE, 15106 Industrial Rd., Omaha, NE, 68144, (402) 330-3400, (1).

Freeborne, William, U.S. DEPARTMENT OF HOUSING AND URBAN DEVELOPMENT (HUD)/OFFICE OF POLICY DEVELOPMENT AND RESEARCH(OPDR)/DIVISION OF INNOVATIVE TECHNOLOGY AND SPECIAL PROJECTS(DITSP), 451 7th St., SW Room 8232, Washington, DC, 20410, (202) 755-0640, (8).

Freed, Ralph R., DELAWARE VALLEY RADON INSPECTIONS/ARSENAL BUSINESS CENTER, 5301 Tacony St. Box 221, Philadelphia, PA, 19137-2309, (215) 289-3300, (215) 947-1169, (1).

Freedman, Steven W., SAFE AND SOUND HOME INSPECTION CONTRACTORS, P.O. Box 195, Cedar Knolls, NJ, 07927, (201) 515-0013, (1).

Freels, Don, MICHIGAN ASSOCIATION OF REALTORS, P.O. Box 40725, Lansing, MI, 48901-7925, (517) 372-8890, (18).

Freeman, Dave, WHIO, 1414 Wilmington, Dayton, OH, 45420, (513) 259-2111

Freeman, Richard, FLORIDA DEPARTMENT OF HEALTH AND REHABILITATIVE SERVICES/DEPUTY SECRETARY FOR HEALTH/OFFICE OF TOXICOLOGY AND HAZARDS ASSESSMENT, 1317 Winewood Blvd. Building 6 Room 466, Tallahassee, FL, 32399-0700, (904) 488-2905, (10).

Freeman, Timothy, AMERICAN RADON TESTING LABS, INC., 21360 Center Ridge Rd. Suite 307, Rocky River, OH, 44116, (216) 331-0200, (1).

Fregelette, Kathy, AFI ENVIRONMENTAL/DIVISION OF BUFFALO CONSULTANTS, INC., 6103 Robinson Rd., Lockport, NY, 14094, (716) 625-8434, (1).

Freile, Herb, H & J FREILE HOME INSPECTIONS, INC., 99 Harrison Ave., North Plainfield, NJ, 07060, (201) 561-7147, (1).

Freitas, John V., ADIRONDACK ENVIRONMENTAL INVESTIGATIONS, P.O. Box 270, Cambridge, NY, 12816, (518) 677-3958, (1).

Fremont, Bob, GILMONT CONSTRUCTION COMPANY, P.O. Box 1268, Exton, PA, 19341, (215) 363-5460, (2).

French, Walter, NEBRASKA WESLEYAN UNIVERSITY/PHYSICS DEPARTMENT/RADON MEASUREMENTS LABORATORY, 5000 St. Paul Ave., Lincoln, NE, 68504, (402) 465-2252, (402) 465-2247, (1).

French, William B., YOUR ENERGY SERVICE, INC., 220 4F Elliston Pl., Nashville, TN, 37203, (615) 329-9747, (1).

Freudan, Neal B., ENVIROSCIENCE CONSULTANTS, INC., 66 Cedar St., Newington, CT, 06111, (203) 666-7167, (1)(3).

Freudenburg, Paul G., HOME INSPECTIONS OF MICHIGAN, 1007 St. Joseph Dr. P. O. Box 936, St. Joseph, MI, 49085, (616) 983-4663, (616) 983-7098, (1).

Freund, Alice, NEW JERSEY DEPARTMENT OF HEALTH/DIVISION OF ENVIRONMENTAL AND OCCUPATIONAL HEALTH/PUBLIC EMPLOYEE OCCUPATIONAL SAFETY AND HEALTH PROGRAM, CN-360, Trenton, NJ, 08625, (609) 984-1863, (10).

Fridg, Jr., Ralph J., R.J. FRIDG COMPANY, INC., 427 Hamill Rd., Indiana, PA, 15701, (412) 463-3225, (1).

Fried, Jean J., EUROPEAN INSTITUTE FOR WATER (EIW), 2, rue Boussingault, F-67000 Strasbourg, France, (011) 33-88-611631, (12).

Friedman, Dan, AMERICAN HOME SERVICE COMPANY, 57 South Grand Ave., Poughkeepsie, NY, 12603, (914) 473-1999, (1).

Friedman, Jay, RTS SYSTEMS, P.O. Box 3070, Gaithersburg, MD, 20878, (301) 258-7890, (1).

Friedman, Stanton T., SCIENCE AND TECHNOLOGY, INC. (SCAT), 79 Pembroke Crescent, Fredericton, NB, E3B 2V1, (506) 457-0232, (1)(3)(5).

Friel, John, MEDRAD, INC., Kappa Manor II 271 Kappa Dr., Pittsburgh, PA, 15238-2870, (412) 767-4600, (1)(3).

Frimerman, Bruce D., REAL ESTATE SUPPORT SERVICES, INC., 7825 Washington Ave., S., Minneapolis, MN, 55435, (612) 941-7558, (1).

Fritchie, Denis, MEMPHIS AND SHELBY COUNTY HEALTH DEPARTMENT/AIR POLLUTION CONTROL BRANCH, 814 Jefferson Ave., Memphis, TN, 38105, (901) 576-7775, (10).

Froehlich, R.C., HELIX ENVIRONMENTAL, INC., 416 Triangle Ave., Dayton, OH, 45419, (513) 298-2990, (1)(3).

Froehlich, Ralph A., HELIX ENVIRONMENTAL, INC., 416 Triangle Ave., Dayton, OH, 45419, (513) 298-2990, (1)(3).

Frohn, Jeffrey M., J.M. FROHN CORPORATION, 35 Bowdoin Dr., Milford, MA, 01757, (508) 478-4342, (1).

Fry, David, BRENTWOOD INDUSTRIES, INC., P.O. Box 605, Reading, PA, 19603, (215) 374-5109, (4).

Fry, Richard M., NORTH CAROLINA DEPARTMENT OF HUMAN RESOURCES/DIVISION OF FACILITY SERVICES/RADIATION PROTECTION SECTION, 701 Barbour Dr., Raleigh, NC, 27603-2008, (919) 733-4283, (1)(10).

Fryzel, Stan, ILLINOIS DEPARTMENT OF COMMERCE AND COMMUNITY AFFAIRS/DIVISION OF INDUSTRIAL SERVICES, State Of Illinois Center 100 West Randolph St. Suite 3-400, Chicago, IL, 60601, (800) 972-4140, (800) 972-4216, (10).

Fuente, Eddie S., MISSISSIPPI STATE DEPARTMENT OF HEALTH/DIVISION OF RADIOLOGICAL HEALTH, 3150 Lawson St. P.O. Box 1700, Jackson, MS, 39215-1700, (601) 354-6657, (10).

Fulp, Robert R., FORSYTH COUNTY ENVIRONMENTAL AFFAIRS DEPARTMENT, 537 North Spruce St., Winston-Salem, NC, 27101, (919) 727-8060, (10).

Funk, Harry, GEOCEL CORPORATION, P.O. Box 398, Elkhart, IN, 46515, (219) 264-0645, (800) 348-7615, (4).

Funkhauser, Michael, U.S. DEPARTMENT OF DEFENSE (DOD)/DEPARTMENT OF THE ARMY(DOA)/BELVOIR RESEARCH, DEVELOPMENT AND ENGINEERING CENTER/MATERIALS, FUELS AND LUBRICANTS LABORATORY, Radiation Research Group, Ft. Belvoir, VA, 22060, (703) 664-5813, (703) 664-4175, (8).

Funkhauser, Rande, CONTAMINATION MANAGEMENT ENGINEERING, P.O. Box 871, Greenbrook, NJ, 08812, (201) 752-0078, (2).

Furman, Richard A., UNIVERSITY OF FLORIDA/COLLEGE OF ARCHITECTURE/SCHOOL OF BUILDING CONSTRUCTION, AFC-101, Gainesville, FL, 32611, (904) 392-5965, (5).

Furniss, William E., BRISTOL-BURLINGTON HEALTH DEPARTMENT, 240 Stafford Ave., Bristol, CT, 06010-4617, (203) 584-7682, (10).

G

Gable, Tom S., NATIONAL SANITATION FOUNDATION (NSF), P.O. Box 1468, Ann Arbor, MI, 48016, (313) 769-8010, (15).

Gabrilson, Tom, RADON TESTING AND MITIGATION SERVICES COMPANY, 5442 Cary Ave., Davenport, IA, 52807, (319) 386-9005, (1)(2).

Gaeta, Neil A., HEALTH AND SAFETY ASSOCIATES, 35 Grove St., Medford, MA, 02155, (617) 488-7081, (1)(3)(6).

Galbraith, Susan, COGITO TECHNICAL SERVICES, INC., 1799 Rte. 13 P.O. Box 86, Sheds, NY, 13151, (315) 662-7529, (3)(6).

Galeota, Anthony, AREA HOME INSPECTIONS, INC., 341 Littleton Rd., Harvard, MA, 01451, (617) 263-7251, (1).

Galeski, James, HARVARD ENGINEERING SERVICES COMPANY, 28 Shaker Rd., Harvard, MA, 01451, (617) 456-3716, (1)(2).

Galeski, Lorelei, HARVARD ENGINEERING SERVICES COMPANY, 28 Shaker Rd., Harvard, MA, 01451, (617) 456-3716, (1)(2).

Galinski, E. Richard, R.I.C. INSPECTION AND ENGINEERING, R.R. 2 P.O. Box 2186-5, Columbia, NJ, 07832, (201) 496-4818, (1).

Gallagher, Charles T., ACCURATE AIR TESTING, 1209 Piney Woods Dr., Friendswood, TX, 77546, (713) 482-5097, (1).

Gallagher, Charles T., NUCLEAR SOURCES AND SERVICES, INC., P.O. Box 34042, Houston, TX, 77234, (713) 641-0391, (1).

Gallagher, Joe, PARAMOUNT PEST CONTROL, 110 Gordon Rd., Falls Church, VA, 22046, (703) 241-9140, (1).

Gallagher, John, GALLAGHER AND ASSOCIATES, INC., 950 Sussex Blvd., Broomall, PA, 19008, (215) 328-2200, (4).

Gallagher, Michael J., GALLAGHER ENGINEERING

AND INSPECTION, 1164 Lincoln Heights Ave., Ephrata, PA, 17522, (717) 733-3734, (1).

Gallagher, R.G., APPLIED HEALTH PHYSICS, INC., 2986 Industrial Blvd., Bethel Park, PA, 15102, (412) 563-2242, (518) 477-7974, (1)(2)(4)(6).

Gallagher, Robert D., ASSAY SERVICES, P.O. Box 1266, Friendswood, TX, 77546, (713) 641-0391, (1).

Galstezer, Tom, FERNCO, INC., 300 S. Dayton, Davison, MI, 48423, (313) 653-9626, (800) 521-1283, (4).

Gamewell, Richard, COLORADO DEPARTMENT OF HEALTH/RADIATION CONTROL DIVISION, 4210 East 11th Ave., Denver, CO, 80220, (303) 331-4812, (10).

Gane, Jim, B-DRY SYSTEM OF DELAWARE AND CHESTER COUNTY, INC., 1054 Saunders Ln., West Chester, PA, 19380, (215) 430-0393, (1).

Gannon, Ray J., MIDWEST RADON ANALYSIS, INC., P.O. Box 33093, Cleveland, OH, 44133, (216) 582-4848, (1).

Gardner, Margaret, NORTHWEST POWER PLANNING COUNCIL (NPPC)/POWER PLANNING DIVISION, 851 SW Sixth Ave. Suite 1100, Portland, OR, 97204-1337, (800) 222-3355, (13).

Gardner, Patricia, NEW JERSEY DEPARTMENT OF ENVIRONMENTAL PROTECTION/DIVISION OF ENVIRONMENTAL QUALITY/BUREAU OF RADIATION AND INORGANIC ANALYTICAL SERVICES, 380 Scotch Rd. CN-411, Trenton, NJ, 08628, (609) 530-4116, (1)(10).

Gardner, Philip, GARDNER, PHILIP, 136B Paoli Pike, Malvern, PA, 19355, (215) 889-0239, (1).

Garey, Carol, LIVINGSTON-STEUBEN-WYOMING BOARD OF COOPERATIVE EDUCATIONAL SERVICES (BOCES), Lackawanna Ave., Mt. Morris, NY, 14510, (716) 658-2291, (10).

Gargaro, Vincent, UCI/ENVIROTEST, 34100 Center Ridge Rd., North Ridgeville, OH, 44039, (216) 327-9432, (3).

Gargaro, Vincent, UCI/ENVIROTEST, 834 Old Thorn Run Rd., Coraopolis, PA, 15108, (412) 264-5891, (1)(3).

Gargaro, Vincent A., ENVIROTEST, P.O. Box 223, North Ridgeville, OH, 44039-0223, (216) 327-9432, (1).

Garland, Robert, KEY TECHNOLOGY, INC., P.O. Box 562, Jonestown, PA, 17038, (717) 274-8310, (800) 523-4964, (1)(4)(6).

Garlington, William D., CRIPPEN CONSULTING ASSOCIATES, INC., 242 North James St. Suite 202, Wilmington, DE, 19804, (302) 994-8000, (1)(2)(3).

Garris, Donna, RDS/RADON DETECTION SERVICES-PENNSYLVANIA OFFICE, 1011 Brookside Rd., Suite 270 P.O. Box 270, Allentown, PA, 18106, (215) 481-9555, (1)(2).

Garstang, M., SIMPSON WEATHER ASSOCIATES, 809 East Jefferson St., Charlottesville, VA, 22902, (804) 979-3571, (1)(2)(3).

Gartrell, Patty, ARIX CORPORATION, 877 North Eighth, W. Suite 8, Riverton, WY, 82501, (307) 858-6506, (1)(2).

Gascoyne, Melvyn, ATOMIC ENERGY OF CANADA, LTD. (AECL)/WASTE MANAGEMENT/APPLIED GEOSCIENCE/HYDROGEOCHEMISTRY SECTION, Tinawa, MB, R0E 1L0, (204) 753-2311, (11).

Gaskill, Jack L., BAYSURE BUILDING INSPECTION, INC., 290 King of Prussia Rd., Radnor, PA, 19087, (215) 688-2572, (1).

Gasparro, Ronald J., RADON SAFETY AND TECHNOLOGY CORPORATION, 4 Charles St., Pleasant Valley, NY, 12569, (914) 635-1138, (800) 552-5257, (1)(2)(3)(4).

Gatner, Jason, PENNSYLVANIA DEPARTMENT OF ENVIRONMENTAL RESOURCES/BUREAU OF RADIATION PROTECTION/DIVISION OF SPECIAL PROJECTS, Fulton Building, 16th Floor Third and Locust Sts., Harrisburg, PA, 17120, (717) 787-2480, (6)(10).

Gatzemeier, Ed, MONTANA DEPARTMENT OF LABOR AND INDUSTRY/BUREAU OF SAFETY AND HEALTH/DIVISION OF WORKERS' COMPENSATION, 5 South Last Chance Gulch, Helena, MT, 59601, (406) 444-6401, (10).

Gaul, Wayne, AMERICAN NUCLEAR SOCIETY/KANSAS-KANSAS SECTION, MS6-02 P.O. Box 2908, Wichita, KS, 67201, (316) 261-6799, (14).

Gaul, Wayne, RADON AND ENVIRONMENTAL PROFESSIONALS, 2768 South Bennett, Wichita, KS, 67217-1812, (316) 945-2570, (1).

Gauthreaux, E.D., TERMINIX INTERNATIONAL, 855 Ridgelake Blvd., Memphis, TN, 38119, (800) 937-3783, (1).

Gavigan, George, SOUTHERN NEW ENGLAND RADON TESTING AND MITIGATION, P.O. Box 1141, Westbrook, CT, 06498, (203) 399-8003, (1)(2).

Gavila, Frank M., F & J SPECIALTY PRODUCTS, INC., P.O. Box 660065, Miami Springs, FL, 33166, (305) 888-0383, (800) 832-5037, (4).

Gavin, Gerard M., GAVIN, GERARD M., 214 East Township Line Rd., Havertown, PA, 19083, (215) 789-2718, (1).

Gay, Kim, CUTTER INFORMATION CORPORATION, 1100 Massachusetts Ave., Arlington, MA, 02174, (617) 648-8700, (3).

Gazzier, Conrad, MISSISSIPPI STATE DEPARTMENT OF NATURAL RESOURCES/BUREAU OF GEOLOGY, P.O. Box 5348, Jackson, MS, 39216, (601) 354-6228, (10).

Geachis, Richard, RADON AIR AND WATER CHECK, INC., 29 Riverside Dr., Binghamton, NY, 13905, (607) 724-5050, (1).

Gearo, Joseph, U.S. ENVIRONMENTAL PROTECTION AGENCY (EPA)/OFFICE OF AIR AND RADIATION(OAR)/OFFICE OF RADIATION PROGRAMS(ORP)/RADON DIVISION (RD), Mitigation Prevention And Quality Assurance Branch ANR-460 401 M St., SW, Washington, DC, 20460, (202) 475-9605, (8).

Gecewizcz, Thomas, COMMUNITY HEALTH ASSOCIATES, INC., 316 Sherman St., Canton, MA, 02021, (617) 828-3123, (1)(2).

Geddis, Robert, AIRREGS, INC./AIR POLLUTION REGULATORY SERVICE, INC., 1640 Powers Ferry Rd. Building 17, Suite 210, Atlanta, GA, 30067, (404) 984-1919, (1)(2)(6).

Gehle, Karen, NEBRASKA REALTORS ASSOCIATION, 1115 K St. Suite 102, Lincoln, NE, 68508, (402) 475-6334, (18).

Gehlke, Scott, ION SYSTEMS, INC., 2546 10th St., Berkeley, CA, 94710, (415) 548-3640, (4).

Geier, Robert J., GEIER CONSTRUCTION COMPANY, INC./DIVISION OF RADON CONTROL, 1506 Gridley Ln., Silver Spring, MD, 20902, (301) 649-2141, (1).

Geiger, Eric L., EBERLINE INSTRUMENT CORPORATION, P.O. Box 2108, Santa Fe, NM, 87504-2108, (505) 471-3232, (1)(4).

Geiger, Jeffrey R., SPENCE-GEIGER ASSOCIATES, INC., 1658 Cole Blvd. #260, Golden, CO, 80401, (303) 233-8577, (1)(3).

Geiger, Kelsie, RADON QC, 2857 Nazareth Rd. Palmer Professional Building Suite 6, Palmer, PA, 18043, (215) 559-1636, (1)(3)(4).

Geisler, Dennis, UCI/ENVIROTEST, 34100 Center Ridge Rd., North Ridgeville, OH, 44039, (216) 327-9432, (3).

Geisler, Dennis, UCI/ENVIROTEST, 834 Old Thorn Run Rd., Coraopolis, PA, 15108, (412) 264-5891, (1)(3).

Geller, Howard, AMERICAN COUNCIL FOR AN ENERGY EFFICIENT ECONOMY (ACEEE), 1001 Connecticut Ave., NW Suite 535, Washington, DC, 20036, (202) 429-8873, (13).

Gemmill, Wayne, EBERLINE INSTRUMENT CORPORATION, P.O. Box 2108, Santa Fe, NM, 87504-2108, (505) 471-3232, (1)(4).

Gentile, Nancy, REALTORS NATIONAL MARKETING INSTITUTE (RNMI), 430 North Michigan Ave., Chicago, IL, 60611, (312) 670-3780, (18).

George, Andreas, U.S. DEPARTMENT OF ENERGY (DOE)/CHICAGO OPERATIONS OFFICE(COO)/ENVIRONMENTAL MEASUREMENTS LABORATORY (EML), 376 Hudson St., New York, NY, 10014-3621, (212) 620-3609, (5)(8).

George, Danny, NATIONAL ASSOCIATION OF HOME BUILDERS OF THE UNITED STATES (NAHB), 15th and M Sts., NW, Washington, DC, 20005, (202) 822-0200, (13).

Georges, Peter W., NATIONWIDE TESTING, RD 1, Box 266B, Annandale, NJ, 08801, (201) 735-2770, (1).

Geren, Charlie, A.T.E.I., 809 East 11th St., Michigan City, IN, 46360, (800) 833-0999, (1)(2)(6).

Gerger, Stuart G., DEGENSHEIN DENKER, 10 South Broadway, Nyack, NY, 10960, (1).

Gerhard, Lee C., KANSAS GEOLOGICAL SURVEY, The University Of Kansas 1930 Constant Ave. Campus West, Lawrence, KS, 66046, (913) 864-3965, (10).

Gerusky, Thomas M., PENNSYLVANIA DEPARTMENT OF ENVIRONMENTAL RESOURCES/BUREAU OF RADIATION PROTECTION, P.O. Box 2063, Harrisburg, PA, 17120, (717) 787-2480, (10).

Gessner, Guy, HOME DETOX SPECIALISTS, 81 Candy Ln., Hamburg, NY, 14075, (716) 648-3321, (1)(2).

Ghahremani, Darioush T., RADON SURVEY SYSTEMS, INC., P.O. Box 22288, Cleveland, OH, 44122, (216) 464-7997, (1)(2)(3)(4)(6).

Ghahremani, Darioush T., TERRADON EXPLORATION TECHNOLOGIES, INC., 3931 White Oak Tr. P.O. Box 22288, Cleveland, OH, 44122, (216) 464-2002, (1)(2)(4)(5)(6).

Ghavi, Mahmoud, QUANTUM TECHNOLOGY, INC., 830 Franklin Ct., Marietta, GA, 30067, (404) 429-8324, (4).

Giardina, Paul, U.S. ENVIRONMENTAL PROTECTION AGENCY REGION 2 (EPA2)/RADIATION DIVISION(ORP), 26 Federal Plaza, New York, NY, 10278, (212) 264-4418, (8).

Gibsson, RADON DETECTION SYSTEMS/GRAND JUNCTION, COLORADO BRANCH, 235 Radcliff Dr., Fruita, CO, 81521, (303) 444-5253, (800) 627-2366, (1).

Gibson, Donald, GEOCHEM, INC., 3200 New Seward Hwy. Suite 201, Anchorage, AK, 99503, (907) 272-2121, (1)(2).

Gibson, Paul, RADON-X CORPORATION, 3 Edinburgh Ln., Mt. Laurel, NJ, 08054, (609) 778-7771, (800) 272-3669, (1)(2)(4).

Giddings, Todd, GIDDINGS AND ASSOCIATES, INC., 3049 Enterprise Dr., State College, PA, 16801, (814) 238-5927, (1)(2).

Giegengack, Robert, UNIVERSITY OF PENNSYLVANIA/DEPARTMENT OF GEOLOGY, 138 West Rose Valley Rd., Wallingford, PA, 19086, (215) 898-5191, (5).

Gilbert, Gary J., GARCO RESEARCH COMPANY, 144A Sutton Rd., Lebanon, NJ, 08833, (201) 832-7895, (1)(2).

Gilbert, John, BOULDER INSPECTION COMPANY, P.O. Box 3369, Boulder, CO, 80307, (303) 444-6657, (1)(4).

Gilchrist, Chris, MINNESOTA DEPARTMENT OF PUBLIC SERVICE/ENERGY INFORMATION CENTER, 900 American Center Building 150 East Kellogg Blvd., St. Paul, MN, 55101, (612) 297-2291, (800) 652-9747, (10).

Gilchrist, James, AMERICAN MINING CONGRESS (AMC), 1920 N St., NW Suite 300, Washington, DC, 20036, (202) 861-2879, (16).

Giles, Jr., Phillip, AMERICAN NUCLEAR SOCIETY/MISSISSIPPI-MISSISSIPPI SECTION, 110 Bottino Dr., Vicksburg, MS, 39180-9721, (601) 437-6214, (14).

Gillespie, Bruce M., CANBERRA INDUSTRIES, INC., 1 State St., Meriden, CT, 06450, (203) 238-2351, (1)(4).

Gillis, Jr., Charles L., FREDERICK COUNTY HEALTH DEPARTMENT, 12 East Church St., Frederick, MD, 21701, (301) 694-1700, (10).

Gilmartin, Thomas F., TOWN OF HAMBURG, S-6100 South Park Ave., Hamburg, NY, 14075, (716) 649-6111, (10).

Gilmore, John, WEST COAST ENVIRONMENTAL SERVICES, INC., P.O. Box 18182, Seattle, WA, 98118, (206) 324-0920, (1).

Gilroy, Daniel, CDM FEDERAL PROGRAMS CORPORATION, 13135 Lee Jackson Memorial Hwy., Fairfax, VA, 22033, (703) 968-0900, (1).

Gilvary, Kenneth, RADON MITIGATORS OF ARIZONA, 6844 East Thunderbird Rd., Scottsdale, AZ, 85254, (602) 483-7506, (1)(2).

Gilvary, Kenneth, RADON MITIGATORS, INC., 16 Hillside Dr., Clinton, NJ, 08809, (201) 730-6596, (1)(2).

Girard, Tom, RADON CONTROL SYSTEMS, 138 Riverview Plantation Dr., Williamsburg, VA, 23185, (804) 564-1893, (2).

Giterman, Michael, GITERMAN ASSOCIATES, 2912 Xenwood, St. Louis Park, MN, 55416, (612) 332-1129, (1).

Glantz, Leif, GAMMADATA/MATTEKNIK AB, Klostergatan 10 Box 1815, Uppsala S-75148, Sweden, 46-18155870, (4).

Glass, Gary, WYOMING GEOLOGICAL SURVEY, P.O. Box 3008 University Station, Laramie, WY, 82071, (307) 742-2054, (5)(10).

Glenn, Robert, U.S. DEPARTMENT OF HEALTH AND HUMAN SERVICES (HHS)/PUBLIC HEALTH SERVICE(PHS)/CENTERS FOR DISEASE CONTROL(CDC)/NATIONAL INSTITUTES FOR OCCUPATIONAL SAFETY AND HEALTH(NIOSH), Division Of Respiratory Disease Studies 944 Chestnut Ridge Rd., Morgantown, WV, 26505, (304) 291-4474, (8).

Glogore, Robert, GLOGORE ENGINEERING COMPANY, 3116 Stanley Rd., Akron, OH, 44313-3254, (216) 836-1880 (800) 702-9905, (1).

Glyde, Toby B., EGAL, INC., P.O. Box 14983 11633 West 83 Terr., Kansas City, KS, 66215, (913) 541-1862, (1).

Goard, Robert W., A & A RADON TESTING SERVICES, INC., 1805 South Bellaire St., Denver, CO, 80222, (303) 753-1044, (1).

Goard, Robert W., COLORADO RADON INSTITUTE, 4675 East Louisiana Suite 114, Denver, CO, 80222, (303) 756-0701, (1).

Goble, Robert L., CLARK UNIVERSITY/ENVIRONMENT, TECHNOLOGY, AND SOCIETY PROGRAM/DEPARTMENT OF PHYSICS, 950 Main St., Worcester, MA, 01610, (617)793-7687, (1)(5).

Godbeer, Barrie C., GODBEER, GODBEER AND ASSOCIATES, P.O. Box 128, New London, NH, 03257, (603) 526-9235, (1).

Goddard, John, ROCKY MOUNTAIN MEDICAL PHYSICS, INC., 7175 West Jefferson Ave. Suite 2150, Lakewood, CO, 80235, (303) 980-7070, (1)(3).

Goeller, Curt, OKLAHOMA CITY-COUNTY HEALTH DEPARTMENT/AIR QUALITY CONTROL SECTION, 921 NE 23rd St., Oklahoma City, OK, 73105, (405) 427-8651, (10).

Gogats, Robert, ENVIRONMENTAL HEALTH INSPECTIONS, INC., 1200 Rte. 9 Cross River Mill Suite 5E, Lakewood, NJ, 08701, (201) 905-8888, (1)(3).

Golack, Janice S., U.S. DEPARTMENT OF HOUSING AND URBAN DEVELOPMENT (HUD)/DEPUTY ASSISTANT SECRETARY FOR PROGRAM DEVELOPMENT, 415 7th St., SW Room 7212, Washington, DC, 20410, (202) 755-5566, (8).

Goldman, Henry, GOLDEN CONSTRUCTION SERVICE, 20 Distillery Rd., Warwick, NY, 10990, (914) 986-5600, (1).

Goldman, Ralph, ASHRAE/ENVIRONMENTAL HEALTH COMMITTEE/INDOOR AIR QUALITY SUBCOMMITTEE, 1791 Tullie Cr., NE, Atlanta, GA, 30329, (404) 636-8400, (13).

Goldnick, Robert, NEW YORK STATE DEPARTMENT OF LABOR/DIVISION OF SAFETY AND HEALTH, Building 12 New York State Office Building Campus, Albany, NY, 12240, (518) 457-2574, (10).

Goldstein, Marvin, BUILDING INSPECTION SERVICES, INC., 1675 Winding Rd., Southampton, PA, 18966, (215) 357-8055, (1).

Goldstein, W. David, AMBIC BUILDING INSPECTION CONSULTANTS, INC., 1200 Rte. 130, Robbinsville, NJ, 08691, (609) 426-1212, (1).

Goldsworthy, John, RADON MANAGEMENT, INC., 474 Easton Rd., Horsham, PA, 19044, (215) 672-8129, (1).

Gollnick, Dan, PACIFIC RADIATION CORPORATION, 2945 Stonehill Dr., Altadena, CA, 91001, (818) 798-8100, (1).

Golub, Saul J., MARTIN BUILDERS COMPANY, INC., THE, 7114 Rising Sun Ave., Philadelphia, PA, 19111, (215) 745-5398, (2).

Goodall, Steve, RADON FREE, 5666 Ashbrook, Toledo, OH, 43614, (419) 865-0129, (1)(2).

Goodell, Grant, UNIVERSITY OF VIRGINIA/DEPARTMENT OF ENVIRONMENTAL SCIENCES, Clark Hall, Charlottesville, VA, 22903, (804) 924-0559, (5).

Goodell, H.G., SIMPSON WEATHER ASSOCIATES, 809 East Jefferson St., Charlottesville, VA, 22902, (804) 979-3571, (1)(2)(3).

Goodman, Penny, CONSTRUCTION SPECIFICATION INSTITUTE (CSI), 601 Madison St., Alexandria, VA, 22314, (703) 684-0300, (13).

Goodwin, Jessica, RADON TESTING AND TREATMENT, INC., 723 North Weber Suite 300, Colorado Springs, CO, 80903, (719) 520-5693, (1)(2)(4).

Goodwin, Richard W., ENVIRONMENTAL CONSULTANTS ASSOCIATES, 14 Ramapo Ln., Upper Saddle River, NJ, 07458, (201) 934-9866, (1).

Gorasi, Donald G., RAD-LAB SERVICES, INC., 1311 Washington Blvd., Williamsport, PA, 17701, (717) 326-2628, (4).

Gordon, Samuel, GORDON, SAMUEL, 146 S. Little Tor Rd., New City, NY, 10956, (914) 634-9331, (3).

Gorham, Michael B., COLORADO REAL ESTATE COMMISSION, 1776 Logan St. 4th Floor, Denver, CO, 80203, (303) 894-2166, (10).

Goriach, Fred, VIRGINIA DEPARTMENT OF HEALTH/DIVISION OF HEALTH HAZARDS CONTROL/BUREAU OF RADIOLOGICAL HEALTH, Northern Regional Office 9504-A Lee Highway, Fairfax, VA, 22031, (703) 430-8609, (10).

Gornto, Gary, REEDY CREEK IMPROVEMENT DISTRICT/POLLUTION CONTROL DEPARTMENT, P.O. Box 36, Lake Buena Vista, FL, 32830, (305) 824-3460, (10).

Goswami, Vinod K., UNIVERSITY HOME HEALTH SERVICES, INC., 670 Centre St. Suite 9, Jamaica Plain, MA, 02130, (617) 524-2149 (617) 522-2793, (1).

Gotchy, Reginald, SCIENCE APPLICATION INTERNATIONAL CORPORATION, 1710 Goodridge Dr., McLean, VA, 22102, (703) 821-4300, (3).

Gotthardt, Peter R., SUNSTREAM DEVELOPMENT CORPORATION, 300 North St., Endicott, NY, 13760, (607) 754-6200, (1)(2)(4).

Gottlieb, S.C., RADON AIR AND WATER CHECK, INC., 29 Riverside Dr., Binghamton, NY, 13905, (607) 724-5050, (1).

Gottschalk, Barry, ALA OF KENTUCKY, P.O. Box 969, Louisville, KY, 40201-0969, (502) 363-2652, (15).

Goy, Terry, GOY, TERRY, 1466 East Lake Dr., Chaska, MN, 55318, (612) 448-3664, (1).

Grafton, Harry E., COLUMBUS HEALTH DEPARTMENT, 1815 Washington Blvd., Columbus, OH, 43215, (614) 222-6992, (10).

Graham, David, AQUATEK LABORATORY/DIVISION OF GRAHAM-MASSEY ANALYTICAL LABORATORY, 2212 Main St., Bridgeport, CT, 06606, (203) 576-0104, (1).

Graham, Ellen H., VERMONT OFFICE OF THE SECRETARY OF STATE/REAL ESTATE COMMISSION, Pavilion Office Building, Montpelier, VT, 05602, (802) 828-3228, (10).

Graham, Thomas C., LANIDEX CORPORATION, 300 Lanidex Plaza, Parsippany, NJ, 07054, (201) 386-5800, (1).

Gralinski, E. Richard, RIC INSPECTION AND ENGINEERING, R.R. 2 P.O. Box 2186-5, Columbia, NJ, 07832, (201) 496-4818, (1).

Grammer, David C., NORTH EAST ENVIRONMENTAL SERVICES, 424 Mine Rd., Asbury, NJ, 08802, (201) 735-6080, (800) 447-2366, (1).

Grandholm, Mark, CONTROL POWER SYSTEMS, INC., 1010 E. 31 St., Hialeah, FL, 33013, (305) 691-2260, (4).

Granlund, Carl, PENNSYLVANIA DEPARTMENT OF ENVIRONMENTAL RESOURCES/BUREAU OF RADIATION PROTECTION/DIVISION OF SPECIAL PROJECTS, Fulton Building, 16th Floor Third and Locust Sts., Harrisburg, PA, 17120, (717) 787-2480, (6)(10).

Grant, Norman, QUOIN PARTNERSHIP, ARCHITECTS AND ENGINEERS, P.O. Box 147, St. Peters, PA, 19470, (215) 469-9595, (2).

Grant, Philip J., WASTREN, INC., 2 Waverly Dr., Hummelstown, PA, 17036, (717) 566-6780, (3).

Grantham, Rhonda, RADIOLOGICAL TESTING SERVICES, HC92, Box2B, Crawford, NE, 69339, (308) 665-1587, (1).

Graser, William F., U.S. DEPARTMENT OF INTERIOR (DOI)/U.S. GEOLOGICAL SURVEY(USGS)/NATIONAL CARTOGRAPHIC INFORMATION CENTER-ROCKY MOUNTAIN(NCIC), Box 25046, Denver Federal Ctr. MS-504, Denver, CO, 80225, (303) 236-5829, (8).

Graves, Paul, NATKIN SERVICE COMPANY, 2241 West Altorfer, Peoria, IL, 61615, (309) 691-9206, (800) 642-8391, (1).

Graves, Thomas, GRAVES RADON CONTRACTING, P.O. Box 266, Terre Hill, PA, 17581, (215) 445-7886, (1)(2).

Gray, Fred O., SPOKANE COUNTY AIR POLLUTION CONTROL AUTHORITY, 1101 West College Ave. Room 230, Spokane, WA, 99201, (509) 456-4727, (10).

Gray, Michael P., IDAHO REAL ESTATE COMMISSION, Statehouse Mail, Boise, ID, 83720, (208) 334-3285, (10).

Gray, Robert, OKLAHOMA REAL ESTATE COMMISSION, 4040 North Lincoln Blvd. Suite 100, Oklahoma City, OK, 73105, (405) 521-3387, (10).

Gray Loveland, David, LEAGUE OF WOMEN VOTERS EDUCATIONAL FUND/NATURAL RESOURCES, 1730 M St., NW, Washington, DC, 20036, (202) 429-1965, (17).

Greco, Joseph M., PRINCETON PLASMA PHYSICS LABORATORY, P.O. Box 451, Princeton, NJ, 08543, (609) 683-3319, (609) 243-3319, (1).

Green, George, GREEN, GEORGE, RD 2, Box 2621 Big Hollow Rd., Lake George, NY, 12845, (518) 668-3233, (1).

Greenaway, Douglas, BUILDING OWNERS AND MANAGERS ASSOCIATION INTERNATIONAL (BOMA), 1250 I St., NW Suite 200, Washington, DC, 20005, (202) 289-7000, (13).

Greenbank, Clinton B., MODOC COUNTY AIR POLLUTION CONTROL DISTRICT, 202 West Fourth St., Alturas, CA, 96101, (916) 233-3939 x 401, (10).

Greenberg, Alan M., WAYNE COUNTY HEALTH DEPARTMENT/AIR POLLUTION CONTROL, 2211 East Jefferson, Detroit, MI, 48207, (313) 567-4100, (10).

Greenberg, Harvey, RADON ENGINEERING/A UNIT OF PSI ENGINEERING, INC., 1 Lethbridge Plaza, Mahwah, NJ, 07430, (201) 529-8300, (1)(2).

Greenberg, Merrill, CAPITAL CONSTRUCTION SERVICES, P.O. Box 568, Troy, NY, 12180, (518) 434-0105, (1).

Greenberg, Stuart, ENVIRONMENTAL HEALTH WATCH OF GREATER CLEVELAND, 4115 Bridge Ave., Cleveland, OH, 44113, (216) 961-4646, (1)(17).

Greene, Jack, U.S. ENVIRONMENTAL PROTECTION AGENCY (EPA)/OFFICE OF RESEARCH AND DEVELOPMENT(ORD)/AIR AND ENERGY ENGINEERING RESEARCH LABORATORY (AEERL), MD-49, Research Triangle Park, NC, 27711, (919) 541-2905, (8).

Greenfield, Richard S., NATIONAL SCIENCE FOUNDATION (NSF)/LOWER ATMOSPHERIC RESEARCH DIVISION, 1800 G St., NW, Washington, DC, 20550, (202) 357-8645, (8).

Greenstein, Jess D., H.P.J. INSPECTION SERVICE, INC., 1303 Woodbridge Commons Way, Iselin, NJ, 08830, (201) 855-9475, (1).

Gregg, L.T., ATLANTA TESTING AND ENGINEERING, INC., 11420 Johns Creek Pkwy., Duluth, GA, 30136, (404) 476-3555, (1).

Grether, John, THRESHOLD TECHNICAL PRODUCTS, 7225 Edington Dr., Cincinnati, OH, 45249, (800) 458-4931, (4).

Grew, Priscilla C., MINNESOTA GEOLOGICAL SURVEY, 2642 University Ave., St. Paul, MN, 55114-1057, (612) 627-4780, (5)(10).

Grigg, David, ENTESCO, 1910 Bobby Ray Rd., Clover, SC, 29710, (803) 222-4130, (1).

Grim, Kenneth, RADON SYSTEMS, INC., P.O. Box 2509, Laurel, MD, 20708, (301) 725-9000, (1)(2).

Grimsrud, David, ELECTRICAL POWER RESEARCH INSTITUTE (EPRI)/LAWRENCE BERKELEY LABORATORY(LBL)/APPLIED SCIENCE DIVISION, University of California One Cyclotran Rd., Berkeley, CA, 94720, (415) 486-4000, (5)(13).

Grisham, Martin, HOMES PLUS RADON SERVICES, INC., 490 Main St. P.O. Box 654, Norwich, CT, 06360, (203) 887-5128, (1)(4).

Grisham, Martin, NEW ENGLAND RADON SERVICES, 95 Vauxhall St., New London, CT, 06320, (203) 442-6675, (1)(3)(6).

Grissley, Michael, AMERICAN RADON TESTERS, 5705 Center Ln. Office 1, Falls Church, VA, 22041, (703) 239-0227, (1).

Groat, Charles, LOUISIANA GEOLOGICAL SURVEY/DEPARTMENT OF NATURAL RESOURCES, Box G University Station, Baton Rouge, LA, 70803, (504) 342-6754, (10).

Grob, A. Kenneth, MEDRAD, INC., 271 Kappa Dr., Pittsburgh, PA, 15238, (412) 782-4600, (3).

Grob, B., TELEDYNE ISOTOPES MIDWEST LABORATORY, 700 Landwehr Rd., Northbrook, IL, 60062, (312) 564-0700, (1).

Grodzins, Lee, MASSACHUSETTS INSTITUTE OF TECHNOLOGY/RADON PROJECT, Building 26 Room 421, Cambridge, MA, 02139, (617) 253-4151, (1)(5).

Grodzins, Lee, NITON CORPORATION, P.O. Box 9153, Lexington, MA, 02173, (617) 229-4929, (800) 451-6467, (1)(4).

Grodzins, Lee, NITONICS CORPORATION, P.O. Box 9153, Lexington, MA, 02173, (617) 229-4929, (800) 451-6467, (4).

Groer, Peter, U.S. DEPARTMENT OF LABOR (DOL)/MINE SAFETY AND HEALTH ADMINISTRATION(MSHA)/OAK RIDGE ASSOCIATED UNIVERSITIES(ORAU)/CENTER FOR EPIDEMILOLOGICAL RESEARCH, P.O. Box 117 210 Badger Rd., Oak Ridge, TN, 37831, (615) 576-3532, (8).

Groff, D., LMS ENGINEERS, 1 Blue Hill Plaza, Pearl River, NY, 10965, (914) 735-8300, (1).

Groff, James, NATIONAL ASSOCIATION OF WATER COMPANIES (NAWC), 1725 K St., NW Suite 1212, Washington, DC, 20006, (202) 833-8383, (14).

Gromicko, Nick, SUNN CORPORATION, Box M, Dept. 213 3520 Forbes Ave., Pittsburgh, PA, 15213, (412) 828-9100, (2)(4).

Groot, Charles, ASSOCIATION OF AMERICAN STATE GEOLOGISTS (AASG), C/O Louisiana Geological Survey P.O. Box G, Baton Rouge, LA, 70893, (504) 342-6754, (14).

Gross, Charlie, CONSTRUCTION SPECIALTIES, INC., 55 Winans Ave., Cranford, NJ, 07016, (201) 272-5200, (4).

Gross, Henry, BRIDGEPORT/DEPARTMENT OF AIR POLLUTION CONTROL/OFFICE OF HUMANE AFFAIRS, 752 East Main St., Bridgeport, CT, 06608, (203) 576-7451, (10).

Grosset, Robert, AAA RADON SERVICES, INC., P.O. Box F, Hellertown, PA, 18055, (215) 868-6025, (215) 821-2938, (1)(2).

Grossman, Susan, AMERICAN SOCIETY OF REAL ESTATE COUNSELORS (ASREC), 430 North Michigan Ave., Chicago, IL, 60611, (312) 329-8429, (18).

Grot, Richard, U.S. DEPARTMENT OF COMMERCE (DOC)/NATIONAL INSTITUTE OF STANDARDS AND TECHNOLOGY(NIST)/CENTER FOR BUILDING TECHNOLOGY(CBT)/BUILDING ENVIRONMENT DIVISION/INDOOR AIR QUALITY GROUP, Building 226 Room A313, Gaithersburg, MD, 20899, (301) 975-6431, (8).

Groth, Edward, CONSUMERS UNION OF UNITED STATES (CU), 256 Washington St., Mt. Vernon, NY, 06608, (914) 667-9400, (17).

Grout, David M., RADIATION HEALTH, 1259 Marlin Dr., Port Orchard, WA, 98366, (206) 871-5454, (1).

Grow, Robert T., U.S. RADON TESTING SERVICE, 820 S. 24th St., Arlington, VA, 22202, (703) 684-4528, (1).

Grubb, Kathleen, ENVIRONMENTAL SCIENCE SERVICES, 235 Promenade St., Providence, RI, 02908, (401) 421-0398, (1).

Grumski, Jr., Joseph, WESTINGHOUSE MATERIAL COMPANY OF OHIO/FEED MATERIALS PRODUCTION CENTER, P.O. Box 398704, Cincinnati, OH, 45239-8704, (513) 738-6200, x6280, (5).

Grundle, James A, U.S. DEPARTMENT OF COMMERCE (DOC)/NATIONAL INSTITUTE OF STANDARDS AND TECHNOLOGY(NIST)/CENTER FOR RADIATION RESEARCH(CRR)/NEUTRON INTERACTIONS AND DOSIMETRY GROUP, Building 245 Room C308, Gaithersburg, MD, 20899, (301) 975-5525, (8).

Guadagno, Janine, U.S. DEPARTMENT OF DEFENSE (DOD)/DEPARTMENT OF THE ARMY(DOA)/BELVOIR RESEARCH, DEVELOPMENT AND ENGINEERING CENTER/MATERIALS, FUELS AND LUBRICANTS LABORATORY, Radiation Research Group, Ft. Belvoir, VA, 22060, (703) 664-5813, (703) 664-4175, (8).

Guadagno, Janine, U.S. DEPARTMENT OF THE ARMY/BELVOIR RESEARCH, DEVELOPMENT AND EDUCATION CENTER, STRBE-VR, Fort Belvoir, VA, 22060-5606, (703) 664-5133, (1).

Guanti, John, AIR VENTILATION SYSTEMS, INC., 3343 Hollins Ferry Rd., Baltimore, MD, 21227, (301) 525-3174, (1).

Guarino, Mark, ENVIRONMENTAL HEALTH INSPECTIONS, INC., 1200 Rte. 9 Cross River Mill Suite 5E, Lakewood, NJ, 08701, (201) 905-8888, (1)(3).

Guest, Gordon, UP-COUNTRY HOME INSPECTORS, INC., P.O. Box 1568, Scarborough, ME, 04074, (207) 883-9876, (1)(2).

Guimond, Richard J., U.S. ENVIRONMENTAL PROTECTION AGENCY (EPA)/OFFICE OF AIR AND RADIATION(OAR)/OFFICE OF RADIATION PROGRAMS(ORP), 401 M St., SW ANR-458, Washington, DC, 20460, (202) 475-9600, (8).

Guinn, Linda, RADON AND ENVIRONMENTAL PROFESSIONALS, P.O. Box 3391, Lawrence, KS, 66046, (913) 749-7420, (1).

Guinn, Otis E., SOUTHEASTERN RADON CONSULTANTS, INC., 161 York Rd., Nolensville, TN, 37135, (615) 776-5569, (1)(2).

Guira, Jose M., SARASOTA COUNTY AIR POLLUTION CONTROL/SARASOTA COUNTY

ENVIRONMENTAL SERVICES DEPARTMENT, 1301 Cattlemen Rd., Sarasota, FL, 34232, (813) 378-6137, (10).

Gullickson, C.W., WESTERN ENVIRONMENTAL, INC., 4320-B Northpark Dr., Colorado Springs, CO, 80907, (719) 528-6132, (1).

Gunderson, Linda, U.S. DEPARTMENT OF INTERIOR (DOI)/U.S. GEOLOGICAL SURVEY(USGS)/GEOLOGIC DIVISION(GD), MS-990, National Ctr., Reston, VA, 22092, (703) 648-4000, (8).

Gunn, John Allen, SCIENTIFIC ANALYSIS, INC., 6012 East Shirley Ln., Montgomery, AL, 36117, (205) 271-0643, (800) 345-2575, (1)(4).

Gunnels, Frederick R., HOLIDAY TERMITE AND PEST CONTROL, 612 3rd St., Herndon, VA, 22070, (703) 471-5533, (1).

Gurney, Glenn, GURNEY KERR CONTRACTORS, INC., Box 753, Bryn Athyn, PA, 19009, (215) 947-7796, (2).

Gutwein, Ed, RADON MONITORING AND SERVICES, 1350 Butter Lane, Mt. Penn, PA, 19606, (215) 779-8314, (1).

Gworek, David, CERT WATER SERVICES, INC., 25 Duval Ln., Plainville, CT, 06062, (203) 793-9142, (1)(4).

H

Haag, Chris, SAFE SHELTER ENVIRONMENTAL, 527 West Lancaster Ave., Frazer, PA, 19355, (215) 640-4839, (1)(2).

Haag, Rick, SAFE SHELTER ENVIRONMENTAL, 527 West Lancaster Ave., Frazer, PA, 19355, (215) 640-4839, (1)(2).

Haarde, Robert, SPARTAN HOME INSPECTION, INC., 54 Woodport Rd., Sparta, NJ, 07871, (201) 729-7972, (1).

Haes, Julius E., WYOMING DEPARTMENT OF HEALTH AND SOCIAL SERVICES/DIVISION OF HEALTH AND MEDICAL SERVICES/RADIOLOGICAL HEALTH SERVICES, Hathaway Building, Cheyenne, WY, 82002-0710, (307) 777-6015, (10).

Hageman, Bruce, RADON DETECTION SYSTEMS/GREENSBORO, NORTH CAROLINA BRANCH, 4324 Cooper Rd., Winston-Salem, NC, 27121, (303) 444-5253, (800) 627-2366, (1).

Hager, Jr., Nat E., NATIONAL INSTITUTE OF BUILDING SCIENCES (NIBS)/INDOOR AIR QUALITY COMMITTEE, C/O Armstrong World Industries 333 Liberty St., Lancaster, PA, 17603, (717) 397-0611, (5)(13).

Hagerty, Frederick J., BAYSTATE ENVIRONMENTAL, 53 Cushing St., Stoughton, MA, 02072, (617) 341-0002, (2).

Hahn, Dennis, MICHIGAN DEPARTMENT OF PUBLIC HEALTH/BUREAU OF ENVIRONMENTAL AND OCCUPATIONAL HEALTH/DIVISION OF RADIOLOGICAL HEALTH, 3423 North Logan St. P.O. Box 30195, Lansing, MI, 48909, (517) 335-8200, (6)(10).

Haines, Philip, MAINE DEPARTMENT OF HUMAN SERVICES/BUREAU OF HEALTH/PUBLIC HEALTH LABORATORY, State House Station 12, Augusta, ME, 04333, (207) 289-2727, (1)(10).

Haines, Rob, NATURAL ENERGY CONSULTANTS, P.O. Box 586, Conifer, CO, 80433, (303) 838-5291, (1)(2)(3).

Hair, Kenneth E., ASSOCIATED PROTECTIVE SYSTEMS, 2511 Winford Ave., Nashville, TN, 37211, (615) 244-7637, (615) 244-7638, (1).

Hajinian, Charles H., ENVIRONMENTAL AUDIT, INC., 1901 West 47th Pl. Suite 310, Westwood, KS, 68205, (913) 384-0000, (1).

Hale, Edmond W., SISKIYOU COUNTY AIR POLLUTION CONTROL DISTRICT, 525 South Foothill Dr., Yreka, CA, 96097, (916) 842-3906, (10).

Hale, Myron, JMX INDUSTRIES, INC., 3959 Wilshire Blvd. Suite 200/B11, Los Angeles, CA, 90010, (213) 871-8054, (1).

Hall, Frances, UNIVERSITY OF NEW HAMPSHIRE/DEPARTMENT OF EARTH SCIENCES, James Hall, Durham, NH, 03824, (603) 862-1234, (5).

Hall, Gary A., RADON CONTROL SYSTEMS, 138 Riverview Plantation Dr., Williamsburg, VA, 23185, (804) 564-1893, (2).

Hall, Ron, EDGELL COMMUNICATIONS, 7500 Old Oak Blvd, Cleveland, OH, 44130, (216) 243-8100,

Hall, Stacey, TECH OPSOL, INC., 2151 East Dublin-Granville Rd. Suite G522, Columbus, OH, 43229, (614) 882-3527, (1).

Hall, Stephen, RADON CONTROL PROFESSIONALS, INC., 11920 Fieldthorn Ct., Reston, VA, 22094, (703) 471-9459, (1)(2)(3).

Hall, Ted, RADON TESTING ASSOCIATION, Rte. 6, 2014 Glenwood Dr., Martinsburg, WV, 25401, (304) 754-7586, (1).

Hall, William, RADON PRODUCTS, INC., 3215 West Hampden Ave., Englewood, CO, 80110, (303) 761-4705, (4).

Halladay, Christopher, APPALACHIAN ENVIRONMENTAL TESTING, INC., 105 South Union St. Suite 326, Danville, VA, 24541, (804) 792-1300, (1).

Hallam, Ellen, U.S. DEPARTMENT OF INTERIOR (DOI)/U.S. GEOLOGICAL SURVEY(USGS)/NATIONAL CARTOGRAPHIC INFORMATION CENTER-ALASKA(NCIC), 4230 University Dr., Anchorage, AK, 99508-4664, (907) 271-4159, (8).

Hallenbach, Tom, HOME OWNERS WARRANTY CORPORATION, 2000 L St., NW Suite 400, Washington, DC, 20036, (202) 463-4600, (18).

Hallisey, Robert, MASSACHUSETTS DEPARTMENT OF PUBLIC HEALTH/OFFICE OF HUMAN SERVICES/RADIATION CONTROL PROGRAM, 150 Tremont St., Boston, MA, 02111, (617) 727-6214, (10).

Hamill, Melanie Ann, ILLINOIS DEPARTMENT OF NUCLEAR SAFETY/OFFICE OF ENVIRONMENTAL SAFETY, 1035 Outer Park Dr., Springfield, IL, 62704, (217) 786-6399, (1)(10).

Hamilton, Richard V., ROANOKE CITY ENGINEERING DEPARTMENT/AIR POLLUTION CONTROL DIVISION, 215 West Campbell Ave., Roanoke, VA, 240111, (703) 981-2731, (10).

Hammitt, Christine, FORESIGHT ENGINEERING, INC., 17 Cook Ave. P.O. Box 621, Madison, NJ, 07940, (201) 377-0600, (1)(2).

Hammitt, F.G., FORESIGHT ENGINEERING, INC., 17 Cook Ave. P.O. Box 621, Madison, NJ, 07940, (201) 377-0600, (1)(2).

Hammond, Douglas, UNIVERSITY OF SOUTHERN CALIFORNIA/GEOLOGICAL SCIENCE DEPARTMENT, University Park, Los Angeles, CA, 90089-0740, (213) 743-6973, (5).

Hancock, Don, SOUTHWEST RESEARCH AND INFORMATION CENTER, P.O. Box 4524, Albuquerque, NM, 87106, (505) 262-1862, (17).

Hancuff, William R., ABIONICS, LTD., 12509 Kings Lake Dr., Reston, VA, 22091, (703) 620-3767, (1).

Haney, Catherine, HEALTH PHYSICS SERVICES, INC., 4 Research Pl. Suite 140, Rockville, MD, 20850, (301) 670-1818, (1).

Haney, Donald, KENTUCKY GEOLOGICAL SURVEY, University Of Kentucky 311 Breckinridge Hall, Lexington, KY, 40506, (606) 257-5863, (10).

Hanmer, Rebecca, U.S. ENVIRONMENTAL PROTECTION AGENCY (EPA)/OFFICE OF WATER(OW), 401 M St., SW WH-556, Washington, DC, 20460, (202) 382-5700, (8).

Hannafin, Laurence L., CONNECTICUT DEPARTMENT OF CONSUMER PROTECTION/REAL ESTATE DIVISION, State Office Building Room G-8 165 Capitol Ave., Hartford, CT, 06106, (203) 566-5130, (10).

Hanneson, Dave, PYLON ELECTRONIC DEVELOPMENT COMPANY, LTD., 147 Colonnade Rd., Nepean, ON, K2E 7L9, (613) 226-7920, (1)(4).

Hannum, William H., AMERICAN NUCLEAR SOCIETY/TENNESSEE-CHATTANOOGA SECTION, 7408 Twinbrook Dr., Chattanooga, TN, 37421, (615) 751-2067, (14).

Hansell, Marty, SKUTTLE MANUFACTURING COMPANY, Rte. 1, Box 16A, Marietta, OH, 45750, (614) 373-9169, (4).

Hansen, Brenda, NEILSON ASSOCIATES, INC., 100 Dickinson Dr. Suites 109-110, Chadds Ford, PA, 19317, (215) 793-1389, (3).

Hansen, Cletus J., WISCONSIN DEPARTMENT OF REGULATION AND LICENSING/BUREAU OF DIRECT LICENSING AND REAL ESTATE, 1400 East Washington P.O. Box 8935, Madison, WI, 53708, (608) 267-7132, (10).

Hanway, Tom, SENSIDYNE, INC., 12345 Starkey Rd. Suite E, Largo, FL, 34643, (813) 530-3602, (800) 451-9444, (4).

Harad, Stanley, HARAD WATERPROOFING, 3301 North Second St., Harrisburg, PA, 17110, (717) 233-6143, (2).

Harbottle, Garman, U.S. DEPARTMENT OF ENERGY (DOE)/CHICAGO OPERATIONS OFFICE(COO)/BROOKHAVEN NATIONAL LABORATORY(BNL), Building 510-A, Upton, NY, 11973, (516) 282-2123, (8).

Hardeman, James C., GEORGIA DEPARTMENT OF NATURAL RESOURCES/PROGRAM COORDINATION BRANCH, 205 Butler St., SE Floyd Towers East, Atlanta, Ga, 30334, (404) 656-4713, (10).

Hardin, Charles, CONFERENCE OF RADIATION CONTROL PROGRAM DIRECTORS (CRCPD), 71 Fountain Pl., Frankfort, KY, 40601, (502) 227-4543, (14).

Harding, Jerry, HARDICO RADON MANAGEMENT, CENTRAL OFFICE, P.O. Box 40, Stevens, PA, 17578, (215) 267-6006, (2).

Harding, Stephen D., TENNESSEE ASSOCIATION OF REALTORS, 1910 Adelicia St., Nashville, TN, 37212, (615) 321-0515, (18).

Harding, Tom, HARDICO RADON MANAGEMENT, GREENVILLE OFFICE, 744 Isaqueena Trail Ridgecrest Apt. #21, Central, SC, 29630, (803) 654-6734, (2).

Hardison, John, NATIONAL SEAL COMPANY, 1255 Monmouth Blvd., Galesburg, IL, 61401, (309) 343-3418, (800) 323-3820, (4).

Hardman, David, CONTEK/CORPORATE HEADQUARTERS, P.O. Box 3065, Newport Beach,

CA, 92663, (714) 759-9226, (1).

Hardy, Jr., C.R., ENVIRONMENTAL HEALTH AND SAFETY, P.O. Box 20052 Suite 211, Brandenton, FL, 34206, (313) 242-1699, (1).

Hare, Robert, HUDSON VALLEY RADON TESTING CORPORATION, P.O. Box 790, Poughkeepsie, NY, 12601, (914) 431-6424, (914) 485-5561, (1).

Hargrove, Gerald, GERALD HARGROVE INSPECTION SERVICE, 5308 Meadowlark Ln., Grovetown, GA, 30813, (404) 860-0797, (1).

Harihar, Subbiah, RADON SAFE, 7596 Wheaton Dr., Canton, MI, 48187, (313) 455-6469, (1).

Harister, Bill, MACKLENBURG COUNTY DEPARTMENT OF ENVIRONMENTAL PROTECTION/AIR QUALITY DIVISION, 1200 Blythe Blvd., Charlotte, NC, 28203, (704) 376-4603, (10).

Harley, Naomi, NEW YORK UNIVERSITY MEDICAL CENTER/INSTITUTE OF ENVIRONMENTAL MEDICINE, 550 1st Ave., New York, NY, 10016, (212) 340-5280, (5).

Harper, Jerry, RADON REDUCTION SERVICES, 5601 East 109th Ter., Kansas City, MO, 64137, (816) 942-7083, (1)(2).

Harpole, Jr., William G., HOME INSPECTION OF KENTUCKIANA, 2416 Frankfort Ave. Suite 206, Louisville, KY, 40206, (502) 897-1450, (1).

Harrigan, Norwell, U.S. VIRGIN ISLANDS/COLLEGE OF THE VIRGIN ISLANDS/CARIBBEAN RESEARCH INSTITUTE, St. Thomas, VI, 809774-9200, (809) 774-9200, (10).

Harrington, Garland H., PLUM GROVE CORPORATION, 33 West Queens Way Mall Old Towne, Hampton, VA, 23669, (804) 722-9841, (1)(2).

Harrington, Ronald G., PROTECTRON, INC., 229 East Main St. Suite E, Westminster, MD, 21157, (301) 857-9116, (1).

Harrington, Scott, HOMES PLUS RADON SERVICES, INC., 490 Main St. P.O. Box 654, Norwich, CT, 06360, (203) 887-5128, (1)(4).

Harris, Bruce, U.S. ENVIRONMENTAL PROTECTION AGENCY (EPA)/OFFICE OF RESEARCH AND DEVELOPMENT(ORD)/AIR AND ENERGY ENGINEERING RESEARCH LABORATORY(AEERL)/RADON MITIGATION PROJECT, MD-60, Research Triangle Park, NC, 27711, (919) 541-2350, (8).

Harris, David, NATIONAL INSTITUTE OF BUILDING SCIENCES (NIBS), 1015 15th St., NW Suite 700, Washington, DC, 20005, (202) 347-5710, (5)(13).

Harris, George H., AMERICAN INSPECTION AGENCY, P.O. Box 2444, Lehigh Valley, PA, 18001, (215) 759-2444, (1).

Harris, Mark E., ENVIROTECH SERVICES, 201 South Paul, Springfield, IL, 62703, (217) 753-1346, (1).

Harris, Terry C., KNOX COUNTY DEPARTMENT OF AIR POLLUTION CONTROL, 400 Main Ave. City-County Building, Room 459, Knoxville, TN, 37902, (615) 521-2488, (10).

Harris, Tony, HARRIS CONSTRUCTION, P.O. Box 1203, Breckenridge, CO, 80424, (303) 453-2045, (1)(2).

Harrison, Bradley, BRADLEY HARRISON ASSOCIATES, INC., 55 Virginia Ave., Freeport, NY, 11520, (516) 546-2283, (516) 546-7760, (1)(3).

Harrison, Jed, U.S. ENVIRONMENTAL PROTECTION AGENCY (EPA)/OFFICE OF AIR AND RADIATION(OAR)/OFFICE OF RADIATION PROGRAMS(ORP)/RADON DIVISION(RD), Policy And Public Information Branch 401 M St., SW ANR-464, Washington, DC, 20460, (202) 475-9605, (8).

Harrison, Steven V., RADON ASSOCIATES MARKETING COMPANY, 1099 18th St. Suite 1830, Denver, CO, 80202, (303) 293-3141, (3).

Harshman, John, APPLIED HEALTH PHYSICS, INC., 2986 Industrial Blvd., Bethel Park, PA, 15102, (412) 563-2242, (518) 477-7974, (1)(2)(4)(6).

Hartsell, Susan, HARTSELL HOME INSPECTION SERVICE, 5501 Executive Center Dr., Charlotte, NC, 28212, (704) 588-3781, (1).

Harvey, Jack, EMERGENCY RADON TESTING, INC., 29 Freeman St., West Orange, NJ, 07052, (201) 481-3992, (201) 674-1570, (1).

Harvey, Melvin, TRI-STATE RADON TESTING AND REDUCTION, 530 Eben Ave., Findlay, OH, 45840, (419) 422-0657, (419) 422-6601, (1).

Hashmi, Zia R., RADON SAFETY AND TECHNOLOGY CORPORATION, 4 Charles St., Pleasant Valley, NY, 12569, (914) 635-1138, (800) 552-5257, (1)(2)(3)(4).

Hatton, Thomas, ALPHA CONCEPTS, INC., P.O. Box 119, Hopatcong, NJ, 07843, (201) 770-3046, (1)(2).

Haugen, Herb, RADON TESTING AND CONTROL, 2910 East Broadway Ave., Bismarck, ND, 58501, (701) 222-1613, (1)(2).

Havewala, J.B., APPLIED EXTRUSION TECHNOLOGIES, INC., P.O. Box 582, Middletown, DE, 19709, (302) 378-8888, (800) 521-6713, (4).

Havlik, C.J., NORLAND CORPORATION, Norland Dr., Ft. Atkinson, WI, 53538, (414) 563-8456, (800) 333-8456, (4).

Hawk, Walter P., TRUESDALE-HAWK INSPECTION CONSULTANTS, INC., P.O. Box 154, Bedminster, NJ, 07921, (201) 234-0323, (1).

Hawks, Jr., Bill, HAWKS INTER-STATE, 814 North Main, Wichita, KS, 67203, (316) 267-8331, (316) 681-0031, (1).

Hawksley, Roland, RADON PROTECTION, INC., The Summit Suite 110, Louisville, KY, 40207, (502) 893-4563, (1)(2).

Hawthorne, Alan, U.S. DEPARTMENT OF ENERGY (DOE)/OAK RIDGE OPERATIONS OFFICE(OROO)/OAK RIDGE NATIONAL LABORATORY(ORNL)/HEALTH STUDIES DIVISION(HSD)/METABOLISM AND DOSIMETRY DIVISION, P.O. Box 2008 Building 7509 MS-6383, Oak Ridge, TN, 37831, (615) 574-6251, (8).

Hayes, Michael J., ILLINOIS ENVIRONMENTAL PROTECTION AGENCY/DIVISON OF AIR POLLUTION CONTROL, 2200 Churchill Rd. P.O. Box 19276, Springfield, IL, 62794-9276, (217) 782-7326, (10).

Hayes, Terry, MAINE REAL ESTATE COMMISSION, State House Station # 35, Augusta, ME, 04333, (207) 783-9303, (10).

Haynes, George F., ADVANCED RADON DETECTION SERVICES, INC., P.O. Box 30458, Knoxville, TN, 37930, (615) 691-9905, (1).

Hayward, Steve, CALIFORNIA DEPARTMENT OF HEALTH SERVICES/AIR AND INDUSTRIAL HYGIENE LABORATORY, 2151 Berkeley Way Room 334, Berkeley, CA, 94704, (415) 540-2134, (10).

Hazic, Gerard M., MICROBAC LABORATORIES, INC./STRAND TESTING DIVISION, 110 Kinzey St., Johnstown, PA, 15904, (814) 266-9548, (1).

Hazle, A.J., COLORADO DEPARTMENT OF HEALTH/RADIATION CONTROL DIVISION, 4210 East 11th Ave., Denver, CO, 80220, (303) 331-4812, (10).

Hazle, Albert J., CONFERENCE OF RADIATION CONTROL PROGRAM DIRECTORS (CRCPD)/COMMITTEE ON RADON POLICY, 71 Fountain Pl., Frankfort, KY, 40601, (502) 227-4543, (14).

Hearon, Robert S., IMC DEVELOPMENT CORPORATION, P.O. Box 7047, Lakeland, FL, 33807, (813) 646-5061, (16).

Heckman, David, NEILSON ASSOCIATES, INC., 100 Dickinson Dr. Suites 109-110, Chadds Ford, PA, 19317, (215) 793-1389, (3).

Hedberg, Michael E., FULL HOUSE CONTRACTING, 1900 South Washington, Casper, WY, 82601, (307) 266-0895, (2).

Hedden, John J., DELAWARE DEPARTMENT OF HEALTH AND SOCIAL SERVICES/BUREAU OF ENVIRONMENTAL HEALTH/OFFICE OF RADIATION CONTROL/DIVISION OF PUBLIC HEALTH, 802 Silver Lake Blvd. and Walker Rd. P.O. Box 637, Dover, DE, 19903, (302) 736-4731, (10).

Hedstrom, Linda, MISSOULA CITY-COUNTY HEALTH DEPARTMENT/ENVIRONMENTAL HEALTH DIVISION, 301 West Adler St., Missoula, MT, 59802, (406) 721-5700, (10).

Heidelberg, Joan C., NORTH AMERICAN ASSOCIATION FOR ENVIRONMENTAL EDUCATION (NAAEE), P.O. Box 400, Troy, OH, 45373, (513) 698-6493, (14).

Heidersdorf, Sidney, ALASKA DEPARTMENT OF HEALTH AND SOCIAL SERVICES/DIVISION OF PUBLIC HEALTH/RADIOLOGICAL HEALTH , Box H, Juneau, AK, 99811-0613, (907) 465-3019, (10).

Heil, James, HEIL COMPANY, P.O. Box 25995, Colorado Springs, CO, 80918, (719) 593-8407, (4).

Heimann, Roger, TAPPAN ZEE DEVELOPMENT, 25 Webb Ln., Highland Falls, NY, 10928, (914) 446-7704, (2).

Heimer, Esther, MATTER RESEARCH, INC., 1923 New York Ave., Huntington Station, NY, 11746, (516) 549-8777, (1).

Heimer, Richard, MATTER RESEARCH, INC., 1923 New York Ave., Huntington Station, NY, 11746, (516) 549-8777, (1).

Heiney, Phil, GILIAN INSTRUMENT CORPORATION, 8 Dawes Hwy., Wayne, NJ, 07470, (201) 831-0440, (4).

Heitzenrater, William, AFI ENVIRONMENTAL/DIVISION OF BUFFALO CONSULTANTS, INC., 6103 Robinson Rd., Lockport, NY, 14094, (716) 625-8434, (1).

Heitzer, Richard, RHH FOAM SYSTEMS, INC., P.O. Box 752 6001 S. Pennsylvania Ave., Cudahy, WI, 53110-0752, (414) 744-6066, (4).

Helbing, Ken, NATIONAL PROPERTY INSPECTIONS, 247 Sunnyhill Ln., Oconomowoc, WI, 53066, (414) 569-9377, (1).

Helf, Samuel, COUNTY COLLEGE OF MORRIS/CHEMISTRY DEPARTMENT, Rte. 10 and Center Grove Rd. A-304, Randolph, NJ, 07869, (201) 361-5000, (1).

Helfrich, Raymond G., MINE INSPECTORS INSTITUTE OF AMERICA (MIIA), 1900 Grant Building, Pittsburgh, PA, 15219, (412) 281-2620, (16).

Hellums, William E., RADON TESTING SERVICES, INC., 4442 18th St., San Francisco, CA, 94114, (415) 552-5496, (1)(3)(4).

Helme, Ned, CENTER FOR CLEAN AIR POLICY (CCAP), 444 North Capitol St., NW, Washington, DC, 20001, (202) 624-7709, (14).

Helmuth, Stephen R., CONSUMER ENERGY

SERVICES, INC., 12 Westerville Sq. Suite 200, Westerville, OH, 43081, (614) 861-0600, (1)(2).

Helweick, Kelly, RPI RADON MEASUREMENT SERVICES, MARYLAND, 10133 Bacon Dr., Beltsville, MD, 20705, (301) 937-4072, (1).

Helwig, John, JANCE ASSOCIATES, INC., P.O. Box 234, East Texas, PA, 18046, (215) 398-0434, (2).

Hemmings, Don, ALTERNATE HEATING, 621 Rothesay Ave., Saint John, NB, E2H 2G9, (506) 633-1500, (4).

Hemschoot, John, FEDERAL HOME LOAN MORTGAGE CORPORATION (FREDDIE MAC), 1776 G St., NW, Washington, DC, 20006, (703) 759-8000, (18).

Hendee, William R., AMERICAN MEDICAL ASSOCIATION (AMA)/COUNCIL ON SCIENTIFIC AFFAIRS, 535 North Dearborn St., Chicago, IL, 60610, (312) 645-5000, (15).

Hendershot, Russell, TEQ CORPORATION, P.O. Box 570, Wilkes-Barre, PA, 18703-0570, (717) 825-3939, (800) 877-2366, (1)(2)(4).

Henderson, Bob, NUTECH ENERGY SYSTEMS, INC., 124 Newbold Ct., London, ON, N6E 1Z7, (519) 686-0797, (4).

Henderson, John E., AIRCARE SYSTEMS, P.O. Box 4942, Lancaster, PA, 17604, (717) 293-9223, (1)(2).

Henderson, Herm, QUALITY ASSURANCE, INC., P.O. Box 4087, Highlands Ranch, CO, 80126, (303) 791-2624, (1).

Hendricks, John, AMC HOME INSPECTION SERVICES, 421 West Union Ave., Bound Brook, NJ, 08805, (201) 469-6050, (1).

Hendricks, Vaughn, U.S. VIRGIN ISLANDS DEPARTMENT OF LABOR/DIVISION OF OCCUPATIONAL SAFETY AND HEALTH, Lagoon St., Frederiksted, VI, 00840, (809) 772-1315, (10).

Hendry, Charles, FLORIDA DEPARTMENT OF NATURAL RESOURCES, Cedars Executive Center 2639 North Monroe Box 20, Building B, Tallahassee, FL, 32399, (904) 488-3177, (10).

Henne, Alfred, B & H RADON SERVICES, P.O. Box 247, Silver Spring, PA, 17575-0247, (717) 684-2612, (717) 426-3723, (1)(2).

Hennecke, Karl, RADONTEST, INC., 4700 East 5th Ave., Columbia, MO, 65201, (314) 875-8045, (1).

Henry, Jim, MELODY HOMES, INC., 11031 Sheridan Sq., Westminster, CO, 80020, (303) 466-1831, (13).

Henry, Mary Jane, AMERICAN RADON TESTING SERVICES, 2302 Basswood Cr., Tustin, CA, 92680, (714) 838-4990, (1).

Henschel, Bruce, U.S. ENVIRONMENTAL PROTECTION AGENCY (EPA)/OFFICE OF RESEARCH AND DEVELOPMENT(ORD)/AIR AND ENERGY ENGINEERING RESEARCH LABORATORY(AEERL)/RADON MITIGATION PROJECT, MD-60, Research Triangle Park, NC, 27711, (919) 541-2745, (8).

Hentgen, Michael, I-T.E.M., LTD., 12850 Middlebrook Rd. Suite 301, Germantown, MD, 20874, (301) 353-0585, (1).

Herra, Jeff, RADON DETECTION SYSTEMS/NORTHERN CALIFORNIA BRANCH, 2567 Greenwich St., San Francisco, CA, 94123, (303) 444-5253, (800) 627-2366, (1).

Herrick, Robert, HERRICK ENGINEERING, INC., 2004 Belvedere Dr., Toledo, OH, 43614, (419) 385-5655, (1)(3).

Herrington, Phil, GEO-CHEK, INC., 2270 Stony Hill Rd., Boulder, CO, 80303, (303) 494-0576, (1).

Hershey, Catherine, NERING, INC., P.O. Box 209, Ridgefield, CT, 06877, (203) 431-3597, (1).

Herz, Norman, RADEX, INC., 1382 Prince Ave. Suite A, Athens, GA, 30606, (404) 548-6960, (1)(2)(3).

Herzog, Scott D., ENVIROSCIENCE CONSULTANTS, INC., 66 Cedar St., Newington, CT, 06111, (203) 666-7167, (1)(3).

Hess, Charles T., UNIVERSITY OF MAINE-ORONO/DEPARTMENT OF PHYSICS, 120 Bennett Hall, Orono, ME, 04469, (207) 581-1016, (5).

Hessling, Peter A., PINELLAS COUNTY DEPARTMENT OF ENVIRONMENTAL MANAGEMENT/AIR QUALITY DIVISION, 16100 Fairchild Dr. Building V102, Clearwater, FL, 34622, (813) 530-6522, (10).

Hester, Norman C., INDIANA GEOLOGICAL SURVEY, 611 North Walnut Grove, Bloomington, IN, 47405, (812) 335-1950, (10).

Hickerson, Glen, HOME CHECK SERVICES, P.O. Box 98, Front Royal, VA, 22630, (703) 635-1274, (1)(2).

Hickey, James E., RHODE ISLAND DEPARTMENT OF HEALTH/DIVISION OF OCCUPATIONAL HEALTH AND RADIATION CONTROL, Cannon Building 75 Davis St., Providence, RI, 02908, (401) 277-2438, (10).

Hicks, Steven R., RADON DETECTION AND CONSULTING, 705 West Fourth St., Lawrence, KS, 66044, (913) 843-7377, (1)(2)(3).

Hiesinger, Ulrich, RADON ALERT® DETECTION CENTER, P.O. Box 323, Flourtown, PA, 19031, (215) 247-9197, (800) 345-6348, (1)(2)(4).

Higgins, Joe, RADON AND ENVIRONMENTAL PROFESSIONALS, P.O. Box 1045, Independence, KS, 67301, (316) 331-5207 , (1).

Hightower, Sandra, AGAPE RADON SERVICES, INC., P.O. Box 969, Alpharetta, GA, 30239-0969, (404) 664-4106, (1).

Hill, A. Alan, EXECUTIVE OFFICE OF THE PRESIDENT (EOP)/COUNCIL ON ENVIRONMENTAL QUALITY(CEQ), 722 Jackson Pl., NW, Washington, DC, 20503, (202) 395-5080, (8).

Hill, Donald R., TEHAMA COUNTY AIR POLLUTION CONTROL DISTRICT, 1760 Walnut St. P.O. Box 38, Red Bluff, CA, 96080, (916) 527-4504, (10).

Hill, Gary, U.S. DEPARTMENT OF INTERIOR (DOI)/U.S. GEOLOGICAL SURVEY(USGS)/ENERGY AND MARINE GEOLOGY(EMG), 12201 Sunrise Valley Dr., Reston, VA, 22092, (703) 648-6472, (8).

Hill, Harry, ENVIRONMENTAL DETECTION ASSOCIATES, INC., P.O. Box 1006 21 Middlemount Rd., Pisgah Forest, NC, 28768-1006, (704) 877-4300, (1)(3)(4).

Hill, Henry, SUBURBAN JERSEY WATERPROOFING, INC., 30-32 Abbett Ave., Morristown, NJ, 07960, (201) 539-1538, (2).

Hill, Jack, RADON MEASURING SYSTEMS, 1129 Wyoming Ave., Scranton, PA, 18509, (717) 961-1666, (1)(2).

Hill, James, U.S. DEPARTMENT OF COMMERCE (DOC)/NATIONAL INSTITUTE OF STANDARDS AND TECHNOLOGY(NIST)/CENTER FOR BUILDING TECHNOLOGY(CBT)/BUILDING ENVIRONMENT DIVISION, Building 226 Room B306, Gaithersburg, MD, 20899, (301) 975-5851, (8).

Hill, M. Arthur, RADON ENVIRONMENTAL TESTING SERVICES, INC., P.O. Box 14613, Dayton, OH, 45414, (513) 890-3361, (1).

Hill, Mae J., H & H DECORATING, 610 Great Falls Rd. Suite 100, Rockville, MD, 20850, (301) 251-6654, (1).

Hill, William, TENNESSEE DEPARTMENT OF CONSERVATION/DIVISION OF GEOLOGY, Customs House 701 Broadway, Nashville, TN, 37219-5237, (615) 742-6691, (10).

Hilley, Edgar C., MARYLAND ASSOCIATION OF REALTORS, 2083 West St. Suite 4-A, Annapolis, MD, 21401, (305) 841-6080, (18).

Hillmann, Joseph, HILLMANN ENVIRONMENTAL COMPANY, 1089 Cedar Ave., Union, NJ, 07083, (201) 686-3335, (1).

Hillmer, Ted, U.S. DEPARTMENT OF INTERIOR (DOI)/NATIONAL PARK SERVICE(NPS)/MID ATLANTIC REGIONAL OFFICE/INDOOR RADON PROGRAM, 143 South Third St., Philadelphia, PA, 19106, (215) 597-3679, (8).

Hills, Richard J., MIDDLESEX COUNTY AIR POLLUTION CONTROL PROGRAM, 280 Hobart St. Room 518, Perth Amboy, NJ, 08861, (201) 826-3100, (10).

Hills, Susan, SCIENTIFIC AND COMMERCIAL SYSTEMS CORPORATION (SCSC), 4651 King St. Suite 200, Alexandria, VA, 22302, (703) 824-8240, (5).

Hiltebrand, David, MALCOLM PIRNIE, INC., 100 Eisenhower Ave., Paramus, NJ, 07652, (201) 845-0400, (14).

Himes, Larry C., MAHONING-TRUMBULL AIR POLLUTION CONTROL AGENCY, 9 West Front St., Youngstown, OH, 44503, (216) 744-1928, (10).

Himmelfarb, Daniel J., DAN HIMMELFARB COMPANY, 7416 Chrome Mine Rd., Gaithersburg, MD, 20879, (301) 744-4004, (1).

Hinger, Bill, MENKE CONSULTING, INC., 6070 Routzong Rd., Greenville, OH, 45331, (513) 447-4225, (1).

Hinkel, Jon, OHIO UNIVERSITY/GEOLOGICAL SCIENCES DEPARTMENT, Athens, OH, 45701, (614) 593-1109, (5).

Hintenlang, David E., UNIVERSITY OF FLORIDA/COLLEGE OF ENGINEERING/NUCLEAR ENGINEERING DEPARTMENT, 202 Nuclear Science Center, Gainesville, FL, 32611, (904) 392-1401, (5).

Hirnak, Michela, CANBERRA INDUSTRIES, INC., 1 State St., Meriden, CT, 06450, (203) 238-2351, (1)(4).

Hitchins, Lynn Ann, U.S. ENVIRONMENTAL PROTECTION AGENCY (EPA)/OFFICE OF RESEARCH AND DEVELOPMENT(ORD)/OFFICE OF ENVIRONMENTAL ENGINEERING AND TECHNOLOGY DEMONSTRATION(OEETD), Radon Mitigation Development And Demonstration Program 401 M St., NW RD 691, Washington, DC, 20460, (202) 382-2583, (8).

Hittinger, Richard, ENVIRONMENTAL SCIENCE SERVICES, 235 Promenade St., Providence, RI, 02908, (401) 421-0398, (1).

Hlavinka, Joanne, STONE RIDGE RADON MITIGATORS, 424 Mine Rd., Asbury, NJ, 08802, (201) 735-6080, (800) 447-2366, (1)(2).

Hoadley, David, SMOOTH-ON, INC., 1000 Valley Rd., Gillette, NJ, 07933, (201) 647-5800, (4).

Hoagland, Douglas C., AIR SENSE, INC., 25 William St., Pawcatuck, CT, 06379-2108, (203) 599-5599, (1)(2).

Hobson, Todd C., ENVIRONMENTAL FIELD SERVICES, INC., 63 Elm St., Topsham, ME, 04086, (207) 725-4112, (1).

Hockstein, Steven, INSPECT-A-HOME, INC., 37 Janet Ln., Springfield, NJ, 07081, (201) 379-4124, (1).

Hodes, Lorraine, RADON TESTING AND CONSULTING, 1771 Post Rd., E. Suite 320, Westport, CT, 06880, (203) 454-3886, (1)(2)(4).

Hodgden, Gary, MIDWEST RADON INSPECTION, INC., 25005 West 129 Ter., Olathe, KS, 66061, (913)

780-2000, (1)(2).

Hodgden, Steve, MIDWEST RADON INSPECTION, INC., 25005 West 129 Ter., Olathe, KS, 66061, (913) 780-2000, (1)(2).

Hodges, Laurent, IOWA STATE UNIVERSITY/BUSINESS AND ENGINEERING EXTENSION/IOWA RADON PROJECT, EES-Building, Room 101 Haber Road, Ames, IA, 50011-0588, (515) 294-0588, (5).

Hodges, Sam, U.S. DEPARTMENT OF AGRICULTURE (USDA)/FARMERS HOME ADMINISTRATION (FMHA), Room 6309 South Building, Washington, DC, 20250, (202) 382-9653, (8).

Hodgson, Michael J., ASHRAE/ENVIRONMENTAL HEALTH COMMITTEE/INDOOR AIR QUALITY SUBCOMMITTEE, Atlanta, GA, 30329, (404) 636-8400, (13).

Hoeck, Charles, FLORIDA DEPARTMENT OF PROFESSIONAL REGULATION/DIVISION OF REAL ESTATE, P.O. Box 1900, Orlando, FL, 32802-1900, (407) 423-6048, (10).

Hoel, David, U.S. DEPARTMENT OF HEALTH AND HUMAN SERVICES (HHS)/PUBLIC HEALTH SERVICE(PHS)/NATIONAL INSTITUTE OF ENVIRONMENTAL HEALTH SCIENCES(NIEHS), P.O. Box 12233, Research Triangle Park, NC, 27711, (919) 541-3441, (8).

Hoff, Richard W., AMERICAN CHEMICAL SOCIETY (ACS), 1155 16th St., NW, Washington, DC, 20036, (202) 872-6062, (14).

Hoffman, Barry, THOMAS L. HOFFMAN, COMPANY, INC., RD 3, Box 173M, Coopersburg, PA, 18036, (215) 433-5887, (2).

Hoffman, Teresa, NEBRASKA REAL ESTATE COMMISSION, 301 Centennial Mall South P.O. Box 94667, Lincoln, NE, 68508, (402) 471-2004, (10).

Hofman, Dirk, INTERSTATE CONFERENCE ON WATER PROBLEMS (ICWP), 2300 M St., NW Suite 800, Washington, DC, 20037, (202) 466-7287, (14).

Hogan, Dorothy, AMERICAN NATIONAL STANDARDS INSTITUTE (ANSI), 1430 Broadway, New York, NY, 10018, (212) 354-3300, (14).

Hogan, Van, TRIANGLE WATERPROOFING AND REPAIR SERVICE, P.O. Box 741 206 North Arendell Ave., Zebulon, NC, 27597, (919) 269-8021, (2).

Hohenemser, Cristoph, CLARK UNIVERSITY/ENVIRONMENT, TECHNOLOGY, AND SOCIETY PROGRAM/DEPARTMENT OF PHYSICS, 950 Main St., Worcester, MA, 01610, (617)793-7687, (1)(5).

Holck, John L., ALA OF NEVADA, P.O. Box 7056, Reno, NV, 89510-7056, (702) 323-5864, (15).

Holecek, Mike, BUILDING DIAGNOSTICS, INC., 380 27th St., Boulder, CO, 80303, (303) 494-2431, (1)(2).

Holland, Mark R., M.R. HOLLAND ASSOCIATES, 70 Woodburn Dr., Methuen, MA, 01844, (508) 681-0821, (3).

Holland, William, BILL EICH CONSTRUCTION COMPANY, Box 63, Spirit Lake, IA, 51360, (712) 366-4438, (1).

Hollandsworth, Gerald, PUEBLO CITY-COUNTY HEALTH DEPARTMENT, 151 Central Main, Pueblo, CO, 81003, (303) 544-8376, (10).

Hollem, Dale C., ALA OF OHIO, P.O. Box 16677, Columbus, OH, 43216-6677, (614) 279-1700, (15).

Hollem, Valeria, JRW INDUSTRIES, INC., 2312 North 2nd St., Harrisburg, PA, 17110, (717) 236-1418, (1)(4).

Holmes, J.J., ATLANTA TESTING AND ENGINEERING, INC., 11420 Johns Creek Pkwy., Duluth, GA, 30136, (404) 476-3555, (1).

Holmes, Monty, INTERMOUNTAIN RADON SERVICE, P.O. Box 3, Salida, CO, 81201, (719) 539-6726, (1)(2)(6).

Holmes, Patrick S., ALPHA SPECTRA OF KENTUCKY, 11518 Main St., Louisville, KY, 40243, (502) 245-5557, (1).

Holmes, Roland, SOLID WASTE AND TOXIC TECHNOLOGY, 23891 Via Fabricante Suite 616, Mission Viejo, CA, 92691, (714) 855-7231, (1).

Holstein, Martha M., NEW YORK CITY BUREAU OF AIR RESOURCES, 295 Lafayette St., New York, NY, 10012, (212) 323-8580, (10).

Holub, R.F., U.S. DEPARTMENT OF INTERIOR (DOI)/BUREAU OF MINES(BOM)/DENVER RESEARCH CENTER DRC)/RADIATION LABORATORY, P.O. Box 25086 Denver Federal Ctr. Building 20, Denver, CO, 80225, (303) 236-0789, (1)(3)(4)(8).

Hon, Rudolph, BOSTON COLLEGE/DEPARTMENT OF GEOLOGY AND GEOPHYSICS, Devlin Hall, Chestnut Hill, MA, 02167, (617) 552-3656, (5).

Hon, Rudolph, RADION LABORATORY, INC., 70 Walnut St., Wellesley, MA, 02181, (617) 239-8066, (1)(2)(3)(4)(5)(6)(13).

Hood, Robert H., RADON ENGINEERING SERVICES, INC., 3129 Overton Rd., Birmingham, AL, 35223, (205) 967-2665, (1)(2).

Hood, Robert H., RADON EVALUATION SERVICES, INC., P.O. Box 13609, Atlanta, GA, 30324, (404) 478-0037, (1)(2).

Hoodes, Liana, ORANGE ENVIRONMENT, INC., RD 2 Maple Ave., Goshen, NY, 10924, (914) 294-5023, (17).

Hook, Robert D., RADIOLOGICAL HEALTH SCIENCES, INC., P.O. Box 4790, Fort Walton Beach, FL, 32549, (904) 863-1191, (1).

Hooper, Phillip M., ROBERTS ENVIRONMENTAL SERVICES, 4110 Old Wake Forest Rd. Suite 109, Raleigh, NC, 27609, (919) 790-9378, (1).

Hoops, Donald L., AMERICAN OCCUPATIONAL MEDICAL ASSOCIATION (AOMA), 2340 South Arlington Heights Rd., Arlington Heights, IL, 60005, (312) 228-6850, (16).

Hoornbeck, John, U.S. ENVIRONMENTAL PROTECTION AGENCY (EPA)/OFFICE OF AIR AND RADIATION(OAR)/OFFICE OF RADIATION PROGRAMS(ORP)/RADON DIVISION (RD), Mitigation Prevention And Quality Assurance Branch ANR-460 401 M St., SW, Washington, DC, 20460, (202) 475-9605, (8).

Hopke, Phil, UNIVERSITY OF ILLINOIS/ENVIRONMENTAL RESEARCH LABORATORY, 1005 West Western Ave., Urbana, IL, 61801, (217) 333-6230, (5).

Hopper, Richard D., U.S. ENVIRONMENTAL PROTECTION AGENCY (EPA)/OFFICE OF AIR AND RADIATION(OAR)/OFFICE OF RADIATION PROGRAMS(ORP)/LAS VEGAS FACILITY(LAF), Evaluation Branch 4220 South Maryland Pkwy. LaPlaza Building C, Room 532, Las Vegas, NV, 89119, (702) 798-2447, (1)(8).

Hoppes, Dale D., U.S. DEPARTMENT OF COMMERCE (DOC)/NATIONAL INSTITUTE OF STANDARDS AND TECHNOLOGY(NIST)/CENTER FOR RADIATION RESEARCH(CRR), Building 245 Room C114, Gaithersburg, MD, 20899, (301) 975-5532, (8).

Horlick, Michael, BLACKHAWK INDUSTRIES, INC., 607 Park St., Regina, SK, S4N 5N1, (306) 924-1551, (4).

Hornberger, Lenny, LDH WATERPROOFING AND RADON CONTROL, 286 Old Delp Rd., Lancaster, PA, 17601, (717) 569-2062, (2).

Horvat, Larry, PROJECT ENVIRONMENTAL GROUP, 10 Duff Rd., Pittsburgh, PA, 15235, (412) 243-3100, (2).

Horvath, John, MAMECO INTERNATIONAL, INC., 4475 East 175th St., Cleveland, OH, 44128, (216) 752-4400, (800) 321-6412, (4).

Hoskins, Donald, PENNSYLVANIA DEPARTMENT OF ENVIRONMENTAL RESOURCES/BUREAU OF TOPOGRAPHIC AND GEOLOGIC SURVEY, P.O. Box 2357, Harrisburg, PA, 17120, (717) 787-2169, (10).

Hough, J. Douglas, HOME RADON SURVEYS, 4209 Olley Ln., Fairfax, VA, 22032, (703) 323-7236, (1).

Houk, Vernon J., U.S. DEPARTMENT OF HEALTH AND HUMAN SERVICES (HHS)/PUBLIC HEALTH SERVICE(PHS)/OFFICE OF THE SURGEON GENERAL(OSG), 200 Independence Ave., SW, Washington, DC, 20201, (202) 245-6467, (8).

Houle, Paul N., EAST STROUDSBURG UNIVERSITY/ PHYSICS DEPARTMENT/RADON PROJECT, East Stroudsburg, PA, 18301, (717) 424-3350, (5).

Householder, Ron, OREGON STATE DEPARTMENT OF ENVIRONMENTAL QUALITY, 811 SW Sixth Ave., Portland, OR, 97204, (503) 229-5696, (10).

Howard, Judith A., AMERICAN ACADEMY OF ENVIRONMENTAL MEDICINE (AAEM), P.O. Box 16106, Denver, CO, 80216, (303) 622-9755, (14).

Howard, Marcia L., INDIANA STATE BOARD OF HEALTH/RADIOLOGICAL HEALTH SECTION, 1330 West Michigan St. Box 1964, Indianapolis, IN, 46206-1964, (317) 633-0153, (10).

Howe, Adrian C., MONTANA DEPARTMENT OF HEALTH AND ENVIRONMENTAL SERVICES/OCCUPATIONAL HEALTH BUREAU/ENVIRONMENTAL SCIENCES DIVISION, Cogswell Building, Helena, MT, 59620, (406) 444-3671, (10).

Howe, Carl, VETERANS ADMINISTRATION CENTRAL OFFICE (VA), 810 Vermont Ave., Washington, DC, 20420, (202) 233-2726, (8).

Howe, Geoffrey, UNIVERSITY OF TORONTO/NATIONAL CANCER INSTITUTE OF CANADA/EPIDEMIOLOGICAL UNIT, 12 Queens Park Crescent West, Toronto , ON, M5S 1A8, (416) 978-4793, (5).

Howell, Terry E., RADON DETECTION AND TESTING, INC., 2222 Cumberland Pl., Smyrna, GA, 30080, (404) 434-7299, (1).

Howell, Terry E., RADON REDUCTION AND TESTING, INC., 2222 Cumberland Pl., Smyrna, GA, 30080, (404) 434-7296, (1).

Howes, Duane, GRANITE STATE RADON SERVICES, P.O. Box 864, Windham, NH, 03087, (603) 437-1820, (1).

Howes, Duane L., SAFE AIR COMPANY, 9 Eastman Dr., Derry, NH, 03038, (603) 437-1425, (1).

Howes, Lewis, GRANITE STATE RADON SERVICES, P.O. Box 864, Windham, NH, 03087, (603) 437-1820, (1).

Howland, David, BERKSHIRE AND PIONEER VALLEY AIR POLLUTION CONTROL DISTRICTS, State House West 436 Dwight St., 4th Fl., Springfield, MA, 01103, (413) 785-5327, (10).

Hoxie, Don, CONFERENCE OF RADIATION CONTROL PROGRAM DIRECTORS (CRCPD)/TASK FORCE FOR ADEQUATE RADIATION CONTROL PROGRAM-RADON, 71 Fountain Pl., Frankfort, KY, 40601, (502) 227-4543, (14).

Hoxie, Don, MAINE DEPARTMENT OF HUMAN SERVICES/BUREAU OF HEALTH/DIVISION OF

HEALTH ENGINEERING, State House Station 10, Augusta, ME, 04333, (207) 289-3826, (10).

Hoyle, Bill, NEW JERSEY DEPARTMENT OF ENVIRONMENTAL PROTECTION/DIVISION OF ENVIRONMENTAL QUALITY/BUREAU OF ENVIRONMENTAL RADIATION, 729 Alexander Rd. CN-415, Trenton, NJ, 08625, (609) 987-6396, (800) 648 0394, (1)(10).

Huber, D. Gravatt, AMERICAN FAN ASSOCIATION (AFA), P.O. Box 888, East Lansing, MI, 48826, No Phone Available, (13).

Huber, Lee, NORTH DAKOTA STATE DEPARTMENT OF HEALTH/DIVISION OF ENVIRONMENTAL ENGINEERING/OCCUPATIONAL HEALTH PROGRAM, 1200 Missouri Ave. Room 304, Bismarck, ND, 58501, (701) 224-2348, (10).

Huber, Michael D., PROFESSIONAL HOME INSPECTIONS, INC., R.D. 1, Fenimore Rd. Box 371-A, Medford, NJ, 08055, (609) 261-7374, (1).

Huber, Stan, ADCO SERVICES, INC., 17650 DuVan Dr., Tinley Park, IL, 60477, (312) 429-1660, (4).

Huber, Stan A., STAN A. HUBER CONSULTANTS, INC., 200 North Cedar Rd., New Lenox, IL, 60451, (815) 485-4433, (1).

Huber, Tom, UNIVERSITY OF COLORADO-COLORADO SPRINGS/PHYSICS DEPARTMENT, 1420 Austin Bluffs Pkwy., Colorado Springs, CO, 80933-7150, (719) 593-3214, (1)(5).

Huebner, L.G., TELEDYNE ISOTOPES MIDWEST LABORATORY, 700 Landwehr Rd., Northbrook, IL, 60062, (312) 564-0700, (1).

Huebner, Terry, RADON DETECTION SPECIALISTS, 7966 South Madison St., Burr Ridge, IL, 60521, (312) 325-4443, (1).

Huey, Stewart, NATIONAL RADON ASSOCIATION, INC., THE (NRA), 2786 North Decatur Rd. Suite 220, Decatur, GA, 30033, (404) 292-1113, (7).

Huff, Robert V., ENVIRONMENTAL RADIATION SERVICES, INC., 305 West Magnolia St. Suite 373, Fort Collins, CO, 80521, (303) 226-5530 , (1).

Huggins, Alan, CAROLINA RADON DETECTION, 209 Tebblewood Dr., Simpsonville, SC, 29681, (803) 282-0494, (1)(2).

Huggins, T. Ruth, WASTE CONTROL INTERNATIONAL, INC., 100 Galleria Pkwy. Suite 400, Atlanta, GA, 30339, (404) 443-2061, (1).

Hughes, Donald R., KENTUCKY DEPARTMENT OF HUMAN RESOURCES/DIVISION OF RADIATION AND PRODUCT SAFETY/RADIATION CONTROL BRANCH, 275 East Main St., Frankfort, KY, 40621, (502) 564-3700, (10).

Hughes, E., ELLIOTT AND ASSOCIATES, 146 North Drexel Ave., La Grange, IL, 60525, (312) 354-1481, (1)(2).

Hughes, James L., ALA OF ALABAMA, P.O. Box 55209, Birmingham, AL, 35255, (205) 933-8821, (15).

Hughes, John, RADON SERVICES ANALYTICAL LABORATORY, 111 Benson Rd., Nashville, TN, 37214, (615) 883-4659, (1).

Hughes, Ronald L., HUGHES, RONALD L., 68 Hanshue Rd., Hummelstown, PA, 17036, (717) 534-1168, (1)(2).

Hughey, Jack W., JH AND ASSOCIATES, INC., 2524 East 71st St., Tulsa, OK, 74136, (918) 455-4984, (1).

Hull, Donald A., OREGON STATE DEPARTMENT OF GEOLOGY AND MINERAL INDUSTRIES, 910 State Office Building, Portland, OR, 97201-5528, (503) 229-5580, (10).

Hull, Gary E., TRUE TEST COMPANY, UTAH, P.O. Box 571, Bountiful, UT, 84010, (801) 298-2256, (1).

Hull, Gary E., TRUE TEST COMPANY, WASHINGTON, 1436-C Brandt Rd., Vancouver, WA, 98661, (206) 696-0957, (1).

Hulsey, John E., ACME EXTERMINATING COMPANY, P.O. Box 85, Gainesville, GA, 30503, (404) 536-3036, (404) 532-4351, (1).

Hung, Min, RADON DETECTION SYSTEMS/PITTSBURGH, PENNSYLVANIA BRANCH, 180 Indionola Rd., Pittsburgh, PA, 15238, (303) 444-5253, (800) 627-2360, (1).

Hungerford, Stan, ALASKA DEPARTMENT OF ENVIRONMENTAL CONSERVATION/AIR AND HAZARDOUS WASTE MANAGEMENT SECTION, Box O, Juneau, AK, 99811-1800, (907) 465-2666, (10).

Hunson, Peter, RADON DETECTION SYSTEMS/SIOUX FALLS, SOUTH DAKOTA BRANCH, 2713 South Prairie, Sioux Falls, SD, 57105, (303) 444-5253, (800) 627-2360, (1).

Hunt, Jim, HUNT'S WATER CONTROL, INC., 607 Tower View Dr., Taylor Mill, KY, 41015, (606) 292-0333, (1)(2).

Hunt, John R., RADON MEASUREMENT AND CONTROL, 7935 Kilbourne Rd., Sunbury, OH, 43074, (614) 866-1197, (1).

Hunt, Phil J., TRC-ECI, 800 Connecticut Blvd., East Hartford, CT, 06108, (203) 289-8631, (1).

Hunt, W.H., HUNT AND ASSOCIATES, 802 Avenue B, Rome, GA, 30161, (404) 232-7653, (404) 232-2034, (1).

Hunter, John, RADON HUNTER, THE, P.O. Box 6009, McLean, VA, 22106, (703) 237-1016, (1).

Hunter, Richard G., WICHITA-SEDGWICK COUNTY DEPARTMENT OF COMMUNITY HEALTH, 1900 East 9th St., Wichita, KS, 67214, (316) 268-8351, (10).

Hunter, Thomas D., RADON TESTING SERVICE, INC., 545 North 33rd, Milwaukee, WI, 53208, (414) 931-0527, (1).

Huntsperger, Gary, HOME CHEK, INC., 234 South Cardington, Wichita, KS, 67209, (316) 267-2324, (1).

Hupman, Janette E., ADVANCED AUTOMATIC ALARM COMPANY, INC., 1722 Mayappleway, Gambrills, MD, 21054, (301) 261-0011, (1).

Hurt, Vern, U.S. DEPARTMENT OF INTERIOR (DOI)/NATIONAL PARK SERVICE(NPS)/NORTH ATLANTIC REGIONAL OFFICE/INDOOR RADON PROGRAM, 15 State St., Boston, MA, 02109, (617) 565-8888, (8).

Hussler, Tim, ABMS, INC., 739 Allston Way, Berkeley, CA, 94710, (415) 548-4300, (1).

Hutchins, David, WASHINGTON STATE DEPARTMENT OF LABOR AND INDUSTY, 814 East 4th P.O. Box 207, Olympia, WA, 98504, (206) 753-6500, (10).

Hutchins, David, WEST VIRGINIA DEPARTMENT OF LABOR, State Capitol Room 415 1900 Washington St., Charleston, WV, 25305, (304) 348-7890, (10).

Hutchinson, Ed, DOW CORNING CORPORATION, Mail Stop CO2306, Midland, MI, 48686-0994, (517) 496-4578, (4).

Hutchinson, J.M. Robin, U.S. DEPARTMENT OF COMMERCE (DOC)/NATIONAL INSTITUTE OF STANDARDS AND TECHNOLOGY(NIST)/CENTER FOR RADIATION RESEARCH(CRR), Building 245 Room C114, Gaithersburg, MD, 20899, (301) 975-5532, (8).

Hyer, Steve A., WASHINGTON ASSOCIATION OF REALTORS, P.O. Box 719, Olympia, WA, 98507, (206) 943-3100, (18).

Hyma, Bowen, ENERGY SAVER IMPORTS, INC., P.O. Box 387 2150 West 6th Ave. Suite E, Broomfield, CO, 80020, (303) 469-1787, (800) 288-3647, (4).

Hymes, Corwin M., HEALTH PHYSICS NORTHWEST, INC., 11535 Southwest 67th Ave., Tigard, OR, 97223, (503) 620-6617, (800) 762-8444, (1)(3)(4).

I

Iachini, Stephen C., RADON DETECTION DEVICES, 5 Lauriann Ct., Parkton, MD, 21120, (301) 329-6306, (1).

Iannone, Pat, UTAH ASSOCIATION OF REALTORS, 5710 South Green St., Murray, UT, 84123, (801) 268-4747, (18).

Ibert, Edward R., GALVESTON COUNTY HEALTH DISTRICT/AIR POLLUTION CONTROL DIVISION, 1207 Oak St. P.O. Box 939, La Marque, TX, 77568, (409) 938-7221, (10).

Ice, Roger, RADON CONTROL PRODUCTS, 511 Industrial Dr., Carmel, IN, 46032, (317) 846-7442, (800) 523-2082, (4).

Iden, Richard L., IMPERIAL POLK COUNTY/BUILDING DIVISION, P.O. Box 726, Bartow, FL, 33830, (813) 534-6573, (10).

Ingham, Bob, CARE SYSTEMS, P.O. Box 122, Brentwood, TN, 37027, (615) 373-4275, (2).

Ingham, Cassandra, ONEIDA COUNTY BOARD OF COOPERATIVE EDUCATIONAL SERVICES (BOCES), Middle Settlement Rd. Box 70, New Hartford, NY, 13413, (315) 768-4583, (10).

Ingrams, Mark, I & M RADON MITIGATION, 529 Browar, Eaton, OH, 45320, (513) 456-6784, (2).

Ip, John, BARRINGER MAGENTA, LTD., 304 Carlingview Dr., Rexdale, ON, M9W 5G2, (416) 675-3870, (1).

Iqbal, M.A., IQBAL CONSULTING ENGINEERS, 5 Ironstone Dr., Reading, PA, 19606, (215) 779-8345, (3).

Irving, George W., U.S. DEPARTMENT OF DEFENSE (DOD)/ARMED FORCES RADIOBIOLOGY RESEARCH INSTITUTE(AFRRI), 8901 Wisconsin Ave. WH-550, Bethesda, MD, 20814, (301) 295-1210, (8).

Irwin, John, KANSAS DEPARTMENT OF HEALTH AND ENVIRONMENT/BUREAU OF AIR QUALITY AND RADIATION CONTROL/ENVIRONMENTAL SURVEILLANCE AND EMERGENCY PLANNING SECTION, Forbes Field Building Suite 740, Topeka, KS, 66620, (913) 296-1567, (1)(10).

Iskow, Milton, RADON REMEDIATION SERVICES, 10400 Connecticut Ave. Suite 504, Kensington, MD, 20895, (301) 933-7900, (2).

Iskow, Milton, RDS/RADON DETECTION SERVICES-MARYLAND OFFICE, 10400 Connecticut Ave. Suite 504, Kensington, MD, 20895, (301) 933-7900, (1)(3).

Isler, John R., AIR ANALYSIS ASSOCIATES, 2063 Stanley Rd., Cazenovia, NY, 13035, (315) 655-8948, (1).

Isler, John R., AIR ANALYSIS ASSOCIATES, P.O. Box 240, Unadilla, NY, 13849, (607) 369-7689, (1).

Isler, Linda W., AIR ANALYSIS ASSOCIATES, 2063 Stanley Rd., Cazenovia, NY, 13035, (315) 655-8948, (1).

Isler, Linda W., AIR ANALYSIS ASSOCIATES, P.O. Box 240, Unadilla, NY, 13849, (607) 369-7689, (1).

Ivancich, Michael, NATIONAL RADON TESTING CENTER, INC., 18 Marshall St., Norwalk, CT, 06854, (203) 853-4777, (203) 853-1219, (1).

Iversen, Tom, HOMEBUYERS INSPECTION SERVICE, INC., 1000 East Chicago Ave., Naperville, IL, 60540,

(312) 357-0903, (1).

J

Jaccarello, John A., HOUSING INSPECTION SERVICES, INC., P.O. Box 373, Murrysville, PA, 15668, (412) 374-1900, (1).

Jackson, J. Wayne, VIRGINIA DRY-BASE, 198 Milnwood Rd. P.O. Box 562, Farmville, VA, 23901, (800) 782-5042, (1).

Jacobi, Warren E., COLORADO DEPARTMENT OF HEALTH/RADIATION CONTROL DIVISION, 4210 East 11th Ave., Denver, CO, 80220, (303) 331-4812, (10).

Jacobs, Bruce, GENERAL PHYSICS CORPORATION, 10650 Hickory Ridge Rd., Columbia, MD, 21044, (301) 290-2300, (301) 997-6241, (1).

Jacobs, Debra J., DJ'S RADIATION PROTECTION, Route 2 Box 332B, St. Pauls, NC, 28384, (919) 425-6680, (1).

Jacobs, Sol J., MINNEAPOLIS POLLUTION CONTROL DIVISION/DEPARTMENT OF INSPECTIONS, 250 South 4th St. Public Health Ctr. Room 300, Minneapolis, MN, 55415, (612) 348-7897, (10).

Jacobsen, Jacob, INTERNATIONAL EXPLORATIONS, 577 Sackettsford Rd., Warminster, PA, 18974, (215) 598-7137, (1)(2).

Jacobsen, Madeleine, RADON RESOURCES, 241 Rumson Rd., Atlanta, GA, 30305, (404) 231-4732, (1).

Jacobson, A.P., UNIVERSITY OF MICHIGAN/SCHOOL OF PUBLIC HEALTH/DEPARTMENT OF ENVIRONMENTAL AND INDUSTRIAL HEALTH/RADON RESOURCE AND TRAINING CENTER, 109 Observatory, Ann Arbor, MI, 48109, (313) 764-0523, (3)(6).

Jacobson, Alan J., OLD COLONY PEST CONTROL COMPANY, P.O. Box 580, Pembroke, MA, 02359, (617) 294-0937, (1).

Jacobson, Dave, GREAT PLAINS NUCLEAR SERVICES, INC., 4502 Leavenworth St., Omaha, NE, 68106, (402) 553-3776, (1).

Jacobson, Robert S., ANNAPOLIS PROPERTY SERVICES, 1680 Yorktown Ct., Crofton, MD, 21114, (301) 721-7177, (1).

Jaffe, Joe, INTERNATIONAL MEDCOM, 7497 Kennedy Rd., Sebastopol, CA, 95472, (707) 823-0336, (4).

James, Gary, RADON DETECTION SYSTEMS/DALLAS, TEXAS BRANCH, 6607 Northwest Hwy., Dallas, TX, 75225, (303) 444-5253, (800) 627-2360, (1).

James, J.Z., AMERICAN NUCLEAR SOCIETY/CALIFORNIA-NORTHERN CALIFORNIA SECTION, C/O James Engineering 6329 Fairmount Ave., El Cerrito, CA, 94530, (415) 525-7350, (14).

James, Myriam, ALA OF THE VIRGIN ISLANDS, P.O. Box 974, St. Thomas, VI, 00801, (809) 774-2077, (15).

James, William R., FORENSIC ANALYTICAL SPECIALTY, 745 Lincoln Ave. Suite 4, Alameda, CA, 94501, (415) 887-8828, (415) 769-1180, (1).

Janeski, Larry, CONNECTICUT BASEMENT DE-WATERING AND RADON SYSTEMS, INC., 1 Clayton St., Milford, CT, 06460, (203) 874-3491, (1)(2).

Jansen, Raymond, U.S. DEPARTMENT OF INTERIOR (DOI)/U.S. GEOLOGICAL SURVEY(USGS)/WATER RESOURCES DIVISION(WRD)/WATER RESOURCES SCIENTIFIC INFORMATION CENTER(WRSIC), 425 National Ctr., Reston, VA, 22092, (703) 648-6821, (8).

Jarnutowski, Robert J., BECKMAN INSTRUMENTS, INC., 2500 Harbor Blvd., Fullerton, CA, 92634, (714) 871-4848, (1)(4).

Jarrett, Paul, GLOBAL RADON TESTING SERVICE TRAINING CENTER, 163 Padgett Dr., Lakeland, FL, 33809, (813) 859-5891, (1)(6).

Jarrett, Robert, U.S. DEPARTMENT OF AGRICULTURE (USDA)/AGRICULTURE RESEARCH SERVICE(ARS)/RADIATION SAFETY(RS), Building 001 Room 226 BARC-West, Beltsville, MD, 20705, (301) 344-4492, (8).

Jarvis, John C., HAZEN RESEARCH, INC., 4601 Indiana St., Golden, CO, 80203, (303) 279-4501, (1).

Jaske, E. Juhani, FMT EQUIPMENT, P.O. Box 6198, Hamden, CT, 06517, (203) 865-2864, (1).

Jaslow, Lee, RESOURCE CONSERVATION TECHNOLOGY, INC., 2633 North Calvert St., Baltimore, MD, 21218, (301) 366-1146, (4).

Javitch, Joel M., ELECTRO ANALYTICAL, INC., 7118 Industrial Park Blvd., Mentor, OH, 44060, (216) 951-3514, (1).

Jaworski, Richard L., AMERICAN NUCLEAR SOCIETY/NEBRASKA-NEBRASKA SECTION, C/O Omaha Public Power District 1623 Harney St., Omaha, NE, 68102, (402) 536-4468, (14).

Jazewsky, Ed, U.S. DEPARTMENT OF ENERGY (DOE)/CHICAGO OPERATIONS OFFICE(COO)/ENVIRONMENTAL SAFETY AND HEALTH(ESH), 9800 South Cass Ave., Argonne, IL, 60439, (312) 972-4800, (8).

Jean, Lucie, SHELBURNE LABORATORIES, INC., P.O. Box 458 Falls Rd., Shelburne, VT, 05482, (802) 985-3379, (1).

Jeck, Richard K., RADON HOME INSPECTION SERVICE, P.O. Box 292, Churchton, MD, 20733, (301) 261-5650, (1).

Jelinek, Robert, RICHARD P. ARBER ASSOCIATES, 44 Cook St., Suite 601, Denver, CO, 80206, (303) 393-6363, (3)(5).

Jelinek, Steve, RADON TESTING AND TREATMENT, INC., 723 North Weber Suite 300, Colorado Springs, CO, 80903, (719) 520-5693, (1)(2)(4).

Jenkins, Peggy, CALIFORNIA AIR RESOURCES BOARD/AIR QUALITY DIVISION, 1102 Q St., Sacramento, CA, 95814, (916) 323-5043, (10).

Jennemann, Paul, MARYLAND RADON LABORATORY, INC., P.O. Box 2578, Columbia, MD, 21045, (301) 997-2366, (1).

Jennemann, Paul V., RADIATION INSTRUMENT COMPANY (RICO), 9165 Helaine Hamlet Way, Columbia, MD, 21045, (301) 997-7426, (4).

Jensen, Larry, U.S. ENVIRONMENTAL PROTECTION AGENCY REGION 5 (EPA5)/RADIATION DIVISION(ORP), 230 S.Dearborn St., Chicago, IL, 60604, (312) 886-6175, (8).

Jensen, Lee, ASSOCIATION OF MAJOR CITY BUILDING OFFICIALS (AMCBO), 481 Carlisle, Herndon, VA, 22070, (703) 437-0100, (13).

Jensen, Mike, U.S. DEPARTMENT OF INTERIOR (DOI)/NATIONAL PARK SERVICE(NPS)/ROCKY MOUNTAIN REGIONAL OFFICE/INDOOR RADON PROGRAM, P.O. Box 25287, Denver, CO, 80225, (303) 969-2503, (8).

Jerrett, Lyle E., FLORIDA DEPARTMENT OF HEALTH AND REHABILITATIVE SERVICES/DEPUTY SECRETARY FOR HEALTH/OFFICE OF RADIATION CONTROL, 1317 Winewood Blvd., Tallahassee, FL, 32399-0700, (904) 487-1004, (10).

Jerszyk, Jeffrey P., FITCHBURG/BOARD OF HEALTH, 718 Main St., Fitchburg, MA, 01420, (617) 342-9843, (10).

Jessup, F. Milton, RADON ENGINEERING SERVICES, INC., 3129 Overton Rd., Birmingham, AL, 35223, (205) 967-2665, (1)(2).

Jessup, F. Milton, RADON EVALUATION SERVICES, INC., P.O. Box 13609, Atlanta, GA, 30324, (404) 478-0037, (1)(2).

Jester, William A., PENNSYLVANIA STATE UNIVERSITY, THE/BREAZEALE NUCLEAR REACTOR/LOW LEVEL RADIATION MONITORING LABORATORY, 101 Academic Projects Building, University Park, PA, 16802, (814) 863-2647, (1)(5).

Jewell, Bill, ALPHA RADON SCREENING AND HOME INSPECTION, INC., 1216 Chinook Way, Boulder, CO, 80303, (303) 494-9533, (1).

Jeziorski, Mark, RADON TESTING CONSULTANTS, INC., P.O. Box 25313, Newark, NJ, 07101, (201) 290-9217, (1)(2)(4).

Jiambalvo, John, POLLENEX/ASSOCIATED MILLS, INC., 165 North Canal St., Chicago, IL, 60606, (312) 454-5400, (4).

Johansen, Jim, INSPEX OF BOULDER, INC., 2955 Valmont Suite 300, Boulder, CO, 80301, (303) 442-8995, (1).

John, Frank, UNITED GILSONITE LABORATORIES, P.O. Box 70, Scranton, PA, 18501, (717) 344-1202, (4).

Johnson, Albert A., NATIONAL ASSOCIATION OF REAL ESTATE BROKERS (REALTIST), P.O. Box 56483, Washington, DC, 20041, (202) 289-6655, (18).

Johnson, Bruce, ACCU-TEST, 1350 Beverly Rd. Suite 115-343, McLean, VA, 22101, (703) 448-8088, (1).

Johnson, Dale L., AIR-N-SOL CORPORATION, 3 Cedar St. P.O. Box 437, Frenchtown, NJ, 08825, (201) 996-2028, (1)(3).

Johnson, Donald, U.S. DEPARTMENT OF COMMERCE (DOC)/NATIONAL INSTITUTE OF STANDARDS AND TECHNOLOGY(NIST)/NATIONAL MEASUREMENTS LABORATORY(NML), Physics Building Room A363, Gaithersburg, MD, 20899, (301) 975-4500, (8).

Johnson, Gordon C., U.S. DEPARTMENT OF HEALTH AND HUMAN SERVICES (HHS)/PUBLIC HEALTH SERVICE PHS)/FOOD AND DRUG ADMINISTRATION FDA)/CENTER FOR DEVICES AND RADIOLOGICAL HEALTH(CDRH), Office Of Health Affairs(OHA) 8757 Georgia Ave., Silver Spring, MD, 20910, (301) 427-7004, (8).

Johnson, Harvey, JOHNSON'S ENERGY SERVICE, 104 Central Dr., Forest City, IA, 50436, (515) 582-2998, (1)(2).

Johnson, Johnny O., ECTOR COUNTY HEALTH DEPARTMENT/AIR POLLUTION CONTROL DIVISION, 221 North Texas, Odessa, TX, 79761, (214) 670-7520, (10).

Johnson, Kirk, JOHNSON PLUMBING AND HEATING, 232 South Main St., Chambersburg, PA, 17201, (717) 264-5028, (1)(2).

Johnson, Mark, RADON TESTING SERVICES, PITTSBURGH, P.O. Box 19425, Pittsburgh, PA, 15213, (412) 687-1533, (1).

Johnson, Neil, U.S. DEPARTMENT OF ENERGY (DOE)/ALBUQUERQUE OPERATIONS OFFICE(AOO)LOVELACE INHALATION TOXICOLOGY INSTITUTE(LITI), P. O. Box 5890, Albuquerque, NM, 87185, (505) 844-2203, (8).

Johnson, Paul E., ENVIRONMENTAL DETECTION SERVICES, 1046 West Peakview Ave., Littleton, CO, 80120, (303) 797-7771, (1).

Johnson, Raymond, COMMUNICATION SCIENCES

INSTITUTE, 16440 Emory Ln., Rockville, MD, 20853, (301) 570-0984, (3).

Johnson, Richard, EXECUTIVE OFFICE OF THE PRESIDENT (EOP)/OFFICE OF SCIENCE AND TECHNOLOGY POLICY(OSTP), New Executive Office Building Room 5005, Washington, DC, 20500, (202) 395-4697, (8).

Johnson, Russell K., NORTHEAST UTILITIES, P.O. Box 270, Hartford, CT, 06141-0270, (203) 665-5000, (13).

Johnson, Steve, CHATAUQUA COUNTY HEALTH DEPARTMENT, Health and Social Services Building, Mayville, NY, 14757, (716) 753-4481, (10).

Johnson, Stewart, RADON TEST SYSTEMS, 1418 Upfield, Carlton, TX, 75006, (214) 242-1020, (1).

Johnson, Terry, NATIONAL ENVIRONMENTAL HEALTH ASSOCIATION (NEHA), 720 South Colorado Blvd. South Tower, Suite 970, Denver, CO, 80222, (303) 756-9090, (15).

Johnson, William B., AMERICAN NUCLEAR SOCIETY/NEW JERSEY-NORTHERN NEW JERSEY SECTION, C/O William B. Johnson And Associates P.O. Box 98 Research Park, Montville, NJ, 07045, (201) 334-9222, (14).

Johnston, Fran, RADON ANALYSIS, INC., P.O. Box 5224, Toms River, NJ, 08754-5224, (201) 929-9575, (1).

Johnston, Laurence, LARIMER COUNTY HEALTH DEPARTMENT/ENVIRONMENTAL HEALTH DIVISION, 363 Jefferson St., Fort Collins, CO, 80524, (303) 221-7497, (10).

Johnston, Robert, BASEMENT DE-WATERING SYSTEMS, INC., 156 Curry Hill Rd., Canonsburg, PA, 15317, (412) 745-5959, (1)(2).

Jokerst, Todd A., RESIDENTIAL RADON MEASUREMENTS, P.O. Box 3051, Fairview Heights, IL, 62208, (618) 397-0099, (1).

Jolley, Richard, J & J QUALITY HOME INSPECTIONS, 909 East Lynda Ln., Marietta, GA, 30060, (404) 425-8868, (1).

Jolly, Henry L., SOUTH CAROLINA REAL ESTATE COMMISSION, 1201 Main St. Suite 1500, Columbia, SC, 29201, (803) 737-0700, (10).

Jones, Dallas, RADON REDUCTION AND TESTING, INC., 2222 Cumberland Pl., Smyrna, GA, 30080, (404) 434-7296, (1).

Jones, Dennis, STRATA ENVIRONMENTAL SERVICE, INC., 401 East Douglas Suite 515, Wichita, KS, 67202, (316) 262-0002, (1)(3).

Jones, Fred W., ABC INSPECTION COMPANY, 9911 Santa Cruz St., Seabrook, MD, 20706, (301) 577-1915, (1).

Jones, Michael J., GALLIPOT, INC., 2401 Pilot Knob Rd., St. Paul, MN, 55120, (612) 681-9517, (800) 423-6967, (1)(4).

Jones, R. Anne, NEW JERSEY INSTITUTE OF TECHNOLOGY/DEPARTMENT OF CIVIL AND ENVIRONMENTAL ENGINEERING, 323 King Blvd., Newark, NJ, 07102, (201) 761-0977, (5).

Jones, Robert P., ALA OF VERMONT, 30 Farrell St., South Burlington, VT, 05401, (802) 863-6817, (15).

Jones, Susan, E.O. CHURCH, INC., 925 East 17th Ave., Denver, CO, 80218, (303) 832-9692, (1).

Jones Kincaide, Holly, DISTRICT OF COLUMBIA DEPARTMENT OF CONSUMER AND REGULATORY AFFAIRS/REAL ESTATE COMMISSION, 614 H St., NW Room 923, Washington, DC, 20013, (202) 727-7468, (202) 727-7469, (10).

Jordan, Bruce, CCI-RADON AND HOME INSPECTION/RADON DETECTION, P.O. Box 1045, Malvern, PA, 19355, (215) 889-1554, (215) 363-9015, (1)(2)(4).

Jordan, David R., INDIANAPOLIS/AIR POLLUTION CONTROL DIVISION, 2700 South Belmont Ave., Indianpolis, IN, 46221, (317) 633-5496, (10).

Jordan, Donna, AIR CONCERNS, 1919 Oakwell Farms Pkwy. Suite 120, San Antonio, TX, 78218, (512) 820-8875, (1).

Jordan, Patricia, RADON SYSTEMS, INC., P.O. Box 2509, Laurel, MD, 20708, (301) 725-9000, (1)(2).

Jordan, Richard, RADON ENVIRONMENTAL SERVICES, INC., 8935 North Meridian St. Suite 110, Indianapolis, IN, 46260, (317) 846-1991, (1)(2).

Jordan, Richard, RADON ANALYTICAL LABS, 9011 North Meridian St., Indianapolis, IN, 46260, (317) 843-0788, (1).

Jordan, Robert, DELAWARE GEOLOGICAL SURVEY, University Of Delaware, Newark, DE, 19716, (302) 452-2833, (10).

Jordan, Vickie, BLUEGRASS RADON SERVICES, 424 Ridgeway Rd., Lexington, KY, 40502, (606) 266-6148, (1).

Jordan, Vickie, RADON ANALYTICAL LABS, 9011 North Meridian St., Indianapolis, IN, 46260, (317) 843-0788, (1).

Joy, Jack, MINE SAFETY APPLIANCES COMPANY, P.O. Box 426, Pittsburgh, PA, 15230, (412) 967-3000, (4).

Joy, Richard, FAIRBANKS NORTH STAR BOROUGH/AIR POLLUTION CONTROL DIVISION, P.O. Box 1267, Fairbanks, AK, 99701, (907) 452-4761, (10).

Joyce, Donald R., JOYCE EXTERMINATING COMPANY, Route 2, Box 249, Ramseur, NC, 27316, (919) 824-2358, (1).

Joyce, Sandra, METRO RADON MEASUREMENT, 1565 Cliff Rd. Suite 3-101, Eagan, MN, 55122, (612) 681-1829, (1)(4).

Joyce, Steven, METRO RADON MEASUREMENT, 1565 Cliff Rd. Suite 3-101, Eagan, MN, 55122, (612) 681-1829, (1)(4).

Ju, Carolyn, NEW YORK STATE HEALTH DEPARTMENT/BUREAU OF ENVIRONMENTAL AND OCCUPATIONAL EPIDEMIOLOGY/ENVIRONMENTAL HEALTH SECTION, Radon Registry, Albany, NY, 12203, (518) 458-6212, (10).

Juliano, Bob, ENVIRONMENTAL SCIENCE LABORATORY, INC., 26 Pearl St. Suite 110, Bellingham, MA, 02019, (508) 966-4344, (800) 343-0707, (1)(4).

Jungblut, David W., JUNGBLUT, DAVID W., GEOLOGIST, Route 12 Box 168, Baptistown, NJ, 08803, (201) 996-3688, (1).

Junghans, Ted, TRACOR JITCO RADON MEASUREMENT SERVICES, 1601 Research Blvd., Rockville, MD, 20850, (301) 984-2800, (1)(4).

Jupa, Michael L., UNION COUNTY BUILDING INSPECTION SERVICE, 502 Twin Oaks Rd., Union, NJ, 07083, (201) 687-6110, (1).

K

Kaganovich, Lev, RADON DETECTION CORPORATION OF MICHIGAN, 4740 Hall St., SE, Kentwood, MI, 49506, (616) 942-0323, (1).

Kahlson, Don, LEWIS AND KAHLSON RADON CONSULTANTS, 459 Dunlin Plaza, Secaucus, NJ, 07094, (201) 866-7219, (1).

Kalka, Jr., James E., KALKA MIDWEST CONSTRUCTION, INC., 1710 South Michigan St., South Bend, IN, 46613, (219) 287-3399, (219) 234-9206, (1).

Kalthoff, Charles, ACCORD CORPORATION, 84 Schuyler St., Belmont, NY, 14813, (716) 268-7605, (1).

Kaminsky, John, ENVIRONMENTAL MAINTENANCE SPECIALISTS COMPANY, INC., 1429 Danby Rd., Ithaca, NY, 14850, (607) 273-5938, (2).

Kance, Aaron, RADON DETECTION SYSTEMS/PORTLAND, OREGON BRANCH, 3710 SE 36th Pl. #27, Portland, OR, 97202, (303) 444-5253, (800) 627-2366, (1).

Kane, Brian, REAL ESTATE SUPPORT SERVICES, INC., 7825 Washington Ave., S., Minneapolis, MN, 55435, (612) 941-7558, (1).

Kane, Margaret E., ALA OF RHODE ISLAND, 10 Abbott Park Pl., Providence, RI, 02903-3703, (401) 421-6487, (15).

Kaplan, Joan, RADON MITIGATORS OF ARIZONA, 6844 East Thunderbird Rd., Scottsdale, AZ, 85254, (602) 483-7506, (1)(2).

Kaplan, Joan, RADON MITIGATORS, INC., 16 Hillside Dr., Clinton, NJ, 08809, (201) 730-6596, (1)(2).

Kaplan, Lawrence, RADON MITIGATORS, INC., 16 Hillside Dr., Clinton, NJ, 08809, (201) 730-6596, (1)(2).

Kaplan, Lawrence H., RADON MITIGATORS OF ARIZONA, 6844 East Thunderbird Rd., Scottsdale, AZ, 85254, (602) 483-7506, (1)(2).

Kapus, Marvin, MASTER BUILDING SERVICE, 245 Huguenot St. Suite 107, New Rochelle, NY, 10801, (914) 235-8282, (1)(2).

Karayianes, George, URETECH, INC., 154 Potomac Ave., Tallmadge, OH, 44278, (216) 633-3586, (4).

Karell Jr., John E., PUTNAM COUNTY HEALTH DEPARTMENT, Putnam County Office Building 110 Old Route Six Center Building #3, Carmel, NY, 10512, (914) 225-2115, (10).

Karim, Jasmin, ENVIRONMENTAL CONTROL SYSTEMS, LTD., 15325 Jones Ln., Gaithersburg, MD, 20878, (301) 421-5150, (1)(3).

Karl, Christy, GARON PRODUCTS, INC., 1974 Hwy. 35 CN20, Wall, NJ, 07719, (201) 499-1776, (800) 631-5380, (4).

Karn, Donald, CONNECTICUT DEPARTMENT OF ENVIRONMENTAL PROTECTION/RADIATION CONTROL UNIT, State Office Building 165 Capital Ave., Hartford, CT, 06106, (203) 566-5668, (10).

Karpel, Richard J., HOME RADON DETECTION, INC., 33488 Lone Rock, Wildwood, IL, 60030, (312) 223-6517, (1).

Karpinski, Gene, U.S. PUBLIC INTEREST RESEARCH GROUP (US PIRG), 215 Pennsylvania Ave., SE, Washington, DC, 20003, (202) 546-9707, (17).

Karr, Janet, HMA HOME INSPECTION AND WARRANTY SERVICE, 13250 Ridge Rd. Unit 7-4, Largo, FL, 32257, (813) 587-0776, (800) 330-0776, (1).

Kasabach, Haig, NEW JERSEY GEOLOGICAL SURVEY/DIVISION OF WATER RESOURCES, CN-029, Trenton, NJ, 08625, (609) 292-1185, (10).

Kasmir, Bernard, BERNELL PROFESSIONAL SERVICES, 361 12th St., Cresskill, NJ, 07626, (201) 568-7083, (1)(3).

Kass, Benny L., NATIONAL HOME BUYERS AND HOME OWNERS ASSOCIATION (NHBHOA), 1050 17th St., NW, Washington, DC, 20036, (202) 659-6500, (17).

Kaufman, Marvin E., AMERICAN GEOLOGICAL INSTITUTE (AGI), 4220 King St., Alexandria, VA, 22310, (703) 379-2480, (14).

Kaufmann, F.B., FORESIGHT ENGINEERING, INC., 17 Cook Ave. P.O. Box 621, Madison, NJ, 07940, (201) 377-0600, (1)(2).

Kaufmann, T.D., FORESIGHT ENGINEERING, INC., 17 Cook Ave. P.O. Box 621, Madison, NJ, 07940, (201) 377-0600, (1)(2).

Kawka, Ganice, CANADIAN INSTITUTE FOR RADIATION SAFETY/RADON IN HOMES COURSE, 595 Bay St. Suite 1050, Toronto, ON, M5G 2C2, (416) 596-1617, (3)(6).

Kay, Jack, DREXEL UNIVERSITY/DEPARTMENT OF CHEMISTRY/ATMOSPHERIC CHEMISTRY PROGRAM, 32nd and Chestnut Sts. 3rd Floor, Philadelphia, PA, 19104, (215) 895-2638, (5).

Kaye, Steve, U.S. DEPARTMENT OF ENERGY (DOE)/OAK RIDGE OPERATIONS OFFICE(OROO)/OAK RIDGE NATIONAL LABORATORY(ORNL)/HEALTH AND SAFETY RESEARCH DIVISION/MEASUREMENTS AND APPLICATIONS SECTION, P.O. Box 2008 Building 7509 South MS-6383, Oak Ridge, TN, 37831, (615) 576-2100, (8).

Kaz, Richard, INDUSTRIAL OPERATIONS, INC./RADON EVALUATION, 3736 South 7th Ave., Phoenix, AZ, 85041, (602) 268-8772, (1).

Kearfott, Kimberly, ARIZONA STATE UNIVERSITY/COLLEGE OF ENGINEERING AND APPLIED SCIENCES/DEPARTMENT OF ELECTRICAL AND COMPUTER ENGINEERING, Department Of Electrical And Computer Engineering, Radiological Imaging Facility, Tempe, AZ, 85287-5706, (602) 965-2727, (5).

Kearney, Philip D., ENVIRONMENTAL RADIATION SERVICES, INC., 305 West Magnolia St. Suite 373, Fort Collins, CO, 80521, (303) 226-5530 , (1).

Keast, Martin, TULARE COUNTY AIR POLLUTION CONTROL DISTRICT/ENVIRONMENTAL HEALTH DIVISION, Health Building County Civic Ctr., Visalia, CA, 93291, (209) 733-6441, (10).

Keaton, Harlan, CONFERENCE OF RADIATION CONTROL PROGRAM DIRECTORS (CRCPD)/COMMITTEE ON RADON POLICY, 71 Fountain Pl., Frankfort, KY, 40601, (502) 227-4543, (14).

Keaton, Harlan, FLORIDA DEPARTMENT OF HEALTH AND REHABILITATIVE SERVICES/DEPUTY SECRETARY FOR HEALTH/OFFICE OF RADIATION CONTROL/ENVIRONMENTAL SECTION, P.O. Box 15490, Orlando, FL, 32858, (407) 628-2274, (407) 297-2097, (1)(10).

Keckler, Norman F., AKRON REGIONAL AIR QUALITY MANAGEMENT DISTRICT/AIR POLLUTION CONTROL DIVISION, 177 South Broadway, Akron, OH, 44308, (216) 375-2480, (10).

Keefe, Lawrence, NEW YORK STATE HEALTH DEPARTMENT/BUREAU OF ENVIRONMENTAL RADIATION PROTECTION/ENVIRONMENTAL RADIATION SECTION, 2 University Place, Albany, NY, 12203, (800) 458-1158, (10).

Keegan, David, RADON DETECTION AGENCY, LTD., 625 5th Ave. Suite 104, New Westminster, BC, V3M 1X4, (604) 525-4455, (1)(4).

Keeley, Kurt M., AMERICAN WATER WORKS ASSOCIATION (AWWA), 6666 West Quincy Ave., Denver, CO, 80235, (303) 794-7711, (14).

Keene, Byron, U.S. ENVIRONMENTAL PROTECTION AGENCY REGION 1 (EPA1)/AIR AND RADIATION DIVISION(OAR), JFK Federal Building Room 2203, Boston, MA, 02203, (617) 565-3234, (8).

Keene, Terrie, ENVIRONMENTAL RADON MEASUREMENTS, 49 North Orange Ave., Orlando, FL, 32801, (407) 648-1969, (800) 237-3065, (1)(2).

Keene, Warren, CATHOLIC UNIVERSITY OF AMERICA, Pangborn Hall Radiation Safety Dept., Washington, DC, 20065, (202) 635-5206, (5).

Keff, Brian, ACCU-TEST, 1350 Beverly Rd. Suite 115-343, McLean, VA, 22101, (703) 448-8088, (1).

Keil, Ronald L., NATURAL RESOURCE LABORATORY, INC., 1100 Simms St., Golden, CO, 80401, (303) 233-8155, (1).

Keiser, Meg, RADIATION RESEARCH SOCIETY (RRS), 1101 Market St. 14th Floor, Philadelphia, PA, 19107, (215) 574-3153, (5)(14).

Keller, Darlene F., FLORIDA DEPARTMENT OF PROFESSIONAL REGULATION/DIVISION OF REAL ESTATE, P.O. Box 1900, Orlando, FL, 32802-1900, (407) 423-6048, (10).

Kelley, William, AMERICAN COUNCIL OF GOVERNMENTAL INDUSTRIAL HYGIENISTS (ACGIH), 6500 Glenway Ave. Building D-7, Cincinnati, OH, 45211, (513) 661-7881, (16).

Kellison, Michael, RADON GASBUSTERS, 892 Green Ln., Union, NJ, 07083, (201) 353-5577, (1).

Kellogg, Ed, KELLOGG ENVIRONMENTAL SERVICES, INC., 816 Anthony Dr., Mechanicsburg, PA, 17055, (717) 732-3364, (717) 732-9263, (1)(2).

Kellogg, Jeanette, GENOVA PRODUCTS, INC., 7034 East Court St., Davison, MI, 48423, (313) 744-4500, (800) 521-7488, (4).

Kelly, Samuel E., MADRONA DIVERSIFIED, INC., 1938 Fairview Ave., E., Seattle, WA, 98102, (206) 324-0920, (1).

Kemner, Jonathan L., NATIONAL MULTI HOUSING COUNCIL (NMHC), 1250 Connecticut Ave., NW Suite 620, Washington, DC, 20036, (202) 659-3381, (18).

Kempski, L.T., ADVANCED BASEMENT DE-WATERING AND RADON CONTROL SYSTEMS, 20 Giblin Dr., West Seneca, NY, 14224, (716) 675-5898, (716) 674-6070, (2).

Kenneweg, Wesley J., NATIONAL DRAEGER, INC., 101 Technology Dr., Pittsburgh, PA, 15275, (412) 787-8383, (4).

Kenney, Kim, CURRENT INDOOR AIR SYSTEMS, INC., P.O. Box 18075, Boulder, CO, 80308, (303) 440-8555, (4).

Kent, Jerard A., PETROLEUM ANALYTICAL LABORATORY, INC., P.O. Drawer AK, Eaton Park, FL, 33840, (813) 665-0028, (1).

Kent, John A., PETROLEUM ANALYTICAL LABORATORY, INC., P.O. Drawer AK, Eaton Park, FL, 33840, (813) 665-0028, (1).

Kent, Paul S., PETROLEUM ANALYTICAL LABORATORY, INC., P.O. Drawer AK, Eaton Park, FL, 33840, (813) 665-0028, (1).

Kerns, Gloria, RADON DETECTION SYSTEMS/SALT LAKE CITY, UTAH BRANCH, 6728 South 1655 East, Salt Lake City, UT, 84121, (303) 444-5253, (800) 627-2360, (1).

Kerr, Robert H., NEW BRUNSWICK DEPARTMENT OF JUSTICE/CONSUMER AFFAIRS BRANCH/REAL ESTATE COUNCIL, P.O. Box 785, Fredericton, NB, E3B 5B4, (506) 455-9733, (11).

Kershner, C.J., FEMTO-TECH, INC., 150 C Industry Dr., Carlisle, OH, 45005, (513) 746-4427, (1)(4)(5).

Khan, M.M. Ali, EAST CHICAGO/DEPARTMENT OF AIR QUALITY CONTROL, 4525 Indianapolis Blvd., East Chicago, IN, 46312, (219) 391-8297, (10).

Khanna, Jogi, SAN JOAQUIN COUNTY AIR POLLUTION CONTROL DISTRICT, 1601 East Hazelton Ave. P.O. Box 2009, Stockton, CA, 95201, (209) 468-3470, (10).

Kilchesty, Al, PECORA CORPORATION, 165 Wambold Rd., Harleysville, PA, 19438, (4).

Kile, Jr., C. Wilson, D-TECH, INC., 7 Laurence Pl., Plymouth Meeting, PA, 19462, (215) 825-4832, (1).

Killingbeck, James, NORTH DAKOTA STATE DEPARTMENT OF HEALTH AND CONSOLIDATED LABORATORIES/DIVISION OF ENVIRONMENTAL ENGINEERING, 1200 Missouri Ave. P.O. Box 5520, Bismarck, ND, 58502-5520, (701) 224-2348, (10).

Kimmel, William, WEATHER SHIELD, INC., 4071 South Salem Church Rd., Dover, PA, 17315, (717) 292-5613, (1)(2).

Kimmell, Mike, RSA CORPORATION/RADON SERVICE AGREEMENT, 918 Woodley Rd., Dayton, OH, 45403, (513) 254-4102, (3).

Kimura, Calvin, HAWAII DEPARTMENT OF COMMERCE AND CONSUMER AFFAIRS/PROFESSIONAL AND VOCATIONAL LICENSING DIVISION, P.O. Box 3469, Honolulu, HI, 96801, (808) 548-7579, (10).

Kincart, Robert O., AMERICAN COMPLIANCE TECHNOLOGIES, INC., 4404 South Florida Ave. Suite 12, Lakeland, FL, 33813, (813) 647-1200, (1)(2).

King, Eleanor, MAINE DEPARTMENT OF HUMAN SERVICES/BUREAU OF HEALTH/DIVISION OF HEALTH ENGINEERING/OCCUPATIONAL AND RESIDENTIAL HEALTH PROGRAM, State House Station 10, Augusta, ME, 04333, (207) 289-3826, (10).

Kingbury, Allan, U.S. DEPARTMENT OF INTERIOR (DOI)/NATIONAL PARK SERVICE(NPS)/ENVIRONMENTAL SANITATION PROGRAM, Box 37127, Washington, DC, 20013, (202) 343-7090, (8).

Kinney, Laurence, SYNERTECH SYSTEMS CORPORATION, 472 South Salina St. Suite 800, Syracuse, NY, 13202, (315) 422-3828, (1).

Kinslow, Roy H., KINSLOW, ROY H., 9935 Coppedge Ln., Dallas, TX, 75220, (214) 351-1446, (3).

Kirk, William P., RPI RADON MEASUREMENT SERVICES, P.O. Box 357, Hershey, PA, 17033, (717) 533-8100, (1)(4).

Kirsch, Laurence S., CADWALADER, WICKERSHAM AND TAFT, 1333 New Hampshire Ave., NW, Washington, DC, 20036, (202) 862-2317, (3).

Kirschbaum, Rebecca, CENTER FOR RADON DETECTION, 12 Bank St., Summit, NJ, 07901, (201) 277-3200, (201) 992-9702, (1).

Kitchens, Jim, RADEX, INC., 1382 Prince Ave. Suite A, Athens, GA, 30606, (404) 548-6960, (1)(2)(3).

Kittinger, David, HOUSE DOCTORS, INC., P.O. Box 10070, Colorado Springs, CO, 80932. (303) 574-6960, (1).

Kittka, Ernest, SOLAR RESOURCE TECHNOLOGY, 128 Market St., South Williamsport, PA, 17701, (717) 323-8919, (1)(2).

Kittrell, G. Ruth, KSE, INC., P.O. Box 368, Amherst, MA, 01004, (413) 549-5506, (1).

Kladder, Douglas L., RADON TESTING AND TREATMENT, INC., 723 North Weber Suite 300, Colorado Springs, CO, 80903, (719) 520-5693, (1)(2)(4).

Klassen, Ken, MANITOBA DEPARTMENT OF ENERGY AND MINES/INFORMATION CENTER/HOME CHECK-UP PROGRAM, 330 Graham Ave. Suite 555, Winnipeg, MB, R3C 4E3,

(204) 945-4154, (800) 282-8069, (11).

Klasterman, James C., RADON INSPECTION SERVICES, 915 6th Ave., North Unit C, Seattle, WA, 98109, (206) 286-0479, (1).

Klatt, Ernst R., EUROPEAN ENVIRONMENTAL BUREAU (EEB), 20, rue du Luxembourg, B-1040 Brussels, Belgium, No Phone Available, (12).

Klein, David, RADON CHECK, 2263 Coppersmith Sq., Reston, VA, 22091, (703) 476-6810, (1).

Klein, Fred, INSPECTIONS UNLIMITED, 514 Ryers Ave., Cheltenham, PA, 19012, (215) 848-0720, (1).

Klein, Gary, SYNERTECH SYSTEMS CORPORATION, 472 South Salina St. Suite 800, Syracuse, NY, 13202, (315) 422-3828, (1).

Klein, Walter, FLORIDA DEPARTMENT OF HEALTH AND REHABILITATIVE SERVICES/DEPUTY SECRETARY FOR HEALTH/OFFICE OF RADIATION CONTROL, 1317 Winewood Blvd., Tallahassee, FL, 32399-0700, (904) 487-1004, (6)(10).

Kline, Stephen, GEORGE MASON UNIVERSITY/CENTER FOR BASIC AND APPLIED SCIENCE, 4400 University Dr., Fairfax, VA, 22030, (703) 323-4318, (5).

Klosterman, James, RADON DETECTION AND REMOVAL SERVICES, 915 6th Ave., N., Seattle, WA, 98109, (206) 282-9156, (1)(2).

Klotz, Judy, NEW JERSEY DEPARTMENT OF HEALTH/DIVISION OF ENVIRONMENTAL AND OCCUPATIONAL HEALTH/ENVIRONMENTAL HEALTH SERVICES/AIR AND RADIATION DIVISION, John Fitch Plaza CN-360, Trenton, NJ, 08625, (609) 633-2043, (10).

Klotz, Norm, N.A.K. PRODUCTION ASSOCIATES, 1317 Apple Ave., Silver Spring, MD, 20910, (301) 565-0355, (3)(4)(6).

Kluk, Tony, U.S. DEPARTMENT OF ENERGY (DOE)/ASSISTANT SECRETARY FOR NUCLEAR ENERGY(ASNE)/OFFICE OF HAZARDOUS WASTE AND REMEDIAL ACTION, DP-14, Germantown, MD, 20874, (301) 353-4971, (8).

Knappmiller, Leonard V., RADON DETECTION SERVICES, P.O. Box 434, West Rutland, VT, 05777, (802) 775-0250, (802) 438-5603, (1).

Kneeland, Craig, NEW YORK STATE ENERGY OFFICE/BUREAU OF RESIDENTIAL CONSERVATION, 2 Rockefeller Plaza 8th Floor, Albany, NY, 12223, (518) 473-7243, (6)(10).

Knessy, Dave, INDIANA DEPARTMENT OF LABOR/BUREAU OF SAFETY, EDUCATION AND TRAINING, 1013 State Office Building, Indianapolis, IN, 46204, (317) 232-2688, (10).

Knickerbocker, Robert T., RTK ENGINEERING, 2939 Newton St., Denver, CO, 80211, (303) 477-0044, (1).

Knight, Ken, EG & G ORTEC, 100 Midland Rd., Oak Ridge, TN, 37831, (615) 482-4411, (800) 251-9750, (1)(4)(6).

Knoff, Patti, NATIONAL ASSOCIATION OF THE REMODELING INDUSTRY (NARI), 1901 North Moore St. Suite 808, Arlington, VA, 22209, (703) 276-7600, (13).

Knote, Elizabeth A., CAPE CHEMICAL COMPANY, 33 North Frederick St., Cape Girardeau, MO, 63701, (314) 334-3002, (1).

Knouse, Carol, SKC, INC., 334 Valley View Rd., Eighty Four, PA, 15330, (412) 941-9701, (800)752-8472, (4).

Knowlton, Blair, SEACOAST LABORATORIES, 55 Village Ln., South Hamilton, MA, 01982, (508) 468-3136, (1).

Knowlton, Stephen, SEACOAST LABORATORIES, 55 Village Ln., South Hamilton, MA, 01982, (508) 468-3136, (1).

Knudsen, Gerald J., THORO SYSTEM PRODUCTS, 7800 Northwest 38th St., Miami, FL, 33166, (305) 592-2081, (4).

Knudson, Mary, AAA DISASTER FORCE, 18219 West 3rd Ave. Suite 3, Golden, CO, 80401, (303) 399-4203, (1)(2).

Knudson, Paula, ACCENT ENGINEERING, INC., 5624 Yukon St., Arvada, CO, 80002, (303) 425-8700, (1).

Kobryn, Alexander Z., GLOBAL RADON TESTING SERVICE TRAINING CENTER, 163 Padgett Dr., Lakeland, FL, 33809, (813) 859-5891, (1)(6).

Koch, Donald A., IOWA GEOLOGICAL SURVEY BUREAU, 123 North Capitol St., Iowa City, IA, 52242, (319) 335-1575, (10).

Koczanowski, Robert S., QUALITY HOME INSPECTIONS, INC., 81 Main St., High Bridge, NJ, 08829, (201) 638-4994, (1).

Koepf, Brian, ALPHA CHECK, 7200 Wisconsin Ave., Bethesda, MD, 20814, (301) 856-3637, (1).

Kolankowski, Thaddeus, NATIONAL FUEL, 2484 Seneca St., Buffalo, NY, 14210, (716) 824-2500, (1).

Kolego, Laurence V., RADON TESTING SERVICES, BALTIMORE, P.O. Box 7999, Baltimore, MD, 21221-0999, (301) 321-8008, (1).

Kolodziej, David, AIR-TIGHT OF FARMINGTON, 28000 Middlebolt, Farmington Hills, MI, 48018, (313) 855-5065, (1).

Komp, Linda L., RADON SERVICES OF NORTH AMERICA, INC., 148 East 632 N., Valparaiso, IN, 46383, (219) 462-5866, (1).

Kondon, Nicholas, TESTING 1.2.3., 50 Freeway Dr., Cranston, RI, 02920, (401) 461-0530, (1).

Koniarczyk, Karl, RADONAIRE, INC., 393 Denrose Dr., Amherst, NY, 14150, (716) 837-2366, (1).

Konkus, Allan, ACCREDITED WATERPROOFING COMPANY, INC., 2504 Church Rd., Cherry Hill, NJ, 08002, (609) 482-2630, (215) 265-2888, (1)(2).

Koop, C. Everett, U.S. DEPARTMENT OF HEALTH AND HUMAN SERVICES (HHS)/PUBLIC HEALTH SERVICE(PHS)/OFFICE OF THE SURGEON GENERAL(OSG), 200 Independence Ave., SW, Washington, DC, 20201, (202) 245-6467, (8).

Koppell, Harry, AMERICAN NUCLEAR SOCIETY/OHIO-NORTHERN OHIO SECTION, C/O Harshaw/Filtrol 14450 Summerfield Rd., University Heights, OH, 44118, (216) 349-6590, (14).

Korchin, Stuart, RADIATION PROTECTION SERVICES, INC., P.O. Box 2395, Darien, CT, 06820, (203) 324-7967, (1)(2).

Kordares, George, MACMILLAN RESEARCH, LTD., 1221 Barclay Cir., Marietta, GA, 30061, (404) 427-3101, (1).

Kordex, Kevin, HOME EXTERMINATING COMPANY, 4608 Bucks Schoolhouse Rd., Baltimore, MD, 21237, (301) 661-7911, (800) 492-5544, (1).

Kordex, Kevin, PARAMOUNT PEST CONTROL, 110 Gordon Rd., Falls Church, VA, 22046, (703) 241-9140, (1).

Korenberg, Paul, W.R. GRACE/CONSTRUCTION PRODUCTS DIVISION, 62 Whittemore Ave., Cambridge, MA, 02140, (617) 876-1400, (4).

Kornbluth, Richard, EN-THERM, INC., 509 West Fayette St., Syracuse, NY, 13204, (315) 474-6549, (1)(2).

Korocij, Roman, HOUSE-CHECK, INC., P.O. Box 731, Valley Forge, PA, 19482, (215) 933-4431, (1).

Kosinski, L.A., INTERNATIONAL GEOGRAPHICAL UNION (IGU), University of Alberta Department of Geology, Edmonton, AB, T6G 2H4, (403) 432-3329, (12).

Koske, Burt, AMERICAN NUCLEAR SOCIETY/NEW MEXICO-TRINITY SECTION, C/O Public Service Company Of New Mexico Alvarado Square, Albuquerque, NM, 87158, (505) 848-2574, (14).

Koslow, James A., YOLO-SOLANO AIR POLLUTION CONTROL DISTRICT, 323 First St. Suite 5 P.O. Box 1006, Woodland, CA, 95695, (916) 666-8146, (10).

Kostige, John V., RMC ENVIRONMENTAL SERVICES/ROYERSFORD DIVISION, 323 Church St., Royersford, PA, 19468, (215) 948-4949, (3).

Kotrappa, Paul, RAD ELEC, INC., 5330 J Spectrum Dr., Frederick, MD, 21701, (301) 694-0011, (1)(4)(5).

Kottlowski, Frank, NEW MEXICO BUREAU OF MINES AND MINERAL RESOURCES, Campus Station, Socorro, NM, 87801, (505) 835-5420, (10).

Kovac, Matt, ELECTRO MECHANICAL CONCEPTS, 28 Woodruff Dr., Carnegie, PA, 15106-3054, (412) 276-2272, (1).

Kovach, Jeff, MAMECO INTERNATIONAL, INC., 4475 East 175th St., Cleveland, OH, 44128, (216) 752-4400, (800) 321-6412, (4).

Kovacs, J., KOVACS CONSULTANTS, 4515 Willard Ave. Suite 1921S, Chevy Chase, MD, 20815, (301) 652-7133, (2).

Kowalski, Donald C., AAA NEW ENGLAND RADON SERVICES, P.O. Box 1096, Westbrook, CT, 06498, (203) 399-5748, (1).

Kowalski, Donald C., SOUTHERN NEW ENGLAND RADON TESTING AND MITIGATION, P.O. Box 1141, Westbrook, CT, 06498, (203) 399-8003, (1)(2).

Kowalski, Paul A., NEW HAVEN HEALTH DEPARTMENT/ENVIRONMENTAL HEALTH DIVISION, One State St., New Haven, CT, 06511, (203) 787-8173, (10).

Kraft, Kirk R., VALLEY HOME INSPECTIONS, INC., P.O. Box 1677, Chandler, AZ, 85244-1677, (602) 988-1375, (1).

Krafthefer, Brian C., HONEYWELL, INC./RESIDENTIAL CONTROLS DIVISION, MN-10-2522 1985 Douglas Dr., N, Golden Valley, MN, 55422, (612) 542-3356, (4).

Kralick, Sharon, ENVIRADON, INC., P.O. Box 13683, Research Triangle Park, NC, 27709, (919) 848-2128, (1).

Kramer, Jeffery, SCWEST, INC., 4950 East Asbury Ave. Suite 5, Denver, CO, 80222, (303) 759-1286, (4).

Kramer, Paul, POLY-AMERICA, INC., 2000 West Marshall Dr., Grand Prairie, TX, 75051, (214) 647-4374, (800) 527-3322, (4).

Kramer, Sidney, MONTGOMERY COUNTY GOVERNMENT, 101 Monroe St., Rockville, MD, 20850, (301) 217-2500, (10).

Kranick, Carl J., SOLAR TESTING LABORATORIES, INC., 9540 Midwest Ave., Garfield Heights, OH, 44125, (216) 663-1800, (1).

Kras, Peter, ALL EASTERN HOME INSPECTION COMPANY, 309 Bloomfield Ave., Caldwell, NJ, 07006, (201) 228-5548, (1).

Krause, Manfred, RADON CHECK, P.O. Box 7882, Greensboro, NC, 27417, (919) 855-1312, (1).

Kreckel, Sylvia, OIL, CHEMICAL AND ATOMIC WORKERS INTERNATIONAL UNION (OCAW), P.O. Box 2812, Denver, CO, 80201, (303) 987-2229, (16).

Kreider, Cheryl, HARDICO RADON MANAGEMENT, D.C. OFFICE, 10720 Main St. Suite 205, Fairfax, VA, 22030, (703) 385-4200, (2).

Kretvix, Bob, O.K. REMS COMPANY, 10 Hillsborough Rd., Flemington, NJ, 08822, (201) 782-3398, (1).

Krisel, Wilfred, WORLD HEALTH ORGANIZATION (WHO)/DIVISION OF ENVIRONMENTAL HEALTH/ENVIRONMENTAL HAZARDS AND FOOD PROTECTION UNIT, 1211 Geneva 27, Switzerland, (011) 4122-34-6061, (12).

Kritchevsky, Evelyn S., ALPHATECH, INC., P.O. Box 254, Haverford, PA, 19041, (215) 525-1490, (1).

Kriviskey, Bruce, AMERICAN INSTITUTE OF ARCHITECTS (AIA), 1735 New York Ave., NW, Washington, DC, 20006, (202) 626-7452, (13).

Krueger, James, AMERICAN RADON CORPORATION, 8300 North Hayden Rd. Suite 100, Scottsdale, AZ, 85258, (602) 967-8029, (1)(2)(6).

Krueger, James W., INTEGRAL RESOURCES, INC., 1916 East El Parque, Tempe, AZ, 85282, (602) 967-8029, (1).

Krug, Harry A., COLUSA COUNTY AIR POLLUTION CONTROL DISTRICT, P.O. Box 1029, Colusa, CA, 95932, (916) 458-5891, (10).

Kruse, James W., AMERICAN RADON TESTING LABORATORIES, 1015 South Birch St. #610, Glendale, CO, 80222, (1).

Kruszka, Mary Jane, TOWN OF HAMBURG, S-6100 South Park Ave., Hamburg, NY, 14075, (716) 649-6111, (10).

Kruthoff, Rob, NORTHWEST FURNACE COMPANY, INC., 3139 5th Ave., S., Fort Dodge, IA, 50105, (515) 944-7262, (1).

Kuchnicki, Richard P., COUNCIL OF AMERICAN BUILDING OFFICIALS (CABO), 5203 Leesburg Pike, Falls Church, VA, 22041, (703) 931-4533, (13).

Kuendig, Betsy, IMPROVED TECHNOLOGY SYSTEMS, P.O. Box 8244, Canton, OH, 44711-8244, (216) 452-1493, (1).

Kuendig, William, IMPROVED TECHNOLOGY SYSTEMS, P.O. Box 8244, Canton, OH, 44711-8244, (216) 452-1493, (1).

Kueser, Paul, AIR POLLUTION CONTROL ASSOCIATION (APCA), P.O. Box 2861, Pittsburgh, PA, 15230, (412) 232-3444, (14).

Kunkel, Harold E., KUNKEL, HAROLD E., 6 Radnor Rd., Newtown Square, PA, 19073, (215) 356-5096, (2).

Kunz, Charles, NEW YORK STATE HEALTH DEPARTMENT/WADSWORTH CENTER FOR LABORATORIES AND RESEARCH/LABORATORY FOR INORGANIC AND NUCLEAR CHEMISTRY, Empire State Plaza Room D366-B, Albany, NY, 12201, (518) 474-6649, (1)(10).

Kunze, Jay F., AMERICAN NUCLEAR SOCIETY/MISSOURI-CENTRAL/EASTERN MISSOURI SECTION, C/O University Of Missouri 1026 Engineering Complex, Rolla, MO, 65211, (314) 882-3550, (14).

Kurtz, Norman R., PROTECTION INSPECTION, INC., Rte. 11 Hill Rd., Knoxville, TN, 37938, (615) 688-9779, (1)(2)(3).

Kuyatt, Chris E., U.S. DEPARTMENT OF COMMERCE (DOC)/NATIONAL INSTITUTE OF STANDARDS AND TECHNOLOGY(NIST)/CENTER FOR RADIATION RESEARCH(CRR), Radiation Physics Building Room C235, Gaithersburg, MD, 20899, (301) 975-6093, (8).

Kyle, Tim, CARROLL COUNTY RADON CONTROL, INC., Box 027, Westminster, MD, 21157, (301) 848-6324, (1).

Kyle, Tim, RAD CON, INC., Box 027, Westminster, MD, 21157, (301) 848-6324, (1).

L

LaBeau, Tammy, SOUTH DAKOTA STATE DEPARTMENT OF HEALTH/OFFICE OF AIR QUALITY AND SOLID WASTE, Joe Foss Office Building 523 East Capital, Pierre, SD, 57501, (605) 773-3329, (10).

Lacker, David K., TEXAS DEPARTMENT OF HEALTH/BUREAU OF RADIATION CONTROL, 1100 West 49th St., Austin, TX, 78756-3189, (512) 835-7000, (10).

Ladnier, William C., PMI CONTRACTORS, INC., P.O. Box 6125, Gulfport, MS, 39506, (601) 863-1988, (601) 863-1998, (1).

Laird, Henry, MISSISSIPPI STATE DEPARTMENT OF HEALTH/DIVISION OF OCCUPATIONAL SAFETY AND HEALTH, P.O. Box 1700, Jackson, MS, 39205, (601) 987-3981, (10).

Laird, William A., MICROBAC LABORATORIES, INC./PITTSBURGH TESTING DIVISION, 4580 McKnight Rd., Pittsburgh, PA, 15237, (412) 931-5851, (1).

Lake, Kerry, ADHESIVE AND SEALANT COUNCIL (ASC), 1627 K St., NW Suite 1000, Arlington, VA, 22209, (202) 452-1500, (13).

Laliberte, Joseph F., ADVANCED HOME INSPECTION AND TESTING, 23 Magna Vista Cr., Tewksbury, MA, 01876, (617) 851-5866, (1).

Lamarre, Bruce L., NORTH EAST ENVIRONMENTAL PRODUCTS, INC., 85 Mechanic St., Lebanon, NH, 03766, (603) 448-3944, (2)(3)(4).

LaMastra, Anthony, A.B.E. RADIATION MEASUREMENTS LABORATORY, RD #1, Box 214, Lenhartsville, PA, 19534, (215) 756-4153, (1).

LaMastra, Anthony, HEALTH PHYSICS ASSOCIATES, INC., RD 1 Box 796, Lenhartsville, PA, 19534, (215) 756-4153, (1).

Lammering, Milton, U.S. ENVIRONMENTAL PROTECTION AGENCY REGION 8 (EPA8)/RADIATION DIVISION, One Denver Pl. 999 18th St. 8HWM-RP, Denver, CO, 80202, (303) 293-1603, (8).

LaMonico, Linda, ESSEX CHEMICAL CORPORATION, 1333 Broad St., Clifton, NJ, 07015, (201) 773-6300, (4).

Lamparelli, Eleanor, MASSACHUSETTS RADON ALERT, INC., 20 Robertson St., Quincy, MA, 02169, (617) 471-9403, (617) 331-7900, (1).

Lamparelli, Josep, MASSACHUSETTS RADON ALERT, INC., 20 Robertson St., Quincy, MA, 02169, (617) 471-9403, (617) 331-7900, (1).

Lampert, Kenneth J., MESA COUNTY HEALTH DEPARTMENT/AIR POLLUTION CONTROL DIVISION, 515 Patterson Rd., Grand Junction, CO, 81501, (303) 244-1750, (10).

Lamson, William M., LAMSON HOME INSPECTIONS, 9 Lougheed Ave., West Caldwell, NJ, 07006, (201) 228-1862, (1).

Lancellotti, Frank, PH SALES COMPANY, P.O. Box 372, Edison, NJ, 08818, (201) 287-5300, (4).

Lane, Peter, RESOURCE CENTER FOR ENVIRONMENTAL PROTECTION, P.O. Box 25, Schuylerville, NY, 12871, (518) 695-3218, (2)(3).

Lane, Ron, CHEMICAL SYSTEMS, INC., P.O. Box 68, Dunbridge, OH, 43414, (419) 833-4492, (4).

Lane, Steve, NATIONAL READY MIXED CONCRETE ASSOCIATION (NRMCA), 900 Spring St., Silver Spring, MD, 20910, (301) 587-1400, (13).

Laney, Stephen E., RECON SYSTEMS, INC., P.O. Box 460, Three Bridges, NJ, 08887, (201) 782-5900, (1).

Lang, Kevin, NORTH EAST ENVIRONMENTAL PRODUCTS, INC., 85 Mechanic St., Lebanon, NH, 03766, (603) 448-3944, (2)(3)(4).

Lange, Francine, U.S. VIRGIN ISLANDS DEPARTMENT OF PLANNING AND NATURAL RESOURCES/ENVIRONMENTAL PROTECTION DIVISION, P.O. Box 3594, St. Thomas, VI, 00801, (809) 776-4844, (10).

Langford, Sharon, TENNESSEE REAL ESTATE COMMISSION, 1808 West End Building 11th Floor, Nashville, TN, 37219-5322, (615) 741-2273, (10).

Langford White, Joan, VIRGINIA DEPARTMENT OF COMMERCE/STATE REAL ESTATE BOARD, 3600 West Broad St. 5th Floor, Richmond, VA, 23230, (804) 367-8552, (10).

Langley, Ron, GLOBAL WHOLESALE, 2410 East Lincolnway, Cheyenne, WY, 82001, (307) 634-7318, (1)(4).

Lanigan, Valerie B., NEW HAMPSHIRE REAL ESTATE COMMISSION, State Office Park South Johnson Hall, 3rd Fl. 107 Pleasant St., Concord, NH, 03301, (603) 271-2701, (10).

Lanksbury, Vicky, HOME ASSURANCE, 17 Birchwood Dr., Naperville, IL, 60540, (312) 961-9154, (312) 355-6792, (1).

Lannus, Arvo, ELECTRICAL POWER RESEARCH INSTITUTE (EPRI)/ENERGY MANAGEMENT AND UTILIZATION DIVISION/RESIDENTIAL AND COMMERCIAL PROGRAMS, 3412 Hillview Ave., Palo Alto, CA, 94304, (415) 855-2000, (5)(13).

Lantz, Carl, LACO, P.O. Box 666, Chesterland, OH, 44026, (216) 729-3034, (4).

Lapin, Byron, CONVENIENCE PRODUCTS, INC., 4205 Forest Park Blvd., St. Louis, MO, 63108-2892, (314) 535-6226, (4).

Lapinski, Frank, RADON X COMPANY, INC., 217 Roxbury St., Keene, NH, 03431, (603) 352-5319, (1)(2).

Lapinski, Ken, BERNER INTERNATIONAL CORPORATION, P.O. Box 5205, New Castle, PA, 16105, (412) 658-1530, (800) 852-5015, (4).

Laprete, Richard, BERGEN HOME INSPECTION SERVICE, 838 Palisades Ave., Teaneck, NJ, 07666, (201) 836-3232, (1).

Lapteff, Alexander, LAPTEFF ASSOCIATES, P.O. Box 4150, Woodbridge, VA, 22194, (703) 491-6700, (1).

LaRiviere, J.W.M., SCIENTIFIC COMMITTEE ON PROBLEMS OF THE ENVIRONMENT (SCOPE), 51 Boulevard De Montmorency, F-75016 Paris, France, (011) 331-45250498, (12).

Larkin, Dave, TREMCO, INC., 3735 Green Rd., Beachwood, OH, 44122, (216) 292-5000, (4).

LaRocca, Thomas J., RADIATION PROTECTION SERVICES, INC., P.O. Box 2395, Darien, CT, 06820, (203) 324-7967, (1)(2).

Larsen, Morella, OREGON REAL ESTATE AGENCY, 158 12th St, NE, Salem, OR, 97310-0240, (503) 378-8414, (10).

Larson, John E., ASTECO, INC., P.O. Box 2204, Niagara Univeristy, NY, 14109, (716) 297-5992, (1).

Lary, Susan, A 'LARY' SERVICE, 686 Somerset Dr., Lawrenceville, GA, 30245, (404) 441-6819, (404) 963-8218, (1).

Lash, Terry, ILLINOIS DEPARTMENT OF NUCLEAR SAFETY/OFFICE OF ENVIRONMENTAL SAFETY, 1035 Outer Park Dr., Springfield, IL, 62704, (217) 786-6399, (1)(10).

Lasher, David, BARRINGER, 15000 West 6th Ave. Suite

300, Golden, CO, 80401, (303) 277-1687, (1)(4).

Lasmanis, Raymond, WASHINGTON STATE DEPARTMENT OF NATURAL RESOURCES/DIVISION OF GEOLOGY AND EARTH RESOURCES DIVISION, MS-PY-12, Olympia, WA, 98504, (206) 459-6372, (10).

Lathrop, Larry, ARAPAHOE HOMES, 4065 North Sinton Street Suite 200, Colorado Springs, CO, 80907, (719) 593-2666, (13).

LaTouche, Y. David, HENRY INTERNATIONAL, INC., 14742 Plaza Dr., Tustin, CA, 92680, (714) 730-8451, (4).

Lau, Thomas, ADVANCE TEX CORPORATION, 204 Plantation Rd., Seneca, SC, 29678, (803) 654-2300, (1).

Lauchert, Jack, TOWN OF HAMBURG, S-6100 South Park Ave., Hamburg, NY, 14075, (716) 649-6111, (10).

Laurer, Gerald, NEW YORK UNIVERSITY MEDICAL CENTER/INSTITUTE OF ENVIRONMENTAL MEDICINE, 550 1st Ave., New York, NY, 10016, (212) 340-5280, (5).

Laurie, Jack, TRI-STATE RADON AND DEWATERING, P.O. Box 922, Dayton, NJ, 08810, (201) 329-8570, (1)(2).

Lawn, Cynthia, FRANKLIN ENVIRONMENTAL ANALYSIS, 716 North Franklin St. Box 1-R, Philadelphia, PA, 19123-2721, (215) 592-8322, (1).

Lawrence, Bill, RADON DETECTION SYSTEMS/CORPORATE HEADQUARTERS, 2450 Central Ave. A-1, Boulder, CO, 80301, (303) 444-5253, (800) 627-2360, (1).

Lawrence, Sherry, CONVENIENCE PRODUCTS, INC., 4205 Forest Park Blvd., St. Louis, MO, 63108-2892, (314) 535-6226, (4).

Leach, Jerry, U.S. ENVIRONMENTAL PROTECTION AGENCY REGION 10 (EPA10)/RADIATION DIVISION(ORP), 1200 Sixth Ave., Seattle, WA, 98101, (206) 442-7299, (8).

Leafe, Edward, LEAFE, EDWARD, 212 Wilson St., Hackensack, NJ, 07601, (201) 343-0780, (1).

Leary, Louise S., NORWALK DEPARTMENT OF HEALTH, 137 East Ave., Norwalk, CT, 06851, (203) 854-7776, (10).

Lebel, Aurel, PYLON ELECTRONIC DEVELOPMENT COMPANY, LTD., 147 Colonnade Rd., Nepean, ON, K2E 7L9, (613) 226-7920, (1)(4).

Ledford, Bud, RADON TEST AND REDUCTION SPECIALIST, 125 Oak Ave. P.O. Box 129, Spruce Pine, NC, 28777, (704) 765-1222, (1)(2).

Ledford, Jerri, RADON TEST AND REDUCTION SPECIALIST, 125 Oak Ave. P.O. Box 129, Spruce Pine, NC, 28777, (704) 765-1222, (1)(2).

Lee, G. Fred, NEW JERSEY INSTITUTE OF TECHNOLOGY/DEPARTMENT OF CIVIL AND ENVIRONMENTAL ENGINEERING, 323 King Blvd., Newark, NJ, 07102, (201) 761-0977, (5).

Lee, James, SABAL INVESTMENTS, 915 Oakfield Drive Suite A, Brandon, FL, 33511, (813) 689-2548, (3).

Lee, Janice, AMERICAN RADON ASSOCIATION (ARA), 4949 South 25-A, Tipp City, OH, 45371, (513) 667-2464, (7).

Lee, Stuart, RADON RESEARCH INTERNATIONAL, INC., 7650 South Tamiami Trail, #9, Sarasota, FL, 34231, (813) 923-7300, (1)(2).

Lefkoff, Kyle, COLORADO VENTURE MANAGEMENT, INC., 2995 Wilderness Pl., Boulder, CO, 80301, (303) 440-4055, (1)(2).

Lefkoff, Kyle, CURRENT INDOOR AIR SYSTEMS, INC., P.O. Box 18075, Boulder, CO, 80308, (303) 440-8555, (4).

Legare, J. M., MINISTERE DE L'ENVIRONMENT DU QUEBEC/DIVISION DE RADIOPROTECTION, 5199 Est, rue Sherbrooke Suite 3860, Montreal, PQ, H1T 3X9, (514) 873-1978, (11).

Lehman, Terry R., ALA OF MASSACHUSETTS, 803 Summer St., South Boston, MA, 02127-1609, (617) 269-9720, (15).

Lehr, Eugene L., U.S. DEPARTMENT OF TRANSPORTATION (DOT)/ENVIRONMENT DIVISION (TED), 400 7th St., SW Nassif Building Room 9217, Washington, DC, 20590, (202) 366-4860, (8).

Lehr, Jay H., ASSOCIATION OF GROUND WATER SCIENTISTS AND ENGINEERS (AGWSE), 6375 Riverside Dr., Dublin, OH, 43017, (614) 761-1711, (14).

Leighton, Morris W., ILLINOIS STATE GEOLOGICAL SURVEY, Natural Resources Building 615 East Peabody Dr. Room 121, Champaign, IL, 61820, (217) 333-5111, (1)(10).

Leisinger, Tom, NATIONAL PROPERTY INSPECTIONS, 1344-S Briar Creek Rd., Charlotte, NC, 28205, (704) 568-6324, (1).

Leister, Kirk, BIOHAZARDOUS MATERIALS CONSULTING, P.O. Box 4376, Ithaca, NY, 14852, (607) 257-5010, (315) 478-0653, (1)(3).

Lembke, Claus H., NORTH DAKOTA ASSOCIATION OF REALTORS, 1910 North 11th St., Bismark, ND, 58501, (701) 258-2361, (18).

Lemen, Richard, U.S. DEPARTMENT OF HEALTH AND HUMAN SERVICES (HHS)/PUBLIC HEALTH SERVICE(PHS)/CENTERS FOR DISEASE CONTROL(CDC)/NATIONAL INSTITUTE FOR OCCUPATIONAL SAFETY AND HEALTH(NIOSH), Division Of Standards Development And Technology Transfer (DSOTT) 4676 Columbia Pkwy., MS-C14, Cincinnati, OH, 45226, (513) 533-8236, (8).

Lennon, Mike, HOMEPRO SYSTEMS, INC., 6715 Lowell Ave. Suite 2, McLean, VA, 22101, (703) 761-1400, (1)(3).

Lents, James, SOUTH COAST AIR QUALITY MANAGEMENT DISTRICT/AIR POLLUTION CONTROL DIVISION, 9150 Flair Dr., El Monte, CA, 91731, (818) 572-6200, (10).

Leonard, Bobby E., ACADEMY OF IRW, INC., The Friary, Winchester-on-Severn 1600 Winchester Rd., Annapolis, MD, 21401, (301) 757-7887, (1).

Leonard, Christopher, LEONARDS HOME INSPECTIONS, 6416 Loring Dr., Columbia, MD, 21045, (301) 381-2240, (1).

Leonard, John A., HOME RADON SURVEYS, 4209 Olley Ln., Fairfax, VA, 22032, (703) 323-7236, (1).

Leonard, Marvin, ACME CHEMICALS AND INSULATION COMPANY, P.O. Box 1404, New Haven, CT, 06505, (203) 562-2171, (4).

Leptich, Ed, AIRXCHANGE, INC., 401 VFW Dr., Rockland, MA, 02370, (617) 871-4816, (4).

Lerman, Alan M., RADON ALERT, P.O. Box 2167, Westminster, MD, 21157-9167, (301) 848-2862, (1).

Lerman, Alan M., RADON REPORTING SYSTEM OF CARROLL COUNTY, P.O. Box 500, Westminster, MD, 21157-9167, (301) 857-1807, (301) 795-0775, (1).

Leslie, Ron, RADON TESTING LAB, 3432 Hollywood, Brookfield, IL, 60513, (312) 485-1133, (1).

Lester, Steven, CITIZEN'S CLEARINGHOUSE FOR HAZARDOUS WASTES (CCHW), P. O. Box 926, Arlington, VA, 22216, (703) 276-7070, (17).

Letourneau, E.G., HEALTH AND WELFARE CANADA/HEALTH PROTECTION BRANCH/BUREAU OF RADIATION AND MEDICAL DEVICES, 775 Brookfield Rd., Ottawa, ON, K1A 1C1, (613) 954-6677, (11).

Leu, Sheng-jia, UNITED SERVICES ALLIANCE, INC., 4900 Leesburg Pike Suite 210, Alexandria, VA, 22302, (703) 845-1324, (1)(2).

Leung, F.C., U.S. DEPARTMENT OF ENERGY (DOE)/RICHLANDS OPERATIONS OFFICE(ROO)/PACIFIC NORTHWEST LABORATORY(PNL)/CELL BIOLOGY DIVISION, M/S K-416, Richland, WA, 99352, (509) 375-2169, (8).

Levin, Michael H., ENVIRONMENTAL RESEARCH ASSOCIATES, P.O. Box 219 490 Darby-Paoli Rd., Villanova, PA, 19085, (215) 687-5629, (1)(3).

Levine, Leslie, NATIONAL INSTITUTE OF BUILDING SCIENCES (NIBS), 1015 15th St., NW Suite 700, Washington, DC, 20005, (202) 347-5710, (5)(13).

Levine, Sumner, HOME INSPECTOR OF NORTH AMERICA, THE, 301 North Main St., Randolph, MA, 02368, (617) 963-0343, (1).

Lewis, Craig, LEWIS HOME INSPECTION, P.O. Box 6149, Lawrenceville, NJ, 08648, (609) 924-1411, (1).

Lewis, Earl K., SKUTTLE MANUFACTURING COMPANY, Rte. 1, Box 16A, Marietta, OH, 45750, (614) 373-9169, (4).

Lewis, Mark, RADON TESTING AND CONSULTING, 1771 Post Rd., E. Suite 320, Westport, CT, 06880, (203) 454-3886, (1)(2)(4).

Libertore, Jr., Larry, LARRY LIBERTORE, INC., 230 West Highland Dr., Lakeland, FL, 33803, (813) 644-2828, (2).

Lightsey, Kevin, RADON REDUCTION EQUIPMENT OF COLORADO, 921 Main Suite C, Louisville, CO, 80027, (303) 665-1776, (4)(6).

Lightsey Pruet, Eleanor, SOUTH CAROLINA REAL ESTATE COMMISSION, 1201 Main St. Suite 1500, Columbia, SC, 29201, (803) 737-0700, (10).

Liming, Glenn, WESTERN CAROLINA UNIVERSITY/DEPARTMENT OF CHEMISTRY AND PHYSICS, Natural Sciences Building, Cullowhee, NC, 28723, (704) 227-7260, (5).

Lindquist, Wayne, TUCSON HOME INSPECTIONS, INC., P.O. Box 50845, Tucson, AZ, 85703-1845, (602) 881-5511, (1).

Lindsey, Terry, NORTH DAKOTA STATE DEPARTMENT OF HEALTH AND CONSOLIDATED LABORATORIES/DIVISION OF ENVIRONMENTAL ENGINEERING, 1200 Missouri Ave. P.O. Box 5520, Bismarck, ND, 58502-5520, (701) 224-2348, (10).

Ling, C. Clifton, UNIVERSITY OF CALIFORNIA-SAN FRANCISCO/DEPARTMENT OF RADIATION ONCOLOGY/PHYSICS DEPARTMENT, L-75, Box 0226, San Francisco, CA, 94143-0226, (415) 476-9000, (5).

Lininger, Rick, LININGER BROTHERS, INC., 12600 Penn Mar Rd., Waynesboro, PA, 17268, (717) 762-4024, (2).

Link, Harry, OXYTHERM, INC., P.O. Box 3467, Greenville, SC, 29602, (803) 246-2055, (2).

Link, John, A & A CONSULTING AND INSPECTION SERVICES, INC., 349 Chenango St., Binghamton, NY, 13901, (607) 723-7968, (1)(3).

Lippert, Ernest L., MONARCH ANALYTICAL LABORATORIES, INC., P.O. Box 2990, Toledo, OH, 43606, (419) 535-1780, (1).

Lipsett, Brian, CITIZEN'S CLEARINGHOUSE FOR HAZARDOUS WASTES (CCHW), P. O. Box 926,

Arlington, VA, 22216, (703) 276-7070, (17).

Lipshutz, A. Arthur, A. ARTHUR LIPSHUTZ ASSOCIATES, 7650 South Tamiami Trail Suite 9, Sarasota, FL, 34231, (813) 923-7300, (1)(2).

Lipshutz, Arthur, RADON RESEARCH INTERNATIONAL, INC., 7650 South Tamiami Trail, #9, Sarasota, FL, 34231, (813) 923-7300, (1)(2).

Litt, Barbara, NEW JERSEY DEPARTMENT OF ENVIRONMENTAL PROTECTION/DIVISION OF SCIENCE AND RESEARCH, 401 East State St. 7th Floor, West Wing, Trenton, NJ, 08625, (609) 984-6070, (1)(10).

Litton Tuck, Ann, TENNESSEE REAL ESTATE COMMISSION, 1808 West End Building 11th Floor, Nashville, TN, 37219-5322, (615) 741-2273, (10).

Livetsberger, Glenn, HERITAGE RADON SERVICES, INC., 6968 Birchwood Rd., Harrisburg, PA, 17112, (717) 599-5060, (1)(2).

Livingston, Joe, RADON READINGS, 9020 Compton Blvd., Bellflower, CA, 90706, (213) 920-2582, (213) 421-8674, (1).

Logan, G.P., ENVIRONMENTAL ENGINEERING AND TESTING, 9185 East Kenyon Ave. Suite 190, Denver, CO, 80237, (303) 722-2112, (1).

Logsdon, Michael E., ENVIRONMENTAL MANAGEMENT GROUP, 9841 Broken Land Pkwy. #117, Columbia, MD, 21046, (301) 290-7078, (1).

Lonchar, Nikola, LONCHAR, NIKOLA, 771 South 20th St., Philadelphia, PA, 19146, (215) 985-1474, (1).

Long, Howard E., ACE PEST CONTROL, Rte. 2, Box 244, Syracuse, IN, 46567, (219) 856-2294, (1).

Longbothum, G. Briggs, BRIGGS LONGBOTHUM THERMINSPECTION, 1440 Prospect Ave., Plainfield, NJ, 07060, (201) 755-8116, (3)(6).

Longenecker, Bud, PENRAD, INC., 54 South Market St., Elizabethtown, PA, 17022, (717) 367-8410, (717) 367-3995, (1)(2).

Longtin, Jon P., U.S. ENVIRONMENTAL PROTECTION AGENCY (EPA)/OFFICE OF WATER(OW)/OFFICE OF DRINKING WATER(ODW)/TECHNICAL SUPPORT DIVISION(TSD), Water Supply Technology Branch 26 West Martin Luther King Dr., Cincinnati, OH, 45268, (513) 569-7919, (8).

Lopez, Nancy, U.S. DEPARTMENT OF INTERIOR (DOI)/U.S. GEOLOGICAL SURVEY(USGS)/OFFICE OF WATER DATA COORDINATION(OWDC), 417 National Ctr., Reston, VA, 22092, (703) 648-5016, (8).

LoPilato, John, RADON-X CORPORATION, 300 Valentine St., Hackettstown, NJ, 07840, (201) 852-6212, (800) 272-3669, (1)(2)(4).

Loreti, Daniel, EXCEL INSPECTION SERVICE, INC., 438 Windmill Way, Somerville, NJ, 08876, (201) 526-7277, (1).

Lori, Ann, TWIN CITY TESTING, 662 Cromwell, St. Paul, MN, 55114, (612) 645-3601, (1).

Loudon, Philip, TECHNORTH, 109 Waterloo Ln., Fairbanks, AK, 99712, (907) 456-8378, (1).

Lovato, Randy Ernest, NEW MEXICO REAL ESTATE COMMISSION, 4125 Carlisle Blvd., NE, Albuquerque, NM, 87107, (505) 841-6524, (10).

Love, Dan, BICRON CORPORATION, 12345 Kinsman Rd., Newbury, OH, 44065, (216) 564-8000, (4).

LoVecchio, Raymond, SHERLOCK HOME INSPECTORS, 6-01 Fair Lawn Ave., Fair Lawn, NJ, 07410, (201) 796-4942, (800) 458-2202, (1).

Lovejoy, Robert, SOUTH CENTRAL ENVIRONMENTAL SERVICES, RD 1, Box 265, Pine City, NY, 14871, (607) 733-1661, (1)(2).

Lovering, Donald, ADVANTAGE HOME INSPECTION, P.O. Box 580, South Dennis, MA, 02660, (617) 398-0861, (1).

Lowe, Leo, SENES CONSULTANTS, 499 McNicholl Ave., Willowdale, ON, M2H 2C9, (416) 499-5030, (1)(3).

Lowry, Jerry, LOWRY ENGINEERING, INC., P.O. Box 536 Back Troy Rd., Unity, ME, 04988, (207) 948-3790, (1)(2)(3)(4)(5).

Lowry, Jerry, UNIVERSITY OF MAINE-ORONO/DEPARTMENT OF ENGINEERING, 455 Aubert Hall, Orono, ME, 04469, (207) 581-2216, (5).

Lowry, Sylvia, LOWRY ENGINEERING, INC., P.O. Box 536 Back Troy Rd., Unity, ME, 04988, (207) 948-3790, (1)(2)(3)(4)(5).

Lubinsky, Jim, OAK HILL RADON TESTING AND REPAIR, 15 1/2 North State St., Westerville, OH, 43081, (614) 895-8365, (1)(2).

Lucas, Henry, ARGONNE NATIONAL LABORATORY, 9700 South Cass Ave., Argonne, IL, 60439, (312) 972-4172, (1).

Lucas, John E., RAMP INDUSTRIES, INC., 1127 West 46th St., Denver, CO, 80210, (1)(2).

Lucas, Ken, FORESIGHT ENGINEERING, INC., P.O. Box 362, Monasquan, NJ, 08736, (201) 223-3316, (1)(2).

Luce, Robert, SCS ENGINEERS, INC., 11260 Roger Bacon Dr. Suite 300, Reston, VA, 22090, (703) 471-6150, (1).

Luciw, George, AMERICAN SOCIETY FOR TESTING AND MATERIALS (ASTM), 1916 Race St., Philadelphia, PA, 19103, (215) 299-5571, (14).

Ludgate, Robert, LUDGATE ENGINEERING CORPORATION, 9A North Fifth St., Reading, PA, 19522, (215) 374-2239, (3).

Ludwig, Gary, NEW JERSEY DEPARTMENT OF HEALTH/DIVISION OF ENVIRONMENTAL AND OCCUPATIONAL HEALTH/PUBLIC EMPLOYEE OCCUPATIONAL SAFETY AND HEALTH PROGRAM, CN-360, Trenton, NJ, 08625, (609) 984-1863, (10).

Ludy, Cheryl B., ENVIRONMENTAL SYSTEMS SERVICE, P.O. Box 512, Culpepper, VA, 22701, (703) 825-6660, (1).

Luke, John H., U.S. GENERAL ACCOUNTING OFFICE (GAO)/RESOURCES, COMMUNITY AND ECONOMIC DEVELOPMENT DIVISION(RCED), 441 G St. NW, Washington, DC, 20548, (202) 426-1780, (8).

Lukenbill, Paul D., LUKENBUILT CONTRACTORS, 965 Dearborn St., Aurora, CO, 80011, (303) 364-9526, (2).

Lukowski, Stephen S., ALBANY COUNTY DEPARTMENT OF HEALTH/DIVISION OF ENVIRONMENTAL HEALTH SERVICES, South Ferry and Green Sts., Albany, NY, 12201, (518) 447-4580, (10).

Lumon, R. Lynn, NEVADA DEPARTMENT OF COMMERCE/ REAL ESTATE DIVISION, 201 South Fall St., Carson City, NV, 89710, (702) 885-4280, (10).

Lundin, Frank, U.S. DEPARTMENT OF HEALTH AND HUMAN SERVICES (HHS)/PUBLIC HEALTH SERVICE(PHS)/FOOD AND DRUG ADMINISTRATION(FDA), 5600 Fishers Ln. HFD-733, Rockville, MD, 20857, (301) 443-3380, (8).

Lussier, Larry, BUILDING INSPECTORS, INC., 282 Harlow St. P.O. Box 2220, Bangor, ME, 04401, (207) 941-0870, (1).

Luxton, Steven, A-1 SHERLOCK HOME INSPECTORS, 6-01 Fairlawn Dr., Fairlawn, NJ, 07410, (201) 796-4942, (1).

Lynch, Kevin, AMERICAN CYANAMID COMPANY, 1 Cyanamid Plaza, Wayne, NJ, 07470, (201) 831-2000, (4).

Lynch, Thomas R., RADON AIRTEST COMPANY, 13800 Bagley Rd., Middleburg Heights, OH, 44130, (216) 845-8050, (1).

Lyngstad, Larry G., SOUTH DAKOTA REAL ESTATE COMMISSION, P.O. Box 490, Pierre, SD, 57501, (605) 773-3600, (10).

Lynn, Charles J., ENCOSYS CORPORATION OF AMERICA, 7202 East 87th St. Suite 109, Indianapolis, IN, 46250, (317) 849-6060, (1).

Lynn, Gary K., DELAWARE DEPARTMENT OF HEALTH AND SOCIAL SERVICES/BUREAU OF ENVIRONMENTAL HEALTH/OFFICE OF RADIATION CONTROL/DIVISION OF PUBLIC HEALTH, 802 Silver Lake Blvd. and Walker Rd. P.O. Box 637, Dover, DE, 19903, (302) 736-4731, (10).

Lynn, Jeffrey N., LYNN PEST MANAGEMENT COMPANY, P.O. Box 274, Jarrettsville, MD, 21084, (301) 557-6839, (800) 327-7134, (1).

Lyon, Bill, RADON DETECTION SYSTEMS/CHICAGO, ILLINOIS BRANCH, 12845 Golf View Ln., Palois Heights, IL, 60463, (303) 444-5253, (800) 627-2360, (1).

Lyon, John W., U.S. DEPARTMENT OF COMMERCE (DOC)/NATIONAL INSTITUTE OF STANDARDS AND TECHNOLOGY(NIST)/NATIONAL ENGINEERING LABORATORY(NEL), Technology Building Room B119, Gaithersburg, MD, 20899, (301) 975-2300, (5)(8).

Lysne, Eric, NORLAND CORPORATION, Norland Dr., Ft. Atkinson, WI, 53538, (414) 563-8456, (800) 333-8456, (4).

M

Mabry, Thomas L., CALIFORNIA DEPARTMENT OF REAL ESTATE, 2201 Broadway P.O. Box 187000, Sacramento, CA, 95818-7000, (916) 739-3721, (10).

Macauley, Louis, QUANTUM INFORMATION, LTD., P.O. Box 467, Westown, PA, 19395-0467, (215) 431-9577, (1).

MacCarter, Katherine, AMERICAN PUBLIC HEALTH ASSOCIATION (APHA), 1015 15th St., NW, Washington, DC, 20005, (202) 789-5663, (15).

MacDaniel, Don, THIRD PARTY INSPECTIONS, P.O. Box 3161, Winchester, VA, 22601, (703) 869-1096, (1).

Macgill, R. Douglas, RADON ENGINEERING SERVICES, INC., 3129 Overton Rd., Birmingham, AL, 35223, (205) 967-2665, (1)(2).

MacGill, R. Douglas, RADON EVALUATION SERVICES, INC., P.O. Box 13609, Atlanta, GA, 30324, (404) 478-0037, (1)(2).

MacGillivray, O.A., SASKATCHEWAN CONSUMER AND COMMERCIAL AFFAIRS, 1871 Smith St., Regina, SK, S4P 3V7, (306) 787-2958, (11).

MacGregor, Ian, NATIONAL SCIENCE FOUNDATION (NSF)/LOWER EARTH SCIENCES DIVISION, 1800 G St., NW, Washington, DC, 20550, (202) 357-7958, (8).

Mack, Thomas E., MACK, E.F., 11 South 8th Ave., Marshalltown, IA, 50158, (515) 753-7328, (1)(2).

MacLean, David A., CON-TEST, INC., 126 Shaker Rd., East Longmeadow, MA, 01028, (413) 525-1198, (1).

Maconaughey, Kirk, U.S. ENVIRONMENTAL

PROTECTION AGENCY (EPA)/OFFICE OF AIR AND RADIATION(OAR)/OFFICE OF RADIATION PROGRAMS(ORP)/RADON DIVISION(RD), Problem Assessment Branch 401 M St. SW ANR-464, Washington, D.C., 20460, (202) 475-9605, (8).

Madigan, John, AMERICAN CANCER SOCIETY (ACS)/NATIONAL PUBLIC AFFAIRS OFFICE, 224 East Capitol St., Washington, DC, 20003, (202) 546-4011, (15).

Maes, Reuben, ROMAES CORPORATION, 6395 Gunpark Dr. Suite C, Boulder, CO, 80301, (303) 530-1637, (4).

Magyar, Lorand, RADON TESTING AND MITIGATION SYSTEMS, P.O. Box 3052, Carmel, IN, 46032, (317) 545-5578, (1)(2).

Maher, Christopher L., THERMO-SCAN, INC., 1137 3rd Ave., SW P.O. Box 705, Carmel, IN, 46032, (317) 846-4655, (1).

Maher, Christopher L., TSI CONSTRUCTION SERVICES, 1137 3rd Ave., SW, Carmel, IN, 46032, (317) 846-4655, (1)(2).

Maher, Ed, U.S. DEPARTMENT OF DEFENSE (DOD)/DEPARTMENT OF THE AIR FORCE(USAF)/OCCUPATIONAL AND ENVIRONMENTAL HEALTH LABORATORY(OEHL)/RADIATION SERVICES DIVISION, USAFOEHL/RZA, Brooks AFB, TX, 78235-5501, (512) 536-2061, (8).

Mahoney, Bob, MAHONEY HEATING AND COOLING, 13620 C St., Omaha, NE, 68144, (402) 895-3333, (1)(2).

Mahoney, Bob, RADON TECHNOLOGIES, INC., 39 Woodrow Ct., Troy, NY, 12180, (518) 447-5834, (1).

Mahoney, Pat, MAHONEY HEATING AND COOLING, 13620 C St., Omaha, NE, 68144, (402) 895-3333, (1)(2).

Maillet, Bruce K., MASSACHUSETTS DIVISION OF AIR QUALITY CONTROL, One Winter St., Boston, MA, 02108, (617) 292-5593, (10).

Majeski, Mark, PRO-TECH RADON, 459 Forestwood Dr., Gahanna, OH, 43230, (614) 475-0023, (1).

Majewski, B., NEW YORK STATE DEPARTMENT OF ENVIRONMENTAL CONSERVATION/BUREAU OF AIR TOXICS, 50 Wolf Rd., Albany, NY, 12223-3250, (518) 457-7688, (10).

Makatura, Elaine, NEW JERSEY DEPARTMENT OF ENVIRONMENTAL PROTECTION/DIVISION OF ENVIRONMENTAL QUALITY/BUREAU OF ENVIRONMENTAL RADIATION, 729 Alexander Rd. CN-415, Trenton, NJ, 08625, (609) 987-6396, (800) 648 0394, (1)(10).

Makofske, William, ORANGE ENVIRONMENT, INC., RD 2 Maple Ave., Goshen, NY, 10924, (914) 294-5023, (17).

Makofske, William, RAMAPO COLLEGE OF NEW JERSEY/INSTITUTE FOR ENVIRONMENTAL STUDIES, 505 Ramapo Valley Rd., Mahwah, NJ, 07430, (201) 529-7742, (3)(5)(6).

Makowski, Gregory, ALLSAFE ENVIRONMENTAL CORPORATION, P.O. Box 89, Bayville, NJ, 08721, (201) 269-6060, (1).

Malbon, Mark, COMPREHENSIVE RADON SERVICES, INC., P.O. Box 548, Manchester, NH, 03105, (603) 644-7301, (1)(2)(4).

Malcolm, Robert, INDIANA DEPARTMENT OF LABOR/BUREAU OF SAFETY, EDUCATION AND TRAINING, 1013 State Office Building, Indianapolis, IN, 46204, (317) 232-2688, (10).

Malec, Mitch, NEW JERSEY DEPARTMENT OF COMMUNITY AFFAIRS/DIVISION OF HOUSING AND DEVELOPMENT/OFFICE OF PLANNING AND OPERATIONS, 101 South Broad St. CN-802, Trenton, NJ, 08625, (609) 633-6060, (10).

Maleck, Wayne, INSPECT, 7737 South Normandy, Burbank, IL, 60459, (312) 599-3553, (1).

Malkasian, William E., WISCONSIN ASSOCIATION OF REALTORS, 4801 Hayes Rd., Madison, WI, 53704, (608) 241-2047, (18).

Mallon, Jr., John B., RADON DETECTION AND CONTROL, P.O. Box 419 153 Jordan St., South Hts., PA, 15081, (412) 375-1700, (1).

Maloney, James C., SUFFOLK COUNTY DEPARTMENT OF HEALTH SERVICES/AIR POLLUTION CONTROL BUREAU, 15 Horse Block Pl., Farmingville, NY, 11738, (516) 451-4637, (10).

Maloney, Walt, NATIONAL ASSOCIATION OF REALTORS (NAR)/PUBLIC AFFAIRS OFFICE, 777 14th St., NW, Washington, DC, 20005, (202) 383-1177, (18).

Maltzie, Scott, NEW HAMPSHIRE GOVERNOR'S ENERGY OFFICE, 2 1/2 Beacon St., Concord, NH, 03301-4498, (603) 271-4498, (10).

Mamak, Joseph T., RADON SAFETY SERVICE, 2204 Frances Dr., Loveland, CO, 80537, (303) 663-3941, (1).

Mamula, Ned, TERRADYNAMICS CORPORATION, Rte. 1, Aldie, VA, 22001, (703) 327-6500, (703) 435-3033, (1).

Mancini, Ernest A., ALABAMA GEOLOGICAL SURVEY, P.O. Box O 420 Hackberry Ln., Tuscaloosa, AL, 35486-9780, (205) 349-2852, (10).

Mandel, Hector, MIDWEST HEALTH PROGRAMS, INC., P.O. Box 3023, Urbana, IL, 61801, (217) 367-5293, (1).

Maniscalco, Peter, R.F. SIMON COMPANY, INC., RD 2, Box 481, Barto, PA, 19504, (215) 754-6455, (1)(2)(3)(4)(6).

Manke, James, MINNESOTA ASSOCIATION OF REALTORS, 5750 Lincoln Dr., Edina, MN, 55436, (612) 935-8313, (18).

Mankin, Charles, OKLAHOMA GEOLOGICAL SURVEY, The University Of Oklahoma 830 Van Vleet Oval Room 163, Norman, OK, 73019, (405) 325-3031, (10).

Mann, James, RADON ENVIRONMENTAL SERVICES, INC., 4320 Parakeet Rd., Toledo, OH, 43612, (419) 447-0209, (1)(2).

Mann, James, RADON ENVIRONMENTAL SERVICES, INC., 8935 North Meridian St. Suite 110, Indianapolis, IN, 46260, (317) 846-1991, (1)(2).

Manning, R., FEMTO-TECH, INC., 150 C Industry Dr., Carlisle, OH, 45005, (513) 746-4427, (1)(4)(5).

Mansfield, Richard, MANSFIELD, PONTICELLI AND LIBBY, 2100 M St., NW, Washington, DC, 20036, (202) 296-1330, (3)(18).

Mansker, William L., INEX, 8704 Gutierrez, NE, Albuquerque, NM, 87111, (505) 292-0805, (2).

Mapes, Royal, OHIO UNIVERSITY/GEOLOGICAL SCIENCES DEPARTMENT, Athens, OH, 45701, (614) 593-1109, (5).

Maraldo, Joseph, RADON REDUCTION SYSTEMS, 169 Old Gage Hill Rd., Pelham, NH, 03076, (603) 893-5549, (2).

Marangos, Paul, TESTING 1.2.3., 50 Freeway Dr., Cranston, RI, 02920, (401) 461-0530, (1).

March, Fred, GEORGIA RADON DETECTION COMPANY, 3400 Elm Creek Dr., Marietta, GA, 30064-1132, (404) 429-0384, (1).

Marchant, Trey, RAD ENVIRONMENTAL SERVICES, INC., 2306 Rodeo Dr., Durham, NC, 27704, (919) 688-9226, (1)(2).

Mardis, Michael, U.S. ENVIRONMENTAL PROTECTION AGENCY (EPA)/OFFICE OF AIR AND RADIATION(OAR)/OFFICE OF RADIATION PROGRAMS(ORP)/RADON DIVISION (RD), Mitigation Prevention And Quality Assurance Branch ANR-460 401 M St., SW, Washington, DC, 20460, (202) 475-9605, (8).

Marguet, Ken, THRESHOLD TECHNICAL PRODUCTS, 7225 Edington Dr., Cincinnati, OH, 45249, (800) 458-4931, (4).

Marhofer, Donald, CLASSIC CONSTRUCTION, 7001 Orchard Lake Rd. Suite 130, West Bloomfield, MI, 48322, (313) 737-8800, (13).

Marion, Richard, CHESTER COUNTY DEVELOPMENT SERVICE, 420 West Market St., West Chester, PA, 19382, (215) 692-4509, (1).

Markels, Michael, VERSAR, INC., 6850 Versar Center, Springfield, VA, 22151, (703) 750-3000, (1)(2)(3).

Markloff, David W., CONCRETE EPOXY TECHNICAL SYSTEMS, INC., P.O. Box F-25 41 Terry Dr., Trevose, PA, 19047, (215) 322-7310, (4).

Marlow, William, TEXAS A & M UNIVERSITY/DEPARTMENT OF NUCLEAR ENGINEERING, 129 Zachry Engineering Center, College Station, TX, 77843-3133, (409) 845-2271, (5).

Marques, Barbara, SUPERIOR SIGNAL, INC., P.O. Box 96, Spotswood, NJ, 08884, (201) 251-0800, (4).

Marquette, Robby, KEY TECHNOLOGY, INC., P.O. Box 562, Jonestown, PA, 17038, (717) 274-8310, (800) 523-4964, (1)(4)(6).

Marr, Sr., J. Robert, JRM ASSOCIATES, Box 602 North Barnstead Rd., Center Barnstead, NH, 03225, (603) 776-6046, (1).

Marriott, Dean, MAINE DEPARTMENT OF ENVIRONMENTAL PROTECTION/BUREAU OF AIR QUALITY CONTROL, State House, Station 17, Augusta, ME, 04333, (207) 289-2437, (10).

Marryott, Theresa R., RADONICS, INC., P.O. 489, Pennington, NJ, 08534, (609) 737-8888, (1)(2).

Marsh, Gordon, AMERICAN RADON TECHNICAL SERVICES, INC., 1941A Nears Pkwy., Margate, FL, 33063, (305) 968-5555, (1)(2).

Marsh, Gordon, WORLDWIDE RADON, INC., 1941A Nears Pkwy., Margate, FL, 33063, (305) 968-5555, (1)(2).

Marsh, Greg, GREGORY K. MARSH AND ASSOCIATES, 7700 West 61st Ave. Suite 12, Arvada, CO, 80004, (303) 421-3383, (1).

Marsh, Ronald W., CONTROL RESOURCE SYSTEMS, INC., 670 Mariner Dr., Michigan City, IN, 46360, (219) 872-5591, (800) 272-3786, (4).

Marshall, Robert E., CALAVERAS COUNTY HEALTH DEPARTMENT/AIR POLLUTION DIVISION, Government Center, San Andreas, CA, 95249, (209) 754-3742, (10).

Marshall, Stanley R., NEVADA DEPARTMENT OF HUMAN RESOURCES/HEALTH DIVISION/RADIOLOGICAL HEALTH SECTION, 505 East King St. Room 202, Carson City, NV, 89710, (702) 885-5394, (10).

Martel, Real, MINISTERE DE LA JUSTICE DU QUEBEC/SERVICE DU COURTAGE IMMOBILIER DU QUEBEC, 220, Grande Allee Est Suite 910, Quebec City, PQ, G1R 2J1, (418) 643-4597, (11).

Martell, Edward, NATIONAL CENTER FOR ATMOSPHERIC RESEARCH/ATMOSPHERIC CHEMISTRY DIVISION, Box 3000, Boulder, CO, 80307, (303) 497-1000, (5).

Marth, II, Olin R., ROBERT A. SWOYER ASSOCIATES,

241 South Fifth St., Reading, PA, 19602, (215) 376-2580, (2).

Martin, Cindy, REAL ESTATE SECURITIES AND SYNDICATION INSTITUTE, 430 North Michigan Ave., Chicago, IL, 60611, (312) 670-6760, (18).

Martin, Dave, BCA LABORATORIES, INC., 1102 South Main St., Bloomington, IL, 61701, (309) 828-7772, (1)(4).

Martin, G. Thomas, PROFESSIONAL HOME INSPECTIONS, INC., 28790 Pine Dr., Evergreen, CO, 80439, (303) 674-4620, (1)(3).

Martin, Gerry, EER PRODUCTS, INC., 3536 East 28th St., Minneapolis, MN, 55406, (612) 721-4231, (4).

Martin, J.D., TELEDYNE ISOTOPES, 50 Van Buren Ave., Westwood, NJ, 07675, (201) 664-7070, (1)(4).

Martin, JoAnne D., DOUGLAS MARTIN AND ASSOCIATES, INC./DMA ANALYSIS GROUP, 520 Chestnut St., Emmaus, PA, 18049, (215) 965-2914, (1)(4).

Martin, Lea, A & N RADON COMPANY, P.O. Box 188, Keene, NH, 03431, (603) 357-0155, (1).

Martin, Rich D., CONSTRUCTION EVALUATION, INC., 102 Hickok Rd., Greenfield Center, NY, 12833, (518) 587-6366, (1).

Martin, Richard, DOUGLAS MARTIN AND ASSOCIATES, INC./DMA ANALYSIS GROUP, 520 Chestnut St., Emmaus, PA, 18049, (215) 965-2914, (1)(4).

Martin, Richard D., DOUGLAS MARTIN AND ASSOCIATES, INC./MARTIN CONSULTING, 520 Chestnut St., Emmaus, PA, 18049, (215) 965-2914, (2)(3)(4)(6).

Martin, Robert G., NOVA SCOTIA DEPARTMENT OF CONSUMER AFFAIRS/CONSUMER SERVICES DIVISION, P.O. Box 998, Halifax, NS, B3J 2X3, (902) 424-4690, (11).

Martin, W.R., W.R. MARTIN COMPANY, 4124-B Clemson Blvd., Anderson, SC, 29621, (803) 225-7010, (1).

Martin, Larry R., LRM, INC., 632 Spooner Rd., Virginia Beach, VA, 23462, (804) 497-6505, (3).

Martini, Silvia L., CONSERVATION ENERGY SYSTEMS, P.O. Box 10146, Minneapolis, MN, 55458, (306) 242-3663, (800) 667-3717, (4).

Martino, Susan, BOULDER COUNTY HEALTH DEPARTMENT/AIR POLLUTION CONTROL DIVISION, 3450 Broadway, Boulder, CO, 80302, (303) 441-1180, (10).

Martonen, Ted, U.S. ENVIRONMENTAL PROTECTION AGENCY (EPA)/OFFICE OF HEALTH RESEARCH(OHR)/HEALTH EFFECTS RESEARCH LABORATORY(HERL), MD-51, Research Triangle Park, NC, 27711, (919) 382-7400, (8).

Maru, Usha, ENVIRONMENTAL ASSOCIATES, 1222 Fairfield Ave., Bridgeport, CT, 06605, (203) 368-6064, (203) 368-6032, (1).

Marx, Andreas, ENVIRONMENTAL QUALITY CONTROL, INC., 2130 South Sawtelle Ave., Los Angeles, CA, 90025, (213) 478-6668, (1).

Mase, David A., DAVID A. MASE REAL ESTATE, 6515 Main St., Trumbull, CT, 06611, (203) 261-8080, (1).

Masich, Jimmie, SAFETY AND HEALTH SERVICES, 3307 North Second St., Harrisburg, PA, 17110, (717) 238-2122, (3).

Masiello, John, RISE, 280 Broadway, Providence, RI, 02903, (401) 272-1040, (1).

Maslany, Thomas, U.S. ENVIRONMENTAL PROTECTION AGENCY REGION 3 (EPA3)/RADIATION DIVISION(ORP)/RADON PROJECT, 841 Chestnut St. MS-3AM12, Philadelphia, PA, 19107, (215) 597-1256, (8).

Mason, Jay, LOUISIANA DEPARTMENT OF ENVIRONMENTAL QUALITY/OFFICE OF AIR QUALITY AND NUCLEAR ENERGY/NUCLEAR ENERGY DIVISION, P.O. Box 14690, Baton Rouge, LA, 70898-4690, (504) 925-4518, (10).

Mast, Jennifer, RADON AND WATER TESTING SERVICE, 10806 Lafayette-Plain City Rd., Plain City, OH, 43064, (614) 873-8821, (1).

Masterson, Pat, ENVIRONMENTAL MANAGEMENT CONTROL, INC., 2408 Belknap Beach, Louisville, KY, 40259, (502) 228-0284, (1).

Matches, Keith, RADIATION DATA, 115 Wall St., Princeton, NJ, 08540, (609) 921-7061, (1)(2).

Mate, C., TECHNORAD, P.O. Box 2144, Rockville, MD, 20852, (301) 881-8293, (1).

Mate, J., TECHNORAD, P.O. Box 2144, Rockville, MD, 20852, (301) 881-8293, (1).

Mateo, Juan Merced, PUERTO RICO/ENVIRONMENTAL QUALITY BOARD, Del Parque St. #204 Corner Pumarada St., Santurce, PR, 00910, (809) 725-5140, (809) 722-0077, (10).

Mateson, Jean F., MATESON CHEMICAL COMPANY, 1025 East Montgomery Ave., Philadelphia, PA, 19125, (215) 358-5453, (4).

Mathieu, Guy, ENVIRONMENTAL RADIOACTIVITY MEASUREMENTS, INC., P.O. Box 226, Piermont, NY, 10968, (914) 534-8679, (201) 768-7651, (1).

Matonti, James, RADON MANAGEMENT LABORATORIES, 2 Riverdale Rd., Pompton Lakes, NJ, 07442, (201) 835-8721, (1).

Matthews, Gerald, FLORIDA ASSOCIATION OF REALTORS, P.O. Box 1231, Orlando, FL, 32802, (407) 849-1400, (18).

Matthias, Darrel, ORGANIZATION OF RESOURCES COUNSELORS (ORC), 1910 Sunderland Pl., NW, Washington, DC, 20036, (202) 737-6300, (16).

Mattix, Doug, GEOCEL CORPORATION, P.O. Box 398, Elkhart, IN, 46515, (219) 264-0645, (800) 348-7615, (4).

Mauger, John, ENERGY RESOURCE TECHNOLOGY, 1494 North Charlotte St., Pottstown, PA, 19464, (215) 327-4641, (2).

Mauro, John, S. COHEN AND ASSOCIATES, INC., 8200 Riding Ridge Pl., McLean, VA, 22102, (703) 893-6592, (3)(6).

Maxwell, Chris, LUDLUM MEASUREMENTS, INC., 501 Oak St. P.O. Box 810, Sweetwater, TX, 79556, (915) 235-5494, (4).

Maxwell, Roger, WILKES COLLEGE/PHYSICS DEPARTMENT, Wilkes-Barre, PA, 18766, (717) 824-4651 ext. 481, (1).

Maxwell, Ron, TENNESSEE VALLEY AUTHORITY (TVA)/DIVISION OF OCCUPATIONAL HEALTH AND SAFETY(DOHS), Radiation Safety 324 MPB, Mussel Shoals, AL, 35660, (205) 386-2767, (8).

May, Michael J., AMERICAN NUCLEAR SOCIETY/ALABAMA-NORTH ALABAMA SECTION, C/O Tennessee Valley Authority Browns Ferry Nuclear Plant/Administrative Bldg. P.O. Box 2000, Decatur, AL, 35602, (205) 259-5677, (14).

Maybee, Lisa, SENECA NATION HEALTH DEPARTMENT, Box 231, Salamanca, NY, 14779, (716) 532-5582, (10).

Mayer, Gary, DAMES AND MOORE, 7101 Wisconsin Ave. Suite 700, Bethesda, MD, 20814, (301) 652-2215, (1).

Mayer, Gregory A., METRO SERVICE LABORATORIES, INC., 6309 Fern Valley Pass, Louisville, KY, 40228, (502) 964-0865, (800) 678-6878, (1).

Mayer, James, KANSAS ASSOCIATION OF REALTORS, 3644 Southwest Burlingamr Road, Topeka, KS, 66611, (913) 267-3610, (18).

Mayer, John, ADVANCED BASEMENT DE-WATERING AND RADON CONTROL SYSTEMS, 20 Giblin Dr., West Seneca, NY, 14224, (716) 675-5898, (716) 674-6070, (2).

Mayer, Sam, CENTRAL HOME INSPECTION, INC., 435 Rte. 18, East Brunswick, NJ, 08816, (201) 545-0300, (1).

Mayer, Sam, CENTRAL RADON TESTING, 435 Rte. 18, East Brunswick, NJ, 08816, (201) 545-0300, (1).

Mazzarella, David, SAFEHOME, P.O. Box 221, Rexford, NY, 12148, (518) 399-0135, (1).

McAllister, Larry A., ALA OF INDIANA, 8777 Purdue Rd. Suite 310, Indianapolis, IN, 46268-3109, (317) 872-9685, (15).

McBeath, William H., WORLD FEDERATION OF PUBLIC HEALTH ASSOCIATIONS (WFPHA), C/O American Public Health Association 1015 15th St., NW, Washington, DC, 20005, (202) 789-5691, (12).

McBryde, Bill, RADON DETECTION OF MISSISSIPPI, P.O. Box 2404, Columbus, MS, 39704, (601) 328-4687, (1).

McCallson, Virgil, ALPINE RADIATION MITIGATION, 35 Kris Ln., Manitou Springs, CO, 80829, (719) 685-1024, (2).

McCandless, Raymond N., VERMONT DEPARTMENT OF HEALTH/DIVISION OF OCCUPATIONAL AND RADIOLOGICAL HEALTH, Administration Building 10 Baldwin St., Montpelier, VT, 05602, (802) 828-2886, (10).

McCarten, Laura, AMERICAN NUCLEAR SOCIETY/MINNESOTA-UPPER MIDWEST SECTION, C/O Northern States Power 414 Nicollet Mall, Minneapolis, MN, 55401, (612) 330-6341, (14).

McCarthy, Gilbert, PHOENIX LABS, L. Rorbes Rd. Route 1, Box 247, Colebrook, NH, 03578, (603) 237-5773, (1).

McCarthy, Kevin, CONNECTICUT DEPARTMENT OF ENVIRONMENTAL PROTECTION/RADIATION CONTROL UNIT, State Office Building 165 Capital Ave., Hartford, CT, 06106, (203) 566-5668, (10).

McCausland, Bob, RADON SERVICES, 1210 Auburn Way N. Suite B-122, Auburn, WA, 98002, (206) 939-3023, (1).

McClenahan, James B., AMADOR COUNTY AIR POLLUTION CONTROL DISTRICT, 108 Court St., Jackson, CA, 95642-2379, (209) 223-6406, (10).

McCloud, Rober, VERMONT DEPARTMENT OF LABOR AND INDUSTRY/DIVISION OF OCCUPATIONAL SAFETY AND HEALTH, 120 State St., Montpelier, VT, 05602, (802) 828-2765, (10).

McCollam, William, EDISON ELECTRIC INSTITUTE (EEI), 1111 19th St., NW, Washington, DC, 20036, (202) 778-6400, (13).

McCollum, Jackie, CHEROKEE DE-WATERING, INC., Rte. 4, Oak Hill Cr., Canton, GA, 30114, (404) 928-5091, (1).

McConnell, Brian R., INDIANA SEAL COMPANY, 12000 Mitchell Rd., Carmel, IN, 46032, (317) 844-8700, (800) 428-5309, (4).

McCray, Kevin, NATIONAL WATER WELL ASSOCIATION (NWWA), 6375 Riverside Dr., Dublin, OH, 43017, (614) 761-1711, (14).

McCreary, Russ, RADON CONSULTANTS, P.O. Box 5209, Sparks, NV, 89432, (800) 327-6826, (1)(4).

McCune, Mark D., RADON GAS DETECTION SYSTEMS, P.O. Box 15189, Hamilton, OH, 45015, (513) 863-6625, (1).

McDaniel, Steven, RSO, INC., P.O. Box 1526 711 Gorman Ave., Laurel, MD, 20707-0953, (301) 953-2482, (4).

McDermott, Dennis, MISSOURI ASSOCIATION OF REALTORS, P.O. Box 1327, Columbia, MO, 65205, (314) 445-8400, (18).

McDonald, R.J., GASOMETRY ASSOCIATES, 29 Pocahontas Trail, Medford, NJ, 08055, (609) 983-3846, (1).

McDonnell, Lawrence J., WISCONSIN DEPARTMENT OF HEALTH AND SOCIAL SERVICES/DIVISION OF HEALTH/RADIATION PROTECTION SECTION, 5708 Odana Rd., Madison, WI, 53719, (608) 273-5180, (10).

McDonnell, William, RE FORMATION, P.O. Box 328, Pine Island, NY, 10969, (914) 258-4711, (3).

McElroy, Jerel K., ENVIRONMENTAL HEALTH TECHNOLOGIES, 566 South Corona St., Denver, CO, 80209, (303) 778-1632, (1).

McFarland, David, MERIT ENVIRONMENTAL SERVICES, INC., P.O. Box 3083, Berea, KY, 40403, (606) 986-7070, (1)(2).

McFarland, David, TEE ENGINEERING COMPANY, 320 Cutters Hill Ct., Lexington, KY, 40509, (606) 263-5350, (1)(2).

McFarland, Sharon, INSTA-FOAM PRODUCTS, INC., 1500 Cedarwood Dr., Joliet, IL, 60435, (800) 435-9359, (4).

McGarvey, Frank, SYBRON CHEMICAL, P.O. Box 66 Birmingham Rd., Birmingham, NJ, 08011, (609) 893-1100, (4).

McGill, Kelly F., AIRMASTERS RADON DIVISION, 1746 Boggs Rd., Forest Hill, MD, 21050, (301) 893-1993, (1)(2).

McGillivary, Terry, ALASKA DIVISION OF OCCUPATIONAL LICENSING, 3601 C St. Suite 722, Anchorage, AK, 99503, (907) 563-2169, (10).

McGreevy, John E., RADONICS, 1764 Old Meadow Ln. Suite 125, McLean, VA, 22102, (703) 734-0050, (1).

McGregor, R.G., HEALTH AND WELFARE CANADA/HEALTH PROTECTION BRANCH/BUREAU OF RADIATION AND MEDICAL DEVICES, 775 Brookfield Rd., Ottawa, ON, K1A 1C1, (613) 954-6677, (11).

McGroddy, Peter, LMS ENGINEERS, 1 Blue Hill Plaza, Pearl River, NY, 10965, (914) 735-8300, (1).

McGuineas, Ann, NITON CORPORATION, P.O. Box 9153, Lexington, MA, 02173, (617) 229-4929, (800) 451-6467, (1)(4).

McGuineas, Ann, NITONICS CORPORATION, P.O. Box 9153, Lexington, MA, 02173, (617) 229-4929, (800) 451-6467, (4).

McGuinness, George D., R/K OCCUPATIONAL AND ENVIRONMENTAL ANALYSIS, INC., 616 Warren St., Alpha, NJ, 08865, (215) 250-5948, (1).

McGuire, Dennis J., GARY/AIR POLLUTION CONTROL OFFICE, 846 Broadway 2nd Floor, Gary, IN, 46402, (219) 882-3000, (10).

McGuire, Robert, MCGUIRE BUILDERS, INC., Rte. 2, Box 1840, Warfordsburg, PA, 17267, (717) 294-3108, (1)(2).

McHard, Dale, OKLAHOMA DEPARTMENT OF HEALTH/RADIATION AND SPECIAL HAZARDS SERVICE, 1000 NW Tenth St. P.O. Box 53551, Oklahoma City, OK, 73152, (405) 271-5221, (10).

McHill, A.D., MCHILL, A.D. AND SONS, INC., 1121 Second St. Pike, Richboro, PA, 18954, (215) 364-2784, (1)(2).

McInerny, T.M., MCINERNY, T.M., 1214 Church St., Elmira, NY, 14905, (607) 733-1994, (2).

McIntyre, Marla, NATIONAL CONFERENCE OF STATES ON BUILDING CODES AND STANDARDS (NCSBCS), 481 Carlisle Dr., Herndon, VA, 22070, (703) 437-0100, (13).

McJunkin, Jerry, EDA INSTRUMENTS, INC., 9200 East Mineral Ave. Suite 370, Englewood, CO, 80112, (303) 790-2541, (1)(4).

McKay, James, HOUSE ANALYSIS AND ASSOCIATES, 747 Hallowell Dr., Huntingdon Valley, PA, 19006, (215) 947-1169, (215) 947-9209, (1).

McKelvey, W., ARIX CORPORATION/A DIVISION OF VERSAR, INC., 1200 East High St., Pottstown, PA, 19464, (215) 327-1140, (1)(2)(6).

McKelvey, William F., ARIX CORPORATION, P.O. Box 2021, Greeley, CO, 80632-2021, (303) 356-3930, (1)(2).

McKenzie, Stewart, MONTGOMERY COUNTY GOVERNMENT/MONTGOMERY COUNTY RADON TASK FORCE, 100 Maryland Ave. 6th Fl., Rockville, MD, 20850, (301) 217-7900, (10).

McKeon, Mike, QUALITY AIR RADON REDUCTION, 5-70 Oakwood Rd., Glastonbury, CT, 06033, (203) 633-3629, (1).

McKillop, D., TELEDYNE ISOTOPES, 50 Van Buren Ave., Westwood, NJ, 07675, (201) 664-7070, (1)(4).

McKinney, Ed, PARADISE COMPANY, 109 Greenleaf, Lumberton, TX, 77656, (409) 755-3524, (1)(2).

McKinney, Linda, PARADISE COMPANY, 109 Greenleaf, Lumberton, TX, 77656, (409) 755-3524, (1)(2).

McKinney, Liz, PARADISE COMPANY, 109 Greenleaf, Lumberton, TX, 77656, (409) 755-3524, (1)(2).

McLain, Milton E., TEXAS A & M UNIVERSITY/DEPARTMENT OF NUCLEAR ENGINEERING/RADON ASSAY LABORATORY, 129 Zachry Engineering Center, College Station, TX, 77843, (409) 845-4107 , (1)(5).

McLane, Lawrence P., SOUTH CENTRAL ENVIRONMENTAL SERVICES, RD 1, Box 265, Pine City, NY, 14871, (607) 733-1661, (1)(2).

McLaughlin, Phyllis A., A & M ASSOCIATES, 1813 Northwest 27th, Portland, OR, 97210, (503) 274-2922, (1).

McLean, Cherl, MINNESOTA COMMERCE DEPARTMENT, 500 Metro Square, St. Paul, MN, 55101, (612) 297-3379, (10).

McLellan, Robert, CENTER FOR ENVIRONMENTAL MEDICINE, 60 Washington Ave., Hamden, CT, 06518, (203) 789-1911, (15).

McLellan, Robert, REMD/RESIDENTIAL ENVIRONMENTAL MEDICAL DIAGNOSTICS, 28 School St. P.O. Box 3285, Stony Creek, CT, 06405, (203) 481-8211, (1).

McLemore, William, GEORGIA GEOLOGICAL SURVEY, 19 Martin Luther King, Jr., Dr., SW Room 400, Atlanta, GA, 30334, (404) 656-3214, (10).

McLeran, David E., RADON DETECTION AGENCY, LTD., 625 5th Ave. Suite 104, New Westminster, BC, V3M 1X4, (604) 525-4455, (1)(4).

McMahon, Robert, MICROBAC LABORATORIES, INC./MASSACHUSETTS TESTING DIVISION, 202 Bussey St., Dedham, MA, 02026, (617) 326-7117, (1).

McMillan, John, W.L. GORE AND ASSOCIATES, 100 Airport Rd. Box 1010, Elkton, MD, 21921, (301) 392-3200, (4).

McMillian, Craig M., MENDOCINO COUNTY AIR POLLUTION CONTROL DISTRICT, Courthouse Sq., Ukiah, CA, 95482, (707) 463-4354, (10).

McNall, Joe, HANCOR, INC., 401 Olive St. P.O. Box 1047, Findlay, OH, 45839, (800) 537-9520, (4).

McNall, Preston, ASHRAE/ENVIRONMENTAL HEALTH COMMITTEE/INDOOR AIR QUALITY SUBCOMMITTEE, Atlanta, GA, 30329, (404) 636-8400, (13).

McNall, Preston, PHOENIX ENGINEERS, 10041 El Monte, Overland Park, KS, 66207, (301) 975-5906, (3).

McNally, William T., POLK COUNTY HEALTH UNIT/RADIOLOGICAL HEALTH SECTION, P.O. Box 1480, Winter Haven, FL, 33882-1480, (813) 294-7481 ext. 283, (1).

McNees, James, ALABAMA DEPARTMENT OF PUBLIC HEALTH/RADIOACTIVE MATERIALS BRANCH/COMPLIANCE SECTION, 434 Monroe St. Room 510, Montgomery, AL, 36130, (205) 261-5315, (10).

McNutt, Gary W., MISSOURI DEPARTMENT OF HEALTH/DIVISION OF ENVIRONMENTAL HEALTH AND EPIDEMIOLOGY SERVICES, 1730 East Elm Plaza P.O. Box 570, Jefferson City, MO, 65102, (314) 751-6083, (10).

McOmber, Ron, CTL/THOMPSON, INC., 1971 West 12th Ave., Denver, CO, 80204, (303) 825-0777, (1).

McPhee, Harvey J., ST. LOUIS PARK INSPECTIONAL SERVICES, 5005 Minnetonka Blvd., St. Louis Park, MN, 55416, (612) 924-2588, (10).

McQueen, Raymond, SAFEHOUSE, INC., 2335 Sigman Industrial Ct., Conyers, GA, 30207, (404) 388-0481, (2).

McRee, Brad, ENVIRADON, INC., P.O. Box 13683, Research Triangle Park, NC, 27709, (919) 848-2128, (1).

McRee, William B., ENVIRADON, INC., 914 Rollingwood Dr., Mt. Holly, NC, 28120, (704) 827-1293, (1).

McReynolds, Bobbie, ALL PRO PEST CONTROL, P.O. Box 1313, Amarillo, TX, 79105-1313, (806) 374-7886, (806) 355-6660, (1).

McSpadden, William H., HOME CHECK SERVICES, P.O. Box 98, Front Royal, VA, 22630, (703) 635-1274, (1)(2).

McWilliams, Alfred R., MCWILLIAMS PEST CONTROL, 7014 McCallum St., Philadelphia, PA, 19119, (215) 849-2223, (1).

Meade, John, A & A CONSULTING AND INSPECTION SERVICES, INC., 349 Chenango St., Binghamton, NY, 13901, (607) 723-7968, (1)(3).

Meade, Jr., Thomas, SUNSTREAM DEVELOPMENT CORPORATION, 300 North St., Endicott, NY, 13760, (607) 754-6200, (1)(2)(4).

Mech, Robert, CANADIAN LUNG ASSOCIATION (CLA), 75 Albert St. Suite 908, Ottawa, ON, K1P 5E7, (613) 237-1208, (11).

Meehan, B. Michael, PROTECT-A-HOME, P.O. Box 584, Newtown, CT, 06470, (203) 426-3000, (1).

Meehan, Chris, BUZZANO CONTRACTING, 62 Mt. Pleasant Rd., Newtown, CT, 06470, (203) 426-7196, (2).

Meehan, Thomas F., SAF-AIR, P.O. Box 1217, Orange, CT, 06477, (203) 799-9609, (1)(2)(6).

Mehta, Dhiro, RADON DETECTION SYSTEMS/NEWARK, NEW JERSEY BRANCH, 6 Earl Ct., Montiville, NJ, 07045, (303) 444-5253, (800) 627-2360, (1).

Mehta, Khurshid, JACKSONVILLE/DUVAL COUNTY BIO-ENVIRONMENTAL SERVICES DIVISION, 515 West 6th St., Jacksonville, FL, 32206, (904) 630-3210, (10).

Meier, R., ARIX CORPORATION/A DIVISION OF VERSAR, INC., 1200 East High St., Pottstown, PA, 19464, (215) 327-1140, (1)(2)(6).

Melgard, Rod, TMA/EBERLINE, LABORATORY, 7021 Pan American Hwy., NE, Albuquerque, NM, 87109, (505) 345-3461, (1).

Meliodon, Bob, RADON TECHNICIANS, INC, 27 Boren Shore Ln., Elkton, MD, 21921, (301) 854-6186, (2).

Mellis, Christopher G., AMERICAN RADON DETECTION COMPANY, INC., 39 Folcroft Ave., Folcroft, PA, 19032, (215) 532-8845, (1).

Mencke, John L., MONTGOMERY COUNTY GOVERNMENT/MONTGOMERY COUNTY RADON TASK FORCE, 100 Maryland Ave. 6th Fl., Rockville, MD, 20850, (301) 217-7900, (10).

Mendelsohn, Eric, MONTGOMERY COUNTY GOVERNMENT/DIVISION OF ENVIRONMENTAL PLANNING AND MONITORING, 101 Monroe St. 6th Fl., Rockville, MD, 20850, (301) 217-2380, (10).

Mendenhall, Carl, U.S. DEPARTMENT OF DEFENSE (DOD)/DEPARTMENT OF THE NAVY(DON)/RADIOLOGICAL CONTROL HEALTH BRANCH(RCHB)/CHIEF OF NAVAL OPERATIONS OFFICE(CNO), Crystal Plaza 5 (Code 455), Washington, DC, 20350, (202) 692-5604, (8).

Mendyka, Jr., Frederick J., PNEUMODYNAMICS, INC., 227 Kasson Rd., Bethlehem, CT, 06751, (203) 266-7786, (1).

Menetrez, Marc Y., RAD ENVIRONMENTAL SERVICES, INC., 2306 Rodeo Dr., Durham, NC, 27704, (919) 688-9226, (1)(2).

Menetrez, Mary, RAD ENVIRONMENTAL SERVICES, INC., 2306 Rodeo Dr., Durham, NC, 27704, (919) 688-9226, (1)(2).

Menke, Tom, MENKE CONSULTING, INC., 6070 Routzong Rd., Greenville, OH, 45331, (513) 447-4225, (1).

Mercer, Richard M., UTAH DEPARTMENT OF BUSINESS REGULATIONS/DIVISION OF REAL ESTATE, P.O. Box 45802 160 East 300 South, Salt Lake City, UT, 84145, (801) 530-6747, (10).

Mercer, Robert R., DUKE UNIVERSITY MEDICAL CENTER/DEPARTMENT OF MEDICINE/DIVISION OF PULMONARY MEDICINE, 353 Bell Building Box 3177, Durham, NC, 27710, (919) 684-2513, (5).

Mercer, Ross L., HEALTH PHYSICS NORTHWEST, INC., 11535 Southwest 67th Ave., Tigard, OR, 97223, (503) 620-6617, (800) 762-8444, (1)(3)(4).

Merion, Richard, DEVELOPMENT SERVICES, LTD., 420 West Market St., West Chester, PA, 19382, (215) 692-4509, (1).

Merkel, Jon, RADON CONSULTING AND TESTING SERVICE, 7449 Highway 44 Box 9, Crystal River, FL, 32629, (904) 563-2141, (904) 563-1771, (1).

Merker, William V., COLORADO FRONT RANGE BUILDING INSPECTION SERVICE, P.O. Box 9614, Ft. Collins, CO, 80525, (303) 482-1976, (800) 248-1976, (1).

Mertz, Charles, ENVIRO-CARE RADON SERVICES, 661 Wyoming St., Buffalo, NY, 14215, (716) 834-1788, (1)(2).

Messa, Jr., John A., J.A. ASSOCIATES/PUBLIC HEALTH CONSULTANTS, 313 Clear Spring Rd., Lansdale, PA, 19446, (215) 855-5592, (1).

Messi, Jeffrey, RADIATION CONSULTANTS OF SOUTHWEST VIRGINIA, P.O. Box 4464, Roanoke, VA, 24015, (703) 985-5778, (703) 981-7732, (1).

Messick, Ray, RAY MESSICK RADON TESTING AND MITIGATION, RDI Box 469L, McConnellsburg, PA, 17233, (202) 332-1798, (717) 485-5456, (1).

Messick, Ray, SCRUB RIDGE RAD TESTING AND REMEDIATION, RDI Box 469L, McConnellsburg, PA, 17233, (202) 332-1798, (717) 485-5456, (1).

Messing, Marc, INFILTEC, D.C., 1105 Park St., NE, Washington, DC, 20002, (202) 547-1611, (1).

Messing, Marc, INFILTEC/MAIN OFFICE, P.O. Box 8007, Falls Church, VA, 22041, (703) 820-7696, (1)(2)(3)(5)(6).

Metzger, Jim, CHATAUQUA COUNTY HEALTH DEPARTMENT, Health and Social Services Building, Mayville, NY, 14757, (716) 753-4481, (10).

Metzger, Robert, RADIATION SAFETY ENGINEERING, 6105 South Ash Suite 1, Tempe, AZ, 85283, (602) 897-9459, (1).

Metzger-Keele, J. Margo, NEW MEXICO DEPARTMENT OF HEALTH AND ENVIRONMENT/COMMUNITY SERVICES BUREAU, P.O. Box 968, Santa Fe, NM, 87504-0968, (505) 827-2948, (1)(10).

Meyer, Bruce F., LANDMARK RADON LABORATORIES, LTD., 3521 West Eisenhower Blvd., Loveland, CO, 80537, (303) 667-6286, (303) 356-6286, (1).

Meyers, Clay, RADON TECHNOLOGY, 11404A East 24 Hwy., Sugar Creek, MO, 64054, (816) 461-0211, (1)(2).

Meyers, Marcee, PENNSYLVANIA DEPARTMENT OF ENVIRONMENTAL RESOURCES/BUREAU OF RADIATION PROTECTION/DIVISION OF SPECIAL PROJECTS, Fulton Building, 16th Floor Third and Locust Sts., Harrisburg, PA, 17120, (717) 787-2480, (6)(10).

Micek, Susanne E., RDS/RADON DETECTION SERVICES, INC.-CORPORATE OFFICE, Rte. 179, Old York Rd. P.O. Box 419, Ringoes, NJ, 08551, (201) 788-3080, (1)(2)(3)(4)(6).

Michael, George A., G.A.M. BUILDING INSPECTION AND RADON MEASUREMENTS AND CONTROL, INC., 17223 East Purdue Pl., Aurora, CO, 80013, (303) 699-2827, (1)(2).

Michael, George A., RAMSES II, LTD., 2355 East Dartmouth Ave., Denver, CO, 80210, (303) 758-5209, (1)(2)(4).

Michael, Richard S., KANSAS CITY,KANSAS-WYANDOTTE COUNTY/ HEALTH DEPARTMENT/AIR POLLUTION CONTROL DIVISION, 619 Ann Ave., Kansas City, KS, 66101, (913) 321-4803, (10).

Michaelson, Ernest C., ECM ENVIRONMENTAL CONSULTANTS, 7425 Hawkins Dr., Hanover, MD, 21076, (301) 761-1136, (1)(2).

Michel, Jacqueline, RPI INTERNATIONAL ENVIRONMENTAL TECHNOLOGY DIVISION, P.O. Box 328, Columbia, SC, 29202, (803) 256-7322, (3)(5).

Micholle, John, REALITY DYNAMICS, 4758 Antelope Trail, Elbert, CO, 80106, (303) 648-3297, (1).

Micka, John, WISCONSIN DEPARTMENT OF HEALTH AND SOCIAL SERVICES/DIVISION OF HEALTH/RADIATION PROTECTION SECTION, 5708 Odana Rd., Madison, WI, 53719, (608) 273-5180, (10).

Mikaelian, Henry, NUWS LABORATORIES, INC., 28 Montgomery St., Rhinebeck, NY, 12572, (914) 876-5666, (1).

Mikucki, Walter, U.S. DEPARTMENT OF DEFENSE (DOD)/ARMY CORPS OF ENGINEERS(ACE)/ENVIRONMENTAL DIVISION(ED)/CONSTRUCTION ENGINEERING RESEARCH LABORATORY(CERL), P.O. Box 4005, Champaign, IL, 61280-1305, (217) 373-7227, (8).

Milewski, M., COLUMBIA RADIOLOGICAL TESTING LABORATORY, 4054 Victoria Ct., Columbia, MO, 65201, (314) 443-3538, (314) 442-6080, (1).

Milhone, John, U.S. DEPARTMENT OF ENERGY (DOE)/OFFICE OF CONSERVATION AND RENEWABLE ENERGY(OCRE)/OFFICE OF BUILDINGS AND COMMUNITY SYSTEMS, 1000 Independence Ave. Room 5E-080, Washington, DC, 20585, (202) 586-1510, (8).

Milici, Robert, VIRGINIA DIVISION OF MINERAL RESOURCES, P.O. Box 3667, Charlottesville, VA, 22903, (804) 293-5121, (10).

Millben, Ann, MICHIGAN DEPARTMENT OF LICENSING AND REGULATION, P.O. Box 30018, Lansing, MI, 48909, (517) 373-0490, (10).

Miller, C. George, MINING ASSOCIATION OF CANADA (MAC), 350 Sparks St. Suite 705, Ottawa, ON, K1R 7S8, (613) 233-9391, (11).

Miller, Charles G., BONDEX INTERNATIONAL, INC., 3616 Scarlet Oak Blvd., St. Louis, MO, 63122, (314) 225-5001, (4).

Miller, David, ENGINEERING ASSOCIATES, 4682 Watch Hill Rd., Manlius, NY, 13104, (315) 682-4389, (1)(2).

Miller, David V., INDOOR POLLUTION CONTROL CENTER, 3921 Sawyer Rd., Sarasota, FL, 34233, (813) 923-3641, (800) 226-2636, (1)(2).

Miller, David W., AMERICAN NUCLEAR SOCIETY/PENNSYLVANIA-NORTHERN PENNSYLVANIA SECTION, 1326 Brassie St., Wescosville, PA, 18106, (215) 770-7893, (14).

Miller, Edward F., ALA OF MAINE, 128 Sewall St., Augusta, ME, 04330, (207) 622-6394, (15).

Miller, Elizabeth C., ALA OF IDAHO, 2621 Camas St., Boise, ID, 83705-2447, (208) 344-6567, (15).

Miller, G. Wade, ASSOCIATION OF STATE DRINKING WATER ADMINISTRATORS (ASDWA), 1911 North Fort Myer Dr., Arlington, VA, 22209, (703) 524-2428, (14).

Miller, James, U.S. DEPARTMENT OF HOUSING AND URBAN DEVELOPMENT (HUD)/COMMUNITY PLANNING AND DEVELOPMENT(CPD)/OFFICE OF ENVIRONMENT AND ENERGY(OEE)/ENVIRONMENTAL PLANNING DIVISION(EPD), 451 7th St., SW Room 7212, Washington, DC, 20410, (202) 755-7225, (8).

Miller, Kennith V., MISSOURI DEPARTMENT OF HEALTH/DIVISION OF ENVIRONMENTAL HEALTH AND EPIDEMIOLOGY SERVICES, 1730 East Elm Plaza P.O. Box 570, Jefferson City, MO, 65102, (314) 751-6083, (10).

Miller, Lee, GARON PRODUCTS, INC., 1974 Hwy. 35 CN20, Wall, NJ, 07719, (201) 499-1776, (800) 631-5380, (4).

Miller, Mike, ENERGY PROFESSIONALS, INC., 387 Irwin St., Lock Haven, PA, 17745, (717) 748-5910, (717) 726-3825, (1)(2).

Miller, Mike, U.S. ENVIRONMENTAL PROTECTION AGENCY REGION 6 (EPA6)/OFFICE OF AIR AND RADIATION(OAR), 1201 Elm St., Dallas, TX, 75270, (214) 655-7208, (8).

Miller, Raymond V., RADON RESEARCH, 700 Winthrop Pl., Orlando, FL, 32803, (813) 377-3740, (1).

Miller, Richard L., LEGIS CORPORATION, 2211 Saxon, Houston, TX, 77018, (713) 680-8863, (1).

Miller, Ronald C., ALA OF GEORGIA, 2452 Spring Rd., Smyrna, GA, 30080, (404) 434-5864, (15).

Miller, Ronald V., INDOOR POLLUTION CONTROL CENTER, 3921 Sawyer Rd., Sarasota, FL, 34233, (813) 923-3641, (800) 226-2636, (1)(2).

Miller, William H., ENVIRONMENTAL INDOOR AIR QUALITY CONSULTANTS, INC. (EIC), 58 Keller Ave., Lancaster, PA, 17601, (717) 393-5774, (800) 367-7223, (1)(2).

Miller Dieppa, Annie, ALA OF PUERTO RICO, G.P.O. 3468, San Juan, PR, 00936, (809) 756-5664, (15).

Mills, David, W.W. GRAINGER, INC. DISTRIBUTION GROUP, 5959 West Howard St., Chicago, IL, 60648, (312) 647-8900, (4).

Mills, William, EXECUTIVE OFFICE OF THE PRESIDENT (EOP)/OFFICE OF SCIENCE AND TECHNOLOGY POLICY(OSTP)/OAK RIDGE ASSOCIATED UNIVERSITIES(ORAU)/COMMITTEE ON INTERAGENCY RADIATION RESEARCH AND POLICY COORDINATION(CIRRPC), 1019 19th St., NW, Washington, DC, 20036, (202) 653-5505, (8).

Milnes, Ian, WRIGHT LABORATORY SERVICES, INC., 3240 Schoolhouse Rd., Middletown, PA, 17057, (717) 944-5541, (1).

Minchik, Richard, MINCHIK AND MINCHIK, 6319 Utah Ave. NW, Washington, DC, 20015, (202) 364-8845, (1).

Minear, Roger, AMERICAN CHEMICAL SOCIETY (ACS), 1155 16th St., NW, Washington, DC, 20036, (202) 872-6062, (14).

Minkin, Sheldon R., KINGS COUNTY AIR POLLUTION CONTROL DISTRICT, 330 Campus Dr., Hanford, CA, 93230, (209) 584-1411, (10).

Minnerly, W. Lee, AMERICAN INSTITUTE OF REAL ESTATE APPRAISERS (AIREA), 430 North Michigan Ave., Chicago, IL, 60611, (312) 329-8437, (18).

Minnick, William, INDUSTRIAL RESOURCE DEVELOPMENT COMPANY, 2490 General Armistead Ave. Suite 303, Norristown, PA, 19403, (215) 630-9150, (1)(2).

Mirabile, William, AIRMASTER RADON DETECTION, INC., P.O. Box 3177, Wayne, NJ, 07474-3177, (201) 794-3733, (1)(2).

Mirkhaef, Vahooman, RANDOLPH AND ASSOCIATES, INC., 8901 North Industrial Rd., Peoria, IL, 61615, (309) 692-4160, (1)(2).

Miscavage, Joe, MISCAVAGE AND COMPANY, 2814 Pearl Ave., Allentown, PA, 18103, (215) 797-3865, (1).

Mitch, Sue F., ACCURATE HOME SERVICE, 928 Weatherbee Pl., Downers Grove, IL, 60516, (312) 971-2662, (1).

Mitchell, Bert, U.S. DEPARTMENT OF INTERIOR (DOI)/NATIONAL PARK SERVICE(NPS)/MIDWEST REGIONAL OFFICE/INDOOR RADON PROGRAM, 1709 Jackson St., Omaha, NE, 68102, (402) 221-3448, (8).

Mitchell, Bob, HARDICO RADON MANAGEMENT, COLUMBUS OFFICE, 95 Cherry St., Groveport, OH, 43215, (614) 836-2788, (2).

Mitchell, Hoyt, RA CONSULTANTS, INC., 1236 North 7th St., Grand Junction, CO, 81501, (303) 245-5237, (1)(2).

Mixson, Karl, VILLAGE OF BEDFORD PARK/AIR POLLUTION CONTROL, 6701 South Archer Rd., Bedford Park, IL, 60501, (312) 458-2975, (10).

Mizrahi, Steve, ACCURATE HOME INSPECTION, INC., P.O. Box 131, Holmdel, NJ, 07733, (201) 264-5485, (1).

Mody, Maria, RADON TESTING AND CONSULTING, 1771 Post Rd., E. Suite 320, Westport, CT, 06880, (203) 454-3886, (1)(2)(4).

Moed, Barbara, GEO-RADON SERVICES, P.O. Box 5985, Berkeley, CA, 94705, (415) 843-8844, (1)(2)(3).

Moeller, Dade W., HARVARD UNIVERSITY/SCHOOL OF PUBLIC HEALTH, 677 Huntington Ave., Boston, MA, 02115, (617) 732-1000, (5).

Moelling, David, AMERICAN NUCLEAR SOCIETY/CONNECTICUT-CONNECTICUT SECTION, C/O Combustion Engineering 9422-2405 1000 Prospect Hill Rd., Windsor, CT, 06095, (203) 285-3881, (14).

Moen, Terry, WISCONSIN DEPARTMENT OF HEALTH AND SOCIAL SERVICES/DIVISION OF HEALTH/BUREAU OF ENVIRONMENTAL HEALTH/SECTION OF OCCUPATIONAL HEALTH, P.O. Box 309, Madison, WI, 53701, (608) 266-0417, (10).

Mohan, Kshitij J., U.S. DEPARTMENT OF HEALTH AND HUMAN SERVICES (HHS)/PUBLIC HEALTH SERVICE(PHS)/FOOD AND DRUG ADMINISTRATION(FDA)/CENTER FOR DEVICE AND RADIOLOGICAL HEALTH, Office Of Device Evaluation (ODE) 8757 Georgia Ave., Silver Spring, MD, 20910, (301) 427-7004, (8).

Mohler, Wayne, U.S. DEPARTMENT OF INTERIOR (DOI)/NATIONAL PARK SERVICE(NPS)/PACIFIC NORTHWEST REGION/INDOOR RADON PROGRAM, 2001 6th Ave., Seattle, WA, 98121, (206) 442-4830, (8).

Mohler, Wayne, U.S. DEPARTMENT OF INTERIOR (DOI)/NATIONAL PARK SERVICE(NPS)REGIONAL OFFICE FOR ALASKA/INDOOR RADON PROJECT, 2525 Gambell St., Anchorage, AK, 99503, (907) 257-2690, (8).

Moline, Donald M., TOLEDO/ENVIRONMENTAL SERVICES DIVISION, 26 Main St., Toledo, OH, 43605, (419) 693-0350, (10).

Monica, Pete, T.R. EDGERTON, INC., 102F Woodwinds Industrial Ct., Cary, NC, 27511, (919) 469-9795, (1).

Monteiro, Richard A., RADON SYSTEMS OF RHODE ISLAND, INC., 156 Earle St., Woonsocket, RI, 02895, (401) 766-7072, (1)(2).

Mood, Robert A., BLOOMINGTON/ENVIRONMENTAL POLLUTION SECTION, 2215 West Old Shakopee Rd., Bloomington, MN, 55431, (612) 881-5811, (10).

Mook, Walter R., SAN BERNARDINO COUNTY AIR POLLUTION CONTROL DISTRICT, 15505 Civic Dr., Victorville, CA, 92392, (619) 243-8147, (10).

Moolgavkar, Suresh H., FRED HUTCHINSON CANCER RESEARCH CENTER/PUBLIC HEALTH SCIENCES DIVISION, 1124 Columbia St., Seattle, WA, 98104, (206) 467-4267, (5).

Mooney, Robert, CONFERENCE OF RADIATION CONTROL PROGRAM DIRECTORS (CRCPD)/TASK FORCE FOR ADEQUATE RADIATION CONTROL PROGRAM-RADON, 71 Fountain Pl., Frankfort, KY, 40601, (502) 227-4543, (14).

Mooney, Robert, WASHINGTON STATE DEPARTMENT OF SOCIAL AND HEALTH SERVICES/OFFICE OF RADIATION PROTECTION, Mail Stop LE-13, Olympia, WA, 98504, (206) 753-5962, (800) 323-9727, (6)(10).

Moore, Earl, I & M RADON MITIGATION, 529 Browar, Eaton, OH, 45320, (513) 456-6784, (2).

Moore, Greg, RADON TECHNOLOGIES, INC., 698 High St., Worthington, OH, 43085, (614) 441-0550, (614) 431-0322, (1).

Moore, Richard C., C SENTINEL ENTERPRISES, 426 Ingel Blvd., Salem, VA, 24153, (703) 380-4828, (1).

Moore, Scott, WATER STORE, THE, P.O. Box 176, Bath, PA, 18014, (215) 837-9660, (3).

Moore, Jr., Dwight G., ENERGY SYSTEMS, INC., 134 Carr Ave., Cookeville, TN, 38501, (615) 528-7247, (1).

Morandi, Larry, NATIONAL CONFERENCE OF STATE LEGISLATURES (NCSL), 1050 17 St. #2100, Denver, CO, 80265, (303) 623-7800, (14).

Morando, Charles A., WEATHER BARRIER SYSTEMS, INC./CAM GROUP, 111 West Second St., Jamestown, NY, 14701, (716) 484-7183, (800) 624-6434, (4).

Morceau, Randy, LOS ANGELES URBAN DEVELOPMENT, 1725 Radcliff St., Los Angeles, CA, 90026-1131, (213) 668-2655, (1)(3).

Moreau, Eugene, MAINE DEPARTMENT OF HUMAN SERVICES/BUREAU OF HEALTH/DIVISION OF HEALTH ENGINEERING/OCCUPATIONAL AND RESIDENTIAL HEALTH PROGRAM, State House Station 10, Augusta, ME, 04333, (207) 289-3826, (10).

Moreno, Jack, BUILDING INSPECTOR OF AMERICA, THE, 129C Gaither Dr., Mt. Laurel, NJ, 08054, (609) 273-8333, (1).

Morgan, Wayne, NORTHERN SIERRA AIR QUALITY MANAGEMENT DISTRICT, 10433 Willow Valley Rd., Nevada City, CA, 95959, (916) 265-1398, (10).

Morony, Ron, U.S. DEPARTMENT OF HOUSING AND URBAN DEVELOPMENT (HUD)/OFFICE OF POLICY DEVELOPMENT AND RESEARCH(OPDR)/DIVISION OF INNOVATIVE TECHNOLOGY AND SPECIAL PROJECTS(DITSP), 451 7th St., SW Room 8232, Washington, DC, 20410, (202) 755-0640, (5)(8).

Morrazzo, Rudolph, U.S. DEPARTMENT OF STATE (DOS)/OFFICE OF MEDICAL SERVICES, 2210 C St., NW Room 4535 A, Washington, DC, 20520, (202) 647-6923, (8).

Morris, Glenn P., RADON CALIBRATIONS, 800 Stiles Ct., Richmond, VA, 23235, (804) 320-3761, (1).

Morrison, Charles, NATIONAL ASSOCIATION OF REALTORS (NAR), Education Division 430 North Michigan Ave., Chicago, IL, 60611, (312) 329-8388, (18).

Morrison, Dave, PHOENIX SAFETY SYSTEMS, 7917 Carnation Dr., Louisville, KY, 40258, (502) 935-8317, (502) 964-0865, (2)(3).

Morrison, David, VIMCO, INC., 300 Hansen Access Rd., King of Prussia, PA, 19406, (215) 768-0500, (4).

Morrison, Robert B., MORRISON WATERPROOFING COMPANY, INC., 7644 L Fullerton Rd., Springfield, VA, 22153, (703) 569-1666, (1).

Morrow, Richard, ATLANTIC ENVIRONMENTAL LABORATORY, 16 Elm St., West Haven, CT, 06516, (203) 932-5808, (1).

Morrow, Robert, RADON PRESS, INC., 500 North Washington St., Alexandria, VA, 22314, (703) 548-2756, (800) 548-1567, (3).

Morrow, Sue, TERRADEX CORPORATION/SUBSIDARY OF TECH/OPS LANDAUER, INC., 2 Science Rd., Glenwood, IL, 60425, (312) 755-7911, (800) 528-8327, (1)(4).

Morse, Ron, NEW ENGLAND DE-WATERING, INC., 70 Heritage Ave. Suite 14 Heritage Pl., Portsmouth, NH, 03801, (800) 535-1211, (2)(4).

Morth, Tom, NORTH DAKOTA STATE DEPARTMENT OF HEALTH AND CONSOLIDATED LABORATORIES/DIVISION OF ENVIRONMENTAL ENGINEERING, 1200 Missouri Ave. P.O. Box 5520, Bismarck, ND, 58502-5520, (701) 224-2348, (10).

Mose, Douglas, GEORGE MASON UNIVERSITY/CENTER FOR BASIC AND APPLIED SCIENCE, 4400 University Dr., Fairfax, VA, 22030, (703) 323-4318, (5).

Moseley, Mitchell, ENERGY UNLIMITED, RFD 1, Box 1090, Mercer, PA, 16137, (412) 662-3591, (1)(2).

Moses, David, U.S. DEPARTMENT OF ENERGY (DOE)/OFFICE OF ENVIRONMENT, SAFETY AND HEALTH(ESH)/ENVIRONMENTAL ANALYSIS(EA), 1000 Independence Ave, SW Room 4G-036 MS-EH22, Washington, DC, 20585, (202) 586-2061, (8).

Moskowitz, Donald A., RADON TECHNOLOGY, INC., 7 Tokanel Dr., Londonderry, NH, 03053, (603) 434-4937, (1).

Mosley, Reynold, U.S. ENVIRONMENTAL PROTECTION AGENCY (EPA)/OFFICE OF RESEARCH AND DEVELOPMENT(ORD)/AIR AND ENERGY ENGINEERING RESEARCH LABORATORY(AEERL)/RADON MITIGATION PROJECT, MD-60, Research Triangle Park, NC, 27711, (919) 541-2745, (8).

Moss, C., BOSTIK, Boston St., Middleton, MA, 01949, (508) 777-0100, (800) 343-8178, (4).

Moss, Doug, POLLENEX/ASSOCIATED MILLS, INC., 165 North Canal St., Chicago, IL, 60606, (312) 454-5400, (4).

Moss, Eugene, U.S. DEPARTMENT OF HEALTH AND HUMAN SERVICES (HHS)/PUBLIC HEALTH SERVICE(PHS)/CENTERS FOR DISEASE CONTROL(CDC)/NATIONAL INSTITUTE FOR OCCUPATIONAL SAFETY AND HEALTH(NIOSH), Ionizing And Nonionizing Radiation Division(INRD)/Hazard And Evaluation Branch 4676 Columbia Pkwy, Cincinnati, OH, 45226, (513) 841-4374, (8).

Moss, Robert, MOSS, ROBERT, 19514 Fisher Ave., Poolesville, MD, 20837, (301) 972-8730, (1).

Moughan, John, RADON TESTING CORPORATION, 4 Cheyenne Ct., Bryan's Road, MD, 20616, (301) 283-6006, (1).

Mount, Craig M., RADON RESEARCH GROUP, INC., P.O. Box 1143, Germantown, MD, 20874, (301) 972-3309, (800) 445-8436, (1)(4).

Mount, Dana K., NORTH DAKOTA STATE DEPARTMENT OF HEALTH AND CONSOLIDATED LABORATORIES/DIVISION OF ENVIRONMENTAL ENGINEERING, 1200 Missouri Ave. P.O. Box 5520, Bismarck, ND, 58502-5520, (701) 224-2348, (10).

Mountel, Bill, GEOMET TECHNOLOGIES, INC., 20251 Century Blvd., Germantown, MD, 20874, (301) 428-9898, (1)(2)(3)(5).

Moyles, Michael O., RADON MONITORING SERVICES, INC., P.O. Box 3730, Plymouth, MA, 02361, (617) 746-8868, (1).

Mu, Robert C., CONSULTING CHEMISTS, 3501 Lakeshore Dr., Kingsport, TN, 37663, (615) 239-5329, (615) 477-8253, (1).

Muchka, Kay, EXPEDITERS, INC., THE, 404 North Summit Ave., Oconomowoc, WI, 53066, (414) 965-3063, (1).

Muchka, Robert, EXPEDITERS, INC., THE, 404 North Summit Ave., Oconomowoc, WI, 53066, (414) 965-3063, (1).

Muhleman, John, ENERMATRIX, INC., P.O. Box 466, Fargo, ND, 58107, (701) 232-3330, (4).

Muir, David, MCMASTER UNIVERSITY/HEALTH SCIENCES CENTER/OCCUPATIONAL HEALTH PROGRAM, 1200 Main St., West Room 3H50, Hamilton, ON, L8N 3Z5, (416) 525-9140, (5).

Mulhearn, Thomas E., RHODE ISLAND REALTORS ASSOCIATION, 120 Lavan St., Warwick, RI, 02888, (401) 785-3650, (18).

Muller, Judy, C.R. LAURENCE, 2503 East Vernon, Los Angeles, CA, 90058, (800) 421-6144, (4).

Mullkoff, Arthur, AMERICAN CONCRETE INSTITUTE (ACI), P.O. Box 19150 22400 West Seven Mile Rd., Detroit, MI, 48219, (313) 532-2600, (13).

Mulready, Marc, W.R. GRACE/ORGANIC CHEMICALS DIVISION, 55 Hayden Ave., Lexington, MA, 02173, (617) 861-6600, (4).

Multer, R., AIDCO MAINE CORPORATION, Grassy Pond House, Rindge, NH, 03461, (603) 899-5166, (1)(2).

Mumma, Stanley A., ASHRAE/ENVIRONMENTAL HEALTH COMMITTEE/INDOOR AIR QUALITY SUBCOMMITTEE, Atlanta, GA, 30329, (404) 636-8400, (13).

Munn, Wanda, AMERICAN NUCLEAR SOCIETY/WASHINGTON-EASTERN WASHINGTON SECTION, C/O Westinghouse Hanford Company P.O. Box 1970, Richland, WA, 99352, (509) 376-4953, (14).

Munoz, Jr., Raul V., EL PASO COUNTY HEALTH UNIT/ENVIRONMENTAL HEALTH SERVICES DIVISION, 222 South Campbell St., El Paso, TX, 79901, (915) 541-4262, (10).

Murane, Dave, U.S. ENVIRONMENTAL PROTECTION AGENCY (EPA)/OFFICE OF AIR AND RADIATION(OAR)/OFFICE OF RADIATION PROGRAMS(ORP)/RADON DIVISION (RD), Mitigation Prevention And Quality Assurance Branch ANR-460 401 M St., SW, Washington, DC, 20460, (202) 475-9605, (8).

Murnick, J.G., DIAGNOSTICS AND DEVICES, INC., 1201 Mt. Kemble Ave., Morristown, NJ, 07960, (201) 766-7111, (1)(4).

Murphy, Gayle, PRINCE EDWARD ISLAND/QUEEN ELIZABETH HOSPITAL, Charlottetown, PE, C1A 8T5, (902) 566-6277, (11).

Murphy, John, AGENTERPRISE, INC., 227 North 7th St., Allentown, PA, 18102, (215) 776-0777, (1)(2).

Murphy, P. Dermot, BRITISH COLUMBIA REAL ESTATE COUNCIL , 626 West Pender St. 5th Floor, Vancouver, BC, V6B 1V9, (604) 683-9664, (11).

Murray, John A., CENTRAL JERSEY RADON, INC., P.O. Box 600, Lincroft, NJ, 07738, (201) 741-0400, (1)(2).

Murray, Stuart, RADON SAFETY TESTING, INC., P.O. Box 277, Gales Ferry, CT, 06335-0277, (203) 886-8327, (1).

Murrell, Andrew W., RADON TESTING AND MITIGATION SYSTEMS, P.O. Box 3052, Carmel, IN, 46032, (317) 545-5578, (1)(2).

Mushrush, George, GEORGE MASON UNIVERSITY/CENTER FOR BASIC AND APPLIED SCIENCE, 4400 University Dr., Fairfax, VA, 22030, (703) 323-4318, (5).

Musilli, Don, CONTROL RESOURCE SYSTEMS, INC., 670 Mariner Dr., Michigan City, IN, 46360, (219) 872-5591, (800) 272-3786, (4).

Musselman, Judy F., COMMONWEALTH ENGINEERING AND TECHNOLOGY, 3113 North Front St., Harrisburg, PA, 17110, (717) 232-2288, (1).

Musser, Ty W., WAYNE MUSSER PROPERTY COMPANY, INC., 517 Broad St. P.O. Box 258, Sewickley, PA, 15143, (412) 741-9517, (1)(4).

Musser, Wayne C., WAYNE MUSSER PROPERTY COMPANY, INC., 517 Broad St. P.O. Box 258, Sewickley, PA, 15143, (412) 741-9517, (1)(4).

Musso, Francis P., MUSSO, FRANCIS P., P.O. Box 651, Bath, NY, 14810, (607) 776-6245, (2).

Musumeci, Robert, AIR RADON SERVICE, INC., 292 Main St. Suite 8, Hackensack, NJ, 07601, (201) 585-2994, (1).

Muth, John B., EL PASO COUNTY HEALTH DEPARTMENT/AIR QUALITY CONTROL SECTION, 501 North Foote Ave., Colorado Springs, CO, 80909-4598, (303) 578-3137, (10).

Myers, Michael J., RADON RESEARCH GROUP, INC., P.O. Box 1143, Germantown, MD, 20874, (301) 972-3309, (800) 445-8436, (1)(4).

Mytelka, Alan I., NEW YORK METROPOLITAN AREA INTERSTATE SANITATION COMMISSION, 311 West 43 St., New York, NY, 10036, (212) 582-0380, (10).

N

Nagda, Niren, VERSAR, INC., 6850 Versar Center, Springfield, VA, 22151, (703) 750-3000, (1)(2)(3).

Nagda, Niren L., GEOMET TECHNOLOGIES, INC., 20251 Century Blvd., Germantown, MD, 20874, (301) 428-9898, (1)(2)(3)(5).

Nagle, Ronald K., EES OF MINNESOTA, INC., 600 Fairview Ave., N., St. Paul, MN, 55104, (612) 646-7991, (1).

Nakiaoka, Bob, RADON DETECTION SYSTEMS/ALBUQUERQUE, NEW MEXICO BRANCH, 4005 Cumino de la Sierra, NE, Albuquerque, NM, 87111, (303) 444-5253, (800) 627-2360, (1).

Nalepa, Dan, NALEPA, DAN, 6 East Courtney St., Dunkirk, NY, 14048, (716) 366-1778, (1).

Namoviev, Matthew, NATIONAL ASSOCIATION OF COUNTY HEALTH OFFICIALS (NACHO), 440 First St., NW, Washington, DC, 20001, (202) 393-6226, (15).

Narushko, Ray, ENVIRONMENTAL AIR CONTROL, INC., 2321 Northwest 30th Ct., Fort Lauderdale, FL, 33311, (305) 739-5159, (1)(2).

Natale, Doris, COMMONWEALTH ENTERPRISES, INC., P.O. Box S187, St. Davids, PA, 19087, (215) 688-7097, (1)(2).

Natkin, Alvin M., CONTINENTAL ENVIRONMENT COMPANY, INC., 34 Maple St., Summit, NJ, 07901, (201) 277-2255, (1)(2).

Nauth, David E., INDIANA STATE BOARD OF HEALTH/RADIOLOGICAL HEALTH SECTION, 1330 West Michigan St. Box 1964, Indianapolis, IN, 46206-1964, (317) 633-0153, (10).

Neal, William A., NEW MEXICO, REALTORS ASSOCIATION OF, P.O. Box 1569, Santa Fe, NM, 87504, (505) 258-2361, (18).

Neblock, David, AIR-TIGHT OF GENESSEE, 2429 North Dort Hwy., Flint, MI, 48506, (313) 235-5010, (1).

Neelley, John W., MISSISSIPPI REAL ESTATE COMMISSION, 1920 Dunbarton Dr., Jackson, MS, 39216-5087, (601) 987-3969, (10).

Neff, Garry, WOODLINE, INC., Box 1170, Frisco, CO, 80443, (303) 668-5603, (2).

Neilson, Arthur, NEILSON ASSOCIATES, INC., 100 Dickinson Dr. Suites 109-110, Chadds Ford, PA, 19317, (215) 793-1389, (3).

Neimeier, Richard, U.S. DEPARTMENT OF HEALTH AND HUMAN SERVICES (HHS)/PUBLIC HEALTH SERVICE(PHS)/CENTERS FOR DISEASE CONTROL(CDC)/NATIONAL INSTITUTE FOR

OCCUPATIONAL SAFETY AND HEALTH(NIOSH), Division Of Standards Development And Technology Transfer (DSOTT) 4676 Columbia Pkwy., MS-C14, Cincinnati, OH, 45226, (513) 533-8236, (8).

Neithercut, Douglas, NEW WAVE ENERGY SYSTEMS, INC., 3645 Pearl St., Boulder, CO, 80301, (303) 449-8628, (2).

Nelson, C. Stuart, NELSON, C. STUART, 1007 River Rd., Quarryville, PA, 17566, (717) 548-3151, (1).

Nelson, Craig L., PROFESSIONAL RESTORATION AND REPAIR, P.O. Box 38932, Colorado Springs, CO, 80937, (719) 632-3996, (2).

Nelson, Gary, MINNEAPOLIS BLOWER DOOR, 920 W. 53rd St., Minneapolis, MN, 55419, (612) 827-1117, (4).

Nelson, Jack, NORTH EAST ENVIRONMENTAL PRODUCTS, INC., 85 Mechanic St., Lebanon, NH, 03766, (603) 448-3944, (2)(3)(4).

Nelson, Jerry, RADON DETECTION SYSTEMS/CASPER, WYOMING BRANCH, 628 East 14th, Casper, WY, 82601, (303) 444-5253, (800) 627-2360, (1).

Nelson, Neal S., U.S. ENVIRONMENTAL PROTECTION AGENCY (EPA)/OFFICE OF AIR AND RADIATION(OAR)/OFFICE OF RADIATION PROGRAMS(ORP) , 401 M St., SW ANR-461, Washington, DC, 20460, (202) 475-9640, (8).

Nelson, IV, Lewis, RADON REMEDIAL SERVICES, P.O. Box 622, Bryn Athyn, PA, 19009, (215) 938-7766, (1)(2).

Nering, Alexander, NERING, INC., P.O. Box 209, Ridgefield, CT, 06877, (203) 431-3597, (1).

Nero, Anthony, U.S. DEPARTMENT OF ENERGY (DOE)/SAN FRANCISCO OPERATIONS OFFICE(SFOO)/LAWRENCE BERKELEY LABORATORY(LBL), 1 Cyclotran Rd. Building 90 Room 3058, Berkeley, CA, 94720, (415) 486-6377, (8).

Nesselbeck, Edward, E-THREE, INC., 4454 Genesee St., Buffalo, NY, 14225, (716) 631-5858, (1).

Neubauer, David, GEOCHEM, INC., 3200 New Seward Hwy. Suite 201, Anchorage, AK, 99503, (907) 272-2121, (1)(2).

Neubauer, Joseph, GEOCHEM, INC., 3200 New Seward Hwy. Suite 201, Anchorage, AK, 99503, (907) 272-2121, (1)(2).

Neuberger, John S., RADON RESEARCH FOUNDATION, INC., 10922 West 101st St., Overland Park, KS, 66214, (913) 492-1723, (1).

Nevin, David, ALPHA-BAN, 3719 Michelle Way, Baltimore, MD, 21208, (301) 484-7120, (1).

Nevins, David H., RADON BAN, 3719 Michelle Way , Baltimore, MD, 21208, (301) 484-7120, (1)(4).

Newell, Vaun A., U.S. ENVIRONMENTAL PROTECTION AGENCY (EPA)/OFFICE OF RESEARCH AND DEVELOPMENT(ORD), 401 M St., SW RD-672, Washington, DC, 20460, (202) 382-7676, (8).

Newman, Bruce, AMERICAN INDEPENDENT LABS, P.O. Box 1069, Sandpoint, ID, 83864, (208) 263-6935, (1).

Newman, Harry J., NORTH CAROLINA DEPARTMENT OF HUMAN RESOURCES/DIVISION OF FACILITY SERVICES/RADIATION PROTECTION SECTION, 701 Barbour Dr., Raleigh, NC, 27603-2008, (919) 733-4283, (1)(10).

Newvine, Ronald G., SURE-WAY SYSTEMS/DIVISION OF E.C.S., INC., 107 Pickard Dr., Syracuse, NY, 13211, (315) 455-2431, (800) 622-1002, (2).

Ney, W. Rosen, NATIONAL COUNCIL ON RADIATION PROTECTION AND MEASUREMENTS (NCRP), 7910 Woodmont Ave. Suite 800, Bethesda, MD, 20814, (301) 657-2652, (5)(14).

Nicholas, John, RADON TESTING AND CONSULTING, 1771 Post Rd., E. Suite 320, Westport, CT, 06880, (203) 454-3886, (1)(2)(4).

Nicholls, George, METROPOLITAN WASHINGTON COUNCIL OF GOVERNMENTS/DEPARTMENT OF ENVIRONMENTAL PROGRAMS, 1875 Eye St., NW Suite 200, Washington, DC, 20006, (202) 223-6800, (14).

Nicholls, Gerald, NEW JERSEY DEPARTMENT OF ENVIRONMENTAL PROTECTION/DIVISION OF ENVIRONMENTAL QUALITY/BUREAU OF ENVIRONMENTAL RADIATION, 729 Alexander Rd. CN-415, Trenton, NJ, 08625, (609) 987-6396, (800) 648 0394, (1)(10).

Nichols, Michael M., CMT, INC., 2813 Rio Vista, Emporia, KS, 66801, (316) 342-2760, (1).

Nichols, Michael M., JH AND ASSOCIATES, INC., 2524 East 71st St., Tulsa, OK, 74136, (918) 455-4984, (1).

Nichols, Jr., Frank, SUPERIOR SIGNAL, INC., P.O. Box 96, Spotswood, NJ, 08884, (201) 251-0800, (4).

Nicolosi, Steve J., BUZZANO CONTRACTING, 62 Mt. Pleasant Rd., Newtown, CT, 06470, (203) 426-7196, (2).

Nied, Gary J., CLEVELAND DEPARTMENT OF PUBLIC HEALTH AND WELFARE/DIVISION OF AIR POLLUTION CONTROL, 2735 Broadway Ave., Cleveland, OH, 44115, (216) 664-3500, (10).

Niedermeyer, Carl, TRUSTY WARNS, INC., 246 East Irving Park Rd., Wood Dale, IL, 60191, (312) 766-9015, (4).

Nielson, Kirk K., ROGERS AND ASSOCIATES ENGINEERING CORPORATION, 515 East 4500 S. Suite G-200, Salt Lake City, UT, 84107, (801) 263-1600, (1)(3)(5).

Niemann, Andrew W., MICROBAC LABORATORIES, INC./MASSACHUSETTS TESTING DIVISION, 202 Bussey St., Dedham, MA, 02026, (617) 326-7117, (1).

Nienaber, Jr., Paul L., HOME ANALYSTS, INC., 748 Considine Ave., Cincinnati, OH, 45205-2336, (513) 244-2860, (1).

Nifong, Ph.D., Gordon, FLORIDA INSTITUTE OF PHOSPHATE RESEARCH, 1855 West Main St., Bartow, FL, 33830, (813) 533-0983, (10).

Nigro, Dan, NIGRO CONSTRUCTION COMPANY, INC., 37 James Rd., Monroe, NY, 10950, (914) 783-4254, (1).

Nitschke, Ian, W.S. FLEMING AND ASSOCIATES, INC., 6310 Fly Rd., East Syracuse, NY, 13057, (315) 437-1869, (1)(3).

Noack, Mary K., ADVANCED RADIATION MONITORING SERVICE, INC. (ARMS), 2219 Prosperity, Portage, MI, 49081, (616) 323-9483, (1)(2)(4).

Noack, Philip B., ADVANCED RADIATION MONITORING SERVICE, INC. (ARMS), 2219 Prosperity, Portage, MI, 49081, (616) 323-9483, (1)(2)(4).

Noble, Michael, ENERGY OUTFITTERS, LTD., 245 North Aldrich Ave., Minneapolis, MN, 55405, (612) 377-6502, (2).

Nocoloff, Robert, WASHINGTON STATE DEPARTMENT OF LICENSING/REAL ESTATE AND ESCROW/PROFESSIONAL PROGRAMS MANAGEMENT DIVISION, P.O. Box 9012, Olympia, WA, 98504, (206) 753-0775, (10).

Nodler, Don, KENTUCKY DEPARTMENT OF HUMAN RESOURCES/DIVISION OF RADIATION AND PRODUCT SAFETY/RADIATION CONTROL BRANCH, 275 East Main St., Frankfort, KY, 40621, (502) 564-3700, (10).

Nolan, Fergal, CANADIAN INSTITUTE FOR RADIATION SAFETY/RADON IN HOMES COURSE, 595 Bay St. Suite 1050, Toronto, ON, M5G 2C2, (416) 596-1617, (3)(6).

Nordberg, Margit Lassen, NORDBERG ASSOCIATES, Clover Hill Farm P.O. Box 520, Alfred, ME, 04002, (207) 490-1105, (1).

Nork, Donald A., PRINCE GEORGE'S COUNTY HEALTH DEPARTMENT/DIVISION OF AIR QUALITY CONTROL, 10210 Greenbelt Rd., Seabrook, MD, 20706, (301) 794-6800, (10).

Norman, Jack C., UNIVERSITY OF WISCONSON-GREEN BAY/COLLEGE OF ENVIRONMENTAL SCIENCES, Natural and Applied Sciences, Green Bay, WI, 54301-7001, (414) 465-2371, (5).

Norman, Jim, AMERICAN SOCIETY OF HEATING, REFRIGERATING AND AIR CONDITIONING ENGINEERS (ASHRAE), 1791 Tullie Circle Rd., NE, Atlanta, GA, 30329, (404) 636-8400, (13).

Norred, Bill, LUDLUM MEASUREMENTS, INC., 501 Oak St. P.O. Box 810, Sweetwater, TX, 79556, (915) 235-5494, (4).

Norris, John G., NATIONAL SOCIETY OF PROFESSIONAL SANITARIANS (NSPS), Missouri Health Department Route 4, Hoffman Dr., Jefferson, MO, 65101, (314) 751-6095, (16).

Norton, Joseph, NORTON CONSERVATION COMPANY, INC., P.O. Box 583, Hackettstown, NJ, 07840, (201) 852-6046, (1)(2).

Norton, Stephen, UNIVERSITY OF MAINE-ORONO/DEPARTMENT OF GEOLOGY, 110 Boardman Hall, Orono, ME, 04469, (207) 581-2156, (5).

Noto, Claude J., RADON SAFETY AND TECHNOLOGY CORPORATION, 4 Charles St., Pleasant Valley, NY, 12569, (914) 635-1138, (800) 552-5257, (1)(2)(3)(4).

Novak, Ronald L., HAMMOND/AIR POLLUTION CONTROL, 5925 Calumet Ave., Hammond, IN, 46320, (219) 853-6306, (10).

Nowakowski, Lou, SECURE-ALL COMPANY, INC., 4300 W. 38th Ave., Denver, CO, 80212, (303) 458-5577, (4).

Nunziato, Steven, S & D CUSTOM ESTATES, 97 Cuba Ave., Staten Island, NY, 10306, (718) 987-2354, (718) 987-8378, (1).

Nyman, Terry L., NORTHWEST AIR POLLUTION AUTHORITY, 207 Pioneer Building, Mount Vernon, WA, 98273, (206) 336-5705, (10).

Nyquist, David, ENERGY ALTERNATIVES UNLIMITED, INC., 8384 Middlecrest Rd., Boulder, CO, 80302, (303) 444-6063, (1)(2)(3).

O

O' Brien, Marc, RADON PRESS, INC., 500 North Washington St., Alexandria, VA, 22314, (703) 548-2756, (800) 548-1567, (3).

O'Brien, David, OBAR SYSTEMS, INC., RR 3, Box 759, Highland Lakes, NJ, 07422, (201) 728-7679, (201) 697-0112, (2)(4).

O'Brien, W., RADON TEST, P.O. Box 5210, San Luis Obispo, CA, 93403-5210, (805) 481-4069, (1).

O'Connor, Catherine, UP-COUNTRY HOME INSPECTORS, INC., P.O. Box 1568, Scarborough, ME, 04074, (207) 883-9876, (1)(2).

O'Connor, Stephen, UP-COUNTRY HOME INSPECTORS, INC., P.O. Box 1568, Scarborough, ME, 04074, (207) 883-9876, (1)(2).

O'Hanlon, Terrence, ENVIRO TECHNOLOGY CORPORATION, 945 Spring Rd., Pelham Manor, NY, 10803, (1).

O'Heir, David, AMERICAN INSULATION, INC., 15 Cherry St. P.O. Box 266, Danvers, MA, 01923, (617) 777-0617, (1)(2).

O'Hornett, Kevin, ECI, 2955 Valmont Suite 300, Boulder, CO, 80301, (303) 442-8995, (1).

O'Hornett, Kevin M., INSPEX OF BOULDER, INC., 2955 Valmont Suite 300, Boulder, CO, 80301, (303) 442-8995, (1).

O'Neill, John T., ARIZONA RADIATION REGULATORY AGENCY, 4814 South 40th St., Phoenix, AZ, 85040, (602) 255-4845, (10).

O'Riordan, Joe, O'RIORDAN INSULATION COMPANY, 8901 North Nebraska Ave., Tampa, FL, 33604, (813) 932-0008, (1)(2).

O'Rourke, Kevin, AAAIRSCAN ENGINEERING CORPORATION, 8 Eric Dr., Granby, CT, 06035, (203) 653-5839, (1).

Oakley, Grayce A., ALASKA DIVISION OF OCCUPATIONAL LICENSING, 3601 C St. Suite 722, Anchorage, AK, 99503, (907) 563-2169, (10).

Odle, Lawrence D., MONTEREY BAY UNIFIED AIR POLLUTION CONTROL DISTRICT, 1164 Monroe St. Suite 10, Salinas, CA, 93906, (408) 443-1135, (10).

Officer, Carl E., EAST ST. LOUIS/AIR POLLUTION CONTROL, 7 Collinsville Ave., East St. Louis, IL, 62201, (618) 874-2115, (10).

Ogden, Albert, TENNESSEE TECH UNIVERSITY/CENTER FOR THE MANAGEMENT, UTILIZATION AND PROTECTION OF WATER RESOURCES, P.O. Box 5082, Cookeville, TN, 38505, (615) 372-3507, (5).

Oge, Margo T., U.S. ENVIRONMENTAL PROTECTION AGENCY (EPA)/OFFICE OF AIR AND RADIATION(OAR)/OFFICE OF RADIATION PROGRAMS(ORP)/RADON DIVISION(RD), Radon Action Program 401 M Street SW ANR-464, Washington, DC, 20460, (202) 475-9605, (8).

Ogilvie, Clayton, MITIGATING CIRCUMSTANCES, 153 Forest Park, Durham, NH, 03824, (603) 868-1557, (2).

Ohanian, Ed, U.S. ENVIRONMENTAL PROTECTION AGENCY (EPA)/OFFICE OF WATER(OW)/OFFICE OF DRINKING WATER(ODW)/CRITERIA AND STANDARDS DIVISION(CSD), Health Effects Branch 401 M St., SW WH-550D, Washington, DC, 20460, (202) 382-7571, (8).

Olcott, Jeffrey C., O.K. REMS CORPORATION, 181 London Rd., Hebron, CT, 06248, (203) 647-7440, (1).

Olcott, Jeffrey C., O.K. REMS CORPORATION, 174 Flock Rd., Mercerville, NJ, 08619, (609) 588-9627, (1).

Olin, Steve, TRACOR JITCO RADON MEASUREMENT SERVICES, 1601 Research Blvd., Rockville, MD, 20850, (301) 984-2800, (1)(4).

Olson, Norman, SOUTH CAROLINA GEOLOGICAL SURVEY, Harbison Forest Rd., Columbia, SC, 29210, (803) 737-9440, (10).

Olson, Richard, GRAND RAPIDS/ENVIRONMENTAL SERVICES DEPARTMENT/AIR POLLUTION CONTROL DIVISION, 1300 Market, SW, Grand Rapids, MI, 49503, (616) 456-3158, (10).

Olson, Wayne, SOUTH DAKOTA ASSOCIATION OF REALTORS, 411 East Capitol St. Box 386, Pierre, SD, 57501, (605) 224-9211, (18).

Olsson III, Theodore A., WATER TEST CORPORATION, 33 South Commercial St., Manchester, NH, 03101, (603) 623-7400, (1).

Olszewski, William, UNIVERSITY OF NEW HAMPSHIRE/DEPARTMENT OF EARTH SCIENCES, James Hall, Durham, NH, 03824, (603) 862-1234, (5).

Opalenik, Jr., John, SOL-AIR, INC., P.O. Box 458, Plantsville, CT, 06479, (203) 621-6211, (1).

Oppenheimer, David, RADON ENVIRONMENTAL TEST EXPERTS, INC., 5271 Holmes Pl., Boulder, CO, 80303, (303) 442-2344, (1).

Oppold, James, NORTH CAROLINA DEPARTMENT OF LABOR/OCCUPATIONAL SAFETY AND HEALTH, Shore Memorial Building 214 West Jones St., Raleigh, NC, 27603, (919) 733-4880, (10).

Ordway, Roderic E., R.J. MOORE AND ASSOCIATES, INC., 1010 West St., Annapolis, MD, 21401, (301) 269-5950, (1).

Orlowsky, Peter, SCIENTIFIC RESEARCH AND DEVELOPMENT COMPANY, P.O. Box 5590, Lakeland, FL, 33807, (813) 644-0404, (3)(4).

Orr, Charles F., CAMPTON DEVELOPMENT CORPORATION, P.O. Box 354, Campton, NH, 03223, (603) 726-3030, (1).

Orr, Stan, OKLAHOMA ASSOCIATION OF REALTORS, 9807 North Broadway Extension, Oklahoma City, OK, 73114, (405) 848-9944, (18).

Ortleib, David, INTERNATIONAL CHEMICAL WORKERS UNION (ICWU), 1655 West Market, Akron, OH, 44313, (216) 867-2444, (16).

Ortolano, Peter, RADON ENVIRONMENTAL CONSTRUCTION SERVICES/RECONS, INC., PA OFFICE, 979 Mt. Eyre Rd., Washington Crossing, PA, 18977, (215) 321-3274, (1)(2).

Ortolano, Jr., Peter J., RADON ENVIRONMENTAL CONSTRUCTION SERVICES/RECONS, INC., MAIN OFFICE, P.O. Box 3058, Princeton, NJ, 08543-3058, (609) 883-1808, (215) 493-0667, (1)(2).

Osage, David A., HOME-TECH ENGINEERING, P.O. Box 43129 51 Upper Montclair Plaza, Upper Montclair, NJ, 07043, (201) 744-6674, (1).

Osborne, Michael C., U.S. ENVIRONMENTAL PROTECTION AGENCY (EPA)/OFFICE OF RESEARCH AND DEVELOPMENT(ORD)/AIR AND ENERGY ENGINEERING RESEARCH LABORATORY(AEERL)/RADON MITIGATION PROJECT, MD-60, Research Triangle Park, NC, 27711, (919) 541-2745, (8).

Osmond, Glenn D., ROBERTS ENVIRONMENTAL SERVICES, 4110 Old Wake Forest Rd. Suite 109, Raleigh, NC, 27609, (919) 790-9378, (1).

Osterberg, Charles L., ENRAD, INC./RADON TESTING AND ENGINEERING, INC., 18705 B North Frederick Rd., Gaithersburg, MD, 20879, (301) 948-8040, (1).

Osterhaus, Craig, TELEDYNE ISOTOPES, 50 Van Buren Ave., Westwood, NJ, 07675, (201) 664-7070, (1)(4).

Ostrom, Meredith, WISCONSIN GEOLOGICAL AND NATURAL HISTORY SURVEY, University of Wisconsin 3817 Mineral Point Rd., Madison, WI, 53705, (608) 262-1705, (10).

Ostroski, Richard J., BLUE ANGEL PUMP COMPANY, 801 Glasgow Ave. P.O. Box 2257, Fort Wayne, IN, 46801, (219) 426-4000, (4).

Oswald, Richard, TERRADEX CORPORATION, 460 North Widget Lane, Walnut Creek, CA, 94598, (415) 938-2545, (4).

Otten, James, U.S. DEPARTMENT OF INTERIOR (DOI)/U.S. GEOLOGICAL SURVEY(USGS)/GEOLOGIC DIVISION(GD), MS-955, National Ctr., Reston, VA, 22092, (703) 648-4000, (8).

Otto, George, REAL ESTATE EDUCATORS ASSOCIATION (REEA), 230 North Michigan Ave. Suite 1200, Chicago, IL, 60601, (312) 372-9800, (18).

Ouellette, R.P., VERSAR, INC., 6850 Versar Center, Springfield, VA, 22151, (703) 750-3000, (1)(2)(3).

Ours, Daniel, ENVIRONMENTAL OIL, INC., P.O. Box 315, Syracuse, NY, 13209, (315) 471-0503, (800) 262-1012, (1).

Outlaw, Larry A., NORTH CAROLINA REAL ESTATE COMMISSION, 1313 Navaho Dr., Raleigh, NC, 27619, (919) 733-9580, (10).

Overman, Ralph, AIR SCIENCES, INC., 508 North Truman Blvd. P.O. Box 320, Crystal City, MO, 63019, (314) 933-3915, (800) 242-8886, (1)(2)(4).

Owczarski, P.C., U.S. DEPARTMENT OF ENERGY (DOE)/RICHLANDS OPERATIONS OFFICE(ROO)/PACIFIC NORTHWEST LABORATORY(PNL)/DEPARTMENT OF ATMOSPHERIC SCIENCES, M/S-K607, Richland, WA, 99352, (509) 376-1701, (8).

Owen, Douglass E., U.S. DEPARTMENT OF INTERIOR (DOI)/U.S. GEOLOGICAL SURVEY(USGS), Denver Federal Center MS-963, Denver, CO, 80225, (303) 236-5400, (8).

Owens, Scott, RA-MED, P.O. Box 4413, Alexandria, VA, 22303, (703) 960-0514, (2).

Owens, Tim, ENVIRADON, INC., P.O. Box 13683, Research Triangle Park, NC, 27709, (919) 848-2128, (1).

Oxburgh, E.R., EUROPEAN UNION OF GEOSCIENCES (EUG)/EUROPEAN SCIENCE FOUNDATION, 1, quai Lezay-Marnesia, F-67000 Strasbourg, France, (011) 33-88-353036, (12).

P

Pagano, Joel A., BUILDING INSPECTOR OF AMERICA, THE, 112 Jockey Hollow Way, Union, NJ, 07083, (201) 473-5656, (201) 579-2734, (1).

Pagano, Joel A., PAGANO ASSSOCIATES, INC., 112 Jockey Hollow Way, Union, NJ, 07083, (201) 686-2939, (1).

Page, Stephen D., U.S. ENVIRONMENTAL PROTECTION AGENCY (EPA)/OFFICE OF AIR AND RADIATION(OAR)/OFFICE OF RADIATION PROGRAMS(ORP)/RADON DIVISION(RD), Policy And Public Information Branch 401 M St., SW ANR-464, Washington, DC, 20460, (202) 475-9605, (8).

Pagello, Paul, ENVIRONMENTAL RADON MEASUREMENTS, 49 North Orange Ave., Orlando, FL, 32801, (407) 648-1969, (800) 237-3065, (1)(2).

Pagello, Paul, ERM, INC., 49 North Orange Ave., Orlando, FL, 32801, (407) 648-1969, (800) 237-3065, (1)(2).

Pai, Hsiang Lin, R.A.D. SERVICE AND INSTRUMENTS, LTD., 50 Silver Star Blvd. Unit 208, Scarborough, ON, M1V 3L3, (416) 298-9200, (1)(4)(5).

Painchard, Phillip A., PRECISION MEASUREMENTS ASSOCIATION (PMA), 9681 Business Center Dr., Rancho Cucomong, CA, 91730, (714) 980-6166, (14).

Palasini, Peter D., INDIANA SEAL COMPANY, 12000 Mitchell Rd., Carmel, IN, 46032, (317) 844-8700, (800) 428-5309, (4).

Pallotta, Arthur J., NATIONAL RADON LABORATORIES/DIVISION OF ANALYTICAL SERVICES CORPORATION, P.O. Box 10681, Arlington, VA, 22210-1681, (703) 525-6672, (1).

Palmer, Chuck, PYLON ELECTRONIC DEVELOPMENT COMPANY, LTD., 147 Colonnade Rd., Nepean, ON, K2E 7L9, (613) 226-7920, (1)(4).

Palmer, Jeffrey A., PALMER TECHNOLOGY, P.O. Box 1072, Glen Burnie, MD, 21061, (301) 766-3176, (1).

Palmer, Laurie E., PALMER TECHNOLOGY, P.O. Box 1072, Glen Burnie, MD, 21061, (301) 766-3176, (1).

Palmer, W.G., BIOMEDICAL TOXICOLOGY ASSOCIATES, P.O. Box 3568, Frederick, MD, 21701, (301) 662-0783, (1)(3).

Palmieri, Christine, RADON TESTING SERVICES, P.O. Box 152227, San Diego, CA, 92115-0870, (619) 229-9002, (619) 287-5317, (1).

Palovcik, Rudolph J., ALBERTA DEPARTMENT OF CONSUMER AND CORPORATE AFFAIRS/REAL ESTATE STANDARDS, 10025 Jasper Ave. 19th Floor, Edmonton, AB, T5J 3Z5, (403) 422-1588, (11).

Palumbo, James, UTILITY ENGINEERS, INC., 37 N. Washington St., Wilkes Barre, PA, 18701, (717) 829-4200, (1).

Pangelinan, Rafael C., GOVERNMENT OF GUAM/DEPARTMENT OF REVENUE AND TAXATION/REAL ESTATE AGENCY, 855 West Marine Dr., Agana, GU, 96910, (671) 477-1040, (10).

Pansire, Richard, SETRA SYSTEMS, INC., 45 Nagog Pk., Acton, MA, 01720, (617) 263-1400, (4).

Pard, David E., RADON REMEDIATION SERVICES, 10400 Connecticut Ave. Suite 504, Kensington, MD, 20895, (301) 933-7900, (2).

Pard, David E., RDS/RADON DETECTION SERVICES-MARYLAND OFFICE, 10400 Connecticut Ave. Suite 504, Kensington, MD, 20895, (301) 933-7900, (1)(3).

Pare, Antonio D., FAST, 929 Gloucester Ave., Bricktown, NJ, 08723, (201) 920-2369, (3).

Parent, Donah, BIO-PURE COMPANY, P.O. Box 20593, Tampa, FL, 33622-0593, (813) 839-8617, (4).

Paris, Ray D., OREGON STATE DEPARTMENT OF HUMAN RESOURCES/HEALTH DIVISION/RADIATION CONTROL SECTION, 1400 SW 5th Ave. P.O. Box 231, Portland, OR, 97201, (503) 229-5797, (10).

Parisi, Jack, SCHENECTADY CITY HEALTH DEPARTMENT, City Hall, Schenectady, NY, 12305, (518) 382-5046, (10).

Parisi, Thomas, HOME TESTING LABS, INC., 299 West Fort Lee Rd., Bogota, NJ, 07603, (201) 343-1199, (1)(3).

Parisi, Thomas, HOME TESTING LABS, INC., 44 Crestview Rd., Phoenixville, PA, 19460, (215) 768-7960, (1)(3).

Parker, Christopher D., NEW YORK STATE HEALTH DEPARTMENT/WADSWORTH CENTER FOR LABORATORIES AND RESEARCH/LABORATORY FOR INORGANIC AND NUCLEAR CHEMISTRY, Empire State Plaza Room D366-B, Albany, NY, 12201, (518) 474-6649, (1)(10).

Parker, Harvey, RADON MONITOR SYSTEMS, P.O. Box 516, Eustis, FL, 32727-0516, (904) 735-2929, (1).

Parkinson, Robert N., RADOMET SERVICES, 7079 East Briarwood Cr., Englewood, CO, 80112, (303) 694-4519, (1).

Parme, Larry, AMERICAN NUCLEAR SOCIETY/CALIFORNIA-SAN DIEGO SECTION, C/O GA Technologies, Inc. P.O. Box 85608, San Diego, CA, 92138, (619) 455-4555, (14).

Parrott, Jack, DAMES AND MOORE, 7101 Wisconsin Ave. Suite 700, Bethesda, MD, 20814, (301) 652-2215, (1).

Parry, Jr., Scott, ALPHA CONTROL, INC., 3509 Cherry Valley Dr., Olney, MD, 20832, (301) 924-0377, (1).

Parsons, Jeff, RADON ANALYSIS, INC., 7 Chamberlain Rd., Flemington, NJ, 08822, (201) 788-0661, (1)(4).

Partington, William M., FLORIDA CONSERVATION FOUNDATION/ENVIRONMENTAL INFORMATION CENTER, 1191 Orange Ave., Winter Park, FL, 32789, (305) 644-5377, (14).

Pasquariello, Thomas, GARDEN STATE HOME INSPECTION SERVICE, 475 Valley Rd., West Orange, NJ, 07052, (201) 678-3333, (1).

Passaro, Jr., Ralph, RESIDENTIAL INSPECTION TECHNOLOGY/RES-I-TEC, 145 D Grassy Plain St., Bethel, CT, 06801, (203) 791-1076, (1).

Passe, Jr., Loel, NATIONAL RADON LABORATORIES/DIVISION OF COLORADO SERVICES CORPORATION, 14677 East Easter Ave. Suite H, Englewood, CO, 80112, (303) 699-7378, (1).

Pastner, Hal, GUNDLE LINING SYSTEMS, INC., 19103 Gundle Rd., Houston, TX, 77073, (713) 443-8564, (800) 435-2008, (4).

Pasupathi, V., BATTELLE MEMORIAL INSTITUTE/NUCLEAR TECHNOLOGY SECTION, 505 King Ave., Columbus, OH, 43201-2693, (614) 879-5192, (1)(5).

Patel, Pramod, BROOKS LABORATORIES, INC., 44 Codfish Ln., Weston, CT, 06883, (203) 226-6384, (1)(2)(3).

Patenaude, Marcel, P.M. WRIGHT, LTD., 1300 Jules Poitras Blvd., St. Lawrence, PQ, H4N 1X8, (514) 337-3331, (4).

Patrek, Walter, BOSSAIRE, INC., 1321 Tyler St., NE, Minneapolis, MN, 55413, (612) 781-0179, (4).

Patterson, Frankie, INFILTEC/FACTORY, 108 South Delphine Ave. P.O. Box 1125, Waynesboro, VA, 22980, (703) 943-2776, (1)(4).

Patterson, Malcolm D., AMERICAN NUCLEAR SOCIETY/DISTRICT OF COLUMBIA-WASHINGTON SECTION, C/O Baltimore Electric And Gas Calvert Cliffs Nuclear Plant P.O. Box 1475, Baltimore, MD, 21203, (301) 254-8529, (14).

Patterson, R.M., ENVIROHEALTH, INC., 1090 General Greene Rd., Wayne, PA, 19087, (215) 964-0668, (1)(3).

Patterson, Robert M., TEMPLE UNIVERSITY RADON PROGRAM, Temple University CECSA 084-53, Philadelphia, PA, 19122, (215) 787-1665, (1)(5)(6).

Paul, Don, D. PAUL BUILDING ENTERPRISE, P.O. Box 1265, Waldorf, MD, 20601, (301) 843-5751, (2).

Paul, John A., DAYTON REGIONAL AIR POLLUTION CONTROL AGENCY, 451 West Third St. P.O. Box 972, Dayton, OH, 45422, (513) 225-4435, (10).

Paul, Joyce, D. PAUL BUILDING ENTERPRISE, P.O. Box 1265, Waldorf, MD, 20601, (301) 843-5751, (2).

Paul, L.G., L.G. PAUL, SR., INC., 333 Union St., Allentown, PA, 18102, (214) 437-3530, (1).

Paul, Roman, AMERICAN HOMECHECK, INC., P.O. Box 565, Wayne, PA, 19087, (215) 259-8707, (1).

Pauley, J. Jackson, WEST VIRGINIA ASSOCIATION OF REALTORS, 814 Virginia St., Charleston, WV, 25301, (304) 342-7600, (18).

Payne, Marvin, U.S. DEPARTMENT OF ENERGY (DOE)/OAK RIDGE OPERATIONS OFFICE(OROO)/OAK RIDGE NATIONAL LABORATORY(ORNL)/DIVISION OF ENVIRONMENTAL RESEARCH(DER), P.O. Box 2008, Oak Ridge, TN, 37831-6113, (615) 574-5898, (8).

Payne, Richard H., CAVIN ANALYTICAL CONSULTANTS, INC., 2165K West Park Ct., Stone Mountain, GA, 30087, (404) 498-8295, (1).

Payne, Jr., Larry D., PAYNE, LARRY D., P.O. Box 1292, Purcellville, VA, 22132, (2).

Peace, Charles E., OLYMPIA AIR POLLUTION CONTROL AUTHORITY, 120 East State Ave., Olympia, WA, 98501, (206) 352-4881, (10).

Peak, Thomas, U.S. ENVIRONMENTAL PROTECTION AGENCY (EPA)/OFFICE OF AIR AND RADIATION(OAR)/OFFICE OF RADIATION PROGRAMS(ORP)/RADON DIVISION(RD), Problem Assessment Branch 401 M St. SW ANR-464, Washington, D.C., 20460, (202) 475-9605, (8).

Pearline, Donald, RADON DETECTION AND REDUCTION LABORATORY/DIVISION OF FORMACHEM, 10834 Ambassador, St. Louis, MO, 63132, (314) 872-9350, (1)(2)(4).

Peck, Robert, SOUTH CAROLINA DEPARTMENT OF LABOR/OCCUPATIONAL SAFETY AND HEALTH/CONSULTATION AND MONITORING, P.O. Box 11329, Columbia, SC, 29211, (803) 734-9599, (10).

Pecor, Robert, A & N RADON COMPANY, P.O. Box 188, Keene, NH, 03431, (603) 357-0155, (1).

Pedersen, Christopher, HOME SENTRY, INC., P.O. Box 555, Midland Park, NJ, 07432, (201) 652-2497, (1).

Pedersen, Mike, WESTERN WATER INTERNATIONAL, INC., 7715 Penn Belt Dr., Forestville, MD, 20747, (301) 568-0200, (4).

Pedersen, Sr., Robert, RADON CONTROL, INC., 337 Route 22 E., Greenbrook, NJ, 08812, (201) 563-4211, (1).

Peer, Denise, STRATA ENVIRONMENTAL SERVICE, INC., 401 East Douglas Suite 515, Wichita, KS, 67202, (316) 262-0002, (1)(3).

Pender, James E., ALA OF WISCONSIN, 10001 West Lisbon Ave., Milwaukee, WI, 53222, (414) 463-3232, (15).

Pennington, Jr., Fred A., A.S. McCREATH AND SONS, INC., P.O. Box 1453, Harrisburg, PA, 17105-1453, (717) 238-9331, (1).

Perez, Betty, ENVIROSERV/RADON DETECTION OF ALABAMA, 751-D Raleigh Ct., Birmingham, AL, 35209, (205) 942-0756, (1).

Perez, Jose, PUERTO RICO/DEPARTMENT OF HEALTH/ RADIOLOGICAL HEALTH DIVISION, G.P.O. Call Box 70184, Rio Piedras, PR, 00936, (809) 767-3563, (10).

Perison, Janice B., PERISON, JANICE B., 13 Beaver Branch Rd., Pennsylvania Furnace, PA, 16865, (814) 234-1871, (1).

Perkins, David C., ENVIRONMENTAL TESTING ASSOCIATES, 6021 Sandy Springs Cr. Suite 205, Atlanta, GA, 30328, (404) 255-3884, (1).

Perkins, Tom, CANBERRA INDUSTRIES, INC., 1 State St., Meriden, CT, 06450, (203) 238-2351, (1)(4).

Perler, Arthur, U.S. ENVIRONMENTAL PROTECTION AGENCY (EPA)/OFFICE OF WATER(OW)/OFFICE OF DRINKING WATER(ODW)/CRITERIA AND STANDARDS DIVISION(CSD), Science And Technology Branch 401 M St. SW WH-550D, Washington, DC, 20460, (202) 382-3022, (8).

Perlman, Daniel, BRANDEIS UNIVERSITY/ROSENTIEL RESEARCH CENTER, 415 South St., Waltham, MA, 02154, (617) 736-2461, (1)(5).

Perlow, Michael, VFC, INC., 1145 Washington St., Allentown, PA, 18102, (215) 437-5010, (1).

Perri, Albert J., A.J. PERRI, INC., 401 Hwy. 35, Red Bank, NJ, 07701, (201) 747-3131, (1).

Perrin, Jr., Alfred C., SUTTER COUNTY AIR POLLUTION CONTROL DISTRICT, 142 Garden Way, Yuba City, CA, 95991, (916) 674-2851, (10).

Perry, Daniel, MILWAUKEE ELECTRIC TOOL, 13135 West Lisbon, Brookfield, WI, 53005, (414) 781-3600, (4).

Perry, George, A.T.E.I., 809 East 11th St., Michigan City, IN, 46360, (800) 833-0999, (1)(2)(6).

Petcovic, Larry, N.A.K. PRODUCTION ASSOCIATES, 1317 Apple Ave., Silver Spring, MD, 20910, (301) 565-0355, (3)(4)(6).

Peters, Fred, HARDICO RADON MANAGEMENT, HARRISBURG OFFICE, 730 Tower Rd., Enola, PA, 17025, (717) 732-9469, (2).

Peters, June E., ENVIRONMENTAL SAFETY PRODUCTS, 8351 Roswell Rd. Suite 325, Atlanta, GA, 30350, (404) 889-3117, (1).

Peters, Tom, BET, INC., 222 South Rolff, Davenport, IA, 52802, (319) 322-8749, (1).

Peterson, Grant, FEDERAL EMERGENCY MANAGEMENT AGENCY (FEMA)/STATE AND LOCAL PROGRAMS AND SUPPORT(SLPS), 500 C St., SW, Washington, DC, 20472, (202) 646-3692, (8).

Peterson, Vernon C., EL DORADO COUNTY AIR POLLUTION CONTROL DISTRICT, 7553 Green Valley Rd., Placerville, CA, 95667, (916) 626-2407, (10).

Petrazzuolo, Gary, BIOMONITORING SERVICES LABORATORY/TECHNICAL RESOURCES, INC., 3202 Monroe St., Rockville, MD, 20852, (301) 231-5250, (1).

Petre, Ion, GEO-ENVIRONMENTAL SERVICES, INC., 130 West Wieuca Rd. Suite 108, Atlanta, GA, 30342, (404) 256-6437, (1).

Petrie, Robert, GLOBAL BUILDING INSPECTORS, 39 East 4th St., Bayonne, NJ, 07002, (201) 858-2386, (800) 321-4677, (1).

Pettit, Pat, RADON LABS OF ATLANTA, 5400 Hwy. 29, Lilburn, GA, 30247, (404) 279-0005, (800) 248-1889, (1).

Pettit, R.A., EASTERN INTEGRATED SERVICES CORPORATION, 3490 U.S. Rte. 1 Building 17-2, Princeton, NJ, 08540, (609) 520-8344, (1).

Petty, Robert C., ASSOCIATED BUILDERS OF THE ROCKIES, INC., P.O. Box 704, Wheat Ridge, CO, 80034, (303) 237-3428, (1).

Peveler, Kenneth, AMERICAN NUCLEAR SOCIETY/IOWA-IOWA SECTION, C/O Iowa Electric Light And Power P.O. Box 351, Cedar Rapids, IA, 52406, (319) 851-5611, (14).

Pfaffenberger, Carl, METROPOLITAN DADE COUNTY DEPARTMENT OF ENVIRONMENTAL RESOURCES MANAGEMENT, Metro Dade Ctr. 111 NW First St. Suite 1310, Miami, FL, 33128, (305) 375-3326, (10).

Pfeifer, William J., ALA OF WESTERN MISSOURI, 2007 Broadway, Kansas City, MO, 64108, (816) 842-5242, (15).

Phelps, M. Jeana, RADON TESTING AND ENVIRONMENTAL CONTROL, 7113 Chippenham Rd., Louisville, KY, 40222, (502) 426-4314, (1).

Philip, Leonard A., SHIELD HOME INSPECTION SERVICE, R.D. 1, Box 274 Summit Dr., Branchville, NJ, 07826, (201) 875-5922, (1).

Phillips, Bob, CONSTRUCTION SPECIALTIES, INC., 55 Winans Ave., Cranford, NJ, 07016, (201) 272-5200, (4).

Phillips, Brian, BRITISH COLUMBIA MINISTRY OF HEALTH/RADIATION PROTECTION SERVICE, 307 West Broadway Suite 200, Vancouver, BC, V5Y 1P9, (604) 660-6630, (11).

Phillips, Bruce C., AIR QUALITY ANALYSTS, INC., 5400 Eisenhower Ave., Alexandria, VA, 22304, (703) 823-7567, (1)(4).

Phillips, Chet, SAFE-AIRE, INC., 162 East Chestnut St. P.O. Box 160, Canton, IL, 61520, (309) 647-0419, (800) 331-2943, (1)(2)(3)(4)(6).

Phillips, Donald D., ENERGY LABORATORIES, INC., 254 North Center St. P.O. Box 3258, Casper, WY, 82602, (307) 235-0515, (1).

Phillips, Jim, AMERICAN MEDICAL MARKETING SERVICES, 2429 West Davie Blvd. Suite 17, Ft. Lauderdale, FL, 33312, (305) 462-2575, (1)(2).

Phillips, John, AMERICAN MONITORING AND ENGINEERING SERVICES, 200 High Tower Blvd. Suite 205, Pittsburgh, PA, 15205, (215) 547-5460, (609) 392-1422, (1).

Phillips, Leroy, APPLIED CONSTRUCTION, INC., The Friary, Winchester-on-Severn 1600 Winchester Rd., Annapolis, MD, 21401, (301) 757-7887, (2).

Piccolo, Richard G., RADON TESTING LABS, P.O. Box 1467, Avon, CT, 06001, (203) 678-8259, (1).

Pickett, Bruce, WASHINGTON STATE DEPARTMENT OF SOCIAL AND HEALTH SERVICES/OFFICE OF RADIATION PROTECTION, Mail Stop LE-13, Olympia, WA, 98504, (206) 753-5962, (800) 323-9727, (6)(10).

Pierce, A.R., HARRIS COUNTY POLLUTION CONTROL DEPARTMENT, 107 North Munger, Pasadena, TX, 77506, (713) 920-2831, (10).

Pierce, Donald, OREGON STATE UNIVERSITY/DEPARTMENT OF STATISTICS, Kidder Hall 44, Corvallis, OR, 97331-4606, (503) 754-0123,

Piersol, Peter, ONTARIO RESEARCH FOUNDATION (ORF), Sheridan Park Research Community, Mississauga, ON, L5K 1B3, (416) 822-4111, (1)(11).

Piette, Greg, ALPHA GUARD/TECTROL, 5613 Yarrow St., Arvada, CO, 80002, (303) 421-7955, (1)(2).

Pitts, Mark R., TRUE TEST COMPANY, UTAH, P.O. Box 571, Bountiful, UT, 84010, (801) 298-2256, (1).

Pizzini, Donald E., CASCADE/CITY-COUNTY AIR POLLUTION CONTROL PROGRAM, 1130 17 Ave. South, Great Falls, MT, 59405, (406) 761-1190, (10).

Platt, James P., SOUTH DAKOTA STATE DEPARTMENT OF HEALTH/OFFICE OF AIR QUALITY AND SOLID WASTE, Joe Foss Office Building 523 East Capital, Pierre, SD, 57501, (605) 773-3329, (10).

Plitnick, Bob, VALLEY DE-WATERING, RD 2, Box 265 A-1, Sugarloaf, PA, 18249, (717) 384-3315, (2).

Pochop, Mike, SOUTH DAKOTA STATE DEPARTMENT OF HEALTH/OFFICE OF AIR QUALITY AND SOLID WASTE, Joe Foss Office Building 523 East Capital, Pierre, SD, 57501, (605) 773-3329, (10).

Poffenroth, Dennis N., RADON CONSULTANTS, INC., P.O. Box 789, Golden, CO, 80402, (303) 279-1358, (1).

Pollackov, Abe, AMERICAN EXCELSIOR COMPANY, 1 Lovellee Ln., Thornton, PA, 19373, (215) 459-1146, (2)(4).

Pollackov, Abe, AURORA INDUSTRIES, P.O. Box 936, Concordville, PA, 19331, (215) 459-1146, (2).

Pollock, Robert, RADON ENVIRONMENTAL MONITORING, INC., Box 5305, Evanston, IL, 60204, (312) 328-6628, (1)(4).

Pompei, Jack, OREGON STATE DEPARTMENT OF WORKERS' COMPENSATION/ACCIDENT PREVENTION DIVISION/CONSULTATIVE SECTION, Labor And Industries Building Room 115, Salem, OR, 97310, (503) 378-3272, (10).

Port, Eli, RADIATION SAFETY SERVICES, INC., 6312 West Oakton St., Morton Grove, IL, 60053, (312) 965-1999, (1)(4)(6).

Porte, Jeff, MICROBAC LABORATORIES, INC./ERIE TESTING DIVISION-MAIN OFFICE, 2411 West 26th St., Erie, PA, 16506, (814) 833-4790, (1)(4).

Porter, Charles R., U.S. ENVIRONMENTAL PROTECTION AGENCY (EPA)/OFFICE OF AIR AND RADIATION(OAR)/OFFICE OF RADIATION PROGRAMS(ORP)/EASTERN ENVIRONMENTAL RADIATION FACILITY(EERF), 1890 Federal Dr., Montgomery, AL, 36109, (205) 272-3402, (8).

Porter, Jr., S.W., PORTER CONSULTANTS, INC., 125 Argyle Rd., Ardmore, PA, 19003, (215) 896-5353, (1)(3)(6).

Portis, Donald E., WEST VIRGINIA REAL ESTATE COMMISSION, 1033 Quarrier St. Suite 400, Charleston, WV, 25301, (304) 348-3555, (10).

Possehl, Robert, RADON TEST COMPANY, P.O. Box 345, Eagle Grove, IA, 50533, (515) 448-4600, (4).

Post, Karen, UTAH DEPARTMENT OF BUSINESS REGULATIONS/DIVISION OF REAL ESTATE, P.O. Box 45802 160 East 300 South, Salt Lake City, UT, 84145, (801) 530-6747, (10).

Poteet, W. Randall, ANCHORAGE AIR POLLUTION CONTROL AGENCY, 825 L St., Anchorage, AK, 99503, (907) 343-4713, (10).

Potochney, Pete, LEHIGH VALLEY HOUSE INSPECTION SERVICE, 200 Delaware Ln., Whitehall, PA, 18052, (215) 435-0208, (1)(2).

Potter, Bruce, ORLEANS-NIAGRA BOARD OF COOPERATIVE EDUCATIONAL SERVICES (BOCES), 4232 Shelby Basin Rd., Medina, NY, 14103, (716) 798-4800, (10).

Potter, Don, RADON DETECTION CENTER OF MICHIGAN, 601 South Saginaw St. Suite 514, Flint, MI, 48502, (313) 767-8630, (1).

Powell, Fred, HOUSE ANALYSIS AND ASSOCIATES, 747 Hallowell Dr., Huntingdon Valley, PA, 19006, (215) 947-1169, (215) 947-9209, (1).

Powell, J.L., RADON GAS TESTING OF CENTRAL OHIO, 5548 Autumn Chase, Columbus, OH, 43232, (614) 864-0191, (1).

Powell, Lori, UNIVERSITY OF MARYLAND/RADIATION SAFETY OFFICE, Chemical-Nuclear Engineering Building Room 2124, College Park, MD, 20742, (301) 454-5294, (1).

Powell, Tyrone, PEPPADYNE RADON DETECTION SERVICE, 126 East 9th St. Suite 25, St. Paul, MN, 55101, (612) 293-0864, (1).

Power, John A., MARICOPA COUNTY DEPARTMENT OF HEALTH SERVICES/ENVIRONMENTAL HEALTH SERVICES DIVISION, 1825 East Roosevelt St., Phoenix, AZ, 85006, (602) 258-6381, (10).

Power, Tom, CATTARAUGUS COUNTY HEALTH DEPARTMENT, 301 Laurens St., Olean, NY, 14760, (716) 375-4121, (10).

Powers, David B., SKUTTLE MANUFACTURING COMPANY, Rte. 1, Box 16A, Marietta, OH, 45750, (614) 373-9169, (4).

Powers, Thomas L., SUN NUCLEAR CORPORATION, 415-C Pineda Ct., Melbourne, FL, 32940, (407) 259-6862, (1)(4).

Powll, William J., HEIMER INSPECTIONS OF NEW JERSEY, INC., 17 Catherine St., Carteret, NJ, 07008, (201) 541-8995, (1).

Prapolsky, Walter, RADON IN-HOUSE DETECTION, Rt. 1, Box 110, Leeds, NY, 12451, (518) 622-3509, (1).

Prentice, Barry, REAL ESTATE INSPECTOR, THE, 2195 Stoney Hill Rd., Boulder, CO, 80303, (303) 499-2688, (1).

Prentice, Jr., R.W., GUILFORD COUNTY DEPARTMENT HEALTH/DIVISON OF ENVIRONMENTAL HEALTH/AIR QUALITY PROGRAM, 301 North Eugene St. P.O. Box 3508, Greensboro, NC, 27401, (919) 373-3771, (10).

Prettyman, Milman, DELAWARE ASSOCIATION OF REALTORS, P.O. Box 1364, Dover, DE, 19903, (302) 734-4444, (18).

Prewitt, Benjamin, FLORIDA DEPARTMENT OF HEALTH AND REHABILITATIVE SERVICES/DEPUTY SECRETARY FOR HEALTH/OFFICE OF RADIATION CONTROL/ENVIRONMENTAL SECTION, P.O. Box 15490, Orlando, FL, 32858, (407) 628-2274, (407) 297-2097, (1)(10).

Price, Evertt, GOVERNMENT OF NEWFOUNDLAND AND LABRADOR/DEPARTMENT OF LABOUR/OCCUPATIONAL HEALTH AND SAFETY DIVISION/MEDICAL AND HYGIENE SERVICES, Beothuck Building Crosbie Pl., St. John's, NF, A1C 5T7, (709) 576-2645, (11).

Price, J., BICRON CORPORATION, 12345 Kinsman Rd., Newbury, OH, 44065, (216) 564-8000, (4).

Price, Norris, OKLAHOMA REAL ESTATE COMMISSION, 4040 North Lincoln Blvd. Suite 100, Oklahoma City, OK, 73105, (405) 521-3387, (10).

Price, Stan, WASHINGTON STATE DEPARTMENT OF ENERGY/ENERGY EXTENSION SERVICE, 914 East Jefferson Suite 300, Seattle, WA, 98122, (509) 624-4180, (10).

Price, Steve, HARDICO RADON MANAGEMENT, ROANOKE OFFICE, P.O. Box 10494, Blacksburg, VA, 24062-0494, (703) 951-0808, (2).

Prichard, Howard M., UNIVERSITY OF TEXAS/SCHOOL OF PUBLIC HEALTH, P.O. Box 20186, Houston, TX, 77225, (713) 792-4431, (1)(5).

Priesmeyer, William L., MANATEE COUNTY PUBLIC HEALTH UNIT/POLLUTION CONTROL DIVISION, 202 Sixth Ave. East, Bradenton, FL, 34208, (813) 748-0666, (10).

Priestner, Pete, SPECIALIZED SERVICES, INC., R.R. 1 7 Wall St., Vernon, NJ, 07462, (201) 827-9055, (1).

Prigel, Don, PRIGEL RADON TESTING, 2866 Travis Ct., York, PA, 17403, (717) 741-3188, (301) 252-1246, (1).

Prikkel, John, ENERGY VENT, INC., 915 Valley St. P.O. Box 14145, Dayton, OH, 45414, (513) 222-7749, (2).

Prikkel, John, JOHN PRIKKEL AND ASSOCIATES, P.O. Box 14145, Dayton, OH, 45414, (513) 222-7749, (513) 278-3176, (1).

Prostka, H., RADON/GEOLOGICAL ASSESSMENT SERVICES, 727 Crown Ridge Dr., Colorado Springs, CO, 80904, (719) 634-5235, (719) 685-4749, (1).

Pruett, Pamela M., ENVIRONMENTAL SYSTEMS SERVICE, P.O. Box 512, Culpepper, VA, 22701, (703) 825-6660, (1).

Pruzan, Ira, POLLENEX/ASSOCIATED MILLS, INC., 165 North Canal St., Chicago, IL, 60606, (312) 454-5400, (4).

Pryor, Gary L., POLK COUNTY PHYSICAL PLANNING, 5895 Northeast 14th St., Des Moines, IA, 50313, (515) 286-3351, (10).

Publis, Morris, U.S. DEPARTMENT OF INTERIOR (DOI)/NATIONAL PARK SERVICE(NPS)/WESTERN REGIONAL OFFICE/INDOOR RADON PROGRAM, 450 Golden Gate Ave. P.O. Box 36063, San Francisco, CA, 94102, (415) 556-5186, (8).

Pukal, Richard, PUKAL CONTRACTING COMPANY, 390 Moorehead Rd., Cabot, PA, 16023, (412) 352-4030, (1)(2).

Purcel, Barbara, BIONOMIC INDUSTRIES, INC., 479 Route 17 N., Mahwah, NJ, 07430, (201) 529-1094, (800) 338-3210, (1)(2).

Purohit, Surendra N., AMERICAN NUCLEAR SOCIETY/NEW YORK- METROPOLITAN NEW YORK SECTION, C/O Con-Edison Company 4 Irving Pl. Room 1014, New York, NY, 10003, (212) 460-6179, (14).

Purtle, B.J., BRYANT AND BRYANT SERVICE COMPANY, 992 1/2 Davidson Dr., Nashville, TN, 37205, (615) 352-2570, (1).

Purvis, Lea Ann, ALA OF COLORADO, 1600 Race St., Denver, CO, 802067-1198, (303) 388-4327, (15).

Pusey, Nancy W., ALPHATECH, INC., P.O. Box 254, Haverford, PA, 19041, (215) 525-1490, (1).

Puskar, Robert, RADON CONSULTANTS, INC., MARYLAND, 4831 Silver Hill Rd., Suitland, MD, 20746, (301) 736-7040, (1)(2).

Pyeatt, Jeri, IDAHO REAL ESTATE COMMISSION, Statehouse Mail, Boise, ID, 83720, (208) 334-3285, (10).

Pyle, B.E., SOUTHERN RESEARCH INSTITUTE, 2000 Ninth Ave., S. P.O. Box 55305, Birmingham, AL, 35255-5305, (205) 581-2000, (1)(5).

Pyles, Michael, PENNSYLVANIA DEPARTMENT OF ENVIRONMENTAL RESOURCES/BUREAU OF RADIATION PROTECTION/DIVISION OF SPECIAL PROJECTS, Fulton Building, 16th Floor Third and Locust Sts., Harrisburg, PA, 17120, (717) 787-2480, (6)(10).

Pyrch, Laurie, NEW JERSEY DEPARTMENT OF HEALTH/DIVISION OF ENVIRONMENTAL AND OCCUPATIONAL HEALTH/ENVIRONMENTAL HEALTH SERVICES/AIR AND RADIATION DIVISION, John Fitch Plaza CN-360, Trenton, NJ, 08625, (609) 633-2043, (10).

Q

Quackenbush, Renee, RADON AWARENESS AND DETECTION, 1408 South Rock Hill Rd., St. Louis, MO, 63119, (314) 961-0400, (1).

Quarles, Ron, ADVANCED MATERIALS ENTERPRISES, INC., P.O. Box 121174, Nashville, TN, 37212, (615) 377-0624, (1).

Quevedo, Lane F., BUILDING INSPECTOR OF AMERICA, THE, 1 Mason Ct. Box 1654, Hawley, PA, 18428, (717) 775-9619, (1).

Quigliano, Lawrence J., L.J. QUIGLIANO, II, ENTERPRISES, 2173 Seneca Ave., Niagara Falls, NY, 14305, (716) 284-4074, (2).

Quillin, Robert M., OHIO DEPARTMENT OF HEALTH/RADIOLOGICAL HEALTH PROGRAM/OHIO RADON PROJECT, 1224 Kinnear Rd., Columbus, OH, 43212, (614) 644-2727, (10).

Quinlan, Charles W., KSE, INC., P.O. Box 368, Amherst, MA, 01004, (413) 549-5506, (1).

Quinlan, Paul, NEBRASKA REAL ESTATE COMMISSION, 301 Centennial Mall South P.O. Box 94667, Lincoln, NE, 68508, (402) 471-2004, (10).

Quitoriano, Vicente D., GOVERNMENT OF GUAM/DEPARTMENT OF PUBLIC HEALTH AND SOCIAL SERVICES/DIVISION OF ENVIRONMENTAL HEALTH, P.O. Box 2816, Agana, GU, 96910, (671) 734-2671, (10).

R

Radcliffe, Jeff, RADON DETECTION SERVICES OF NEBRASKA, 16506 Dorcas St., Omaha, NE, 68130, (402) 330-8968, (1).

Radke, Vincent J., STAMFORD HEALTH DEPARTMENT, 229 North St., Stamford, CT, 06902, (203) 358-4396, (10).

Rahon, Theodore E., NDL ORGANIZATION, INC., THE, P.O. Box 791, Peekskill, NY, 10566, (914) 737-7200, (1).

Rains, Gloria, MANASOTA 88, 5314 Bay State Rd., Palmetto, FL, 34221, (813) 722-7413, (17).

Raisanen, David, RAYDOT, INC., 145 Jackson Ave., Cokato, MN, 55321, (612) 286-2103, (800) 328-3813, (4).

Raisanen, Niilo, RAYDOT, INC., 145 Jackson Ave., Cokato, MN, 55321, (612) 286-2103, (800) 328-3813, (4).

Raju, Mudindi, U.S. DEPARTMENT OF ENERGY (DOE)/ALBUQUERQUE OPERATIONS OFFICE(AOO)/LOS ALAMOS NATIONAL LABORATORY(LANL), P.O. Box 1663 M 888, Los Alamos, NM, 87545, (505) 667-5061, (8).

Rakowski, Bruce, CONSHOHOCKEN RADON TESTING SERVICES, 476 West Elm St., Conshohocken, PA, 19428, (215) 828-2059, (215) 272-3694, (1).

Rammacher, Richard E., BIOSOURCES, LTD., 18 Sevilla Rd., Andover, MA, 01810, (508) 698-8707, (1).

Rammien, Dale, AIR MOVEMENT AND CONTROL ASSOCIATION (AMCA)/HOME VENTILATING INSTITUTE(HVI), 30 West University Dr., Arlington Heights, IL, 60004, (312) 394-0150, (6)(13).

Ramsden, Ann Marie, CLEAN AIR ENGINEERING, INC., 207 North Woodwork Ln., Palatine, IL, 60067, (312) 991-3300, (1).

Ramsey, Paul R., RTS SYSTEMS, P.O. Box 3070, Gaithersburg, MD, 20878, (301) 258-7890, (1).

Ramsey, Robert W., RAD ELEC, INC., 5330 J Spectrum Dr., Frederick, MD, 21701, (301) 694-0011, (1)(4)(5).

Randolph, Roger D., TULSA CITY-COUNTY HEALTH DEPARTMENT, 4616 East 15th St., Tulsa, OK, 74112, (918) 744-1000, (10).

Ranieri, Ernie, IDAHO DEPARTMENT OF HEALTH AND WELFARE/DIVISION OF ENVIRONMENTAL QUALITY/BUREAU OF HAZARDOUS MATERIALS/RADIATION CONTROL DIVISION, 450 West State St., Boise, ID, 83720, (208) 334-5879, (10).

Rankin, Dorothy M., PINAL-GILA COUNTIES AIR QUALITY CONTROL DISTRICT, P.O. Box 1076, Florence, AZ, 85232, (602) 685-5801 x486, (10).

Rao, S., NEW YORK STATE DEPARTMENT OF ENVIRONMENTAL CONSERVATION/BUREAU OF AIR RESEARCH, 50 Wolf Rd., Albany, NY, 12223-3250, (518) 457-3200, (10).

Ratner, Marty, NUCLEAR ASSOCIATES/DIVISION OF VICTOREEN, 100 Voice Rd., Carle Place, NY, 11514, (516) 741-6360, (4).

Ratte, Charles, VERMONT GEOLOGICAL SURVEY, 103 South Main St. Center Building, Waterbury, VT, 05676, (802) 244-5164, (10).

Ravenhold, Otto, CLARK COUNTY HEALTH DISTRICT/AIR POLLUTION CONTROL DIVISION, 625 Shadow Ln., Las Vegas, NV, 89127, (702) 383-1276, (10).

Ray, Earl D., ST. LOUIS/DIVISION OF AIR POLLUTION CONTROL, City Hall Room 419, St.

S

Index of Names

1014, Manchester, NH, 03105, (603) 669-2411, (15).

Smith, Tim, RADON DETECTION SYSTEMS/LOS ANGELES, CALIFORNIA BRANCH, 313 8th St. Apt. E, Seal Beach, CA, 90740, (213) 596-1948, (800) 627-2366, (1).

Smith, Tony, RADON DETECTION SYSTEMS/DURANGO, COLORADO BRANCH, 77 Roach Ranch Rd., Durango, CO, 81302, (303) 444-5253, (800) 627-2360, (1).

Smith, Trudy Y., ENVIRONMENTAL SOLUTIONS, INC., P.O. Box 11323, St. Louis, MO, 63105, (314) 863-9801, (1)(3)(4).

Snapp, Charles E., ACI CHEMICAL COMPANY, INC., P. O. Box 306, Walnut Ridge, AR, 72476, (501) 886-6625, (1).

Snethen, Donna, MAINE ASSOCIATION OF REALTORS, 21 Bangor St., Augusta, ME, 04330, (207) 622-7501, (18).

Snider, Nelson, OHIO DEPARTMENT OF COMMERCE/DIVISION OF REAL ESTATE, Two Nationwide Plaza 5th Floor, Columbus, OH, 43266-0547, (614) 466-4100, (10).

Snow, Charles T., AMERICAN NUCLEAR SOCIETY/VIRGINIA-VIRGINIA SECTION, 10108 Heritage Ln., Glen Arlen, VA, 23060, (804) 771-4595, (14).

Snyder, Carole A., ALA OF NEW YORK CITY/BROOKLYN, 165 Cadman Plaza, E. Room 210, Brooklyn, NY, 11201, (718) 624-8531, (15).

Snyder, David M., TIMBERLINE RADON SERVICE, INC., P.O. Box 3815, Evergreen, CO, 80439, (303) 674-2828, (1).

Soby, Dan, RHH FOAM SYSTEMS, INC., P.O. Box 752 6001 S. Pennsylvania Ave., Cudahy, WI, 53110-0752, (414) 744-6066, (4).

Socolow, R.H., PRINCETON UNIVERSITY/SCHOOL OF ENGINEERING AND APPLIED SCIENCES/CENTER FOR ENGINEERING AND APPLIED STUDIES, Engineering Quadrangle, Princeton, NJ, 08544, (609) 452-5445, (5).

Solberg, Richard, NATURAL PANTRY, THE, 300 West 36th Ave., Anchorage, AK, 99503, (907) 563-2727, (1).

Solge, Joseph, FRANKLIN ENVIRONMENTAL ANALYSIS, 716 North Franklin St. Box 1-R, Philadelphia, PA, 19123-2721, (215) 592-8322, (1).

Solini, John A., ENVIRONMENTAL QUALITY ASSOCIATES, INC., 308 Harrison St., Fulton, NY, 13069, (315) 593-2085, (800) 727-2366, (1).

Sommerville, R.J., SAN DIEGO COUNTY AIR POLLUTION CONTROL DISTRICT, 9150 Chesapeake Dr., San Diego, CA, 92123-1095, (619) 694-3307, (10).

Soos, John, SOOS RADON SYSTEMS, INC., 25 Pheasant Run Dr., Skillman, NJ, 08558, (201) 359-0356, (2).

Sorensen, Peter F., SORENSEN ENTERPRISES, INC., 886 West Hunt Rd., Alcoa, TN, 37701, (615) 984-1376, (1).

Sorg, Tom, U.S. ENVIRONMENTAL PROTECTION AGENCY (EPA)/OFFICE OF WATER(OW)/OFFICE OF DRINKING WATER(ODW)/WATER ENGINEERING RESEARCH LABORATORY(WERL), Drinking Water Research Division (DWRD) 26 West Martin Luther King Dr., Cincinnati, OH, 45268, (513) 569-7370, (8).

Sotelo, Joe, ARIZONA DEPARTMENT OF REAL ESTATE, 202 East Earll Dr., Phoenix, AZ, 85012, (602) 255-3481, (10).

Sowby, F.D., INTERNATIONAL COMMISSION ON RADIOLOGICAL PROTECTION (ICRP), Clifton Ave., Sutton (Surrey), UK, SM2 5PU, (011) 44-642-4680, (12).

Spady, Gary D., SPADY, GARY D., 1305 Court St. Box 4 RR #2, Imperial, NE, 69033, (308) 882-5198, (1).

Spangler, Bob, COUNCIL OF AMERICAN BUILDING OFFICIALS (CABO), 5203 Leesburg Pike, Falls Church, VA, 22041, (703) 931-4533, (13).

Sparks, Glendora, RADON SERVICES OF MID-AMERICA, 935B Southeast 3rd St., Lee's Summit, MO, 64063, (816) 525-7412, (1).

Spaulding, Jim D., UNIVERSITY OF GEORGIA/CENTER FOR APPLIED ISOTOPE STUDY, 120 Riverbend Rd., Athens, GA, 30605, (404) 542-1395, (1)(5).

Spear, Michael, R.E. WRIGHT ASSOCIATES, INC./RADONIX GROUP, 3240 Schoolhouse Rd., Middletown, PA, 17057, (717) 944-5520, (1)(2)(4).

Spears, John, NATIONAL ASSOCIATION OF HOME BUILDERS (NAHB)/NATIONAL RESEARCH CENTER, 400 Prince Georges Center Blvd., Upper Marlboro, MD, 20772, (301) 249-2000, (1)(5)(6)(13).

Speedy, Ron, U.S. DEPARTMENT OF INTERIOR (DOI)/NATIONAL PARK SERVICE(NPS)/SOUTHEAST REGIONAL OFFICE/INDOOR RADON PROGRAM, Richard B. Russell Federal Building And U.S. Courthouse 75 Spring St., SW, Atlanta, GA, 30303, (404) 331-5187, (8).

Spell, William H., LOUISIANA DEPARTMENT OF ENVIRONMENTAL QUALITY/OFFICE OF AIR QUALITY AND NUCLEAR ENERGY/NUCLEAR ENERGY DIVISION, P.O. Box 14690, Baton Rouge, LA, 70898-4690, (504) 925-4518, (10).

Spence, Hugh, ATOMIC ENERGY CONTROL BOARD (AECB)/TECHNICAL INFORMATION DIVISION/OFFICE OF PUBLIC INFORMATION, P.O. Box 1046, Ottawa, ON, K1P 559, (613) 992-9206, (11).

Spencer, John, TELEDYNE WATER PIK, 1730 East Prospect, Fort Collins, CO, 80525, (303) 484-1352, (4).

Spencer, John S., AMERICAN NUCLEAR SOCIETY/ILLINOIS-CENTRAL ILLINOIS SECTION, C/O Illinois Power Company 500 South 27th St., Decatur, IL, 62525, (217) 424-8301, (14).

Spevak, Frank, ALNOR INSTRUMENT COMPANY, 7555 N. Linder Ave., Skokie, IL, 60076, (312) 677-3500, (4).

Spier, Lou, RADON REDUCTION EQUIPMENT OF COLORADO, 921 Main Suite C, Louisville, CO, 80027, (303) 665-1776, (4)(6).

Spiker, Harold, KANSAS DEPARTMENT OF HEALTH AND ENVIRONMENT/BUREAU OF AIR QUALITY AND RADIATION CONTROL/ENVIRONMENTAL SURVEILLANCE AND EMERGENCY PLANNING SECTION, Forbes Field Building Suite 740, Topeka, KS, 66620, (913) 296-1567, (1)(10).

Spilhaus, Fred, AMERICAN GEOPHYSICAL UNION (AGU), 2000 Florida Ave., Washington, DC, 20009, (202) 462-6903, (14).

Spivey, R. Jack, CHAPPARAL ENTERPRISES, INC., P.O. Box 601, Elberton, GA, 30635, (404) 283-7611, (1).

Sprague, Kent, UNIVERSITY RADON TESTING ASSOCIATES, 145 A Sycamore Ln., Athens, GA, 30605, (404) 549-8057, (1).

Sprague, Kent , MCCRONE ENVIRONMENTAL SERVICES, INC., 1412 Oakbrook Dr. Suite 100, Norcross, GA, 30093, (404) 368-9600, (1).

Sprehe, Steve, CONVENIENCE PRODUCTS, INC., 4205 Forest Park Blvd., St. Louis, MO, 63108-2892, (314) 535-6226, (4).

Springel, Stan, QUADREX, 1700 Dell Ave., Campbell, CA, 95008, (408) 866-4510, (2).

Springer, David, SPRINGER COMPANY, 629 Lancaster Ave., Malvern, PA, 19355, (215) 344-9800, (1)(2).

Sroka, John, SHEET METAL AND AIR CONDITIONING CONTRACTORS NATIONAL ASSOCIATION (SMACCNA), P.O. Box 70, Merrifield, VA, 22116, (703) 790-9890, (13).

St. Clair, Kevin, AIR SOLUTIONS, 1004 Dennis Ave., Silver Spring, MD, 20901, (301) 681-9242, (1)(2).

Stablein, Paul F., KANSAS CITY, MISSOURI/AIR QUALITY SECTION, 21st Floor, City Hall 414 East 12th St., Kansas City, MO, 64106, (816) 274-2501, (10).

Stafford, C. Fred, NATIONAL RADON LABORATORIES, INC., 1000 Johnson Ferry Rd. Suite B-145, Marietta, GA, 30068, (404) 565-3886, (1).

Stafford, C. Fred, SOUTHERN RADON SERVICES, INC., 1000 Johnson Ferry Rd. Suite B-145, Marietta, GA, 30068, (404) 565-3886, (800) 537-2366, (1)(3).

Stafford, Michael T., MCWILLIAMS PEST CONTROL, 7014 McCallum St., Philadelphia, PA, 19119, (215) 849-2223, (1).

Staib, Bob, ENVIRONMENTAL HEALTH WATCH OF GREATER CLEVELAND, 4115 Bridge Ave., Cleveland, OH, 44113, (216) 961-4646, (1)(17).

Staing, Fred, RESOURCE CENTER FOR ENVIRONMENTAL PROTECTION, P.O. Box 25, Schuylerville, NY, 12871, (518) 695-3218, (2)(3).

Staley, Thomas, RADON SCREENING SERVICE, INC., P.O. Box 37007, Denver, CO, 80237, (303) 741-3207, (1)(2).

Stallings, Freeman, RADON REDUCTION AND TESTING COMPANY, 605 South Sherman Suite 705 J, Richardson, TX, 75081, (214) 680-2296, (1)(2).

Stallings, Phillip, RADON REDUCTION AND TESTING COMPANY, 605 South Sherman Suite 705 J, Richardson, TX, 75081, (214) 680-2296, (1)(2).

Stang, Peter, U.S. DEPARTMENT OF AGRICULTURE (USDA)/AGRICULTURE RESEARCH SERVICE(ARS)/RADIATION SAFETY(RS), Building 001 Room 226 BARC-West, Beltsville, MD, 20705, (301) 344-4492, (8).

Stangle, Scott A., HOME-TECH, P.O. Box 423, Arnold, MD, 21012, (301) 757-8458, (1).

Stanizzo, Jack, ENVIRO TECH SYSTEMS, 430 Fernhill Ave., Pittsburgh, PA, 15226, (412) 561-2100, (1)(2)(4).

Staro, Charles M., NEW YORK STATE ASSOCIATION, 107 Washington Ave. Box 122, Albany, NY, 12260, (518) 462-9563, (18).

Starr, Martin, STAR LABORATORIES, P.O. Box 1666, Plymouth, MA, 02360, (617) 747-5227, (617) 947-4940, (1).

Stascavage, Paul, EASTERN ANALYTICAL SERVICES, 4 Westchester Plaza, Elmsford, NY, 10523, (914) 939-6992, (800) 327-0121, (1).

Staw, Sandy, INSURANCE HEALTH SERVICES, 35 Stony Brook Rd., Westport, CT, 06880, (203) 226-3444, (1).

Steadman, Bernadette, RESIDENTIAL SECURITY ADVISORS, 265 Kenny Brook Ln., Elyria, OH, 44035, (216) 329-2365, (4).

Steadman, James, RESIDENTIAL SECURITY ADVISORS, 265 Kenny Brook Ln., Elyria, OH, 44035, (216) 329-2365, (4).

Stear, David, LAKE BUILDERS, INC., Star Rte. 2, Box 39 Rte. 6, Hawley, PA, 18428, (717) 226-1850, (2).

Stebbings, J.H., U.S. DEPARTMENT OF ENERGY (DOE)/CHICAGO OPERATIONS OFFICE(COO)/ARGONNE NATIONAL LABORATORY(ANL)/DIVISION OF BIOLOGICAL, ENVIRONMENTAL AND MEDICAL RESEARCH/ENVIRONMENTAL HEALTH SECTION, 900 South Cass Ave., Argonne, IL, 60439, (312) 972-7678, (8).

Steckowych, Virginia, BURNS AND ASSOCIATES, 265 Kearsarge Way, Portsmouth, NH, 03801, (603) 431-3400, (1)(2).

Stecy, Peter, LAKE COUNTY HEALTH DEPARTMENT, 2293 North Main St., Crown Point, IN, 46307, (219) 738-2020, (10).

Steding, W., FOMO PRODUCTS, INC., 2775 Barber Rd., Norton, OH, 44203, (216) 753-4585, (800) 321-5585, (4).

Steele, D.F., AIRXCHANGE, INC., 401 VFW Dr., Rockland, MA, 02370, (617) 871-4816, (4).

Stees, Don, SAFE-AIRE, INC., 162 East Chestnut St. P.O. Box 160, Canton, IL, 61520, (309) 647-0419, (800) 331-2943, (1)(2)(3)(4)(6).

Steeves, Vaughan M., SOUTHEASTERN MASSACHUSETTS AIR POLLUTION CONTROL DISTRICT, Lakeville Hospital, Lakeville, MA, 02346, (617) 947-1231 x680, (10).

Steffen, Joe, LYONS SAFETY, North 92 W. 14350 Anthony Ave., Menomonee Falls, WI, 53051, (800) 558-8900, (4).

Steichen, John, RADON/SOILS AND WATER TESTING SERVICE, 7080 Price-Hilliard Rd., Plain City, OH, 43064, (614) 873-8821, (1)(4).

Steiger, Bill, ENVIRONMENTAL TESTING ASSOCIATES, 240 Harvest Ln., Altoona, WI, 54720, (715) 832-8585, (1).

Steiger, Don, ENVIRONMENTAL TESTING ASSOCIATES, 9 White Pine Trail, Madison, WI, 53717, (608) 833-4073, (1)(2).

Stein, Charles A., ENVIRONMENTAL TECHNOLOGIES, INC., P.O. Box 1300, Monticello, MN, 55362, (612) 295-3008, (1).

Stein, Ellen, VERITECH, 47 Carey Ave., Butler, NJ, 07405, (201) 492-8744, (1).

Stein, Ronald, RONALD STEIN CONSTRUCTION, INC., 21 Davis Ave., Poughkeepsie, NY, 12603, (914) 473-0064, (1)(2).

Steinberger, Paul W., GEOCENTRIC ENGINEERING CORPORATION, P.O. Box 277, Lorton, VA, 22070, (703) 550-9253, (1).

Steinmeyer, Paul, RADIATION SAFETY ASSOCIATES, INC., 10 Pendleton Dr. P.O. Box 107, Hebron, CT, 06248, (203) 228-0487, (1).

Stemmer, John, RADON BUSTERS OF CENTRAL NEW YORK, 3218 Valley Pl., Sauquoit, NY, 13456, (315) 737-7290, (1).

Stempler, Samuel, NEW YORK CITY BUREAU OF AIR RESOURCES, 295 Lafayette St., New York, NY, 10012, (212) 323-8580, (10).

Stenhouse, Mike, BATTELLE MEMORIAL INSTITUTE/NUCLEAR TECHNOLOGY SECTION, 505 King Ave., Columbus, OH, 43201-2693, (614) 879-5192, (1)(5).

Stephens, James, RADON LABS OF ATLANTA, 5400 Hwy. 29, Lilburn, GA, 30247, (404) 279-0005, (800) 248-1889, (1).

Stephenson, Charlie, MECHANICAL EQUIPMENT SERVICES CORPORATION, 3122 49th St., Des Moines, IA, 50310, (515) 276-9131, (1)(2).

Stephenson, Charlie, RADON CONTROL SERVICES CORPORATION, 3122 49th St., Des Moines, IA, 50310, (515) 276-9131, (1)(2).

Stephenson, Douglas, JOHN L. RENSHAW, INC., 10738 Tucker St., Beltsville, MD, 20705, (301) 937-4440, (2).

Stephenson, Thomas R., CERTUS LABORATORIES, P.O. Box 105633, Atlanta, GA, 30348, (404) 762-5411, (1)(6).

Sterbel, Sylvester, STERBAL, SYLVESTER, 16 Rose Ct., Staten Island, NY, 10301, (718) 273-2422, (1).

Stern, Richard J., POLLENEX/ASSOCIATED MILLS, INC., 165 North Canal St., Chicago, IL, 60606, (312) 454-5400, (4).

Stern, Robert, NEW JERSEY DEPARTMENT OF ENVIRONMENTAL PROTECTION/DIVISION OF ENVIRONMENTAL QUALITY/BUREAU OF ENVIRONMENTAL RADIATION, 729 Alexander Rd. CN-415, Trenton, NJ, 08625, (609) 987-6396, (800) 648 0394, (1)(10).

Stesis, Harold, U.S. TOXIC SUBSTANCE TESTING BUREAU, 1024 Cottman Ave., Philadelphia, PA, 19111, (215) 745-7778, (800) 535-8378, (1)(4).

Steva, Deborah, OHIO DEPARTMENT OF HEALTH/RADIOLOGICAL HEALTH PROGRAM/OHIO RADON PROJECT, 1224 Kinnear Rd., Columbus, OH, 43212, (614) 644-2727, (10).

Stevens, Dwayne, LUDLUM MEASUREMENTS, INC., 501 Oak St. P.O. Box 810, Sweetwater, TX, 79556, (915) 235-5494, (4).

Stevens, Gregg, ENVIROSCIENCE CONSULTANTS, INC., 66 Cedar St., Newington, CT, 06111, (203) 666-7167, (1)(3).

Stevens, William, ASSOCIATION FOR IMPROVEMENT IN MOHEGAN, INC., East Main St. P.O. Box 103, Mohegan Lake, NY, 10547, (914) 528-3300, (1)(6)(17).

Stevenson, Richard D., ALA OF EASTERN MISSOURI, 1118 Hampton Ave., St. Louis, MO, 63139-3147, (314) 645-5505, (15).

Stewart, Bob, ION SYSTEMS, INC., 2546 10th St., Berkeley, CA, 94710, (415) 548-3640, (4).

Stewart, Mary Beth, U.S. GENERAL SERVICES ADMINISTRATION REGIONAL HEADQUARTERS(GSAR), 7th and D Sts. Room 1126, Washington, DC, 20407, (202) 453-5236, (8).

Stewart, Norma, DISTRICT OF COLUMBIA DEPARTMENT OF CONSUMER AND REGULATORY AFFAIRS/SERVICE FACILITY REGULATION ADMINISTRATION/PHARMACEUTICAL AND MEDICAL DEVICES CONTROL DIVISION, 614 H St. Room 1014, Washington, DC, 20004, (202) 727-7190, (10).

Stewart, Sam, SAM STEWART TESTING, P.O. Box 7304, Fairfax Station, VA, 22039, (703) 250-9203, (1).

Stewart, William Barry, APPLIED ANTENNA TECHNOLOGIES/RADON MEASUREMENTS DIVISION, 2270 Honeycomb Ct., Duluth, GA, 30136, (404) 623-0403, (1).

Stklovsky, Garry, RADIOACTIVITY ALERT DETECTION SERVICES, INC., P.O. Box L-227, Langhorne, PA, 19047, (215) 357-0574, (1)(2).

Stocknoff, Michael S., CENTER FOR RADON SERVICES, P.O. Box 2250, Cherry Hill, NJ, 08034, (609) 424-7191, (1).

Stocks, Hal S., INDIANA STATE BOARD OF HEALTH/RADIOLOGICAL HEALTH SECTION, 1330 West Michigan St. Box 1964, Indianapolis, IN, 46206-1964, (317) 633-0153, (10).

Stoebner, Tom, RAVEN INDUSTRIES, INC., P.O. Box 1007, Sioux Falls, SD, 57117, (605) 336-2750, (800) 227-2836, (4).

Stokes, Allan, IOWA DEPARTMENT OF NATURAL RESOURCES/ENVIRONMENTAL PROTECTION DIVISION, Henry A. Wallace Building, Des Moines, IA, 50319-0034, (515) 281-5145, (10).

Stokes, Lee, IDAHO OSHA ONSITE CONSULTATION PROGRAM, Boise State University Community And Environmental Health 1910 University Drive, Boise, ID, 83725, (208) 385-3929, (10).

Stoll, Jeff, WELD COUNTY HEALTH DEPARTMENT/AIR POLLUTION CONTROL DIVISION, 1516 Hospital Rd., Greeley, CO, 80631, (303) 353-0635, (10).

Stoller, Jim, STO-COTE, INC., P.O. Drawer 310, Richmond, IL, 60071, (815) 675-2358, (4).

Stolwijk, Jan A.J., ASHRAE/ENVIRONMENTAL HEALTH COMMITTEE/INDOOR AIR QUALITY SUBCOMMITTEE, 1791 Tullie Cr., NE, Atlanta, GA, 30329, (404) 636-8400, (13).

Stopher, Susan G., KENTUCKY REAL ESTATE COMMISSION, 222 South First St. Suite 300, Louisville, KY, 40202, (502) 588-4462, (10).

Stoss, Frederick W., CENTER FOR ENVIRONMENTAL INFORMATION, INC. (CEI), 33 South Washington St., Rochester, NY, 14608, (716) 546-3796, (14).

Strachen, David, WASHINGTON DC ASSOCIATION, 777 14th St., NW Suite 200, Washington, DC, 20005, (202) 628-4494, (18).

Stranberg, R.W., CALIFORNIA DEPARTMENT OF INDUSTRIAL RELATIONS/DIVISION OF OCCUPATIONAL SAFETY AND HEALTH, P.O. Box 603, San Francisco, CA, 94101-0603, (415) 557-2870, (10).

Strausburg, Herbert T., PRE-PURCHASE HOME INSPECTION, 3512 Breeland Ave., Louisvile, KY, 40241, (502) 425-6810, (1).

Strawder, Kay, U.S. DEPARTMENT OF HEALTH AND HUMAN SERVICES (HHS)/PUBLIC HEALTH SERVICE(PHS)/OFFICE OF THE SURGEON GENERAL(OSG), 5600 Fisher Ln. Parklawn Building Room 1867, Washington, DC, 20857, (301) 443-6496, (8).

Strobel, Craig, HOME EXTERMINATING COMPANY, 4608 Bucks Schoolhouse Rd., Baltimore, MD, 21237, (301) 661-7911, (800) 492-5544, (1).

Strobel, Craig, PARAMOUNT PEST CONTROL, 110 Gordon Rd., Falls Church, VA, 22046, (703) 241-9140, (1).

Strobel, Gary, FLORIDA DEPARTMENT OF LABOR AND EMPLOYMENT SECURITY/BUREAU OF INDUSTRIAL SAFETY AND HEALTH, Lafayette Building Room 204 2551 Executive Center Circle West, Talahassee, FL, 32301, (904) 488-3044, (10).

Strong, Dave, HEIL COMPANY, P.O. Box 25995, Colorado Springs, CO, 80918, (719) 593-8407, (4).

Stroud, Thomas W., HOUSE MASTER OF AMERICA, 9989 West 60th Ave. Suite 201A, Arvada, CO, 80004, (303) 431-5383, (1).

Strout, Ty, ARIZONA ASSOCIATION OF REALTORS, 4414 North 19th Ave. Suite R, Phoenix, AZ, 85015, (602) 248-7785, (18).

Stuart, Tom, ALASKA DEPARTMENT OF LABOR/ OCCUPATIONAL SAFETY AND HEALTH, 3301 Eagle St. Suite 303 Pouch 7-022, Anchorage, AK, 99510, (907) 276-5013, (10).

Stubbolo, Phillip, SPECTRUM INDUSTRIES, 1128 Forrest St., Dover, DE, 19901, (302) 736-0792, (1)(2).

Sturk, W.E., MADERA COUNTY AIR POLLUTION CONTROL DISTRICT, 135 West Yosemite Ave., Madera, CA, 93637, (209) 675-7823, (10).

Sturm, Avery E., MARIPOSA COUNTY HEALTH

DEPARTMENT, 4988 11th St. P.O. Box 5, Mariposa, CA, 95338, (209) 966-3689, (10).

Suben, Victor, AIR HAZARDS ASSESSMENT, INC., 415 Main St. Suite 3, Laurel, MD, 20707, (301) 792-7111, (1).

Suda, J.R., REDICORP PROTECTIVE MATERIALS, 212 Dominion Rd., NE, Vienna, VA, 22180, (703) 938-2928, (4).

Sudakar, Ash, FCA DEVELOPMENT CORPORATION, P.O. Box 1268, Exton, PA, 19341, (215) 363-5460, (2).

Sullivan, David G., ZANUS CORPORATION, 1259 El Camino Real Suite 134, Menlo Park, CA, 94025, (415) 591-7783, (1).

Sullivan, Eugene, SULLIVAN CONSULTANTS, INC., 302 Forest Dr., Union, NJ, 07083, (201) 964-5316, (1).

Sullivan, Mike, WYOMING DEPARTMENT OF OCCUPATIONAL HEALTH AND SAFETY/COMMISSIONER OF LABOR, 604 East 25th St., Cheyenne, WY, 82002, (307) 777-7786, (10).

Sullivan, Ray, APPLIED ULTRASONICS, INC., 145 C Grassy Plains St., Bethel, CT, 06801, (203) 790-1102, (4).

Sullivan, Toni, ARIX CORPORATION, 2155 North 200 West Suite 100, Provo, UT, 84604, (801) 373-2749, (1)(2).

Sullivan, Robert, SULLIVAN, ROBERT, Rte. 17, Box 11, Lowman, NY, 14861, (607) 732-8617, (2).

Sumares, John, RADON LABS, P.O. Box 2243, Woburn, MA, 01888-0443, (617) 664-0316, (1).

Summers, James, PROFESSIONAL RADON ASSESSMENT, 405 Valley Dr., Oliver Springs, TN, 37840, (615) 435-6432, (615) 690-2208, (1).

Sunseri, Matthew, AMERICAN NUCLEAR SOCIETY/TEXAS-NORTH TEXAS SECTION, C/O TU Electric Box 2300, Glen Rose, TX, 76043, (817) 897-5688, (14).

Surace, Joseph, RADIOLOGICAL PHYSICS SERVICES/HEALTH EAST, INC., P.O. Box 13367, Roanoke, VA, 24033, (703) 981-7379, (1).

Surace, Joseph L., RADIATION CONSULTANTS OF SOUTHWEST VIRGINIA, P.O. Box 4464, Roanoke, VA, 24015, (703) 985-5778, (703) 981-7732, (1).

Survil, John, QDT, LTD., 1000 Singleton Blvd., Dallas, TX, 75212, (214) 741-1993, (4).

Susman, Harold, RADON DETECTION AND REDUCTION LABORATORY/DIVISION OF FORMACHEM, 10834 Ambassador, St. Louis, MO, 63132, (314) 872-9350, (1)(2)(4).

Suter, Edward, KEY TECHNOLOGY, INC., P.O. Box 562, Jonestown, PA, 17038, (717) 274-8310, (800) 523-4964, (1)(4)(6).

Sutherland, Arthur, ROGERS AND ASSOCIATES ENGINEERING CORPORATION, 515 East 4500 S. Suite G-200, Salt Lake City, UT, 84107, (801) 263-1600, (1)(3)(5).

Sutter, John, BONDEX INTERNATIONAL, INC., 3616 Scarlet Oak Blvd., St. Louis, MO, 63122, (314) 225-5001, (4).

Svehlak, Hank, U.S. DEPARTMENT OF INTERIOR (DOI)/U.S. GEOLOGICAL SURVEY(USGS)/NATIONAL CARTOGRAPHIC INFORMATION CENTER-NSTL STATION(NCIC), Building 3101, NSTL Station, MS, 39529, (601)688-3544, (8).

Swale, William, SWALE, INC., 57 Indian Hill Rd., Collinsville, CT, 06022, (203) 693-2146, (1).

Swankin, David, CONSUMER FEDERATION OF AMERICA (CFA), 1424 16th St., NW Suite 604, Washington, DC, 20036, (202) 387-6121, (17).

Swanson, Norma, BOSSAIRE, INC., 1321 Tyler St., NE, Minneapolis, MN, 55413, (612) 781-0179, (4).

Swartz, James, CAPITAL MATERIALS TESTING, INC., P.O. Box 81 5 Hemlock St., Latham, NY, 12110, (518) 783-3150, (800) 541-4109, (1)(2)(4).

Swatski, Ronald J., U.S. DEPARTMENT OF THE ARMY/ARMY ENVIRONMENTAL HYGIENE AGENCY, HSHB-ML-RR, Aberdeen Proving Ground, MD, 21010-5422, (301) 671-2619, (301) 671-2637, (1).

Sweet, Charles E., CONFERENCE OF PUBLIC HEALTH LABORATORIANS (COPHL), P.O. Box 9083, Austin, TX, 78766, (512) 458-7318, (15).

Swible, George, MASSACHUSETTS DEPARTMENT OF PUBLIC HEALTH/OFFICE OF HUMAN SERVICES/RADIATION CONTROL PROGRAM, 150 Tremont St., Boston, MA, 02111, (617) 727-6214, (10).

Swift, David L., JOHNS HOPKINS UNIVERSITY/DEPARTMENT OF ENVIRONMENTAL HEALTH SCIENCES/DIVISION OF ENVIRONMENTAL HEALTH ENGINEERING, 615 Wolfe St., Baltimore, MD, 21205, (301) 955-3720, (5).

Swirsky, Larry, GARNER CONSULTANTS INSPECTION SERVICE, 1686 Fifth St., Trenton, NJ, 08638, (609) 883-2021, (1).

Swotinsky, Robert, WASHINGTON OCCUPATIONAL HEALTH ASSOCIATES, INC., 1120 19th St., NW Suite 410, Washington, DC, 20036, (202) 463-6698, (16).

Swoyer, Robert A., ROBERT A. SWOYER ASSOCIATES, 241 South Fifth St., Reading, PA, 19602, (215) 376-2580, (2).

Swyers, Warrick, GOVERNMENT OF NEWFOUNDLAND AND LABRADOR/DEPARTMENT OF HEALTH/PUBLIC HEALTH INSPECTION DIVISION, P.O. Box 4750, St. John's, NF, A1C 5T7, (709) 576-2679, (11).

Sykes, III, J., PORTER CONSULTANTS, INC., 125 Argyle Rd., Ardmore, PA, 19003, (215) 896-5353, (1)(3)(6).

Sylvester, George H., NATURAL RESOURCE TECHNOLOGIES, 32131 Steven Way, Conifer, CO, 80433, (303) 850-9335, (1).

Sythe, Dan, INTERNATIONAL MEDCOM, 7497 Kennedy Rd., Sebastopol, CA, 95472, (707) 823-0336, (4).

Szabo, Anthony, ACS CONSULTANTS, INC., 129 Maitland Ave., Paramus, NJ, 07652, (201) 845-8278, (1).

Szabo, Gustav, PROGRESS UNLIMITED, INC., 200 Madison Ave. Suite 1806, New York, NY, 10016, (212) 689-7030, (4).

Szabo, Tom, PROGRESS UNLIMITED, INC., 200 Madison Ave. Suite 1806, New York, NY, 10016, (212) 689-7030, (4).

Szalinski, Paul, CALIFORNIA DEPARTMENT OF HEALTH SERVICES/RADIOLOGICAL HEALTH BRANCH, 714 P St. Building #8, Sacramento, CA, 95814, (916) 322-2040, (10).

Szczepanski, Stanley J., STANDOR RADON DETECTION, 598 South Edgewood Ave., Jacksonville, FL, 32205, (904) 388-0284, (1)(2).

Szonntagh, Thomas G., AMERICAN NUCLEAR SOCIETY/PENNSYLVANIA- DELAWARE VALLEY SECTION, 6327 Fairfield Dr., Flourtown, PA, 19031, (215) 327-1200, (14).

T

Tafuri, Bryan, NORTON CONSERVATION COMPANY, INC., P.O. Box 583, Hackettstown, NJ, 07840, (201) 852-6046, (1)(2).

Tagomori, Manabu, HAWAII DIVISION OF WATER AND LAND DEVELOPMENT, P. O. Box 373, Honolulu, HI, 96809, (808) 548-7533, (10).

Tai-Pow, John, ONTARIO MINISTRY OF LABOUR/RADIATION PROTECTION SERVICES, 81 Resources Rd., Weston, ON, M9P 3T1, (416) 235-5922, (11).

Talabac, Gail, LYONS SAFETY, North 92 W. 14350 Anthony Ave., Menomonee Falls, WI, 53051, (800) 558-8900, (4).

Talbot, John, U.S. DEPARTMENT OF ENERGY (DOE)/OFFICE OF CONSERVATION AND RENEWABLE ENERGY(OCRE)/OFFICE OF BUILDINGS AND COMMUNITY SYSTEMS(OBCS)/BUILDING SYSTEMS DIVISION(BSD), CE-131, Washington, DC, 20585, (202) 586-9445, (8).

Taliefero, Michael, MORTGAGE BANKERS ASSOCIATION OF AMERICA (MBAA), 1125 15th St., NW, Washington, DC, 20005, (202) 861-6500, (18).

Tallent, Jeffrey L., RAD-TEST, INC., P.O. Box 22576, Louisville, KY, 40222, (502) 368-5996, (1).

Tangney, Louis, HOME CHEK, INC., 234 South Cardington, Wichita, KS, 67209, (316) 267-2324, (1).

Tangye, William J., SOUTHERN BUILDING CODE CONGRESS INTERNATIONAL (SBCCI), 900 Montclair Rd., Birmingham, AL, 35213, (205) 591-1853, (13).

Tanner, Alan, U.S. DEPARTMENT OF INTERIOR (DOI)/U.S. GEOLOGICAL SURVEY(USGS)/GEOLOGIC DIVISION(GD), MS-955, National Ctr., Reston, VA, 22092, (703) 648-4000, (8).

Tantleff, Richard, INTERNATIONAL PEST CONTROL, INC., P.O. Box 5922, Edison, NJ, 08818, (201) 494-9313, (1).

Tapert, Allan C., DELAWARE DEPARTMENT OF HEALTH AND SOCIAL SERVICES/BUREAU OF ENVIRONMENTAL HEALTH/OFFICE OF RADIATION CONTROL/DIVISION OF PUBLIC HEALTH, 802 Silver Lake Blvd. and Walker Rd. P.O. Box 637, Dover, DE, 19903, (302) 736-4731, (10).

Tappan, J., VERSAR, INC., 6850 Versar Center, Springfield, VA, 22151, (703) 750-3000, (1)(2)(3).

Tappan, J. Tell, ARIX CORPORATION, 760 Horizon Dr., Grand Junction, CO, 81506, (303) 242-6203, (1)(2).

Taranik, James, NEVADA BUREAU OF MINES AND GEOLOGY, University Of Nevada-Reno, Reno, NV, 89557-0088, (702) 784-6691, (10).

Tarantola, Anthony, A.D.T. INSPECTIONS, INC., P.O. Box 425, Keyport, NJ, 07735, (201) 566-4079, (1).

Tarnowski, Richard, ENVIRO TESTING, 97 Greenridge St., Johnson City, NY, 13790, (607) 770-9098, (607)797-3665, (1).

Tate, Earnest E., MINNESOTA DEPARTMENT OF HEALTH/ENVIRONMENTAL HEALTH DIVISION/SECTION OF RADIATION CONTROL, 717 Delaware St., SE P.O. Box 9441, Minneapolis, MN, 55440, (612) 623-5351, (10).

Taylor, Dave, ION SYSTEMS, INC., 2546 10th St., Berkeley, CA, 94710, (415) 548-3640, (4).

Taylor, Eugene, RADON ANALYSIS, INC., 4635 B South 28 Rd., Arlington, VA, 22206, (703) 379-8614, (1).

Taylor, Frank, PALCO LININGS, INC., 2500-B Hamilton Blvd., South Plainfield, NJ, 07080, (201) 753-6262, (800) 457-2526, (4).

Taylor, John, ENVIRONMENTAL CHEMISTRY, INC., 1233 Ogden Suite 504, Denver, CO, 80218, (303) 831-0437, (1)(4).

Taylor, Richard, JPS ELASTOMERICS CORPORATION, 395 Pleasant St., Northampton, MA, 01060, (413) 586-8750, (4).

Taylor, Robert B., TENNESSEE DEPARTMENT OF LABOR/TENNESSEE OCCUPATIONAL SAFETY AND HEALTH, 501 Union Building 3rd Floor, Nashville, TN, 37219, (615) 741-2793, (10).

Taylor, William C., RADON ENGINEERING/A UNIT OF PSI ENGINEERING, INC., 1 Lethbridge Plaza, Mahwah, NJ, 07430, (201) 529-8300, (1)(2).

Taylor, Zachary S., ENCOSYS CORPORATION OF AMERICA, 7202 East 87th St. Suite 109, Indianapolis, IN, 46250, (317) 849-6060, (1).

Tedesco, Michael, HOUSEMASTERS OF AMERICA, 2935 Hawk, Carson City, NV, 89701, (702) 849-2229, (1).

Tedford, Charles, ARIZONA RADIATION REGULATORY AGENCY, 4814 South 40th St., Phoenix, AZ, 85040, (602) 255-4845, (10).

Tefft, Diane, NEW HAMPSHIRE DEPARTMENT OF HEALTH AND HUMAN SERVICES/DIVISION OF PUBLIC HEALTH SERVICES/BUREAU OF RADIOLOGICAL HEALTH, P.O. Box 148, Concord, NH, 03301, (603) 271-4588, (10).

Teichman, Kevin, U.S. ENVIRONMENTAL PROTECTION AGENCY (EPA)/OFFICE OF RESEARCH AND DEVELOPMENT(ORD)/OFFICE OF TECHNOLOGY TRANSFER AND REGULATORY SUPPORT (OTTRS)/AIR TEAM, 401 M St., SW RD-672, Washington, DC, 20460, (202) 382-7669, (8).

Tengler, Robert J., RADON SAFETY SERVICE, 2204 Frances Dr., Loveland, CO, 80537, (303) 663-3941, (1).

Terpilak, Michael S., NUCLEON LECTERN ASSOCIATES, INC., 1916 Grayslake Dr., Silver Spring, MD, 20906, (301) 774-3301, (1).

Terpilak, Michael S., RDS/RADON DETECTION SERVICES, INC.-CORPORATE OFFICE, Rte. 179, Old York Rd. P.O. Box 419, Ringoes, NJ, 08551, (201) 788-3080, (1)(2)(3)(4)(6).

Terrill, Wallace, WALDON ASSOCIATES, 80 Hoadley St., Naugatuck, CT, 06770, (203) 729-5045, (1).

Terry, David B., RADON INSPECTION SERVICE, INC., 789 East Glen Ave., Ridgewood, NJ, 07450, (201) 670-8821, (1)(2).

Terry, Duane O., RADON APPLIANCES, INC., P.O. Box 1461, Glen Burnie, MD, 21061, (301) 766-5711, (4).

Terry, Patti, ARIX CORPORATION, P.O. Box 2021, Greeley, CO, 80632-2021, (303) 356-3930, (1)(2).

Terry, R.D., BICRON CORPORATION, 12345 Kinsman Rd., Newbury, OH, 44065, (216) 564-8000, (4).

Testa, Gene, FOMO PRODUCTS, INC., 2775 Barber Rd., Norton, OH, 44203, (216) 753-4585, (800) 321-5585, (4).

Thayer, Richard, NEW HAMPSHIRE DEPARTMENT OF ENVIRONMENTAL SERVICES/WATER SUPPLY AND POLLUTION CONTROL DIVISION/BUREAU OF WATER SUPPLY AND ENGINEERING, 6 Hazen Dr. P.O. Box 95, Concord, NH, 03301-6527, (603) 271-3139, (10).

Thelin, Ellen, RADON ASSOCIATES, INC., 126 Greenlawn Ave., Clifton, NJ, 07013, (201) 773-4955, (1).

Thelin, Lowell C., RADON ASSOCIATES, INC., 126 Greenlawn Ave., Clifton, NJ, 07013, (201) 773-4955, (1).

Theroux, Dennis R., THEROUX ENGINEERING, P.O. Box 4096, Hamden, CT, 06514, (203) 932-1313, (1)(2).

Thibeault, William, COMMUNITY HEALTH ASSOCIATES, INC., 316 Sherman St., Canton, MA, 02021, (617) 828-3123, (1)(2).

Thomas, Dolly, AMERICAN HOME INSPECTION SERVICE, 684 Main St., Wakefield, MA, 01880, (617) 536-6510, (1).

Thomas, Donald, UNIVERSITY OF HAWAII-MANOA/HAWAII INSTITUTE OF GEOPHYSICS/DIVISION OF EARTH SCIENCES, 2525 Correa, Honolulu, HI, 96822, (808) 948-8855, (5).

Thomas, Earl W., ALA OF MONTANA, Christmas Seal Building 825 Helena Ave., Helena, MT, 59601, (406) 442-6556, (15).

Thomas, Hugo, CONNECTICUT GEOLOGICAL AND NATURAL HISTORY SURVEY/NATURAL RESOURCES CENTER, 165 Capitol Ave. Room 553, Hartford, CT, 06106, (203) 566-3540, (10).

Thomas, John, SUFFOLK II BOARD OF COOPERATIVE EDUCATIONAL SERVICES (BOCES), 201 Sunrise Hwy., Patchogue, NY, 11772, (516) 289-2200, (10).

Thomas, Lee M., U.S. ENVIRONMENTAL PROTECTION AGENCY (EPA), 401 M St., SW, Washington, DC, 20460, (202) 382-4700, (8).

Thomas, Lee W., CRIPPEN CONSULTING ASSOCIATES, INC., 242 North James St. Suite 202, Wilmington, DE, 19804, (302) 994-8000, (1)(2)(3).

Thomas, Thomas M., HARFORD COUNTY HEALTH DEPARTMENT, P.O. Box 191, Bel Air, MD, 21014-0191, (301) 838-1500 x246, (10).

Thomas, Greg, SYRACUSE ENERGY SERVICE, 100 New Street, Syracuse, NY, 13202, (315) 424-0679, (1).

Thomassen, David, U.S. DEPARTMENT OF ENERGY (DOE)/ALBUQUERQUE OPERATIONS OFFICE(AOO)LOVELACE INHALATION TOXICOLOGY INSTITUTE(LITI), P. O. Box 5890, Albuquerque, NM, 87185, (505) 844-2203, (8).

Thompson, Corey, I-T.E.M., LTD., 12850 Middlebrook Rd. Suite 301, Germantown, MD, 20874, (301) 353-0585, (1).

Thompson, Franklin, ADVANCE PEST CONTROL, P.O. Box 749, Franklin, NC, 28734, (704) 524-8061, (1).

Thompson, George, FORUM FOR SCIENTIFIC EXCELLENCE, INC., THE, P.O. Box 682, Sparta, NJ, 07871, (201) 729-4823, (5).

Thompson, John, TRI-COUNTY RADON, INC., 2796 North Constitution Rd., Camden, NJ, 08104, (609) 963-1533, (1)(2).

Thompson, Jonathan, EXECUTIVE OFFICE OF THE PRESIDENT (EOP)/FEDERAL COORDINATING COUNCIL FOR SCIENCE, ENGINEERING AND TECHNOLOGY POLICY(FCSETP), New Executive Office Building, Washington, DC, 20506, (202) 456-6272, (8).

Thompson, Linda J., REALTORS LAND INSTITUTE (RLI), 430 North Michigan Ave., Chicago, IL, 60611, (312) 329-8440, (18).

Thompson, Mary, NEVADA DEPARTMENT OF HUMAN RESOURCES/HEALTH DIVISION/RADIOLOGICAL HEALTH SECTION, 505 East King St. Room 202, Carson City, NV, 89710, (702) 885-5394, (10).

Thompson, Meredith, CONFERENCE OF STATE HEALTH AND ENVIRONMENTAL MANAGERS (CSHEM), P.O. Box 5254, Hudson, FL, 34674, (15).

Thompson, R. Stanley, RADON MEASUREMENT AND SERVICES, 13131 West Cedar Dr., Lakewood, CO, 80228, (303) 980-5086, (303) 988-3033, (1)(2).

Thompson, Robert, ALABAMA OCCUPATIONAL SAFETY AND HEALTH CONSULTATION PROGRAM, P.O. Box 6005, University, AL, 35486, (205) 348-7136, (10).

Thompson, Sharon, RITE WAY WATERPROOFING, 7337 Lake Dr., Lino Lakes, MN, 55014, (612) 786-0550, (2).

Thompson, W.A., EASTERN INTEGRATED SERVICES CORPORATION, 3490 U.S. Rte. 1 Building 17-2, Princeton, NJ, 08540, (609) 520-8344, (1).

Thomson, Ian, THOMSON AND NIELSEN ELECTRONICS, LTD., 4019 Carling Ave. Suite 303, Kanata, ON, K2K 2A3, (613) 592-3019, (1)(4)(5).

Thor, Phillip, U.S. DEPARTMENT OF ENERGY (DOE)/BONNEVILLE POWER ADMINISTRATION (BPA)/OFFICE OF ENERGY RESOURCES(OE)/DIVISION OF RESOURCE MANAGEMENT/RESIDENTIAL TECHNOLOGY SECTION, 905 Northeast 11th St., Portland, OR, 97208, (503) 230-5488, (8).

Thornburg, D.R., QUALITY RADON TESTING CORPORATION, P.O. Box 11980, Chicago, IL, 60611, (312) 280-4908, (1).

Thornell, Mike, NIMCO/NESSI, P.O. Box 356, Menasha, WI, 54952-0356, (414) 725-0357, (1).

Tierny, Timothy, MINNESOTA DEPARTMENT OF LABOR AND INDUSTY, 444 Lafayette Rd. 5th Floor, St. Paul, MN, 55101, (612) 297-1953, (10).

Tilbrook, Roger, AMERICAN NUCLEAR SOCIETY/ILLINOIS-CHICAGO SECTION, C/O Argonne National Laboratory 9700 South Cass Ave., Argonne, IL, 60439, (312) 972-8352, (14).

Tillet, Tom, RADON ANALYTICAL LABS, 9011 North Meridian St., Indianapolis, IN, 46260, (317) 843-0788, (1).

Tilley, Walter, PARAMOUNT PEST CONTROL, 110 Gordon Rd., Falls Church, VA, 22046, (703) 241-9140, (1).

Tillman, J. Edward, TARGET RADON SERVICES, INC., Oakland Ctr. 8940A Rte. 108, Columbia, MD, 21045, (301) 992-6622, (1)(3).

Timblin, Frank, F.H. TIMBLIN COMPANY, P.O. Box 254, Industry, PA, 15052, (412) 643-1755, (4).

Timmerman, R.L., OMAHA/AIR QUALITY CONTROL, 1819 Farnam, Omaha, NE, 68183, (402) 444-5379, (10).

Timmons, K.C., TISCO, 2615 Skyline Dr., College Park, GA, 30337, (404) 768-2517, (1).

Tipton, Merlin, SOUTH DAKOTA DEPARTMENT OF WATER AND NATURAL RESOURCES/SOUTH DAKOTA GEOLOGICAL SURVEY, Science Center University of South Dakota, Vermillion, SD, 57069, (605) 677-5227, (10).

Tironi, John O., NATIONAL CONFERENCE OF LOCAL ENVIRONMENTAL HEALTH ADMINISTRATORS (NCLEHA), 3415 28th St., Port Huron, MI, 48060, (313) 987-5306, (15).

Tkatch, David P., RADON X COMPANY, INC., 217 Roxbury St., Keene, NH, 03431, (603) 352-5319, (1)(2).

To, Dominic, KANSAS DEPARTMENT OF HEALTH AND ENVIRONMENT/BUREAU OF AIR QUALITY AND RADIATION CONTROL/ENVIRONMENTAL SURVEILLANCE AND EMERGENCY PLANNING SECTION, Forbes Field Building Suite 740, Topeka,

KS, 66620, (913) 296-1567, (1)(10).

Tobin, Tom E., HOME RADON SURVEYS, 4209 Olley Ln., Fairfax, VA, 22032, (703) 323-7236, (1).

Todd, Sami, C & D HEATING AND CONSTRUCTION, 1111 South Sheridan Blvd., Lakewood, CO, 80226, (303) 935-6461, (1)(2).

Tofteberg, Harlan L., RADON TEST COMPANY, P.O. Box 345, Eagle Grove, IA, 50533, (515) 448-4600, (4).

Toivonen, Pentti, FINNEX (RADON), 23 North Ave., Cincinnati, OH, 45215, (513) 821-7019 (915) 757-2068, (1).

Tolley, Marc T., SUN DISTRIBUTING GROUP, 813 East Bloomingdale Ave. Suite 313, Brandon, FL, 33511, (813) 681-0350, (1).

Tolmasoff, Michael W., NORTHERN SONOMA COUNTY AIR POLLUTION CONTROL DISTRICT, 109 North St., Healdsburg, CA, 95448, (707) 433-5911 (707) 433-5742, (10).

Tomasetti, Frank J., DK ROCKWELL, 101 East Cornell St., Somerdale, NJ, 08083, (609) 627-2166, (1).

Tombach, Tom, AEROVIRONMENT, INC., 825 Myrtle Ave., Monrovia, CA, 91016-3424, (813) 357-9983, (4).

Tompkins, Carolyn, SAFE AIR, P.O. Box 120064, Nashville, TN, 37212, (615) 865-5325, (1).

Tonelli, Mario G., COOK COUNTY DEPARTMENT OF ENVIRONMENTAL CONTROL, Maybrook Civic Center 1500 Maybrook Dr., Maywood, IL, 60153, (312) 865-6165, (10).

Toney, Leroy, TRI-COURT SAFETY/SECURITY, P.O. Box 1257, Sunset Beach, CA, 90742, (213) 335-3290, (714) 846-4346, (1).

Tooley, Jr., John J., NATURAL FLORIDA RETROFIT, INC., P.O. Box 301, Montverde, FL, 32756, (305) 469-2173, (305) 841-4376, (1).

Toombs, George L., OREGON STATE DEPARTMENT OF HUMAN RESOURCES/HEALTH DIVISION/RADIATION CONTROL SECTION, 1400 SW 5th Ave. P.O. Box 231, Portland, OR, 97201, (503) 229-5797, (10).

Topikian, Harry, P.M. WRIGHT, LTD., 1300 Jules Poitras Blvd., St. Lawrence, PQ, H4N 1X8, (514) 337-3331, (4).

Toppan, W. Clough, MAINE DEPARTMENT OF HUMAN SERVICES/BUREAU OF HEALTH/DIVISION OF HEALTH ENGINEERING/RADIOLOGICAL HEALTH PROGRAM, State House Station 10, Augusta, ME, 04333, (207) 289-5676, (10).

Tousimis, C., TECHNORAD, P.O. Box 2144, Rockville, MD, 20852, (301) 881-8293, (1).

Tousimis, J., TECHNORAD, P.O. Box 2144, Rockville, MD, 20852, (301) 881-8293, (1).

Townsend, Douglas W., RADON APPLIANCES, INC., P.O. Box 1461, Glen Burnie, MD, 21061, (301) 766-5711, (4).

Tracey, Dick, RYAN HOMES/PRODUCTS DIVISION, 100 Ryan Court, Pittsburgh, PA, 15205, (412) 276-8000, (13).

Tracy, Hank, RADON DETECTION SYSTEMS/BOSTON, MASSACHUSETTS BRANCH, 91 Westland Ave. Suite 606, Boston, MA, 02115, (303) 444-5253, (800) 627-2366, (1).

Trainor, Diane, ENVIROHAZARDS MANAGEMENT, INC., P.O. Box 1054, South Orange, NJ, 07079, (201) 761-7226, (1).

Trapani, Charles, MICRO-TRENDS, 25 Caton Terr., Caldwell, NJ, 07006, (201) 228-5907, (3).

Traudt, Jon, APPLIED ENERGY COMPANY, 3316 Augusta Ave., Omaha, NE, 68144, (402) 334-5881, (1).

Trenga, Eric, TFC NUCLEAR ASSOCIATES, INC., 425 Bridgeboro Rd., Moorestown, NJ, 08057, (609) 778-4529, (1)(2).

Trischuk, Howard, RADON DETECTION SERVICES, 323 Neusch Way, Saskatoon, SK, S7K 7V2, (306) 931-7729, (1).

Trivelpiece, Alvin, AMERICAN ASSOCIATION FOR THE ADVANCEMENT OF SCIENCE (AAAS), 1333 H St., NW, Washington, DC, 20005, (202) 326-6400, (14).

Trost, Paul E., ST. JOSEPH COUNTY AIR POLLUTION CONTROL, County-City Building 8th & 9th Floors, South Bend, IN, 46601, (219) 284-9721, (10).

Troy, Joyce, MINISTRY OF CONCERN FOR PUBLIC HEALTH (MCPH), 5495 Main St. Suite 147, Buffalo, NY, 14221, No Phone Available, (17).

Truex, Bryan, SENSIDYNE, INC., 12345 Starkey Rd. Suite E, Largo, FL, 34643, (813) 530-3602, (800) 451-9444, (4).

Trujillo, Paul, PLASTICRAFTS, 600 West Bayaud Ave., Denver, CO, 80223, (303) 744-3700, (4).

Tseng, Eddie, AIR RADON SERVICE, INC., 292 Main St. Suite 8, Hackensack, NJ, 07601, (201) 585-2994, (1).

Tseng, Tom T., NUCLEAR RESEARCH CORPORATION, P.O. Box H, Warrington, PA, 18976, (215) 343-5900, (1).

Tu, William W., UNITED SERVICES ALLIANCE, INC., 4900 Leesburg Pike Suite 210, Alexandria, VA, 22302, (703) 845-1324, (1)(2).

Tuck, Frederick, RADEX, INC., 1382 Prince Ave. Suite A, Athens, GA, 30606, (404) 548-6960, (1)(2)(3).

Tucker, Brian, CALIFORNIA DEPARTMENT OF CONSERVATION/DIVISION OF MINES, 1416 Ninth St. Room 1341, Sacramento, CA, 95814, (916) 445-1923, (10).

Tucker, Kathleen M., HEALTH AND ENERGY INSTITUTE, 236 Massachusetts Ave., NE Suite 506, Washington, DC, 20002, (202) 543-1070, (15).

Tucker, Robert, NEW JERSEY DEPARTMENT OF ENVIRONMENTAL PROTECTION/DIVISION OF SCIENCE AND RESEARCH, 401 East State St. 7th Floor, West Wing, Trenton, NJ, 08625, (609) 984-6070, (1)(10).

Tucker, Rosemary Cash, WYOMING ASSOCIATION OF REALTORS, P.O. Box 2312, Casper, WY, 82602-2312, (307) 237-4085, (18).

Tucker, W. Gene, U.S. ENVIRONMENTAL PROTECTION AGENCY (EPA)/OFFICE OF RESEARCH AND DEVELOPMENT(ORD)/AIR AND ENGINEERING RESEARCH LABORATORY(AEERL)/INDOOR AIR BRANCH, MD-54, Research Triangle Park, NC, 27711, (919) 541-2746, (8).

Tuhoy, Mary Beth, AMERICAN MEDICAL ASSOCIATION (AMA)/COUNCIL ON SCIENTIFIC AFFAIRS, 535 North Dearborn St., Chicago, IL, 60610, (312) 645-5000, (15).

Tully, J. Charles, AMERICAN HI-TECH, 263-265 Leighton Ave., Red Bank, NJ, 07701, (201) 842-9477, (201) 671-5887, (1).

Turbett, John A., ADVANCED RADON SERVICES, P.O. Box 729, Parker, CO, 80134, (303) 841-6702, (1).

Turekion, Carl, YALE UNIVERSITY/DEPARTMENT OF GEOLOGY AND GEOPHYSICS, Kline Geolab P.O. Box 6666, New Haven, CT, 06511, (203) 432-3161,

Turner, Ann, NATIONAL CONFERENCE ON WEIGHTS AND MEASURES (NCWM), P.O. Box 3137, Gaithersburg, MD, 20870, (301) 975-4004, (13).

Turner, Dea, ALASKA ASSOCIATION OF REALTORS, 1836 West Northern Lights Blvd., Anchorage, AK, 99517, (907) 276-0655, (18).

Turner, J.B., RADON BUSTERS, P.O. Box R247, Radnor, PA, 19087, (215) 566-3059, (1).

Turner, Joe, ELECTRO ANALYTICAL, INC., 7118 Industrial Park Blvd., Mentor, OH, 44060, (216) 951-3514, (1).

Turner, L. David, RADON MEASUREMENT AND CONTROL OF NEW MEXICO, 1527 Granite St., NW, Albuquerque, NM, 87104, (505) 843-7643, (1)(4).

Turner, William, HARRIMAN ASSOCIATES, 292 Court St., Auburn, ME, 04210, (207) 784-5728, (1).

Turner, William M., RADON MEASUREMENT AND CONTROL OF NEW MEXICO, 1527 Granite St., NW, Albuquerque, NM, 87104, (505) 843-7643, (1)(4).

Tyler, Don, MICROBAC LABORATORIES, INC./ERIE TESTING DIVISION-MAIN OFFICE, 2411 West 26th St., Erie, PA, 16506, (814) 833-4790, (1)(4).

Tyler, Michael, NEVADA DIVISION OF OCCUPATIONAL SAFETY AND HEALTH, Capitol Complex 1370 South Curry St., Carson City, NV, 89710, (702) 885-5240, (10).

Tyree, Paul M., BIOSOURCES, LTD., 7 Bailey St., Nashua, NH, 03063, (603) 880-6004, (1).

Tyrrell, Les, NATIONAL ASSOCIATION OF REAL ESTATE LICENSE LAW OFFICIALS (NARELLO), P.O. Box 129, Centerville, UT, 84010-0129, (801) 531-8202, (18).

Tyson, Brent, HOUSE DOCTORS OF LAKEWOOD, INC., P.O. Box 489, Englewood, CO, 80151-0489, (303) 841-2443, (1).

U

Ulbrich, D. Todd, ENVIROSERV/RADON DETECTION OF OHIO, 781 Mount Vernon Ave., Marion, OH, 43302, (614) 382-2354, (1).

Ulbrich, Donald, ENVIROSERV/MAIN OFFICE, 15 Buckley Hill Rd., Morristown, NJ, 07960, (201) 285-1065, (1).

Ulbrich, Douglas, ENVIROSERV/RADON DETECTION OF ALABAMA, 751-D Raleigh Ct., Birmingham, AL, 35209, (205) 942-0756, (1).

Unser, Al, INDIANA ASSOCIATION OF REALTORS, 7301 North Shadeland, Indianapolis, IN, 46250, (317) 842-0890, (18).

Updike, Alan K., INDIANA REAL ESTATE COMMISSION, 100 North Senate Ave. 1021 State Office Building, Indianapolis, IN, 46204, (317) 232-2980, (10).

Upton, Arthur, NEW YORK UNIVERSITY MEDICAL CENTER/INSTITUTE OF ENVIRONMENTAL MEDICINE, 550 1st Ave., New York, NY, 10016, (212) 340-5280, (5).

Urman, Steve, U.S. DEPARTMENT OF STATE (DOS)/SAFETY OFFICE, 2210 C St., NW Room 2728, Washington, DC, 20520, (202) 647-4302, (8).

Usiskin, Clive, PRINCETON HOME INSPECTION SERVICE, 197 Wall St., Princeton, NJ, 08540, (609) 921-3775, (1)(2).

Uslan, Paul, AA HOME INSPECTIONS, P.O. Box 248, Springfield, NJ, 07081, (201) 635-6060, (1)(3).

Utgard, Russell, RADON DETECTION SYSTEMS/COLUMBUS, OHIO BRANCH, 3487 Colchester Rd., Columbus, OH, 43221, (303) 444-5253, (800) 627-2366, (1).

Uyesugi, George S., CALIFORNIA DEPARTMENT OF HEALTH SERVICES/SANITATION AND

RADIATION LABORATORY, 2151 Berkeley Way Room 465, Berkeley, CA, 94704, (415) 540-2201, (1)(10).

V

Vackoc, Misha, U.S. ENVIRONMENTAL PROTECTION AGENCY REGION 10 (EPA10)/RADIATION DIVISION(ORP), 1200 Sixth Ave., Seattle, WA, 98101, (206) 442-7299, (8).

Vaillancourt, Sr., Gerald O., TRU-KIL SERVICE, INC., P.O. Box 87156, College Park, GA, 30337, (404) 768-4533, (1).

Vakili, Moe, ENVIRONMENTAL CONTROL SYSTEMS, LTD., 15325 Jones Ln., Gaithersburg, MD, 20878, (301) 421-5150, (1)(3).

Valastro, James, K & J SOLAR SERVICE, 1142 Fillmore St., Denver, CO, 80206, (303) 393-8372, (1).

Valdec, David, PUBLIC CITIZEN LITIGATION GROUP (PCLG)/HEALTH RESEARCH GROUP, 2000 P St., NW Suite 700, Washington, DC, 20036, (202) 872-0320, (17).

Valenti, JoAnn Myer, UNIVERSITY OF FLORIDA/COLLEGE OF JOURNALISM AND COMMUNICATION/COMMUNICATION RESEARCH CENTER, 2042 Weimer, Gainesville, FL, 32611, (904) 392-6660, (5).

Van der Werss, Maria, CONNECTICUT DEPARTMENT OF HEALTH SERVICES/TOXIC HAZARDS SECTION, 150 Washington St., Hartford, CT, 06106, (203) 566-3122, (10).

Van House, Kerry, AIR SCIENCES, INC., 508 North Truman Blvd. P.O. Box 320, Crystal City, MO, 63019, (314) 933-3915, (800) 242-8886, (1)(2)(4).

Van Kampen, Gerald, RADON DETECTION SYSTEMS/BUTTE, MONTANA BRANCH, 1123 West Park, Butte, MT, 59701, (303) 444-5253, (800) 627-2360, (1).

Van Laningham, Bob, NATIONAL CONCRETE MASONRY ASSOCIATION (NCMA), P.O. Box 781 2302 Horse Pen Rd., Herndon, VA, 22070, (703) 435-4900, (13).

Van Leeuwen, Thomas, RADIATION DATA/DIVISION OF PRINCETON GEOPHYSICAL SERVICES, 1200 State Rd., Princeton, NJ, 08540, (609) 921-7061, (1)(2).

Van Pelt, Wesley R., WESLEY R. VAN PELT ASSOCIATES, INC., 773 Paramus Rd., Paramus, NJ, 07652, (201) 445-5124, (3).

Van Praag, Kenneth, RENSSELAER COUNTY DEPARTMENT OF HEALTH/DIVISION OF ENVIRONMENTAL HEALTH, County Office Building 1600 Seventh Ave., Troy, NY, 12180, (518) 270-2664, (10).

Van Schaick, Richard, R.K. VAN SCHAICK AND ASSOCIATES, R.R. 4 Box 4038, Greenwich, NY, 12834, (518) 692-2015, (1).

Van Steenkiste, Dwight, GENOVA PRODUCTS, INC., 7034 East Court St., Davison, MI, 48423, (313) 744-4500, (800) 521-7488, (4).

Van Zandt, Arnold, COLORADO ASSOCIATION OF REALTORS, 6558 South Yosemite Cr., Englewood, CO, 80111, (303) 770-6550, (18).

Vanags, Ulois, NORTHERN ENVIRONMENTAL LABORATORY, INC., RR 1, Box 1810, Dixmont, ME, 04932-9740, (207) 234-2387, (1).

Vanderbeek, Thomas, LMS ENGINEERS, 1 Blue Hill Plaza, Pearl River, NY, 10965, (914) 735-8300, (1).

Vandine, Dale, PARADISE COMPANY, 109 Greenleaf, Lumberton, TX, 77656, (409) 755-3524, (1)(2).

Vann, Thad I., MISSISSIPPI ASSOCIATION OF REALTORS, P.O. Box 2454, Jackson, MS, 39205, (601) 355-0216, (18).

Varadi, Laslo, LV SAFETY AND HEALTH CONSULTANTS, 446 Timothy Dr., Bethlehem, PA, 18017, (215) 758-9072, (1)(2).

Varani, Frederick, ENVIRONMENTAL DATA COMPANY, INC., 2700 Youngfield, Lakewood, CO, 80215, (303) 238-1104, (1).

Vargo, George J., AMERICAN NUCLEAR SOCIETY/NEW YORK-NIAGARA FINGER LAKES SECTION, C/O New York Power Authority P.O. Box 41, Lycoming, NY, 13093, (416) 449-6706, (14).

Vargo, George J., ENVIRONMENTAL QUALITY ASSOCIATES, INC., 308 Harrison St., Fulton, NY, 13069, (315) 593-2085, (800) 727-2366, (1).

Vartanian, Rafik, CTL ENVIRONMENTAL SERVICES, 2905 East Century Blvd., South Gate, CA, 90280, (213) 564-2641, (1).

Vass, David T., BONDEX INTERNATIONAL, INC., 3616 Scarlet Oak Blvd., St. Louis, MO, 63122, (314) 225-5001, (4).

Vasys, Al, CAMBRIDGE ASSOCIATES, LTD., 11 Harwood Ln. Trowbridge Complex, Clementon, NJ, 08021, (609) 435-2626, (215) 563-6611, (1).

Vaughan, I.N., HUNTSVILLE/AIR POLLUTION CONTROL DEPARTMENT, 2033 City Airport Rd., Huntsville, AL, 35802, (205) 881-7803, (10).

Vaughan, Maurice, NIAGARA COUNTY HEALTH DEPARTMENT/AIR RESOURCES BUREAU, 10th and East Falls Sts., Niagara Falls, NY, 14302, (716) 284-3128, (10).

Vaughan, William M., ENVIRONMENTAL SOLUTIONS, INC., P.O. Box 11323, St. Louis, MO, 63105, (314) 863-9801, (1)(3)(4).

Vegh, Cindy, FTLC SERVICES, INC., P.O. Box 563, Amherst, OH, 44001, (216) 960-2550, (1).

Vegh, John, FTLC SERVICES, INC., P.O. Box 563, Amherst, OH, 44001, (216) 960-2550, (1).

Venable, C. Scott, ALA OF NORTH CAROLINA, P.O. Box 27985, Raleigh, NC, 27611-7985, (919) 832-8326, (15).

Veradi, Laslo, GARCO RESEARCH COMPANY, 144A Sutton Rd., Lebanon, NJ, 08833, (201) 832-7895, (1)(2).

Verhagen, Nicholas, AAA RADON TESTING, 26 Flynn Pl., Wayne, NJ, 07470, (201) 772-7166, (1)(2).

Verhagen, Nicholas, VER-GON BUILDERS, 193 Christie Ave., Clifton, NJ, 07011, (201) 772-4040, (2).

Vernetson, William G., AMERICAN NUCLEAR SOCIETY/FLORIDA-FLORIDA SECTION, C/O University of Florida Nuclear Engineering Room 202 NSC, Gainesville, FL, 32611, (904) 392-1408, (14).

Vetter, William, MCWILLIAMS PEST CONTROL, 7014 McCallum St., Philadelphia, PA, 19119, (215) 849-2223, (1).

Vick, Robert I., VICK ENTERPRISES, 2060 East Easter Ave., Littleton, CO, 80122, (303) 798-7918, (2).

Viles, Frederick M., RADON CHECK/DIVISION OF HORIZON SYSTEMS, P.O. Box 721, Harvard, MA, 01451, (617) 456-8811, (617) 890-2334, (1).

Villforth, John C., U.S. DEPARTMENT OF HEALTH AND HUMAN SERVICES (HHS)/PUBLIC HEALTH SERVICE(PHS)/FOOD AND DRUG ADMINISTRATION(FDA)/CENTER FOR DEVICES AND RADIOLOGICAL HEALTH(CDRH), 5600 Fishers Ln., Rockville, MD, 20857, (301) 443-4690, (8).

Viola, Ralph, DPR BUILDERS, 21 Stacy Dr., Belle Mead, NJ, 08502, (201) 874-3419, (1)(2).

Vitale, Thomas J., SURE-WAY SYSTEMS/DIVISION OF E.C.S., INC., 107 Pickard Dr., Syracuse, NY, 13211, (315) 455-2431, (800) 622-1002, (2).

Vittitow, Lee, ANALYTICAL MANAGEMENT, INC., P.O. Box 11279, Lexington, KY, 40574, (606) 233-7774, (1).

Vitz, Edward, KUTZTOWN UNIVERSITY/DEPARTMENT OF PHYSICAL SCIENCES/RADON PROJECT, 110 Grim Science Building, Kutztown, PA, 19530, (215) 683-4443, (5).

Vivian, Robert, RADON TESTING SERVICES, 4096 Scenic Dr., Provo, UT, 84604, (801) 227-0179, (800) 777-2320, (1).

Voccio, Valerie, RHODE ISLAND DEPARTMENT OF BUSINESS REGULATION, LICENSING AND CONSUMER PROTECTION, 100 North Main St., Providence, RI, 02903, (401) 277-2255, (10).

Voightman, Tom, PROFESSIONAL HOUSE DOCTORS, 599 Short Ave., Marengo, IA, 52301, (319) 642-7133, (1)(2).

Voll, Edwin, TFC NUCLEAR ASSOCIATES, INC., 425 Bridgeboro Rd., Moorestown, NJ, 08057, (609) 778-4529, (1)(2).

Vom Eigen, Ann, MORTGAGE BANKERS ASSOCIATION OF AMERICA (MBAA), 1125 15th St., NW, Washington, DC, 20005, (202) 861-6500, (18).

Von Weizsacker, Ernst, INSTITUTE FOR EUROPEAN ENVIRONMENTAL POLICY (IEEP), Aloys-Schulte-Strasse 6, D-5300 Bonn 1, West Germany, (011) 49-228-213810, (12).

Vories, David H., REMSEARCH, INC., 160 Bleeker St. Suite 5BE, New York, NY, 10012, (212) 982-1115, (1).

Vukas, Ronald, INSTITUTE OF REAL ESTATE MANAGEMENT, 430 North Michigan Ave., Chicago, IL, 60611, (312) 661-1930, (18).

W

Wacks, Margaret, ECODEX, 1 Aaron Rd., Lexington, MA, 02173, (617) 862-3440, (1)(2).

Wacks, Margaret, HOME TESTING SERVICE, 1 Aaron Rd., Lexington, MA, 02173, (617) 862-3440, (1).

Wagner, Carol, MORTGAGE BANKERS ASSOCIATION OF AMERICA (MBAA), 1125 15th St., NW, Washington, DC, 20005, (202) 861-6500, (18).

Wagner, Frank, SANI-DUCT SERVICE OF THE INLAND NORTHWEST, 16015 East Sprague Ave., Spokane, WA, 99037, (509) 924-8472, (1).

Wagner, William D., U.S. DEPARTMENT OF HEALTH AND HUMAN SERVICES (HHS)/PUBLIC HEALTH SERVICE(PHS)/CENTERS FOR DISEASE CONTROL(CDC)/NATIONAL INSTITUTE FOR OCCUPATIONAL SAFETY AND HEALTH(NIOSH), Division Of Standards Development And Technology Transfer (DSOTT) 4676 Columbia Pkwy., MS-C14, Cincinnati, OH, 45226, (513) 533-8236, (8).

Wahl, Douglas F., RMC ENVIRONMENTAL SERVICES, 919 Edgewood Rd., Havertown, PA, 19083, (215) 841-5268, (3).

Waite, David, ENVIROSPHERE, 10900 Northeast Eighth St., Bellevue, WA, 98004, (206) 451-4600, (3).

Waite, Joyce C., ALA OF MARYLAND, Heaver Plaza 1301 York Rd. Suite 705, Lutherville, MD, 21093-6010, (301) 494-1100, (15).

Wakamo, Chuck, U.S. ENVIRONMENTAL PROTECTION AGENCY REGION 4 (EPA4)/RADIATION DIVISION(ORP), 345 Courtland

St. NE, Atlanta, GA, 30365, (404) 347-3907, (8).

Waligora, Stanley, TMA/EBERLINE, HEALTH PHYSICS, 5635 Jefferson Blvd., NE, Albuquerque, NM, 87109, (505) 345-9931, (1)(2)(3)(4).

Walk, John, TRIANGLE WATERPROOFING AND REPAIR SERVICE, P.O. Box 741 206 North Arendell Ave., Zebulon, NC, 27597, (919) 269-8021, (2).

Walker, Barrett P., RDS/RADON DETECTION SERVICES-GEORGIA OFFICE, 403 West Ponce de Leon Ave. P.O. Box 1826, Decatur, GA, 30031, (404) 377-2366, (1)(2).

Walker, Cathleen, RADON TESTING AND CONSULTING SERVICES, 34 Jones River Dr., Kingston, MA, 02364, (617) 585-9924, (1)(2)(3).

Walker, Daniel C., RADON TESTING AND CONSULTING SERVICES, 34 Jones River Dr., Kingston, MA, 02364, (617) 585-9924, (1)(2)(3).

Walker, Peggy, RDS/RADON DETECTION SERVICES-GEORGIA OFFICE, 403 West Ponce de Leon Ave. P.O. Box 1826, Decatur, GA, 30031, (404) 377-2366, (1)(2).

Walker, Randall R., TOPEKA-SHAWNEE COUNTY HEALTH AGENCY, 1615 West Eighth St., Topeka, KS, 66606, (913) 233-8961, (10).

Walker, Steven W., ALBUQUERQUE/ENVIRONMENTAL HEALTH DEPARTMENT/AIR POLLUTON CONTROL DIVISION, P.O. Box 1293, Albuquerque, NM, 87103, (505) 768-2600, (10).

Walker, Thomas, ALABAMA ASSOCIATION OF REALTORS, P.O. Box 4070, Montgomery, AL, 36103, (205) 262-3808, (18).

Walker, Jr., Donald G., DONALD WALKER COMPANY, 60591 Mount Vernon, Rochester, MI, 48064, (313) 652-1504, (1).

Walker, Jr., Paris R., ETE CONSULTING ENGINEERING, INC., P.O. Box 3471, Oak Ridge, TN, 37831, (615) 482-4503, (615) 546-0154, (1).

Walker, Jr., Vance F., RADON MANAGEMENT OF KENTUCKY, 12203 Old Shelbyville Rd., Middletown, KY, 40243, (502) 244-0517, (1)(2).

Walkiewicz, Thomas A., EDINBORO SCIENTIFIC COMPANY, P.O. Box 589 Wetsell Ridge Rd., Edinboro, PA, 16412, (814) 734-4648, (1)(3)(4).

Walkiewicz, Thomas A., PROTECTION INSPECTION, INC., Rte. 11 Hill Rd., Knoxville, TN, 37938, (615) 688-9779, (1)(2)(3).

Wall, Douglas R., ENVIRONMENTAL SERVICES OF ILLINOIS, INC., 302 West State St., O'Fallon, IL, 62269, (618) 624-8079, (314) 421-4374, (1)(2).

Wall, Elliott, ELLIOTT AND ASSOCIATES, 146 North Drexel Ave., La Grange, IL, 60525, (312) 354-1481, (1)(2).

Wall, G.R., ENVIRONMENTAL SERVICES OF ILLINOIS, INC., 302 West State St., O'Fallon, IL, 62269, (618) 624-8079, (314) 421-4374, (1)(2).

Wallace, Douglas, RADON CORRECTIONS INSTITUTE, INC., P.O. Box 314, Longmont, CO, 80502, (303) 651-6190, (800) 648-9281, (2)(3).

Walling, Patricia P., OHIO ENVIRONMENTAL PROTECTION AGENCY/DIVISION OF AIR POLLUTION CONTROL, 1800 Watermark Dr., Columbus, OH, 43266-0149, (614) 644-3020, (10).

Walsh, Jack, POLYMERIC SYSTEMS, INC., Wheatland and Mason Sts., Phoenixville, PA, 19460, (215) 935-1170, (800) 228-5548, (4).

Walsh, James, RADON HOME SEARCH, 349 Maxim Dr., Hopatcong, NJ, 07843, (201) 398-4106, (1)(2).

Warchol, Mark, ENVIRONMENTAL SERVICES OF ILLINOIS, INC., 302 West State St., O'Fallon, IL, 62269, (618) 624-8079, (314) 421-4374, (1)(2).

Ward, Carolyn, RADON REDUCTION, INC., 1125 Mechanic St., Jeffersonville, IN, 47130, (812) 284-4466, (2).

Ward, Gil, HOME ENVIRONMENT COMPANY, P.O. Box 24744, Speedway, IN, 46224, (317) 247-1881, (1)(2)(4).

Ward, Jim, RADON REDUCTION, INC., 1125 Mechanic St., Jeffersonville, IN, 47130, (812) 284-4466, (2).

Ward, John F., UNIVERSITY OF CALIFORNIA-SAN DIEGO/DEPARTMENT OF RADIOLOGY/RADIATION BIOLOGY, M-010, La Jolla, CA, 92093, (619) 534-4916, (5).

Ward, Thomas, TMW MARKETING, 1660 North LaSalle Suite 3301, Chicago, IL, 60614, (312) 951-7213, (1).

Warden, Steven, CANCERSCREEN, INC., 1120 First Colonial Rd., Virginia Beach, VA, 23454, (804) 481-4742, (1).

Warner, Ann, AIR CONDITIONING CONTRACTORS OF AMERICA (ACCA), 1513 16th St., NW, Washington, DC, 20036, (202) 296-7610, (13).

Warner, Norman, CCI-RADON AND HOME INSPECTION/RADON DETECTION, P.O. Box 1045, Malvern, PA, 19355, (215) 889-1554, (215) 363-9015, (1)(2)(4).

Warren, Gooden T., WARREN WOLFE ASSOCIATES, 212 South Queen St., Dover, DE, 19901, (302) 734-8275, (1).

Warren, Thomas E., RADON TECHNOLOGIES, INC., 698 High St., Worthington, OH, 43085, (614) 441-0550, (614) 431-0322, (1).

Warren, Willliam, COLORADO FRONT RANGE BUILDING INSPECTION SERVICE, P.O. Box 9614, Ft. Collins, CO, 80525, (303) 482-1976, (800) 248-1976, (1).

Washburn, Timothy A., RADON LABORATORIES, INC., 3 Deer Tr., Tabernacle, NJ, 08088, (800) 527-4520, (609) 859-0944, (1).

Wasick, Bruce, WASICK, BRUCE, 1888 S. Jacoby Rd., Copley, OH, 44321, (216) 666-8078, (1).

Wass, Alan R., ALA OF NEBRASKA, 8901 Indian Hills Dr. Suite 107, Omaha, NE, 68114-4057, (402) 393-2222, (15).

Wasson, Larry, AFFILIATED BUILDING INSPECTORS, INC., P.O. Box 15770, Chevy Chase, MD, 20815, (301) 986-8866, (1)(4).

Waterdale, Jerry F., TEXAS DEPARTMENT OF HEALTH/DIVISION OF OCCUPATIONAL SAFETY AND STATE SAFETY ENGINEER, 1100 West 49th St., Austin, TX, 78756, (512) 458-7287, (10).

Watkins, Andrew, WATKINS ENVIRONMENTAL SCIENCES, 207 McLennan Dr., Fayetteville, NY, 13066, (315) 637-4077, (1).

Watkins, Robert T., MASSACHUSETTS DEPARTMENT OF PUBLIC HEALTH/OFFICE OF HUMAN SERVICES/RADIATION CONTROL PROGRAM, 150 Tremont St., Boston, MA, 02111, (617) 727-6214, (10).

Watson, Mark, NEW YORK STATE ENERGY OFFICE/BUREAU OF RESIDENTIAL CONSERVATION, 2 Rockefeller Plaza 8th Floor, Albany, NY, 12223, (518) 473-7243, (6)(10).

Watson, William J., STRATFORD DEPARTMENT OF HEALTH, 2730 Main St., Stratford, CT, 06497, (203) 385-4090, (10).

Watson, Jr., James E., UNIVERSITY OF NORTH CAROLINA/SCHOOL OF PUBLIC HEALTH/DEPARTMENT OF ENVIRONMENTAL SCIENCES AND ENGINEERING, Room 201H, Chapel Hill, NC, 27599-7400, (919) 966-3840, (1)(5).

Watt, Flint C., MICHIGAN DEPARTMENT OF PUBLIC HEALTH/DIVISION OF OCCUPATIONAL HEALTH/SPECIAL PROGRAMS SECTION, 3500 North Logan, Lansing, MI, 48909, (517) 322-1809, (10).

Watters, Robert L., ENRAD, INC./RADON TESTING AND ENGINEERING, INC., 18705 B North Frederick Rd., Gaithersburg, MD, 20879, (301) 948-8040, (1).

Way, Anthony, LUBBOCK CITY HEALTH DEPARTMENT/AIR POLLUTION CONTROL DIVISION, 1902 Texas Ave. P.O. Box 2548, Lubbock, TX, 79408, (806) 762-6411 x2918, (10).

Way, J.T., ANTI-PEST COMPANY OF SHREVEPORT, P.O. Box 4266, Shreveport, LA, 71104, (318) 221-6181, (1).

Way, James T., ANTI-PEST COMPANY, P.O. Box 8082, Longview, TX, 75607, (214) 758-0169, (1).

Wayne, Michele, WAYNE, MICHELE, Box 266, Terre Hille, PA, 17581, (215) 745-0129, (1).

Wayne, Jr., Charles R., RADON DIAGNOSTICS, Box 761, Andover, NJ, 07821, (201) 729-1251, (1)(2).

Waynick, Jackie, TENNESSEE DEPARTMENT OF HEALTH AND ENVIRONMENT/DIVISION OF AIR POLLUTION CONTROL, 701 Broadway 4th Floor, Nashville, TN, 37219-5403, (615) 741-4634, (615) 741-3931, (10).

Weaver, Kenneth, MARYLAND GEOLOGICAL SURVEY, 2300 St. Paul St., Baltimore, MD, 21218, (301) 554-5500, (10).

Weaver, Steven, RADON CONTROL PRODUCTS, 511 Industrial Dr., Carmel, IN, 46032, (317) 846-7442, (800) 523-2082, (4).

Weaver, Steven, WEAVER WERKS, 511 Industrial Dr., Carmel, IN, 46032, (317) 846-7442, (800) 523-2082, (4).

Weaver, William, THREE W RADON SERVICES, INC., 1619 Springhill Dr., Hummelstown, PA, 17036, (717) 566-0499, (1).

Webb, Jim, C.R. LAURENCE, 2503 East Vernon, Los Angeles, CA, 90058, (800) 421-6144, (4).

Webb, Larry, NUCLEUS, THE, P.O. Box 2561, Oak Ridge, TN, 37830-2561, (615) 482-4041, (1)(4).

Webb, Sparty M., CLEVELAND COUNTY HEALTH DEPARTMENT/ENVIRONMENTAL HEALTH DIVISION, 315 Grover St., Shelby, NC, 28150, (704) 487-1138, (10).

Weber, Brent, W.R. MEADOWS, INC., P.O. Box 543, Elgin, IL, 60121, (312) 683-4500, (4).

Weber, Charles E., CCI-RADON AND HOME INSPECTION/RADON DETECTION, P.O. Box 1045, Malvern, PA, 19355, (215) 889-1554, (215) 363-9015, (1)(2)(4).

Weber, Karl, RADON DETECTION SPECIALISTS, 7966 South Madison St., Burr Ridge, IL, 60521, (312) 325-4443, (1).

Weber, Ronald E., WEBER AND ASSOCIATES, 8657 Fallbrook Way, Sacramento, CA, 95826, (916) 386-8745, (1).

Webster, Stephen, SASKATCHEWAN DEPARTMENT OF HUMAN RESOURCES, LABOUR AND EMPLOYMENT, Saskatchewan Place 1870 Albert St., Regina, SK, S4P 3V4, (306) 787-4486, (11).

Webster, Stephen A., U.S. RADON TESTING/REDUCTION OF MARYLAND, P.O. Box 3204, Gaithersburg, MD, 20878, (301) 330-4187, (1).

Weeden, Paul, CONFERENCE OF RADIATION CONTROL PROGRAM DIRECTORS (CRCPD)/COMMITTEE ON RADON POLICY, 71

Fountain Pl., Frankfort, KY, 40601, (502) 227-4543, (14).

Weeden, Paul, CONFERENCE OF RADIATION CONTROL PROGRAM DIRECTORS (CRCPD)/TASK FORCE FOR ADEQUATE RADIATION CONTROL PROGRAM-RADON, 71 Fountain Pl., Frankfort, KY, 40601, (502) 227-4543, (14).

Weeden, Paul J., ARIZONA RADIATION REGULATORY AGENCY/ENVIRONMENTAL SURVEILLANCE LABORATORY, 4814 South 40th St., Phoenix, AZ, 85040, (602) 255-4845, (1)(10).

Weel, Judy, AMERICAN SOCIETY OF SAFETY ENGINEERS (ASSE), 1800 East Oakton, Des Plaines, IL, 60018, (312) 692-4121, (16).

Weems, Bill, ALABAMA OCCUPATIONAL SAFETY AND HEALTH CONSULTATION PROGRAM, P.O. Box 6005, University, AL, 35486, (205) 348-7136, (10).

Weidlein, John L., NATIONAL HOMEOWNERS ASSOCIATION (NHA), 1906 Sunderland Pl., NW, Washington, DC, 20036, (202) 223-1453, (18).

Weiffenbach, Conrad, CORNELL COLLEGE, First St. West, Mt. Vernon, IA, 52314, (319) 895-8811, (1).

Weinert, Donald G., NATIONAL SOCIETY OF PROFESSIONAL ENGINEERS (NSPE), 1420 King St., Alexandria, VA, 22314, (703) 684-2800, (13).

Weinress, Stephen, L.H. FRIEND AND COMPANY, 1875 Century Park, E., Los Angeles, CA, 90210, (213) 556-5488, (800) 351-9923, (3).

Weisenbaum, Jack, ERIE BOARD OF COOPERATIVE EDUCATIONAL SERVICES (BOCES), 2 Pleasant Ave., West Lancaster, NY, 14086, (716) 686-2094, (10).

Weiss, David, NATIONAL ASSOCIATION OF REALTORS (NAR)/GOVERNMENT AFFAIRS OFFICE, 777 14th St., NW, Washington, DC, 20005, (202) 383-1087, (18).

Weitzel, Leon A., LAKE COUNTY GENERAL HEALTH DISTRICT/AIR POLLUTION CONTROL DIVISION, 105 Main St. P.O. Box 490, Painesville, OH, 44077, (216) 357-2543, (10).

Welch, Ashton, NEW HAMPSHIRE ASSOCIATION, 115 A. Airport Rd. Box 515A, Concord, NH, 03301, (603) 225-5549, (18).

Welch, Richard, MERCED COUNTY AIR POLLUTION CONTROL DISTRICT, 385 East13th St. P.O. Box 471, Merced, CA, 95341, (209) 385-7391, (10).

Welke, Rick, IOWA DEPARTMENT OF PUBLIC HEALTH/DIVISION OF DISEASE PREVENTION/BUREAU OF RADIOLOGICAL HEALTH, Lucas State Office Building, Des Moines, IA, 50319, (515) 281-7781, (10).

Welling, William, GROUNDWATER TECHNOLOGY, INC., 12 Walker Way, Albany, NY, 12205, (518) 456-2444, (1)(2).

Wenz, Harry W., CONNECTICUT ASSOCIATION OF REALTORS, 316 Farmington Ave., Hartford, CT, 06105, (203) 522-7255, (18).

Wenz, Thomas W., HAMILTON COUNTY-SOUTHWESTERN OHIO/AIR POLLUTION CONTROL AGENCY, 2400 Beekman St., Cincinnati, OH, 45214, (513) 251-8777, (10).

Werner, Erhard, ASAFE-HOUSE, INC., 260 Litchfield Rd., New Preston, CT, 06777, (203) 868-0512, (1).

Werner, Jack, DISTRICT OF COLUMBIA/ENERGY OFFICE, 613 G St., NW Suite 500, Washington, DC, 20001, (202) 724-2100, (10).

Werner, James M., AMERICAN IMPROVEMENT COMPANY, 2101 South Branch Ave., Altoona, PA, 16602, (814) 946-1664, (800) 922-5857, (2).

Werner, Norman E., RADON HOME PROTECTION SERVICES, 48 Woodruff Ave., Thomaston, CT, 06787, (203) 283-4078, (1).

Werner, Pete, RADON REDUCTION EQUIPMENT OF COLORADO, 921 Main Suite C, Louisville, CO, 80027, (303) 665-1776, (4)(6).

Werner, Robert F., RADON TESTING OF MICHIANA, Colfax Ctr. 914 Lincolnway W., South Bend, IN, 46616, (219) 282-2435, (1).

Werper, Larry, PUTNAM COUNTY HEALTH DEPARTMENT, Putnam County Office Building 110 Old Route Six Center Building #3, Carmel, NY, 10512, (914) 225-2115, (10).

Weshnak, Ron, GARDEN STATE RADON DEWATERING SERVICES, INC., 2 Edgewater Pl. Suite 1A, New Brunswick, NJ, 08901, (201) 846-1948, (1)(2).

Wesselhoft, Prudence, U.S. VIRGIN ISLANDS DEPARTMENT OF LICENSING AND CONSUMER SERVICES, Sub Base #1 Property and Procurement Building, St. Thomas, VI, 00802, (809) 774-3130, (10).

Wessinger, Hugh J., U.S. GENERAL ACCOUNTING OFFICE (GAO)/ENVIRONMENTAL PROGRAMS(EP), 441 G St. NW, Washington, DC, 20548, (202) 275-5489, (8).

West, Charles, WEST ENVIRO SYSTEMS, INC., 45 Sycamore Ln., Skillman, NJ, 08558, (609) 924-0680, (1)(2).

West, Dan, INSUL-TECH, INC., 5559 Westerville Rd., Westerville, OH, 43081, (614) 891-9555, (1)(2).

West, Fred W., WEST ENVIRO SYSTEMS, INC., 45 Sycamore Ln., Skillman, NJ, 08558, (609) 924-0680, (1)(2).

West, H.E., WEST ENVIRO SYSTEMS, INC., 45 Sycamore Ln., Skillman, NJ, 08558, (609) 924-0680, (1)(2).

West, Philip, GRANITE STATE ANALYTICAL, INC., 16 Manning St. Unit 107, Derry, NH, 03038, (603) 432-3044, (1)(4).

West, Raymond E., MERIDEN DEPARTMENT OF HEALTH/HUMAN SERVICES, 165 Miller St., Meriden, CT, 06450, (203) 634-0003, (10).

West, Tad, SOUTHWEST ABATEMENT, INC., 4200 Broadway, SE, Albuquerque, NM, 87102, (505) 873-2957, (1).

Westby, Steve, RADON DETECTION SYSTEMS/COLORADO SPRINGS, COLORADO BRANCH, 2010 Wooten Ln., Colorado Springs, CO, 80907, (303) 444-5253, (800) 627-2366, (1).

Westdal, Harold, DRANJER CORPORATION, 1441 Pembina Hwy., Winnipeg, MB, R3T 2C4, (204) 474-0451, (4).

Westfall, Carl, RADON SAFE, P.O. Box 631 Highway 62 W., Boonville, IN, 47601, (812) 897-1700, (1).

Westrick, James, U.S. ENVIRONMENTAL PROTECTION AGENCY (EPA)/OFFICE OF WATER(OW)/OFFICE OF DRINKING WATER(ODW)/TECHNICAL SUPPORT DIVISION(TSD), Water Supply Technology Branch 26 West Martin Luther King Dr., Cincinnati, OH, 45268, (513) 569-7919, (8).

Whalen, Dan, SUNDANCE ENTERPRISES, P.O. Box 3607, Dillon, CO, 80435, (303) 468-3911, (303) 468-8369, (1).

Wheeland, Jeff C., RAD-LAB SERVICES, INC., 1311 Washington Blvd., Williamsport, PA, 17701, (717) 326-2628, (4).

Wheeler, Robert, TECH/OPS LANDAUER, INC., 2 Science Rd., Glenwood, IL, 60425-1586, (312) 755-7000, (4).

Wheeler, Shirley, KANSAS REAL ESTATE COMMISSION, Landon State Office Building 900 Jackson St. Room 501, Topeka, KS, 66612-1220, (913) 296-3411, (10).

Wherley, Bill, SAFE-AIRE, INC., 162 East Chestnut St. P.O. Box 160, Canton, IL, 61520, (309) 647-0419, (800) 331-2943, (1)(2)(3)(4)(6).

Whicker, F. W., COLORADO STATE UNIVERSITY/DEPARTMENT OF RADIOLOGY AND RADIATION BIOLOGY, Ft. Collins, CO, 80523, (303) 491-5222, (5).

Whigham, William, DELAWARE DEPARTMENT OF LABOR/DIVISION OF INDUSTRIAL AFFAIRS, 820 North French St. 6th Fl., Wilmington, DE, 19801, (302) 571-3908, (10).

Whitaker, T.J., U.S. DEPARTMENT OF ENERGY (DOE)/RICHLANDS OPERATIONS OFFICE(ROO)/PACIFIC NORTHWEST LABORATORY(PNL)/CHEMICAL SCIENCE DEPARTMENT, Atomic And Molecular Chemistry Section M/S P7-07, Richland, WA, 99352, (509) 376-3294, (8).

White, Charles J., ALA OF PENNSYLVANIA, P.O. Box 4213, Harrisburg, PA, 17111-0213, (717) 564-4850, (15).

White, Gregory, UNIVERSITY OF MAINE-ORONO/LAND AND WATER RESOURCES CENTER, 11 Coburn Hall, Orono, ME, 04469, (207) 581-1490, (5).

White, Lorenzo, DISTRICT OF COLUMBIA DEPARTMENT OF EMPLOYMENT SERVICES/OCCUPATIONAL SAFETY AND HEALTH DIVISION, 950 Upshur St., NW, Washington, DC, 20011, (202) 576-6651, (10).

White, Ron, AMERICAN LUNG ASSOCIATION (ALA)/NATIONAL HEADQUARTERS, 1740 Broadway, New York, NY, 10019, (212) 315-8700, (15).

White, William F., ALA OF FLORIDA, P.O. Box 8127, Jacksonville, FL, 32239-8127, (904) 743-2933, (15).

Wible, Robert C., NATIONAL CONFERENCE OF STATES ON BUILDING CODES AND STANDARDS (NCSBCS), 481 Carlisle Dr., Herndon, VA, 22070, (703) 437-0100, (13).

Wiblin, Claude, KEY TECHNOLOGY, INC., P.O. Box 562, Jonestown, PA, 17038, (717) 274-8310, (800) 523-4964, (1)(4)(6).

Wicks, Doug, CONSERVATION ENERGY SYSTEMS, P.O. Box 10146, Minneapolis, MN, 55458, (306) 242-3663, (800) 667-3717, (4).

Widmayer, Michele, RADON SYSTEMS, INC., P.O. Box 2509, Laurel, MD, 20708, (301) 725-9000, (1)(2).

Widmer, Janet M., ALA OF DISTRICT OF COLUMBIA, 475 H St., NW, Washington, DC, 20001, (202) 682-5864, (15).

Wiebers, Kathleen, ALA OF SOUTH DAKOTA, 208 East 13th St., Sioux Falls, SD, 57102, (605) 336-7222, (15).

Wiedmann, Donald A., CALIFORNIA ASSOCIATION OF REALTORS, P.O. Box 76917, Los Angeles, CA, 90067, (213) 739-8201, (18).

Wiggers, Ken, RADON DETECTION SYSTEMS/DES MOINES, IOWA BRANCH, Box 93 East 57 Rd., Kelly, IA, 50134, (303) 444-5253, (800) 627-2366, (1).

Wiggers, Kenneth D., AMERICAN RADON SERVICES, LTD., 303 I.S.I.S. Center Iowa State University, Ames, IA, 50011-2130, (515) 296-9926, (1)(2).

Wiggers, Kenneth D., EMMI-RADON CHECK LABORATORY, E-57 Rd. Box 93, Kelley, IA, 50134-8724, (515) 769-2400, (1).

Index of Organizations

This section lists all companies, organizations, and associations listed in the Directory. Each entry is followed by the appropriate address, telephone number, and the sections where the listing can be located. Please note that Section 8 is not included in this index.

A

Index of Organizations

AMERICAN NUCLEAR SOCIETY/KANSAS-KANSAS SECTION, MS6-02 P.O. Box 2908, Wichita, KS, 67201, (316) 261-6799, (14).

AMERICAN NUCLEAR SOCIETY/LOUISIANA-LOUISIANA SECTION, C/O GSU-River Bend P.O. Box 220, St. Francisville, LA, 70775, (504) 381-4622, (14).

AMERICAN NUCLEAR SOCIETY/MAINE, MASSACHUSETTS, NEW HAMPSHIRE, RHODE ISLAND, VERMONT-NORTHEASTERN SECTION, 32 Knight Rd., Framingham, MA, 01701, (617) 872-8100, (14).

AMERICAN NUCLEAR SOCIETY/MICHIGAN-MICHIGAN SECTION, 5340 Deerridge Rd., Jackson, MI, 49201, (14).

AMERICAN NUCLEAR SOCIETY/MINNESOTA-UPPER MIDWEST SECTION, C/O Northern States Power 414 Nicollet Mall, Minneapolis, MN, 55401, (612) 330-6341, (14).

AMERICAN NUCLEAR SOCIETY/MISSISSIPPI-MISSISSIPPI SECTION, 110 Bottino Dr., Vicksburg, MS, 39180-9721, (601) 437-6214, (14).

AMERICAN NUCLEAR SOCIETY/MISSOURI-CENTRAL/EASTERN MISSOURI SECTION, C/O University Of Missouri 1026 Engineering Complex, Rolla, MO, 65211, (314) 882-3550, (14).

AMERICAN NUCLEAR SOCIETY/MISSOURI-MISSOURI/KANSAS SECTION, C/O Kansas City Power And Light P.O. Box 670, Kansas City, MO, 64141, (816) 556-2942, (14).

AMERICAN NUCLEAR SOCIETY/NEBRASKA-NEBRASKA SECTION, C/O Omaha Public Power District 1623 Harney St., Omaha, NE, 68102, (402) 536-4468, (14).

AMERICAN NUCLEAR SOCIETY/NEVADA-NEVADA SECTION, 7164 South Abbeyville Dr., Las Vegas, NV, 89119, (702) 794-7639, (14).

AMERICAN NUCLEAR SOCIETY/NEW JERSEY-NORTHERN NEW JERSEY SECTION, C/O William B. Johnson And Associates P.O. Box 98 Research Park, Montville, NJ, 07045, (201) 334-9222, (14).

AMERICAN NUCLEAR SOCIETY/NEW MEXICO-TRINITY SECTION, C/O Public Service Company Of New Mexico Alvarado Square, Albuquerque, NM, 87158, (505) 848-2574, (14).

AMERICAN NUCLEAR SOCIETY/NEW YORK-METROPOLITAN NEW YORK SECTION, C/O Con-Edison Company 4 Irving Pl. Room 1014, New York, NY, 10003, (212) 460-6179, (14).

AMERICAN NUCLEAR SOCIETY/NEW YORK-LONG ISLAND SECTION, C/O Brookhaven National Laboratory Building 475 B, Upton, NY, 11973, (516) 282-2146, (14).

AMERICAN NUCLEAR SOCIETY/NEW YORK-NIAGARA FINGER LAKES SECTION, C/O New York Power Authority P.O. Box 41, Lycoming, NY, 13093, (416) 449-6706, (14).

AMERICAN NUCLEAR SOCIETY/NEW YORK-NORTHEASTERN NEW YORK SECTION, C/O Knolls Atomic Power Laboratory P.O. Box 1072 Building E6, Schenectady, NY, 12301, (518) 393-6611, (14).

AMERICAN NUCLEAR SOCIETY/NORTH CAROLINA-EASTERN CAROLINAS SECTION, C/O Carolina Power And Light P.O. Box 1551, Raleigh, NC, 27602, (919) 836-6628, (14).

AMERICAN NUCLEAR SOCIETY/NORTH CAROLINA-PIEDMONT CAROLINAS SECTION, C/O Duke Power Comapny Nuclear Production P.O. Box 33189, Charlotte, NC, 28242, (704) 373-4504, (14).

AMERICAN NUCLEAR SOCIETY/OHIO-NORTHERN OHIO SECTION, C/O Harshaw/Filtrol 14450 Summerfield Rd., University Heights, OH, 44118, (216) 349-6590, (14).

AMERICAN NUCLEAR SOCIETY/OHIO-SOUTHWESTERN OHIO SECTION, 2520 Westmont Blvd., Columbus, OH, 43221, (614) 422-0445, (14).

AMERICAN NUCLEAR SOCIETY/OREGON-OREGON SECTION, C/O Portland General Electric 121 Southwest Salmon St., Portland, OR, 97204, (503) 226-8126, (14).

AMERICAN NUCLEAR SOCIETY/PENNSYLVANIA-DELAWARE VALLEY SECTION, 6327 Fairfield Dr., Flourtown, PA, 19031, (215) 327-1200, (14).

AMERICAN NUCLEAR SOCIETY/PENNSYLVANIA-CENTRAL PENNSYLVANIA SECTION, 447 Lovell Ct., Hummelstown, PA, 17036, (717) 948-1043, (14).

AMERICAN NUCLEAR SOCIETY/PENNSYLVANIA-NORTHERN PENNSYLVANIA SECTION, 1326 Brassie St., Wescosville, PA, 18106, (215) 770-7893, (14).

AMERICAN NUCLEAR SOCIETY/PENNSYLVANIA-PITTSBURGH SECTION, C/O Westinghouse Electric Corporation P.O. Box 355, Pittsburgh, PA, 15230-0355, (412) 374-3632, (14).

AMERICAN NUCLEAR SOCIETY/SOUTH CAROLINA-COLUMBIA SECTION, 335 Tram Rd., Columbia, SC, 29210, (803) 345-5209, (14).

AMERICAN NUCLEAR SOCIETY/SOUTH CAROLINA-SAVANNAH RIVER SECTION, C/O Georgia Power Route 2 Box 299A, Waynesboro, GA, 30830, (404) 554-9961, (14).

AMERICAN NUCLEAR SOCIETY/TENNESSEE-CHATTANOOGA SECTION, 7408 Twinbrook Dr., Chattanooga, TN, 37421, (615) 751-2067, (14).

AMERICAN NUCLEAR SOCIETY/TENNESSEE-OAK RIDGE/KNOXVILLE SECTION, C/O Oak Ridge National Laboratory P.O. Box X, Oak Ridge, TN, 37831-6089, (615) 574-6852, (14).

AMERICAN NUCLEAR SOCIETY/TEXAS-NORTH TEXAS SECTION, C/O TU Electric Box 2300, Glen Rose, TX, 76043, (817) 897-5688, (14).

AMERICAN NUCLEAR SOCIETY/TEXAS-SOUTH TEXAS SECTION, C/O Houston Lighting And Power Company STP-5400 P.O. Box 1700, Houston, TX, 77001, (713) 993-1335, (14).

AMERICAN NUCLEAR SOCIETY/VIRGINIA-VIRGINIA SECTION, 10108 Heritage Ln., Glen Arlen, VA, 23060, (804) 771-4595, (14).

AMERICAN NUCLEAR SOCIETY/WASHINGTON-EASTERN WASHINGTON SECTION, C/O Westinghouse Hanford Company P.O. Box 1970, Richland, WA, 99352, (509) 376-4953, (14).

AMERICAN NUCLEAR SOCIETY/WASHINGTON-PUGET SOUND SECTION, University Of Washington/Nuclear Engineering Department, Seattle, WA, 98105, (206) 543-2754, (14).

AMERICAN NUCLEAR SOCIETY/WISCONSIN-WISCONSIN SECTION, C/O Wisconsin Power And Light P.O. Box 192, Madison, WI, 53701-0192, (608) 252-3345, (14).

AMERICAN OCCUPATIONAL MEDICAL ASSOCIATION (AOMA), 2340 South Arlington Heights Rd., Arlington Heights, IL, 60005, (312) 228-6850, (16).

AMERICAN PEST MANAGEMENT, 3929 Martin Way Suite D, Olympia, WA, 98506, (206) 491-1809, (1)(4).

AMERICAN PUBLIC HEALTH ASSOCIATION (APHA), 1015 15th St., NW, Washington, DC, 20005, (202) 789-5663, (15).

AMERICAN RADON ASSOCIATION (ARA), 4949 South 25-A, Tipp City, OH, 45371, (513) 667-2464, (7).

AMERICAN RADON CONSULTANTS, INC., 109 Ladbrook Rd., Knoxville, TN, 37922, (615) 891-9581, (1).

AMERICAN RADON CONTROL, INC., 2758 S. Langley Ct., Denver, CO, 80210, (303) 758-6580, (1).

AMERICAN RADON CORPORATION, 8300 North Hayden Rd. Suite 100, Scottsdale, AZ, 85258, (602) 967-8029, (1)(2)(6).

AMERICAN RADON DETECTION COMPANY, INC., 39 Folcroft Ave., Folcroft, PA, 19032, (215) 532-8845, (1).

AMERICAN RADON SERVICES, LTD., 303 I.S.I.S. Center Iowa State University, Ames, IA, 50011-2130, (515) 296-9926, (1)(2).

AMERICAN RADON TECHNICAL SERVICES, INC., 1941A Nears Pkwy., Margate, FL, 33063, (305) 968-5555, (1)(2).

AMERICAN RADON TESTERS, 5705 Center Ln. Office 1, Falls Church, VA, 22041, (703) 239-0227, (1).

AMERICAN RADON TESTING LABORATORIES, 1015 South Birch St. #610, Glendale, CO, 80222, (1).

AMERICAN RADON TESTING LABS, INC., 21360 Center Ridge Rd. Suite 307, Rocky River, OH, 44116, (216) 331-0200, (1).

AMERICAN RADON TESTING SERVICES, 2302 Basswood Cr., Tustin, CA, 92680, (714) 838-4990, (1).

AMERICAN RADON TESTING SERVICES, INC., P.O. Box 70, Vandalia, OH, 45377, (513) 898-1949, (1).

AMERICAN RADON TESTING, INC., P.O. Box 60, Trexlertown, PA, 18087, (215) 778-2167, (1).

AMERICAN REAL ESTATE INSPECTIONS, P.O. Box 61, Blairstown, NJ, 07825, (201) 850-7435, (1).

AMERICAN SAMOA/ENVIRONMENTAL QUALITY COMMISSION/GOVERNOR'S OFFICE, Pago Pago, American Samoa, AS, 96799, (684) 633-4116, (684) 633-2304, (10).

AMERICAN SOCIETY FOR TESTING AND MATERIALS (ASTM), 1916 Race St., Philadelphia, PA, 19103, (215) 299-5571, (14).

AMERICAN SOCIETY OF HEATING, REFRIGERATING AND AIR CONDITIONING ENGINEERS (ASHRAE), 1791 Tullie Circle Rd., NE, Atlanta, GA, 30329, (404) 636-8400, (13).

AMERICAN SOCIETY OF HOME INSPECTORS (ASHI), 3299 K St., NW Suite 700, Washington, DC, 20007, (202) 842-3096, (18).

AMERICAN SOCIETY OF REAL ESTATE COUNSELORS (ASREC), 430 North Michigan Ave., Chicago, IL, 60611, (312) 329-8429, (18).

AMERICAN SOCIETY OF SAFETY ENGINEERS (ASSE), 1800 East Oakton, Des Plaines, IL, 60018, (312) 692-4121, (16).

AMERICAN WATER WORKS ASSOCIATION (AWWA), 6666 West Quincy Ave., Denver, CO, 80235, (303) 794-7711, (14).

AMERICAN WATER WORKS ASSOCIATION RESEARCH FOUNDATION (AWWARF), 6666 West Quincy Ave., Denver, CO, 80235, (303) 794-7711, (5)(14).

AMES ASSOCIATES, INC., 6935 Wisconsin Ave., Chevy Chase, MD, 20815, (301) 656-5222, (1).

ANALYTICAL MANAGEMENT, INC., P.O. Box 11279, Lexington, KY, 40574, (606) 233-7774, (1).

ANCHORAGE AIR POLLUTION CONTROL AGENCY, 825 L St., Anchorage, AK, 99503, (907) 343-4713,

(10).

ANDERSON AIR POLLUTION CONTROL DEPARTMENT, P.O. Box 2100, Anderson, IN, 46018, (317) 646-9835, (10).

ANGELO, MARC, 110 Woods Ln., Radnor, PA, 19087, (215) 687-6183, (1).

ANNAPOLIS PROPERTY SERVICES, 1680 Yorktown Ct., Crofton, MD, 21114, (301) 721-7177, (1).

ANNE ARUNDEL COUNTY DEPARTMENT OF HEALTH/DIVISON OF ENVIRONMENTAL HEALTH/AIR QUALITY CONTROL SECTION, 3 Harry S. Truman Pkwy., Annapolis, MD, 21401, (301) 244-7360, (10).

ANTI-PEST COMPANY, P.O. Box 8082, Longview, TX, 75607, (214) 758-0169, (1).

ANTI-PEST COMPANY OF SHREVEPORT, P.O. Box 4266, Shreveport, LA, 71104, (318) 221-6181, (1).

APPALACHIAN ENVIRONMENTAL TESTING, INC., 105 South Union St. Suite 326, Danville, VA, 24541, (804) 792-1300, (1).

APPLIED ANALYTICAL TECHNOLOGIES, INC., P.O. Box 668, Narragansett, RI, 02882, (401) 782-8188, (1).

APPLIED ANTENNA TECHNOLOGIES/RADON MEASUREMENTS DIVISION, 2270 Honeycomb Ct., Duluth, GA, 30136, (404) 623-0403, (1).

APPLIED CONSTRUCTION, INC., The Friary, Winchester-on-Severn 1600 Winchester Rd., Annapolis, MD, 21401, (301) 757-7887, (2).

APPLIED ENERGY COMPANY, 3316 Augusta Ave., Omaha, NE, 68144, (402) 334-5881, (1).

APPLIED EXTRUSION TECHNOLOGIES, INC., P.O. Box 582, Middletown, DE, 19709, (302) 378-8888, (800) 521-6713, (4).

APPLIED HEALTH PHYSICS, INC., 2986 Industrial Blvd., Bethel Park, PA, 15102, (412) 563-2242, (518) 477-7974, (1)(2)(4)(6).

APPLIED RADON, INC., 52205 Warehime Rd., Cumberland, OH, 43732-9804, (614) 638-2078, (800) 638-2078, (1)(2)(4).

APPLIED TECHNICAL SERVICES, INC., 1190 Atlanta Industrial Dr., Marietta, GA, 30066, (404) 423-1400, (800) 451-3405, (1)(4).

APPLIED ULTRASONICS, INC., 145 C Grassy Plains St., Bethel, CT, 06801, (203) 790-1102, (4).

AQUATEK LABORATORY/DIVISION OF GRAHAM-MASSEY ANALYTICAL LABORATORY, 2212 Main St., Bridgeport, CT, 06606, (203) 576-0104, (1).

ARAPAHOE HOMES, 4065 North Sinton Street Suite 200, Colorado Springs, CO, 80907, (719) 593-2666, (13).

ARDEX, INC., 630 Stoops Ferry Rd., Corapolis, PA, 15108, (4).

AREA HOME INSPECTIONS, INC., 341 Littleton Rd., Harvard, MA, 01451, (617) 263-7251, (1).

ARGONNE NATIONAL LABORATORY, 9700 South Cass Ave., Argonne, IL, 60439, (312) 972-4172, (1).

ARID BASEMENT COMPANY, 237 East Utica, Buffalo, NY, 14208, (716) 885-2162, (2).

ARIX CORPORATION, 325 East Southern Ave. Suite 18, Tempe, AZ, 85282, (602) 968-0052, (1)(2).

ARIX CORPORATION, 760 Horizon Dr., Grand Junction, CO, 81506, (303) 242-6203, (1)(2).

ARIX CORPORATION, P.O. Box 2021, Greeley, CO, 80632-2021, (303) 356-3930, (1)(2).

ARIX CORPORATION, 2155 North 200 West Suite 100, Provo, UT, 84604, (801) 373-2749, (1)(2).

ARIX CORPORATION, 877 North Eighth, W. Suite 8, Riverton, WY, 82501, (307) 858-6506, (1)(2).

ARIX CORPORATION/A DIVISION OF VERSAR, INC., 1200 East High St., Pottstown, PA, 19464, (215) 327-1140, (1)(2)(6).

ARIZONA ASSOCIATION OF REALTORS, 4414 North 19th Ave. Suite R, Phoenix, AZ, 85015, (602) 248-7785, (18).

ARIZONA BUREAU OF GEOLOGY AND MINERAL TECHNOLOGY, 845 North Park Ave., Tucson, AZ, 85719, (602) 621-7906, (10).

ARIZONA DEPARTMENT OF REAL ESTATE, 202 East Earll Dr., Phoenix, AZ, 85012, (602) 255-3481, (10).

ARIZONA DIVISION OF OCCUPATIONAL SAFETY AND HEALTH/CONSULTATION AND TRAINING, P.O. Box 19070 1624 West Adams, Phoenix, AZ, 85005, (602) 255-5795, (10).

ARIZONA LUNG ASSOCIATION, 102 West McDowell Rd., Phoenix, AZ, 85003, (602) 258-7505, (15).

ARIZONA RADIATION REGULATORY AGENCY, 4814 South 40th St., Phoenix, AZ, 85040, (602) 255-4845, (10).

ARIZONA RADIATION REGULATORY AGENCY/ENVIRONMENTAL SURVEILLANCE LABORATORY, 4814 South 40th St., Phoenix, AZ, 85040, (602) 255-4845, (1)(10).

ARIZONA STATE UNIVERSITY/COLLEGE OF ENGINEERING AND APPLIED SCIENCES/DEPARTMENT OF ELECTRICAL AND COMPUTER ENGINEERING, Department Of Electrical And Computer Engineering, Radiological Imaging Facility, Tempe, AZ, 85287-5706, (602) 965-2727, (5).

ARKANSAS DEPARTMENT OF HEALTH/RADIATION CONTROL AND EMERGENCY MANAGEMENT/NUCLEAR AND ENVIRONMENTAL SAFETY SECTION, 4815 West Markham St., Little Rock, AR, 72201, (501) 661-2301, (10).

ARKANSAS DEPARTMENT OF LABOR/OSHA CONSULTATION, 1022 High St., Little Rock, AR, 72202, (501) 375-8442, (10).

ARKANSAS GEOLOGICAL COMMISSION, 3815 West Roosevelt Rd., Little Rock, AR, 72204, (501) 371-1488, (10).

ARKANSAS REAL ESTATE COMMISSION, One Riverfront Pl. Suite 660, North Little Rock, AR, 72114, (501) 371-1247, (10).

ARKANSAS REALTORS ASSOCIATION, Realtors Building 201 Executive Ct., Suite 100, Little Rock, AR, 72205, (602) 248-7785, (18).

ARTHUR SCOTT AND ASSOCIATES/AMERICAN ATCON, 2150 Burnhamthorpe Rd., W. Box 164, Mississauga, ON, L5L 3A2, (416) 828-2389, (2)(3)(5).

ARTHUR W. MATTES, INC., 134 Washington Ave., Bethlehem, PA, 18018, (215) 867-5268, (2).

ASAFE-HOUSE, INC., 260 Litchfield Rd., New Preston, CT, 06777, (203) 868-0512, (1).

ASBESTOS AND RADON CONSULTANTS QUALI-PRO-TECH SERVICES, INC., 204 Devon Dr., Exiom, PA, 19341, (215) 363-7268, (1).

ASHRAE/ENVIRONMENTAL HEALTH COMMITTEE/INDOOR AIR QUALITY SUBCOMMITTEE, 1791 Tullie Cr., NE, Atlanta, GA, 30329, (404) 636-8400, (13).

ASSAY SERVICES, P.O. Box 1266, Friendswood, TX, 77546, (713) 641-0391, (1).

ASSOCIATED BUILDERS OF THE ROCKIES, INC., P.O. Box 704, Wheat Ridge, CO, 80034, (303) 237-3428, (1).

ASSOCIATED DESERT DRY WATERPROOFING COMPANY, INC., 504 South Van Dorn St., Alexandria, VA, 22304, (703) 360-7100, (2).

ASSOCIATED PROTECTIVE SYSTEMS, 2511 Winford Ave., Nashville, TN, 37211, (615) 244-7637, (615) 244-7638, (1).

ASSOCIATED RADON TESTING SERVICES, INC., 303 South Ave., Media, PA, 19063, (215) 566-2274, (1).

ASSOCIATION FOR IMPROVEMENT IN MOHEGAN, INC., East Main St. P.O. Box 103, Mohegan Lake, NY, 10547, (914) 528-3300, (1)(6)(17).

ASSOCIATION OF AMERICAN STATE GEOLOGISTS (AASG), C/O Louisiana Geological Survey P.O. Box G, Baton Rouge, LA, 70893, (504) 342-6754, (14).

ASSOCIATION OF GROUND WATER SCIENTISTS AND ENGINEERS (AGWSE), 6375 Riverside Dr., Dublin, OH, 43017, (614) 761-1711, (14).

ASSOCIATION OF MAJOR CITY BUILDING OFFICIALS (AMCBO), 481 Carlisle, Herndon, VA, 22070, (703) 437-0100, (13).

ASSOCIATION OF STATE AND INTERSTATE WATER POLLUTION CONTROL ADMINISTRATORS (ASIWPCA), 444 North Capitol St. Suite 330, Washington, DC, 20001, (202) 624-7782, (14).

ASSOCIATION OF STATE AND TERRITORIAL HEALTH OFFICIALS (ASTHO), 1311 Dolly Madison Blvd., McLean, VA, 22101, (703) 556-9222, (15).

ASSOCIATION OF STATE DRINKING WATER ADMINISTRATORS (ASDWA), 1911 North Fort Myer Dr., Arlington, VA, 22209, (703) 524-2428, (14).

ASTECO, INC., P.O. Box 2204, Niagara Univeristy, NY, 14109, (716) 297-5992, (1).

ASTLE CORPORATION, 146 Old Farm Rd., Leominister, MA, 01453-2029, (508) 534-0979, (1)(2)(4).

ATESYS, INC., 1 Pheasant Ln., Westport, CT, 06880, (203) 261-0528, (203) 226-5173, (1).

ATLANTA TESTING AND ENGINEERING, INC., 11420 Johns Creek Pkwy., Duluth, GA, 30136, (404) 476-3555, (1).

ATLANTIC ELECTRIC, 65 W. Highland Ave., Atlantic Highlands, NJ, 07716, (201) 291-3468, (1).

ATLANTIC ENVIRONMENTAL LABORATORY, 16 Elm St., West Haven, CT, 06516, (203) 932-5808, (1).

ATLANTIC NUCLEAR, 6 Village Gate Rd., Canton, MA, 02021, (617) 828-9118, (4).

ATLAS WATERPROOFING COMPANY, INC., P.O. Box 14953, Reading, PA, 19612-4953, (215) 374-6454, (1)(2).

ATOMETRICS, 911 Veterans Blvd. Suite 202, Metairie, LA, 70005, (504) 837-1367, (1).

ATOMIC ENERGY CONTROL BOARD (AECB)/TECHNICAL INFORMATION DIVISION/OFFICE OF PUBLIC INFORMATION, P.O. Box 1046, Ottawa, ON, K1P 5S9, (613) 992-9206, (11).

ATOMIC ENERGY OF CANADA, LTD. (AECL)/ADMINISTRATIVE DIVISION/RADIATION INDUSTRIAL SAFETY BRANCH, Tinawa, MB, R0E 1L0, (204) 753-2720, (11).

ATOMIC ENERGY OF CANADA, LTD. (AECL)/WASTE MANAGEMENT/APPLIED GEOSCIENCE/HYDROGEOCHEMISTRY SECTION, Tinawa, MB, R0E 1L0, (204) 753-2311, (11).

AUDUBON CONTROLS, INC., P.O. Box 477, Audubon, PA, 19403, (215) 566-7089, (1).

AUGUSTUS W. SMITH, INC., 1 Short Hills Ave. P.O. Box 164, Short Hills, NJ, 07078, (201) 376-7740, (1).

AURORA INDUSTRIES, P.O. Box 936, Concordville, PA, 19331, (215) 459-1146, (2).

B

B & H RADON SERVICES, P.O. Box 247, Silver Spring, PA, 17575-0247, (717) 684-2612, (717) 426-3723, (1)(2).

B & I SERVICES, P.O. Box 291, Edmond, KS, 67636, (913) 622-4341, (1).

B & W SALES AND MARKETING, P.O. Box 130, Chiefland, FL, 32626, (904) 528-3600, (1).

B & W SALES AND MARKETING CORPORATION, 4045-A Jonesboro Rd., Forest Park, GA, 30050, (404) 366-0957, (800) 331-1449, (1).

B-DRY SYSTEM OF DELAWARE AND CHESTER COUNTY, INC., 1054 Saunders Ln., West Chester, PA, 19380, (215) 430-0393, (1).

BAIN PEST CONTROL SERVICE, INC., 1320 Middlesex St., Lowell, MA, 01851, (617) 453-7792, (1).

BALTIMORE COUNTY DEPARTMENT OF ENVIRONMENTAL PROTECTION AND RESOURCE MANAGEMENT, 401 Bosley Ave., Towson, MD, 21204, (301) 494-3775, (10).

BALTIMORE COUNTY HEALTH DEPARTMENT/BUREAU OF INDUSTRIAL HYGIENE/AIR POLLUTION CONTROL, 303 East Fayette St. Room 400, Baltimore, MD, 21202, (301) 396-4427, (10).

BALTIMORE GAS AND ELECTRIC, P.O. Box 1475, Baltimore, MD, 21203, (301) 281-3781, (1)(13).

BAR-TEL INFRASCAN, 748 Kirkland Cir. F-102, Kirkland, WA, 98033, (206) 822-1482, (1).

BARBARA L. SALTHOUSE, 251 Crown St., Stratford, CT, 06497, (203) 375-5837, (1).

BARNES AND CONE, INC., 5894 Court St., Syracuse, NY, 13206, (315) 437-0305, (3).

BARRINGER, 15000 West 6th Ave. Suite 300, Golden, CO, 80401, (303) 277-1687, (1)(4).

BARRINGER MAGENTA, LTD., 304 Carlingview Dr., Rexdale, ON, M9W 5G2, (416) 675-3870, (1).

BASEMENT DE-WATERING SYSTEMS, INC., 156 Curry Hill Rd., Canonsburg, PA, 15317, (412) 745-5959, (1)(2).

BATEMAN HOME IMPROVEMENTS, 17024 Sioux Ln., Gaithersburg, MD, 20878, (301) 330-1528, (2).

BATTELLE MEMORIAL INSTITUTE/NUCLEAR TECHNOLOGY SECTION, 505 King Ave., Columbus, OH, 43201-2693, (614) 879-5192, (1)(5).

BAUBIOLOGIE BY BANTA, P.O. Box 51250 Suite 125, Pacific Grove, CA, 93950, (408) 372-8626, (1).

BAY AREA AIR QUALITY MANAGEMENT DISTRICT, 939 Eillis St., San Francisco, CA, 94109, (415) 771-6000, (10).

BAYSTATE ENVIRONMENTAL, 53 Cushing St., Stoughton, MA, 02072, (617) 341-0002, (2).

BAYSURE BUILDING INSPECTION, INC., 290 King of Prussia Rd., Radnor, PA, 19087, (215) 688-2572, (1).

BBC & M ENGINEERING, 195 Chittenden Ave., Columbus, OH, 43201, (614) 294-3745, (1).

BCA LABORATORIES, INC., 1102 South Main St., Bloomington, IL, 61701, (309) 828-7772, (1)(4).

BCT, 12 Magnolia Ave., Denville, NJ, 07834, (201) 627-2802, (1).

BEAVER INDUSTRIES, 890 Hersey St., St. Paul, MN, 55114, (612) 644-9933, (4).

BECKMAN INSTRUMENTS, INC., 2500 Harbor Blvd., Fullerton, CA, 92634, (714) 871-4848, (1)(4).

BENCHMARK INSPECTION SERVICES, 939 South 26th St., Arlington, VA, 22202, (703) 548-5690, (1)(3).

BENSENVILLE AIR POLLUTION CONTROL DISTRICT, 1253 Roosevelt Ave., Glenview, IL, 60025, (312) 498-9840, (10).

BENTON-FRANKLIN-WALLA WALLA COUNTIES AIR POLLUTION AUTHORITY/AIR POLLUTION CONTROL DIVISION, 650 George Washington Way, Richland, WA, 99352, (509) 545-2354, (10).

BERG GROUP, 1401 West Hamburg, Baltimore, MD, 21230, (301) 837-5525, (1)(2).

BERGEN HOME INSPECTION SERVICE, 838 Palisades Ave., Teaneck, NJ, 07666, (201) 836-3232, (1).

BERKSHIRE AND PIONEER VALLEY AIR POLLUTION CONTROL DISTRICTS, State House West 436 Dwight St., 4th Fl., Springfield, MA, 01103, (413) 785-5327, (10).

BERNELL PROFESSIONAL SERVICES, 361 12th St., Cresskill, NJ, 07626, (201) 568-7083, (1)(3).

BERNER INTERNATIONAL CORPORATION, P.O. Box 5205, New Castle, PA, 16105, (412) 658-1530, (800) 852-5015, (4).

BET, INC., 222 South Rolff, Davenport, IA, 52802, (319) 322-8749, (1).

BGI, INC., 58 Guinan St., Waltham, MA, 02154, (617) 891-9380, (4).

BICRON CORPORATION, 12345 Kinsman Rd., Newbury, OH, 44065, (216) 564-8000, (4).

BIEDERER CONSTRUCTION SERVICE, 421 Hillandale, Round Lake, IL, 60073, (312) 546-3890, (2).

BILL EICH CONSTRUCTION COMPANY, Box 63, Spirit Lake, IA, 51360, (712) 366-4438, (1).

BIO-PURE COMPANY, P.O. Box 20593, Tampa, FL, 33622-0593, (813) 839-8617, (4).

BIOHAZARDOUS MATERIALS CONSULTING, P.O. Box 4376, Ithaca, NY, 14852, (607) 257-5010, (315) 478-0653, (1)(3).

BIOMEDICAL TOXICOLOGY ASSOCIATES, P.O. Box 3568, Frederick, MD, 21701, (301) 662-0783, (1)(3).

BIOMONITORING SERVICES LABORATORY/TECHNICAL RESOURCES, INC., 3202 Monroe St., Rockville, MD, 20852, (301) 231-5250, (1).

BIONOMIC INDUSTRIES, INC., 479 Route 17 N., Mahwah, NJ, 07430, (201) 529-1094, (800) 338-3210, (1)(2).

BIOSOURCES, LTD., 18 Sevilla Rd., Andover, MA, 01810, (508) 698-8707, (1).

BIOSOURCES, LTD., 7 Bailey St., Nashua, NH, 03063, (603) 880-6004, (1).

BIOSPHERICS, INC., 4928 Wyaconda Rd., Rockville, MD, 20852, (301) 770-7700, (1).

BLACKHAWK INDUSTRIES, INC., 607 Park St., Regina, SK, S4N 5N1, (306) 924-1551, (4).

BLOOMINGTON/ENVIRONMENTAL POLLUTION SECTION, 2215 West Old Shakopee Rd., Bloomington, MN, 55431, (612) 881-5811, (10).

BLUE ANGEL PUMP COMPANY, 801 Glasgow Ave. P.O. Box 2257, Fort Wayne, IN, 46801, (219) 426-4000, (4).

BLUEGRASS RADON SERVICES, 424 Ridgeway Rd., Lexington, KY, 40502, (606) 266-6148, (1).

BOHNSACK GROUP, THE, 4329 Tamalga, South Euclid, OH, 44121, (216) 291-8720, (2).

BONDAR-CLEGG, INC., 12980 West Cedar Dr., Lakewood, CO, 80228, (303) 989-1404, (1)(2).

BONDEX INTERNATIONAL, INC., 3616 Scarlet Oak Blvd., St. Louis, MO, 63122, (314) 225-5001, (4).

BORO RADON TESTING SERVICE, INC., 501 Sharp Ave., Glenolden, PA, 19036, (215) 532-8378, (215) 586-5640, (1).

BOSSAIRE, INC., 1321 Tyler St., NE, Minneapolis, MN, 55413, (612) 781-0179, (4).

BOSTIK, Boston St., Middleton, MA, 01949, (508) 777-0100, (800) 343-8178, (4).

BOSTON AIR POLLUTION CONTROL COMMISSION, Boston City Hall Room 805, Boston, MA, 02201, (617) 725-4416, (10).

BOSTON COLLEGE/DEPARTMENT OF GEOLOGY AND GEOPHYSICS, Devlin Hall, Chestnut Hill, MA, 02167, (617) 552-3656, (5).

BOSTON UNIVERSITY MEDICAL CENTER/RADIATION PROTECTION OFFICE, 80 East Concord, Boston, MA, 02118, (617) 638-7052, (1).

BOULDER COUNTY HEALTH DEPARTMENT/AIR POLLUTION CONTROL DIVISION, 3450 Broadway, Boulder, CO, 80302, (303) 441-1180, (10).

BOULDER INSPECTION COMPANY, P.O. Box 3369, Boulder, CO, 80307, (303) 444-6657, (1)(4).

BOXFORD RADON LABORATORY, 52 Wildmeadow Rd., Boxford, MA, 01921, (617) 887-9641, (1).

BRADLEY HARRISON ASSOCIATES, INC., 55 Virginia Ave., Freeport, NY, 11520, (516) 546-2283, (516) 546-7760, (1)(3).

BRAILSFORD AND COMPANY, INC., 670 Milton Rd., Rye, NY, 10580, (914) 967-1820, (4).

BRAMA CONSTRUCTION CORPORATION, 47 Buena Vista Ave., Yonkers, NY, 10701, (914) 968-4700, (1).

BRANDEIS UNIVERSITY/ROSENTIEL RESEARCH CENTER, 415 South St., Waltham, MA, 02154, (617) 736-2461, (1)(5).

BRANDYWINE ENVIRONMENTAL SERVICES, P.O. Box 3343 206-B East Ayre St., Newport, DE, 19804, (302) 995-7565, (1)(2).

BRENT, DAVID J., 3000 South Jamaica Ct. Suite 130, Aurora, CO, 80014, (303) 755-7012, (1).

BRENTWOOD INDUSTRIES, INC., P.O. Box 605, Reading, PA, 19603, (215) 374-5109, (4).

BRIDGEPORT/DEPARTMENT OF AIR POLLUTION CONTROL/OFFICE OF HUMANE AFFAIRS, 752 East Main St., Bridgeport, CT, 06608, (203) 576-7451, (10).

BRIGGS LONGBOTHUM THERMINSPECTION, 1440 Prospect Ave., Plainfield, NJ, 07060, (201) 755-8116, (3)(6).

BRISTOL-BURLINGTON HEALTH DEPARTMENT, 240 Stafford Ave., Bristol, CT, 06010-4617, (203) 584-7682, (10).

BRITISH COLUMBIA MINISTRY OF HEALTH/RADIATION PROTECTION SERVICE, 307 West Broadway Suite 200, Vancouver, BC, V5Y 1P9, (604) 660-6630, (11).

BRITISH COLUMBIA REAL ESTATE COUNCIL , 626 West Pender St. 5th Floor, Vancouver, BC, V6B 1V9, (604) 683-9664, (11).

BROOKS LABORATORIES, INC., 44 Codfish Ln., Weston, CT, 06883, (203) 226-6384, (1)(2)(3).

BROWARD COUNTY ENVIRONMENTAL QUALITY CONTROL BOARD, 500 SW 14th Ct., Fort

Lauderdale, FL, 33315, (305) 765-5881, (10).

BRUN-REUSS ASSOCIATES, P.O. Box 0842, Lionville, PA, 19353-0482, (800) 522-0072, (1)(2).

BRYANT AND BRYANT SERVICE COMPANY, 992 1/2 Davidson Dr., Nashville, TN, 37205, (615) 352-2570, (1).

BUILDERS EQUIPMENT AND SUPPLY, 1500 South Hanover St., Pottstown, PA, 19464, (215) 323-3320, (4).

BUILDING DIAGNOSTICS, INC., 380 27th St., Boulder, CO, 80303, (303) 494-2431, (1)(2).

BUILDING INSPECTION SERVICES, INC., 1675 Winding Rd., Southampton, PA, 18966, (215) 357-8055, (1).

BUILDING INSPECTOR OF AMERICA, THE, 243 Reed Rd., Westport, MA, 02790, (508) 771-2477, (508) 636-6767, (1).

BUILDING INSPECTOR OF AMERICA, THE, 129C Gaither Dr., Mt. Laurel, NJ, 08054, (609) 273-8333, (1).

BUILDING INSPECTOR OF AMERICA, THE, 1 Mason Ct. Box 1654, Hawley, PA, 18428, (717) 775-9619, (1).

BUILDING INSPECTOR OF AMERICA, THE, 112 Jockey Hollow Way, Union, NJ, 07083, (201) 473-5656, (201) 579-2734, (1).

BUILDING INSPECTORS, INC., 282 Harlow St. P.O. Box 2220, Bangor, ME, 04401, (207) 941-0870, (1).

BUILDING OFFICIALS AND CODE ADMINISTRATORS INTERNATIONAL (BOCA), 4051 West Flossmore Rd., Country Club Hills, IL, 60477, (312) 799-2300, (13).

BUILDING OWNERS AND MANAGERS ASSOCIATION INTERNATIONAL (BOMA), 1250 I St., NW Suite 200, Washington, DC, 20005, (202) 289-7000, (13).

BULAVA ENVIRONMENTAL, INC., 13 Hunt Club Rd., Belle Mead, NJ, 08502, (201) 874-6207, (1).

BURENS ENVIRONMENTAL SERVICES CORPORATION, P.O. Box 96, Wallingford, PA, 19086, (215) 565-7455, (1).

BURNS AND ASSOCIATES, 265 Kearsarge Way, Portsmouth, NH, 03801, (603) 431-3400, (1)(2).

BUTTE COUNTY AIR POLLUTION CONTROL DISTRICT, 316 Nelson Ave. P.O. Box 1229, Oroville, CA, 95965, (916) 534-4383, (10).

BUYERS UP RADON PROJECT, Box 53005, Washington, DC, 20036, (202) 328-3800, (17).

BUZZANO CONTRACTING, 62 Mt. Pleasant Rd., Newtown, CT, 06470, (203) 426-7196, (2).

C

C & D ASSOCIATES HOME INSPECTIONS, 542 Union Ave., Middlesex, NJ, 08846, (201) 469-0263, (1).

C & D HEATING AND CONSTRUCTION, 1111 South Sheridan Blvd., Lakewood, CO, 80226, (303) 935-6461, (1)(2).

C SENTINEL ENTERPRISES, 426 Ingel Blvd., Salem, VA, 24153, (703) 380-4828, (1).

C-P LABORATORY, 783 State Rd. Suite 174, Plymouth, MA, 02360, (617) 477-8128, (1).

C. KENT CHAMBERLAIN AND ASSOCIATES, 6579 West David Dr., Littleton, CO, 80123, (303) 979-3916, (1)(2).

C.R. ADAMS COMPANY, INC., 17228 West 16th St., Golden, CO, 80401, (303) 279-5967, (1).

C.R. LAURENCE, 2503 East Vernon, Los Angeles, CA, 90058, (800) 421-6144, (4).

CADWALADER, WICKERSHAM AND TAFT, 1333 New Hampshire Ave., NW, Washington, DC, 20036, (202) 862-2317, (3).

CALAVERAS COUNTY HEALTH DEPARTMENT/AIR POLLUTION DIVISION, Government Center, San Andreas, CA, 95249, (209) 754-3742, (10).

CALBAR, INC., 2626 North Martha, Philadelphia, PA, 19125, (215) 739-9141, (4).

CALIFORNIA AIR RESOURCES BOARD/AIR QUALITY DIVISION, 1102 Q St., Sacramento, CA, 95814, (916) 323-5043, (10).

CALIFORNIA ASSOCIATION OF REALTORS, P.O. Box 76917, Los Angeles, CA, 90067, (213) 739-8201, (18).

CALIFORNIA DEPARTMENT OF CONSERVATION/DIVISION OF MINES, 1416 Ninth St. Room 1341, Sacramento, CA, 95814, (916) 445-1923, (10).

CALIFORNIA DEPARTMENT OF HEALTH SERVICES/AIR AND INDUSTRIAL HYGIENE LABORATORY, 2151 Berkeley Way Room 334, Berkeley, CA, 94704, (415) 540-2134, (10).

CALIFORNIA DEPARTMENT OF HEALTH SERVICES/RADIOLOGICAL HEALTH BRANCH, 714 P St. Building #8, Sacramento, CA, 95814, (916) 322-2040, (10).

CALIFORNIA DEPARTMENT OF HEALTH SERVICES/SANITATION AND RADIATION LABORATORY, 2151 Berkeley Way Room 465, Berkeley, CA, 94704, (415) 540-2201, (1)(10).

CALIFORNIA DEPARTMENT OF INDUSTRIAL RELATIONS/DIVISION OF OCCUPATIONAL SAFETY AND HEALTH, P.O. Box 603, San Francisco, CA, 94101-0603, (415) 557-2870, (10).

CALIFORNIA DEPARTMENT OF REAL ESTATE, 2201 Broadway P.O. Box 187000, Sacramento, CA, 95818-7000, (916) 739-3721, (10).

CALOIERO RADON TECHNOLOGIES, RD5, Box 84 North Main St., Williamstown, NJ, 08094, (609) 728-9257, (1)(2).

CAMBRIDGE ASSOCIATES, LTD., 11 Harwood Ln. Trowbridge Complex, Clementon, NJ, 08021, (609) 435-2626, (215) 563-6611, (1).

CAMP DRESSER AND MCKEE, INC., Plaza 1 Raritan Ctr., Raritan, NJ, 08818, (201) 225-7000, (1).

CAMPAGNA BUILDERS, 398B Union Blvd., Totowa, NJ, 07512, (201) 942-9661, (2).

CAMPTON DEVELOPMENT CORPORATION, P.O. Box 354, Campton, NH, 03223, (603) 726-3030, (1).

CAMRODEN ASSOCIATES, RFD 4, Box 62, Rome, NY, 13440, (315) 865-4269, (1)(2)(3)(5)(6).

CANADA MORTGAGE AND HOUSING CORPORATION (CMHC)/RESEARCH DIVISION, Montreal Rd., Ottawa, ON, K1A 0P7, (613) 748-2306, (11).

CANADIAN HOMEBUILDERS ASSOCIATION (CHA), 200 Elgin St. Suite 502, Ottawa, ON, K2P 1L5, (613) 230-3060, (11).

CANADIAN INSTITUTE FOR RADIATION SAFETY/ELLIOTT LAKE LABORATORY, 7 Timmins Rd. Suite 7-15, Elliott Lake, ON, P5A 2R7, (705) 848-3687, (1)(4)(5)(6)(11).

CANADIAN INSTITUTE FOR RADIATION SAFETY/RADON IN HOMES COURSE, 595 Bay St. Suite 1050, Toronto, ON, M5G 2C2, (416) 596-1617, (3)(6).

CANADIAN LUNG ASSOCIATION (CLA), 75 Albert St. Suite 908, Ottawa, ON, K1P 5E7, (613) 237-1208, (11).

CANBERRA INDUSTRIES, INC., 1 State St., Meriden, CT, 06450, (203) 238-2351, (1)(4).

CANCERSCREEN, INC., 1120 First Colonial Rd., Virginia Beach, VA, 23454, (804) 481-4742, (1).

CANNONDALE HEATING AND AIR CONDITIONING, P.O. Box 458, Wilton, CT, 06897, (203) 834-0875, (1).

CANTON CITY HEALTH DEPARTMENT/AIR POLLUTION CONTROL DIVISION, City Hall, Canton, OH, 44702, (216) 489-3385, (10).

CAPE CHEMICAL COMPANY, 33 North Frederick St., Cape Girardeau, MO, 63701, (314) 334-3002, (1).

CAPE COD HOME INSPECTIONS, 243 Reed Rd., Westport, MA, 02790, (508) 771-2477, (800) 828-1919, (1).

CAPITAL CONSTRUCTION SERVICES, P.O. Box 568, Troy, NY, 12180, (518) 434-0105, (1).

CAPITAL HOME INSPECTION CORPORATION, P.O. Box 2541, Manassas, VA, 22110, (703) 369-4206, (1).

CAPITAL MATERIALS TESTING, INC., P.O. Box 81 5 Hemlock St., Latham, NY, 12110, (518) 783-3150, (800) 541-4109, (1)(2)(4).

CARE SYSTEMS, P.O. Box 122, Brentwood, TN, 37027, (615) 373-4275, (2).

CARL HENRY AND COMPANY, INC., Benson East Suite 207B, Jenkintown, PA, 19046, (215) 576-7766, (1).

CARLSON HEATING, COOLING, AND ELECTRIC COMPANY, 3340 West Lake Ave., Glenview, IL, 60025-1297, (312) 729-0123, (1).

CARLSON INSPECTION ASSOCIATES, INC., 51 R Dudley St., Arlington, MA, 02174, (617) 641-0600, (1).

CAROLINA RADON CONSULTANTS, P.O. Box 2000, Etowah, NC, 28729, (704) 891-9559, (1)(2).

CAROLINA RADON CONSULTANTS, P.O. Box 6683, Columbia, SC, 29260, (803) 779-2043, (1)(2).

CAROLINA RADON DETECTION, 209 Tebblewood Dr., Simpsonville, SC, 29681, (803) 282-0494, (1)(2).

CARPENTER, CHARLES D., 13462 West Florida Dr., Lakewood, CO, 80228, (303) 988-6818, (1).

CARROLL COUNTY RADON CONTROL, INC., Box 027, Westminster, MD, 21157, (301) 848-6324, (1).

CASCADE/CITY-COUNTY AIR POLLUTION CONTROL PROGRAM, 1130 17 Ave. South, Great Falls, MT, 59405, (406) 761-1190, (10).

CASE WESTERN RESERVE UNIVERSITY/SCHOOL OF MEDICINE, 2048 Adelbert Rd.., Cleveland, OH, 44186, (216) 844-3530, (5).

CATHOLIC UNIVERSITY OF AMERICA, Pangborn Hall Radiation Safety Dept., Washington, DC, 20065, (202) 635-5206, (5).

CATTARAUGUS COUNTY HEALTH DEPARTMENT, 301 Laurens St., Olean, NY, 14760, (716) 375-4121, (10).

CAVALIER CORPORATION, 11516 East Sprague, Spokane, WA, 99206, (509) 926-6217, (1)(2)(4)(6).

CAVIN ANALYTICAL CONSULTANTS, INC., 2165K West Park Ct., Stone Mountain, GA, 30087, (404) 498-8295, (1).

CCI-RADON AND HOME INSPECTION/RADON DETECTION, P.O. Box 1045, Malvern, PA, 19355, (215) 889-1554, (215) 363-9015, (1)(2)(4).

CDM FEDERAL PROGRAMS CORPORATION, 13135 Lee Jackson Memorial Hwy., Fairfax, VA, 22033, (703) 968-0900, (1).

CENTARUS CORPORATION, P.O. Box 513, Glasgow, KY, 42141, (502) 651-2396, (1).

CENTER FOR CLEAN AIR POLICY (CCAP), 444 North Capitol St., NW, Washington, DC, 20001, (202) 624-7709, (14).

CENTER FOR ENVIRONMENTAL INFORMATION, INC. (CEI), 33 South Washington St., Rochester, NY, 14608, (716) 546-3796, (14).

CENTER FOR ENVIRONMENTAL MEDICINE, 60 Washington Ave., Hamden, CT, 06518, (203) 789-1911, (15).

CENTER FOR RADON DETECTION, 12 Bank St., Summit, NJ, 07901, (201) 277-3200, (201) 992-9702, (1).

CENTER FOR RADON SERVICES, P.O. Box 2250, Cherry Hill, NJ, 08034, (609) 424-7191, (1).

CENTRAL HOME INSPECTION, INC., 435 Rte. 18, East Brunswick, NJ, 08816, (201) 545-0300, (1).

CENTRAL JERSEY RADON, INC., P.O. Box 600, Lincroft, NJ, 07738, (201) 741-0400, (1)(2).

CENTRAL RADON TESTING, 435 Rte. 18, East Brunswick, NJ, 08816, (201) 545-0300, (1).

CERT WATER SERVICES, INC., 25 Duval Ln., Plainville, CT, 06062, (203) 793-9142, (1)(4).

CERTIFIED TESTING SERVICES, INC., 474 Easton Rd., Horsham, PA, 19044, (215) 672-8100, (1)(2).

CERTUS LABORATORIES, P.O. Box 105633, Atlanta, GA, 30348, (404) 762-5411, (1)(6).

CHAPPARAL ENTERPRISES, INC., P.O. Box 601, Elberton, GA, 30635, (404) 283-7611, (1).

CHATAUQUA COUNTY HEALTH DEPARTMENT, Health and Social Services Building, Mayville, NY, 14757, (716) 753-4481, (10).

CHATTANOOGA-HAMILTON COUNTY/AIR POLLUTION CONTROL BUREAU, 3511 Rossville Blvd., Chattanooga, TN, 37407, (615) 867-4321, (10).

CHEM-SAFE LABORATORIES, North 131 Grand Ave. P.O. Box 546, Pullman, WA, 99163-0546, (509) 334-0922, (1)(4).

CHEMICAL SYSTEMS, INC., P.O. Box 68, Dunbridge, OH, 43414, (419) 833-4492, (4).

CHENANGO ENVIRONMENTAL LAB, INC., 35 State St., Binghamton, NY, 13901, (607) 723-8175, (1).

CHEROKEE DE-WATERING, INC., Rte. 4, Oak Hill Cr., Canton, GA, 30114, (404) 928-5091, (1).

CHESTER COUNTY DEVELOPMENT SERVICE, 420 West Market St., West Chester, PA, 19382, (215) 692-4509, (1).

CHICAGO/DEPARTMENT OF CONSUMER SERVICES/ENERGY AND ENVIRONMENTAL PROGRAMS DIVISION, 121 North LaSalle St. Room 810, Chicago, IL, 60602, (312) 744-4034, (10).

CHIEF HOME AND BUILDING INSPECTION SERVICES, Box 284, Towaco, NJ, 07082, (201) 299-9087, (1).

CHILTON ASSOCIATES (RADON GROUP), 144 Valley Rd., Haworth, NJ, 07641, (201) 385-1660, (1).

CITIZEN'S CLEARINGHOUSE FOR HAZARDOUS WASTES (CCHW), P. O. Box 926, Arlington, VA, 22216, (703) 276-7070, (17).

CITIZENS FOR A BETTER ENVIRONMENT (CBE), 506 South Wabash, Chicago, IL, 60605, (312) 939-1530, (17).

CLANCY & COMPANY, 405 West Rockrimmon, Colorado Springs, CO, 80919, (303) 599-4660, (13).

CLARK COUNTY HEALTH DISTRICT/AIR POLLUTION CONTROL DIVISION, 625 Shadow Ln., Las Vegas, NV, 89127, (702) 383-1276, (10).

CLARK UNIVERSITY/ENVIRONMENT, TECHNOLOGY, AND SOCIETY PROGRAM/DEPARTMENT OF PHYSICS, 950 Main St., Worcester, MA, 01610, (617)793-7687, (1)(5).

CLASSIC CONSTRUCTION, 7001 Orchard Lake Rd. Suite 130, West Bloomfield, MI, 48322, (313) 737-8800, (13).

CLASSIC RENOVATIONS AND BUILDERS, 6762 Riverview Rd., Peninsula, OH, 44264, (216) 468-3119, (1).

CLEAN AIR ENGINEERING, INC., 207 North Woodwork Ln., Palatine, IL, 60067, (312) 991-3300, (1).

CLEAN HARBORS ANALYTICAL SERVICES, INC., P.O. Box 327 325 Wood Rd., Braintree, MA, 02184, (617) 849-6070, (1).

CLEMSON UNIVERSITY/ENVIRONMENTAL SYSTEMS ENGINEERING, 501 Rhodes Engineering Research Center, Clemson, SC, 29634-0919, (803) 656-5569, (5).

CLEVELAND COUNTY HEALTH DEPARTMENT/ENVIRONMENTAL HEALTH DIVISION, 315 Grover St., Shelby, NC, 28150, (704) 487-1138, (10).

CLEVELAND DEPARTMENT OF PUBLIC HEALTH AND WELFARE/DIVISION OF AIR POLLUTION CONTROL, 2735 Broadway Ave., Cleveland, OH, 44115, (216) 664-3500, (10).

CMT, INC., 2813 Rio Vista, Emporia, KS, 66801, (316) 342-2760, (1).

COCONINO COUNTY AIR POLLUTION CONTROL DISTRICT/ENVIRONMENTAL HEALTH DIVISION, 2500 North Fort Valley Rd., Flagstaff, AZ, 86001, (602) 779-5164, (10).

COGITO TECHNICAL SERVICES, INC., 1799 Rte. 13 P.O. Box 86, Sheds, NY, 13151, (315) 662-7529, (3)(6).

COLE-PARMER INSTRUMENT COMPANY, 7425 N. Oak Park Ave., Chicago, IL, 60648, (312) 647-7600, (800) 323-4340, (4).

COLORADO ASSOCIATION OF REALTORS, 6558 South Yosemite Cr., Englewood, CO, 80111, (303) 770-6550, (18).

COLORADO DEPARTMENT OF HEALTH/RADIATION CONTROL DIVISION, 4210 East 11th Ave., Denver, CO, 80220, (303) 331-4812, (10).

COLORADO FRONT RANGE BUILDING INSPECTION SERVICE, P.O. Box 9614, Ft. Collins, CO, 80525, (303) 482-1976, (800) 248-1976, (1).

COLORADO GEOLOGICAL SURVEY, 1313 Sherman St. Room 715, Denver, CO, 80203, (303) 866-2611, (10).

COLORADO HOME PROTECTION, 6318 South Gallop Ct., Littleton, CO, 80120, (303) 797-2007, (1).

COLORADO INSTITUTE OF ENVIRONMENTAL HEALTH/OCCUPATIONAL SAFETY AND HEALTH SECTION, 110 Veterinary Science Bldg. Colorado State University, Fort Collins, CO, 80523, (303) 491-6151, (10).

COLORADO RADON DETECTION COMPANY, 8824 West Ontario Ave., Littleton, CO, 80123, (303) 973-3261, (1).

COLORADO RADON ENGINEERING, INC., 5104 Greenview Ct., Ft. Collins, CO, 80525, (303) 223-9755, (1).

COLORADO RADON INSTITUTE, 4675 East Louisiana Suite 114, Denver, CO, 80222, (303) 756-0701, (1).

COLORADO REAL ESTATE COMMISSION, 1776 Logan St. 4th Floor, Denver, CO, 80203, (303) 894-2166, (10).

COLORADO REAL ESTATE INSPECTION SERVICES, P.O. Box 24962, Denver, CO, 80224-0962, (303) 753-0543, (1).

COLORADO SAFE-N-DRY, INC., 7950 East Prentice Ave. Suite 201, Englewood, CO, 80111, (303) 771-2212, (1)(2).

COLORADO STATE UNIVERSITY/DEPARTMENT OF RADIOLOGY AND RADIATION BIOLOGY, Ft. Collins, CO, 80523, (303) 491-5222, (5).

COLORADO VENTURE MANAGEMENT, INC., 2995 Wilderness Pl., Boulder, CO, 80301, (303) 440-4055, (1)(2).

COLUMBIA RADIOLOGICAL TESTING LABORATORY, 4054 Victoria Ct., Columbia, MO, 65201, (314) 443-3538, (314) 442-6080, (1).

COLUMBIA/DEPARTMENT OF HEALTH/AIR POLLUTION CONTROL DIVISION, 1225 Laurel St., Columbia, SC, 29217, (803) 733-8320, (10).

COLUMBUS HEALTH DEPARTMENT, 1815 Washington Blvd., Columbus, OH, 43215, (614) 222-6992, (10).

COLUSA COUNTY AIR POLLUTION CONTROL DISTRICT, P.O. Box 1029, Colusa, CA, 95932, (916) 458-5891, (10).

COMFORT SYSTEMS, 4100 Nebraska Ave., Nashville, TN, 37209, (615) 298-1855, (1).

COMMISSION OF THE EUROPEAN COMMUNITIES (CEC)/RADIATION PROTECTION PROGRAM, rue de la Loi, 200, B-1049 Brussels, Belgium, (011) 322-235-5211, (011) 322-235-1111, (12).

COMMONWEALTH ENGINEERING AND TECHNOLOGY, 3113 North Front St., Harrisburg, PA, 17110, (717) 232-2288, (1).

COMMONWEALTH ENTERPRISES, INC., P.O. Box S187, St. Davids, PA, 19087, (215) 688-7097, (1)(2).

COMMONWEALTH RADON SPECIALISTS, 2204 Grundy Rd., Woodbridge, VA, 22191, (703) 494-6241, (1).

COMMUNICATION SCIENCES INSTITUTE, 16440 Emory Ln., Rockville, MD, 20853, (301) 570-0984, (3).

COMMUNITY HEALTH ASSOCIATES, INC., 316 Sherman St., Canton, MA, 02021, (617) 828-3123, (1)(2).

COMMUNITY RADON TESTING/ DIVISION OF SANTIAGO ENTERPRISES, 4702 Kilburn Ct., Woodbridge, VA, 22193, (703) 690-4282, (800) 451-8164, (1).

COMPREHENSIVE MANAGEMENT SERVICES, 650 Division St., Harrisburg, PA, 17110, (717) 238-1010, (1).

COMPREHENSIVE RADON SERVICES, INC., P.O. Box 548, Manchester, NH, 03105, (603) 644-7301, (1)(2)(4).

CON-TEST, INC., 126 Shaker Rd., East Longmeadow, MA, 01028, (413) 525-1198, (1).

CONCEPT 90, INC., 12843 West Jewell Dr., Lakewood, CO, 80228, (303) 985-5355, (1).

CONCRETE EPOXY TECHNICAL SYSTEMS, INC., P.O. Box F-25 41 Terry Dr., Trevose, PA, 19047, (215) 322-7310, (4).

CONFERENCE OF PUBLIC HEALTH LABORATORIANS (COPHL), P.O. Box 9083, Austin, TX, 78766, (512) 458-7318, (15).

CONFERENCE OF RADIATION CONTROL PROGRAM DIRECTORS (CRCPD), 71 Fountain Pl., Frankfort, KY, 40601, (502) 227-4543, (14).

CONFERENCE OF STATE HEALTH AND ENVIRONMENTAL MANAGERS (CSHEM), P.O. Box 5254, Hudson, FL, 34674, (15).

CONNECTICUT ASSOCIATION OF REALTORS, 316 Farmington Ave., Hartford, CT, 06105, (203) 522-7255, (18).

CONNECTICUT BASEMENT DE-WATERING AND RADON SYSTEMS, INC., 1 Clayton St., Milford, CT, 06460, (203) 874-3491, (1)(2).

CONNECTICUT DEPARTMENT OF CONSUMER PROTECTION/REAL ESTATE DIVISION, State Office Building Room G-8 165 Capitol Ave., Hartford, CT, 06106, (203) 566-5130, (10).

CONNECTICUT DEPARTMENT OF ENVIRONMENTAL PROTECTION/RADIATION CONTROL UNIT, State Office Building 165 Capital Ave., Hartford, CT, 06106, (203) 566-5668, (10).

CONNECTICUT DEPARTMENT OF HEALTH SERVICES/TOXIC HAZARDS SECTION, 150 Washington St., Hartford, CT, 06106, (203) 566-3122, (10).

CONNECTICUT DEPARTMENT OF LABOR/DIVISION OF OCCUPATIONAL SAFETY AND HEALTH, 200 Folly Brook Rd., Wethersfield, CT, 06109, (203) 566-4550, (10).

CONNECTICUT ENVIRONMENTAL TESTING, 6 East Main St., Westport, CT, 06880, (203) 454-1544, (1).

CONNECTICUT GEOLOGICAL AND NATURAL HISTORY SURVEY/NATURAL RESOURCES CENTER, 165 Capitol Ave. Room 553, Hartford, CT, 06106, (203) 566-3540, (10).

CONSERVATION ENERGY SYSTEMS, P.O. Box 10146, Minneapolis, MN, 55458, (306) 242-3663, (800) 667-3717, (4).

CONSHOHOCKEN RADON TESTING SERVICES, 476 West Elm St., Conshohocken, PA, 19428, (215) 828-2059, (215) 272-3694, (1).

CONSTRUCTION EVALUATION, INC., 102 Hickok Rd., Greenfield Center, NY, 12833, (518) 587-6366, (1).

CONSTRUCTION SERVICES OF INDIANA, INC., 9341 Sargeant Rd., Indianapolis, IN, 46256, (317) 849-2686, (2).

CONSTRUCTION SPECIALTIES, INC., 55 Winans Ave., Cranford, NJ, 07016, (201) 272-5200, (4).

CONSTRUCTION SPECIFICATION INSTITUTE (CSI), 601 Madison St., Alexandria, VA, 22314, (703) 684-0300, (13).

CONSULTING CHEMISTS, 3501 Lakeshore Dr., Kingsport, TN, 37663, (615) 239-5329, (615) 477-8253, (1).

CONSUMER ENERGY SERVICES, INC., 12 Westerville Sq. Suite 200, Westerville, OH, 43081, (614) 861-0600, (1)(2).

CONSUMER FEDERATION OF AMERICA (CFA), 1424 16th St., NW Suite 604, Washington, DC, 20036, (202) 387-6121, (17).

CONSUMERS UNION OF UNITED STATES (CU), 256 Washington St., Mt. Vernon, NY, 06608, (914) 667-9400, (17).

CONTAMINATION MANAGEMENT ENGINEERING, P.O. Box 871, Greenbrook, NJ, 08812, (201) 752-0078, (2).

CONTEK/CORPORATE HEADQUARTERS, P.O. Box 3065, Newport Beach, CA, 92663, (714) 759-9226, (1).

CONTINENTAL ENVIRONMENT COMPANY, INC., 34 Maple St., Summit, NJ, 07901, (201) 277-2255, (1)(2).

CONTROL POWER SYSTEMS, INC., 1010 E. 31 St., Hialeah, FL, 33013, (305) 691-2260, (4).

CONTROL RESOURCE SYSTEMS, INC., 670 Mariner Dr., Michigan City, IN, 46360, (219) 872-5591, (800) 272-3786, (4).

CONVENIENCE PRODUCTS, INC., 4205 Forest Park Blvd., St. Louis, MO, 63108-2892, (314) 535-6226, (4).

COOK COUNTY DEPARTMENT OF ENVIRONMENTAL CONTROL, Maybrook Civic Center 1500 Maybrook Dr., Maywood, IL, 60153, (312) 865-6165, (10).

CORE LABORATORIES, INC., 420 West First St., Casper, WY, 82601, (307) 235-5741, (800) 666-0306, (1).

CORNELL COLLEGE, First St. West, Mt. Vernon, IA, 52314, (319) 895-8811, (1).

CORNELL UNIVERSITY/INSTITUTE FOR COMPARATIVE AND ENVIRONMENTAL TOXICOLOGY, Fernow Hall, Ithaca, NY, 14850, (607) 255-9107, (5).

CORNERSTONES, 1770 Central St., Yorktown Heights, NY, 10598, (914) 962-7037, (1).

CORPUS CHRISTI-NUECES COUNTY DEPARTMENT OF PUBLIC HEALTH/ENVIRONMENTAL CONTROL DIVISION, 1702 Horne Rd., Corpus Christi, TX, 78416, (512) 851-7226, (10).

COTHERN, C. RICHARD, 4732 Merivale Rd., Chevy Chase, MD, 20815, (301) 652-2863, (3).

COUNCIL OF AMERICAN BUILDING OFFICIALS (CABO), 5203 Leesburg Pike, Falls Church, VA, 22041, (703) 931-4533, (13).

COUNTY COLLEGE OF MORRIS/CHEMISTRY DEPARTMENT, Rte. 10 and Center Grove Rd. A-304, Randolph, NJ, 07869, (201) 361-5000, (1).

COUNTY HOME INSPECTIONS, P.O. Box 391, Wilton, CT, 06897, (203) 762-2608, (1).

COURTNEY CONSULTANTS, INC., 520 Carriage Dr., Atlanta, GA, 30328, (404) 256-2487, (1).

CRIPPEN CONSULTING ASSOCIATES, INC., 242 North James St. Suite 202, Wilmington, DE, 19804, (302) 994-8000, (1)(2)(3).

CTL ENVIRONMENTAL SERVICES, 2905 East Century Blvd., South Gate, CA, 90280, (213) 564-2641, (1).

CTL/THOMPSON, INC., 1971 West 12th Ave., Denver, CO, 80204, (303) 825-0777, (1).

CUMBERLAND ANALYTICAL LABS, INC., 56 North 2nd St., Chambersburg, PA, 17201, (717) 263-5943, (1).

CUMBERLAND COUNTY/E. NEWTON SMITH PUBLIC HEALTH CENTER/ENVIRONMENTAL HEALTH DIVISION, 227 Fountainhead Lane, Fayetteville, NC, 28301, (919) 483-9046, (10).

CURRENT INDOOR AIR SYSTEMS, INC., P.O. Box 18075, Boulder, CO, 80308, (303) 440-8555, (4).

CUTTER INFORMATION CORPORATION, 1100 Massachusetts Ave., Arlington, MA, 02174, (617) 648-8700, (3).

D

D-TECH, INC., 7 Laurence Pl., Plymouth Meeting, PA, 19462, (215) 825-4832, (1).

D. PAUL BUILDING ENTERPRISE, P.O. Box 1265, Waldorf, MD, 20601, (301) 843-5751, (2).

DADE COUNTY CONSUMER ADVOCATES OFFICE, 140 West Flagler St. Room 1605, Miami, FL, 33130, (305) 375-4206, (10).

DALLAS/HEALTH AND HUMAN SERVICES/ENVIRONMENTAL HEALTH DIVISION/AIR POLLUTION CONTROL SECTION, 400 South Zang Blvd., Dallas, TX, 75208, (214) 670-7520, (10).

DAMES AND MOORE, 7101 Wisconsin Ave. Suite 700, Bethesda, MD, 20814, (301) 652-2215, (1).

DAN HIMMELFARB COMPANY, 7416 Chrome Mine Rd., Gaithersburg, MD, 20879, (301) 744-4004, (1).

DAP, INC., P.O. Box 277, Dayton, OH, 45401, (513) 667-4461, (4).

DAVID A. MASE REAL ESTATE, 6515 Main St., Trumbull, CT, 06611, (203) 261-8080, (1).

DAVID E. WOLFE CONSULTANTS, INC., 6728 Memorial Hwy. Suite 255, Tampa, FL, 33615, (813) 885-5887, (813) 887-8164, (1).

DAYTON REGIONAL AIR POLLUTION CONTROL AGENCY, 451 West Third St. P.O. Box 972, Dayton, OH, 45422, (513) 225-4435, (10).

DEAN ARRINGTON ASSOCIATES, 89A Haytown Rd., Lebanon, NJ, 08833, (201) 236-9697, (1).

DECISION RESEARCH, 1201 Oak, Eugene, OR, 97401, (503) 485-2400, (3)(17).

DEFLECTO CORPORATION, 7035 East 86th St., Indianapolis, IN, 46250, (317) 849-9555, (4).

DEGENSHEIN DENKER, 10 South Broadway, Nyack, NY, 10960, (1).

DELAWARE ASSOCIATION OF REALTORS, P.O. Box 1364, Dover, DE, 19903, (302) 734-4444, (18).

DELAWARE DEPARTMENT OF HEALTH AND SOCIAL SERVICES/BUREAU OF ENVIRONMENTAL HEALTH/OFFICE OF RADIATION CONTROL/DIVISION OF PUBLIC HEALTH, 802 Silver Lake Blvd. and Walker Rd. P.O. Box 637, Dover, DE, 19903, (302) 736-4731, (10).

DELAWARE DEPARTMENT OF LABOR/DIVISION OF INDUSTRIAL AFFAIRS, 820 North French St. 6th Fl., Wilmington, DE, 19801, (302) 571-3908, (10).

DELAWARE DIVISION OF PROFESSIONAL REGULATION, Margaret O'Neill Building P.O. Box 1401, Dover, DE, 19903, (302) 736-4522, (10).

DELAWARE GEOLOGICAL SURVEY, University Of Delaware, Newark, DE, 19716, (302) 452-2833, (10).

DELAWARE VALLEY GEOTECHNICAL ASSOCIATES, P.O. Box 98, Washington Crossing, PA, 18977, (215) 357-0907, (1)(2).

DELAWARE VALLEY RADON INSPECTIONS/ARSENAL BUSINESS CENTER, 5301 Tacony St. Box 221, Philadelphia, PA, 19137-2309, (215) 289-3300, (215) 947-1169, (1).

DEMAR PEST CONTROL COMPANY, 3522 Kensington Ave., Philadelphia, PA, 19134, (215) 744-0336, (1).

DEMPSEY, JOSEPH H., P.O. Box 247, Eldorado Springs, CO, 80025, (303) 499-7578, (2).

DENVER/ CITY-COUNTY DEPARTMENT OF HEALTH AND HOSPITALS/ENVIRONMENTAL HEALTH SERVICE, 605 Bannock St., Denver, CO, 80204, (303) 893-6243, (10).

DES CHAMPS LABORATORIES, INC., P.O. Box 440 17 Farinella Dr., East Hanover, NJ, 07936, (201) 994-4660, (4).

DESERT-DRY WATERPROOFING, 904 Rydal Rd., Rydal, PA, 19046, (215) 886-6140, (703) 490-4010, (1).

DEVELOPMENT SERVICES, LTD., 420 West Market St., West Chester, PA, 19382, (215) 692-4509, (1).

DIAGNOSTICS AND DEVICES, INC., 1201 Mt. Kemble Ave., Morristown, NJ, 07960, (201) 766-7111, (1)(4).

DIAMOND BACK SERVICES, INC., 7396 Reynolds Dr.,

Sedalia, CO, 80135, (303) 791-1984, (1)(2).

DIAMOND REALTY INSPECTIONS, INC., 211 Monocacy Creek Rd., Birdsboro, PA, 19508, (215) 582-2070, (1).

DINOFF ASSOCIATES, INC., 112 South Decatur, Strasburg, PA, 17579, (717) 687-0938, (2).

DISTRICT OF COLUMBIA DEPARTMENT OF CONSUMER AND REGULATORY AFFAIRS/REAL ESTATE COMMISSION, 614 H St., NW Room 923, Washington, DC, 20013, (202) 727-7468, (202) 727-7469, (10).

DISTRICT OF COLUMBIA DEPARTMENT OF CONSUMER AND REGULATORY AFFAIRS/SERVICE FACILITY REGULATION ADMINISTRATION/PHARMACEUTICAL AND MEDICAL DEVICES CONTROL DIVISION, 614 H St. Room 1014, Washington, DC, 20004, (202) 727-7190, (10).

DISTRICT OF COLUMBIA DEPARTMENT OF EMPLOYMENT SERVICES/OCCUPATIONAL SAFETY AND HEALTH DIVISION, 950 Upshur St., NW, Washington, DC, 20011, (202) 576-6651, (10).

DISTRICT OF COLUMBIA/ENERGY OFFICE, 613 G St., NW Suite 500, Washington, DC, 20001, (202) 724-2100, (10).

DITTMAR INSPECTION COMPANY, INC., 20 Redding Pl., Towaco, NJ, 07082, (201) 263-4745, (1).

DJ'S RADIATION PROTECTION, Route 2 Box 332B, St. Pauls, NC, 28384, (919) 425-6680, (1).

DK ROCKWELL, 101 East Cornell St., Somerdale, NJ, 08083, (609) 627-2166, (1).

DMC ASSOCIATES, P.O. Box 11, Bergenfield, NJ, 07621, (201) 384-6060, (1).

DONALD WALKER COMPANY, 60591 Mount Vernon, Rochester, MI, 48064, (313) 652-1504, (1).

DOSIMETER CORPORATION, Box 897, Fayetteville, GA, 30214, (404) 461-5639, (4).

DOUGLAS MARTIN AND ASSOCIATES, INC./DMA ANALYSIS GROUP, 520 Chestnut St., Emmaus, PA, 18049, (215) 965-2914, (1)(4).

DOUGLAS MARTIN AND ASSOCIATES, INC./MARTIN CONSULTING, 520 Chestnut St., Emmaus, PA, 18049, (215) 965-2914, (2)(3)(4)(6).

DOW CORNING CORPORATION, Mail Stop CO2306, Midland, MI, 48686-0994, (517) 496-4578, (4).

DPR BUILDERS, 21 Stacy Dr., Belle Mead, NJ, 08502, (201) 874-3419, (1)(2).

DR. RADON TESTING AND REDUCTION, 687 North Cary St., Brockton, MA, 02402, (508) 583-2950, (1).

DRANJER CORPORATION, 1441 Pembina Hwy., Winnipeg, MB, R3T 2C4, (204) 474-0451, (4).

DREWRY HOME INSPECTIONS, INC., 2625 Oakton Glen Dr., Vienna, VA, 22180, (703) 938-1044, (1)(2).

DREXEL UNIVERSITY/DEPARTMENT OF CHEMISTRY/ATMOSPHERIC CHEMISTRY PROGRAM, 32nd and Chestnut Sts. 3rd Floor, Philadelphia, PA, 19104, (215) 895-2638, (5).

DUDLEY, PAUL, 140 Railroad Mills Rd., Pittsford, NY, 14534, (716) 385-2368, (1)(3).

DUKE UNIVERSITY MEDICAL CENTER/DEPARTMENT OF MEDICINE/DIVISION OF PULMONARY MEDICINE, 353 Bell Building Box 3177, Durham, NC, 27710, (919) 684-2513, (5).

DUOVENT INDUSTRIES, LTD., 1375 Bl. Charest Ouest Suite 6, Quebec, PQ, G1N 2E7, (418) 688-8805, (4).

DUPAGE COUNTY HEALTH DEPARTMENT/ENVIRONMENTAL HEALTH DIVISION, 111 North County Farm Rd., Wheaton, IL, 60187, (312) 682-7400, (10).

DUTCHESS COUNTY HEALTH DEPARTMENT, County Office Building 22 Market St., Poughkeepsie, NY, 12601, (914) 294-7961, (10).

DWYER INSTRUMENTS, INC., P.O. Box 373, Michigan City, IN, 46360, (219) 872-9141, (4).

E

E-THREE, INC., 4454 Genesee St., Buffalo, NY, 14225, (716) 631-5858, (1).

E. VERNON HILL, INC., P.O. Box 7053, Corte Madera, CA, 94925, (415) 924-6837, (4).

E.C. CHAMBERLAIN, RADIOLOGICAL CONSULTANT, 3304 Susileen Dr., Reno, NV, 89509, (702) 329-5638, (702) 323-8341, (1).

E.D.A., INC., P.O. Box 4854, Austin, TX, 78765, (512) 322-9013, (1).

E.O. CHURCH, INC., 925 East 17th Ave., Denver, CO, 80218, (303) 832-9692, (1).

EASON AND EASON CHARTERED/ATTORNEYS AT LAW, 4423 Montgomery Ave., Bethesda, MD, 20814-4491, (301) 652-2666, (3).

EAST CHICAGO/DEPARTMENT OF AIR QUALITY CONTROL, 4525 Indianapolis Blvd., East Chicago, IN, 46312, (219) 391-8297, (10).

EAST ST. LOUIS/AIR POLLUTION CONTROL, 7 Collinsville Ave., East St. Louis, IL, 62201, (618) 874-2115, (10).

EAST STROUDSBURG UNIVERSITY/ PHYSICS DEPARTMENT/RADON PROJECT, East Stroudsburg, PA, 18301, (717) 424-3350, (5).

EASTERN ANALYTICAL SERVICES, 4 Westchester Plaza, Elmsford, NY, 10523, (914) 939-6992, (800) 327-0121, (1).

EASTERN FASTENERS, INC., 2 Commerce Dr., Moorestown, NJ, 08057, (800) 257-5830, (4).

EASTERN INTEGRATED SERVICES CORPORATION, 3490 U.S. Rte. 1 Building 17-2, Princeton, NJ, 08540, (609) 520-8344, (1).

EAU CLAIRE/CITY-COUNTY HEALTH DEPARTMENT/ENVIRONMENTAL HEALTH DIVISION, 720 2nd Ave., Eau Claire, WI, 54701, (715) 839-4718, (10).

EBERLINE INSTRUMENT CORPORATION, P.O. Box 2108, Santa Fe, NM, 87504-2108, (505) 471-3232, (1)(4).

ECI, 2955 Valmont Suite 300, Boulder, CO, 80301, (303) 442-8995, (1).

ECM ENVIRONMENTAL CONSULTANTS, 7425 Hawkins Dr., Hanover, MD, 21076, (301) 761-1136, (1)(2).

ECODEX, 1 Aaron Rd., Lexington, MA, 02173, (617) 862-3440, (1)(2).

ECOLOGICAL INTERACTIONS, INC., 96 East Main St., Rockaway, NJ, 07866, (201) 625-7700, (800) 543-2116, (1)(2)(4).

ECONOMY HEATING AND AIR CONDITIONING, 519 C Pulaski Hwy., Joppa, MD, 21085, (301) 679-3000, (1).

ECTOR COUNTY HEALTH DEPARTMENT/AIR POLLUTION CONTROL DIVISION, 221 North Texas, Odessa, TX, 79761, (214) 670-7520, (10).

EDA INSTRUMENTS, INC., 9200 East Mineral Ave. Suite 370, Englewood, CO, 80112, (303) 790-2541, (1)(4).

EDGELL COMMUNICATIONS, 7500 Old Oak Blvd, Cleveland, OH, 44130, (216) 243-8100,.

EDINBORO SCIENTIFIC COMPANY, P.O. Box 589 Wetsell Ridge Rd., Edinboro, PA, 16412, (814) 734-4648, (1)(3)(4).

EDISON ELECTRIC INSTITUTE (EEI), 1111 19th St., NW, Washington, DC, 20036, (202) 778-6400, (13).

EER PRODUCTS, INC., 3536 East 28th St., Minneapolis, MN, 55406, (612) 721-4231, (4).

EES OF MINNESOTA, INC., 600 Fairview Ave., N., St. Paul, MN, 55104, (612) 646-7991, (1).

EG & G ORTEC, 100 Midland Rd., Oak Ridge, TN, 37831, (615) 482-4411, (800) 251-9750, (1)(4)(6).

EGAL, INC., P.O. Box 14983 11633 West 83 Terr., Kansas City, KS, 66215, (913) 541-1862, (1).

EKS RADTECH, INC., 1000 Herald Sq., Ashton, PA, 19014, (215) 358-9425, (800) 682-5378, (1)(4).

EL DORADO COUNTY AIR POLLUTION CONTROL DISTRICT, 7553 Green Valley Rd., Placerville, CA, 95667, (916) 626-2407, (10).

EL PASO COUNTY HEALTH DEPARTMENT/AIR QUALITY CONTROL SECTION, 501 North Foote Ave., Colorado Springs, CO, 80909-4598, (303) 578-3137, (10).

EL PASO COUNTY HEALTH UNIT/ENVIRONMENTAL HEALTH SERVICES DIVISION, 222 South Campbell St., El Paso, TX, 79901, (915) 541-4262, (10).

ELECTRICAL POWER RESEARCH INSTITUTE (EPRI)/ENERGY MANAGEMENT AND UTILIZATION DIVISION/RESIDENTIAL AND COMMERCIAL PROGRAMS, 3412 Hillview Ave., Palo Alto, CA, 94304, (415) 855-2000, (5)(13).

ELECTRICAL POWER RESEARCH INSTITUTE (EPRI)/LAWRENCE BERKELEY LABORATORY(LBL)/APPLIED SCIENCE DIVISION, University of California One Cyclotran Rd., Berkeley, CA, 94720, (415) 486-4000, (5)(13).

ELECTRO ANALYTICAL, INC., 7118 Industrial Park Blvd., Mentor, OH, 44060, (216) 951-3514, (1).

ELECTRO MECHANICAL CONCEPTS, 28 Woodruff Dr., Carnegie, PA, 15106-3054, (412) 276-2272, (1).

ELIZABETH/DEPARTMENT OF HEALTH, WELFARE AND HOUSING, 60 Winfield Scott Plaza, Elizabeth, NJ, 07201, (201) 820-4068, (10).

ELLIOTT AND ASSOCIATES, 146 North Drexel Ave., La Grange, IL, 60525, (312) 354-1481, (1)(2).

EMERGENCY RADON TESTING, INC., 29 Freeman St., West Orange, NJ, 07052, (201) 481-3992, (201) 674-1570, (1).

EMMI-RADON CHECK LABORATORY, E-57 Rd. Box 93, Kelley, IA, 50134-8724, (515) 769-2400, (1).

EMPAC, INC., 25 Science Pk., New Haven, CT, 06511, (203) 786-5420, (800) 872-3662, (1)(4).

EMPIRE INSPECTION, INC., 1020 Springfield Ave., Mountainside, NJ, 07092, (201) 654-3111, (1)(4).

EMPLOYEE RELOCATION COUNCIL (ERC), 1720 N St., NW, Washington, DC, 20036, (202) 857-0857, (18).

EN-THERM, INC., 509 West Fayette St., Syracuse, NY, 13204, (315) 474-6549, (1)(2).

ENCOSYS CORPORATION OF AMERICA, 7202 East 87th St. Suite 109, Indianapolis, IN, 46250, (317) 849-6060, (1).

ENERGY ALTERNATIVES UNLIMITED, INC., 8384 Middlecrest Rd., Boulder, CO, 80302, (303) 444-6063, (1)(2)(3).

ENERGY CONSERVATION SERVICES, 6290 North Vine Valley Rd., Rushville, NY, 14544, (716) 554-3994, (1)(2)(3).

ENERGY INDEPENDENCE/THE RADON RESPONSE TEAM, 9940 East 36th St., Independence, MO, 64052, (816) 353-3755, (1)(2)(4).

ENERGY LABORATORIES, INC., 254 North Center St. P.O. Box 3258, Casper, WY, 82602, (307) 235-0515, (1).

ENERGY MARKETING MIDWEST, INC., East 57 Rd. Box 93, Kelley, IA, 50134-8724, (515) 769-2400, (1).

ENERGY OUTFITTERS, LTD., 245 North Aldrich Ave., Minneapolis, MN, 55405, (612) 377-6502, (2).

ENERGY PROFESSIONALS, INC., 387 Irwin St., Lock Haven, PA, 17745, (717) 748-5910, (717) 726-3825, (1)(2).

ENERGY RESOURCE CENTER, P.O. Box 267, Colorado Springs, CO, 80901, (719) 636-2447, (1)(2)(3)(4)(6).

ENERGY RESOURCE TECHNOLOGY, 1494 North Charlotte St., Pottstown, PA, 19464, (215) 327-4641, (2).

ENERGY SAVER IMPORTS, INC., P.O. Box 387 2150 West 6th Ave. Suite E, Broomfield, CO, 80020, (303) 469-1787, (800) 288-3647, (4).

ENERGY SAVINGS WAREHOUSE, INC., 3490 Hwy. 22 W., Somerville, NJ, 08876, (201) 534-6922, (2)(4).

ENERGY SYSTEMS, INC., 134 Carr Ave., Cookeville, TN, 38501, (615) 528-7247, (1).

ENERGY UNLIMITED, RFD 1, Box 1090, Mercer, PA, 16137, (412) 662-3591, (1)(2).

ENERGY VENT, INC., 915 Valley St. P.O. Box 14145, Dayton, OH, 45414, (513) 222-7749, (2).

ENERMATRIX, INC., P.O. Box 466, Fargo, ND, 58107, (701) 232-3330, (4).

ENGINEERING ASSOCIATES, 4682 Watch Hill Rd., Manlius, NY, 13104, (315) 682-4389, (1)(2).

ENGINEERING DEVELOPMENT, INC., 4850 Northpark, Colorado Springs, CO, 80918, (717) 599-9080, (4).

ENRAD, INC./RADON TESTING AND ENGINEERING, INC., 18705 B North Frederick Rd., Gaithersburg, MD, 20879, (301) 948-8040, (1).

ENSA, INC., 4550 Bailey Ave., Buffalo, NY, 14226, (716) 834-3111, (1).

ENTESCO, 1910 Bobby Ray Rd., Clover, SC, 29710, (803) 222-4130, (1).

ENTROPY ENVIRONMENTALISTS, INC., P.O. Box 12291, Research Triangle Park, NC, 27709, (919) 781-3550, (1).

ENV SERVICES, INC., 216 Goddard Blvd. Suite 117, King of Prussia, PA, 19406, (215) 337-8222, (800) 345-6094, (1).

ENVIRADON, INC., P.O. Box 13683, Research Triangle Park, NC, 27709, (919) 848-2128, (1).

ENVIRADON, INC., 914 Rollingwood Dr., Mt. Holly, NC, 28120, (704) 827-1293, (1).

ENVIRO SCIENCES, INC, 1548 Valwood Pkwy. Suite 118, Carrollton, TX, 75066, (214) 406-8100 (817) 792-3434, (1).

ENVIRO TECH SYSTEMS, 430 Fernhill Ave., Pittsburgh, PA, 15226, (412) 561-2100, (1)(2)(4).

ENVIRO TECHNOLOGY CORPORATION, 945 Spring Rd., Pelham Manor, NY, 10803, (1).

ENVIRO TESTING, 97 Greenridge St., Johnson City, NY, 13790, (607) 770-9098, (607)797-3665, (1).

ENVIRO-CARE RADON SERVICES, 661 Wyoming St., Buffalo, NY, 14215, (716) 834-1788, (1)(2).

ENVIROHAZARDS MANAGEMENT, INC., P.O. Box 1054, South Orange, NJ, 07079, (201) 761-7226, (1).

ENVIROHEALTH, INC., 1090 General Greene Rd., Wayne, PA, 19087, (215) 964-0668, (1)(3).

ENVIRONMENT AIR, LTD., P.O. Box 459, Bouctouche, NB, E0A 1G0, (506) 743-8901, (4).

ENVIRONMENT LIAISON CENTRE (ELC), P.O. Box 72461, Nairobi, Kenya, (011) 254-224-770, (12).

ENVIRONMENTAL AIR CONTROL, INC., 2321 Northwest 30th Ct., Fort Lauderdale, FL, 33311, (305) 739-5159, (1)(2).

ENVIRONMENTAL ASPECS, INC., 1527 North Dale Mabry Suite 103, Lutz, FL, 33549, (813) 948-1387, (1).

ENVIRONMENTAL ASSOCIATES, 1222 Fairfield Ave., Bridgeport, CT, 06605, (203) 368-6064, (203) 368-6032, (1).

ENVIRONMENTAL AUDIT, INC., 1901 West 47th Pl. Suite 310, Westwood, KS, 68205, (913) 384-0000, (1).

ENVIRONMENTAL CHEMISTRY, INC., 1233 Ogden Suite 504, Denver, CO, 80218, (303) 831-0437, (1)(4).

ENVIRONMENTAL CONSULTANTS ASSOCIATES, 14 Ramapo Ln., Upper Saddle River, NJ, 07458, (201) 934-9866, (1).

ENVIRONMENTAL CONSULTING LABORATORY, 1353 Boston Post Rd. Suite 11, Madison, CT, 06443, (203) 245-7039, (1).

ENVIRONMENTAL CONTROL SYSTEMS, LTD., 15325 Jones Ln., Gaithersburg, MD, 20878, (301) 421-5150, (1)(3).

ENVIRONMENTAL CONTROL TECHNOLOGIES, 400 Amherst St., Nashua, NH, 03063, (603) 881-7133, (1).

ENVIRONMENTAL DATA COMPANY, INC., 2700 Youngfield, Lakewood, CO, 80215, (303) 238-1104, (1).

ENVIRONMENTAL DEFENSE FUND (EDF)/NATIONAL HEADQUARTERS, 257 Park Avenue South, New York, NY, 10010, (212) 505-2100, (17).

ENVIRONMENTAL DETECTION AGENCY, INC., P.O. Box 4854, Austin, TX, 78765, (512) 322-9013, (1)(4).

ENVIRONMENTAL DETECTION ASSOCIATES, INC., P.O. Box 1006 21 Middlemount Rd., Pisgah Forest, NC, 28768-1006, (704) 877-4300, (1)(3)(4).

ENVIRONMENTAL DETECTION MANAGEMENT, P.O. Box 12316, Research Triangle Park, NC, 27709, (919) 469-8005, (1).

ENVIRONMENTAL DETECTION SERVICES, 1046 West Peakview Ave., Littleton, CO, 80120, (303) 797-7771, (1).

ENVIRONMENTAL ENGINEERING AND TESTING, 9185 East Kenyon Ave. Suite 190, Denver, CO, 80237, (303) 722-2112, (1).

ENVIRONMENTAL ENGINEERING LABORATORIES, INC., P.O. Box 55304, Grand Junction, CO, 81505, (303) 245-2500, (1).

ENVIRONMENTAL ENTERPRISES, INC., P.O. Box 16847, Atlanta, GA, 30321, (404) 766-5551, (1)(2)(4).

ENVIRONMENTAL FIELD SERVICES, INC., 63 Elm St., Topsham, ME, 04086, (207) 725-4112, (1).

ENVIRONMENTAL HAZARD CONTROL, 590 Whitney Ave., New Haven, CT, 06511, (203) 787-1871, (1).

ENVIRONMENTAL HEALTH AND SAFETY, P.O. Box 20052 Suite 211, Brandenton, FL, 34206, (313) 242-1699, (1).

ENVIRONMENTAL HEALTH CONSULTANTS, INC., 1505 North Prospect, Champaign, IL, 61820, (217) 356-4797, (1).

ENVIRONMENTAL HEALTH INSPECTIONS, INC., 1200 Rte. 9 Cross River Mill Suite 5E, Lakewood, NJ, 08701, (201) 905-8888, (1)(3).

ENVIRONMENTAL HEALTH TECHNOLOGIES, 566 South Corona St., Denver, CO, 80209, (303) 778-1632, (1).

ENVIRONMENTAL HEALTH WATCH OF GREATER CLEVELAND, 4115 Bridge Ave., Cleveland, OH, 44113, (216) 961-4646, (1)(17).

ENVIRONMENTAL HOME SERVICES, INC., 12354 Howard Lodge Dr., Sykesville, MD, 21784, (301) 442-5861, (1)(2).

ENVIRONMENTAL INDOOR AIR QUALITY CONSULTANTS, INC. (EIC), 58 Keller Ave., Lancaster, PA, 17601, (717) 393-5774, (800) 367-7223, (1)(2).

ENVIRONMENTAL MAINTENANCE SPECIALISTS COMPANY, INC., 1429 Danby Rd., Ithaca, NY, 14850, (607) 273-5938, (2).

ENVIRONMENTAL MANAGEMENT CONTROL, INC., 2408 Belknap Beach, Louisville, KY, 40259, (502) 228-0284, (1).

ENVIRONMENTAL MANAGEMENT GROUP, 9841 Broken Land Pkwy. #117, Columbia, MD, 21046, (301) 290-7078, (1).

ENVIRONMENTAL OIL, INC., P.O. Box 315, Syracuse, NY, 13209, (315) 471-0503, (800) 262-1012, (1).

ENVIRONMENTAL POLICY INSTITUTE (EPI)/RADIATION HEALTH INFORMATION PROJECT, 218 D St., SE, Washington, DC, 20003, (202) 544-2600, (14).

ENVIRONMENTAL POLLUTION CONTROL, 195 Cliffwood Dr., Atlanta, GA, 30328, (400) 425-6454, (1)(2).

ENVIRONMENTAL PRO-TECH SERVICES, INC., 13 Edgar Dr. P.O. Box 599, West Milford, NJ, 07480, (201) 728-2666, (1).

ENVIRONMENTAL PRODUCTS ALLIANCE, LTD., 218 North 4th St., Suite 21 P.O. Box 2156, Bismarck, ND, 58502, (701) 258-5050, (1)(4).

ENVIRONMENTAL QUALITY ASSOCIATES, INC., 308 Harrison St., Fulton, NY, 13069, (315) 593-2085, (800) 727-2366, (1).

ENVIRONMENTAL QUALITY CONTROL, INC., 2130 South Sawtelle Ave., Los Angeles, CA, 90025, (213) 478-6668, (1).

ENVIRONMENTAL RADIATION GROUP, 4900 Northwest 19th Pl., Gainesville, FL, 32605, (904) 378-5581, (3).

ENVIRONMENTAL RADIATION SERVICES, INC., 305 West Magnolia St. Suite 373, Fort Collins, CO, 80521, (303) 226-5530 , (1).

ENVIRONMENTAL RADIOACTIVITY MEASUREMENTS, INC., P.O. Box 226, Piermont, NY, 10968, (914) 534-8679, (201) 768-7651, (1).

ENVIRONMENTAL RADON MEASUREMENTS, 49 North Orange Ave., Orlando, FL, 32801, (407) 648-1969, (800) 237-3065, (1)(2).

ENVIRONMENTAL RESEARCH ASSOCIATES, P.O. Box 219 490 Darby-Paoli Rd., Villanova, PA, 19085, (215) 687-5629, (1)(3).

ENVIRONMENTAL RISK CONSULTANTS, INC., 800 Cluster, Evanston, IL, 60202, (312) 864-1255, (1).

ENVIRONMENTAL SAFETY PRODUCTS, 8351 Roswell Rd. Suite 325, Atlanta, GA, 30350, (404) 889-3117, (1).

ENVIRONMENTAL SCIENCE AND ENGINEERING, INC., P.O. Box ESE, Gainesville, FL, 32602-3053, (904) 332-3318, (1).

ENVIRONMENTAL SCIENCE LABORATORY, INC.,

26 Pearl St. Suite 110, Bellingham, MA, 02019, (508) 966-4344, (800) 343-0707, (1)(4).

ENVIRONMENTAL SCIENCE SERVICES, 235 Promenade St., Providence, RI, 02908, (401) 421-0398, (1).

ENVIRONMENTAL SERVICES AND TECHNOLOGY, INC., 22 Oakland Ave., Atlantic Highlands, NJ, 07716, (201) 291-5525, (1)(2).

ENVIRONMENTAL SERVICES OF ILLINOIS, INC., 302 West State St., O'Fallon, IL, 62269, (618) 624-8079, (314) 421-4374, (1)(2).

ENVIRONMENTAL SOLUTIONS, INC., P.O. Box 11323, St. Louis, MO, 63105, (314) 863-9801, (1)(3)(4).

ENVIRONMENTAL SYSTEMS SERVICE, P.O. Box 512, Culpepper, VA, 22701, (703) 825-6660, (1).

ENVIRONMENTAL SYSTEMS SERVICE, LTD., P.O. Box 220, College Park, MD, 20740, (301) 779-0606, (1).

ENVIRONMENTAL TECHNOLOGIES, INC., P.O. Box 1300, Monticello, MN, 55362, (612) 295-3008, (1).

ENVIRONMENTAL TESTING, 133 Defense Hwy. Suite 105, Annapolis, MD, 21401, (301) 266-8700, (1).

ENVIRONMENTAL TESTING AND CONSULTING, 1316 Gress St., Manville, NJ, 08835, (201) 722-5293, (1).

ENVIRONMENTAL TESTING ASSOCIATES, 6021 Sandy Springs Cr. Suite 205, Atlanta, GA, 30328, (404) 255-3884, (1).

ENVIRONMENTAL TESTING ASSOCIATES, 240 Harvest Ln., Altoona, WI, 54720, (715) 832-8585, (1).

ENVIRONMENTAL TESTING ASSOCIATES, 9 White Pine Trail, Madison, WI, 53717, (608) 833-4073, (1)(2).

ENVIRONMENTAL TESTING ASSOCIATES, 2201 East Ridge Rd., Beloit, WI, 53511, (608) 365-2774, (1).

ENVIRONMENTAL TESTING ASSOCIATES, INC., P.O. Box 142, Media, PA, 19050, (215) 565-5665, (1).

ENVIROSCAPE, LTD., 4500 19th St., Suite 172, Boulder, CO, 80302, (303) 444-9076, (1).

ENVIROSCIENCE CONSULTANTS, INC., 66 Cedar St., Newington, CT, 06111, (203) 666-7167, (1)(3).

ENVIROSERV/MAIN OFFICE, 15 Buckley Hill Rd., Morristown, NJ, 07960, (201) 285-1065, (1).

ENVIROSERV/RADON DETECTION OF ALABAMA, 751-D Raleigh Ct., Birmingham, AL, 35209, (205) 942-0756, (1).

ENVIROSERV/RADON DETECTION OF OHIO, 781 Mount Vernon Ave., Marion, OH, 43302, (614) 382-2354, (1).

ENVIROSPHERE, 10900 Northeast Eighth St., Bellevue, WA, 98004, (206) 451-4600, (3).

ENVIROTECH, 132 Owsley Ave., Lexington, KY, 40502, (606) 269-4228, (1).

ENVIROTECH SERVICES, 201 South Paul, Springfield, IL, 62703, (217) 753-1346, (1).

ENVIROTEST, P.O. Box 223, North Ridgeville, OH, 44039-0223, (216) 327-9432, (1).

ERIE BOARD OF COOPERATIVE EDUCATIONAL SERVICES (BOCES), 2 Pleasant Ave., West Lancaster, NY, 14086, (716) 686-2094, (10).

ERM, INC., 49 North Orange Ave., Orlando, FL, 32801, (407) 648-1969, (800) 237-3065, (1)(2).

ERNEST W. BLOCHINGER, INC., 2 Burnbrae Ln., Sparta, NJ, 07871, (201) 729-6638, (1).

ERO RESOURCES CORPORATION, 2801 Youngfield Suite 121, Golden, CO, 80401, (303) 231-9022, (1).

ESSEX CHEMICAL CORPORATION, 1333 Broad St., Clifton, NJ, 07015, (201) 773-6300, (4).

ETE CONSULTING ENGINEERING, INC., P.O. Box 3471, Oak Ridge, TN, 37831, (615) 482-4503, (615) 546-0154, (1).

ETS ENVIRONMENTAL TESTING SERVICE, Rte. 2, Box 54, Mechanicsville, MD, 20659, (301) 884-4489, (301) 475-2169, (1).

EUROPEAN ENVIRONMENTAL BUREAU (EEB), 20, rue du Luxembourg, B-1040 Brussels, Belgium, No Phone Available, (12).

EUROPEAN INSTITUTE FOR WATER (EIW), 2, rue Boussingault, F-67000 Strasbourg, France, (011) 33-88-611631, (12).

EUROPEAN UNION OF GEOSCIENCES (EUG)/EUROPEAN SCIENCE FOUNDATION, 1, quai Lezay-Marnesia, F-67000 Strasbourg, France, (011) 33-88-353036, (12).

EVANSHIRE COMPANY, LTD., THE, Rte. 4, Box 57, Rome, NY, 13440, (315) 865-5941, (800) 962-5941, (1)(2)(4).

EVANSTON/DEPARTMENT OF BUILDING AND ZONING, 2100 Ridge Ave., Evanston, IL, 60204, (312) 328-2100, (10).

EVANSVILLE/AIR POLLUTION CONTROL OFFICER, Administration Building Room 207, Evansville, IN, 47708, (812) 426-5597, (10).

EXCEL INSPECTION SERVICE, INC., 438 Windmill Way, Somerville, NJ, 08876, (201) 526-7277, (1).

EXECUTIVE HOME CONSULTANTS, 19 Grove St., Waldwick, NJ, 07403, (201) 444-9571, (1).

EXPANDED SHALE, CLAY AND SLATE INSTITUTE (ESCSI), 6218 Montrose Rd., Rockville, MD, 20852, (301) 231-9497, (13).

EXPEDITERS, INC., THE, 404 North Summit Ave., Oconomowoc, WI, 53066, (414) 965-3063, (1).

F

F & J SPECIALTY PRODUCTS, INC., P.O. Box 660065, Miami Springs, FL, 33166, (305) 888-0383, (800) 832-5037, (4).

F.H. TIMBLIN COMPANY, P.O. Box 254, Industry, PA, 15052, (412) 643-1755, (4).

FAIRBANKS NORTH STAR BOROUGH/AIR POLLUTION CONTROL DIVISION, P.O. Box 1267, Fairbanks, AK, 99701, (907) 452-4761, (10).

FAIRFAX CITY DEPARTMENT.OF HEALTH, 10777 Main St. Suite 100, Fairfax, VA, 22030, (703) 246-2411, (10).

FAIRFAX COUNTY AIR POLLUTION CONROL, 10777 Main St. Suite 100A, Fairfax, VA, 22030, (703) 246-2541, (10).

FANTECH, INC., 2225 Industrial Blvd., Sarasota, FL, 34234, (813) 351-2947, (4).

FAST, 929 Gloucester Ave., Bricktown, NJ, 08723, (201) 920-2369, (3).

FAYTEK, INC., P.O. Box 2229, Coeur D'Alene, ID, 83814, (208) 667-3263, (1).

FCA DEVELOPMENT CORPORATION, P.O. Box 1268, Exton, PA, 19341, (215) 363-5460, (2).

FEDERAL HOME LOAN MORTGAGE CORPORATION (FREDDIE MAC), 1776 G St., NW, Washington, DC, 20006, (703) 759-8000, (18).

FEDERAL NATIONAL MORTGAGE ASSOCIATION (FANNIE MAE), 3900 Wisconson Ave., NW, Washington, DC, 20016, (202) 537-7000, (18).

FEDERAL NUCLIDE MONITORS, 2023 Park Rd., NW, Washington, DC, 20010, (202) 265-0429, (1).

FEDERAL RADON, INC., 206 Third St. P.O. Box 374, Cameron, MO, 64429, (816) 632-1318, (1)(4).

FEDERAL RADON, INC., 93 Parmalee Ave., Hawthorne, NJ, 07506, (201) 427-9391, (1).

FEMTO-TECH, INC., 150 C Industry Dr., Carlisle, OH, 45005, (513) 746-4427, (1)(4)(5).

FERNCO, INC., 300 S. Dayton, Davison, MI, 48423, (313) 653-9626, (800) 521-1283, (4).

FGHB ASSOCIATES, 1141 Hohlfelder Rd., Glencoe, IL, 60022, (312) 835-4433, (3).

FINN CONSULTANT ENGINEERS, INC., 227 North St., Dalton, MA, 01226, (413) 684-2019, (1)(2)(3).

FINNEX (RADON), 23 North Ave., Cincinnati, OH, 45215, (513) 821-7019 (915) 757-2068, (1).

FITCHBURG/BOARD OF HEALTH, 718 Main St., Fitchburg, MA, 01420, (617) 342-9843, (10).

FLORIDA ASSOCIATION OF REALTORS, P.O. Box 1231, Orlando, FL, 32802, (407) 849-1400, (18).

FLORIDA CONSERVATION FOUNDATION/ENVIRONMENTAL INFORMATION CENTER, 1191 Orange Ave., Winter Park, FL, 32789, (305) 644-5377, (14).

FLORIDA DEFENDERS OF THE ENVIRONMENT (FDE), 1523 Northwest 4th St., Gainesville, FL, 32608, (904) 372-6965, (17).

FLORIDA DEPARTMENT OF NATURAL RESOURCES/BUREAU OF MINE RECLAMATION, Innovation Park 2051 East Dirac Dr., Tallahassee, FL, 32304-3760, (904) 488-8217, (10).

FLORIDA DEPARTMENT OF COMMUNITY AFFAIRS/DIVISION OF HOUSING AND COMMUNITY DEVELOPMENT/CODES AND STANDARDS SECTION, 2740 Centerview Dr. The Rhyne Building, Tallahassee, FL, 32399, (904) 487-1824, (10).

FLORIDA DEPARTMENT OF HEALTH AND REHABILITATIVE SERVICES/DEPUTY SECRETARY FOR HEALTH/OFFICE OF RADIATION CONTROL, 1317 Winewood Blvd., Tallahassee, FL, 32399-0700, (904) 487-1004, (10).

FLORIDA DEPARTMENT OF HEALTH AND REHABILITATIVE SERVICES/DEPUTY SECRETARY FOR HEALTH/OFFICE OF RADIATION CONTROL, 1317 Winewood Blvd., Tallahassee, FL, 32399-0700, (904) 487-1004, (6)(10).

FLORIDA DEPARTMENT OF HEALTH AND REHABILITATIVE SERVICES/DEPUTY SECRETARY FOR HEALTH/OFFICE OF RADIATION CONTROL/ENVIRONMENTAL SECTION, P.O. Box 15490, Orlando, FL, 32858, (407) 628-2274, (407) 297-2097, (1)(10).

FLORIDA DEPARTMENT OF HEALTH AND REHABILITATIVE SERVICES/DEPUTY SECRETARY FOR HEALTH/OFFICE OF TOXICOLOGY AND HAZARDS ASSESSMENT, 1317 Winewood Blvd. Building 6 Room 466, Tallahassee, FL, 32399-0700, (904) 488-2905, (10).

FLORIDA DEPARTMENT OF LABOR AND EMPLOYMENT SECURITY/BUREAU OF INDUSTRIAL SAFETY AND HEALTH, Lafayette Building Room 204 2551 Executive Center Circle West, Talahassee, FL, 32301, (904) 488-3044, (10).

FLORIDA DEPARTMENT OF NATURAL RESOURCES, Cedars Executive Center 2639 North Monroe Box 20, Building B, Tallahassee, FL, 32399, (904) 488-3177, (10).

FLORIDA DEPARTMENT OF PROFESSIONAL REGULATION/DIVISION OF REAL ESTATE, P.O.

Box 1900, Orlando, FL, 32802-1900, (407) 423-6048, (10).

FLORIDA INSTITUTE OF PHOSPHATE RESEARCH, 1855 West Main St., Bartow, FL, 33830, (813) 533-0983, (10).

FLORIDA POWER AND LIGHT COMPANY/ENVIRONMENTAL AFFAIRS DEPARTMENT, P.O. Box 14000, Juno Beach, FL, 33408, (305) 694-3651, (13).

FLORIDA RADON CONSULTANTS, INC., 5448 Hoffner Ave. Suite 305, Orlando, FL, 32812, (407) 281-0339, (1).

FLORIDA STATE UNIVERSITY/DEPARTMENT OF GEOLOGY, 305 Carroway Bldg., Tallahassee, FL, 32306, (904) 644-5784, (5).

FMT EQUIPMENT, P.O. Box 6198, Hamden, CT, 06517, (203) 865-2864, (1).

FOAM ENTERPRISES, INC., 13630 Watertower Cir., Minneapolis, MN, 55441, (612) 559-3266, (800) 888-3342, (4).

FOMO PRODUCTS, INC., 2775 Barber Rd., Norton, OH, 44203, (216) 753-4585, (800) 321-5585, (4).

FORCHETTI PSI, P.O. Box 1234, Morrisville, PA, 19067, (215) 493-1646, (609) 695-1229, (1)(2).

FORENSIC ANALYTICAL SPECIALTY, 745 Lincoln Ave. Suite 4, Alameda, CA, 94501, (415) 887-8828, (415) 769-1180, (1).

FORESIGHT ENGINEERING, P.O. Box 349, Pittstown, NJ, 08867, (201) 996-3090, (1)(2).

FORESIGHT ENGINEERING, INC., P.O. Box 3111, Wayne, NJ, 07474-3111, (201) 942-3061, (1)(2).

FORESIGHT ENGINEERING, INC., P.O. Box 362, Monasquan, NJ, 08736, (201) 223-3316, (1)(2).

FORESIGHT ENGINEERING, INC., 17 Cook Ave. P.O. Box 621, Madison, NJ, 07940, (201) 377-0600, (1)(2).

FORESIGHT ENGINEERING, INC., VIRGINIA, P.O. Box 7333, Arlington, VA, 22207-0333, (703) 525-4790, (1).

FORSYTH COUNTY ENVIRONMENTAL AFFAIRS DEPARTMENT, 537 North Spruce St., Winston-Salem, NC, 27101, (919) 727-8060, (10).

FORT WORTH AIR POLLUTION CONTROL/DEPARTMENT OF PUBLIC HEALTH, 1800 University Dr., Fort Worth, TX, 76107, (817) 870-7280, (10).

FORUM FOR SCIENTIFIC EXCELLENCE, INC., THE, P.O. Box 682, Sparta, NJ, 07871, (201) 729-4823, (5).

FRANKLIN ENVIRONMENTAL ANALYSIS, 716 North Franklin St. Box 1-R, Philadelphia, PA, 19123-2721, (215) 592-8322, (1).

FRANKS, TERRY M., P.O. Box 5224, Louisville, KY, 40205, (603) 437-1425, (1).

FRAZIER COMPANY, THE, 15106 Industrial Rd., Omaha, NE, 68144, (402) 330-3400, (1).

FRED HUTCHINSON CANCER RESEARCH CENTER/PUBLIC HEALTH SCIENCES DIVISION, 1124 Columbia St., Seattle, WA, 98104, (206) 467-4267, (5).

FREDERICK COUNTY HEALTH DEPARTMENT, 12 East Church St., Frederick, MD, 21701, (301) 694-1700, (10).

FRESNO COUNTY AIR POLLUTION CONTROL DISTRICT, 1221 Fulton Mall, Fresno, CA, 93710, (209) 445-3239, (10).

FTLC SERVICES, INC., P.O. Box 563, Amherst, OH, 44001, (216) 960-2550, (1).

FULL HOUSE CONTRACTING, 1900 South Washington, Casper, WY, 82601, (307) 266-0895, (2).

FUTURA MEDICAL SYSTEMS, INC., 387 Regency Ridge, Centerville, OH, 45459, (513) 436-2215, (1).

G

G.A.M. BUILDING INSPECTION AND RADON MEASUREMENTS AND CONTROL, INC., 17223 East Purdue Pl., Aurora, CO, 80013, (303) 699-2827, (1)(2).

G.K. YUILL AND ASSOCIATES, LTD., 1441 Pembina Hwy., Winnipeg, MB, R3T 2C4, (204) 474-2461, (5).

G.M.D. CONSTRUCTION, INC., P.O. Box 204, Lisbon, MD, 21765, (301) 854-6186, (2).

G.S. TECHNOLOGIES, INC., 108 Hooten Rd., Mt. Laurel, NJ, 08054, (609) 234-2605, (1).

GALLAGHER AND ASSOCIATES, INC., 950 Sussex Blvd., Broomall, PA, 19008, (215) 328-2200, (4).

GALLAGHER ENGINEERING AND INSPECTION, 1164 Lincoln Heights Ave., Ephrata, PA, 17522, (717) 733-3734, (1).

GALLIPOT, INC., 2401 Pilot Knob Rd., St. Paul, MN, 55120, (612) 681-9517, (800) 423-6967, (1)(4).

GALSON TECHNICAL SERVICES, 6601 Kirkville Rd., East Syracuse, NY, 13057, (315) 432-0506, (1).

GALVESTON COUNTY HEALTH DISTRICT/AIR POLLUTION CONTROL DIVISION, 1207 Oak St. P.O. Box 939, La Marque, TX, 77568, (409) 938-7221, (10).

GAMMADATA/MATTEKNIK AB, Klostergatan 10 Box 1815, Uppsala S-75148, Sweden, 46-18155870, (4).

GARCO RESEARCH COMPANY, 144A Sutton Rd., Lebanon, NJ, 08833, (201) 832-7895, (1)(2).

GARDEN STATE HOME INSPECTION SERVICE, 475 Valley Rd., West Orange, NJ, 07052, (201) 678-3333, (1).

GARDEN STATE RADON DEWATERING SERVICES, INC., 2 Edgewater Pl. Suite 1A, New Brunswick, NJ, 08901, (201) 846-1948, (1)(2).

GARDNER, PHILIP, 136B Paoli Pike, Malvern, PA, 19355, (215) 889-0239, (1).

GARDWELL MANUFACTURING CORPORATION, 602 John F. Kennedy St. P.O. Box 7, Wampum, PA, 16157, (412) 535-4779, (4).

GARNER CONSULTANTS INSPECTION SERVICE, 1686 Fifth St., Trenton, NJ, 08638, (609) 883-2021, (1).

GARNET HOMES, INC., 3900 Germantown Rd. Suite 400, Fairfax, VA, 22030, (703) 591-4663, (13).

GARON PRODUCTS, INC., 1974 Hwy. 35 CN20, Wall, NJ, 07719, (201) 499-1776, (800) 631-5380, (4).

GARY/AIR POLLUTION CONTROL OFFICE, 846 Broadway 2nd Floor, Gary, IN, 46402, (219) 882-3000, (10).

GAS RESEARCH INSTITUTE (GRI)/INDOOR AIR QUALITY RESEARCH PROGRAM, 8600 West Bryn Mawr Ave., Chicago, IL, 60631, (312) 399-8100, (5)(13).

GASOMETRY ASSOCIATES, 29 Pocahontas Trail, Medford, NJ, 08055, (609) 983-3846, (1).

GAVIN, GERARD M., 214 East Township Line Rd., Havertown, PA, 19083, (215) 789-2718, (1).

GE SILICONES, 260 Hudson River Rd., Waterford, NY, 12188, (518) 237-3330, (800) 255-8886, (4).

GEIER CONSTRUCTION COMPANY, INC./DIVISION OF RADON CONTROL, 1506 Gridley Ln., Silver Spring, MD, 20902, (301) 649-2141, (1).

GENE CALVERT AND ASSOCIATES, P.O. Box 217, Waxahachie, TX, 75165, (214) 937-2223, (1).

GENERAL ENVIRONMENTAL MONITORING, 3312 Commercial Ave., Northbrook, IL, 60062-1909, (312) 291-0772, (1).

GENERAL HEALTH PHYSICS, INC., 7217 Lockport Pl., Lorton, VA, 22079, (703) 550-7525, (1).

GENERAL PHYSICS CORPORATION, 10650 Hickory Ridge Rd., Columbia, MD, 21044, (301) 290-2300, (301) 997-6241, (1).

GENOVA PRODUCTS, INC., 7034 East Court St., Davison, MI, 48423, (313) 744-4500, (800) 521-7488, (4).

GENUS CORPORATION, THE, 2400 Morris Ave., Union, NJ, 07083, (201) 687-5000, (1).

GEO-CHEK, INC., 2270 Stony Hill Rd., Boulder, CO, 80303, (303) 494-0576, (1).

GEO-ENVIRONMENTAL SERVICES, INC., 130 West Wieuca Rd. Suite 108, Atlanta, GA, 30342, (404) 256-6437, (1).

GEO-ENVIRONMENTAL SERVICES, INC., 444 North Frederick Ave. Suite L-148, Gaithersburg, MD, 20877-2432, (301) 353-0338, (1).

GEO-RADON SERVICES, P.O. Box 5985, Berkeley, CA, 94705, (415) 843-8844, (1)(2)(3).

GEOCEL CORPORATION, P.O. Box 398, Elkhart, IN, 46515, (219) 264-0645, (800) 348-7615, (4).

GEOCENTRIC ENGINEERING CORPORATION, P.O. Box 277, Lorton, VA, 22070, (703) 550-9253, (1).

GEOCHEM, INC., 3200 New Seward Hwy. Suite 201, Anchorage, AK, 99503, (907) 272-2121, (1)(2).

GEOHAZARDS, INC., P.O. Box 14956, Gainesville, FL, 32604, (904) 371-7243, (1)(2)(3).

GEOLOGICAL SOCIETY OF AMERICA (GSA), P.O. Box 9140 3300 Penrose Pl., Boulder, CO, 80301, (303) 447-2020, (14).

GEOMET TECHNOLOGIES, INC., 20251 Century Blvd., Germantown, MD, 20874, (301) 428-9898, (1)(2)(3)(5).

GEOPHYSICAL INSTRUMENT AND SUPPLY COMPANY, INC./GISCO, 4665 Joliet St., Denver, CO, 80239, (303) 371-1940, (4).

GEORGE MASON UNIVERSITY/CENTER FOR BASIC AND APPLIED SCIENCE, 4400 University Dr., Fairfax, VA, 22030, (703) 323-4318, (5).

GEORGE WASHINGTON UNIVERSITY/DEPARTMENT OF GEOLOGY, 2029 G St., NW, Washington, DC, 20052, (202) 994-6194, (5).

GEORGIA ASSOCIATION OF REALTORS, 3200 Presidential Dr., Atlanta, GA, 30340, (404) 451-1831, (18).

GEORGIA DEPARTMENT OF NATURAL RESOURCES/PROGRAM COORDINATION BRANCH, 205 Butler St., SE Floyd Towers East, Atlanta, Ga, 30334, (404) 656-4713, (10).

GEORGIA GEOLOGICAL SURVEY, 19 Martin Luther King, Jr., Dr., SW Room 400, Atlanta, GA, 30334, (404) 656-3214, (10).

GEORGIA INSTITUTE OF TECHNOLOGY/NUCLEAR ENGINEERING AND HEALTH PHYSICS PROGRAM, ESM Building, Atlanta, GA, 30332, (404) 894-3720, (5).

GEORGIA ONSITE CONSULTATION PROGRAM, Georgia Institute Of Technology EES/EDL/OSHG, Atlanta, GA, 30332, (404) 894-3806, (10).

GEORGIA RADON DETECTION COMPANY, 3400 Elm Creek Dr., Marietta, GA, 30064-1132, (404) 429-0384, (1).

GEORGIA REAL ESTATE COMMISSION, 40 Pryor St., SW, Atlanta, GA, 30303-3184, (404) 656-3916, (10).

GEOTHERMAL ENERGY COMPANY, LTD., P.O. Box 59, Wolfeboro, NH, 03894, (603) 569-2519, (4).

GERALD HARGROVE INSPECTION SERVICE, 5308 Meadowlark Ln., Grovetown, GA, 30813, (404) 860-0797, (1).

GIDDINGS AND ASSOCIATES, INC., 3049 Enterprise Dr., State College, PA, 16801, (814) 238-5927, (1)(2).

GILIAN INSTRUMENT CORPORATION, 8 Dawes Hwy., Wayne, NJ, 07470, (201) 831-0440, (4).

GILMONT CONSTRUCTION COMPANY, P.O. Box 1268, Exton, PA, 19341, (215) 363-5460, (2).

GITERMAN ASSOCIATES, 2912 Xenwood, St. Louis Park, MN, 55416, (612) 332-1129, (1).

GLENN COUNTY AIR POLLUTION CONTROL DISTRICT, 720 North Colusa St., Willows, CA, 95988, (916) 934-4651, (10).

GLOBAL BUILDING INSPECTORS, 39 East 4th St., Bayonne, NJ, 07002, (201) 858-2386, (800) 321-4677, (1).

GLOBAL RADON TESTING SERVICE TRAINING CENTER, 163 Padgett Dr., Lakeland, FL, 33809, (813) 859-5891, (1)(6).

GLOBAL WHOLESALE, 2410 East Lincolnway, Cheyenne, WY, 82001, (307) 634-7318, (1)(4).

GLOGORE ENGINEERING COMPANY, 3116 Stanley Rd., Akron, OH, 44313-3254, (216) 836-1880 (800) 702-9905, (1).

GMAC HOME IMPROVEMENT, 111 East Broadway St., Campbellsville, KY, 42718, (800) 222-5008, (2).

GODBEER, GODBEER AND ASSOCIATES, P.O. Box 128, New London, NH, 03257, (603) 526-9235, (1).

GOLDEN CONSTRUCTION SERVICE, 20 Distillery Rd., Warwick, NY, 10990, (914) 986-5600, (1).

GORDON, SAMUEL, 146 S. Little Tor Rd., New City, NY, 10956, (914) 634-9331, (3).

GOVERNMENT OF GUAM/DEPARTMENT OF LABOR, P.O. Box 23548 Guam Main Facility, Agana, GU, 96921, (671) 477-9821, (10).

GOVERNMENT OF GUAM/DEPARTMENT OF PUBLIC HEALTH AND SOCIAL SERVICES/DIVISION OF ENVIRONMENTAL HEALTH, P.O. Box 2816, Agana, GU, 96910, (671) 734-2671, (10).

GOVERNMENT OF GUAM/DEPARTMENT OF REVENUE AND TAXATION/REAL ESTATE AGENCY, 855 West Marine Dr., Agana, GU, 96910, (671) 477-1040, (10).

GOVERNMENT OF GUAM/ENVIRONMENTAL PROTECTION AGENCY, P.O. Box 2999, Agana, GU, 96910, (671) 646-8863, (10).

GOVERNMENT OF NEWFOUNDLAND AND LABRADOR/DEPARTMENT OF HEALTH/PUBLIC HEALTH INSPECTION DIVISION, P.O. Box 4750, St. John's, NF, A1C 5T7, (709) 576-2679, (11).

GOVERNMENT OF NEWFOUNDLAND AND LABRADOR/DEPARTMENT OF LABOUR/OCCUPATIONAL HEALTH AND SAFETY DIVISION/MEDICAL AND HYGIENE SERVICES, Beothuck Building Crosbie Pl., St. John's, NF, A1C 5T7, (709) 576-2645, (11).

GOY, TERRY, 1466 East Lake Dr., Chaska, MN, 55318, (612) 448-3664, (1).

GRAND RAPIDS/ENVIRONMENTAL SERVICES DEPARTMENT/AIR POLLUTION CONTROL DIVISION, 1300 Market, SW, Grand Rapids, MI, 49503, (616) 456-3158, (10).

GRANITE STATE ANALYTICAL, INC., 16 Manning St. Unit 107, Derry, NH, 03038, (603) 432-3044, (1)(4).

GRANITE STATE RADON SERVICES, P.O. Box 864, Windham, NH, 03087, (603) 437-1820, (1).

GRAVES RADON CONTRACTING, P.O. Box 266, Terre Hill, PA, 17581, (215) 445-7886, (1)(2).

GREAT BASIN UNIFIED AIR POLLUTION CONTROL DISTRICT, 157 Short St. Suite 6, Bishop, CA, 93514, (619) 872-8211, (10).

GREAT PLAINS NUCLEAR SERVICES, INC., 4502 Leavenworth St., Omaha, NE, 68106, (402) 553-3776, (1).

GREAT TRACERS, 3 Schoenbeck Rd., Prospect Heights, IL, 60070-1435, (312) 255-0436, (1)(4).

GREEN, GEORGE, RD 2, Box 2621 Big Hollow Rd., Lake George, NY, 12845, (518) 668-3233, (1).

GREENE COUNTY-CITY OF SPRINGFIELD/AIR POLLUTION CONTROL AUTHORITY, 227 East Chestnut Expressway, Springfield, MO, 65802, (417) 864-1662, (10).

GREENWICH HEALTH DEPARTMENT/DIVISON OF ENVIRONMENTAL HEALTH AND LABORATORY, Town Hall Field Point Rd., Greenwich, CT, 06830, (203) 622-7843, (10).

GREGORY K. MARSH AND ASSOCIATES, 7700 West 61st Ave. Suite 12, Arvada, CO, 80004, (303) 421-3383, (1).

GROUNDWATER TECHNOLOGY, INC., Meadowbrook Industrial Park, Milford, NH, 03055, (603) 672-5303, (3).

GROUNDWATER TECHNOLOGY, INC., 12 Walker Way, Albany, NY, 12205, (518) 456-2444, (1)(2).

GS ANALYTICS, INC., 108 Hooten Rd., Mt. Laurel, NJ, 08054, (609) 722-5557, (1).

GTS CONSULTANTS, 5599 North Maplewood Pl., Littleton, CO, 80123, (303) 794-1153, (3).

GUILFORD COUNTY DEPARTMENT HEALTH/DIVISON OF ENVIRONMENTAL HEALTH/AIR QUALITY PROGRAM, 301 North Eugene St. P.O. Box 3508, Greensboro, NC, 27401, (919) 373-3771, (10).

GUNDLE LINING SYSTEMS, INC., 19103 Gundle Rd., Houston, TX, 77073, (713) 443-8564, (800) 435-2008, (4).

GUNTER EXTERMINATING COMPANY, 220 West 72nd St., Kansas City, MO, 64114, (816) 523-0777, (1).

GURNEY KERR CONTRACTORS, INC., Box 753, Bryn Athyn, PA, 19009, (215) 947-7796, (2).

H

H & H DECORATING, 610 Great Falls Rd. Suite 100, Rockville, MD, 20850, (301) 251-6654, (1).

H & J FREILE HOME INSPECTIONS, INC., 99 Harrison Ave., North Plainfield, NJ, 07060, (201) 561-7147, (1).

H.B. FULLER, 315 South Hicks, Palatine, IL, 60067, (800) 323-7407, (4).

H.M. COOK, JR., PROFESSIONAL ENGINEER, 14632 Lufthansa Cr., Chantilly, VA, 22021, (703) 378-0114, (703) 248-3809, (1).

H.P.J. INSPECTION SERVICE, INC., 1303 Woodbridge Commons Way, Iselin, NJ, 08830, (201) 855-9475, (1).

HABITABILITY ENGINEERING CONSULTANTS (HECO), 2108 Southeast 12th St., Ocala, FL, 32671, (904) 622-3873, (1).

HALL-KIMBRELL ENVIRONMENTAL SERVICES, 5319 Southwest Westgate, Portland, OR, 97221, (503) 292-9406, (1)(2)(3).

HALLTECH, INC., 465 Coronation Dr., West Hill, ONT, M1E 2K2, (416) 284-6111, (4).

HAMILTON COUNTY-SOUTHWESTERN OHIO/AIR POLLUTION CONTROL AGENCY, 2400 Beekman St., Cincinnati, OH, 45214, (513) 251-8777, (10).

HAMMOND/AIR POLLUTION CONTROL, 5925 Calumet Ave., Hammond, IN, 46320, (219) 853-6306, (10).

HANCOR, INC., 401 Olive St. P.O. Box 1047, Findlay, OH, 45839, (800) 537-9520, (4).

HANDY ANDY, 90 Corona St. #1102, Denver, CO, 80218, (303) 733-2903, (1).

HARAD WATERPROOFING, 3301 North Second St., Harrisburg, PA, 17110, (717) 233-6143, (2).

HARDICO RADON MANAGEMENT, CENTRAL OFFICE, P.O. Box 40, Stevens, PA, 17578, (215) 267-6006, (2).

HARDICO RADON MANAGEMENT, COLUMBUS OFFICE, 95 Cherry St., Groveport, OH, 43215, (614) 836-2788, (2).

HARDICO RADON MANAGEMENT, D.C. OFFICE, 10720 Main St. Suite 205, Fairfax, VA, 22030, (703) 385-4200, (2).

HARDICO RADON MANAGEMENT, GREENVILLE OFFICE, 744 Isaqueena Trail Ridgecrest Apt. #21, Central, SC, 29630, (803) 654-6734, (2).

HARDICO RADON MANAGEMENT, HARRISBURG OFFICE, 730 Tower Rd., Enola, PA, 17025, (717) 732-9469, (2).

HARDICO RADON MANAGEMENT, PITTSBURGH OFFICE, 216 West Sycamore St., Pittsburgh, PA, 15211, (412) 381-6155, (2).

HARDICO RADON MANAGEMENT, ROANOKE OFFICE, P.O. Box 10494, Blacksburg, VA, 24062-0494, (703) 951-0808, (2).

HARFORD COUNTY HEALTH DEPARTMENT, P.O. Box 191, Bel Air, MD, 21014-0191, (301) 838-1500 x246, (10).

HARRIMAN ASSOCIATES, 292 Court St., Auburn, ME, 04210, (207) 784-5728, (1).

HARRIS CONSTRUCTION, P.O. Box 1203, Breckenridge, CO, 80424, (303) 453-2045, (1)(2).

HARRIS COUNTY POLLUTION CONTROL DEPARTMENT, 107 North Munger, Pasadena, TX, 77506, (713) 920-2831, (10).

HARTSELL HOME INSPECTION SERVICE, 5501 Executive Center Dr., Charlotte, NC, 28212, (704) 588-3781, (1).

HARVARD ENGINEERING SERVICES COMPANY, 28 Shaker Rd., Harvard, MA, 01451, (617) 456-3716, (1)(2).

HARVARD UNIVERSITY/SCHOOL OF PUBLIC HEALTH, 677 Huntington Ave., Boston, MA, 02115, (617) 732-1000, (5).

HAWAII ASSOCIATION OF REALTORS, 505 Ward Ave., Honolulu, HI, 96814, (808) 538-3641, (18).

HAWAII DEPARTMENT OF COMMERCE AND CONSUMER AFFAIRS/PROFESSIONAL AND VOCATIONAL LICENSING DIVISION, P.O. Box 3469, Honolulu, HI, 96801, (808) 548-7579, (10).

HAWAII DEPARTMENT OF HEALTH/ENVIRONMENTAL PROTECTION AND HEALTH SERVICES DIVISION/NOISE AND RADIATION BRANCH, 591 Ala Moana Blvd., Honolulu, HI, 96813, (808) 548-4383, (10).

HAWAII DEPARTMENT OF LABOR AND INDUSTRIAL RELATIONS/DIVISION OF

OCCUPATIONAL SAFETY AND HEALTH, 677 Ala Moana Suite 910, Honolulu, HI, 96813, (808) 548-2511, (10).

HAWAII DIVISION OF WATER AND LAND DEVELOPMENT, P. O. Box 373, Honolulu, HI, 96809, (808) 548-7533, (10).

HAWKS INTER-STATE, 814 North Main, Wichita, KS, 67203, (316) 267-8331, (316) 681-0031, (1).

HAYNES REALTY, 335 Nashville Pike, Gallatin, TN, 37066, (615) 452-7500, (1).

HAZEN RESEARCH, INC., 4601 Indiana St., Golden, CO, 80203, (303) 279-4501, (1).

HEALTH AND ENERGY INSTITUTE, 236 Massachusetts Ave., NE Suite 506, Washington, DC, 20002, (202) 543-1070, (15).

HEALTH AND SAFETY ASSOCIATES, 35 Grove St., Medford, MA, 02155, (617) 488-7081, (1)(3)(6).

HEALTH AND WELFARE CANADA/HEALTH PROTECTION BRANCH/BUREAU OF RADIATION AND MEDICAL DEVICES, 775 Brookfield Rd., Ottawa, ON, K1A 1C1, (613) 954-6677, (11).

HEALTH HAZARDS DETECTION, Route 3, Box 372, Albany, KY, 42602, (606) 387-7021, (1).

HEALTH PHYSICS ASSOCIATES, INC., RD 1 Box 796, Lenhartsville, PA, 19534, (215) 756-4153, (1).

HEALTH PHYSICS ASSOCIATES, LTD., 3312 Commercial Ave., Northbook, IL, 60062-1909, (312) 480-1900, (1)(3)(4).

HEALTH PHYSICS NORTHWEST, INC., 11535 Southwest 67th Ave., Tigard, OR, 97223, (503) 620-6617, (800) 762-8444, (1)(3)(4).

HEALTH PHYSICS SERVICES, INC., 4 Research Pl. Suite 140, Rockville, MD, 20850, (301) 670-1818, (1).

HEALTH PHYSICS SOCIETY (HPS), 800 Westpark Dr. Suite 400, McLean, VA, 22102, (703) 790-1745, (15).

HEALTH SCIENCE SERVICES CORPORATION, P.O. Box 102, Marlboro, NJ, 07746, (201) 536-2066, (1).

HEIL COMPANY, P.O. Box 25995, Colorado Springs, CO, 80918, (719) 593-8407, (4).

HEIMER INSPECTIONS OF NEW JERSEY, INC., 17 Catherine St., Carteret, NJ, 07008, (201) 541-8995, (1).

HELIX ENVIRONMENTAL, INC., 416 Triangle Ave., Dayton, OH, 45419, (513) 298-2990, (1)(3).

HENIGAR AND RAY ENGINEERING ASSOCIATES, INC., 640 East Hwy. 44, Crystal River, FL, 32629, (904) 795-6551, (904) 447-2889, (3).

HENRY INTERNATIONAL, INC., 14742 Plaza Dr., Tustin, CA, 92680, (714) 730-8451, (4).

HERITAGE RADON SERVICES, INC., 6968 Birchwood Rd., Harrisburg, PA, 17112, (717) 599-5060, (1)(2).

HERRICK ENGINEERING, INC., 2004 Belvedere Dr., Toledo, OH, 43614, (419) 385-5655, (1)(3).

HI-TEK POLYMERS, 9808 Bluegrass Pkwy., Louisville, KY, 40299, (502) 499-4300, (4).

HILLMANN ENVIRONMENTAL COMPANY, 1089 Cedar Ave., Union, NJ, 07083, (201) 686-3335, (1).

HMA HOME INSPECTION AND WARRANTY SERVICE, 13250 Ridge Rd. Unit 7-4, Largo, FL, 32257, (813) 587-0776, (800) 330-0776, (1).

HOECHST-CELANESE, INC., 1250 West Mockingbird Ln., Dallas, TX, 75247, (214) 689-4000, (4).

HOLIDAY TERMITE AND PEST CONTROL, 612 3rd St., Herndon, VA, 22070, (703) 471-5533, (1).

HOLMES INSPECTION SERVICE, 159 Liberty St., Little Ferry, NJ, 07643, (201) 641-8377, (800) 642-4603, (1).

HOME ANALYSTS, INC., 748 Considine Ave., Cincinnati, OH, 45205-2336, (513) 244-2860, (1).

HOME ASSURANCE, 17 Birchwood Dr., Naperville, IL, 60540, (312) 961-9154, (312) 355-6792, (1).

HOME BUYERS EDGE, 880 Benton St., Lakewood, CO, 80214, (303) 238-6275, (1).

HOME CHECK SERVICES, P.O. Box 98, Front Royal, VA, 22630, (703) 635-1274, (1)(2).

HOME CHEK, INC., 234 South Cardington, Wichita, KS, 67209, (316) 267-2324, (1).

HOME DETOX SPECIALISTS, 81 Candy Ln., Hamburg, NY, 14075, (716) 648-3321, (1)(2).

HOME ENVIRONMENT COMPANY, P.O. Box 24744, Speedway, IN, 46224, (317) 247-1881, (1)(2)(4).

HOME EXTERMINATING COMPANY, 4608 Bucks Schoolhouse Rd., Baltimore, MD, 21237, (301) 661-7911, (800) 492-5544, (1).

HOME INSPECTION OF KENTUCKIANA, 2416 Frankfort Ave. Suite 206, Louisville, KY, 40206, (502) 897-1450, (1).

HOME INSPECTION ONE, P.O. Box 1177 2 Valley Rd., Denville, NJ, 07834, (201) 586-2730, (1).

HOME INSPECTION SERVICE, INC., 1916 Southland Ave., Highland Park, IL, 60035, (312) 831-4545, (1)(2).

HOME INSPECTION SERVICE, INC., RR 1 Box 225 Walsh Rd., La Grangeville, NY, 12540, (914) 227-5417, (1).

HOME INSPECTION SERVICES, P.O. Box 455, Boulder, CO, 80306, (303) 443-9015, (1).

HOME INSPECTIONS BY CHANEL, INC., 20 Redding Pl., Towaco, NJ, 07082, (201) 263-9163, (1).

HOME INSPECTIONS OF MICHIGAN, 1007 St. Joseph Dr. P. O. Box 936, St. Joseph, MI, 49085, (616) 983-4663, (616) 983-7098, (1).

HOME INSPECTOR OF NORTH AMERICA, THE, 301 North Main St., Randolph, MA, 02368, (617) 963-0343, (1).

HOME OWNERS WARRANTY CORPORATION, 2000 L St., NW Suite 400, Washington, DC, 20036, (202) 463-4600, (18).

HOME RADON ASSOCIATES, 490 Main St., Norwich, CT, 06360, (203) 889-8104 (203) 887-1528, (1).

HOME RADON DETECTION COMPANY, 5824 El Monte, Fairway, KS, 66205, (913) 262-1555, (1).

HOME RADON DETECTION, INC., 33488 Lone Rock, Wildwood, IL, 60030, (312) 223-6517, (1).

HOME RADON SURVEYS, 4209 Olley Ln., Fairfax, VA, 22032, (703) 323-7236, (1).

HOME SENTRY, INC., P.O. Box 555, Midland Park, NJ, 07432, (201) 652-2497, (1).

HOME TESTING LABS, INC., 299 West Fort Lee Rd., Bogota, NJ, 07603, (201) 343-1199, (1)(3).

HOME TESTING LABS, INC., 44 Crestview Rd., Phoenixville, PA, 19460, (215) 768-7960, (1)(3).

HOME TESTING SERVICE, 1 Aaron Rd., Lexington, MA, 02173, (617) 862-3440, (1).

HOME-TECH, P.O. Box 423, Arnold, MD, 21012, (301) 757-8458, (1).

HOME-TECH ENGINEERING, P.O. Box 43129 51 Upper Montclair Plaza, Upper Montclair, NJ, 07043, (201) 744-6674, (1).

HOMEAIR TESTERS, 1234 Old Town South Dr., Indianapolis, IN, 46260, (317) 846-5711, (1).

HOMEBUYERS INSPECTION SERVICE OF NORTH WEST FLORIDA, INC., P.O. Box 9433, Pensacola, FL, 32513-9433, (904) 479-3276, (1).

HOMEBUYERS INSPECTION SERVICE OF SOUTH WEST FLORIDA, 2062 Stella St., Ft. Myers, FL, 33901, (813) 334-6116, (1).

HOMEBUYERS INSPECTION SERVICE, INC., 1000 East Chicago Ave., Naperville, IL, 60540, (312) 357-0903, (1).

HOMEPRO SYSTEMS, INC., 6715 Lowell Ave. Suite 2, McLean, VA, 22101, (703) 761-1400, (1)(3).

HOMERAD, 908 Wild Forest Dr., Gaithersburg, MD, 20879, (301) 889-5769, (1).

HOMES PLUS RADON SERVICES, INC., 490 Main St. P.O. Box 654, Norwich, CT, 06360, (203) 887-5128, (1)(4).

HOMESAFE RADON ENGINEERING, P.O. Box 6, Livingston, NJ, 07039, (201) 992-8060, (201) 342-2246, (1).

HOMESTEAD INSPECTION SERVICE, 7 Sunny View Dr., Succasunna, NJ, 07876, (201) 927-8343, (1).

HOMESURE INSPECTIONS, INC., 444 North Frederick Ave. Suite L348, Gaithersburg, MD, 20877-2432, (301) 428-7672, (1).

HONEYWELL, INC./RESIDENTIAL CONTROLS DIVISION, MN-10-2522 1985 Douglas Dr., N, Golden Valley, MN, 55422, (612) 542-3356, (4).

HORIZON ENGINEERING AND ENVIRONMENTAL SERVICES, INC., 599 Waldron Rd., LaVergne, TN, 37086, (615) 793-7993, (1).

HOUSE ANALYSIS AND ASSOCIATES, 747 Hallowell Dr., Huntingdon Valley, PA, 19006, (215) 947-1169, (215) 947-9209, (1).

HOUSE CHECK, INC./MARYLAND, 2025 Maryland Ave., Baltimore, MD, 21218, (301) 727-5553, (703) 550-7525, (1).

HOUSE CHEK, P.O. Box 209, Port Monmouth, NJ, 07758, (201) 495-0303, (1).

HOUSE DETECTIVES, INC., 2927 Champa St., Denver, CO, 80205, (303) 399-6557, (1).

HOUSE DOCTORS OF LAKEWOOD, INC., P.O. Box 489, Englewood, CO, 80151-0489, (303) 841-2443, (1).

HOUSE DOCTORS, INC., P.O. Box 10070, Colorado Springs, CO, 80932, (303) 574-6960, (1).

HOUSE MASTER OF AMERICA, 9989 West 60th Ave. Suite 201A, Arvada, CO, 80004, (303) 431-5383, (1).

HOUSE-CHECK, INC., P.O. Box 731, Valley Forge, PA, 19482, (215) 933-4431, (1).

HOUSEHOLD LABORATORIES, INC., P. O. Box 2247, Glenview, IL, 60025, (312) 724-6330, (1).

HOUSEHOLD RADON TESTERS, INC., 4 Seabury Ln., Downingtown, PA, 19335, (215) 525-8427, (215) 458-8730, (1).

HOUSEMASTER, 601 N. Olive St., Media, PA, 19063, (302) 658-2880, (1).

HOUSEMASTERS OF AMERICA, 2935 Hawk, Carson City, NV, 89701, (702) 849-2229, (1).

HOUSING CONSULTANTS, INC., P.O. Box 4838, Lousiville, KY, 40204, (502) 459-0772, (1).

HOUSING INSPECTION SERVICES, INC., P.O. Box 373, Murrysville, PA, 15668, (412) 374-1900, (1).

HOUSTON/BUREAU OF AIR QUALITY CONTROL, 7411 Park Place Blvd., Houston, TX, 77087, (713) 640-4200, (10).

HOWARD COUNTY HEALTH DEPARTMENT/BUREAU OF ENVIRONMENTAL HEALTH, P.O. Box 476, Ellicott City, MD, 21043, (301) 461-9955, (10).

HUDSON REGIONAL HEALTH COMMISSION, 215 Harrison Ave., Harrison, NJ, 07029, (201) 485-7001,

(10).

HUDSON VALLEY RADON TESTING CORPORATION, P.O. Box 790, Poughkeepsie, NY, 12601, (914) 431-6424, (914) 485-5561, (1).

HUGHES, RONALD L., 68 Hanshue Rd., Hummelstown, PA, 17036, (717) 534-1168, (1)(2).

HUNT AND ASSOCIATES, 802 Avenue B, Rome, GA, 30161, (404) 232-7653, (404) 232-2034, (1).

HUNT'S WATER CONTROL, INC., 607 Tower View Dr., Taylor Mill, KY, 41015, (606) 292-0333, (1)(2).

HUNTER-DON PEST CONTROL, 777 Mountain View Rd., Asbury, NJ, 08802, (201) 638-4024, (2).

HUNTSVILLE/AIR POLLUTION CONTROL DEPARTMENT, 2033 City Airport Rd., Huntsville, AL, 35802, (205) 881-7803, (10).

HURT AND PROFFITT, PROFESSIONAL ENGINEERS, 2524 Langhorne Rd., Lynchburg, VA, 24501, (804) 847-7796, (1).

HUTCHISON PEST CONTROL, 2662 North Freeway, Pueblo, CO, 81003, (719) 544-7378, (719) 633-7378, (1).

I

I & M RADON MITIGATION, 529 Browar, Eaton, OH, 45320, (513) 456-6784, (2).

I-T.E.M., LTD., 12850 Middlebrook Rd. Suite 301, Germantown, MD, 20874, (301) 353-0585, (1).

IDAHO ASSOCIATION OF REALTORS, 1450 West Bannock, Boise, ID, 83702, (208) 342-3585, (18).

IDAHO DEPARTMENT OF HEALTH AND WELFARE/DIVISION OF ENVIRONMENTAL QUALITY/BUREAU OF HAZARDOUS MATERIALS/RADIATION CONTROL DIVISION, 450 West State St., Boise, ID, 83720, (208) 334-5879, (10).

IDAHO GEOLOGICAL SURVEY, University Of Idaho Morrill Hall, Room 332, Moscow, ID, 83843, (208) 885-7991, (10).

IDAHO OSHA ONSITE CONSULTATION PROGRAM, Boise State University Community And Environmental Health 1910 University Drive, Boise, ID, 83725, (208) 385-3929, (10).

IDAHO REAL ESTATE COMMISSION, Statehouse Mail, Boise, ID, 83720, (208) 334-3285, (10).

ILLINOIS ASSOCIATION OF REALTORS, Illinois Realtors Pl. Suite 400, Springfield, IL, 62703, (217) 529-2600, (18).

ILLINOIS DEPARTMENT OF COMMERCE AND COMMUNITY AFFAIRS/DIVISION OF INDUSTRIAL SERVICES, State Of Illinois Center 100 West Randolph St. Suite 3-400, Chicago, IL, 60601, (800) 972-4140, (800) 972-4216, (10).

ILLINOIS DEPARTMENT OF NUCLEAR SAFETY/OFFICE OF ENVIRONMENTAL SAFETY, 1035 Outer Park Dr., Springfield, IL, 62704, (217) 786-6399, (1)(10).

ILLINOIS DEPARTMENT OF PROFESSIONAL REGULATION, 320 West Washington St. 3rd Floor, Springfield, IL, 62786, (217) 785-0800, (10).

ILLINOIS ENVIRONMENTAL PROTECTION AGENCY/DIVISON OF AIR POLLUTION CONTROL, 2200 Churchill Rd. P.O. Box 19276, Springfield, IL, 62794-9276, (217) 782-7326, (10).

ILLINOIS STATE GEOLOGICAL SURVEY, Natural Resources Building 615 East Peabody Dr. Room 121, Champaign, IL, 61820, (217) 333-5111, (1)(10).

ILMASTI ELECTRONICS, LTD., P.B. 50, Vantaa SF-01711, Finland, 011-358-0-840711, (4).

IMC DEVELOPMENT CORPORATION, P.O. Box 7047, Lakeland, FL, 33807, (813) 646-5061, (16).

IMPERIAL POLK COUNTY/BUILDING DIVISION, P.O. Box 726, Bartow, FL, 33830, (813) 534-6573, (10).

IMPROVED TECHNOLOGY SYSTEMS, P.O. Box 8244, Canton, OH, 44711-8244, (216) 452-1493, (1).

INDEPENDENCE PLANNING DEPARTMENT/AIR POLLUTION CONTROL DIVISION, 111 East Maple Ave., Independence, MO, 64050, (816) 836-8300 x363, (10).

INDEPENDENT HOME INSPECTION SERVICE, INC., 51 Forest Rd., Demarest, NJ, 07627, (201) 767-3449, (1).

INDIANA ASSOCIATION OF REALTORS, 7301 North Shadeland, Indianapolis, IN, 46250, (317) 842-0890, (18).

INDIANA DEPARTMENT OF LABOR/BUREAU OF SAFETY, EDUCATION AND TRAINING, 1013 State Office Building, Indianapolis, IN, 46204, (317) 232-2688, (10).

INDIANA GEOLOGICAL SURVEY, 611 North Walnut Grove, Bloomington, IN, 47405, (812) 335-1950, (10).

INDIANA REAL ESTATE COMMISSION, 100 North Senate Ave. 1021 State Office Building, Indianapolis, IN, 46204, (317) 232-2980, (10).

INDIANA SEAL COMPANY, 12000 Mitchell Rd., Carmel, IN, 46032, (317) 844-8700, (800) 428-5309, (4).

INDIANA STATE BOARD OF HEALTH/RADIOLOGICAL HEALTH SECTION, 1330 West Michigan St. Box 1964, Indianapolis, IN, 46206-1964, (317) 633-0153, (10).

INDIANAPOLIS/AIR POLLUTION CONTROL DIVISION, 2700 South Belmont Ave., Indianpolis, IN, 46221, (317) 633-5496, (10).

INDOOR POLLUTION CONTROL CENTER, 3921 Sawyer Rd., Sarasota, FL, 34233, (813) 923-3641, (800) 226-2636, (1)(2).

INDUSTRIAL OPERATIONS, INC./RADON EVALUATION, 3736 South 7th Ave., Phoenix, AZ, 85041, (602) 268-8772, (1).

INDUSTRIAL RESOURCE DEVELOPMENT COMPANY, 2490 General Armistead Ave. Suite 303, Norristown, PA, 19403, (215) 630-9150, (1)(2).

INDUSTRIAL SAFETY EQUIPMENT ASSOCIATION (ISEA), 1901 North Moore St., Arlington, VA, 22209, (703) 525-1695, (16).

INEX, 8704 Gutierrez, NE, Albuquerque, NM, 87111, (505) 292-0805, (2).

INFILTEC, D.C., 1105 Park St., NE, Washington, DC, 20002, (202) 547-1611, (1).

INFILTEC/FACTORY, 108 South Delphine Ave. P.O. Box 1125, Waynesboro, VA, 22980, (703) 943-2776, (1)(4).

INFILTEC/MAIN OFFICE, P.O. Box 8007, Falls Church, VA, 22041, (703) 820-7696, (1)(2)(3)(5)(6).

INFOTECH HOME INSPECTION SERVICES, 984 Route 9, Parlin, NJ, 08859, (201) 727-5335, (1).

INK FILTERING SYSTEMS, 3402 Habersham Club Dr., Cumming, GA, 30130, (404) 889-1061, (4).

INSPECT, 7737 South Normandy, Burbank, IL, 60459, (312) 599-3553, (1).

INSPECT-A-HOME OF UTAH, 455 East 400 S., Salt Lake City, UT, 84111, (801) 532-4663, (1).

INSPECT-A-HOME, INC., 37 Janet Ln., Springfield, NJ, 07081, (201) 379-4124, (1).

INSPECTIONS UNLIMITED, 514 Ryers Ave., Cheltenham, PA, 19012, (215) 848-0720, (1).

INSPEX OF BOULDER, INC., 2955 Valmont Suite 300, Boulder, CO, 80301, (303) 442-8995, (1).

INSTA-FOAM PRODUCTS, INC., 1500 Cedarwood Dr., Joliet, IL, 60435, (800) 435-9359, (4).

INSTITUTE FOR EUROPEAN ENVIRONMENTAL POLICY (IEEP), Aloys-Schulte-Strasse 6, D-5300 Bonn 1, West Germany, (011) 49-228-213810, (12).

INSTITUTE OF ENVIRONMENTAL HEALTH/OCCUPATIONAL SAFETY AND HEALTH SECTION, Colorado State University 110 Veterinary Science Building, Fort Collins, CO, 80523, (303) 491-6151, (10).

INSTITUTE OF REAL ESTATE MANAGEMENT, 430 North Michigan Ave., Chicago, IL, 60611, (312) 661-1930, (18).

INSUL-TECH, INC., 5559 Westerville Rd., Westerville, OH, 43081, (614) 891-9555, (1)(2).

INSULATING SYSTEMS, 340 Gargrave Rd., West Carrollton, OH, 45449, (513) 435-0079, (1)(2).

INSURANCE HEALTH SERVICES, 35 Stony Brook Rd., Westport, CT, 06880, (203) 226-3444, (1).

INTEGRAL RESOURCES, INC., 1916 East El Parque, Tempe, AZ, 85282, (602) 967-8029, (1).

INTEGRATED PEST MANAGEMENT, Box 93, RD #2 White Horse Pike, Hammonton, NJ, 08037, (609) 567-0100, (1).

INTERMOUNTAIN RADON SERVICE, P.O. Box 3, Salida, CO, 81201, (719) 539-6726, (1)(2)(6).

INTERNATIONAL ASSOCIATION OF HYDROGEOLOGISTS (IAH), Postbus 9090, NL-6800 Arnhem, Netherlands, (011) 3185-340396, (12).

INTERNATIONAL ASSOCIATION OF HYDROLOGICAL SCIENCES (IAHS)/INSTITUTE OF HYDROLOGY, Maclean Building Crowmarsh Gifford Wallingford, Oxfordshire, UK, 0X10 8BB, (011) 44-491-8800, (12).

INTERNATIONAL ASSOCIATION OF RADIATION RESEARCH (IARR)/RADIOBIOLOGICAL INSTITUTE TNO, Lange Kleiweg 151, NL-2288 GJ Rijswijk, Netherlands, (011) 3115-136940, (12).

INTERNATIONAL CHEMICAL WORKERS UNION (ICWU), 1655 West Market, Akron, OH, 44313, (216) 867-2444, (16).

INTERNATIONAL COMMISSION ON RADIATION UNITS AND MEASUREMENTS (ICRU), 7910 Woodmont Ave. Suite 1016, Bethesda, MD, 20814, (301) 657-2652, (12).

INTERNATIONAL COMMISSION ON RADIOLOGICAL PROTECTION (ICRP), Clifton Ave., Sutton (Surrey), UK, SM2 5PU, (011) 44-642-4680, (12).

INTERNATIONAL EXPLORATIONS, 577 Sackettsford Rd., Warminster, PA, 18974, (215) 598-7137, (1)(2).

INTERNATIONAL EXTERMINATOR CORPORATION, 320 South Freeway, Fort Worth, TX, 76104, (817) 335-5531, (800) 666-7287, (1).

INTERNATIONAL GEOGRAPHICAL UNION (IGU), University of Alberta Department of Geology, Edmonton, AB, T6G 2H4, (403) 432-3329, (12).

INTERNATIONAL GEOLOGICAL CORRELATION PROGRAM (IGCP), C/O UNESCO Maison de l' UNESCO 7 Place de Fontenoy, F-75700 Paris, France, (011) 331-45684123, (12).

INTERNATIONAL MEDCOM, 7497 Kennedy Rd., Sebastopol, CA, 95472, (707) 823-0336, (4).

INTERNATIONAL PEST CONTROL, INC., P.O. Box 5922, Edison, NJ, 08818, (201) 494-9313, (1).

INTERNATIONAL RADIATION PROTECTION ASSOCIATION (IRPA), BP N Q 33, F-92260 Fontenay-aux-Roses, France, (011) 331-746547139, (12).

INTERNATIONAL RADON TECHNOLOGIES, 2417 Regency Rd., Lexington, KY, 40503, (606) 278-0104, (800) 537-5066, (1)(4).

INTERNATIONAL UNION AGAINST TUBERCULOSIS AND LUNG DISEASE (IUATLD), 68, Boulevard Saint-Michel, 75006 Paris, France, No Phone Available, (12).

INTERNATIONAL UNION OF GEOLOGICAL SCIENCES (IUGS), Postboks 3006, N-7002 Tronheim, Norway, (12).

INTERSTATE CONFERENCE ON WATER PROBLEMS (ICWP), 2300 M St., NW Suite 800, Washington, DC, 20037, (202) 466-7287, (14).

ION SYSTEMS, INC., 2546 10th St., Berkeley, CA, 94710, (415) 548-3640, (4).

IOWA ASSOCIATION OF REALTORS, 999 Oakridge Dr., Des Moines, IA, 50308, (515) 244-2294, (18).

IOWA BUREAU OF LABOR/IOWA OCCUPATIONAL SAFETY AND HEALTH ADMINISTRATION, 307 East Seventh St., Des Moines, IA, 50319, (515) 281-5352, (10).

IOWA DEPARTMENT OF COMMERCE/PROFESSIONAL LICENSING AND REGULATION DIVISION, 1918 SE Hulsizer, Ankeny, IA, 50021, (515) 281-3183, (10).

IOWA DEPARTMENT OF NATURAL RESOURCES/ENVIRONMENTAL PROTECTION DIVISION, Henry A. Wallace Building, Des Moines, IA, 50319-0034, (515) 281-5145, (10).

IOWA DEPARTMENT OF PUBLIC HEALTH/DIVISION OF DISEASE PREVENTION, Lucas State Office Building, Des Moines, IA, 50319, (515) 281-6549, (10).

IOWA DEPARTMENT OF PUBLIC HEALTH/DIVISION OF DISEASE PREVENTION/BUREAU OF RADIOLOGICAL HEALTH, Lucas State Office Building, Des Moines, IA, 50319, (515) 281-7781, (10).

IOWA GEOLOGICAL SURVEY BUREAU, 123 North Capitol St., Iowa City, IA, 52242, (319) 335-1575, (10).

IOWA STATE UNIVERSITY/BUSINESS AND ENGINEERING EXTENSION/IOWA RADON PROJECT, EES-Building, Room 101 Haber Road, Ames, IA, 50011-0588, (515) 294-0588, (5).

IQBAL CONSULTING ENGINEERS, 5 Ironstone Dr., Reading, PA, 19606, (215) 779-8345, (3).

J

J & J QUALITY HOME INSPECTIONS, 909 East Lynda Ln., Marietta, GA, 30060, (404) 425-8868, (1).

J. BECK AND ASSOCIATES, INC., 35 Meetinghouse Ln., Bradford Woods, PA, 15015, (412) 935-6711, (1)(2).

J.A. ASSOCIATES/PUBLIC HEALTH CONSULTANTS, 313 Clear Spring Rd., Lansdale, PA, 19446, (215) 855-5592, (1).

J.J. LAY COMPANY, 3130 South Platte River Dr. P.O. Box 1745, Englewood, CO, 80150, (303) 761-2021, (1).

J.M. FROHN CORPORATION, 35 Bowdoin Dr., Milford, MA, 01757, (508) 478-4342, (1).

J.M. MONTGOMERY ENGINEERS, 555 East Walnut St., Pasadena, CA, 91101, (818) 796-9141, (1).

J.M.C. ENTERPRISES, 700 Ferndale Rd., Johnson City, TN, 37604, (615) 282-2136, (1).

JACKSONVILLE/DUVAL COUNTY BIO-ENVIRONMENTAL SERVICES DIVISION, 515 West 6th St., Jacksonville, FL, 32206, (904) 630-3210, (10).

JANCE ASSOCIATES, INC., P.O. Box 234, East Texas, PA, 18046, (215) 398-0434, (2).

JB TESTING COMPANY, P.O. Box 1303, Frederick, MD, 21701, (301) 662-7043, (1).

JBK INSPECTION SERVICE, INC., 225B Cottrell Rd., Matawan, NJ, 07747, (201) 525-9410, (1).

JEFFERSON COUNTY DEPARTMENT OF HEALTH, 1400 Sixth Avenue South, Birmingham, AL, 35233, (205) 933-9110, (10).

JEFFERSON COUNTY HEALTH DEPARTMENT, 260 South Kipling, Lakewood, CO, 80226, (303) 232-6301, (10).

JEFFERSON COUNTY/AIR POLLUTION CONTROL, 914 East Broadway, Louisville, KY, 40204, (502) 584-8151, (10).

JH AND ASSOCIATES, INC., 2524 East 71st St., Tulsa, OK, 74136, (918) 455-4984, (1).

JMX INDUSTRIES, INC., 3959 Wilshire Blvd. Suite 200/B11, Los Angeles, CA, 90010, (213) 871-8054, (1).

JOHN A. FIXARY, INC., P.O. Box 490, Riverside, CT, 06878, (203) 622-6661, (1).

JOHN L. RENSHAW, INC., 10738 Tucker St., Beltsville, MD, 20705, (301) 937-4440, (2).

JOHN PRIKKEL AND ASSOCIATES, P.O. Box 14145, Dayton, OH, 45414, (513) 222-7749, (513) 278-3176, (1).

JOHNS HOPKINS UNIVERSITY/DEPARTMENT OF ENVIRONMENTAL HEALTH SCIENCES/DIVISION OF ENVIRONMENTAL HEALTH ENGINEERING, 615 Wolfe St., Baltimore, MD, 21205, (301) 955-3720, (5).

JOHNSON PLUMBING AND HEATING, 232 South Main St., Chambersburg, PA, 17201, (717) 264-5028, (1)(2).

JOHNSON'S ENERGY SERVICE, 104 Central Dr., Forest City, IA, 50436, (515) 582-2998, (1)(2).

JOYCE EXTERMINATING COMPANY, Route 2, Box 249, Ramseur, NC, 27316, (919) 824-2358, (1).

JP HOME INSPECTION, INC., 100 Polifly Rd. Suite 4I, Hackensack, NJ, 07601, (201) 488-0417, (1).

JPS ELASTOMERICS CORPORATION, 395 Pleasant St., Northampton, MA, 01060, (413) 586-8750, (4).

JRM ASSOCIATES, Box 602 North Barnstead Rd., Center Barnstead, NH, 03225, (603) 776-6046, (1).

JRW INDUSTRIES, INC., 2312 North 2nd St., Harrisburg, PA, 17110, (717) 236-1418, (1)(4).

JUNGBLUT, DAVID W., GEOLOGIST, Route 12 Box 168, Baptistown, NJ, 08803, (201) 996-3688, (1).

K

K & J SOLAR SERVICE, 1142 Fillmore St., Denver, CO, 80206, (303) 393-8372, (1).

K & S ASSOCIATES, 1854 Airlane Dr., Nashville, TN, 37210, (615) 883-9760, (3)(4).

KALKA MIDWEST CONSTRUCTION, INC., 1710 South Michigan St., South Bend, IN, 46613, (219) 287-3399, (219) 234-9206, (1).

KANSAS ASSOCIATION OF REALTORS, 3644 Southwest Burlingamr Road, Topeka, KS, 66611, (913) 267-3610, (18).

KANSAS CITY, MISSOURI/AIR QUALITY SECTION, 21st Floor, City Hall 414 East 12th St., Kansas City, MO, 64106, (816) 274-2501, (10).

KANSAS CITY, KANSAS-WYANDOTTE COUNTY/HEALTH DEPARTMENT/AIR POLLUTION CONTROL DIVISION, 619 Ann Ave., Kansas City, KS, 66101, (913) 321-4803, (10).

KANSAS DEPARTMENT OF HEALTH AND ENVIRONMENT/BUREAU OF AIR QUALITY AND RADIATION CONTROL/ENVIRONMENTAL SURVEILLANCE AND EMERGENCY PLANNING SECTION, Forbes Field Building Suite 740, Topeka, KS, 66620, (913) 296-1567, (1)(10).

KANSAS DEPARTMENT OF HUMAN RESOURCES/DIVISION OF INDUSTRIAL SAFETY AND HEALTH, 512 West Sixth St., Topeka, KS, 66603, (913) 296-4386, (10).

KANSAS GEOLOGICAL SURVEY, The University Of Kansas 1930 Constant Ave. Campus West, Lawrence, KS, 66046, (913) 864-3965, (10).

KANSAS REAL ESTATE COMMISSION, Landon State Office Building 900 Jackson St. Room 501, Topeka, KS, 66612-1220, (913) 296-3411, (10).

KANSAS TESTING SERVICES, 957 North Coolidge, Wichita, KS, 67203, (316) 267-8378, (1).

KBC TECHNOLOGY, Box 170 RR 2, Raymond, IL, 62580, (217) 229-4437, (1).

KELLOGG ENVIRONMENTAL SERVICES, INC., 816 Anthony Dr., Mechanicsburg, PA, 17055, (717) 732-3364, (717) 732-9263, (1)(2).

KENMAR, INC., 2626 North Martha St., Philadelphia, PA, 19125-1493, (215) 739-1245, (4).

KENTUCKY ASSOCIATION OF REALTORS, 2224 Young Dr., Lexington, KY, 40505, (606) 266-2185, (18).

KENTUCKY DEPARTMENT OF HUMAN RESOURCES/DIVISION OF RADIATION AND PRODUCT SAFETY/RADIATION CONTROL BRANCH, 275 East Main St., Frankfort, KY, 40621, (502) 564-3700, (10).

KENTUCKY DEPARTMENT OF LABOR/EDUCATION AND TRAINING, OCCUPATIONAL SAFETY AND HEALTH, U.S. Highway 127 South, Frankfort, KY, 40601, (502) 564-6895, (10).

KENTUCKY GEOLOGICAL SURVEY, University Of Kentucky 311 Breckinridge Hall, Lexington, KY, 40506, (606) 257-5863, (10).

KENTUCKY REAL ESTATE COMMISSION, 222 South First St. Suite 300, Louisville, KY, 40202, (502) 588-4462, (10).

KERN COUNTY AIR POLLUTION CONTROL DISTRICT, 2700 M St., Bakersfield, CA, 93305-4198, (805) 861-3682, (10).

KEY TECHNOLOGY, INC., P.O. Box 562, Jonestown, PA, 17038, (717) 274-8310, (800) 523-4964, (1)(4)(6).

KINGS COUNTY AIR POLLUTION CONTROL DISTRICT, 330 Campus Dr., Hanford, CA, 93230, (209) 584-1411, (10).

KINSLOW, ROY H., 9935 Coppedge Ln., Dallas, TX, 75220, (214) 351-1446, (3).

KITTREDGE CONSTRUCTION COMPANY, 2758 South Langley Ct., Denver, CO, 80210, (303) 825-4940, (2).

KNOX COUNTY DEPARTMENT OF AIR POLLUTION CONTROL, 400 Main Ave. City-County Building, Room 459, Knoxville, TN, 37902, (615) 521-2488, (10).

KOVACS CONSULTANTS, 4515 Willard Ave. Suite

1921S, Chevy Chase, MD, 20815, (301) 652-7133, (2).

KSE, INC., P.O. Box 368, Amherst, MA, 01004, (413) 549-5506, (1).

KUNKEL, HAROLD E., 6 Radnor Rd., Newtown Square, PA, 19073, (215) 356-5096, (2).

KUTZTOWN UNIVERSITY/DEPARTMENT OF PHYSICAL SCIENCES/RADON PROJECT, 110 Grim Science Building, Kutztown, PA, 19530, (215) 683-4443, (5).

L

L. CLARK AND SONS, INC., 986 Fisher Rd., Fitchburg, MA, 01420, (508) 343-7849, (1)(2).

L.G. PAUL, SR., INC., 333 Union St., Allentown, PA, 18102, (214) 437-3530, (1).

L.H. FRIEND AND COMPANY, 1875 Century Park, E., Los Angeles, CA, 90210, (213) 556-5488, (800) 351-9923, (3).

L.J. QUIGLIANO, II, ENTERPRISES, 2173 Seneca Ave., Niagara Falls, NY, 14305, (716) 284-4074, (2).

LACO, P.O. Box 666, Chesterland, OH, 44026, (216) 729-3034, (4).

LAKE BUILDERS, INC., Star Rte. 2, Box 39 Rte. 6, Hawley, PA, 18428, (717) 226-1850, (2).

LAKE COUNTY AIR QUALITY MANAGEMENT DISTRICT, 883 Lakeport Blvd., Lakeport, CA, 95453, (707) 263-7000, (10).

LAKE COUNTY GENERAL HEALTH DISTRICT/AIR POLLUTION CONTROL DIVISION, 105 Main St. P.O. Box 490, Painesville, OH, 44077, (216) 357-2543, (10).

LAKE COUNTY HEALTH DEPARTMENT, 2293 North Main St., Crown Point, IN, 46307, (219) 738-2020, (10).

LAMSON HOME INSPECTIONS, 9 Lougheed Ave., West Caldwell, NJ, 07006, (201) 228-1862, (1).

LANDMARK RADON LABORATORIES, LTD., 3521 West Eisenhower Blvd., Loveland, CO, 80537, (303) 667-6286, (303) 356-6286, (1).

LANE REGIONAL AIR POLLUTION AUTHORITY, 225 North 5th St. Suite 501, Springfield, OR, 97477, (503) 726-2514, (10).

LANIDEX CORPORATION, 300 Lanidex Plaza, Parsippany, NJ, 07054, (201) 386-5800, (1).

LAPTEFF ASSOCIATES, P.O. Box 4150, Woodbridge, VA, 22194, (703) 491-6700, (1).

LARIMER COUNTY HEALTH DEPARTMENT/ENVIRONMENTAL HEALTH DIVISION, 363 Jefferson St., Fort Collins, CO, 80524, (303) 221-7497, (10).

LARRY LIBERTORE, INC., 230 West Highland Dr., Lakeland, FL, 33803, (813) 644-2828, (2).

LASSEN COUNTY AIR POLLUTION CONTROL DISTRICT, 175 Russell Ave., Susanville, CA, 96130, (916) 257-8311, (10).

LAW ENVIRONMENTAL, INC., P.O. Box 23184, Tampa, FL, 33623, (813) 877-9182, (1).

LDH WATERPROOFING AND RADON CONTROL, 286 Old Delp Rd., Lancaster, PA, 17601, (717) 569-2062, (2).

LEAFE, EDWARD, 212 Wilson St., Hackensack, NJ, 07601, (201) 343-0780, (1).

LEAGUE OF WOMEN VOTERS EDUCATIONAL FUND/NATURAL RESOURCES, 1730 M St., NW, Washington, DC, 20036, (202) 429-1965, (17).

LEGIS CORPORATION, 2211 Saxon, Houston, TX, 77018, (713) 680-8863, (1).

LEHIGH VALLEY HOUSE INSPECTION SERVICE, 200 Delaware Ln., Whitehall, PA, 18052, (215) 435-0208, (1)(2).

LEONARDS HOME INSPECTIONS, 6416 Loring Dr., Columbia, MD, 21045, (301) 381-2240, (1).

LEWIS AND KAHLSON RADON CONSULTANTS, 459 Dunlin Plaza, Secaucus, NJ, 07094, (201) 866-7219, (1).

LEWIS HOME INSPECTION, P.O. Box 6149, Lawrenceville, NJ, 08648, (609) 924-1411, (1).

LIFELINE ENVIRONMENTAL TESTING, 2522 Stuart Ave., Richmond, VA, 23220, (804) 358-4192, (1).

LINCOLN DEVORE, INC., 1000 West Fillmore St., Colorado Springs, CO, 80907, (303) 632-3593, (1).

LINCOLN-LANCASTER COUNTY/AIR POLLUTION CONTROL AGENCY, 2200 St. Mary's Ave., Lincoln, NE, 68502, (402) 471-8021, (10).

LININGER BROTHERS, INC., 12600 Penn Mar Rd., Waynesboro, PA, 17268, (717) 762-4024, (2).

LINN COUNTY HEALTH DEPARTMENT/AIR POLLUTION CONTROL DIVISION, 751 Center Point Rd., NE, Cedar Rapids, IA, 52402-4692, (318) 398-3551, (10).

LIVINGSTON-STEUBEN-WYOMING BOARD OF COOPERATIVE EDUCATIONAL SERVICES (BOCES), Lackawanna Ave., Mt. Morris, NY, 14510, (716) 658-2291, (10).

LMS ENGINEERS, 1 Blue Hill Plaza, Pearl River, NY, 10965, (914) 735-8300, (1).

LONCHAR, NIKOLA, 771 South 20th St., Philadelphia, PA, 19146, (215) 985-1474, (1).

LOS ANGELES COUNTY DEPARTMENT OF HEALTH SERVICES/RADIATION MANAGEMENT, 2615 South Grand Ave., Los Angeles, CA, 90007, (213) 744-3244, (10).

LOS ANGELES URBAN DEVELOPMENT, 1725 Radcliff St., Los Angeles, CA, 90026-1131, (213) 668-2655, (1)(3).

LOUISIANA DEPARTMENT OF ENVIRONMENTAL QUALITY/OFFICE OF AIR QUALITY AND NUCLEAR ENERGY/NUCLEAR ENERGY DIVISION, P.O. Box 14690, Baton Rouge, LA, 70898-4690, (504) 925-4518, (10).

LOUISIANA GEOLOGICAL SURVEY/DEPARTMENT OF NATURAL RESOURCES, Box G University Station, Baton Rouge, LA, 70803, (504) 342-6754, (10).

LOUISIANA REAL ESTATE COMMISSION, P.O. Box 14785, Baton Rouge, LA, 70898, (504) 925-4771, (10).

LOUISIANA REALTORS ASSOCIATION, 4639 Bennington, Baton Rouge, LA, 70808, (504) 923-2210, (18).

LOWRY ENGINEERING, INC., P.O. Box 536 Back Troy Rd., Unity, ME, 04988, (207) 948-3790, (1)(2)(3)(4)(5).

LRM, INC., 632 Spooner Rd., Virginia Beach, VA, 23462, (804) 497-6505, (3).

LUBBOCK CITY HEALTH DEPARTMENT/AIR POLLUTION CONTROL DIVISION, 1902 Texas Ave. P.O. Box 2548, Lubbock, TX, 79408, (806) 762-6411 x2918, (10).

LUDGATE ENGINEERING CORPORATION, 9A North Fifth St., Reading, PA, 19522, (215) 374-2239, (3).

LUDLUM MEASUREMENTS, INC., 501 Oak St. P.O. Box 810, Sweetwater, TX, 79556, (915) 235-5494, (4).

LUKENBUILT CONTRACTORS, 965 Dearborn St., Aurora, CO, 80011, (303) 364-9526, (2).

LV SAFETY AND HEALTH CONSULTANTS, 446 Timothy Dr., Bethlehem, PA, 18017, (215) 758-9072, (1)(2).

LW ENVIRONMENTAL SERVICES, 2111 Jennings St., San Francisco, CA, 94124, (415) 822-4555, (1)(2).

LYNN PEST MANAGEMENT COMPANY, P.O. Box 274, Jarrettsville, MD, 21084, (301) 557-6839, (800) 327-7134, (1).

LYONS SAFETY, North 92 W. 14350 Anthony Ave., Menomonee Falls, WI, 53051, (800) 558-8900, (4).

M

M.A.D. EXTERMINATORS, INC., 3116 Methacton Ave., Norristown, PA, 19403, (215) 539-7647, (1).

M.L. DOWNING AND ASSOCIATES, 2806 Farm Rd., Alexandria, VA, 22314, (703) 548-8607, (1).

M.R. HOLLAND ASSOCIATES, 70 Woodburn Dr., Methuen, MA, 01844, (508) 681-0821, (3).

M.V.E., INC., 1911 Lelaray St., Colorado Springs, CO, 80909, (303) 635-5736, (1).

MACK, E.F., 11 South 8th Ave., Marshalltown, IA, 50158, (515) 753-7328, (1)(2).

MACKLENBURG COUNTY DEPARTMENT OF ENVIRONMENTAL PROTECTION/AIR QUALITY DIVISION, 1200 Blythe Blvd., Charlotte, NC, 28203, (704) 376-4603, (10).

MACLAREN ENGINEERS, INC., 33 Yonge St., Toronto, ON, M5E 1E7, (416) 756-4919, (1)(2).

MACMILLAN RESEARCH, LTD., 1221 Barclay Cir., Marietta, GA, 30061, (404) 427-3101, (1).

MADERA COUNTY AIR POLLUTION CONTROL DISTRICT, 135 West Yosemite Ave., Madera, CA, 93637, (209) 675-7823, (10).

MADISON DEPARTMENT OF PUBLIC HEALTH/ENVIRONMENTAL HEALTH AND LABORATORY, 210 Martin Luther King, Jr. Blvd., Madison, WI, 53710, (608) 266-4821, (10).

MADRONA DIVERSIFIED, INC., 1938 Fairview Ave., E., Seattle, WA, 98102, (206) 324-0920, (1).

MAHONEY HEATING AND COOLING, 13620 C St., Omaha, NE, 68144, (402) 895-3333, (1)(2).

MAHONING-TRUMBULL AIR POLLUTION CONTROL AGENCY, 9 West Front St., Youngstown, OH, 44503, (216) 744-1928, (10).

MAINE ASSOCIATION OF REALTORS, 21 Bangor St., Augusta, ME, 04330, (207) 622-7501, (18).

MAINE DEPARTMENT OF ENVIRONMENTAL PROTECTION/BUREAU OF AIR QUALITY CONTROL, State House, Station 17, Augusta, ME, 04333, (207) 289-2437, (10).

MAINE DEPARTMENT OF HUMAN SERVICES/BUREAU OF HEALTH/DIVISION OF HEALTH ENGINEERING, State House Station 10, Augusta, ME, 04333, (207) 289-3826, (10).

MAINE DEPARTMENT OF HUMAN SERVICES/BUREAU OF HEALTH/DIVISION OF HEALTH ENGINEERING/OCCUPATIONAL AND RESIDENTIAL HEALTH PROGRAM, State House Station 10, Augusta, ME, 04333, (207) 289-3826, (10).

MAINE DEPARTMENT OF HUMAN SERVICES/BUREAU OF HEALTH/DIVISION OF HEALTH ENGINEERING/RADIOLOGICAL HEALTH PROGRAM, State House Station 10, Augusta, ME, 04333, (207) 289-5676, (10).

MAINE DEPARTMENT OF HUMAN SERVICES/BUREAU OF HEALTH/PUBLIC HEALTH LABORATORY, State House Station 12,

Augusta, ME, 04333, (207) 289-2727, (1)(10).

MAINE DEPARTMENT OF LABOR/DIVISION OF INDUSTRIAL SAFETY, Labor Station 82 283 State St., Augusta, ME, 04333, (207) 289-2591, (10).

MAINE GEOLOGICAL SURVEY/DEPARTMENT OF CONSERVATION, State House Station 22, Augusta, ME, 04333, (207) 289-2801, (10).

MAINE REAL ESTATE COMMISSION, State House Station # 35, Augusta, ME, 04333, (207) 783-9303, (10).

MALCOLM PIRNIE, INC., 100 Eisenhower Ave., Paramus, NJ, 07652, (201) 845-0400, (5)(14).

MAMECO INTERNATIONAL, INC., 4475 East 175th St., Cleveland, OH, 44128, (216) 752-4400, (800) 321-6412, (4).

MANASOTA 88, 5314 Bay State Rd., Palmetto, FL, 34221, (813) 722-7413, (17).

MANATEE COUNTY PUBLIC HEALTH UNIT/POLLUTION CONTROL DIVISION, 202 Sixth Ave. East, Bradenton, FL, 34208, (813) 748-0666, (10).

MANITOBA DEPARTMENT OF ENERGY AND MINES/INFORMATION CENTER/HOME CHECK-UP PROGRAM, 330 Graham Ave. Suite 555, Winnipeg, MB, R3C 4E3, (204) 945-4154, (800) 282-8069, (11).

MANSFIELD, PONTICELLI AND LIBBY, 2100 M St., NW, Washington, DC, 20036, (202) 296-1330, (3)(18).

MARICOPA COUNTY DEPARTMENT OF HEALTH SERVICES/ENVIRONMENTAL HEALTH SERVICES DIVISION, 1825 East Roosevelt St., Phoenix, AZ, 85006, (602) 258-6381, (10).

MARIPOSA COUNTY HEALTH DEPARTMENT, 4988 11th St. P.O. Box 5, Mariposa, CA, 95338, (209) 966-3689, (10).

MARTIN BUILDERS COMPANY, INC., THE, 7114 Rising Sun Ave., Philadelphia, PA, 19111, (215) 745-5398, (2).

MARTIN J. EGAN ASSOCIATES, 37 Waldron Ave., Summit, NJ, 07901, (201) 522-0931, (1).

MARTIN MARIETTA ENERGY SYSTEMS, P.O. Box 1410 Hobbs Rd., Paducah, KY, 42001, (502) 444-6311, (1).

MARYLAND ASSOCIATION OF REALTORS, 2083 West St. Suite 4-A, Annapolis, MD, 21401, (305) 841-6080, (18).

MARYLAND DEPARTMENT OF THE ENVIRONMENT/AIR MANAGEMENT ADMINISTRATION, 201 West Preston St., Baltimore, MD, 21201, (301) 225-5235, (10).

MARYLAND DEPARTMENT OF THE ENVIRONMENT/OFFICE OF TOXICS, ENVIRONMENTAL SAFETY AND HEALTH, 201 West Preston St. 7th Floor Mail Room, Baltimore, MD, 21201, (301) 333-3130, (800) 872-3666, (10).

MARYLAND DIVISION OF LABOR AND INDUSTRY/CONSULTATION PROGRAM, 501 St. Paul Pl. 3rd Fl., Baltimore, MD, 21202, (301) 659-4218, (10).

MARYLAND GEOLOGICAL SURVEY, 2300 St. Paul St., Baltimore, MD, 21218, (301) 554-5500, (10).

MARYLAND RADON LABORATORY, INC., P.O. Box 2578, Columbia, MD, 21045, (301) 997-2366, (1).

MARYLAND REAL ESTATE COMMISSION, 501 St. Paul Pl. Suite 804, Baltimore, MD, 21202-2272, (301) 333-6230, (10).

MASSACHUSETTS ASSOCIATION OF REALTORS, P.O. Box 650, Waltham, MA, 02254-0650, (617) 890-3700, (18).

MASSACHUSETTS BOARD OF REGISTRATION OF REAL ESTATE BROKERS AND SALESMEN, 100 Cambridge St. Room 1518, Boston, MA, 02202, (617) 727-7376, (10).

MASSACHUSETTS DEPARTMENT OF LABOR AND INDUSTRIES/DIVISION OF INDUSTRIAL SAFETY, 100 Cambridge St., Boston, MA, 02202, (617) 727-3463, (10).

MASSACHUSETTS DEPARTMENT OF PUBLIC HEALTH/OFFICE OF HUMAN SERVICES/RADIATION CONTROL PROGRAM, 150 Tremont St., Boston, MA, 02111, (617) 727-6214, (10).

MASSACHUSETTS DIVISION OF AIR QUALITY CONTROL, One Winter St., Boston, MA, 02108, (617) 292-5593, (10).

MASSACHUSETTS INSTITUTE OF TECHNOLOGY/RADON PROJECT, Building 26 Room 421, Cambridge, MA, 02139, (617) 253-4151, (1)(5).

MASSACHUSETTS RADON ALERT, INC., 20 Robertson St., Quincy, MA, 02169, (617) 471-9403, (617) 331-7900, (1).

MASSACHUSETTS STATE GEOLOGIST/COASTAL ZONE MANAGEMENT, 100 Cambridge St. 20th Floor, Boston, MA, 02202, (617) 292-5690, (10).

MASTER BUILDING SERVICE, 245 Huguenot St. Suite 107, New Rochelle, NY, 10801, (914) 235-8282, (1)(2).

MATESON CHEMICAL COMPANY, 1025 East Montgomery Ave., Philadelphia, PA, 19125, (215) 358-5453, (4).

MATTER RESEARCH, INC., 1923 New York Ave., Huntington Station, NY, 11746, (516) 549-8777, (1).

MAYA-TEK, INC., P.O. Box 1092, Chadds Ford, PA, 19317, (215) 358-5453, (1)(2).

MBI RADON TECHNOLOGY, P.O. Box 21-157, St. Paul, MN, 55121, (612) 454-4534, (1).

MCCRONE ENVIRONMENTAL SERVICES, INC., 1412 Oakbrook Dr. Suite 100, Norcross, GA, 30093, (404) 368-9600, (1).

MCGUIRE BUILDERS, INC., Rte. 2, Box 1840, Warfordsburg, PA, 17267, (717) 294-3108, (1)(2).

MCHILL, A.D. AND SONS, INC., 1121 Second St. Pike, Richboro, PA, 18954, (215) 364-2784, (1)(2).

MCINERNY, T.M., 1214 Church St., Elmira, NY, 14905, (607) 733-1994, (2).

MCMASTER UNIVERSITY/HEALTH SCIENCES CENTER/DEPARTMENT OF EPIDEMIOLOGY, 1200 Main St., West, Hamilton, ON, L8N 3Z5, (416) 525-3892, (5).

MCMASTER UNIVERSITY/HEALTH SCIENCES CENTER/OCCUPATIONAL HEALTH PROGRAM, 1200 Main St., West Room 3H50, Hamilton, ON, L8N 3Z5, (416) 525-9140, (5).

MCWILLIAMS PEST CONTROL, 7014 McCallum St., Philadelphia, PA, 19119, (215) 849-2223, (1).

MECHANICAL EQUIPMENT SERVICES CORPORATION, 3122 49th St., Des Moines, IA, 50310, (515) 276-9131, (1)(2).

MECKLENBURG PURIFIER, P.O. Box 404, Newell, NC, 28126, (704) 537-8503, (1).

MEDLAB ENVIRONMENTAL TESTING, INC., P.O. Box 2045, Wilmington, DE, 19899, (302) 655-5227, (800) 633-5221, (1)(4).

MEDRAD, INC., Kappa Manor II 271 Kappa Dr., Pittsburgh, PA, 15238-2870, (412) 767-4600, (1).

MEDRAD, INC., 271 Kappa Dr., Pittsburgh, PA, 15238, (412) 782-4600, (3).

MELODY HOMES, INC., 11031 Sheridan Sq., Westminster, CO, 80020, (303) 466-1831, (13).

MEMPHIS AND SHELBY COUNTY HEALTH DEPARTMENT/AIR POLLUTION CONTROL BRANCH, 814 Jefferson Ave., Memphis, TN, 38105, (901) 576-7775, (10).

MEMPHREMAGOG HEAT EXCHANGERS, INC., P.O. Box 456, Newport, VT, 05855, (802) 334-5412, (4).

MENDOCINO COUNTY AIR POLLUTION CONTROL DISTRICT, Courthouse Sq., Ukiah, CA, 95482, (707) 463-4354, (10).

MENKE CONSULTING, INC., 6070 Routzong Rd., Greenville, OH, 45331, (513) 447-4225, (1).

MERCED COUNTY AIR POLLUTION CONTROL DISTRICT, 385 East13th St. P.O. Box 471, Merced, CA, 95341, (209) 385-7391, (10).

MERIDEN DEPARTMENT OF HEALTH/HUMAN SERVICES, 165 Miller St., Meriden, CT, 06450, (203) 634-0003, (10).

MERIT ENVIRONMENTAL SERVICES, INC., P.O. Box 3083, Berea, KY, 40403, (606) 986-7070, (1)(2).

MERRIMACK VALLEY AND METROPOLITAN BOSTON AIR POLLUTION CONTROL DISTRICTS, 5 Commonwealth Ave., Woburn, MA, 01801, (617) 935-2160, (10).

MESA COUNTY HEALTH DEPARTMENT/AIR POLLUTION CONTROL DIVISION, 515 Patterson Rd., Grand Junction, CO, 81501, (303) 244-1750, (10).

METRO RADON MEASUREMENT, 1565 Cliff Rd. Suite 3-101, Eagan, MN, 55122, (612) 681-1829, (1)(4).

METRO SERVICE LABORATORIES, INC., 6309 Fern Valley Pass, Louisville, KY, 40228, (502) 964-0865, (800) 678-6878, (1).

METROPOLITAN DADE COUNTY DEPARTMENT OF ENVIRONMENTAL RESOURCES MANAGEMENT, Metro Dade Ctr. 111 NW First St. Suite 1310, Miami, FL, 33128, (305) 375-3326, (10).

METROPOLITAN RADON SERVICE, 4011 Newbys Bridge Rd., Chesterfield, VA, 23832, (804) 745-4702, (1)(2).

METROPOLITAN REAL ESTATE ANALYSIS, 1810 South Josephine, Denver, CO, 80210, (303) 777-7594, (1).

METROPOLITAN WASHINGTON COUNCIL OF GOVERNMENTS/DEPARTMENT OF ENVIRONMENTAL PROGRAMS, 1875 Eye St., NW Suite 200, Washington, DC, 20006, (202) 223-6800, (14).

MICHIGAN ASSOCIATION OF REALTORS, P.O. Box 40725, Lansing, MI, 48901-7925, (517) 372-8890, (18).

MICHIGAN DEPARTMENT OF LICENSING AND REGULATION, P.O. Box 30018, Lansing, MI, 48909, (517) 373-0490, (10).

MICHIGAN DEPARTMENT OF NATURAL RESOURCES, P.O. Box 30028, Lansing, MI, 48909, (517) 334-6923, (10).

MICHIGAN DEPARTMENT OF PUBLIC HEALTH/BUREAU OF ENVIRONMENTAL AND OCCUPATIONAL HEALTH/DIVISION OF RADIOLOGICAL HEALTH, 3423 North Logan St. P.O. Box 30195, Lansing, MI, 48909, (517) 335-8200, (6)(10).

MICHIGAN DEPARTMENT OF PUBLIC HEALTH/DIVISION OF OCCUPATIONAL HEALTH/SPECIAL PROGRAMS SECTION, 3500 North Logan, Lansing, MI, 48909, (517) 322-1809, (10).

MICRO-TRENDS, 25 Caton Terr., Caldwell, NJ, 07006,

(201) 228-5907, (3).

MICROBAC ENVIRONMENTAL LABORATORY, P.O. Box 49, Fayetteville, NC, 28303, (919) 864-1920, (1).

MICROBAC LABORATORIES, INC./ERIE TESTING DIVISION-MAIN OFFICE, 2411 West 26th St., Erie, PA, 16506, (814) 833-4790, (1)(4).

MICROBAC LABORATORIES, INC./HANOVER DIVISION, 701 3rd St., Hanover, PA, 17331, (717) 633-6011, (1).

MICROBAC LABORATORIES, INC./KENTUCKY TESTING DIVISION, 1121 West Broadway, Louisville, KY, 40203, (502) 583-5256, (502) 589-6019, (1).

MICROBAC LABORATORIES, INC./MASSACHUSETTS TESTING DIVISION, 202 Bussey St., Dedham, MA, 02026, (617) 326-7117, (1).

MICROBAC LABORATORIES, INC./NEW CASTLE TESTING DIVISION, R. D. #3 Pulaski Rd., New Castle, PA, 16105, (412) 854-4212, (1)(4).

MICROBAC LABORATORIES, INC./PITTSBURGH TESTING DIVISION, 4580 McKnight Rd., Pittsburgh, PA, 15237, (412) 931-5851, (1).

MICROBAC LABORATORIES, INC./SCHILLER DIVISION, 449 Rochester Rd., Pittsburgh, PA, 15237-1733, (412) 369-4830, (1).

MICROBAC LABORATORIES, INC./SEAWAY DIVISION, 542-544 Conkey St., Hammond, IN, 46324, (219) 932-1770, (1).

MICROBAC LABORATORIES, INC./STRAND TESTING DIVISION, 110 Kinzey St., Johnstown, PA, 15904, (814) 266-9548, (1).

MICROBAC MID-ATLANTIC, INC., 5808 Jefferson Ave., Newport News, VA, 23605, (804) 245-8613, (1).

MID AMERICA RADON TESTING, INC., P.O. Box 25643, Overland Park, KS, 66225, (913) 469-1997, (1).

MIDDLESEX COUNTY AIR POLLUTION CONTROL PROGRAM, 280 Hobart St. Room 518, Perth Amboy, NJ, 08861, (201) 826-3100, (10).

MIDWEST HEALTH PROGRAMS, INC., P.O. Box 3023, Urbana, IL, 61801, (217) 367-5293, (1).

MIDWEST RADON ANALYSIS, INC., P.O. Box 33093, Cleveland, OH, 44133, (216) 582-4848, (1).

MIDWEST RADON INSPECTION, INC., 25005 West 129 Ter., Olathe, KS, 66061, (913) 780-2000, (1)(2).

MILWAUKEE COUNTY DEPARTMENT OF PUBLIC WORKS/PROFESSIONAL SERVICES DIVISION/ENGINEERS ENVIRONMENTAL AND ENERGY SERVICES SECTION, 901 North 9th St., Milwaukee, WI, 53233, (414) 278-4874, (10).

MILWAUKEE ELECTRIC TOOL, 13135 West Lisbon, Brookfield, WI, 53005, (414) 781-3600, (4).

MINCHIK AND MINCHIK, 6319 Utah Ave. NW, Washington, DC, 20015, (202) 364-8845, (1).

MINE INSPECTORS INSTITUTE OF AMERICA (MIIA), 1900 Grant Building, Pittsburgh, PA, 15219, (412) 281-2620, (16).

MINE SAFETY APPLIANCES COMPANY, P.O. Box 426, Pittsburgh, PA, 15230, (412) 967-3000, (4).

MINING ASSOCIATION OF CANADA (MAC), 350 Sparks St. Suite 705, Ottawa, ON, K1R 7S8, (613) 233-9391, (11).

MINISTERE DE L'ENVIRONMENT DU QUEBEC/DIVISION DE RADIOPROTECTION, 5199 Est, rue Sherbrooke Suite 3860, Montreal, PQ, H1T 3X9, (514) 873-1978, (11).

MINISTERE DE LA JUSTICE DU QUEBEC/SERVICE DU COURTAGE IMMOBILIER DU QUEBEC, 220, Grande Allee Est Suite 910, Quebec City, PQ, G1R 2J1, (418) 643-4597, (11).

MINISTRY OF CONCERN FOR PUBLIC HEALTH (MCPH), 5495 Main St. Suite 147, Buffalo, NY, 14221, No Phone Available, (17).

MINNEAPOLIS BLOWER DOOR, 920 W. 53rd St., Minneapolis, MN, 55419, (612) 827-1117, (4).

MINNEAPOLIS ENERGY OFFICE, City Hall Room 330, Minneapolis, MN, 55415, (612) 348-4830, (10).

MINNEAPOLIS POLLUTION CONTROL DIVISION/DEPARTMENT OF INSPECTIONS, 250 South 4th St. Public Health Ctr. Room 300, Minneapolis, MN, 55415, (612) 348-7897, (10).

MINNESOTA ASSOCIATION OF REALTORS, 5750 Lincoln Dr., Edina, MN, 55436, (612) 935-8313, (18).

MINNESOTA COMMERCE DEPARTMENT, 500 Metro Square, St. Paul, MN, 55101, (612) 297-3379, (10).

MINNESOTA DEPARTMENT OF HEALTH/ENVIRONMENTAL HEALTH DIVISION/SECTION OF RADIATION CONTROL, 717 Delaware St., SE P.O. Box 9441, Minneapolis, MN, 55440, (612) 623-5351, (10).

MINNESOTA DEPARTMENT OF LABOR AND INDUSTY, 444 Lafayette Rd. 5th Floor, St. Paul, MN, 55101, (612) 297-1953, (10).

MINNESOTA DEPARTMENT OF PUBLIC SERVICE/ENERGY INFORMATION CENTER, 900 American Center Building 150 East Kellogg Blvd., St. Paul, MN, 55101, (612) 297-2291, (800) 652-9747, (10).

MINNESOTA GEOLOGICAL SURVEY, 2642 University Ave., St. Paul, MN, 55114-1057, (612) 627-4780, (10).

MINNESOTA POLLUTION CONTROL AGENCY/DIVISION OF AIR QUALITY, 520 Lafayette Rd., St. Paul, MN, 55155, (612) 296-7802, (10).

MINUTEMAN HOME INSPECTORS, INC., 90 Conz St., Northampton, MA, 01060, (413) 586-1009, (413) 549-1800, (1).

MISCAVAGE AND COMPANY, 2814 Pearl Ave., Allentown, PA, 18103, (215) 797-3865, (1).

MISSISSIPPI ASSOCIATION OF REALTORS, P.O. Box 2454, Jackson, MS, 39205, (601) 355-0216, (18).

MISSISSIPPI REAL ESTATE COMMISSION, 1920 Dunbarton Dr., Jackson, MS, 39216-5087, (601) 987-3969, (10).

MISSISSIPPI STATE DEPARTMENT OF HEALTH/DIVISION OF OCCUPATIONAL SAFETY AND HEALTH, P.O. Box 1700, Jackson, MS, 39205, (601) 987-3981, (10).

MISSISSIPPI STATE DEPARTMENT OF HEALTH/DIVISION OF RADIOLOGICAL HEALTH, 3150 Lawson St. P.O. Box 1700, Jackson, MS, 39215-1700, (601) 354-6657, (10).

MISSISSIPPI STATE DEPARTMENT OF NATURAL RESOURCES/BUREAU OF GEOLOGY, P.O. Box 5348, Jackson, MS, 39216, (601) 354-6228, (10).

MISSOULA CITY-COUNTY HEALTH DEPARTMENT/ENVIRONMENTAL HEALTH DIVISION, 301 West Adler St., Missoula, MT, 59802, (406) 721-5700, (10).

MISSOURI ASSOCIATION OF REALTORS, P.O. Box 1327, Columbia, MO, 65205, (314) 445-8400, (18).

MISSOURI DEPARTMENT OF HEALTH/DIVISION OF ENVIRONMENTAL HEALTH AND EPIDEMIOLOGY SERVICES, 1730 East Elm Plaza P.O. Box 570, Jefferson City, MO, 65102, (314) 751-6083, (10).

MISSOURI DEPARTMENT OF LABOR AND INDUSTRIAL RELATIONS, 1001-D Southwest Blvd. P.O. Box 449, Jefferson City, MO, 65102, (314) 751-3403, (10).

MISSOURI DEPARTMENT OF NATURAL RESOURCES/DIVISION OF GEOLOGY AND LAND SURVEY, 111 Fairgrounds Rd. P.O. Box 250, Rolla, MO, 65401, (314) 364-1752, (10).

MISSOURI REAL ESTATE COMMISSION, P.O. Box 1339, Jefferson City, MO, 65102, (314) 751-2334, (10).

MISSOURI TESTING LABORATORIES, INC., P.O. Box 1, Trenton, MO, 64683, (816) 359-6833, (1).

MITIGATING CIRCUMSTANCES, 153 Forest Park, Durham, NH, 03824, (603) 868-1557, (2).

MIXLEE ENVIRONMENTAL RESEARCH CORPORATION, 684-B Commerce Dr., Fairfield, CT, 06430, (203) 336-1211, (800) 545-1211, (1).

MOBILE EMISSIONS TESTING CORPORATION, 1780 Maple St. Suite 21, Northfield, IL, 60093, (312) 501-2525, (1).

MOBILE RADON DIAGNOSTICS, INC., 1018 Delta Ave. Suite 302, Cincinnati, OH, 45208, (513) 321-5300, (1).

MODOC COUNTY AIR POLLUTION CONTROL DISTRICT, 202 West Fourth St., Alturas, CA, 96101, (916) 233-3939 x 401, (10).

MONARCH ANALYTICAL LABORATORIES, INC., P.O. Box 2990, Toledo, OH, 43606, (419) 535-1780, (1).

MONTANA ASSOCIATION OF REALTORS, 910 Helena Pk., Helena, MT, 59601, (406) 443-4032, (18).

MONTANA BOARD OF REALTY REGULATION, 1424 9th Ave., Helena, MT, 59620, (406) 444-2961, (10).

MONTANA BUREAU OF MINES AND GEOLOGY, Montana College of Mineral Science and Technology, Butte, MT, 59701, (406) 496-4180, (10).

MONTANA DEPARTMENT OF HEALTH AND ENVIRONMENTAL SERVICES/OCCUPATIONAL HEALTH BUREAU/ENVIRONMENTAL SCIENCES DIVISION, Cogswell Building, Helena, MT, 59620, (406) 444-3671, (10).

MONTANA DEPARTMENT OF LABOR AND INDUSTRY/BUREAU OF SAFETY AND HEALTH/DIVISION OF WORKERS' COMPENSATION, 5 South Last Chance Gulch, Helena, MT, 59601, (406) 444-6401, (10).

MONTEREY BAY UNIFIED AIR POLLUTION CONTROL DISTRICT, 1164 Monroe St. Suite 10, Salinas, CA, 93906, (408) 443-1135, (10).

MONTGOMERY COUNTY GOVERNMENT, 101 Monroe St., Rockville, MD, 20850, (301) 217-2500, (10).

MONTGOMERY COUNTY GOVERNMENT/DIVISION OF ENVIRONMENTAL PLANNING AND MONITORING, 101 Monroe St. 6th Fl., Rockville, MD, 20850, (301) 217-2380, (10).

MONTGOMERY COUNTY GOVERNMENT/MONTGOMERY COUNTY RADON TASK FORCE, 100 Maryland Ave. 6th Fl., Rockville, MD, 20850, (301) 217-7900, (10).

MONTGOMERY LABORATORIES, 555 East Walnut, Pasadena, CA, 91109, (818) 796-9141, (1).

MORRISON WATERPROOFING COMPANY, INC., 7644 L Fullerton Rd., Springfield, VA, 22153, (703) 569-1666, (1).

MORTGAGE BANKERS ASSOCIATION OF AMERICA (MBAA), 1125 15th St., NW, Washington, DC, 20005, (202) 861-6500, (18).

MOSS, ROBERT, 19514 Fisher Ave., Poolesville, MD, 20837, (301) 972-8730, (1).

MOUNTAIN ENERGY AND RESOURCES, INC., 15800

W. 6th Ave., Golden, CO, 80401, (303) 279-4971, (4).

MR. RENOVATOR, 395 South End Ave. Apt. 27P, New York, NY, 10280, (212) 321-9129, (1).

MRC HOME INSPECTIONS, P.O. Box 157, Wyckoff, NJ, 07481, (201) 891-7057, (1).

MUSSO, FRANCIS P., P.O. Box 651, Bath, NY, 14810, (607) 776-6245, (2).

N

N.A.K. PRODUCTION ASSOCIATES, 1317 Apple Ave., Silver Spring, MD, 20910, (301) 565-0355, (3)(4)(6).

NAHAJO NATION/OCCUPATIONAL SAFETY AND HEALTH, P.O. Box 1690, Window Rock, 86515, (614) 644-2223, (800) 282-1425, (10).

NALEPA, DAN, 6 East Courtney St., Dunkirk, NY, 14048, (716) 366-1778, (1).

NASHVILLE AND DAVIDSON COUNTYMETROPOLITAN HEALTH DEPARTMENT/POLLUTION CONTROL DIVISION, 311 23rd Ave. North, Nashville, TN, 37203, (615) 327-9313, (10).

NASSAU COUNTY DEPARTMENT OF HEALTH/DIVISION OF ENVIRONMENTAL HEALTH SERVICES/BUREAU OF AIR QUALITY MANAGEMENT, 240 Old County Rd., Mineola, NY, 11501, (516) 535-3232, (10).

NATIONAL ACADEMY OF SCIENCES (NAS)/NATIONAL RESEARCH COUNCIL (NRC)/FEDERAL CONSTRUCTION COUNCIL, 2101 Constitution Ave., NW, Washington, DC, 20418, (202) 334-3378, (5)(13).

NATIONAL ACADEMY OF SCIENCES (NAS)/NATIONAL RESEARCH COUNCIL (NRC)/MEDICAL FOLLOW-UP AGENCY, 2101 Constitution Ave., NW, Washington, DC, 20418, (202) 334-2000, (15).

NATIONAL APARTMENT ASSOCIATION (NAA), 1101 14th St., NW Suite 804, Washington, DC, 20005, (202) 842-4050, (13).

NATIONAL ASSOCIATION OF COUNTY HEALTH OFFICIALS (NACHO), 440 First St., NW, Washington, DC, 20001, (202) 393-6226, (15).

NATIONAL ASSOCIATION OF ENVIRONMENTAL PROFESSIONALS (NAEP), P.O. Box 9400, Washington, DC, 20016, (202) 683-3746, (14).

NATIONAL ASSOCIATION OF HOME BUILDERS (NAHB)/NATIONAL RESEARCH CENTER, 400 Prince Georges Center Blvd., Upper Marlboro, MD, 20772, (301) 249-2000, (1)(5)(6)(13).

NATIONAL ASSOCIATION OF HOME BUILDERS OF THE UNITED STATES (NAHB), 15th and M Sts., NW, Washington, DC, 20005, (202) 822-0200, (13).

NATIONAL ASSOCIATION OF REAL ESTATE BROKERS (REALTIST), P.O. Box 56483, Washington, DC, 20041, (202) 289-6655, (18).

NATIONAL ASSOCIATION OF REAL ESTATE LICENSE LAW OFFICIALS (NARELLO), P.O. Box 129, Centerville, UT, 84010-0129, (801) 531-8202, (18).

NATIONAL ASSOCIATION OF REALTORS (NAR), Education Division 430 North Michigan Ave., Chicago, IL, 60611, (312) 329-8388, (18).

NATIONAL ASSOCIATION OF REALTORS (NAR)/GOVERNMENT AFFAIRS OFFICE, 777 14th St., NW, Washington, DC, 20005, (202) 383-1087, (18).

NATIONAL ASSOCIATION OF REALTORS (NAR)/PUBLIC AFFAIRS OFFICE, 777 14th St., NW, Washington, DC, 20005, (202) 383-1177, (18).

NATIONAL ASSOCIATION OF THE REMODELING INDUSTRY (NARI), 1901 North Moore St. Suite 808, Arlington, VA, 22209, (703) 276-7600, (13).

NATIONAL ASSOCIATION OF WATER COMPANIES (NAWC), 1725 K St., NW Suite 1212, Washington, DC, 20006, (202) 833-8383, (14).

NATIONAL CENTER FOR ATMOSPHERIC RESEARCH/ATMOSPHERIC CHEMISTRY DIVISION, Box 3000, Boulder, CO, 80307, (303) 497-1000, (5).

NATIONAL CLEAN AIR COALITION (NCAC), 801 Pennsylvania Ave., SE, Washington, DC, 20003, (202) 543-8200, (17).

NATIONAL CONCRETE MASONRY ASSOCIATION (NCMA), P.O. Box 781 2302 Horse Pen Rd., Herndon, VA, 22070, (703) 435-4900, (13).

NATIONAL CONFERENCE OF LOCAL ENVIRONMENTAL HEALTH ADMINISTRATORS (NCLEHA), 3415 28th St., Port Huron, MI, 48060, (313) 987-5306, (15).

NATIONAL CONFERENCE OF STATE LEGISLATURES (NCSL), 1050 17 St. #2100, Denver, CO, 80265, (303) 623-7800, (14).

NATIONAL CONFERENCE OF STATES ON BUILDING CODES AND STANDARDS (NCSBCS), 481 Carlisle Dr., Herndon, VA, 22070, (703) 437-0100, (13).

NATIONAL CONFERENCE ON WEIGHTS AND MEASURES (NCWM), P.O. Box 3137, Gaithersburg, MD, 20870, (301) 975-4004, (13).

NATIONAL COUNCIL ON RADIATION PROTECTION AND MEASUREMENTS (NCRP), 7910 Woodmont Ave. Suite 800, Bethesda, MD, 20814, (301) 657-2652, (5)(14).

NATIONAL DRAEGER, INC., 101 Technology Dr., Pittsburgh, PA, 15275, (412) 787-8383, (4).

NATIONAL ENVIRONMENTAL ENGINEERING AND DIAGNOSTICS, Box 519, Rockland, MA, 02370, (617) 871-4426, (1).

NATIONAL ENVIRONMENTAL HEALTH ASSOCIATION (NEHA), 720 South Colorado Blvd. South Tower, Suite 970, Denver, CO, 80222, (303) 756-9090, (15).

NATIONAL FUEL, 2484 Seneca St., Buffalo, NY, 14210, (716) 824-2500, (1).

NATIONAL FUEL, 800 State St., Erie, PA, 16501, (814) 871-8000, (1).

NATIONAL GOVERNORS ASSOCIATION (NGA)/COMMITTEE ON ENERGY AND ENVIRONMENT, Hall of States, Suite 250 444 North Capitol St., NW, Washington, DC, 20001, (202) 624-5300, (14).

NATIONAL HOME BUYERS AND HOME OWNERS ASSOCIATION (NHBHOA), 1050 17th St., NW, Washington, DC, 20036, (202) 659-6500, (17).

NATIONAL HOMEOWNERS ASSOCIATION (NHA), 1906 Sunderland Pl., NW, Washington, DC, 20036, (202) 223-1453, (18).

NATIONAL INSTITUTE OF BUILDING SCIENCES (NIBS), 1015 15th St., NW Suite 700, Washington, DC, 20005, (202) 347-5710, (5)(13).

NATIONAL INSTITUTE OF BUILDING SCIENCES (NIBS)/INDOOR AIR QUALITY COMMITTEE, C/O Armstrong World Industries 333 Liberty St., Lancaster, PA, 17603, (717) 397-0611, (5)(13).

NATIONAL MULTI HOUSING COUNCIL (NMHC), 1250 Connecticut Ave., NW Suite 620, Washington, DC, 20036, (202) 659-3381, (18).

NATIONAL PROPERTY INSPECTIONS, 7917 North Anita Dr., Kansas City, MO, 64151, (816) 587-9595, (1).

NATIONAL PROPERTY INSPECTIONS, 1344-S Briar Creek Rd., Charlotte, NC, 28205, (704) 568-6324, (1).

NATIONAL PROPERTY INSPECTIONS, 11212 Davenport St. Suite 15, Omaha, NE, 68154, (402) 330-1074, (1).

NATIONAL PROPERTY INSPECTIONS, 247 Sunnyhill Ln., Oconomowoc, WI, 53066, (414) 569-9377, (1).

NATIONAL PROPERTY INSPECTIONS, INC., 224 South 108th Ave. Suite 2, Omaha, NE, 68154, (402) 333-9807, (800) 333-9807, (1).

NATIONAL RADON ASSOCIATION, INC., THE (NRA), 2786 North Decatur Rd. Suite 220, Decatur, GA, 30033, (404) 292-1113, (7).

NATIONAL RADON CONTROL, INC., P.O. Box 5342, North Branch, NJ, 08876, (201) 231-0844, (1).

NATIONAL RADON LABORATORIES, INC., 1000 Johnson Ferry Rd. Suite B-145, Marietta, GA, 30068, (404) 565-3886, (1).

NATIONAL RADON LABORATORIES/DIVISION OF ANALYTICAL SERVICES CORPORATION, P.O. Box 10681, Arlington, VA, 22210-1681, (703) 525-6672, (1).

NATIONAL RADON LABORATORIES/DIVISION OF COLORADO SERVICES CORPORATION, 14677 East Easter Ave. Suite H, Englewood, CO, 80112, (303) 699-7378, (1).

NATIONAL RADON SERVICE, P.O. Box 3596, Gaithersburg, MD, 20878-0596, (301) 953-4598, (1).

NATIONAL RADON TEST COMPANY, P.O. Box 50468, Austin, TX, 78763, (512) 495-9330, (1).

NATIONAL RADON TESTING CENTER, INC., 18 Marshall St., Norwalk, CT, 06854, (203) 853-4777, (203) 853-1219, (1).

NATIONAL RADON TESTING SERVICE, P.O. Box 266, Maple Glen, PA, 19002, (215) 628-8738, (1).

NATIONAL READY MIXED CONCRETE ASSOCIATION (NRMCA), 900 Spring St., Silver Spring, MD, 20910, (301) 587-1400, (13).

NATIONAL SANITATION FOUNDATION (NSF), P.O. Box 1468, Ann Arbor, MI, 48016, (313) 769-8010, (15).

NATIONAL SEAL COMPANY, 1255 Monmouth Blvd., Galesburg, IL, 61401, (309) 343-3418, (800) 323-3820, (4).

NATIONAL SOCIETY OF PROFESSIONAL ENGINEERS (NSPE), 1420 King St., Alexandria, VA, 22314, (703) 684-2800, (13).

NATIONAL SOCIETY OF PROFESSIONAL SANITARIANS (NSPS), Missouri Health Department Route 4, Hoffman Dr., Jefferson, MO, 65101, (314) 751-6095, (16).

NATIONAL STONE ASSOCIATION (NSA), 1415 Elliot Pl., NW, Washington, DC, 20007, (202) 342-1100, (13).

NATIONAL WATER WELL ASSOCIATION (NWWA), 6375 Riverside Dr., Dublin, OH, 43017, (614) 761-1711, (14).

NATIONWIDE TESTING, RD 1, Box 266B, Annandale, NJ, 08801, (201) 735-2770, (1).

NATKIN SERVICE COMPANY, 2241 West Altorfer, Peoria, IL, 61615, (309) 691-9206, (800) 642-8391, (1).

NATLSCO, Rte. 22 and Kemper Dr., Long Grove, IL, 60049, (312) 540-2488, (1).

NATURAL ENERGY CONSULTANTS, P.O. Box 586,

Conifer, CO, 80433, (303) 838-5291, (1)(2)(3).

NATURAL FLORIDA RETROFIT, INC., P.O. Box 301, Montverde, FL, 32756, (305) 469-2173, (305) 841-4376, (1).

NATURAL PANTRY, THE, 300 West 36th Ave., Anchorage, AK, 99503, (907) 563-2727, (1).

NATURAL RESOURCE LABORATORY, INC., 1100 Simms St., Golden, CO, 80401, (303) 233-8155, (1).

NATURAL RESOURCE TECHNOLOGIES, 32131 Steven Way, Conifer, CO, 80433, (303) 850-9335, (1).

NAVAJO NATION/OCCUPATIONAL HEALTH AND SAFETY, P.O. Box 1690, Window Rock, 86515, (602) 871-6335, (10).

NDL ORGANIZATION, INC., THE, P.O. Box 791, Peekskill, NY, 10566, (914) 737-7200, (1).

NEA, INC., 10950 Southwest 5th St. Suite 380, Beaverton, OR, 97005, (503) 643-4661, (1).

NEBRASKA DEPARTMENT OF ENVIRONMENTAL CONTROL/AIR POLLUTION CONTROL DIVISION, 301 Centennial Mall South P.O. Box 94877, Lincoln, NE, 68509, (402) 471-2186, (10).

NEBRASKA DEPARTMENT OF HEALTH/DIVISION OF RADIOLOGICAL HEALTH, 301 Centennial Mall South P.O. Box 93007, Lincoln, NE, 68509, (402) 471-2168, (10).

NEBRASKA DEPARTMENT OF LABOR, State Office Building 301 Centennial Mall, South, Lincoln, NE, 68509, (402) 471-2239, (10).

NEBRASKA INSTITUTE OF AGRICULTURE AND NATURAL RESOURCES/CONSERVATION AND SURVEY DIVISION, 113 Nebraska Hall The University Of Nebraska, Lincoln, NE, 68588-0517, (402) 472-3471, (10).

NEBRASKA REAL ESTATE COMMISSION, 301 Centennial Mall South P.O. Box 94667, Lincoln, NE, 68508, (402) 471-2004, (10).

NEBRASKA REALTORS ASSOCIATION, 1115 K St. Suite 102, Lincoln, NE, 68508, (402) 475-6334, (18).

NEBRASKA WESLEYAN UNIVERSITY/PHYSICS DEPARTMENT/RADON MEASUREMENTS LABORATORY, 5000 St. Paul Ave., Lincoln, NE, 68504, (402) 465-2252, (402) 465-2247, (1).

NEILSON ASSOCIATES, INC., 100 Dickinson Dr. Suites 109-110, Chadds Ford, PA, 19317, (215) 793-1389, (3).

NELSON, C. STUART, 1007 River Rd., Quarryville, PA, 17566, (717) 548-3151, (1).

NEOTRONICS, P.O. Box 370, Gainesville, GA, 30501, (404) 535-0600, (4).

NERING, INC., P.O. Box 209, Ridgefield, CT, 06877, (203) 431-3597, (1).

NEVADA ASSOCIATION OF REALTORS, P.O. Box 7338, Reno, NV, 89510, (702) 329-6648, (18).

NEVADA BUREAU OF MINES AND GEOLOGY, University Of Nevada-Reno, Reno, NV, 89557-0088, (702) 784-6691, (10).

NEVADA DEPARTMENT OF COMMERCE/ REAL ESTATE DIVISION, 201 South Fall St., Carson City, NV, 89710, (702) 885-4280, (10).

NEVADA DEPARTMENT OF HUMAN RESOURCES/HEALTH DIVISION/RADIOLOGICAL HEALTH SECTION, 505 East King St. Room 202, Carson City, NV, 89710, (702) 885-5394, (10).

NEVADA DIVISION OF OCCUPATIONAL SAFETY AND HEALTH, Capitol Complex 1370 South Curry St., Carson City, NV, 89710, (702) 885-5240, (10).

NEVADA STATE AIR QUALITY DEPARTMENT/DIVISION OF ENVIRONMENTAL PROTECTION, 201 South Fall St., Carson City, NV, 89710, (702) 885-5065, (10).

NEW BRUNSWICK DEPARTMENT OF HEALTH AND COMMUNITY SERVICES/RADIATION PROTECTION SERVICES, P.O. Box 5100, Frederickton, NB, E3B 5G8, (506) 453-2067, (11).

NEW BRUNSWICK DEPARTMENT OF JUSTICE/CONSUMER AFFAIRS BRANCH/REAL ESTATE COUNCIL, P.O. Box 785, Fredericton, NB, E3B 5B4, (506) 455-9733, (11).

NEW ENGLAND DE-WATERING, INC., 70 Heritage Ave. Suite 14 Heritage Pl., Portsmouth, NH, 03801, (800) 535-1211, (2)(4).

NEW ENGLAND RADON SERVICES, 95 Vauxhall St., New London, CT, 06320, (203) 442-6675, (1)(3)(6).

NEW ENGLAND RADON, LTD., 373 Main St., Salem, NH, 03079, (603) 893-4260, (800) 637-2366, (1)(2)(4).

NEW HAMPSHIRE ASSOCIATION, 115 A. Airport Rd. Box 515A, Concord, NH, 03301, (603) 225-5549, (18).

NEW HAMPSHIRE DEPARTMENT OF ENVIRONMENTAL SERVICES/WATER SUPPLY AND POLLUTION CONTROL DIVISION/BUREAU OF WATER SUPPLY AND ENGINEERING, 6 Hazen Dr. P.O. Box 95, Concord, NH, 03301-6527, (603) 271-3139, (10).

NEW HAMPSHIRE DEPARTMENT OF HEALTH AND HUMAN SERVICES/DIVISION OF PUBLIC HEALTH SERVICES/BUREAU OF RADIOLOGICAL HEALTH, P.O. Box 148, Concord, NH, 03301, (603) 271-4588, (10).

NEW HAMPSHIRE DEPARTMENT OF LABOR/ONSITE CONSULTATION PROGRAM, 19 Pillsbury St., Concord, NH, 03301, (603) 271-3170, (10).

NEW HAMPSHIRE DEPARTMENT OF RESOURCES AND ECONOMIC DEVELOPMENT/ENVIRONMENTAL SERVICES/GEOLOGY UNIT, University Of New Hampshire 117 James Hall, Durham, NH, 03824, (603) 862-3160, (10).

NEW HAMPSHIRE GOVERNOR'S ENERGY OFFICE, 2 1/2 Beacon St., Concord, NH, 03301-4498, (603) 271-4498, (10).

NEW HAMPSHIRE REAL ESTATE COMMISSION, State Office Park South Johnson Hall, 3rd Fl. 107 Pleasant St., Concord, NH, 03301, (603) 271-2701, (10).

NEW HAVEN HEALTH DEPARTMENT/ENVIRONMENTAL HEALTH DIVISION, One State St., New Haven, CT, 06511, (203) 787-8173, (10).

NEW JERSEY ASSOCIATION OF REALTORS, P.O. Box 2098, Edison, NJ, 08818, (201) 494-5616, (18).

NEW JERSEY DEPARTMENT OF COMMUNITY AFFAIRS/DIVISION OF HOUSING AND DEVELOPMENT/OFFICE OF PLANNING AND OPERATIONS, 101 South Broad St. CN-802, Trenton, NJ, 08625, (609) 633-6060, (10).

NEW JERSEY DEPARTMENT OF ENVIRONMENTAL PROTECTION/DIVISION OF ENVIRONMENTAL PROTECTION/BUREAU OF ENVIRONMENTAL RADIATION/RADON PROGRAM, 380 Scotch Rd. CN-411, Trenton, NJ, 08625, (201) 879-2612, (6)(10).

NEW JERSEY DEPARTMENT OF ENVIRONMENTAL PROTECTION/DIVISION OF ENVIRONMENTAL QUALITY, 401 East State St. 2nd Floor, East Wing, Trenton, NJ, 08625, (609) 292-5383, (10).

NEW JERSEY DEPARTMENT OF ENVIRONMENTAL PROTECTION/DIVISION OF ENVIRONMENTAL QUALITY/BUREAU OF ENVIRONMENTAL RADIATION, 729 Alexander Rd. CN-415, Trenton, NJ, 08625, (609) 987-6396, (800) 648 0394, (1)(10).

NEW JERSEY DEPARTMENT OF ENVIRONMENTAL PROTECTION/DIVISION OF ENVIRONMENTAL QUALITY/BUREAU OF ENVIRONMENTAL RADIATION/RADIOLOGICAL ENVIRONMENTAL ASSESSMENT, 380 Scotch Rd. CN-411, Trenton, NJ, 08625, (609) 530-4051, (10).

NEW JERSEY DEPARTMENT OF ENVIRONMENTAL PROTECTION/DIVISION OF ENVIRONMENTAL QUALITY/BUREAU OF RADIATION AND INORGANIC ANALYTICAL SERVICES, 380 Scotch Rd. CN-411, Trenton, NJ, 08628, (609) 530-4116, (1)(10).

NEW JERSEY DEPARTMENT OF ENVIRONMENTAL PROTECTION/DIVISION OF SCIENCE AND RESEARCH, 401 East State St. 7th Floor, West Wing, Trenton, NJ, 08625, (609) 984-6070, (1)(10).

NEW JERSEY DEPARTMENT OF HEALTH/DIVISION OF ENVIRONMENTAL AND OCCUPATIONAL HEALTH/ENVIRONMENTAL HEALTH SERVICES/AIR AND RADIATION DIVISION, John Fitch Plaza CN-360, Trenton, NJ, 08625, (609) 633-2043, (10).

NEW JERSEY DEPARTMENT OF HEALTH/DIVISION OF ENVIRONMENTAL AND OCCUPATIONAL HEALTH/PUBLIC EMPLOYEE OCCUPATIONAL SAFETY AND HEALTH PROGRAM, CN-360, Trenton, NJ, 08625, (609) 984-1863, (10).

NEW JERSEY DEPARTMENT OF HEALTH/DIVISION OF EPIDEMIOLOGY AND DISEASE CONTROL/CANCER EPIDEMIOLOGICAL PROGRAM, CN-369, Trenton, NJ, 08625, (609) 588-3500, (10).

NEW JERSEY DEPARTMENT OF LABOR AND INDUSTRY/DIVISION OF WORKPLACE STANDARDS, 36 West State Room 315, Trenton, NJ, 08625, (609) 984-3507, (10).

NEW JERSEY GEOLOGICAL SURVEY/DIVISION OF WATER RESOURCES, CN-029, Trenton, NJ, 08625, (609) 292-1185, (10).

NEW JERSEY HOME SERVICE, INC., 16 Westwood Dr., N., West Orange, NJ, 07052, (201) 669-8166, (1).

NEW JERSEY INSTITUTE OF TECHNOLOGY/DEPARTMENT OF CIVIL AND ENVIRONMENTAL ENGINEERING, 323 King Blvd., Newark, NJ, 07102, (201) 761-0977, (5).

NEW JERSEY REAL ESTATE COMMISSION, Commerce Building 20 West State St. CN 325, Trenton, NJ, 08625, (609) 987-2010, (10).

NEW MEXICO BUREAU OF MINES AND MINERAL RESOURCES, Campus Station, Socorro, NM, 87801, (505) 835-5420, (10).

NEW MEXICO DEPARTMENT OF HEALTH AND ENVIRONMENT/COMMUNITY SERVICES BUREAU, P.O. Box 968, Santa Fe, NM, 87504-0968, (505) 827-2948, (1)(10).

NEW MEXICO DEPARTMENT OF HEALTH AND ENVIRONMENT/OCCUPATIONAL HEALTH AND SAFETY BUREAU, 1476 St. Francis Dr., Santa Fe, NM, 87504, (505) 827-2877, (10).

NEW MEXICO INSTITUTE OF MINING AND TECHNOLOGY/DEPARTMENT OF PHYSICS/RADON LABORATORY, Weir 104, Socorro, NM, 87801, (505) 835-5328, (5).

NEW MEXICO INSTITUTE OF MINING AND TECHNOLOGY/GEOSCIENCE DEPARTMENT, Socorro, NM, 87801, (505) 835-5308, (5).

NEW MEXICO REAL ESTATE COMMISSION, 4125 Carlisle Blvd., NE, Albuquerque, NM, 87107, (505) 841-6524, (10).

NEW MEXICO, REALTORS ASSOCIATION OF, P.O.

Box 1569, Santa Fe, NM, 87504, (505) 258-2361, (18).

NEW WAVE ENERGY SYSTEMS, INC., 3645 Pearl St., Boulder, CO, 80301, (303) 449-8628, (2).

NEW YORK CITY BUREAU OF AIR RESOURCES, 295 Lafayette St., New York, NY, 10012, (212) 323-8580, (10).

NEW YORK DEPARTMENT OF STATE/DIVISION OF LICENSING SERVICES, 162 Washington Ave., Albany, NY, 12231-0001, (518) 473-2419, (10).

NEW YORK METROPOLITAN AREA INTERSTATE SANITATION COMMISSION, 311 West 43 St., New York, NY, 10036, (212) 582-0380, (10).

NEW YORK STATE ASSOCIATION, 107 Washington Ave. Box 122, Albany, NY, 12260, (518) 462-9563, (18).

NEW YORK STATE DEPARTMENT OF ENVIRONMENTAL CONSERVATION/BUREAU OF AIR RESEARCH, 50 Wolf Rd., Albany, NY, 12223-3250, (518) 457-3200, (10).

NEW YORK STATE DEPARTMENT OF ENVIRONMENTAL CONSERVATION/BUREAU OF AIR TOXICS, 50 Wolf Rd., Albany, NY, 12223-3250, (518) 457-7688, (10).

NEW YORK STATE DEPARTMENT OF ENVIRONMENTAL CONSERVATION/BUREAU OF TOXICS AIR SAMPLING, 50 Wolf Rd., Albany, NY, 12223-3250, (518) 457-7454, (10).

NEW YORK STATE DEPARTMENT OF ENVIRONMENTAL CONSERVATION/DIVISION OF AIR RESOURCES, 50 Wolf Rd., Albany, NY, 12223-3250, (518) 457-7230, (10).

NEW YORK STATE DEPARTMENT OF LABOR/DIVISION OF SAFETY AND HEALTH, Building 12 New York State Office Building Campus, Albany, NY, 12240, (518) 457-2574, (10).

NEW YORK STATE ENERGY OFFICE/BUREAU OF RESIDENTIAL CONSERVATION, 2 Rockefeller Plaza 8th Floor, Albany, NY, 12223, (518) 473-7243, (6)(10).

NEW YORK STATE ENERGY RESEARCH AND DEVELOPMENT AUTHORITY/ENERGY EFFICIENCY AND ECONOMIC DEVELOPMENT/BUILDING ENERGY SYSTEMS PROGRAM, 2 Rockefeller Plaza 14th Floor, Albany, NY, 12223, (518) 465-6251, (10).

NEW YORK STATE GEOLOGICAL SURVEY, 3136 Cultural Education Center Empire State Plaza, Albany, NY, 12230, (518) 474-5816, (10).

NEW YORK STATE HEALTH DEPARTMENT/BUREAU OF ENVIRONMENTAL AND OCCUPATIONAL EPIDEMIOLOGY/ENVIRONMENTAL HEALTH SECTION, Radon Registry, Albany, NY, 12203, (518) 458-6212, (10).

NEW YORK STATE HEALTH DEPARTMENT/BUREAU OF ENVIRONMENTAL RADIATION PROTECTION, 2 University Place, Albany, NY, 12203, (518) 458-6461, (10).

NEW YORK STATE HEALTH DEPARTMENT/BUREAU OF ENVIRONMENTAL RADIATION PROTECTION/ENVIRONMENTAL RADIATION SECTION, 2 University Place, Albany, NY, 12203, (800) 458-1158, (10).

NEW YORK STATE HEALTH DEPARTMENT/WADSWORTH CENTER FOR LABORATORIES AND RESEARCH/LABORATORY FOR INORGANIC AND NUCLEAR CHEMISTRY, Empire State Plaza Room D366-B, Albany, NY, 12201, (518) 474-6649, (1)(10).

NEW YORK UNIVERSITY MEDICAL CENTER/INSTITUTE OF ENVIRONMENTAL MEDICINE, 550 1st Ave., New York, NY, 10016, (212) 340-5280, (5).

NEW YORK UNIVERSITY MEDICAL CENTER/INSTITUTE OF ENVIRONMENTAL MEDICINE, 550 1st Ave., New York, NY, 10016, (212) 340-5280, (5).

NIAGARA COUNTY HEALTH DEPARTMENT/AIR RESOURCES BUREAU, 10th and East Falls Sts., Niagara Falls, NY, 14302, (716) 284-3128, (10).

NIGRO CONSTRUCTION COMPANY, INC., 37 James Rd., Monroe, NY, 10950, (914) 783-4254, (1).

NIMCO/NESSI, P.O. Box 356, Menasha, WI, 54952-0356, (414) 725-0357, (1).

NITE HAWK ENTERPRISES, INC., 17 Lincoln Ave., White Plains, NY, 10606, (914) 949-3207, (1)(2).

NITON CORPORATION, P.O. Box 9153, Lexington, MA, 02173, (617) 229-4929, (800) 451-6467, (1)(4).

NITONICS CORPORATION, P.O. Box 9153, Lexington, MA, 02173, (617) 229-4929, (800) 451-6467, (4).

NODAR, INC., Route 5 Box 333R, Athens, AL, 35611, (206) 729-1900, (800) 445-3297, (1).

NORDBERG ASSOCIATES, Clover Hill Farm P.O. Box 520, Alfred, ME, 04002, (207) 490-1105, (1).

NORLAND CORPORATION, Norland Dr., Ft. Atkinson, WI, 53538, (414) 563-8456, (800) 333-8456, (4).

NORTH AMERICAN ASSOCIATION FOR ENVIRONMENTAL EDUCATION (NAAEE), P.O. Box 400, Troy, OH, 45373, (513) 698-6493, (14).

NORTH CAROLINA ASSOCIATION OF REALTORS, P.O. Box 7918, Greensboro, NC, 27417, (919) 294-1415, (18).

NORTH CAROLINA DEPARTMENT OF HUMAN RESOURCES/DIVISION OF FACILITY SERVICES/RADIATION PROTECTION SECTION, 701 Barbour Dr., Raleigh, NC, 27603-2008, (919) 733-4283, (1)(10).

NORTH CAROLINA DEPARTMENT OF LABOR/OCCUPATIONAL SAFETY AND HEALTH, Shore Memorial Building 214 West Jones St., Raleigh, NC, 27603, (919) 733-4880, (10).

NORTH CAROLINA DEPARTMENT OF NATURAL RESOURCES AND COMMUNITY DEVELOPMENT/DIVISION OF LAND RESOURCES, P.O. Box 27687, Raleigh, NC, 27611, (919) 733-3833, (10).

NORTH CAROLINA REAL ESTATE COMMISSION, 1313 Navaho Dr., Raleigh, NC, 27619, (919) 733-9580, (10).

NORTH DAKOTA ASSOCIATION OF REALTORS, 1910 North 11th St., Bismark, ND, 58501, (701) 258-2361, (18).

NORTH DAKOTA GEOLOGICAL SURVEY, University Station, Grand Forks, ND, 58202-8156, (701) 777-5771, (10).

NORTH DAKOTA STATE DEPARTMENT OF HEALTH AND CONSOLIDATED LABORATORIES/DIVISION OF ENVIRONMENTAL ENGINEERING, 1200 Missouri Ave. P.O. Box 5520, Bismarck, ND, 58502-5520, (701) 224-2348, (10).

NORTH DAKOTA STATE DEPARTMENT OF HEALTH/DIVISION OF ENVIRONMENTAL ENGINEERING/OCCUPATIONAL HEALTH PROGRAM, 1200 Missouri Ave. Room 304, Bismarck, ND, 58501, (701) 224-2348, (10).

NORTH DAKOTA STATE REAL ESTATE COMMISSION, 314 East Thayer Ave., Bismarck, ND, 58502, (701) 224-2749, (701) 224-2737, (10).

NORTH EAST ENVIRONMENTAL PRODUCTS, INC., 85 Mechanic St., Lebanon, NH, 03766, (603) 448-3944, (2)(3)(4).

NORTH EAST ENVIRONMENTAL SERVICES, 424 Mine Rd., Asbury, NJ, 08802, (201) 735-6080, (800) 447-2366, (1).

NORTH OHIO VALLEY AIR AUTHORITY, 814 Adams St., Steubenville, OH, 43952, (614) 282-3908, (10).

NORTHEAST RADON CONSULTANTS, 180 Zachary Rd., Manchester, NH, 03103, (603) 622-9391, (1).

NORTHEAST RADON TESTING SERVICE, INC., P.O. Box 94, Stormville, NY, 12582, (914) 221-0236, (1).

NORTHEAST UTILITIES, P.O. Box 270, Hartford, CT, 06141-0270, (203) 665-5000, (13).

NORTHEAST VECTOR MANAGEMENT, INC., P.O. Box 754, Freehold, NJ, 07728, (201) 431-9563, (1).

NORTHEASTERN RADON SERVICE, 21 Rte. 202, Montville, NJ, 07045, (201) 402-1030, (1)(2).

NORTHERN ENVIRONMENTAL LABORATORY, INC., RR 1, Box 1810, Dixmont, ME, 04932-9740, (207) 234-2387, (1).

NORTHERN NECK RADON SERVICE, Rte. 1, Box 259-E, Hague, VA, 22469, (804) 472-4252, (1)(2).

NORTHERN SIERRA AIR QUALITY MANAGEMENT DISTRICT, P.O. Box 5, Sierra City, CA, 96125, (916) 862-1233, (10).

NORTHERN SIERRA AIR QUALITY MANAGEMENT DISTRICT, 10433 Willow Valley Rd., Nevada City, CA, 95959, (916) 265-1398, (10).

NORTHERN SONOMA COUNTY AIR POLLUTION CONTROL DISTRICT, 109 North St., Healdsburg, CA, 95448, (707) 433-5911 (707) 433-5742, (10).

NORTHWEST AIR POLLUTION AUTHORITY, 207 Pioneer Building, Mount Vernon, WA, 98273, (206) 336-5705, (10).

NORTHWEST FURNACE COMPANY, INC., 3139 5th Ave., S., Fort Dodge, IA, 50105, (515) 944-7262, (1).

NORTHWEST POWER PLANNING COUNCIL (NPPC)/POWER PLANNING DIVISION, 851 SW Sixth Ave. Suite 1100, Portland, OR, 97204-1337, (800) 222-3355, (13).

NORTON CONSERVATION COMPANY, INC., P.O. Box 583, Hackettstown, NJ, 07840, (201) 852-6046, (1)(2).

NORWALK DEPARTMENT OF HEALTH, 137 East Ave., Norwalk, CT, 06851, (203) 854-7776, (10).

NOVA SCOTIA DEPARTMENT OF CONSUMER AFFAIRS/CONSUMER SERVICES DIVISION, P.O. Box 998, Halifax, NS, B3J 2X3, (902) 424-4690, (11).

NOVA SCOTIA DEPARTMENT OF HEALTH, P.O. Box 488, Halifax, NS, B3H 2R8, (902) 424-4077, (11).

NUCHEM CORPORATION, P.O. Box 3369 2900 William Penn Hwy., Palmer, PA, 18043, (215) 252-9050, (1)(2).

NUCLEAR ASSOCIATES/DIVISION OF VICTOREEN, 100 Voice Rd., Carle Place, NY, 11514, (516) 741-6360, (4).

NUCLEAR MEASUREMENTS CORPORATION, 2460 North Arlington Ave., Indianapolis, IN, 46218, (317) 543-4419, (317) 546-2415, (1).

NUCLEAR RESEARCH CORPORATION, P.O. Box H, Warrington, PA, 18976, (215) 343-5900, (1).

NUCLEAR SOURCES AND SERVICES, INC., P.O. Box 34042, Houston, TX, 77234, (713) 641-0391, (1).

NUCLEAR TECHNOLOGY, INC./RADON CHEK, 20 H. South State St., Westerville, OH, 43081, (614) 899-2435, (1)(4).

NUCLEON LECTERN ASSOCIATES, INC., 1916 Grayslake Dr., Silver Spring, MD, 20906, (301) 774-

3301, (1).

NUCLEUS, THE, P.O. Box 2561, Oak Ridge, TN, 37830-2561, (615) 482-4041, (1)(4).

NUTECH ENERGY SYSTEMS, INC., 124 Newbold Ct., London, ON, N6E 1Z7, (519) 686-0797, (4).

NUWS LABORATORIES, INC., 28 Montgomery St., Rhinebeck, NY, 12572, (914) 876-5666, (1).

O

O'RIORDAN INSULATION COMPANY, 8901 North Nebraska Ave., Tampa, FL, 33604, (813) 932-0008, (1)(2).

O.K. REMS COMPANY, 10 Hillsborough Rd., Flemington, NJ, 08822, (201) 782-3398, (1).

O.K. REMS CORPORATION, 181 London Rd., Hebron, CT, 06248, (203) 647-7440, (1).

O.K. REMS CORPORATION, 174 Flock Rd., Mercerville, NJ, 08619, (609) 588-9627, (1).

OAK HILL RADON TESTING AND REPAIR, 15 1/2 North State St., Westerville, OH, 43081, (614) 895-8365, (1)(2).

OBAR SYSTEMS, INC., RR 3, Box 759, Highland Lakes, NJ, 07422, (201) 728-7679, (201) 697-0112, (2)(4).

OCCUPATIONAL AND ENVIRONMENTAL HEALTH CONSULTING SERVICES, 635 Harding Rd., Hinsdale, IL, 60521, (312) 325-2083, (1)(2).

OHIO ASSOCIATION OF REALTORS, 200 East Town St., Columbus, OH, 43215, (614) 228-6675, (18).

OHIO DEPARTMENT OF COMMERCE/DIVISION OF REAL ESTATE, Two Nationwide Plaza 5th Floor, Columbus, OH, 43266-0547, (614) 466-4100, (10).

OHIO DEPARTMENT OF HEALTH/RADIOLOGICAL HEALTH PROGRAM/OHIO RADON PROJECT, 1224 Kinnear Rd., Columbus, OH, 43212, (614) 644-2727, (10).

OHIO DEPARTMENT OF INDUSTRIAL RELATIONS/OCCUPATIONAL SAFETY AND HEALTH/DIVISION OF ONSITE CONSULTATION, P.O. Box 825 2323 Fifth Ave., Columbus, OH, 43216, (614) 644-2223, (800) 282-1425, (10).

OHIO DEPARTMENT OF NATURAL RESOURCES/DIVISION OF GEOLOGICAL SURVEY, Fountain Square Building B, Columbus, OH, 43224, (614) 265-6605, (10).

OHIO ENVIRONMENTAL PROTECTION AGENCY/DIVISION OF AIR POLLUTION CONTROL, 1800 Watermark Dr., Columbus, OH, 43266-0149, (614) 644-3020, (10).

OHIO UNIVERSITY/GEOLOGICAL SCIENCES DEPARTMENT, Athens, OH, 45701, (614) 593-1109, (5).

OIL, CHEMICAL AND ATOMIC WORKERS INTERNATIONAL UNION (OCAW), P.O. Box 2812, Denver, CO, 80201, (303) 987-2229, (16).

OKLAHOMA ASSOCIATION OF REALTORS, 9807 North Broadway Extension, Oklahoma City, OK, 73114, (405) 848-9944, (18).

OKLAHOMA CITY-COUNTY HEALTH DEPARTMENT/AIR QUALITY CONTROL SECTION, 921 NE 23rd St., Oklahoma City, OK, 73105, (405) 427-8651, (10).

OKLAHOMA DEPARTMENT OF HEALTH/RADIATION AND SPECIAL HAZARDS SERVICE, 1000 NW Tenth St. P.O. Box 53551, Oklahoma City, OK, 73152, (405) 271-5221, (10).

OKLAHOMA GEOLOGICAL SURVEY, The University Of Oklahoma 830 Van Vleet Oval Room 163, Norman, OK, 73019, (405) 325-3031, (10).

OKLAHOMA REAL ESTATE COMMISSION, 4040 North Lincoln Blvd. Suite 100, Oklahoma City, OK, 73105, (405) 521-3387, (10).

OLD COLONY PEST CONTROL COMPANY, P.O. Box 580, Pembroke, MA, 02359, (617) 294-0937, (1).

OLYMPIA AIR POLLUTION CONTROL AUTHORITY, 120 East State Ave., Olympia, WA, 98501, (206) 352-4881, (10).

OMAHA/AIR QUALITY CONTROL, 1819 Farnam, Omaha, NE, 68183, (402) 444-5379, (10).

ONEIDA COUNTY BOARD OF COOPERATIVE EDUCATIONAL SERVICES (BOCES), Middle Settlement Rd. Box 70, New Hartford, NY, 13413, (315) 768-4583, (10).

ONTARIO MINISTRY OF CONSUMER AND COMMERCIAL RELATIONS/REAL ESTATE AND BUSINESS BROKERS ACT, Toronto, ON, M7A 2H6, (416) 965-0406, (11).

ONTARIO MINISTRY OF HEALTH/DIVISION OF COMMUNITY HEALTH/PUBLIC HEALTH BRANCH/DISEASE CONTROL AND EPIDEMIOLOGICAL SERVICES, 15 Orelea Blvd., Toronto, ON, M4H 1A9, (416) 963-2238, (11).

ONTARIO MINISTRY OF LABOUR/RADIATION PROTECTION SERVICES, 81 Resources Rd., Weston, ON, M9P 3T1, (416) 235-5922, (11).

ONTARIO RESEARCH FOUNDATION (ORF), Sheridan Park Research Community, Mississauga, ON, L5K 1B3, (416) 822-4111, (1)(11).

ORANGE COUNTY HEALTH DEPARTMENT, 124 Main St., Goshen, NY, 10924, (914) 294-7961, (10).

ORANGE ENVIRONMENT, INC., RD 2 Maple Ave., Goshen, NY, 10924, (914) 294-5023, (17).

OREGON ASSOCIATION OF REALTORS, P.O. Box 351, Salem, OR, 97308, (503) 362-3645, (18).

OREGON REAL ESTATE AGENCY, 158 12th St, NE, Salem, OR, 97310-0240, (503) 378-8414, (10).

OREGON STATE DEPARTMENT OF ENVIRONMENTAL QUALITY, 811 SW Sixth Ave., Portland, OR, 97204, (503) 229-5696, (10).

OREGON STATE DEPARTMENT OF GEOLOGY AND MINERAL INDUSTRIES, 910 State Office Building, Portland, OR, 97201-5528, (503) 229-5580, (10).

OREGON STATE DEPARTMENT OF HUMAN RESOURCES/HEALTH DIVISION/RADIATION CONTROL SECTION, 1400 SW 5th Ave. P.O. Box 231, Portland, OR, 97201, (503) 229-5797, (10).

OREGON STATE DEPARTMENT OF WORKERS' COMPENSATION/ACCIDENT PREVENTION DIVISION/CONSULTATIVE SECTION, Labor And Industries Building Room 115, Salem, OR, 97310, (503) 378-3272, (10).

OREGON STATE UNIVERSITY/DEPARTMENT OF STATISTICS, Kidder Hall 44, Corvallis, OR, 97331-4606, (503) 754-0123.

ORGANIZATION OF RESOURCES COUNSELORS (ORC), 1910 Sunderland Pl., NW, Washington, DC, 20036, (202) 737-6300, (16).

ORKIN ENVIRONMENTAL SERVICES, P.O. Box 647, Atlanta, GA, 30301, (404) 888-2801, (1).

ORLEANS-NIAGRA BOARD OF COOPERATIVE EDUCATIONAL SERVICES (BOCES), 4232 Shelby Basin Rd., Medina, NY, 14103, (716) 798-4800, (10).

ORR-DUVALL AND ASSOCIATES, 22346 Metamora, Birmingham, MI, 48010, (313) 644-0706, (1).

OXYTHERM, INC., P.O. Box 3467, Greenville, SC, 29602, (803) 246-2055, (2).

P

P.C. ENGINEERING, Rte. 1, Box 347-B, Lexington, VA, 24450, (703) 463-9116, (2).

P.M. WRIGHT, LTD., 1300 Jules Poitras Blvd., St. Lawrence, PQ, H4N 1X8, (514) 337-3331, (4).

PACIFIC RADIATION CORPORATION, 2945 Stonehill Dr., Altadena, CA, 91001, (818) 798-8100, (1).

PACKARD INSTRUMENT COMPANY, 2200 Warrenville Rd., Downers Grove, IL, 60515, (312) 969-6000, (800) 323-1891, (1)(4).

PACKARD INSTRUMENT COMPANY, 1640 Worcester Rd. Unit 61D, Framingham, MA, 01701, (617) 872-5415, (800) 638-4058, (4).

PAGANO ASSSOCIATES, INC., 112 Jockey Hollow Way, Union, NJ, 07083, (201) 686-2939, (1).

PALCO LININGS, INC., 2500-B Hamilton Blvd., South Plainfield, NJ, 07080, (201) 753-6262, (800) 457-2526, (4).

PALM BEACH COUNTY HEALTH DEPARTMENT/AIR POLLUTION CONTROL PROGRAM, 901 Evernia St., West Palm Beach, FL, 33401, (305) 820-3070, (10).

PALMER TECHNOLOGY, P.O. Box 1072, Glen Burnie, MD, 21061, (301) 766-3176, (1).

PARADISE COMPANY, 109 Greenleaf, Lumberton, TX, 77656, (409) 755-3524, (1)(2).

PARAMOUNT PEST CONTROL, 110 Gordon Rd., Falls Church, VA, 22046, (703) 241-9140, (1).

PAYNE, LARRY D., P.O. Box 1292, Purcellville, VA, 22132, (2).

PECORA CORPORATION, 165 Wambold Rd., Harleysville, PA, 19438, (4).

PELICAN ENVIRONMENTAL CORPORATION, 42 Lehigh Rd., Wellesley, MA, 02181, (617) 235-5997, (1)(2)(3).

PENN NUCLEAR SERVICES, 3190 Industrial Blvd., Bethel Park, PA, 15102, (412) 833-5300, (2).

PENNSYLVANIA ASSOCIATION OF REALTORS, 4501 Chambers Hill Rd., Harrisburg, PA, 17111, (717) 561-1303, (18).

PENNSYLVANIA BUILDING INSPECTORS, INC., 1273 Wilton Crescent, Yardley, PA, 19067, (215) 493-9426, (1).

PENNSYLVANIA DEPARTMENT OF ENVIRONMENTAL RESOURCES/BUREAU OF RADIATION PROTECTION, P.O. Box 2063, Harrisburg, PA, 17120, (717) 787-2480, (10).

PENNSYLVANIA DEPARTMENT OF ENVIRONMENTAL RESOURCES/BUREAU OF RADIATION PROTECTION/DIVISION OF ENVIRONMENTAL RADIATION, Fulton Building, 16th Fl. Third and Locust Sts., Harrisburg, PA, 17120, (717) 787-3479, (10).

PENNSYLVANIA DEPARTMENT OF ENVIRONMENTAL RESOURCES/BUREAU OF RADIATION PROTECTION/DIVISION OF SPECIAL PROJECTS, Fulton Building, 16th Floor Third and Locust Sts., Harrisburg, PA, 17120, (717) 787-2480, (6)(10).

PENNSYLVANIA DEPARTMENT OF ENVIRONMENTAL RESOURCES/BUREAU OF TOPOGRAPHIC AND GEOLOGIC SURVEY, P.O. Box 2357, Harrisburg, PA, 17120, (717) 787-2169, (10).

PENNSYLVANIA DEPARTMENT OF STATE/BUREAU OF PROFESSIONAL AND OCCUPATIONAL AFFAIRS/REAL ESTATE AND VEHICLE

DIVISION, P.O. Box 2649, Harrisburg, PA, 17105-3658, (717) 783-3658, (10).

PENNSYLVANIA RADON CENTER, 21 Hillbrook Cr., Malvern, PA, 19355, (215) 647-9999, (1)(2).

PENNSYLVANIA STATE UNIVERSITY, THE/BREAZEALE NUCLEAR REACTOR/LOW LEVEL RADIATION MONITORING LABORATORY, 101 Academic Projects Building, University Park, PA, 16802, (814) 863-2647, (1)(5).

PENNSYLVANIA STATE UNIVERSITY, THE/COLLEGE OF EARTH AND MINERAL SCIENCE/GEOSCIENCES DEPARTMENT, 332 Deike Building, University Park, PA, 16802, (814) 865-7261, (5).

PENNSYLVANIA STATE UNIVERSITY, THE/DEPARTMENT OF CHEMISTRY, 152 Davey Laboratory, University Park, PA, 16802, (814) 865-6553, (5).

PENNSYLVANIA STATE UNIVERSITY, THE/RADON DETECTION AND MITIGATION CERTIFICATION TRAINING PROGRAM, 231 Sackett Building Energy Technology Projects, University Park, PA, 16802, (814) 863-2133, (3)(6).

PENRAD, INC., 54 South Market St., Elizabethtown, PA, 17022, (717) 367-8410, (717) 367-3995, (1)(2).

PEPPADYNE RADON DETECTION SERVICE, 126 East 9th St. Suite 25, St. Paul, MN, 55101, (612) 293-0864, (1).

PERISON, JANICE B., 13 Beaver Branch Rd., Pennsylvania Furnace, PA, 16865, (814) 234-1871, (1).

PERMA TREAT, INC., P.O. Box 8260, Fredericksburg, VA, 22404, (703) 373-6686, (1).

PEST CONTROL TECHNICIANS, INC., P.O. Box 1951, Norristown, PA, 19401, (215) 275-7378, (1).

PEST DOCTOR, P.O. Box 32, Lithia Springs, GA, 30057, (404) 920-2288, (1).

PETROLEUM ANALYTICAL LABORATORY, INC., P.O. Drawer AK, Eaton Park, FL, 33840, (813) 665-0028, (1).

PH SALES COMPANY, P.O. Box 372, Edison, NJ, 08818, (201) 287-5300, (4).

PHILADELPHIA DEPARTMENT OF PUBLIC HEALTH/AIR MANAGEMENT SERVICES, 500 South Broad St., Philadelphia, PA, 19146, (215) 875-5623, (10).

PHOENIX ENGINEERS, 10041 El Monte, Overland Park, KS, 66207, (301) 975-5906, (3).

PHOENIX LABS, L. Rorbes Rd. Route 1, Box 247, Colebrook, NH, 03578, (603) 237-5773, (1).

PHOENIX LABS, P.O. Box 472, Hollis, NH, 03049, (603) 465-7868, (1).

PHOENIX SAFETY SYSTEMS, 7917 Carnation Dr., Louisville, KY, 40258, (502) 935-8317, (502) 964-0865, (2)(3).

PIMA COUNTY HEALTH DEPARTMENT/AIR QUALITY CONTROL DISTRICT, 150 West Congress St., Tucson, AZ, 85701, (602) 792-8803, (10).

PINAL-GILA COUNTIES AIR QUALITY CONTROL DISTRICT, P.O. Box 1076, Florence, AZ, 85232, (602) 685-5801 x486, (10).

PINELLAS COUNTY DEPARTMENT OF ENVIRONMENTAL MANAGEMENT/AIR QUALITY DIVISION, 16100 Fairchild Dr. Building V102, Clearwater, FL, 34622, (813) 530-6522, (10).

PLACER COUNTY AIR POLLUTION CONTROL DISTRICT, 11484 B Ave., Auburn, CA, 95603, (916) 823-4443, (10).

PLASTICRAFTS, 600 West Bayaud Ave., Denver, CO, 80223, (303) 744-3700, (4).

PLUM GROVE CORPORATION, 33 West Queens Way Mall Old Towne, Hampton, VA, 23669, (804) 722-9841, (1)(2).

PLUMAS COUNTY DEPARTMENT OF ENVIRONMENTAL HEALTH/AIR POLLUTION CONTROL DIVISION, P.O. Box 480, Quincy, CA, 95971, (916) 283-1255, (10).

PMI CONTRACTORS, INC., P.O. Box 6125, Gulfport, MS, 39506, (601) 863-1988, (601) 863-1998, (1).

PNEUMODYNAMICS, INC., 227 Kasson Rd., Bethlehem, CT, 06751, (203) 266-7786, (1).

POLK COUNTY HEALTH UNIT/RADIOLOGICAL HEALTH SECTION, P.O. Box 1480, Winter Haven, FL, 33882-1480, (813) 294-7481 ext. 283, (1).

POLK COUNTY PHYSICAL PLANNING, 5895 Northeast 14th St., Des Moines, IA, 50313, (515) 286-3351, (10).

POLLENEX/ASSOCIATED MILLS, INC., 165 North Canal St., Chicago, IL, 60606, (312) 454-5400, (4).

POLY-AMERICA, INC., 2000 West Marshall Dr., Grand Prairie, TX, 75051, (214) 647-4374, (800) 527-3322, (4).

POLYMERIC SYSTEMS, INC., Wheatland and Mason Sts., Phoenixville, PA, 19460, (215) 935-1170, (800) 228-5548, (4).

PORTER CONSULTANTS, INC., 125 Argyle Rd., Ardmore, PA, 19003, (215) 896-5353, (1)(3)(6).

PORTLAND CEMENT ASSOCIATION (PCA), 5420 Old Orchard Rd., Skokie, IL, 60077-4321, (312) 966-6200, (13).

PORTSMOUTH LOCAL AIR AGENCY/AIR POLLUTION CONTROL DIVISION, 740 Second St., Portsmouth, OH, 45662, (614) 353-5156, (10).

POTOMAC AIR TESTING, P.O. Box 30091, Bethesda, MD, 20814, (301) 657-4391, (1).

PRE-PURCHASE HOME INSPECTION, 3512 Breeland Ave., Louisvile, KY, 40241, (502) 425-6810, (1).

PRECISION MEASUREMENTS ASSOCIATION (PMA), 9681 Business Center Dr., Rancho Cucomong, CA, 91730, (714) 980-6166, (14).

PRECISION RADON TESTING, 302 Leidy Rd., Souderton, PA, 18964, (215) 721-1958, (800) 545-8378, (1).

PRIGEL RADON TESTING, 2866 Travis Ct., York, PA, 17403, (717) 741-3188, (301) 252-1246, (1).

PRINCE EDWARD ISLAND/QUEEN ELIZABETH HOSPITAL, Charlottetown, PE, C1A 8T5, (902) 566-6277, (11).

PRINCE GEORGE'S COUNTY HEALTH DEPARTMENT/DIVISION OF AIR QUALITY CONTROL, 10210 Greenbelt Rd., Seabrook, MD, 20706, (301) 794-6800, (10).

PRINCETON HOME INSPECTION SERVICE, 197 Wall St., Princeton, NJ, 08540, (609) 921-3775, (1)(2).

PRINCETON PLASMA PHYSICS LABORATORY, P.O. Box 451, Princeton, NJ, 08543, (609) 683-3319, (609) 243-3319, (1).

PRINCETON TESTING LABORATORY, Princeton Service Center 3490 U.S. Rte. 1, Princeton, NJ, 08540, (609) 452-2037, (1).

PRINCETON UNIVERSITY/SCHOOL OF ENGINEERING AND APPLIED SCIENCES/CENTER FOR ENGINEERING AND APPLIED STUDIES, Engineering Quadrangle, Princeton, NJ, 08544, (609) 452-5445, (5).

PRO CHEK, INC., Box 2000, Arden, NC, 28704, (800) 522-2435, (1).

PRO-GUARD BUILDING INSPECTIONS, INC., RD 1, Box 193B, Auburn, PA, 17922, (717) 754-7673, (1)(2).

PRO-TEC HOME INSPECTORS, INC., P.O. Box 132, Franklin, NJ, 07416, (201) 827-4996, (1).

PRO-TECH RADON, 459 Forestwood Dr., Gahanna, OH, 43230, (614) 475-0023, (1).

PROFESSIONAL HOME ANALYSIS, 1025 Northeast Sanchez Ave., Ocala, FL, 32670, (904) 694-3982, (1)(2).

PROFESSIONAL HOME INSPECTIONS, INC., 28790 Pine Dr., Evergreen, CO, 80439, (303) 674-4620, (1)(3).

PROFESSIONAL HOME INSPECTIONS, INC., R.D. 1, Fenimore Rd. Box 371-A, Medford, NJ, 08055, (609) 261-7374, (1).

PROFESSIONAL HOUSE DOCTORS, 599 Short Ave., Marengo, IA, 52301, (319) 642-7133, (1)(2).

PROFESSIONAL HOUSE DOCTORS, INC., 1406 East 14th St., Des Moines, IA, 50316, (515) 265-6667, (1)(2)(4)(6).

PROFESSIONAL RADON ASSESSMENT, 405 Valley Dr., Oliver Springs, TN, 37840, (615) 435-6432, (615) 690-2208, (1).

PROFESSIONAL RADON MEASUREMENTS, P.O. Box 791, Normal, IL, 61761, (309) 827-3377, (309)452-9904, (1).

PROFESSIONAL RESTORATION AND REPAIR, P.O. Box 38932, Colorado Springs, CO, 80937, (719) 632-3996, (2).

PROGRESS UNLIMITED, INC., 200 Madison Ave. Suite 1806, New York, NY, 10016, (212) 689-7030, (4).

PROJECT ENVIRONMENTAL GROUP, 10 Duff Rd., Pittsburgh, PA, 15235, (412) 243-3100, (2).

PROTECT-A-HOME, P.O. Box 584, Newtown, CT, 06470, (203) 426-3000, (1).

PROTECTION INSPECTION, INC., Rte. 11 Hill Rd., Knoxville, TN, 37938, (615) 688-9779, (1)(2)(3).

PROTECTRON, INC., 229 East Main St. Suite E, Westminster, MD, 21157, (301) 857-9116, (1).

PUBLIC CITIZEN LITIGATION GROUP (PCLG)/HEALTH RESEARCH GROUP, 2000 P St., NW Suite 700, Washington, DC, 20036, (202) 872-0320, (17).

PUBLIC HEALTH FOUNDATION (PHF), 1220 L St., NW Suite 350, Washington, DC, 20005, (202) 898-5600, (15).

PUEBLO CITY-COUNTY HEALTH DEPARTMENT, 151 Central Main, Pueblo, CO, 81003, (303) 544-8376, (10).

PUERTO RICO DEPARTMENT OF LABOR AND HUMAN RESOURCES/OCCUPATIONAL SAFETY AND HEALTH, 505 Munoz Rivera Ave. 21st Fl., Hato Rey, PR, 00918, (809) 754-2134, (809) 754-2171, (10).

PUERTO RICO GEOLOGICALL SURVEY/DEPARTMENT DE RECURSOS NATURALES, Apartado 5887 Puerta de Tierra, San Juan, PR, 00906, (809) 723-2716, (10).

PUERTO RICO/DEPARTMENT OF HEALTH/ RADIOLOGICAL HEALTH DIVISION, G.P.O. Call Box 70184, Rio Piedras, PR, 00936, (809) 767-3563, (10).

PUERTO RICO/ENVIRONMENTAL QUALITY BOARD, Del Parque St. #204 Corner Pumarada St., Santurce, PR, 00910, (809) 725-5140, (809) 722-0077, (10).

PUGET SOUND AIR POLLUTION CONTROL AGENCY, 200 West Mercer St. Room 205 P.O. Box 9863, Seattle, WA, 98119-3958, (206) 344-7330, (10).

PUKAL CONTRACTING COMPANY, 390 Moorehead

Rd., Cabot, PA, 16023, (412) 352-4030, (1)(2).

PUTNAM COUNTY HEALTH DEPARTMENT, Putnam County Office Building 110 Old Route Six Center Building #3, Carmel, NY, 10512, (914) 225-2115, (10).

PYLON ELECTRONIC DEVELOPMENT COMPANY, LTD., 147 Colonnade Rd., Nepean, ON, K2E 7L9, (613) 226-7920, (1)(4).

PYRAMID ENVIRONMENTAL SYSTEMS, INC., P.O. Box 548, Sparta, NJ, 07871, (201) 729-9375, (1)(2).

Q

QDT, LTD., 1000 Singleton Blvd., Dallas, TX, 75212, (214) 741-1993, (4).

QUADREX, 1700 Dell Ave., Campbell, CA, 95008, (408) 866-4510, (2).

QUALITY AIR RADON REDUCTION, 5-70 Oakwood Rd., Glastonbury, CT, 06033, (203) 633-3629, (1).

QUALITY ASSURANCE, INC., P.O. Box 4087, Highlands Ranch, CO, 80126, (303) 791-2624, (1).

QUALITY ENVIRONMENTS, INC., P.O. Box 3689, Chapel Hill, NC, 27515, (919) 493-7033, (1).

QUALITY HOME INSPECTIONS, INC., 81 Main St., High Bridge, NJ, 08829, (201) 638-4994, (1).

QUALITY RADON TESTING CORPORATION, P.O. Box 11980, Chicago, IL, 60611, (312) 280-4908, (1).

QUANTEC, 212 Lucia St., Middlesex, NJ, 08846, (201) 469-9461, (1).

QUANTUM INFORMATION, LTD., P.O. Box 467, Westown, PA, 19395-0467, (215) 431-9577, (1).

QUANTUM TECHNOLOGY, INC., 830 Franklin Ct., Marietta, GA, 30067, (404) 429-8324, (4).

QUOIN PARTNERSHIP, ARCHITECTS AND ENGINEERS, P.O. Box 147, St. Peters, PA, 19470, (215) 469-9595, (2).

R

R. BERNT AND SONS, 96 Laural Ave. P.O. Box 721, Woodacre, CA, 94973, (415) 488-9106, (1).

R.A.D. SERVICE AND INSTRUMENTS, LTD., 50 Silver Star Blvd. Unit 208, Scarborough, ON, M1V 3L3, (416) 298-9200, (1)(4)(5).

R.B. KANALFLAKT, INC., 1121 Lewis Ave., Sarasota, FL, 34237, (813) 366-7505, (813) 365-4118, (4).

R.E. WRIGHT ASSOCIATES, INC./RADONIX GROUP, 3240 Schoolhouse Rd., Middletown, PA, 17057, (717) 944-5520, (1)(2)(4).

R.F. SIMON COMPANY, INC., RD 2, Box 481, Barto, PA, 19504, (215) 754-6455, (1)(2)(3)(4)(6).

R.I.C. INSPECTION AND ENGINEERING, R.R. 2 P.O. Box 2186-5, Columbia, NJ, 07832, (201) 496-4818, (1).

R.J. FRIDG COMPANY, INC., 427 Hamill Rd., Indiana, PA, 15701, (412) 463-3225, (1).

R.J. MOORE AND ASSOCIATES, INC., 1010 West St., Annapolis, MD, 21401, (301) 269-5950, (1).

R.K. OCCUPATIONAL AND ENVIRONMENTAL ANALYSIS, INC., 19 Burrows Ave., Bernardsville, NJ, 07924, (201) 766-1737, (1).

R.K. VAN SCHAICK AND ASSOCIATES, R.R. 4 Box 4038, Greenwich, NY, 12834, (518) 692-2015, (1).

R.T.C., P.O. Box 832, Linbrook, NY, 11563, (516) 887-1371, (1).

R/K OCCUPATIONAL AND ENVIRONMENTAL ANALYSIS, INC., 616 Warren St., Alpha, NJ, 08865, (215) 250-5948, (1).

RA CONSULTANTS, INC., 1236 North 7th St., Grand Junction, CO, 81501, (303) 245-5237, (1)(2).

RA-MED, P.O. Box 4413, Alexandria, VA, 22303, (703) 960-0514, (2).

RA-TEK/DIVISION OF SHAMROCK, INC., P.O. Box 496 800 East Main St., Marshall, MN, 56258, (507) 532-7313, (1)(4).

RACHECK, INC., P.O. Box 548, Southampton, PA, 18966, (215) 322-5309, (1).

RAD ALERT, LTD., P.O. Box 9238, Springfield, IL, 62791, (217) 546-9549, (1)(2).

RAD CHEK SERVICES COMPANY, P.O. Box 173, North Granby, CT, 06060-0173, (203) 653-4159, (1).

RAD CHEK SERVICES COMPANY, P.O. Box 279, Livingston, NJ, 07039, (201) 535-6459, (1).

RAD CON, INC., Box 027, Westminster, MD, 21157, (301) 848-6324, (1).

RAD ELEC, INC., 5330 J Spectrum Dr., Frederick, MD, 21701, (301) 694-0011, (1)(4)(5).

RAD ENVIRONMENTAL SERVICES, INC., 2306 Rodeo Dr., Durham, NC, 27704, (919) 688-9226, (1)(2).

RAD ENVIRONMENTAL SERVICES, INC., 30 Basswood Ct., Red Bank, NJ, 07701, (201) 758-9324, (1)(2).

RAD SYSTEMS, INC., 21 East Main St., Westborough, MA, 01581, (508) 366-5051, (4).

RAD-LAB SERVICES, INC., 1311 Washington Blvd., Williamsport, PA, 17701, (717) 326-2628, (4).

RAD-TEST, INC., P.O. Box 22576, Louisville, KY, 40222, (502) 368-5996, (1).

RADALYTIC LABS, INC., P.O. Box 202, McConnells, SC, 29726, (803) 684-9859, (1)(2).

RADEX, 265 Stone St., Walpole, MA, 02081, (617) 660-1548, (1).

RADEX, INC., 1382 Prince Ave. Suite A, Athens, GA, 30606, (404) 548-6960, (1)(2)(3).

RADIAN CORPORATION/ENGINEERING DIVISION, P.O. Box 1300, Research Triangle Park, NC, 27709, (919) 541-9100, (5)(14).

RADIATION CONSULTANTS OF SOUTHWEST VIRGINIA, P.O. Box 4464, Roanoke, VA, 24015, (703) 985-5778, (703) 981-7732, (1).

RADIATION CONSULTANTS, INC., P.O. Box 34012, Houston, TX, 77234, (713) 476-0072, (1).

RADIATION DATA, 115 Wall St., Princeton, NJ, 08540, (609) 921-7061, (1)(2).

RADIATION DATA/DIVISION OF PRINCETON GEOPHYSICAL SERVICES, 37 Castleman Dr., Southington, CT, 06489, (203) 621-0678, (1)(2).

RADIATION DATA/DIVISION OF PRINCETON GEOPHYSICAL SERVICES, 1200 State Rd., Princeton, NJ, 08540, (609) 921-7061, (1)(2).

RADIATION HEALTH, 1259 Marlin Dr., Port Orchard, WA, 98366, (206) 871-5454, (1).

RADIATION INSTRUMENT COMPANY (RICO), 9165 Helaine Hamlet Way, Columbia, MD, 21045, (301) 997-7426, (4).

RADIATION MEASUREMENT SERVICES, P.O. Box 3681, Crofton, MD, 21114, (301) 261-6966, (1).

RADIATION PROTECTION SERVICES, INC., P.O. Box 2395, Darien, CT, 06820, (203) 324-7967, (1)(2).

RADIATION PROTECTION SERVICES, INC., 646 Rte. 18 Suite 215, East Brunswick, NJ, 08816, (201) 390-7360, (1)(2)(6).

RADIATION RESEARCH SOCIETY (RRS), 1101 Market St. 14th Floor, Philadelphia, PA, 19107, (215) 574-3153, (5)(14).

RADIATION SAFETY ASSOCIATES, INC., 10 Pendleton Dr. P.O. Box 107, Hebron, CT, 06248, (203) 228-0487, (1).

RADIATION SAFETY ENGINEERING, 6105 South Ash Suite 1, Tempe, AZ, 85283, (602) 897-9459, (1).

RADIATION SAFETY SERVICES, INC., 6312 West Oakton St., Morton Grove, IL, 60053, (312) 965-1999, (1)(4)(6).

RADIATION SERVICE ORGANIZATION, INC., P.O. Box 1526, Laurel, MD, 20707-7444, (301) 953-2482, (1).

RADIATION SERVICES, INC., 3412 Doriswood Ct., Antioch, TN, 37013, (615) 361-6492, (1).

RADIATION SURVEILLANCE ASSOCIATES, INC., 1230 Princeton Ave., Salt Lake City, UT, 84105, (801) 583-6919, (801) 467-1030, (1).

RADIATION TESTING DESIGN, 279 County Rd., 84W, Allenspark, CO, 80510, (303) 747-2702, (1).

RADIOACTIVITY ALERT DETECTION SERVICES, INC., P.O. Box L-227, Langhorne, PA, 19047, (215) 357-0574, (1)(2).

RADIOLOGICAL CONTROL ASSOCIATES, 665 Robbins Ave. Unit 2, Dracut, MA, 01826, (617) 453-5268, (1)(2).

RADIOLOGICAL HEALTH SCIENCES, INC., P.O. Box 4790, Fort Walton Beach, FL, 32549, (904) 863-1191, (1).

RADIOLOGICAL PHYSICS SERVICES/HEALTH EAST, INC., P.O. Box 13367, Roanoke, VA, 24033, (703) 981-7379, (1).

RADIOLOGICAL TESTING SERVICES, HC92, Box2B, Crawford, NE, 69339, (308) 665-1587, (1).

RADION LABORATORY, INC., 70 Walnut St., Wellesley, MA, 02181, (617) 239-8066, (1)(2)(3)(4)(5)(6)(13).

RADOMET SERVICES, 7079 East Briarwood Cr., Englewood, CO, 80112, (303) 694-4519, (1).

RADON AIR AND WATER CHECK, INC., 29 Riverside Dr., Binghamton, NY, 13905, (607) 724-5050, (1).

RADON AIRTEST COMPANY, 13800 Bagley Rd., Middleburg Heights, OH, 44130, (216) 845-8050, (1).

RADON ALERT, P.O. Box 2167, Westminster, MD, 21157-9167, (301) 848-2862, (1).

RADON ALERT SYSTEMS, P.O. Box 1247, Regina, SK, S4P 3B8, (306) 949-0445, (1).

RADON ALERT, INC., P.O. Box 131378, Tyler, TX, 75713, (214) 592-4077, (1)(6).

RADON ALERT® DETECTION CENTER, P.O. Box 323, Flourtown, PA, 19031, (215) 247-9197, (800) 345-6348, (1)(2)(4).

RADON AMERICA, P.O. Box 13284, Pensacola, FL, 32591, (1).

RADON ANALYSIS AND DETECTION OF NEW ENGLAND, INC., 111 Jefferson St., Laconia, NH, 03246, (603) 528-5772, (1).

RADON ANALYSIS, INC., 2520 15th Ave., Longmont, CO, 80501, (303) 651-7314, (1).

RADON ANALYSIS, INC., 7 Chamberlain Rd., Flemington, NJ, 08822, (201) 788-0661, (1)(4).

RADON ANALYSIS, INC., P.O. Box 5224, Toms River, NJ, 08754-5224, (201) 929-9575, (1).

RADON ANALYSIS, INC., 4635 B South 28 Rd., Arlington, VA, 22206, (703) 379-8614, (1).

RADON ANALYSTS, INC., P.O. Box 509, Livingston

Manor, NY, 12758, (914) 439-5111, (914) 292-2277, (1).

RADON ANALYTICAL LABS, 9011 North Meridian St., Indianapolis, IN, 46260, (317) 843-0788, (1).

RADON AND DEWATERING SYSTEMS, INC. OF MARYLAND, Box 1814, Silver Spring, MD, 20902, (301) 593-1861, (4).

RADON AND ENVIRONMENTAL PROFESSIONALS, P.O. Box 1045, Independence, KS, 67301, (316) 331-5207 , (1).

RADON AND ENVIRONMENTAL PROFESSIONALS, P.O. Box 3391, Lawrence, KS, 66046, (913) 749-7420, (1).

RADON AND ENVIRONMENTAL PROFESSIONALS, 2768 South Bennett, Wichita, KS, 67217-1812, (316) 945-2570, (1).

RADON AND RADIATION CONSULTANTS, 3575 Peninsula Dr., Akron, OH, 44319, (216) 644-7664, (1).

RADON AND WATER TESTING SERVICE, 10806 Lafayette-Plain City Rd., Plain City, OH, 43064, (614) 873-8821, (1).

RADON APPLIANCES, INC., P.O. Box 1461, Glen Burnie, MD, 21061, (301) 766-5711, (4).

RADON ASSOCIATES MARKETING COMPANY, 1099 18th St. Suite 1830, Denver, CO, 80202, (303) 293-3141, (3).

RADON ASSOCIATES, INC., 126 Greenlawn Ave., Clifton, NJ, 07013, (201) 773-4955, (1).

RADON AWARENESS AND DETECTION, 1408 South Rock Hill Rd., St. Louis, MO, 63119, (314) 961-0400, (1).

RADON AWARENESS, INC., 415 East Galena P.O. Box 566, Hailey, ID, 83333, (208) 788-2634, (1)(2).

RADON BAN, 3719 Michelle Way , Baltimore, MD, 21208, (301) 484-7120, (1)(4).

RADON BUSTERS, 1345 Main Ave. P.O. Box 2426, Fargo, ND, 58108, (701) 293-5370, (1)(2).

RADON BUSTERS, P.O. Box R247, Radnor, PA, 19087, (215) 566-3059, (1).

RADON BUSTERS OF CENTRAL NEW YORK, 3218 Valley Pl., Sauquoit, NY, 13456, (315) 737-7290, (1).

RADON CALIBRATIONS, 800 Stiles Ct., Richmond, VA, 23235, (804) 320-3761, (1).

RADON CHECK, 2524 South El Paradiso Suite 50, Mesa, AZ, 85202, (602) 730-5559, (1).

RADON CHECK, P.O. Box 7882, Greensboro, NC, 27417, (919) 855-1312, (1).

RADON CHECK, 2263 Coppersmith Sq., Reston, VA, 22091, (703) 476-6810, (1).

RADON CHECK/DIVISION OF HORIZON SYSTEMS, P.O. Box 721, Harvard, MA, 01451, (617) 456-8811, (617) 890-2334, (1).

RADON CONSULTANTS, P.O. Box 5209, Sparks, NV, 89432, (800) 327-6826, (1)(4).

RADON CONSULTANTS, INC., P.O. Box 789, Golden, CO, 80402, (303) 279-1358, (1).

RADON CONSULTANTS, INC., MARYLAND, 4831 Silver Hill Rd., Suitland, MD, 20746, (301) 736-7040, (1)(2).

RADON CONSULTANTS, INC., VIRGINIA, 4012 North 35th St., Arlington, VA, 22207, (703) 522-6669, (301) 736-7040, (1)(2).

RADON CONSULTING AND TESTING SERVICE, 7449 Highway 44 Box 9, Crystal River, FL, 32629, (904) 563-2141, (904) 563-1771, (1).

RADON CONTROL PRODUCTS, 511 Industrial Dr., Carmel, IN, 46032, (317) 846-7442, (800) 523-2082, (4).

RADON CONTROL PROFESSIONALS, INC., 11920 Fieldthorn Ct., Reston, VA, 22094, (703) 471-9459, (1)(2)(3).

RADON CONTROL SERVICES CORPORATION, 3122 49th St., Des Moines, IA, 50310, (515) 276-9131, (1)(2).

RADON CONTROL SYSTEMS, 138 Riverview Plantation Dr., Williamsburg, VA, 23185, (804) 564-1893, (2).

RADON CONTROL, INC., 337 Route 22 E., Greenbrook, NJ, 08812, (201) 563-4211, (1).

RADON CORRECTIONS INSTITUTE, INC., P.O. Box 314, Longmont, CO, 80502, (303) 651-6190, (800) 648-9281, (2)(3).

RADON DECON, INC., P.O. Box 28, Muncy Valley, PA, 17758, (717) 482-2076, (1)(2).

RADON DETECTION OF MISSISSIPPI, P.O. Box 2404, Columbus, MS, 39704, (601) 328-4687, (1).

RADON DETECTION AGENCY, LTD., 625 5th Ave. Suite 104, New Westminster, BC, V3M 1X4, (604) 525-4455, (1)(4).

RADON DETECTION AND CONSULTING, 705 West Fourth St., Lawrence, KS, 66044, (913) 843-7377, (1)(2)(3).

RADON DETECTION AND CONTROL, P.O. Box 419 153 Jordan St., South Hts., PA, 15081, (412) 375-1700, (1).

RADON DETECTION AND REDUCTION LABORATORY/DIVISION OF FORMACHEM, 10834 Ambassador, St. Louis, MO, 63132, (314) 872-9350, (1)(2)(4).

RADON DETECTION AND REMOVAL SERVICES, 915 6th Ave., N., Seattle, WA, 98109, (206) 282-9156, (1)(2).

RADON DETECTION AND TESTING, INC., 2222 Cumberland Pl., Smyrna, GA, 30080, (404) 434-7299, (1).

RADON DETECTION CENTER OF MICHIGAN, 601 South Saginaw St. Suite 514, Flint, MI, 48502, (313) 767-8630, (1).

RADON DETECTION CORPORATION OF MICHIGAN, 4740 Hall St., SE, Kentwood, MI, 49506, (616) 942-0323, (1).

RADON DETECTION DEVICES, 5 Lauriann Ct., Parkton, MD, 21120, (301) 329-6306, (1).

RADON DETECTION SERVICE OF KANSAS CITY, 8405 Booth, Raytown, MO, 64138, (816) 356-9677, (1).

RADON DETECTION SERVICE OF MONTANA, P.O. Box 1114, Kalispell, MT, 59901, (406) 755-9477, (406) 752-8404, (1).

RADON DETECTION SERVICES, 323 Neusch Way, Saskatoon, SK, S7K 7V2, (306) 931-7729, (1).

RADON DETECTION SERVICES, P.O. Box 434, West Rutland, VT, 05777, (802) 775-0250, (802) 438-5603, (1).

RADON DETECTION SERVICES OF NEBRASKA, 16506 Dorcas St., Omaha, NE, 68130, (402) 330-8968, (1).

RADON DETECTION SERVICES, LTD., 1500 Harbour Oaks Dr., Tucker, GA, 30084, (404) 621-9300, (1).

RADON DETECTION SPECIALISTS, 7966 South Madison St., Burr Ridge, IL, 60521, (312) 325-4443, (1).

RADON DETECTION SYSTEMS OF VERMONT, P.O. Box 383, Randolph, VT, 05060, (802) 728-5190, (1).

RADON DETECTION SYSTEMS, INC., OHIO, P.O. Box 473, Aurora, OH, 44202, (216) 562-4647, (1).

RADON DETECTION SYSTEMS/ALBUQUERQUE, NEW MEXICO BRANCH, 4005 Cumino de la Sierra, NE, Albuquerque, NM, 87111, (303) 444-5253, (800) 627-2360, (1).

RADON DETECTION SYSTEMS/ARIZONA BRANCH, 1301 East Dava Dr., Tempe, AZ, 85283, (303) 444-5253, (800) 627-2360, (1).

RADON DETECTION SYSTEMS/ATLANTA, GEORGIA BRANCH, 2965 Atterbury Ct., Decatur, GA, 30033, (303) 444-5253, (800) 627-2366, (1).

RADON DETECTION SYSTEMS/BLOOMINGTON-NORMAL BRANCH, 1207 Stephens Rd., Normal, IL, 61761, (303) 444-5253, (800) 627-2366, (1).

RADON DETECTION SYSTEMS/BOSTON, MASSACHUSETTS BRANCH, 91 Westland Ave. Suite 606, Boston, MA, 02115, (303) 444-5253, (800) 627-2366, (1).

RADON DETECTION SYSTEMS/BUTTE, MONTANA BRANCH, 1123 West Park, Butte, MT, 59701, (303) 444-5253, (800) 627-2360, (1).

RADON DETECTION SYSTEMS/CASPER, WYOMING BRANCH, 628 East 14th, Casper, WY, 82601, (303) 444-5253, (800) 627-2360, (1).

RADON DETECTION SYSTEMS/CHARLESTON, WEST VIRGINIA BRANCH, 2300 McCorkle Ave., SE, Charleston, WV, 25304, (303) 444-5253, (800) 627-2366, (1).

RADON DETECTION SYSTEMS/CHICAGO, ILLINOIS BRANCH, 12845 Golf View Ln., Palois Heights, IL, 60463, (303) 444-5253, (800) 627-2360, (1).

RADON DETECTION SYSTEMS/COLORADO SPRINGS, COLORADO BRANCH, 2010 Wooten Ln., Colorado Springs, CO, 80907, (303) 444-5253, (800) 627-2366, (1).

RADON DETECTION SYSTEMS/COLUMBUS, OHIO BRANCH, 3487 Colchester Rd., Columbus, OH, 43221, (303) 444-5253, (800) 627-2366, (1).

RADON DETECTION SYSTEMS/CORPORATE HEADQUARTERS, 2450 Central Ave. A-1, Boulder, CO, 80301, (303) 444-5253, (800) 627-2360, (1).

RADON DETECTION SYSTEMS/DALLAS, TEXAS BRANCH, 6607 Northwest Hwy., Dallas, TX, 75225, (303) 444-5253, (800) 627-2360, (1).

RADON DETECTION SYSTEMS/DES MOINES, IOWA BRANCH, Box 93 East 57 Rd., Kelly, IA, 50134, (303) 444-5253, (800) 627-2366, (1).

RADON DETECTION SYSTEMS/DETROIT, MICHIGAN BRANCH, 6051 Dunn Rd., Howell, MI, 48843, (303) 444-5253, (800) 627-2366, (1).

RADON DETECTION SYSTEMS/DURANGO, COLORADO BRANCH, 77 Roach Ranch Rd., Durango, CO, 81302, (303) 444-5253, (800) 627-2360, (1).

RADON DETECTION SYSTEMS/FARGO, NORTH DAKOTA BRANCH, 1225 Oak St., North, Fargo, ND, 58102, (303) 444-5253, (800) 627-2360, (1).

RADON DETECTION SYSTEMS/GRAND JUNCTION, COLORADO BRANCH, 235 Radcliff Dr., Frulta, CO, 81521, (303) 444-5253, (800) 627-2366, (1).

RADON DETECTION SYSTEMS/GREENSBORO, NORTH CAROLINA BRANCH, 4324 Cooper Rd., Winston-Salem, NC, 27121, (303) 444-5253, (800) 627-2366, (1).

RADON DETECTION SYSTEMS/LOS ANGELES, CALIFORNIA BRANCH, 313 8th St. Apt. E, Seal Beach, CA, 90740, (213) 596-1948, (800) 627-2366, (1).

RADON DETECTION SYSTEMS/MIAMI, FLORIDA

BRANCH, 17001 SW 109 Ave., Miami, FL, 33157, (303) 444-5253, (800) 627-2366, (1).

RADON DETECTION SYSTEMS/MILWAUKEE BRANCH, 2630 North Newhall St., Milwaukee, WI, 53211, (303) 444-5253, (800) 627-2366, (1).

RADON DETECTION SYSTEMS/MINNEAPOLIS-ST. PAUL BRANCH, 548 Frontenae, St. Paul, MN, 55104, (303) 444-5253, (800) 627-2366, (1).

RADON DETECTION SYSTEMS/NASHVILLE, TENNESSEE BRANCH, 3102 Bellwood Ave., Nashville, TN, 37203, (303) 444-5253, (800) 627-2366, (1).

RADON DETECTION SYSTEMS/NEWARK, NEW JERSEY BRANCH, 6 Earl Ct., Montiville, NJ, 07045, (303) 444-5253, (800) 627-2360, (1).

RADON DETECTION SYSTEMS/NORTHERN CALIFORNIA BRANCH, 2567 Greenwich St., San Francisco, CA, 94123, (303) 444-5253, (800) 627-2366, (1).

RADON DETECTION SYSTEMS/PITTSBURGH, PENNSYLVANIA BRANCH, 180 Indionola Rd., Pittsburgh, PA, 15238, (303) 444-5253, (800) 627-2360, (1).

RADON DETECTION SYSTEMS/PORTLAND, OREGON BRANCH, 3710 SE 36th Pl. #27, Portland, OR, 97202, (303) 444-5253, (800) 627-2366, (1).

RADON DETECTION SYSTEMS/SALT LAKE CITY, UTAH BRANCH, 6728 South 1655 East, Salt Lake City, UT, 84121, (303) 444-5253, (800) 627-2360, (1).

RADON DETECTION SYSTEMS/SIOUX FALLS, SOUTH DAKOTA BRANCH, 2713 South Prairie, Sioux Falls, SD, 57105, (303) 444-5253, (800) 627-2360, (1).

RADON DETECTION SYSTEMS/SPOKANE, WASHINGTON BRANCH, 1115 Lambert Ln., Coer D'Alene, ID, 83814, (303) 444-5253, (800) 627-2360, (1).

RADON DETECTION, INC., 131 Amherst St. P.O. Box 900, Amherst, NH, 03031-0900, (603) 882-7340, (1).

RADON DETECTORS, RD 1, Box 177, Van Etten, NY, 14889, (607) 589-6223, (1)(2).

RADON DIAGNOSTIC LABORATORIES, 231 Front St. P.O. Box 2890, South Portland, ME, 04106, (207) 767-2818, (1)(2).

RADON DIAGNOSTICS, Box 761, Andover, NJ, 07821, (201) 729-1251, (1)(2).

RADON DOCTOR, INC., THE, Alpha Building Suite 904, Easton, PA, 18042, (215) 252-4545, (1)(2).

RADON ENGINEERING SERVICES, INC., 3129 Overton Rd., Birmingham, AL, 35223, (205) 967-2665, (1)(2).

RADON ENGINEERING/A UNIT OF PSI ENGINEERING, INC., 1 Lethbridge Plaza, Mahwah, NJ, 07430, (201) 529-8300, (1)(2).

RADON ENVIRONMENTAL CONSTRUCTION SERVICES/RECONS, INC., MAIN OFFICE, P.O. Box 3058, Princeton, NJ, 08543-3058, (609) 883-1808, (215) 493-0667, (1)(2).

RADON ENVIRONMENTAL CONSTRUCTION SERVICES/RECONS, INC., PA OFFICE, 979 Mt. Eyre Rd., Washington Crossing, PA, 18977, (215) 321-3274, (1)(2).

RADON ENVIRONMENTAL MONITORING, INC., Box 5305, Evanston, IL, 60204, (312) 328-6628, (1)(4).

RADON ENVIRONMENTAL SERVICES, INC., 8935 North Meridian St. Suite 110, Indianapolis, IN, 46260, (317) 846-1991, (1)(2).

RADON ENVIRONMENTAL SERVICES, INC., 901 Mason Headley Rd., Lexington, KY, 40502, (606) 255-6375, (1)(2).

RADON ENVIRONMENTAL SERVICES, INC., 4320 Parakeet Rd., Toledo, OH, 43612, (419) 447-0209, (1)(2).

RADON ENVIRONMENTAL TEST EXPERTS, INC., 5271 Holmes Pl., Boulder, CO, 80303, (303) 442-2344, (1).

RADON ENVIRONMENTAL TESTING SERVICE INC., 6229 North Dixie Dr., Dayton, OH, 45414, (513) 890-3361, (1).

RADON ENVIRONMENTAL TESTING SERVICES, INC., P.O. Box 14613, Dayton, OH, 45414, (513) 890-3361, (1).

RADON EVALUATION DIVISION/EVALDO, LTD., 64 Sunset Hill Dr., Branford, CT, 06405, (203) 488-4715, (1).

RADON EVALUATION SERVICES, INC., P.O. Box 13609, Atlanta, GA, 30324, (404) 478-0037, (1)(2).

RADON FINDERS, 1118 Millington Rd., Schenectady, NY, 12309, (518) 372-8395, (1).

RADON FREE, 5666 Ashbrook, Toledo, OH, 43614, (419) 865-0129, (1)(2).

RADON GAS DETECTION SYSTEMS, P.O. Box 15189, Hamilton, OH, 45015, (513) 863-6625, (1).

RADON GAS TESTING OF CENTRAL OHIO, 5548 Autumn Chase, Columbus, OH, 43232, (614) 864-0191, (1).

RADON GASBUSTERS, 892 Green Ln., Union, NJ, 07083, (201) 353-5577, (1).

RADON HOME INSPECTION SERVICE, P.O. Box 292, Churchton, MD, 20733, (301) 261-5650, (1).

RADON HOME PROTECTION SERVICES, 48 Woodruff Ave., Thomaston, CT, 06787, (203) 283-4078, (1).

RADON HOME SEARCH, 349 Maxim Dr., Hopatcong, NJ, 07843, (201) 398-4106, (1)(2).

RADON HUNTER, THE, P.O. Box 6009, McLean, VA, 22106, (703) 237-1016, (1).

RADON IN-HOUSE DETECTION, Rt. 1, Box 110, Leeds, NY, 12451, (518) 622-3509, (1).

RADON INFORMATION AND DETECTION SERVICE, Rte. 2, Isbill Rd., Madisonville, TN, 37354, (615) 442-5849, (1).

RADON INFORMATION AND TESTING SERVICE, 29 Appleton St., Manchester, NH, 03104, (603) 627-7261, (1).

RADON INSPECTION SERVICE, INC., 789 East Glen Ave., Ridgewood, NJ, 07450, (201) 670-8821, (1)(2).

RADON INSPECTION SERVICES, 915 6th Ave., North Unit C, Seattle, WA, 98109, (206) 286-0479, (1).

RADON INSTRUMENTAL DETECTION, 8000 West 110th Suite 115, Overland Park, KS, 66210, (913) 451-9270, (1).

RADON INTERVENTION, P.O. Box 827, Three Rivers, TX, 78071, (512) 786-4276, (1).

RADON LABORATORIES, P.O. Box 2054 Harvard Sq., Cambridge, MA, 02138, (617) 923-6454, (1).

RADON LABORATORIES, P.O. Box 829, Middletown, MD, 21769, (301) 371-5867, (1)(3)(4).

RADON LABORATORIES, INC., 33 Kensington Pkwy. P.O. Box 867, Abingdon, MD, 21009, (301) 679-6494, (1).

RADON LABORATORIES, INC., 3 Deer Tr., Tabernacle, NJ, 08088, (800) 527-4520, (609) 859-0944, (1).

RADON LABS, P.O. Box 2054 Harvard Square, Cambridge, MA, 02238, (617) 923-6454, (1)(2).

RADON LABS, P.O. Box 2243, Woburn, MA, 01888-0443, (617) 664-0316, (1).

RADON LABS OF ATLANTA, 5400 Hwy. 29, Lilburn, GA, 30247, (404) 279-0005, (800) 248-1889, (1).

RADON MANAGEMENT COMPANY, 4332 Gingham Ct., Alexandria, VA, 22310, (703) 960-4407, (1).

RADON MANAGEMENT LABORATORIES, 2 Riverdale Rd., Pompton Lakes, NJ, 07442, (201) 835-8721, (1).

RADON MANAGEMENT LABORATORIES, Overlook Terr., Chester, NJ, 07930, (201) 879-2220, (201)835-8721, (1).

RADON MANAGEMENT OF KENTUCKY, 12203 Old Shelbyville Rd., Middletown, KY, 40243, (502) 244-0517, (1)(2).

RADON MANAGEMENT, INC., 474 Easton Rd., Horsham, PA, 19044, (215) 672-8129, (1).

RADON MANAGEMENT/DIVISION OF PHYSICS CONSULTANTS, INC., P.O. Box 3250, Auburn, ME, 04210, (207) 795-2459, (1).

RADON MEASURE COMPANY, THE, P.O. Box 16243, St. Louis Park, MN, 55416, (612) 339-9961, (612) 374-3849, (1).

RADON MEASUREMENT AND CONTROL, 7935 Kilbourne Rd., Sunbury, OH, 43074, (614) 866-1197, (1).

RADON MEASUREMENT AND CONTROL OF NEW MEXICO, 1527 Granite St., NW, Albuquerque, NM, 87104, (505) 843-7643, (1)(4).

RADON MEASUREMENT AND SERVICES, 13131 West Cedar Dr., Lakewood, CO, 80228, (303) 980-5086, (303) 988-3033, (1)(2).

RADON MEASUREMENT SERVICE, 495 North Maple St., Ephrata, PA, 17522, (717) 738-1219, (1).

RADON MEASUREMENTS, INC., 2916 West T.C. Jester, Houston, TX, 77018, (713) 683-8227, (1).

RADON MEASURING SYSTEMS, 1129 Wyoming Ave., Scranton, PA, 18509, (717) 961-1666, (1)(2).

RADON MITIGATION SERVICE, 421 Hillandale, Round Lake, IL, 60073, (312) 546-3890, (2).

RADON MITIGATION SERVICES, 4044 Southwest 13th St. Suite 153, Gainesville, FL, 32608, (904) 375-5656, (1).

RADON MITIGATORS OF ARIZONA, 6844 East Thunderbird Rd., Scottsdale, AZ, 85254, (602) 483-7506, (1)(2).

RADON MITIGATORS, INC., 16 Hillside Dr., Clinton, NJ, 08809, (201) 730-6596, (1)(2).

RADON MONITOR SYSTEMS, P.O. Box 516, Eustis, FL, 32727-0516, (904) 735-2929, (1).

RADON MONITORING AND SERVICES, 1350 Butter Lane, Mt. Penn, PA, 19606, (215) 779-8314, (1).

RADON MONITORING SERVICES, P.O. Box 9083, Colorado Springs, CO, 80932, (719) 577-9662, (1).

RADON MONITORING SERVICES, INC., P.O. Box 3730, Plymouth, MA, 02361, (617) 746-8868, (1).

RADON PRESS, INC., 500 North Washington St., Alexandria, VA, 22314, (703) 548-2756, (800) 548-1567, (3).

RADON PRODUCTS, INC., 3215 West Hampden Ave., Englewood, CO, 80110, (303) 761-4705, (4).

RADON PROJECT, THE, P.O. Box 90069, Pittsburgh, PA, 15224, (412) 687-2886, (1)(4).

RADON PROTECTION SERVICES, P.O. Box 158, Hummelstown, PA, 17036, (717) 566-5609, (1)(2)(6).

RADON PROTECTION, INC., The Summit Suite 110, Louisville, KY, 40207, (502) 893-4563, (1)(2).

RADON QC, 2857 Nazareth Rd. Palmer Professional Building Suite 6, Palmer, PA, 18043, (215) 559-1636, (1)(3)(4).

RADON RANGERS, 7979 Old Georgetown Rd. Suite 1100, Bethesda, MD, 20814, (301) 543-4300, (1).

RADON READINGS, 9020 Compton Blvd., Bellflower, CA, 90706, (213) 920-2582, (213) 421-8674, (1).

RADON REDUCTION AND TESTING COMPANY, 605 South Sherman Suite 705 J, Richardson, TX, 75081, (214) 680-2296, (1)(2).

RADON REDUCTION AND TESTING, INC., 2222 Cumberland Pl., Smyrna, GA, 30080, (404) 434-7296, (1).

RADON REDUCTION ASSOCIATES, INC., 815 Saratoga Rd., Burnt Hills, NY, 12027, (518) 399-6560, (1).

RADON REDUCTION ENGINEERING AND LABORATORY, 211 Mill St., NE, Vienna, VA, 22180, (703) 281-7865, (703) 281-3730, (1)(2).

RADON REDUCTION EQUIPMENT OF COLORADO, 921 Main Suite C, Louisville, CO, 80027, (303) 665-1776, (4)(6).

RADON REDUCTION SERVICES, 5601 East 109th Ter., Kansas City, MO, 64137, (816) 942-7083, (1)(2).

RADON REDUCTION SERVICES, INC., P.O. Box 627, Mt. Airy, MD, 21771, (301) 831-0228, (1)(2).

RADON REDUCTION SYSTEMS, 169 Old Gage Hill Rd., Pelham, NH, 03076, (603) 893-5549, (2).

RADON REDUCTION, INC., 1125 Mechanic St., Jeffersonville, IN, 47130, (812) 284-4466, (2).

RADON REDUCTION, INC., 363 Massachusetts Ave., Lexington, MA, 02173, (617) 862-8462, (2).

RADON REFEREEING/DIVISION OF HOME QUALITY HEADQUARTERS, 3 Dogwood Rd., Tolland, CT, 06084, (203) 871-7433, (1).

RADON REGISTRY, 819 2nd St. B-178, Manchester, NH, 03102, (603) 645-6780, (1).

RADON REMEDIAL SERVICES, P.O. Box 622, Bryn Athyn, PA, 19009, (215) 938-7766, (1)(2).

RADON REMEDIATION COMPANY, INC., 38 Whippoorwill Way, Belle Mead, NJ, 08502, (201) 359-0050, (2).

RADON REMEDIATION SERVICES, 10400 Connecticut Ave. Suite 504, Kensington, MD, 20895, (301) 933-7900, (2).

RADON REMEDIES, INC., 2700 West Broad St. P.O. Box 44234, Columbus, OH, 43204, (614) 272-7772, (1)(3)(4).

RADON REMOVAL, INC., Box 283, Milltown, NJ, 08850, (201) 545-2000, (1)(2)(4).

RADON REPORTING SYSTEM OF CARROLL COUNTY, P.O. Box 500, Westminster, MD, 21157-9167, (301) 857-1807, (301) 795-0775, (1).

RADON RESEARCH, 700 Winthrop Pl., Orlando, FL, 32803, (813) 377-3740, (1).

RADON RESEARCH FOUNDATION, INC., 10922 West 101st St., Overland Park, KS, 66214, (913) 492-1723, (1).

RADON RESEARCH GROUP, INC., P.O. Box 1143, Germantown, MD, 20874, (301) 972-3309, (800) 445-8436, (1)(4).

RADON RESEARCH INTERNATIONAL, INC., 7650 South Tamiami Trail, #9, Sarasota, FL, 34231, (813) 923-7300, (1)(2).

RADON RESIDENTIAL SYSTEMS, INC., 1474 Rte. 23 N. Suite 3, Wayne, NJ, 07470, (201) 633-3976, (2).

RADON RESOURCE, INC., 49 John St., Westminster, MD, 21157, (301) 857-9330, (1)(2).

RADON RESOURCES, 241 Rumson Rd., Atlanta, GA, 30305, (404) 231-4732, (1).

RADON RESOURCES, INC., 55 Constitution Dr., Bedford, NH, 03102, (603) 472-5955, (1).

RADON SAFE, P.O. Box 631 Highway 62 W., Boonville, IN, 47601, (812) 897-1700, (1).

RADON SAFE, 7596 Wheaton Dr., Canton, MI, 48187, (313) 455-6469, (1).

RADON SAFETY AND TECHNOLOGY CORPORATION, 4 Charles St., Pleasant Valley, NY, 12569, (914) 635-1138, (800) 552-5257, (1)(2)(3)(4).

RADON SAFETY SERVICE, 2204 Frances Dr., Loveland, CO, 80537, (303) 663-3941, (1).

RADON SAFETY SERVICES, INC., P.O. Box 441, New Providence, NJ, 07974, (201) 825-8267, (201) 665-1188, (1).

RADON SAFETY TESTING, INC., P.O. Box 277, Gales Ferry, CT, 06335-0277, (203) 886-8327, (1).

RADON SCREENING SERVICE, INC., P.O. Box 37007, Denver, CO, 80237, (303) 741-3207, (1)(2).

RADON SCREENING, INC., 7 Brookcrest Ct., Fredericksburg, VA, 22401, (703) 786-4320, (1).

RADON SEARCH-USA, 2701 North Rocky Point Rd. Suite 800, Tampa, FL, 33607, (813) 960-1360, (1).

RADON SERVICES, 1210 Auburn Way N. Suite B-122, Auburn, WA, 98002, (206) 939-3023, (1).

RADON SERVICES ANALYTICAL LABORATORY, 111 Benson Rd., Nashville, TN, 37214, (615) 883-4659, (1).

RADON SERVICES OF MID-AMERICA, 935B Southeast 3rd St., Lee's Summit, MO, 64063, (816) 525-7412, (1).

RADON SERVICES OF NORTH AMERICA, INC., 148 East 632 N., Valparaiso, IN, 46383, (219) 462-5866, (1).

RADON SERVICES OF WYOMING, 1714 Rollins, Cheyenne, WY, 82001, (307) 634-9505, (1)(2).

RADON SERVICES, INC., 2950 Banksville Rd., Pittsburgh, PA, 15216, (412) 563-0553, (1).

RADON SPECIALIST OF AMERICA, 12628 Gray Eagle Ct. Suite 12, Germantown, MD, 20874, (301) 231-4091, (1).

RADON SPECIALISTS, 1840 Folsom St. Suite 255, Boulder, CO, 80302, (303) 673-0494, (1).

RADON SURVEY SYSTEMS, INC., P.O. Box 22288, Cleveland, OH, 44122, (216) 464-7997, (1)(2)(3)(4)(6).

RADON SURVEYS, 40 Pinetree, Seymour, IL, 61875, (217) 687-2611, (1).

RADON SURVEYS/DIVISION OF COPE TECHNICAL SERVICES, 3506 87th Pl., Kenosha, WI, 53142, (414) 694-3730, (1).

RADON SYSTEMATICS OF CALIFORNIA, 1500 San Remo Dr., Pacific Palisades, CA, 90272, (213) 459-9146, (213) 825-1934, (1).

RADON SYSTEMS OF RHODE ISLAND, INC., 156 Earle St., Woonsocket, RI, 02895, (401) 766-7072, (1)(2).

RADON SYSTEMS, INC., P.O. Box 2509, Laurel, MD, 20708, (301) 725-9000, (1)(2).

RADON TECHNICAL SYSTEMS, INC., 37 Tree Brook Dr., Rochester, NY, 14625, (716) 385-4255, (1).

RADON TECHNICIANS, INC, 27 Boren Shore Ln., Elkton, MD, 21921, (301) 854-6186, (2).

RADON TECHNOLOGIES, INC., 39 Woodrow Ct., Troy, NY, 12180, (518) 447-5834, (1).

RADON TECHNOLOGIES, INC., 698 High St., Worthington, OH, 43085, (614) 441-0550, (614) 431-0322, (1).

RADON TECHNOLOGIES, INC., 800 Forest Ave., Norristown, PA, 19401, (215) 279-5188, (1).

RADON TECHNOLOGIES, INC., 12200 Sunrise Valley Dr. Suite 320, Reston, VA, 22091, (703) 620-4004, (800) 237-5024, (1)(4).

RADON TECHNOLOGY, 11404A East 24 Hwy., Sugar Creek, MO, 64054, (816) 461-0211, (1)(2).

RADON TECHNOLOGY AND CONSULTANTS, RFD 2 Louisquisset Pike, North Smithfield, RI, 02895, (401) 769-2129, (401) 568-0296, (1)(2).

RADON TECHNOLOGY, INC., 19595 Merriman Rd., Romulus, MI, 48174, (313) 753-4211, (1).

RADON TECHNOLOGY, INC., 7 Tokanel Dr., Londonderry, NH, 03053, (603) 434-4937, (1).

RADON TEST, P.O. Box 5210, San Luis Obispo, CA, 93403-5210, (805) 481-4069, (1).

RADON TEST AND REDUCTION SPECIALIST, 125 Oak Ave. P.O. Box 129, Spruce Pine, NC, 28777, (704) 765-1222, (1)(2).

RADON TEST CENTER, P.O. Box 1542, Ellicott City, MD, 21043, (301) 997-2366, (1).

RADON TEST COMPANY, P.O. Box 345, Eagle Grove, IA, 50533, (515) 448-4600, (4).

RADON TEST COMPANY, P.O. Box 800, Worcester, MA, 01602, (617) 853-7809, (617) 793-7687, (1).

RADON TEST SYSTEMS, 1418 Upfield, Carlton, TX, 75006, (214) 242-1020, (1).

RADON TESTING AND CONSULTING, 1771 Post Rd., E. Suite 320, Westport, CT, 06880, (203) 454-3886, (1)(2)(4).

RADON TESTING AND CONSULTING SERVICES, 34 Jones River Dr., Kingston, MA, 02364, (617) 585-9924, (1)(2)(3).

RADON TESTING AND CONTROL, 2910 East Broadway Ave., Bismarck, ND, 58501, (701) 222-1613, (1)(2).

RADON TESTING AND ENVIRONMENTAL CONTROL, 7113 Chippenham Rd., Louisville, KY, 40222, (502) 426-4314, (1).

RADON TESTING AND MITIGATION OF TEXAS, INC., P.O. Box 34268, Houston, TX, 77234-4268, (800) 920-1663, (1)(2).

RADON TESTING AND MITIGATION SERVICES COMPANY, 5442 Cary Ave., Davenport, IA, 52807, (319) 386-9005, (1)(2).

RADON TESTING AND MITIGATION SYSTEMS, P.O. Box 3052, Carmel, IN, 46032, (317) 545-5578, (1)(2).

RADON TESTING AND REDUCTION, P.O. Box 437, Howell, MI, 48844, (517) 548-2497, (1)(2).

RADON TESTING AND TREATMENT, INC., 723 North Weber Suite 300, Colorado Springs, CO, 80903, (719) 520-5693, (1)(2)(4).

RADON TESTING ASSOCIATION, Rte. 6, 2014 Glenwood Dr., Martinsburg, WV, 25401, (304) 754-7586, (1).

RADON TESTING COMPANY OF NEW YORK, 1790 Meadowbrook Rd., Merrick, NY, 11566, (516) 868-1778, (1).

RADON TESTING CONSULTANTS, INC., P.O. Box 25313, Newark, NJ, 07101, (201) 290-9217, (1)(2)(4).

RADON TESTING CORPORATION, 4 Cheyenne Ct., Bryan's Road, MD, 20616, (301) 283-6006, (1).

RADON TESTING CORPORATION OF AMERICA, 12 West Main St., Elmsford, NY, 10523, (914) 347-5010, (800) 457-2366, (1)(4)(6).

RADON TESTING GROUP, P.O. Box 23759,

Jacksonville, FL, 32241, (904) 260-2752, (1).

RADON TESTING LAB, 3432 Hollywood, Brookfield, IL, 60513, (312) 485-1133, (1).

RADON TESTING LABORATORIES, INC., 10390 Tuscany Ln., Ellicott City, MD, 21043, (301) 465-1093, (1).

RADON TESTING LABORATORY OF NEW ENGLAND, P.O. Box 4508, Boston, MA, 02101, (617) 426-9224, (1).

RADON TESTING LABORATORY OF NEW JERSEY, 1303 Birch Ave., Ocean Township, NJ, 07712-4514, (201) 502-0727, (1).

RADON TESTING LABS, P.O. Box 1467, Avon, CT, 06001, (203) 678-8259, (1).

RADON TESTING OF MICHIANA, Colfax Ctr. 914 Lincolnway W., South Bend, IN, 46616, (219) 282-2435, (1).

RADON TESTING SERVICE, INC., P.O. Box 1130, Englewood, CO, 80110, (303) 762-8598, (1).

RADON TESTING SERVICE, INC., Rte. 212, Lake Hill, NY, 12488, (914) 679-6835, (1)(2).

RADON TESTING SERVICE, INC., 545 North 33rd, Milwaukee, WI, 53208, (414) 931-0527, (1).

RADON TESTING SERVICES, P.O. Box 152227, San Diego, CA, 92115-0870, (619) 229-9002, (619) 287-5317, (1).

RADON TESTING SERVICES, 4096 Scenic Dr., Provo, UT, 84604, (801) 227-0179, (800) 777-2320, (1).

RADON TESTING SERVICES OF MISSOURI, P.O. Box 362, Trenton, MO, 64683, (816) 359-4442, (1).

RADON TESTING SERVICES, BALTIMORE, P.O. Box 7999, Baltimore, MD, 21221-0999, (301) 321-8008, (1).

RADON TESTING SERVICES, CHERRY HILL, P.O. Box 3013, Cherry Hill, NJ, 08034, (609) 665-9228, (1).

RADON TESTING SERVICES, HARRISBURG, P.O. Box 3013, Cherry Hill, NJ, 08034, (717) 232-4616, (1).

RADON TESTING SERVICES, INC., 4442 18th St., San Francisco, CA, 94114, (415) 552-5496, (1)(3)(4).

RADON TESTING SERVICES, INC., 45 Streit Ave., Poughkeepsie, NY, 12603, (914) 485-3060, (1).

RADON TESTING SERVICES, PITTSBURGH, P.O. Box 19425, Pittsburgh, PA, 15213, (412) 687-1533, (1).

RADON TESTING SERVICES, PRINCETON, P.O. Box 3311, Princeton, NJ, 08543, (609) 737-8818, (1).

RADON TESTING SERVICES, SILVER SPRING, 1774 Featherwood St., Silver Spring, MD, 20904, (301) 236-9434, (1).

RADON TESTING SYSTEMS, P.O. Box 243, Newtown, PA, 18940, (215) 639-1933, (1).

RADON TESTING, INC., P.O. Box 7304, Fairfax Station, VA, 22039, (703) 591-1111, (1)(2).

RADON TRAC, INC., P.O. Box 4015, Sarasota, FL, 34230, (813) 955-5426, (813) 923-5196, (1).

RADON TRACKING SERVICE, 1915 Whitney Rd. Suite 1, Clearwater, FL, 34230, (904) 260-2752, (1).

RADON WATCH, INC., 472 North Wooster Way, Lancaster, OH, 43130, (614) 837-5748, (1).

RADON X COMPANY, INC., 217 Roxbury St., Keene, NH, 03431, (603) 352-5319, (1)(2).

RADON, INC., 1637 Downtown West Blvd. Suite 9, Knoxville, TN, 37919, (615) 694-4080, (1).

RADON, LTD., 411 North Weinbach Ave., Evansville, IN, 47711, (812) 479-3110, (800) 634-2901, (1)(2)(4).

RADON-ONE LABORATORY SERVICES, 4575 Chandlers Forde, Sarasota, FL, 34235, (813) 377-3740, (1).

RADON-ONE LABORATORY SERVICES, 4700 East Fifth Ave., Columbus, OH, 43219, (614) 239-7171, (1).

RADON-X CORPORATION, 300 Valentine St., Hackettstown, NJ, 07840, (201) 852-6212, (800) 272-3669, (1)(2)(4).

RADON-X CORPORATION, 3 Edinburgh Ln., Mt. Laurel, NJ, 08054, (609) 778-7771, (800) 272-3669, (1)(2)(4).

RADON/GEOLOGICAL ASSESSMENT SERVICES, 727 Crown Ridge Dr., Colorado Springs, CO, 80904, (719) 634-5235, (719) 685-4749, (1).

RADON/SOILS AND WATER TESTING SERVICE, 7080 Price-Hilliard Rd., Plain City, OH, 43064, (614) 873-8821, (1)(4).

RADONAIRE, P.O. Box 3702, Wayne, NJ, 07474, (201) 790-3252, (1)(2).

RADONAIRE, INC., 393 Denrose Dr., Amherst, NY, 14150, (716) 837-2366, (1).

RADONICS, 1764 Old Meadow Ln. Suite 125, McLean, VA, 22102, (703) 734-0050, (1).

RADONICS, INC., P.O. 489, Pennington, NJ, 08534, (609) 737-8888, (1)(2).

RADONIX, LTD., P.O. Box R, Lake Grove, NY, 11755, (516) 467-3780, (1).

RADONTECH, INC., 1616 Walnut St. Suite 2200, Philadelphia, PA, 19103, (215) 546-7328, (1).

RADONTEST, INC., 4700 East 5th Ave., Columbia, MO, 65201, (314) 875-8045, (1).

RAF ENGINEERING, Hillside Rd., Rockaway, NJ, 07866, (201) 983-9041, (1).

RAM EXTERMINATORS, 97 Frank St., Somerset, NJ, 08873, (201) 937-6900, (1).

RAMAPO COLLEGE OF NEW JERSEY/INSTITUTE FOR ENVIRONMENTAL STUDIES, 505 Ramapo Valley Rd., Mahwah, NJ, 07430, (201) 529-7742, (3)(5)(6).

RAMP INDUSTRIES, INC., 1127 West 46th St., Denver, CO, 80210, (1)(2).

RAMSES II, LTD., 2355 East Dartmouth Ave., Denver, CO, 80210, (303) 758-5209, (1)(2)(4).

RANDOLPH AND ASSOCIATES, INC., 8901 North Industrial Rd., Peoria, IL, 61615, (309) 692-4160, (1)(2).

RAVEN INDUSTRIES, INC., P.O. Box 1007, Sioux Falls, SD, 57117, (605) 336-2750, (800) 227-2836, (4).

RAY MESSICK RADON TESTING AND MITIGATION, RDI Box 469L, McConnellsburg, PA, 17233, (202) 332-1798, (717) 485-5456, (1).

RAYDOT, INC., 145 Jackson Ave., Cokato, MN, 55321, (612) 286-2103, (800) 328-3813, (4).

RCI SERVICES, 11412 Hounds Way, Rockville, MD, 20852, (301) 231-5134, (1)(2).

RDS/RADON DETECTION SERVICES, INC.-CORPORATE OFFICE, Rte. 179, Old York Rd. P.O. Box 419, Ringoes, NJ, 08551, (201) 788-3080, (1)(2)(3)(4)(6).

RDS/RADON DETECTION SERVICES-GEORGIA OFFICE, 403 West Ponce de Leon Ave. P.O. Box 1826, Decatur, GA, 30031, (404) 377-2366, (1)(2).

RDS/RADON DETECTION SERVICES-MARYLAND OFFICE, 10400 Connecticut Ave. Suite 504, Kensington, MD, 20895, (301) 933-7900, (1)(3).

RDS/RADON DETECTION SERVICES-PENNSYLVANIA OFFICE, 1011 Brookside Rd., Suite 270 P.O. Box 270, Allentown, PA, 18106, (215) 481-9555, (1)(2).

RE FORMATION, P.O. Box 328, Pine Island, NY, 10969, (914) 258-4711, (3).

REAL ESTATE EDUCATORS ASSOCIATION (REEA), 230 North Michigan Ave. Suite 1200, Chicago, IL, 60601, (312) 372-9800, (18).

REAL ESTATE INSPECTION SERVICE, P.O. Box 3621 CRS, Johnson City, TN, 37602, (615) 282-4216, (1).

REAL ESTATE INSPECTOR, THE, 2195 Stoney Hill Rd., Boulder, CO, 80303, (303) 499-2688, (1).

REAL ESTATE SECURITIES AND SYNDICATION INSTITUTE, 430 North Michigan Ave., Chicago, IL, 60611, (312) 670-6760, (18).

REAL ESTATE SUPPORT SERVICES, INC., 7825 Washington Ave., S., Minneapolis, MN, 55435, (612) 941-7558, (1).

REALITY DYNAMICS, 4758 Antelope Trail, Elbert, CO, 80106, (303) 648-3297, (1).

REALTORS LAND INSTITUTE (RLI), 430 North Michigan Ave., Chicago, IL, 60611, (312) 329-8440, (18).

REALTORS NATIONAL MARKETING INSTITUTE (RNMI), 430 North Michigan Ave., Chicago, IL, 60611, (312) 670-3780, (18).

REALTY REPAIR AND DEVELOPMENT, 8 Mount Hope Rd., Rockaway, NJ, 07866, (201) 625-0030, (2).

RECON SYSTEMS, INC., P.O. Box 460, Three Bridges, NJ, 08887, (201) 782-5900, (1).

REDICORP PROTECTIVE MATERIALS, 212 Dominion Rd., NE, Vienna, VA, 22180, (703) 938-2928, (4).

REDIHOME INSPECTIONS, INC., 18 Kennedy Ter., Westwood, NJ, 07675, (201) 358-6888, (1).

REEDY CREEK IMPROVEMENT DISTRICT/POLLUTION CONTROL DEPARTMENT, P.O. Box 36, Lake Buena Vista, FL, 32830, (305) 824-3460, (10).

RELIABLE BASEMENT WATERPROOF COMPANY, 2811 Westerville Rd., Columbus, OH, 43224, (614) 267-4477, (1)(2).

REMD/RESIDENTIAL ENVIRONMENTAL MEDICAL DIAGNOSTICS, 28 School St. P.O. Box 3285, Stony Creek, CT, 06405, (203) 481-8211, (1).

REMEDIAL TECHNOLOGIES, INC., 10110 Kile Rd., Chardon, OH, 44024, (216) 765-2442, (2)(4).

REMSEARCH, INC., 160 Bleeker St. Suite 5BE, New York, NY, 10012, (212) 982-1115, (1).

RENSSELAER COUNTY DEPARTMENT OF HEALTH/DIVISION OF ENVIRONMENTAL HEALTH, County Office Building 1600 Seventh Ave., Troy, NY, 12180, (518) 270-2664, (10).

RENSSELAER POLYTECHNICAL INSTITUTE, 110 Eighth Street, Troy, NY, 12180, (518) 276-8237, (1)(2).

RESIDENCE RADON, 28077 West 85th Ter., Desoto, KS, 66018, (913) 383-4610, (913) 585-1094, (1)(4).

RESIDENTIAL INSPECTION TECHNOLOGY/RES-I-TEC, 145 D Grassy Plain St., Bethel, CT, 06801, (203) 791-1076, (1).

RESIDENTIAL RADON MEASUREMENTS, P.O. Box 3051, Fairview Heights, IL, 62208, (618) 397-0099, (1).

RESIDENTIAL RADON TEST SERVICE, 126 Old Oaken Bucket Rd., Scituate, MA, 02066, (617) 545-9160, (1).

RESIDENTIAL SECURITY ADVISORS, 265 Kenny Brook Ln., Elyria, OH, 44035, (216) 329-2365, (4).

RESOURCE CENTER FOR ENVIRONMENTAL PROTECTION, P.O. Box 25, Schuylerville, NY, 12871, (518) 695-3218, (2)(3).

RESOURCE CONSERVATION TECHNOLOGY, INC.,

2633 North Calvert St., Baltimore, MD, 21218, (301) 366-1146, (4).

RESTORED ENVIRONMENTS COMPANY, INC., 4940 Nutmeg Knoll, Cincinnati, OH, 45244, (513) 831-0743, (1)(2).

RETROTEC, P.O. Box 939, Ogdensburg, NY, 13669, (613) 723-2453, (4).

REYNOLDS RADIOLOGICAL SERVICES, INC., RD #1, Box 2715 Governor Stable Rd., Bainbridge, PA, 17502, (717) 299-6429, (717) 367-7829, (1).

RHH FOAM SYSTEMS, INC., P.O. Box 752 6001 S. Pennsylvania Ave., Cudahy, WI, 53110-0752, (414) 744-6066, (4).

RHODE ISLAND DEPARTMENT OF BUSINESS REGULATION, LICENSING AND CONSUMER PROTECTION, 100 North Main St., Providence, RI, 02903, (401) 277-2255, (10).

RHODE ISLAND DEPARTMENT OF GEOLOGY, Green Hall The University of Rhode Island, Kingston, RI, 02881, (401) 792-2265, (10).

RHODE ISLAND DEPARTMENT OF HEALTH/DIVISION OF OCCUPATIONAL HEALTH AND RADIATION CONTROL, Cannon Building 75 Davis St., Providence, RI, 02908, (401) 277-2438, (10).

RHODE ISLAND REALTORS ASSOCIATION, 120 Lavan St., Warwick, RI, 02888, (401) 785-3650, (18).

RIC INSPECTION AND ENGINEERING, R.R. 2 P.O. Box 2186-5, Columbia, NJ, 07832, (201) 496-4818, (1).

RICHARD P. ARBER ASSOCIATES, 44 Cook St., Suite 601, Denver, CO, 80206, (303) 393-6363, (3)(5).

RICHFIELD/SUPPORT SERVICES DIVISION, 6700 Portland Ave., Richfield, MN, 55423, (612) 869-7521, (10).

RIEHS AND RIEHS, 501 George St., New Bern, NC, 28560, (919) 637-9752, (2)(4).

RISE, 280 Broadway, Providence, RI, 02903, (401) 272-1040, (1).

RITE WAY WATERPROOFING, 7337 Lake Dr., Lino Lakes, MN, 55014, (612) 786-0550, (2).

RMC ENVIRONMENTAL SERVICES, 919 Edgewood Rd., Havertown, PA, 19083, (215) 841-5268, (3).

RMC ENVIRONMENTAL SERVICES/ROYERSFORD DIVISION, 323 Church St., Royersford, PA, 19468, (215) 948-4949, (3).

RMG/PC ENGINEERING, INC., 617 North 17th St. Suite 201, Colorado Springs, CO, 80904, (719) 475-9600, (1).

ROANOKE CITY ENGINEERING DEPARTMENT/AIR POLLUTION CONTROL DIVISION, 215 West Campbell Ave., Roanoke, VA, 240111, (703) 981-2731, (10).

ROANOKE COUNTY AND CITY OF SALEM/AIR POLLUTION CONTROL, P.O. Box 1074, Salem, VA, 24153, (703) 387-6224, (10).

ROBERT A. SWOYER ASSOCIATES, 241 South Fifth St., Reading, PA, 19602, (215) 376-2580, (2).

ROBERTS ENVIRONMENTAL SERVICES, 4110 Old Wake Forest Rd. Suite 109, Raleigh, NC, 27609, (919) 790-9378, (1).

ROCKLAND COUNTY HEALTH DEPARTMENT, Sanatorium Rd. Building D, Pomona, NY, 10970, (914) 354-0250, (10).

ROCKY MOUNTAIN HOME INSPECTIONS, 1571 East Mineral Pl., Littleton, CO, 80122, (303) 730-0922, (1)(2).

ROCKY MOUNTAIN MEDICAL PHYSICS, INC., 7175 West Jefferson Ave. Suite 2150, Lakewood, CO, 80235, (303) 980-7070, (1)(3).

ROCKY MOUNTAIN RADON TESTING, 6820 South Brook Forest Dr., Evergreen, CO, 80439, (303) 674-7029, (1).

ROGERS AND ASSOCIATES ENGINEERING CORPORATION, 515 East 4500 S. Suite G-200, Salt Lake City, UT, 84107, (801) 263-1600, (1)(3)(5).

ROMA RADON, P.O. Box 90875, Lakeland, FL, 33804-0875, (813) 859-5001, (1).

ROMAES CORPORATION, 6395 Gunpark Dr. Suite C, Boulder, CO, 80301, (303) 530-1637, (4).

RONALD BARTA AND ASSOCIATES, 345 Meadow View Dr., Evergreen, CO, 80439, (303) 674-3025, (1).

RONALD STEIN CONSTRUCTION, INC., 21 Davis Ave., Poughkeepsie, NY, 12603, (914) 473-0064, (1)(2).

ROSS RADON TESTING, 4200 Chimney Lake Dr., Roswell, GA, 30075, (404) 993-8270, (1).

ROSS SYSTEMS, INC., 955 Connecticut Ave., Bridgeport, CT, 06607, (203) 366-2210, (1)(2).

ROSS SYSTEMS, INC., 21 Main St. Box 670, Blairstown, NJ, 07825, (201) 362-5571, (1)(2)(3)(4)(6).

ROY F. WESTON, INC., Building 4-2 1 Weston Way, West Chester, PA, 19380, (215) 692-3030, (1).

RPI INTERNATIONAL ENVIRONMENTAL TECHNOLOGY DIVISION, P.O. Box 328, Columbia, SC, 29202, (803) 256-7322, (3)(5).

RPI RADON MEASUREMENT SERVICES, P.O. Box 357, Hershey, PA, 17033, (717) 533-8100, (1)(4).

RPI RADON MEASUREMENT SERVICES, MARYLAND, 10133 Bacon Dr., Beltsville, MD, 20705, (301) 937-4072, (1).

RSA CORPORATION/RADON SERVICE AGREEMENT, 918 Woodley Rd., Dayton, OH, 45403, (513) 254-4102, (3).

RSO, INC., P.O. Box 1526 711 Gorman Ave., Laurel, MD, 20707-0953, (301) 953-2482, (4).

RTK ENGINEERING, 2939 Newton St., Denver, CO, 80211, (303) 477-0044, (1).

RTS SYSTEMS, P.O. Box 3070, Gaithersburg, MD, 20878, (301) 258-7890, (1).

RUECK ASSOCIATES, 103 Sage Rd., Silver Lake, KS, 66539-9461, (913) 582-4050, (1)(3)(4).

RUTGERS UNIVERSITY/DEPARTMENT OF COMMUNICATIONS/ENVIRONMENTAL COMMUNICATIONS RESEARCH PROGRAM, Cook College Ryders Lane, New Brunswick, NJ, 08903, (201) 932-8795, (5)(6).

RYAN HOMES/PRODUCTS DIVISION, 100 Ryan Court, Pittsburgh, PA, 15205, (412) 276-8000, (13).

RYAN NUCLEAR LABORATORIES, P.O. Box 88, West Jefferson, OH, 43162, (614) 848-4414, (1).

S

S & D CUSTOM ESTATES, 97 Cuba Ave., Staten Island, NY, 10306, (718) 987-2354, (718) 987-8378, (1).

S & W ENVIRONMENTAL TESTING, Route 1, Box 185-D, Telephone, TX, 75488, (214) 664-3105, (214) 980-4121, (1).

S. COHEN AND ASSOCIATES, INC., 8200 Riding Ridge Pl., McLean, VA, 22102, (703) 893-6592, (3)(6).

SABAL INVESTMENTS, 915 Oakfield Drive Suite A, Brandon, FL, 33511, (813) 689-2548, (3).

SACCHETTI, R. MICHAEL, 103 Intervale St., Quincy, MA, 02169, (617) 479-2111, (1)(2).

SACRAMENTO COUNTY AIR POLLUTION CONTROL DISTRICT, 9323 Tech Center Dr. Suite 800, Sacramento, CA, 95826, (916) 366-2107, (10).

SAF-AIR, P.O. Box 1217, Orange, CT, 06477, (203) 799-9609, (1)(2)(6).

SAFE AIR, P.O. Box 120064, Nashville, TN, 37212, (615) 865-5325, (1).

SAFE AIR COMPANY, 9 Eastman Dr., Derry, NH, 03038, (603) 437-1425, (1).

SAFE AND SOUND HOME INSPECTION CONTRACTORS, P.O. Box 195, Cedar Knolls, NJ, 07927, (201) 515-0013, (1).

SAFE SHELTER ENVIRONMENTAL, 527 West Lancaster Ave., Frazer, PA, 19355, (215) 640-4839, (1)(2).

SAFE-AIR, 2599 East Main St. Suite 525, Bexley, OH, 43209, (614) 447-8300, (800) 553-3117, (2).

SAFE-AIRE, INC., 162 East Chestnut St. P.O. Box 160, Canton, IL, 61520, (309) 647-0419, (800) 331-2943, (1)(2)(3)(4)(6).

SAFEHOME, P.O. Box 221, Rexford, NY, 12148, (518) 399-0135, (1).

SAFEHOUSE, INC., 2335 Sigman Industrial Ct., Conyers, GA, 30207, (404) 388-0481, (2).

SAFETY AND HEALTH SERVICES, 3307 North Second St., Harrisburg, PA, 17110, (717) 238-2122, (3).

SAGE RESOURCES, /INDOOR AIR QUALITY DIVISION, P.O. Box 480559, Los Angeles, CA, 90048, (213) 825-1934, (213) 651-0155, (1).

SALAMON, PAUL, 2621 North Knoxville Ave., Peoria, IL, 61604, (309) 686-9392, (309) 686-0046, (2).

SAM STEWART TESTING, P.O. Box 7304, Fairfax Station, VA, 22039, (703) 250-9203, (1).

SAN BERNARDINO COUNTY AIR POLLUTION CONTROL DISTRICT, 15505 Civic Dr., Victorville, CA, 92392, (619) 243-8147, (10).

SAN DIEGO COUNTY AIR POLLUTION CONTROL DISTRICT, 9150 Chesapeake Dr., San Diego, CA, 92123-1095, (619) 694-3307, (10).

SAN JOAQUIN COUNTY AIR POLLUTION CONTROL DISTRICT, 1601 East Hazelton Ave. P.O. Box 2009, Stockton, CA, 95201, (209) 468-3470, (10).

SAN LUIS OBISPO COUNTY AIR POLLUTION CONTROL DISTRICT, 2156 Sierra Way Suite B, San Luis Obispo, CA, 93401, (805) 549-5912, (10).

SANI-DUCT SERVICE OF THE INLAND NORTHWEST, 16015 East Sprague Ave., Spokane, WA, 99037, (509) 924-8472, (1).

SANTA BARBARA COUNTY AIR POLLUTION CONTROL DISTRICT, 5540 Ekwill St. Suite B, Santa Barbara, CA, 93111, (805) 964-8111, (10).

SARASOTA COUNTY AIR POLLUTION CONTROL/SARASOTA COUNTY ENVIRONMENTAL SERVICES DEPARTMENT, 1301 Cattlemen Rd., Sarasota, FL, 34232, (813) 378-6137, (10).

SASKATCHEWAN CONSUMER AND COMMERCIAL AFFAIRS, 1871 Smith St., Regina, SK, S4P 3V7, (306) 787-2958, (11).

SASKATCHEWAN DEPARTMENT OF HUMAN RESOURCES, LABOUR AND EMPLOYMENT, Saskatchewan Place 1870 Albert St., Regina, SK, S4P 3V4, (306) 787-4486, (11).

SCHENECTADY CITY HEALTH DEPARTMENT, City Hall, Schenectady, NY, 12305, (518) 382-5046, (10).

SCHMELTZER, APTAKER AND SHEPPARD, 1800 Massachusetts Ave., NW Suite 500, Washington, DC, 20036, (202) 828-1000, (3)(16).

SCHNABEL ENGINEERING ASSOCIATES/GEOLOGICAL SERVICES, 4909 Cordell Ave., Bethesda, MD, 20814, (301) 652-8922, (3).

SCHOHARIE COUNTY HEALTH DEPARTMENT, 342 Main St., Schoharie, NY, 12157, (518) 295-8182, (10).

SCHUCK HOME SERVICES, P.O. Box 15581, Ft. Wayne, IN, 46885, (219) 486-7945, (2).

SCHWABE, WALTER A., Rd. 3 Box 533, Branchville, NJ, 07826, (201) 948-7500, (1).

SCIENCE AND TECHNOLOGY, INC. (SCAT), 79 Pembroke Crescent, Fredericton, NB, E3B 2V1, (506) 457-0232, (1)(3)(5).

SCIENCE APPLICATION INTERNATIONAL CORPORATION, 1710 Goodridge Dr., McLean, VA, 22102, (703) 821-4300, (3).

SCIENCE MANAGEMENT SERVICES, LTD., 25 Eshelman Rd., Lancaster, PA, 17601, (717) 392-1425, (1).

SCIENTIFIC ANALYSIS, INC., 6012 East Shirley Ln., Montgomery, AL, 36117, (205) 271-0643, (800) 345-2575, (1)(4).

SCIENTIFIC AND COMMERCIAL SYSTEMS CORPORATION (SCSC), 4651 King St. Suite 200, Alexandria, VA, 22302, (703) 824-8240, (5).

SCIENTIFIC COMMITTEE ON PROBLEMS OF THE ENVIRONMENT (SCOPE), 51 Boulevard De Montmorency, F-75016 Paris, France, (011) 331-45250498, (12).

SCIENTIFIC RESEARCH AND DEVELOPMENT COMPANY, P.O. Box 5590, Lakeland, FL, 33807, (813) 644-0404, (3)(4).

SCIENTIFIC TESTING ASSOCIATES, LTD., Star Route Box 124, Topping, VA, 23169, (804) 758-5728, (1)(4).

SCINTA, INC., 2421 Homestead Dr., Silver Spring, MD, 20902, (301) 593-9478, (3).

SCOTT HOME INSPECTION, INC., 87 Beach Ave., Milford, CT, 06460, (203) 874-2414, (1).

SCRUB RIDGE RAD TESTING AND REMEDIATION, RDI Box 469L, McConnellsburg, PA, 17233, (202) 332-1798, (717) 485-5456, (1).

SCS ENGINEERS, INC., 11260 Roger Bacon Dr. Suite 300, Reston, VA, 22090, (703) 471-6150, (1).

SCWEST, INC., 4950 East Asbury Ave. Suite 5, Denver, CO, 80222, (303) 759-1286, (4).

SEACOAST LABORATORIES, 55 Village Ln., South Hamilton, MA, 01982, (508) 468-3136, (1).

SEALANT AND WATERPROOFERS INSTITUTE (SWI), 3101 Broadway Suite 300, Kansas City, MO, 6411, (816) 561-8230, (13).

SEBRING HOME INSPECTION COMPANY, 73 Mulwal Dr., Painesville, OH, 44077, (216) 354-8768, (1).

SECONDO, THOMAS A., 50 Spruce St., Hornell, NY, 14843, (607) 324-1986, (1).

SECURE-ALL COMPANY, INC., 4300 W. 38th Ave., Denver, CO, 80212, (303) 458-5577, (4).

SELSER, ROBERT C., 442 Broadway, Hillsdale, NJ, 07642, (201) 666-4060, (3).

SENECA NATION HEALTH DEPARTMENT, Box 231, Salamanca, NY, 14779, (716) 532-5582, (10).

SENES CONSULTANTS, 499 McNicholl Ave., Willowdale, ON, M2H 2C9, (416) 499-5030, (1)(3).

SENSIDYNE, INC., 12345 Starkey Rd. Suite E, Largo, FL, 34643, (813) 530-3602, (800) 451-9444, (4).

SEPTECH, 7625 North Howard, Spokane, WA, 99208, (509) 467-6274, (1).

SETRA SYSTEMS, INC., 45 Nagog Pk., Acton, MA, 01720, (617) 263-1400, (4).

SHANNON HEATING, COOLING AND ELECTRIC, 1701 North 74th St., Omaha, NE, 68114, (513) 831-0743, (1).

SHAPIRO, MARK, 7709 Morgan Ln., Laverock, PA, 19118, (215) 886-3060, (2).

SHASTA COUNTY AIR QUALITY MANAGEMENT DISTRICT, 1855 Placer St., Redding, CA, 96001, (916) 225-5674, (10).

SHEET METAL AND AIR CONDITIONING CONTRACTORS NATIONAL ASSOCIATION (SMACCNA), P.O. Box 70, Merrifield, VA, 22116, (703) 790-9890, (13).

SHELBURNE LABORATORIES, INC., P.O. Box 458 Falls Rd., Shelburne, VT, 05482, (802) 985-3379, (1).

SHELTERWORKS, INC., 53 New St., Dover, NJ, 07801, (201) 328-2912, (1).

SHERLOCK HOME INSPECTORS, 6-01 Fair Lawn Ave., Fair Lawn, NJ, 07410, (201) 796-4942, (800) 458-2202, (1).

SHEVENELL, GALLEN, AND ASSOCIATES, P. O. Box 433 195 Commerce Way, Portsmouth, NH, 03801, (603) 436-1490, (1)(3).

SHIELD HOME INSPECTION SERVICE, R.D. 1, Box 274 Summit Dr., Branchville, NJ, 07826, (201) 875-5922, (1).

SHINN COMPANY, 145 Wolf Creek Trail, Broomfield, CO, 80020, (303) 465-1331, (13).

SHORTRIDGE INSTRUMENTS, INC., 7855 E. Redfield, Scottsdale, AZ, 85260, (602) 991-6744, (4).

SHOTWELL ASSOCIATES, Box 83, Budd Lake, NJ, 07828, (201) 691-9037, (1).

SIGMA ENVIRONMENTAL SYSTEMS, INC., P.O. Box 229, Wappingers Falls, NY, 12590, (914) 297-4000, (2).

SIKA CORPORATION, 201 Polito Ave., Lyndhurst, NJ, 07071, (201) 933-8800, (800) 631-7270, (4).

SILVER SPRINGS HOME INSPECTION, P.O. Box 182, Landing, NJ, 07850, (201) 770-0026, (1).

SIMPSON WEATHER ASSOCIATES, 809 East Jefferson St., Charlottesville, VA, 22902, (804) 979-3571, (1)(2)(3).

SISKIYOU COUNTY AIR POLLUTION CONTROL DISTRICT, 525 South Foothill Dr., Yreka, CA, 96097, (916) 842-3906, (10).

SKC, INC., 334 Valley View Rd., Eighty Four, PA, 15330, (412) 941-9701, (800)752-8472, (4).

SKUTTLE MANUFACTURING COMPANY, Rte. 1, Box 16A, Marietta, OH, 45750, (614) 373-9169, (4).

SMALL UNITED LABORATORIES, 7606 Patapsco Dr., Sykesville, MD, 21784, (301) 781-6636, (301) 876-7636, (1).

SMALLEY AND COMPANY, 861 South Jason, Denver, CO, 80223, (303) 777-3435, (4).

SMITH, STANLEY, 2901 26th St., W. Suite 702, Bradenton, FL, 34205, (813) 756-3635, (1)(2).

SMOOTH-ON, INC., 1000 Valley Rd., Gillette, NJ, 07933, (201) 647-5800, (4).

SOCIETY OF INDUSTRIAL AND OFFICE REALTORS (SIRO), 777 14th St., NW, Washington, DC, 20005, (202) 383-1150, (18).

SOL-AIR, INC., P.O. Box 458, Plantsville, CT, 06479, (203) 621-6211, (1).

SOLAPLEXUS, R.D. 3, Box 507, New Paultz, NY, 12561, (914) 384-6798, (6).

SOLAPLEXUS/PRODUCT FILE DIVISION, 21 Tamarack Rd., Fishkill, NY, 12524, (914) 896-4796, (914) 384-6798, (3).

SOLAR RESOURCE TECHNOLOGY, 128 Market St., South Williamsport, PA, 17701, (717) 323-8919, (1)(2).

SOLAR TESTING LABORATORIES, INC., 9540 Midwest Ave., Garfield Heights, OH, 44125, (216) 663-1800, (1).

SOLID STATE RADON DETECTION AND ELIMINATION SERVICES, P.O. Box 107, Glenwood, NJ, 07418, (201) 764-8555, (1).

SOLID WASTE AND TOXIC TECHNOLOGY, 23891 Via Fabricante Suite 616, Mission Viejo, CA, 92691, (714) 855-7231, (1).

SOOS RADON SYSTEMS, INC., 25 Pheasant Run Dr., Skillman, NJ, 08558, (201) 359-0356, (2).

SORENSEN ENTERPRISES, INC., 886 West Hunt Rd., Alcoa, TN, 37701, (615) 984-1376, (1).

SOUTH CAROLINA ASSOCIATION OF REALTORS, 3780 Fernandina Rd. Box 21827, Columbia, SC, 29221, (803) 772-5206, (18).

SOUTH CAROLINA DEPARTMENT OF HEALTH AND ENVIRONMENTAL CONTROL/BUREAU OF RADIOLOGICAL HEALTH, 2600 Bull St., Columbia, SC, 29201, (803) 734-4700, (1)(10).

SOUTH CAROLINA DEPARTMENT OF LABOR/OCCUPATIONAL SAFETY AND HEALTH/CONSULTATION AND MONITORING, P.O. Box 11329, Columbia, SC, 29211, (803) 734-9599, (10).

SOUTH CAROLINA ELECTRIC AND GAS, P.O. Box 88, Jenkinsville, SC, 29065, (803) 345-1915, (1)(13).

SOUTH CAROLINA GEOLOGICAL SURVEY, Harbison Forest Rd., Columbia, SC, 29210, (803) 737-9440, (10).

SOUTH CAROLINA REAL ESTATE COMMISSION, 1201 Main St. Suite 1500, Columbia, SC, 29201, (803) 737-0700, (10).

SOUTH CENTRAL ENVIRONMENTAL SERVICES, RD 1, Box 265, Pine City, NY, 14871, (607) 733-1661, (1)(2).

SOUTH COAST AIR QUALITY MANAGEMENT DISTRICT/AIR POLLUTION CONTROL DIVISION, 9150 Flair Dr., El Monte, CA, 91731, (818) 572-6200, (10).

SOUTH DAKOTA ASSOCIATION OF REALTORS, 411 East Capitol St. Box 386, Pierre, SD, 57501, (605) 224-9211, (18).

SOUTH DAKOTA DEPARTMENT OF WATER AND NATURAL RESOURCES/SOUTH DAKOTA GEOLOGICAL SURVEY, Science Center University of South Dakota, Vermillion, SD, 57069, (605) 677-5227, (10).

SOUTH DAKOTA REAL ESTATE COMMISSION, P.O. Box 490, Pierre, SD, 57501, (605) 773-3600, (10).

SOUTH DAKOTA STATE DEPARTMENT OF HEALTH/OFFICE OF AIR QUALITY AND SOLID WASTE, Joe Foss Office Building 523 East Capital, Pierre, SD, 57501, (605) 773-3329, (10).

SOUTH DAKOTA STATE UNIVERSITY/S.T.A.T.E. ENGINEERING DIVISION, Box 2218, Brookings, SD, 57007, (605) 688-4101, (10).

SOUTH PARK URBAN RENEWAL CORPORATION, 1938 South Park Ave., Buffalo, NY, 14220, (716) 873-4763, (1).

SOUTHEASTERN MASSACHUSETTS AIR POLLUTION CONTROL DISTRICT, Lakeville Hospital, Lakeville, MA, 02346, (617) 947-1231 x680, (10).

SOUTHEASTERN RADON CONSULTANTS, INC., 161

York Rd., Nolensville, TN, 37135, (615) 776-5569, (1)(2).

SOUTHERN BUILDING CODE CONGRESS INTERNATIONAL (SBCCI), 900 Montclair Rd., Birmingham, AL, 35213, (205) 591-1853, (13).

SOUTHERN MECHANICAL, 1721 Lee St., Decatur, GA, 30035, (404) 289-9637 , (1).

SOUTHERN NEW ENGLAND RADON TESTING AND MITIGATION, P.O. Box 1141, Westbrook, CT, 06498, (203) 399-8003, (1)(2).

SOUTHERN OHIO RADON PROTECTION COMPANY, 4940 Nutmeg Knoll, Cincinnati, OH, 45244, (513) 831-0743, (1).

SOUTHERN RADON SERVICES, INC., 1000 Johnson Ferry Rd. Suite B-145, Marietta, GA, 30068, (404) 565-3886, (800) 537-2366, (1)(3).

SOUTHERN RESEARCH INSTITUTE, 2000 Ninth Ave., S. P.O. Box 55305, Birmingham, AL, 35255-5305, (205) 581-2000, (1)(5).

SOUTHFACE ENERGY INSTITUTE, P.O. Box 5506, Atlanta, GA, 30307, (404) 525-7657, (17).

SOUTHWEST ABATEMENT, INC., 4200 Broadway, SE, Albuquerque, NM, 87102, (505) 873-2957, (1).

SOUTHWEST AIR POLLUTION CONTROL AUTHORITY, 1306 NE 134th St. Suite D, Vancouver, WA, 98685, (206) 574-3058, (10).

SOUTHWEST RESEARCH AND INFORMATION CENTER, P.O. Box 4524, Albuquerque, NM, 87106, (505) 262-1862, (17).

SPADY, GARY D., 1305 Court St. Box 4 RR #2, Imperial, NE, 69033, (308) 882-5198, (1).

SPARTAN HOME INSPECTION, RFD 3, Box 533, Branchville, NJ, 07826, (201) 948-7500, (1).

SPARTAN HOME INSPECTION, INC., 54 Woodport Rd., Sparta, NJ, 07871, (201) 729-7972, (1).

SPECIAL ENVIRONMENTAL SERVICES, INC., P.O. Box 20058, Atlanta, GA, 30325, (404) 355-5600, (1).

SPECIALIZED SERVICES, INC., R.R. 1 7 Wall St., Vernon, NJ, 07462, (201) 827-9055, (1).

SPECTRUM INDUSTRIES, 1128 Forrest St., Dover, DE, 19901, (302) 736-0792, (1)(2).

SPENCE-GEIGER ASSOCIATES, INC., 1658 Cole Blvd. #260, Golden, CO, 80401, (303) 233-8577, (1)(3).

SPOKANE COUNTY AIR POLLUTION CONTROL AUTHORITY, 1101 West College Ave. Room 230, Spokane, WA, 99201, (509) 456-4727, (10).

SPRINGER COMPANY, 629 Lancaster Ave., Malvern, PA, 19355, (215) 344-9800, (1)(2).

ST. JOSEPH COUNTY AIR POLLUTION CONTROL, County-City Building 8th & 9th Floors, South Bend, IN, 46601, (219) 284-9721, (10).

ST. LOUIS COUNTY AIR POLLUTION CONTROL BRANCH, 801 South Brentwood Blvd., Clayton, MO, 63105, (314) 854-6921, (10).

ST. LOUIS PARK INSPECTIONAL SERVICES, 5005 Minnetonka Blvd., St. Louis Park, MN, 55416, (612) 924-2588, (10).

ST. LOUIS/DIVISION OF AIR POLLUTION CONTROL, City Hall Room 419, St. Louis, MO, 63103, (314) 622-3334, (10).

STAFFORD, C. FRED, 1000 Johnson Ferry Rd. Suite B145, Marietta, GA, 30068, (404) 565-3886, (1).

STAMFORD HEALTH DEPARTMENT, 229 North St., Stamford, CT, 06902, (203) 358-4396, (10).

STAN A. HUBER CONSULTANTS, INC., 200 North Cedar Rd., New Lenox, IL, 60451, (815) 485-4433, (1).

STANDOR RADON DETECTION, 598 South Edgewood Ave., Jacksonville, FL, 32205, (904) 388-0284, (1)(2).

STANFORD UNIVERSITY/DEPARTMENT OF CIVIL ENGINEERING, Stanford, CA, 94306, (415) 723-0861, (5).

STANISLAUS COUNTY AIR POLLUTION CONTROL DISTRICT, 1716 Morgan Rd., Modesto, CA, 95351, (209) 571-6908, (10).

STAR LABORATORIES, P.O. Box 1666, Plymouth, MA, 02360, (617) 747-5227, (617) 947-4940, (1).

STATE AND TERRITORIAL AIR POLLUTION PROGRAM ADMINISTRATORS (STAPPA)/ASSOCIATION OF LOCAL AIR POLLUTION CONTROL OFFICIALS (ALAPCO), 444 North Capitol St., NW, Washington, DC, 20001, (202) 624-7864, (14).

STERBAL, SYLVESTER, 16 Rose Ct., Staten Island, NY, 10301, (718) 273-2422, (1).

STO-COTE, INC., P.O. Drawer 310, Richmond, IL, 60071, (815) 675-2358, (4).

STONE RIDGE RADON MITIGATORS, 424 Mine Rd., Asbury, NJ, 08802, (201) 735-6080, (800) 447-2366, (1)(2).

STRATA ENVIRONMENTAL SERVICE, INC., 401 East Douglas Suite 515, Wichita, KS, 67202, (316) 262-0002, (1)(3).

STRATFORD DEPARTMENT OF HEALTH, 2730 Main St., Stratford, CT, 06497, (203) 385-4090, (10).

STUART HUNT AND ASSOCIATES/OCCUPATIONAL AND RADIATION SAFETY SERVICES, LTD., 26 Raybom Crescent, St. Albert, AB, T8N 5D8, (403) 458-0291, (1).

SUBURBAN JERSEY WATERPROOFING, INC., 30-32 Abbett Ave., Morristown, NJ, 07960, (201) 539-1538, (2).

SUBURBAN REGIONAL HEALTH COMMISSION, 141 South Harrison St. Suite 11, East Orange, NJ, 07018, (201) 675-1774, (10).

SUFFOLK COUNTY DEPARTMENT OF HEALTH SERVICES/AIR POLLUTION CONTROL BUREAU, 15 Horse Block Pl., Farmingville, NY, 11738, (516) 451-4637, (10).

SUFFOLK II BOARD OF COOPERATIVE EDUCATIONAL SERVICES (BOCES), 201 Sunrise Hwy., Patchogue, NY, 11772, (516) 289-2200, (10).

SULLIVAN CONSULTANTS, INC., 302 Forest Dr., Union, NJ, 07083, (201) 964-5316, (1).

SULLIVAN, ROBERT, Rte. 17, Box 11, Lowman, NY, 14861, (607) 732-8617, (2).

SUMP AND SEWAGE PUMP MANUFACTURERS ASSOCIATION (SSPMA), 560 West Washington St. Suite 301, Chicago, IL, 60606, (312) 332-4146, (13).

SUN DISTRIBUTING GROUP, 813 East Bloomingdale Ave. Suite 313, Brandon, FL, 33511, (813) 681-0350, (1).

SUN NUCLEAR CORPORATION, 415-C Pineda Ct., Melbourne, FL, 32940, (407) 259-6862, (1)(4).

SUNDANCE ENTERPRISES, P.O. Box 3607, Dillon, CO, 80435, (303) 468-3911, (303) 468-8369, (1).

SUNN CORPORATION, Box M, Dept. 213 3520 Forbes Ave., Pittsburgh, PA, 15213, (412) 828-9100, (2)(4).

SUNSTREAM DEVELOPMENT CORPORATION, 300 North St., Endicott, NY, 13760, (607) 754-6200, (1)(2)(4).

SUPERIOR SIGNAL, INC., P.O. Box 96, Spotswood, NJ, 08884, (201) 251-0800, (4).

SURE-WAY SYSTEMS/DIVISION OF E.C.S., INC., 107 Pickard Dr., Syracuse, NY, 13211, (315) 455-2431, (800) 622-1002, (2).

SUTTER COUNTY AIR POLLUTION CONTROL DISTRICT, 142 Garden Way, Yuba City, CA, 95991, (916) 674-2851, (10).

SWALE, INC., 57 Indian Hill Rd., Collinsville, CT, 06022, (203) 693-2146, (1).

SWAT TERMITE AND PEST CONTROL, 1797 Roswell Rd., Marietta, GA, 30062, (404) 565-7928, (1).

SYBRON CHEMICAL, P.O. Box 66 Birmingham Rd., Birmingham, NJ, 08011, (609) 893-1100, (4).

SYNERTECH SYSTEMS CORPORATION, 472 South Salina St. Suite 800, Syracuse, NY, 13202, (315) 422-3828, (1).

SYRACUSE ENERGY SERVICE, 100 New Street, Syracuse, NY, 13202, (315) 424-0679, (1).

T

T.R. EDGERTON, INC., 102F Woodwinds Industrial Ct., Cary, NC, 27511, (919) 469-9795, (1).

TAPPAN ZEE DEVELOPMENT, 25 Webb Ln., Highland Falls, NY, 10928, (914) 446-7704, (2).

TARGET RADON SERVICES, INC., Oakland Ctr. 8940A Rte. 108, Columbia, MD, 21045, (301) 992-6622, (1)(3).

TCS INDUSTRIES, INC., 3805 Paxton St., Harrisburg, PA, 17111, (717) 558-8206, (1)(4).

TECH OPSOL, INC., 2151 East Dublin-Granville Rd. Suite G522, Columbus, OH, 43229, (614) 882-3527, (1).

TECH/OPS LANDAUER, INC., 2 Science Rd., Glenwood, IL, 60425-1586, (312) 755-7000, (4).

TECHNICAL ANALYSIS CORPORATION, 6723 Whittier Ave. Suite 202, McLean, VA, 22101, (703) 883-3700, (3).

TECHNORAD, P.O. Box 2144, Rockville, MD, 20852, (301) 881-8293, (1).

TECHNORTH, 109 Waterloo Ln., Fairbanks, AK, 99712, (907) 456-8378, (1).

TECHTRAC, P.O. Box 5050 Suite 303, Lewisville, TX, 75067, (214) 221-1047, (1).

TECON, 2220 Plainfield Pike, Cranston, RI, 02920, (401) 946-3720, (1).

TECTONIC INSPECTION COMPANY, P.O. Box 278, Denville, NJ, 07834, (201) 625-2233, (1).

TEE ENGINEERING COMPANY, 320 Cutters Hill Ct., Lexington, KY, 40509, (606) 263-5350, (1)(2).

TEHAMA COUNTY AIR POLLUTION CONTROL DISTRICT, 1760 Walnut St. P.O. Box 38, Red Bluff, CA, 96080, (916) 527-4504, (10).

TEK ENGINEERING, 7150 East Hampden Ave. #101, Denver, CO, 80224, (303) 758-3173, (1).

TELEDYNE ISOTOPES, 50 Van Buren Ave., Westwood, NJ, 07675, (201) 664-7070, (1)(4).

TELEDYNE ISOTOPES MIDWEST LABORATORY, 700 Landwehr Rd., Northbrook, IL, 60062, (312) 564-0700, (1).

TELEDYNE WATER PIK, 1730 East Prospect, Fort Collins, CO, 80525, (303) 484-1352, (4).

TEMPLE UNIVERSITY RADON PROGRAM, Temple University CECSA 084-53, Philadelphia, PA, 19122, (215) 787-1665, (1)(5)(6).

TENNESSEE ASSOCIATION OF REALTORS, 1910 Adelicia St., Nashville, TN, 37212, (615) 321-0515, (18).

TENNESSEE DEPARTMENT OF

CONSERVATION/DIVISION OF GEOLOGY, Customs House 701 Broadway, Nashville, TN, 37219-5237, (615) 742-6691, (10).

TENNESSEE DEPARTMENT OF HEALTH AND ENVIRONMENT/DIVISION OF AIR POLLUTION CONTROL, 701 Broadway 4th Floor, Nashville, TN, 37219-5403, (615) 741-4634, (615) 741-3931, (10).

TENNESSEE DEPARTMENT OF LABOR/TENNESSEE OCCUPATIONAL SAFETY AND HEALTH, 501 Union Building 3rd Floor, Nashville, TN, 37219, (615) 741-2793, (10).

TENNESSEE REAL ESTATE COMMISSION, 1808 West End Building 11th Floor, Nashville, TN, 37219-5322, (615) 741-2273, (10).

TENNESSEE TECH UNIVERSITY/CENTER FOR THE MANAGEMENT, UTILIZATION AND PROTECTION OF WATER RESOURCES, P.O. Box 5082, Cookeville, TN, 38505, (615) 372-3507, (5).

TEQ CORPORATION, P.O. Box 570, Wilkes-Barre, PA, 18703-0570, (717) 825-3939, (800) 877-2366, (1)(2)(4).

TERMINIX INTERNATIONAL, 855 Ridgelake Blvd., Memphis, TN, 38119, (800) 937-3783, (1).

TERRADEX CORPORATION, 460 North Widget Lane, Walnut Creek, CA, 94598, (415) 938-2545, (4).

TERRADEX CORPORATION/SUBSIDARY OF TECH/OPS LANDAUER, INC., 2 Science Rd., Glenwood, IL, 60425, (312) 755-7911, (800) 528-8327, (1)(4).

TERRADON EXPLORATION TECHNOLOGIES, INC., 3931 White Oak Tr. P.O. Box 22288, Cleveland, OH, 44122, (216) 464-2002, (1)(2)(4)(5)(6).

TERRADYNAMICS CORPORATION, Rte. 1, Aldie, VA, 22001, (703) 327-6500, (703) 435-3033, (1).

TESTING 1.2.3., 50 Freeway Dr., Cranston, RI, 02920, (401) 461-0530, (1).

TESTWELL CRAIG LABORATORIES, 11553 U.S. Hwy. 41 South, Gibsonton, FL, 33534, (813) 677-0242, (1).

TEXAS A & M UNIVERSITY/DEPARTMENT OF NUCLEAR ENGINEERING, 129 Zachry Engineering Center, College Station, TX, 77843-3133, (409) 845-2271, (5).

TEXAS A & M UNIVERSITY/DEPARTMENT OF NUCLEAR ENGINEERING/RADON ASSAY LABORATORY, 129 Zachry Engineering Center, College Station, TX, 77843, (409) 845-4107 , (1)(5).

TEXAS ASSOCIATION OF REALTORS, P.O. Box 14488, Austin, TX, 78761, (512) 836-8630, (18).

TEXAS BUREAU OF ECONOMIC GEOLOGY, The University Of Texas, Austin Balcones Research Center University Station, Box X, Austin, TX, 78713-7508, (512) 471-1534, (10).

TEXAS DEPARTMENT OF HEALTH/BUREAU OF RADIATION CONTROL, 1100 West 49th St., Austin, TX, 78756-3189, (512) 835-7000, (10).

TEXAS DEPARTMENT OF HEALTH/DIVISION OF OCCUPATIONAL SAFETY AND STATE SAFETY ENGINEER, 1100 West 49th St., Austin, TX, 78756, (512) 458-7287, (10).

TEXAS REAL ESTATE COMMISSION, P.O. Box 12188 Capitol Station, Austin, TX, 78711, (512) 465-3980, (10).

TFC NUCLEAR ASSOCIATES, INC., 425 Bridgeboro Rd., Moorestown, NJ, 08057, (609) 778-4529, (1)(2).

THERMAL INSULATION MANUFACTURERS ASSOCIATION (TIMA), 8341 Sangre De Cristo Rd., Littleton, CO, 80127, (303) 933-9774, (13).

THERMO ENVIRONMENTAL CORPORATION, 101 1st Ave. P.O. Box 9046, Waltham, MA, 02254, (617) 622-1103, (505) 345-9931, (1)(2)(4).

THERMO-SCAN, INC., 1137 3rd Ave., SW P.O. Box 705, Carmel, IN, 46032, (317) 846-4655, (1).

THEROUX ENGINEERING, P.O. Box 4096, Hamden, CT, 06514, (203) 932-1313, (1)(2).

THIRD PARTY INSPECTIONS, P.O. Box 3161, Winchester, VA, 22601, (703) 869-1096, (1).

THOMAS L. HOFFMAN, COMPANY, INC., RD 3, Box 173M, Coopersburg, PA, 18036, (215) 433-5887, (2).

THOMSON AND NIELSEN ELECTRONICS, LTD., 4019 Carling Ave. Suite 303, Kanata, ON, K2K 2A3, (613) 592-3019, (1)(4)(5).

THORO SYSTEM PRODUCTS, 7800 Northwest 38th St., Miami, FL, 33166, (305) 592-2081, (4).

THOROUGH CHECK, INC., 2606 Rte. 37 East, Toms River, NJ, 08753, (201) 506-0200, (800) 852-0116, (1).

THREE W RADON SERVICES, INC., 1619 Springhill Dr., Hummelstown, PA, 17036, (717) 566-0499, (1).

THRESHOLD TECHNICAL PRODUCTS, 7225 Edington Dr., Cincinnati, OH, 45249, (800) 458-4931, (4).

TIMBERLINE RADON SERVICE, INC., P.O. Box 3815, Evergreen, CO, 80439, (303) 674-2828, (1).

TISCO, 2615 Skyline Dr., College Park, GA, 30337, (404) 768-2517, (1).

TMA/EBERLINE, HEALTH PHYSICS, 5635 Jefferson Blvd., NE, Albuquerque, NM, 87109, (505) 345-9931, (1)(2)(3)(4).

TMA/EBERLINE, LABORATORY, 7021 Pan American Hwy., NE, Albuquerque, NM, 87109, (505) 345-3461, (1).

TMB COMPANY, 3900 Parkside Dr., Baltimore, MD, 21206, (301) 485-8734, (1).

TMW MARKETING, 1660 North LaSalle Suite 3301, Chicago, IL, 60614, (312) 951-7213, (1).

TODD GIDDINGS AND ASSOCIATES, 3049 Enterprise Dr., State College, PA, 16801, (814) 238-5927, (1).

TODOL PRODUCTS, INC., P.O. Box 398, Natick, MA, 01760, (508) 879-7741, (4).

TOLEDO/ENVIRONMENTAL SERVICES DIVISION, 26 Main St., Toledo, OH, 43605, (419) 693-0350, (10).

TOM SCURRY ASSOCIATES, 1976 Linden Dr., Boulder, CO, 80302, (303) 444-4802, (4).

TOPEKA-SHAWNEE COUNTY HEALTH AGENCY, 1615 West Eighth St., Topeka, KS, 66606, (913) 233-8961, (10).

TOTAL HOME TECH, 808 Sunnyside Ave., Audubon, PA, 19403, (215) 631-9313, (1).

TOWN OF HAMBURG, S-6100 South Park Ave., Hamburg, NY, 14075, (716) 649-6111, (10).

TRA-CON, INC., 55 North St., Medford, MA, 02155, (617) 391-5550, (800) 872-2661, (4).

TRACOR JITCO RADON MEASUREMENT SERVICES, 1601 Research Blvd., Rockville, MD, 20850, (301) 984-2800, (1)(4).

TRADON ENVIRONMENTAL SERVICES, 41 Old Pascack Rd., Woodcliff Lake, NJ, 07675, (201) 307-1063, (201) 307-1728, (1).

TRC-ECI, 800 Connecticut Blvd., East Hartford, CT, 06108, (203) 289-8631, (1).

TREMCO, INC., 3735 Green Rd., Beachwood, OH, 44122, (216) 292-5000, (4).

TRI-CITIES RADON, 30 Clarke St., Binghamton, NY, 13905, (607) 724-4088, (1).

TRI-COUNTY RADON, INC., 2796 North Constitution Rd., Camden, NJ, 08104, (609) 963-1533, (1)(2).

TRI-COURT SAFETY/SECURITY, P.O. Box 1257, Sunset Beach, CA, 90742, (213) 335-3290, (714) 846-4346, (1).

TRI-STATE RADON AND DEWATERING, P.O. Box 922, Dayton, NJ, 08810, (201) 329-8570, (1)(2).

TRI-STATE RADON REDUCTION, INC., 816 Ainsworth Dr., Linden, NJ, 07036, (609) 275-0155, (201) 925-8472, (1)(2).

TRI-STATE RADON TESTING AND REDUCTION, 530 Eben Ave., Findlay, OH, 45840, (419) 422-0657, (419) 422-6601, (1).

TRIAD BASEMENT DE-WATERING SYSTEM, 1443 Spring Hill Cr., Kernersville, NC, 27284, (919) 996-7870, (919) 996-1685, (1).

TRIANGLE WATERPROOFING AND REPAIR SERVICE, P.O. Box 741 206 North Arendell Ave., Zebulon, NC, 27597, (919) 269-8021, (2).

TRIPLE SEAL, INC., 2425 Patterson Rd., Kettering, OH, 45420, (513) 258-8255, (1).

TROCAL, 10 Link Drive, Rockleigh, NJ, 07647, (201) 767-1660, (4).

TRU-KIL SERVICE, INC., P.O. Box 87156, College Park, GA, 30337, (404) 768-4533, (1).

TRUE TEST COMPANY, UTAH, P.O. Box 571, Bountiful, UT, 84010, (801) 298-2256, (1).

TRUE TEST COMPANY, WASHINGTON, 1436-C Brandt Rd., Vancouver, WA, 98661, (206) 696-0957, (1).

TRUESDALE-HAWK INSPECTION CONSULTANTS, INC., P.O. Box 154, Bedminster, NJ, 07921, (201) 234-0323, (1).

TRUSTY WARNS, INC., 246 East Irving Park Rd., Wood Dale, IL, 60191, (312) 766-9015, (4).

TSI CONSTRUCTION SERVICES, 1137 3rd Ave., SW, Carmel, IN, 46032, (317) 846-4655, (1)(2).

TUCSON HOME INSPECTIONS, INC., P.O. Box 50845, Tucson, AZ, 85703-1845, (602) 881-5511, (1).

TULANE UNIVERSITY MEDICAL CENTER/OCCUPATIONAL HEALTH AND SAFETY SERVICES, 1430 Tulane Ave., New Orleans, LA, 70112-2699, (504) 588-5486, (1).

TULARE COUNTY AIR POLLUTION CONTROL DISTRICT/ENVIRONMENTAL HEALTH DIVISION, Health Building County Civic Ctr., Visalia, CA, 93291, (209) 733-6441, (10).

TULSA CITY-COUNTY HEALTH DEPARTMENT, 4616 East 15th St., Tulsa, OK, 74112, (918) 744-1000, (10).

TUOLUMNE COUNTY AIR POLLUTION CONTROL DISTRICT, 2 South Green St., Sonora, CA, 95730, (209) 533-5698, (10).

TWIN CITY TESTING, 662 Cromwell, St. Paul, MN, 55114, (612) 645-3601, (1).

U

U.S. ASSETS RECOVERY INVESTIGATIONS, INC., P.O. Box 6953, Washington, DC, 20032, (202) 362-4764, (1).

U.S. DEPARTMENT OF INTERIOR (DOI)/BUREAU OF MINES(BOM)/DENVER RESEARCH CENTER DRC)/RADIATION LABORATORY, P.O. Box 25086 Denver Federal Ctr. Building 20, Denver, CO, 80225, (303) 236-0789, (1)(3)(4)(5).

U.S. DEPARTMENT OF THE ARMY/ARMY ENVIRONMENTAL HYGIENE AGENCY, HSHB-ML-RR, Aberdeen Proving Ground, MD, 21010-5422, (301) 671-2619, (301) 671-2637, (1).

U.S. DEPARTMENT OF THE ARMY/BELVOIR RESEARCH, DEVELOPMENT AND EDUCATION CENTER, STRBE-VR, Fort Belvoir, VA, 22060-5606, (703) 664-5133, (1).

U.S. HOME INSPECTION SERVICES, 2603 Berwyn St., Richmond, VA, 23234, (804) 231-6262, (804) 423-0730, (1).

U.S. PUBLIC INTEREST RESEARCH GROUP (US PIRG), 215 Pennsylvania Ave., SE, Washington, DC, 20003, (202) 546-9707, (17).

U.S. RADON TESTING SERVICE, 820 S. 24th St., Arlington, VA, 22202, (703) 684-4528, (1).

U.S. RADON TESTING/REDUCTION OF MARYLAND, P.O. Box 3204, Gaithersburg, MD, 20878, (301) 330-4187, (1).

U.S. TOXIC SUBSTANCE TESTING BUREAU, 1024 Cottman Ave., Philadelphia, PA, 19111, (215) 745-7778, (800) 535-8378, (1)(4).

U.S. VIRGIN ISLANDS DEPARTMENT OF CONSERVATION AND CULTURAL AFFAIRS/DIVISION OF NATURAL RESOURCES MANAGEMENT/AIR POLLUTION PROGRAM, P.O. Box 4399, St. Thomas, VI, 00801-4399, (809) 774-3320, (10).

U.S. VIRGIN ISLANDS DEPARTMENT OF LABOR/DIVISION OF OCCUPATIONAL SAFETY AND HEALTH, Lagoon St., Frederiksted, VI, 00840, (809) 772-1315, (10).

U.S. VIRGIN ISLANDS DEPARTMENT OF LICENSING AND CONSUMER SERVICES, Sub Base #1 Property and Procurement Building, St. Thomas, VI, 00802, (809) 774-3130, (10).

U.S. VIRGIN ISLANDS DEPARTMENT OF PLANNING AND NATURAL RESOURCES/ENVIRONMENTAL PROTECTION DIVISION, P.O. Box 3594, St. Thomas, VI, 00801, (809) 776-4844, (10).

U.S. VIRGIN ISLANDS/COLLEGE OF THE VIRGIN ISLANDS/CARIBBEAN RESEARCH INSTITUTE, St. Thomas, VI, 809774-9200, (809) 774-9200, (10).

UCI/ENVIROTEST, 34100 Center Ridge Rd., North Ridgeville, OH, 44039, (216) 327-9432, (3).

UCI/ENVIROTEST, 834 Old Thorn Run Rd., Coraopolis, PA, 15108, (412) 264-5891, (1)(3).

UNC GEOTECH, 2594 B 314 Road P.O. Box 14000, Grand Junction, CO, 81502-5504, (303) 248-6702, (5).

UNION COUNTY BUILDING INSPECTION SERVICE, 502 Twin Oaks Rd., Union, NJ, 07083, (201) 687-6110, (1).

UNION OF CONCERNED SCIENTISTS (UCS), 26 Church St., Cambridge, MA, 02238, (617) 547-5552, (17).

UNITED GILSONITE LABORATORIES, P.O. Box 70, Scranton, PA, 18501, (717) 344-1202, (4).

UNITED NATIONS SCIENTIFIC COMMITTEE ON THE EFFECTS OF ATOMIC RADIATION (UNSCEAR), P.O. Box 500, Vienna, Austria A1440, (011) 43-222-26310, (12).

UNITED SERVICES ALLIANCE, INC., 4900 Leesburg Pike Suite 210, Alexandria, VA, 22302, (703) 845-1324, (1)(2).

UNIVERSAL HOME INSPECTION, INC., 5 Boar Ct., Suffern, NY, 10901, (914) 354-1119, (1).

UNIVERSITY HOME HEALTH SERVICES, INC., 670 Centre St. Suite 9, Jamaica Plain, MA, 02130, (617) 524-2149 (617) 522-2793, (1).

UNIVERSITY OF ALASKA/COOPERATIVE EXTENSION SERVICE, 207D Eielson Building, Fairbanks, AK, 99775-5200, (907) 474-7201, (5).

UNIVERSITY OF CALIFORNIA-SAN DIEGO/DEPARTMENT OF RADIOLOGY/RADIATION BIOLOGY, M-010, La Jolla, CA, 92093, (619) 534-4916, (5).

UNIVERSITY OF CALIFORNIA-SAN FRANCISCO/DEPARTMENT OF RADIATION ONCOLOGY/PHYSICS DEPARTMENT, L-75, Box 0226, San Francisco, CA, 94143-0226, (415) 476-9000, (5).

UNIVERSITY OF CALIFORNIA-SAN FRANCISCO/LABORATORY OF RADIOBIOLOGY AND ENVIRONMENTAL HEALTH, San Francisco, CA, 94143-0750, (415) 476-1636, (5).

UNIVERSITY OF CALIFORNIA/LAWRENCE BERKELEY LABORATORY/DONNER LABORATORY BIOMEDICAL SCIENCES, One Cyclotran Rd./Donner Pavilion Cowell Hospital, Room 420, Berkeley, CA, 94720, (415) 486-4000, (5).

UNIVERSITY OF CHICAGO MEDICAL CENTER/DEPARTMENT OF RADIATION ONCOLOGY, 5841 South Maryland Ave., Chicago, IL, 60637, (312) 702-6860, (5).

UNIVERSITY OF COLORADO-COLORADO SPRINGS/PHYSICS DEPARTMENT, 1420 Austin Bluffs Pkwy., Colorado Springs, CO, 80933-7150, (719) 593-3214, (1)(5).

UNIVERSITY OF FLORIDA INDOOR RADON TASK FORCE, A.P. Black Hall, Gainesville, FL, 32611, (904) 392-0841, (5).

UNIVERSITY OF FLORIDA/COLLEGE OF ARCHITECTURE/SCHOOL OF BUILDING CONSTRUCTION, AFC-101, Gainesville, FL, 32611, (904) 392-5965, (5).

UNIVERSITY OF FLORIDA/COLLEGE OF ENGINEERING/ENVIRONMENTAL ENGINEERING DEPARTMENT, 217 Black Hall, Gainesville, FL, 32611, (904) 392-0841, (5).

UNIVERSITY OF FLORIDA/COLLEGE OF ENGINEERING/NUCLEAR ENGINEERING DEPARTMENT, 202 Nuclear Science Center, Gainesville, FL, 32611, (904) 392-1401, (5).

UNIVERSITY OF FLORIDA/COLLEGE OF JOURNALISM AND COMMUNICATION/COMMUNICATION RESEARCH CENTER, 2042 Weimer, Gainesville, FL, 32611, (904) 392-6660, (5).

UNIVERSITY OF FLORIDA/COLLEGE OF LIBERAL ARTS AND SCIENCES/DEPARTMENT OF GEOLOGY, 1112 Turlington Hall, Gainesville, FL, 32611, (904) 392-2231, (5).

UNIVERSITY OF GEORGIA/CENTER FOR APPLIED ISOTOPE STUDY, 120 Riverbend Rd., Athens, GA, 30605, (404) 542-1395, (1)(5).

UNIVERSITY OF HAWAII-MANOA/HAWAII INSTITUTE OF GEOPHYSICS/DIVISION OF EARTH SCIENCES, 2525 Correa, Honolulu, HI, 96822, (808) 948-8855, (5).

UNIVERSITY OF HOUSTON/DEPARTMENT OF CIVIL AND ENVIRONMENTAL ENGINEERING, 4800 Calhoun, Houston, TX, 77204-4791, (713) 749-4481, (5).

UNIVERSITY OF ILLINOIS/ENVIRONMENTAL RESEARCH LABORATORY, 1005 West Western Ave., Urbana, IL, 61801, (217) 333-6230, (5).

UNIVERSITY OF IOWA/UNIVERSITY HYGIENIC LABORATORY, Oakdale Campus, Iowa City, IA, 52242, (319) 335-4500, (5).

UNIVERSITY OF LOWELL/RADIOLOGICAL SCIENCES DEPARTMENT, 1 University Ave., Lowell, MA, 01854, (617) 453-1045, (1)(5).

UNIVERSITY OF MAINE-ORONO/COOPERATIVE EXTENSION SERVICE, 100 Winslow Hall, Orono, ME, 04469, (207) 581-3188, (14).

UNIVERSITY OF MAINE-ORONO/DEPARTMENT OF CIVIL ENGINEERING, 103 Boardman, Orono, ME, 04469, (207) 581-2171, (5).

UNIVERSITY OF MAINE-ORONO/DEPARTMENT OF ENGINEERING, 455 Aubert Hall, Orono, ME, 04469, (207) 581-2216, (5).

UNIVERSITY OF MAINE-ORONO/DEPARTMENT OF GEOLOGY, 110 Boardman Hall, Orono, ME, 04469, (207) 581-2156, (5).

UNIVERSITY OF MAINE-ORONO/DEPARTMENT OF PHYSICS, 120 Bennett Hall, Orono, ME, 04469, (207) 581-1016, (5).

UNIVERSITY OF MAINE-ORONO/LAND AND WATER RESOURCES CENTER, 11 Coburn Hall, Orono, ME, 04469, (207) 581-1490, (5).

UNIVERSITY OF MARYLAND/RADIATION SAFETY OFFICE, Chemical-Nuclear Engineering Building Room 2124, College Park, MD, 20742, (301) 454-5294, (1).

UNIVERSITY OF MICHIGAN/SCHOOL OF PUBLIC HEALTH/DEPARTMENT OF ENVIRONMENTAL AND INDUSTRIAL HEALTH/RADON RESOURCE AND TRAINING CENTER, 109 Observatory, Ann Arbor, MI, 48109, (313) 764-0523, (3)(6).

UNIVERSITY OF NEW HAMPSHIRE/DEPARTMENT OF EARTH SCIENCES, James Hall, Durham, NH, 03824, (603) 862-1234, (5).

UNIVERSITY OF NEW MEXICO SCHOOL OF MEDICINE/CANCER RESEARCH AND TREATMENT CENTER/NEW MEXICO TUMOR REGISTRY, 900 Camino Dey Salud, NE, Albuquerque, NM, 87131, (505) 277-5541, (5).

UNIVERSITY OF NEW MEXICO/DEPARTMENT OF GEOLOGY, Albuquerque, NM, 87131, (505) 277-2310, (505) 277-4204, (1)(5).

UNIVERSITY OF NORTH CAROLINA/SCHOOL OF PUBLIC HEALTH/DEPARTMENT OF ENVIRONMENTAL SCIENCES AND ENGINEERING, Room 201H, Chapel Hill, NC, 27599-7400, (919) 966-3840, (1)(5).

UNIVERSITY OF PENNSYLVANIA/DEPARTMENT OF GEOLOGY, 138 West Rose Valley Rd., Wallingford, PA, 19086, (215) 898-5191, (5).

UNIVERSITY OF PITTSBURGH/PHYSICS DEPARTMENT, 100 Allen Hall, Pittsburgh, PA, 15260, (412) 624-9245, (1)(5).

UNIVERSITY OF SOUTHERN CALIFORNIA/GEOLOGICAL SCIENCE DEPARTMENT, University Park, Los Angeles, CA, 90089-0740, (213) 743-6973, (5).

UNIVERSITY OF TEXAS/SCHOOL OF PUBLIC HEALTH, P.O. Box 20186, Houston, TX, 77225, (713) 792-4431, (1)(5).

UNIVERSITY OF TORONTO/NATIONAL CANCER INSTITUTE OF CANADA/EPIDEMIOLOGICAL UNIT, 12 Queens Park Crescent West, Toronto , ON, M5S 1A8, (416) 978-4793, (5).

UNIVERSITY OF UTAH/SCHOOL OF MEDICINE/RADIOBIOLOGICAL LABORATORY, Building 351, Salt Lake City, UT, 84112, (801) 581-7201, (5).

UNIVERSITY OF VIRGINIA/DEPARTMENT OF ENVIRONMENTAL SCIENCES, Clark Hall, Charlottesville, VA, 22903, (804) 924-0559, (5).

UNIVERSITY OF WASHINGTON/DEPARTMENT OF ENVIRONMENTAL HEALTH/RADIATION ECOLOGY LABORATORY, LRE-WH-10, Seattle, WA, 98195, (206) 543-4259, (1)(5).

UNIVERSITY OF WISCONSON-GREEN BAY/COLLEGE OF ENVIRONMENTAL SCIENCES, Natural and Applied Sciences, Green Bay, WI, 54301-7001, (414) 465-2371, (5).

UNIVERSITY RADON TESTING ASSOCIATES, 145 A Sycamore Ln., Athens, GA, 30605, (404) 549-8057, (1).

UP-COUNTRY HOME INSPECTORS, INC., P.O. Box 1568, Scarborough, ME, 04074, (207) 883-9876, (1)(2).

URETECH, INC., 154 Potomac Ave., Tallmadge, OH, 44278, (216) 633-3586, (4).

UTAH ASSOCIATION OF REALTORS, 5710 South Green St., Murray, UT, 84123, (801) 268-4747, (18).

UTAH DEPARTMENT OF BUSINESS REGULATIONS/DIVISION OF REAL ESTATE, P.O. Box 45802 160 East 300 South, Salt Lake City, UT, 84145, (801) 530-6747, (10).

UTAH DEPARTMENT OF HEALTH/BUREAU OF RADIATION CONTROL, 288 North 1460 West P.O. Box 16690, Salt Lake City, UT, 84116-0690, (801) 538-6734, (10).

UTAH GEOLOGICAL AND MINERAL SURVEY, 606 Black Hawk Way, Salt Lake City, UT, 84108-1280, (801) 244-5164, (10).

UTILITY ENGINEERS, INC., 37 N. Washington St., Wilkes Barre, PA, 18701, (717) 829-4200, (1).

V

VACU-RAY CORPORATION, P.O. Box 291, Whitehouse Station, NJ, 08889, (201) 534-5471, (4).

VALLEN SAFETY SUPPLY COMPANY, 13333 Northwest Fwy., Houston, TX, 77040, (713) 462-8700, (4).

VALLEY DE-WATERING, RD 2, Box 265 A-1, Sugarloaf, PA, 18249, (717) 384-3315, (2).

VALLEY HOME INSPECTIONS, INC., P.O. Box 1677, Chandler, AZ, 85244-1677, (602) 988-1375, (1).

VALLEY RADON TESTING, INC., 26 Oak Valley Rd., Huntington, CT, 06484, (203) 929-4109, (1).

VENABLE, BAETJER AND HOWARD, 2 Hopkins Plaza, Baltimore, MD, 21201-2978, (301) 244-7772, (3).

VENTURA COUNTY AIR POLLUTION CONTROL DISTRICT, 800 South Victoria Ave., Ventura, CA, 93009, (805) 654-2806, (10).

VER-GON BUILDERS, 193 Christie Ave., Clifton, NJ, 07011, (201) 772-4040, (2).

VERITECH, 47 Carey Ave., Butler, NJ, 07405, (201) 492-8744, (1).

VERMONT ASSOCIATION OF REALTORS, P.O. Box 1074, Montpelier, VT, 05602, (802) 229-0513, (18).

VERMONT DEPARTMENT OF HEALTH/DIVISION OF OCCUPATIONAL AND RADIOLOGICAL HEALTH, Administration Building 10 Baldwin St., Montpelier, VT, 05602, (802) 828-2886, (10).

VERMONT DEPARTMENT OF LABOR AND INDUSTRY/DIVISION OF OCCUPATIONAL SAFETY AND HEALTH, 120 State St., Montpelier, VT, 05602, (802) 828-2765, (10).

VERMONT GEOLOGICAL SURVEY, 103 South Main St. Center Building, Waterbury, VT, 05676, (802) 244-5164, (10).

VERMONT OFFICE OF THE SECRETARY OF STATE/REAL ESTATE COMMISSION, Pavilion Office Building, Montpelier, VT, 05602, (802) 828-3228, (10).

VERSAR, INC., 6850 Versar Center, Springfield, VA, 22151, (703) 750-3000, (1)(2)(3).

VFC, INC., 1145 Washington St., Allentown, PA, 18102, (215) 437-5010, (1).

VICK ENTERPRISES, 2060 East Easter Ave., Littleton, CO, 80122, (303) 798-7918, (2).

VICTOREEN, INC., 6000 Cochran Rd., Cleveland, OH, 44139, (216) 248-9300, (4).

VIGO COUNTY AIR POLLUTION CONTROL, 201 Cherry St., Terre Haute, IN, 47807, (812) 238-8433, (10).

VIKING HOME IMPROVEMENT, INC., 15 South 15th St., Fargo, ND, 58103, (701) 237-6300, (1)(2)(4).

VILLAGE OF BEDFORD PARK/AIR POLLUTION CONTROL, 6701 South Archer Rd., Bedford Park, IL, 60501, (312) 458-2975, (10).

VIMCO, INC., 300 Hansen Access Rd., King of Prussia, PA, 19406, (215) 768-0500, (4).

VINCE SCHAEFER CONSTRUCTION, 3220 Beaver Rd., Des Moines, IA, 50310, (515) 279-4600, (1)(2).

VIRGINIA ASSOCIATION OF REALTORS, P.O. Box 15719, Richmond, VA, 23227, (804) 264-5033, (18).

VIRGINIA DEPARTMENT OF AGRICULTURE AND CONSUMER SERVICES/OFFICE OF CONSUMER AFFAIRS, Washington Building 1100 Bank St., Richmond, VA, 23219, (800) 552-9963, (10).

VIRGINIA DEPARTMENT OF COMMERCE/STATE REAL ESTATE BOARD, 3600 West Broad St. 5th Floor, Richmond, VA, 23230, (804) 367-8552, (10).

VIRGINIA DEPARTMENT OF HEALTH/DIVISION OF HEALTH HAZARDS CONTROL, 109 Governor St. Room 700, Richmond, VA, 23219, (804) 786-6029, (10).

VIRGINIA DEPARTMENT OF HEALTH/DIVISION OF HEALTH HAZARDS CONTROL/BUREAU OF RADIOLOGICAL HEALTH, Northern Regional Office 9504-A Lee Highway, Fairfax, VA, 22031, (703) 430-8609, (10).

VIRGINIA DEPARTMENT OF HEALTH/DIVISION OF HEALTH HAZARDS CONTROL/BUREAU OF RADIOLOGICAL HEALTH, 109 Governor St., Richmond, VA, 23219, (804) 786-6029, (10).

VIRGINIA DIVISION OF MINERAL RESOURCES, P.O. Box 3667, Charlottesville, VA, 22903, (804) 293-5121, (10).

VIRGINIA DRY-BASE, 198 Milnwood Rd. P.O. Box 562, Farmville, VA, 23901, (800) 782-5042, (1).

VIRGINIA RADON SERVICE, INC., Box 332-B Route 1, Roseland, VA, 22967, (804) 277-5491, (1).

VISUAL BUILDING INSPECTIONS, 15 B Sickletown Rd., Orangeburg, NY, 10962, (914) 358-2282, (1).

VMS AND ASSOCIATES, Route 1, Box 472A, Madison Heights, VA, 24572, (804) 595-2413, (1).

W

W.L. GORE AND ASSOCIATES, 100 Airport Rd. Box 1010, Elkton, MD, 21921, (301) 392-3200, (4).

W.R. GRACE/CONSTRUCTION PRODUCTS DIVISION, 62 Whittemore Ave., Cambridge, MA, 02140, (617) 876-1400, (4).

W.R. GRACE/ORGANIC CHEMICALS DIVISION, 55 Hayden Ave., Lexington, MA, 02173, (617) 861-6600, (4).

W.R. MARTIN COMPANY, 4124-B Clemson Blvd., Anderson, SC, 29621, (803) 225-7010, (1).

W.R. MEADOWS, INC., P.O. Box 543, Elgin, IL, 60121, (312) 683-4500, (4).

W.S. FLEMING AND ASSOCIATES, INC., 6310 Fly Rd., East Syracuse, NY, 13057, (315) 437-1869, (1)(3).

W.W. GRAINGER, INC. DISTRIBUTION GROUP, 5959 West Howard St., Chicago, IL, 60648, (312) 647-8900, (4).

WALDON ASSOCIATES, 80 Hoadley St., Naugatuck, CT, 06770, (203) 729-5045, (1).

WARREN WOLFE ASSOCIATES, 212 South Queen St., Dover, DE, 19901, (302) 734-8275, (1).

WASHINGTON ASSOCIATION OF REALTORS, P.O. Box 719, Olympia, WA, 98507, (206) 943-3100, (18).

WASHINGTON DC ASSOCIATION, 777 14th St., NW Suite 200, Washington, DC, 20005, (202) 628-4494, (18).

WASHINGTON OCCUPATIONAL HEALTH ASSOCIATES, INC., 1120 19th St., NW Suite 410, Washington, DC, 20036, (202) 463-6698, (16).

WASHINGTON STATE DEPARTMENT OF ENERGY/ENERGY EXTENSION SERVICE, 914 East Jefferson Suite 300, Seattle, WA, 98122, (509) 624-4180, (10).

WASHINGTON STATE DEPARTMENT OF LABOR AND INDUSTY, 814 East 4th P.O. Box 207, Olympia, WA, 98504, (206) 753-6500, (10).

WASHINGTON STATE DEPARTMENT OF LICENSING/REAL ESTATE AND ESCROW/PROFESSIONAL PROGRAMS MANAGEMENT DIVISION, P.O. Box 9012, Olympia, WA, 98504, (206) 753-0775, (10).

WASHINGTON STATE DEPARTMENT OF NATURAL RESOURCES/DIVISION OF GEOLOGY AND EARTH RESOURCES DIVISION, MS-PY-12, Olympia, WA, 98504, (206) 459-6372, (10).

WASHINGTON STATE DEPARTMENT OF SOCIAL AND HEALTH SERVICES/OFFICE OF RADIATION PROTECTION, Mail Stop LE-13, Olympia, WA, 98504, (206) 753-5962, (800) 323-9727, (6)(10).

WASHINGTON STATE UNIVERSITY/NUCLEAR RADIATION CENTER, Pullman, WA, 99164-1300, (509) 335-8917, (509) 335-8641, (1).

WASHOE COUNTY DISTRICT HEALTH DEPARTMENT, 1001 East Ninth, Reno, NV, 89520, (702) 328-2400, (10).

WASICK, BRUCE, 1888 S. Jacoby Rd., Copley, OH, 44321, (216) 666-8078, (1).

WASTE CONTROL INTERNATIONAL, INC., 100 Galleria Pkwy. Suite 400, Atlanta, GA, 30339, (404) 443-2061, (1).

WASTREN, INC., 2 Waverly Dr., Hummelstown, PA, 17036, (717) 566-6780, (3).

WATER POLLUTION CONTROL FEDERATION (WPCF), 601 Wyeth St., Alexandria, VA, 22314-1994, (703) 684-2400, (14).

WATER QUALITY ASSOCIATION (WQA), 4151 Naperville Rd., Lisle, IL, 60532, (312) 369-1600, (14).

WATER STORE, THE, P.O. Box 176, Bath, PA, 18014, (215) 837-9660, (3).

WATER TEST CORPORATION, 33 South Commercial St., Manchester, NH, 03101, (603) 623-7400, (1).

WATKINS ENVIRONMENTAL SCIENCES, 207 McLennan Dr., Fayetteville, NY, 13066, (315) 637-4077, (1).

WAYNE COUNTY HEALTH DEPARTMENT/AIR POLLUTION CONTROL, 2211 East Jefferson, Detroit, MI, 48207, (313) 567-4100, (10).

WAYNE MUSSER PROPERTY COMPANY, INC., 517 Broad St. P.O. Box 258, Sewickley, PA, 15143, (412) 741-9517, (1)(4).

WAYNE, MICHELE, Box 266, Terre Hille, PA, 17581, (215) 745-0129, (1).

WB CONTRACTORS, 3209 Morrison St., NW, Washington, DC, 20015, (202) 966-6312, (1).

WEATHER BARRIER SYSTEMS, INC./CAM GROUP, 111 West Second St., Jamestown, NY, 14701, (716) 484-7183, (800) 624-6434, (4).

WEATHER SHIELD, INC., 4071 South Salem Church Rd., Dover, PA, 17315, (717) 292-5613, (1)(2).

WEAVER WERKS, 511 Industrial Dr., Carmel, IN, 46032, (317) 846-7442, (800) 523-2082, (4).

WEBER AND ASSOCIATES, 8657 Fallbrook Way, Sacramento, CA, 95826, (916) 386-8745, (1).

WELD COUNTY HEALTH DEPARTMENT/AIR POLLUTION CONTROL DIVISION, 1516 Hospital Rd., Greeley, CO, 80631, (303) 353-0635, (10).

WESLEY R. VAN PELT ASSOCIATES, INC., 773 Paramus Rd., Paramus, NJ, 07652, (201) 445-5124, (3).

WEST COAST ENVIRONMENTAL SERVICES, INC., P.O. Box 18182, Seattle, WA, 98118, (206) 324-0920, (1).

WEST ENVIRO SYSTEMS, INC., 45 Sycamore Ln., Skillman, NJ, 08558, (609) 924-0680, (1)(2).

WEST VIRGINIA ASSOCIATION OF REALTORS, 814 Virginia St., Charleston, WV, 25301, (304) 342-7600, (18).

WEST VIRGINIA DEPARTMENT OF HEALTH/INDUSTRIAL HYGIENE DIVISION, 151 11th Ave., South Charleston, WV, 25303, (304) 348-3526, (6)(10).

WEST VIRGINIA DEPARTMENT OF LABOR, State Capitol Room 415 1900 Washington St., Charleston, WV, 25305, (304) 348-7890, (10).

WEST VIRGINIA GEOLOGICAL SURVEY, Mt. Chateau Research Center P.O. Box 879, Morgantown, WV, 26507-0879, (304) 594-2331, (10).

WEST VIRGINIA REAL ESTATE COMMISSION, 1033 Quarrier St. Suit 400, Charleston, WV, 25301, (304) 348-3555, (10).

WESTCHESTER COUNTY HEALTH DEPARTMENT, County Office Building #2, White Plains, NY, 10601, (914) 285-1500, (10).

WESTERN CAROLINA UNIVERSITY/DEPARTMENT OF CHEMISTRY AND PHYSICS, Natural Sciences Building, Cullowhee, NC, 28723, (704) 227-7260, (5).

WESTERN ENVIRONMENTAL, INC., 4320-B Northpark Dr., Colorado Springs, CO, 80907, (719) 528-6132, (1).

WESTERN KENTUCKY UNIVERSITY/DEPARTMENT OF GEOGRAPHY AND GEOLOGY, WKU EST-304, Bowling Green, KY, 42101, (502) 745-5989, (5).

WESTERN NORTH CAROLINA REGIONAL AIR POLLUTION CONTROL BOARD, Buncombe County Courthouse, Asheville, NC, 28802-3569, (704) 255-5655, (10).

WESTERN TECHNOLOGY, INC., 18708 Woodway Dr., Derwood, MD, 20855, (301) 869-2628 , (1).

WESTERN WATER INTERNATIONAL, INC., 7715 Penn Belt Dr., Forestville, MD, 20747, (301) 568-0200, (4).

WESTINGHOUSE ELECTRIC CORPORATION, P.O. Box 355, Pittsburgh, PA, 15230, (412) 374-5716, (4).

WESTINGHOUSE MATERIAL COMPANY OF OHIO/FEED MATERIALS PRODUCTION CENTER, P.O. Box 398704, Cincinnati, OH, 45239-8704, (513) 738-6200, x6280, (5).

WICHITA-SEDGWICK COUNTY DEPARTMENT OF COMMUNITY HEALTH, 1900 East 9th St., Wichita, KS, 67214, (316) 268-8351, (10).

WILKES COLLEGE/PHYSICS DEPARTMENT, Wilkes-Barre, PA, 18766, (717) 824-4651 ext. 481, (1).

WILLIAM B. JOHNSON AND ASSOCIATES, INC., P.O. Box 98 Research Park, Montville, NJ, 07045, (201) 334-9222, (4).

WILLIAM G. CHANEY, PE, 1708 Park Ave., SW, Albuquerque, NM, 87104, (505) 247-2087, (1)(3).

WILSON JR., WALTER E., 700 Harston Ln., Erdenheim, PA, 19118, (215) 836-7257, (1)(2).

WISCONSIN ASSOCIATION OF REALTORS, 4801 Hayes Rd., Madison, WI, 53704, (608) 241-2047, (18).

WISCONSIN DEPARTMENT OF HEALTH AND SOCIAL SERVICES/DIVISION OF HEALTH/BUREAU OF ENVIRONMENTAL HEALTH/SECTION OF OCCUPATIONAL HEALTH, P.O. Box 309, Madison, WI, 53701, (608) 266-0417, (10).

WISCONSIN DEPARTMENT OF HEALTH AND SOCIAL SERVICES/DIVISION OF HEALTH/RADIATION PROTECTION SECTION, 5708 Odana Rd., Madison, WI, 53719, (608) 273-5180, (10).

WISCONSIN DEPARTMENT OF REGULATION AND LICENSING/BUREAU OF DIRECT LICENSING AND REAL ESTATE, 1400 East Washington P.O. Box 8935, Madison, WI, 53708, (608) 267-7132, (10).

WISCONSIN GEOLOGICAL AND NATURAL HISTORY SURVEY, University of Wisconsin 3817 Mineral Point Rd., Madison, WI, 53705, (608) 262-1705, (10).

WJLA TV, 4461 Connecticut Ave., Washington, DC, 20008, (212) 364-7777, (3).

WOODLINE, INC., Box 1170, Frisco, CO, 80443, (303) 668-5603, (2).

WORLD FEDERATION OF PUBLIC HEALTH ASSOCIATIONS (WFPHA), C/O American Public Health Association 1015 15th St., NW, Washington, DC, 20005, (202) 789-5691, (12).

WORLD HEALTH ORGANIZATION (WHO)/DIVISION OF ENVIRONMENTAL HEALTH/ENVIRONMENTAL HAZARDS AND FOOD PROTECTION UNIT, 1211 Geneva 27, Switzerland, (011) 4122-34-6061, (12).

WORLD HEALTH ORGANIZATION (WHO)/WASHINGTON OFFICE ENVIRONMENTAL HEALTH PROGRAM, 525 23rd St., NW, Washington, DC, 20037, (202) 861-3311, (12).

WORLD INFORMATION SERVICE ON ENERGY (WISE)/WASHINGTON OFFICE, P.O. Box 53373, Washington, DC, 20009, (202) 745-7467, (12).

WORLDWIDE RADON, INC., 1941A Nears Pkwy., Margate, FL, 33063, (305) 968-5555, (1)(2).

WORTH ENGINEERING, 650 Palm Ave., Hialeah, FL, 33010, (305) 885-4515, (1)(2).

WPB ENTERPRISES, INC., RD 1, Box 209, Riegelsville, PA, 18077, (215) 346-8004, (1)(2)(6).

WRIGHT LABORATORY SERVICES, INC., 3240 Schoolhouse Rd., Middletown, PA, 17057, (717) 944-5541, (1).

WYOMING ASSOCIATION OF REALTORS, P.O. Box 2312, Casper, WY, 82602-2312, (307) 237-4085, (18).

WYOMING DEPARTMENT OF HEALTH AND SOCIAL SERVICES/DIVISION OF HEALTH AND MEDICAL SERVICES/RADIOLOGICAL HEALTH SERVICES, Hathaway Building, Cheyenne, WY, 82002-0710, (307) 777-6015, (10).

WYOMING DEPARTMENT OF OCCUPATIONAL HEALTH AND SAFETY/COMMISSIONER OF LABOR, 604 East 25th St., Cheyenne, WY, 82002, (307) 777-7786, (10).

WYOMING GEOLOGICAL SURVEY, P.O. Box 3008 University Station, Laramie, WY, 82071, (307) 742-2054, (10).

WYOMING REAL ESATE COMMISSION, 4301 Herschler Building, Cheyenne, WY, 82002, (307) 777-7141, (10).

X

Y

YAKIMA COUNTY CLEAN AIR AUTHORITY, County Courthouse, Yakima, WA, 98901, (509) 575-4116, (10).

YALE UNIVERSITY/DEPARTMENT OF GEOLOGY AND GEOPHYSICS, Kline Geolab P.O. Box 6666, New Haven, CT, 06511, (203) 432-3161.

YOLO-SOLANO AIR POLLUTION CONTROL DISTRICT, 323 First St. Suite 5 P.O. Box 1006, Woodland, CA, 95695, (916) 666-8146, (10).

YOUR ENERGY SERVICE, INC., 220 4F Elliston Pl., Nashville, TN, 37203, (615) 329-9747, (1).

YUBA COUNTY AIR POLLUTION CONTROL DISTRICT, 938 14th St., Marysville, CA, 95901, (916) 741-6484, (10).

Z

ZANUS CORPORATION, 1259 El Camino Real Suite 134, Menlo Park, CA, 94025, (415) 591-7783, (1).

It's Like Having The Entire Radon Industry Delivered To Your Door!

Everything You Need To Know. Everyone You Want To Reach.

Finally there's a reference that contains all the information on the radon issue — everything you need to know — everyone you want to reach!

The Radon Industry Directory is an indispensable sourcebook for anyone concerned about radon.

Edited for a wide and varied audience, it serves as a frequently consulted reference to the rapidly emerging radon industry.

Thoroughly up-to-date, with over 3,000 listings, **The Radon Industry Directory** gives you instant access to thousands of organizations and radon professionals — radon detection and mitigation firms, product manufacturers, federal, state, and local government agencies, special interest groups, workshops, conferences, indexes, and much, much more. In all, 30 vital categories are concisely listed, described, and indexed. There's even a section providing the reader a complete overview of radon — its origins, testing for radon, and radon mitigation.

The Radon Industry Directory puts the names, numbers, and addresses of key radon people and organizations right in your hands — everything you need to know, everyone you want to reach! Published annually and available on standing order.

$75.00 ea.

ISBN: 0-929840-00-3
ISSN: 0899-9023

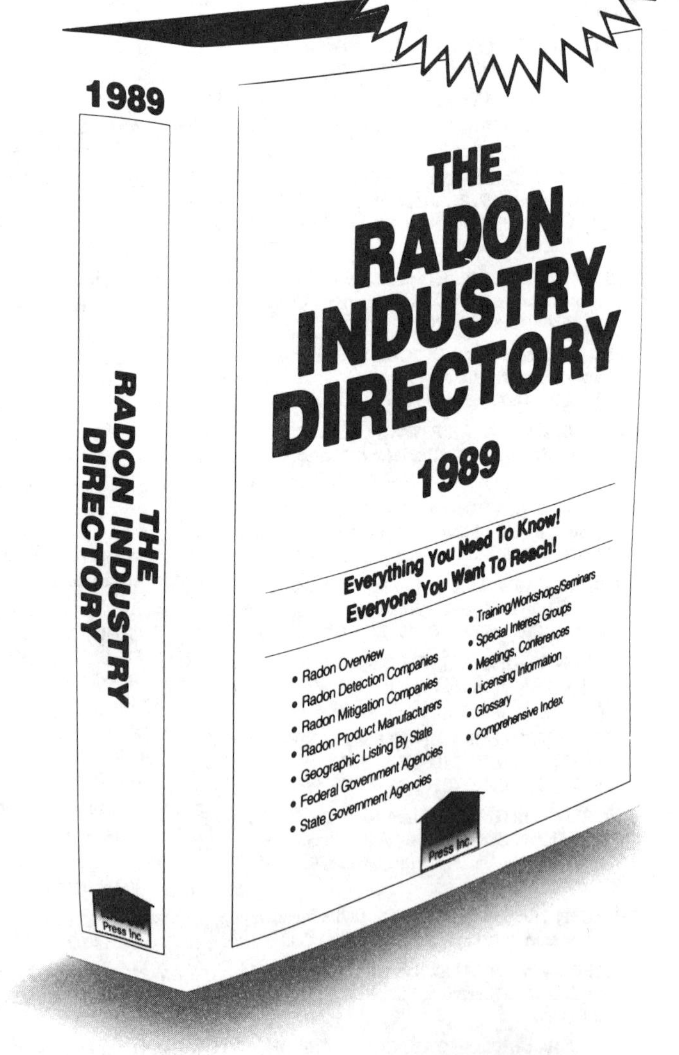

To Order
Call 1-800-548-1567 or fill out and return the order form at the back of the directory.

Satisfaction Guaranteed

You must be completely satisfied with The 1989 Radon Industry Directory or simply return it within 10 days of receipt for a prompt refund.

RADON
Press Inc.

The Radon Industry Directory Listing Form

Here is your opportunity to be included in the next **Radon Industry Directory.** In one comprehensive reference, this defintive directory lists radon related services, radon detection companies, radon mitigation companies, manufacturers, distributors, government agencies, seminars, workshops, training, meetings, conferences, licensing, proficiency tests, research publications, prominent individuals, geographic listings by state, and much more.

Your First Listing Is Free!

Entries will be listed alphabetically by category. Each listing will include organization name, address, phone number, description of product, service, purpose and names of key personnel. **Your first listing in one category is free.** Additional listings in other categories will be charged at $20 per listing. If you are involved in the radon industry, you simply must be represented in the next **Radon Industry Directory.**

Please fill out each section completely and accurately—thank you.

1 Please list my organization/agency/institution as follows (if branch offices please list on additional copies of this form):

Company name: ____________________

Address ____________________

City ____________________ State __________ Zip ____________________

Phone number (area code) ____________________ (800) ____________________

2 Please describe your organization/agency/facility in the space below. (Describe your product, service, purpose. What do you provide? For whom?):

____

____

____

If appropriate, briefly describe any licensing, credentials, qualifications applicable to your organization.

____

____

3 Is your service/product available: ❑ Nationwide ❑ Statewide

Specific States ____________________

4 Please list names and titles of personnel to be listed:

Name	Title	Name	Title

Please turn page over to complete then detach page and mail

The Radon Industry Directory Listing Form—Page 2

5 Please check each of the following categories that describes your organization. Remember you first listing is FREE, listings in additional categories are $20 each:

- ❑ Radon Detection Company
- ❑ Radon Mitigation Company
- ❑ Other Professional/Consultant
- ❑ Radon Detection Product Mfg.
- ❑ Radon Detection Product Distributor
- ❑ Radon Mitigation Product Manufacturer
- ❑ Radon Mitigation Product Distributor
- ❑ Federal Government Agency
- ❑ State Government Agency
- ❑ Local Government Agency
- ❑ Radon Course/Workshop/Training
- ❑ Research Facility
- ❑ Buildings/Construction
- ❑ Health Services
- ❑ Real Estate Industry
- ❑ Environmental
- ❑ Public Interest

6 If you checked any **manufacturer** or **distributor** listings, please indicate which products you are involved with:

- ❑ Sealers
- ❑ Membranes
- ❑ Air Flow Measurement Equipment
- ❑ Sump Pump Equipment
- ❑ Ventilation Equipment
- ❑ Protective Equipment
- ❑ Instrument Repair
- ❑ Other__________________

7 Please check the last RMP Program you have completed: ❑ 4 ❑ 5 ❑ 6 ❑ 7 ❑ 8

For which instruments?: ____

8 **Listing Fee:** Your **first** listing in Section **Five** is **FREE.** Listings in additional categories are $20 each. Please Check the number of additional categories you wish to be listed in as indicated in Section Five.

Example

1 Free listing+ 4 additional listings @ $20 each = $80 Total Due

1 Free listing+ ______ Additional listing(s) @$20 each = $______________ Total Due

Payment method: ❑ Check ❑ P.O. Number______________

9 Mail Completed Questionaire and Payment (Check, Money Order, Purchase Order) To:

RADON Press, Inc.
Attn: Editor
P.O. Box 25551
Alexandria, Virgina 22313

For Additional Information:

- ❑ Please send me information on **reserving** my copy of **The 1989 Radon Industry Directory.**
- ❑ Please send me information on **display advertising** in **The 1990 Radon Industry Directory.**
- ❑ Please send me information on **subscribing** to **The Radon Hotline Letter.**
- ❑ Please send me information on **Radon Industry Mailing Lists.**

It's Like Having The Entire Radon Industry Delivered To Your Door!

ISBN: 0-929840-00-3 ISSN: 0899-9023

Satisfaction Guaranteed

You must be completely satisfied with The 1989 Radon Industry Directory or simply return it within 10 days of receipt for a prompt refund.

For Faster Service, and Standing Order Information, Call Toll-Free
1-800-548-1567

RADON Press Inc.

ORDER FORM

The RADON INDUSTRY DIRECTORY

Yes! Please send me _____ copies of the Directory for only $75 each — Total Price $________. All invoices payable U.S. dollars.

☐ Bill Me ☐ Bill Company ☐ P.O. Enclosed #________ ☐ Check Enclosed
☐ Begin a Standing Order (Standing order qualifies for a 10% discount).

Name____________________ Title____________________
Company____
Address____
City____________________ State ______ Zip__________
Phone____________________ Ext.____________________

Just Drop In The Mail. Postage Is Prepaid. DB1001

ORDER FORM

The RADON INDUSTRY DIRECTORY

Yes! Please send me _____ copies of the Directory for only $75 each — Total Price $________. All invoices payable U.S. dollars.

☐ Bill Me ☐ Bill Company ☐ P.O. Enclosed #________ ☐ Check Enclosed
☐ Begin a Standing Order (Standing order qualifies for a 10% discount).

Name____________________ Title____________________
Company____
Address____
City____________________ State ______ Zip__________
Phone____________________ Ext.____________________

Just Drop In The Mail. Postage Is Prepaid. DB1002

ORDER FORM

The RADON INDUSTRY DIRECTORY

Yes! Please send me _____ copies of the Directory for only $75 each — Total Price $________. All invoices payable U.S. dollars.

☐ Bill Me ☐ Bill Company ☐ P.O. Enclosed #________ ☐ Check Enclosed
☐ Begin a Standing Order (Standing order qualifies for a 10% discount)

Name____________________ Title____________________
Company____
Address____
City____________________ State ______ Zip__________
Phone____________________ Ext.____________________

Just Drop In The Mail. Postage Is Prepaid. DB1003

NO POSTAGE
NECESSARY
IF MAILED
IN THE
UNITED STATES

BUSINESS REPLY MAIL
FIRST-CLASS MAIL PERMIT NO. 8873 ALEXANDRIA, VIRGINIA
POSTAGE WILL BE PAID BY ADDRESSEE

RADON PRESS, INC.
P.O. BOX 25551
ALEXANDRIA, VA. 22313

NO POSTAGE
NECESSARY
IF MAILED
IN THE
UNITED STATES

BUSINESS REPLY MAIL
FIRST-CLASS MAIL PERMIT NO. 8873 ALEXANDRIA, VIRGINIA
POSTAGE WILL BE PAID BY ADDRESSEE

RADON PRESS, INC.
P.O. BOX 25551
ALEXANDRIA, VA. 22313

NO POSTAGE
NECESSARY
IF MAILED
IN THE
UNITED STATES

BUSINESS REPLY MAIL
FIRST-CLASS MAIL PERMIT NO. 8873 ALEXANDRIA, VIRGINIA
POSTAGE WILL BE PAID BY ADDRESSEE

RADON PRESS, INC.
P.O. BOX 25551
ALEXANDRIA, VA. 22313